MyStatLab Enhancements for *Stats: Data and Models*, Third Edition

- **Increased exercise coverage** provides even more opportunities to practice and gain mastery of statistics.

- **Reading assessment questions** allow students to check their understanding of the concepts.

- **Algorithmically generated data sets** can be copied from MyStatLab exercises into external applications, such as StatCrunch or Excel,® for in-depth analysis.

- MyStatLab now includes access to **StatCrunch.com.** Log on today to learn more.

Video Lectures are scripted and presented by the authors themselves, helping students review the important points in each chapter. Different video presenters also work through examples from the text.

StatCrunch
Data analysis on the Web

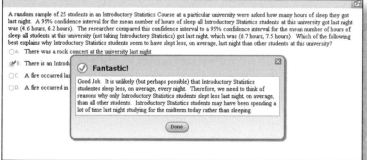

A **1,000 Conceptual Question Library** in MyStatLab helps students develop strong analytical skills and conceptual understanding.

Stats

Data and Models

THIRD EDITION

Stats
Data and Models

EDITION 3

Richard D. De Veaux
Williams College

Paul F. Velleman
Cornell University

David E. Bock
Cornell University

Addison-Wesley

Boston Columbus Indianapolis New York San Francisco Upper Saddle River
Amsterdam Cape Town Dubai London Madrid Milan Munich Paris Montréal Toronto
Delhi Mexico City São Paulo Sydney Hong Kong Seoul Singapore Taipei Tokyo

Editor in Chief	Deirdre Lynch
Acquisitions Editor	Christopher Cummings
Senior Content Editor	Chere Bemelmans
Assistant Editors	Dana Jones Bettez and Christina Lepre
Senior Managing Editor	Karen Wernholm
Associate Managing Editor	Tamela Ambush
Senior Production Project Manager	Sheila Spinney
Digital Assets Manager	Marianne Groth
Supplements Production Coordinator	Kerri McQueen
Senior Media Producer	Christine Stavrou
Software Development	Robert Carroll and Marty Wright
Marketing Manager	Alex Gay
Marketing Coordinator	Kathleen DeChavez
Senior Author Support/Technology Specialist	Joe Vetere
Rights and Permissions Advisor	Michael Joyce
Image Manager	Rachel Youdelman
Senior Manufacturing Buyer	Carol Melville
Senior Media Buyer	Ginny Michaud
Design Manager	Andrea Nix
Cover Designer	Beth Paquin
Text Design	The Davis Group, Inc.
Production Coordination, Composition, and Illustrations	PreMediaGlobal
Cover Image	Internet mapping project Patent(s) Pending & Copyright © Lumeta Corporation 2009–2010. All Rights Reserved.

Library of Congress Cataloging-in-Publication Data
De Veaux, Richard D.
  Stats : data and models / Richard D. De Veaux, Paul F. Velleman,
  David E. Bock. — 3rd ed.
    p. cm.
  ISBN 13: 978-0-321-69260-3 (instructor's edition)   ISBN 13: 978-0-321-69255-9 (student edition)
  ISBN 10: 0-321-69260-8 (instructor's edition)   ISBN 10: 0-321-69255-1 (student edition)
  1. Statistics—Textbooks. 2. Mathematical statistics—Textbooks.   I. Velleman,
Paul F., 1949-   II. Bock, David E.   III. Title.
QA276.12.D417 2012
519.5—dc22   2010005111

5 6 7 8 9 10—V057—14

Addison-Wesley
is a imprint of

www.pearsonhighered.com

ISBN 13: 978-0-321-69255-9
ISBN 10: 0-321-69255-1

To Sylvia, who has helped me in more ways than she'll ever know,
and to Nicholas, Scyrine, Frederick, and Alexandra,
who make me so proud in everything that they are and do
—Dick

To my sons, David and Zev, from whom I've learned so much,
and to my wife, Sue, for taking a chance on me
—Paul

To Greg and Becca, great fun as kids and great friends as adults,
and especially to my wife and best friend, Joanna, for her
understanding, encouragement, and love
—Dave

Meet the Authors

Richard D. De Veaux is an internationally known educator and consultant. He has taught at the Wharton School and the Princeton University School of Engineering, where he won a "Lifetime Award for Dedication and Excellence in Teaching." Since 1994, he has been Professor of Statistics at Williams College. Dick has won both the Wilcoxon and Shewell awards from the American Society for Quality. He is a fellow of the American Statistical Association. Dick is also well known in industry, where for the past 20 years he has consulted for such companies as Hewlett-Packard, Alcoa, DuPont, Pillsbury, General Electric, and Chemical Bank. He has also sometimes been called the "Official Statistician for the Grateful Dead." His real-world experiences and anecdotes illustrate many of this book's chapters.

Dick holds degrees from Princeton University in Civil Engineering (B.S.E.) and Mathematics (A.B.) and from Stanford University in Dance Education (M.A.) and Statistics (Ph.D.), where he studied with Persi Diaconis. His research focuses on the analysis of large data sets and data mining in science and industry.

In his spare time he is an avid cyclist and swimmer. He also is the founder and bass for the "Diminished Faculty," an a cappella Doo-Wop quartet at Williams College. Dick is the father of four children.

Paul F. Velleman has an international reputation for innovative Statistics education. He is the author and designer of the multimedia statistics CD-ROM *ActivStats*, for which he was awarded the EDUCOM Medal for innovative uses of computers in teaching Statistics, and the ICTCM Award for Innovation in Using Technology in College Mathematics. He also developed the award-winning statistics program, Data Desk, and the Internet site Data And Story Library (DASL) (http://dasl.datadesk.com), which provides data sets for teaching Statistics. Paul's understanding of using and teaching with technology informs much of this book's approach.

Paul has taught Statistics at Cornell University since 1975. He holds an A.B. from Dartmouth College in Mathematics and Social Science, and M.S. and Ph.D. degrees in Statistics from Princeton University, where he studied with John Tukey. His research often deals with statistical graphics and data analysis methods. Paul co-authored (with David Hoaglin) *ABCs of Exploratory Data Analysis*. Paul is a Fellow of the American Statistical Association and of the American Association for the Advancement of Science.

Outside of class, Paul sings baritone in a barbershop quartet. He is the father of two boys.

David E. Bock taught mathematics at Ithaca High School for 35 years. He has taught Statistics at Ithaca High School, Tompkins-Cortland Community College, Ithaca College, and Cornell University. Dave has won numerous teaching awards, including the MAA's Edyth May Sliffe Award for Distinguished High School Mathematics Teaching (twice), Cornell University's Outstanding Educator Award (three times), and has been a finalist for New York State Teacher of the Year.

Dave holds degrees from the University at Albany in Mathematics (B.A.) and Statistics/Education (M.S.). Dave has been a reader and table leader for the AP Statistics exam, serves as a Statistics consultant to the College Board, and leads workshops and institutes for AP Statistics teachers. He has recently served as K-12 Education and Outreach Coordinator and as a senior lecturer for the Mathematics Department at Cornell University. His understanding of how students learn informs much of this book's approach.

Dave relaxes by biking and hiking. He and his wife have enjoyed many days camping across Canada and through the Rockies. They have a son, a daughter, and three grandchildren.

Contents

Preface

W e've been thrilled with the feedback we've received from instructors and students using *Stats: Data and Models*. If there is a single hallmark of this book it is that students actually read it. We have reports from every level—from high school to graduate school—that students find our books easy and even enjoyable to read. We strive for a conversational, approachable style, and introduce anecdotes to maintain students' interest. And it works. Instructors report their amazement that students are voluntarily reading ahead of their assignments. Students write to tell us (to their amazement) that they actually enjoyed the book.

Unlike any other introductory book, *Stats: Data and Models* is written with the understanding that Statistics is practiced with technology. This insight informs everything from our choice of forms for equations (favoring intuitive forms over calculation forms) to our extensive use of real data. Most important, it allows us to focus on teaching Statistical Thinking rather than calculation. The questions that motivate each of our hundreds of examples are not "how do you find the answer?" but "how do you think about the answer?"

What's New in This Edition

The third edition of *Stats: Data and Models* continues and extends the successful innovations pioneered in our books, teaching Statistics and statistical thinking as it is practiced today. We've rewritten sections throughout the book to make them clearer and more interesting. We've introduced new up-to-the-minute motivating examples throughout. And, we've added a number of new features, each with the goal of making it even easier for students to put the concepts of Statistics together into a coherent whole.

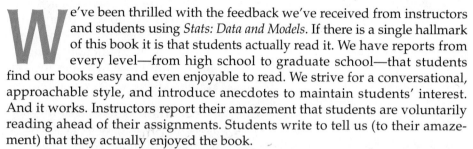

▶ *For Example.* In every chapter, you'll find new worked examples that illustrate how to apply new concepts and methods—**more than 100 new illustrative examples**. But these aren't isolated examples. We carry the discussion through the chapter with each *For Example*, picking up the story and moving it forward as students learn to apply each new concept.

▶ *Exercises.* We've added **hundreds of new exercises,** including more single-concept exercises at the beginning of each set so students can be sure they have a clear understanding of each important topic before they're asked to tie them all together in more comprehensive exercises. Continuing exercises have been **updated with the most recent data.** Whenever possible, the data are on the DVD and the book's website so students can explore them further.

▶ *Step-by-Step Worked Examples.* More than a third of the chapters have new or updated Think/Show/Tell Step-by-Step examples.

▶ *ActivStats Pointers.* In the third edition, the *ActivStats* pointers have been revised for clarity and now indicate exactly what they are pointing to—activity, video, simulation, or animation—paralleling the book's discussions to enhance learning.

▶ *Data Sources.* Most of the data used in examples and exercises are from recent news stories, research articles, and other real-world sources. We've listed more of those sources in this edition.

▶ *New Motivating Examples.* Each chapter starts with a real-world example, which we follow up with data analyses. We've updated and replaced many of these examples, introducing new stories (and data) about earthquakes, hurricane path prediction, penguin dives, belief in ghosts, motorcycle helmet compliance, whether men are less likely than women to wear seat belts, and Olympic speedskating.

▶ *Simulation.* We've improved the discussion of simulation in Chapter 11 so it relates more easily to discussions of experimental design and probability. The simulations included in the *ActivStats* multimedia software on the book's DVD carry those ideas forward in a student-friendly fashion.

▶ *Instructor's Podcasts.* (10 points in 10 minutes). Created and presented by the authors, these podcasts focus on key points in each chapter to help instructors prepare for class. They can be easily downloaded from the IRC.

Continuing Features

▶ *Think, Show, Tell.* The worked examples repeat the mantra of *Think, Show,* and *Tell* in every chapter. They emphasize the importance of thinking about a Statistics question (What do we know? What do we hope to learn? Are the assumptions and conditions satisfied?) and reporting our findings (the *Tell* step). The *Show* step contains the mechanics of calculating results and conveys our belief that it is only one part of the process. This rubric is highlighted in the *Step-by-Step* examples that guide the students through the process of analyzing the problem with the general explanation on the left and the worked-out problem on the right. The result is a better understanding of the concept, not just number crunching.

▶ *Just Checking.* Within each chapter, we ask students to pause and think about what they've just read. These questions are designed to be a quick check that they understand the material. Answers are at the end of the exercise sets in each chapter so students can easily check themselves.

▶ *Math Boxes.* In many chapters we present the mathematical underpinnings of the statistical methods and concepts. By setting these proofs, derivations, and justifications apart from the narrative, we allow the student to continue to follow the logical development of the topic at hand, yet also refer to the underlying mathematics for greater depth.

▶ *What Can Go Wrong?* Each chapter still contains our innovative *What Can Go Wrong?* sections that highlight the most common errors people make and the misconceptions they have about Statistics. Our goals are to help students avoid these pitfalls, and to arm them with the tools to detect statistical errors and to debunk misuses of statistics, whether intentional or not. In this spirit, some of our exercises probe the understanding of such failures.

WHAT HAVE WE LEARNED?

REALITY CHECK

NOTATION ALERT

ON THE COMPUTER

▶ *What Have We Learned?* These chapter-ending summaries are great study guides, providing complete overviews that highlight the new concepts, define the new terms, and list the skills that the student should have acquired in the chapter.

▶ *Exercises.* Throughout, we've maintained the pairing of examples so that each odd-numbered exercise (with an answer in the back of the book) is followed by an even-numbered exercise on the same concept. Exercises are still ordered by level of difficulty.

▶ *Reality Check.* We regularly remind students that Statistics is about understanding the world with data. Results that make no sense are probably wrong, no matter how carefully we think we did the calculations. Mistakes are often easy to spot with a little thought, so we ask students to stop for a reality check before interpreting their result.

▶ *Notation Alert.* Throughout this book we emphasize the importance of clear communication, and proper notation is part of the vocabulary of Statistics. We've found that it helps students when we call attention to the letters and symbols statisticians use to mean very specific things.

▶ *Connections.* Each chapter has a *Connections* section to link key terms and concepts with previous discussions and to point out continuing themes, helping students fit newly learned concepts into a growing understanding of Statistics.

▶ *On the Computer.* In the real world, Statistics is practiced with computers. We prefer not to choose a particular Statistics program. Instead, at the end of each chapter, we summarize what students can find in the most common packages, often with an annotated example. Computer output appearing in the book and in exercises is often generic, resembling all of the common packages to some degree.

Our Goal: Read This Book!

The best text in the world is of little value if students don't read it. Here are some of the ways we have made *Stats: Data and Models*, Third Edition even more approachable:

- *Readability.* You'll see immediately that this book doesn't read like other Statistics texts. The style is both colloquial and informative, engaging students to actually read the book to see what it says.
- *Informality.* Our informal diction doesn't mean that the subject matter is covered lightly or informally. We have tried to be precise and, wherever possible, to offer deeper explanations and justifications than those found in most introductory texts.
- *Focused lessons.* The chapters are shorter than in most other texts, to make it easier to focus on one topic at a time.
- *Consistency.* We've worked hard to avoid the "do what we say, not what we do" trap. From the very start we teach the importance of plotting data and checking assumptions and conditions, and we have been careful to model that behavior right through the rest of the book.
- *The need to read.* Students who plan just to skim the book may find our presentation a bit frustrating. The important concepts, definitions, and sample solutions don't sit in little boxes. This is a book that needs to be read, so we've tried to make the reading experience enjoyable.

Coverage

Textbooks are often defined more by what they choose not to cover than by what they do cover. We've been guided in the choice and order of topics by several fundamental principles. First, we have tried to ensure that each new topic fits into the growing structure of understanding that we hope students will build. Several topic orders can support this goal. We explain our reasons for the topic order of the chapters in the supplement Printed Test Bank and Resource Guide (also available for download at pearsonhighered.com/irc).

GAISE Guidelines. We have worked to provide materials to help each class, in its own way, follow the guidelines of the GAISE (Guidelines for Assessment and Instruction in Statistics Education) project sponsored by the American Statistical Association. That report urges that Statistics education should

1. Emphasize Statistical literacy and develop Statistical thinking.
2. Use real data.
3. Stress conceptual understanding rather than mere knowledge of procedures.
4. Foster active learning.
5. Use technology for developing concepts and analyzing data.
6. Make assessment a part of the learning process.

Mathematics

Mathematics traditionally appears in Statistics texts in several roles:

1. It can provide a concise, clear statement of important concepts.
2. It can describe calculations to be performed with data.
3. It can embody proofs of fundamental results.

Of these, we emphasize the first. Mathematics can make discussions of Statistics concepts, probability, and inference clear and concise. We have tried to be sensitive to those who are discouraged by equations by also providing verbal descriptions and numerical examples.

This book is not concerned with proving theorems about Statistics. Some of these theorems are quite interesting, and many are important. Often, though, their proofs are not enlightening to introductory Statistics students, and can distract the audience from the concepts we want them to understand. However, we have not shied away from the mathematics where we believed that it helped clarify without intimidating. You will find some important proofs, derivations, and justifications in Math Boxes that accompany the development of many topics.

Nor do we concentrate on calculations. Although statistics calculations are generally straightforward, they are also usually tedious. And, more to the point, they are often unnecessary. Today, virtually all statistics are calculated with technology, so there is little need for students to work by hand. The equations we use have been selected for their focus on understanding concepts and methods.

Technology and Data

To experience the real world of Statistics, it's best to explore real data sets using modern technology.

▶ *Technology.* We assume that you are using some form of technology in your Statistics course. That could be a calculator, a spreadsheet, or a statistics package. Rather than adopt any particular software, we discuss generic computer output. The DVD includes *ActivStats* and the software package

Data Desk. Also, at the end of each chapter in On the Computer, we offer general guidance to help students get started on common software platforms (Excel, Minitab, Data Desk, JMP, and SPSS), and TI-83/84 and TI-89 calculators.

▶ *Data.* Because we use technology for computing, we don't limit ourselves to small, artificial data sets. In addition to including some small data sets, we have built examples and exercises on real data with a moderate number of cases—usually more than you would want to enter by hand into a program or calculator. These data are included on the DVD as well as on the book's website.

On the DVD

The DVD holds a number of supporting materials, including *ActivStats*, the *Data Desk* statistics package, an Excel add-in (*DDXL*), and all large data sets from the text formatted for the most popular technologies.

ActivStats (for Data Desk). The award-winning *ActivStats* multimedia program supports learning chapter by chapter. It complements the book with videos of real-world stories, worked examples, animated expositions of each of the major Statistics topics, and tools for performing simulations, visualizing inference, and learning to use statistics software. The new version of *ActivStats* includes

- improved navigation and a cleaner design that makes it easier to find and us tools such as the Index and Glossary
- more than **1000 homework exercises,** including many new exercises, plus answers to the "odd numbered" exercises. Many are from the text, providing the data already set up for calculations, and some are unique to *ActivStats.* Many exercises link to data files for each statistics package.
- **17** short **video clips,** many new and updated
- **70 animated activities**
- **117 teaching applets**
- more than **300 data sets**

Supplements

Student Supplements

Stats: Data and Models, Third Edition, for-sale student edition (ISBN-13: 978-0-321-69255-9; ISBN-10: 0-321-69255-1)

Student Solutions Manual for Stats: Data and Models, Third Edition, by William Craine, provides detailed, worked-out solutions to odd-numbered exercises. (ISBN-13: 978-0-321-69349-5; ISBN-10: 0-321-69349-3)

Graphing Calculator Manual, by Patricia Humphrey (Georgia Southern University) is organized to follow the sequence of topics in the text, and is an easy-to-follow, step-by-step guide on how to use the TI-83/84 Plus and TI-89 graphing calculators. It provides worked-out examples to help students fully understand and use the graphing calculator. (ISBN-13: 978-0-321-49943-1; ISBN-10: 0-321-49943-3)

Statistics Study Card is a resource for students containing important formulas, definitions, and tables that correspond precisely to the De Veaux/Velleman/Bock Statistics series. This card can work as a reference for completing homework assignments or as an aid in studying. (ISBN-13: 978-0-321-46370-8; ISBN-10: 0-321-46370-6)

Study Cards for Statistics Software. Technology Study Cards for Statistics are a convenient resource for students, with instructions and screenshots for using the most popular technologies. The following Study Cards are available in print (8-page fold-out cards) and within MyStatLab: SPSS (0-321-58979-3), Excel 2007 w/DDXL (0-321-59280-8), Minitab (0-321-59282-4), R (0-321-59283-2), Graphing Calculator (0-321-57077-4), StatCrunch (0-321-62892-6), and JMP (0-321-59281-6).

Graphing Calculator Tutorial for Statistics will guide students through the keystrokes needed to most efficiently use their graphing calculator. Although based on the TI-84 Plus Silver Edition, operating system 2.30, the keystrokes for this calculator are identical to those on the TI-84 Plus, and very similar to the TI-83 and TI-83 Plus. This tutorial should be helpful to students using any of these calculators, though there may be differences in some lessons. The tutorial is organized by topic. (ISBN-13: 978-0-321-41382-6; ISBN-10: 0-321-41382-2)

Instructor Supplements

Instructor's Edition contains answers to all exercises. (ISBN-13: 978-0-321-69260-3; ISBN-10: 0-321-69260-8)

Printed Test Bank and Resource Guide, by William Craine, contains chapter-by-chapter comments on the major concepts, tips on presenting topics (and what to avoid), teaching examples, suggested assignments, Web links and lists of other resources, as well as chapter quizzes, unit tests, investigative tasks, and suggestions for projects. (ISBN-13: 978-0-321-69257-3; ISBN-10: 0-321-69257-8). The Printed Test Bank and Resource Guide is also available to download from www.pearsonhighered.com/irc.

Instructor's Solutions Manual, by William Craine, contains detailed solutions to all of the exercises. (ISBN-13: 978-0-321-69900-8; ISBN-10: 0-321-69900-9)

TestGen® (www.pearsonhighered.com/testgen) enables instructors to build, edit, print, and administer tests using a computerized bank of questions developed to cover all the objectives of the text. TestGen is algorithmically based, allowing instructors to create multiple but equivalent versions of the same question or test with the click of a button. Instructors can also modify test bank questions or add new questions. Tests can be printed or administered online. The software and test bank are available for download from Pearson Education's online catalog.

PowerPoint Lecture Slides provide an outline to use in a lecture setting, presenting definitions, key concepts, and figures from the text. These slides are available for download from www.pearsonhighered.com/irc.

Technology Resources

A multimedia program on DVD designed to support learning chapter by chapter is bundled with student books or may be purchased separately. It is available per student or as a lab version (per work station). The DVD holds a number of supporting materials, including:

- *ActivStats® for Data Desk.* The award-winning *ActivStats* multimedia program supports learning chapter by chapter with the book. It complements the book with videos of real-word stories, worked examples, animated expositions of each of the major Statistics topics, and tools for performing simulations, visualizing inference, and learning to use statistics software. The new version of *ActivStats* includes 17 short video clips; 170 animated activities and teaching applets; 300 data sets; 1,000 homework exercises, many with links to Data Desk files; interactive graphs, simulations, visualization tools, and much more.
- *Data Desk* statistics package.
- *DDXL*, an Excel add-in, adds sound statistics and statistical graphics capabilities to Excel. DDXL adds, among other capabilities, boxplots, histograms, statistical scatterplots, normal probability plots, and statistical inference procedures not available in Excel's Data Analysis pack.
- *Data.* Data for exercises marked **T** are available on the DVD and website formatted for the most popular technologies and as text files suitable for these and virtually all other statistics software.

ActivStats® The award-winning ActivStats multimedia program supports learning chapter by chapter with the book and is available as a stand-alone DVD. It complements the book with videos of real-word stories, worked examples, animated expositions of each of the major Statistics topics, and tools for performing simulations, visualizing inference, and learning to use statistics software. The new version of ActivStats includes 17 short video clips; 170 animated activities and teaching applets; 300 data sets; 1,000 homework exercises, many with links to Data Desk files; interactive graphs, simulations, visualization tools, and much more. ActivStats (Mac and PC) is available in an all-in-one version for Data Desk, Excel, JMP, MINITAB, and SPSS. This DVD also includes Data Desk statistical software. (ISBN-13: 978-0-321-50014-4; ISBN-10: 0-321-50014-8)

MyStatLab™ Online Course (access code required) MyStatLab is a series of text-specific, easily customizable online courses for Pearson Education's textbooks in statistics. For students, MyStatLab™ provides students with a personalized interactive learning environment that adapts to each student's learning style and gives them immediate feedback and help. Because MyStatLab is delivered over the Internet, students can learn at their own pace and work whenever they want. MyStatLab provides instructors with a rich and flexible set of text-specific resources, including course management tools, to support online, hybrid, or traditional courses. MyStatLab is available to qualified adopters and includes access to StatCrunch, a web-based statistics software. For more information, visit www.mystatlab.com or contact your Pearson representative.

StatCrunch.com access is now included with MyStatLab. StatCrunch.com is the leading web-based statistics software designed for teaching statistics. Users can perform complex analyses, share data sets, and generate compelling reports. The vibrant online community offers more than 12,000 data sets for students to analyze.

MathXL® for Statistics Online Course (access code required) MathXL® for Statistics is a powerful online homework, tutorial, and assessment system that accompanies Pearson textbooks in statistics. With MathXL for Statistics, instructors can:

- Create, edit, and assign online homework and tests using algorithmically generated exercises correlated at the objective level to the textbook.
- Create and assign their own online exercises and import TestGen tests for added flexibility.
- Maintain records of all student work, tracked in MathXL's online gradebook.

With MathXL for Statistics, students can:

- Take chapter tests in MathXL and receive personalized study plans and/or personalized homework assignments based on their test results.
- Use the study plan and/or the homework to link directly to tutorial exercises for the objectives they need to study.
- Students can also access supplemental animations and video clips directly from selected exercises.

MathXL for Statistics is available to qualified adopters. For more information, visit our website at www.mathxl.com, or contact your Pearson representative.

Companion Web site www.pearsonhighered.com/deveaux provides additional resources for instructors and students.

The Student Edition of MINITAB is a condensed edition of the Professional release of MINITAB statistical software that offers the full range of statistical methods and graphical capabilities, along with worksheets that can include up to 10,000 data points. Individual copies of the software can be bundled with the text. (ISBN 13: 978-013-143661-9; ISBN-10: 0-13-143661-9)

JMP Student Edition is an easy-to-use, streamlined version of JMP desktop statistical discovery software from SAS Institute, Inc. and is available for bundling with the text. (ISBN 13: 978-0-321-67212-4; ISBN-10: 0-321-67212-7)

SPSS Student Version, a statistical and data management software package, is also available for bundling with the text. (ISBN-13: 978-0-321-67537-8; ISBN-10: 0-321-67537-1)

Acknowledgments

Many people have contributed to this book in all of its editions. This edition never would have seen the light of day without the assistance of the incredible team at Addison-Wesley. Our Editor, Chris Cummings, was central to the genesis, development, and realization of the book from day one. Chere Bemelmans, Senior Content Editor, kept us on task as much as humanly possible with much needed humor and grace. Sheila Spinney, Senior Production Project Manager, kept the cogs from getting into the wheels where they often wanted to wander. Senior Marketing Manager Alex Gay made sure the word got out. Dana Jones, Associate Editor, and Kathleen DeChavez, Marketing Assistant, were essential in managing all of the behind-the-scenes work that needed to be done. Christine Stavrou, Senior Media Producer, put together a top-notch media package for this book. Geri Davis is responsible for the wonderful way the book looks. Evelyn Beaton, Manufacturing Manager, and Ginny Michaud, Manufacturing Buyer, worked miracles to get this book and CD in your hands, and Greg Tobin, President, was supportive and good-humored throughout all aspects of the project.

A special thanks goes out to Laura Hakala, Senior Project Manager at PreMediaGlobal, for her close attention to detail and her amazing editorial skills.

We'd also like to thank our accuracy checkers whose monumental task was to make sure we said what we thought we were saying. They are Elaine McDonald, Sonoma State University; Dave Bregenzer, Utah State University; Stan Seltzer, Ithaca College; and Michael Zwilling, Mt. Union College.

We extend our sincere thanks for the suggestions and contributions made by the following reviewers of this edition:

Jeff Kolath,
Oregon State University

Scott Nickleach,
Sonoma State University

John Verzani,
College of Staten Island

Engin Sungur,
University of Minnesota Morris

Ali Arab,
Georgetown University

Michael Zwilling,
Mt. Union College

John D. Emerson,
Middlebury College

We extend our sincere thanks for the suggestions and contributions made by the following reviewers, focus group participants, and class-testers of the previous edition:

Jon Angellotti, *Cornell University*
Sanjib Basu, *Northern Illinois University*
James Bearden, *SUNY Geneseo*
Peter Blaskiewicz, *McLennan Community College*
Steven Bogart, *Shoreline Community College*
Dana Calland, *Maysville Community College*
Ann Cannon, *Cornell College*
David G. Caraballo, *Georgetown University*
Grace Cascio-Houston, *Louisiana State University–Eunice*
Smiley Cheng, *University of Manitoba*
Crista Lynn Coles, *Elon University*
Jon Cryer, *University of Iowa*
Carolyn Cuff, *Westminster College*
Nasser Dastrange, *Buena Vista University*
Mary Ellen Davis, *Georgia Perimeter College*
Rick Denman, *Southwestern University*
Scott Desposato, *University of California, San Diego*
Jody DeVoe, *Valencia Community College–East Campus*
Jeffrey Eldridge, *Edmonds Community College*
David Elesh, *Temple University*
Karen Estes, *St. Petersburg Junior College*

John W. Emerson, *Yale University*
Hans Engler, *Georgetown University*
Russell Euler, *Northwest Missouri State University*
Amy Fisher, *Miami University, Middletown*
William Fox, *Francis Marion University*
Richard Friary
John Gabrosek, *Grand Valley State University*
Jinadasa Gamage, *Illinois State University*
Dermot Gately, *New York University*
James Gehrmann, *California State University–Sacramento*
Paramjit Gill, *Okanagan University College*
Martha Goshaw, *Seminole Community College*
Kimberly Goyette, *Temple University*
Robert Gould, *University of California, Los Angeles*
Ken Grace, *Anoka-Ramsey Community College*
Jonathan Graham, *University of Montana*
David Graves, *Elmira College*
Richard Greene, *Temple University*
Scott Greene, *University of Oklahoma*
Hasan Hamdan, *James Madison University*
Josephine Hamer, *Western Connecticut State University*

Mary Hartz, *Mohawk Valley Community College*
Nancy Heckman, *University of British Columbia*
James Helreich, *Marist College*
Susan Herring, *Sonoma State University*
Robert Hollister, *Jacksonville University*
Patricia Humphrey, *Georgia Southern University*
Debra L. Hydorn, *Mary Washington College*
Debra Ingram, *Arkansas State University*
Coleen Jacobson, *Elmira College*
Lloyd R. Jaisingh, *Moorehead State University*
Rebecka Jornsten, *Rutgers University*
Mohammed Kazemi, *University of North Carolina-Charlotte*
John Khoury, *Brevard Community College–Melbourne*
Jeff Kollath, *Oregon State University*
Catherine Kong, *Carson–Newman College*
Karole Kurnow, *Dominican University of California*
Christopher Lacke, *Rowan University*
James Lang, *Valencia Community College–East Campus*
Sheila Lawrence, *Rutgers University*
Julie Legler, *St. Olaf College*
Michael Lichter, *State University of New York–Buffalo*
Pamela Lockwood, *Western Texas A & M University*
Wei-Yin Loh, *University of Wisconsin-Madison*
Catherine Matos, *Clayton College & State University*
Elaine McDonald, *Sonoma State University*
Amy McElroy, *San Diego State University*
Josiah (Si) Meyer, *Elmira College*
Donald Miller, *St. Mary's College*

Jackie Miller, *The Ohio State University*
Panagis Moschopoulos, *University of Texas at El Paso*
Weston I. Nathanson, *California State University*
Sondra Perdue, *University of Washington–Tacoma*
William Peterson, *Middlebury College*
Kimberley Polly, *Parkland College*
Anne Puciloski, *Stonehill College*
Shane Redmond, *Southeastern Louisiana University*
Gina Reed, *Gainesville College*
Jerry Reiter, *Duke University*
Mary Richardson, *Grand Valley State University*
Scott Richter, *Western Kentucky University*
William Roberts
Kim Robinson, *Clayton College & State University*
Richard Rogers, *University of Massachusetts–Amherst*
Edith Seier, *East Tennessee State University*
Nagambal Shah, *Spelman College*
Therese Shelton, *Southwestern University*
Sounny Slitine, *Palo Alto College*
Jeffrey Stuart, *Pacific Lutheran University*
Sharon Testone, *Onondaga Community College*
Theresa Vecchiarelli, *Nassau Community College*
Anita Wah, *Chabot College*
John Walker, *California Polytechnic State University–SLO*
Chamont Wang, *The College of New Jersey*
Edward Welsh, *Westfield State College*
Janit M. Winter-Becker, *Penn State Berks Lehigh Valley College*
Kenny Ye, *SUNY at Stony Brook*

Index of Applications

BE = Boxed Example; E = Exercises; IE = In-Text Example; JC = Just Checking

Accounting

Audits and Taxes (E), 303, 351; (IE), 294, 536
Budgets (E), 520
Casualty Actuarial Society (BE), 381
Company Assets, Profit, and Revenue (BE), 162, 192; (E), 106–107, 149, 265, 707–708; (IE), 239–241
Society of Actuaries (BE), 381

Advertising

Ads (IE), 477
Branding (E), 523, 607–608
Free Products (E), 279
Product Claims (E), 779–780; (JC), 61, 310–312
Sexual Images in Advertising (E), 626, 630–631
Target Audience (E), 549
Text Messaging (JC), 460
Truth in Advertising (E), 631

Agriculture

American Angus Association (IE), 552
Beef and Livestock (BE), 552; (E), 136–137, 139, 574, 632
Crop Damage (E), 329, 473; (JC), 646–647
Farmers' Markets (E), 402
Fertilizers (E), 747, 782–783
Pesticides (E), 861
Salinity Experiment (E), 778–780
Seeds (E), 401, 452, 497
Vineyards (E), 16, 72, 103, 258–259, 334

Banking

Bank Tellers (E), 745
Credit Card Charges (BE), 50, 58, 60–61; (E), 575
Credit Card Companies (BE), 61, 320; (E), 422, 450–451
Credit Card Customers (BE), 50, 58, 60–61
Credit Card Debt (E), 475–476; (JC), 488, 509, 515
Credit Card Fraud (E), 143

Credit Card Offers (BE), 468; (E), 474–475, 496; (IE), 467
Interest Rates (E), 175, 205, 233–234, 254; (IE), 90–91
Loan Approval (E), 452, 521
World Bank (E), 236

Business (General)

Awards and Recognition (E), 43
Chief Executives (BE), 637–640; (E), 453–454, 575; (IE), 91–92, 130, 148, 448, 633–635
Company Databases (E), 77, 138–139
Contract Bids (E), 400
Forbes 500 Companies (BE), 149; (IE), 239
Fortune 500 Companies (BE), 723; (IE), 633, 635
Small Business (IE), 221
Women-Led Businesses (E), 429, 497, 666

Company Names

Amazon.com (IE), 6–11
Arby's (E), 16
AT&T (BE), 9
Buick (E), 173
Burger King (BE), 182, 186–187; (E), 207–208, 812; (IE), 178–180, 182, 185, 189–193, 196, 817–818
Cleveland Casting Plant (E), 16
Coca-Cola (E), 520; (IE), 477
CUPPS (BE), 85–86
First USA (E), 496
Ford (E), 173
Ford Motor Company (E), 16
GfK Roper (E), 301, 376
GlaxoSmithKline (BE), 514, 517
Guinness Brewery (IE), 553
Holes-R-Us (E), 77, 745
The Home Depot (IE), 221
Honda (BE), 391; (E), 173
Hostess Company (IE), 284–285
Hummer (IE), 238–239
Husqvarna (BE), 224
ISA Babcock Company (E), 547
Kroll, Inc. (E), 853

Masterfoods (BE), 345; (E), 353
Nabisco Company (E), 578
Nambe Mills (E), 851
Nissan (BE), 62–63, 85–86; (IE), 129–130
PepsiCo (E), 354, 520; (IE), 477
Pew Research Center (BE), 284; (E), 38, 148, 327, 334, 352–353, 540, 544, 546; (IE), 281, 457; (JC), 345, 416, 460
Pontiac (E), 173
Preusser Group (IE), 499
Progressive Insurance (E), 143–144
SIGG (BE), 85–86
Starbucks (BE), 85–86
Toyota (E), 173, 207, 704
United Parcel Service (UPS) (IE), 1, 11
Veritas Software (E), 853

Consumers

Consumer Price Index (CPI) (E), 236
Consumer Spending (BE), 50, 58
Consumers Union (E), 301
Customer Databases (BE), 10; (E), 138–139, 745; (IE), 6–12
Customer Service (E), 135, 422–423; (IE), 6–7

Demographics

Age (BE), 283, 618–620; (E), 102, 174, 233–235, 577, 704; (IE), 181, 355, 618
American Community Survey (JC), 557
Birth and Death Rates (BE), 83, 797–799, 829–835; (E), 38, 41, 73, 101–102, 141–142, 144, 210–211, 259, 263–264, 329, 497, 542–543, 545, 606–607, 668, 710–711, 811, 845, 851, 853, 857–858, 860; (IE), 249, 795–796, 801, 837
Census-Taking (JC), 557
Eye Color (E), 28, 103, 421, 744–745, 862
Family Size/History (E), 279, 350, 399; (IE), 283, 293, 355
Income (BE), 59, 283; (E), 174–175, 204–205, 668, 811, 845, 864–865; (IE), 537; (JC), 61
Left-handedness (BE), 438–439; (E), 351, 422–423, 427, 857
Licensed Drivers (E), 42

Stats

Data and Models

THIRD EDITION

Stats Starts Here[1]

Where are we going?

Statistics gets no respect. People say things like "You can prove anything with Statistics." People will write off a claim based on data as "just a statistical trick." And Statistics courses don't have the reputation of being students' first choice for a fun elective.

But Statistics *is* fun. That's probably not what you heard on the street, but it's true. Statistics is about how to think clearly with data. A little practice thinking statistically is all it takes to start seeing the world more clearly and accurately.

Q: What is Statistics?
A: Statistics is a way of reasoning, along with a collection of tools and methods, designed to help us understand the world.
Q: What are statistics?
A: Statistics (plural) are particular calculations made from data.
Q: So what is data?
A: You mean, "what *are* data?" Data is the plural form. The singular is datum.
Q: OK, OK, so what are data?
A: Data are values along with their context.

"But where shall I begin?" asked Alice. "Begin at the beginning,"
the King said gravely, "and go on till you come to the end: then stop."

–Lewis Carroll,
Alice's Adventures in Wonderland

So, What Is (Are?) Statistics?

It seems every time we turn around, someone is collecting data on us, from every purchase we make in the grocery store, to every click of our mouse as we surf the Web. The United Parcel Service (UPS) tracks every package it ships from one place to another around the world and stores these records in a giant database. You can access part of it if you send or receive a UPS package. The database is about 17 terabytes big—about the same size as a database that contained every book in the Library of Congress would be. (But, we suspect, not *quite* as interesting.) What can anyone hope to do with all these data?

Statistics plays a role in making sense of the complex world in which we live today. Statisticians assess the risk of genetically engineered foods or of a new drug being considered by the Food and Drug Administration (FDA). They predict the number of new cases of AIDS by regions of the country or the number of customers likely to respond to a sale at the mall. And statisticians help scientists and social scientists understand how unemployment is related to environmental controls, whether enriched early education affects later performance of school children, and whether vitamin C really prevents illness. Whenever there are data and a need for understanding the world, you need Statistics.

---

[1] This chapter might have been called "Introduction," but nobody reads the introduction, and we wanted you to read this. We feel safe admitting this here, in the footnote, because nobody reads footnotes either.

So our objectives in this book are to help you develop the insights to think clearly about the questions, use the tools to show what the data are saying, and acquire the skills to tell clearly what it all means.

The ads say, "Don't drink and drive; you don't want to be a statistic." But you can't be a statistic.

We say: "Don't be a datum."

FRAZZ reprinted by permission of United Feature Syndicate, Inc.

Statistics in a Word

It can be fun, and sometimes useful, to summarize a discipline in only a few words. So,

Statistics is about variation. Data vary because we don't see everything and because even what we do see and measure, we measure imperfectly.

So, in a very basic way, Statistics is about the real, imperfect world in which we live.

Economics is about . . . *Money (and why it is good).*

Psychology: *Why we think what we think (we think).*

Biology: *Life.*

Anthropology: *Who?*

History: *What, where, and when?*

Philosophy: *Why?*

Engineering: *How?*

Accounting: *How much?*

In such a caricature, Statistics is about . . . **Variation**.

Data vary. People are different. We can't see everything, let alone measure it all. And even what we do measure, we measure imperfectly. So the data we wind up looking at and basing our decisions on provide, at best, an imperfect picture of the world. This fact lies at the heart of what Statistics is all about. How to make sense of it is a central challenge of Statistics.

So, How Will This Book Help?

A fair question. Most likely, this book will not turn out to be quite what you expected.

What's different?

Close your eyes and open the book to a page at random. Is there a graph or table on that page? Do that again, say, 10 times. We'll bet you saw data displayed in many ways, even near the back of the book and in the exercises.

We can better understand everything we do with data by making pictures. This book leads you through the entire process of thinking about a problem, finding and showing results, and telling others about what you have discovered. At each of these steps, we display data for better understanding and insight.

You looked at only a few randomly selected pages to get an impression of the entire book. We'll see soon that doing so was sound Statistics practice and reasoning.

Next, pick a chapter and read the first two sentences. (Go ahead; we'll wait.)

We'll bet you didn't see anything about Statistics. Why? Because the best way to understand Statistics is to see it at work. In this book, chapters usually start by presenting a story and posing questions. That's when Statistics really gets down to work.

There are three simple steps to doing Statistics right: *think, show,* and *tell*:

Think first. Know where you're headed and why. It will save you a lot of work.

Show is what most folks think Statistics is about. The *mechanics* of calculating statistics and making displays is important, but not the most important part of Statistics.

Tell what you've learned. Until you've explained your results so that someone else can understand your conclusions, the job is not done.

The best way to learn new skills is to take them out for a spin. In **For Example** boxes you'll see brief ways to apply new ideas and methods as you learn them. You'll also find more comprehensive worked examples called **Step-by-Steps**. These show you fully worked solutions side by side with commentary and discussion, modeling the way statisticians attack and solve problems. They illustrate how to think about the problem, what to show, and how to tell what it all means. These step-by-step examples will show you how to produce the kind of solutions instructors hope to see.

Sometimes, in the middle of the chapter, we've put a section called **Just Checking.** . . . There you'll find a few short questions you can answer without much calculation—a quick way to check to see if you've understood the basic ideas in the chapter. You'll find the answers at the end of the chapter's exercises.

MATH BOX

Knowing where the formulas and procedures of Statistics come from and why they work will help you understand the important concepts. We'll provide brief, clear explanations of the mathematics that supports many of the statistical methods in **Math Boxes** like this.

A S **You'll find ActivStats** parallels the chapters in this book and includes expanded lessons and activities to increase your understanding of the material covered in the text.

From time to time, you'll see an icon like this in the margin to signal that the *ActivStats* multimedia materials on the DVD in the back of the book have an activity that you might find helpful at this point. Typically, we've flagged simulations and interactive activities because they're the most fun and will probably help you see how things work best. The chapters in *ActivStats* are the same as those in the text—just look for the named activity in the corresponding chapter.

"Get your facts first, and then you can distort them as much as you please. (Facts are stubborn, but statistics are more pliable.)"

–Mark Twain

One of the interesting challenges of Statistics is that, unlike in some math and science courses, there can be more than one right answer. This is why two statisticians can testify honestly on opposite sides of a court case. And it's why some people think that you can prove anything with statistics. But that's not true. People make mistakes using statistics, sometimes on purpose in order to mislead others. Most of the unintentional mistakes people make, though, are avoidable. We're not talking about arithmetic. More often, the mistakes come from using a method in the wrong situation or misinterpreting the results. Each chapter has a section called **What Can Go Wrong?** to help you avoid some of the most common mistakes.

Although we'll show you all the formulas you need to understand the calculations, you will most often use a calculator or computer to perform the mechanics of a Statistics problem. But there are times when you may want to see how it's done by hand too. So we'll show you how.

> **Time out.** From time to time, we'll take time out to discuss an interesting or important side issue. We indicate these by setting them apart like this.[2]

A S **Introduction to (Your Statistics Package).** *ActivStats* launches your statistics package (such as Data Desk) automatically. Try it now.

There are a number of statistics packages available for computers, and they differ widely in the details of how to use them and in how they present their results. But they all work from the same basic information and find the same results. Rather than adopt one package for this book, we present generic output and point out common features that you should look for. The . . . **on the Computer** section of most chapters (just before the exercises) holds this information. We also give a table of instructions to get you started on any of several commonly used packages.

At the end of each chapter, you'll see a brief summary of the important concepts you've covered in a section called **What Have We Learned?** That section includes a list of the **Terms** and a summary of the important **Skills** you've acquired in the chapter. You won't be able to learn the material from these summaries, but you can use them to check your knowledge of the important ideas in the chapter. If you have the skills, know the terms, and understand the concepts, you should be well prepared for the exam—and ready to use Statistics!

You'll find all sorts of stuff in margin notes, such as stories and quotations. For example:

"Computers are useless. They can only give you answers."

–Pablo Picasso

While Picasso underestimated the value of good statistics software, he did know that creating a solution requires more than just *Showing* an answer—it means you have to *Think* and *Tell*, too!

Beware: No one can learn Statistics just by reading or listening. The only way to learn it is to do it. So, of course, at the end of each chapter (except this one) you'll find **Exercises** designed to help you learn to use the Statistics you've just read about.

Some exercises are marked with an orange **T**. You'll find the data for these exercises on the DVD in the back of the book or on the book's website at http://www.pearsonhighered.com/bock.

We've paired up the exercises, putting similar ones together. So, if you're having trouble doing an exercise, you will find a similar one either just before or just after it. You'll find answers to the odd-numbered exercises at the back of the book. But these are only "answers" and not complete "solutions." Huh? What's the difference? The answers are sketches of the complete solutions. For most problems, your solution should follow the model of the Step-by-Step Examples. If your calculations match the numerical parts of the "answer" and your argument contains the elements shown in the answer, you're on the right track. Your complete solution should explain the context, show your reasoning and calculations, and state your conclusions. Don't fret too much if your numbers don't match the printed answers to every decimal place. Statistics is more about getting the reasoning correct—pay more attention to how you interpret a result than what the digit in the third decimal place was.

"Far too many scientists have only a shaky grasp of the statistical techniques they are using. They employ them as an amateur chef employs a cookbook, believing the recipes will work without understanding why. A more cordon bleu attitude . . . might lead to fewer statistical soufflés failing to rise."

–*The Economist*, June 3, 2004, "Sloppy stats shame science"

---

[2] Or in a footnote.

In the real world, problems don't come with chapters attached. So, in addition to the exercises at the ends of chapters, we've also collected a variety of problems at the end of each part of the text to make it more like the real world. This should help you to see whether you can sort out which methods to use when. If you can do that successfully, then you'll know you understand Statistics.

*Optional Sections and Chapters

Some sections and chapters of this book are marked with an asterisk(*). These are optional in the sense that subsequent material does not depend on them directly. We hope you'll read them anyway, as you did this section.

Onward!

It's only fair to warn you: You can't get there by just picking out the highlighted sentences and the summaries. This book is different. It's not about memorizing definitions and learning equations. It's deeper than that. And much more fun. But . . .

You have to read the book![3]

---

[3] So turn the page.

Where are we going?

This is a book about understanding the world by using data. So we'd better start by understanding data. There's more to that than you might have thought.

"Data is king at Amazon. Clickstream and purchase data are the crown jewels at Amazon. They help us build features to personalize the website experience."

–Ronny Kohavi,
Director of Data Mining and
Personalization, Amazon.com

Many years ago, most stores in small towns knew their customers personally. If you walked into the hobby shop, the owner might tell you about a new bridge that had come in for your Lionel train set. The tailor knew your dad's size, and the hairdresser knew how your mom liked her hair. There are still some stores like that around today, but we're increasingly likely to shop at large stores, by phone, or on the Internet. Even so, when you phone an 800 number to buy new running shoes, customer service representatives may call you by your first name or ask about the socks you bought 6 weeks ago. Or the company may send an e-mail in October offering new head warmers for winter running. This company has millions of customers, and you called without identifying yourself. How did the sales rep know who you are, where you live, and what you had bought?

The answer is data. Collecting data on their customers, transactions, and sales lets companies track their inventory and helps them predict what their customers prefer. These data can help them predict what their customers may buy in the future so they know how much of each item to stock. The store can use the data and what it learns from the data to improve customer service, mimicking the kind of personal attention a shopper had 50 years ago.

Amazon.com opened for business in July 1995, billing itself as "Earth's Biggest Bookstore." By 1997, Amazon had a catalog of more than 2.5 million book titles and had sold books to more than 1.5 million customers in 150 countries. In 2007, the company's revenue reached $14.8 billion. Amazon has expanded into selling a wide selection of merchandise, from $400,000 necklaces[1] to yak cheese from Tibet to the largest book in the world.

---

[1] Please get credit card approval before purchasing online.

Amazon is constantly monitoring and evolving its website to serve its customers better and maximize sales performance. To decide which changes to make to the site, the company experiments, collecting data and analyzing what works best. When you visit the Amazon website, you may encounter a different look or different suggestions and offers. Amazon statisticians want to know whether you'll follow the links offered, purchase the items suggested, or even spend a longer time browsing the site. As Ronny Kohavi, former director of Data Mining and Personalization, said, "Data trumps intuition. Instead of using our intuition, we experiment on the live site and let our customers tell us what works for them."

But What *Are* Data?

We bet you thought you knew this instinctively. Think about it for a minute. What exactly *do* we mean by "data"?

Do data have to be numbers? The amount of your last purchase in dollars is numerical data, but some data record names or other labels. The names in Amazon.com's database are data, but not numerical.

Sometimes, data can have values that look like numerical values but are just numerals serving as labels. This can be confusing. For example, the ASIN (Amazon Standard Item Number) of a book, like 0321692551, may have a numerical value, but it's really just another name for *Stats: Data and Models*.

Data values, no matter what kind, are useless without their context. Newspaper journalists know that the lead paragraph of a good story should establish the "Five W's": *Who, What, When, Where*, and (if possible) *Why*. Often we add *How* to the list as well. Answering these questions can provide the **context** for data values. The answers to the first two questions are essential. If you can't answer *Who* and *What*, you don't have **data**, and you don't have any useful information.

THE W'S:
WHO
WHAT
and in what units
WHEN
WHERE
WHY
HOW

Data Tables

Here are some data Amazon might collect:

B0000001OAA	10.99	Chris G.	902	15783947	15.98	Kansas	Illinois	Boston
Canada	Samuel P.	Orange County	N	B000068ZVQ	Bad Blood	Nashville	Katherine H.	N
Mammals	10783489	Ohio	N	Chicago	12837593	11.99	Massachusetts	16.99
312	Monique D.	10675489	413	B00000I5Y6	440	B000002BK9	Let Go	Y

A S **Activity: What Is (Are) Data?** Do you really know what's data and what's just numbers?

Try to guess what they represent. Why is that hard? Because these data have no *context*. If we don't know *Who* they're about or *What* they measure, these values are meaningless. We can make the meaning clear if we organize the values into a **data table** such as this one:

Purchase Order	Name	Ship to State/Country	Price	Area Code	Previous CD Purchase	Gift?	ASIN	Artist
10675489	Katharine H.	Ohio	10.99	440	Nashville	N	B00000I5Y6	Kansas
10783489	Samuel P.	Illinois	16.99	312	Orange County	Y	B000002BK9	Boston
12837593	Chris G.	Massachusetts	15.98	413	Bad Blood	N	B000068ZVQ	Chicago
15783947	Monique D.	Canada	11.99	902	Let Go	N	B0000001OAA	Mammals

Now we can see that these are four purchase records, relating to CD orders from Amazon. The column titles tell *What* has been recorded. The rows tell us *Who*. But be careful. Look at all the variables to see *Who* the variables are about. Even if people are involved, they may not be the *Who* of the data. For example, the *Who* here are the purchase orders (not the people who made the purchases). A common place to find the *Who* of the table is the leftmost column. The other W's might have to come from the company's database administrator.[2]

Who

In general, the rows of a data table correspond to individual **cases** about *Whom* (or about which—if they're not people) we record some characteristics. These cases go by different names, depending on the situation. Individuals who answer a survey are referred to as *respondents*. People on whom we experiment are *subjects* or (in an attempt to acknowledge the importance of their role in the experiment) *participants,* but animals, plants, websites, and other inanimate subjects are often just called *experimental units*. In a database, rows are called *records*—in this example, purchase records. Perhaps the most generic term is **cases.** In the Amazon table, the cases are the individual CD orders.

Sometimes people just refer to data values as *observations*, without being clear about the *Who*. Be sure you know the *Who* of the data, or you may not know what the data say.

Often, the cases are a **sample** of cases selected from some larger **population** that we'd like to understand. Amazon certainly cares about its customers, but also wants to know how to attract all those other Internet users who may never have made a purchase from Amazon's site. To be able to generalize from the sample of cases to the larger population, we'll want the sample to be *representative* of that population—a kind of snapshot image of the larger world.

A S
Activity: **Consider the Context . . .**
Can you tell who's *Who* and what's *What*?
And *Why*? This activity offers real-world
examples to help you practice identifying
the context.

FOR EXAMPLE   Identifying the "Who"

In March 2009, *Consumer Reports* published an evaluation of large-screen, high-definition television sets (HDTVs). The magazine purchased and tested 116 different models from a variety of manufacturers.

QUESTION: Describe the population of interest, the sample, and the *Who* of this study.

The magazine is interested in the performance of all HDTVs currently being offered for sale. It tested a sample of 116 sets, the "Who" for these data. Each HDTV set represents all similar sets offered by that manufacturer.

What and Why

The characteristics recorded about each individual are called **variables.** These are usually shown as the columns of a data table, and they should have a name that identifies *What* has been measured. Variables may seem simple, but to really understand your variables, you must *Think* about what you want to know.

Although area codes are numbers, do we use them that way? Is 610 twice 305? Of course it is, but is that the question? Why would we want to know

---

[2] In database management, this kind of information is called "metadata."

whether Allentown, PA (area code 610), is twice Key West, FL (305)? Variables play different roles, and you can't tell a variable's role just by looking at it.

Some variables just tell us what group or category each individual belongs to. Are you male or female? Pierced or not? . . . What kinds of things can we learn about variables like these? A natural start is to *count* how many cases belong in each category. (Are you listening to music while reading this? We could count the number of students in the class who were and the number who weren't.) We'll look for ways to compare and contrast the sizes of such categories.

Some variables have measurement **units**. Units tell how each value has been measured. But, more importantly, units such as yen, cubits, carats, angstroms, nanoseconds, miles per hour, or degrees Celsius tell us the *scale* of measurement. The units tell us how much of something we have or how far apart two values are. Without units, the values of a measured variable have no meaning. It does little good to be promised a raise of 5000 a year if you don't know whether it will be paid in euros, dollars, yen, or Estonian krooni.

What kinds of things can we learn about measured variables? We can do a lot more than just counting categories. We can look for patterns and trends. (How much did you pay for your last movie ticket? What is the range of ticket prices available in your town? How has the price of a ticket changed over the past 20 years?)

When a variable names categories and answers questions about how cases fall into those categories, we call it a **categorical variable**.[3] When a measured variable with units answers questions about the quantity of what is measured, we call it a **quantitative variable**. These types can help us decide what to do with a variable, but they are really more about what we hope to learn from a variable than about the variable itself. It's the questions we ask a variable (the *Why* of our analysis) that shape how we think about it and how we treat it.

Some variables can answer questions only about categories. If the values of a variable are words rather than numbers, it's a good bet that it is categorical. But some variables can answer both kinds of questions. Amazon could ask for your *Age* in years. That seems quantitative, and would be if the company wanted to know the average age of those customers who visit their site after 3 a.m. But suppose Amazon wants to decide which CD to offer you in a special deal—one by Raffi, Kings of Leon, James Taylor, or Barry Manilow—and needs to be sure to have adequate supplies on hand to meet the demand. Then thinking of your age in one of the categories—child, teen, adult, or senior—might be more useful. If it isn't clear whether a variable is categorical or quantitative, think about *Why* you are looking at it and what you want it to tell you.

A typical course evaluation survey asks, "How valuable do you think this course will be to you?": 1 = Worthless; 2 = Slightly; 3 = Middling; 4 = Reasonably; 5 = Invaluable. Is *Educational Value* categorical or quantitative? Once again, we'll look to the *Why*. A teacher might just count the number of students who gave each response for her course, treating *Educational Value* as a categorical

---

By international agreement, the International System of Units links together all systems of weights and measures. There are seven base units from which all other physical units are derived:

- Distance · · · · · Meter
- Mass · · · · · Kilogram
- Time · · · · · Second
- Electric current · · · · · Ampere
- Temperature · · · · · °Kelvin
- Amount of substance · · · · · Mole
- Intensity of light · · · · · Candela

It is wise to be careful. The *What* and *Why* of area codes are not as simple as they may first  seem. When area codes were first introduced, AT&T was still the source of all telephone equipment, and phones had dials.

To reduce wear and tear on the dials, the area codes with the lowest digits (for which the dial would have to spin least) were assigned to the most populous regions—those with the most phone numbers and thus the area codes most likely to be dialed. New York City was assigned 212, Chicago 312, and Los Angeles 213, but rural upstate New York was given 607, Joliet was 815, and San Diego 619. For that reason, at one time the numerical value of an area code could be used to guess something about the population of its region. Now that phones have push buttons, area codes have finally become just categories.

A S

Activity: Recognize variables measured in a variety of ways. This activity shows examples of the many ways to measure data.

A S

Activities: Variables. Several activities show you how to begin working with data in your statistics package.

---

[3] You may also see it called a *qualitative variable*.

One tradition that hangs on in some quarters is to name variables with cryptic abbreviations written in uppercase letters. This can be traced back to the 1960s, when the very first statistics computer programs were controlled with instructions punched on cards. The earliest punch card equipment used only uppercase letters, and the earliest statistics programs limited variable names to six or eight characters, so variables were called things like PRSRF3. Modern programs do not have such restrictive limits, so there is no reason for variable names that you wouldn't use in an ordinary sentence.

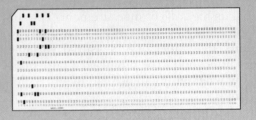

variable. When she wants to see whether the course is improving, she might treat the responses as the *amount* of perceived value—in effect, treating the variable as quantitative. But what are the units? There is certainly an *order* of perceived worth: Higher numbers indicate higher perceived worth. A course that averages 4.5 seems more valuable than one that averages 2, but we should be careful about treating *Educational Value* as purely quantitative. To treat it as quantitative, she'll have to imagine that it has "educational value units" or some similar arbitrary construction. Because there are no natural units, she should be cautious. Variables like this that report order without natural units are often called "ordinal" variables. But saying "that's an ordinal variable" doesn't get you off the hook. You must still look to the *Why* of your study to decide whether to treat it as categorical or quantitative.

FOR EXAMPLE | Identifying "What" and "Why" of HDTVs

RECAP: A *Consumer Reports* article about 116 HDTVs lists each set's manufacturer, cost, screen size, type (LCD or plasma or rear projection), and overall performance score (0–100).

QUESTION: Are these variables categorical or quantitative? Include units where appropriate, and describe the "Why" of this investigation.

The "What" of this article includes the following variables:

- manufacturer (categorical);
- cost (in dollars, quantitative);
- screen size (in inches, quantitative);
- type (categorical);
- performance score (quantitative).

The magazine hopes to help consumers pick a good HDTV set.

Counts Count

In Statistics, we often count things. When Amazon considers a special offer of free shipping to customers, it might first analyze how purchases are shipped. They'd probably start by counting the number of purchases shipped by ground transportation, by second-day air, and by overnight air. Counting is a natural way to summarize the categorical variable *Shipping Method.* So every time we see counts, does that mean the variable is categorical? Actually, no.

We also use counts to measure the amounts of things. How many songs are on your digital music player? How many classes are you taking this semester? To measure these quantities, we'd naturally count. The variables (*Songs, Classes*) would be quantitative, and we'd consider the units to be "number of . . ." or, generically, just "counts" for short.

So we use counts in two different ways. When we count the cases in each category of a categorical variable, the category labels are the *What* and the individuals counted are the *Who* of our data. The counts themselves are not the data, but are something we summarize about the data. Amazon counts the number of purchases in each category of the categorical variable *Shipping Method*. For this purpose (the *Why*), the *What* is shipping method and the *Who* is purchases.

Shipping Method	Number of Purchases
Ground	20,345
Second-day	7,890
Overnight	5,432

Other times our focus is on the amount of something, which we measure by counting. Amazon might record the number of teenage customers visiting their site each month to track customer growth and forecast CD sales (the *Why*). Now the *What* is *Teens,* the *Who* is *Months,* and the units are *Number of Teenage Customers. Teen* was a category when we looked at the categorical variable *Age.* But now it is a quantitative variable in its own right whose amount is measured by counting the number of customers.

Month	Number of Teenage Customers
January	123,456
February	234,567
March	345,678
April	456,789
May	. . .
. . .	. . .

Identifying Identifiers

What's your student ID number? It is numerical, but is it a quantitative variable? No, it doesn't have units. Is it categorical? Yes, but it is a special kind. Look at how many categories there are and at how many individuals are in each. There are as many categories as individuals and only one individual in each category. While it's easy to count the totals for each category, it's not very interesting. Amazon wants to know who you are when you sign in again and doesn't want to confuse you with some other customer. So it assigns you a unique identifier.

Identifier variables themselves don't tell us anything useful about the categories because we know there is exactly one individual in each. However, they are crucial in this age of large data sets. They make it possible to combine data from different sources, to protect confidentiality, and to provide unique labels. The variables *UPS Tracking Number, Social Security Number,* and Amazon's *ASIN* are all examples of identifier variables.

You'll want to recognize when a variable is playing the role of an identifier so you won't be tempted to analyze it. There's probably a list of unique ID numbers for students in a class (so they'll each get their own grade confidentially), but you might worry about the professor who keeps track of the average of these numbers from class to class. Even though this year's average ID number happens to be higher than last's, it doesn't mean that the students are better.

Where, When, and How

We must know *Who*, *What*, and *Why* to analyze data. Without knowing these three, we don't have enough to start. Of course, we'd always like to know more. The more we know about the data, the more we'll understand about the world.

If possible, we'd like to know the **When** and **Where** of data as well. Values recorded in 1803 may mean something different from similar values recorded last year. Values measured in Tanzania may differ in meaning from similar measurements made in Mexico.

How the data are collected can make the difference between insight and nonsense. As we'll see later, data that come from a voluntary survey on the Internet are almost always worthless. One primary concern of Statistics, to be discussed in Part III, is the design of sound methods for collecting data.

Throughout this book, whenever we introduce data, we'll provide a margin note listing the W's (and H) of the data. It's a habit we recommend. The first step of any data analysis is to know why you are examining the data (what you want to know), whom each row of your data table refers to, and what the variables (the columns of the table) record. These are the *Why*, the *Who*, and the *What*. Identifying them is a key part of the *Think* step of any analysis. Make sure you know all three before you proceed to *Show* or *Tell* anything about the data.

JUST CHECKING

In the 2003 Tour de France, Lance Armstrong averaged 40.94 kilometers per hour (km/h) for the entire course, making it the fastest Tour de France in its 100-year history. In 2004, he made history again by winning the race for an unprecedented sixth time. In 2005, he became the only 7-time winner and once again set a new record for the fastest average speed. You can find data on all the Tour de France races on the DVD. Here are the first three and last ten lines of the data set. Keep in mind that the entire data set has nearly 100 entries.

1. List as many of the W's as you can for this data set.

2. Classify each variable as categorical or quantitative; if quantitative, identify the units.

Year	Winner	Country of Origin	Total Time (h/min/s)	Avg. Speed (km/h)	Stages	Total Distance Ridden (km)	Starting Riders	Finishing Riders
1903	Maurice Garin	France	94.33.00	25.3	6	2428	60	21
1904	Henri Cornet	France	96.05.00	24.3	6	2388	88	23
1905	Louis Trousselier	France	112.18.09	27.3	11	2975	60	24
⋮								
1999	Lance Armstrong	USA	91.32.16	40.30	20	3687	180	141
2000	Lance Armstrong	USA	92.33.08	39.56	21	3662	180	128
2001	Lance Armstrong	USA	86.17.28	40.02	20	3453	189	144
2002	Lance Armstrong	USA	82.05.12	39.93	20	3278	189	153
2003	Lance Armstrong	USA	83.41.12	40.94	20	3427	189	147
2004	Lance Armstrong	USA	83.36.02	40.53	20	3391	188	147
2005	Lance Armstrong	USA	86.15.02	41.65	21	3608	189	155
2006	Óscar Periero	Spain	89.40.27	40.78	20	3657	176	139
2007	Alberto Contador	Spain	91.00.26	38.97	20	3547	189	141
2008	Carlos Sastre	Spain	87.52.52	40.50	21	3559	199	145
2009	Alberto Contador	Spain	85.48.35	40.32	21	3460	180	156

There's a world of data on the Internet. These days, one of the richest sources of data is the Internet. With a bit of practice, you can learn to find data on almost any subject. Many of the data sets we use in this book were found in this way. The Internet has both advantages and disadvantages as a source of data. Among the advantages are the fact that often you'll be able to find even more current data than those we present. The disadvantage is that references to Internet addresses can "break" as sites evolve, move, and die.

Our solution to these challenges is to offer the best advice we can to help you search for the data, wherever they may be residing. We usually point you to a website. We'll sometimes suggest search terms and offer other guidance.

Some words of caution, though: Data found on Internet sites may not be formatted in the best way for use in statistics software. Although you may see a data table in standard form, an attempt to copy the data may leave you with a single column of values. You may have to work in your favorite statistics or spreadsheet program to reformat the data into variables. You will also probably want to remove commas from large numbers and such extra symbols as money indicators ($, ¥, £); few statistics packages can handle these.

What Can Go Wrong?

- **Don't label a variable as categorical or quantitative without thinking about the question you want it to answer.** The same variable can sometimes take on different roles.

- **Just because your variable's values are numbers, don't assume that it's quantitative.** Categories are often given numerical labels. Don't let that fool you into thinking they have quantitative meaning. Look at the context.

- **Always be skeptical.** One reason to analyze data is to discover the truth. Even when you are told a context for the data, it may turn out that the truth is a bit (or even a lot) different. The context colors our interpretation of the data, so those who want to influence what you think may slant the context. A survey that seems to be about all students may in fact report just the opinions of those who visited a fan website. The question that respondents answered may have been posed in a way that influenced their responses.

WHAT HAVE WE LEARNED?

We've learned that data are information in a context.

▶ The W's help nail down the context: *Who, What, Why, Where, When,* and *hoW.*

▶ We must know at least the *Who, What,* and *Why* to be able to say anything useful based on the data. The *Who* are the *cases.* The *What* are the *variables.* A variable gives information about each of the cases. The *Why* helps us decide which way to treat the variables.

We treat variables in two basic ways: as *categorical* or *quantitative*.

▶ Categorical variables identify a category for each case. Usually, we think about the counts of cases that fall into each category. (An exception is an identifier variable that just names each case.)

▶ Quantitative variables record measurements or amounts of something; they must have *units*.

▶ Sometimes we treat a variable as categorical or quantitative depending on what we want to learn from it, which means that some variables can't be pigeonholed as one type or the other. That's an early hint that in Statistics we can't always pin things down precisely.

Terms

Context	The context ideally tells *Who* was measured, *What* was measured, *How* the data were collected, *Where* the data were collected, and *When* and *Why* the study was performed (p. 7).
Data	Systematically recorded information, whether numbers or labels, together with its context (p. 7).
Data table	An arrangement of data in which each row represents a case and each column represents a variable (p. 7).
Case	A case is an individual about whom or which we have data (p. 8).
Sample	The cases we actually examine in seeking to understand the much larger population (p. 8).
Population	All the cases we wish we knew about (p. 8).
Variable	A variable holds information about the same characteristic for many cases (p. 8).
Units	A quantity or amount adopted as a standard of measurement, such as dollars, hours, or grams (p. 9).
Categorical variable	A variable that names categories (whether with words or numerals) is called categorical (p. 9).
Quantitative variable	A variable in which the numbers act as numerical values is called quantitative. Quantitative variables always have units (p. 9).
Identifier variable	A variable holding a unique name, ID number, or other identification for a case. Identifiers are particularly useful in matching data from two different databases or relations (p. 11).

Skills

▶ Be able to identify the *Who, What, When, Where, Why,* and *How* of data, or recognize when some of this information has not been provided.

▶ Be able to identify the cases and variables in any data set.

▶ Be able to identify the population from which a sample was chosen.

▶ Be able to classify a variable as categorical or quantitative, depending on its use.

▶ For any quantitative variable, be able to identify the units in which the variable has been measured (or note that they have not been provided).

▶ Be able to describe a variable in terms of its *Who, What, When, Where, Why,* and *How* (and be prepared to remark when that information is not provided).

DATA ON THE COMPUTER

A S *Activity: Examine the Data.* Take a look at your own data from your experiment (p. 11) and get comfortable with your statistics package as you find out about the experiment test results.

Most often we find statistics on a computer using a program, or *package*, designed for that purpose. There are many different statistics packages, but they all do essentially the same things. If you understand what the computer needs to know to do what you want and what it needs to show you in return, you can figure out the specific details of most packages pretty easily.

For example, to get your data into a computer statistics package, you need to tell the computer:

- Where to find the data. This usually means directing the computer to a file stored on your computer's disk or to data on a database. Or it might just mean that you have copied the data from a spreadsheet program or Internet site and it is currently on your computer's clipboard. Usually, the data should be in the form of a data table. Most computer statistics packages prefer the *delimiter* that marks the division between elements of a data table to be a tab character and the delimiter that marks the end of a case to be a return character.

- Where to put the data. (Usually this is handled automatically.)

- What to call the variables. Some data tables have variable names as the first row of the data, and often statistics packages can take the variable names from the first row automatically.

EXERCISES

1. **Voters.** A February 2007 Gallup Poll question asked, "In politics, as of today, do you consider yourself a Republican, a Democrat, or an Independent?" The possible responses were "Democrat", "Republican", "Independent", "Other", and "No Response". What kind of variable is the response?

2. **Mood.** A January 2007 Gallup Poll question asked, "In general, do you think things have gotten better or gotten worse in this country in the last five years?" Possible answers were "Better", "Worse", "No Change", "Don't Know", and "No Response". What kind of variable is the response?

3. **Medicine.** A pharmaceutical company conducts an experiment in which a subject takes 100 mg of a substance orally. The researchers measure how many minutes it takes for half of the substance to exit the bloodstream. What kind of variable is the company studying?

4. **Stress.** A medical researcher measures the increase in heart rate of patients under a stress test. What kind of variable is the researcher studying?

(Exercises 5–12) For each description of data, identify Who and What were investigated and the population of interest.

5. **The news.** Find a newspaper or magazine article in which some data are reported. For the data discussed in the article, answer the same questions as for Exercises 1–4. Include a copy of the article with your report.

6. **The Internet.** Find an Internet source that reports on a study and describes the data. Print out the description and answer the same questions as for Exercises 1–4.

7. **Bicycle safety.** Ian Walker, a psychologist at the University of Bath, wondered whether drivers treat bicycle riders differently when they wear helmets. He rigged his bicycle with an ultrasonic sensor that could measure how close each car was that passed him. He then rode on alternating days with and without a helmet. Out of 2500 cars passing him, he found that when he wore his helmet, motorists passed 3.35 inches closer to him, on average, than when his head was bare. [*NY Times*, Dec. 10, 2006]

8. **Investments.** Some companies offer 401(k) retirement plans to employees, permitting them to shift part of their before-tax salaries into investments such as mutual funds. Employers typically match 50% of the employees' contribution up to about 6% of salary. One company, concerned with what it believed was a low employee participation rate in its 401(k) plan, sampled 30 other companies with similar plans and asked for their 401(k) participation rates.

9. **Honesty.** Coffee stations in offices often just ask users to leave money in a tray to pay for their coffee, but many people cheat. Researchers at Newcastle University alternately taped two posters over the coffee station. During one week, it was a picture of flowers; during the other, it was a pair of staring eyes. They found that the average contribution was significantly higher when the eyes poster was up than when the flowers were there. Apparently, the mere feeling of being watched—even by eyes that were not real—was enough to encourage people to behave more honestly. [*NY Times,* Dec. 10, 2006]

10. **Movies.** Some motion pictures are profitable and others are not. Understandably, the movie industry would like to know what makes a movie successful. Data from 120 first-run movies released in 2005 suggest that longer movies actually make *less* profit.

11. **Fitness.** Are physically fit people less likely to die of cancer? An article in the May 2002 issue of *Medicine and Science in Sports and Exercise* reported results of a study that followed 25,892 men aged 30 to 87 for 10 years. The most physically fit men had a 55% lower risk of death from cancer than the least fit group.

12. **Molten iron.** The Cleveland Casting Plant is a large, highly automated producer of gray and nodular iron automotive castings for Ford Motor Company. The company is interested in keeping the pouring temperature of the molten iron (in degrees Fahrenheit) close to the specified value of 2550 degrees. Cleveland Casting measured the pouring temperature for 10 randomly selected crankshafts.

(Exercises 13–26) For each description of data, identify the W's, name the variables, specify for each variable whether its use indicates that it should be treated as categorical or quantitative, and, for any quantitative variable, identify the units in which it was measured (or note that they were not provided).

13. **Weighing bears.** Because of the difficulty of weighing a bear in the woods, researchers caught and measured 54 bears, recording their weight, neck size, length, and sex. They hoped to find a way to estimate weight from the other, more easily determined quantities.

14. **Schools.** The State Education Department requires local school districts to keep these records on all students: age, race or ethnicity, days absent, current grade level, standardized test scores in reading and mathematics, and any disabilities or special educational needs.

15. **Arby's menu.** A listing posted by the Arby's restaurant chain gives, for each of the sandwiches it sells, the type of meat in the sandwich, the number of calories, and the serving size in ounces. The data might be used to assess the nutritional value of the different sandwiches.

16. **Age and party.** The Gallup Poll conducted a representative telephone survey of 1180 American voters during the first quarter of 2007. Among the reported results were the voter's region (Northeast, South, etc.), age, party affiliation, and whether or not the person had voted in the 2006 midterm congressional election.

17. **Babies.** Medical researchers at a large city hospital investigating the impact of prenatal care on newborn health collected data from 882 births during 1998–2000. They kept track of the mother's age, the number of weeks the pregnancy lasted, the type of birth (cesarean, induced, natural), the level of prenatal care the mother had (none, minimal, adequate), the birth weight and sex of the baby, and whether the baby exhibited health problems (none, minor, major).

18. **Flowers.** In a study appearing in the journal *Science*, a research team reports that plants in southern England are flowering earlier in the spring. Records of the first flowering dates for 385 species over a period of 47 years show that flowering has advanced an average of 15 days per decade, an indication of climate warming, according to the authors.

19. **Herbal medicine.** Scientists at a major pharmaceutical firm conducted an experiment to study the effectiveness of an herbal compound to treat the common cold. They exposed each patient to a cold virus, then gave them either the herbal compound or a sugar solution known to have no effect on colds. Several days later they assessed each patient's condition, using a cold severity scale ranging from 0 to 5. They found no evidence of benefits of the compound.

20. **Vineyards.** Business analysts hoping to provide information helpful to American grape growers compiled these data about vineyards: size (acres), number of years in existence, state, varieties of grapes grown, average case price, gross sales, and percent profit.

21. **Streams.** In performing research for an ecology class, students at a college in upstate New York collect data on streams each year. They record a number of biological, chemical, and physical variables, including the stream name, the substrate of the stream (limestone, shale, or mixed), the acidity of the water (pH), the temperature (°C), and the BCI (a numerical measure of biological diversity).

22. **Fuel economy.** The Environmental Protection Agency (EPA) tracks fuel economy of automobiles based on information from the manufacturers (Ford, Toyota, etc.). Among the data the agency collects are the manufacturer, vehicle type (car, SUV, etc.), weight, horsepower, and gas mileage (mpg) for city and highway driving.

23. **Refrigerators.** In 2006, *Consumer Reports* published an article evaluating refrigerators. It listed 41 models, giving the brand, cost, size (cu ft), type (such as top freezer), estimated annual energy cost, an overall rating (good, excellent, etc.), and the repair history for that brand (percentage requiring repairs over the past 5 years).

24. **Walking in circles.** People who get lost in the desert, mountains, or woods often seem to wander in circles rather than walk in straight lines. To see whether people naturally walk in circles in the absence of visual clues, researcher Andrea Axtell tested 32 people on a football field. One at a time, they stood at the center of one goal line, were blindfolded, and then tried to walk to the other goal line. She recorded each individual's sex, height, handedness, the number of yards each was able to walk before going out of bounds, and whether each wandered off course to the left or the right. No one made it all the way to the far end of the field without crossing one of the sidelines. [*STATS* No. 39, Winter 2004]

T 25. **Horse race 2008.** The Kentucky Derby is a horse race that has been run every year since 1875 at Churchill Downs, Louisville, Kentucky. The race started as a 1.5-mile race, but in 1896, it was shortened to 1.25 miles because experts felt that 3-year-old horses shouldn't run such a long race that early in the season. (It has been run in May every year but one—1901—when it took place on April 29). Here are the data for the first four and several recent races.

Date	Winner	Margin (lengths)	Jockey	Winner's Payoff ($)	Duration (min:sec)	Track Condition
May 17, 1875	Aristides	2	O. Lewis	2850	2:37.75	Fast
May 15, 1876	Vagrant	2	B. Swim	2950	2:38.25	Fast
May 22, 1877	Baden-Baden	2	W. Walker	3300	2:38.00	Fast
May 21, 1878	Day Star	1	J. Carter	4050	2:37.25	Dusty
. . . . . .						
May 1, 2004	Smarty Jones	2 3/4	S. Elliott	854800	2:04.06	Sloppy
May 7, 2005	Giacomo	1/2	M. Smith	5854800	2:02.75	Fast
May 6, 2006	Barbaro	6 1/2	E. Prado	1453200	2:01.36	Fast
May 5, 2007	Street Sense	2 1/4	C. Borel	1450000	2:02.17	Fast
May 3, 2008	Big Brown	4 3/4	K. Desormeaux	1451800	2:01.82	Fast

T 26. **Indy 2009.** The 2.5-mile Indianapolis Motor Speedway has been the home to a race on Memorial Day nearly every year since 1911. Even during the first race, there were controversies. Ralph Mulford was given the checkered flag first but took three extra laps just to make sure he'd completed 500 miles. When he finished, another driver, Ray Harroun, was being presented with the winner's trophy, and Mulford's protests were ignored. Harroun averaged 74.6 mph for the 500 miles. In 2009, the winner, Hélio Castroneves, averaged 150.318 mph.

Here are the data for the first five races and five recent Indianapolis 500 races.

Year	Driver	Time	Speed (mph)
1911	Ray Harroun	0.279260	74.602
1912	Joe Dawson	0.264654	78.719
1913	Jules Goux	0.274365	75.933
1914	René Thomas	0.252605	82.474
2005	Dan Wheldon	0.132188	157.603
2006	Sam Hornish, Jr.	0.132625	157.085
2007	Dario Franchitti	0.113930	151.774
2008	Scott Dixon	0.145112	143.567
2009	Hélio Castroneves	0.138595	150.318

ANSWERS

1. Who—Tour de France races; What—year, winner, country of origin, total time, average speed, stages, total distance ridden, starting riders, finishing riders; How—official statistics at race; Where—France (for the most part); When—1903 to 2009; Why—not specified (To see progress in speeds of cycling racing?)

2.

Variable	Type	Units
Year	Quantitative or Categorical	Years
Winner	Categorical	
Country of Origin	Categorical	
Total Time	Quantitative	Hours/minutes/seconds
Average Speed	Quantitative	Kilometers per hour
Stages	Quantitative	Counts (stages)
Total Distance	Quantitative	Kilometers
Starting Riders	Quantitative	Counts (riders)
Finishing Riders	Quantitative	Counts (riders)

Displaying and Describing Categorical Data

Where are we going?

What is your class: Freshman, Sophomore, Junior, or Senior? What is your blood type? Which candidate do you plan to vote for?

Data are not just numbers—the answers to all these questions are categories. We'll study categorical variables like these, but things get even more exciting when we look at how categorical variables work together. Are men or women more likely to be Democrats? Are people with blue eyes more likely to be left-handed?

You often see categorical data displayed in pie charts or bar charts, or summarized in tables, sometimes in confusing ways. But, with a little skill, it's not hard to do it right.

WHO	People on the *Titanic*
WHAT	Survival status, age, sex, ticket class
WHEN	April 14, 1912
WHERE	North Atlantic
HOW	A variety of sources and Internet sites
WHY	Historical interest

A S *Video:* **The Incident** tells the story of the *Titanic*, and includes rare film footage.

What happened on the *Titanic* at 11:40 on the night of April 14, 1912, is well known. Frederick Fleet's cry of "Iceberg, right ahead" and the three accompanying pulls of the crow's nest bell signaled the beginning of a nightmare that has become legend. By 2:15 a.m., the *Titanic*, thought by many to be unsinkable, had sunk, leaving more than 1500 passengers and crew members on board to meet their icy fate.

Below are some data about the passengers and crew aboard the *Titanic*. Each case (row) of the data table represents a person on board the ship. The variables are the person's *Survival* status (Dead or Alive), *Age* (Adult or Child), *Sex* (Male or Female), and ticket *Class* (First, Second, Third, or Crew).

The problem with a data table like this—and in fact with all data tables—is that you can't *see* what's going on. And seeing is just what we want to do. We need ways to show the data so that we can see patterns, relationships, trends, and exceptions.

Survival	Age	Sex	Class
Dead	Adult	Male	Third
Dead	Adult	Male	Crew
Dead	Adult	Male	Third
Dead	Adult	Male	Crew
Dead	Adult	Male	Crew
Dead	Adult	Male	Crew
Alive	Adult	Female	First
Dead	Adult	Male	Third
Dead	Adult	Male	Crew

TABLE 3.1
Part of a data table showing four variables for nine people aboard the *Titanic*.

The Three Rules of Data Analysis

So, what should we do with data like these? There are three things you should always do first with data:

1. **Make a picture.** A display of your data will reveal things you are not likely to see in a table of numbers and will help you to *Think* clearly about the patterns and relationships that may be hiding in your data.
2. **Make a picture.** A well-designed display will *Show* the important features and patterns in your data. A picture will also show you the things you did not expect to see: the extraordinary (possibly wrong) data values or unexpected patterns.
3. **Make a picture.** The best way to *Tell* others about your data is with a well-chosen picture.

These are the three rules of data analysis. There are pictures of data throughout the book, and new kinds keep showing up. These days, technology makes drawing pictures of data easy, so there is no reason not to follow the three rules.

FIGURE 3.1
A picture to tell a story.

Florence Nightingale (1820–1910), a founder of modern nursing, was also a pioneer in health management, statistics, and epidemiology. She was the first female member of the British Statistical Society and was granted honorary membership in the newly formed American Statistical Association.

To argue forcefully for better hospital conditions for soldiers, she and her colleague, Dr. William Farr, invented this display, which showed that in the Crimean War, far more soldiers died of illness and infection than of battle wounds. Her campaign succeeded in improving hospital conditions and nursing for soldiers.

Florence Nightingale went on to apply statistical methods to a variety of important health issues and published more than 200 books, reports, and pamphlets during her long and illustrious career.

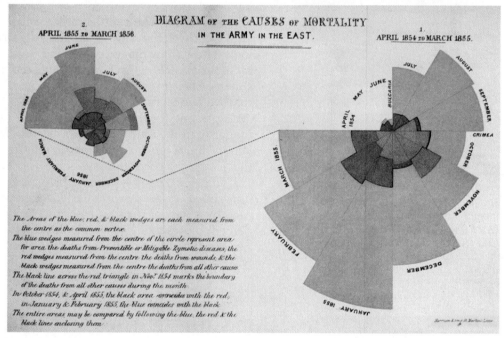

A S
Activity: **Make and examine a table of counts.** Even data on something as simple as hair color can reveal surprises when you organize it in a data table.

TABLE 3.2
A frequency table of the *Titanic* passengers.

Class	Count
First	325
Second	285
Third	706
Crew	885

Frequency Tables: Making Piles

To make a picture of data, the first thing we have to do is to make piles. Making piles is the beginning of understanding about data. We pile together things that seem to go together, so we can see how the cases distribute across different categories. For categorical data, piling is easy. We just count the number of cases corresponding to each category and pile them up.

One way to put all 2201 people on the *Titanic* into piles is by ticket *Class*, counting up how many had each kind of ticket. We can organize these counts into a **frequency table**, which records the totals and the category names.

Even when we have thousands of cases, a variable like ticket *Class*, with only a few categories, has a frequency table that's easy to read. A frequency table with dozens or hundreds of categories would be much harder to read. We use the names of the categories to label each row in the frequency table. For ticket *Class*, these are "First," "Second," "Third," and "Crew."

Class	%
First	14.77
Second	12.95
Third	32.08
Crew	40.21

TABLE 3.3
A relative frequency
table for the same data.

Counts are useful, but sometimes we want to know the fraction or **proportion** of the data in each category, so we divide the counts by the total number of cases. Usually we multiply by 100 to express these proportions as **percentages**. A **relative frequency table** displays the *percentages*, rather than the counts, of the values in each category. Both types of tables show how the cases are distributed across the categories. In this way, they describe the **distribution** of a categorical variable because they name the possible categories and tell how frequently each occurs.

The Area Principle

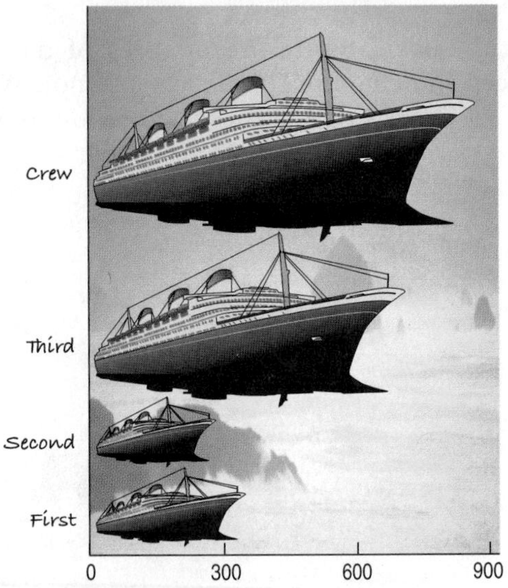

Now that we have the frequency table, we're ready to follow the three rules of data analysis and make a picture of the data. But a bad picture can distort our understanding rather than help it. Here's a graph of the *Titanic* data. What impression do you get about who was aboard the ship?

It sure looks like most of the people on the *Titanic* were crew members, with a few passengers along for the ride. That doesn't seem right. What's wrong? The lengths of the ships *do* match the totals in the table. (You can check the scale at the bottom.) However, experience and psychological tests show that our eyes tend to be more impressed by the *area* than by other aspects of each ship image. So, even though the *length* of each ship matches up with one of the totals, it's the associated *area* in the image that we notice. Since there were about 3 times as many crew as second-class passengers, the ship depicting the number of crew is about 3 times longer than the ship depicting second-class passengers, but it occupies about 9 times the area. As you can see from the frequency table (Table 3.2), that just isn't a correct impression.

The best data displays observe a fundamental principle of graphing data called the **area principle**. The area principle says that the area occupied by a part of the graph should correspond to the magnitude of the value it represents. Violations of the area principle are a common way to lie (or, since most mistakes are unintentional, we should say err) with Statistics.

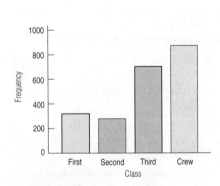

FIGURE 3.2
How many people were in each class on the *Titanic*? From this display, it looks as though the service must have been great, since most aboard were crew members. Although the length of each ship here corresponds to the correct number, the impression is all wrong. In fact, only about 40% were crew.

Bar Charts

Here's a chart that obeys the area principle. It's not as visually entertaining as the ships, but it does give an *accurate* visual impression of the distribution. The height of each bar shows the count for its category. The bars are the same width, so their heights determine their areas, and the areas are proportional to the counts in each class. Now it's easy to see that the majority of people on board were *not* crew, as the ships picture led us to believe. We can also see that there were about 3 times as many crew as second-class passengers. And there were more than twice as many third-class passengers as either first- or second-class passengers, something you may have missed in the frequency table. Bar charts make these kinds of comparisons easy and natural.

A **bar chart** displays the distribution of a categorical variable, showing the counts for each category next to each other for easy comparison. Bar charts should have small spaces between the bars to indicate that these are freestanding bars that could be rearranged into any order. The bars are lined up along a common base.

FIGURE 3.3

People on the *Titanic* by ticket *Class*.

With the area principle satisfied, we can see the true distribution more clearly.

Usually they stick up like this but sometimes they run sideways

like this  .

For some reason, some computer programs give the name "bar chart" to any graph that uses bars. And others use different names according to whether the bars are horizontal or vertical. Don't be misled. "Bar Chart" is the term for a *display of counts of a categorical variable* with bars.

If we really want to draw attention to the relative *proportion* of passengers falling into each of these classes, we could replace the counts with percentages and use a **relative frequency bar chart**.

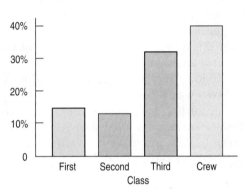

FIGURE 3.4
The relative frequency bar chart looks the same as the bar chart (Figure 3.3) but shows the proportion of people in each category rather than the counts.

Pie Charts

Another common display that shows how a whole group breaks into several categories is a pie chart. **Pie charts** show the whole group of cases as a circle. They slice the circle into pieces whose sizes are proportional to the fraction of the whole in each category.

Pie charts give a quick impression of how a whole group is partitioned into smaller groups. Because we're used to cutting up pies into 2, 4, or 8 pieces, pie charts are good for seeing relative frequencies near 1/2, 1/4, or 1/8. For example, you may be able to tell that the purple slice, representing the second-class passengers, is very close to 1/8 of the total. It's harder to see that there were about twice as many third-class as first-class passengers. Which category had the most passengers? Were there more crew or more third-class passengers? Comparisons such as these are easier in a bar chart.

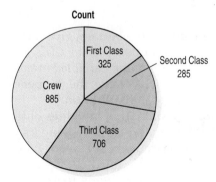

FIGURE 3.5
Number of *Titanic* passengers in each class.

Think before you draw. Our first rule of data analysis is *Make a picture*. But what kind of picture? We don't have a lot of options—yet. There's more to Statistics than pie charts and bar charts, and knowing when to use each type of graph is a critical first step in data analysis. That decision depends in part on what type of data we have.

It's important to check that the data are appropriate for whatever method of analysis you choose. Before you make a bar chart or a pie chart, always check the **Categorical Data Condition:** The data are counts or percentages of individuals in categories.

If you want to make a relative frequency bar chart or a pie chart, you'll need to also make sure that the categories don't overlap so that no individual is counted twice. If the categories do overlap, you can still make a bar chart, but the percentages won't add up to 100%. For the *Titanic* data, either kind of display is appropriate because the categories don't overlap.

Throughout this course, you'll see that doing Statistics right means selecting the proper methods. That means you have to *Think* about the situation at hand. An important first step, then, is to check that the type of analysis you plan is appropriate. The Categorical Data Condition is just the first of many such checks.

Contingency Tables: Children and First-Class Ticket Holders First?

A S *Activity:* **Children at Risk.** This activity looks at the fates of children aboard the *Titanic*; the subsequent activity shows how to make such tables on a computer.

We know how many tickets of each class were sold on the *Titanic*, and we know that only about 32% of all those aboard the *Titanic* survived. After looking at the distribution of each variable by itself, it's natural and more interesting to ask how they relate. Was there a relationship between the kind of ticket a passenger held and the passenger's chances of making it into the lifeboat? To answer this question, we need to look at the two categorical variables *Class* and *Survival* together.

To look at two categorical variables together, we often arrange the counts in a two-way table. Here is a two-way table of those aboard the *Titanic*, classified according to the class of ticket and whether the ticket holder survived or didn't. Because the table shows how the individuals are distributed along each variable, contingent on the value of the other variable, such a table is called a **contingency table**.

TABLE 3.4
Contingency table of ticket *Class* and *Survival*.
The bottom line of "Totals" is the same as the previous frequency table.

		Class				
		First	Second	Third	Crew	Total
Survival	Alive	203	118	178	212	711
	Dead	122	167	528	673	1490
	Total	325	285	706	885	2201

The margins of the table, both on the right and at the bottom, give totals. The bottom line of the table is just the frequency distribution of ticket *Class*. The right column of the table is the frequency distribution of the variable *Survival*. When presented like this, in the margins of a contingency table, the frequency distribution of one of the variables is called its **marginal distribution**.

Each **cell** of the table gives the count for a combination of values of the two variables. If you look down the column for second-class passengers to the first cell, you can see that 118 second-class passengers survived. Looking at the third-class passengers, you can see that more third-class passengers (178) survived. Were second-class passengers more likely to survive? Questions like this are easier to address by using percentages. The 118 survivors in second class were 41.4% of the total 285 second-class passengers, while the 178 surviving third-class passengers were only 25.2% of that class's total.

We know that 118 second-class passengers survived. We could display this number as a percentage—but as a percentage of what? The total number of passengers? (118 is 5.4% of the total: 2201.) The number of second-class passengers? (118 is 41.4% of the 285 second-class passengers.) The number of survivors? (118 is 16.6% of the 711 survivors.) All of these are possibilities, and all are potentially useful or interesting. You'll probably wind up calculating (or letting your technology calculate) lots of percentages. Most statistics programs

A bell-shaped artifact from the Titanic.

offer a choice of total percent, row percent, or column percent for contingency tables. Unfortunately, they often put them all together with several numbers in each cell of the table. The resulting table holds lots of information, but it can be hard to understand:

TABLE 3.5

Another contingency table of ticket _Class_. This time we see not only the counts for each combination of _Class_ and _Survival_ (in bold) but the percentages these counts represent. For each count, there are three choices for the percentage: by row, by column, and by table total. There's probably too much information here for this table to be useful.

			Class				
			First	Second	Third	Crew	Total
Survival	Alive	Count	**203**	**118**	**178**	**212**	**711**
		% of Row	28.6%	16.6%	25.0%	29.8%	100%
		% of Column	62.5%	41.4%	25.2%	24.0%	32.3%
		% of Table	9.2%	5.4%	8.1%	9.6%	32.3%
	Dead	Count	**122**	**167**	**528**	**673**	**1490**
		% of Row	8.2%	11.2%	35.4%	45.2%	100%
		% of Column	37.5%	58.6%	74.8%	76.0%	67.7%
		% of Table	5.5%	7.6%	24.0%	30.6%	67.7%
	Total	Count	**325**	**285**	**706**	**885**	**2201**
		%of Row	14.8%	12.9%	32.1%	40.2%	100%
		% of Column	100%	100%	100%	100%	100%
		% of Table	14.8%	12.9%	32.1%	40.2%	100%

To simplify the table, let's first pull out the percent of table values:

TABLE 3.6

A contingency table of _Class_ by _Survival_ with only the table percentages.

		Class				
		First	Second	Third	Crew	Total
Survival	Alive	9.2%	5.4%	8.1%	9.6%	32.3%
	Dead	5.5%	7.6%	24.0%	30.6%	67.7%
	Total	14.8%	12.9%	32.1%	40.2%	100%

These percentages tell us what percent of _all_ passengers belong to each combination of column and row category. For example, we see that although 8.1% of the people aboard the _Titanic_ were surviving third-class ticket holders, only 5.4% were surviving second-class ticket holders. Is this fact useful? Comparing these percentages, you might think that the chances of surviving were better in third class than in second. But be careful. There were many more third-class than second-class passengers on the _Titanic,_ so there were more third-class survivors. That group is a larger percentage of the passengers, but is that really what we want to know?

Percent of what?   The English language can be tricky when we talk about percentages. If you're asked "What percent _of the survivors_ were in second class?" it's pretty clear that we're interested only in survivors. It's as if we're restricting the _Who_ in the question to the survivors, so we should look at the number of second-class passengers among all the survivors—in other words, the row percent.

But if you're asked "What percent were second-class passengers who survived?" you have a different question. Be careful; here, the _Who_ is everyone on board, so 2201 should be the denominator, and the answer is the table percent.

And if you're asked "What percent of the second-class passengers survived?" you have a third question. Now the *Who* is the second-class passengers, so the denominator is the 285 second-class passengers, and the answer is the column percent.

Always be sure to ask "percent of what?" That will help you to know the *Who* and whether we want *row, column,* or *table* percentages.

FOR EXAMPLE   **Finding Marginal Distributions**

In January 2007, a Gallup poll asked 1008 Americans age 18 and over whether they planned to watch the upcoming Super Bowl. The pollster also asked those who planned to watch whether they were looking forward more to seeing the football game or the commercials. The results are summarized in the table:

QUESTION: What's the marginal distribution of the responses?

To determine the percentages for the three responses, divide the count for each response by the total number of people polled:

		Sex		
		Male	Female	Total
Response	Game	279	200	**479**
	Commercial	81	156	**237**
	Won't Watch	132	160	**292**
	Total	**492**	**516**	**1008**

$$\frac{479}{1008} = 47.5\% \qquad \frac{237}{1008} = 23.5\% \qquad \frac{292}{1008} = 29.0\%$$

According to the poll, 47.5% of American adults were looking forward to watching the Super Bowl game, 23.5% were looking forward to watching the commercials, and 29% didn't plan to watch at all.

Conditional Distributions

The more interesting questions are *contingent*. We'd like to know, for example, what percentage of *second-class passengers* survived and how that compares with the survival rate for third-class passengers.

It's more interesting to ask whether the chance of surviving the *Titanic* sinking *depended* on ticket class. We can look at this question in two ways. First, we could ask how the distribution of ticket *Class* changes between survivors and nonsurvivors. To do that, we look at the *row percentages:*

TABLE 3.7
The conditional distribution of ticket *Class* conditioned on each value of *Survival: Alive* and *Dead.*

		Class				
		First	Second	Third	Crew	Total
Survival	**Alive**	**203**	**118**	**178**	**212**	**711**
		28.6%	16.6%	25.0%	29.8%	**100%**
	Dead	**122**	**167**	**528**	**673**	**1490**
		8.2%	11.2%	35.4%	45.2%	**100%**

By focusing on each row separately, we see the distribution of class under the *condition* of surviving or not. The sum of the percentages in each row is 100%, and we divide that up by ticket class. In effect, we temporarily restrict the *Who* first to survivors and make a pie chart for them. Then we refocus the *Who* on the nonsurvivors and make their pie chart. These pie charts show the distribution of ticket classes *for each row* of the table: survivors and nonsurvivors. The distributions we create this way are called **conditional distributions**, because they show the distribution of one variable for just those cases that satisfy a condition on another variable.

FIGURE 3.6
Pie charts of the conditional distributions of ticket *Class* for the survivors and nonsurvivors, **separately.** Do the distributions appear to be the same? We're primarily concerned with percentages here, so pie charts are a reasonable choice.

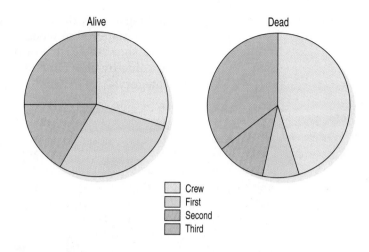

Alive   Dead

☐ Crew
☐ First
☐ Second
☐ Third

FOR EXAMPLE | **Finding Conditional Distributions**

RECAP: The table shows results of a poll asking adults whether they were looking forward to the Super Bowl game, looking forward to the commercials, or didn't plan to watch.

QUESTION: How do the conditional distributions of interest in the commercials differ for men and women?

Look at the group of people who responded "Commercials" and determine what percent of them were male and female:

		Sex	
		Male	Female	Total
Response	Game	279	200	**479**
	Commercial	81	156	**237**
	Won't Watch	132	160	**292**
	Total	**492**	**516**	**1008**

$$\frac{81}{237} = 34.2\% \quad \frac{156}{237} = 65.8\%$$

Women make up a sizable majority of the adult Americans who look forward to seeing Super Bowl commercials more than the game itself. Nearly 66% of people who voiced a preference for the commercials were women, and only 34% were men.

But we can also turn the question around. We can look at the distribution of *Survival* for each category of ticket *Class*. To do this, we look at the *column percentages*. Those show us whether the chance of surviving was roughly the same *for each of the four classes*. Now the percentages in each column add to 100%, because we've restricted the *Who*, in turn, to each of the four ticket classes:

TABLE 3.8
A contingency table of *Class* by *Survival* with only counts and column percentages. Each column represents the conditional distribution of *Survival* for a given category of ticket *Class*.

			Class				
			First	Second	Third	Crew	Total
Survival	Alive	Count	**203**	**118**	**178**	**212**	**711**
		% of Column	62.5%	41.4%	25.2%	24.0%	**32.3%**
	Dead	Count	**122**	**167**	**528**	**673**	**1490**
		% of Column	37.5%	58.6%	74.8%	76.0%	**67.7%**
	Total	Count	**325**	**285**	**706**	**885**	**2201**
			100%	100%	100%	100%	100%

Looking at how the percentages change across each row, it sure looks like ticket class mattered in whether a passenger survived. To make it more vivid, we could show the distribution of *Survival* for each ticket class in a display. Here's a side-by-side bar chart showing percentages of surviving and not for each category:

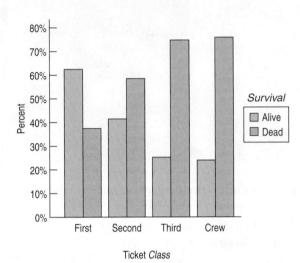

FIGURE 3.7
Side-by-side bar chart showing the conditional distribution of *Survival* for each category of ticket *Class*. The corresponding pie charts would have only two categories in each of four pies, so bar charts seem the better alternative.

These bar charts are simple because, for the variable *Survival*, we have only two alternatives: Alive and Dead. When we have only two categories, we really need to know only the percentage of one of them. Knowing the percentage that survived tells us the percentage that died. We can use this fact to simplify the display even more by dropping one category. Here are the percentages of dying *across the classes* displayed in one chart:

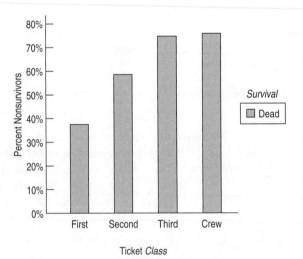

FIGURE 3.8
Bar chart showing just nonsurvivor percentages for each value of ticket *Class*. Because we have only two values, the second bar doesn't add any information. Compare this chart to the side-by-side bar chart shown earlier.

Now it's easy to compare the risks. Among first-class passengers, 37.5% perished, compared to 58.6% for second-class ticket holders, 74.8% for those in third class, and 76.0% for crew members.

If the risk had been about the same across the ticket classes, we would have said that survival was *independent* of class. But it's not. The differences we see among these conditional distributions suggest that survival may have depended on ticket class. You may find it useful to consider conditioning on each variable in a contingency table in order to explore the dependence between them.

It is interesting to know that *Class* and *Survival* are associated. That's an important part of the *Titanic* story. And we know how important this is because the margins show us the actual numbers of people involved.

Variables can be associated in many ways and to different degrees. The best way to tell whether two variables are associated is to ask whether they are *not*.[1] In a contingency table, when the distribution of *one* variable is the same for all categories of another, we say that the variables are **independent**. That tells us there's no association between these variables. We'll see a way to check for independence formally later in the book. For now, we'll just compare the distributions.

FOR EXAMPLE | **Looking for Associations between Variables**

RECAP: The table shows results of a poll asking adults whether they were looking forward to the Super Bowl game, looking forward to the commercials, or didn't plan to watch.

QUESTION: Does it seem that there's an association between interest in Super Bowl TV coverage and a person's sex?

	Male	Female	Total
Game	279	200	**479**
Commercials	81	156	**237**
Won't watch	132	160	**292**
Total	**492**	**516**	**1008**

First find the distribution of the three responses for the men (the column percentages):

$$\frac{279}{492} = 56.7\% \quad \frac{81}{492} = 16.5\% \quad \frac{132}{492} = 26.8\%$$

Then do the same for the women who were polled, and display the two distributions with a side-by-side bar chart:

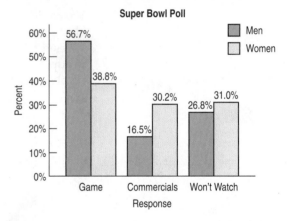

Based on this poll it appears that women were only slightly less interested than men in watching the Super Bowl telecast: 31% of the women said they didn't plan to watch, compared to just under 27% of men. Among those who planned to watch, however, there appears to be an association between the viewer's sex and what the viewer is most looking forward to. While more women are interested in the game (38.8%) than the commercials (30.2%), the margin among men is much wider: 56.7% of men said they were looking forward to seeing the game, compared to only 16.5% who cited the commercials.

---

[1]This kind of "backwards" reasoning shows up surprisingly often in science—and in Statistics. We'll see it again.

JUST CHECKING

A Statistics class reports the following data on Sex and Eye Color for students in the class:

		Eye Color			
		Blue	Brown	Green/Hazel/Other	Total
Sex	Males	6	20	6	32
	Females	4	16	12	32
	Total	10	36	18	64

1. What percent of females are brown-eyed?

2. What percent of brown-eyed students are female?

3. What percent of students are brown-eyed females?

4. What's the distribution of Eye Color?

5. What's the conditional distribution of Eye Color for the males?

6. Compare the percent who are female among the blue-eyed students to the percent of all students who are female.

7. Does it seem that Eye Color and Sex are independent? Explain.

Segmented Bar Charts

We could display the *Titanic* information by dividing up bars rather than circles. The resulting **segmented bar chart** treats each bar as the "whole" and divides it proportionally into segments corresponding to the percentage in each group. We can clearly see that the distributions of ticket *Class* are different, indicating again that survival was not independent of ticket *Class*.

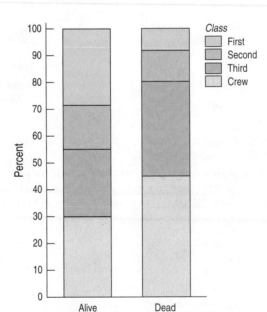

FIGURE 3.9

A segmented bar chart for *Class by Survival*.

Notice that although the totals for survivors and nonsurvivors are quite different, the bars are the same height because we have converted the numbers to *percentages*. Compare this display with the side-by-side pie charts of the same data in Figure 3.6.

STEP-BY-STEP EXAMPLE | Examining Contingency Tables

Medical researchers followed 6272 Swedish men for 30 years to see if there was any association between the amount of fish in their diet and prostate cancer ("Fatty Fish Consumption and Risk of Prostate Cancer," *Lancet*, June 2001). Their results are summarized in this table:

	Prostate Cancer	
Fish Consumption	No	Yes
Never/seldom	110	14
Small part of diet	2420	201
Moderate part	2769	209
Large part	507	42

TABLE 3.9

We asked for a picture of a man eating fish. This is what we got.

Question: Is there an association between fish consumption and prostate cancer?

THINK

Plan Be sure to state what the problem is about.

I want to know if there is an association between fish consumption and prostate cancer.

Variables Identify the variables and report the W's.

The individuals are 6272 Swedish men followed by medical researchers for 30 years. The variables record their fish consumption and whether or not they were diagnosed with prostate cancer.

Be sure to check the appropriate condition.

✓ **Categorical Data Condition:** I have counts for both fish consumption and cancer diagnosis. The categories of diet do not overlap, and the diagnoses do not overlap. It's okay to draw pie charts or bar charts.

SHOW

Mechanics It's a good idea to check the marginal distributions first before looking at the two variables together.

	Prostate Cancer		
Fish Consumption	No	Yes	Total
Never/seldom	110	14	**124 (2.0%)**
Small part of diet	2420	201	**2621 (41.8%)**
Moderate part	2769	209	**2978 (47.5%)**
Large part	507	42	**549 (8.8%)**
Total	**5806 (92.6%)**	**466 (7.4%)**	**6272 (100%)**

Two categories of the diet are quite small, with only 2.0% Never/seldom eating fish and 8.8% in the "Large part" category. Overall, 7.4% of the men in this study had prostate cancer.

Then, make appropriate displays to see whether there is a difference in the relative proportions. These pie charts compare fish consumption for men who have prostate cancer to fish consumption for men who don't.

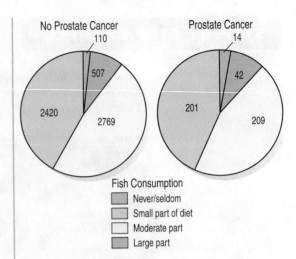

It's hard to see much difference in the pie charts. So, I made a display of the row percentages. Because there are only two alternatives, I chose to display the risk of prostate cancer for each group:

Both pie charts and bar charts can be used to compare conditional distributions. Here we compare prostate cancer rates based on differences in fish consumption.

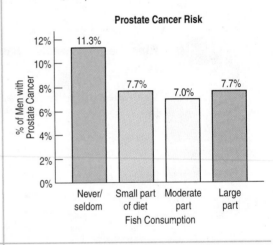

Conclusion Interpret the patterns in the table and displays in context. If you can, discuss possible real-world consequences. Be careful not to overstate what you see. The results may not generalize to other situations.

Overall, there is a 7.4% rate of prostate cancer among men in this study. Most of the men (89.3%) ate fish either as a moderate or small part of their diet. From the pie charts, it's hard to see a difference in cancer rates among the groups. But in the bar chart, it looks like the cancer rate for those who never/seldom ate fish may be somewhat higher.

However, only 124 of the 6272 men in the study fell into this category, and only 14 of them developed prostate cancer. More study would probably be needed before we would recommend that men change their diets.[2]

---

[2] The original study actually used pairs of twins, which enabled the researchers to discern that the risk of cancer for those who never ate fish actually *was* substantially greater. Using pairs is a special way of gathering data. We'll discuss such study design issues and how to analyze the data in the later chapters.

This study is an example of looking at a sample of data to learn something about a larger population. We care about more than these particular 6272 Swedish men. We hope that learning about their experiences will tell us something about the value of eating fish in general. That raises the interesting question of what population we think this sample might represent. Do we hope to learn about all Swedish men? About all men? About the value of eating fish for all adult humans?[3] Often, it can be hard to decide just which population our findings may tell us about, but that also is how researchers decide what to look into in future studies.

What Can Go Wrong?

- **Don't violate the area principle.** This is probably the most common mistake in a graphical display. It is often made in the cause of artistic presentation. Here, for example, are two displays of the pie chart of the *Titanic* passengers by class:

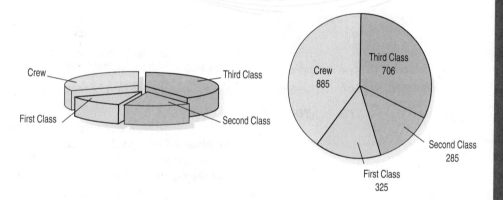

The one on the left looks pretty, doesn't it? But showing the pie on a slant violates the area principle and makes it much more difficult to compare fractions of the whole made up of each class—the principal feature that a pie chart ought to show.

- **Keep it honest.** Here's a pie chart that displays data on the percentage of high school students who engage in specified dangerous behaviors as reported by the Centers for Disease Control and Prevention. What's wrong with this plot?

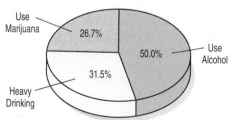

Try adding up the percentages. Or look at the 50% slice. Does it look right? Then think: What are these percentages of? Is there a "whole" that has been sliced up? In a pie chart, the proportions shown by each slice of the pie must add up to 100% and each individual must fall into only one category. Of course, showing the pie on a slant makes it even harder to detect the error.

(continued)

---
[3] Probably not, since we're looking only at prostate cancer risk.

Here's another. This bar chart shows the number of airline passengers searched in security screening, by year:

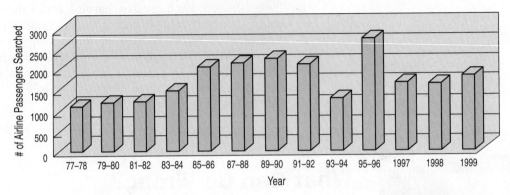

Looks like things didn't change much in the final years of the 20th century—until you read the bar labels and see that the last three bars represent single years while all the others are for *pairs* of years. Of course, the false depth makes it harder to see the problem.

■ **Don't confuse similar-sounding percentages.** These percentages sound similar but are different:

		Class			
	First	**Second**	**Third**	**Crew**	**Total**
Alive	203	118	178	212	**711**
Dead	122	167	528	673	**1490**
Total	**325**	**285**	**706**	**885**	**2201**

(Survival)

- ■ The percentage of the passengers who were both in first class and survived: This would be 203/2201, or 9.2%.
- ■ The percentage of the first-class passengers who survived: This is 203/325, or 62.5%.
- ■ The percentage of the survivors who were in first class: This is 203/711, or 28.6%.

In each instance, pay attention to the *Who* implicitly defined by the phrase. Often there is a restriction to a smaller group (all aboard the *Titanic*, those in first class, and those who survived, respectively) before a percentage is found. Your discussion of results must make these differences clear.

■ **Don't forget to look at the variables separately, too.** When you make a contingency table or display a conditional distribution, be sure you also examine the marginal distributions. It's important to know how many cases are in each category.

■ **Be sure to use enough individuals.** When you consider percentages, take care that they are based on a large enough number of individuals. Take care not to make a report such as this one:

We found that 66.67% of the rats improved their performance with training. The other rat died.

■ **Don't overstate your case.** Independence is an important concept, but it is rare for two variables to be *entirely* independent. We can't conclude that one variable has no effect whatsoever on another. Usually, all we know is that little effect was observed in our study. Other studies of other groups under other circumstances could find different results.

Simpson's Paradox

■ **Don't use unfair or silly averages.** Sometimes averages can be misleading. Sometimes they just don't make sense at all. Be careful when averaging different variables that the quantities you're averaging are comparable. The Centerville sign says it all.

When using averages of proportions across several different groups, it's important to make sure that the groups really are comparable.

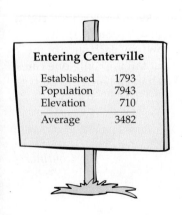

Entering Centerville

Established	1793
Population	7943
Elevation	710
Average	3482

It's easy to make up an example showing that averaging across very different values or groups can give absurd results. Here's how that might work: Suppose there are two pilots, Moe and Jill. Moe argues that he's the better pilot of the two, since he managed to land 83% of his last 120 flights on time compared with Jill's 78%. But let's look at the data a little more closely. Here are the results for each of their last 120 flights, broken down by the time of day they flew:

TABLE 3.10

On-time flights by *Time of Day* and *Pilot*. Look at the percentages within each *Time of Day* category. Who has a better on-time record during the day? At night? Who is better overall?

		Time of Day		
		Day	Night	Overall
Pilot	Moe	90 out of 100 90%	10 out of 20 50%	100 out of 120 83%
	Jill	19 out of 20 95%	75 out of 100 75%	94 out of 120 78%

One famous example of Simpson's paradox arose during an investigation of admission rates for men and women at the University of California at Berkeley's graduate schools. As reported in an article in *Science*, about 45% of male applicants were admitted, but only about 30% of female applicants got in. It looked like a clear case of discrimination. However, when the data were broken down by school (Engineering, Law, Medicine, etc.), it turned out that, within each school, the women were admitted at nearly the same or, in some cases, much *higher* rates than the men. How could this be? Women applied in large numbers to schools with very low admission rates (Law and Medicine, for example, admitted fewer than 10%). Men tended to apply to Engineering and Science. Those schools have admission rates above 50%. When the *average* was taken, the women had a much lower *overall* rate, but the average didn't really make sense.

Look at the daytime and nighttime flights separately. For day flights, Jill had a 95% on-time rate and Moe only a 90% rate. At night, Jill was on time 75% of the time and Moe only 50%. So Moe is better "overall," but Jill is better both during the day and at night. How can this be?

What's going on here is a problem known as **Simpson's paradox**, named for the statistician who described it in the 1960s. It comes up rarely in real life, but there have been several well-publicized cases. As we can see from the pilot example, the problem is *unfair averaging* over different groups. Jill has mostly night flights, which are more difficult, so her *overall average* is heavily influenced by her nighttime average. Moe, on the other hand, benefits from flying mostly during the day, with its high on-time percentage. With their very different patterns of flying conditions, taking an overall average is misleading. It's not a fair comparison.

The moral of Simpson's paradox is to be careful when you average across different levels of a second variable. It's always better to compare percentages or other averages *within* each level of the other variable. The overall average may be misleading.

CONNECTIONS

All of the methods of this chapter work with *categorical variables.* You must know the *Who* of the data to know who is counted in each category and the *What* of the variable to know where the categories come from.

WHAT HAVE WE LEARNED?

We've learned that we can summarize categorical data by counting the number of cases in each category, sometimes expressing the resulting distribution as percents. We can display the distribution in a bar chart or a pie chart. When we want to see how two categorical variables are related, we put the counts (and/or percentages) in a two-way table called a contingency table.

▶ We look at the marginal distribution of each variable (found in the margins of the table).

▶ We also look at the conditional distribution of a variable within each category of the other variable.

▶ We can display these conditional and marginal distributions by using bar charts or pie charts.

▶ If the conditional distributions of one variable are (roughly) the same for every category of the other, the variables are independent.

Terms

Frequency table (Relative frequency table)	A frequency table lists the categories in a categorical variable and gives the count (or percentage) of observations for each category (p. 19).
Distribution	The distribution of a variable gives ▶ the possible values of the variable and ▶ the relative frequency of each value (p. 20).
Area principle	In a statistical display, each data value should be represented by the same amount of area (p. 20).
Bar chart (Relative frequency bar chart)	Bar charts show a bar whose area represents the count (or percentage) of observations for each category of a categorical variable (p. 20).
Pie chart	Pie charts show how a "whole" divides into categories by showing a wedge of a circle whose area corresponds to the proportion in each category (p. 21).
Categorical data condition	The methods in this chapter are appropriate for displaying and describing categorical data. Be careful not to use them with quantitative data (p. 21).
Contingency table	A contingency table displays counts and, sometimes, percentages of individuals falling into named categories on two or more variables. The table categorizes the individuals on all variables at once to reveal possible patterns in one variable that may be contingent on the category of the other (p. 22).
Marginal distribution	In a contingency table, the distribution of either variable alone is called the marginal distribution. The counts or percentages are the totals found in the margins (last row or column) of the table (p. 22).
Conditional distribution	The distribution of a variable restricting the *Who* to consider only a smaller group of individuals is called a conditional distribution (p. 24).
Independence	Variables are said to be independent if the conditional distribution of one variable is the same for each category of the other. We'll show how to check for independence in a later chapter (p. 27).
Segmented bar chart	A segmented bar chart displays the conditional distribution of a categorical variable within each category of another variable (p. 28).
Simpson's paradox	When averages are taken across different groups, they can appear to contradict the overall averages. This is known as "Simpson's paradox" (p. 33).

Skills

THINK

▶ Be able to recognize when a variable is categorical and choose an appropriate display for it.

▶ Understand how to examine the association between categorical variables by comparing conditional and marginal percentages.

SHOW

▶ Be able to summarize the distribution of a categorical variable with a frequency table.

▶ Be able to display the distribution of a categorical variable with a bar chart or pie chart.

▶ Know how to make and examine a contingency table.

▶ Know how to make and examine displays of the conditional distributions of one variable for two or more groups.

▶ Be able to describe the distribution of a categorical variable in terms of its possible values and relative frequencies.

▶ Know how to describe any anomalies or extraordinary features revealed by the display of a variable.

▶ Be able to describe and discuss patterns found in a contingency table and associated displays of conditional distributions.

DISPLAYING CATEGORICAL DATA ON THE COMPUTER

Although every package makes a slightly different bar chart, they all have similar features:

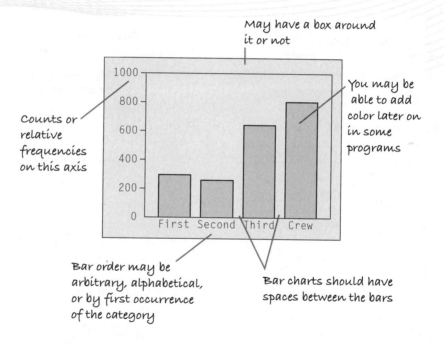

Sometimes the count or a percentage is printed above or on top of each bar to give some additional information. You may find that your statistics package sorts category names in annoying orders by default. For example, many packages sort categories alphabetically or by the order the categories are seen in the data set. Often, neither of these is the best choice.

DATA DESK

To make a bar chart or pie chart, select the variable. In the **Plot** menu, choose **Bar Chart** or **Pie Chart.** To make a frequency table, in the **Calc** menu choose **Frequency Table**. Data Desk cannot make stacked bar charts.

COMMENTS

These commands treat data as categorical even if they are numerals. If you select a quantitative variable by mistake, you'll see an error message warning of too many categories. The **Replicate Y by X** command can generate variables from summary counts with values for each case, and thus suitable for these commands.

EXCEL

First make a pivot table (Excel's name for a frequency table). From the **Data** menu, choose **Pivot Table** and **Pivot Chart Report**.

When you reach the Layout window, drag your variable to the row area and drag your variable again to the data area. This tells Excel to count the occurrences of each category. Once you have an Excel pivot table, you can construct bar charts and pie charts.

Click inside the Pivot Table.

Click the Pivot Table Chart Wizard button. Excel creates a bar chart.

A longer path leads to a pie chart; see your Excel documentation.

COMMENTS

Excel uses the pivot table to specify the category names and find counts within each category. If you already have that information, you can proceed directly to the Chart Wizard.

EXCEL 2007

To make a bar chart:

- Select the variable in Excel you want to work with.
- Choose the **Column** command from the Insert tab in the Ribbon.
- Select the appropriate chart from the drop-down dialog.

To change the bar chart into a pie chart:

- Right-click the chart and select **Change Chart Type...** from the menu. The Chart type dialog opens.
- Select a pie chart type.
- Click the **OK** button. Excel changes your bar chart into a pie chart.

JMP

JMP makes a bar chart and frequency table together. From the **Analyze** menu, choose **Distribution**.

In the Distribution dialog, drag the name of the variable into the empty variable window beside the label "Y, Columns"; click **OK**.

To make a pie chart, choose **Chart** from the **Graph** menu. In the Chart dialog, select the variable name from

the Columns list, click on the button labeled "Statistics," and select "N" from the drop-down menu.

Click the "**Categories, X, Levels**" button to assign the same variable name to the x-axis.

Under Options, click on the **second** button—labeled "**Bar Chart**"—and select "Pie" from the drop-down menu.

MINITAB

To make a bar chart, choose **Bar Chart** from the **Graph** menu. Select "Counts of unique values" in the first menu, and select "Simple" for the type of graph. Click **OK**.

In the Chart dialog, enter the name of the variable that you wish to display in the box labeled "Categorical variables." Click **OK**.

SPSS

To make a bar chart, open the **Chart Builder** from the **Graphs** menu.

Click the **Gallery** tab.

Choose **Bar Chart** from the list of chart types.

Drag the appropriate bar chart onto the canvas.

Drag a categorical variable onto the x-axis drop zone.

Click **OK**.

COMMENT

A similar path makes a pie chart by choosing **Pie chart** from the list of chart types.

TI-83/84 PLUS

The TI-83 won't do displays for categorical variables.

TI-89

The TI-89 won't do displays for categorical variables.

EXERCISES

1. **Graphs in the news.** Find a bar chart of categorical data from a newspaper, a magazine, or the Internet.
   a) Is the graph clearly labeled?
   b) Does it violate the area principle?
   c) Does the accompanying article tell the W's of the variable?
   d) Do you think the article correctly interprets the data? Explain.

2. **Graphs in the news II.** Find a pie chart of categorical data from a newspaper, a magazine, or the Internet.
   a) Is the graph clearly labeled?
   b) Does it violate the area principle?
   c) Does the accompanying article tell the W's of the variable?
   d) Do you think the article correctly interprets the data? Explain.

3. **Tables in the news.** Find a frequency table of categorical data from a newspaper, a magazine, or the Internet.
   a) Is it clearly labeled?
   b) Does it display percentages or counts?
   c) Does the accompanying article tell the W's of the variable?
   d) Do you think the article correctly interprets the data? Explain.

4. **Tables in the news II.** Find a contingency table of categorical data from a newspaper, a magazine, or the Internet.
   a) Is it clearly labeled?
   b) Does it display percentages or counts?
   c) Does the accompanying article tell the W's of the variables?
   d) Do you think the article correctly interprets the data? Explain.

5. **Movie genres.** The pie chart summarizes the genres of 120 first-run movies released in 2005.
   a) Is this an appropriate display for the genres? Why/why not?
   b) Which genre was least common?

Genre2
- Action/Adventure
- Comedy
- Drama
- Thriller/Horror

6. **Movie ratings.** The pie chart shows the ratings assigned to 120 first-run movies released in 2005.
   a) Is this an appropriate display for these data? Explain.
   b) Which was the most common rating?

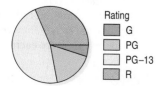

Rating
- G
- PG
- PG–13
- R

7. **Genres again.** Here is a bar chart summarizing the 2005 movie genres, as seen in the pie chart in Exercise 5.
   a) Which genre was most common?
   b) Is it easier to see that in the pie chart or the bar chart? Explain.

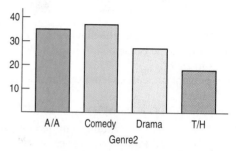

8. **Ratings again.** Here is a bar chart summarizing the 2005 movie ratings, as seen in the pie chart in Exercise 6.
   a) Which was the least common rating?
   b) An editorial claimed that there's been a growth in PG-13 rated films that, according to the writer, "have too much sex and violence," at the expense of G-rated films that offer "good, clean fun." The writer offered the bar chart below as evidence to support his claim. Does the bar chart support his claim? Explain.

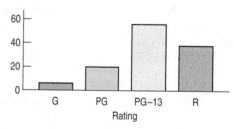

9. **Magnet schools.** An article in the Winter 2003 issue of *Chance* magazine reported on the Houston Independent School District's magnet schools programs. Of the 1755 qualified applicants, 931 were accepted, 298 were wait-listed, and 526 were turned away for lack of space. Find the relative frequency distribution of the decisions made, and write a sentence describing it.

10. **Magnet schools again.** The *Chance* article about the Houston magnet schools program described in Exercise 9 also indicated that 517 applicants were black or Hispanic, 292 Asian, and 946 white. Summarize the relative frequency distribution of ethnicity with a sentence or two (in the proper context, of course).

11. **Causes of death 2006.** The Centers for Disease Control and Prevention (www.cdc.gov) lists causes of death in the United States during 2006:

Cause of Death	Percent
Heart disease	26.6
Cancer	22.8
Circulatory diseases and stroke	5.9
Respiratory diseases	5.3
Accidents	4.8

a) Is it reasonable to conclude that heart or respiratory diseases were the cause of approximately 32% of U.S. deaths in 2006?
b) What percent of deaths were from causes not listed here?
c) Create an appropriate display for these data.

12. **Plane crashes.** An investigation compiled information about recent nonmilitary plane crashes (www.planecrashinfo.com). The causes, to the extent that they could be determined, are summarized in the table.

Cause	Percent
Pilot error	40
Other human error	5
Weather	6
Mechanical failure	14
Sabotage	6

a) Is it reasonable to conclude that the weather or mechanical failures caused only about 20% of recent plane crashes?
b) In what percent of crashes were the causes not determined?
c) Create an appropriate display for these data.

13. **Oil spills 2008.** Data from the International Tanker Owners Pollution Federation Limited (www.itopf.com) give the cause of spillage for 319 large oil tanker accidents from 1974–2008. Here are displays.

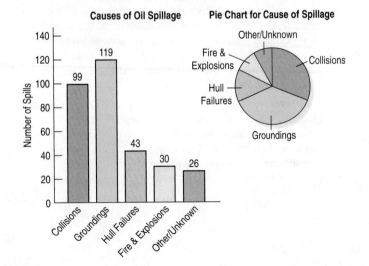

a) Write a brief report interpreting what the displays show.
b) Is a pie chart an appropriate display for these data? Why or why not?

14. **Winter Olympics 2006.** Twenty-six countries won medals in the 2006 Winter Olympics. The table lists them, along with the total number of medals each won:

Country	Medals	Country	Medals
Germany	29	Finland	9
United States	25	Czech Republic	4
Canada	24	Estonia	3
Austria	23	Croatia	3
Russia	22	Australia	2
Norway	19	Poland	2
Sweden	14	Ukraine	2
Switzerland	14	Japan	1
South Korea	11	Belarus	1
Italy	11	Bulgaria	1
China	11	Great Britain	1
France	9	Slovakia	1
Netherlands	9	Latvia	1

a) Try to make a display of these data. What problems do you encounter?
b) Can you find a way to organize the data so that the graph is more successful?

15. **Global warming.** The Pew Research Center for the People and the Press (http://people-press.org) has asked a representative sample of U.S. adults about global warming, repeating the question over time. In January 2007, the responses reflected an increased belief that global warming is real and due to human activity. Here's a display of the percentages of respondents choosing each of the major alternatives offered:

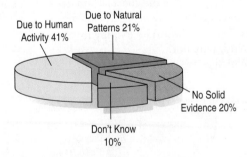

List the errors in this display.

16. **Modalities.** A survey of athletic trainers (Scott F. Nadler, Michael Prybicien, Gerard A. Malanga, and Dan Sicher, "Complications from Therapeutic Modalities: Results of a National Survey of Athletic Trainers." *Archives of Physical Medical Rehabilitation* 84 [June 2003]) asked what modalities (treatment methods such as ice, whirlpool, ultrasound, or exercise) they commonly use to treat injuries. Respondents were each asked to list

three modalities. The article included the following figure reporting the modalities used:

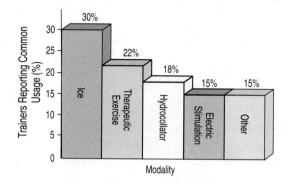

a) What problems do you see with the graph?
b) Consider the percentages for the named modalities. Do you see anything odd about them?

17. **Teen smokers.** The organization Monitoring the Future (www.monitoringthefuture.org) asked 2048 eighth graders who said they smoked cigarettes what brands they preferred. The table below shows brand preferences for two regions of the country. Write a few sentences describing the similarities and differences in brand preferences among eighth graders in the two regions listed.

Brand Preference	South	West
Marlboro	58.4%	58.0%
Newport	22.5%	10.1%
Camel	3.3%	9.5%
Other (over 20 brands)	9.1%	9.5%
No usual brand	6.7%	12.9%

18. **Handguns.** In an effort to reduce the number of gun-related homicides, some cities have run buyback programs in which the police offer cash (often $50) to anyone who turns in an operating handgun. *Chance* magazine looked at results from a four-year period in Milwaukee. The table below shows what types of guns were turned in and what types were used in homicides during a four-year period. Write a few sentences comparing the two distributions.

Caliber of Gun	Buyback	Homicide
Small (.22, .25, .32)	76.4%	20.3%
Medium (.357, .38, 9 mm)	19.3%	54.7%
Large (.40, .44, .45)	2.1%	10.8%
Other	2.2%	14.2%

19. **Movies by genre and rating.** Here's a table that classifies movies released in 2005 by genre and MPAA rating:
a) The table gives column percents. How could you tell that from the table itself?
b) What percentage of these movies were comedies?
c) What percentage of the PG-rated movies were comedies?

	G	PG	PG-13	R	Total
Action/Adventure	66.7	25	30.4	23.7	**29.2**
Comedy	33.3	60.0	35.7	10.5	**31.7**
Drama	0	15.0	14.3	44.7	**23.3**
Thriller/Horror	0	0	19.6	21.1	**15.8**
Total	**100%**	**100%**	**100%**	**100%**	**100%**

d) Which of the following can you learn from this table? Give the answer if you can find it from the table.
i) The percentage of PG-13 movies that were comedies
ii) The percentage of dramas that were R-rated
iii) The percentage of dramas that were G-rated
iv) The percentage of 2005 movies that were PG-rated comedies

20. **The last picture show.** Here's another table showing information about 120 movies released in 2005. This table gives percentages of the table total:

	G	PG	PG-13	R	Total
Action/Adventure	3.33%	4.17	14.2	7.50	**29.2**
Comedy	1.67	10	16.7	3.33	**31.7**
Drama	0	2.50	6.67	14.2	**23.3**
Thriller/Horror	0	0	9.17	6.67	**15.8**
Total	**5**	**16.7**	**46.7**	**31.7**	**100%**

a) How can you tell that this table holds table percentages (rather than row or column percentages)?
b) What was the most common genre/rating combination in 2005 movies?
c) How many of these movies were PG-rated comedies?
d) How many were G-rated?
e) An editorial about the movies noted, "More than three-quarters of the movies made today can be seen only by patrons 13 years old or older." Does this table support that assertion? Explain.

21. **Seniors.** Prior to graduation, a high school class was surveyed about its plans. The following table displays the results for white and minority students (the "Minority" group included African-American, Asian, Hispanic, and Native American students):

SENIORS		
	White	Minority
4-year college	198	44
2-year college	36	6
Military	4	1
Employment	14	3
Other	16	3

Plans

a) What percent of the seniors are white?
b) What percent of the seniors are planning to attend a 2-year college?
c) What percent of the seniors are white and planning to attend a 2-year college?
d) What percent of the white seniors are planning to attend a 2-year college?
e) What percent of the seniors planning to attend a 2-year college are white?

22. **Politics.** Students in an Intro Stats course were asked to describe their politics as "Liberal," "Moderate," or "Conservative." Here are the results:

		Politics		
		L	M	C	Total
Sex	Female	35	36	6	77
	Male	50	44	21	115
	Total	85	80	27	192

a) What percent of the class is male?
b) What percent of the class considers themselves to be "Conservative"?
c) What percent of the males in the class consider themselves to be "Conservative"?
d) What percent of all students in the class are males who consider themselves to be "Conservative"?

23. **More about seniors.** Look again at the table of post-graduation plans for the senior class in Exercise 21.
a) Find the conditional distributions (percentages) of plans for the white students.
b) Find the conditional distributions (percentages) of plans for the minority students.
c) Create a graph comparing the plans of white and minority students.
d) Do you see any important differences in the post-graduation plans of white and minority students? Write a brief summary of what these data show, including comparisons of conditional distributions.

24. **Politics revisited.** Look again at the table of political views for the Intro Stats students in Exercise 22.
a) Find the conditional distributions (percentages) of political views for the females.
b) Find the conditional distributions (percentages) of political views for the males.
c) Make a graphical display that compares the two distributions.
d) Do the variables *Politics* and *Sex* appear to be independent? Explain.

25. **Magnet schools revisited.** The *Chance* magazine article described in Exercise 9 further examined the impact of an applicant's ethnicity on the likelihood of admission to the Houston Independent School District's magnet schools programs. Those data are summarized in the table below.
a) What percent of all applicants were Asian?
b) What percent of the students accepted were Asian?

		Admission Decision			
		Accepted	Wait-listed	Turned away	Total
Ethnicity	Black/Hispanic	485	0	32	517
	Asian	110	49	133	292
	White	336	251	359	946
	Total	931	300	524	1755

c) What percent of Asians were accepted?
d) What percent of all students were accepted?

26. **More politics.** Look once more at the table summarizing the political views of Intro Stats students in Exercise 22.
a) Produce a graphical display comparing the conditional distributions of males and females among the three categories of politics.
b) Comment briefly on what you see from the display in a.

27. **Back to school.** Examine the table about ethnicity and acceptance for the Houston Independent School District's magnet schools program, shown in Exercise 25. Does it appear that the admissions decisions are made independent of the applicant's ethnicity? Explain.

28. **Parking lots.** A survey of autos parked in student and staff lots at a large university classified the brands by country of origin, as seen in the table.

		Driver	
		Student	Staff
Origin	American	107	105
	European	33	12
	Asian	55	47

a) What percent of all the cars surveyed were foreign?
b) What percent of the American cars were owned by students?
c) What percent of the students owned American cars?
d) What is the marginal distribution of origin?
e) What are the conditional distributions of origin by driver classification?
f) Do you think that the origin of the car is independent of the type of driver? Explain.

T 29. **Weather forecasts.** Just how accurate are the weather forecasts we hear every day? The following table compares the daily forecast with a city's actual weather for a year:

		Actual Weather	
		Rain	No rain
Forecast	Rain	27	63
	No rain	7	268

a) On what percent of days did it actually rain?
b) On what percent of days was rain predicted?

c) What percent of the time was the forecast correct?

d) Do you see evidence of an association between the type of weather and the ability of forecasters to make an accurate prediction? Write a brief explanation, including an appropriate graph.

T 30. **Twin births.** In 2000, the *Journal of the American Medical Association (JAMA)* published a study that examined pregnancies that resulted in the birth of twins. Births were classified as preterm with intervention (induced labor or cesarean), preterm without procedures, or term/post-term. Researchers also classified the pregnancies by the level of prenatal medical care the mother received (inadequate, adequate, or intensive). The data, from the years 1995–1997, are summarized in the table below. Figures are in thousands of births. (*JAMA* 284 [2000]:335–341)

TWIN BIRTHS 1995–1997 (IN THOUSANDS)

	Preterm (induced or cesarean)	Preterm (without procedures)	Term or post-term	Total
Intensive	18	15	28	**61**
Adequate	46	43	65	**154**
Inadequate	12	13	38	**63**
Total	**76**	**71**	**131**	**278**

(Level of Prenatal Care)

a) What percent of these mothers received inadequate medical care during their pregnancies?

b) What percent of all twin births were preterm?

c) Among the mothers who received inadequate medical care, what percent of the twin births were preterm?

d) Create an appropriate graph comparing the outcomes of these pregnancies by the level of medical care the mother received.

e) Write a few sentences describing the association between these two variables.

T 31. **Blood pressure.** A company held a blood pressure screening clinic for its employees. The results are summarized in the table below by age group and blood pressure level:

		Age		
		Under 30	30–49	Over 50
Blood Pressure	**Low**	27	37	31
	Normal	48	91	93
	High	23	51	73

a) Find the marginal distribution of blood pressure level.

b) Find the conditional distribution of blood pressure level within each age group.

c) Compare these distributions with a segmented bar graph.

d) Write a brief description of the association between age and blood pressure among these employees.

e) Does this prove that people's blood pressure increases as they age? Explain.

32. **Obesity and exercise.** The Centers for Disease Control and Prevention (CDC) has estimated that 19.8% of Americans over 15 years old are obese. The CDC conducts a survey on obesity and various behaviors. Here is a table on self-reported exercise classified by body mass index (BMI):

	Body Mass Index		
	Normal (%)	Overweight (%)	Obese (%)
Inactive	23.8	26.0	35.6
Irregularly active	27.8	28.7	28.1
Regular, not intense	31.6	31.1	27.2
Regular, intense	16.8	14.2	9.1

(Physical Activity)

a) Are these percentages column percentages, row percentages, or table percentages?

b) Use graphical displays to show different percentages of physical activities for the three BMI groups.

c) Do these data prove that lack of exercise causes obesity? Explain.

33. **Anorexia.** Hearing anecdotal reports that some patients undergoing treatment for the eating disorder anorexia seemed to be responding positively to the antidepressant Prozac, medical researchers conducted an experiment to investigate. They found 93 women being treated for anorexia who volunteered to participate. For one year, 49 randomly selected patients were treated with Prozac and the other 44 were given an inert substance called a placebo. At the end of the year, patients were diagnosed as healthy or relapsed, as summarized in the table:

	Prozac	Placebo	Total
Healthy	35	32	67
Relapse	14	12	26
Total	**49**	**44**	**93**

Do these results provide evidence that Prozac might be helpful in treating anorexia? Explain.

34. **Antidepressants and bone fractures.** For a period of five years, physicians at McGill University Health Center followed more than 5000 adults over the age of 50. The researchers were investigating whether people taking a certain class of antidepressants (SSRIs) might be at greater risk of bone fractures. Their observations are summarized in the table:

	Taking SSRI	No SSRI	Total
Experienced fractures	14	244	258
No fractures	123	4627	4750
Total	**137**	**4871**	**5008**

Do these results suggest there's an association between taking SSRI antidepressants and experiencing bone fractures? Explain.

T 35. Drivers' licenses 2006. The following table shows the number of licensed U.S. drivers by age and by sex (www.dot.gov):

Age	Male Drivers (number)	Female Drivers (number)	Total
(19 and under)	4,972,491	4,755,025	9,727,516
20	1,661,381	1,601,090	3,262,471
21	1,686,255	1,654,810	3,341,065
22	1,697,270	1,672,630	3,369,900
23	1,760,214	1,738,319	3,498,533
24	1,799,077	1,790,905	3,589,982
(20–24)	8,604,197	8,457,754	17,061,951
25–29	8,942,559	8,894,582	17,837,141
30–34	8,901,311	8,756,497	17,657,808
35–39	9,790,487	9,623,673	19,414,160
40–44	10,328,973	10,255,683	20,584,656
45–49	10,535,359	10,589,323	21,124,682
50–54	9,616,108	9,721,217	19,337,325
55–59	8,517,397	8,625,943	17,143,340
60–64	6,360,556	6,425,837	12,786,393
65–69	4,726,657	4,819,918	9,546,575
70–74	3,661,284	3,801,952	7,463,236
75–79	2,922,713	3,176,789	6,099,502
80–84	1,964,351	2,267,787	4,232,138
85 AND OVER	1,271,839	1,522,176	2,794,015
Total	101,116,282	101,694,156	219,872,389

a) What percent of total drivers are under 20?
b) What percent of total drivers are male?
c) Write a few sentences comparing the number of male and female licensed drivers in each age group.
d) Do a driver's age and sex appear to be independent? Explain.

T 36. Tattoos. A study by the University of Texas Southwestern Medical Center examined 626 people to see if an increased risk of contracting hepatitis C was associated with having a tattoo. If the subject had a tattoo, researchers asked whether it had been done in a commercial tattoo parlor or elsewhere. Write a brief description of the association between tattooing and hepatitis C, including an appropriate graphical display.

	Tattoo done in commercial parlor	Tattoo done elsewhere	No tattoo
Has hepatitis C	17	8	18
No hepatitis C	35	53	495

37. Hospitals. Most patients who undergo surgery make routine recoveries and are discharged as planned. Others suffer excessive bleeding, infection, or other postsurgical complications and have their discharges from the hospital delayed. Suppose your city has a large hospital and a small hospital, each performing major and minor surgeries. You collect data to see how many surgical patients have their discharges delayed by postsurgical complications, and you find the results shown in the following table.

	Discharge Delayed	
	Large hospital	Small hospital
Major surgery	120 of 800	10 of 50
Minor surgery	10 of 200	20 of 250

a) Overall, for what percent of patients was discharge delayed?
b) Were the percentages different for major and minor surgery?
c) Overall, what were the discharge delay rates at each hospital?
d) What were the delay rates at each hospital for each kind of surgery?
e) The small hospital advertises that it has a lower rate of postsurgical complications. Do you agree?
f) Explain, in your own words, why this confusion occurs.

38. Delivery service. A company must decide which of two delivery services it will contract with. During a recent trial period, the company shipped numerous packages with each service and kept track of how often deliveries did not arrive on time. Here are the data:

Delivery Service	Type of Service	Number of Deliveries	Number of Late Packages
Pack Rats	Regular	400	12
	Overnight	100	16
Boxes R Us	Regular	100	2
	Overnight	400	28

a) Compare the two services' overall percentage of late deliveries.
b) On the basis of the results in part a, the company has decided to hire Pack Rats. Do you agree that Pack Rats delivers on time more often? Explain.
c) The results here are an instance of what phenomenon?

39. Graduate admissions. A 1975 article in the magazine *Science* examined the graduate admissions process at Berkeley for evidence of sex discrimination. The table below shows the number of applicants accepted to each of four graduate programs:

		Males accepted (of applicants)	Females accepted (of applicants)
Program	1	511 of 825	89 of 108
	2	352 of 560	17 of 25
	3	137 of 407	132 of 375
	4	22 of 373	24 of 341
	Total	1022 of 2165	262 of 849

a) What percent of total applicants were admitted?
b) Overall, was a higher percentage of males or females admitted?
c) Compare the percentage of males and females admitted in each program.
d) Which of the comparisons you made do you consider to be the most valid? Why?

40. **Be a Simpson!** Can you design a Simpson's paradox? Two companies are vying for a city's "Best Local Employer" award, to be given to the company most committed to hiring local residents. Although both employers hired 300 new people in the past year, Company A brags that it deserves the award because 70% of its new jobs went to local residents, compared to only 60% for Company B. Company B concedes that those percentages are correct, but points out that most of its new jobs were full-time, while most of Company A's were part-time. Not only that, says Company B, but a higher percentage of its full-time jobs went to local residents than did Company A's, and the same was true for part-time jobs. Thus, Company B argues, it's a better local employer than Company A.

Show how it's possible for Company B to fill a higher percentage of both full-time and part-time jobs with local residents, even though Company A hired more local residents overall.

ANSWERS

1. 50.0%

2. 44.4%

3. 25.0%

4. 15.6% Blue, 56.3% Brown, 28.1% Green/Hazel/Other

5. 18.8% Blue, 62.5% Brown, 18.8% Green/Hazel/Other

6. 40% of the blue-eyed students are female, while 50% of all students are female.

7. Since blue-eyed students appear less likely to be female, it seems that *Sex* and *Eye Color* may not be independent. (But the numbers are small.)

Displaying and Summarizing Quantitative Data

Where are we going?

If someone asked you to summarize a variable, what would you say? You might start by making a picture. For quantitative data, that first picture would probably be a histogram. We've all looked at histograms, but what should we look *for*? We'll describe the histogram, and we'll often do more—reporting numerical summaries of the center and spread as well. Spread measures are a bit less common, but in Statistics they are even more important. This chapter is where we'll first encounter the single most important calculated value in all of Statistics.

Tsunamis are potentially destructive waves that can occur when the sea floor is suddenly and abruptly deformed. They are most often caused by earthquakes beneath the sea that shift the earth's crust, displacing a large mass of water.

The tsunami of December 26, 2004, with its epicenter off the west coast of Sumatra, was caused by an earthquake of magnitude 9.0 on the Richter scale. It killed an estimated 297,248 people, making it the most disastrous tsunami on record. But was the earthquake that caused it truly extraordinary, or did it just happen at an unlucky place and time? The U.S. National Geophysical Data Center[1] has information on more than 2400 tsunamis dating back to 2000 B.C.E., and we have estimates of the magnitude of the underlying earthquake for 1240 of them. What can we learn from these data?

Histograms

Let's start with a picture. For categorical variables, it is easy to draw the distribution because each category is a natural "pile." But for quantitative variables, there's no obvious way to choose piles. So, usually, we slice up all the possible values into equal-width bins. We then count the number of cases that fall into each bin. The bins, together with these counts, give the **distribution** of the quantitative variable and provide the building blocks for the histogram. By representing the counts as bars and plotting them against the bin values, the **histogram** displays the distribution at a glance.

---

[1] www.ngdc.noaa.gov.

WHO	1240 earthquakes known to have caused tsunamis for which we have data or good estimates
WHAT	Magnitude (Richter scale[2]), depth (m), date, location, and other variables
WHEN	From 2000 B.C.E. to the present
WHERE	All over the earth

For example, here are the *Magnitudes* (on the Richter scale) of the 1240 earthquakes in the NGDC data:

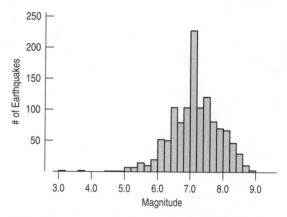

FIGURE 4.1
A histogram of earthquake magnitudes shows the number of earthquakes with magnitudes (in Richter scale units) in each bin.

Like a bar chart, a histogram plots the bin counts as the heights of bars. In this histogram of earthquake magnitudes, each bin has a width of 0.2, so, for example, the height of the tallest bar says that there were about 230 earthquakes with magnitudes between 7.0 and 7.2. In this way, the histogram displays the entire distribution of earthquake magnitudes.

How Do Histograms Work? If you make a histogram by hand, you'll need to decide the endpoints of the bins. Usually, it helps to make them come out to "nice" numbers that are easy to think about. The standard rule for a value that falls exactly on a bin boundary is to put it into the next higher bin, so if a bin spans magnitudes 5.0 to 5.2, and the next goes from 5.2 to 5.4, you'd put an earthquake with magnitude 5.2 into the higher bin.

Different features of the distribution may appear more obvious at different bin width choices. When you use technology, it's usually easy to vary the bin width interactively so you can make sure that a feature you think you see isn't a consequence of a certain bin width choice.

Does the distribution look as you expected? It is often a good idea to *imagine* what the distribution might look like before you make the display. That way you'll be less likely to be fooled by errors in the data or when you accidentally graph the wrong variable.

From the histogram, we can see that these earthquakes typically have magnitudes around 7. Most are between 5.5 and 8.5, and some are as small as 3 and as big as 9. Now we can answer the question about the Sumatra tsunami. With a value of 9.0 it's clear that the earthquake that caused it was an extraordinarily powerful earthquake—one of the largest on record.[3]

The bar charts of categorical variables we saw in Chapter 3 had spaces between the bars to separate the counts of different categories. But in a

---

[2] Technically, Richter scale values are in units of log dyne-cm. But the Richter scale is so common now that usually the units are assumed. The U.S. Geological Survey gives the background details of Richter scale measurements on its website www.usgs.gov.
[3] Some experts now estimate the magnitude at between 9.1 and 9.3.

One surprising feature of the earthquake magnitudes is the spike around magnitude 7.0. Only one other bin holds even half that many earthquakes. These values include historical data for which the magnitudes were estimated by experts and not measured by modern seismographs. Perhaps the experts thought 7 was a typical and reasonable value for a tsunami-causing earthquake when they lacked detailed information. That would explain the overabundance of magnitudes right at 7.0 rather than spread out near that value.

histogram, the bins slice up *all the values* of the quantitative variable, so any spaces in a histogram are actual **gaps** in the data, indicating a region where there are no values.

Sometimes it is useful to make a **relative frequency histogram**, replacing the counts on the vertical axis with the *percentage* of the total number of cases falling in each bin. Of course, the shape of the histogram is exactly the same; only the vertical scale is different.

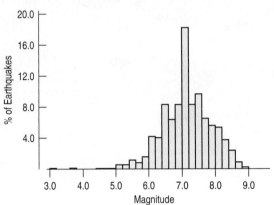

FIGURE 4.2
A relative frequency histogram looks just like a frequency histogram except for the labels on the *y*-axis, which now show the percentage of earthquakes in each bin.

Stem-and-Leaf Displays

Histograms provide an easy-to-understand summary of the distribution of a quantitative variable, but they don't show the data values themselves. Here's a histogram of the pulse rates of 24 women, taken by a researcher at a health clinic:

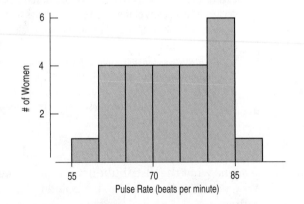

FIGURE 4.3
The pulse rates of 24 women at a health clinic.

The story seems pretty clear. We can see the entire span of the data and can easily see what a typical pulse rate might be. But is that all there is to these data?

A **stem-and-leaf display** is like a histogram, but it shows the individual values. It's also easier to make by hand. Here's a stem-and-leaf display of the same data:

The Stem-and-Leaf display was devised by John W. Tukey, one of the greatest statisticians of the 20th century. It is called a "Stemplot" in some texts and computer programs, but we prefer Tukey's original name for it.

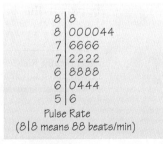

A S ***Activity:* Stem-and-Leaf Displays.** As you might expect of something called "stem-and leaf," these displays grow as you consider each data value.

Turn the stem-and-leaf on its side (or turn your head to the right) and squint at it. It should look roughly like the histogram of the same data. Does it? Well, it's backwards because now the higher values are on the left, but other than that, it has the same shape.[4]

What does the line at the top of the display that says 8 | 8 mean? It stands for a pulse of 88 beats per minute (bpm). We've taken the tens place of the number and made that the "stem." Then we sliced off the ones place and made it a "leaf." The next line down is 8 | 000044. That shows that there were four pulse rates of 80 and two of 84 bpm.

Stem-and-leaf displays are especially useful when you make them by hand for batches of fewer than a few hundred data values. They are a quick way to display—and even to record—numbers. Because the leaves show the individual values, we can sometimes see even more in the data than the distribution's shape. Take another look at all the leaves of the pulse data. See anything unusual? At a glance you can see that they are all even. With a bit more thought you can see that they are all multiples of 4—something you couldn't possibly see from a histogram. How do you think the nurse took these pulses? Counting beats for a full minute or counting for only 15 seconds and multiplying by 4?

How do stem-and-leaf displays work? Stem-and-leaf displays work like histograms, but they show more information. They use part of the number itself (called the stem) to name the bins. To make the "bars," they use the next digit of the number. For example, if we had a test score of 83, we could write it 8 | 3, where 8 serves as the stem and 3 as the leaf. Then, to display the scores 83, 76, and 88 together, we would write

```
8 | 38
7 | 6
```

For the pulse data, we have

```
8 | 0000448
7 | 22226666
6 | 04448888
5 | 6
```

Pulse Rate
(5 | 6 means 56 beats/min)

This display is OK, but a little crowded. A histogram might split each line into two bars. With a stem-and-leaf, we can do the same by putting the leaves 0-4 on one line and 5-9 on another, as we saw above:

```
8 | 8
8 | 000044
7 | 6666
7 | 2222
6 | 8888
6 | 0444
5 | 6
```

Pulse Rate
(8 | 8 means 88 beats/min)

(continued)

[4] You could make the stem-and-leaf with the higher values on the bottom. Usually, though, higher on the top makes sense.

For numbers with three or more digits, you'll often decide to truncate (or round) the number to two places, using the first digit as the stem and the second as the leaf. So, if you had 432, 540, 571, and 638, you might display them as shown below with an indication that 6 | 3 means 630–639.

```
6 | 3
5 | 4 7
4 | 3
```

When you make a stem-and-leaf by hand, make sure to give each digit the same width, in order to preserve the area principle. (That can lead to some fat 1's and thin 8's—but it makes the display honest.)

Dotplots

A S

Activity: **Dotplots.** Click on points to see their values and even drag them around.

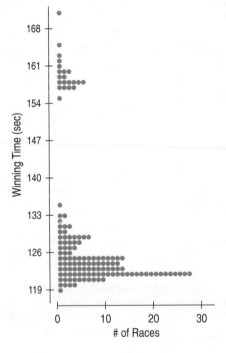

FIGURE 4.4
A dotplot of Kentucky Derby winning times plots each race as its own dot. We can see two distinct groups corresponding to the two different race distances.

A **dotplot** is a simple display. It just places a dot along an axis for each case in the data. It's like a stem-and-leaf display, but with dots instead of digits for all the leaves. Dotplots are a great way to display a small data set (especially if you forget how to write the digits from 0 to 9). Here's a dotplot of the time (in seconds) that the winning horse took to win the Kentucky Derby in each race between the first Derby in 1875 and the 2008 Derby.

Dotplots show basic facts about the distribution. We can find the slowest and quickest races by finding times for the topmost and bottommost dots. It's also clear that there are two clusters of points, one just below 160 seconds and the other at about 122 seconds. Something strange happened to the Derby times. Once we know to look for it, we can find out that in 1896 the distance of the Derby race was changed from 1.5 miles to the current 1.25 miles. That explains the two clusters of winning times.

Some dotplots stretch out horizontally, with the counts on the vertical axis, like a histogram. Others, such as the one shown here, run vertically, like a stem-and-leaf display. Some dotplots place points next to each other when they would otherwise overlap. Others just place them on top of one another. Newspapers sometimes offer dotplots with the dots made up of little pictures.

Think Before You Draw

Suddenly, we face a lot more options when it's time to invoke our first rule of data analysis and make a picture. You'll need to *Think* carefully to decide which type of graph to make. In the previous chapter you learned to check the Categorical Data Condition before making a pie chart or a bar chart. Now, before making a stem-and-leaf display, a histogram, or a dotplot, you need to check the

Quantitative Data Condition: The data are values of a quantitative variable whose units are known.

Although a bar chart and a histogram may look somewhat similar, they're not the same display. You can't display categorical data in a histogram or quantitative data in a bar chart. Always check the condition that confirms what type of data you have before proceeding with your display.

Step back from a histogram or stem-and-leaf display. What can you say about the distribution? When you describe a distribution, you should always tell about three things: its **shape, center,** and **spread**.

The Shape of a Distribution

1. *Does the histogram have a single, central hump or several separated humps?* These humps are called **modes**.[5] The earthquake magnitudes have a single mode at just about 7. A histogram with one peak, such as the earthquake magnitudes, is dubbed **unimodal**; histograms with two peaks are **bimodal**, and those with three or more are called **multimodal**.[6] For example, here's a bimodal histogram.

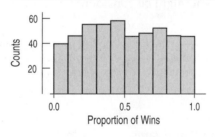

FIGURE 4.5
A bimodal histogram has two apparent peaks.

A histogram that doesn't appear to have any mode and in which all the bars are approximately the same height is called **uniform**.

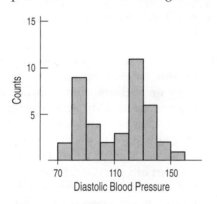

FIGURE 4.6
In a uniform histogram, the bars are all about the same height. The histogram doesn't appear to have a mode.

2. *Is the histogram **symmetric?*** Can you fold it along a vertical line through the middle and have the edges match pretty closely, or are more of the values on one side?

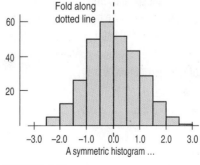

FIGURE 4.7

The **mode** is sometimes defined as the single value that appears most often. That definition is fine for categorical variables because all we need to do is count the number of cases for each category. For quantitative variables, the mode is more ambiguous. What is the mode of the Kentucky Derby times? Well, seven races were timed at 122.2 seconds—more than any other race time. Should that be the mode? Probably not. For quantitative data, it makes more sense to use the term "mode" in the more general sense of the peak of the histogram rather than as a single summary value. In this sense, the important feature of the Kentucky Derby races is that there are two distinct modes, representing the two different versions of the race and warning us to consider those two versions separately.

You've heard of pie à la mode. Is there a connection between pie and the mode of a distribution? Actually, there is! The mode of a distribution is a *popular* value near which a lot of the data values gather. And "à la mode" means "in style"—*not* "with ice cream." That just happened to be a *popular* way to have pie in Paris around 1900.

[5] Well, technically, it's the value on the horizontal axis of the histogram that is the mode, but anyone asked to point to the mode would point to the hump.
[6] Apparently, statisticians don't like to count past two.

The (usually) thinner ends of a distribution are called the **tails**. If one tail stretches out farther than the other, the histogram is said to be **skewed** to the side of the longer tail.

Activity: Attributes of Distribution Shape. This activity and the others on this page show off aspects of distribution shape through animation and example, then let you make and interpret histograms with your statistics package.

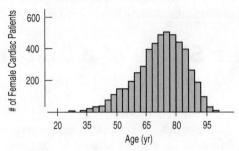
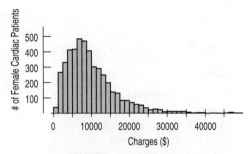

FIGURE 4.8
Two skewed histograms showing data on two variables for all female heart attack patients in New York State in one year. The blue one (age in years) is skewed to the left. The purple one (charges in $) is skewed to the right.

3. *Do any unusual features stick out?* Often such features tell us something interesting or exciting about the data. You should always mention any stragglers, or **outliers**, that stand away from the body of the distribution. If you're collecting data on nose lengths and Pinocchio is in the group, you'd probably notice him, and you'd certainly want to mention it.

   Outliers can affect almost every method we discuss in this course. So we'll always be on the lookout for them. An outlier can be the most informative part of your data. Or it might just be an error. But don't throw it away without comment. Treat it specially and discuss it when you tell about your data. Or find the error and fix it if you can. Be sure to look for outliers. Always.

In the next chapter you'll learn a handy rule of thumb for deciding when a point might be considered an outlier.

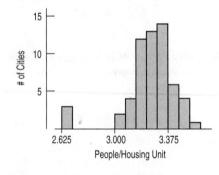

FIGURE 4.9
A histogram with outliers. There are three cities in the leftmost bar.

FOR EXAMPLE   **Describing Histograms**

A credit card company wants to see how much customers in a particular segment of their market use their credit card. They have provided you with data[7] on the amount spent by 500 selected customers during a 3-month period and have asked you to summarize the expenditures. Of course, you begin by making a histogram.

QUESTION: Describe the shape of this distribution.

The distribution of expenditures is unimodal and skewed to the high end. There is an extraordinarily large value at about $7000, and some of the expenditures are negative.

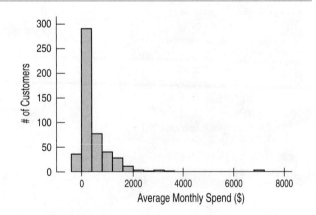

---

[7] These data are real, but cannot be further identified for obvious privacy reasons.

Are there any gaps in the distribution? The Kentucky Derby data that we saw in the dotplot on page 48 has a large gap between two groups of times, one near 120 seconds and one near 160. Gaps help us see multiple modes and encourage us to notice when the data may come from different sources or contain more than one group.

Toto, I've a feeling we're not in math class anymore . . .   When Dorothy and her dog Toto land in Oz, everything is more vivid and colorful, but also more dangerous and exciting. Dorothy has new choices to make. She can't always rely on the old definitions, and the yellow brick road has many branches. You may be coming to a similar realization about Statistics.

When we summarize data, our goal is usually more than just developing a detailed knowledge of the data we have at hand. Scientists generally don't care about the particular guinea pigs they've treated, but rather about what their reactions say about how animals (and, perhaps, humans) would respond.

When you look at data, you want to know what the data say about the world, so you'd like to know whether the patterns you see in histograms and summary statistics generalize to other individuals and situations. You'll want to calculate summary statistics accurately, but then you'll also want to think about what they may say beyond just describing the data. And your knowledge about the world matters when you think about the overall meaning of your analysis.

It may surprise you that many of the most important concepts in Statistics are not defined as precisely as most concepts in mathematics. That's done on purpose, to leave room for judgment.

Because we want to see broader patterns rather than focus on the details of the data set we're looking at, we deliberately leave some statistical concepts a bit vague. Whether a histogram is symmetric or skewed, whether it has one or more modes, whether a point is far enough from the rest of the data to be considered an outlier—these are all somewhat vague concepts. And they all require judgment. You may be used to finding a single correct and precise answer, but in Statistics, there may be more than one interpretation. That may make you a little uncomfortable at first, but soon you'll see that this room for judgment brings you enormous power and responsibility. It means that using your own knowledge and judgment and supporting your findings with statistical evidence and justifications entitles you to your own opinions about what you see.

JUST CHECKING

It's often a good idea to think about what the distribution of a data set might look like before we collect the data. What do you think the distribution of each of the following data sets will look like? Be sure to discuss its shape. Where do you think the center might be? How spread out do you think the values will be?

1. Number of miles run by Saturday morning joggers at a park.

2. Hours spent by U.S. adults watching football on Thanksgiving Day.

3. Amount of winnings of all people playing a particular state's lottery last week.

4. Ages of the faculty members at your school.

5. Last digit of phone numbers on your campus.

The Center of the Distribution: The Median

Let's return to the tsunami earthquakes. But this time, let's look at just 25 years of data: 176 earthquakes that occurred from 1981 through 2005. These should be more accurately measured than prehistoric quakes because seismographs were in wide use. Try to put your finger on the histogram at the value you think is typical. (Read the value from the horizontal axis and remember it.) When we think of a typical value, we usually look for the **center** of the distribution. Where do you think the center of this distribution is? For a unimodal, symmetric distribution such as these earthquake data, it's easy. We'd all agree on the center of symmetry, where we would fold the histogram to match the two sides. But when the distribution is skewed or possibly multimodal, it's not immediately clear what we even mean by the center.

One natural choice of typical value is the value that is literally in the middle, with half the values below it and half above it.

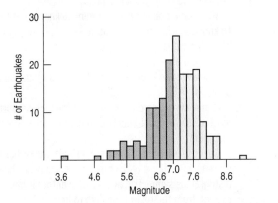

FIGURE 4.10
Tsunami-causing earthquakes (1981–2005).

The median splits the histogram into two halves of equal area.

Histograms follow the area principle, and each half of the data has about 88 earthquakes, so each colored region has the same area in the display. The middle value that divides the histogram into two equal areas is called the **median**.

The median has the same units as the data. Be sure to include the units whenever you discuss the median.

For the recent tsunamis, there are 176 earthquakes, so the median is found at the $(176 + 1)/2 = 88.5$th place in the sorted data. That ".5" just says to average the two values on either side: the 88th and the 89th. The median earthquake magnitude is 7.0.

NOTATION ALERT

We always use n to indicate the number of values. Some people even say, "How big is the n?" when they mean the number of data values.

How do medians work?   Finding the median of a batch of n numbers is easy as long as you remember to order the values first. If n is odd, the median is the middle value. Counting in from the ends, we find this value in the $\dfrac{n+1}{2}$ position.

When n is even, there are two middle values. So, in this case, the median is the average of the two values in positions $\dfrac{n}{2}$ and $\dfrac{n}{2} + 1$.

Here are two examples:
Suppose the batch has these values: 14.1, 3.2, 25.3, 2.8, −17.5, 13.9, 45.8.
First we order the values: −17.5, 2.8, 3.2, 13.9, 14.1, 25.3, 45.8.

Since there are 7 values, the median is the $(7 + 1)/2 = 4$th value, counting from the top or bottom: 13.9. Notice that 3 values are lower, 3 higher.

Suppose we had the same batch with another value at 35.7. Then the ordered values are −17.5, 2.8, 3.2, 13.9, 14.1, 25.3, 35.7, 45.8.

The median is the average of the 8/2 or 4th, and the (8/2) + 1, or 5th, values. So the median is $(13.9 + 14.1)/2 = 14.0$. Four data values are lower, and four higher.

The median is one way to find the center of the data. But there are many others. We'll look at an even more important measure later in this chapter.

Knowing the median, we could say that in recent years, a typical tsunami-causing earthquake has been about 7.0 on the Richter scale. How much does that really say? How well does the median describe the data? After all, not every earthquake has a Richter scale value of 7.0. Whenever we find the center of data, the next step is always to ask how well it actually summarizes the data.

Spread: Home on the Range

Statistics pays close attention to what we *don't* know as well as what we do know. Understanding how spread out the data are is a first step in understanding what a summary *cannot* tell us about the data. It's the beginning of telling us what we don't know.

If every earthquake that caused a tsunami registered 7.0 on the Richter scale, then knowing the median would tell us everything about the distribution of earthquake magnitudes. The more the data vary, however, the less the median alone can tell us. So we need to measure how much the data values vary around the center. In other words, how spread out are they? When we describe a distribution numerically, we always report a measure of its **spread** along with its center.

How should we measure the spread? We could simply look at the extent of the data. How far apart are the two extremes? The **range** of the data is defined as the *difference* between the maximum and minimum values:

$$Range = max - min.$$

Notice that the range is a *single number, not* an interval of values, as you might think from its use in common speech. The maximum magnitude of these earthquakes is 9.0 and the minimum is 3.7, so the *range* is $9.0 - 3.7 = 5.3$.

The range has the disadvantage that a single extreme value can make it very large, giving a value that doesn't really represent the data overall.

Spread: The Interquartile Range

A better way to describe the spread of a variable might be to ignore the extremes and concentrate on the middle of the data. We could, for example, find the range of just the middle half of the data. What do we mean by the middle half? Divide the data in half at the median. Now divide both halves in half again, cutting the data into four quarters. We call these new dividing points **quartiles**. One quarter of the data lies below the **lower quartile**, and one quarter of the data lies above the **upper quartile**, so half the data lies between them. The quartiles border the middle half of the data.

For any percentage there is a corresponding **percentile** that cuts off that percentage of the data below it. The 10th and 90th percentiles, for example, identify the values below which 10% and 90% (respectively) of the data lie. The median, of course, is the 50th percentile.

How do quartiles work? A simple way to find the quartiles is to start by splitting the batch into two halves at the median. (When n is odd, some statisticians include the median in both halves; others omit it.) The lower quartile is the median of the lower half, and the upper quartile is the median of the upper half.

Here are our two examples again.

The ordered values of the first batch were -17.5, 2.8, 3.2, 13.9, 14.1, 25.3, and 45.8, with a median of 13.9. Notice that 7 is odd; we'll include the median in both halves to get -17.5, 2.8, 3.2, 13.9 and 13.9, 14.1, 25.3, 45.8.

Each half has 4 values, so the median of each is the average of its 2nd and 3rd values. So, the lower quartile is $(2.8 + 3.2)/2 = 3.0$ and the upper quartile is $(14.1 + 25.3)/2 = 19.7$.

The second batch of data had the ordered values -17.5, 2.8, 3.2, 13.9, 14.1, 25.3, 35.7, and 45.8. Here n is even, so the two halves of 4 values are -17.5, 2.8, 3.2, 13.9 and 14.1, 25.3, 35.7, 45.8. Now the lower quartile is $(2.8 + 3.2)/2 = 3.0$ and the upper quartile is $(25.3 + 35.7)/2 = 30.5$.

OTHER PERCENTILE DIFFERENCES

Could we use other percentiles besides the quartiles to measure the spread? Sure, but the IQR is the most commonly used percentile difference and the one you're most likely to see in practice.

The difference between the quartiles tells us how much territory the middle half of the data covers and is called the **interquartile range**. It's commonly abbreviated IQR (and pronounced "eye-cue-are"):

$$IQR = upper\ quartile - lower\ quartile.$$

For the earthquakes, there are 88 values below the median and 88 values above the median. The midpoint of the lower half is the average of the 44th and 45th values in the ordered data; that turns out to be 6.6. In the upper half we average the 132nd and 133rd values, finding a magnitude of 7.6 as the third quartile. The *difference* between the quartiles gives the IQR:

$$IQR = 7.6 - 6.6 = 1.0.$$

Now we know that the middle half of the earthquake magnitudes extends across a (interquartile) range of 1.0 Richter scale units. This seems like a reasonable summary of the spread of the distribution, as we can see from this histogram:

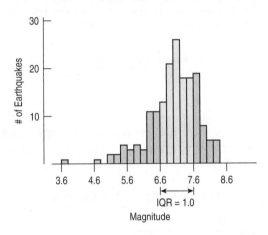

FIGURE 4.11
The quartiles bound the middle 50% of the values of the distribution. This gives a visual indication of the spread of the data. Here we see that the IQR is 1.0 Richter scale units.

The IQR is almost always a reasonable summary of the spread of a distribution. Even if the distribution itself is skewed or has some outliers, the IQR should provide useful information. The one exception is when the data are strongly bimodal. For example, remember the dotplot of winning times in the Kentucky Derby (page 48)? Because the race distance was changed, we really have data on two different races, and they shouldn't be summarized together.

So, what is a quartile anyway?   Finding the quartiles sounds easy, but surprisingly, the quartiles are not well-defined. It's not always clear how to find a value such that exactly one quarter of the data lies above or below that value. We offered a simple rule for finding quartiles in the box on page 53: Find the median of each half of the data split by the median. When *n* is odd, we include the median with each of the halves. Some other texts omit the median from each half before finding the quartiles. Both methods are commonly used. If you are willing to do a bit more calculating, there are several other methods that locate a quartile somewhere between adjacent data values. We know of at least six different rules for finding quartiles. Remarkably, each one is in use in some software package or calculator.

So don't worry too much about getting the "exact" value for a quartile. All of the methods agree pretty closely when the data set is large. When the data set is small, different rules will disagree more, but in that case there's little need to summarize the data anyway.

Remember, Statistics is about understanding the world, not about calculating the right number. The "answer" to a statistical question is a sentence about the issue raised in the question.

The lower and upper quartiles are also known as the 25th and 75th **percentiles** of the data, respectively, since the lower quartile falls above 25% of the data and the upper quartile falls above 75% of the data. If we count this way, the median is the 50th percentile. We could, of course, define and calculate any percentile that we want. For example, the 10th percentile would be the number that falls above the lowest 10% of the data values.

5-Number Summary

The **5-number summary** of a distribution reports its median, quartiles, and extremes (maximum and minimum). The 5-number summary for the recent tsunami earthquake *Magnitudes* looks like this:

We always use Q1 to label the lower (25%) quartile and Q3 to label the upper (75%) quartile. We skip the number 2 because the median would, by this system, naturally be labeled Q2— but we don't usually call it that.

Max	9.0
Q3	7.6
Median	7.0
Q1	6.6
Min	3.7

It's a good idea to report the number of data values and the identity of the cases (the *Who*). Here there are 176 earthquakes.

The 5-number summary provides a good overview of the distribution of magnitudes of these tsunami-causing earthquakes. For a start, we can see that the median magnitude is 7.0. Because the IQR is only 7.6 − 6.6 = 1, we see that many quakes are close to the median magnitude. Indeed, the quartiles show us that the middle half of these earthquakes had magnitudes between 6.6 and 7.6. One quarter of the earthquakes had magnitudes above 7.6, although one tsunami was caused by a quake measuring only 3.7 on the Richter scale.

STEP-BY-STEP EXAMPLE | **Shape, Center, and Spread: Flight Cancellations**

The U.S. Bureau of Transportation Statistics (www.bts.gov) reports data on airline flights. Let's look at data giving the percentage of flights cancelled each month between 1995 and 2006.

Question: How often are flights cancelled?

WHO	Months
WHAT	Percentage of flights cancelled at U.S. airports
WHEN	1995–2006
WHERE	United States

Variable: Identify the *variable,* and decide how you wish to display it.

To identify a variable, report the W's.

I want to learn about the monthly percentage of flight cancellations at U.S airports.

I have data from the U.S. Bureau of Transportation Statistics giving the percentage of flights cancelled at U.S. airports each month between 1995 and 2006.

Select an appropriate display based on the nature of the data and what you want to know.

✔ **Quantitative Data Condition:** Percentages are quantitative. A histogram and numerical summaries would be appropriate.

(continued)

Mechanics: We usually make histograms with a computer or graphing calculator.

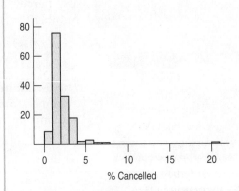

The histogram shows a distribution skewed to the high end and one extreme outlier, a month in which more than 20% of flights were cancelled.

 It's always a good idea to think about what you expect to see so that you can check whether the histogram looks like what you expected.

In most months, fewer than 5% of flights are cancelled and usually only about 2% or 3%. That seems reasonable.

With 144 cases, we probably have more data than you'd choose to work with by hand. The results given here are from technology.

Count	144
Max	20.24
Q3	2.525
Median	1.740
Q1	1.445
Min	0.770
IQR	1.080

TELL **Interpretation:** Describe the shape, center, and spread of the distribution. Report on the symmetry, number of modes, and any gaps or outliers. You should also mention any concerns you may have about the data.

The distribution of cancellations is skewed to the right, and this makes sense: The values can't fall below 0%, but can increase almost arbitrarily due to bad weather or other events.

The median is 1.74% and the IQR is 1.08%. The low IQR indicates that in most months the cancellation rate is close to the median. In fact, it's between 1.4% and 2.5% in the middle 50% of all months, and in only 1/4 of the months were more than 2.5% of flights cancelled.

There is one extraordinary value: 20.2%. Looking it up, I find that the extraordinary month was September 2001. The attacks of September 11 shut down air travel for several days, accounting for this outlier.

Summarizing Symmetric Distributions: The Mean

Medians do a good job of summarizing the center of a distribution, even when the shape is skewed or when there is an outlier, as with the flight cancellations. But when we have symmetric data, there's another alternative. You probably already know how to average values. In fact, to find the median when n is even, we said you should average the two middle values, and you didn't even flinch.

The earthquake magnitudes are pretty close to symmetric, so we can also summarize their center with a mean. The mean tsunami earthquake magnitude is 6.96—about what we might expect from the histogram. You already know how to average values, but this is a good place to introduce notation that we'll use throughout the book. We use the Greek capital letter sigma, Σ, to mean "sum" (sigma is "S" in Greek), and we'll write:

$$\bar{y} = \frac{Total}{n} = \frac{\Sigma y}{n}.$$

The formula says to add up all the values of the variable and divide that sum by the number of data values, n—just as you've always done.[8]

Once we've averaged the data, you'd expect the result to be called the *average*, but that would be too easy. Informally, we speak of the "average person" but we don't add up people and divide by the number of people. A median is also a kind of average. To make this distinction, the value we calculated is called the mean, $\bar{y}$, and pronounced "*y*-bar."

The **mean** feels like the center because it is the point where the histogram balances:

In everyday language, sometimes "average" *does* mean what we want it to mean. We don't talk about your grade point mean or a baseball player's batting mean or the Dow Jones industrial mean. So we'll continue to say "average" when that seems most natural. When we do, though, you may assume that what we mean is the mean.

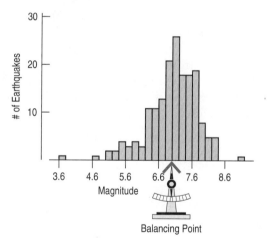

FIGURE 4.12
The mean is located at the *balancing point* of the histogram.

Mean or Median?

Using the center of balance makes sense when the data are symmetric. But data are not always this well behaved. If the distribution is skewed or has outliers, the center is not so well defined and the mean may not be what we want. For example, the mean of the flight cancellations doesn't give a very good idea of the typical percentage of cancellations.

---

[8] You may also see the variable called x and the equation written $\bar{x} = \frac{Total}{n} = \frac{\Sigma x}{n}$. Don't let that throw you. You are free to name the variable anything you want, but we'll generally use y for variables like this that we want to summarize, model, or predict. (Later we'll talk about variables that are used to explain, model, or predict y. We'll call them x.)

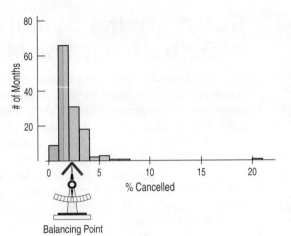

FIGURE 4.13
The median splits the area of the histogram in half at 1.755%. Because the distribution is skewed to the right, the mean (2.28%) is higher than the median. The points at the right have pulled the mean toward them away from the median.

 A S   *Activity:* **The Center of a Distribution.** Compare measures of center by dragging points up and down and seeing the consequences. Another activity shows how to find summaries with your statistics package.

The mean is 2.28%, but nearly 70% of months had cancellation rates below that, so the mean doesn't feel like a good overall summary. Why is the balancing point so high? The large outlying value pulls it to the right. For data like these, the median is a better summary of the center.

Because the median considers only the order of the values, it is **resistant** to values that are extraordinarily large or small; it simply notes that they are one of the "big ones" or the "small ones" and ignores their distance from the center.

For the tsunami earthquake magnitudes, it doesn't seem to make much difference—the mean is 6.96; the median is 7.0. When the data are symmetric, the mean and median will be close, but when the data are skewed, the median is likely to be a better choice. So, why not just use the median? Well, for one, the median can go overboard. It's not just resistant to occasional outliers, but can be unaffected by changes in up to half the data values. By contrast, the mean includes input from each data value and gives each one equal weight. It's also easier to work with, so when the distribution is unimodal and symmetric, we'll use the mean.

Of course, to choose between mean and median, we'll start by looking at the data. If the histogram is symmetric and there are no outliers, we'll prefer the mean. However, if the histogram is skewed or has outliers, we're usually better off with the median. If you're not sure, report both and discuss why they might differ.

FOR EXAMPLE   Describing Center

RECAP: You want to summarize the expenditures of 500 credit card company customers, and have looked at a histogram.

QUESTION: You have found the mean expenditure to be $478.19 and the median to be $216.28. Which is the more appropriate measure of center, and why?

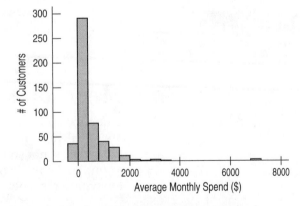

Because the distribution of expenditures is skewed, the median is the more appropriate measure of center. Unlike the mean, it's not affected by the large outlying value or by the skewness. Half of these credit card customers had average monthly expenditures less than $216.28 and half more.

When to expect skewness.   Even without making a histogram, we can expect some variables to be skewed. When values of a quantitative variable are bounded on one side but not the other, the distribution may be skewed. For example, incomes and waiting times can't be less than zero, so they are often skewed to the right. Amounts of things (dollars, employees) are often skewed to the right for the same reason. If a test is too easy, the distribution will be skewed to the left because many scores will bump against 100%. And combinations of things are often skewed. In the case of the cancelled flights, flights are more likely to be cancelled in January (due to snowstorms) and in August (thunderstorms). Combining values across months leads to a skewed distribution.

What About Spread? The Standard Deviation

The IQR is always a reasonable summary of spread, but because it uses only the two quartiles of the data, it ignores much of the information about how individual values vary. A more powerful approach uses the **standard deviation**, which takes into account how far *each* value is from the mean. Like the mean, the standard deviation is appropriate only for symmetric data.

One way to think about spread is to examine how far each data value is from the mean. This difference is called a *deviation*. We could just average the deviations, but the positive and negative differences always cancel each other out. So the average deviation is always zero—not very helpful.

To keep them from canceling out, we *square* each deviation. Squaring always gives a positive value, so the sum won't be zero. That's great. Squaring also emphasizes larger differences—a feature that turns out to be both good and bad.

When we add up these squared deviations and find their average (almost), we call the result the **variance**:

$$s^2 = \frac{\sum (y - \bar{y})^2}{n - 1}.$$

Why almost? It *would* be a mean if we divided the sum by n. Instead, we divide by $n - 1$. Why? The simplest explanation is "to drive you crazy." But there are good technical reasons, some of which we'll see later.

The variance will play an important role later in this book, but it has a problem as a measure of spread. Whatever the units of the original data are, the variance is in *squared* units. We want measures of spread to have the same units as the data. And we probably don't want to talk about squared dollars or mpg². So, to get back to the original units, we take the square root of s^2. The result, s, is the **standard deviation**.

Putting it all together, the standard deviation of the data is found by the following formula:

$$s = \sqrt{\frac{\sum (y - \bar{y})^2}{n - 1}}.$$

WHO	52 adults
WHAT	Resting heart rates
UNITS	Beats per minute

You will almost always rely on a calculator or computer to do the calculating.

Understanding what the standard deviation really means will take some time, and we'll revisit the concept in later chapters. For now, have a look at this histogram of resting pulse rates. The distribution is roughly symmetric, so it's okay to choose the mean and standard deviation as our summaries of center and spread. The mean pulse rate is 72.7 beats per minute, and we can see that's a typical heart rate. We also see that some heart rates are higher and some lower—but how much? Well, the standard deviation of 6.5 beats per minute indicates that, on average, we might expect people's heart rates to differ from the

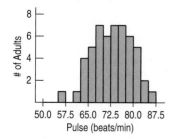

mean rate by about 6.5 beats per minute. Looking at the histogram, we can see that 6.5 beats above or below the mean appears to be a typical deviation.

A S *Activity:* **The Spread of a Distribution.** What happens to measures of spread when some of the data values change may not be quite what you expect.

How does standard deviation work?   To find the standard deviation, start with the mean, $\bar{y}$. Then find the *deviations* by taking $\bar{y}$ from each value: $(y - \bar{y})$. Square each deviation: $(y - \bar{y})^2$.

Now just add up the squared deviations and divide by $n - 1$. That gives you the variance, s^2. To find the standard deviation, s, take the square root. Here we go:

Suppose the batch of values is 14, 13, 20, 22, 18, 19, and 13.

The mean is $\bar{y} = 17$. So the deviations are found by subtracting 17 from each value:

Original Values	Deviations	Squared Deviations
14	$14 - 17 = -3$	$(-3)^2 = 9$
13	$13 - 17 = -4$	$(-4)^2 = 16$
20	$20 - 17 = 3$	9
22	$22 - 17 = 5$	25
18	$18 - 17 = 1$	1
19	$19 - 17 = 2$	4
13	$13 - 17 = -4$	16

Add up the squared deviations: $9 + 16 + 9 + 25 + 1 + 4 + 16 = 80$.

Now divide by $n - 1$: $80/6 = 13.33$.

Finally, take the square root: $s = \sqrt{13.33} = 3.65$

Thinking About Variation

A S *Activity:* **Displaying Spread.** What does the standard deviation look like on a histogram? How about the IQR?

Statistics is about variation, so spread is an important fundamental concept in Statistics. Measures of spread help us to be precise about what we *don't* know. If many data values are scattered far from the center, the IQR and the standard deviation will be large. If the data values are close to the center, then these measures of spread will be small. If all our data values were exactly the same, we'd have no question about summarizing the center, and all measures of spread would be zero—and we wouldn't need Statistics. You might think this would be a big plus, but it would make for a boring world. Fortunately (at least for Statistics), data do vary.

Measures of spread tell how well other summaries describe the data. That's why we always (always!) report a spread along with any summary of the center.

FOR EXAMPLE   **Describing Spread**

RECAP: The histogram has shown you that the distribution of credit card expenditures is skewed, and you have used the median to describe the center. The quartiles are $73.84 and $624.80.

QUESTION: What is the IQR and why is it a suitable measure of spread?

For these data, the interquartile range (IQR) is $624.80 − $73.84 = $550.96. Like the median, the IQR is not affected by the outlying value or by the skewness of the distribution, so it is an appropriate measure of spread for the given expenditures.

JUST CHECKING

6. The U.S. Census Bureau reports the median family income in its summary of census data. Why do you suppose they use the median instead of the mean? What might be the disadvantages of reporting the mean?

7. You've just bought a new car that claims to get a highway fuel efficiency of 31 miles per gallon. Of course, your mileage will "vary." If you had to guess, would you expect the IQR of gas mileage attained by all cars like yours to be 30 mpg, 3 mpg, or 0.3 mpg? Why?

8. A company selling a new MP3 player advertises that the player has a mean lifetime of 5 years. If you were in charge of quality control at the factory, would you prefer that the standard deviation of life spans of the players you produce be 2 years or 2 months? Why?

What to *Tell* About a Quantitative Variable

A S *Activity:* **Playing with Summaries.** Here's a Statistics game about summaries that even some experienced statisticians find . . . well, challenging. Your intuition may be better. Give it a try!

What should you *Tell* about a quantitative variable?

- Start by making a histogram or stem-and-leaf display, and discuss the shape of the distribution.
- Next, discuss the center and spread.
  - Always pair the median with the IQR and the mean with the standard deviation. It's not useful to report one without the other. Reporting a center without a spread is dangerous. You may think you know more than you do about the distribution. Reporting only the spread leaves us wondering where we are.
  - If the shape is skewed, report the median and IQR. You may want to include the mean and standard deviation as well, but you should point out why the mean and median differ.
  - If the shape is symmetric, report the mean and standard deviation and possibly the median and IQR as well. For unimodal symmetric data, the IQR is usually a bit larger than the standard deviation. If that's not true of your data set, look again to make sure that the distribution isn't skewed and there are no outliers.
- Also, discuss any unusual features.
  - If there are multiple modes, try to understand why. If you can identify a reason for separate modes (for example, women and men typically have heart attacks at different ages), it may be a good idea to split the data into separate groups.
  - If there are any clear outliers, you should point them out. If you are reporting the mean and standard deviation, report them with the outliers present and with the outliers omitted. The differences may be revealing. (Of course, the median and IQR won't be affected very much by the outliers.)

HOW "ACCURATE" SHOULD WE BE?

Don't think you should report means and standard deviations to a zillion decimal places; such implied accuracy is really meaningless. Although there is no ironclad rule, statisticians commonly report summary statistics to one or two decimal places more than the original data have.

FOR EXAMPLE | **Choosing Summary Statistics**

RECAP: You have provided the credit card company's board of directors with a histogram of customer expenditures, and you have summarized the center and spread with the median and IQR. Knowing a little Statistics, the directors now insist on having the mean and standard deviation as summaries of the spending data.

QUESTION: Although you know that the mean is $478.19 and the standard deviation is $741.87, you need to explain to them why these are not suitable summary statistics for these expenditures data. What would you give as reasons?

The high outlier at $7000 pulls the mean up substantially and inflates the standard deviation. Locating the mean value on the histogram shows that it is not a typical value at all, and the standard deviation suggests that expenditures vary much more than they do. The median and IQR are more resistant to the presence of skewness and outliers, giving more realistic descriptions of center and spread.

STEP-BY-STEP EXAMPLE | **Summarizing a Distribution**

One of the authors owned a 1989 Nissan Maxima for 8 years. Being a statistician, he recorded the car's fuel efficiency (in mpg) each time he filled the tank. He wanted to know what fuel efficiency to expect as "ordinary" for his car. (Hey, he's a statistician. What would you expect?[9]) Knowing this, he was able to predict when he'd need to fill the tank again and to notice if the fuel efficiency suddenly got worse, which could be a sign of trouble.

Question: How would you describe the distribution of *Fuel efficiency* for this car?

THINK **Plan** State what you want to find out.	I want to summarize the distribution of Nissan Maxima fuel efficiency.
Variable Identify the variable and report the W's.	The data are the fuel efficiency values in miles per gallon for the first 100 fill-ups of a 1989 Nissan Maxima between 1989 and 1992.
Be sure to check the appropriate condition.	✔ **Quantitative Data Condition:** The fuel efficiencies are quantitative with units of miles per gallon. Histograms are appropriate displays for displaying the distribution. Numerical summaries are appropriate as well.

 SHOW **Mechanics** Make a histogram and boxplot. Based on the shape, choose appropriate numerical summaries.

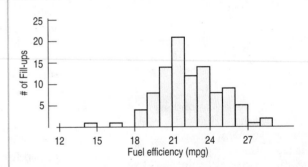

REALITY CHECK A value of 22 mpg seems reasonable for such a car. The spread is reasonable, although the range looks a bit large.

A histogram of the data shows a fairly symmetric distribution with a low outlier.

Count	100
Mean	22.4 mpg
StdDev	2.45
Q1	20.8
Median	22.0
Q3	24.0
IQR	3.2

The mean and median are close, so the outlier doesn't seem to be a problem. I can use the mean and standard deviation.

---

[9] He also recorded the time of day, temperature, price of gas, and phase of the moon. (OK, maybe not phase of the moon.) His data are on the DVD.

Conclusion Summarize and interpret your findings in context. Be sure to discuss the distribution's shape, center, spread, and unusual features (if any).

The distribution of mileage is unimodal and roughly symmetric with a mean of 22.4 mpg. There is a low outlier that should be investigated, but it does not influence the mean very much. The standard deviation suggests that from tankful to tankful, I can expect the car's fuel economy to differ from the mean by an average of about 2.45 mpg.

Are my statistics "right"? When you calculate a mean, the computation is clear: You sum all the values and divide by the sample size. You may round your answer less or more than someone else (we recommend one more decimal place than the data), but all books and technologies agree on how to find the mean. Some statistics, however, are more problematic. For example, we've already pointed out that methods of finding quartiles differ.

Differences in numeric results can also arise from decisions in the middle of calculations. For example, if you round off your value for the mean before you calculate the sum of squared deviations, your standard deviation probably won't agree with a computer program that calculates using many decimal places. (We do recommend that you use as many digits as you can during the calculation and round only when you are done to minimize this effect.)

Don't be overly concerned with these discrepancies, especially if the differences are small. They don't mean that your answer is "wrong," and they usually won't change any conclusion you might draw about the data. Sometimes (in footnotes and in the answers in the back of the book) we'll note alternative results, but we could never list all the possible values, so we'll rely on your common sense to focus on the meaning rather than on the digits. Remember: Answers are sentences!

What Can Go Wrong?

A data display should tell a story about the data. To do that, it must speak in a clear language, making plain what variable is displayed, what any axis shows, and what the values of the data are. And it must be consistent in those decisions.

A display of quantitative data can go wrong in many ways. The most common failures arise from only a few basic errors:

- **Don't make a histogram of a categorical variable.** Just because the variable contains numbers doesn't mean that it's quantitative. Here's a histogram of the insurance policy numbers of some workers.

FIGURE 4.14
It's not appropriate to display these data with a histogram.

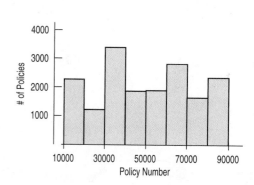

It's not very informative because the policy numbers are just labels. A histogram or stem-and-leaf display of a categorical variable makes no sense. A bar chart or pie chart would be more appropriate.

■ **Don't look for shape, center, and spread of a bar chart.** A bar chart showing the sizes of the piles displays the distribution of a categorical variable, but the bars could be arranged in any order left to right. Concepts like symmetry, center, and spread make sense only for quantitative variables.

■ **Don't use bars in every display—save them for histograms and bar charts.** In a bar chart, the bars indicate how many cases of a categorical variable are piled in each category. Bars in a histogram indicate the number of cases piled in each interval of a quantitative variable. In both bar charts and histograms, the bars represent counts of data values. Some people create other displays that use bars to represent individual data values. Beware: Such graphs are neither bar charts nor histograms. For example, a student was asked to make a histogram from data showing the number of juvenile bald eagles seen during each of the 13 weeks in the winter of 2003–2004 at a site in Rock Island, IL. Instead, he made this plot:

FIGURE 4.15
This isn't a histogram or a bar chart. It's an ill-conceived graph that uses bars to represent individual data values (number of eagles sighted) week by week.

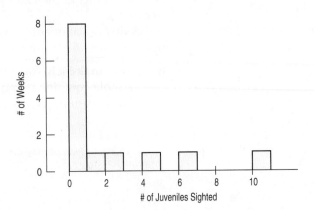

Look carefully. That's not a histogram. A histogram shows *What* we've measured along the horizontal axis and counts of the associated *Who*'s represented as bar heights. This student has it backwards: He used bars to show counts of birds for each week.[10] We need counts of weeks. A correct histogram should have a tall bar at "0" to show there were many weeks when no eagles were seen, like this:

FIGURE 4.16
A histogram of the eagle-sighting data shows the number of weeks in which different counts of eagles occurred. This display shows the distribution of juvenile-eagle sightings.

---

[10] Edward Tufte, in his book *The Visual Display of Quantitative Information*, proposes that graphs should have a high data-to-ink ratio. That is, we shouldn't waste a lot of ink to display a single number when a dot would do the job.

- **Choose a bin width appropriate to the data.** Computer programs usually do a pretty good job of choosing histogram bin widths. Often there's an easy way to adjust the width, sometimes interactively. Here are the tsunami earthquakes with two (rather extreme) choices for the bin size:

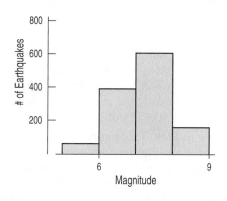 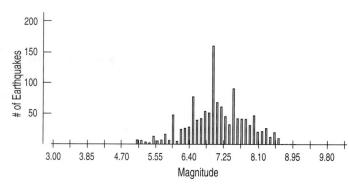

The task of summarizing a quantitative variable is relatively simple, and there is a simple path to follow. However, you need to watch out for certain features of the data that make summarizing them with a number dangerous. Here's some advice:

- **Don't forget to do a reality check.** Don't let the computer or calculator do your thinking for you. Make sure the calculated summaries make sense. For example, does the mean look like it is in the center of the histogram? Think about the spread: An IQR of 50 mpg would clearly be wrong for gas mileage. And no measure of spread can be negative. The standard deviation can take the value 0, but only in the very unusual case that all the data values equal the same number. If you see an IQR or standard deviation equal to 0, it's probably a sign that something's wrong with the data.

- **Don't forget to sort the values before finding the median or percentiles.** It seems obvious, but when you work by hand, it's easy to forget to sort the data first before counting in to find medians, quartiles, or other percentiles. Don't report that the median of the five values 194, 5, 1, 17, and 893 is 1 just because 1 is the middle number.

- **Don't worry about small differences when using different methods.** Finding the 10th percentile or the lower quartile in a data set sounds easy enough. But it turns out that the definitions are not exactly clear. If you compare different statistics packages or calculators, you may find that they give slightly different answers for the same data. These differences, though, are unlikely to be important in interpreting the data, the quartiles, or the IQR, so don't let them worry you.

Gold Card Customers—Regions National Banks		
Month	**April 2007**	**May 2007**
Average Zip Code	45,034.34	38,743.34

- **Don't compute numerical summaries of a categorical variable.** Neither the mean zip code nor the standard deviation of social security numbers is meaningful. If the variable is categorical, you should instead report summaries such as percentages of individuals in each category. It is easy to make this mistake when using technology to do the summaries for you. After all, the computer doesn't care what the numbers mean.

- **Don't report too many decimal places.** Statistical programs and calculators often report a ridiculous number of digits. A general rule for numerical summaries is to report one or two more digits than the number of digits in the data. For example, earlier we saw a dotplot of the Kentucky

Derby race times. The mean and standard deviation of those times could be reported as:

$$\bar{y} = 130.63401639344262 \text{ sec} \qquad s = 13.66448201942662 \text{ sec}$$

But we knew the race times only to the nearest quarter second, so the extra digits are meaningless.

- **Don't round in the middle of a calculation.** Don't *report* too many decimal places, but it's best not to do any rounding until the end of your calculations. Even though you might report the mean of the earthquake magnitudes as 7.08, it's really 7.08339. Use the more precise number in your calculations if you're finding the standard deviation by hand—or be prepared to see small differences in your final result.

- **Watch out for multiple modes.** The summaries of the Kentucky Derby times are meaningless for another reason. As we saw in the dotplot, the Derby was initially a longer race. It would make much more sense to report that the old 1.5 mile Derby had a mean time of 159.6 seconds, while the current Derby has a mean time of 124.6 seconds. If the distribution has multiple modes, consider separating the data into different groups and summarizing each group separately.

- **Beware of outliers.** The median and IQR are resistant to outliers, but the mean and standard deviation are not. To help spot outliers . . .

- **Don't forget to: Make a picture (make a picture, make a picture).** The sensitivity of the mean and standard deviation to outliers is one reason you should always make a picture of the data. Summarizing a variable with its mean and standard deviation when you have not looked at a histogram or dotplot to check for outliers or skewness invites disaster. You may find yourself drawing absurd or dangerously wrong conclusions about the data. And, of course, you should demand no less of others. Don't accept a mean and standard deviation blindly without some evidence that the variable they summarize is unimodal, symmetric, and free of outliers.

CONNECTIONS

Distributions of quantitative variables, like those of categorical variables, show the possible values and their relative frequencies. A histogram shows the distribution of values in a quantitative variable with adjacent bars. Don't confuse histograms with bar charts, which display categorical variables. For categorical data, the mode is the category with the biggest count. For quantitative data, modes are peaks in the histogram.

The shape of the distribution of a quantitative variable is an important concept in most of the subsequent chapters. We will be especially interested in distributions that are unimodal and symmetric.

In addition to their shape, we summarize distributions with center and spread, usually pairing a measure of center with a measure of spread: median with IQR and mean with standard deviation. We favor the mean and standard deviation when the shape is unimodal and symmetric, but choose the median and IQR for skewed distributions or when there are outliers we can't otherwise set aside.

WHAT HAVE WE LEARNED?

We've learned how to make a picture of quantitative data to help us see the story the data have to *Tell*.

▶ We can display the distribution of quantitative data with a *histogram*, a *stem-and-leaf* display, or a *dotplot*.

▶ We *Tell* what we see about the distribution by talking about *shape, center, spread*, and any *unusual features*.

We've learned how to summarize distributions of quantitative variables numerically.

▶ Measures of center for a distribution include the median and the mean.

We write the formula for the mean as $\bar{y} = \dfrac{\sum y}{n}$.

▶ Measures of spread include the range, IQR, and standard deviation.

The standard deviation is computed as $s = \sqrt{\dfrac{\sum (y - \bar{y})^2}{n - 1}}$.

The median and IQR are not usually given as formulas.

▶ We'll report the median and IQR when the distribution is skewed. If it's symmetric, we'll summarize the distribution with the mean and standard deviation (and possibly the median and IQR as well). Always pair the median with the IQR and the mean with the standard deviation.

We've learned to *Think* about the type of variable we're summarizing.

▶ All the methods of this chapter assume that the data are quantitative.

▶ The **Quantitative Data Condition** serves as a check that the data are, in fact, quantitative. One good way to be sure is to know the measurement units. You'll want those as part of the *Think* step of your answers.

Terms	
Distribution	The distribution of a quantitative variable slices up all the possible values of the variable into equal-width bins and gives the number of values (or counts) falling into each bin (p. 44).
Histogram (relative frequency histogram)	A histogram uses adjacent bars to show the distribution of a quantitative variable. Each bar represents the frequency (or relative frequency) of values falling in each bin (p. 44).
Gap	A region of the distribution where there are no values (p. 46).
Stem-and-leaf display	A stem-and-leaf display shows quantitative data values in a way that sketches the distribution of the data. It's best described in detail by example (p. 46).
Dotplot	A dotplot graphs a dot for each case against a single axis (p. 48).
Shape	To describe the shape of a distribution, look for (p. 49)
	▶ single vs. multiple modes.
	▶ symmetry vs. skewness.
	▶ outliers and gaps.
Mode	A hump or local high point in the shape of the distribution of a variable. The apparent location of modes can change as the scale of a histogram is changed (p. 49).
Unimodal (Bimodal)	Having one mode. This is a useful term for describing the shape of a histogram when it's generally mound-shaped. Distributions with two modes are called **bimodal**. Those with more than two are **multimodal** (p. 49).
Uniform	A distribution that's roughly flat is said to be uniform (p. 49).

Symmetric	A distribution is symmetric if the two halves on either side of the center look approximately like mirror images of each other (p. 49).
Tails	The tails of a distribution are the parts that typically trail off on either side. Distributions can be characterized as having long tails (if they straggle off for some distance) or short tails (if they don't) (p. 50).
Skewed	A distribution is skewed if it's not symmetric and one tail stretches out farther than the other. Distributions are said to be **skewed left** when the longer tail stretches to the left, and **skewed right** when it goes to the right (p. 50).
Outliers	Outliers are extreme values that don't appear to belong with the rest of the data. They may be unusual values that deserve further investigation, or they may be just mistakes; there's no obvious way to tell. Don't delete outliers automatically—you have to think about them. Outliers can affect many statistical analyses, so you should always be alert for them (p. 50).
Center	The place in the distribution of a variable that you'd point to if you wanted to attempt the impossible by summarizing the entire distribution with a single number. Measures of center include the mean and median (p. 52).
Median	The median is the middle value, with half of the data above and half below it. If n is even, it is the average of the two middle values. It is usually paired with the IQR (p. 52).
Spread	A numerical summary of how tightly the values are clustered around the center. Measures of spread include the IQR and standard deviation (p. 53).
Range	The difference between the lowest and highest values in a data set (p. 53). $Range = max - min.$
Quartile	The lower quartile (Q1) is the value with a quarter of the data below it. The upper quartile (Q3) has three quarters of the data below it. The median and quartiles divide data into four parts with equal numbers of data values (p. 53).
Percentile	The ith percentile is the number that falls above i% of the data (p. 54).
Interquartile range (IQR)	The IQR is the difference between the first and third quartiles. $IQR = Q3 - Q1$. It is usually reported along with the median (p. 54).
5-Number Summary	The 5-number summary of a distribution reports the minimum value, Q1, the median, Q3, and the maximum value (p. 55).
Mean	The mean is found by summing all the data values and dividing by the count: $$\bar{y} = \frac{Total}{n} = \frac{\sum y}{n}.$$ It is usually paired with the standard deviation (p. 57).
Resistant	A calculated summary is said to be resistant if outliers have only a small effect on it (p. 58).
Variance	The variance is the sum of squared deviations from the mean, divided by the count minus 1: $$s^2 = \frac{\sum (y - \bar{y})^2}{n - 1}.$$ It is useful in calculations later in the book (p. 59).
Standard deviation	The standard deviation is the square root of the variance: $$s = \sqrt{\frac{\sum (y - \bar{y})^2}{n - 1}}$$ It is usually reported along with the mean (p. 59).

Skills

THINK

▸ Be able to identify an appropriate display for any quantitative variable.

▸ Be able to guess the shape of the distribution of a variable by knowing something about the data.

▸ Be able to select a suitable measure of center and a suitable measure of spread for a variable based on information about its distribution.

▶ Know the basic properties of the median: The median divides the data into the half of the data values that are below the median and the half that are above.

▶ Know the basic properties of the mean: The mean is the point at which the histogram balances.

▶ Know that the standard deviation summarizes how spread out all the data are around the mean.

▶ Understand that the median and IQR resist the effects of outliers, while the mean and standard deviation do not.

▶ Understand that in a skewed distribution, the mean is pulled in the direction of the skewness (toward the longer tail) relative to the median.

SHOW ▶ Know how to display the distribution of a quantitative variable with a stem-and-leaf display (drawn by hand for smaller data sets), a dotplot, or a histogram (made by computer for larger data sets).

▶ Know how to compute the mean and median of a set of data.

▶ Know how to compute the standard deviation and IQR of a set of data.

TELL ▶ Be able to describe the distribution of a quantitative variable in terms of its shape, center, and spread.

▶ Be able to describe any anomalies or extraordinary features revealed by the display of a variable.

▶ Know how to describe summary measures in a sentence. In particular, know that the common measures of center and spread have the same units as the variable that they summarize, and should be described in those units.

▶ Be able to describe the distribution of a quantitative variable with a description of the shape of the distribution, a numerical measure of center, and a numerical measure of spread. Be sure to note any unusual features, such as outliers, too.

DISPLAYING AND SUMMARIZING QUANTITATIVE VARIABLES ON THE COMPUTER

Almost any program that displays data can make a histogram, but some will do a better job of determining where the bars should start and how they should partition the span of the data.

The vertical scale may be counts or proportions. Sometimes it isn't clear which. But the shape of the histogram is the same either way.

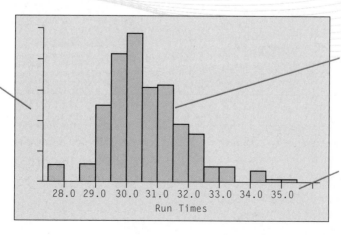

Most packages choose the number of bars for you automatically. Often you can adjust that choice.

The axis should be clearly labeled so you can tell what "pile" each bar represents. You should be able to tell the lower and upper bounds of each bar.

Many statistics packages offer a prepackaged collection of summary measures. The result might look like this:

Variable: Weight
N = 234
Mean = 143.3   Median = 139
St. Dev = 11.1   IQR = 14

Alternatively, a package might make a table for several variables and summary measures:

Case Study: Describing Distribution Shapes. Who's safer in a crash–passengers or the driver? Investigate with your statistics package.

Variable	N	mean	median	stdev	IQR
Weight	234	143.3	139	11.1	14
Height	234	68.3	68.1	4.3	5
Score	234	86	88	9	5

You should be able to read the summary statistics produced by any computer package. It is usually easy to read the results and identify each computed summary.

Packages often provide many more summary statistics than you need. Of course, some of these may not be appropriate when the data are skewed or have outliers. It is your responsibility to check a histogram or stem-and-leaf display and decide which summary statistics to use.

It is common for packages to report summary statistics to many decimal places of "accuracy." Of course, it is rare data that have such accuracy in the original measurements. Just because a package calculates to six or seven digits beyond the decimal point doesn't mean that those digits have any meaning. Generally it's a good idea to round these values, allowing perhaps one more digit of precision than was given in the original data.

Displays and summaries of quantitative variables are among the simplest things you can do in most statistics packages.

DATA DESK

To make a histogram:

- Select the variable to display.
- In the **Plot** menu, choose **Histogram**.

To calculate summaries:

- In the **Calc** menu, open the **summaries** submenu. **Options** offer separate tables, a single unified table, and other formats.

EXCEL

Excel cannot make histograms or dotplots without an add-in.

To calculate summaries,

Click on an empty cell. Type an equals sign and choose "**Average**" from the pop-up list of functions that appears to the left of the text editing box. Enter the data range in the box that says "**Number 1.**" Click the **OK** button.

To compute the standard deviation of a column of data directly, use the **STDEV** from the pop-up list of functions in the same way.

COMMENTS

Excel's Data Analysis add-in offers a way to compute a histogram, but you will have to adjust the bar edges, eliminate space between the bars, and interpret the bar labels carefully. The DDXL add-in provided on our DVD adds these and other capabilities to Excel.

EXCEL 2007

In Excel 2007 there is another way to find some of the standard summary statistics. For example, to compute the mean:

- Click on an empty cell.
- Go to the Formulas tab in the Ribbon. Click on the drop down arrow next to "AutoSum" and choose "**Average.**"

- Enter the data range in the formula displayed in the empty box you selected earlier.
- Press **Enter**. This computes the mean for the values in that range.

(continued)

To compute the standard deviation:

- Click on an empty cell.
- Go to the Formulas tab in the Ribbon and click the drop down arrow next to "AutoSum" and select **"More functions..."**
- In the dialog window that opens, select **"STDEV"** from the list of functions and click **OK**. A new dialog

window opens. Enter a range of fields into the text fields and click **OK**.

Excel 2007 computes the standard deviation for the values in that range and places it in the specified cell of the spreadsheet.

JMP

To make a histogram and find summary statistics:

- Choose **Distribution** from the **Analyze** menu.
- In the **Distribution** dialog, drag the name of the variable that you wish to analyze into the empty window beside the label "**Y, Columns.**"

- Click **OK**. JMP computes standard summary statistics along with displays of the variables.

MINITAB

To make a histogram:

- Choose **Histogram** from the **Graph** menu.
- Select "Simple" for the type of graph and click **OK**.
- Enter the name of the quantitative variable you wish to display in the box labeled "Graph variables." Click **OK**.

To calculate summary statistics:

- Choose **Basic statistics** from the **Stat** menu. From the **Basic Statistics** submenu, choose **Display Descriptive Statistics**.
- Assign variables from the variable list box to the Variables box. MINITAB makes a Descriptive Statistics table.

SPSS

To make a histogram in SPSS open the Chart Builder from the Graphs menu.

- Click the **Gallery** tab.
- Choose **Histogram** from the list of chart types.
- Drag the histogram onto the canvas.
- Drag a scale variable to the y-axis drop zone.
- Click **OK**.

To calculate summary statistics:

- Choose **Explore** from the **Descriptive Statistics** submenu of the **Analyze** menu. In the Explore dialog, assign one or more variables from the source list to the Dependent List and click the **OK** button.

TI-83/84 PLUS

Choose **1-VarStats** from the **STAT CALC** menu and specify the List where the data are stored. You must scroll down to see the 5-number summary.

To make a boxplot, set up a **STAT PLOT** using the boxplot icon.

COMMENTS

Note that the standard deviation is identified as Sx; don't use σ_x by mistake, because it divides by n instead of $n - 1$.

TI-89

- To compute summary statistics press F4 (Calc). Input the name of the list using VAR-LINK. Press ENTER. Use the down arrow to scroll through the output.
- To create a boxplot, press F2 (Plots) then ENTER. Select a plot to define and press F1. Select either 3: **Box Plot** or 4: **Mod Box Plot** (to identify outliers). Select the mark type of your choice (for outliers). Press ENTER to finish. Press F5 to display the graph.

COMMENTS

If the data are stored as a frequency table (say, with data values in list1 and frequencies in list2), use VAR-LINK to select list2 as the frequency variable in 1-Var Stats.

For the plot, change Use Freq and Categories to YES and use VAR-LINK to select list2 as the frequency variable on the plot definition screen.

EXERCISES

1. **Histogram.** Find a histogram that shows the distribution of a variable in a newspaper, a magazine, or the Internet.
   a) Does the article identify the W's?
   b) Discuss whether the display is appropriate.
   c) Discuss what the display reveals about the variable and its distribution.
   d) Does the article accurately describe and interpret the data? Explain.

2. **Not a histogram.** Find a graph other than a histogram that shows the distribution of a quantitative variable in a newspaper, a magazine, or the Internet.
   a) Does the article identify the W's?
   b) Discuss whether the display is appropriate for the data.
   c) Discuss what the display reveals about the variable and its distribution.
   d) Does the article accurately describe and interpret the data? Explain.

3. **In the news.** Find an article in a newspaper, a magazine, or the Internet that discusses an "average."
   a) Does the article discuss the W's for the data?
   b) What are the units of the variable?
   c) Is the average used the median or the mean? How can you tell?
   d) Is the choice of median or mean appropriate for the situation? Explain.

4. **In the news II.** Find an article in a newspaper, a magazine, or the Internet that discusses a measure of spread.
   a) Does the article discuss the W's for the data?
   b) What are the units of the variable?
   c) Does the article use the range, IQR, or standard deviation?
   d) Is the choice of measure of spread appropriate for the situation? Explain.

5. **Thinking about shape.** Would you expect distributions of these variables to be uniform, unimodal, or bimodal? Symmetric or skewed? Explain why.
   a) The number of speeding tickets each student in the senior class of a college has ever had.
   b) Players' scores (number of strokes) at the U.S. Open golf tournament in a given year.
   c) Weights of female babies born in a particular hospital over the course of a year.
   d) The length of the average hair on the heads of students in a large class.

6. **More shapes.** Would you expect distributions of these variables to be uniform, unimodal, or bimodal? Symmetric or skewed? Explain why.
   a) Ages of people at a Little League game.
   b) Number of siblings of people in your class.
   c) Pulse rates of college-age males.
   d) Number of times each face of a die shows in 100 tosses.

7. **Sugar in cereals.** The histogram displays the sugar content (as a percent of weight) of 49 brands of breakfast cereals.

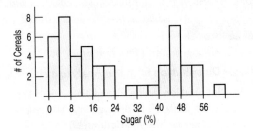

   a) Describe this distribution.
   b) What do you think might account for this shape?

8. **Singers.** The display shows the heights of some of the singers in a chorus, collected so that the singers could be positioned on stage with shorter ones in front and taller ones in back.

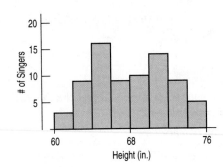

   a) Describe the distribution.
   b) Can you account for the features you see here?

9. **Vineyards.** The histogram shows the sizes (in acres) of 36 vineyards in the Finger Lakes region of New York.

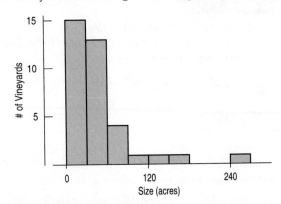

   a) Approximately what percentage of these vineyards are under 60 acres?
   b) Write a brief description of this distribution (shape, center, spread, unusual features).

10. **Run times.** One of the authors collected the times (in minutes) it took him to run 4 miles on various courses during a 10-year period. Here is a histogram of the times.

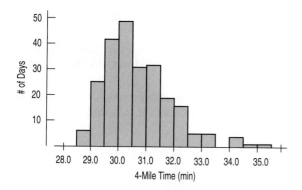

Describe the distribution and summarize the important features. What is it about running that might account for the shape you see?

T 11. Heart attack stays. The histogram shows the lengths of hospital stays (in days) for all the female patients admitted to hospitals in New York during one year with a primary diagnosis of acute myocardial infarction (heart attack).

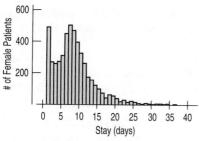

a) From the histogram, would you expect the mean or median to be larger? Explain.
b) Write a few sentences describing this distribution (shape, center, spread, unusual features).
c) Which summary statistics would you choose to summarize the center and spread in these data? Why?

T 12. E-mails. A university teacher saved every e-mail received from students in a large Introductory Statistics class during an entire term. He then counted, for each student who had sent him at least one e-mail, how many e-mails each student had sent.

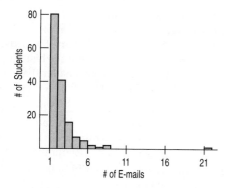

a) From the histogram, would you expect the mean or the median to be larger? Explain.
b) Write a few sentences describing this distribution (shape, center, spread, unusual features).
c) Which summary statistics would you choose to summarize the center and spread in these data? Why?

13. Super Bowl points. How many points do football teams score in the Super Bowl? Here are the total numbers of points scored by both teams in each of the first 43 Super Bowl games:

45, 47, 23, 30, 29, 27, 21, 31, 22, 38, 46, 37, 66, 50, 37, 47, 44, 47, 54, 56, 59, 52, 36, 65, 39, 61, 69, 43, 75, 44, 56, 55, 53, 39, 41, 37, 69, 61, 45, 31, 46, 31, 50

a) Find the median.
b) Find the quartiles.
c) Write a description based on the 5-number summary.

14. Super Bowl wins. In the Super Bowl, by how many points does the winning team outscore the losers? Here are the winning margins for the first 43 Super Bowl games:

25, 19, 9, 16, 3, 21, 7, 17, 10, 4, 18, 17, 4, 12, 17, 5, 10, 29, 22, 36, 19, 32, 4, 45, 1, 13, 35, 17, 23, 10, 14, 7, 15, 7, 27, 3, 27, 3, 3, 11, 12, 3, 4

a) Find the median.
b) Find the quartiles.
c) Write a description based on the 5-number summary.

15. Summaries. Here are costs of 10 electric smoothtop ranges rated very good or excellent by *Consumer Reports* in August 2002.

$850 900 1400 1200 1050 1000 750 1250 1050 565

Find these statistics by *hand* (no calculator!):
a) mean
b) median and quartiles
c) range and IQR

16. Tornadoes 2008. Here are the annual numbers of deaths from tornadoes in the United States from 1998 through 2008 (Source: NOAA):

130   94   40   40   555   54   35   38   67   81   125

Find these statistics *by hand* (no calculator!):
a) mean
b) median and quartiles
c) range and IQR

17. Mistake. A clerk entering salary data into a company spreadsheet accidentally put an extra "0" in the boss's salary, listing it as $2,000,000 instead of $200,000. Explain how this error will affect these summary statistics for the company payroll:

a) measures of center: median and mean.
b) measures of spread: range, IQR, and standard deviation.

18. Sick days. During contract negotiations, a company seeks to change the number of sick days employees may take, saying that the annual "average" is 7 days of absence per employee. The union negotiators counter that the "average" employee misses only 3 days of work each year. Explain how both sides might be correct, identifying the measure of center you think each side is using and why the difference might exist.

19. Standard deviation I. For each lettered part, a through c, examine the two given sets of numbers. Without doing any calculations, decide which set has the larger standard deviation and explain why. Then check by finding the standard deviations *by hand*.

Set 1	Set 2
a) 3, 5, 6, 7, 9	2, 4, 6, 8, 10
b) 10, 14, 15, 16, 20	10, 11, 15, 19, 20
c) 2, 6, 6, 9, 11, 14	82, 86, 86, 89, 91, 94

20. **Standard deviation II.** For each lettered part, a through c, examine the two given sets of numbers. Without doing any calculations, decide which set has the larger standard deviation and explain why. Then check by finding the standard deviations *by hand*.

Set 1	Set 2
a) 4, 7, 7, 7, 10	4, 6, 7, 8, 10
b) 100, 140, 150, 160, 200	10, 50, 60, 70, 110
c) 10, 16, 18, 20, 22, 28	48, 56, 58, 60, 62, 70

21. **Pizza prices.** The histogram shows the distribution of the prices of plain pizza slices (in $) for 156 weeks in Dallas, TX.

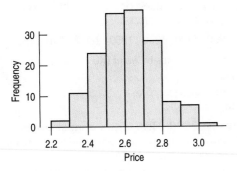

Which summary statistics would you choose to summarize the center and spread in these data? Why?

22. **Neck size.** The histogram shows the neck sizes (in inches) of 250 men recruited for a health study in Utah.

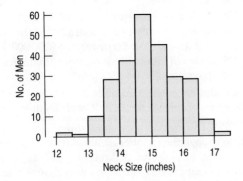

Which summary statistics would you choose to summarize the center and spread in these data? Why?

23. **Pizza prices again.** Look again at the histogram of the pizza prices in Exercise 21.
   a) Is the mean closer to $2.40, $2.60, or $2.80? Why?
   b) Is the standard deviation closer to $0.15, $0.50, or $1.00? Explain.

24. **Neck sizes again.** Look again at the histogram of men's neck sizes in Exercise 22.
   a) Is the mean closer to 14, 15, or 16 inches? Why?
   b) Is the standard deviation closer to 1 inch, 3 inches, or 5 inches? Explain.

25. **Movie lengths.** The histogram shows the running times in minutes of 122 feature films released in 2005.

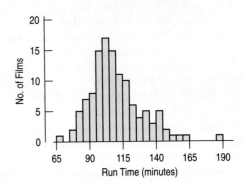

   a) You plan to see a movie this weekend. Based on these movies, how long do you expect a typical movie to run?
   b) Would you be surprised to find that your movie ran for $2\frac{1}{2}$ hours (150 minutes)?
   c) Which would you expect to be higher: the mean or the median run time for all movies? Why?

26. **Golf drives.** The display shows the average drive distance (in yards) for 202 professional golfers on the men's PGA tour.

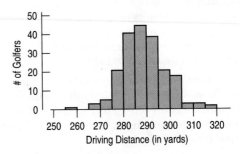

   a) Describe this distribution.
   b) Approximately what proportion of professional male golfers drive, on average, less than 280 yards?
   c) Estimate the mean by examining the histogram.
   d) Do you expect the mean to be smaller than, approximately equal to, or larger than the median? Why?

27. **Movie lengths II.** Exercise 25 looked at the running times of movies released in 2005. The standard deviation of these running times is 19.6 minutes, and the quartiles are $Q_1 = 97$ minutes and $Q_3 = 119$ minutes.
   a) Write a sentence or two describing the spread in running times based on
      i) the quartiles.
      ii) the standard deviation.
   b) Do you have any concerns about using either of these descriptions of spread? Explain.

T 28. Golf drives II. Exercise 26 looked at distances PGA golfers can hit the ball. The standard deviation of these average drive distances is 9.3 yards, and the quartiles are Q_1 = 282 yards and Q_3 = 294 yards.

a) Write a sentence or two describing the spread in distances based on
   i) the quartiles.
   ii) the standard deviation.
b) Do you have any concerns about using either of these descriptions of spread? Explain.

T 29. Movie budgets. The histogram shows the budgets (in millions of dollars) of major release movies in 2005.

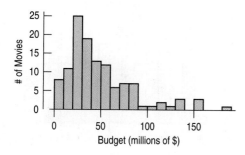

An industry publication reports that the average movie costs $35 million to make, but a watchdog group concerned with rising ticket prices says that the average cost is $46.8 million. What statistic do you think each group is using? Explain.

30. Cold weather. A meteorologist preparing a talk about global warming compiled a list of weekly low temperatures (in degrees Fahrenheit) he observed at his southern Florida home last year. The coldest temperature for any week was 36°F, but he inadvertently recorded the Celsius value of 2°. Assuming that he correctly listed all the other temperatures, explain how this error will affect these summary statistics:

a) measures of center: mean and median.
b) measures of spread: range, IQR, and standard deviation.

31. Payroll. A small warehouse employs a supervisor at $1200 a week, an inventory manager at $700 a week, six stock boys at $400 a week, and four drivers at $500 a week.

a) Find the mean and median wage.
b) How many employees earn more than the mean wage?
c) Which measure of center best describes a typical wage at this company: the mean or the median?
d) Which measure of spread would best describe the payroll: the range, the IQR, or the standard deviation? Why?

T 32. Singers. The frequency table shows the heights (in inches) of 130 members of a choir.

Height	Count	Height	Count
60	2	69	5
61	6	70	11
62	9	71	8
63	7	72	9
64	5	73	4
65	20	74	2
66	18	75	4
67	7	76	1
68	12		

a) Find the median and IQR.
b) Find the mean and standard deviation.
c) Display these data with a histogram.
d) Write a few sentences describing the distribution.

33. Gasoline. In March 2006, 16 gas stations in Grand Junction, CO, posted these prices for a gallon of regular gasoline:

2.22	2.21	2.45	2.24
2.27	2.28	2.27	2.23
2.26	2.46	2.29	2.32
2.36	2.38	2.33	2.27

a) Make a stem-and-leaf display of these gas prices. Use split stems; for example, use two 2.2 stems—one for prices between $2.20 and $2.24 and the other for prices from $2.25 to $2.29.
b) Describe the shape, center, and spread of this distribution.
c) What unusual feature do you see?

34. The Great One. During his 20 seasons in the NHL, Wayne Gretzky scored 50% more points than anyone who ever played professional hockey. He accomplished this amazing feat while playing in 280 fewer games than Gordie Howe, the previous record holder. Here are the number of games Gretzky played during each season:

79, 80, 80, 80, 74, 80, 80, 79, 64, 78, 73, 78, 74, 45, 81, 48, 80, 82, 82, 70

a) Create a stem-and-leaf display for these data, using split stems.
b) Describe the shape of the distribution.
c) Describe the center and spread of this distribution.
d) What unusual feature do you see? What might explain this?

35. States. The stem-and-leaf display shows populations of the 50 states and Washington, DC, in millions of people, according to the 2000 census.

```
3 | 4
2 |
2 | 1
1 | 69
1 | 0122
0 | 5555666667888
0 | 11111111111112222233333334444
```
State Populations (1| 2 means 12 million)

a) What measures of center and spread are most appropriate?

b) Without doing any calculations, which must be larger: the median or the mean? Explain how you know.

c) From the stem-and-leaf display, find the median and the interquartile range.

d) Write a few sentences describing this distribution.

36. **Wayne Gretzky.** In Exercise 34, you examined the number of games played by hockey great Wayne Gretzky during his 20-year career in the NHL.

a) Would you use the median or the mean to describe the center of this distribution? Why?

b) Find the median.

c) Without actually finding the mean, would you expect it to be higher or lower than the median? Explain.

37. **A-Rod 2009.** Alex Rodriguez (known to fans as A-Rod) was the youngest player ever to hit 500 home runs. Here is a stem-and-leaf display of the number of home runs hit by A-Rod during the 1994–2009 seasons. Describe the distribution, mentioning its shape and any unusual features.

```
5 | 247
4 | 1 2278
3 | 05566
2 | 03
1 |
0 | 05
```
(5|2 means 52)

38. **Bird species.** The Cornell Lab of Ornithology holds an annual Christmas Bird Count (www.birdsource.org), in which bird watchers at various locations around the country see how many different species of birds they can spot. Here are some of the counts reported from sites in Texas during the 1999 event:

```
228  178  186  162  206  166  163
183  181  206  177  175  167  162
160  160  157  156  153  153  152
```

a) Create a stem-and-leaf display of these data.

b) Write a brief description of the distribution. Be sure to discuss the overall shape as well as any unusual features.

39. **Hurricanes 2006.** The data below give the number of hurricanes classified as major hurricanes in the Atlantic Ocean each year from 1944 through 2006, as reported by NOAA (www.nhc.noaa.gov):

3, 2, 1, 2, 4, 3, 7, 2, 3, 3, 2, 5, 2, 2, 4, 2, 2, 6, 0, 2, 5, 1, 3, 1, 0, 3, 2, 1, 0, 1, 2, 3, 2, 1, 2, 2, 2, 3, 1, 1, 1, 3, 0, 1, 3, 2, 1, 2, 1, 1, 0, 5, 6, 1, 3, 5, 3, 3, 2, 3, 6, 7, 2

a) Create a dotplot of these data.

b) Describe the distribution.

40. **Horsepower.** Create a stem-and-leaf display for these horsepowers of autos reviewed by *Consumer Reports* one year, and describe the distribution:

155	103	130	80	65
142	125	129	71	69
125	115	138	68	78
150	133	135	90	97
68	105	88	115	110
95	85	109	115	71
97	110	65	90	
75	120	80	70	

41. **A-Rod again.** Students were asked to make a histogram of the number of home runs hit by Alex Rodriguez from 1995 to 2009 (see Exercise 37). One student submitted the following display:

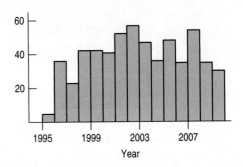

a) Comment on this graph.

b) Create your own histogram of the data.

42. **Return of the birds.** Students were given the assignment to make a histogram of the data on bird counts reported in Exercise 38. One student submitted the following display:

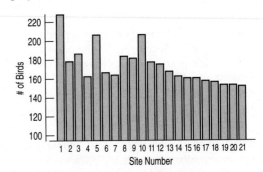

a) Comment on this graph.

b) Create your own histogram of the data.

43. **Acid rain.** Two researchers measured the pH (a scale on which a value of 7 is neutral and values below 7 are acidic) of water collected from rain and snow over a 6-month period in Allegheny County, PA. Describe their data with a graph and a few sentences:

```
4.57  5.62  4.12  5.29  4.64  4.31  4.30  4.39  4.45
5.67  4.39  4.52  4.26  4.26  4.40  5.78  4.73  4.56
5.08  4.41  4.12  5.51  4.82  4.63  4.29  4.60
```

44. **Marijuana 2007.** In 2007 the Council of Europe published a report entitled *The European School Survey Project on Alcohol and Other Drugs* (www.espad.org). Among other issues, the survey investigated the percentages of 16-year-olds who had used marijuana. Shown here are the results for 20 European countries. Create an appropriate graph of these data, and describe the distribution.

Country	Cannabls	Country	Cannabls
Armenia	3	Italy	23
Austria	17	Latvia	18
Belgium	24	Lithuania	18
Bulgaria	22	Malta	13
Croatia	18	Monaco	28
Cyprus	5	Netherlands	28
Czech Republic	45	Norway	6
Estonia	26	Poland	16
Faroe Islands	6	Portugal	13
Finland	8	Romania	4
France	31	Russia	19
Germany	20	Slovak Republic	32
Greece	6	Slovenia	22
Hungary	13	Sweden	7
Iceland	9	Switzerland	33
Ireland	20	Ukraine	14
Isle of Man	34	United Kingdom	29

45. Final grades. A professor (of something other than Statistics!) distributed the following histogram to show the distribution of grades on his 200-point final exam. Comment on the display.

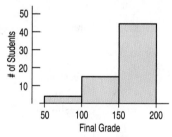

46. Final grades revisited. After receiving many complaints about his final-grade histogram from students currently taking a Statistics course, the professor from Exercise 45 distributed the following revised histogram:

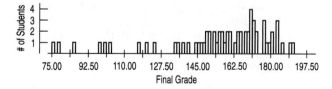

a) Comment on this display.
b) Describe the distribution of grades.

47. Zip codes. Holes-R-Us, an Internet company that sells piercing jewelry, keeps transaction records on its sales. At a recent sales meeting, one of the staff presented a histogram of the zip codes of the last 500 customers, so that the staff might understand where sales are coming from. Comment on the usefulness and appropriateness of the display.

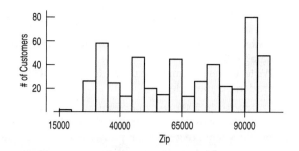

48. Zip codes revisited. Here are some summary statistics to go with the histogram of the zip codes of 500 customers from the Holes-R-Us Internet Jewelry Salon that we saw in Exercise 47:

Count	500
Mean	64,970.0
StdDev	23,523.0
Median	64,871
IQR	44,183
Q1	46,050
Q3	90,233

What can these statistics tell you about the company's sales?

T 49. Math scores 2005. The National Center for Education Statistics (http://nces.ed.gov/nationsreportcard/) reported 2005 average mathematics achievement scores for eighth graders in all 50 states:

State	Score	State	Score
Alabama	225	Montana	241
Alaska	236	Nebraska	238
Arizona	230	Nevada	230
Arkansas	236	New Hampshire	246
California	230	New Jersey	244
Colorado	239	New Mexico	224
Connecticut	242	New York	238
Delaware	240	North Carolina	241
Florida	239	North Dakota	243
Georgia	234	Ohio	242
Hawaii	230	Oklahoma	234
Idaho	242	Oregon	238
Illinois	233	Pennsylvania	241
Indiana	240	Rhode Island	233
Iowa	240	South Carolina	238
Kansas	246	South Dakota	242
Kentucky	231	Tennessee	232
Louisiana	230	Texas	242
Maine	241	Utah	239
Maryland	238	Vermont	244
Massachusetts	247	Virginia	240
Michigan	238	Washington	242
Minnesota	246	West Virginia	231
Mississippi	227	Wisconsin	241
Missouri	235	Wyoming	243

a) Find the median, the IQR, the mean, and the standard deviation of these state averages.
b) Which summary statistics would you report for these data? Why?
c) Write a brief summary of the performance of eighth graders nationwide.

T 50. **Boomtowns.** In 2006, *Inc.* magazine (www.inc.com) listed its choice of "boomtowns" in the United States— larger cities that are growing rapidly. Here is the magazine's top 20, along with their job growth percentages:

City	1-Year Job Growth (%)
Las Vegas, NV	7.5
Fort Lauderdale, FL	4.2
Orlando, FL	4.5
West Palm Beach-Boca Raton, FL	3.4
San Bernadino-Riverside, CA	1.9
Phoenix, AZ	4.4
Northern Virginia, VA	3.1
Washington, DC-Arlington-Alexandria, VA	3.2
Tampa-St. Petersburg, FL	2.6
Camden-Burlington Counties, NJ	2.6
Jacksonville, FL	2.6
Charlotte, NC	3.3
Raleigh-Cary, NC	2.8
Richmond, VA	2.9
Salt Lake City, UT	3.3
Putnam-Rockland-Westchester counties, New York	2.3
Santa Ana-Anaheim-Irvine, CA	1.7
Miami-Miami Beach, FL	2.2
Sacramento, CA	1.5
San Diego, CA	1.4

a) Make a suitable display of the growth rates.
b) Summarize the typical growth rate among these cities with a median and mean. Why do they differ?
c) Given what you know about the distribution, which of the measures in part b does the better job of summarizing the growth rates? Why?
d) Summarize the spread of the growth rate distribution with a standard deviation and with an IQR.
e) Given what you know about the distribution, which of the measures in part d does the better job of summarizing the growth rates? Why?
f) Suppose we subtract from each of the preceding growth rates the predicted U.S. average growth rate of 1.20%, so that we can look at how much these growth rates exceed the U.S. rate. How would this change the values of the summary statistics you calculated above? (*Hint:* You need not recompute any of the summary statistics from scratch.)
g) If we were to omit Las Vegas from the data, how would you expect the mean, median, standard deviation, and IQR to change? Explain your expectations for each.
h) Write a brief report about all of these growth rates.

T 51. **Gasoline usage 2004.** The California Energy Commission (www.energy.ca.gov/gasoline/) collects data on the amount of gasoline sold in each state. The following data show the per capita (gallons used per person) consumption in the year 2004. Using appropriate graphical displays and summary statistics, write a report on the gasoline use by state in the year 2004.

State	Gallons per Capita	State	Gallons per Capita
Alabama	529.4	Nebraska	470.1
Alaska	461.7	Nevada	367.9
Arizona	381.9	New Hampshire	544.4
Arkansas	512.0	New Jersey	488.2
California	414.4	New Mexico	508.8
Colorado	435.7	New York	293.4
Connecticut	435.7	North Carolina	505.0
Delaware	541.6	North Dakota	553.7
Florida	496.0	Ohio	451.1
Georgia	537.1	Oklahoma	614.2
Hawaii	358.7	Oregon	418.4
Idaho	454.8	Pennsylvania	386.8
Illinois	408.3	Rhode Island	454.6
Indiana	491.7	South Carolina	578.6
Iowa	555.1	South Dakota	564.4
Kansas	511.8	Tennessee	552.5
Kentucky	526.6	Texas	532.7
Maine	576.3	Utah	460.6
Maryland	447.5	Vermont	545.5
Massachusetts	458.5	Virginia	526.9
Michigan	482.0	Washington	423.6
Minnesota	527.7	West Virginia	426.7
Mississippi	558.5	Wisconsin	449.8
Missouri	550.5	Wyoming	615.0
Montana	544.4		

T 52. **Prisons 2005.** A report from the U.S. Department of Justice (www.ojp.usdoj.gov/bjs/) reported the percent changes in federal prison populations in 21 northeastern and midwestern states during 2005. Using appropriate graphical displays and summary statistics, write a report on the changes in prison populations.

State	Percent Change	State	Percent Change
Connecticut	−0.3	Iowa	2.5
Maine	0.0	Kansas	1.1
Massachusetts	5.5	Michigan	1.4
New Hampshire	3.3	Minnesota	6.0
New Jersey	2.2	Missouri	−0.8
New York	−1.6	Nebraska	7.9
Pennsylvania	3.5	North Dakota	4.4
Rhode Island	6.5	Ohio	2.3
Vermont	5.6	South Dakota	11.9
Illinois	2.0	Wisconsin	−1.0
Indiana	1.9		

ANSWERS

(Thoughts will vary.)

1. Roughly symmetric, slightly skewed to the right. Center around 3 miles? Few over 10 miles.

2. Bimodal. Center between 1 and 2 hours? Many people watch no football; others watch most of one or more games. Probably only a few values over 5 hours.

3. Strongly skewed to the right, with almost everyone at $0; a few small prizes, with the winner an outlier.

4. Fairly symmetric, somewhat uniform, perhaps slightly skewed to the right. Center in the 40s? Few ages below 25 or above 70.

5. Uniform, symmetric. Center near 5. Roughly equal counts for each digit 0–9.

6. Incomes are probably skewed to the right and not symmetric, making the median the more appropriate measure of center. The mean will be influenced by the high end of family incomes and not reflect the "typical" family income as well as the median would. It will give the impression that the typical income is higher than it is.

7. An IQR of 30 mpg would mean that only 50% of the cars get gas mileages in an interval 30 mpg wide. Fuel economy doesn't vary that much. 3 mpg is reasonable. It seems plausible that 50% of the cars will be within about 3 mpg of each other. An IQR of 0.3 mpg would mean that the gas mileage of half the cars varies little from the estimate. It's unlikely that cars, drivers, and driving conditions are that consistent.

8. We'd prefer a standard deviation of 2 months. Making a consistent product is important for quality. Customers want to be able to count on the MP3 player lasting somewhere close to 5 years, and a standard deviation of 2 years would mean that life spans were highly variable.

CHAPTER 5

Understanding and Comparing Distributions

Where are we going?

We can answer much more interesting questions about variables when we compare distributions for different groups. Are heart attack rates the same for men and women? Is that expensive beverage container really worth the price? Are wind patterns the same throughout the year? These are the kinds of questions where Statistics can really help. Some simple graphical displays and summaries can start us thinking about patterns, trends, and models—something we'll do throughout the rest of this book.

WHO	Days during 1989
WHAT	Average daily wind speed (mph), Average barometric pressure (mb), Average daily temperature (deg Celsius)
WHEN	1989
WHERE	Hopkins Forest, in Western Massachusetts
WHY	Long-term observations to study ecology and climate

The Hopkins Memorial Forest is a 2500-acre reserve in Massachusetts, New York, and Vermont managed by the Williams College Center for Environmental Studies (CES). As part of their mission, CES monitors forest resources and conditions over the long term. They post daily measurements at their website.[1] You can go there, download, and analyze data for any range of days. We'll focus for now on 1989. As we'll see, some interesting things happened that year.

One of the variables measured in the forest is wind speed. Three remote anemometers generate far too much data to report, so, as summaries, you'll find minimum, maximum, and average wind speed (in mph) for each day.

Wind is caused as air flows from areas of high pressure to areas of low pressure. Centers of low pressure often accompany storms, so both high winds and low pressure are associated with some of the fiercest storms. Wind speeds can vary greatly during a day and from day to day, but if we step back a bit farther, we can see patterns. By modeling these patterns, we can understand things about *Average Wind Speed* that we may not have known.

In Chapter 3 we looked at the association between two categorical variables using contingency tables and displays. Here we'll explore different ways of examining the relationship between two variables when one is quantitative, and the other is categorical and indicates groups to compare. We are given wind speed averages for each day of 1989. But we can collect the days together into different size groups and compare the wind speeds among them. If we consider *Time* as a categorical variable in this way, we'll gain enormous flexibility

---

[1] www.williams.edu/CES/hopkins.htm

for our analysis and for our understanding. We'll discover new insights as we change the granularity of the grouping variable—from viewing the whole year's data at one glance, to comparing seasons, to looking for patterns across months, and, finally, to looking at the data day by day.

The Big Picture

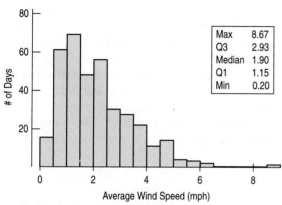

Let's start with the "big picture." Here's a histogram and 5-number summary of the *Average Wind Speed* for every day in 1989. Because of the skewness, we'll report the median and IQR. We can see that the distribution of *Average Wind Speed* is unimodal and skewed to the right. Median daily wind speed is about 1.90 mph, and on half of the days, the average wind speed is between 1.15 and 2.93 mph. We also see a rather windy 8.67-mph day. Was that unusually windy or just the windiest day of the year? To answer that, we'll need to work with the summaries a bit more.

FIGURE 5.1

A histogram of daily *Average Wind Speed* for 1989. It is unimodal and skewed to the right, with a possible high outlier.

Boxplots

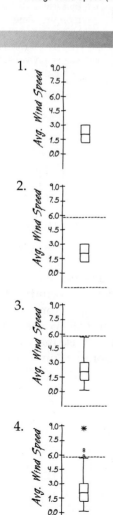

Once we have a 5-number summary of a (quantitative) variable, we can display that information in a **boxplot**. To make a boxplot of the average wind speeds, follow these steps:

1. Draw a single vertical axis spanning the extent of the data.[2] Draw short horizontal lines at the lower and upper quartiles and at the median. Then connect them with vertical lines to form a box. The box can have any width that looks OK.[3]

2. To help us construct the boxplot, we erect "fences" around the main part of the data. We place the upper fence 1.5 IQRs above the upper quartile and the lower fence 1.5 IQRs below the lower quartile. For the wind speed data, we compute

$$Upper\ fence = Q3 + 1.5\ IQR = 2.93 + 1.5 \times 1.78 = 5.60\ mph$$

and

$$Lower\ fence = Q1 - 1.5\ IQR = 1.15 - 1.5 \times 1.78 = -1.52\ mph$$

The fences are just for construction and are not part of the display. We show them here with dotted lines for illustration. You should never include them in your boxplot.

3. We use the fences to grow "whiskers." Draw lines from the ends of the box up and down to *the most extreme data values found within the fences*. If a data value falls outside one of the fences, we do *not* connect it with a whisker.

4. Finally, we add the **outliers** by displaying any data values beyond the fences with special symbols. (We often use a different symbol for "**far outliers**"— data values farther than 3 IQRs from the quartiles.)

---

[2] The axis could also run horizontally.

[3] Some computer programs draw wider boxes for larger data sets. That can be useful when comparing groups.

A boxplot is just a graphical representation of a 5-number summary. The box shows the middle half of the data—the values that lie between the quartiles. Because the top and bottom of the box are the 3rd and 1st quartiles, the height of the box is equal to the IQR. If the median is roughly centered between the quartiles, then the middle half of the data is roughly symmetric. If the median is not centered, the distribution is skewed. The whiskers show skewness as well if they are not roughly the same length. Any outliers are displayed individually, both to keep them out of the way for judging skewness and to encourage you to give them special attention. They may be mistakes, or they may be the most interesting cases in your data.

For the Hopkins Forest data, the central box contains each day whose *Average Wind Speed* is between 1.15 and 2.93 miles per hour (see Figure 5.2). From the shape of the box, it looks like the central part of the distribution of wind speeds is roughly symmetric, but the longer upper whisker indicates that the distribution stretches out at the upper end. We also see a few very windy days. Boxplots are particularly good at pointing out outliers. These extraordinarily windy days may deserve more attention. We'll give them that extra attention shortly.

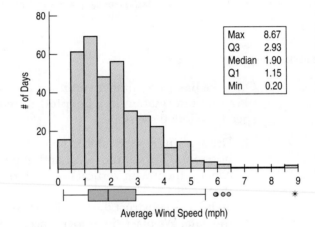

FIGURE 5.2
By turning the boxplot and putting it on the same scale as the histogram, we can compare both displays of the daily wind speeds and see how each represents the distribution.

Max	8.67
Q3	2.93
Median	1.90
Q1	1.15
Min	0.20

Comparing Groups with Histograms

It is almost always more interesting to compare groups. Is it windier in the winter or the summer? Are any months particularly windy? Are weekends a special problem? Let's split the year into two groups: April through September (Spring/Summer) and October through March (Fall/Winter). To compare the groups, we create two histograms, being careful to use the same scale. Here are displays of the average daily wind speed for Spring/Summer (on the left) and Fall/Winter (on the right):

FIGURE 5.3
Histograms of *Average Wind Speed* for days in Spring/Summer (left) and Fall/Winter (right) show very different patterns.

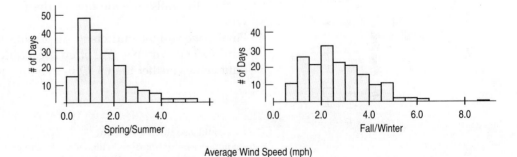

The shapes, centers, and spreads of these two distributions are strikingly different. During spring and summer (histogram on the left), the distribution is skewed to the right. A typical day during these warmer months has an average wind speed of only 1 to 2 mph, and few have average speeds above 3 mph. In the colder months (histogram on the right), however, the shape is less strongly skewed and more spread out. The typical wind speed is higher, and days with average wind speeds above 3 mph are not unusual. There are several noticeable high values.

Summaries for *Average Wind Speed* by Season				
Group	Mean	StdDev	Median	IQR
Fall/Winter	2.71	1.36	2.47	1.87
Spring/Summer	1.56	1.01	1.34	1.32

FOR EXAMPLE   Comparing Groups with Stem-and-Leaf Displays

In 2004 the infant death rate in the United States was 6.8 deaths per 1000 live births. The Kaiser Family Foundation collected data from all 50 states and the District of Columbia, allowing us to look at different regions of the country. Since there are only 51 data values, a back-to-back stem-and-leaf plot is an effective display. Here's one comparing infant death rates in the Northeast and Midwest to those in the South and West. In this display the stems run down the middle of the plot, with the leaves for the two regions to the left or right. Be careful when you read the values on the left: 4|11| means a rate of 11.4 deaths per 1000 live births for one of the southern or western states.

QUESTION: How do infant death rates compare for these regions?

Infant Death Rates (by state) 2004

South and West		North and Midwest
4	11	
3 0	10	
0 0	9	
0 4 1 6 9 5 8	8	1 0
0 5 0 3	7	5 8 0 7 4 1
4 1 0 4 9 1 1 6 4	6	3 1 5 4 4
6 3 6 2	5	8 4 0 6
	4	8 8 9 7
	3	

(4 |11| means 11.4 deaths per 1000 live births)

Infant death rates were generally higher for states in the South and West than in the Northeast and Midwest. The distribution for the northeastern and midwestern states is roughly uniform, varying from a low of 4.8 to a high of 8.1 deaths per 1000 live births. Ten southern and western states had higher infant death rates than any in the Northeast or Midwest, with one state over 11. Rates varied more widely in the South and West, where the distribution is skewed to the right and possibly bimodal. We should investigate further to see which states represent the cluster of high death rates.

A S
Video: **Can Diet Prolong Life?** Here's a subject that's been in the news: Can you live longer by eating less? (Or would it just seem longer?) Look at the data in subsequent activities, and you'll find that you can learn a lot by comparing two groups with boxplots.

Comparing Groups with Boxplots

Are some months windier than others? Even residents may not have a good idea which parts of the year are the most windy. (Do you know for your hometown?) We're not interested just in the centers, but also in the spreads. Are wind speeds equally variable from month to month, or do some months show more variation?

Earlier, we compared histograms of the wind speeds for two halves of the year. To look for seasonal trends, though, we'll group the daily observations by month. Histograms or stem-and-leaf displays are a fine way to look at one distribution or two. But it would be hard to see patterns by comparing 12 histograms. Boxplots offer an ideal balance of information and simplicity, hiding the details while displaying the overall summary information. So we often plot them side by side for groups or categories we wish to compare.

By placing boxplots side by side, we can easily see which groups have higher medians, which have the greater IQRs, where the central 50% of the data is located in each group, and which have the greater overall range. And, when the boxes are in an order, we can get a general idea of patterns in both the centers and the spreads. Equally important, we can see past any outliers in making these comparisons because they've been displayed separately.

Here are boxplots of the *Average Daily Wind Speed* by month:

FIGURE 5.4
Boxplots of the *Average Daily Wind Speed* for each month show seasonal patterns in both the centers and spreads.

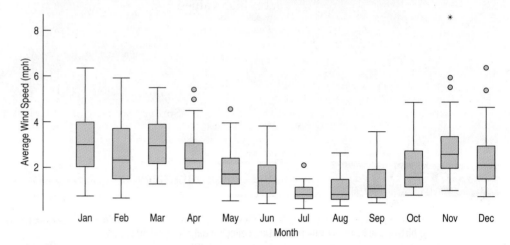

Here we see that wind speeds tend to decrease in the summer. The months in which the winds are both strongest and most variable are November through March. And there was one remarkably windy day in November.

When we looked at a boxplot of wind speeds for the entire year, there were only 5 outliers. Now, when we group the days by *Month*, the boxplots display more days as outliers and call out one in November as a far outlier. The boxplots show different outliers than before because some days that seemed ordinary when placed against the entire year's data look like outliers for the month that they're in. That windy day in July certainly wouldn't stand out in November or December, but for July, it was remarkable.

FOR EXAMPLE   Comparing Distributions

Roller coasters[4] are a thrill ride in many amusement parks worldwide. And thrill seekers want a coaster that goes fast. There are two main types of roller coasters: those with wooden tracks and those with steel tracks. Do they typically run at different speeds? Here are boxplots:

QUESTION: Compare the speeds of wood and steel roller coasters.

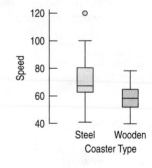

Overall, wooden-track roller coasters are slower than steel-track coasters. In fact, the fastest half of the steel coasters are faster than three quarters of the wooden coasters. Although the IQRs of the two groups are similar, the range of speeds among steel coasters is larger than the range for wooden coasters. The distribution of speeds of wooden coasters appears to be roughly symmetric, but the speeds of the steel coasters are skewed to the right, and there is a high outlier at 120 mph. We should look into why that steel coaster is so fast.

---

[4] See the Roller Coaster Data Base at www.rcdb.com.

STEP-BY-STEP EXAMPLE | **Comparing Groups**

Of course, we can compare groups even when they are not in any particular order. Most scientific studies compare two or more groups. It is almost always a good idea to start an analysis of data from such studies by comparing boxplots for the groups. Here's an example:

For her class project, a student compared the efficiency of various coffee containers. For her study, she decided to try four different containers and to test each of them eight different times. Each time, she heated water to 180°F, poured it into a container, and sealed it. (We'll learn the details of how to set up experiments in Chapter 13.) After 30 minutes, she measured the temperature again and recorded the difference in temperature. Because these are temperature differences, smaller differences mean that the liquid stayed hot—just what we would want in a coffee mug.

Question: What can we say about the effectiveness of these four mugs?

Plan State what you want to find out.

Variables Identify the *variables* and report the W's.

Be sure to check the appropriate condition.

I want to compare the effectiveness of four different brands of mugs in maintaining temperature. I have eight measurements of *Temperature Change* for a single example of each of the brands.

✓ **Quantitative Data Condition:** The *Temperature Changes* are quantitative, with units of °F. Boxplots are appropriate displays for comparing the groups. Numerical summaries of each group are appropriate as well.

Mechanics Report the 5-number summaries of the four groups. Including the IQR is a good idea as well.

Make a picture. Because we want to compare the distributions for four groups, boxplots are an appropriate choice.

	Min	Q1	Median	Q3	Max	IQR
CUPPS	6°F	6	8.25	14.25	18.50	8.25
Nissan	0	1	2	4.50	7	3.50
SIGG	9	11.50	14.25	21.75	24.50	10.25
Starbucks	6	6.50	8.50	14.25	17.50	7.75

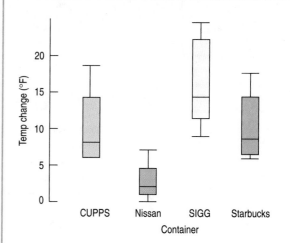

Conclusion Interpret what the boxplots and summaries say about the ability of these mugs to retain heat. Compare the shapes, centers, and spreads, and note any outliers.

The individual distributions of temperature changes are all slightly skewed to the high end. The Nissan cup does the best job of keeping liquids hot, with a median loss of only 2°F, and the SIGG cup does the worst, typically losing 14°F. The difference is large enough to be important: A coffee drinker would be likely to notice a 14° drop in temperature. And the mugs are clearly different: 75% of the Nissan tests showed less heat loss than any of the other mugs in the study. The IQR of results for the Nissan cup is also the smallest of these test cups, indicating that it is a consistent performer.

JUST CHECKING

The Bureau of Transportation Statistics of the U.S. Department of Transportation collects and publishes statistics on airline travel (www.transtats.bts.gov). Here are three displays of the % of flights arriving late each month from 1995 through 2005:

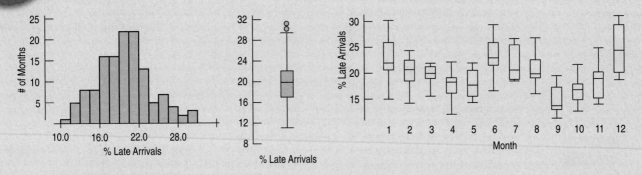

1. Describe what the histogram says about late arrivals.

2. What does the boxplot of late arrivals suggest that you can't see in the histogram?

3. Describe the patterns shown in the boxplots by month. At what time of year are flights least likely to be late? Can you suggest reasons for this pattern?

Outliers

When we looked at boxplots for the *Average Wind Speed* by *Month*, we noticed that several days stood out as possible outliers and that one very windy day in November seemed truly remarkable. What should we do with such outliers?

Cases that stand out from the rest of the data almost always deserve our attention. An outlier is a value that doesn't fit with the rest of the data, but exactly how different it should be to receive special treatment is a judgment call. Boxplots provide a rule of thumb to highlight these unusual points, but that rule doesn't tell you what to do with them.

So, what *should* we do with outliers? The first thing to do is to try to understand them in the context of the data. A good place to start is with a histogram. Histograms show us more detail about a distribution than a boxplot can, so they give us a better idea of how the outlier fits (or doesn't fit) in with the rest of the data.

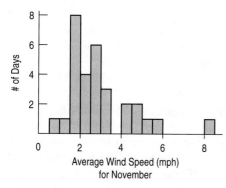

FIGURE 5.5

The *Average Wind Speed* in November is slightly skewed with a high outlier.

A histogram of the *Average Wind Speed* in November shows a slightly skewed main body of data and that very windy day clearly set apart from the other days. When considering whether a case is an outlier, we often look at the gap between that case and the rest of the data. A large gap suggests that the case really is quite different. But a case that just happens to be the largest or smallest value at the end of a possibly stretched-out tail may be best thought of as just . . . the largest or smallest value. After all, *some* case has to be the largest or smallest.

Some outliers are simply unbelievable. If a class survey includes a student who claims to be 170 inches tall (about 14 feet, or 4.3 meters), you can be pretty sure that's an error.

Once you've identified likely outliers, you should always investigate them. Some outliers are just errors. A decimal point may have been misplaced, digits transposed, or digits repeated or omitted. The units may be wrong. (Was that outlying height reported in centimeters rather than in inches [170 cm = 65 in.]?) Or a number may just have been transcribed incorrectly, perhaps copying an adjacent value on the original data sheet. If you can identify the correct value, then you should certainly fix it. One important reason to look into outliers is to correct errors in your data.

14-year-old widowers?   Careful attention to outliers can often reveal problems in data collection and management. Two researchers, Ansley Coale and Fred Stephan, looking at data from the 1950 census noticed that the number of widowed 14-year-old boys had increased from 85 in 1940 to a whopping 1600 in 1950. The number of divorced 14-year-old boys had increased, too, from 85 to 1240. Oddly, the number of teenaged widowers and divorcees *decreased* for every age group after 14, from 15 to 19. When Coale and Stephan also noticed a large increase in the number of young Native Americans in the Northeast United States, they began to look for data problems. Data in the 1950 census were recorded on computer cards. (For a picture of a computer card, see p. 10.) Cards are hard to read and mistakes are easy to make. It turned out that data punches had been shifted to the right by one column on hundreds of cards. Because each card column meant something different, the shift turned 43-year-old widowed males into 14-year-olds, 42-year-old divorcees into 14-year-olds, and children of white parents into Native Americans. Not all outliers have such a colorful (or famous) story, but it is always worthwhile to investigate them. And, as in this case, the explanation is often surprising. (A. Coale and F. Stephan, "The case of the Indians and the teen-age widows," *J. Am. Stat. Assoc.* 57 [Jun 1962]: 338-347.).

Many outliers are not wrong; they're just different. Such cases often repay the effort to understand them. You can learn more from the extraordinary cases than from summaries of the overall data set.

What about that windy November day? Was it really that windy, or could there have been a problem with the anemometers? A quick Internet search for weather on November 21, 1989, finds that there was a severe storm:

WIND, SNOW, COLD GIVE N.E. A TASTE OF WINTER

Published on November 22, 1989
Author: Andrew Dabilis, Globe Staff

An intense storm roared like the Montreal Express through New England yesterday, bringing frigid winds of up to 55 m.p.h., 2 feet of snow in some parts of Vermont and a preview of winter after weeks of mild weather. Residents throughout the region awoke yesterday to an icy vortex that lifted an airplane off the runway in Newark and made driving dangerous in New England because of rapidly shifting winds that seemed to come from all directions.

When we have outliers, we need to decide what to *Tell* about the data. If we can correct an error, we'll just summarize the corrected data (and note the correction). But if we see no way to correct an outlying value, or if we confirm that it is correct, our best path is to report summaries and analyses with *and* without the outlier. In this way a reader can judge for him- or herself what influence the outlier has and decide what to think about the data.

There are two things we should *never* do with outliers. The first is to silently leave an outlier in place and proceed as if nothing were unusual. Analyses of data with outliers are very likely to be influenced by those outliers—sometimes to a large and misleading degree. The other is to drop an outlier from the analysis without comment just because it's unusual. If you want to exclude an outlier, you must discuss your decision and, to the extent you can, justify your decision.

A S *Case Study:* **Are passengers or drivers safer in a crash?** Practice the skills of this chapter by comparing these two groups.

FOR EXAMPLE | **Checking Out the Outliers**

RECAP: We've looked at the speeds of roller coasters and found a difference between steel- and wooden-track coasters. We also noticed an extraordinary value.

QUESTION: The fastest coaster in this collection turns out to be the "Top Thrill Dragster" at Cedar Point amusement park. What might make this roller coaster unusual? You'll have to do some research, but that's often what happens with outliers.

The Top Thrill Dragster is easy to find in an Internet search. We learn that it is a "hydraulic launch" coaster. That is, it doesn't get its remarkable speed just from gravity, but rather from a kick-start by a hydraulic piston. That could make it different from the other roller coasters.

(You might also discover that it is no longer the fastest roller coaster in the world.)

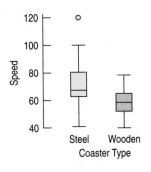

Timeplots: Order, Please!

The Hopkins Forest wind speeds are reported as daily averages. Previously, we grouped the days into months or seasons, but we could look at the wind speed values day by day. Whenever we have data measured over time, it is a good idea to look for patterns by plotting the data in time order. Here are the daily average wind speeds plotted over time:

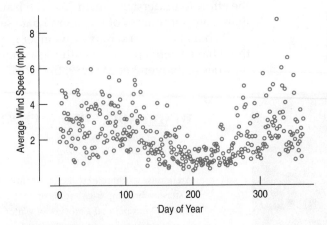

FIGURE 5.6
A timeplot of *Average Wind Speed* shows the overall pattern and changes in variation.

A display of values against time is sometimes called a **timeplot**. This timeplot reflects the pattern that we saw when we plotted the wind speeds by month. But without the arbitrary divisions between months, we can see a calm period during the summer, starting around day 200 (the middle of July), when the wind is relatively mild and doesn't vary greatly from day to day. We can also see that the wind becomes both more variable and stronger during the early and late parts of the year.

*Smoothing Timeplots

Timeplots often show a great deal of point-to-point variation, as this one does. We usually want to see past this variation to understand any underlying smooth trends, and then also think about how the values vary around that trend—the timeplot version of center and spread. You could take your pencil, squint at the plot a bit to blur things, and sketch a smooth curve. And you'd probably do a pretty good job. But we can also ask a computer to do it for us.

There are many ways for computers to run a smooth trace through a timeplot. Some follow local bumps, others emphasize long-term trends. Some provide an equation that gives a typical value for any given time point, others just offer a smooth trace. For our purposes here, it doesn't matter all that much. You can try the methods offered by your favorite program and pick the one that looks best for the timeplot you are looking at and the questions you'd like to answer.

You'll often see timeplots drawn with all the points connected, especially in financial publications. This trace through the data falls at one extreme, following every bump as the data unfold. For the wind speeds, the daily fluctuations are so large that connecting the points doesn't help us a great deal:

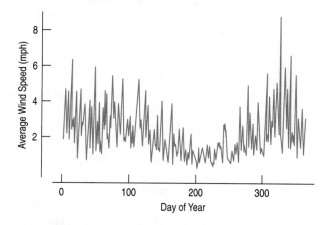

FIGURE 5.7
The *Average Wind Speeds* of Figure 5.6, drawn by connecting all the points. Sometimes this can help in seeing the underlying pattern, but here there is too much daily variation for this to be very useful.

A smooth trace can highlight long-term patterns and help us see them through the more local variation. Here are the daily average wind speed values with a smooth trace found by a method called *lowess,* available in many statistics programs:

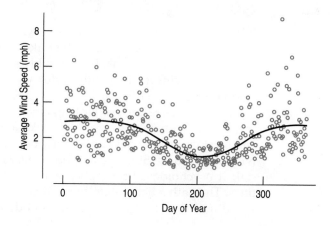

FIGURE 5.8
The *Average Wind Speeds* of Figure 5.6, with a smooth trace added to help your eye see the long-term pattern.

With the smooth trace, it's a bit easier to see a pattern. The trace helps our eye follow the main trend and alerts us to points that don't fit the overall pattern.

FOR EXAMPLE   Timeplots and Smoothing

RECAP: We have looked at the current speeds of the world's fastest roller coasters. Have coasters been getting faster? Our data include coasters going back to 1909. Here's a timeplot with a *lowess* smooth trace:

QUESTION: What does the timeplot say about the trend in roller coaster speeds?

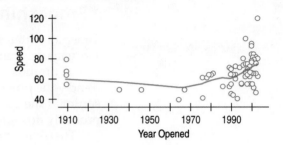

Roller coasters do seem to have been getting faster. Speeds were stagnant, or even declining, until 1970, but have increased (on average) since then. The increase seems to have been more rapid since 1990, pushed along by a few outliers.

HOW DO SMOOTHERS WORK?

If you planned to go camping in the Hopkins Forest around July 15 and wanted to know how windy it was likely to be, you'd probably look at a typical value for mid-July. Maybe you'd consider all the wind speeds for July, or maybe only those from July 10th to July 20th. That's just the kind of thing we do when we smooth a timeplot.

One simple way to smooth is with a **moving average**. To find a smooth value for a particular time, we average the values around that point in an interval called the "window." To find the value for the next point, we *move* the window by one point in time and take the new *average*. The size of the window you choose affects how smooth the resulting trace will be. For the Hopkins Forest winds, we might use a 5-day moving average. Stock analysts often use a 50- or 200-day moving average to help them (attempt) to see the underlying pattern in stock price movements.

Can we use smoothing to predict the future? We have only the values of a stock in the past, but (unfortunately) none of the future values. We could use the recent past as our window and take the simple average of those values. A more sophisticated method, **exponential smoothing**, gives more weight to the recent past values and less and less weight to values as they recede into the past.

Looking into the Future

It is always tempting to try to extend what we see in a timeplot into the future. Sometimes that makes sense. Most likely, the Hopkins Forest climate follows regular seasonal patterns. It's probably safe to predict a less windy June next year and a windier November. But we certainly wouldn't predict another storm on November 21.

Other patterns are riskier to extend into the future. If a stock has been rising, will it continue to go up? No stock has ever increased in value indefinitely, and no stock analyst has consistently been able to forecast when a stock's value will turn around. Stock prices, unemployment rates, and other economic, social, or psychological concepts are much harder to predict than physical quantities. The path a ball will follow when thrown from a certain height at a given speed and direction is well understood. The path interest rates will take is much less clear. Unless we have strong (nonstatistical) reasons for doing otherwise, we should resist the temptation to think that any trend we see will continue, even into the near future.

Statistical models often tempt those who use them to think beyond the data. We'll pay close attention later in this book to understanding when, how, and how much we can justify doing that.

Re-expressing Data: A First Look

Re-expressing to Improve Symmetry

When data are skewed, it can be hard to summarize them simply with a center and spread, and hard to decide whether the most extreme values are outliers or just part of the stretched-out tail. How can we say anything useful about such data? The secret is to *re-express* the data by applying a simple function to each value.

Many relationships and "laws" in the sciences and social sciences include functions such as logarithms, square roots, and reciprocals. Similar relationships often show up in data. Here's a simple example:

In 1980 large companies' chief executive officers (CEOs) made, on average, about 42 times what workers earned. In the next two decades, CEO compensation soared when compared to the average worker. By 2000 that multiple had jumped[5] to 525. What does the distribution of the compensation of Fortune 500 companies' CEOs look like? Here's a histogram and boxplot for 2005 compensation:

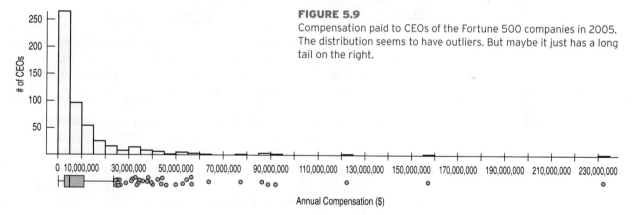

FIGURE 5.9

Compensation paid to CEOs of the Fortune 500 companies in 2005. The distribution seems to have outliers. But maybe it just has a long tail on the right.

We have 500 CEOs and about 48 possible histogram bins, most of which are empty—but don't miss the tiny bars straggling out to the right. The boxplot indicates that some CEOs received extraordinarily high compensations, while the majority received relatively "little." But look at the values of the bins. The first bin, with about half the CEOs, covers compensations of $0 to $5,000,000. Imagine receiving a salary survey with these categories:

What is your income?
a) $0 to $5,000,000
b) $5,000,001 to $10,000,000
c) $10,000,001 to $15,000,000
d) More than $15,000,000

The reason that the histogram seems to leave so much of the area blank is that the salaries are spread all along the axis from about $15,000,000 to

---

[5]*Sources:* United for a Fair Economy, *Business Week* annual CEO pay surveys, Bureau of Labor Statistics, "Average Weekly Earnings of Production Workers, Total Private Sector." Series ID: EEU00500004.

$240,000,000. After $50,000,000 there are so few for each bin that it's very hard to see the tiny bars. What we *can* see from this histogram and boxplot is that this distribution is highly skewed to the right.

It can be hard to decide what we mean by the "center" of a skewed distribution, so it's hard to pick a typical value to summarize the distribution. What would you say was a typical CEO total compensation? The mean value is $10,307,000, while the median is "only" $4,700,000. Each tells us something different about the data.

One approach is to **re-express**, or **transform**, the data by applying a simple function to make the skewed distribution more symmetric. For example, we could take the square root or logarithm of each compensation value. Taking logs works pretty well for the CEO compensations, as you can see:

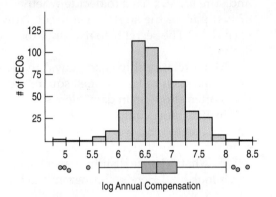

FIGURE 5.10
The logarithms of 2005 CEO compensations are much more nearly symmetric.

The histogram of the logs of the total CEO compensations is much more nearly symmetric, so we can see that a typical log compensation is between 6, which corresponds to $1,000,000, and 7, corresponding to $10,000,000. And it's easier to talk about a typical value for the logs. The mean log compensation is 6.73, while the median is 6.67. (That's $5,370,317 and $4,677,351, respectively.) Notice that nearly all the values are between 6.0 and 8.0—in other words, between $1,000,000 and $100,000,000 a year, but who's counting?

Against the background of a generally symmetric main body of data, it's easier to decide whether the largest compensations are outliers. In fact, the three most highly compensated CEOs are identified as outliers by the boxplot rule of thumb even after this re-expression. It's perhaps impressive to be an outlier CEO in annual compensation. It's even more impressive to be an outlier in the log scale!

Variables that are skewed to the right often benefit from a re-expression by square roots, logs, or reciprocals. Those skewed to the left may benefit from squaring the data. Because computers and calculators can do the calculating, re-expressing data is quite easy. Consider re-expression as a helpful tool whenever you have skewed data.

> **Dealing with logarithms**   You have probably learned about logs in math courses and seen them in psychology or science classes. In this book, we use them only for making data behave better. Base 10 logs are the easiest to understand, but natural logs are often used as well. (Either one is fine.) You can think of base 10 logs as roughly one less than the number of digits you need to write the number. So 100, which is the smallest number to require three digits, has a $\log_{10}$ of 2. And 1000 has a $\log_{10}$ of 3. The $\log_{10}$ of 500 is between 2 and 3, but you'd need a calculator to find that it's approximately 2.7. All salaries of "six figures" have $\log_{10}$ between 5 and 6. Logs are incredibly useful for making skewed data more symmetric. But don't worry—nobody does logs without technology and neither should you. Often, remaking a histogram or other display of the data is as easy as pushing another button.

Re-expressing to Equalize Spread Across Groups

Researchers measured the concentration (nanograms per milliliter) of cotinine in the blood of three groups of people: nonsmokers who have not been exposed to smoke, nonsmokers who have been exposed to smoke (ETS), and smokers. Cotinine is left in the blood when the body metabolizes nicotine, so this measure gives a direct measurement of the effect of passive smoke exposure. The boxplots of the cotinine levels of the three groups tell us that the smokers have higher cotinine levels, but if we want to compare the levels of the passive smokers to those of the nonsmokers, we're in trouble, because on this scale, the cotinine levels for both nonsmoking groups are too low to be seen.

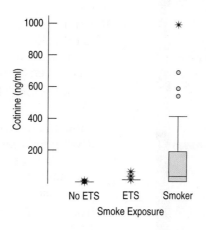

FIGURE 5.11
Cotinine levels (nanograms per milliliter) for three groups with different exposures to tobacco smoke. Can you compare the ETS (exposed to smoke) and No-ETS groups?

Re-expressing can help alleviate the problem of comparing groups that have very different spreads. For measurements like the cotinine data, whose values can't be negative and whose distributions are skewed to the high end, a good first guess at a re-expression is the logarithm.

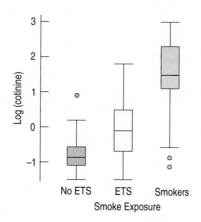

FIGURE 5.12
Blood cotinine levels after taking logs. What a difference a log makes!

After taking logs, we can compare the groups and see that the nonsmokers exposed to environmental smoke (the ETS group) do show increased levels of (log) cotinine, although not the high levels found in the blood of smokers.

Notice that the same re-expression has also improved the symmetry of the cotinine distribution for smokers and pulled in most of the apparent outliers in all of the groups. It is not unusual for a re-expression that improves one aspect of data to improve others as well. We'll talk about other ways to re-express data as the need arises throughout the book. We'll explore some common re-expressions more thoroughly in Chapter 10.

What Can Go Wrong?

- **Avoid inconsistent scales.** Parts of displays should be mutually consistent—no fair changing scales in the middle or plotting two variables on different scales but on the same display. When comparing two groups, be sure to compare them on the same scale.

  - **Label clearly.** Variables should be identified clearly and axes labeled so a reader knows what the plot displays.

    Here's a remarkable example of a plot gone wrong. It illustrated a news story about rising college costs. It uses timeplots, but it gives a misleading impression. First think about the story you're being told by this display. Then try to figure out what has gone wrong.

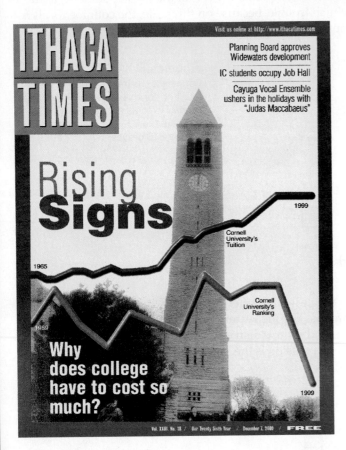

What's wrong? Just about everything.

- The horizontal scales are inconsistent. Both lines show trends over time, but exactly for what years? The tuition sequence starts in 1965, but rankings are graphed from 1989. Plotting them on the same (invisible) scale makes it seem that they're for the same years.

- The vertical axis isn't labeled. That hides the fact that it's inconsistent. Does it graph dollars (of tuition) or ranking (of Cornell University)?

  This display violates three of the rules. And it's even worse than that: It violates a rule that we didn't even bother to mention.

- The two inconsistent scales for the vertical axis don't point in the same direction! The line for Cornell's rank shows that it has "plummeted" from 15th place to 6th place in academic rank. Most of us think that's an *improvement*, but that's not the message of this graph.

- **Beware of outliers.** If the data have outliers and you can correct them, you should do so. If they are clearly wrong or impossible, you should remove them and report on them. Otherwise, consider summarizing the data both with and without the outliers.

CONNECTIONS

We discussed the value of summarizing a distribution with shape, center, and spread in Chapter 4, and we developed several ways to measure these attributes. Now we've seen the value of comparing distributions for different groups and of looking at patterns in a quantitative variable measured over time. Although it can be interesting to summarize a single variable for a single group, it is almost always more interesting to compare groups and look for patterns across several groups and over time. We'll continue to make comparisons like these throughout the rest of our work.

WHAT HAVE WE LEARNED?

▶ We've learned the value of comparing groups and looking for patterns among groups and over time.

▶ We've seen that boxplots are very effective for comparing groups graphically. When we compare groups, we discuss their shape, center, and spreads, and any unusual features.

▶ We've experienced the value of identifying and investigating outliers. And we've seen that when we group data in different ways, it can allow different cases to emerge as possible outliers.

▶ We've graphed data that have been measured over time against a time axis and looked for long-term trends both by eye and with a data smoother.

Terms

Boxplot
A boxplot displays the 5-number summary as a central box with whiskers that extend to the non-outlying data values. Boxplots are particularly effective for comparing groups and for displaying outliers (p. 81).

Outlier
Any point more than 1.5 IQR from either end of the box in a boxplot is nominated as an outlier (p. 81).

Far Outlier
If a point is more than 3.0 IQR from either end of the box in a boxplot, it is nominated as a *far outlier* (p. 81).

Comparing distributions
When comparing the distributions of several groups using histograms or stem-and-leaf displays, consider their:

▶ Shape

▶ Center

▶ Spread (p. 83).

Comparing boxplots
When comparing groups with boxplots:

▶ Compare the shapes. Do the boxes look symmetric or skewed? Are there differences between groups?

▶ Compare the medians. Which group has the higher center? Is there any pattern to the medians?

▶ Compare the IQRs. Which group is more spread out? Is there any pattern to how the IQRs change?

▶ Using the IQRs as a background measure of variation, do the medians seem to be different, or do they just vary much as you'd expect from the overall variation?

▶ Check for possible outliers. Identify them if you can and discuss why they might be unusual. Of course, correct them if you find that they are errors (p. 83).

Timeplot
A timeplot displays data that change over time. Often, successive values are connected with lines to show trends more clearly. Sometimes a smooth curve is added to the plot to help show long-term patterns and trends (p. 88).

Skills

THINK

▶ Be able to select a suitable display for comparing groups. Understand that histograms show distributions well, but are difficult to use when comparing more than two or three groups. Boxplots are more effective for comparing several groups, in part because they show much less information about the distribution of each group.

▶ Understand that how you group data can affect what kinds of patterns and relationships you are likely to see. Know how to select groupings to show the information that is important for your analysis.

▶ Be aware of the effects of skewness and outliers on measures of center and spread. Know how to select appropriate measures for comparing groups based on their displayed distributions.

▶ Understand that outliers can emerge at different groupings of data and that, whatever their source, they deserve special attention.

▶ Recognize when it is appropriate to make a timeplot. Be able to choose between connecting consecutive points in a timeplot to show the sequence clearly and adding a smooth trace, which helps long-term trends emerge from the background scatter.

SHOW

▶ Know how to make side-by-side histograms on comparable scales to compare the distributions of two groups.

▶ Know how to make side-by-side boxplots to compare the distributions of two or more groups.

▶ Know how to describe differences among groups in terms of patterns and changes in their center, spread, shape, and unusual values.

▶ Know how to make a timeplot of data that have been measured over time.

TELL

▶ Know how to compare the distributions of two or more groups by comparing their shapes, centers, and spreads. Be prepared to explain your choice of measures of center and spread for comparing the groups.

▶ Be able to describe trends and patterns in the centers and spreads of groups—especially if there is a natural order to the groups, such as a time order.

▶ Be prepared to discuss patterns in a timeplot in terms of both the general trend of the data and the changes in how spread out the pattern is. Be able to use a smooth trace as a general guide to the long-term patterns.

▶ Be cautious about assuming that trends over time will continue into the future.

▶ Be able to describe the distribution of a quantitative variable in terms of its shape, center, and spread.

▶ Be able to describe any anomalies or extraordinary features revealed by the display of a variable.

▶ Know how to compare the distributions of two or more groups by comparing their shapes, centers, and spreads.

▶ Know how to describe patterns over time shown in a timeplot.

▶ Be able to discuss any outliers in the data, noting how they deviate from the overall pattern of the data.

COMPARING DISTRIBUTIONS ON THE COMPUTER

Most programs for displaying and analyzing data can display plots to compare the distributions of different groups. Typically, these are boxplots displayed side by side.

Side-by-side boxplots should be on the same y-axis scale so they can be compared.

Some programs offer a graphical way to assess how much the medians differ by drawing a band around the median or by "notching" the boxes.

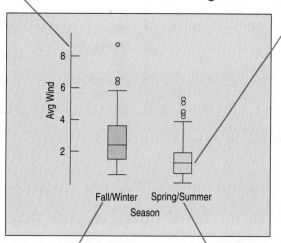

Boxes are typically labeled with a group name. Often they are placed in alphabetical order by group name—not the most useful order.

There are two ways to organize data when we want to compare groups. Each group can be in its own variable (or list, on a calculator). In this form, the experiment comparing coffee cups would have four variables, one for each type of cup:

CUPPS	SIGG	Nissan	Starbucks
6	2	12	13
6	1.5	16	7
6	2	9	7
18.5	3	23	17.5
10	0	11	10
17.5	7	20.5	15.5
11	0.5	12.5	6
6.5	6	24.5	6

But there's another way to think about and organize the data. What is the variable of interest (the *What*) in this experiment? It's the number of degrees lost by the water in each cup. And the *Who* is each time she tested a cup. We could gather all of the temperature values into one variable and put the names of the cups in a second variable listing the individual results, one on each row. Now, the *Who* is clearer—it's an experimental run, one row of the table. Most statistics packages prefer data on groups organized in this way.

That's actually the way we've thought about the wind speed data in this chapter, treating wind speeds as one variable and the groups (whether seasons, months, or days) as a second variable.

Container	Temperature Container	Container	Temperature Container
Difference	Difference	Difference	Difference
CUPPS	6	SIGG	12
CUPPS	6	SIGG	16
CUPPS	6	SIGG	9
CUPPS	18.5	SIGG	23
CUPPS	10	SIGG	11
CUPPS	17.5	SIGG	20.5
CUPPS	11	SIGG	12.5
CUPPS	6.5	SIGG	24.5
Nissan	2	Starbucks	13
Nissan	1.5	Starbucks	7
Nissan	2	Starbucks	7
Nissan	3	Starbucks	17.5
Nissan	0	Starbucks	10
Nissan	7	Starbucks	15.5
Nissan	0.5	Starbucks	6
Nissan	6	Starbucks	6

DATA DESK

If the data are in separate variables, select the variables and choose **Boxplot side by side** from the **Plot** menu. The boxes will appear in the order in which the variables were selected.

If the data are a single quantitative variable and a second variable holding group names, select the

quantitative variable as Y and the group variable as X. Then choose **Boxplot y by x** from the **Plot** menu. The boxes will appear in alphabetical order by group name.

Data Desk offers options for assessing whether any pair of medians differ.

EXCEL

Excel cannot make boxplots.

COMMENT

The DDXL add-on provided on the DVD adds the ability to make boxplots to Excel.

JMP

Choose **Fit y by x.** Assign a continuous response variable to Y, **Response** and a nominal group variable holding the group names to X, **Factor,** and click **OK.** JMP will offer (among other things) dotplots of the

data. Click the red triangle and, under **Display Options,** select Boxplots. *Note:* if the variables are of the wrong type, the display options might not offer boxplots.

MINITAB

Choose **Boxplot...** from the **Graph** menu. If your data are in the form of one quantitative variable and one group variable, choose **One Y** and **with Groups.** If your

data are in separate columns of the worksheet, choose **Multiple Y's.**

SPSS

To make a boxplot in SPSS, open the **Chart Builder** from the Graphs menu.

Click the **Gallery** tab.

Choose **Boxplot** from the list of chart types.

Drag a single or 2-D (side-by-side) boxplot onto the canvas.

Drag a scale variable to the y-axis drop zone.

To make side-by-side boxplots, drag a categorical variable to the x-axis drop zone.

Click **OK.**

(continued)

TI-83/84 PLUS

To make a boxplot, set up a **STAT PLOT** by using the boxplot icon.

To compare groups with boxplots, enter the data in lists, for example in L1 and L2.

Set up STATPLOT's Plot1 to make a boxplot of the L1 data:

- Turn the plot On;
- Choose the first boxplot icon (so the plot will indicate outliers);
- Specify Xlist:L1 and Freq:1, and select the Mark you want the calculator to use for displaying any outliers.

Use ZoomStat to display the boxplot for L1. You can now TRACE to see the statistics in the 5-number summary.

Then set up Plot2 to display the L2 data. This time when you use ZoomStat with both plots turned on, the display shows the boxplots in parallel.

TI-89

For the plot, change Use Freq and Categories to YES and use VAR-LINK to select list2 as the frequency variable on the plot definition screen.

To create a boxplot, press [F2] (**Plots**), then [ENTER]. Select a plot to define and press [F1]. Select either 3: **Box Plot** or 4: **Mod Box Plot** (to identify outliers). Select the mark type of your choice (for outliers). Press [ENTER] to finish. Press [F5] to display the graph.

EXERCISES

1. **In the news.** Find an article in a newspaper, magazine, or the Internet that compares two or more groups of data.
   a) Does the article discuss the W's?
   b) Is the chosen display appropriate? Explain.
   c) Discuss what the display reveals about the groups.
   d) Does the article accurately describe and interpret the data? Explain.

2. **In the news.** Find an article in a newspaper, magazine, or the Internet that shows a time plot.
   a) Does the article discuss the W's?
   b) Is the timeplot appropriate for the data? Explain.
   c) Discuss what the timeplot reveals about the variable.
   d) Does the article accurately describe and interpret the data? Explain.

3. **Time on the Internet.** Find data on the Internet (or elsewhere) that give results recorded over time. Make an appropriate display and discuss what it shows.

4. **Groups on the Internet.** Find data on the Internet (or elsewhere) for two or more groups. Make appropriate displays to compare the groups, and interpret what you find.

5. **Pizza prices.** A company that sells frozen pizza to stores in four markets in the United States (Denver, Baltimore, Dallas, and Chicago) wants to examine the prices that the stores charge for pizza slices. Here are boxplots comparing data from a sample of stores in each market:

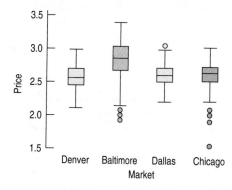

a) Do prices appear to be the same in the four markets? Explain.
b) Does the presence of any outliers affect your overall conclusions about prices in the four markets?

6. Costs. To help travelers know what to expect, researchers collected the prices of commodities in 16 cities throughout the world. Here are boxplots comparing the prices of a ride on public transportation, a newspaper, and a cup of coffee in the 16 cities (prices are all in $US).

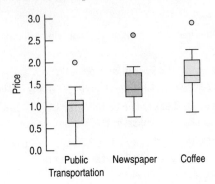

a) On average, which commodity is the most expensive?
b) Is a newspaper always more expensive than a ride on public transportation? Explain.
c) Does the presence of outliers affect your conclusions in part a or b?

7. Rock concert accidents. Crowd Management Strategies monitors accidents at rock concerts. In their database, they list the names and other variables of victims whose deaths were attributed to "crowd crush" at rock concerts. Here are the histogram and boxplot of the victims' ages for data from 1999 to 2000:

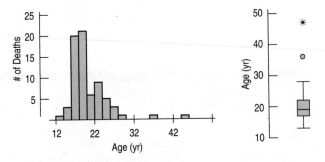

a) What features of the distribution can you see in both the histogram and the boxplot?
b) What features of the distribution can you see in the histogram that you could not see in the boxplot?
c) What summary statistic would you choose to summarize the center of this distribution? Why?
d) What summary statistic would you choose to summarize the spread of this distribution? Why?

8. Slalom times. The **Men's Combined** skiing event consists of a downhill and a slalom. Two displays of the slalom times in the Men's Combined at the 2006 Winter Olympics are shown below.

a) What features of the distribution can you see in both the histogram and the boxplot?
b) What features of the distribution can you see in the histogram that you could not see in the boxplot?
c) What summary statistic would you choose to summarize the center of this distribution? Why?
d) What summary statistic would you choose to summarize the spread of this distribution? Why?

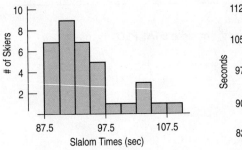
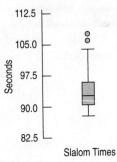

9. Cereals. Sugar is a major ingredient in many breakfast cereals. The histogram displays the sugar content as a percentage of weight for 49 brands of cereal. The boxplot compares sugar content for adult and children's cereals.

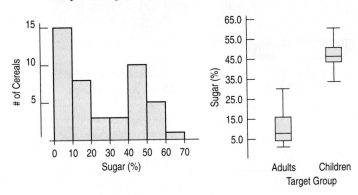

a) What is the range of the sugar contents of these cereals?
b) Describe the shape of the distribution.
c) What aspect of breakfast cereals might account for this shape?
d) Are all children's cereals higher in sugar than adult cereals?
e) Which group of cereals varies more in sugar content? Explain.

10. Tendon transfers. People with spinal cord injuries may lose function in some, but not all, of their muscles. The ability to push oneself up is particularly important for shifting position when seated and for transferring into and out of wheelchairs. Surgeons compared two operations to restore the ability to push up in children. The histogram shows scores rating pushing strength two years after surgery and boxplots compare results for the two surgical methods. (Mulcahey, Lutz, Kozen, Betz,

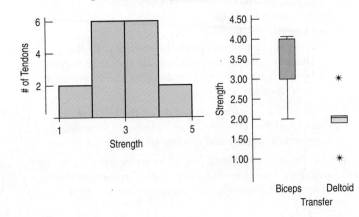

"Prospective Evaluation of Biceps to Triceps and Deltoid to Triceps for Elbow Extension in Tetraplegia," *Journal of Hand Surgery*, 28, 6, 2003)

a) Describe the shape of this distribution.
b) What is the range of the strength scores?
c) What fact about results of the two procedures is hidden in the histogram?
d) Which method had the higher (better) median score?
e) Was that method always best?
f) Which method produced the most consistent results? Explain.

(T) 11. Population growth. Here is a "back-to-back" stem-and-leaf display that shows two data sets at once—one going to the left, one to the right. The display compares the percent change in population for two regions of the United States (based on census figures for 1990 and 2000). The fastest growing states were Nevada at 66% and Arizona at 40%. To show the distributions better, this display breaks each stem into two lines, putting leaves 0–4 on one stem and leaves 5–9 on the other.

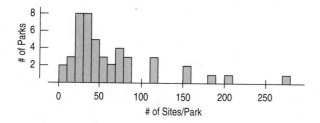

```
 NE/MW States     | S/W States
               6  | 6
               6  |
               5  |
               5  |
               4  |
               4  | 0
               3  |
               3  | 001
               2  | 6
               2  | 001134
               1  | 578
          2100 1  | 001134444
  99998876655 0  | 6999
          4431 0  | 1
```

Population Growth rate
(|6| 6 means 66%)

a) Use the data displayed in the stem-and-leaf display to construct comparative boxplots.
b) Write a few sentences describing the difference in growth rates for the two regions of the United States.

12. Camp sites. Shown below are the histogram and summary statistics for the number of camp sites at public parks in Vermont.

a) Which statistics would you use to identify the center and spread of this distribution? Why?
b) How many parks would you classify as outliers? Explain.

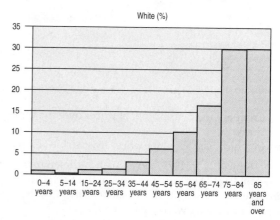

Count	46
Mean	62.8 sites
Median	43.5
StdDev	56.2
Min	0
Max	275
Q1	28
Q3	78

c) Create a boxplot for these data.
d) Write a few sentences describing the distribution.

13. Hospital stays. The U.S. National Center for Health Statistics compiles data on the length of stay by patients in short-term hospitals and publishes its findings in *Vital and Health Statistics*. Data from a sample of 39 male patients and 35 female patients on length of stay (in days) are displayed in the histograms below.

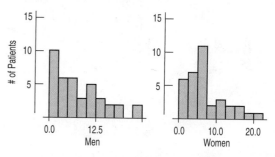

a) What would you suggest be changed about these histograms to make them easier to compare?
b) Describe these distributions by writing a few sentences comparing the duration of hospitalization for men and women.
c) Can you suggest a reason for the peak in women's length of stay?

14. Deaths 2003. A National Vital Statistics Report (www.cdc.gov/nchs/) indicated that nearly 300,000 black Americans died in 2003, compared with just over 2 million white Americans. Below are histograms displaying the distributions of their ages at death.

a) Describe the overall shapes of these distributions.
b) How do the distributions differ?
c) Look carefully at the bar definitions. Where do these plots violate the rules for statistical graphs?

White (%)

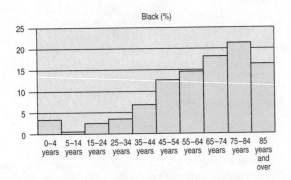

Black (%)

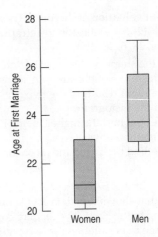

15. **Women's basketball.** Here are boxplots of the points scored during the first 10 games of the season for both Scyrine and Alexandra:

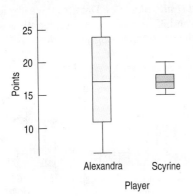

a) Summarize the similarities and differences in their performance so far.
b) The coach can take only one player to the state championship. Which one should she take? Why?

16. **Gas prices.** Here are boxplots of weekly gas prices at a service station in the midwestern United States (prices in $ per gallon):

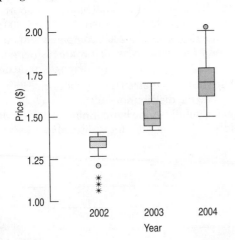

a) Compare the distribution of prices over the three years.
b) In which year were the prices least stable? Explain.

17. **Marriage age.** In 1975, did men and women marry at the same age? Here are boxplots of the age at first marriage for a sample of U.S. citizens then. Write a brief report discussing what these data show.

18. **Fuel economy.** Describe what these boxplots tell you about the relationship between the number of cylinders a car's engine has and the car's fuel economy (mpg):

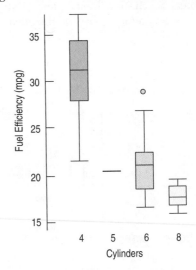

19. **Fuel economy II.** The Environmental Protection Agency provides fuel economy and pollution information on over 2000 car models. Here is a boxplot of *Combined Fuel Economy* (using an average of driving conditions) in *miles per gallon* by vehicle *Type* (car, van, or SUV). Summarize what you see about the fuel economies of the three vehicle types.

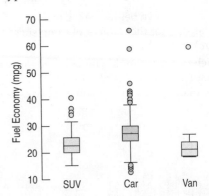

20. **Ozone.** Ozone levels (in parts per billion, ppb) were recorded at sites in New Jersey monthly between 1926

and 1971. Here are boxplots of the data for each month (over the 46 years), lined up in order (January = 1):

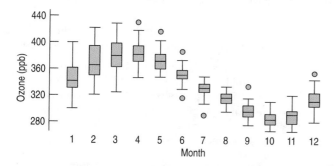

a) In what month was the highest ozone level ever recorded?
b) Which month has the largest IQR?
c) Which month has the smallest range?
d) Write a brief comparison of the ozone levels in January and June.
e) Write a report on the annual patterns you see in the ozone levels.

21. **Test scores.** Three Statistics classes all took the same test. Histograms and boxplots of the scores for each class are shown below. Match each class with the corresponding boxplot.

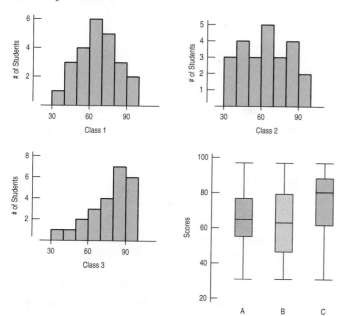

22. **Eye and hair color.** A survey of 1021 school-age children was conducted by randomly selecting children from several large urban elementary schools. Two of the questions concerned eye and hair color. In the survey, the following codes were used:

Hair Color	Eye Color
1 = Blond	1 = Blue
2 = Brown	2 = Green
3 = Black	3 = Brown
4 = Red	4 = Grey
5 = Other	5 = Other

The Statistics students analyzing the data were asked to study the relationship between eye and hair color. They produced this plot:

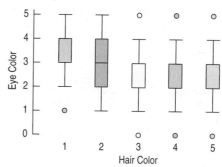

Is their graph appropriate? If so, summarize the findings. If not, explain why not.

23. **Graduation?** A survey of major universities asked what percentage of incoming freshmen usually graduate "on time" in 4 years. Use the summary statistics given to answer the questions that follow.

	% on Time
Count	48
Mean	68.35
Median	69.90
StdDev	10.20
Min	43.20
Max	87.40
Range	44.20
25th %tile	59.15
75th %tile	74.75

a) Would you describe this distribution as symmetric or skewed? Explain.
b) Are there any outliers? Explain.
c) Create a boxplot of these data.
d) Write a few sentences about the graduation rates.

24. **Vineyards.** Here are summary statistics for the sizes (in acres) of Finger Lakes vineyards:

Count	36
Mean	46.50 acres
StdDev	47.76
Median	33.50
IQR	36.50
Min	6
Q1	18.50
Q3	55
Max	250

a) Would you describe this distribution as symmetric or skewed? Explain.
b) Are there any outliers? Explain.
c) Create a boxplot of these data.
d) Write a few sentences about the sizes of the vineyards.

25. Caffeine. A student study of the effects of caffeine asked volunteers to take a memory test 2 hours after drinking soda. Some drank caffeine-free cola, some drank regular cola (with caffeine), and others drank a mixture of the two (getting a half-dose of caffeine). Here are the 5-number summaries for each group's scores (number of items recalled correctly) on the memory test:

	n	Min	Q1	Median	Q3	Max
No caffeine	15	16	20	21	24	26
Low caffeine	15	16	18	21	24	27
High caffeine	15	12	17	19	22	24

a) Describe the W's for these data.
b) Name the variables and classify each as categorical or quantitative.
c) Create side-by-side boxplots to display these results as best you can with this information.
d) Write a few sentences comparing the performances of the three groups.

26. SAT scores. Here are the summary statistics for Verbal SAT scores for a high school graduating class:

	n	Mean	Median	SD	Min	Max	Q1	Q3
Male	80	590	600	97.2	310	800	515	650
Female	82	602	625	102.0	360	770	530	680

a) Create side-by-side boxplots comparing the scores of boys and girls as best you can from the information given.
b) Write a brief report on these results. Be sure to discuss the shape, center, and spread of the scores.

27. Derby speeds 2007. How fast do horses run? Kentucky Derby winners top 30 miles per hour, as shown in this graph. The graph shows the percentage of Derby winners that have run *slower* than each given speed. Note that few have won running less than 33 miles per hour, but about 86% of the winning horses have run less than 37 miles per hour. (A cumulative frequency graph like this is called an "ogive.")

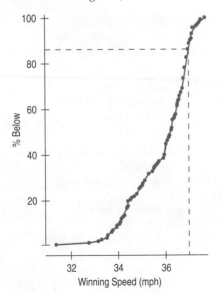

Winning Speed (mph)

a) Estimate the median winning speed.
b) Estimate the quartiles.
c) Estimate the range and the IQR.
d) Create a boxplot of these speeds.
e) Write a few sentences about the speeds of the Kentucky Derby winners.

28. Framingham. The Framingham Heart Study recorded the cholesterol levels of more than 1400 men. Here is an ogive of the distribution of these cholesterol measures. (An ogive shows the percentage of cases at or below a certain value.) Construct a boxplot for these data, and write a few sentences describing the distribution.

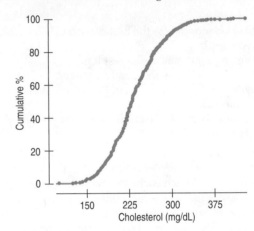

29. Reading scores. A class of fourth graders takes a diagnostic reading test, and the scores are reported by reading grade level. The 5-number summaries for the 14 boys and 11 girls are shown:

Boys:	2.0	3.9	4.3	4.9	6.0
Girls:	2.8	3.8	4.5	5.2	5.9

a) Which group had the highest score?
b) Which group had the greater range?
c) Which group had the greater interquartile range?
d) Which group's scores appear to be more skewed? Explain.
e) Which group generally did better on the test? Explain.
f) If the mean reading level for boys was 4.2 and for girls was 4.6, what is the overall mean for the class?

30. Cloud seeding. In an experiment to determine whether seeding clouds with silver iodide increases rainfall, 52 clouds were randomly assigned to be seeded or not. The amount of rain they generated was then measured (in acre-feet). Here are the summary statistics:

	n	Mean	Median	SD	IQR	Q1	Q3
Unseeded	26	164.59	44.20	278.43	138.60	24.40	163
Seeded	26	441.98	221.60	650.79	337.60	92.40	430

a) Which of the summary statistics are most appropriate for describing these distributions. Why?
b) Do you see any evidence that seeding clouds may be effective? Explain.

31. Industrial experiment. Engineers at a computer production plant tested two methods for accuracy in drilling holes into a PC board. They tested how fast they could set the drilling machine by running 10 boards at each of two different speeds. To assess the results, they measured the distance (in inches) from the center of a target on the board to the center of the hole. The data and summary statistics are shown in the table:

Distance (in.)	Speed	Distance (in.)	Speed
0.000101	Fast	0.000098	Slow
0.000102	Fast	0.000096	Slow
0.000100	Fast	0.000097	Slow
0.000102	Fast	0.000095	Slow
0.000101	Fast	0.000094	Slow
0.000103	Fast	0.000098	Slow
0.000104	Fast	0.000096	Slow
0.000102	Fast	0.975600	Slow
0.000102	Fast	0.000097	Slow
0.000100	Fast	0.000096	Slow

Mean	0.000102	Mean	0.097647
StdDev	0.000001	StdDev	0.308481

Write a report summarizing the findings of the experiment. Include appropriate visual and verbal displays of the distributions, and make a recommendation to the engineers if they are most interested in the accuracy of the method.

32. Cholesterol. A study examining the health risks of smoking measured the cholesterol levels of people who had smoked for at least 25 years and people of similar ages who had smoked for no more than 5 years and then stopped. Create appropriate graphical displays for both groups, and write a brief report comparing their cholesterol levels. Here are the data:

Smokers				Ex-Smokers		
225	211	209	284	250	134	300
258	216	196	288	249	213	310
250	200	209	280	175	174	328
225	256	243	200	160	188	321
213	246	225	237	213	257	292
232	267	232	216	200	271	227
216	243	200	155	238	163	263
216	271	230	309	192	242	249
183	280	217	305	242	267	243
287	217	246	351	217	267	218
200	280	209		217	183	228

33. MPG. A consumer organization compared gas mileage figures for several models of cars made in the United States with autos manufactured in other countries. The data are shown in the table.

a) Create graphical displays for these two groups.
b) Write a few sentences comparing the distributions.

Gas Mileage (mpg)	Country	Gas Mileage (mpg)	Country
16.9	U.S.	26.8	U.S.
15.5	U.S.	33.5	U.S.
19.2	U.S.	34.2	U.S.
18.5	U.S.	16.2	Other
30.0	U.S.	20.3	Other
30.9	U.S.	31.5	Other
20.6	U.S.	30.5	Other
20.8	U.S.	21.5	Other
18.6	U.S.	31.9	Other
18.1	U.S.	37.3	Other
17.0	U.S.	27.5	Other
17.6	U.S.	27.2	Other
16.5	U.S.	34.1	Other
18.2	U.S.	35.1	Other
26.5	U.S.	29.5	Other
21.9	U.S.	31.8	Other
27.4	U.S.	22.0	Other
28.4	U.S.	17.0	Other
28.8	U.S.	21.6	Other

34. Baseball 2008. American League baseball teams play their games with the designated hitter rule, meaning that pitchers do not bat. The League believes that replacing the pitcher, typically a weak hitter, with another player in the batting order produces more runs and generates more interest among fans. Following are the average number of runs scored in American League and National League stadiums for the first half of the 2008 season:

American	Runs/game	National	Runs/game
Texas	12.29	Colorado	10.25
Detroit	10.74	Chicago Cubs	9.79
Baltimore	10.51	Atlanta	9.74
Chicago Sox	9.98	Cincinnati	9.59
Boston	9.85	Pittsburgh	9.45
Cleveland	9.64	Arizona	9.35
NY Yankees	9.54	Florida	9.32
Minnesota	9.07	Washington	9.27
LA Angels	9.07	Philadelphia	9.25
Seattle	8.82	Houston	9.20
Kansas City	8.75	NY Mets	9.07
Tampa Bay	8.71	St. Louis	9.01
Toronto	8.00	San Francisco	8.82
Oakland	7.93	Milwaukee	8.62

a) Create an appropriate graphical display of these data.
b) Write a few sentences comparing the average number of runs scored per game in the two leagues. (Remember: shape, center, spread, unusual features!)
c) The Texas Rangers ballpark in Arlington, TX, was built in a "retro" style recalling older style parks. It is relatively small and has a reputation for home runs. Do you see evidence that the number of runs scored there is unusually high? Explain.

T 35. **Fruit flies.** Researchers tracked a population of 1,203,646 fruit flies, counting how many died each day for 171 days. Here are three timeplots offering different views of these data. One shows the number of flies alive on each day, one the number who died that day, and the third the mortality rate—the fraction of the number alive who died. On the last day studied, the last 2 flies died, for a mortality rate of 1.0.

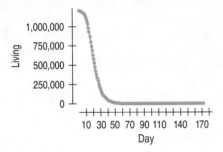

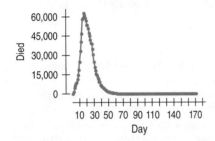

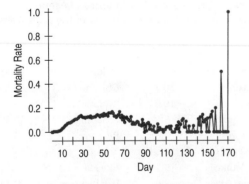

a) On approximately what day did the most flies die?
b) On what day during the first 100 days did the largest *proportion* of flies die?
c) When did the number of fruit flies alive stop changing very much from day to day?

T 36. **Drunk driving 2007.** Accidents involving drunk drivers account for about 40% of all deaths on the nation's highways. The table tracks the number of alcohol-related fatalities for 26 years. (www.alcoholalert.com)

a) Create a stem-and-leaf display or a histogram of these data.
b) Create a timeplot.
c) Using features apparent in the stem-and-leaf display (or histogram) and the timeplot, write a few sentences about deaths caused by drunk driving.

Year	Deaths (thousands)	Year	Deaths (thousands)
1982	26.2	1995	17.7
1983	24.6	1996	17.7
1984	24.8	1997	16.7
1985	23.2	1998	16.7
1986	25.0	1999	16.6
1987	24.1	2000	17.4
1988	23.8	2001	17.4
1989	22.4	2002	17.5
1990	22.6	2003	17.1
1991	20.2	2004	16.9
1992	18.3	2005	16.9
1993	17.9	2006	15.8
1994	17.3	2007	15.4

T 37. **Assets.** Here is a histogram of the assets (in millions of dollars) of 79 companies chosen from the *Forbes* list of the nation's top corporations:

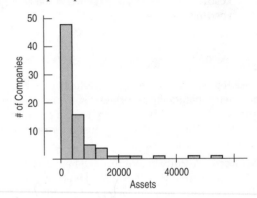

a) What aspect of this distribution makes it difficult to summarize, or to discuss, center and spread?
b) What would you suggest doing with these data if we want to understand them better?

T 38. **Music library.** Students were asked how many songs they had in their digital music libraries. Here's a display of the responses:

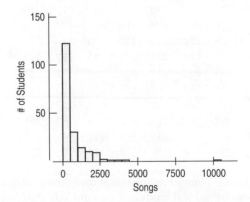

a) What aspect of this distribution makes it difficult to summarize, or to discuss, center and spread?
b) What would you suggest doing with these data if we want to understand them better?

T 39. **Assets again.** Here are the same data you saw in Exercise 37 after re-expressions as the square root of assets and the logarithm of assets:

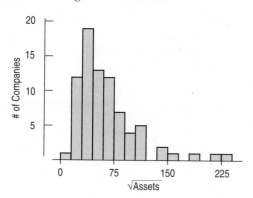

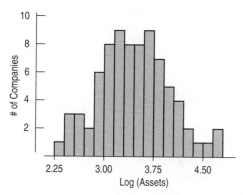

a) Which re-expression do you prefer? Why?
b) In the square root re-expression, what does the value 50 actually indicate about the company's assets?
c) In the logarithm re-expression, what does the value 3 actually indicate about the company's assets?

T 40. **Rainmakers.** The table lists the amount of rainfall (in acre-feet) from the 26 clouds seeded with silver iodide discussed in Exercise 30:

2745	703	302	242	119	40	7
1697	489	274	200	118	32	4
1656	430	274	198	115	31	
978	334	255	129	92	17	

a) Why is acre-feet a good way to measure the amount of precipitation produced by cloud seeding?
b) Plot these data, and describe the distribution.
c) Create a re-expression of these data that produces a more advantageous distribution.
d) Explain what your re-expressed scale means.

T 41. **Stereograms.** Stereograms appear to be composed entirely of random dots. However, they contain separate images that a viewer can "fuse" into a three-dimensional (3D) image by staring at the dots while defocusing the eyes. An experiment was performed to determine whether knowledge of the embedded image affected the time required for subjects to fuse the images. One group

of subjects (group NV) received no information or just verbal information about the shape of the embedded object. A second group (group VV) received both verbal information and visual information (specifically, a drawing of the object). The experimenters measured how many seconds it took for the subject to report that he or she saw the 3D image.

a) What two variables are discussed in this description?
b) For each variable, is it quantitative or categorical? If quantitative, what are the units?
c) The boxplots compare the fusion times for the two treatment groups. Write a few sentences comparing these distributions. What does the experiment show?

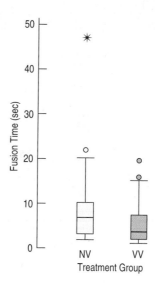

T 42. **Stereograms, revisited.** Because of the skewness of the distributions of fusion times described in Exercise 41, we might consider a re-expression. Here are the boxplots of the *log* of fusion times. Is it better to analyze the original fusion times or the log fusion times? Explain.

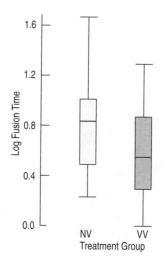

JUST CHECKING

ANSWERS

1. The % late arrivals have a unimodal, symmetric distribution centered at about 20%. In most months between 16% and 23% of the flights arrived late.

2. The boxplot of % late arrivals makes it easier to see that the median is just below 20%, with quartiles at about 17% and 22%. It nominates two months as high outliers.

3. The boxplots by month show a strong seasonal pattern. Flights are more likely to be late in the winter and summer and less likely to be late in the spring and fall. One likely reason for the pattern is snowstorms in the winter and thunderstorms in the summer.

The Standard Deviation as a Ruler and the Normal Model

Where are we going?

A college admissions officer is looking at the files of two candidates, one with a total SAT score of 1500, another with an ACT score of 21. Which candidate scored better? How do we compare things when they're measured on different scales?

To answer a question like this, we need to standardize the results. To do that, we need to know two things. First, we need a base value for comparison—we'll often use the mean for that. Next, we need to know how far away we are from the mean. So, we'll need some sort of ruler. Fortunately, the standard deviation is just the thing we need.

The idea of measuring distances from means by counting standard deviations shows up throughout Statistics, and it starts right here.

The women's heptathlon in the Olympics consists of seven track and field events: the 200-m and 800-m runs, 100-m high hurdles, shot put, javelin, high jump, and long jump. To determine who should get the gold medal, somehow the performances in all seven events have to be combined into one score. How can performances in such different events be compared? They don't even have the same units; the races are recorded in minutes and seconds and the throwing and jumping events in meters. In the 2004 Olympics, Austra Skujytė of Lithuania put the shot 16.4 meters, about 3 meters farther than the average of all contestants. Carolina Klüft won the long jump with a 6.78-m jump, about a meter better than the average. Which performance deserves more points? Even though both events are measured in meters, it's not clear how to compare them. The solution to the problem of how to compare scores turns out to be a useful method for comparing all sorts of values whether they have the same units or not.

The Standard Deviation as a Ruler

The trick in comparing very different-looking values is to use standard deviations. The standard deviation tells us how the whole collection of values varies, so it's a natural ruler for comparing an individual value to the group. Over and over during this course, we will ask questions such as "How far is this value from the mean?" or "How different are these two statistics?" The answer in every case will be to measure the distance or difference in standard deviations.

GRADING ON A CURVE

If you score 79% on an exam, what grade should you get? One teaching philosophy looks only at the raw percentage, 79, and bases the grade on that alone. Another looks at your *relative* performance and bases the grade on how you did compared with the rest of the class. Teachers and students still debate which method is better.

The concept of the standard deviation as a ruler is not special to this course. You'll find statistical distances measured in standard deviations throughout Statistics, up to the most advanced levels.[1] This approach is one of the basic tools of statistical thinking.

In order to compare the two events, let's start with a picture. This time we'll use stem-and-leaf displays so we can see the individual distances.

Long Jump Stem	Leaf	Shot Put Stem	Leaf
67	8	16	4
66		15	
65	1	15	
64	2	14	56778
63	0566	14	24
62	11235	13	5789
61	0569	13	012234
60	2223	12	55
59	0278	12	0144
58	4	11	59
57	0	11	23

67|8 means 6.78 m 16|4 means 16.4 m

FIGURE 6.1
Stem-and-leaf displays for both the long jump and the shot put in the 2004 Olympic Heptathlon. Carolina Klüft (green scores) won the long jump, and Austra Skujyté (red scores) won the shot put. Which heptathlete did better for both events *combined*?

	Event	
	Long Jump	Shot Put
Mean (all contestants)	6.16 m	13.29 m
SD	0.23 m	1.24 m
n	26	28
Klüft	6.78 m	14.77 m
Skujyté	6.30 m	16.40 m

The two winning performances on the top of each stem-and-leaf display appear to be about the same distance from the center of the pack. But look again carefully. What do we mean by the *same distance*? The two displays have different scales. Each line in the stem-and-leaf for the shot put represents half a meter, but for the long jump each line is only a tenth of a meter. It's only because our eyes naturally adjust the scales and use the standard deviation as the ruler that we see each as being about the same distance from the center of the data. How can we make this hunch more precise? Let's see how many standard deviations each performance is from the mean.

Klüft's 6.78-m long jump is 0.62 meter longer than the mean jump of 6.16 m. How many *standard deviations* better than the mean is that? The standard deviation for this event was 0.23 m, so her jump was $(6.78 - 6.16)/0.23 = 0.62/0.23 = 2.70$ standard deviations better than the mean. Skujyté's winning shot put was $16.40 - 13.29 = 3.11$ meters longer than the mean shot put distance, and that's $3.11/1.24 = 2.51$ standard deviations better than the mean. That's a great performance but not quite as impressive as Klüft's long jump, which was farther above the mean, as measured in standard deviations.

Standardizing with *z*-Scores

NOTATION ALERT

There goes another letter. We always use the letter z to denote values that have been standardized with the mean and standard deviation.

To compare these athletes' performances, we determined how many standard deviations from the event's mean each was.

Expressing the distance in standard deviations ***standardizes*** the performances. To standardize a value, we simply subtract the mean performance in that event and then divide this difference by the standard deviation. We can

[1] Other measures of spread could be used as well, but the standard deviation is the most common measure, and it is almost always used as the ruler.

write the calculation as

$$z = \frac{y - \bar{y}}{s}.$$

These values are called **standardized values**, and are commonly denoted with the letter z. Usually, we just call them **z-scores**.

Standardized values have *no units*. z-scores measure the distance of each data value from the mean in standard deviations. A z-score of 2 tells us that a data value is 2 standard deviations above the mean. It doesn't matter whether the original variable was measured in inches, dollars, or seconds. Data values below the mean have negative z-scores, so a z-score of −1.6 means that the data value was 1.6 standard deviations below the mean. Of course, regardless of the direction, the farther a data value is from the mean, the more unusual it is, so a z-score of −1.3 is more extraordinary than a z-score of 1.2. Looking at the z-scores, we can see that even though both were winning scores, Klüft's long jump with a z-score of 2.70 is slightly more impressive than Skujyté's shot put with a z-score of 2.51.

FOR EXAMPLE   Standardizing Skiing Times

The men's combined skiing event in the winter Olympics consists of two races: a downhill and a slalom. Times for the two events are added together, and the skier with the lowest total time wins. In the 2006 Winter Olympics, the mean slalom time was 94.2714 seconds with a standard deviation of 5.2844 seconds. The mean downhill time was 101.807 seconds with a standard deviation of 1.8356 seconds. Ted Ligety of the United States, who won the gold medal with a combined time of 189.35 seconds, skied the slalom in 87.93 seconds and the downhill in 101.42 seconds.

QUESTION: On which race did he do better compared with the competition?

For the slalom, Ligety's z-score is found by subtracting the mean time from his time and then dividing by the standard deviation:

$$z_{Slalom} = \frac{87.93 - 94.2714}{5.2844} = -1.2$$

Similarly, his z-score for the downhill is:

$$z_{Downhill} = \frac{101.42 - 101.807}{1.8356} = -0.21$$

The z-scores show that Ligety's time in the slalom is farther below the mean than his time in the downhill. His performance in the slalom was more remarkable.

By using the standard deviation as a ruler to measure statistical distance from the mean, we can compare values that are measured on different variables, with different scales, with different units, or for different individuals. To determine the winner of the heptathlon, the judges must combine performances on seven very different events. Because they want the score to be absolute, and *not* dependent on the particular athletes in each Olympics, they use predetermined tables, but they could combine scores by standardizing each, and then adding the z-scores together to reach a total score. The only trick is that they'd have to switch the sign of the z-score for running events, because unlike throwing and jumping, it's better to have a running time below the mean (with a negative z-score).

To combine the scores Skujyté and Klüft earned in the long jump and the shot put, we standardize both events as shown in the table. That gives Klüft her 2.70

		Event	
		Long Jump	Shot Put
	Mean	6.16 m	13.29 m
	SD	0.23 m	1.24 m
Klüft	Performance	6.78 m	14.77 m
	z-score	$\dfrac{6.78 - 6.16}{0.23} = 2.70$	$\dfrac{14.77 - 13.29}{1.24} = 1.19$
	Total z-score	\multicolumn{2}{c}{$2.70 + 1.19 = 3.89$}
Skujyté	Performance	6.30 m	16.40 m
	z-score	$\dfrac{6.30 - 6.16}{0.23} = 0.61$	$\dfrac{16.40 - 13.29}{1.24} = 2.51$
	Total z-score	\multicolumn{2}{c}{$0.61 + 2.51 = 3.12$}

z-score in the long jump and a 1.19 in the shot put, for a total of 3.89. Skujyté's shot put gave her a 2.51, but her long jump was only 0.61 SDs above the mean, so her total is 3.12.

Is this the result we wanted? Yes. Each won one event, but Klüft's shot put was second best, while Skujyté's long jump was seventh. The z-scores measure how far each result is from the event mean in standard deviation units. And because they are both in standard deviation units, we can combine them. Not coincidentally, Klüft went on to win the gold medal for the entire seven-event heptathlon, while Skujyté got the silver.

FOR EXAMPLE — Combining z-Scores

In the 2006 winter Olympics men's combined event, Ivica Kostelić of Croatia skied the slalom in 89.44 seconds and the downhill in 100.44 seconds. He thus beat Ted Ligety in the downhill, but not in the slalom. Maybe he should have won the gold medal.

QUESTION: Considered in terms of standardized scores, which skier did better?

Kostelić's z-scores are:

$$z_{Slalom} = \frac{89.44 - 94.2714}{5.2844} = -0.91 \quad \text{and} \quad z_{Downhill} = \frac{100.44 - 101.807}{1.8356} = -0.74$$

The sum of his z-scores is approximately −1.65. Ligety's z-score sum is only about −1.41. Because the standard deviation of the downhill times is so much smaller, Kostelić's better performance there means that he would have won the event if standardized scores were used.

When we standardize data to get a z-score, we do two things. First, we shift the data by subtracting the mean. Then, we rescale the values by dividing by their standard deviation. We often shift and rescale data. What happens to a grade distribution if *everyone* gets a five-point bonus? Everyone's grade goes up, but does the shape change? (*Hint:* Has anyone's distance from the mean changed?) If we switch from feet to meters, what happens to the distribution of heights of students in your class? Even though your intuition probably tells you the answers to these questions, we need to look at exactly how shifting and rescaling work.

JUST CHECKING

1. Your Statistics teacher has announced that the lower of your two tests will be dropped. You got a 90 on test 1 and an 80 on test 2. You're all set to drop the 80 until she announces that she grades "on a curve." She standardized the scores in order to decide which is the lower one. If the mean on the first test was 88 with a standard deviation of 4 and the mean on the second was 75 with a standard deviation of 5,

   a) Which one will be dropped?
   b) Does this seem "fair"?

Shifting Data

WHO	80 male participants of the NHANES survey between the ages of 19 and 24 who measured between 68 and 70 inches tall
WHAT	Their weights
UNIT	Kilograms
WHEN	2001–2002
WHERE	United States
WHY	To study nutrition, and health issues and trends
HOW	National survey

Since the 1960s, the Centers for Disease Control and Prevention's National Center for Health Statistics has been collecting health and nutritional information on people of all ages and backgrounds. A recent survey, the National Health and Nutrition Examination Survey (NHANES) 2001–2002,[2] measured a wide variety of variables, including body measurements, cardiovascular fitness, blood chemistry, and demographic information on more than 11,000 individuals.

Included in this group were 80 men between 19 and 24 years old of average height (between 5'8" and 5'10" tall). Here are a histogram and boxplot of their weights:

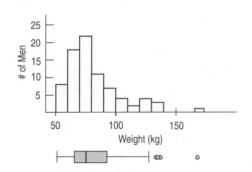

FIGURE 6.2
Histogram and boxplot for the men's weights. The shape is skewed to the right with several high outliers.

Their mean weight is 82.36 kg. For this age and height group, the National Institutes of Health recommends a maximum healthy weight of 74 kg, but we can see that some of the men are heavier than the recommended weight. To compare their weights to the recommended maximum, we could subtract 74 kg from each of their weights. What would that do to the center, shape, and spread of the histogram? Here's the picture:

A S
Activity: **Changing the Baseline.** What happens when we shift data? Do measures of center and spread change?

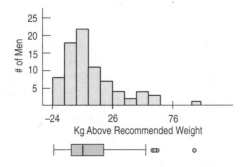

FIGURE 6.3
Subtracting 74 kilograms shifts the entire histogram down but leaves the spread and the shape exactly the same.

---

[2] www.cdc.gov/nchs/nhanes.htm

SHIFTING IDEAL HEIGHTS
Doctors' height and weight charts sometimes give ideal weights for various heights that include 2-inch heels. If the mean height of adult women is 66 inches including 2-inch heels, what is the mean height of women without shoes? Each woman is shorter by 2 inches when barefoot, so the mean is decreased by 2 inches, to 64 inches.

On average, they weigh 82.36 kg, so on average they're 8.36 kg overweight. And, after subtracting 74 from each weight, the mean of the new distribution is $82.36 - 74 = 8.36$ kg. In fact, when we **shift** the data by adding (or subtracting) a constant to each value, all measures of position (center, percentiles, min, max) will increase (or decrease) by the same constant.

What about the spread? What does adding or subtracting a constant value do to the spread of the distribution? Look at the two histograms again. Adding or subtracting a constant changes each data value equally, so the entire distribution just shifts. Its shape doesn't change and neither does the spread. None of the measures of spread we've discussed—not the range, not the IQR, not the standard deviation—changes.

Adding (or subtracting) a constant to every data value adds (or subtracts) the same constant to measures of position, but leaves measures of spread unchanged.

Rescaling Data

Not everyone thinks naturally in metric units. Suppose we want to look at the weights in pounds instead. We'd have to **rescale** the data. Because there are about 2.2 pounds in every kilogram, we'd convert the weights by multiplying each value by 2.2. Multiplying or dividing each value by a constant changes the measurement units. Here are histograms of the two weight distributions, plotted on the same scale, so you can see the effect of multiplying:

FIGURE 6.4
Men's weights in both kilograms and pounds. How do the distributions and numerical summaries change?

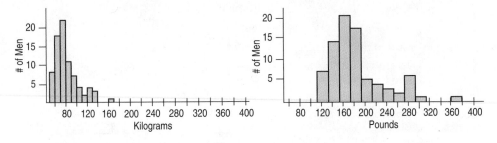

A S *Simulation:* **Changing the Units.**
Change the center and spread values for a distribution and watch the summaries change (or not, as the case may be).

What happens to the shape of the distribution? Although the histograms don't look exactly alike, we see that the shape really hasn't changed: Both are unimodal and skewed to the right.

What happens to the mean? Not too surprisingly, it gets multiplied by 2.2 as well. The men weigh 82.36 kg on average, which is 181.19 pounds. As the boxplots and 5-number summaries show, all measures of position act the same way. They all get multiplied by this same constant.

What happens to the spread? Take a look at the boxplots. The spread in pounds (on the right) is larger. How much larger? If you guessed 2.2 times, you've figured out how measures of spread get rescaled.

FIGURE 6.5
The boxplots (drawn on the same scale) show the weights measured in kilograms (on the left) and pounds (on the right). Because 1 kg is 2.2 lb, all the points in the right box are 2.2 times larger than the corresponding points in the left box. So each measure of position and spread is 2.2 times as large when measured in pounds rather than kilograms.

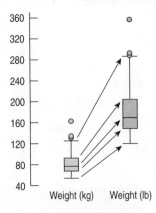

	Weight (kg)	Weight (lb)
Min	54.3	119.46
Q1	67.3	148.06
Median	76.85	169.07
Q3	92.3	203.06
Max	161.5	355.30
IQR	25	55
SD	22.27	48.99

When we multiply (or divide) all the data values by any constant, all measures of position (such as the mean, median, and percentiles) and measures of spread (such as the range, the IQR, and the standard deviation) are multiplied (or divided) by that same constant.

FOR EXAMPLE | Rescaling the Slalom

RECAP: The times in the men's combined event at the winter Olympics are reported in minutes and seconds. Previously, we converted these to seconds and found the mean and standard deviation of the slalom times to be 94.2714 seconds and 5.2844 seconds, respectively.

QUESTION: Suppose instead that we had reported the times in minutes—that is, that each individual time was divided by 60. What would the resulting mean and standard deviation be?

Dividing all the times by 60 would divide both the mean and the standard deviation by 60:

Mean = 94.2714/60 = 1.5712 minutes;   SD = 5.2844/60 = 0.0881 minutes.

JUST CHECKING

2. In 1995 the Educational Testing Service (ETS) adjusted the scores of SAT tests. Before ETS recentered the SAT Verbal test, the mean of all test scores was 450.

   a) How would adding 50 points to each score affect the mean?
   b) The standard deviation was 100 points. What would the standard deviation be after adding 50 points?
   c) Suppose we drew boxplots of test takers' scores a year before and a year after the recentering. How would the boxplots of the two years differ?

3. A company manufactures wheels for in-line skates. The diameters of the wheels have a mean of 3 inches and a standard deviation of 0.1 inch. Because so many of their customers use the metric system, the company decided to report their production statistics in millimeters (1 inch = 25.4 mm). They report that the standard deviation is now 2.54 mm. A corporate executive is worried about this increase in variation. Should he be concerned? Explain.

Back to z-Scores

A S   *Activity: Standardizing.* What if we both shift and rescale? The result is so nice that we give it a name.

Standardizing data into z-scores is just shifting them by the mean and rescaling them by the standard deviation. Now we can see how standardizing affects the distribution. When we subtract the mean of the data from every data value, we shift the mean to zero. As we have seen, such a shift doesn't change the standard deviation.

When we *divide* each of these shifted values by s, however, the standard deviation should be divided by s as well. Since the standard deviation was s to start with, the new standard deviation becomes 1.

How, then, does standardizing affect the distribution of a variable? Let's consider the three aspects of a distribution: the shape, center, and spread.

z-scores have mean 0 and standard deviation 1.

- *Standardizing into z-scores does not change the **shape** of the distribution of a variable.*
- *Standardizing into z-scores changes the **center** by making the mean 0.*
- *Standardizing into z-scores changes the **spread** by making the standard deviation 1.*

STEP-BY-STEP EXAMPLE | Working with Standardized Variables

Many colleges and universities require applicants to submit scores on standardized tests such as the SAT Writing, Math, and Critical Reading (Verbal) tests. The college your little sister wants to apply to says that while there is no minimum score required, the middle 50% of their students have combined SAT scores between 1530 and 1850. You'd feel confident if you knew her score was in their top 25%, but unfortunately she took the ACT test, an alternative standardized test.

Question: How high does her ACT need to be to make it into the top quarter of equivalent SAT scores?

To answer that question you'll have to standardize all the scores, so you'll need to know the mean and standard deviations of scores for some group on both tests. The college doesn't report the mean or standard deviation for their applicants on either test, so we'll use the group of all test takers nationally. For college-bound seniors, the average combined SAT score is about 1500 and the standard deviation is about 250 points. For the same group, the ACT average is 20.8 with a standard deviation of 4.8.

Plan State what you want to find out.

Variables Identify the variables and report the W's (if known).

Check the appropriate conditions.

I want to know what ACT score corresponds to the upper-quartile SAT score. I know the mean and standard deviation for both the SAT and ACT scores based on all test takers, but I have no individual data values.

✔ **Quantitative Data Condition:** Scores for both tests are quantitative but have no meaningful units other than points.

Mechanics Standardize the variables.

The middle 50% of SAT scores at this college fall between 1530 and 1850 points. To be in the top quarter, my sister would have to have a score of at least 1850. That's a z-score of

$$z = \frac{(1850 - 1500)}{250} = 1.40$$

So an SAT score of 1850 is 1.40 standard deviations above the mean of all test takers.

The y-value we seek is z standard deviations above the mean.

For the ACT, 1.40 standard deviations above the mean is $20.8 + 1.40(4.8) = 27.52$.

Conclusion Interpret your results in context.

To be in the top quarter of applicants in terms of combined SAT score, she'd need to have an ACT score of at least 27.52.

When Is a *z*-Score BIG?

IS NORMAL NORMAL?

Don't be misled. The name "Normal" doesn't mean that these are the *usual* shapes for histograms. The name follows a tradition of positive thinking in Mathematics and Statistics in which functions, equations, and relationships that are easy to work with or have other nice properties are called "normal," "common," "regular," "natural," or similar terms. It's as if by calling them ordinary, we could make them actually occur more often and simplify our lives.

"All models are wrong—but some are useful."

—George Box, famous statistician

NOTATION ALERT

$N(\mu, \sigma)$ always denotes a Normal model. The μ, pronounced "mew," is the Greek letter for "m" and always represents the mean in a model. The σ, sigma, is the lowercase Greek letter for "s" and always represents the standard deviation in a model.

IS THE STANDARD NORMAL A STANDARD?

Yes. We call it the "Standard Normal" because it models standardized values. It is also a "standard" because this is the particular Normal model that we almost always use.

A *z*-score gives us an indication of how unusual a value is because it tells us how far it is from the mean. If the data value sits right at the mean, it's not very far at all and its *z*-score is 0. A *z*-score of 1 tells us that the data value is 1 standard deviation above the mean, while a *z*-score of −1 tells us that the value is 1 standard deviation below the mean. How far from 0 does a *z*-score have to be to be interesting or unusual? There is no universal standard, but the larger the score is (negative or positive), the more unusual it is. We know that 50% of the data lie between the quartiles. For symmetric data, the standard deviation is usually a bit smaller than the IQR, and it's not uncommon for at least half of the data to have *z*-scores between −1 and 1. But no matter what the shape of the distribution, a *z*-score of 3 (plus or minus) or more is rare, and a *z*-score of 6 or 7 shouts out for attention.

To say more about how big we expect a *z*-score to be, we need to *model* the data's distribution. A model will let us say much more precisely how often we'd be likely to see *z*-scores of different sizes. Of course, like all models of the real world, the model will be wrong—wrong in the sense that it can't match reality exactly. But it can still be useful. Like a physical model, it's something we can look at and manipulate in order to learn more about the real world.

Models help our understanding in many ways. Just as a model of an airplane in a wind tunnel can give insights even though it doesn't show every rivet,[3] models of data give us summaries that we can learn from and use, even though they don't fit each data value exactly. It's important to remember that they're only *models* of reality and not reality itself. But without models, what we can learn about the world at large is limited to only what we can say about the data we have at hand.

There is no universal standard for *z*-scores, but there is a model that shows up over and over in Statistics. You may have heard of "bell-shaped curves." Statisticians call them Normal models. **Normal models** are appropriate for distributions whose shapes are unimodal and roughly symmetric. For these distributions, they provide a measure of how extreme a *z*-score is. Fortunately, there is a Normal model for every possible combination of mean and standard deviation. We write $N(\mu, \sigma)$ to represent a Normal model with a mean of μ and a standard deviation of σ. Why the Greek? Well, *this* mean and standard deviation are not numerical summaries of data. They are part of the model. They don't come from the data. Rather, they are numbers that we choose to help specify the model. Such numbers are called **parameters** of the model.

We don't want to confuse the parameters with summaries of the data such as $\bar{y}$ and s, so we use special symbols. In Statistics, we almost always use Greek letters for parameters. By contrast, summaries of data are called **statistics** and are usually written with Latin letters.

If we model data with a Normal model and standardize them using the corresponding μ and σ, we still call the standardized value a **z-score**, and we write

$$z = \frac{y - \mu}{\sigma}.$$

Usually it's easier to standardize data first (using its mean and standard deviation). Then we need only the model $N(0,1)$. The Normal model with mean 0 and standard deviation 1 is called the **standard Normal model** (or the **standard Normal distribution**).

---

[3] In fact, the model is useful *because* it doesn't have every rivet. It is because models offer a simpler view of reality that they are so useful as we try to understand reality.

But be careful. You shouldn't use a Normal model for just any data set. Remember that standardizing won't change the shape of the distribution. If the distribution is not unimodal and symmetric to begin with, standardizing won't make it Normal.

When we use the Normal model, we assume that the distribution of the data is, well, Normal. Practically speaking, there's no way to check whether this **Normality Assumption** is true. In fact, it almost certainly is not true. Real data don't behave like mathematical models. Models are idealized; real data are real. The good news, however, is that to use a Normal model, it's sufficient to check the following condition:

> **Nearly Normal Condition.** The shape of the data's distribution is unimodal and symmetric. Check this by making a histogram (or a Normal probability plot, which we'll explain later).

Don't model data with a Normal model without checking whether the condition is satisfied.

All models make **assumptions**. Whenever we model—and we'll do that often—we'll be careful to point out the assumptions that we're making. And, what's even more important, we'll check the associated **conditions** in the data to make sure that those assumptions are reasonable.

> **A S** *Activity:* **Working with Normal Models.** Learn more about the Normal model and see what data drawn at random from a Normal model might look like.

The 68-95-99.7 Rule

Normal models give us an idea of how extreme a value is by telling us how likely it is to find one that far from the mean. We'll soon show how to find these numbers precisely—but one simple rule is usually all we need.

It turns out that in a Normal model, about 68% of the values fall within 1 standard deviation of the mean, about 95% of the values fall within 2 standard deviations of the mean, and about 99.7%—almost all—of the values fall within 3 standard deviations of the mean. These facts are summarized in a rule that we call (let's see . . .) the **68–95–99.7 Rule**.[4]

> **ONE IN A MILLION**
>
> These magic 68, 95, 99.7 values come from the Normal model. As a model, it can give us corresponding values for any z-score. For example, it tells us that fewer than 1 out of a million values have z-scores smaller than −5.0 or larger than +5.0. So if someone tells you you're "one in a million," they must really admire your z-score.

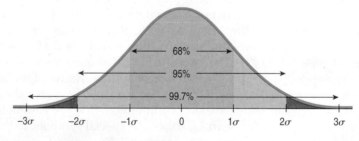

FIGURE 6.6
Reaching out one, two, and three standard deviations on a Normal model gives the 68-95-99.7 Rule, seen as proportions of the area under the curve.

FOR EXAMPLE   Using the 68–95–99.7 Rule

QUESTION: In the 2006 Winter Olympics men's combined event, Jean-Baptiste Grange of France skied the slalom in 88.46 seconds–about 1 standard deviation faster than the mean. If a Normal model is useful in describing slalom times, about how many of the 35 skiers finishing the event would you expect skied the slalom *faster* than Jean-Baptiste?

From the 68–95–99.7 Rule, we expect 68% of the skiers to be within one standard deviation of the mean. Of the remaining 32%, we expect half on the high end and half on the low end. 16% of 35 is 5.6, so, conservatively, we'd expect about 5 skiers to do better than Jean-Baptiste.

---

[4] This rule is also called the "Empirical Rule" because it originally came from observation. The rule was first published by Abraham de Moivre in 1733, 75 years before the Normal model was discovered. Maybe it should be called "de Moivre's Rule," but that wouldn't help us remember the important numbers, 68, 95, and 99.7.

JUST CHECKING

4. As a group, the Dutch are among the tallest people in the world. The average Dutch man is 184 cm tall—just over 6 feet (and the average Dutch woman is 170.8 cm tall—just over 5′7″). If a Normal model is appropriate and the standard deviation for men is about 8 cm, what percentage of all Dutch men will be over 2 meters (6′6″) tall?

5. Suppose it takes you 20 minutes, on average, to drive to school, with a standard deviation of 2 minutes. Suppose a Normal model is appropriate for the distributions of driving times.

   a) How often will you arrive at school in less than 22 minutes?
   b) How often will it take you more than 24 minutes?
   c) Do you think the distribution of your driving times is unimodal and symmetric?
   d) What does this say about the accuracy of your predictions? Explain.

The First Three Rules for Working with Normal Models

A S

Activity: **Working with Normal Models.** Well, actually playing with them. This interactive tool lets you do what this chapter's figures can't do, move them!

1. Make a picture.
2. Make a picture.
3. Make a picture.

    Although we're thinking about models, not histograms of data, the three rules don't change. To help you think clearly, a simple hand-drawn sketch is all you need. Even experienced statisticians sketch pictures to help them think about Normal models. You should too.

    Of course, when we have data, we'll also need to make a histogram to check the **Nearly Normal Condition** to be sure we can use the Normal model to model the data's distribution. Other times, we may be told that a Normal model is appropriate based on prior knowledge of the situation or on theoretical considerations.

A S

Activity: **Normal Models.** Normal models have several interesting properties—see them here.

How to sketch a normal curve that looks normal   To sketch a good Normal curve, you need to remember only three things:

▶ The Normal curve is bell-shaped and symmetric around its mean. Start at the middle, and sketch to the right and left from there.

▶ Even though the Normal model extends forever on either side, you need to draw it only for 3 standard deviations. After that, there's so little left that it isn't worth sketching.

▶ The place where the bell shape changes from curving downward to curving back up—the *inflection point*—is exactly one standard deviation away from the mean.

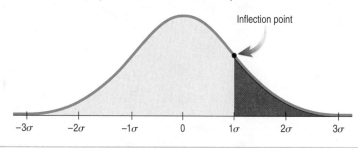

STEP-BY-STEP EXAMPLE Working with the 68–95–99.7 Rule

The SAT Reasoning Test has three parts: Writing, Math, and Critical Reading (Verbal). Each part has a distribution that is roughly unimodal and symmetric and is designed to have an overall mean of about 500 and a standard deviation of 100 for all test takers. In any one year, the mean and standard deviation may differ from these target values by a small amount, but they are a good overall approximation.

Question: Suppose you earned a 600 on one part of your SAT. Where do you stand among all students who took that test?

You could calculate your z-score and find out that it's $z = (600 - 500)/100 = 1.0$, but what does that tell you about your percentile? You'll need the Normal model and the 68–95–99.7 Rule to answer that question.

	Plan State what you want to know.	I want to see how my SAT score compares with the scores of all other students. To do that, I'll need to model the distribution.
	Variables Identify the variable and report the W's.	Let y = my SAT score. Scores are quantitative but have no meaningful units other than points.
	Be sure to check the appropriate conditions.	✓ **Nearly Normal Condition:** If I had data, I would check the histogram. I have no data, but I am told that the SAT scores are roughly unimodal and symmetric.
	Specify the parameters of your model.	I will model SAT score with a $N(500, 100)$ model.
	Mechanics Make a picture of this Normal model. (A simple sketch is all you need.)	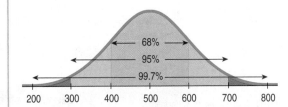
	Locate your score.	My score of 600 is 1 standard deviation above the mean. That corresponds to one of the points of the 68–95–99.7 Rule.
	Conclusion Interpret your result in context.	About 68% of those who took the test had scores that fell no more than 1 standard deviation from the mean, so 100% − 68% = 32% of all students had scores more than 1 standard deviation away. Only half of those were on the high side, so about 16% (half of 32%) of the test scores were better than mine. My score of 600 is higher than about 84% of all scores on this test.

The bounds of SAT scoring at 200 and 800 can also be explained by the 68–95–99.7 Rule. Since 200 and 800 are three standard deviations from 500, it hardly pays to extend the scoring any farther on either side. We'd get more information only on $100 - 99.7 = 0.3\%$ of students.

The worst-case scenario: Tchebycheff's Inequality* Suppose we encounter an observation that's 5 standard deviations above the mean. Should we be surprised? We've just seen that when a Normal model is appropriate, such a value is exceptionally rare. After all, 99.7% of all the values should be within 3 standard deviations of the mean, so anything farther away would be unusual indeed.

But our handy 68-95-99.7 Rule applies only to Normal models, and the Normal is such a *nice* shape. What if we're dealing with a distribution that's strongly skewed (like the CEO salaries), or one that is uniform or bimodal or something really strange? A Normal model has 68% of its observations within one standard deviation of the mean, but a bimodal distribution could even be entirely empty in the middle. In that case could we still say anything at all about an observation 5 standard deviations above the mean?

Remarkably, even with really weird distributions, the worst case can't get all that bad. A Russian mathematician named Pafnuty Tchebycheff[5] answered the question by proving this theorem:

In any distribution, at least $1 - \dfrac{1}{k^2}$ *of the values must lie within* $\pm k$ *standard deviations of the mean.*

What does that mean?

▶ For $k = 1, 1 - \dfrac{1}{1^2} = 0$; if the distribution is far from Normal, it's possible that none of the values are within 1 standard deviation of the mean. We should be really cautious about saying anything about 68% unless we think a Normal model is justified. (Tchebycheff's theorem really is about the worst case; it tells us nothing about the middle; only about the extremes.)

▶ For $k = 2, 1 - \dfrac{1}{2^2} = \dfrac{3}{4}$; no matter how strange the shape of the distribution, at least 75% of the values must be within 2 standard deviations of the mean. Normal models may expect 95% in that 2-standard-deviation interval, but even in a worst-case scenario it can never go lower than 75%.

▶ For $k = 3, 1 - \dfrac{1}{3^2} = \dfrac{8}{9}$; in any distribution, at least 89% of the values lie within 3 standard deviations of the mean.

What we see is that values beyond 3 standard deviations from the mean are uncommon, Normal model or not. Tchebycheff tells us that at least 96% of all values must be within 5 standard deviations of the mean. While we can't always apply the 68-95-99.7 Rule, we can be sure that the observation we encountered 5 standard deviations above the mean is unusual.

---

[5] He may have made the worst case for deviations clear, but the English spelling of his name is not. You'll find his first name spelled Pavnutii or Pavnuty and his last name spelled Chebsheff, Cebysev, and other creative versions.

Finding Normal Percentiles

An SAT score of 600 is easy to assess, because we can think of it as one standard deviation above the mean. If your score was 680, though, where do you stand among the rest of the people tested? Your z-score is 1.80, so you're somewhere between 1 and 2 standard deviations above the mean. We figured out that no more than 16% of people score better than 600. By the same logic, no more than 2.5% of people score better than 700. Can we be more specific than "between 16% and 2.5%"?

When the value doesn't fall exactly 1, 2, or 3 standard deviations from the mean, we can look it up in a table of **Normal percentiles** or use technology.[6] Either way, we first convert our data to z-scores before using the table. Your SAT score of 680 has a z-score of $(680 - 500)/100 = 1.80$.

FIGURE 6.7
A table of Normal percentiles (Table Z in Appendix D) lets us find the percentage of individuals in a Standard Normal distribution falling below any specified z-score value.

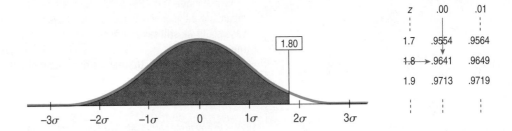

In the piece of the table shown, we find your z-score by looking down the left column for the first two digits, 1.8, and across the top row for the third digit, 0. The table gives the percentile as 0.9641. That means that 96.4% of the z-scores are less than 1.80. Only 3.6% of people, then, scored better than 680 on the SAT.

Most of the time, though, you'll do this with a calculator, Internet tool, or computer package.

Finding Normal Percentiles Using Technology

These days, finding percentiles from a Normal probability table is a "desert island" method—something we might do if we desperately needed a Normal percentile and were stranded miles from the mainland with only a Normal probability table. (Of course, you might feel just that way during a Statistics exam, so it's a good idea to know how to do it.) Fortunately, most of the time, we can use a calculator, a computer, or the Internet. Graphing calculators, such as the TI-84, find Normal percentiles and offer to draw the picture as well. Most statistics programs have functions to find Normal percentiles, and values of z-scores are easy to find on the Internet (search for *z-scores*, or *normal probability tables*).

The *ActivStats* Multimedia Assistant provided on the DVD that accompanies this book offers two methods. The "Normal Model Tool" introduced along with the Normal model makes it easy to see how areas under parts of the

---

[6] See Table Z in Appendix D, if you're curious. But many statistics computer packages do this, too—and more easily!

A S
Activity: **The Normal Table.** Table Z just sits there, but this version of the Normal table changes so it always Makes a Picture that fits. A great way to learn to use the table.

Normal model correspond to particular cut points. The tool is especially useful for problems in which you want to find the area *between* two z-scores.

The tool also allows you to work in the original units if you wish, without converting everything to z-scores. This can be particularly helpful for understanding statements about relative frequencies.

ActivStats also offers a Normal table in which the picture of the Normal model is interactive.[7] Grab the z-score cut point with your mouse and drag it to the value you are interested in. The table will adjust accordingly.

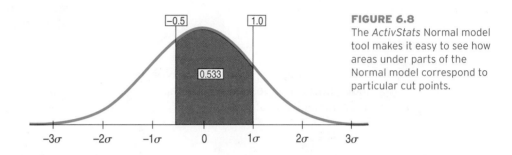

FIGURE 6.8
The *ActivStats* Normal model tool makes it easy to see how areas under parts of the Normal model correspond to particular cut points.

A S
Normal Models. Normal models have several interesting properties–see them here.

With a tool like the *ActivStats* Normal model tool, it's easy to answer a question like what percentage of all test takers score between 950 and 1080 on the combined SAT. You can leave the units as they are and just enter the mean and standard deviation on the Normal model tool, sliding the cut points to 950 and 1080. Alternatively, you could convert both values to z-scores and slide the area to match those scores on the Standard Normal model tool. In either case, you would read the percentage right off the picture.

Other Models

Of course, the Normal is not the only model for data. There are models for skewed data (watch out for the χ^2 and F later in the book) and we'll see models for variables that can take on only a few values (Binomial and Poisson, to name two). But the Normal will return in an important and surprising way.

STEP-BY-STEP EXAMPLE   **Working with Normal Models Part I**

The Normal model is our first model for data. It's the first in a series of modeling situations where we step away from the data at hand to make more general statements about the world. We'll become more practiced in thinking about and learning the details of models as we progress through the book. To give you some practice in thinking about the Normal model, here are several problems that ask you to find percentiles in detail.

Question: What proportion of SAT scores fall between 450 and 600?

[7] Now it's time to open the DVD that accompanies the book if you haven't done so already. You'll find the instructions for using the tools there. You can also get to the tools directly in the Appendix following the last chapter of the *ActivStats* Lesson Book.

Plan State the problem.

I want to know the proportion of SAT scores between 450 and 600.

Variables Name the variable.

Let y = SAT score.

Check the appropriate conditions and specify which Normal model to use.

✔ **Nearly Normal Condition:** We are told that SAT scores are nearly Normal.

I'll model SAT scores with a N(500, 100) model, using the mean and standard deviation specified for them.

Mechanics Make a picture of this Normal model. Locate the desired values and shade the region of interest.

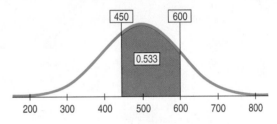

Find z-scores for the cut points 450 and 600. Use technology to find the desired proportions, represented by the area under the curve.

Standardizing the two scores, I find that

$$z = \frac{(y - \mu)}{\sigma} = \frac{(600 - 500)}{100} = 1.00$$

and

$$z = \frac{(450 - 500)}{100} = -0.50$$

So,

Area $(450 < y < 600)$ = Area $(-0.5 < z < 1.0)$
$= 0.5328$

(If you use a table, then you need to subtract the two areas to find the area *between* the cut points.)

(**OR:** From Table Z, the area $(z < 1.0) = 0.8413$ and area $(z < -0.5) = 0.3085$, so the proportion of z-scores between them is $0.8413 - 0.3085 = 0.5328$, or 53.28%.)

Conclusion Interpret your result in context.

The Normal model estimates that about 53.3% of SAT scores fall between 450 and 600.

From Percentiles to Scores: z in Reverse

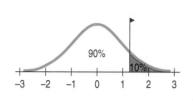

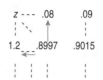

Finding areas from z-scores is the simplest way to work with the Normal model. But sometimes we start with areas and are asked to work backward to find the corresponding z-score or even the original data value. For instance, what z-score cuts off the top 10% in a Normal model?

Make a picture like the one shown, shading the rightmost 10% of the area. Notice that this is the 90th percentile. Look in Table Z for an area of 0.900. The exact area is not there, but 0.8997 is pretty close. That shows up in the table with 1.2 in the left margin and .08 in the top margin. The z-score for the 90th percentile, then, is approximately $z = 1.28$.

Computers and calculators will determine the cut point more precisely (and more easily).

STEP-BY-STEP EXAMPLE | **Working with Normal Models Part II**

Question: Suppose a college says it admits only people with SAT Verbal test scores among the top 10%. How high a score does it take to be eligible?

 THINK

Plan State the problem.

How high an SAT Verbal score do I need to be in the top 10% of all test takers?

Variable Define the variable.

Let y = my SAT score.

Check to see if a Normal model is appropriate, and specify which Normal model to use.

✓ **Nearly Normal Condition:** I am told that SAT scores are nearly Normal. I'll model them with N(500, 100).

 SHOW

Mechanics Make a picture of this Normal model. Locate the desired percentile approximately by shading the rightmost 10% of the area.

The college takes the top 10%, so its cutoff score is the 90th percentile. Find the corresponding z-score using Table Z as shown on p. 124.

Convert the z-score back to the original units.

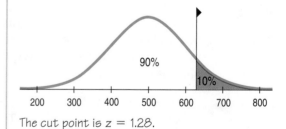

The cut point is z = 1.28.

A z-score of 1.28 is 1.28 standard deviations above the mean. Since the SD is 100, that's 128 SAT points. The cutoff is 128 points above the mean of 500, or 628.

 TELL

Conclusion Interpret your results in the proper context.

Because the school wants SAT Verbal scores in the top 10%, the cutoff is 628. (Actually, since SAT scores are reported only in multiples of 10, I'd have to score at least a 630.)

STEP-BY-STEP EXAMPLE   More Working with Normal Models

Working with Normal percentiles can be a little tricky, depending on how the problem is stated. Here are a few more worked examples of the kind you're likely to see.

A cereal manufacturer has a machine that fills the boxes. Boxes are labeled "16 ounces," so the company wants to have that much cereal in each box, but since no packaging process is perfect, there will be minor variations. If the machine is set at exactly 16 ounces and the Normal model applies (or at least the distribution is roughly symmetric), then about half of the boxes will be underweight, making consumers unhappy and exposing the company to bad publicity and possible lawsuits. To prevent underweight boxes, the manufacturer has to set the mean a little higher than 16.0 ounces.

Based on their experience with the packaging machine, the company believes that the amount of cereal in the boxes fits a Normal model with a standard deviation of 0.2 ounce. The manufacturer decides to set the machine to put an average of 16.3 ounces in each box. Let's use that model to answer a series of questions about these cereal boxes.

Question 1: What fraction of the boxes will be underweight?

 THINK

Plan State the problem.

Variable Name the variable.

Check to see if a Normal model is appropriate.

Specify which Normal model to use.

What proportion of boxes weigh less than 16 ounces?

Let y = weight of cereal in a box.

✓ **Nearly Normal Condition:** I have no data, so I cannot make a histogram, but I am told that the company believes the distribution of weights from the machine is Normal.

I'll use a N(16.3, 0.2) model.

SHOW

Mechanics Make a picture of this Normal model. Locate the value you're interested in on the picture, label it, and shade the appropriate region.

REALITY CHECK Estimate from the picture the percentage of boxes that are underweight. (This will be useful later to check that your answer makes sense.) It looks like a low percentage. Less than 20% for sure.

Convert your cutoff value into a z-score.

Find the area with your calculator (or use the Normal table).

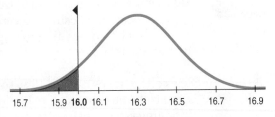

15.7   15.9 **16.0** 16.1   16.3   16.5   16.7   16.9

I want to know what fraction of the boxes will weigh less than 16 ounces.

$$z = \frac{y - \mu}{\sigma} = \frac{16 - 16.3}{0.2} = -1.50$$

Area $(y < 16)$ = Area $(z < -1.50)$ = 0.0668

TELL ▶ **Conclusion** State your conclusion, and check that it's consistent with your earlier guess. It's below 20%—seems okay.

I estimate that approximately 6.7% of the boxes will contain less than 16 ounces of cereal.

Question 2: The company's lawyers say that 6.7% is too high. They insist that no more than 4% of the boxes can be underweight. So the company needs to set the machine to put a little more cereal in each box. What mean setting do they need?

THINK ▶ **Plan** State the problem.

What mean weight will reduce the proportion of underweight boxes to 4%?

Variable Name the variable.

Let *y* = weight of cereal in a box.

Check to see if a Normal model is appropriate.

✔ **Nearly Normal Condition:** I am told that a Normal model applies.

Specify which Normal model to use. This time you are not given a value for the mean!

I don't know μ, the mean amount of cereal. The standard deviation for this machine is 0.2 ounce. The model is $N(\mu, 0.2)$.

REALITY CHECK ▶ We found out earlier that setting the machine to $\mu = 16.3$ ounces made 6.7% of the boxes too light. We'll need to raise the mean a bit to reduce this fraction.

No more than 4% of the boxes can be below 16 ounces.

SHOW ▶ **Mechanics** Make a picture of this Normal model. Center it at μ (since you don't know the mean), and shade the region below 16 ounces.

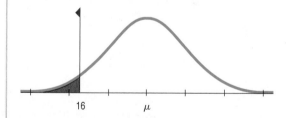

Using your calculator (or the Normal table), find the *z*-score that cuts off the lowest 4%.

The z-score that has 0.04 area to the left of it is z = −1.75.

Use this information to find μ. It's located 1.75 standard deviations to the right of 16. Since σ is 0.2, that's 1.75×0.2, or 0.35 ounces more than 16.

For 16 to be 1.75 standard deviations below the mean, the mean must be

$$16 + 1.75(0.2) = 16.35 \text{ ounces.}$$

TELL ▶ **Conclusion** Interpret your result in context.
(This makes sense; we knew it would have to be just a bit higher than 16.3.)

The company must set the machine to average 16.35 ounces of cereal per box.

Question 3: The company president vetoes that plan, saying the company should give away less free cereal, not more. Her goal is to set the machine no higher than 16.2 ounces and still have only 4% underweight boxes. The only way to accomplish this is to reduce the standard deviation. What standard deviation must the company achieve, and what does that mean about the machine?

 THINK

Plan State the problem.

What standard deviation will allow the mean to be 16.2 ounces and still have only 4% of boxes underweight?

Variable Name the variable.

Let y = weight of cereal in a box.

Check conditions to be sure that a Normal model is appropriate.

✔ **Nearly Normal Condition:** The company believes that the weights are described by a Normal model.

Specify which Normal model to use. This time you don't know σ.

I know the mean, but not the standard deviation, so my model is N(16.2, σ).

REALITY CHECK ▷ We know the new standard deviation must be less than 0.2 ounce.

 SHOW

Mechanics Make a picture of this Normal model. Center it at 16.2, and shade the area you're interested in. We want 4% of the area to the left of 16 ounces.

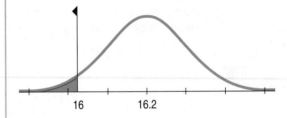

Find the z-score that cuts off the lowest 4%.

I know that the z-score with 4% below it is z = −1.75.

Solve for σ. (We need 16 to be 1.75 σ's below 16.2, so 1.75 σ must be 0.2 ounce. You could just start with that equation.)

$$z = \frac{y - \mu}{\sigma}$$

$$-1.75 = \frac{16 - 16.2}{\sigma}$$

$$1.75\,\sigma = 0.2$$

$$\sigma = 0.114$$

 TELL

Conclusion Interpret your result in context.

As we expected, the standard deviation is lower than before—actually, quite a bit lower.

The company must get the machine to box cereal with a standard deviation of only 0.114 ounce. This means the machine must be more consistent (by nearly a factor of 2) in filling the boxes.

Are You Normal? Find Out with a Normal Probability Plot

In the examples we've worked through, we've assumed that the underlying data distribution was roughly unimodal and symmetric, so that using a Normal model makes sense. When you have data, you must *check* to see whether a Normal model is reasonable. How? Make a picture, of course! Drawing a histogram of the data and looking at the shape is one good way to see if a Normal model might work.

There's a more specialized graphical display that can help you to decide whether the Normal model is appropriate: the **Normal probability plot**. If the distribution of the data is roughly Normal, the plot is roughly a diagonal straight line. Deviations from a straight line indicate that the distribution is not Normal. This plot is usually able to show deviations from Normality more clearly than the corresponding histogram, but it's usually easier to understand *how* a distribution fails to be Normal by looking at its histogram.

Some data on a car's fuel efficiency provide an example of data that are nearly Normal. The overall pattern of the Normal probability plot is straight. The two trailing low values correspond to the values in the histogram that trail off the low end. They're not quite in line with the rest of the data set. The Normal probability plot shows us that they're a bit lower than we'd expect of the lowest two values in a Normal model.

FIGURE 6.9
Histogram and Normal probability plot for gas mileage (mpg) recorded by one of the authors over the 8 years he owned a 1989 Nissan Maxima. The vertical axes are the same, so each dot on the probability plot would fall into the bar on the histogram immediately to its left.

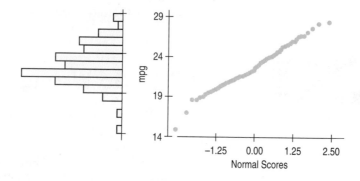

By contrast, the Normal probability plot of the men's *Weight*s from the NHANES Study is far from straight. The weights are skewed to the high end, and the plot is curved. We'd conclude from these pictures that approximations using the 68–95–99.7 Rule for these data would not be very accurate.

FIGURE 6.10
Histogram and Normal probability plot for men's weights. Note how a skewed distribution corresponds to a bent probability plot.

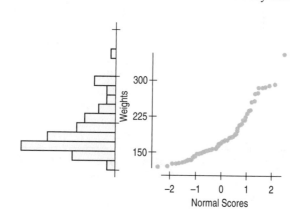

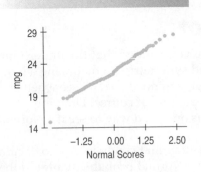

How Does a Normal Probability Plot Work?

Why does the Normal probability plot work like that? We looked at 100 fuel efficiency measures for the author's Nissan car. The smallest of these has a z-score of −3.16. The Normal model can tell us what value to expect for the smallest z-score in a batch of 100 if a Normal model were appropriate. That turns out to be −2.58. So our first data value is smaller than we would expect from the Normal.

We can continue this and ask a similar question for each value. For example, the 14th-smallest fuel efficiency has a z-score of almost exactly −1, and that's just what we should expect (well, −1.1 to be exact). A Normal probability plot takes each data value and plots it against the z-score you'd expect that point to have if the distribution were perfectly Normal.[8]

When the values match up well, the line is straight. If one or two points are surprising from the Normal's point of view, they don't line up. When the entire distribution is skewed or different from the Normal in some other way, the values don't match up very well at all and the plot bends.

It turns out to be tricky to find the values we expect. They're called *Normal scores*, but you can't easily look them up in the tables. That's why probability plots are best made with technology and not by hand.

The best advice on using Normal probability plots is to see whether they are straight. If so, then your data look like data from a Normal model. If not, make a histogram to understand how they differ from the model.

A S   *Activity:* **Assessing Normality.** This activity guides you through the process of checking the Nearly Normal Condition using your statistics package.

What Can Go Wrong?

■ **Don't use a Normal model when the distribution is not unimodal and symmetric.** Normal models are so easy and useful that it is tempting to use them even when they don't describe the data very well. That can lead to wrong conclusions. Don't use a Normal model without first checking the **Nearly Normal Condition**. Look at a picture of the data to check that it is unimodal and symmetric. A histogram, or a Normal probability plot, can help you tell whether a Normal model is appropriate.

The CEOs (p. 91) had a mean total compensation of $10,307,311.87 with a standard deviation of $17,964,615.16. Using the Normal model rule, we should expect about 68% of the CEOs to have compensations between −$7,657,303.29 and $28,271,927.03. In fact, more than 90% of the CEOs have annual compensations in this range. What went wrong? The distribution is skewed, not symmetric. Using the 68–95–99.7 Rule for data like these will lead to silly results.

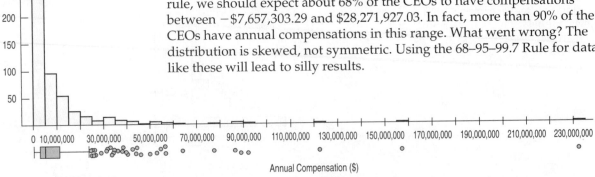

----

[8] Sometimes a Normal probability plot is drawn with the two axes switched, putting the data on the x-axis and the z-scores on the y-axis.

- **Don't use the mean and standard deviation when outliers are present.** Both means and standard deviations can be distorted by outliers, and no model based on distorted values will do a good job. A z-score calculated from a distribution with outliers may be misleading. It's always a good idea to check for outliers. How? Make a picture.

- **Don't round your results in the middle of a calculation.** We reported the mean of the heptathletes' long jump as 6.16 meters. More precisely, it was 6.16153846153846 meters.

    You should use all the precision available in the data for all the intermediate steps of a calculation. Using the more precise value for the mean (and also carrying 15 digits for the SD), the z-score calculation for Klüft's long jump comes out to

$$z = \frac{6.78 - 6.16153846153846}{0.2297597407326585} = 2.691775053755667700$$

We'd report that as 2.692, as opposed to the rounded-off value of 2.70 we got earlier from the table.

- **Do what we say, not what we do.** When we showed the z-score calculations for Kluft, we rounded the mean to 6.16 m and the SD to 0.23 m. Then to make the story clearer we used *those values* in the displayed calculation. But that gave a z-score of 2.70, not 2.692.

    We'll continue to show simplified calculations in the book to make the story simpler. When you calculate with full precision, your results may differ slightly from ours. So, we also advise . . .

- **Don't worry about minor differences in results.** Because various calculators and programs may carry different precision in calculations, your answers may differ slightly from those we show in the text and in the Step-By-Steps, or even from the values given in the answers in the back of the book. Those differences aren't anything to worry about. They're not the main story Statistics tries to tell.

CONNECTIONS

Changing the center and spread of a variable is equivalent to changing its *units*. Indeed, the only part of the data's context changed by standardizing is the units. All other aspects of the context do not depend on the choice or modification of measurement units. This fact points out an important distinction between the numbers the data provide for calculation and the meaning of the variables and the relationships among them. Standardizing can make the numbers easier to work with, but it does not alter the meaning.

Another way to look at this is to note that standardizing may change the center and spread values, but it does not affect the *shape* of a distribution. A histogram or boxplot of standardized values looks just the same as the histogram or boxplot of the original values except, perhaps, for the numbers on the axes.

When we summarized *shape, center,* and *spread* for histograms, we compared them to unimodal, symmetric shapes. You couldn't ask for a nicer example than the Normal model. And if the shape *is* like a Normal, we'll use the the mean and standard deviation to standardize the values.

WHAT HAVE WE LEARNED?

We've learned that the story data can tell may be easier to understand after shifting or rescaling the data.

▶ Shifting data by adding or subtracting the same amount from each value affects measures of center and position but not measures of spread.

▶ Rescaling data by multiplying or dividing every value by a constant changes all the summary statistics—center, position, and spread.

We've learned the power of standardizing data.

▶ Standardizing uses the standard deviation as a ruler to measure distance from the mean, creating z-scores.

▶ Using these z-scores, we can compare apples and oranges—values from different distributions or values based on different units.

▶ And a z-score can identify unusual or surprising values among data.

We've learned that the 68-95-99.7 Rule can be a useful rule of thumb for understanding distributions.

▶ For data that are unimodal and symmetric, about 68% fall within 1 SD of the mean, 95% fall within 2 SDs of the mean, and 99.7% fall within 3 SDs of the mean (see p. 118).

Again we've seen the importance of *Thinking* about whether a method will work.

▶ **Normality Assumption:** We sometimes work with Normal tables (Table Z). Those tables are based on the Normal model.

▶ Data can't be exactly Normal, so we check the **Nearly Normal Condition** by making a histogram (is it unimodal, symmetric, and free of outliers?) or a Normal probability plot (is it straight enough?). (See p. 129.)

Terms

Standardizing	We standardize to eliminate units. Standardized values can be compared and combined even if the original variables had different units and magnitudes (p. 110).
Standardized value	A value found by subtracting the mean and dividing by the standard deviation (p. 111).
Shifting	Adding a constant to each data value adds the same constant to the mean, the median, and the quartiles, but does not change the standard deviation or IQR (p. 114).
Rescaling	Multiplying each data value by a constant multiplies both the measures of position (mean, median, and quartiles) and the measures of spread (standard deviation and IQR) by that constant (p. 114).
Normal model	A useful family of models for unimodal, symmetric distributions (p. 117).
Parameter	A numerically valued attribute of a model. For example, the values of μ and σ in a $N(\mu, \sigma)$ model are parameters (p. 117).
Statistic	A value calculated from data to summarize aspects of the data. For example, the mean, $\bar{y}$ and standard deviation, s, are statistics (p. 117).
z-score	A z-score tells how many standard deviations a value is from the mean; z-scores have a mean of 0 and a standard deviation of 1 (pp. 111, 117).

When working with data, use the statistics $\bar{y}$ and s:

$$z = \frac{y - \bar{y}}{s}$$

When working with models, use the parameters μ and σ:

$$z = \frac{y - \mu}{\sigma}$$

Standard Normal model	A Normal model, $N(\mu, \sigma)$ with mean $\mu = 0$ and standard deviation $\sigma = 1$. Also called the **standard Normal distribution** (p. 117).
Nearly Normal Condition	A distribution is nearly Normal if it is unimodal and symmetric. We can check by looking at a histogram or a Normal probability plot (p. 118).
68–95–99.7 Rule	In a Normal model, about 68% of values fall within 1 standard deviation of the mean, about 95% fall within 2 standard deviations of the mean, and about 99.7% fall within 3 standard deviations of the mean (p. 118).
Normal percentile	The Normal percentile corresponding to a z-score gives the percentage of values in a standard Normal distribution found at that z-score or below (p. 122).
Normal probability plot	A display to help assess whether a distribution of data is approximately Normal. If the plot is nearly straight, the data satisfy the **Nearly Normal Condition** (p. 129).

Skills

▸ Understand how adding (subtracting) a constant or multiplying (dividing) by a constant changes the center and/or spread of a variable.

▸ Recognize when standardization can be used to compare values.

▸ Understand that standardizing uses the standard deviation as a ruler.

▸ Recognize when a Normal model is appropriate.

▸ Know how to calculate the z-score of an observation.

▸ Know how to compare values of two different variables using their z-scores.

▸ Be able to use Normal models and the 68-95-99.7 Rule to estimate the percentage of observations falling within 1, 2, or 3 standard deviations of the mean.

▸ Know how to find the percentage of observations falling below any value in a Normal model using a Normal table or appropriate technology.

▸ Know how to check whether a variable satisfies the **Nearly Normal Condition** by making a Normal probability plot or a histogram.

▸ Know what z-scores mean.

▸ Be able to explain how extraordinary a standardized value may be by using a Normal model.

NORMAL PROBABILITY PLOTS ON THE COMPUTER

The best way to tell whether your data can be modeled well by a Normal model is to make a picture or two. We've already talked about making histograms. Normal probability plots are almost never made by hand because the values of the Normal scores are tricky to find. But most statistics software make Normal plots, though various packages call the same plot by different names and array the information differently.

DATA DESK

To make a "Normal Probability Plot" in Data Desk,

• Select the Variable.

• Choose **Normal Prob Plot** from the **Plot** menu.

COMMENTS

Data Desk places the ordered data values on the vertical axis and the Normal scores on the horizontal axis.

EXCEL

Excel offers a "Normal probability plot" as part of the Regression command in the Data Analysis extension, but (as of this writing) it is not a correct Normal probability plot and should not be used.

JMP

To make a "Normal Quantile Plot" in JMP,

- Make a histogram using **Distributions** from the **Analyze** menu.
- Click on the drop-down menu next to the variable name.
- Choose **Normal Quantile Plot** from the drop-down menu.
- JMP opens the plot next to the histogram.

COMMENTS

JMP places the ordered data on the vertical axis and the Normal scores on the horizontal axis. The vertical axis aligns with the histogram's axis, a useful feature.

MINITAB

To make a "Normal Probability Plot" in MINITAB,

- Choose **Probability Plot** from the **Graph** menu.
- Select "Single" for the type of plot. Click **OK.**
- Enter the name of the variable in the "Graph variables" box. Click **OK.**

COMMENTS

MINITAB places the ordered data on the horizontal axis and the Normal scores on the vertical axis.

SPSS

To make a Normal "P-P plot" in SPSS,

- Choose **P-P** from the **Graphs** menu.
- Select the variable to be displayed in the source list.
- Click the arrow button to move the variable into the target list.
- Click the **OK** button.

COMMENTS

SPSS places the ordered data on the horizontal axis and the Normal scores on the vertical axis. You may safely ignore the options in the P-P dialog.

TI-83/84 PLUS

To create a "Normal Percentile Plot" on the TI-83,

- Set up a **STAT PLOT** using the last of the Types.
- Specify your datalist, and the axis you choose to represent the data.

Although most people wouldn't open a statistics package just to find a Normal model value they could find in a table, you *would* use a calculator for that function. So . . . To find what percent of a Normal model lies between two z-scores, choose **normalcdf** from the **DISTRibutions** menu and enter the command **normalcdf(zLeft, zRight)**. To find the z-score that corresponds to a given percentile in a Normal model, choose **invNorm** from the **DISTRibutions** menu and enter the command **invNorm(percentile)**.

COMMENTS

We often want to find Normal percentages from a certain z-score to infinity. On the calculator, indicate "infinity" as a very large *z*-score, say, 99. For example, the percentage of a Normal model over 2 standard deviations above the mean can be evaluated with **normalcdf(2, 99)**.

To make a Normal Probability plot:

- Turn a STATPLOT On.
- Tell it to make a Normal probability plot by choosing the last of the icons.
- Specify your datalist and which axis you want the data on. (Use Y to make the plot look like those here.)
- Specify the Mark you want the plot to use.
- Now ZoomStat does the rest.

TI-89

- To create a "Normal Prob Plot," press F2 and select choice 2: **Norm Prob Plot**. Select a plot number and use VAR-LINK to enter the data list. Select X or Y for the data axis. Press ENTER to calculate the z-scores.

- Press F2 and select choice 1: **Plot Setup**. Turn off any undesired plots (either F3 (**Clear**) or F4 (√)). Press F5 to display the plot.

- To find what percent of a Normal model lies between two z-scores, press F5 (**Distr**). Then select 4: **Normal Cdf**. Enter the lower and upper z-scores, specify mean 0 and standard deviation 1, and press ENTER.

- To find the z-score for a given percentile, press F5 (**Distr**). Then arrow down to 2: **Inverse** press the right arrow to see the sub menu and select 1: **Inverse Normal**. Enter the area to the left of the desired point, mean 0 and standard deviation 1, and press ENTER.

COMMENTS

Normal models strictly go to infinity on either end, which is 1EE99 on the calculator. In practice, any "large" number will work. For example, the percentage of the Normal model over two standard deviations above the mean can use Lower Value 2 and Upper Value 99. To find area more than 2 standard deviations below the mean, use Lower Value −99, and Upper Value −2.

EXERCISES

1. **Shipments.** A company selling clothing on the Internet reports that the packages it ships have a median weight of 68 ounces and an IQR of 40 ounces.

   a) The company plans to include a sales flyer weighing 4 ounces in each package. What will the new median and IQR be?

   b) If the company recorded the shipping weights of these new packages in pounds instead of ounces, what would the median and IQR be? (1 lb. = 16 oz.)

2. **Hotline.** A company's customer service hotline handles many calls relating to orders, refunds, and other issues. The company's records indicate that the median length of calls to the hotline is 4.4 minutes with an IQR of 2.3 minutes.

   a) If the company were to describe the duration of these calls in seconds instead of minutes, what would the median and IQR be?

   b) In an effort to speed up the customer service process, the company decides to streamline the series of push-button menus customers must navigate, cutting the time by 24 seconds. What will the median and IQR of the length of hotline calls become?

3. **Payroll.** Here are the summary statistics for the weekly payroll of a small company: lowest salary = $300, mean salary = $700, median = $500, range = $1200, IQR = $600, first quartile = $350, standard deviation = $400.

   a) Do you think the distribution of salaries is symmetric, skewed to the left, or skewed to the right? Explain why.

   b) Between what two values are the middle 50% of the salaries found?

   c) Suppose business has been good and the company gives every employee a $50 raise. Tell the new value of each of the summary statistics.

   d) Instead, suppose the company gives each employee a 10% raise. Tell the new value of each of the summary statistics.

4. **Hams.** A specialty foods company sells "gourmet hams" by mail order. The hams vary in size from 4.15 to 7.45 pounds, with a mean weight of 6 pounds and standard deviation of 0.65 pounds. The quartiles and median weights are 5.6, 6.2, and 6.55 pounds.

   a) Find the range and the IQR of the weights.

   b) Do you think the distribution of the weights is symmetric or skewed? If skewed, which way? Why?

   c) If these weights were expressed in ounces (1 pound = 16 ounces) what would the mean, standard deviation, quartiles, median, IQR, and range be?

   d) When the company ships these hams, the box and packing materials add 30 ounces. What are the mean, standard deviation, quartiles, median, IQR, and range of weights of boxes shipped (in ounces)?

   e) One customer made a special order of a 10-pound ham. Which of the summary statistics of part d might *not* change if that data value were added to the distribution?

5. **SAT or ACT?** Each year thousands of high school students take either the SAT or the ACT, standardized tests used in the college admissions process. Combined SAT Math and Verbal scores go as high as 1600, while the maximum ACT composite score is 36. Since the two exams use very different scales, comparisons of performance are difficult. A convenient rule of thumb is $SAT = 40 \times ACT + 150$; that is, multiply an ACT score by 40 and add 150 points to estimate the equivalent SAT score. An admissions officer reported the following statistics about the ACT scores of 2355 students who applied to her college one year. Find the summaries of equivalent SAT scores.

   Lowest score = 19   Mean = 27   Standard deviation = 3
   Q3 = 30   Median = 28 IQR = 6

6. **Cold U?** A high school senior uses the Internet to get information on February temperatures in the town where he'll be going to college. He finds a website with some statistics, but they are given in degrees Celsius. The conversion formula is $°F = 9/5°C + 32$. Determine the Fahrenheit equivalents for the summary information below.

   Maximum temperature = 11°C   Range = 33°
   Mean = 1°   Standard deviation = 7°
   Median = 2°   IQR = 16°

7. **Stats test.** Suppose your Statistics professor reports test grades as z-scores, and you got a score of 2.20 on an exam. Write a sentence explaining what that means.

8. **Checkup.** One of the authors has an adopted grandson whose birth family members are very short. After examining him at his 2-year checkup, the boy's pediatrician said that the z-score for his height relative to American 2-year-olds was −1.88. Write a sentence explaining what that means.

9. **Stats test, part II.** The mean score on the Stats exam was 75 points with a standard deviation of 5 points, and Gregor's z-score was −2. How many points did he score?

10. **Mensa.** People with z-scores above 2.5 on an IQ test are sometimes classified as geniuses. If IQ scores have a mean of 100 and a standard deviation of 16 points, what IQ score do you need to be considered a genius?

11. **Temperatures.** A town's January high temperatures average 36°F with a standard deviation of 10°, while in July the mean high temperature is 74° and the standard deviation is 8°. In which month is it more unusual to have a day with a high temperature of 55°? Explain.

12. **Placement exams.** An incoming freshman took her college's placement exams in French and mathematics. In French, she scored 82 and in math 86. The overall results on the French exam had a mean of 72 and a standard deviation of 8, while the mean math score was 68, with a standard deviation of 12. On which exam did she do better compared with the other freshmen?

13. **Combining test scores.** The first Stats exam had a mean of 65 and a standard deviation of 10 points; the second had a mean of 80 and a standard deviation of 5 points. Derrick scored an 80 on both tests. Julie scored a 70 on the first test and a 90 on the second. They both totaled 160 points on the two exams, but Julie claims that her total is better. Explain.

14. **Combining scores again.** The first Stat exam had a mean of 80 and a standard deviation of 4 points; the second had a mean of 70 and a standard deviation of 15 points. Reginald scored an 80 on the first test and an 85 on the second. Sara scored an 88 on the first but only a 65 on the second. Although Reginald's total score is higher, Sara feels she should get the higher grade. Explain her point of view.

15. **Final exams.** Anna, a language major, took final exams in both French and Spanish and scored 83 on each. Her roommate Megan, also taking both courses, scored 77 on the French exam and 95 on the Spanish exam. Overall, student scores on the French exam had a mean of 81 and a standard deviation of 5, and the Spanish scores had a mean of 74 and a standard deviation of 15.

    a) To qualify for language honors, a major must maintain at least an 85 average for all language courses taken. So far, which student qualifies?

    b) Which student's overall performance was better?

16. **MP3s.** Two companies market new batteries targeted at owners of personal music players. DuraTunes claims a mean battery life of 11 hours, while RockReady advertises 12 hours.

    a) Explain why you would also like to know the standard deviations of the battery lifespans before deciding which brand to buy.

    b) Suppose those standard deviations are 2 hours for DuraTunes and 1.5 hours for RockReady. You are headed for 8 hours at the beach. Which battery is most likely to last all day? Explain.

    c) If your beach trip is all weekend, and you probably will have the music on for 16 hours, which battery is most likely to last? Explain.

17. **Cattle.** The Virginia Cooperative Extension reports that the mean weight of yearling Angus steers is 1152 pounds. Suppose that weights of all such animals can be described by a Normal model with a standard deviation of 84 pounds.

    a) How many standard deviations from the mean would a steer weighing 1000 pounds be?

    b) Which would be more unusual, a steer weighing 1000 pounds or one weighing 1250 pounds?

18. **Car speeds.** John Beale of Stanford, CA, recorded the speeds of cars driving past his house, where the speed limit read 20 mph. The mean of 100 readings was 23.84 mph, with a standard deviation of 3.56 mph. (He actually recorded every car for a two-month period. These are 100 representative readings.)

    a) How many standard deviations from the mean would a car going under the speed limit be?

    b) Which would be more unusual, a car traveling 34 mph or one going 10 mph?

19. More cattle. Recall that the beef cattle described in Exercise 17 had a mean weight of 1152 pounds, with a standard deviation of 84 pounds.

 a) Cattle buyers hope that yearling Angus steers will weigh at least 1000 pounds. To see how much over (or under) that goal the cattle are, we could subtract 1000 pounds from all the weights. What would the new mean and standard deviation be?

 b) Suppose such cattle sell at auction for 40 cents a pound. Find the mean and standard deviation of the sale prices for all the steers.

T 20. Car speeds again. For the car speed data of Exercise 18, recall that the mean speed recorded was 23.84 mph, with a standard deviation of 3.56 mph. To see how many cars are speeding, John subtracts 20 mph from all speeds.

 a) What is the mean speed now? What is the new standard deviation?

 b) His friend in Berlin wants to study the speeds, so John converts all the original miles-per-hour readings to kilometers per hour by multiplying all speeds by 1.609 (km per mile). What is the mean now? What is the new standard deviation?

21. Cattle, part III. Suppose the auctioneer in Exercise 19 sold a herd of cattle whose minimum weight was 980 pounds, median was 1140 pounds, standard deviation 84 pounds, and IQR 102 pounds. They sold for 40 cents a pound, and the auctioneer took a $20 commission on each animal. Then, for example, a steer weighing 1100 pounds would net the owner 0.40 (1100) − 20 = $420. Find the minimum, median, standard deviation, and IQR of the net sale prices.

22. Caught speeding. Suppose police set up radar surveillance on the Stanford street described in Exercise 18. They handed out a large number of tickets to speeders going a mean of 28 mph, with a standard deviation of 2.4 mph, a maximum of 33 mph, and an IQR of 3.2 mph. Local law prescribes fines of $100, plus $10 per mile per hour over the 20 mph speed limit. For example, a driver convicted of going 25 mph would be fined 100 + 10(5) = $150. Find the mean, maximum, standard deviation, and IQR of all the potential fines.

23. Professors. A friend tells you about a recent study dealing with the number of years of teaching experience among current college professors. He remembers the mean but can't recall whether the standard deviation was 6 months, 6 years, or 16 years. Tell him which one it must have been, and why.

24. Rock concerts. A popular band on tour played a series of concerts in large venues. They always drew a large crowd, averaging 21,359 fans. While the band did not announce (and probably never calculated) the standard deviation, which of these values do you think is most likely to be correct: 20, 200, 2000, or 20,000 fans? Explain your choice.

25. Guzzlers? Environmental Protection Agency (EPA) fuel economy estimates for automobile models tested recently predicted a mean of 24.8 mpg and a standard deviation of 6.2 mpg for highway driving. Assume that a Normal model can be applied.

 a) Draw the model for auto fuel economy. Clearly label it, showing what the 68–95–99.7 Rule predicts.

 b) In what interval would you expect the central 68% of autos to be found?

 c) About what percent of autos should get more than 31 mpg?

 d) About what percent of cars should get between 31 and 37.2 mpg?

 e) Describe the gas mileage of the worst 2.5% of all cars.

26. IQ. Some IQ tests are standardized to a Normal model, with a mean of 100 and a standard deviation of 16.

 a) Draw the model for these IQ scores. Clearly label it, showing what the 68–95–99.7 Rule predicts.

 b) In what interval would you expect the central 95% of IQ scores to be found?

 c) About what percent of people should have IQ scores above 116?

 d) About what percent of people should have IQ scores between 68 and 84?

 e) About what percent of people should have IQ scores above 132?

27. Small steer. In Exercise 17 we suggested the model $N(1152, 84)$ for weights in pounds of yearling Angus steers. What weight would you consider to be unusually low for such an animal? Explain.

28. High IQ. Exercise 26 proposes modeling IQ scores with $N(100, 16)$. What IQ would you consider to be unusually high? Explain.

29. Trees. A forester measured 27 of the trees in a large woods that is up for sale. He found a mean diameter of 10.4 inches and a standard deviation of 4.7 inches. Suppose that these trees provide an accurate description of the whole forest and that a Normal model applies.

 a) Draw the Normal model for tree diameters.

 b) What size would you expect the central 95% of all trees to be?

 c) About what percent of the trees should be less than an inch in diameter?

 d) About what percent of the trees should be between 5.7 and 10.4 inches in diameter?

 e) About what percent of the trees should be over 15 inches in diameter?

30. Rivets. A company that manufactures rivets believes the shear strength (in pounds) is modeled by $N(800, 50)$.

 a) Draw and label the Normal model.

 b) Would it be safe to use these rivets in a situation requiring a shear strength of 750 pounds? Explain.

 c) About what percent of these rivets would you expect to fall below 900 pounds?

 d) Rivets are used in a variety of applications with varying shear strength requirements. What is the maximum shear strength for which you would feel comfortable approving this company's rivets? Explain your reasoning.

31. **Trees, part II.** Later on, the forester in Exercise 29 shows you a histogram of the tree diameters he used in analyzing the woods that was for sale. Do you think he was justified in using a Normal model? Explain, citing some specific concerns.

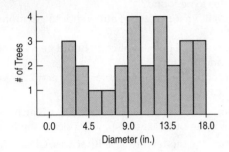

32. **Car speeds, the picture.** For the car speed data of Exercise 18, here is the histogram, boxplot, and Normal probability plot of the 100 readings. Do you think it is appropriate to apply a Normal model here? Explain.

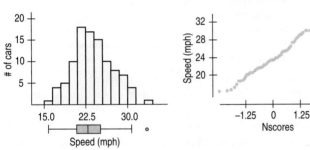

33. **Winter Olympics 2006 downhill.** Fifty-three men qualified for the men's alpine downhill race in Torino. The gold medal winner finished in 1 minute, 48.8 seconds. All competitors' times (in seconds) are found in the following list:

108.80	109.52	109.82	109.88	109.93	110.00
110.04	110.12	110.29	110.33	110.35	110.44
110.45	110.64	110.68	110.70	110.72	110.84
110.88	110.88	110.90	110.91	110.98	111.37
111.48	111.51	111.55	111.70	111.72	111.93
112.17	112.55	112.87	112.90	113.34	114.07
114.65	114.70	115.01	115.03	115.73	116.10
116.58	116.81	117.45	117.54	117.56	117.69
118.77	119.24	119.41	119.79	120.93	

a) The mean time was 113.02 seconds, with a standard deviation of 3.24 seconds. If the Normal model is appropriate, what percent of times will be less than 109.78 seconds?

b) What is the actual percent of times less than 109.78 seconds?

c) Why do you think the two percentages don't agree?

d) Create a histogram of these times. What do you see?

34. **Check the model.** The mean of the 100 car speeds in Exercise 20 was 23.84 mph, with a standard deviation of 3.56 mph.

a) Using a Normal model, what values should border the middle 95% of all car speeds?

b) Here are some summary statistics.

Percentile		Speed
100%	Max	34.060
97.5%		30.976
90.0%		28.978
75.0%	Q3	25.785
50.0%	Median	23.525
25.0%	Q1	21.547
10.0%		19.163
2.5%		16.638
0.0%	Min	16.270

From your answer in part a, how well does the model do in predicting those percentiles? Are you surprised? Explain.

35. **Receivers.** NFL data from the 2006 football season reported the number of yards gained by each of the league's 167 wide receivers:

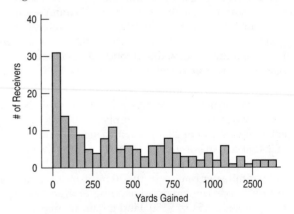

The mean is 435 yards, with a standard deviation of 384 yards.

a) According to the Normal model, what percent of receivers would you expect to gain fewer yards than 2 standard deviations below the mean number of yards?

b) For these data, what does that mean?

c) Explain the problem in using a Normal model here.

36. **Customer database.** A large philanthropic organization keeps records on the people who have contributed to their cause. In addition to keeping records of past giving, the organization buys demographic data on neighborhoods from the U.S. Census Bureau. Eighteen of these variables concern the ethnicity of the neighborhood of the donor.

Here are a histogram and summary statistics for the percentage of whites in the neighborhoods of 500 donors:

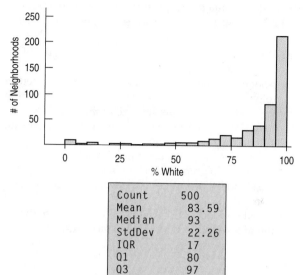

Count	500
Mean	83.59
Median	93
StdDev	22.26
IQR	17
Q1	80
Q3	97

a) Which is a better summary of the percentage of white residents in the neighborhoods, the mean or the median? Explain.

b) Which is a better summary of the spread, the IQR or the standard deviation? Explain.

c) From a Normal model, about what percentage of neighborhoods should have a percent white within one standard deviation of the mean?

d) What percentage of neighborhoods actually have a percent white within one standard deviation of the mean?

e) Explain the discrepancy between parts c and d.

37. **Normal cattle.** Using $N(1152, 84)$, the Normal model for weights of Angus steers in Exercise 17, what percent of steers weigh

a) over 1250 pounds?

b) under 1200 pounds?

c) between 1000 and 1100 pounds?

38. **IQs revisited.** Based on the Normal model $N(100, 16)$ describing IQ scores, what percent of people's IQs would you expect to be

a) over 80?

b) under 90?

c) between 112 and 132?

39. **More cattle.** Based on the model $N(1152, 84)$ describing Angus steer weights, what are the cutoff values for

a) the highest 10% of the weights?

b) the lowest 20% of the weights?

c) the middle 40% of the weights?

40. **More IQs.** In the Normal model $N(100, 16)$, what cutoff value bounds

a) the highest 5% of all IQs?

b) the lowest 30% of the IQs?

c) the middle 80% of the IQs?

41. **Cattle, finis.** Consider the Angus weights model $N(1152, 84)$ one last time.

a) What weight represents the 40th percentile?

b) What weight represents the 99th percentile?

c) What's the IQR of the weights of these Angus steers?

42. **IQ, finis.** Consider the IQ model $N(100, 16)$ one last time.

a) What IQ represents the 15th percentile?

b) What IQ represents the 98th percentile?

c) What's the IQR of the IQs?

43. **Cholesterol.** Assume the cholesterol levels of adult American women can be described by a Normal model with a mean of 188 mg/dL and a standard deviation of 24.

a) Draw and label the Normal model.

b) What percent of adult women do you expect to have cholesterol levels over 200 mg/dL?

c) What percent of adult women do you expect to have cholesterol levels between 150 and 170 mg/dL?

d) Estimate the IQR of the cholesterol levels.

e) Above what value are the highest 15% of women's cholesterol levels?

44. **Tires.** A tire manufacturer believes that the treadlife of its snow tires can be described by a Normal model with a mean of 32,000 miles and standard deviation of 2500 miles.

a) If you buy a set of these tires, would it be reasonable for you to hope they'll last 40,000 miles? Explain.

b) Approximately what fraction of these tires can be expected to last less than 30,000 miles?

c) Approximately what fraction of these tires can be expected to last between 30,000 and 35,000 miles?

d) Estimate the IQR of the treadlives.

e) In planning a marketing strategy, a local tire dealer wants to offer a refund to any customer whose tires fail to last a certain number of miles. However, the dealer does not want to take too big a risk. If the dealer is willing to give refunds to no more than 1 of every 25 customers, for what mileage can he guarantee these tires to last?

45. **Kindergarten.** Companies that design furniture for elementary school classrooms produce a variety of sizes for kids of different ages. Suppose the heights of kindergarten children can be described by a Normal model with a mean of 38.2 inches and standard deviation of 1.8 inches.

a) What fraction of kindergarten kids should the company expect to be less than 3 feet tall?

b) In what height interval should the company expect to find the middle 80% of kindergarteners?

c) At least how tall are the biggest 10% of kindergarteners?

46. **Body temperatures.** Most people think that the "normal" adult body temperature is 98.6°F. That figure, based on a 19th-century study, has recently been challenged. In a 1992 article in the *Journal of the American Medical Association*, researchers reported that a more accurate figure may be 98.2°F. Furthermore, the standard deviation appeared to be around 0.7°F. Assume that a Normal model is appropriate.

a) In what interval would you expect most people's body temperatures to be? Explain.

b) What fraction of people would be expected to have body temperatures above 98.6°F?

c) Below what body temperature are the coolest 20% of all people?

47. **Eggs.** Hens usually begin laying eggs when they are about 6 months old. Young hens tend to lay smaller eggs, often weighing less than the desired minimum weight of 54 grams.

    a) The average weight of the eggs produced by the young hens is 50.9 grams, and only 28% of their eggs exceed the desired minimum weight. If a Normal model is appropriate, what would the standard deviation of the egg weights be?

    b) By the time these hens have reached the age of 1 year, the eggs they produce average 67.1 grams, and 98% of them are above the minimum weight. What is the standard deviation for the appropriate Normal model for these older hens?

    c) Are egg sizes more consistent for the younger hens or the older ones? Explain.

48. **Tomatoes.** Agricultural scientists are working on developing an improved variety of Roma tomatoes. Marketing research indicates that customers are likely to bypass Romas that weigh less than 70 grams. The current variety of Roma plants produces fruit that averages 74 grams, but 11% of the tomatoes are too small. It is reasonable to assume that a Normal model applies.

    a) What is the standard deviation of the weights of Romas now being grown?

    b) Scientists hope to reduce the frequency of undersized tomatoes to no more than 4%. One way to accomplish this is to raise the average size of the fruit. If the standard deviation remains the same, what target mean should they have as a goal?

    c) The researchers produce a new variety with a mean weight of 75 grams, which meets the 4% goal. What is the standard deviation of the weights of these new Romas?

    d) Based on their standard deviations, compare the tomatoes produced by the two varieties.

JUST CHECKING

ANSWERS

1. **a)** On the first test, the mean is 88 and the SD is 4, so $z = (90 - 88)/4 = 0.5$. On the second test, the mean is 75 and the SD is 5, so $z = (80 - 75)/5 = 1.0$. The first test has the lower z-score, so it is the one that will be dropped.

   b) No. The second test is 1 standard deviation above the mean, farther away than the first test, so it's the better score relative to the class.

2. **a)** The mean would increase to 500.

   b) The standard deviation is still 100 points.

   c) The two boxplots would look nearly identical (the shape of the distribution would remain the same), but the later one would be shifted 50 points higher.

3. The standard deviation is now 2.54 millimeters, which is the same as 0.1 inch. Nothing has changed. The standard deviation has "increased" only because we're reporting it in millimeters now, not inches.

4. The mean is 184 centimeters, with a standard deviation of 8 centimeters. 2 meters is 200 centimeters, which is 2 standard deviations above the mean. We expect 5% of the men to be more than 2 standard deviations below or above the mean, so half of those, 2.5%, are likely to be above 2 meters.

5. **a)** We know that 68% of the time we'll be within 1 standard deviation (2 min) of 20. So 32% of the time we'll arrive in less than 18 or more than 22 minutes. Half of those times (16%) will be greater than 22 minutes, so 84% will be less than 22 minutes.

   b) 24 minutes is 2 standard deviations above the mean. Because of the 95% rule, we know 2.5% of the times will be more than 24 minutes.

   c) Traffic incidents may occasionally increase the time it takes to get to school, so the driving times may be skewed to the right, and there may be outliers.

   d) If so, the Normal model would not be appropriate and the percentages we predict would not be accurate.

Exploring and Understanding Data

Quick Review

It's time to put it all together. Real data don't come tagged with instructions for use. So let's step back and look at how the key concepts and skills we've seen work together. This brief list and the review exercises that follow should help you check your understanding of Statistics so far.

▸ We treat data two ways: as categorical and as quantitative.

▸ To describe categorical data:

- Make a picture. Bar graphs work well for comparing counts in categories.
- Summarize the distribution with a table of counts or relative frequencies (percents) in each category.
- Pie charts and segmented bar charts display divisions of a whole.
- Compare distributions with plots side by side.
- Look for associations between variables by comparing marginal and conditional distributions.

▸ To describe quantitative data:

- Make a picture. Use histograms, boxplots, stem-and-leaf displays, or dotplots. Stem-and-leafs are great when working by hand and good for small data sets. Histograms are a good way to see the distribution. Boxplots are best for comparing several distributions.
- Describe distributions in terms of their shape, center, and spread, and note any unusual features such as gaps or outliers.
- The shape of most distributions you'll see will likely be uniform, unimodal, or bimodal. It may be multimodal. If it is unimodal, then it may be symmetric or skewed.
- A 5-number summary makes a good numerical description of a distribution: min, Q1, median, Q3, and max.

- If the distribution is skewed, be sure to include the median and interquartile range (IQR) when you describe its center and spread.
- A distribution that is severely skewed may benefit from re-expressing the data. If it is skewed to the high end, taking logs often works well.
- If the distribution is unimodal and symmetric, describe its center and spread with the mean and standard deviation.
- Use the standard deviation as a ruler to tell how unusual an observed value may be, or to compare or combine measurements made on different scales.
- Shifting a distribution by adding or subtracting a constant affects measures of position but not measures of spread. Rescaling by multiplying or dividing by a constant affects both.
- When a distribution is roughly unimodal and symmetric, a Normal model may be useful. For Normal models, the 68–95–99.7 Rule is a good rule of thumb.
- If the Normal model fits well (check a histogram or Normal probability plot), then Normal percentile tables or functions found in most statistics technology can provide more detailed values.

Need more help with some of this? It never hurts to reread sections of the chapters! And in the following pages we offer you more opportunities[1] to review these concepts and skills.

The exercises that follow use the concepts and skills you've learned in the first six chapters. To be more realistic and more useful for your review, they don't tell you which of the concepts or methods you need. But neither will the exam.

_____

[1] If you doubted that we are teachers, this should convince you. Only a teacher would call additional homework exercises "opportunities."

REVIEW EXERCISES

1. **Bananas.** Here are the prices (in cents per pound) of bananas reported from 15 markets surveyed by the U.S. Department of Agriculture.

51	52	45
48	53	52
50	49	52
48	43	46
45	42	50

a) Display these data with an appropriate graph.
b) Report appropriate summary statistics.
c) Write a few sentences about this distribution.

2. **Prenatal care.** Results of a 1996 American Medical Association report about the infant mortality rate for twins carried for the full term of a normal pregnancy are shown on the next page, broken down by the level of prenatal care the mother had received.

Full-Term Pregnancies, Level of Prenatal Care	Infant Mortality Rate Among Twins (deaths per thousand live births)
Intensive	5.4
Adequate	3.9
Inadequate	6.1
Overall	5.1

a) Is the overall rate the average of the other three rates? Should it be? Explain.
b) Do these results indicate that adequate prenatal care is important for pregnant women? Explain.
c) Do these results suggest that a woman pregnant with twins should be wary of seeking too much medical care? Explain.

3. Singers. The boxplots shown display the heights (in inches) of 130 members of a choir.

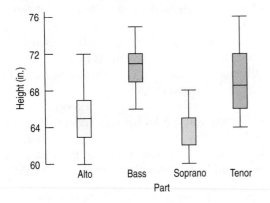

a) It appears that the median height for sopranos is missing, but actually the median and the upper quartile are equal. How could that happen?
b) Write a few sentences describing what you see.

4. Dialysis. In a study of dialysis, researchers found that "of the three patients who were currently on dialysis, 67% had developed blindness and 33% had their toes amputated." What kind of display might be appropriate for these data? Explain.

5. Beanstalks. Beanstalk Clubs are social clubs for very tall people. To join, a man must be over 6'2" tall, and a woman over 5'10". The National Health Survey suggests that heights of adults may be Normally distributed, with mean heights of 69.1" for men and 64.0" for women. The respective standard deviations are 2.8" and 2.5."

a) You are probably not surprised to learn that men are generally taller than women, but what does the greater standard deviation for men's heights indicate?
b) Who are more likely to qualify for Beanstalk membership, men or women? Explain.

6. Bread. Clarksburg Bakery is trying to predict how many loaves to bake. In the last 100 days, they have sold between 95 and 140 loaves per day. Here is a histogram of the number of loaves they sold for the last 100 days.

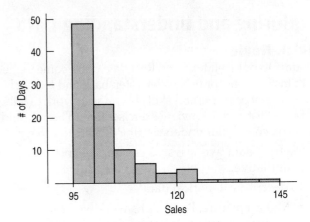

a) Describe the distribution.
b) Which should be larger, the mean number of sales or the median? Explain.
c) Here are the summary statistics for Clarksburg Bakery's bread sales. Use these statistics and the histogram above to create a boxplot. You may approximate the values of any outliers.

Summary of Sales	
Median	100
Min	95
Max	140
25th %tile	97
75th %tile	105.5

d) For these data, the mean was 103 loaves sold per day, with a standard deviation of 9 loaves. Do these statistics suggest that Clarksburg Bakery should expect to sell between 94 and 112 loaves on about 68% of the days? Explain.

7. State University. Public relations staff at State U. phoned 850 local residents. After identifying themselves, the callers asked the survey participants their ages, whether they had attended college, and whether they had a favorable opinion of the university. The official report to the university's directors claimed that, in general, people had very favorable opinions about the university.

a) Identify the W's of these data.
b) Identify the variables, classify each as categorical or quantitative, and specify units if relevant.
c) Are you confident about the report's conclusion? Explain.

8. Acid rain. Based on long-term investigation, researchers have suggested that the acidity (pH) of rainfall in the

Shenandoah Mountains can be described by the Normal model $N(4.9, 0.6)$.

a) Draw and carefully label the model.
b) What percent of storms produce rainfall with pH over 6?
c) What percent of storms produce rainfall with pH under 4?
d) The lower the pH, the more acidic the rain. What is the pH level for the most acidic 20% of all storms?
e) What is the pH level for the least acidic 5% of all storms?
f) What is the IQR for the pH of rainfall?

9. **Fraud detection.** A credit card bank is investigating the incidence of fraudulent card use. The bank suspects that the type of product bought may provide clues to the fraud. To examine this situation, the bank looks at the Standard Industrial Code (SIC) of the business related to the transaction. This is a code that was used by the U.S. Census Bureau and Statistics Canada to identify the type of every registered business in North America.[2] For example, 1011 designates Meat and Meat Products (except Poultry), 1012 is Poultry Products, 1021 is Fish Products, 1031 is Canned and Preserved Fruits and Vegetables, and 1032 is Frozen Fruits and Vegetables.

A company intern produces the following histogram of the SIC codes for 1536 transactions:

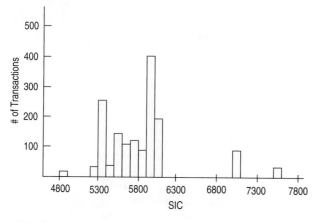

He also reports that the mean SIC is 5823.13 with a standard deviation of 488.17.

a) Comment on any problems you see with the use of the mean and standard deviation as summary statistics.
b) How well do you think the Normal model will work on these data? Explain.

10. **Streams.** As part of the course work, a class at an upstate NY college collects data on streams each year. Students record a number of biological, chemical, and physical variables, including the stream name, the substrate of the stream (*limestone, shale,* or *mixed*), the pH, the temperature (°C), and the BCI, a measure of biological diversity.

Group	Count	%
Limestone	77	44.8
Mixed	26	15.1
Shale	69	40.1

a) Name each variable, indicating whether it is categorical or quantitative, and give the units if available.
b) These streams have been classified according to their substrate—the composition of soil and rock over which they flow—as summarized in the table. What kind of graph might be used to display these data?

11. **Cramming.** One Thursday, researchers gave students enrolled in a section of basic Spanish a set of 50 new vocabulary words to memorize. On Friday the students took a vocabulary test. When they returned to class the following Monday, they were retested—without advance warning. Both sets of test scores for the 28 students are shown below.

Fri	Mon	Fri	Mon
42	36	50	47
44	44	34	34
45	46	38	31
48	38	43	40
44	40	39	41
43	38	46	32
41	37	37	36
35	31	40	31
43	32	41	32
48	37	48	39
43	41	37	31
45	32	36	41
47	44		

a) Create a graphical display to compare the two distributions of scores.
b) Write a few sentences about the scores reported on Friday and Monday.
c) Create a graphical display showing the distribution of the *changes* in student scores.
d) Describe the distribution of changes.

12. **Computers and Internet.** A U.S. Census Bureau report (August 2000, *Current Population Survey*) found that 51.0% of homes had a personal computer and 41.5% had access to the Internet. A newspaper concluded that 92.5% of homes had either a computer or access to the Internet. Do you agree? Explain.

13. **Let's play cards.** You pick a card from a deck (see description in Chapter 11) and record its denomination (7, say) and its suit (maybe spades).

a) Is the variable *suit* categorical or quantitative?
b) Name a game you might be playing for which you would consider the variable *denomination* to be categorical. Explain.
c) Name a game you might be playing for which you would consider the variable *denomination* to be quantitative. Explain.

14. **Accidents.** In 2001, Progressive Insurance asked customers who had been involved in auto accidents how far they were from home when the accident happened. The data are summarized in the table.

---

[2] Since 1997 the SIC has been replaced by the NAICS, a code of six letters.

Miles from Home	% of Accidents
Less than 1	23
1 to 5	29
6 to 10	17
11 to 15	8
16 to 20	6
Over 20	17

a) Create an appropriate graph of these data.
b) Do these data indicate that driving near home is particularly dangerous? Explain.

T 15. Hard water. In an investigation of environmental causes of disease, data were collected on the annual mortality rate (deaths per 100,000) for males in 61 large towns in England and Wales. In addition, the water hardness was recorded as the calcium concentration (parts per million, ppm) in the drinking water.

a) What are the variables in this study? For each, indicate whether it is quantitative or categorical and what the units are.
b) Here are histograms of calcium concentration and mortality. Describe the distributions of the two variables.

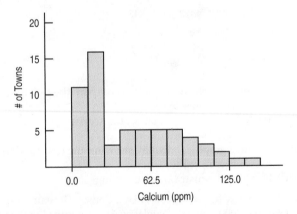

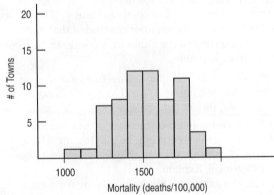

T 16. Hard water II. The data set from England and Wales also notes for each town whether it was south or north of Derby. Here are some summary statistics and a comparative boxplot for the two regions.

Summary of Mortality				
Group	Count	Mean	Median	StdDev
North	34	1631.59	1631	138.470
South	27	1388.85	1369	151.114

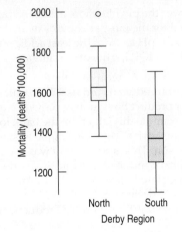

a) What is the overall mean mortality rate for the two regions?
b) Do you see evidence of a difference in mortality rates? Explain.

17. Seasons. Average daily temperatures in January and July for 60 large U.S. cities are graphed in the histograms below.

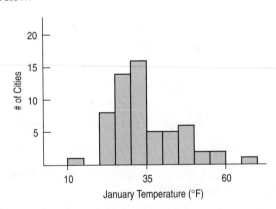

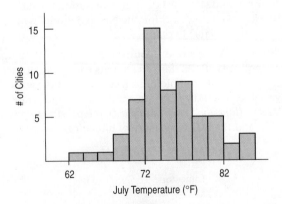

a) What aspect of these histograms makes it difficult to compare the distributions?
b) What differences do you see between the distributions of January and July average temperatures?

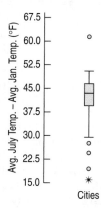

Cities

c) Differences in temperatures (July–January) for each of the cities are displayed in the boxplot above. Write a few sentences describing what you see.

18. Old Faithful. It is a common belief that Yellowstone's most famous geyser erupts once an hour at very predictable intervals. The histogram below shows the time gaps (in minutes) between 222 successive eruptions. Describe this distribution.

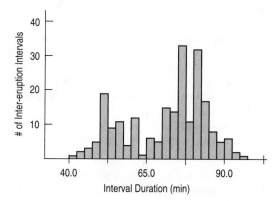

Interval Duration (min)

19. Old Faithful? Does the duration of an eruption have an effect on the length of time that elapses before the next eruption?

a) The histogram below shows the duration (in minutes) of those 222 eruptions. Describe this distribution.

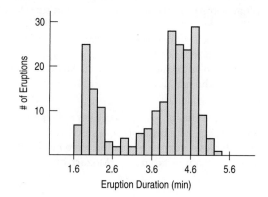

Eruption Duration (min)

b) Explain why it is not appropriate to find summary statistics for this distribution.
c) Let's classify the eruptions as "long" or "short," depending upon whether or not they last at least 3 minutes. Describe what you see in the comparative boxplots.

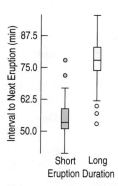

Short Long
Eruption Duration

20. Teen drivers. In its *Traffic Safety Facts 2005*, the National Highway Traffic Safety Administration reported that 6.3% of licensed drivers were between the ages of 15 and 20, yet this age group was behind the wheel in 15.9% of all fatal crashes. Use these statistics to explain the concept of independence.

21. Liberty's nose. Is the Statue of Liberty's nose too long? Her nose measures 4'6", but she is a large statue, after all. Her arm is 42 feet long. That means her arm is 42/45 = 9.3 times as long as her nose. Is that a reasonable ratio? Shown in the table are arm and nose lengths of 18 girls in a Statistics class, and the ratio of arm-to-nose length for each.

Arm (cm)	Nose (cm)	Arm/Nose Ratio
73.8	5.0	14.8
74.0	4.5	16.4
69.5	4.5	15.4
62.5	4.7	13.3
68.6	4.4	15.6
64.5	4.8	13.4
68.2	4.8	14.2
63.5	4.4	14.4
63.5	5.4	11.8
67.0	4.6	14.6
67.4	4.4	15.3
70.7	4.3	16.4
69.4	4.1	16.9
71.7	4.5	15.9
69.0	4.4	15.7
69.8	4.5	15.5
71.0	4.8	14.8
71.3	4.7	15.2

a) Make an appropriate plot and describe the distribution of the ratios.
b) Summarize the ratios numerically, choosing appropriate measures of center and spread.
c) Is the ratio of 9.3 for the Statue of Liberty unrealistically low? Explain.

T 22. Winter Olympics 2006 speed skating. The top 25 women's 500-m speed skating times are listed in the table:

Skater	Country	Time
Svetlana Zhurova	Russia	76.57
Wang Manli	China	76.78
Hui Ren	China	76.87
Tomomi Okazaki	Japan	76.92
Lee Sang-Hwa	South Korea	77.04
Jenny Wolf	Germany	77.25
Wang Beixing	China	77.27
Sayuri Osuga	Japan	77.39
Sayuri Yoshii	Japan	77.43
Chiara Simionato	Italy	77.68
Jennifer Rodriguez	United States	77.70
Annette Gerritsen	Netherlands	78.09
Xing Aihua	China	78.35
Sanne van der Star	Netherlands	78.59
Yukari Watanabe	Japan	78.65
Shannon Rempel	Canada	78.85
Amy Sannes	United States	78.89
Choi Seung-Yong	South Korea	79.02
Judith Hesse	Germany	79.03
Kim You-Lim	South Korea	79.25
Kerry Simpson	Canada	79.34
Krisy Myers	Canada	79.43
Elli Ochowicz	United States	79.48
Pamela Zoellner	Germany	79.56
Lee Bo-Ra	South Korea	79.73

a) The mean finishing time was 78.21 seconds, with a standard deviation of 1.03 seconds. If the Normal model is appropriate, what percent of the times should be within 0.5 second of 78.21?

b) What percent of the times actually fall within this interval?

c) Explain the discrepancy between a and b.

23. **Sample.** A study in South Africa focusing on the impact of health insurance identified 1590 children at birth and then sought to conduct follow-up health studies 5 years later. Only 416 of the original group participated in the 5-year follow-up study. This made researchers concerned that the follow-up group might not accurately resemble the total group in terms of health insurance. The table in the next column summarizes the two groups by race and by presence of medical insurance when the child was born. Carefully explain how this study demonstrates Simpson's paradox. (*Birth to Ten Study*, Medical Research Council, South Africa)

		Number (%) Insured	
		Follow-Up	Not Traced
Race	Black	36 of 404 (8.9%)	91 of 1048 (8.7%)
	White	10 of 12 (83.3%)	104 of 126 (82.5%)
	Overall	46 of 416 (11.1%)	195 of 1174 (16.6%)

24. **Sluggers.** Babe Ruth was the first great "slugger" in baseball. His record of 60 home runs in one season held for 34 years until Roger Maris hit 61 in 1961. Mark McGwire (with the aid of steroids) set a new standard of 70 in 1998. Listed below are the home run totals for each season McGwire played. Also listed are Babe Ruth's home run totals, whose record of 60 was the mark that Maris broke.

McGwire: 3*, 49, 32, 33, 39, 22, 42, 9*, 9*, 39, 52, 58, 70, 65, 32*, 29*

Ruth: 54, 59, 35, 41, 46, 25, 47, 60, 54, 46, 49, 46, 41, 34, 22

a) Find the 5-number summary for McGwire's career.

b) Do any of his seasons appear to be outliers? Explain.

c) McGwire played in only 18 games at the end of his first big league season, and missed major portions of some other seasons because of injuries to his back and knees. Those seasons might not be representative of his abilities. They are marked with asterisks in the list above. Omit these values and make parallel boxplots comparing McGwire's career to Babe Ruth's.

d) Write a few sentences comparing the two sluggers.

e) Create a side-by-side stem-and-leaf display comparing the careers of the two players.

f) What aspects of the distributions are apparent in the stem-and-leaf displays that did not clearly show in the boxplots?

25. **Be quick!** Avoiding an accident when driving can depend on reaction time. That time, measured from the moment the driver first sees the danger until he or she steps on the brake pedal, is thought to follow a Normal model with a mean of 1.5 seconds and a standard deviation of 0.18 second.

a) Use the 68–95–99.7 Rule to draw the Normal model.

b) Write a few sentences describing driver reaction times.

c) What percent of drivers have a reaction time less than 1.25 seconds?

d) What percent of drivers have reaction times between 1.6 and 1.8 seconds?

e) What is the interquartile range of reaction times?

f) Describe the reaction times of the slowest 1/3 of all drivers.

26. **Music and memory.** Is it a good idea to listen to music when studying for a big test? In a study conducted by some Statistics students, 62 people were randomly assigned to listen to rap music, Mozart, or no music

while attempting to memorize objects pictured on a page. They were then asked to list all the objects they could remember. Here are the 5-number summaries for each group:

	n	Min	Q1	Median	Q3	Max
Rap	29	5	8	10	12	25
Mozart	20	4	7	10	12	27
None	13	8	9.5	13	17	24

a) Describe the W's for these data: *Who, What, Where, Why, When, How.*
b) Name the variables and classify each as categorical or quantitative.
c) Create parallel boxplots as best you can from these summary statistics to display these results.
d) Write a few sentences comparing the performances of the three groups.

27. **Mail.** Here are the number of pieces of mail received at a school office for 36 days.

123	70	90	151	115	97
80	78	72	100	128	130
52	103	138	66	135	76
112	92	93	143	100	88
118	118	106	110	75	60
95	131	59	115	105	85

a) Plot these data.
b) Find appropriate summary statistics.
c) Write a brief description of the school's mail deliveries.
d) What percent of the days actually lie within one standard deviation of the mean? Comment.

28. **Birth order.** Is your birth order related to your choice of major? A Statistics professor at a large university polled his students to find out what their majors were and what position they held in the family birth order. The results are summarized in the table.

a) What percent of these students are oldest or only children?
b) What percent of Humanities majors are oldest children?
c) What percent of oldest children are Humanities students?
d) What percent of the students are oldest children majoring in the Humanities?

		Birth Order*				
		1	2	3	4+	Total
Major	Math/Science	34	14	6	3	57
	Agriculture	52	27	5	9	93
	Humanities	15	17	8	3	43
	Other	12	11	1	6	30
	Total	113	69	20	21	223

*1 = oldest or only child

29. **Herbal medicine.** Researchers for the Herbal Medicine Council collected information on people's experiences with a new herbal remedy for colds. They went to a store that sold natural health products. There they asked 100 customers whether they had taken the cold remedy and, if so, to rate its effectiveness (on a scale from 1 to 10) in curing their symptoms. The Council concluded that this product was highly effective in treating the common cold.

a) Identify the W's of these data.
b) Identify the variables, classify each as categorical or quantitative, and specify units if relevant.
c) Are you confident about the Council's conclusion? Explain.

30. **Birth order revisited.** Consider again the data on birth order and college majors in Exercise 28.

a) What is the marginal distribution of majors?
b) What is the conditional distribution of majors for the oldest children?
c) What is the conditional distribution of majors for the children born second?
d) Do you think that college major appears to be independent of birth order? Explain.

31. **Engines.** One measure of the size of an automobile engine is its "displacement," the total volume (in liters or cubic inches) of its cylinders. Summary statistics for several models of new cars are shown. These displacements were measured in cubic inches.

Summary of Displacement	
Count	38
Mean	177.29
Median	148.5
StdDev	88.88
Range	275
25th %tile	105
75th %tile	231

a) How many cars were measured?
b) Why might the mean be so much larger than the median?
c) Describe the center and spread of this distribution with appropriate statistics.
d) Your neighbor is bragging about the 227-cubic-inch engine he bought in his new car. Is that engine unusually large? Explain.
e) Are there any engines in this data set that you would consider to be outliers? Explain.
f) Is it reasonable to expect that about 68% of car engines measure between 88 and 266 cubic inches? (That's 177.289 ± 88.8767.) Explain.
g) We can convert all the data from cubic inches to cubic centimeters (cc) by multiplying by 16.4. For example, a 200-cubic-inch engine has a displacement of 3280 cc. How would such a conversion affect each of the summary statistics?

32. **Engines, again.** Horsepower is another measure commonly used to describe auto engines. Here are the summary statistics and histogram displaying horsepowers of the same group of 38 cars discussed in Exercise 31.

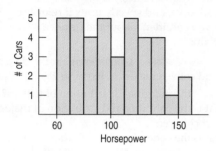

Summary of Horsepower	
Count	38
Mean	101.7
Median	100
StdDev	26.4
Range	90
25th %tile	78
75th %tile	125

a) Describe the shape, center, and spread of this distribution.
b) What is the interquartile range?
c) Are any of these engines outliers in terms of horsepower? Explain.
d) Do you think the 68–95–99.7 Rule applies to the horsepower of auto engines? Explain.
e) From the histogram, make a rough estimate of the percentage of these engines whose horsepower is within one standard deviation of the mean.
f) A fuel additive boasts in its advertising that it can "add 10 horsepower to any car." Assuming that is true, what would happen to each of these summary statistics if this additive were used in all the cars?

33. **Age and party 2007.** The Pew Research Center conducts surveys regularly asking respondents which political party they identify with. Among their results is the following table relating preferred political party and age. (http://people-press.org/reports/)

		Party		
		Republican	Democrat	Others	Total
	18–29	2636	2738	4765	**10139**
	30–49	6871	6442	8160	**21473**
Age	50–64	3896	4286	4806	**12988**
	65+	3131	3718	2934	**9784**
	Total	**16535**	**17183**	**20666**	**54384**

a) What percent of people surveyed were Republicans?
b) Do you think this might be a reasonable estimate of the percentage of all voters who are Republicans? Explain.
c) What percent of people surveyed were under 30 or over 65?
d) What percent of people were classified as "Other" and under the age of 30?

e) What percent of the people classified as "Other" were under 30?
f) What percent of people under 30 were classified as "Other"?

34. **Pay.** According to the *2006 National Occupational Employment and Wage Estimates for Management Occupations*, the mean hourly wage for Chief Executives was $69.52 and the median hourly wage was "over $70.00." By contrast, for General and Operations Managers, the mean hourly wage was $47.73 and the median was $40.97. Are these wage distributions likely to be symmetric, skewed left, or skewed right? Explain.

35. **Age and party II.** Consider again the Pew Research Center results on age and political party in Exercise 33.
a) What is the marginal distribution of party affiliation?
b) Create segmented bar graphs displaying the conditional distribution of party affiliation for each age group.
c) Summarize these poll results in a few sentences that might appear in a newspaper article about party affiliation in the United States.
d) Do you think party affiliation is independent of the voter's age? Explain.

36. **Bike safety.** The Bicycle Helmet Safety Institute website includes a report on the number of bicycle fatalities per year in the United States. The table below shows the counts for the years 1994–2003.

Year	Bicycle Fatalities
1994	796
1995	828
1996	761
1997	811
1998	757
1999	750
2000	689
2001	729
2002	663
2003	619

a) What are the W's for these data?
b) Display the data in a stem-and-leaf display.
c) Display the data in a timeplot.
d) What is apparent in the stem-and-leaf display that is hard to see in the timeplot?
e) What is apparent in the timeplot that is hard to see in the stem-and-leaf display?
f) Write a few sentences about bicycle fatalities in the United States.

37. **Some assembly required.** A company that markets build-it-yourself furniture sells a computer desk that is advertised with the claim "less than an hour to assemble." However, through postpurchase surveys the company has learned that only 25% of its customers succeeded in building the desk in under an hour. The mean time was 1.29 hours. The company assumes that consumer assembly time follows a Normal model.

a) Find the standard deviation of the assembly time model.

b) One way the company could solve this problem would be to change the advertising claim. What assembly time should the company quote in order that 60% of customers succeed in finishing the desk by then?

c) Wishing to maintain the "less than an hour" claim, the company hopes that revising the instructions and labeling the parts more clearly can improve the 1-hour success rate to 60%. If the standard deviation stays the same, what new lower mean time does the company need to achieve?

d) Months later, another postpurchase survey shows that new instructions and part labeling did lower the mean assembly time, but only to 55 minutes. Nonetheless, the company did achieve the 60%-in-an-hour goal, too. How was that possible?

T 38. **Profits.** Here is a stem-and-leaf display showing profits as a percent of sales for 29 of the *Forbes* 500 largest U.S. corporations. The stems are split; each stem represents a span of 5%, from a loss of 9% to a profit of 25%.

```
 -0 | 9 9
 -0 | 1 2 3 4
  0 | 1 1 1 1 2 3 4 4 4
  0 | 5 5 5 5 6 7 9
  1 | 0 0 1 1 3
  1 |
  2 | 2
  2 | 5
```

Profits (% of sales)
(−0|3 means a loss of 3%)

a) Find the 5-number summary.

b) Draw a boxplot for these data.

c) Find the mean and standard deviation.

d) Describe the distribution of profits for these corporations.

CHAPTER 7

Scatterplots, Association, and Correlation

Where are we going?

Is the price of sneakers related to how long they last? Is your alertness in class related to how much (or little) sleep you got the night before?

   In this chapter we'll look at relationships between two quantitative variables. We'll start by looking at scatterplots and describing the essence of what we see—the direction, form, and strength of the association. Then, as we did for histograms, we'll find a quantitative summary of what we learned from the display. We'll use the correlation to measure the strength of the association we see in the scatterplot.

WHO	Years 1970–2007
WHAT	Mean error in the position of Atlantic hurricanes as predicted 72 hours ahead by the NHC and Year.
UNITS	nautical miles and years
WHEN	1970–2007
WHERE	Atlantic and Gulf of Mexico
WHY	NHC wants to improve prediction models

Hurricane Katrina killed 1836 people[1] and caused well over 100 billion dollars in damage—the most ever recorded. Much of the damage caused by Katrina was due to its almost perfectly deadly aim at New Orleans.

   Where will a hurricane go? People want to know if a hurricane is coming their way, and the National Hurricane Center (NHC) of the National Oceanic and Atmospheric Administration (NOAA) tries to predict the path a hurricane will take. But hurricanes tend to wander around aimlessly and are pushed by fronts and other weather phenomena in their area, so they are notoriously difficult to predict. Even relatively small changes in a hurricane's track can make big differences in the damage it causes.

   To improve hurricane prediction, NOAA[2] relies on sophisticated computer models, and has been working for decades to improve them. How well are they doing? Have predictions improved in recent years? Has the improvement been consistent? Here's a timeplot of the mean error, in nautical miles, of the NHC's 72-hour predictions of Atlantic hurricanes since 1970:

---

[1] In addition, 705 are still listed as missing.

[2] www.nhc.noaa.gov

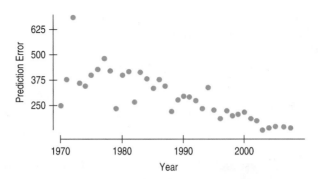

FIGURE 7.1
A scatterplot of the average error in nautical miles of the predicted position of Atlantic hurricanes for predictions made by the National Hurricane Center of NOAA, plotted against the *Year* in which the predictions were made.

Clearly, predictions have improved. The plot shows a fairly steady decline in the average error, from almost 500 nautical miles in the late 1970s to about 140 nautical miles in 2007. We can also see a few years when predictions were unusually good and that 1972 was a really bad year for predicting hurricane tracks.

This timeplot is an example of a more general kind of display called a **scatterplot**. Scatterplots may be the most common displays for data. By just looking at them, you can see patterns, trends, relationships, and even the occasional extraordinary value sitting apart from the others. As the great philosopher Yogi Berra[3] once said, "You can observe a lot by watching."[4] Scatterplots are the best way to start observing the relationship between two *quantitative* variables.

Relationships between variables are often at the heart of what we'd like to learn from data:

- Are grades actually higher now than they used to be?
- Do people tend to reach puberty at a younger age than in previous generations?
- Does applying magnets to parts of the body relieve pain? If so, are stronger magnets more effective?
- Do students learn better with the use of computer technology?

Questions such as these relate two quantitative variables and ask whether there is an **association** between them. Scatterplots are the ideal way to *picture* such associations.

Looking at Scatterplots

How would you describe the association of hurricane *Prediction Error* and *Year*? Everyone looks at scatterplots. But, if asked, many people would find it hard to say what to look for in a scatterplot. What do *you* see? Try to describe the scatterplot of *Prediction Error* against *Year*.

You might say that the **direction** of the association is important. Over time, the NHC's prediction errors have decreased. A pattern like this that runs from the upper left to the lower right is said to be **negative**. A pattern running the other way is called **positive**.

The second thing to look for in a scatterplot is its **form**. If there is a straight line relationship, it will appear as a cloud or swarm of points stretched out in a generally consistent, straight form. For example, the scatterplot of *Prediction Error* vs. *Year* has such an underlying **linear** form, although some points stray away from it.

---

[3] Hall of Fame catcher and manager of the New York Mets and Yankees.
[4] But then he also said "I really didn't say everything I said." So we can't really be sure.

Scatterplots can reveal many kinds of patterns. Often they will not be straight, but straight line patterns are both the most common and the most useful for statistics.

> Look for **Form**: straight, curved, something exotic, or no pattern?

If the relationship isn't straight, but curves gently, while still increasing or decreasing steadily, ⠿⠶⠄⠄⠄ , we can often find ways to make it more nearly straight. But if it curves sharply up and down, for example like this: , there is much less we can say about it with the methods of this book.

The third feature to look for in a scatterplot is how strong the relationship is.

> Look for **Strength**: how much scatter?

At one extreme, do the points appear tightly clustered in a single stream (whether straight, curved, or bending all over the place)? Or, at the other extreme, does the swarm of points seem to form a vague cloud through which we can barely discern any trend or pattern? The *Prediction error* vs. *Year* plot shows moderate scatter around a generally straight form. This indicates that the linear trend of improving prediction is pretty consistent and moderately strong.

> Look for **Unusual Features**: are there outliers or subgroups?

Finally, always look for the unexpected. Often the most interesting thing to see in a scatterplot is something you never thought to look for. One example of such a surprise is an **outlier** standing away from the overall pattern of the scatterplot. Such a point is almost always interesting and always deserves special attention. In the scatterplot of prediction errors, the year 1972 stands out as a year with very high prediction errors. An Internet search shows that it was a relatively quiet hurricane season. However, it included the very unusual—and deadly—Hurricane Agnes, which combined with another low-pressure center to ravage the northeastern United States, killing 122 and causing 1.3 billion 1972 dollars in damage. Possibly, Agnes was also unusually difficult to predict.

You should also look for clusters or subgroups that stand away from the rest of the plot or that show a trend in a different direction. Deviating groups should raise questions about why they are different. They may be a clue that you should split the data into subgroups instead of looking at them all together.

FOR EXAMPLE   Describing the scatterplot of hurricane winds and pressure

Hurricanes develop low pressure at their centers. This pulls in moist air, pumps up their rotation, and generates high winds. Standard sea-level pressure is around 1013 millibars (mb), or 29.9 inches of mercury. Hurricane Katrina had a central pressure of 920 mb and sustained winds of 110 knots.

Here's a scatterplot of *Maximum Wind Speed* (kts) vs. *Central Pressure* (mb) for 163 hurricanes that have hit the United States since 1851.

QUESTION: Describe what this plot shows.

The scatterplot shows a negative direction; in general, lower central pressure is found in hurricanes that have higher maximum wind speeds. This association is linear and moderately strong.

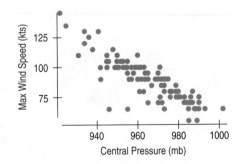

Roles for Variables

Which variable should go on the *x*-axis and which on the *y*-axis? What we want to know about the relationship can tell us how to make the plot. We often have questions such as:

- Do baseball teams that score more runs sell more tickets to their games?
- Do older houses sell for less than newer ones of comparable size and quality?
- Do students who score higher on their SAT tests have higher grade point averages in college?
- Can we estimate a person's percent body fat more simply by just measuring waist or wrist size?

In these examples, the two variables play different roles. We'll call the variable of interest the **response variable** and the other the **explanatory** or **predictor variable**.[5] We'll continue our practice of naming the variable of interest *y*. Naturally we'll plot it on the *y*-axis and place the explanatory variable on the *x*-axis. Sometimes, we'll call them the ***x*- and *y*-variables**. When you make a scatterplot, you can assume that those who view it will think this way, so choose which variables to assign to which axes carefully.

The roles that we choose for variables are more about how we *think* about them than about the variables themselves. Just placing a variable on the *x*-axis doesn't necessarily mean that it explains or predicts *anything*. And the variable on the *y*-axis may not respond to it in any way. We plotted prediction error on the *y*-axis against year on the *x*-axis because the National Hurricane Center is interested in how their predictions have changed over time. Could we have plotted them the other way? In this case, it's hard to imagine reversing the roles—knowing the prediction error and wanting to guess in what year it happened. But for some scatterplots, it can make sense to use either choice, so you have to think about how the choice of role helps to answer the question you have.

Correlation

Data collected from students in Statistics classes included their *Height* (in inches) and *Weight* (in pounds). It's no great surprise to discover that there is a positive association between the two. As you might suspect, taller students tend to weigh more. (If we had reversed the roles and chosen height as the explanatory variable, we might say that heavier students tend to be taller.)[6] And the form of the scatterplot is fairly straight as well, although there seems to be a high outlier, as the plot shows.

NOTATION ALERT

So *x* and *y* are reserved letters as well, but not just for labeling the axes of a scatterplot. In Statistics, the assignment of variables to the *x*- and *y*-axes (and the choice of notation for them in formulas) often conveys information about their roles as predictor or response variable.

A S   *Self-Test:* **Scatterplot Check.** Can you identify a scatterplot's direction, form, and strength?

WHO	Students
WHAT	Height (inches), weight (pounds)
WHERE	Ithaca, NY
WHY	Data for class
HOW	Survey

FIGURE 7.2
Weight vs. Height of Statistics students.

Plotting *Weight* vs. *Height* in different units doesn't change the shape of the pattern.

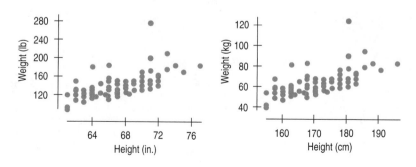

5 The *x*- and *y*-variables have sometimes been referred to as the *independent* and *dependent* variables, respectively. The idea was that the *y*-variable depended on the *x*-variable and the *x*-variable acted independently to make *y* respond. These names, however, conflict with other uses of the same terms in Statistics.

6 The son of one of the authors, when told (as he often was) that he was tall for his age, used to point out that, actually, he was young for his height.

The pattern in the scatterplots looks straight and is clearly a positive association, but how strong is it? If you had to put a number (say, between 0 and 1) on the strength, what would it be? Whatever measure you use shouldn't depend on the choice of units for the variables. After all, if we measure heights and weights in centimeters and kilograms instead, it doesn't change the direction, form, or strength, so it shouldn't change the number.

Since the units shouldn't matter to our measure of strength, we can remove them by standardizing each variable. Now, for each point, instead of the values (x, y) we'll have the standardized coordinates (z_x, z_y). Remember that to standardize values, we subtract the mean of each variable and then divide by its standard deviation:

$$(z_x, z_y) = \left(\frac{x - \bar{x}}{s_x}, \frac{y - \bar{y}}{s_y} \right).$$

Because standardizing makes the means of both variables 0, the center of the new scatterplot is at the origin. The scales on both axes are now standard deviation units.

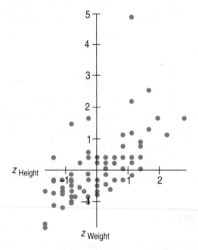

FIGURE 7.3
A scatterplot of standardized heights and weights.

Standardizing shouldn't affect the appearance of the plot. Does the plot of z-scores (Figure 7.3) look like the previous plots? Well, no. The underlying linear pattern seems steeper in the standardized plot. That's because the scales of the axes are now the same, so the length of one standard deviation is the same vertically and horizontally. When we worked in the original units, we were free to make the plot as tall and thin
or as squat and wide

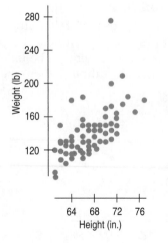

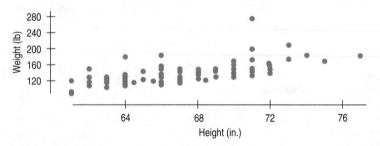

as we wanted to, but that can change the impression the plot gives. By contrast, equal scaling gives a neutral way of drawing the scatterplot and a fairer impression of the strength of the association.[7]

---

[7] When we draw a scatterplot, what often looks best is to make the length of the x-axis slightly larger than the length of the y-axis. This is an aesthetic choice, probably related to the Golden Ratio of the Greeks.

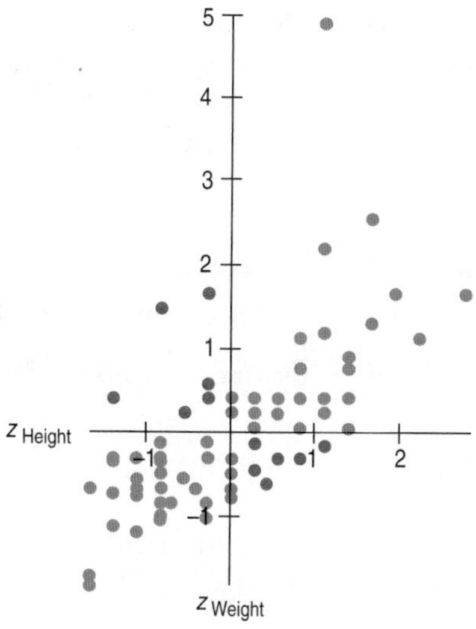

FIGURE 7.4
In this scatterplot of *z*-scores, points are colored according to how they affect the association: green for positive, red for negative, and blue for neutral.

Which points in the scatterplot of the *z*-scores give the impression of a positive association? In a positive association, *y* tends to increase as *x* increases. So, the points in the upper right and lower left (colored green) strengthen that impression. For these points, z_x and z_y have the same sign, so the product $z_x z_y$ is positive. Points far from the origin (which make the association look more positive) have bigger products.

The red points in the upper left and lower right quadrants tend to weaken the positive association (or support a negative association). For these points, z_x and z_y have opposite signs. So the product $z_x z_y$ for these points is negative. Points far from the origin (which make the association look more negative) have a negative product even larger in magnitude.

Points with *z*-scores of zero on either variable don't vote either way, because $z_x z_y = 0$. They're colored blue.

To turn these products into a measure of the strength of the association, just add up the $z_x z_y$ products for every point in the scatterplot:

$$\sum z_x z_y.$$

This summarizes the direction *and* strength of the association for all the points. If most of the points are in the green quadrants, the sum will tend to be positive. If most are in the red quadrants, it will tend to be negative.

But the *size* of this sum gets bigger the more data we have. To adjust for this, the natural (for statisticians anyway) thing to do is to divide the sum by $n - 1$.[8] The ratio is the famous **correlation coefficient**:

$$r = \frac{\sum z_x z_y}{n - 1}.$$

For the students' heights and weights, the correlation is 0.644. There are a number of alternative formulas for the correlation coefficient, using *x* and *y* in their original units, that you may encounter. Here are two of the most common:

$$r = \frac{\sum (x - \bar{x})(y - \bar{y})}{\sqrt{\sum (x - \bar{x})^2 (y - \bar{y})^2}} = \frac{\sum (x - \bar{x})(y - \bar{y})}{(n - 1)s_x s_y}.$$

----

[8] Yes, the same $n - 1$ as in the standard deviation calculation. And we offer the same promise to explain it later.

A S
Activity: Correlation and Relationship Strength. What does a correlation of 0.8 look like? How about 0.3?

NOTATION ALERT

The letter *r* is always used for correlation, so you can't use it for anything else in Statistics. Whenever you see an *r*, it's safe to assume it's a correlation.

These formulas can be more convenient for computing correlation by hand. But the form using z-scores is best for understanding what correlation means. If you want to see how to go from the formula using z-scores to these, just look at the Math Box for details.

MATH BOX **Are the Formulas for *r* Really Equivalent?**

Standardizing the variables first gives us an easy-to-understand expression for the correlation.

$$r = \frac{\sum z_x z_y}{n - 1}$$

But sometimes you'll see other formulas. Just remembering how standardizing works gets us from one formula to the other.

Since

$$z_x = \frac{x - \bar{x}}{s_x}$$

and

$$z_y = \frac{y - \bar{y}}{s_y},$$

we can just substitute these and get

$$r = \left(\frac{1}{n-1}\right)\sum z_x z_y = \left(\frac{1}{n-1}\right)\sum \frac{(x - \bar{x})}{s_x}\frac{(y - \bar{y})}{s_y} = \sum \frac{(x - \bar{x})(y - \bar{y})}{(n-1)s_x s_y}.$$

That's one version. And since we know the formula for standard deviation:

$$s_y = \sqrt{\frac{\sum (y - \bar{y})^2}{n - 1}},$$

we could (if we really wanted to) write:

$$r = \left(\frac{1}{n-1}\right)\sum \frac{(x - \bar{x})}{s_x}\frac{(y - \bar{y})}{s_y}$$

$$= \left(\frac{1}{n-1}\right)\frac{\sum (x - \bar{x})(y - \bar{y})}{\sqrt{\dfrac{\sum (x - \bar{x})^2}{n - 1}}\sqrt{\dfrac{\sum (y - \bar{y})^2}{n - 1}}}$$

$$= \left(\frac{1}{n-1}\right)\frac{\sum (x - \bar{x})(y - \bar{y})}{\left(\dfrac{1}{n-1}\right)\sqrt{\sum (x - \bar{x})^2}\sqrt{\sum (y - \bar{y})^2}}$$

$$= \frac{\sum (x - \bar{x})(y - \bar{y})}{\sqrt{\sum (x - \bar{x})^2 \sum (y - \bar{y})^2}}$$

That's the other common version. If you ever have to compute the correlation by hand, it's easier to start with one of these. But if you want to remember how correlation works, stick with the first formula in this Math Box.

FINDING THE CORRELATION COEFFICIENT BY HAND

To find the correlation coefficient by hand, we'll use a formula in original units, rather than z-scores. This will save us the work of having to standardize each individual data value first. Start with the summary statistics for both variables: $\bar{x}, \bar{y}, s_x$, and s_y. Then find the deviations as we did for the standard deviation, but now in both x and y: $(x - \bar{x})$ and $(y - \bar{y})$. For each data pair, multiply these deviations together: $(x - \bar{x})(y - \bar{y})$. Add the products up for all data pairs. Finally, divide the sum by the product of $(n - 1) \times s_x \times s_y$ to get the correlation coefficient.

Here we go:
Suppose the data pairs are:

x	6	10	14	19	21
y	5	3	7	8	12

Then $\bar{x} = 14, \bar{y} = 7, s_x = 6.20$, and $s_y = 3.39$

Deviations in x	Deviations in y	Product
$6 - 14 = -8$	$5 - 7 = -2$	$-8 \times -2 = 16$
$10 - 14 = -4$	$3 - 7 = -4$	16
$14 - 14 = 0$	$7 - 7 = 0$	0
$19 - 14 = 5$	$8 - 7 = 1$	5
$21 - 14 = 7$	$12 - 7 = 5$	35

Add up the products: $16 + 16 + 0 + 5 + 35 = 72$
Finally, we divide by $(n - 1) \times s_x \times s_y = (5 - 1) \times 6.20 \times 3.39 = 84.07$
The ratio is the correlation coefficient:

$$r = 72/84.07 = 0.856$$

Correlation Conditions

Correlation measures the strength of the *linear* association between two *quantitative* variables. Before you use correlation, you must check several *conditions:*

- **Quantitative Variables Condition:** Correlation applies only to quantitative variables. Don't apply correlation to categorical data masquerading as quantitative. Check that you know the variables' units and what they measure.
- **Straight Enough Condition:** Sure, you can *calculate* a correlation coefficient for any pair of variables. But correlation measures the strength only of the *linear* association, and will be misleading if the relationship is not linear. What is "straight enough"? How non-straight would the scatterplot have to be to fail the condition? This is a judgment call that you just have to think about. Do you think that the underlying relationship is curved? If so, then summarizing its strength with a correlation would be misleading.
- **Outlier Condition:** Outliers can distort the correlation dramatically. An outlier can make an otherwise weak correlation look big or hide a strong correlation. It can even give an otherwise positive association a negative correlation coefficient (and vice versa). When you see an outlier, it's often a good idea to report the correlation with and without the point.

Each of these conditions is easy to check with a scatterplot. Many correlations are reported without supporting data or plots. Nevertheless, you should still think about the conditions. And you should be cautious in interpreting (or accepting others' interpretations of) the correlation when you can't check the conditions for yourself.

A S **Simulation: Correlation and Linearity.** How much does straightness matter? See for yourself as you bend the scatterplot.

A S **Case Study: Mortality and Education.** Is the mortality rate lower in cities with higher median education levels?

FOR EXAMPLE | Correlating wind speed and pressure

RECAP: We looked at the scatterplot displaying hurricane wind speeds and central pressures.

The correlation coefficient for these wind speeds and pressures is $r = -0.879$.

QUESTION: Check the conditions for using correlation. If you feel they are satisfied, interpret this correlation.

- Quantitative Variables Condition: Both *Wind Speed* and *Central Pressure* are quantitative variables, measured (respectively) in knots and millibars.
- Straight Enough Condition: The pattern in the scatterplot is quite straight.
- Outlier Condition: A few hurricanes seem to straggle away from the main pattern, but they don't appear to be extreme enough to be called outliers. It may be worthwhile to check on them, however.

The conditions for using correlation are satisfied. The correlation coefficient of $r = -0.879$ indicates quite a strong negative linear association between the wind speeds of hurricanes and their central pressures.

JUST CHECKING

Your Statistics teacher tells you that the correlation between the scores (points out of 50) on Exam 1 and Exam 2 was 0.75.

1. Before answering any questions about the correlation, what would you like to see? Why?

2. If she adds 10 points to each Exam 1 score, how will this change the correlation?

3. If she standardizes scores on each exam, how will this affect the correlation?

4. In general, if someone did poorly on Exam 1, are they likely to have done poorly or well on Exam 2? Explain.

5. If someone did poorly on Exam 1, can you be sure that they did poorly on Exam 2 as well? Explain.

STEP-BY-STEP EXAMPLE | Looking at Association

When your blood pressure is measured, it is reported as two values: systolic blood pressure and diastolic blood pressure.

Questions: How are these variables related to each other? Do they tend to be both high or both low? How strongly associated are they?

THINK **Plan** State what you are trying to investigate.	I'll examine the relationship between two measures of blood pressure.

Variables Identify the two quantitative variables whose relationship we wish to examine. Report the W's, and be sure both variables are recorded for the same individuals.

The variables are systolic and diastolic blood pressure (SBP and DBP), recorded in millimeters of mercury (mm Hg) for each of 1406 participants in the Framingham Heart Study, a famous health study in Framingham, MA.[9]

Plot Make the scatterplot. Use a computer program or graphing calculator if you can.

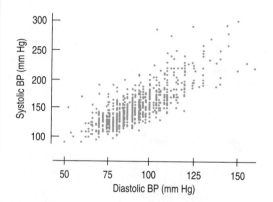

Check the conditions.

✔ **Quantitative Variables Condition:** Both SBP and DBP are quantitative and measured in mm Hg.

✔ **Straight Enough Condition:** The scatterplot looks straight.

✔ **Outlier Condition:** There are a few straggling points, but none far enough from the body of the data to be called outliers.

 REALITY CHECK Looks like a strong positive linear association. We shouldn't be surprised if the correlation coefficient is positive and fairly large.

I have two quantitative variables that satisfy the conditions, so correlation is a suitable measure of association.

 SHOW **Mechanics** We usually calculate correlations with technology. Here we have 1406 cases, so we'd never try it by hand.

The correlation coefficient is r = 0.792.

TELL **Conclusion** Describe the direction, form, and strength you see in the plot, along with any unusual points or features. Be sure to state your interpretations in the proper context.

The scatterplot shows a positive direction, with higher SBP going with higher DBP. The plot is generally straight, with a moderate amount of scatter. The correlation of 0.792 is consistent with what I saw in the scatterplot. A few cases stand out with unusually high SBP compared with their DBP. It seems far less common for the DBP to be high by itself.

[9] www.nhlbi.nih.gov/about/framingham

A S   *Activity: Construct Scatterplots with a Given Correlation.* Try to make a scatterplot that has a given correlation. How close can you get?

HEIGHT AND WEIGHT, AGAIN

We could have measured the students' weights in stones. In the now outdated UK system of measures, a stone is a measure equal to 14 pounds. And we could have measured heights in hands. Hands are still commonly used to measure the heights of horses. A hand is 4 inches. But no matter what *units* we use to measure the two variables, the *correlation* stays the same.

Correlation Properties

Here's a useful list of facts about the correlation coefficient:

- The sign of a correlation coefficient gives the direction of the association.
- Correlation is always between -1 and $+1$. Correlation *can* be exactly equal to -1.0 or $+1.0$, but these values are unusual in real data because they mean that all the data points fall *exactly* on a single straight line. Unlike the other properties here, this one is not readily apparent from the definitions, but it can be shown with simple algebra—as we'll do in the next chapter.
- Correlation treats x and y symmetrically. The correlation of x with y is the same as the correlation of y with x.
- Correlation has no units. This fact can be especially appropriate when the data's units are somewhat vague to begin with (IQ score, personality index, socialization, and so on). Correlation is sometimes given as a percentage, but you probably shouldn't do that because it suggests a percentage of *something*—and correlation, lacking units, has no "something" of which to be a percentage.
- Correlation is not affected by changes in the center or scale of either variable. Changing the units or baseline of either variable has no effect on the correlation coefficient. Correlation depends only on the z-scores, and they are unaffected by changes in center or scale.
- Correlation measures the strength of the *linear* association between the two variables. Variables can be strongly associated but still have a small correlation if the association isn't linear.
- Correlation is sensitive to outliers. A single outlying value can make a small correlation large or make a large one small.

> **How strong is strong?** You'll often see correlations characterized as "weak," "moderate," or "strong," but be careful. There's no agreement on what those terms mean. The same numerical correlation might be strong in one context and weak in another. You might be thrilled to discover a correlation of 0.7 between the new summary of the economy you've come up with and stock market prices, but you'd consider it a design failure if you found a correlation of "only" 0.7 between two tests intended to measure the same skill. Deliberately vague terms like "weak," "moderate," or "strong" that describe a linear association can be useful additions to the numerical summary that correlation provides. But be sure to include the correlation and show a scatterplot, so others can judge for themselves.

FOR EXAMPLE   Changing scales

RECAP: We found a correlation of $r = -0.879$ between hurricane wind speeds in knots and their central pressures in millibars.

QUESTION: Suppose we wanted to consider the wind speeds in miles per hour (1 mile per hour = 0.869 knot) and central pressures in inches of mercury (1 inch of mercury = 33.86 millibars). How would that conversion affect the conditions, the value of r, and our interpretation of the correlation coefficient?

Not at all! Correlation is based on standardized values (z-scores), so the conditions, the value of r, and the proper interpretation are all unaffected by changes in units.

Warning: Correlation ≠ Causation

Whenever we have a strong correlation, it's tempting to try to explain it by imagining that the predictor variable has *caused* the response to change. Humans are like that; we tend to see causes and effects in everything.

Sometimes this tendency can be amusing. A scatterplot of the human population (*y*) of Oldenburg, Germany, in the beginning of the 1930s plotted against the number of storks nesting in the town (*x*) shows a tempting pattern.

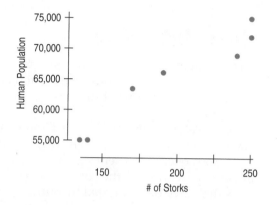

FIGURE 7.5
The number of storks in Oldenburg, Germany, plotted against the population of the town for 7 years in the 1930s. The association is clear. How about the causation? (*Ornithologishe Monatsberichte*, 44, no. 2)

Anyone who has seen the beginning of the movie *Dumbo* remembers Mrs. Jumbo anxiously waiting for the stork to bring her new baby. Even though you know it's silly, you can't help but think for a minute that this plot shows that storks are the culprits. The two variables are obviously related to each other (the correlation is 0.97!), but that doesn't prove that storks bring babies.

It turns out that storks nest on house chimneys. More people means more houses, more nesting sites, and so more storks. The causation is actually in the *opposite* direction, but you can't tell from the scatterplot or correlation. You need additional information—not just the data—to determine the real mechanism.

A scatterplot of the damage (in dollars) caused to a house by fire would show a strong correlation with the number of firefighters at the scene. Surely the damage doesn't cause firefighters. And firefighters do seem to cause damage, spraying water all around and chopping holes. Does that mean we shouldn't call the fire department? Of course not. There is an underlying variable that leads to both more damage and more firefighters: the size of the blaze.

A hidden variable that stands behind a relationship and determines it by simultaneously affecting the other two variables is called a **lurking variable**. You can often debunk claims made about data by finding a lurking variable behind the scenes.

Scatterplots and correlation coefficients *never* prove causation. That's one reason it took so long for the U.S. Surgeon General to get warning labels on cigarettes. Although there was plenty of evidence that increased smoking was *associated* with increased levels of lung cancer, it took years to provide evidence that smoking actually *causes* lung cancer.

Does cancer cause smoking? Even if the correlation of two variables is due to a causal relationship, the correlation itself cannot tell us what causes what.

Sir Ronald Aylmer Fisher (1890-1962) was one of the greatest statisticians of the 20th century. Fisher testified in court (in testimony paid for by the tobacco companies) that a causal relationship might underlie the correlation of smoking and cancer:

"Is it possible, then, that lung cancer . . . is one of the causes of smoking cigarettes? I don't think it can be excluded . . . the pre-cancerous condition is one involving a certain amount of slight chronic inflammation . . . ."

A slight cause of irritation . . . is commonly accompanied by pulling out a cigarette, and getting a little compensation for life's minor ills in that way. And . . . is not unlikely to be associated with smoking more frequently."

Ironically, the proof that smoking indeed is the cause of many cancers came from experiments conducted following the principles of experiment design and analysis that Fisher himself developed—and that we'll see in Chapter 13.

Correlation Tables

It is common in some fields to compute the correlations between every pair of variables in a collection of variables and arrange these correlations in a table. The rows and columns of the table name the variables, and the cells hold the correlations.

Correlation tables are compact and give a lot of summary information at a glance. They can be an efficient way to start to look at a large data set, but a dangerous one. By presenting all of these correlations without any checks for linearity and outliers, the correlation table risks showing truly small correlations that have been inflated by outliers, truly large correlations that are hidden by outliers, and correlations of any size that may be meaningless because the underlying form is not linear.

Sir Maurice Kendall (1907–1983) was a prominent statistician, writer, and teacher. He made significant contributions to the study of randomness (developing tests for randomness that are still used), the theory of statistics, and rank-based statistics. He also served as director of the World Fertility Study, for which the United Nations awarded him their Peace Medal.

	Assets	Sales	Market Value	Profits	Cash Flow	Employees
Assets	1.000					
Sales	0.746	1.000				
Market Value	0.682	0.879	1.000			
Profits	0.602	0.814	0.968	1.000		
Cash Flow	0.641	0.855	0.970	0.989	1.000	
Employees	0.594	0.924	0.818	0.762	0.787	1.000

TABLE 7.1

A correlation table of data reported by *Forbes* magazine for large companies. From this table, can you be sure that the variables are linearly associated and free from outliers?

The diagonal cells of a correlation table always show correlations of exactly 1. (Can you see why?) Correlation tables are commonly offered by statistics packages on computers. These same packages often offer simple ways to make all the scatterplots that go with these correlations.

*Measuring Trend: Kendall's Tau

Survey researchers often ask people to rank their opinions or attitudes on a scale that lacks units. (Common scales run from 1 to 5, 1 to 7, or 1 to 10.) For example, we might ask:

"How would you assess the pace of your Statistics course so far?"

1 = Way too slow   2 = A little too slow   3 = About right
4 = A little too fast   5 = Way too fast

Scales of this sort that attempt to measure attitudes numerically are called Likert scales, after their developer, Rensis Likert. Likert scales have order. The higher the number, the faster you think the course has gone, but a 4 doesn't necessarily mean *twice* as fast as 2. So we may want to be careful when applying methods such as correlation that require the Straight Enough Condition. That

PRONUNCIATION NOTE:

Many people pronounce Likert's name with a long "i" as in "like." But he actually pronounced it with a short "i" as in "lick."

A MONOTONOUS RELATIONSHIP?

You probably think of monotonous as *boring*. A monotone relationship shares the same root, but it's only the *sameness* that's important. A monotone relationship is one that increases or decreases consistently (but it doesn't have to be boring).

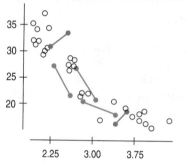

FIGURE 7.6
For each pair of points, Kendall's tau records whether the slope between them is positive (red), negative (blue), or zero.

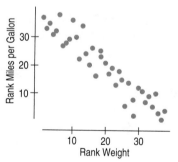

FIGURE 7.7
Spearman's rho finds the correlation of the *ranks* of the two variables.

may be too strong a requirement when all we may want to ask of the data is whether there is a consistent trend.

Suppose we collected data from lots of courses and tried to relate perceived pace to the number of students in the courses. We might wonder if students perceive the course as being faster in larger courses, but it isn't necessarily clear what it would mean for *pace* to be related to *class size* in a linear way. So the correlation coefficient might not be the appropriate measure. Still, we might want to know if the speed of the course and its size are associated. To answer this question, we can use an alternative measure of association, Kendall's tau.

Kendall's tau is a statistic designed to assess how close the relationship between two variables is to being *monotone*. A monotone relationship is one that consistently increases or decreases, but not necessarily in a linear fashion. That is often all we are interested in. And, of course, if we only care about a consistent trend, we don't have to worry that one or both of our variables might not be on a perfectly quantitative scale.

Kendall's tau measures monotonicity directly. For each pair of points in a scatterplot, it records only whether the slope of a line between those two points is positive, negative, or zero. (If the points have the same x-value, the slope between them is ignored.) In a monotone plot, these pairwise slopes would all be positive or all be negative. In a non-monotone plot, the counts of positive and negative slopes will tend to balance out. Tau (often written with the Greek letter τ) is the difference between the number of positive slopes and the number of negative slopes divided by the total number of slopes between pairs.

Obviously, tau can take values between -1.0 and $+1.0$. If every pairwise slope is negative, then tau will equal -1.0, and the plot will show a single stream of points trending consistently downward. If every pairwise slope is positive, then tau will equal $+1.0$, and the plot will show a single stream of points trending consistently upward. In a generally horizontal plot or one that goes up and then down, the counts of positive and negative slopes will balance out and tau will be near zero.

*Nonparametric Association: Spearman's Rho

One of the problems we have seen with the correlation coefficient is that it is very sensitive to violations of the Straight Enough Condition. Both outliers and bends in the data make it impossible to interpret correlation. **Spearman's rho** (often denoted by the Greek letter ρ) can deal with both of these problems. Rho replaces the original data values with their *ranks* within each variable. That is, it replaces the lowest value in x by the number 1, the next lowest value by 2, and so on, until the highest value is assigned the value n. The same ranking method is applied to the y-variable. A scatterplot of the y-ranks against the x-ranks shows the same general *trend* as the original data, going up whenever the data went up and down whenever the data went down. If the original scatterplot shows a consistent but bent trend, however, the scatterplot of the ranks is likely to be more nearly linear. And if either variable has an extreme outlying value, the ranking process keeps it from being extreme by just counting it as the highest or lowest value, ignoring how extreme it really is.

Spearman's rho is the correlation of the two rank variables. Because this is a correlation coefficient, it must be between -1.0 and 1.0.

Both Kendall's tau and Spearman's rho have advantages over the correlation coefficient r. For one, they can be used even when we know only the ranks. Also, they measure the consistency of the trend between the variables without insisting that the trend be linear.

Charles Edward Spearman, FRS (1863–1945) was an English psychologist known for work in Statistics, as a pioneer of factor analysis, and for Spearman's rank correlation coefficient. He also did fundamental work on measures of human intelligence.

They have the added advantage that they are not much affected by outliers. Spearman's rho limits the outlier to the value of its rank, and Kendall's tau cares only about the sign of each slope between points, not how *steep* it might be. Neither tau nor rho is changed at all by re-expressing variables by functions that don't alter the order of values (that is, by just the sort of functions we might use to straighten a scatterplot). But, unlike the correlation coefficient, neither statistic can be used as a base for more advanced or complex methods, so they tend to be specialized methods used when we care primarily about consistent trend between two variables.

Both of these measures are examples of what are called *nonparametric* or *distribution-free* methods. We first discussed parameters in Chapter 6. They are the constants in a model. The correlation coefficient attempts to estimate a particular parameter in the Normal model for two quantitative variables. Kendall's tau and Spearman's rho are less specific. They measure association, but there is no parameter that they are tied to and no specific model they require.

We'll run into more examples of nonparametric methods throughout the book. Because they tend to be somewhat specialized methods, we'll continue to put them in starred sections in the chapters where they appear.

Straightening Scatterplots

Correlation is a suitable measure of strength for straight relationships only. When a scatterplot shows a bent form that consistently increases or decreases, we can often straighten the form of the plot by re-expressing one or both variables.

Some camera lenses have an adjustable aperture, the hole that lets the light in. The size of the aperture is expressed in a mysterious number called the f/stop. Each increase of one f/stop number corresponds to a halving of the light that is allowed to come through. The f/stops of one digital camera are

f/stop:	2.8	4	5.6	8	11	16	22	32

When you halve the shutter speed, you cut down the light, so you have to open the aperture one notch. We could experiment to find the best f/stop value for each shutter speed. A table of recommended shutter speeds and f/stops for a camera lists the relationship like this:

Shutter speed:	1/1000	1/500	1/250	1/125	1/60	1/30	1/15	1/8
f/stop:	2.8	4	5.6	8	11	16	22	32

The correlation of these shutter speeds and f/stops is 0.979. That sounds pretty high. You might assume that there must be a strong linear relationship. But when we check the scatterplot (we *always* check the scatterplot), it shows that something is not quite right:

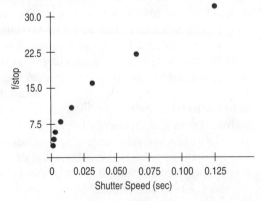

FIGURE 7.8
A scatterplot of *f/stop* vs. *Shutter Speed* shows a bent relationship.

We can see that the f/stop is not *linearly* related to the shutter speed. Can we find a transformation of *f/stop* that straightens out the line? What if we look at the *square* of the f/stop against the shutter speed?

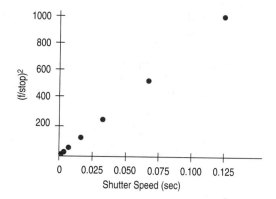

FIGURE 7.9
Re-expressing *f/stop* by squaring
straightens the plot.

The second plot looks much more nearly straight. In fact, the correlation is now 0.998, but the increase in correlation is not important. (The original value of 0.979 should please almost anyone who sought a large correlation.) What is important is that the *form* of the plot is now straight, so the correlation is now an appropriate measure of association.[10]

We can often find transformations that straighten a scatterplot's form. Here, we found the square. Chapter 10 discusses simple ways to find a good re-expression.

What Can Go Wrong?

■ **Don't say "correlation" when you mean "association."** How often have you heard the word "correlation"? Chances are pretty good that when you've heard the term, it's been misused. When people want to sound scientific, they often say "correlation" when talking about the relationship between two variables. It's one of the most widely misused Statistics terms, and given how often statistics are misused, that's saying a lot. One of the problems is that many people use the specific term *correlation* when they really mean the more general term *association*. "Association" is a deliberately vague term describing the relationship between two variables.

"Correlation" is a precise term that measures the strength and direction of the linear relationship between quantitative variables.

■ **Don't correlate categorical variables.** People who misuse the term "correlation" to mean "association" often fail to notice whether the variables they discuss are quantitative. Be sure to check the **Quantitative Variables Condition.**

■ **Don't confuse correlation with causation.** One of the most common mistakes people make in interpreting statistics occurs when they observe a high correlation between two variables and jump to the perhaps tempting conclusion that one thing must be causing the other. Scatterplots and correlations *never* demonstrate causation. At best, these statistical tools can only reveal an association between variables, and that's a far cry from establishing cause and effect. While it's true that some associations may be causal, the nature and direction of the causation can be very hard to establish, and there's always the risk of overlooking lurking variables.

[10]Sometimes we can do a "reality check" on our choice of re-expression. In this case, a bit of research reveals that f/stops are related to the diameter of the open shutter. Since the amount of light that enters is determined by the *area* of the open shutter, which is related to the diameter by squaring, the square re-expression seems reasonable. Not all re-expressions have such nice explanations, but it's a good idea to think about them.

■ **Make sure the association is linear.** Not all associations between quantitative variables are linear. Correlation can miss even a strong nonlinear association. A student project evaluating the quality of brownies baked at different temperatures reports a correlation of −0.05 between judges' scores and baking temperature. That seems to say there is no relationship—until we look at the scatterplot:

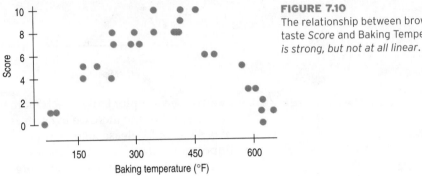

FIGURE 7.10
The relationship between brownie taste *Score* and Baking Temperature *is strong, but not at all linear.*

There is a strong association, but the relationship is not linear. Don't forget to check the Straight Enough Condition.

■ **Don't assume the relationship is linear just because the correlation coefficient is high.** Recall that the correlation of f/stops and shutter speeds is 0.979 and yet the relationship is clearly not straight. Although the relationship must be straight for the correlation to be an appropriate measure, a high correlation is no guarantee of straightness. Nor is it safe to use correlation to judge the best re-expression. It's always important to look at the scatterplot.

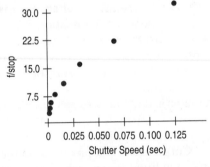

FIGURE 7.11
A scatterplot of *f/stop* vs. *Shutter Speed* shows a bent relationship even though the correlation is *r* = 0.979.

■ **Beware of outliers.** You can't interpret a correlation coefficient safely without a background check for outliers. Here's a silly example:

The relationship between IQ and shoe size among comedians shows a surprisingly strong positive correlation of 0.50. To check assumptions, we look at the scatterplot:

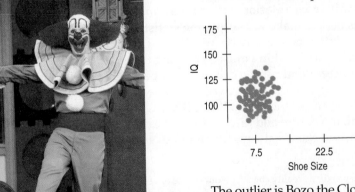

FIGURE 7.12
A scatterplot of *IQ* vs. *Shoe Size*. From this "study," what is the relationship between the two? The correlation is 0.50. Who does that point (the green x) in the upper right-hand corner belong to?

The outlier is Bozo the Clown, known for his large shoes, and widely acknowledged to be a comic "genius." Without Bozo, the correlation is near zero.

Even a single outlier can dominate the correlation value. That's why you need to check the Outlier Condition.

CONNECTIONS

Scatterplots are the basic tool for examining the relationship between two quantitative variables. We start with a picture when we want to understand the distribution of a single variable, and we always make a scatterplot to begin to understand the relationship between two quantitative variables.

We used z-scores as a way to measure the statistical distance of data values from their means. Now we've seen the z-scores of *x* and *y* working together to build the correlation coefficient. Correlation is a summary statistic like the mean and standard deviation—only it summarizes the strength of a linear relationship. And we interpret it as we did z-scores, using the standard deviations as our rulers in both *x* and *y*.

WHAT HAVE WE LEARNED?

In recent chapters we learned how to listen to the story told by data from a single variable. Now we've turned our attention to the more complicated (and more interesting) story we can discover in the association between two quantitative variables.

We've learned to begin our investigation by looking at a scatterplot. We're interested in the *direction* of the association, the *form* it takes, and its *strength*.

We've learned that, although not every relationship is linear, when the scatterplot is straight enough, the *correlation coefficient* is a useful numerical summary.

▶ The sign of the correlation tells us the direction of the association.

▶ The magnitude of the correlation tells us the *strength* of a linear association. Strong associations have correlations near −1 or +1 and very weak associations near 0.

▶ Correlation has no units, so shifting or scaling the data, standardizing, or even swapping the variables has no effect on the numerical value.

A S
Simulation: Correlation, Center, and Scale. If you have any lingering doubts that shifting and rescaling the data won't change the correlation, watch nothing happen right before your eyes!

Once again we've learned that doing Statistics right means we have to *Think* about whether our choice of methods is appropriate.

▶ The correlation coefficient is appropriate only if the underlying relationship is linear.

▶ We'll check the **Straight Enough Condition** by looking at a scatterplot.

▶ And, as always, we'll watch out for outliers!

Finally, we've learned not to make the mistake of assuming that a high correlation or strong association is evidence of a cause-and-effect relationship. Beware of lurking variables!

Terms

Scatterplots   A scatterplot shows the relationship between two quantitative variables measured on the same cases (p. 151).

Association   ▶ **Direction:** A positive direction or association means that, in general, as one variable increases, so does the other. When increases in one variable generally correspond to decreases in the other, the association is negative (p. 151).

▶ **Form:** The form we care about most is straight, but you should certainly describe other patterns you see in scatterplots (p. 151).

▶ **Strength:** A scatterplot is said to show a strong association if there is little scatter around the underlying relationship (p. 152).

Outlier   A point that does not fit the overall pattern seen in the scatterplot (p. 152).

Response variable, Explanatory variable, *x*-variable, *y*-variable	In a scatterplot, you must choose a role for each variable. Assign to the *y*-axis the response variable that you hope to predict or explain. Assign to the *x*-axis the explanatory or predictor variable that accounts for, explains, predicts, or is otherwise responsible for the *y*-variable (p. 153).
Correlation Coefficient	The correlation coefficient is a numerical measure of the direction and strength of a linear association (p. 155).

$$r = \frac{\sum z_x z_y}{n - 1}.$$

Lurking variable	A variable other than *x* and *y* that simultaneously affects both variables, accounting for the correlation between the two (p. 161).

Skills

 THINK

▸ Recognize when interest in the pattern of a possible relationship between two quantitative variables suggests making a scatterplot.

▸ Know how to identify the roles of the variables and that you should place the response variable on the *y*-axis and the explanatory variable on the *x*-axis.

▸ Know the conditions for correlation and how to check them.

▸ Know that correlations are between −1 and +1, and that each extreme indicates a perfect linear association.

▸ Understand how the magnitude of the correlation reflects the strength of a linear association as viewed in a scatterplot.

▸ Know that correlation has no units.

▸ Know that the correlation coefficient is not changed by changing the center or scale of either variable.

▸ Understand that causation cannot be demonstrated by a scatterplot or correlation.

 SHOW

▸ Know how to make a scatterplot by hand (for a small set of data) or with technology.

▸ Know how to compute the correlation of two variables.

▸ Know how to read a correlation table produced by a statistics program.

 TELL

▸ Be able to describe the direction, form, and strength of a scatterplot.

▸ Be prepared to identify and describe points that deviate from the overall pattern.

▸ Be able to use correlation as part of the description of a scatterplot.

▸ Be alert to misinterpretations of correlation.

▸ Understand that finding a correlation between two variables does not indicate a causal relationship between them. Beware the dangers of suggesting causal relationships when describing correlations.

SCATTERPLOTS AND CORRELATION ON THE COMPUTER

Statistics packages generally make it easy to look at a scatterplot to check whether the correlation is appropriate. Some packages make this easier than others.

Many packages allow you to modify or enhance a scatterplot, altering the axis labels, the axis numbering, the plot symbols, or the colors used. Some options, such as color and symbol choice, can be used to display additional information on the scatterplot.

DATA DESK

To make a scatterplot of two variables, click to select one variable as Y, shift-click to select the other as X, and choose **Scatterplot** from the **Plot** menu. Then find the correlation by choosing **Correlation** from the scatterplot's HyperView menu.

Alternatively, select the two variables and choose **Pearson Product-Moment** from the **Correlations** submenu of the **Calc** menu.

COMMENTS

We prefer that you look at the scatterplot first and then find the correlation. But if you've found the correlation first, click on the correlation value to drop down a menu that offers to make the scatterplot.

EXCEL

To make a scatterplot with the Excel Chart Wizard,

- Click on the **Chart Wizard** Button in the menu bar. Excel opens the Chart Wizard's Chart Type Dialog window.

- Make sure the **Standard Types** tab is selected, and select **XY (Scatter)** from the choices offered.

- Specify the **scatterplot without lines** from the choices offered in the Chart subtype selections. The **Next** button takes you to the Chart Source Data dialog.

- If it is not already frontmost, click on the **Data Range** tab, and enter the data range in the space provided.

- By convention, we always represent variables in columns. The Chart Wizard refers to variables as Series. Be sure the **Column** option is selected.

- Excel places the leftmost column of those you select on the x-axis of the scatterplot. If the column you wish to see on the x-axis is not the leftmost column in your spreadsheet, click on the **Series** tab and edit the specification of the individual axis series.

- Click the **Next** button. The Chart Options dialog appears.

- Select the **Titles** tab. Here you specify the title of the chart and names of the variables displayed on each axis.

- Type the chart title in the **Chart title:** edit box.

- Type the x-axis variable name in the **Value (X) Axis:** edit box. Note that you must name the columns correctly here. Naming another variable will not alter the plot, only mislabel it.

- Type the y-axis variable name in the **Value (Y) Axis:** edit box.

- Click the **Next** button to open the chart location dialog.

- Select the **As new sheet:** option button.

- Click the **Finish** button.

Often, the resulting scatterplot will not be useful. By default, Excel includes the origin in the plot even when the data are far from zero. You can adjust the axis scales.

To change the scale of a plot axis in Excel,

- Double-click on the axis. The **Format Axis Dialog** appears.

- If the **scale tab** is not the frontmost, select it.

- Enter new minimum or new maximum values in the spaces provided. You can drag the dialog box over the scatterplot as a straightedge to help you read the maximum and minimum values on the axes.

- Click the **OK** button to view the rescaled scatterplot.

- Follow the same steps for the x-axis scale.

Compute a correlation in Excel with the **CORREL** function from the drop-down menu of functions. If CORREL is not on the menu, choose **More Functions** and find it among the statistical functions in the browser.

In the dialog that pops up, enter the range of cells holding one of the variables in the space provided.

Enter the range of cells for the other variable in the space provided.

EXCEL 2007

To make a scatterplot in Excel 2007:

- Select the columns of data to use in the scatterplot. You can select more than one column by holding down the control key while clicking.

- In the Insert tab, click on the **Scatter** button and select the **Scatter with only Markers** chart from the menu.

Unfortunately, the plot this creates is often statistically useless. To make the plot useful, we need to change the display:

- With the chart selected click on the **Gridlines** button in the Layout tab to cause the Chart Tools tab to appear.

- Within Primary Horizontal Gridlines, select **None**. This will remove the gridlines from the scatterplot.

(continued)

- To change the axis scaling, click on the numbers of each axis of the chart, and click on the **Format Selection** button in the Layout tab.
- Select the **Fixed** option instead of the Auto option, and type a value more suited for the scatterplot. You can use the pop-up dialog window as a straight-edge to approximate the appropriate values.

Excel 2007 automatically places the leftmost of the two columns you select on the x-axis, and the rightmost one on the y-axis. If that's not what you'd prefer for your plot, you'll want to switch them.

To switch the X- and Y-variables:

- Click the chart to access the **Chart Tools** tabs.
- Click on the **Select Data** button in the Design tab.
- In the pop-up window's Legend Entries box, click on **Edit.**
- Highlight and delete everything in the Series X Values line, and select new data from the spreadsheet. (Note that selecting the column would inadvertently select the title of the column, which would not work well here.)
- Do the same with the Series Y Values line.
- Press **OK**, then press **OK** again.

JMP

To make a scatterplot and compute correlation, choose **Fit Y by X** from the **Analyze** menu.

In the Fit Y by X dialog, drag the Y variable into the **"Y, Response"** box, and drag the X variable into the **"X, Factor"** box. Click the **OK** button.

Once JMP has made the scatterplot, click on the red triangle next to the plot title to reveal a menu of options. Select **Density Ellipse** and select .95. JMP draws an ellipse around the data and reveals the **Correlation** tab. Click the blue triangle next to Correlation to reveal a table containing the correlation coefficient.

MINITAB

To make a scatterplot, choose **Scatterplot** from the **Graph** menu. Choose "Simple" for the type of graph. Click **OK**. Enter variable names for the Y-variable and X-variable into the table. Click **OK.**

To compute a correlation coefficient, choose **Basic Statistics** from the **Stat** menu. From the Basic Statistics sub-menu, choose **Correlation.** Specify the names of at least two quantitative variables in the "Variables" box. Click **OK** to compute the correlation table.

SPSS

To make a scatterplot in SPSS, open the Chart Builder from the Graphs menu. Then

- Click the Gallery tab.
- Choose Scatterplot from the list of chart types.
- Drag the scatterplot onto the canvas.
- Drag a scale variable you want as the response variable to the y-axis drop zone.
- Drag a scale variable you want as the factor or predictor to the x-axis drop zone.
- Click OK.

To compute a correlation coefficient, choose **Correlate** from the **Analyze** menu. From the Correlate submenu, choose **Bivariate.** In the Bivariate Correlations dialog, use the arrow button to move variables between the source and target lists.

Make sure the **Pearson** option is selected in the Correlation Coefficients field.

TI-83/84 PLUS

To create a scatterplot, set up the **STAT PLOT** by choosing the **scatterplot** icon (the first option). Specify the lists where the data are stored as Xlist and Ylist. Set the graphing **WINDOW** to the appropriate scale and **GRAPH** (or take the easy way out and just ZoomStat!).

To find the correlation, go to **STAT CALC** menu and select **8: LinReg(a + bx).** Then specify the lists where the data are stored. The final command you will enter should look like **LinReg(a + bx) L1, L2.**

COMMENTS

Notice that if you **TRACE** the scatterplot, the calculator will tell you the x- and y-value at each point.

If the calculator does not tell you the correlation after you enter a LinReg command, try this: Hit **2nd CATALOG.** You now see a list of everything the calculator knows how to do. Scroll down until you find **DiagnosticOn.** Hit **ENTER** twice. (It should say Done.) Now and forevermore (or until you change batteries), you can find a correlation by using your calculator.

TI-89

To create a scatterplot, press F2 (Plots). Select choice 1: **Plot Setup.** Select a plot to define and press F1. Select **Plot Type 1: Scatter.** Select a mark type. Specify the lists where the data are stored as Xlist and Ylist, using VAR-LINK. Press ENTER to finish. Press F5 to display the plot.

To find the correlation, press F4 (CALC), then arrow to **3: Regressions,** press the right arrow, and select **1:LinReg(a + bx).** Then specify the lists where the data are stored. You can also select a y-function to store the equation of the line.

COMMENTS

Notice that if you **TRACE** (press F3) the scatterplot, the calculator will tell you the x- and y-value at each point.

EXERCISES

1. **Association.** Suppose you were to collect data for each pair of variables. You want to make a scatterplot. Which variable would you use as the explanatory variable and which as the response variable? Why? What would you expect to see in the scatterplot? Discuss the likely direction, form, and strength.
   a) Apples: weight in grams, weight in ounces
   b) Apples: circumference (inches), weight (ounces)
   c) College freshmen: shoe size, grade point average
   d) Gasoline: number of miles you drove since filling up, gallons remaining in your tank

2. **Association.** Suppose you were to collect data for each pair of variables. You want to make a scatterplot. Which variable would you use as the explanatory variable and which as the response variable? Why? What would you expect to see in the scatterplot? Discuss the likely direction, form, and strength.
   a) T-shirts at a store: price each, number sold
   b) Scuba diving: depth, water pressure
   c) Scuba diving: depth, visibility
   d) All elementary school students: weight, score on a reading test

3. **Association.** Suppose you were to collect data for each pair of variables. You want to make a scatterplot. Which variable would you use as the explanatory variable and which as the response variable? Why? What would you expect to see in the scatterplot? Discuss the likely direction, form, and strength.
   a) When climbing mountains: altitude, temperature
   b) For each week: ice cream cone sales, air-conditioner sales
   c) People: age, grip strength
   d) Drivers: blood alcohol level, reaction time

4. **Association.** Suppose you were to collect data for each pair of variables. You want to make a scatterplot. Which variable would you use as the explanatory variable and which as the response variable? Why? What would you expect to see in the scatterplot? Discuss the likely direction, form, and strength.
   a) Long-distance calls: time (minutes), cost
   b) Lightning strikes: distance from lightning, time delay of the thunder
   c) A streetlight: its apparent brightness, your distance from it
   d) Cars: weight of car, age of owner

5. **Scatterplots.** Which of these scatterplots show
   a) little or no association?
   b) a negative association?
   c) a linear association?
   d) a moderately strong association?
   e) a very strong association?

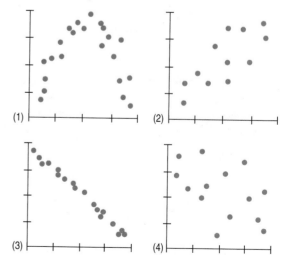

6. **Scatterplots.** Which of these scatterplots show
   a) little or no association?
   b) a negative association?
   c) a linear association?
   d) a moderately strong association?
   e) a very strong association?

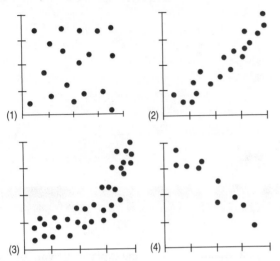

(1) (2) (3) (4)

7. **Performance IQ scores vs. brain size.** A study examined brain size (measured as pixels counted in a digitized magnetic resonance image [MRI] of a cross section of the brain) and IQ (4 Performance scales of the Weschler IQ test) for college students. The scatterplot shows the Performance IQ scores vs. the brain size. Comment on the association between brain size and IQ as seen in the scatterplot.

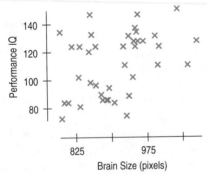

8. **Kentucky Derby 2007.** The fastest horse in Kentucky Derby history was Secretariat in 1973. The scatterplot shows speed (in miles per hour) of the winning horses each year.

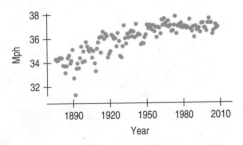

What do you see? In most sporting events, performances have improved and continue to improve, so surely we anticipate a positive direction. But what of the form? Has the performance increased at the same rate throughout the last 130 years?

9. **Firing pottery.** A ceramics factory can fire eight large batches of pottery a day. Sometimes a few of the pieces break in the process. In order to understand the problem better, the factory records the number of broken pieces in each batch for 3 days and then creates the scatterplot shown.

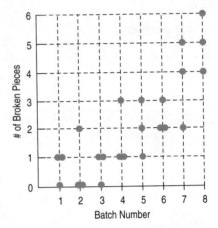

a) Make a histogram showing the distribution of the number of broken pieces in the 24 batches of pottery examined.
b) Describe the distribution as shown in the histogram. What feature of the problem is more apparent in the histogram than in the scatterplot?
c) What aspect of the company's problem is more apparent in the scatterplot?

10. **Coffee sales.** Owners of a new coffee shop tracked sales for the first 20 days and displayed the data in a scatterplot (by day).

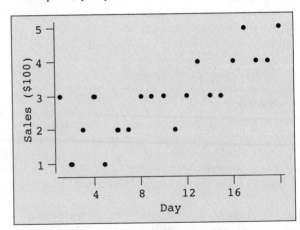

a) Make a histogram of the daily sales since the shop has been in business.
b) State one fact that is obvious from the scatterplot, but not from the histogram.
c) State one fact that is obvious from the histogram, but not from the scatterplot.

11. **Matching.** Here are several scatterplots. The calculated correlations are −0.923, −0.487, 0.006, and 0.777. Which is which?

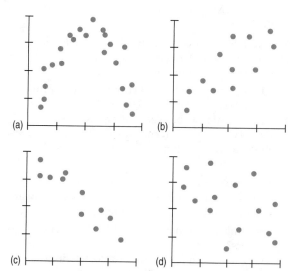

12. **Matching.** Here are several scatterplots. The calculated correlations are −0.977, −0.021, 0.736, and 0.951. Which is which?

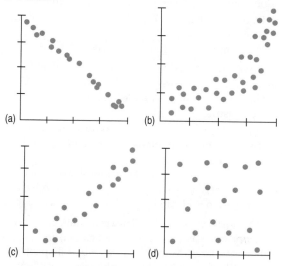

13. **Politics.** A candidate for office claims that "there is a correlation between television watching and crime." Criticize this statement on statistical grounds.

14. **Car thefts.** The National Insurance Crime Bureau reports that Honda Accords, Honda Civics, and Toyota Camrys are the cars most frequently reported stolen, while Ford Tauruses, Pontiac Vibes, and Buick LeSabres are stolen least often. Is it reasonable to say that there's a correlation between the type of car you own and the risk that it will be stolen?

15. **Roller coasters.** Roller coasters get all their speed by dropping down a steep initial incline, so it makes sense that the height of that drop might be related to the speed of the coaster. Here's a scatterplot of top *Speed* and largest *Drop* for 75 roller coasters around the world.

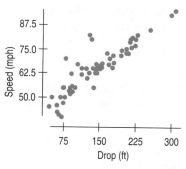

a) Does the scatterplot indicate that it is appropriate to calculate the correlation? Explain.
b) In fact, the correlation of Speed and Drop is 0.91. Describe the association.

16. **Antidepressants.** A study compared the effectiveness of several antidepressants by examining the experiments in which they had passed the FDA requirements. Each of those experiments compared the active drug with a placebo, an inert pill given to some of the subjects. In each experiment some patients treated with the placebo had improved, a phenomenon called the *placebo effect*. Patients' depression levels were evaluated on the Hamilton Depression Rating Scale, where larger numbers indicate greater improvement. (The Hamilton scale is a widely accepted standard that was used in each of the independently run studies.) The scatterplot below compares mean improvement levels for the antidepressants and placebos for several experiments.

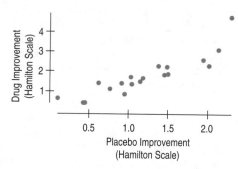

a) Is it appropriate to calculate the correlation? Explain.
b) The correlation is 0.898. Explain what we have learned about the results of these experiments.

17. **Streams and hard water.** In a study of streams in the Adirondack Mountains, the following relationship was found between the water's pH and its hardness (measured in grains):

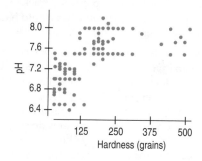

a) Is it appropriate to summarize the strength of association with a correlation? Explain.

*b) Would Kendall's tau be an appropriate measure? Explain.

18. **Traffic headaches.** A study of traffic delays in 68 U.S. cities found the following relationship between total delays (in total hours lost) and mean highway speed:

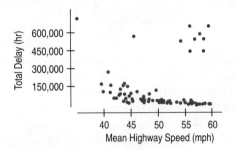

a) Is it appropriate to summarize the strength of association with a correlation? Explain.

*b) Would Spearman's correlation be a better choice of summary? Explain.

19. **Cold nights.** Is there an association between time of year and the nighttime temperature in North Dakota? A researcher assigned the numbers 1–365 to the days January 1–December 31 and recorded the temperature at 2:00 a.m. for each. What might you expect the correlation between *DayNumber* and *Temperature* to be? Explain.

20. **Association.** A researcher investigating the association between two variables collected some data and was surprised when he calculated the correlation. He had expected to find a fairly strong association, yet the correlation was near 0. Discouraged, he didn't bother making a scatterplot. Explain to him how the scatterplot could still reveal the strong association he anticipated.

21. **Prediction units.** The errors in predicting hurricane tracks (examined in this chapter) were given in nautical miles. A statutory mile is 0.86898 nautical mile. Most people living on the Gulf Coast of the United States would prefer to know the prediction errors in statutory miles rather than nautical miles. Explain why converting the errors to miles would not change the correlation between *Prediction Error* and *Year*.

22. **More predictions.** Hurricane Katrina's hurricane force winds extended 120 miles from its center. Katrina was a big storm, and that affects how we think about the prediction errors. Suppose we add 120 miles to each error to get an idea of how far from the predicted track we might still find damaging winds. Explain what would happen to the correlation between *Prediction Error* and *Year*, and why.

23. **Correlation errors.** Your Economics instructor assigns your class to investigate factors associated with the gross domestic product (*GDP*) of nations. Each student examines a different factor (such as *Life Expectancy*, *Literacy Rate*, etc.) for a few countries and reports to the class. Apparently, some of your classmates do not understand Statistics very well because you know several of their conclusions are incorrect. Explain the mistakes in their statements at the top of the next column.

a) "My very low correlation of −0.772 shows that there is almost no association between *GDP* and *Infant Mortality Rate*."

b) "There was a correlation of 0.44 between *GDP* and *Continent*."

24. **More correlation errors.** Students in the Economics class discussed in Exercise 23 also wrote these conclusions. Explain the mistakes they made.

a) "There was a very strong correlation of 1.22 between *Life Expectancy* and *GDP*."

b) "The correlation between *Literacy Rate* and *GDP* was 0.83. This shows that countries wanting to increase their standard of living should invest heavily in education."

25. **Height and reading.** A researcher studies children in elementary school and finds a strong positive linear association between height and reading scores.

a) Does this mean that taller children are generally better readers?

b) What might explain the strong correlation?

26. **Cellular telephones and life expectancy.** A survey of the world's nations in 2004 shows a strong positive correlation between percentage of the country using cell phones and life expectancy in years at birth.

a) Does this mean that cell phones are good for your health?

b) What might explain the strong correlation?

27. **Correlation conclusions I.** The correlation between *Age* and *Income* as measured on 100 people is $r = 0.75$. Explain whether or not each of these possible conclusions is justified:

a) When *Age* increases, *Income* increases as well.

b) The form of the relationship between *Age* and *Income* is straight.

c) There are no outliers in the scatterplot of *Income* vs. *Age*.

d) Whether we measure *Age* in years or months, the correlation will still be 0.75.

28. **Correlation conclusions II.** The correlation between *Fuel Efficiency* (as measured by miles per gallon) and *Price* of 150 cars at a large dealership is $r = -0.34$. Explain whether or not each of these possible conclusions is justified:

a) The more you pay, the lower the fuel efficiency of your car will be.

b) The form of the relationship between *Fuel Efficiency* and *Price* is moderately straight.

c) There are several outliers that explain the low correlation.

d) If we measure *Fuel Efficiency* in kilometers per liter instead of miles per gallon, the correlation will increase.

29. **Baldness and heart disease.** Medical researchers followed 1435 middle-aged men for a period of 5 years, measuring the amount of *Baldness* present (none = 1, little = 2, some = 3, much = 4, extreme = 5) and presence of *Heart Disease* (No = 0, Yes = 1). They found a correlation of 0.089 between the two variables. Comment on their conclusion that this shows that baldness is not a possible cause of heart disease.

30. **Sample survey.** A polling organization is checking its database to see if the two data sources it used sampled the same zip codes. The variable *Datasource* = 1 if the data source is MetroMedia, 2 if the data source is DataQwest, and 3 if it's RollingPoll. The organization finds that the correlation between five-digit zip code and *Datasource* is −0.0229. It concludes that the correlation is low enough to state that there is no dependency between *Zip Code* and *Source of Data*. Comment.

(T) 31. **Income and housing.** The Office of Federal Housing Enterprise Oversight (www.ofheo.gov) collects data on various aspects of housing costs around the United States. Here is a scatterplot of the *Housing Cost Index* versus the *Median Family Income* for each of the 50 states. The correlation is 0.65.

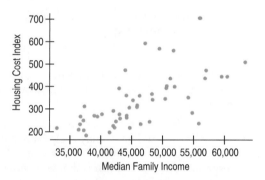

a) Describe the relationship between the *Housing Cost Index* and the *Median Family Income* by state.
b) If we standardized both variables, what would the correlation coefficient between the standardized variables be?
c) If we had measured *Median Family Income* in thousands of dollars instead of dollars, how would the correlation change?
d) Washington, DC, has a Housing Cost Index of 548 and a median income of about $45,000. If we were to include DC in the data set, how would that affect the correlation coefficient?
e) Do these data provide proof that by raising the median income in a state, the Housing Cost Index will rise as a result? Explain.
*f) For these data Kendall's tau is 0.51. Does that provide proof that by raising the median income in a state, the Housing Cost Index will rise as a result? Explain what Kendall's tau says and does not say.

(T) 32. **Interest rates and mortgages.** Since 1980, average mortgage interest rates have fluctuated from a low of under 6% to a high of over 14%. Is there a relationship between the amount of money people borrow and the interest rate that's offered? Here is a scatterplot of *Total Mortgages* in the United States (in millions of 2005 dollars) versus *Interest Rate* at various times over the past 26 years. The correlation is −0.84.

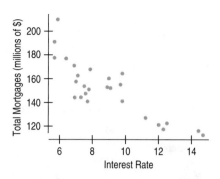

a) Describe the relationship between *Total Mortgages* and *Interest Rate*.
b) If we standardized both variables, what would the correlation coefficient between the standardized variables be?
c) If we were to measure *Total Mortgages* in thousands of dollars instead of millions of dollars, how would the correlation coefficient change?
d) Suppose in another year, interest rates were 11% and mortgages totaled $250 million. How would including that year with these data affect the correlation coefficient?
e) Do these data provide proof that if mortgage rates are lowered, people will take out more mortgages? Explain.
*f) For these data Kendall's tau is −0.61. Does that provide proof that if mortgage rates are lowered, people will take out more mortgages? Explain what Kendall's tau says and does not say.

(T) 33. **Fuel economy 2007.** Here are advertised horsepower ratings and expected gas mileage for several 2007 vehicles. (http://www.kbb.com)

Vehicle	Horsepower	Highway Gas Mileage (mpg)
Audi A4	200	32
BMW 328	230	30
Buick LaCrosse	200	30
Chevy Cobalt	148	32
Chevy TrailBlazer	291	22
Ford Expedition	300	20
GMC Yukon	295	21
Honda Civic	140	40
Honda Accord	166	34
Hyundai Elantra	138	36
Lexus IS 350	306	28
Lincoln Navigator	300	18
Mazda Tribute	212	25
Toyota Camry	158	34
Volkswagen Beetle	150	30

a) Make a scatterplot for these data.
b) Describe the direction, form, and strength of the plot.
c) Find the correlation between horsepower and miles per gallon.
d) Write a few sentences telling what the plot says about fuel economy.

34. Drug abuse. A survey was conducted in the United States and 10 countries of Western Europe to determine the percentage of teenagers who had used marijuana and other drugs. The results are summarized in the table.

Country	Percent Who Have Used	
	Marijuana	Other Drugs
Czech Rep.	22	4
Denmark	17	3
England	40	21
Finland	5	1
Ireland	37	16
Italy	19	8
No. Ireland	23	14
Norway	6	3
Portugal	7	3
Scotland	53	31
USA	34	24

a) Create a scatterplot.
b) What is the correlation between the percent of teens who have used marijuana and the percent who have used other drugs?
c) Write a brief description of the association.
d) Do these results confirm that marijuana is a "gateway drug," that is, that marijuana use leads to the use of other drugs? Explain.

35. Burgers. Fast food is often considered unhealthy because much of it is high in both fat and sodium. But are the two related? Here are the fat and sodium contents of several brands of burgers.

Fat (g)	19	31	34	35	39	39	43
Sodium (mg)	920	1500	1310	860	1180	940	1260

a) Analyze the association between fat content and sodium using correlation and scatterplots.
*b) Find Spearman's rho for these data. Compare it with the Pearson correlation. Comment.

36. Burgers II. In the previous exercise you analyzed the association between the amounts of fat and sodium in fast food hamburgers. What about fat and calories? Here are data for the same burgers:

Fat (g)	19	31	34	35	39	39	43
Calories	410	580	590	570	640	680	660

a) Analyze the association between fat content and calories using correlation and scatterplots.
*b) Repeat your analysis using Spearman's rho. Explain any differences you see.

37. Attendance 2006. American League baseball games are played under the designated hitter rule, meaning that pitchers, often weak hitters, do not come to bat. Baseball owners believe that the designated hitter rule means more runs scored, which in turn means higher

attendance. Is there evidence that more fans attend games if the teams score more runs? Data collected from American League games during the 2006 season indicate a correlation of 0.667 between runs scored and the number of people at the game. (http://mlb.mlb.com)

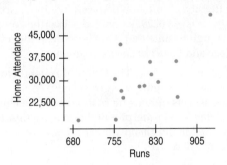

a) Does the scatterplot indicate that it's appropriate to calculate a correlation? Explain.
b) Describe the association between attendance and runs scored.
c) Does this association prove that the owners are right that more fans will come to games if the teams score more runs?

38. Second inning 2006. Perhaps fans are just more interested in teams that win. The displays below are based on American League teams for the 2006 season. (http://espn.go.com) Are the teams that win necessarily those which score the most runs?

	Correlation		
	Wins	Runs	Attend
Wins	1.000		
Runs	0.605	1.000	
Attend	0.697	0.667	1.000

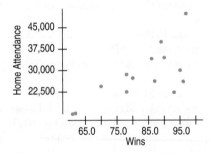

a) Do winning teams generally enjoy greater attendance at their home games? Describe the association.
b) Is attendance more strongly associated with winning or scoring runs? Explain.
c) How strongly is scoring more runs associated with winning more games?

39. Thrills. People who responded to a July 2004 Discovery Channel poll named the 10 best roller coasters in the United States. The table below shows the length of the initial drop (in feet) and the duration of the ride (in seconds). What do these data indicate about the height of a roller coaster and the length of the ride you can expect?

Roller Coaster	State	Drop (ft)	Duration (sec)
Incredible Hulk	FL	105	135
Millennium Force	OH	300	105
Goliath	CA	255	180
Nitro	NJ	215	240
Magnum XL-2000	OH	195	120
The Beast	OH	141	65
Son of Beast	OH	214	140
Thunderbolt	PA	95	90
Ghost Rider	CA	108	160
Raven	IN	86	90

T 40. Vehicle weights. The Minnesota Department of Transportation hoped that they could measure the weights of big trucks without actually stopping the vehicles by using a newly developed "weight-in-motion" scale. To see if the new device was accurate, they conducted a calibration test. They weighed several stopped trucks (static weight) and assumed that this weight was correct. Then they weighed the trucks again while they were moving to see how well the new scale could estimate the actual weight. Their data are given in the table below.

Weights (1000s of lbs)	
Weight-in-Motion	Static Weight
26.0	27.9
29.9	29.1
39.5	38.0
25.1	27.0
31.6	30.3
36.2	34.5
25.1	27.8
31.0	29.6
35.6	33.1
40.2	35.5

a) Make a scatterplot for these data.
b) Describe the direction, form, and strength of the plot.
c) Write a few sentences telling what the plot says about the data. (*Note*: The sentences should be about weighing trucks, not about scatterplots.)
d) Find the correlation.
e) If the trucks were weighed in kilograms, how would this change the correlation? (1 kilogram = 2.2 pounds)
f) Do any points deviate from the overall pattern? What does the plot say about a possible recalibration of the weight-in-motion scale?

T 41. Planets (more or less). On August 24, 2006, the International Astronomical Union voted that Pluto is not a planet. Some members of the public have been reluctant to accept that decision. Let's look at some of the data. (We'll see more in the next chapter.) Is there any pattern to the locations of the planets? The table shows the average distance of each of the traditional nine planets from the sun.

Planet	Position Number	Distance from Sun (million miles)
Mercury	1	36
Venus	2	67
Earth	3	93
Mars	4	142
Jupiter	5	484
Saturn	6	887
Uranus	7	1784
Neptune	8	2796
Pluto	9	3666

a) Make a scatterplot and describe the association. (Remember: direction, form, and strength!)
b) Why would you not want to talk about the correlation between a planet's *Position* and *Distance* from the sun?
c) Make a scatterplot showing the logarithm of *Distance* vs. *Position*. What is better about this scatterplot?
*d) Looking only at the scatterplot, what is Kendall's tau for these data? Explain.

T 42. Flights. The number of flights by U.S. Airlines has grown rapidly. Here are the number of flights flown in each year from 1995 to 2005.

a) Find the correlation of *Flights* with *Year*.
b) Make a scatterplot and describe the trend.
c) Note two reasons that the correlation you found in part a is not a suitable summary of the strength of the association. Can you account for these violations of the conditions?
*d) Find Kendall's tau for these data. Would that be an appropriate measure for summarizing their relationship? Explain.

Year	Flights
1995	5,327,435
1996	5,351,983
1997	5,411,843
1998	5,384,721
1999	5,527,884
2000	5,683,047
2001	5,967,780
2002	5,271,359
2003	6,488,539
2004	7,129,270
2005	7,140,596

JUST CHECKING

ANSWERS

1. We know the scores are quantitative. We should check to see if the Straight Enough Condition and the Outlier Condition are satisfied by looking at a scatterplot of the two scores.

2. It won't change.

3. It won't change.

4. They are likely to have done poorly. The positive correlation means that low scores on Exam 1 are associated with low scores on Exam 2 (and similarly for high scores).

5. No. The general association is positive, but individual performances may vary.

Linear Regression

Where are we going?

We know that lower pressure in storms is associated with higher wind speeds. In fact, for hurricanes, the correlation is −0.88. If we know that a hurricane has a central pressure of 860 mb, how high would we expect the winds to be? The correlation alone won't tell us the answer.

We need a model to be able to use one variable to predict another. Using linear regression, we can understand the relationship between two quantitative variables better and make predictions.

WHO	Items on the Burger King menu
WHAT	Protein content and total fat content
UNITS	Grams of protein Grams of fat
HOW	Supplied by BK on request or at their website

A S *Video:* **Manatees and Motorboats.** Are motorboats killing more manatees in Florida? Here's the story on video.

The Whopper™ has been Burger King's signature sandwich since 1957. One Double Whopper with cheese provides 53 grams of protein—all the protein you need in a day. It also supplies 1020 calories and 65 grams of fat. The Daily Value (based on a 2000-calorie diet) for fat is 65 grams. So after a Double Whopper you'll want the rest of your calories that day to be fat-free.[1]

Of course, the Whopper isn't the only item Burger King sells. How are fat and protein related on the entire BK menu? The scatterplot of the *Fat* (in grams) versus the *Protein* (in grams) for foods sold at Burger King shows a positive, moderately strong, linear relationship.

FIGURE 8.1
Total *Fat* versus *Protein* for 30 items on the BK menu. The Double Whopper is in the upper right corner. It's extreme, but is it out of line?

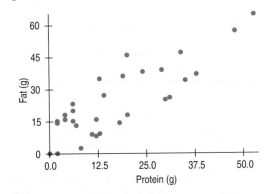

If you want 25 grams of protein in your lunch, how much fat should you expect to consume at Burger King? The correlation between *Fat* and *Protein*

---

[1] Sorry about the fries.

A S *Activity:* **Linear Equations.** For a quick review of linear equations, view this activity and play with the interactive tool.

"Statisticians, like artists, have the bad habit of falling in love with their models."

—George Box, famous statistician

A S *Activity:* **Residuals.** Residuals are the basis for fitting lines to scatterplots. See how they work.

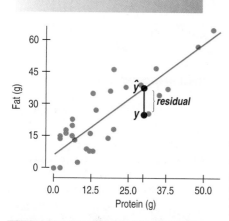

residual = observed value − predicted value.
A *negative* residual means the predicted value is too big—an **overestimate**. And a *positive* residual shows that the model makes an **underestimate**. These may seem backwards until you think about them.

A S *Activity:* **The Least Squares Criterion.** Does your sense of "best fit" look like the least squares line?

is 0.83, a sign that the linear association seen in the scatterplot is fairly strong. But *strength* of the relationship is only part of the picture. The correlation says, "The linear association between these two variables is fairly strong," but it doesn't tell us *what the line is*.

Of course, we *can* say more. We can **model** the relationship with a line and give its equation. The equation will let us predict the fat content for any Burger King food, given its amount of protein.

We met our first model in Chapter 6. We saw there that we can specify a Normal model with two parameters: its mean (μ) and standard deviation (σ).

For the Burger King foods, we'd choose a linear model to describe the relationship between *Protein* and *Fat*. The **linear model** is just an equation of a straight line through the data. Of course, no line can go through all the points, but a linear model can summarize the general pattern with only a couple of parameters. Like all models of the real world, the line will be wrong—wrong in the sense that it can't match reality *exactly*. But it can help us understand how the variables are associated.

Residuals

Not only can't we draw a line through all the points, the best line might not even hit *any* of the points. Then how can it be the "best" line? We want to find the line that somehow comes *closer* to all the points than any other line. Some of the points will be above the line and some below. For example, the line might suggest that a BK Broiler chicken sandwich with 30 grams of protein should have 36 grams of fat when, in fact, it actually has only 25 grams of fat. We call the estimate made from a model the **predicted value**, and write it as $\hat{y}$ (called *y-hat*) to distinguish it from the true value, *y*. The difference between the observed value and its associated predicted value is called the **residual.** The residual value tells us how far off the model's prediction is at that point. The BK Broiler chicken residual would be $y - \hat{y} = 25 - 36 = -11g$ of fat.

To find the residuals, we always subtract the predicted value from the observed one. The negative residual tells us that the actual fat content of the BK Broiler chicken is about 11 grams *less* than the model predicts for a typical Burger King menu item with 30 grams of protein.

Our challenge now is how to find the right line.

"Best Fit" Means Least Squares

When we draw a line through a scatterplot, some residuals are positive and some negative. We can't assess how well the line fits by adding up all the residuals—the positive and negative ones would just cancel each other out. We faced the same issue when we calculated a standard deviation to measure spread. And we deal with it the same way here: by squaring the residuals. Squaring makes them all positive. Now we can add them up. Squaring also emphasizes the large residuals. After all, points near the line are consistent with the model, but we're more concerned about points far from the line. When we add all the squared residuals together, that sum indicates how well the line we drew fits the data—the smaller the sum, the better the fit. A different line will produce a different sum, maybe bigger, maybe smaller. The **line of best fit** is the line for which the sum of the squared residuals is smallest, the **least squares** line.

You might think that finding this line would be pretty hard. Surprisingly, it's not, although it was an exciting mathematical discovery when Legendre published it in 1805:

WHO'S ON FIRST

In 1805, the French mathematician Adrien-Marie Legendre was the first to publish the "least squares" solution to the problem of fitting a line to data when the points don't all fall exactly on the line. The main challenge was how to distribute the errors "fairly." After considerable thought, he decided to minimize the sum of the squares of what we now call the residuals. When Legendre published his paper, though, the German mathematician Carl Friedrich Gauss claimed he had been using the method since 1795. Gauss later referred to the "least squares" solution as "*our* method" (*principium nostrum*), which certainly didn't help his relationship with Legendre.

The Linear Model

You may remember from Algebra that a straight line can be written as

$$y = mx + b.$$

We'll use this form for our linear model, but in Statistics we use slightly different notation:

$$\hat{y} = b_0 + b_1 x.$$

We write $\hat{y}$ (y-hat) to emphasize that the points that satisfy this equation are just our *predicted* values, not the actual data values (which scatter around the line). If the model is a good one, the data values will scatter closely around it.

We write b_1 and b_0 for the slope and intercept of the line. The b's are called the **coefficients** of the linear model. The coefficient b_1 is the **slope**, which tells how rapidly $\hat{y}$ changes with respect to x. The coefficient b_0 is the **intercept**, which tells where the line hits (intercepts) the y-axis.[2]

For the Burger King menu items, the best fit line is

$$\widehat{Fat} = 6.8 + 0.97\, Protein.$$

What does this mean? The slope, 0.97, says that a Burger King item with one more gram of protein can be expected, on average, to have 0.97 more grams of fat. Less formally, we might say that Burger King foods pack about 0.97 grams of fat per gram of protein. Slopes are always expressed in y-units per x-unit. They tell how the y-variable changes (in its units) for a one-unit change in the x-variable. When you see a phrase like "students per teacher" or "kilobytes per second," think slope.

How about the intercept, 6.8? Algebraically, that's the value the line takes when x is zero. Here, our model predicts that even a BK item with no protein would have, on average, about 6.8 grams of fat. Is that reasonable? Well, the apple pie, with 2 grams of protein, has 14 grams of fat, so it's not impossible. But often 0 is not a plausible value for x (the year 0, a baby born weighing 0 grams, . . .). Then the intercept serves only as a starting value for our predictions, and we don't interpret it as a meaningful predicted value.

NOTATION ALERT

"Putting a hat on it" is standard Statistics notation to indicate that something has been predicted by a model. Whenever you see a hat over a variable name or symbol, you can assume it is the predicted version of that variable or symbol (and look around for the model).

In a linear model, we use b_1 for the slope and b_0 for the y-intercept.

A S   *Simulation: Interpreting Equations.* This activity demonstrates how to use and interpret linear equations.

---

[2] We change from $mx + b$ to $b_0 + b_1 x$ for a reason: Eventually we'll want to add more x's to the model to make it more realistic, and we don't want to use up the entire alphabet. What would we use after m? The next letter is n and that one's already taken. o? See our point? Sometimes subscripts are the best approach.

A Linear Model for Hurricanes

In Chapter 7 we looked at the relationship between the central pressure and maximum wind speed of Atlantic hurricanes. We saw that the scatterplot was straight enough, and then found a correlation of -0.879, but we had no model to describe how these two important variables are related or to allow us to predict wind speed from pressure. Since the conditions we need to check to find a linear model are the same ones we checked before, we can use technology to find the model. It looks like this:

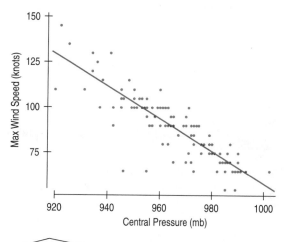

$$\widehat{MaxWindSpeed} = 955.27 - 0.897\,CentralPressure$$

QUESTION: Interpret this model. What does the slope mean in this context? Does the intercept have a meaningful interpretation?

The negative slope says that as CentralPressure falls, MaxWindSpeed increases. That makes sense from our general understanding of how hurricanes work: Low central pressure pulls in moist air, driving the rotation and the resulting destructive winds. The slope's value says that, on average, the maximum wind speed increases by about 0.897 knots for every 1-millibar drop in central pressure.

It's not meaningful, however, to interpret the intercept as the wind speed predicted for a central pressure of 0—that would be a vacuum. Instead, it is merely a starting value for the model.

The Least Squares Line

How do we find the actual values of slope and intercept for the least squares line? The formulas are simple. The model is built from the summary statistics you already know: the correlation (to tell us the strength of the linear association), the standard deviations (to give us the units), and the means (to tell us where to put the line). Remarkably, that's all you need to find the value of the slope:

$$b_1 = r\,\frac{s_y}{s_x}.$$

We've already seen that the correlation tells us the *sign* as well as the strength of the relationship, so it's no surprise that the slope inherits this sign as well.

Correlations don't have units, but slopes do. How x and y are measured—what units they have—doesn't affect their correlation, but can change their slope. If children grow an average of 3 inches per year, that's the same as 0.21 millimeters per day. For the slope, it matters whether you express age in days or years and whether you measure height in inches or millimeters.

SLOPE

$$b_1 = r\,\frac{s_y}{s_x}$$

UNITS OF Y PER UNIT OF X

Get into the habit of identifying the units by writing down "y-units per x-unit," with the unit names put in place. You'll find it'll really help you to Tell about the line in context.

Changing the units of x and y affects their standard deviations directly. And that's how the slope gets its units—from the ratio of the two standard deviations. So, the units of the slope are a ratio too. The units of the slope are always the units of y *per* unit of x.

What about the intercept? If you had to predict the y-value for a data point whose x-value was average, what would you say? The best fit line always predicts $\bar{y}$ for points whose x-value is $\bar{x}$. Putting that into our equation, we find

$$\bar{y} = b_0 + b_1\bar{x},$$

or, rearranging the terms,

$$b_0 = \bar{y} - b_1\bar{x}.$$

Knowing the slope and the fact that the line goes through the point $(\bar{x}, \bar{y})$ tells us how to find the intercept.

> **INTERCEPT**
>
> $$b_0 = \bar{y} - b_1\bar{x}$$

Here are summary statistics for the Burger King data:

PROTEIN	FAT
$\bar{x} = 17.2\,g$	$\bar{y} = 23.5\,g$
$s_x = 14.0\,g$	$s_y = 16.4\,g$
	$r = 0.83$

FOR EXAMPLE   Finding the Regression Equation

Let's try out the slope and intercept formulas on the "BK data" and see how to find the line. We checked the conditions when we calculated the correlation. We'll need the summary statistics (located conveniently in the margin). So, the slope is

$$b_1 = r\frac{s_y}{s_x} = 0.83 \times \frac{16.4\,g\ fat}{14.0\,g\ protein} = 0.97 \text{ grams of fat } per \text{ gram of protein.}$$

The intercept is

$$b_0 = \bar{y} - b_1\bar{x} = 23.5\,g\ fat - 0.97\frac{g\ fat}{g\ protein} \times 17.2\,g\ protein = 6.8\,g\ fat.$$

Putting these results back into the equation gives

$$\widehat{Fat} = 6.8\,g\ fat + 0.97\frac{g\ fat}{g\ protein}\ Protein,$$

or more simply,

$$\widehat{Fat} = 6.8 + 0.97\ Protein.$$

With an estimated linear model, it's easy to predict fat content for any menu item we want. For example, for the BK Broiler chicken sandwich with 30 grams of protein, we can plug in 30 grams for the amount of protein and see that the predicted fat content is $6.8 + 0.97(30) = 35.9$ grams of fat. Because the BK Broiler chicken sandwich actually has 25 grams of fat, its residual is

$$Fat - \widehat{Fat} = 25 - 35.9 = -10.9\,g.$$

Least squares lines are commonly called regression lines. In a few pages we'll see that this name is an accident of history. For now, you just need to know that "regression" almost always means "the linear model fit by least squares."

To use a regression model, we should check the same conditions for regressions as we did for correlation: the **Quantitative Variables Condition**, the **Straight Enough Condition**, and the **Outlier Condition**.

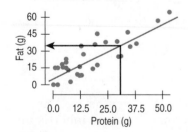

FIGURE 8.2
Burger King menu items with the regression line.

STEP-BY-STEP EXAMPLE   Calculating a Regression Equation

According to the director of the American Association of State Highway and Transportation Officials, "We just haven't been keeping up with the maintenance and preservation" of our highways and bridges. Concern for bridge safety mounted after some prominent bridge failures. Each state is responsible for inspecting its own bridges, and they use different standards and make available different amounts of information.

New York State uses a scale that runs from 1 to 7. Bridge inspectors evaluate, assign a condition score, and document the condition of up to 47 structural elements for each bridge. The Department of Transportation (DOT) combines these ratings of individual components using a formula that assigns greater weights to the ratings of the bridge elements having the greatest structural importance and lesser weights for minor structural and nonstructural elements. If a bridge has multiple spans, each element common to the spans is rated and the lowest individual span element rating is used in the condition rating formula. A bridge with a rating less than 5.0 is labeled "deficient."

New York has more than 17,000 bridges. Scores are reported by counties. We have available data on the 194 bridges of Tompkins County (coincidentally, the home county of two of our coauthors.)

One natural concern is the age of a bridge. A model that relates age to score might help the DOT to focus inspector efforts where they are most needed.

Question: Is there a relationship between the age of a bridge and its safety rating?

THINK	**Plan** State the problem.	I want to know whether there is a relationship between the age of a bridge in Tompkins County, NY, and its safety rating.
	Variables Identify the variables and report the W's.	I have data giving the *Condition*, year of inspection, and year of construction for 194 bridges constructed since 1865. From the year of construction and the year of inspection, I will calculate the *Age* when the bridge was inspected.

Just as we did for correlation, check the conditions for a regression by making a picture. Never fit a regression without looking at the scatterplot first.

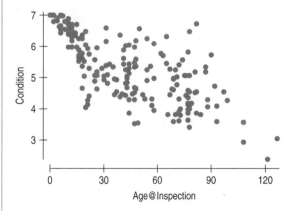

✔ **Quantitative Variables Condition:** Both the *Condition* rating and the *Age* are quantitative. The condition values are quantitative, but have no units because they combine ratings of various bridge components.

✔ **Straight Enough Condition:** The scatterplot shows a moderately strong linear relationship with a negative association.

✔ **Outlier Condition:** No outliers are evident in this scatterplot.

Because these conditions are satisfied, it is OK to model the relationship with a regression model.

Mechanics Find the equation of the regression line. Summary statistics give the building blocks of the calculation.

(We generally report summary statistics to one more digit of accuracy than the data. We do the same for intercept and predicted values, but for slopes we usually report an additional digit. Remember, though, not to round off until you finish computing an answer.)[3]

Find the slope, b_1.

Find the intercept, b_0.

Write the equation of the model, using meaningful variable names.

Age

$\bar{x} = 44.9$

$s_x = 30.75$

Condition Index

$\bar{y} = 5.2779$

$s_y = 1.0297$

Correlation:

$r = -0.681$

$$b_1 = r\frac{s_y}{s_x} = -0.681\frac{1.0297}{30.75} = -0.0228$$

$$\text{condition points per year}$$

$$b_0 = \bar{y} - b_1\bar{x} = 5.2779 - (-0.0228)44.9$$
$$= 6.30$$

The least squares line is

$$\hat{y} = 6.30 - 0.0228x$$

or

$$\widehat{Condition} = 6.30 - 0.0228\,Age$$

Conclusion Interpret what you have found in the context of the question. Discuss in terms of the variables and their units.

The condition of the bridges in Tompkins County, NY, decreases with the age of the bridges at the time of inspection. Bridge condition declines by about 0.023 on the condition scale from 1 to 7 for each year of age. The model uses a base of 6.3, which is quite reasonable because a new bridge (0 years of age) should have a condition near the maximum of 7.

Because I have only data from one county, I can't tell from these data whether this model would apply to bridges in other counties of New York or in other locations.

A S

Activity: Find a Regression Equation.
Now that we've done it by hand, try it with technology using the statistics package paired with your version of *ActivStats*.

---

[3] We warned you in Chapter 6 that we'll round in the intermediate steps of a calculation to show the steps more clearly. If you repeat these calculations yourself on a calculator or statistics program, you may get somewhat different results. When calculated with more precision, the intercept is 6.40495 and the slope is −0.02565.

Correlation and the Line

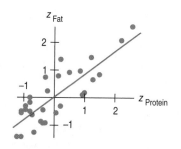

FIGURE 8.3
The Burger King scatterplot in z-scores.

In Chapter 7, we learned a lot about how correlation worked by looking at a scatterplot of the standardized variables. Let's see what else standardizing can tell us. Here's a scatterplot for the BK items of z_y (standardized *Fat*) vs. z_x (standardized *Protein*) along with their least squares line. What's the slope of this line?

We know that the slope $b_1 = r\frac{s_y}{s_x}$, but here we are working with standardized variables whose standard deviations are both 1. So for the regression of z_y on z_x, $b_1 = r$.

What about the intercept? Look at the plot. We know that $b_0 = \bar{y} - b_1\bar{x}$. But standardized variables have zero means, so we just get $b_0 = 0$; the line goes through the origin.

Wow! This line has an equation that's about as simple as we could possibly hope for:

$$\hat{z}_y = rz_x.$$

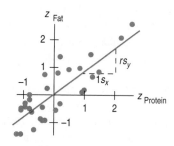

FIGURE 8.4
Standardized *Fat* vs. standardized *Protein* with the regression line. Each one-standard-deviation change in *Protein* results in a predicted change of r standard deviations in *Fat*.

Great. It's simple, but what does it tell us? It says that in moving one standard deviation from the mean in x, we can expect to move about r standard deviations away from the mean in y. Now that we're thinking about regression lines, the correlation is more than just a vague measure of strength of association: It's a great way to think about what regression tells us.

Let's be more specific. For the BK menu items, the correlation is 0.83. If we standardize both protein and fat, we can write

$$\hat{z}_{Fat} = 0.83 z_{Protein}.$$

We ordinarily don't standardize variables to perform a regression, but it's useful to think about what this equation says. It tells us that for every standard deviation above (or below) the mean a menu item is in protein, we'd predict that its fat content is 0.83 standard deviations above (or below) the mean fat content.

A double hamburger has 31 grams of protein, about 1 SD above the mean. How much fat do you expect it to have? Putting 1.0 in for z_x in the model gives a $\hat{z}_y$ value of 0.83. If you trust the model, you'd expect the fat content to be about 0.83 fat SDs above the mean fat level. In general, menu items that are one standard deviation away from the mean in x are, on average, r standard deviations away from the mean in y.

If $r = 0$, there's no linear relationship. No matter how many standard deviations you move in x, the predicted value for y doesn't change. On the other hand, if $r = 1.0$ or -1.0, there's a perfect linear association. In that case, moving any number of standard deviations in x moves the prediction exactly the same number of standard deviations in y.

JUST CHECKING

A scatterplot of house *Price* (in thousands of dollars) vs. house *Size* (in thousands of square feet) for houses sold recently in Saratoga, NY, shows a relationship that is straight, with only moderate scatter and no outliers. The correlation between house *Price* and house *Size* is 0.77, and the equation of the regression model is

$$\widehat{Price} = -3.117 + 94.45\,Size.$$

1. What does the slope of 94.45 mean?

2. What are the units of the slope?

3. Your house in Saratoga is 2000 sq ft bigger than your neighbor's house. How much more do you expect it to be worth?

How Big Can Predicted Values Get?

Suppose you were told that a new male student was about to join the class, and you were asked to guess his height in inches. What would be your guess? A good guess would be the mean height of male students. Now suppose you are also told that this student has a grade point average (*GPA*) of 3.9—about 2 SDs above the mean *GPA*. Would that change your guess? Probably not. There's no association between *GPA* and *Height*, so knowing the *GPA* value doesn't tell you anything and doesn't move your guess. (And the formula tells us that as well, since it says that we should guess 0×2 SDs from the mean.)

On the other hand, if you were told that, measured in centimeters, the student's height was 2 SDs above the mean, you'd know his height in inches. There's a perfect correlation between *Height in Inches* and *Height in Centimeters*, so you know he's 2 SDs above mean height in inches as well. (The formula would tell us to guess 1.0×2 SDs from the mean.)

What if you're told that the student is 2 SDs above the mean in *Shoe Size*? Would you still guess that he's of average *Height*? You might guess that he's taller than average, since there's a positive correlation between *Height* and *Shoe Size*. But would you guess that he's 2 SDs above the mean? When there was no correlation, we didn't move away from the mean at all. With a perfect correlation, we moved our guess the full 2 SDs. Any correlation between these extremes should lead us to guess somewhere between 0 and 2 SDs above the mean. (To be exact, the formula tells us to guess $r \times 2$ standard deviations away from the mean.)

You won't ever guess more than 2 SDs away, since r can't be bigger than 1.0. So, each predicted y tends to be closer to its mean (in standard deviations) than its corresponding x was. This property of the linear model is called **regression to the mean**, and *that's* where we got the term **regression line**.

Sir Francis Galton was the first to speak of "regression," although others had fit lines to data by the same method.

THE FIRST REGRESSION

Sir Francis Galton related the heights of sons to the heights of their fathers with a regression line. The slope of his line was less than 1. That is, sons of tall fathers were tall, but not as much above the average height as their fathers had been above their mean. Sons of short fathers were short, but generally not as far from their mean as their fathers. Galton interpreted the slope correctly as indicating a "regression" toward the mean height—and "regression" stuck as a description of the method he had used to find the line.

MATH BOX

Where does the equation of the line of best fit come from? And what does it really mean?

We can write the equation of any line if we know a point on the line and the slope. Finding a point on the line isn't hard. For the BK sandwiches, it's logical to predict that a sandwich with average protein will have average fat. In fact, it's fairly easy to show that the best fit line *always* passes through the point $(\bar{x}, \bar{y})$.

To find the slope, we'll again convert both x and y to z-scores. We need to keep a few facts in mind:

1. The mean of any set of z-scores is 0. Since we know the line passes through $(\bar{x}, \bar{y})$ this tells us that the line of best fit for z-scores passes through the origin $(0, 0)$.

2. The standard deviation of a set of z-scores is 1, and so, of course, are the variances. This means, in particular, that

$$\frac{\sum (z_y - \bar{z}_y)^2}{(n-1)} = \frac{\sum (z_y - 0)^2}{(n-1)} = \frac{\sum z_y^2}{(n-1)} = 1,$$

a fact that will be important soon.

3. Remember from Chapter 7, the correlation is $r = \dfrac{\sum z_x z_y}{(n-1)}$, also important soon.

We want to find the slope of the best fit line. Because it passes through the origin, its equation will be of the simple form $\hat{z}_y = m z_x$. We want to find the value of m that minimizes the sum of the squared residuals $\sum (z_y - \hat{z}_y)^2$. Because it actually makes things simpler, we'll divide that sum by $(n-1)$ and minimize this quantity instead.

Minimize: $\dfrac{\sum (z_y - \hat{z}_y)^2}{(n-1)}$

Since $\hat{z}_y = m z_x$, we are minimizing $\dfrac{\sum (z_y - m z_x)^2}{(n-1)}$

We can expand this and write it as: $\dfrac{\sum z_y^2}{(n-1)} - 2m \dfrac{\sum z_x z_y}{(n-1)} + m^2 \dfrac{\sum z_x^2}{(n-1)}$

4. Notice that the first term is 1 (because it's the variance of the z_y), the second term is $-2mr$ and the last term is m^2 (because the variance of z_x is 1 as well). So to minimize the sum of squared residuals we only have to minimize: $1 - 2mr + m^2$. This last expression is quadratic, which, as you may remember, describes a parabola. And a parabola of the form $y = ax^2 + bx + c$ reaches its minimum at its turning point, which occurs when $x = -\dfrac{b}{2a}$. So, for this particular quadratic, we can minimize the sum of the squared residuals by choosing $m = -\dfrac{-(2r)}{2(1)} = r$.

This is really an amazing fact and explains to some extent the importance of the correlation as a summary of the strength of the linear relationship between x and y. To say it again: The slope of the best fit line for the z-scores (the standardized values) of any two variables x and y is their correlation, r.

We've converted to z-scores to make the mathematics easier, but what does this result mean for any two variables x and y?

- A slope of r for the best fit line for z-scores means that if we move 1 standard deviation in z_x, we expect to move r standard deviations in z_y. So a BK item that is one standard deviation above the mean in protein would be expected to be r standard deviations above the mean in fat. And a sandwich that is 2 standard deviations higher than the mean in protein would be expected to be $2 \times r$ standard deviations higher than the mean in fat. If the correlation were 1, we'd expect it to be a full 2 standard deviations higher in fat. On the other hand, if the correlation were 0, we'd predict it to have the mean fat content. For the Burger King items, the correlation actually is 0.83, so for a sandwich 2 standard deviations higher than the mean in protein, we'd predict its fat content to be 1.66 standard deviations higher than the average fat content.

  Rather than just saying that we'd predict a BK sandwich to be r standard deviations above the mean in fat for each standard deviation above the mean in protein, we can write the relationship in the original units by multiplying by the ratio of the standard deviations. The slope of the regression line is then $b = r\dfrac{s_y}{s_x}$.

- We know that choosing $m = r$ minimizes the sum of the squared residuals, but how small does that sum get? Equation 4 told us that the sum of the squared residuals divided by $(n-1)$ is $1 - 2mr + m^2$. When $m = r$, this quantity is $1 - 2r \times r + r^2 = 1 - r^2$. Because it's a sum of squares it has to be positive, so $1 - r^2 \geq 0$. That means that $r^2 \leq 1$, and therefore, $-1 \leq r \leq 1$. In other words, it shows that the correlation is always between -1 and $+1$.

The quantity r^2 is also important in its own right. Look back at the expression $\dfrac{\sum(z_y - \hat{z}_y)^2}{(n-1)}$. Compare it to the variance of the z_y: $Var(z_y) = \dfrac{\sum(z_y - \bar{z}_y)^2}{(n-1)}$. Almost the same, aren't they? The variance of z_y measures its variation from the mean. The sum of the residuals measures variation from the predicted values, from the regression line. The variance of z_y is 1. Now that we've minimized the variation around the regression line it's only $1 - r^2$. In other words, the regression line has removed (or accounted for) r^2 of the original variance in y. Knowing how much variance the regression removes helps us assess the strength of our model. We'll use r^2 to help us do just that.

JUST CHECKING

Remember the homes in Saratoga? Our analysis found a correlation of $r = 0.77$ between the house *Price* (in thousands of dollars) and the house *Size* (in thousands of square feet).

4. You go to an open house and find that the house is 1 standard deviation above the mean in size. What would you guess about its price?

5. You read an ad for a house priced 2 standard deviations below the mean. What would you guess about its size?

6. A friend tells you about a house whose size in square meters (he's European) is 1.5 standard deviations above the mean. What would you guess about its size in square feet?

Residuals Revisited

The linear model we are using assumes that the relationship between the two variables is a perfect straight line. The residuals are the part of the data that *hasn't* been modeled. We can write

$$Data = Model + Residual$$

or, equivalently,

$$Residual = Data - Model.$$

Or, in symbols,

$$e = y - \hat{y}.$$

When we want to know how well the model fits, we can ask instead what the model missed. To see that, we look at the residuals.

WHY *e* FOR "RESIDUAL"?

The flip answer is that *r* is already taken, but the truth is that *e* stands for "error." No, that doesn't mean it's a mistake. Statisticians often refer to variability not explained by a model as error.

FOR EXAMPLE | **Katrina's Residual**

RECAP: The linear model relating hurricanes' wind speeds to their central pressures was

$$\widehat{MaxWindSpeed} = 955.27 - 0.897\,CentralPressure$$

Let's use this model to make predictions and see how those predictions do.

QUESTION: Hurricane Katrina had a central pressure measured at 920 millibars. What does our regression model predict for her maximum wind speed? How good is that prediction, given that Katrina's actual wind speed was measured at 110 knots?

Substituting 920 for the central pressure in the regression model equation gives

$$\widehat{MaxWindSpeed} = 955.27 - 0.897(920) = 130.03$$

(continued)

The regression model predicts a maximum wind speed of 130 knots for Hurricane Katrina.

The residual for this prediction is the observed value minus the predicted value:

$$110 - 130 = -20kts.$$

In the case of Hurricane Katrina, the model predicts a wind speed 20 knots higher than was actually observed.

Residuals help us to see whether the model makes sense. When a regression model is appropriate, it should model the underlying relationship. Nothing interesting should be left behind. So after we fit a regression model, we usually plot the residuals in the hope of finding... nothing.

FIGURE 8.5
The residuals for the BK menu regression look appropriately boring.

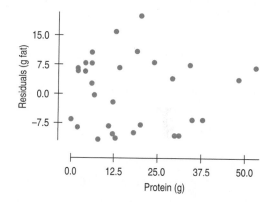

A scatterplot of the residuals versus the *x*-values should be the most boring scatterplot you've ever seen. It shouldn't have any interesting features, like a direction or shape. It should stretch horizontally, with about the same amount of scatter throughout. It should show no bends, and it should have no outliers. If you see any of these features, find out what the regression model missed.

Most computer statistics packages plot the residuals against the predicted values $\hat{y}$, rather than against x. When the slope is negative, the two versions are mirror images. When the slope is positive, they're virtually identical except for the axis labels. Since all we care about is the patterns (or, better, lack of patterns) in the plot, it really doesn't matter which way we plot the residuals.

The Residual Standard Deviation

If the residuals show no interesting pattern when we plot them against x, we can look at how big they are. After all, we're trying to make them as small as possible. Since their mean is always zero, though, it's only sensible to look at how much they vary. The standard deviation of the residuals, s_e, gives us a measure of how much the points spread around the regression line. Of course, for this summary to make sense, the residuals should all share the same underlying spread. That's why we check to make sure that the residual plot has about the same amount of scatter throughout.

This gives us a new assumption: the **Equal Variance Assumption**. The associated condition to check is the **Does the Plot Thicken? Condition**. We check to make sure that the spread is about the same throughout. We can check that either in the original scatterplot of y against x or in the scatterplot of residuals.

We estimate the standard deviation of the residuals in almost the way you'd expect:

$$s_e = \sqrt{\frac{\Sigma e^2}{n - 2}}.$$

We don't need to subtract the mean because the mean of the residuals $\bar{e} = 0$.

Why $n - 2$ rather than $n - 1$? We used $n - 1$ for s when we estimated the mean. Now we're estimating both a slope and an intercept. Looks like a pattern—and it is. We subtract one more for each parameter we estimate.

For the Burger King foods, the standard deviation of the residuals is 9.2 grams of fat. That looks about right in the scatterplot of residuals. The residual for the BK Broiler chicken was −11 grams, just over one standard deviation.

It's a good idea to make a histogram of the residuals. If we see a unimodal, symmetric histogram, then we can apply the 68–95–99.7 Rule to see how well the regression model describes the data. In particular, we know that 95% of the residuals should be no larger in size than $2s_e$. The Burger King residuals look like this:

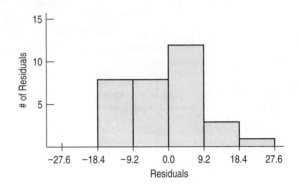

Sure enough, most are less than 2(9.2), or 18.4, g of fat in size.

R^2–The Variation Accounted For

The variation in the residuals is the key to assessing how well the model fits. Let's compare the variation of the response variable with the variation of the residuals. The total *Fat* has a standard deviation of 16.4 grams. The standard deviation of the residuals is 9.2 grams. If the correlation were 1.0 and the model predicted the *Fat* values perfectly, the residuals would all be zero and have no variation. We couldn't possibly do any better than that.

On the other hand, if the correlation were zero, the model would simply predict 23.5 grams of *Fat* (the mean) for all menu items. The residuals from that prediction would just be the observed *Fat* values minus their mean. These residuals would have the same variability as the original data because, as we know, just subtracting the mean doesn't change the spread.

A S *Simulation:* **Interpreting Equations.** This demonstrates how to use and interpret linear equations.

FIGURE 8.6
Compare the variability of total *Fat* with the residuals from the regression. The means have been subtracted to make it easier to compare spreads. The variation left in the residuals is unaccounted for by the model, but it's less than the variation in the original data.

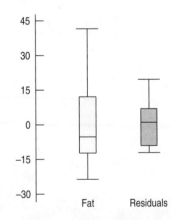

How well does the BK regression model do? Look at the boxplots. The variation in the residuals is smaller than in the data, but certainly bigger than zero. That's nice to know, but how much of the variation is still left in the residuals? If you had to put a number between 0% and 100% on the fraction of the variation left in the residuals, what would you say?

All regression models fall somewhere between the two extremes of zero correlation and perfect correlation. We'd like to gauge where our model falls. Can we use the correlation to do that? Well, a regression model with correlation −0.5 is doing as well as one with correlation +0.5. They just have different directions. But if we *square* the correlation coefficient, we'll get a value between 0 and 1, and the direction won't matter. The squared correlation, r^2, gives the fraction of the data's variation accounted for by the model, and $1 − r^2$ is the fraction of the original variation left in the residuals. For the Burger King model, $r^2 = 0.83^2 = 0.69$, and $1 − r^2$ is 0.31, so 31% of the variability in total *Fat* has been left in the residuals. How close was that to your guess?

All regression analyses include this statistic, although by tradition, it is written with a capital letter, $\boldsymbol{R^2}$, and pronounced "R-squared." An R^2 of 0 means that none of the variance in the data is in the model; all of it is still in the residuals. It would be hard to imagine using that model for anything.

Because R^2 is a fraction of a whole, it is often given as a percentage.[4] For the Burger King data, R^2 is 69%.

When interpreting a regression model, you need to *Tell* what R^2 means. According to our linear model, 69% of the variability in the fat content of Burger King sandwiches is accounted for by variation in the protein content.

> Is a correlation of 0.80 twice as strong as a correlation of 0.40? Not if you think in terms of R^2. A correlation of 0.80 means an R^2 of $0.80^2 = 64\%$. A correlation of 0.40 means an R^2 of $0.40^2 = 16\%$—only a quarter as much of the variability accounted for. A correlation of 0.80 gives an R^2 *four* times as strong as a correlation of 0.40 and accounts for four times as much of the variability.

> **How can we see that R^2 is really the fraction of variance accounted for by the model?** It's a simple calculation. The variance of the fat content of the Burger King foods is $16.4^2 = 268.96$. If we treat the residuals as data, the variance of the residuals is 83.195.[5] As a fraction, that's $83.195/268.96 = 0.31$, or 31%. That's the fraction of the variance that is *not* accounted for by the model. The fraction that is accounted for is $100\% − 31\% = 69\%$, just the value we got for R^2.

FOR EXAMPLE   Interpreting R^2

RECAP: Our regression model that predicts maximum wind speed in hurricanes based on the storm's central pressure has $R^2 = 77.3\%$.

QUESTION: What does that say about our regression model?

An R^2 of 77.3% indicates that 77.3% of the variation in maximum wind speed can be accounted for by the hurricane's central pressure. Other factors, such as temperature and whether the storm is over water or land, may account for some of the remaining variation.

JUST CHECKING

Back to our regression of house *Price* (in thousands of $) on house *Size* (in thousands of square feet):

$$\widehat{Price} = -3.117 + 94.45\,Size.$$

The R^2 value is reported as 59.5%, and the standard deviation of the residuals is 53.79.

7. What does the R^2 value mean about the relationship of price and size?

8. Is the correlation of price and size positive or negative? How do you know?

9. You find that your house is worth $100,000 more than the regression model predicts. Should you be very surprised (as well as pleased)?

---

[4] By contrast, we usually give correlation coefficients as decimal values between −1.0 and 1.0.
[5] This isn't quite the same as squaring the s_e that we discussed on the previous pages, but it's very close. We'll deal with the distinction in Chapter 27.

How Big Should R^2 Be?

R^2 is always between 0% and 100%. But what's a "good" R^2 value? The answer depends on the kind of data you are analyzing and on what you want to do with it. Just as with correlation, there is no value for R^2 that automatically determines that the regression is "good." Data from scientific experiments often have R^2 in the 80% to 90% range and even higher. Data from observational studies and surveys, though, often show relatively weak associations because it's so difficult to measure responses reliably. An R^2 of 50% to 30% or even lower might be taken as evidence of a useful regression. The standard deviation of the residuals can give us more information about the usefulness of the regression by telling us how much scatter there is around the line.

As we've seen, an R^2 of 100% is a perfect fit, with no scatter around the line. The s_e would be zero. All of the variance is accounted for by the model and none is left in the residuals at all. This sounds great, but it's too good to be true for real data.[6]

Along with the slope and intercept for a regression, you should always report R^2 so that readers can judge for themselves how successful the regression is at fitting the data. Statistics is about variation, and R^2 measures the success of the regression model in terms of the fraction of the variation of y accounted for by the regression. R^2 is the first part of a regression that many people look at because, along with the scatterplot, it tells whether the regression model is even worth thinking about.

A Tale of Two Regressions

Regression slopes may not behave exactly the way you'd expect at first. Our regression model for the Burger King sandwiches was $\widehat{Fat} = 6.8 + 0.97\,Protein$. That equation allowed us to estimate that a sandwich with 30 grams of protein would have 35.9 grams of fat. Suppose, though, that we knew the fat content and wanted to predict the amount of protein. It might seem natural to think that by solving our equation for *Protein* we'd get a model for predicting *Protein* from *Fat*. But that doesn't work.

Our original model is $\hat{y} = b_0 + b_1 x$, but the new one needs to evaluate an $\hat{x}$ based on a value of y. We don't have y in our original model, only $\hat{y}$, and that makes all the difference. Our model doesn't fit the BK data values perfectly, and the least squares criterion focuses on the *vertical* (y) errors the model makes in using x to model y—not on *horizontal* errors related to x.

A quick look at the equations reveals why. Simply solving our equation for x would give a new line whose slope is the reciprocal of ours. To model y in terms of x, our slope is $b_1 = r\frac{s_y}{s_x}$. To model x in terms of y, we'd need to use the slope $b_1 = r\frac{s_x}{s_y}$. That's *not* the reciprocal of ours.

Sure, if the correlation, r, were 1.0 or -1.0 the slopes *would* be reciprocals, but that would happen only if we had a perfect fit. Real data don't follow perfect straight lines, so in the real world y and $\hat{y}$ aren't the same, r is a fraction, and the slopes of the two models are not simple reciprocals of one another. Also, if the standard deviations were equal—for example, if we standardize both variables—the two slopes would be *the same*. Far from being reciprocals, both would be equal to the correlation—but we already knew that the correlation of x with y is the same as the correlation of y with x.

Otherwise, slopes of the two lines will not be reciprocals, so we can't derive one equation from the other. If we want to predict *Protein* from *Fat*, we

Protein	Fat
$\bar{x} = 17.2\,g$	$\bar{y} = 23.5\,g$
$s_x = 14.0\,g$	$s_y = 16.4\,g$
$r = 0.83$	

[6] If you see an R^2 of 100%, it's a good idea to figure out what happened. You may have discovered a new law of Physics, but it's much more likely that you accidentally regressed two variables that measure the same thing.

need to create that model. The slope is $b_1 = 0.83 \frac{14.0}{16.4} = 0.709$ grams of protein per gram of fat. The equation turns out to be $\widehat{Protein} = 0.55 + 0.709\ Fat$, so we'd predict that a sandwich with 35.9 grams of fat should have 26.0 grams of protein—not the 30 grams that we used in the first equation.

Moral of the story: *Think.* (Where have you heard *that* before?) Decide which variable you want to use (x) to predict values for the other (y). Then find the model that does that. If, later, you want to make predictions in the other direction, you'll need to start over and create the other model from scratch.

Regression Assumptions and Conditions

The linear regression model is perhaps the most widely used model in all of Statistics. It has everything we could want in a model: two easily estimated parameters, a meaningful measure of how well the model fits the data, and the ability to predict new values. It even provides a self-check in plots of the residuals, to help us avoid silly mistakes.

Like all models, though, linear models don't apply all the time, so we'd better think about whether they're reasonable. It makes no sense to make a scatterplot of categorical variables, and even less to perform a regression on them. Always check the **Quantitative Variables Condition** to be sure a regression is appropriate.

The linear model makes several assumptions. First, and foremost, is the **Linearity Assumption**—that the relationship between the variables is, in fact, linear. You can't verify an assumption, but you can check the associated condition. A quick look at the scatterplot will help you check the **Straight Enough Condition**. You don't need a *perfectly* straight plot, but it must be straight enough for the linear model to make sense. If you try to model a curved relationship with a straight line, you'll usually get exactly what you deserve.

If the scatterplot is not straight enough, stop here. You can't use a linear model for *any* two variables, even if they are related. They must have a *linear* association, or the model won't mean a thing.

For the standard deviation of the residuals to summarize the scatter, all the residuals should share the same spread. The most common pattern in the residual spread is for the residuals to spread out more for larger values of $\hat{y}$. We can examine the scatterplot of y vs. x and plot the residuals against $\hat{y}$. Then we check for such a "thickening" with the **Does the Plot Thicken? Condition**.

Check the **Outlier Condition**. Outlying points can dramatically change a regression model. Outliers can even change the sign of the slope, misleading us about the underlying relationship between the variables. We'll see examples in the next chapter.

Even though we've checked the conditions in the scatterplot of the data, a scatterplot of the residuals can sometimes help us see any violations even more clearly. And examining the residuals is the best way to look for additional patterns and interesting quirks in the data.

MAKE A PICTURE

To use regression, first check that

- the scatterplot is straight enough.

After you've fit the regression, make a residual plot and check that there are no obvious patterns. In particular, check that

- there are no obvious bends,
- the spread of the residuals is about the same throughout, and
- there are no obvious outliers.

STEP-BY-STEP EXAMPLE   **Regression**

If you hit the fast-food joints for lunch, you should have a good breakfast. Nutritionists, concerned about "empty calories" in breakfast cereals, recorded facts about 77 cereals, including their *Calories* per serving and *Sugar* content (in grams).

Question: How are calories and sugar content related in breakfast cereals?

Plan State the problem.

I am interested in the relationship between sugar content and calories in cereals.

Variables Name the variables and report the W's.

✔ **Quantitative Variables Condition:** I have two quantitative variables, Calories and Sugar content per serving, measured on 77 breakfast cereals. The units of measurement are calories and grams of sugar, respectively.

Check the conditions for a regression by making a picture. Never fit a regression without looking at the scatterplot first.

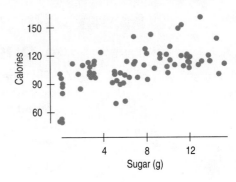

✔ **Outlier Condition:** There are no obvious outliers or groups.

✔ The **Straight Enough Condition** is satisfied; I will fit a regression model to these data.

✔ The **Does the Plot Thicken? Condition** is satisfied. The spread in the x-y scatterplot around the line looks about the same throughout, but I'll check it again in the residuals.

Mechanics If there are no clear violations of the conditions, fit a straight line model of the form $\hat{y} = b_0 + b_1 x$ to the data. Summary statistics give the building blocks of the calculation.

Calories
$$\bar{y} = 107.0 \text{ calories}$$
$$s_y = 19.5 \text{ calories}$$

Sugar
$$\bar{x} = 7.0 \text{ grams}$$
$$s_x = 4.4 \text{ grams}$$

Correlation
$$r = 0.564$$

Find the slope.

$$b_1 = r\frac{s_y}{s_x} = 0.564\frac{19.5}{4.4}$$
$$= 2.50 \text{ calories per gram of sugar.}$$

Find the intercept.

$$b_0 = \bar{y} - b_1\bar{x} = 107 - 2.50(7) = 89.5 \text{ calories.}$$

So the least squares line is
$$\hat{y} = 89.5 + 2.50\,x,$$

Write the equation, using meaningful variable names.

or $\widehat{Calories} = 89.5 + 2.50 \text{ Sugar.}$

State the value of R^2.

Squaring the correlation gives
$$R^2 = 0.564^2 = 0.318 \text{ or } 31.8\%.$$

Conclusion Describe what the model says in words and numbers. Be sure to use the names of the variables and their units.

The key to interpreting a regression model is to start with the phrase "b_1 y-units per x-unit," substituting the estimated value of the slope for b_1 and the names of the respective units. The intercept is then a starting or base value.

R^2 gives the fraction of the variability of y accounted for by the linear regression model.

Find the standard deviation of the residuals, s_e, and compare it to the original, s_y.

The scatterplot shows a positive, linear relationship and no outliers. The least squares regression line fit through these data has the equation

$$\widehat{Calories} = 89.5 + 2.50\ Sugar.$$

The slope says that cereals have about 2.50 more *Calories* per gram of *Sugar*.

The intercept predicts that sugar-free cereals would average about 89.5 calories.

The R^2 says that 31.8% of the variability in *Calories* is accounted for by variation in *Sugar* content.

$s_e = 16.2$ calories. That's smaller than the original SD of 19.5, but still fairly large.

Check Again Even though we looked at the scatterplot *before* fitting a regression model, a plot of the residuals is essential to any regression analysis because it is the best check for additional patterns and interesting quirks in the data.

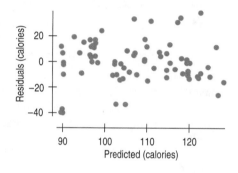

The residuals show a horizontal direction, a shapeless form, and roughly equal scatter for all predicted values. The linear model appears to be appropriate.

Reality Check: Is the Regression Reasonable?

Statistics don't come out of nowhere. They are based on data. The results of a statistical analysis should reinforce your common sense, not fly in its face. If the results are surprising, then either you've learned something new about the world or your analysis is wrong.

Whenever you perform a regression, think about the coefficients and ask whether they make sense. Is a slope of 2.5 calories per gram of sugar reasonable? That's hard to say right off. We know from the summary statistics that a typical cereal has about 100 calories and 7 grams of sugar per serving. A gram of sugar contributes some calories (actually, 4, but you don't need to know that), so calories should go up with increasing sugar. The direction of the slope seems right.

To see if the *size* of the slope is reasonable, a useful trick is to consider its order of magnitude. We'll start by asking if deflating the slope by a factor of 10 seems reasonable. Is 0.25 calories per gram of sugar enough? The 7 grams of sugar found in the average cereal would contribute less than 2 calories. That seems too small.

Now let's try inflating the slope by a factor of 10. Is 25 calories per gram reasonable? Then the average cereal would have 175 calories from sugar alone.

The average cereal has only 100 calories per serving, though, so that slope seems too big.

We have tried inflating the slope by a factor of 10 and deflating it by 10 and found both to be unreasonable. So, like Goldilocks, we're left with the value in the middle that's just right. And an increase of 2.5 calories per gram of sugar is certainly *plausible*.

The small effort of asking yourself whether the regression equation is plausible is repaid whenever you catch errors or avoid saying something silly or absurd about the data. It's too easy to take something that comes out of a computer at face value and assume that it makes sense.

Always be skeptical and ask yourself if the answer is reasonable.

FOR EXAMPLE Causation and Regression

RECAP: Our regression model predicting hurricane wind speeds from the central pressure was reasonably successful. The negative slope indicates that, in general, storms with lower central pressures have stronger winds.

QUESTION: Can we conclude that lower central barometric pressure *causes* the higher wind speeds in hurricanes?

No. While it may be true that lower pressure causes higher winds, a regression model for observed data such as these cannot demonstrate causation. Perhaps higher wind speeds reduce the barometric pressure, or perhaps both pressure and wind speed are both driven by some other variable we have not observed.

(As it happens, in hurricanes it is reasonable to say that the low central pressure at the eye is responsible for the high winds because it draws moist, warm air into the center of the storm, where it swirls around, generating the winds. But as is often the case, things aren't quite that simple. The winds themselves contribute to lowering the pressure at the center of the storm as it becomes organized into a hurricane. The lesson is that to understand causation in hurricanes, we must do more than just model the relationship of two variables; we must study the mechanism itself.)

What Can Go Wrong?

There are many ways in which data that appear at first to be good candidates for regression analysis may be unsuitable. And there are ways that people use regression that can lead them astray. Here's an overview of the most common problems. We'll discuss them at length in the next chapter.

- **Don't fit a straight line to a nonlinear relationship.** Linear regression is suited only to relationships that are, well, *linear*. Fortunately, we can often improve the linearity easily by using re-expression. We'll come back to that topic in Chapter 10.

- **Beware of extraordinary points.** Data points can be extraordinary in a regression in two ways: They can have y-values that stand off from the linear pattern suggested by the bulk of the data, or extreme x-values. Both kinds of extraordinary points require attention.

- **Don't invert the regression.** The BK regression model was $\widehat{Fat} = 6.8 + 0.97\ Protein$. Knowing protein content, we can predict the amount of fat. But that doesn't let us switch the regression around. We can't use this model to predict protein values from fat values. To model y from x, the least squares slope is $b_1 = r\frac{s_y}{s_x}$. To model x in terms of y, we'd find $b_1 = r\frac{s_x}{s_y}$. That's not the reciprocal of the first slope (unless the correlation is 1.0). To swap the predictor–response roles of the variables in a regression (which can sometimes make sense), we must fit a new regression equation.

- **Don't extrapolate beyond the data.** A linear model will often do a reasonable job of summarizing a relationship in the narrow range of observed x-values. Once we have a working model for the relationship, it's tempting to use it. But beware of predicting y-values for x-values that lie outside the range of the original data. The model may no longer hold there, so such *extrapolations* too far from the data are dangerous.

- **Don't infer that x causes y just because there is a good linear model for their relationship.** When two variables are strongly correlated, it is often tempting to assume a causal relationship between them. Putting a regression line on a scatterplot tempts us even further, but it doesn't make the assumption of causation any more valid.

- **Don't choose a model based on R^2 alone.** Although R^2 measures the *strength* of the linear association, a high R^2 does not demonstrate the *appropriateness* of the regression. A single outlier, or data that separate into two groups rather than a single cloud of points, can make R^2 seem quite large when, in fact, the linear regression model is simply inappropriate. Conversely, a low R^2 value may be due to a single outlier as well. It may be that most of the data fall roughly along a straight line, with the exception of a single point. Always look at the scatterplot.

> R^2 does **not** mean that protein accounts for 69% of the fat in a BK food item. It is the *variation* in fat content that is accounted for by the linear model.

CONNECTIONS

We've talked about the importance of models before, but have seen only the Normal model as an example. The linear model is one of the most important models in Statistics. Chapter 7 talked about the assignment of variables to the y- and x-axes. That didn't matter to correlation, but it does matter to regression because y is predicted by x in the regression model.

The connection of R^2 to correlation is obvious, although it may not be immediately clear that just by squaring the correlation we can learn the fraction of the variability of y accounted for by a regression on x. We'll return to this in subsequent chapters.

We made a big fuss about knowing the units of your quantitative variables. We didn't need units for correlation, but without the units we can't define the slope of a regression. A regression makes no sense if you don't know the *Who*, the *What*, and the *Units* of both your variables.

We've summed squared deviations before when we computed the standard deviation and variance. That's not coincidental. They are closely connected to regression.

When we first talked about models, we noted that deviations away from a model were often interesting. Now we have a formal definition of these deviations as residuals.

WHAT HAVE WE LEARNED?

We've learned that when the relationship between quantitative variables is fairly straight, a linear model can help summarize that relationship and give us insights about it:

- The regression (best fit) line doesn't pass through all the points, but it is the best compromise in the sense that the sum of squares of the residuals is the smallest possible.

We've learned several things the correlation, r, tells us about the regression:

- The slope of the line is based on the correlation, adjusted for the units of x and y:

$$b_1 = r \frac{s_y}{s_x}.$$

▶ For each SD of x that a case is away from the x mean, we expect it to be r SDs of y away from the y mean.

▶ Because r is always between -1 and $+1$, each predicted y is fewer SDs away from its mean than the corresponding x was—from its mean—a phenomenon called regression to the mean.

▶ The square of the correlation coefficient, R^2, gives us the fraction of the variation of the response accounted for by the regression model. The remaining $1 - R^2$ of the variation is left in the residuals.

The residuals also reveal how well the model works:

▶ If a plot of residuals against predicted values shows a pattern, we should re-examine the data to see why.

▶ The standard deviation of the residuals

$$s_e = \sqrt{\frac{\Sigma e^2}{n - 2}}$$

quantifies the amount of scatter around the line.

Of course, the linear model makes no sense unless the **Linearity Assumption** is satisfied. We check the **Straight Enough Condition** and **Outlier Condition** with a scatterplot, as we did for correlation, and also with a plot of residuals against either the x or the predicted values. For the standard deviation of the residuals to make sense as a summary, we have to make the **Equal Variance Assumption**. We check it by looking at both the original scatterplot and the residual plot for the **Does the Plot Thicken? Condition.**

Terms

Model
An equation or formula that simplifies and represents reality (p. 179).

Linear model
A linear model is an equation of the form

$$\hat{y} = b_0 + b_1 x.$$

To interpret a linear model, we need to know the variables (along with their W's) and their units (p. 179).

Predicted value
The value of $\hat{y}$ found for a given x-value in the data. A predicted value is found by substituting the x-value in the regression equation. The predicted values are the values on the fitted line; the points $(x, \hat{y})$ all lie exactly on the fitted line (p. 179).
The predicted values are found from the linear model that we fit:

$$\hat{y} = b_0 + b_1 x.$$

Residuals
Residuals are the differences between data values and the corresponding values predicted by the regression model—or, more generally, values predicted by any model (p. 179).

$$\text{Residual} = \text{observed value} - \text{predicted value} = y - \hat{y}$$

Regression line (Line of best fit)
The particular linear equation

$$\hat{y} = b_0 + b_1 x$$

that satisfies the least squares criterion is called the least squares regression line. Casually, we often just call it the regression line, or the line of best fit (p. 180).

Least squares
The least squares criterion specifies the unique line that minimizes the variance of the residuals or, equivalently, the sum of the squared residuals (p. 180).

Slope
The slope, b_1, gives a value in "y-units per x-unit." Changes of one unit in x are associated with changes of b_1 units in predicted values of y (p. 180).
The slope can be found by

$$b_1 = r \frac{s_y}{s_x}.$$

Intercept	The intercept, b_0, gives a starting value in y-units. It's the $\hat{y}$-value when x is 0 (p. 180). You can find the intercept from $b_0 = \bar{y} - b_1\bar{x}$.
Regression to the mean	Because the correlation is always less than 1.0 in magnitude, each predicted $\hat{y}$ tends to be fewer standard deviations from its mean than its corresponding x was from its mean (p. 186).
s_e	The standard deviation of the residuals is found by $s_e = \sqrt{\dfrac{\Sigma e^2}{n-2}}$. When the assumptions and conditions are met, the residuals can be well described by using this standard deviation and the 68–95–99.7 Rule (p. 189).
R^2	R^2 is the square of the correlation between y and x (p. 190).

▸ R^2 gives the fraction of the variability of y accounted for by the least squares linear regression on x.

▸ R^2 is an overall measure of how successful the regression is in linearly relating y to x.

Skills

THINK

▸ Be able to identify response (y) and explanatory (x) variables in context.

▸ Understand how a linear equation summarizes the relationship between two variables.

▸ Recognize when a regression should be used to summarize a linear relationship between two quantitative variables.

▸ Be able to judge whether the slope of a regression makes sense.

▸ Know how to examine your data for violations of the **Straight Enough Condition** that would make it inappropriate to compute a regression.

▸ Understand that the least squares slope is easily affected by extreme values.

▸ Know that residuals are the differences between the data values and the corresponding values predicted by the regression line and that the *least squares criterion* finds the line that minimizes the sum of the squared residuals.

SHOW

▸ Know how to use a plot of residuals against predicted values to check the **Straight Enough Condition,** the **Does the Plot Thicken? Condition,** and the **Outlier Condition.**

▸ Understand that the standard deviation of the residuals, s_e, measures variability around the line. A large s_e means the points are widely scattered; a small s_e means they lie close to the line.

▸ Know how to find a regression equation from the summary statistics for each variable and the correlation between the variables.

▸ Know how to find a regression equation using your statistics software and how to find the slope and intercept values in the regression output table.

▸ Know how to use regression to predict a value of y for a given x.

▸ Know how to compute the residual for each data value and how to display the residuals.

TELL

▸ Be able to write a sentence explaining what a linear equation says about the relationship between y and x, basing it on the fact that the slope is given in *y-units per x-unit*.

▸ Understand how the correlation coefficient and the regression slope are related. Know how R^2 describes how much of the variation in y is accounted for by its linear relationship with x.

▸ Be able to describe a prediction made from a regression equation, relating the predicted value to the specified x-value.

▸ Be able to write a sentence interpreting s_e as representing typical errors in predictions–the amounts by which actual y-values differ from the $\hat{y}$'s estimated by the model.

REGRESSION ON THE COMPUTER

All statistics packages make a table of results for a regression. These tables may differ slightly from one package to another, but all are essentially the same—and all include much more than we need to know for now. Every computer regression table includes a section that looks something like this:

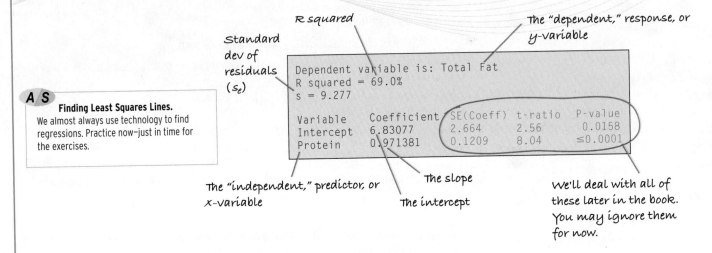

R squared

The "dependent," response, or y-variable

Standard dev of residuals (s_e)

A S

Finding Least Squares Lines.
We almost always use technology to find regressions. Practice now—just in time for the exercises.

```
Dependent variable is: Total Fat
R squared = 69.0%
s = 9.277

Variable    Coefficient   SE(Coeff)   t-ratio    P-value
Intercept   6.83077       2.664       2.56       0.0158
Protein     0.971381      0.1209      8.04       ≤0.0001
```

The "independent," predictor, or x-variable

The slope

The intercept

We'll deal with all of these later in the book. You may ignore them for now.

The slope and intercept coefficient are given in a table such as this one. Usually the slope is labeled with the name of the x-variable, and the intercept is labeled "Intercept" or "Constant." So the regression equation shown here is

$$\widehat{Fat} = 6.83077 + 0.97138 \ Protein.$$

It is not unusual for statistics packages to give many more digits of the estimated slope and intercept than could possibly be estimated from the data. (The original data were reported to the nearest gram.) Ordinarily, you should round most of the reported numbers to one digit more than the precision of the data, and the slope to two. We will learn about the other numbers in the regression table later in the book. For now, all you need to be able to do is find the coefficients, the s_e, and the R^2 value.

DATA DESK

Select the y-variable and the x-variable. In the **Plot** menu choose **Scatterplot**. From the scatterplot Hyper-View menu, choose **Add Regression Line** to display the line. From the HyperView menu, choose **Regression** to compute the regression. To plot the residuals, Click on the **HyperView** menu on the **Regression** output table.

A menu drops down that offers scatterplots of residuals against predicted values (as well as other options).

COMMENTS

Alternatively, find the regression first with the **Regression** command in the **Calc** menu. Click on the x-variable's name to open a menu that offers the scatterplot.

EXCEL

Make a scatterplot of the data. With the scatterplot front-most, select **Add Trendline . . .** from the **Chart** menu. Click the **Options** tab and select **Display Equation on Chart**. Click OK.

COMMENTS

The computer section for Chapter 7 shows how to make a scatterplot. We don't repeat those steps here.

(continued)

EXCEL 2007

- Click on a blank cell in the spreadsheet.
- Go to the **Formulas** tab in the Ribbon and click **More Functions → Statistical**.
- Choose the **CORREL** function from the drop-down menu of functions.
- In the dialog that pops up, enter the range of one of the variables in the space provided.
- Enter the range of the other variable in the space provided.
- Click **OK**.

COMMENTS

The correlation is computed in the selected cell. Correlations computed this way will update if any of the data values are changed.

Before you interpret a correlation coefficient, always make a scatterplot to check for nonlinearity and outliers. If the variables are not linearly related, the correlation coefficient cannot be interpreted.

JMP

Choose **Fit Y by X** from the **Analyze** menu. Specify the y-variable in the Select Columns box and click the **Y, Response** button. Specify the x-variable and click the **X, Factor** button. Click **OK** to make a scatterplot. In the scatterplot window, click on the red triangle beside the

heading labeled **"Bivariate Fit ..."** and choose **Fit Line**. JMP draws the least squares regression line on the scatterplot and displays the results of the regression in tables below the plot.

MINITAB

Choose **Regression** from the **Stat** menu. From the Regression submenu, choose **Fitted Line Plot**. In the Fitted Line Plot dialog, click in the **Response Y** box, and assign the y-variable from the variable list. Click in the

Predictor X box, and assign the x-variable from the Variable list. Make sure that the Type of Regression Model is set to Linear. Click the **OK** button.

SPSS

Choose **Interactive** from the **Graphs** menu. From the interactive Graphs submenu, choose **Scatterplot**. In the Create Scatterplot dialog, drag the y-variable into the

y-axis target, and the x-variable into the **x-axis target**. Click on the **Fit** tab. Choose **Regression** from the **Method** popup menu. Click the **OK** button.

TI-83/84 PLUS

To find the equation of the regression line (add the line to a scatterplot), choose **LinReg(a+bx)**, tell it the list names, and then add a comma to specify a function name (from VARS Y-Vars 1:Function). The final command looks like

$$\text{LinReg (a+bx) L1, L2, Y1.}$$

To make a residuals plot, set up a **STATPLOT** as a scatterplot. Specify your explanatory data list as Xlist. For Ylist, import the name RESID from the **LIST NAMES** menu. **ZoomStat** will now create the residuals plot.

COMMENTS

Each time you execute a **LinReg** command, the calculator automatically computes the residuals and stores them in a data list named RESID. If you want to see them, go to **STAT EDIT**. Space through the names of the lists until you find a blank. Import RESID from the LIST NAMES menu. Now every time you have the calculator compute a regression analysis, it will show you the residuals.

TI-89

To find the equation of the regression line (and add the line to a scatterplot), choose **LinReg(a+bx)** from the **Calc Regressions** menu and tell it the list names and a function to store the equation. To make a residuals plot, define a **PLOT** as a scatterplot.

Specify your explanatory data list as Xlist. For Ylist, find the list name **resid** from VAR-LINK by arrowing to the STATVARS portion. Then press F2 (r) and locate the list. Press ENTER to finish the plot definition and F5 to display the plot.

COMMENTS

Each time you execute a **LinReg** command, the calculator automatically computes the residuals and stores them in a data list named RESID. If you don't want to see this (or any other calculator-generated list) anymore, press F1 (Tools) and select choice 3: Setup Editor. Leaving the box for lists to display blank will reset the calculator to show only lists 1 through 6.

EXERCISES

1. **Cereals.** For many people, breakfast cereal is an important source of fiber in their diets. Cereals also contain potassium, a mineral shown to be associated with maintaining a healthy blood pressure. An analysis of the amount of fiber (in grams) and the potassium content (in milligrams) in servings of 77 breakfast cereals produced the regression model $\widehat{Potassium} = 38 + 27Fiber$. If your cereal provides 9 grams of fiber per serving, how much potassium does the model estimate you will get?

2. **Horsepower.** In Chapter 7's Exercise 33 we examined the relationship between the fuel economy (mpg) and horsepower for 15 models of cars. Further analysis produces the regression model $\widehat{mpg} = 46.87 - 0.084HP$. If the car you are thinking of buying has a 200-horsepower engine, what does this model suggest your gas mileage would be?

3. **More cereal.** Exercise 1 describes a regression model that estimates a cereal's potassium content from the amount of fiber it contains. In this context, what does it mean to say that a cereal has a negative residual?

4. **Horsepower, again.** Exercise 2 describes a regression model that uses a car's horsepower to estimate its fuel economy. In this context, what does it mean to say that a certain car has a positive residual?

5. **Another bowl.** In Exercise 1, the regression model $\widehat{Potassium} = 38 + 27Fiber$ relates fiber (in grams) and potassium content (in milligrams) in servings of breakfast cereals. Explain what the slope means.

6. **More horsepower.** In Exercise 2, the regression model $\widehat{mpg} = 46.87 - 0.084HP$ relates cars' horsepower to their fuel economy (in mpg). Explain what the slope means.

7. **Cereal, again.** The correlation between a cereal's fiber and potassium contents is $r = 0.903$. What fraction of the variability in potassium is accounted for by the variation in the amount of fiber that servings contain?

8. **Another car.** The correlation between a car's horsepower and its fuel economy (in mpg) is $r = -0.869$. What fraction of the variability in fuel economy is accounted for by the horsepower?

9. **Last bowl!** For Exercise 1's regression model predicting potassium content (in milligrams) from the amount of fiber (in grams) in breakfast cereals, $s_e = 30.77$. Explain in this context what that means.

10. **Last tank!** For Exercise 2's regression model predicting fuel economy (in mpg) from the car's horsepower, $s_e = 3.287$. Explain in this context what that means.

11. **Regression equations.** Fill in the missing information in the following table.

	$\bar{x}$	s_x	$\bar{y}$	s_y	r	$\hat{y} = b_0 + b_1 x$
a)	10	2	20	3	0.5	
b)	2	0.06	7.2	1.2	-0.4	
c)	12	6			-0.8	$\hat{y} = 200 - 4x$
d)		2.5	1.2		100	$\hat{y} = -100 + 50x$

12. **More regression equations.** Fill in the missing information in the following table.

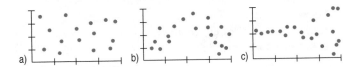

	$\bar{x}$	s_x	$\bar{y}$	s_y	r	$\hat{y} = b_0 + b_1 x$
a)	30	4	18	6	-0.2	
b)	100	18	60	10	0.9	
c)		0.8	50	15		$\hat{y} = -10 + 15x$
d)			18	4	-0.6	$\hat{y} = 30 - 2x$

13. **Residuals.** Tell what each of the residual plots below indicates about the appropriateness of the linear model that was fit to the data.

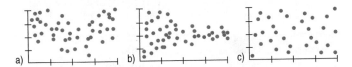

a)   b)   c)

14. **Residuals.** Tell what each of the residual plots below indicates about the appropriateness of the linear model that was fit to the data.

a)   b)   c)

15. **Real estate.** A random sample of records of sales of homes from Feb. 15 to Apr. 30, 1993, from the files maintained by the Albuquerque Board of Realtors gives the *Price* and *Size* (in square feet) of 117 homes. A regression to predict *Price* (in thousands of dollars) from *Size* has an R^2 of 71.4%. The residuals plot indicated that a linear model is appropriate.
   a) What are the variables and units in this regression?
   b) What units does the slope have?
   c) Do you think the slope is positive or negative? Explain.

T 16. **Roller coaster.** People who responded to a July 2004 Discovery Channel poll named the 10 best roller coasters in the United States. A table in the previous chapter's exercises shows the length of the initial drop (in feet) and the duration of the ride (in seconds). A regression to predict *Duration* from *Drop* has $R^2 = 12.4\%$.
   a) What are the variables and units in this regression?
   b) What units does the slope have?
   c) Do you think the slope is positive or negative? Explain.

17. **What slope?** If you create a regression model for predicting the *Weight* of a car (in pounds) from its *Length* (in feet), is the slope most likely to be 3, 30, 300, or 3000? Explain.

18. **What slope?** If you create a regression model for estimating the *Height* of a pine tree (in feet) based on the *Circumference* of its trunk (in inches), is the slope most likely to be 0.1, 1, 10, or 100? Explain.

19. Real estate, again. The regression of *Price* on *Size* of homes in Albuquerque had $R^2 = 71.4\%$, as described in Exercise 15. Write a sentence (in context, of course) summarizing what the R^2 says about this regression.

T 20. Coasters, again. Exercise 16 examined the association between the *Duration* of a roller coaster ride and the height of its initial *Drop*, reporting that $R^2 = 12.4\%$. Write a sentence (in context, of course) summarizing what the R^2 says about this regression.

21. Misinterpretations. A Biology student who created a regression model to use a bird's *Height* when perched for predicting its *Wingspan* made these two statements. Assuming the calculations were done correctly, explain what is wrong with each interpretation.

a) My R^2 of 93% shows that this linear model is appropriate.

b) A bird 10 inches tall will have a wingspan of 17 inches.

22. More misinterpretations. A Sociology student investigated the association between a country's *Literacy Rate* and *Life Expectancy*, then drew the conclusions listed below. Explain why each statement is incorrect. (Assume that all the calculations were done properly.)

a) The R^2 of 64 means that the *Literacy Rate* determines 64% of the *Life Expectancy* for a country.

b) The slope of the line shows that an increase of 5% in *Literacy Rate* will produce a 2-year improvement in *Life Expectancy*.

23. Real estate redux. The regression of *Price* on *Size* of homes in Albuquerque had $R^2 = 71.4\%$, as described in Exercise 15.

a) What is the correlation between *Size* and *Price*?

b) What would you predict about the *Price* of a home 1 SD above average in *Size*?

c) What would you predict about the *Price* of a home 2 SDs below average in *Size*?

T 24. Another ride. The regression of *Duration* of a roller coaster ride on the height of its initial *Drop*, described in Exercise 16, had $R^2 = 12.4\%$.

a) What is the correlation between *Drop* and *Duration*?

b) What would you predict about the *Duration* of the ride on a coaster whose initial *Drop* was 1 standard deviation below the mean *Drop*?

c) What would you predict about the *Duration* of the ride on a coaster whose initial *Drop* was 3 standard deviations above the mean *Drop*?

25. ESP. People who claim to "have ESP" participate in a screening test in which they have to guess which of several images someone is thinking of. You and a friend both took the test. You scored 2 standard deviations above the mean, and your friend scored 1 standard deviation below the mean. The researchers offer everyone the opportunity to take a retest.

a) Should you choose to take this retest? Explain.

b) Now explain to your friend what his decision should be and why.

26. *SI* jinx. Players in any sport who are having great seasons, turning in performances that are much better than anyone might have anticipated, often are pictured on the cover of *Sports Illustrated*. Frequently, their performances then falter somewhat, leading some athletes to believe in a "*Sports Illustrated* jinx." Similarly, it is common for phenomenal rookies to have less stellar second seasons—the so-called "sophomore slump." While fans, athletes, and analysts have proposed many theories about what leads to such declines, a statistician might offer a simpler (statistical) explanation. Explain.

27. More real estate. Consider the Albuquerque home sales from Exercise 15 again. The regression analysis gives the model $\widehat{Price} = 47.82 + 0.061\ Size$.

a) Explain what the slope of the line says about housing prices and house size.

b) What price would you predict for a 3000-square-foot house in this market?

c) A real estate agent shows a potential buyer a 1200-square-foot home, saying that the asking price is $6000 less than what one would expect to pay for a house of this size. What is the asking price, and what is the $6000 called?

T 28. Last ride. Consider the roller coasters described in Exercise 16 again. The regression analysis gives the model $\widehat{Duration} = 91.033 + 0.242\ Drop$.

a) Explain what the slope of the line says about how long a roller coaster ride may last and the height of the coaster.

b) A new roller coaster advertises an initial drop of 200 feet. How long would you predict the rides last?

c) Another coaster with a 150-foot initial drop advertises a 2-minute ride. Is this longer or shorter than you'd expect? By how much? What's that called?

T 29. Cigarettes. Is the nicotine content of a cigarette related to the "tars"? A collection of data (in milligrams) on 29 cigarettes produced the scatterplot, residuals plot, and regression analysis shown:

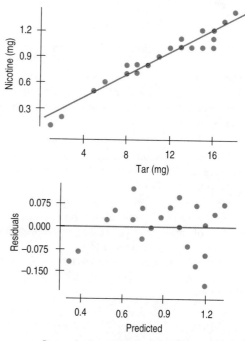

Dependent variable is: nicotine
R-squared = 92.4%

Variable	Coefficient
Constant	0.154030
Tar	0.065052

a) Do you think a linear model is appropriate here? Explain.

b) Explain the meaning of R^2 in this context.

30. Attendance 2006. In the previous chapter you looked at the relationship between the number of wins by American League baseball teams and the average attendance at their home games for the 2006 season. Here are the scatterplot, the residuals plot, and part of the regression analysis:

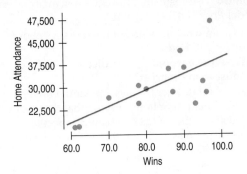

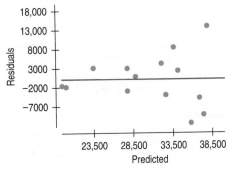

Dependent variable is: Home Attendance
R-squared = 48.5%

Variable	Coefficient
Constant	−14364.5
Wins	538.915

a) Do you think a linear model is appropriate here? Explain.

b) Interpret the meaning of R^2 in this context.

c) Do the residuals show any pattern worth remarking on?

d) The point in the upper right of the plots is the New York Yankees. What can you say about the residual for the Yankees?

31. Another cigarette. Consider again the regression of *Nicotine* content on *Tar* (both in milligrams) for the cigarettes examined in Exercise 29.

a) What is the correlation between *Tar* and *Nicotine*?

b) What would you predict about the average *Nicotine* content of cigarettes that are 2 standard deviations below average in *Tar* content?

c) If a cigarette is 1 standard deviation above average in *Nicotine* content, what do you suspect is true about its *Tar* content?

32. Second inning 2006. Consider again the regression of *Average Attendance* on *Wins* for the baseball teams examined in Exercise 30.

a) What is the correlation between *Wins* and *Average Attendance*?

b) What would you predict about the *Average Attendance* for a team that is 2 standard deviations above average in *Wins*?

c) If a team is 1 standard deviation below average in attendance, what would you predict about the number of games the team has won?

33. Last cigarette. Take another look at the regression analysis of tar and nicotine content of the cigarettes in Exercise 29.

a) Write the equation of the regression line.

b) Estimate the *Nicotine* content of cigarettes with 4 milligrams of *Tar*.

c) Interpret the meaning of the slope of the regression line in this context.

d) What does the *y*-intercept mean?

e) If a new brand of cigarette contains 7 milligrams of tar and a nicotine level whose residual is −0.5 mg, what is the nicotine content?

34. Last inning 2006. Refer again to the regression analysis for average attendance and games won by American League baseball teams, seen in Exercise 30.

a) Write the equation of the regression line.

b) Estimate the *Average Attendance* for a team with 50 *Wins*.

c) Interpret the meaning of the slope of the regression line in this context.

d) In general, what would a negative residual mean in this context?

e) The St. Louis Cardinals, the 2006 World Champions, are not included in these data because they are a National League team. During the 2006 regular season, the Cardinals won 83 games and averaged 42,588 fans at their home games. Calculate the residual for this team, and explain what it means.

35. Income and housing revisited. In Chapter 7, Exercise 31, we learned that the Office of Federal Housing Enterprise Oversight (OFHEO) collects data on various aspects of housing costs around the United States. Here's a scatterplot (by state) of the *Housing Cost Index* (HCI) versus the *Median Family Income* (MFI) for the 50 states. The correlation is $r = 0.65$. The mean HCI is 338.2, with a standard deviation of 116.55. The mean MFI is $46,234, with a standard deviation of $7072.47.

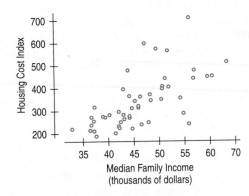

a) Is a regression analysis appropriate? Explain.

b) What is the equation that predicts Housing Cost Index from median family income?

c) For a state with MFI = $44,993, what would be the predicted HCI?

d) Washington, DC, has an MFI of $44,993 and an HCI of 548.02. How far off is the prediction in part b from the actual HCI?

e) If we standardized both variables, what would be the regression equation that predicts standardized HCI from standardized MFI?

f) If we standardized both variables, what would be the regression equation that predicts standardized MFI from standardized HCI?

T 36. Interest rates and mortgages again. In Chapter 7, Exercise 32, we saw a plot of total mortgages in the United States (in millions of 2005 dollars) versus the interest rate at various times over the past 26 years. The correlation is $r = -0.84$. The mean mortgage amount is $151.9 million and the mean interest rate is 8.88%. The standard deviations are $23.86 million for mortgage amounts and 2.58% for the interest rates.

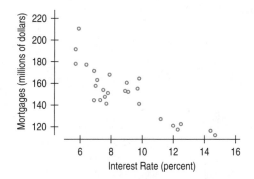

a) Is a regression model appropriate for predicting mortgage amount from interest rates? Explain.

b) What is the equation that predicts mortgage amount from interest rates?

c) What would you predict the mortgage amount would be if the interest rates climbed to 20%?

d) Do you have any reservations about your prediction in part c?

e) If we standardized both variables, what would be the regression equation that predicts standardized mortgage amount from standardized interest rates?

f) If we standardized both variables, what would be the regression equation that predicts standardized interest rates from standardized mortgage amount?

37. Online clothes. An online clothing retailer keeps track of its customers' purchases. For those customers who signed up for the company's credit card, the company also has information on the customer's *Age* and *Income*. A random sample of 500 of these customers shows the following scatterplot of *Total Yearly Purchases* by *Age*:

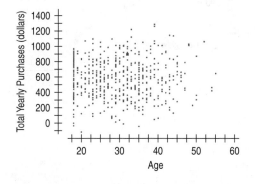

The correlation between *Total Yearly Purchases* and *Age* is $r = 0.037$. Summary statistics for the two variables are:

	Mean	SD
Age	29.67 yrs	8.51 yrs
Total Yearly Purchase	$572.52	$253.62

a) What is the linear regression equation for predicting *Total Yearly Purchase* from *Age*?

b) Do the assumptions and conditions for regression appear to be met?

c) What is the predicted average *Total Yearly Purchase* for an 18-year-old? For a 50-year-old?

d) What percent of the variability in *Total Yearly Purchases* is accounted for by this model?

e) Do you think the regression might be a useful one for the company? Explain.

38. Online clothes II. For the online clothing retailer discussed in the previous problem, the scatterplot of *Total Yearly Purchases* by *Income* looks like this:

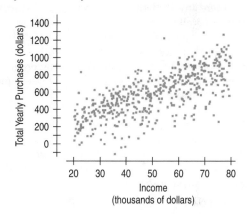

The correlation between *Total Yearly Purchases* and *Income* is 0.722. Summary statistics for the two variables are:

	Mean	SD
Income	$50,343.40	$16,952.50
Total Yearly Purchase	$572.52	$253.62

a) What is the linear regression equation for predicting *Total Yearly Purchase* from *Income*?

b) Do the assumptions and conditions for regression appear to be met?

c) What is the predicted average *Total Yearly Purchase* for someone with a yearly *Income* of $20,000? For someone with an annual *Income* of $80,000?

d) What percent of the variability in *Total Yearly Purchases* is accounted for by this model?

e) Do you think the regression might be a useful one for the company? Comment.

T 39. SAT scores. The SAT is a test often used as part of an application to college. SAT scores are between 200 and 800, but have no units. Tests are given in both Math and Verbal areas. Doing the SAT-Math problems also involves the ability to read and understand the questions, but can a person's verbal score be used to predict the math score?

Verbal and math SAT scores of a high school graduating class are displayed in the scatterplot, with the regression line added.

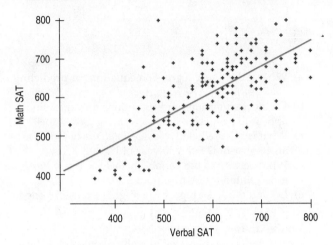

a) Describe the relationship.
b) Are there any students whose scores do not seem to fit the overall pattern?
c) For these data, $r = 0.685$. Interpret this statistic.
d) These verbal scores averaged 596.3, with a standard deviation of 99.5, and the math scores averaged 612.2, with a standard deviation of 96.1. Write the equation of the regression line.
e) Interpret the slope of this line.
f) Predict the math score of a student with a verbal score of 500.
g) Every year some student scores a perfect 1600. Based on this model, what would that student's residual be for her math score?

40. **Success in college.** Colleges use SAT scores in the admissions process because they believe these scores provide some insight into how a high school student will perform at the college level. Suppose the entering freshmen at a certain college have mean combined *SAT Scores* of 1222, with a standard deviation of 123. In the first semester these students attained a mean *GPA* of 2.66, with a standard deviation of 0.56. A scatterplot showed the association to be reasonably linear, and the correlation between *SAT* score and *GPA* was 0.47.

a) Write the equation of the regression line.
b) Explain what the *y*-intercept of the regression line indicates.
c) Interpret the slope of the regression line.
d) Predict the GPA of a freshman who scored a combined 1400.
e) Based upon these statistics, how effective do you think SAT scores would be in predicting academic success during the first semester of the freshman year at this college? Explain.
f) As a student, would you rather have a positive or a negative residual in this context? Explain.

41. **SAT, take 2.** Suppose we wanted to use SAT math scores to estimate verbal scores based on the information in Exercise 39.

a) What is the correlation?
b) Write the equation of the line of regression predicting verbal scores from math scores.
c) In general, what would a positive residual mean in this context?
d) A person tells you her math score was 500. Predict her verbal score.
e) Using that predicted verbal score and the equation you created in Exercise 39, predict her math score.
f) Why doesn't the result in part e come out to 500?

42. **Success, part 2.** Based on the statistics for college freshmen given in Exercise 40, what SAT score might be expected among freshmen who attained a first-semester GPA of 3.0?

43. **Wildfires 2008.** The National Interagency Fire Center (www.nifc.gov) reports statistics about wildfires. Here's an analysis of the number of wildfires between 1985 and 2007.

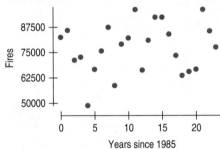

Dependent variable is: Fires
R-squared = 2.9%
s = 12323

Variable	Coefficient
Intercept	73790.7
Years since 1985	296.665

a) Is a linear model appropriate for these data? Explain.
b) Interpret the slope in this context.
c) Can we interpret the intercept? Why or why not?
d) What does the value of s_e say about the size of the residuals? What does it say about the effectiveness of the model?
e) What does R^2 mean in this context?

44. **Wildfire size 2008.** We saw in Exercise 43 that the number of fires was nearly constant. But has the damage they cause remained constant as well? Here's a regression that examines the trend in *Acres per Fire*, together with some supporting plots:

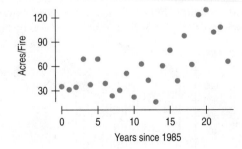

Dependent variable is: Acres/fire
R-squared = 44.2%
s = 24.89

Variable	Coefficient
Intercept	24.8069
Years since 1985	3.06530

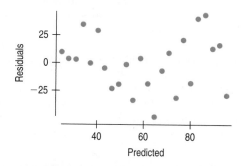

a) Is the regression model appropriate for these data? Explain.
b) What interpretation (if any) can you give for the R^2 in the regression table?

T 45. Used cars 2007. Classified ads in the *Ithaca Journal* offered several used Toyota Corollas for sale. Listed below are the ages of the cars and the advertised prices.

Age (yr)	Price Advertised ($)
1	13,990
1	13,495
3	12,999
4	9500
4	10,495
5	8995
5	9495
6	6999
7	6950
7	7850
8	6999
8	5995
10	4950
10	4495
13	2850

a) Make a scatterplot for these data.
b) Describe the association between *Age* and *Price* of a used Corolla.
c) Do you think a linear model is appropriate?
d) Computer software says that R^2 = 94.4%. What is the correlation between *Age* and *Price*?
e) Explain the meaning of R^2 in this context.
f) Why doesn't this model explain 100% of the variability in the price of a used Corolla?

T 46. Drug abuse. In the exercises of the last chapter you examined results of a survey conducted in the United States and 10 countries of Western Europe to determine the percentage of teenagers who had used marijuana and other drugs. Below is the scatterplot. Summary statistics showed that the mean percent that had used marijuana was 23.9%, with a standard deviation of 15.6%. An average of 11.6% of teens had used other drugs, with a standard deviation of 10.2%.

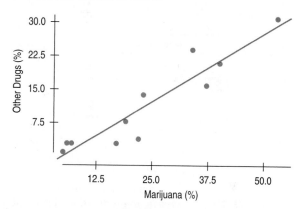

a) Do you think a linear model is appropriate? Explain.
b) For this regression, R^2 is 87.3%. Interpret this statistic in this context.
c) Write the equation you would use to estimate the percentage of teens who use other drugs from the percentage who have used marijuana.
d) Explain in context what the slope of this line means.
e) Do these results confirm that marijuana is a "gateway drug," that is, that marijuana use leads to the use of other drugs?

T 47. More used cars 2007. Use the advertised prices for Toyota Corollas given in Exercise 45 to create a linear model for the relationship between a car's *Age* and its *Price*.

a) Find the equation of the regression line.
b) Explain the meaning of the slope of the line.
c) Explain the meaning of the *y*-intercept of the line.
d) If you want to sell a 7-year-old Corolla, what price seems appropriate?
e) You have a chance to buy one of two cars. They are about the same age and appear to be in equally good condition. Would you rather buy the one with a positive residual or the one with a negative residual? Explain.
f) You see a "For Sale" sign on a 10-year-old Corolla stating the asking price as $3500. What is the residual?
g) Would this regression model be useful in establishing a fair price for a 20-year-old car? Explain.

48. Veggie burgers. Recently Burger King introduced a meat-free burger. The nutrition label is shown here:

Nutrition Facts

Calories	330
Fat	10g*
Sodium	760g
Sugars	5g
Protein	14g
Carbohydrates	43g
Dietary Fiber	4g
Cholesterol	0

* (2 grams of saturated fat)

RECOMMENDED DAILY VALUES
(based on a 2,000-calorie/day diet)

Iron	20%
Vitamin A	10%
Vitamin C	10%
Calcium	6%

a) Use the regression model created in this chapter, $\widehat{Fat} = 6.8 + 0.97\ Protein$, to predict the fat content of this burger from its protein content.

b) What is its residual? How would you explain the residual?

c) Write a brief report about the *Fat* and *Protein* content of this new menu item. Be sure to talk about the variables by name and in the correct units.

49. **Burgers.** In the last chapter, you examined the association between the amounts of *Fat* and *Calories* in fast-food hamburgers. Here are the data:

Fat (g)	19	31	34	35	39	39	43
Calories	410	580	590	570	640	680	660

a) Create a scatterplot of *Calories* vs. *Fat*.

b) Interpret the value of R^2 in this context.

c) Write the equation of the line of regression.

d) Use the residuals plot to explain whether your linear model is appropriate.

e) Explain the meaning of the *y*-intercept of the line.

f) Explain the meaning of the slope of the line.

g) A new burger containing 28 grams of fat is introduced. According to this model, its residual for calories is +33. How many calories does the burger have?

50. **Chicken.** Chicken sandwiches are often advertised as a healthier alternative to beef because many are lower in fat. Tests on 11 brands of fast-food chicken sandwiches produced the following summary statistics and scatterplot from a graphing calculator:

	Fat (g)	Calories
Mean	20.6	472.7
St. Dev.	9.8	144.2
Correlation		0.947

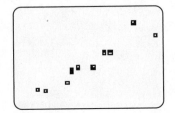

a) Do you think a linear model is appropriate in this situation?

b) Describe the strength of this association.

c) Write the equation of the regression line.

d) Explain the meaning of the slope.

e) Explain the meaning of the *y*–intercept.

f) What does it mean if a certain sandwich has a negative residual?

g) If a chicken sandwich and a burger each advertised 35 grams of fat, which would you expect to have more calories? (See Exercise 49.)

h) McDonald's Filet-O-Fish sandwich has 26 grams of fat and 470 calories. Does the fat–calorie relationship in this sandwich appear to be very different from that found in chicken sandwiches or in burgers (see Exercise 49)? Explain.

51. **A second helping of burgers.** In Exercise 49 you created a model that can estimate the number of *Calories* in a burger when the *Fat* content is known.

a) Explain why you cannot use that model to estimate the fat content of a burger with 600 calories.

b) Using an appropriate model, estimate the fat content of a burger with 600 calories.

52. **Cost of living 2008.** The *Worldwide Cost of Living Survey City Rankings* determine the cost of living in the 25 most expensive cities in the world. (www.finfacts.com/costofliving.htm) These rankings scale New York City as 100, and express the cost of living in other cities as a percentage of the New York cost. For example, the table indicates that in Tokyo the cost of living was 22.1% higher than New York in 2007, and increased to 27.0% higher in 2008.

City	2007	2008
Moscow	134.4	142.4
Tokyo	122.1	127.0
London	126.3	125.0
Oslo	105.8	118.3
Seoul	122.4	117.7
Hong Kong	119.4	117.6
Copenhagen	110.2	117.2
Geneva	109.8	115.8
Zurich	107.6	112.7
Milan	104.4	111.3
Osaka	108.4	110.0
Paris	101.4	109.4
Singapore	100.4	109.1
Tel Aviv	97.7	105.0
Sydney	94.9	104.1
Dublin	99.6	103.9
Rome	97.6	103.9
St. Petersburg	103.0	103.1
Vienna	96.9	102.3
Beijing	95.9	101.9
Helsinki	93.3	101.1
New York City	100.0	100.0
Istanbul	87.7	99.4
Shanghai	92.1	98.3
Amsterdam	92.2	97.0

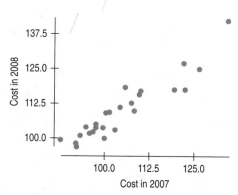

Year	Halloween Candy Sales (millions of dollars)
1995	1.474
1996	1.660
1997	1.708
1998	1.787
1999	1.896
2000	1.985
2001	2.035

a) Here is a scatterplot. Describe the association between costs of living in 2007 and 2008.

b) The correlation is 0.938. Find and interpret the value of R^2.

c) The regression equation predicting the 2008 cost of living from the 2007 figure is $\widehat{Cost08} = 21.75 + 0.84\,Cost07$. Use this equation to find the residual for Oslo.

d) Explain what the residual means.

53. New York bridges. We saw in this chapter that in Tompkins County, NY, older bridges were in worse condition than newer ones. Tompkins is a rural area. Is this relationship true in New York City as well? Here are data on the *Condition* (as measured by the state Department of Transportation Condition Index) and *Age at Inspection* for bridges in New York City.

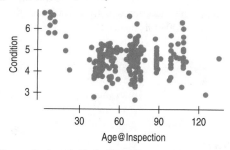

Dependent variable is: Condition
R-squared = 2.6%
s = 0.6708

Variable	Coefficient
Intercept	4.95147
Age@Inspection	−0.00481

a) New York State defines any bridge with a condition score less than 5 as *deficient*. What does this model predict for the condition scores of New York City bridges?

b) Our earlier model found that the condition of bridges in Tompkins County was decreasing at about 0.025 per year. What does this model say about New York City bridges?

c) How much faith would you place in this model? Explain.

54. Candy 2006. The table shows the increase in Halloween candy sales over a 7-year period as reported by the National Confectioners Association (www.ecandy.com). Using these data, estimate the amount of candy sold in 2006. Discuss the appropriateness of your model and your faith in the estimate. Then comment on the fact that NCA reported 2006 sales of $2.136 million. (Enter *Year* as 95, 96, . . . , 101.)

55. Climate change. The climate on earth is getting warmer. The most common theory relates an increase in atmospheric levels of carbon dioxide (CO_2), a greenhouse gas, to increases in temperature. Here is a scatterplot showing the mean annual CO_2 concentration in the atmosphere, measured in parts per million (ppm) at the top of Mauna Loa in Hawaii, and the mean annual air temperature over both land and sea across the globe, in degrees Celsius (°C).

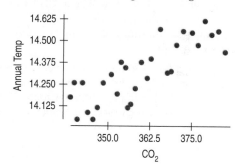

A regression predicting *Temperature* from CO_2 produces the following output table (in part):

Dependent variable is: Temperature
R-squared = 67.8%

Variable	Coefficient
Intercept	10.71
CO_2	0.010

a) What is the correlation between CO_2 and *Temperature*?
b) Explain the meaning of R-squared in this context.
c) Give the regression equation.
d) What is the meaning of the slope in this equation?
e) What is the meaning of the y-intercept of this equation?
f) Here is a scatterplot of the residuals vs. CO_2. Does this plot show evidence of the violation of any assumptions behind the regression? If so, which ones?

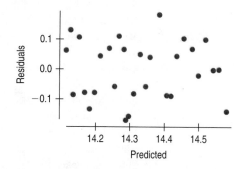

g) CO_2 levels will probably reach 390 ppm by 2013. What mean *Temperature* does the regression predict from that information?

T 56. Birthrates 2005. The table shows the number of live births per 1000 women aged 15–44 years in the United States, starting in 1965. (National Center for Health Statistics, www.cdc.gov/nchs/)

Year	1965	1970	1975	1980	1985	1990	1995	2000	2005
Rate	19.4	18.4	14.8	15.9	15.6	16.4	14.8	14.4	14.0

a) Make a scatterplot and describe the general trend in *Birthrates*. (Enter *Year* as years since 1900: 65, 70, 75, etc.)
b) Find the equation of the regression line.
c) Check to see if the line is an appropriate model. Explain.
d) Interpret the slope of the line.
e) The table gives rates only at 5-year intervals. Estimate what the rate was in 1978.
f) In 1978 the birthrate was actually 15.0. How close did your model come?
g) Predict what the *Birthrate* will be in 2010. Comment on your faith in this prediction.
h) Predict the *Birthrate* for 2025. Comment on your faith in this prediction.

T 57. Body fat. It is difficult to determine a person's body fat percentage accurately without immersing him or her in water. Researchers hoping to find ways to make a good estimate immersed 20 male subjects, then measured their waists and recorded their weights.

Waist (in.)	Weight (lb)	Body Fat (%)	Waist (in.)	Weight (lb)	Body Fat (%)
32	175	6	33	188	10
36	181	21	40	240	20
38	200	15	36	175	22
33	159	6	32	168	9
39	196	22	44	246	38
40	192	31	33	160	10
41	205	32	41	215	27
35	173	21	34	159	12
38	187	25	34	146	10
38	188	30	44	219	28

a) Create a model to predict *%Body Fat* from *Weight*.
b) Do you think a linear model is appropriate? Explain.
c) Interpret the slope of your model.
d) Is your model likely to make reliable estimates? Explain.
e) What is the residual for a person who weighs 190 pounds and has 21% body fat?

T 58. Body fat, again. Would a model that uses the person's *Waist* size be able to predict the *%Body Fat* more accurately than one that uses *Weight*? Using the data in Exercise 57, create and analyze that model.

T 59. Heptathlon 2004. We discussed the women's 2004 Olympic heptathlon in Chapter 6. Here are the results from the high jump, 800-meter run, and long jump for the 26 women who successfully completed all three events in the 2004 Olympics (www.espn.com):

Name	Country	High Jump (m)	800-m (sec)	Long Jump (m)
Carolina Klüft	SWE	1.91	134.15	6.51
Austra Skujyte	LIT	1.76	135.92	6.30
Kelly Sotherton	GBR	1.85	132.27	6.51
Shelia Burrell	USA	1.70	135.32	6.25
Yelena Prokhorova	RUS	1.79	131.31	6.21
Sonja Kesselschlaeger	GER	1.76	135.21	6.42
Marie Collonville	FRA	1.85	133.62	6.19
Natalya Dobrynska	UKR	1.82	137.01	6.23
Margaret Simpson	GHA	1.79	137.72	6.02
Svetlana Sokolova	RUS	1.70	133.23	5.84
J J Shobha	IND	1.67	137.28	6.36
Claudia Tonn	GER	1.82	130.77	6.35
Naide Gomes	POR	1.85	140.05	6.10
Michelle Perry	USA	1.70	133.69	6.02
Aryiro Strataki	GRE	1.79	137.90	5.97
Karin Ruckstuhl	NED	1.85	133.95	5.90
Karin Ertl	GER	1.73	138.68	6.03
Kylie Wheeler	AUS	1.79	137.65	6.36
Janice Josephs	RSA	1.70	138.47	6.21
Tiffany Lott Hogan	USA	1.67	145.10	6.15
Magdalena Szczepanska	POL	1.76	133.08	5.98
Irina Naumenko	KAZ	1.79	134.57	6.16
Yuliya Akulenko	UKR	1.73	142.58	6.02
Soma Biswas	IND	1.70	132.27	5.92
Marsha Mark-Baird	TRI	1.70	141.21	6.22
Michaela Hejnova	CZE	1.70	145.68	5.70

Let's examine the association among these events. Perform a regression to predict high-jump performance from the 800-meter results.

a) What is the regression equation? What does the slope mean?
b) What percent of the variability in high jumps can be accounted for by differences in 800-m times?
c) Do good high jumpers tend to be fast runners? (Be careful—low times are good for running events and high distances are good for jumps.)
d) What does the residuals plot reveal about the model?
e) Do you think this is a useful model? Would you use it to predict high-jump performance? (Compare the residual standard deviation to the standard deviation of the high jumps.)

T 60. **Heptathlon 2004 again.** We saw the data for the women's 2004 Olympic heptathlon in Exercise 59. Are the two jumping events associated? Perform a regression of the long-jump results on the high-jump results.

a) What is the regression equation? What does the slope mean?
b) What percentage of the variability in long jumps can be accounted for by high-jump performances?
c) Do good high jumpers tend to be good long jumpers?
d) What does the residuals plot reveal about the model?
e) Do you think this is a useful model? Would you use it to predict long-jump performance? (Compare the residual standard deviation to the standard deviation of the long jumps.)

T 61. **Hard water.** In an investigation of environmental causes of disease, data were collected on the annual mortality rate (deaths per 100,000) for males in 61 large towns in England and Wales. In addition, the water hardness was recorded as the calcium concentration (parts per million, ppm) in the drinking water. The following display shows the relationship between *Mortality* and *Calcium* concentration for these towns:

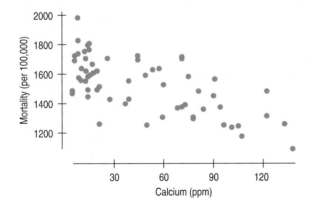

a) Describe what you see in this scatterplot, in context.
b) Here is the regression analysis of *Mortality* and *Calcium* concentration. What is the regression equation?

Dependent variable is: Mortality
R-squared = 43%
s = 143.0

Variable	Coefficient	SE(Coeff)	t-ratio	P-value
Intercept	1676	29.30	57.2	<0.0001
Calcium	−3.23	0.48	26.66	<0.0001

c) Interpret the slope and *y*-intercept of the line, in context.
d) The largest residual, with a value of −348.6, is for the town of Exeter. Explain what this value means.
e) The hardness of Derby's municipal water is about 100 ppm of calcium. Use this equation to predict the mortality rate in Derby.
f) Explain the meaning of *R*-squared in this situation.

62. **Gators.** Wildlife researchers monitor many wildlife populations by taking aerial photographs. Can they estimate the weights of alligators accurately from the air? Here is a regression analysis of the *Weight* of alligators (in pounds) and their *Length* (in inches) based on data collected about captured alligators.

Dependent variable is: Weight
R-squared = 83.6%
s = 54.01

Variable	Coefficient	SE(Coeff)	t-ratio	P-value
Intercept	−393	47.53	−8.27	<0.0001
Length	5.9	0.5448	10.8	<0.0001

a) Did they choose the correct variable to use as the dependent variable and the predictor? Explain.
b) What is the correlation between an alligator's length and weight?
c) Write the regression equation.
d) Interpret the slope of the equation in this context.
e) Do you think this equation will allow the scientists to make accurate predictions about alligators? What part of the regression analysis indicates this? What additional concerns do you have?

63. **Least squares.** Consider the four points (10,10), (20,50), (40,20), and (50,80). The least squares line is $\hat{y} = 7.0 + 1.1x$. Explain what "least squares" means, using these data as a specific example.

64. **Least squares.** Consider the four points (200,1950), (400,1650), (600,1800), and (800,1600). The least squares line is $\hat{y} = 1975 − 0.45x$. Explain what "least squares" means, using these data as a specific example.

ANSWERS

1. An increase in home size of 1000 square feet is associated with an increase in price of $94,450, on average.

2. Units are thousands of dollars per thousand square feet.

3. About $188,900, on average.

4. You should expect the price to be 0.77 standard deviations above the mean.

5. You should expect the size to be $2(0.77) = 1.54$ standard deviations below the mean.

6. The home is 1.5 standard deviations above the mean in size no matter how size is measured.

7. Differences in the size of houses account for about 59.5% of the variation in the house prices.

8. It's positive. The correlation and the slope have the same sign.

9. No, the standard deviation of the residuals is 53.79 thousand dollars. We shouldn't be surprised by any residual smaller than 2 standard deviations, and a residual of $100,000 is less than $2 \times (\$53,790)$.

Regression Wisdom

Where are we going?

What happens when we fit a regression model to data that aren't straight? How bad will the predictions be? How can we tell if the model is appropriate or not? Questions like these are as important as fitting the model itself. In this chapter we'll see how to tell whether a regression model is sensible and what to do if it isn't.

A S *Activity:* **Construct a Plot with a Given Slope.** How's your feel for regression lines? Can you make a scatterplot that has a specified slope?

Regression may be the most widely used Statistics method. It is used every day throughout the world to predict customer loyalty, numbers of admissions at hospitals, sales of automobiles, and many other things. Because regression is so widely used, it's also widely abused and misinterpreted. This chapter presents examples of regressions in which things are not quite as simple as they may have seemed at first and shows how you can still use regression to discover what the data have to say.

Getting the "Bends": When the Residuals Aren't Straight

No regression analysis is complete without a display of the residuals to check that the linear model is reasonable. Because the residuals are what is "left over" after the model describes the relationship, they often reveal subtleties that were not clear from a plot of the original data. Sometimes these are additional details that help confirm or refine our understanding. Sometimes they reveal violations of the regression conditions that require our attention.

The fundamental assumption in working with a linear model is that the relationship you are modeling is, in fact, linear. That sounds obvious, but when you fit a regression, you can't take it for granted. Often it's hard to tell from the scatterplot you looked at before you fit the regression model. Sometimes you can't see a bend in the relationship until you plot the residuals.

We can't *know* whether the **Linearity Assumption** is true, but we can see if it's *plausible* by checking the **Straight Enough Condition**.

Jessica Meir and Paul Ponganis study emperor penguins at the Scripps Institution of Oceanography's Center for Marine Biotechnology and Biomedicine at the University of California at San Diego. Says Jessica:

Emperor penguins are the most accomplished divers among birds, making routine dives of 5–12 minutes, with the longest recorded dive over 27 minutes. These birds can also dive to depths of over 500 meters! Since air-breathing animals like penguins must hold their breath while submerged, the duration of any given dive depends on how much oxygen is in the bird's body at the beginning of the dive, how quickly that oxygen gets used, and the lowest level of oxygen the bird can tolerate. The rate of oxygen depletion is primarily determined by the penguin's heart rate. Consequently, studies of heart rates during dives can help us understand how these animals regulate their oxygen consumption in order to make such impressive dives.

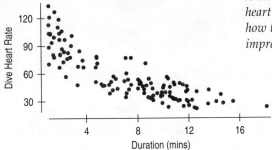

FIGURE 9.1

The scatterplot of *Dive Heart Rate* in beats per minute (bpm) vs. *Duration* (minutes) shows a strong, roughly linear, negative association.

The researchers equip emperor penguins with devices that record their heart rates during dives. Here's a scatterplot of the *Dive Heart Rate* (beats per minute) and the *Duration* (minutes) of dives by these high-tech penguins.

The scatterplot looks fairly linear with a moderately strong negative association ($R^2 = 71.5\%$). The linear regression equation

$$\widehat{DiveHeartRate} = 96.9 - 5.47\,Duration$$

says that for longer dives, the average *Dive Heart Rate* is lower by about 5.47 beats per dive minute, starting from a value of 96.9 beats per minute.

The scatterplot of the residuals against *Duration* holds a surprise. The Linearity Assumption says we should not see a pattern, but instead there's a bend, starting high on the left, dropping down in the middle of the plot, and rising again at the right. Graphs of residuals often reveal patterns such as this that were easy to miss in the original scatterplot.

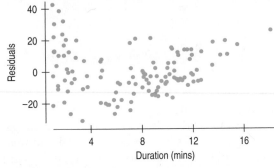

FIGURE 9.2

Plotting the residuals against *Duration* reveals a bend. It was also in the original scatterplot, but here it's easier to see.

Now looking back at the original scatterplot, you may see that the scatter of points isn't really straight. There's a slight bend to that plot, but the bend is much easier to see in the residuals. Even though it means rechecking the Straight Enough Condition *after* you find the regression, it's always a good idea to check your scatterplot of the residuals for bends that you might have overlooked in the original scatterplot.

Sifting Residuals for Groups

In the Step-By-Step analysis in Chapter 8 to predict *Calories* from *Sugar* content in breakfast cereals, we examined a scatterplot of the residuals. Our first impression was that it had no particular structure—a conclusion that supported using the regression model. But let's look again.

Here's a histogram of the residuals. How would you describe its shape? It looks like there might be small modes on both sides of the central body of the data. One group of cereals seems to stand out as having large negative residuals, with fewer calories than our regression model predicted, and another stands out with large positive residuals. The calories in these cereals were overestimated by the model. Whenever we suspect multiple modes, we ask whether they are somehow different.

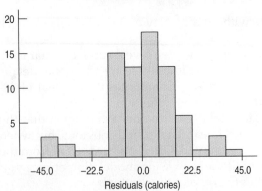

FIGURE 9.3

A histogram of the regression residuals shows small modes both above and below the central large mode. These may be worth a second look.

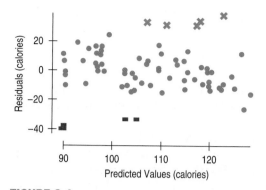

FIGURE 9.4
A scatterplot of the residuals vs. predicted values for the cereal regression. The green "**x**" points are cereals whose calorie content is higher than the linear model predicts. The red "**-**" points show cereals with fewer calories than the model predicts. Is there something special about these cereals?

Let's look more carefully at those points in the residual plot. (See Figure 9.4.). Now we can see that those two groups stand away from the central pattern in the scatterplot. The high-residual cereals are Just Right Fruit & Nut; Muesli Raisins, Dates & Almonds; Peaches & Pecans; Mueslix Crispy Blend; and Nutri-Grain Almond Raisin. Do these cereals seem to have something in common? They all present themselves as "healthy." This might be surprising, but in fact, "healthy" cereals often contain more fat, and therefore more calories, than we might expect from looking at their sugar content alone.

The low-residual cereals are Puffed Rice, Puffed Wheat, three bran cereals, and Golden Crisps. You might not have grouped these cereals together before. What they have in common is a low calorie count *relative to their sugar content*—even though their sugar contents are quite different.

These observations may not lead us to question the overall linear model, but they do help us understand that other factors may be part of the story. An examination of residuals often leads us to discover groups of observations that are different from the rest.

When we discover that there is more than one group in a regression, we may decide to analyze the groups separately, using a different model for each group. Or we can stick with the original model and simply note that there are groups that are a little different. Either way, the model will be wrong, but useful, so it will improve our understanding of the data.

Subsets

> Here's an important unstated condition for fitting models:
> **All the data must come from the same population.**

Cereal manufacturers aim cereals at different segments of the market. Supermarkets and cereal manufacturers try to attract different customers by placing different types of cereals on certain shelves. Cereals for kids tend to be on the "kid's shelf," at their eye level. Toddlers wouldn't be likely to grab a box from this shelf and beg, "Mom, can we please get this All-Bran with Extra Fiber?"

Should we take this extra information into account in our analysis? Figure 9.5 shows a scatterplot of *Calories* and *Sugar*, colored according to the shelf on which the cereals were found and with a separate regression line fit for each. The top shelf is clearly different. We might want to report two regressions, one for the top shelf and one for the bottom two shelves.[1]

FIGURE 9.5
Calories and *Sugar* colored according to the shelf on which the cereal was found in a supermarket, with regression lines fit for each shelf individually. Do these data appear homogeneous? That is, do all the cereals seem to be from the same population of cereals? Or are there different kinds of cereals that we might want to consider separately?

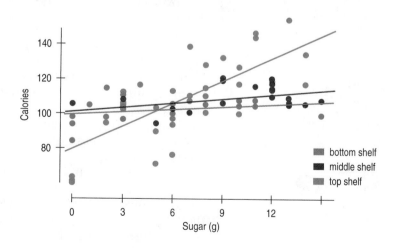

---

[1] More complex models can take into account both sugar content and shelf information. This kind of *multiple regression* model is a natural extension of the model we're using here. We'll see such models in Chapters 30 and 31.

Extrapolation: Reaching Beyond the Data

Case Study: Predicting Manatee Kills. Can we use regression to predict the number of manatees that will be killed by power boats this year?

"Prediction is difficult, especially about the future."

–Niels Bohr,
Danish physicist

Linear models give a predicted value for each case in the data. Put a new x-value into the equation, and it gives a predicted value, $\hat{y}$, to go with it. But when the new x-value lies far from the data we used to build the regression, how trustworthy is the prediction?

The simple answer is that the farther the new x-value is from $\bar{x}$, the less trust we should place in the predicted value. Once we venture into new x territory, such a prediction is called an **extrapolation**. Extrapolations are dubious because they require the very questionable assumption that nothing about the relationship between x and y changes even at extreme values of x and beyond.

Extrapolation is a good way to see just where the limits of our model may be. But it requires caution. When the x-variable is *Time*, extrapolation becomes an attempt to peer into the future. People have always wanted to see into the future, and it doesn't take a crystal ball to foresee that they always will. In the past, seers, oracles, and wizards were called on to predict the future. Today mediums, fortune-tellers, astrologers, and Tarot card readers still find many customers.

FOXTROT © 2002 Bill Amend. Reprinted with permission of UNIVERSAL PRESS SYNDICATE. All rights reserved.

Those with a more scientific outlook may use a linear model as their digital crystal ball. Some physical phenomena do exhibit a kind of "inertia" that allows us to guess that current systematic behavior will continue, but be careful in counting on that kind of regularity in phenomena such as stock prices, sales figures, hurricane tracks, or public opinion.

Extrapolating from current trends is so tempting that even professional forecasters sometimes expect too much from their models—and sometimes the errors are striking. In the mid-1970s, oil prices surged and long lines at gas stations were common. In 1970, oil cost about $17 a barrel (in 2005 dollars)—about what it had cost for 20 years or so. But then, within just a few years, the price surged to over $40. In 1975, a survey of 15 top econometric forecasting models (built by groups that included Nobel prize–winning economists) found predictions for 1985 oil prices that ranged from $300 to over $700 a barrel (in 2005 dollars). How close were these forecasts?

WHEN THE DATA ARE YEARS . . .

. . . we usually don't enter them as four-digit numbers. here we used 0 for 1970, 10 for 1980, and so on. Or we may simply enter two digits, using 82 for 1982, for instance. Rescaling years like this often makes calculations easier and equations simpler. We recommend you do it, too. But be careful: if 1982 is 82, then 2004 is 104 (not 4), right?

Here's a scatterplot of oil prices from 1971 to 1981 (in 2005 dollars).

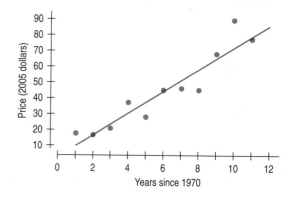

FIGURE 9.6
The scatterplot shows an average *increase* in the price of a barrel of oil of over $7 per year from 1971 to 1981.

The regression model

$$\widehat{Price} = 3.08 + 6.90\ Years\ since\ 1970$$

says that prices had been going up 6.90 dollars per year, or nearly $69 in 10 years. If you assume that they would *keep going up*, it's not hard to imagine almost any price you want.

So, how did the forecasters do? Well, in the period from 1982 to 1998 oil prices didn't exactly continue that steady increase. In fact, they went down so much that by 1998, prices (adjusted for inflation) were the lowest they'd been since before World War II.

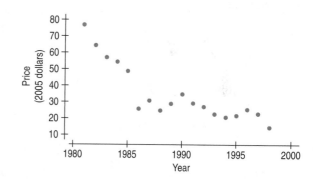

FIGURE 9.7
This scatterplot of oil prices from 1981 to 1998 shows a fairly constant *decrease* of about $3 per barrel per year.

Not one of the experts' models predicted that.

Of course, these decreases clearly couldn't continue, or oil would be free by now. The Energy Information Administration offered two *different* 20-year forecasts for oil prices after 1998, and both called for relatively modest increases in oil prices. So, how accurate have *these* forecasts been? Here's a timeplot of the EIA's predictions and the actual prices (in 2005 dollars).

Oops! They seemed to have missed the sharp run-up in oil prices between 2004 and 2008. And they also missed the sharp drop in prices at the beginning of 2009 back to about $60 per barrel.

Where do you think oil prices will go in the next decade? *Your* guess may be as good as anyone's!

Of course, knowing that extrapolation requires thought and caution doesn't stop people. The temptation to see into the future is hard to resist. So our more realistic advice is this:

If you extrapolate into the future, at least don't believe blindly that the prediction will come true.

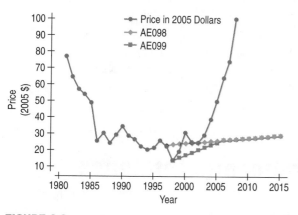

FIGURE 9.8
Here are the EIA forecasts with the actual prices from 1981 to 2009. Neither forecast predicted the sharp run-up in the past few years.

Extrapolation: Reaching Beyond the Data

The U.S. Census Bureau reports the median age at first marriage for men and women. Here's a regression of median *Age* (at first marriage) for men against *Year* (since 1890) at every census from 1890 to 1940:

R-squared = 92.6%
s = 0.2417

Variable	Coefficient
Intercept	25.7
Year	−0.04

The regression equation is

$$\widehat{Age} = 25.7 - 0.04\ Year.$$

QUESTION: What would this model predict as the age at first marriage for men in the year 2000?

When *Year* counts from 0 in 1890, the year 2000 is "110." Substituting 110 for *Year*, we find that the model predicts a first marriage *Age* of 25.7 − 0.04 × 110 = 21.3 years old.

QUESTION: In the year 2000, the median *Age* at first marriage for men was almost 27 years. What's gone wrong?

Here's a scatterplot of the median *Age* at first marriage for men for all the data from 1890 to 2007:

The average age at which men married fell at the rate of about a year every 25 years from 1890 to 1940.

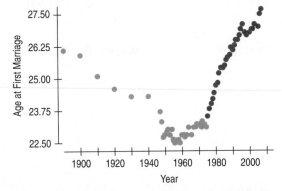

Median *Age* at first marriage (years of age) for men in the United States vs. *Year*. The regression line is fit only to the first 50 years of the data (shown in blue), which looked nicely linear. But the linear pattern could not have continued, and in fact it changed in direction, steepness, and strength.

Now we can see why the extrapolation failed. Although the trend in *Age* at first marriage was linear for parts of the century, it did not follow the same linear pattern over the entire century.

Outliers, Leverage, and Influence

The outcome of the 2000 U.S. presidential election was determined in Florida amid much controversy. The main race was between George W. Bush and Al Gore, but two minor candidates played a significant role. To the political right of the main party candidates was Pat Buchanan, while to the political left was

"*Nature is nowhere accustomed more openly to display her secret mysteries than in cases where she shows traces of her workings apart from the beaten path.*"

—William Harvey
(1657)

Ralph Nader. Generally, Nader earned more votes than Buchanan throughout the state. We would expect counties with larger vote totals to give more votes to each candidate. Here's a regression relating *Buchanan*'s vote totals by county in the state of Florida to *Nader*'s:

Dependent variable is: Buchanan
R-squared = 42.8%

Variable	Coefficient
Intercept	50.3
Nader	0.14

The regression model,

$$\widehat{Buchanan} = 50.3 + 0.14\,Nader,$$

says that, in each county, Buchanan received about 0.14 times (or 14% of) the vote Nader received, starting from a base of 50.3 votes.

This seems like a reasonable regression, with an R^2 of almost 43%. But we've violated all three Rules of Data Analysis by going straight to the regression table without making a picture.

Here's a scatterplot that shows the vote for Buchanan in each county of Florida plotted against the vote for Nader. The striking **outlier** is Palm Beach County.

FIGURE 9.9
Votes received by Buchanan against votes for Nader in all Florida counties in the presidential election of 2000. The red "x" point is Palm Beach County, home of the "butterfly ballot."

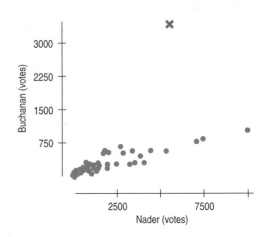

The so-called "butterfly ballot," used only in Palm Beach County, was a source of controversy. It has been claimed that the format of this ballot confused voters so that some who intended to vote for the Democrat, Al Gore, punched the wrong hole next to his name and, as a result, voted for Buchanan.

The scatterplot shows a strong, positive, linear association, and one striking point. With Palm Beach removed from the regression, the R^2 jumps from 42.8% to 82.1% and the slope of the line changes to 0.1, suggesting that Buchanan received only about 10% of the vote that Nader received. With more than 82% of the variability of the Buchanan vote accounted for, the model when Palm Beach is omitted certainly fits better. Palm Beach County now stands out, not as a Buchanan stronghold, but rather as a clear violation of the model that begs for explanation.

One of the great values of models is that, by establishing an idealized behavior, they help us to see when and how data values are unusual. In regression, a point can stand out in two different ways. First, a data value can have a large residual, as Palm Beach County

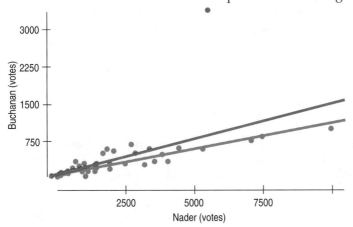

FIGURE 9.10
The red line shows the effect that one unusual point can have on a regression.

"Give me a place to stand and I will move the Earth."

–Archimedes
(287–211 B.C.E.)

A S **Activity: Leverage.** You may be surprised to see how sensitive to a single influential point a regression line is.

"For whoever knows the ways of Nature will more easily notice her deviations; and, on the other hand, whoever knows her deviations will more accurately describe her ways."

–Francis Bacon
(1561–1626)

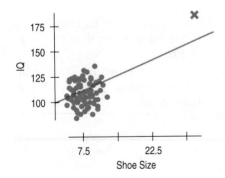

FIGURE 9.11
Bozo's extraordinarily large shoes give his data point high leverage in the regression. Wherever Bozo's *IQ* falls, the regression line will follow.

does in this example. Because they seem to be different from the other cases, points whose residuals are large always deserve special attention.

A data point can also be unusual if its *x*-value is far from the mean of the *x*-values. Such a point is said to have high **leverage**. The physical image of a lever is exactly right. We know the line must pass through $(\bar{x}, \bar{y})$, so you can picture that point as the fulcrum of the lever. Just as sitting farther from the hinge on a see-saw gives you more leverage to pull it your way, points with values far from $\bar{x}$ pull more strongly on the regression line.

A point with high leverage has the potential to change the regression line. But it doesn't always use that potential. If the point lines up with the pattern of the other points, then including it doesn't change our estimate of the line. By sitting so far from $\bar{x}$, though, it may strengthen the relationship, inflating the correlation and R^2. How can you tell if a high-leverage point actually changes the model? Just fit the linear model twice, both with and without the point in question. We say that a point is **influential** if omitting it from the analysis gives a very different model.[2]

Influence depends on both leverage and residual; a case with high leverage whose *y*-value sits right on the line fit to the rest of the data is not influential. Removing that case won't change the slope, even if it does affect R^2. A case with modest leverage but a very large residual (such as Palm Beach County) can be influential. Of course, if a point has enough leverage, it can pull the line right to it. Then it's highly influential, but its residual is small. The only way to be sure is to fit both regressions.

Unusual points in a regression often tell us more about the data and the model than any other points. We face a challenge: The best way to identify unusual points is against the background of a model, but a model dominated by a single case is unlikely to be useful for identifying unusual cases. (That insight's at least 400 years old. See the quote in the margin.) Don't give in to the temptation to simply delete points that don't fit the line. You can set aside cases and discuss what the model looks like with and without them, but arbitrarily deleting cases can give a false sense of how well the model fits the data. Your goal should be understanding the data, not making R^2 as big as you can.

In 2000, George W. Bush won Florida (and thus the presidency) by only a few hundred votes, so Palm Beach County's residual is big enough to be meaningful. It's the rare unusual point that determines a presidency, but all are worth examining and trying to understand.

A point with so much influence that it pulls the regression line close to it can make its residual deceptively small. Influential points like that can have a shocking effect on the regression. Here's a plot of *IQ* against *Shoe Size*, again from the fanciful study of intelligence and foot size in comedians we saw in Chapter 7. The linear regression output shows

Dependent variable is: IQ
R-squared = 24.8%

Variable	Coefficient
Intercept	93.3265
Shoe size	2.08318

Although this is a silly example, it illustrates an important and common potential problem: Almost all of the variance accounted for ($R^2 = 24.8\%$) is due to *one* point, namely, Bozo. Without Bozo, there is little correlation

---

[2] Some textbooks use the term *influential point* for any observation that influences the slope, intercept, or R^2. We'll reserve the term for points that influence the slope.

between *Shoe Size* and *IQ*. Look what happens to the regression when we take him out:

Dependent variable is: IQ
R-squared 0.7%

Variable	Coefficient
Intercept	105.458
Shoe size	−0.460194

The R^2 value is now 0.7%—a very weak linear relationship (as one might expect!). One single point exhibits a great influence on the regression analysis.

What would have happened if Bozo hadn't shown his comic genius on IQ tests? Suppose his measured *IQ* had been only 50. The slope of the line would then drop from 0.96 IQ points/shoe size to −0.69 IQ points/shoe size. No matter where Bozo's *IQ* is, the line tends to follow it because his *Shoe Size*, being so far from the mean *Shoe Size*, makes this a high-leverage point.

Even though this example is far-fetched, similar situations occur all the time in real life. For example, a regression of sales against floor space for hardware stores that looked primarily at small-town businesses could be dominated in a similar way if The Home Depot were included.

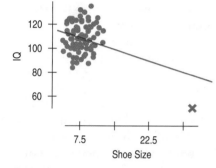

FIGURE 9.12
If Bozo's IQ were low, the regression slope would change from positive to negative. A single influential point can change a regression model drastically.

Warning: Influential points can hide in plots of residuals. Points with high leverage pull the line close to them, so they often have small residuals. You'll see influential points more easily in scatterplots of the original data or by finding a regression model with and without the points.

Each of these scatterplots shows an unusual point. For each, tell whether the point is a high-leverage point, would have a large residual, or is influential.

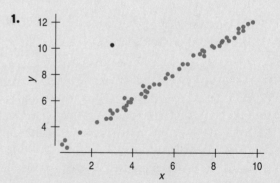

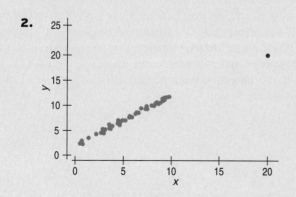

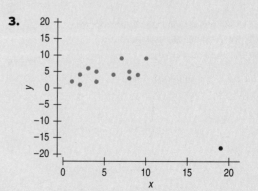

Lurking Variables and Causation

One common way to interpret a regression slope is to say that "a change of 1 unit in x results in a change of b_1 units in y." This way of saying things encourages causal thinking. Beware.

In Chapter 7, we tried to make it clear that no matter how strong the correlation is between two variables, there's no simple way to show that one variable causes the other. Putting a regression line through a cloud of points just increases the temptation to think and to say that the x-variable *causes* the y-variable. Just to make sure, let's repeat the point again: No matter how strong the association, no matter how large the R^2 value, no matter how straight the line, there is no way to conclude from a regression alone that one variable *causes* the other. There's always the possibility that some third variable is driving both of the variables you have observed. With observational data, as opposed to data from a designed experiment, there is no way to be sure that a **lurking variable** is not the cause of any apparent association.[3]

Here's an example: The scatterplot shows the *Life Expectancy* (average of men and women, in years) for each of 40 countries of the world, plotted against the square root of the number of *Doctors per person* in the country. (The square root is here to make the relationship satisfy the Straight Enough Condition, as we saw back in Chapter 7.)

The strong positive association ($R^2 = 62.9\%$) seems to confirm our expectation that more *Doctors per person* improves health care, leading to longer lifetimes and a greater *Life Expectancy*. The strength of the association would *seem* to argue that we should send more doctors to developing countries to increase life expectancy.

That conclusion is about the consequences of a change. Would sending more doctors increase life expectancy? Specifically, do doctors *cause* greater life expectancy? Perhaps, but these are observed data, so there may be another explanation for the association.

In Figure 9.14, the similar-looking scatterplot's x-variable is the square root of the number of *Televisions* per person in each country. The positive association in this scatterplot is even *stronger* than the association in the previous plot ($R^2 = 72.5\%$). We can fit the linear model, and quite possibly use the number of TVs as a way to predict life expectancy. Should we conclude that increasing the number of TVs actually extends lifetimes? If so, we should send TVs instead of doctors to developing countries. Not only is the correlation with life expectancy higher, but TVs are much cheaper than doctors.

What's wrong with this reasoning? Maybe we were a bit hasty earlier when we concluded that doctors *cause* longer lives. Maybe there's a lurking variable here. Countries with higher standards of living have both longer life expectancies *and* more doctors (and more TVs). Could higher living standards cause changes in the other variables? If so, then improving living standards might be expected to prolong lives, increase the number of doctors, and increase the number of TVs.

From this example, you can see how easy it is to fall into the trap of mistakenly inferring causality from a regression. For all we know, doctors (or TVs!) *do* increase life expectancy. But we can't tell that from data like these, no matter how much we'd like to. Resist the temptation to conclude that x causes y from a regression, no matter how obvious that conclusion seems to you.

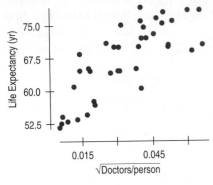

FIGURE 9.13
The relationship between *Life Expectancy* (years) and availability of *Doctors* (measured as $\sqrt{\text{doctors per person}}$) for countries of the world is strong, positive, and linear.

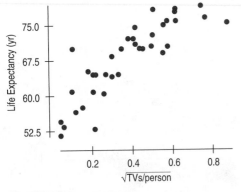

FIGURE 9.14
To increase life expectancy, don't send doctors, send TVs; they're cheaper and more fun. Or maybe that's not the right interpretation of this scatterplot of *life expectancy* against availability of TVs (as $\sqrt{\text{TVs per person}}$).

---

[3] Chapter 13 discusses observational data and experiments at greater length.

Working with Summary Values

Scatterplots of statistics summarized over groups tend to show less variability than we would see if we measured the same variable on individuals. This is because the summary statistics themselves vary less than the data on the individuals do—a fact we will make more specific in coming chapters.

In Chapter 7 we looked at the heights and weights of individual students. There we saw a correlation of 0.644, so R^2 is 41.5%.

FIGURE 9.15
Weight (lb) against *Height* (in.) for a sample of men. There's a strong, positive, linear association.

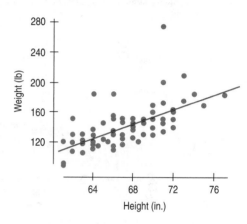

Suppose, instead of data on individuals, we knew only the mean weight for each height value. The scatterplot of mean weight by height would show less scatter. And the R^2 would increase to 80.1%.

FIGURE 9.16
Mean Weight (lb) shows a stronger linear association with *Height* than do the weights of individuals. Means vary less than individual values.

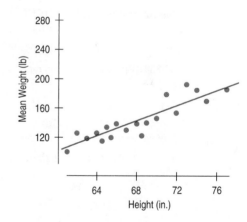

To think about lurking variables you must think "outside the box." What variables are *not* in your data that ought to be?

Scatterplots of summary statistics show less scatter than the baseline data on individuals and can give a false impression of how well a line summarizes the data. There's no simple correction for this phenomenon. Once we're given summary data, there's no simple way to get the original values back.

In the life expectancy and TVs example, we have no good measure of exposure to doctors or to TV on an individual basis. But if we did, we should expect the scatterplot to show more variability and the corresponding R^2 to be smaller. The bottom line is that you should be a bit suspicious of conclusions based on regressions of summary data. They may look better than they really are.

FOR EXAMPLE Using Several of These Methods Together

Motorcycles designed to run off-road, often known as dirt bikes, are specialized vehicles. We have data on 104 dirt bikes available for sale in 2005. Some cost as little as $3000, while others are substantially more expensive. Let's investigate how the size and type of engine contribute to the cost of a dirt bike. As always, we start with a scatterplot.

Here's a scatterplot of the manufacturer's suggested retail price (*MSRP*) in dollars against the engine *Displacement* (in cubic centimeters), along with a regression analysis:

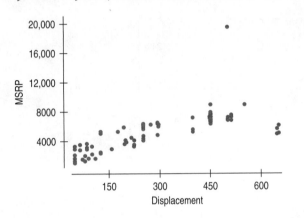

Dependent variable is: MSRP
R-squared = 49.9%   s = 1737

Variable	Coefficient
Intercept	2273.67
Displacement	10.0297

QUESTION: What do you see in the scatterplot?

There is a strong positive association between the engine displacement of dirt bikes and the manufacturer's suggested retail price. One of the dirt bikes is an outlier; its price is more than double that of any other bike.

The outlier is the Husqvarna TE 510 Centennial. Most of its components are handmade exclusively for this model, including extensive use of carbon fiber throughout. That may explain its $19,500 price tag! Clearly, the TE 510 is not like the other bikes. We'll set it aside for now and look at the data for the remaining dirt bikes.

QUESTION: What effect will removing this outlier have on the regression? Describe how the slope, R^2, and s_e will change.

The TE 510 was an influential point, tilting the regression line upward. With that point removed, the regression slope will get smaller. With that dirt bike omitted, the pattern becomes more consistent, so the value of R^2 should get larger and the standard deviation of the residuals, s_e, should get smaller.

With the outlier omitted, here's the new regression and a scatterplot of the residuals:

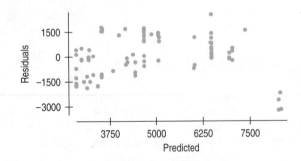

Dependent variable is: MSRP
R-squared = 61.3%   s = 1237

Variable	Coefficient
Intercept	2411.02
Displacement	9.05450

QUESTION: What do you see in the residuals plot?

The points at the far right don't fit well with the other dirt bikes. Overall, there appears to be a bend in the relationship, so a linear model may not be appropriate.

Let's try a re-expression. Here's a scatterplot showing *MSRP* against the cube root of *Displacement* to make the relationship closer to straight. (Since displacement is measured in cubic centimeters, its cube root has the simple units of centimeters.) In addition, we've colored the plot according to the cooling method used in the bike's engine: liquid or air. Each group is shown with its own regression line, as we did for the cereals on different shelves.

QUESTION: What does this plot say about dirt bikes?

There appears to be a positive, linear relationship between *MSRP* and the cube root of *Displacement*. In general, the larger the engine a bike has, the higher the suggested price. Liquid-cooled dirt bikes, however, typically cost more than air-cooled bikes with comparable displacement. A few liquid-cooled bikes appear to be much less expensive than we might expect, given their engine displacements (but without separating the groups we might have missed that because they look more like air-cooled bikes.)

[Jiang Lu, Joseph B. Kadane, and Peter Boatwright, "The Dirt on Bikes: An Illustration of CART Models for Brand Differentiation," provided data on 2005-model bikes.]

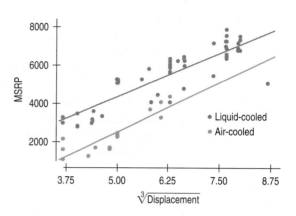

What Can Go Wrong?

This entire chapter has held warnings about things that can go wrong in a regression analysis. So let's just recap. When you make a linear model:

- **Make sure the relationship is straight.** Check the Straight Enough Condition. Always examine the residuals for evidence that the Linearity Assumption has failed. It's often easier to see deviations from a straight line in the residuals plot than in the scatterplot of the original data. Pay special attention to the most extreme residuals because they may have something to add to the story told by the linear model.

- **Be on guard for different groups in your regression.** Check for evidence that the data consist of separate subsets. If you find subsets that behave differently, consider fitting a different linear model to each subset.

- **Beware of extrapolating.** Beware of extrapolation beyond the *x*-values that were used to fit the model. Although it's common to use linear models to extrapolate, the practice is dangerous.

- **Beware especially of extrapolating into the future!** Be especially cautious about extrapolating into the future with linear models. To predict the future, you must assume that future changes will continue at the same rate you've observed in the past. Predicting the future is particularly tempting and particularly dangerous.

- **Look for unusual points.** Unusual points always deserve attention and may well reveal more about your data than the rest of the points combined. Always look for them and try to understand why they stand apart. A scatterplot of the data is a good way to see high-leverage and influential points. A scatterplot of the residuals against the predicted values is a good tool for finding points with large residuals.

- **Beware of high-leverage points and especially of those that are influential.** Influential points can alter the regression model a great deal. The resulting model may say more about one or two points than about the overall relationship.

- **Consider comparing two regressions.** To see the impact of outliers on a regression, it's often wise to run two regressions, one with and one without the extraordinary points, and then to discuss the differences.

- **Treat unusual points honestly.** If you remove enough carefully selected points, you will eventually get a regression with a high R^2, but it won't give you much understanding. Some variables are not related in a way that's simple enough for a linear model to fit very well. When that happens, report the failure and stop.

- **Beware of lurking variables.** Think about lurking variables before interpreting a linear model. It's particularly tempting to explain a strong regression by thinking that the *x*-variable *causes* the *y*-variable. A linear model alone can never demonstrate such causation, in part because it cannot eliminate the chance that a lurking variable has caused the variation in both *x* and *y*.

- **Watch out when dealing with data that are summaries.** Be cautious in working with data values that are themselves summaries, such as means or medians. Such statistics are less variable than the data on which they are based, so they tend to inflate the impression of the strength of a relationship.

CONNECTIONS

We should always be alert to things that could go wrong if we were to use statistics without thinking carefully. Regression opens new vistas of potential problems. But each one relates to issues we've thought about before.

It is always important that our data be from a single homogeneous group and not made up of disparate groups. We looked for multiple modes in single variables. Now we check scatterplots for evidence of subgroups in our data. As with modes, it's often best to split the data and analyze the groups separately.

Our concern with unusual points and their potential influence also harks back to our earlier concern with outliers in histograms and boxplots—and for many of the same reasons. As we've seen here, regression offers such points new scope for mischief.

The risks of interpreting linear models as causal or predictive arose in Chapters 7 and 8. And they're important enough to mention again in later chapters.

WHAT HAVE WE LEARNED?

We've learned that there are many ways in which a data set may be unsuitable for a regression analysis.

- ▶ Watch out for more than one group hiding in your regression analysis. If you find subsets of the data that behave differently, consider fitting a different regression model to each subset.

- ▶ The **Straight Enough Condition** says that the relationship should be reasonably straight to fit a regression. Somewhat paradoxically, sometimes it's easier to see that the relationship is not straight *after* fitting the regression by examining the residuals. The same is true of outliers.

- ▶ The **Outlier Condition** actually means two things: Points with large residuals or high leverage (especially both) can influence the regression model significantly. It's a good idea to perform the regression analysis with and without such points to see their impact.

And we've learned that even a good regression doesn't mean we should believe that the model says more than it really does.

- ▶ Extrapolation far from $\bar{x}$ can lead to silly and useless predictions.

- ▶ Even an R^2 near 100% doesn't indicate that *x* causes *y* (or the other way around). Watch out for lurking variables that may affect both *x* and *y*.

- ▶ Be careful when you interpret regressions based on *summaries* of the data sets. These regressions tend to look stronger than the regression based on all the individual data.

Terms

Extrapolation   Although linear models provide an easy way to predict values of y for a given value of x, it is unsafe to predict for values of x far from the ones used to find the linear model equation. Such extrapolation may pretend to see into the future, but the predictions should not be trusted (p. 216).

Outlier   Any data point that stands away from the others can be called an outlier. In regression, outliers can be extraordinary in two ways: by having a large residual or by having high leverage (p. 218).

Leverage   Data points whose x-values are far from the mean of x are said to exert leverage on a linear model. High-leverage points pull the line close to them, and so they can have a large effect on the line, sometimes completely determining the slope and intercept. With high enough leverage, their residuals can be deceptively small (p. 220).

Influential point   If omitting a point from the data results in a very different regression model, then that point is called an influential point (p. 220).

Lurking variable   A variable that is not explicitly part of a model but affects the way the variables in the model appear to be related is called a lurking variable. Because we can never be certain that observational data are not hiding a lurking variable that influences both x and y, it is never safe to conclude that a linear model demonstrates a causal relationship, no matter how strong the linear association (p. 222).

Skills

THINK

▶ Understand that we cannot fit linear models or use linear regression if the underlying relationship between the variables is not itself linear.

▶ Understand that data used to find a model must be homogeneous. Look for subgroups in data before you find a regression, and analyze each separately.

▶ Know the danger of extrapolating beyond the range of the x-values used to find the linear model, especially when the extrapolation tries to predict into the future.

▶ Understand that points can be unusual by having a large residual or by having high leverage.

▶ Understand that an influential point can change the slope and intercept of the regression line.

▶ Look for lurking variables whenever you consider the association between two variables. Understand that a strong association does not mean that the variables are causally related.

▶ Know how to display residuals from a linear model by making a scatterplot of residuals against predicted values or against the x-variable, and know what patterns to look for in the picture.

SHOW

▶ Know how to look for high-leverage and influential points by examining a scatterplot of the data and how to look for points with large residuals by examining a scatterplot of the residuals against the predicted values or against the x-variable. Understand how fitting a regression line with and without influential points can add to your understanding of the regression model.

▶ Know how to look for high-leverage points by examining the distribution of the x-values or by recognizing them in a scatterplot of the data, and understand how they can affect a linear model.

TELL

▶ Include diagnostic information such as plots of residuals and leverages as part of your report of a regression.

▶ Report any high-leverage points.

▶ Report any outliers. Consider reporting analyses with and without outliers, to assess their influence on the regression.

▶ Include appropriate cautions about extrapolation when reporting predictions from a linear model.

▶ Discuss possible lurking variables.

REGRESSION DIAGNOSIS ON THE COMPUTER

Most statistics technology offers simple ways to check whether your data satisfy the conditions for regression. We have already seen that these programs can make a simple scatterplot. They can also help us check the conditions by plotting residuals.

DATA DESK

Click on the **HyperView** menu on the **Regression** output table. A menu drops down to offer scatterplots of residuals against predicted values, Normal probability plots of residuals, or just the ability to save the residuals and predicted values.

Click on the name of a predictor in the regression table to be offered a scatterplot of the residuals against that predictor.

COMMENTS

If you change any of the variables in the regression analysis, Data Desk will offer to update the plots of residuals.

EXCEL

The Data Analysis add-in for Excel includes a Regression command. The dialog box it shows offers to make plots of residuals.

COMMENTS

Do not use the Normal probability plot offered in the regression dialog. It is not what it claims to be and is wrong.

JMP

From the **Analyze** menu, choose **Fit Y by X**. Select **Fit Line**. Under Linear Fit, select **Plot Residuals**. You can also choose to **Save Residuals**. Subsequently, from the

Distribution menu, choose **Normal quantile plot** or **histogram** for the residuals.

MINITAB

From the **Stat** menu, choose **Regression**. From the **Regression** submenu, select **Regression** again. In the Regression dialog, enter the response variable name in the "Response" box and the predictor variable name in the "Predictor" box. To specify saved results, in the Regression dialog, click **Storage**. Check "Residuals" and

"Fits." Click **OK**. To specify displays, in the Regression dialog, click **Graphs**. Under "Residual Plots," select "Individual plots" and check "Residuals versus fits." Click **OK**. Now back in the Regression dialog, click **OK**. Minitab computes the regression and the requested saved values and graphs.

SPSS

From the **Analyze** menu, choose **Regression**. From the Regression submenu, choose **Linear**. After assigning variables to their roles in the regression, click the "Plots ..." button.

In the Plots dialog, you can specify a Normal probability plot of residuals and scatterplots of various versions of standardized residuals and predicted values.

COMMENTS

A plot of ***ZRESID** against ***PRED** will look most like the residual plots we've discussed. SPSS standardizes the residuals by dividing by their standard deviation. (There's no need to subtract their mean; it must be zero.) The standardization doesn't affect the scatterplot.

TI-83/84 PLUS

To make a residuals plot, set up a **STATPLOT** as a scatterplot. Specify your explanatory data list as **Xlist**. For **Ylist**, import the name **RESID** from the **LIST NAMES** menu. **ZoomStat** will now create the residuals plot.

COMMENTS

Each time you execute a **LinReg** command, the calculator automatically computes the residuals and stores them in a data list named **RESID**. If you want to see them, go to **STAT EDIT**. Space through the names of the lists until you find a blank. Import **RESID** from the **LIST NAMES** menu. Now every time you have the calculator compute a regression analysis, it will show you the residuals.

TI-89

To make a residuals plot, define a Plot as a scatterplot. Specify your explanatory data list as **Xlist.** For **Ylist,** find the list name **resid** from **VAR-LINK** by arrowing to the **STATVARS** portion. Then press ② **(r)** and locate the list. Press ⟨ENTER⟩ to finish the plot definition and ⟨F5⟩ to display the plot.

COMMENTS

Each time you execute a **LinReg** command, the calculator automatically computes the residuals and stores them in a data list named **RESID**. If you don't want to see this (or any other calculator-generated list) anymore, press ⟨F1⟩ (Tools) and select choice 3: Setup Editor. Leaving the box for lists to display blank will reset the calculator to show only lists 1 through 6.

EXERCISES

T 1. **Marriage age 2007.** Is there evidence that the age at which women get married has changed over the past 100 years? The scatterplot shows the trend in age at first marriage for American women (www.census.gov).

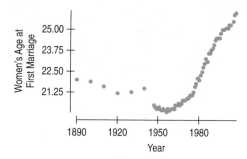

a) Is there a clear pattern? Describe the trend.
b) Is the association strong?
c) Is the correlation high? Explain.
d) Is a linear model appropriate? Explain.

T 2. **Smoking 2006.** The Centers for Disease Control and Prevention track cigarette smoking in the United States. How has the percentage of people who smoke changed since the danger became clear during the last half of the 20th century? The scatterplot shows percentages of smokers among men 18–24 years of age, as estimated by surveys, from 1965 through 2006 (http://www.cdc.gov/nchs/products/pubs/pubd/hus/healthrisk.htm).

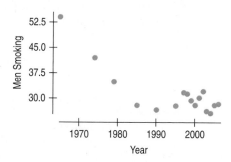

a) Is there a clear pattern? Describe the trend.
b) Is the association strong?
c) Is a linear model appropriate? Explain.

T 3. **Human Development Index.** The United Nations Development Programme (UNDP) uses the Human Development Index (HDI) in an attempt to summarize in one number the progress in health, education, and economics of a country. In 2006, the HDI was as high as 0.965 for Norway and as low as 0.331 for Niger. The gross domestic product per capita (GDPPC), by contrast, is often used to summarize the *overall* economic strength of a country. Is the HDI related to the GDPPC? Here is a scatterplot of *HDI* against *GDPPC*.

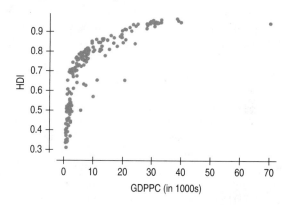

a) Explain why fitting a linear model to these data might be misleading.
b) If you fit a linear model to the data, what do you think a scatterplot of residuals versus predicted *HDI* will look like?
c) There is an outlier (Luxembourg) with a *GDPPC* of around $70,000. Will setting this point aside improve the model substantially? Explain.

T 4. **HDI revisited.** The United Nations Development Programme (UNDP) uses the Human Development Index (HDI) in an attempt to summarize in one number the progress in health, education, and economics of a country. The number of cell phone subscribers per 1000 people is positively associated with economic progress in a country. Can the number of cell phone subscribers be used to predict the HDI? Here is a scatterplot of HDI against cell phone subscribers:

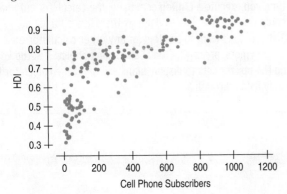

a) Explain why fitting a linear model to these data might be misleading.
b) If you fit a linear model to the data, what do you think a scatterplot of residuals vs. predicted *HDI* will look like?

5. **Good model?** In justifying his choice of a model, a student wrote, "I know this is the correct model because $R^2 = 99.4\%$."

a) Is this reasoning correct? Explain.
b) Does this model allow the student to make accurate predictions? Explain.

6. **Bad model?** A student who has created a linear model is disappointed to find that her R^2 value is a very low 13%.

a) Does this mean that a linear model is not appropriate? Explain.
b) Does this model allow the student to make accurate predictions? Explain.

(T) 7. **Movie dramas.** Here's a scatterplot of the production budgets (in millions of dollars) vs. the running time (in minutes) for major release movies in 2005. Dramas are plotted as red x's and all other genres are plotted as blue dots. (The re-make of *King Kong* is plotted as a black "-". At the time it was the most expensive movie ever made, and not typical of any genre.) A separate least squares regression line has been fitted to each group. For the following questions, just examine the plot:

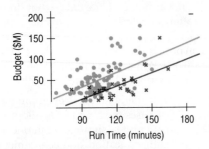

a) What are the units for the slopes of these lines?
b) In what way are dramas and other movies similar with respect to this relationship?
c) In what way are dramas different from other genres of movies with respect to this relationship?

(T) 8. **Smoking 2006, women and men.** In Exercise 2 we examined the percentage of men aged 18–24 who

smoked from 1965 to 2006 according to the Centers for Disease Control and Prevention. How about women? Here's a scatterplot showing the corresponding percentages for both men and women:

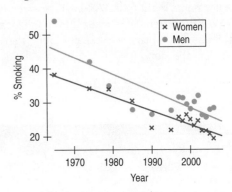

a) In what ways are the trends in smoking behavior similar for men and women?
b) How do the smoking rates for women differ from those for men?
c) Viewed alone, the trend for men may have seemed to violate the Linearity Condition. How about the trend for women? Does the consistency of the two patterns encourage you to think that a linear model for the trend in men might be appropriate? (Note: there is no correct answer to this question; it is raised for you to think about.)

(T) 9. **Oakland passengers.** The scatterplot below shows the number of passengers departing from Oakland (CA) airport month by month since the start of 1997. Time is shown as years since 1990, with fractional years used to represent each month. (Thus, June of 1997 is 7.5—halfway through the 7th year after 1990.)
www.oaklandairport.com

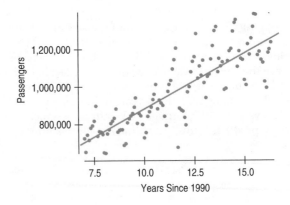

Here's a regression and the residuals plot:

Dependent variable is: Passengers
R-squared = 71.1%   s = 104330

Variable	Coefficient
Constant	282584
Year - 1990	59704.4

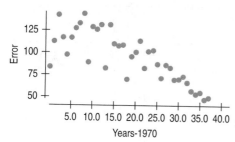

a) Interpret the slope and intercept of the model.
b) What does the value of R^2 say about the model?
c) Interpret s_e in this context.
d) Would you use this model to predict the numbers of passengers in 2010 (*Years Since 1990* = 20)? Explain.
e) There's a point near the middle of this time span with a large negative residual. Can you explain this outlier?

T 10. Tracking hurricanes 2007. In a previous chapter, we saw data on the errors (in nautical miles) made by the National Hurricane Center in predicting the path of hurricanes. The scatterplot shows the trend in the 24-hour tracking errors since 1970 (www.nhc.noaa.gov).

Dependent variable is: Error
R-squared = 61.9%   s = 17.02

Variable	Coefficient
Intercept	131.257
Years - 1970	−1.92572

a) Interpret the slope and intercept of the model.
b) Interpret s_e in this context.
c) The Center had a stated goal of achieving an average tracking error of 125 nautical miles in 2009. Will they make it? Why do you think so?
d) What if their goal were an average tracking error of 90 nautical miles?
e) What cautions would you state about your conclusion?

11. Unusual points. Each of the four scatterplots that follow shows a cluster of points and one "stray" point. For each, answer these questions:

1) In what way is the point unusual? Does it have high leverage, a large residual, or both?
2) Do you think that point is an influential point?
3) If that point were removed, would the correlation become stronger or weaker? Explain.
4) If that point were removed, would the slope of the regression line increase or decrease? Explain.

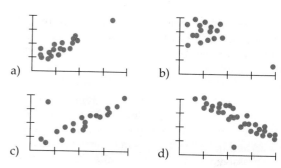

12. More unusual points. Each of the following scatterplots shows a cluster of points and one "stray" point. For each, answer these questions:

1) In what way is the point unusual? Does it have high leverage, a large residual, or both?
2) Do you think that point is an influential point?
3) If that point were removed, would the correlation become stronger or weaker? Explain.
4) If that point were removed, would the slope of the regression line increase or decrease? Explain.

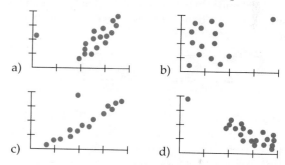

13. The extra point. The scatterplot shows five blue data points at the left. Not surprisingly, the correlation for these points is $r = 0$. Suppose *one* additional data point is added at one of the five positions suggested below in green. Match each point (a–e) with the correct new correlation from the list given.

1) −0.90   4) 0.05
2) −0.40   5) 0.75
3) 0.00

14. The extra point, revisited. The original five points in Exercise 13 produce a regression line with slope 0. Match each of the green points (a–e) with the slope of the line after that one point is added:

1) −0.45   4) 0.05
2) −0.30   5) 0.85
3) 0.00

15. **What's the cause?** Suppose a researcher studying health issues measures blood pressure and the percentage of body fat for several adult males and finds a strong positive association. Describe three different possible cause-and-effect relationships that might be present.

16. **What's the effect?** A researcher studying violent behavior in elementary school children asks the children's parents how much time each child spends playing computer games and has their teachers rate each child on the level of aggressiveness they display while playing with other children. Suppose that the researcher finds a moderately strong positive correlation. Describe three different possible cause-and-effect explanations for this relationship.

17. **Reading.** To measure progress in reading ability, students at an elementary school take a reading comprehension test every year. Scores are measured in "grade-level" units; that is, a score of 4.2 means that a student is reading at slightly above the expected level for a fourth grader. The school principal prepares a report to parents that includes a graph showing the mean reading score for each grade. In his comments he points out that the strong positive trend demonstrates the success of the school's reading program.

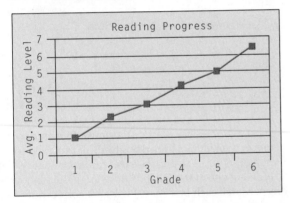

a) Does this graph indicate that students are making satisfactory progress in reading? Explain.
b) What would you estimate the correlation between *Grade* and *Average Reading Level* to be?
c) If, instead of this plot showing average reading levels, the principal had produced a scatterplot of the reading levels of all the individual students, would you expect the correlation to be the same, higher, or lower? Explain.
d) Although the principal did not do a regression analysis, someone as statistically astute as you might do that. (But don't bother.) What value of the slope of that line would you view as demonstrating acceptable progress in reading comprehension? Explain.

18. **Grades.** A college admissions officer, defending the college's use of SAT scores in the admissions process, produced the following graph. It shows the mean GPAs for last year's freshmen, grouped by SAT scores. How strong is the evidence that *SAT Score* is a good predictor of *GPA*? What concerns you about the graph, the statistical methodology or the conclusions reached?

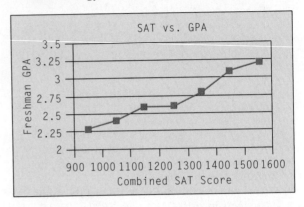

19. **Heating.** After keeping track of his heating expenses for several winters, a homeowner believes he can estimate the monthly cost from the average daily Fahrenheit temperature by using the model $\widehat{Cost} = 133 - 2.13\ Temp$. Here is the residuals plot for his data:

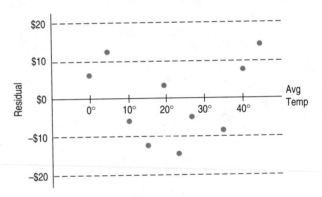

a) Interpret the slope of the line in this context.
b) Interpret the *y*-intercept of the line in this context.
c) During months when the temperature stays around freezing, would you expect cost predictions based on this model to be accurate, too low, or too high? Explain.
d) What heating cost does the model predict for a month that averages 10°?
e) During one of the months on which the model was based, the temperature did average 10°. What were the actual heating costs for that month?
f) Should the homeowner use this model? Explain.
g) Would this model be more successful if the temperature were expressed in degrees Celsius? Explain.

20. **Speed.** How does the speed at which you drive affect your fuel economy? To find out, researchers drove a compact car for 200 miles at speeds ranging from 35 to 75 miles per hour. From their data, they created the

model $\widehat{Fuel\ Efficiency} = 32 - 0.1\ Speed$ and created this residual plot:

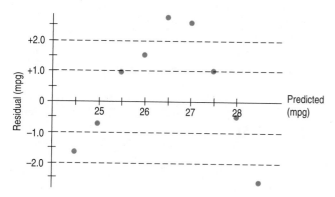

a) Interpret the slope of this line in context.
b) Explain why it's silly to attach any meaning to the y-intercept.
c) When this model predicts high *Fuel Efficiency*, what can you say about those predictions?
d) What *Fuel Efficiency* does the model predict when the car is driven at 50 mph?
e) What was the actual *Fuel Efficiency* when the car was driven at 45 mph?
f) Do you think there appears to be a strong association between *Speed* and *Fuel Efficiency*? Explain.
g) Do you think this is the appropriate model for that association? Explain.

T 21. Interest rates. Here's a plot showing the federal rate on 3-month Treasury bills from 1950 to 1980, and a regression model fit to the relationship between the *Rate* (in %) and *Years since 1950* (www.gpoaccess.gov/eop/).

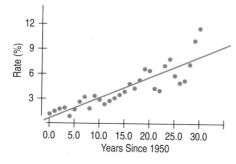

Dependent variable is: Rate
R-squared = 77.4%   s = 1.239

Variable	Coefficient
Intercept	0.640282
Year - 1950	0.247637

a) What is the correlation between *Rate* and *Year*?
b) Interpret the slope and intercept.
c) What does this model predict for the interest rate in the year 2000?
d) Would you expect this prediction to have been accurate? Explain.

T 22. Ages of couples 2007. The graph shows the ages of both men and women at first marriage (www.census.gov).

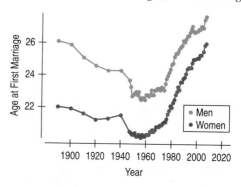

Clearly, the patterns for men and women are similar. But are the two lines getting closer together?

Here's a timeplot showing the *difference* in average age (men's age − women's age) at first marriage, the regression analysis, and the associated residuals plot.

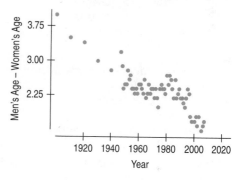

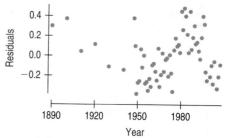

Dependent variable is: Age Difference
R-squared = 77.4%   s = 0.2334

Variable	Coefficient
Intercept	36.2639
Year	−0.01718

a) What is the correlation between *Age Difference* and *Year*?
b) Interpret the slope of this line.
c) Predict the average age difference in 2015.
d) Describe reasons why you might not place much faith in that prediction.

T 23. Interest rates revisited. In Exercise 21 you investigated the federal rate on 3-month Treasury bills between 1950 and 1980. The scatterplot below shows that the trend changed dramatically after 1980.

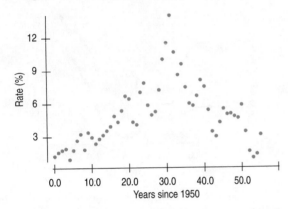

Years since 1950

Here's a regression model for the data since 1950.

Dependent variable is: Rate
R-squared = 74.5%   s = 1.630

Variable	Coefficient
Intercept	21.0688
Year - 1950	−0.356578

a) How does this model compare to the one in Exercise 21?
b) What does this model estimate the interest rate to have been in 2000? How does this compare to the rate you predicted in Exercise 21?
c) Do you trust this newer predicted value? Explain.
d) Given these two models, what would you predict the interest rate on 3-month Treasury bills will be in 2020?

T 24. Ages of couples, again. Has the trend of decreasing difference in age at first marriage seen in Exercise 22 gotten stronger recently? The scatterplot and residual plot for the data from 1980 through 2007, along with a regression for just those years, are below.

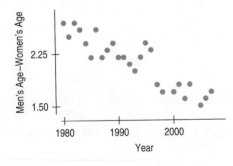

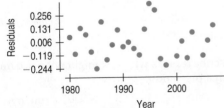

Dependent variable is: Men − Women
R-squared = 82.2%   s = 0.1583

Variable	Coefficient
Intercept	83.3416
Year	−0.04075

a) Is this linear model appropriate for the post-1980 data? Explain.
b) What does the slope say about marriage ages?
c) Explain why it's not reasonable to interpret the y-intercept.

T 25. Gestation. For humans, pregnancy lasts about 280 days. In other species of animals, the length of time from conception to birth varies. Is there any evidence that the gestation period is related to the animal's lifespan? The first scatterplot shows *Gestation Period* (in days) vs. *Life Expectancy* (in years) for 18 species of mammals. The highlighted point at the far right represents humans.

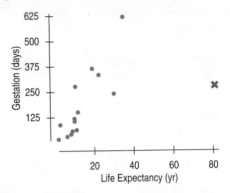

a) For these data, $r = 0.54$, not a very strong relationship. Do you think the association would be stronger or weaker if humans were removed? Explain.
b) Is there reasonable justification for removing humans from the data set? Explain.
c) Here are the scatterplot and regression analysis for the 17 nonhuman species. Comment on the strength of the association.

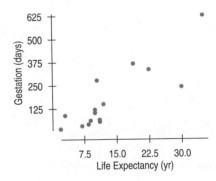

Dependent variable is: Gestation
R-squared = 72.2%

Variable	Coefficient
Constant	−39.5172
LifExp	15.4980

d) Interpret the slope of the line.
e) Some species of monkeys have a life expectancy of about 20 years. Estimate the expected gestation period of one of these monkeys.

T 26. Swim the lake 2008. People swam across Lake Ontario 45 times between 1974 and 2008 (www.soloswims.com). We might be interested in whether they are getting any faster or slower. Here are the regression of the crossing

Times (minutes) against the *Year* of the crossing and the residuals plot:

Dependent variable is: Time
R-squared = 2.8%   s = 461.3

Variable	Coefficient
Intercept	−13464.9
Year	7.41683

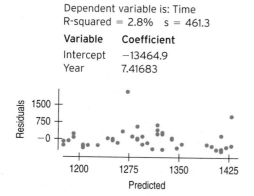

a) What does the R^2 mean for this regression?
b) Are the swimmers getting faster or slower? Explain.
c) The outlier seen in the residuals plot is a crossing by Vicki Keith in 1987 in which she swam a round trip, north to south, and then back again. Clearly, this swim doesn't belong with the others. Would removing it change the model a lot? Explain.

27. **Elephants and hippos.** We removed humans from the scatterplot in Exercise 25 because our species was an outlier in life expectancy. The resulting scatterplot (below) shows two points that now may be of concern. The point in the upper right corner of this scatterplot is for elephants, and the other point at the far right is for hippos.

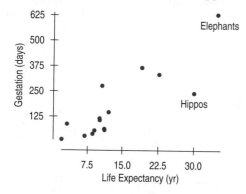

a) By removing one of these points, we could make the association appear to be stronger. Which point? Explain.
b) Would the slope of the line increase or decrease?
c) Should we just keep removing animals to increase the strength of the model? Explain.
d) If we remove elephants from the scatterplot, the slope of the regression line becomes 11.6 days per year. Do you think elephants were an influential point? Explain.

28. **Another swim 2006.** In Exercise 26 we saw that Vicki Keith's round-trip swim of Lake Ontario was an obvious outlier among the other one-way times. Here is the new regression after this unusual point is removed:

Dependent variable is: Time
R-Squared = 4.1%   s = 292.6

Variable	Coefficient
Intercept	−11048.7
Year	6.17091

a) In this new model, the value of s_e is much smaller. Explain what that means in this context.
b) Now would you be willing to say that the Lake Ontario swimmers are getting faster (or slower)?

29. **Marriage age 2007, revisited.** Suppose you wanted to predict the trend in marriage age for American women into the early part of this century.
a) How could you use the data graphed in Exercise 1 to get a good prediction? Marriage ages in selected years starting in 1900 are listed below. Use all or part of these data to create an appropriate model for predicting the average age at which women will first marry in 2015.

1900–1950 (10-yr intervals): 21.9, 21.6, 21.2, 21.3, 21.5, 20.3
1955–2005 (5-yr intervals): 20.2, 20.2, 20.6, 20.8, 21.1, 22.0, 23.3, 23.9, 24.5, 25.1, 25.9

b) How much faith do you place in this prediction? Explain.
c) Do you think your model would produce an accurate prediction about your grandchildren, say, 50 years from now? Explain.

30. **Bridges covered.** In Chapter 8 we found a relationship between the age of a bridge in Tompkins County, NY, and its condition as found by inspection. But we considered only bridges built since 1880. Tompkins County is the home of the oldest covered bridge in daily use in New York State. Built in 1853, it is judged to have a condition of 4.523.

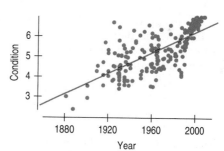

Dependent variable is: Condition
R-squared = 51.8%   s = 0.7192

Variable	Coefficient
Intercept	−44.9905
Year	0.025601

Suppose we were to add this bridge to the data.
a) If we use this regression to predict the condition of the covered bridge, what would its residual be?
b) If we add the covered bridge to the data, what would you expect to happen to the regression slope? Explain.
c) If we add the covered bridge to the data, what would you expect to happen to the R^2? Explain.
d) The bridge was extensively restored in 1972. If we use that date instead, do you find the condition of the bridge remarkable?

31. Life expectancy 2004. Data from the World Bank for 26 Western Hemisphere countries can be used to examine the association between female *Life Expectancy* and the average *Number of Children* women give birth to (http://devdata.worldbank.org/data-query/).

Country	Births/ Woman	Life Exp.	Country	Births/ Woman	Life Exp.
Argentina	2.3	74.6	Guatemala	4.4	67.6
Bahamas	2.3	70.5	Honduras	3.6	68.2
Barbados	1.7	75.4	Jamaica	2.4	70.8
Belize	3.0	71.9	Mexico	2.2	75.1
Bolivia	3.7	64.5	Nicaragua	3.2	70.1
Brazil	2.3	70.9	Panama	2.6	75.1
Canada	1.5	79.8	Paraguay	3.7	71.2
Chile	2.0	78.0	Peru	2.8	70.4
Colombia	2.4	72.6	Puerto Rico	1.9	77.5
Costa Rica	24.9	78.7	United States	2.0	77.4
Dominican			Uruguay	2.1	75.2
Republic	2.8	67.8	Venezuela	2.7	73.7
Ecuador	2.7	74.5	Virgin Islands	2.2	78.6
El Salvador	2.8	71.1			

a) Create a scatterplot relating these two variables, and describe the association.
b) Are there any countries that do not seem to fit the overall pattern?
c) Find the correlation, and interpret the value of R^2.
d) Find the equation of the regression line.
e) Is the line an appropriate model? Describe what you see in the residuals plot.
f) Interpret the slope and the *y*-intercept of the line.
g) If government leaders wanted to increase life expectancy in their country, should they encourage women to have fewer children? Explain.

32. Tour de France 2009. We met the Tour de France data set in Chapter 2 (in Just Checking). One hundred years ago, the fastest rider finished the course at an average speed of about 25.3 kph (around 15.8 mph). In 2005, Lance Armstrong averaged 41.65 kph (25.88 mph) for the fastest average winning speed in history.

a) Make a scatterplot of *Avg Speed* against *Year*. Describe the relationship of *Avg Speed* by *Year*, being careful to point out any unusual features in the plot.
b) Find the regression equation of *Avg Speed* on *Year*.
c) Are the conditions for regression met? Comment.

33. Inflation 2006. The Consumer Price Index (CPI) tracks the prices of consumer goods in the United States, as shown in the following table (ftp://ftp.bis.gov). It indicates, for example, that the average item costing $17.70 in 1926 cost $201.60 in the year 2006.

Year	CPI	Year	CPI
1914	10.0	1962	30.2
1918	15.1	1966	32.4
1922	16.8	1970	38.8
1926	17.7	1974	49.3
1930	16.7	1978	65.2
1934	13.4	1982	96.5
1938	14.1	1986	109.6
1942	16.3	1990	130.7
1946	19.5	1994	148.2
1950	24.1	1998	163.0
1954	26.9	2002	179.9
1958	28.9	2006	201.6

a) Make a scatterplot showing the trend in consumer prices. Describe what you see.
b) Be an economic forecaster: Project increases in the cost of living over the next decade. Justify decisions you make in creating your model.

34. Second stage 2009. Look once more at the data from the Tour de France. In Exercise 32 we looked at the whole history of the race, but now let's consider just the post–World War II era.

a) Find the regression of *Avg Speed* by *Year* only for years from 1947 to the present. Are the conditions for regression met?
b) Interpret the slope.
c) In 1979 Bernard Hinault averaged 39.8 kph, while in 2005 Lance Armstrong averaged 41.65 kph. Which was the more remarkable performance and why?

JUST CHECKING

ANSWERS

1. Not high leverage, not influential, large residual

2. High leverage, not influential, small residual

3. High leverage, influential, not large residual

Re-expressing Data: Get It Straight!

Where are we going?

What should we do when the data don't satisfy one of the conditions we're checking? What can we do if they are really skewed, or fail to satisfy the Straight Enough Condition?

Re-expressing the data–replacing the original by a simple function of it such as a logarithm or reciprocal–can often improve things. That makes the methods in this book far more useful because they can be applied to many more kinds of data.

And technology makes it all pretty easy.

A S
Activity: **Re-expressing Data.** Should you re-express data? Actually, you already do.

Scan through any Physics book. Most equations have powers, reciprocals, or logs.

How fast can you go on a bicycle? If you measure your speed, you probably do it in miles per hour or kilometers per hour. In a 12-mile-long time trial in the 2005 Tour de France, Dave Zabriskie *averaged* nearly 35 mph (54.7 kph), beating Lance Armstrong by 2 seconds. You probably realize that's a tough act to follow. It's fast. You can tell that at a glance because you have no trouble thinking in terms of distance covered per time.

OK, then, if you averaged 12.5 mph (20.1 kph) for a mile *run,* would *that* be fast? Would it be fast for a 100-m dash? Even if you run the mile often, you probably have to stop and calculate. Running a mile in under 5 minutes (12 mph) is fast. A mile at 16 mph would be a world record (that's a 3-minute, 45-second mile). There's no single *natural* way to measure speed. Sometimes we use time over distance; other times we use the *reciprocal,* distance over time. Neither one is *correct.* We're just used to thinking that way in each case.

So, how does this insight help us understand data? All quantitative data come to us measured in some way, with units specified. But maybe those units aren't the best choice. It's not that meters are better (or worse) than fathoms or leagues. What we're talking about is re-expressing the data another way by applying a function, such as a square root, log, or reciprocal. You already use some of them, even though you may not know it. For example, the Richter scale of earthquake strength (logs), the decibel scale for sound intensity (logs), the f/stop scale for camera aperture openings (squares), and the gauges of shotguns (square roots) all include simple functions of this sort.

Why bother? As with speeds, some expressions of the data may be easier to think about. And some may be much easier to analyze with statistical methods. We've seen that symmetric distributions are easier to summarize and

237

straight scatterplots are easier to model with regressions. We often look to re-express our data if doing so makes them more suitable for our methods.

Straight to the Point

We know from common sense and from Physics that heavier cars need more fuel, but exactly how does a car's weight affect its fuel efficiency? Here are the scatterplot of *Weight* (in pounds) and *Fuel Efficiency* (in miles per gallon) for 38 cars, and the residuals plot:

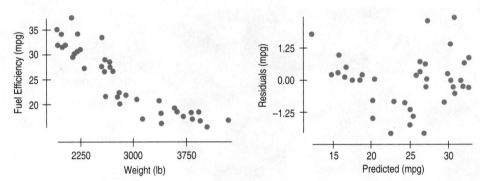

FIGURE 10.1

Fuel Efficiency (mpg) vs. *Weight* for 38 cars as reported by *Consumer Reports*.

The scatterplot shows a negative direction, roughly linear shape, and strong relationship. However, the residuals from a regression of *Fuel Efficiency* on *Weight* reveal a bent shape when plotted against the predicted values. Looking back at the original scatterplot, you may be able to see the bend.

Hmm. . . . Even though R^2 is 81.6%, the residuals don't show the random scatter we were hoping for. The shape is clearly bent. Looking back at the first scatterplot, you can probably see the slight bending. Think about the regression line through the points. How heavy would a car have to be to have a predicted gas mileage of 0? It looks like the *Fuel Efficiency* would go negative at about 6000 pounds. A Hummer H2 weighs about 6400 pounds. The H2 is hardly known for fuel efficiency, but it does get more than the *minus* 5 mpg this regression predicts. Extrapolation always requires caution, but it can go dangerously wrong when your model is wrong, because wrong models tend to do even worse the farther you get from the middle of the data.

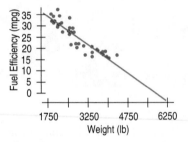

FIGURE 10.2

Extrapolating the regression line gives an absurd answer for vehicles that weigh as little as 6000 pounds.

The bend in the relationship between *Fuel Efficiency* and *Weight* is the kind of failure to satisfy the conditions for an analysis that we can repair by re-expressing the data. Instead of looking at miles per gallon, we could take the reciprocal and work with gallons per hundred miles.[1]

> **"Gallons per hundred miles—what an absurd way to measure fuel efficiency! Who would ever do it that way?"** Not all re-expressions are easy to understand, but in this case the answer is "Everyone except U.S. drivers." Most of the world measures fuel efficiency in liters per 100 kilometers (L/100 km). This is the same reciprocal form (fuel amount per distance driven) and differs from gallons per 100 miles only by a constant multiple of about 2.38. It has been suggested that most of the world says, "I've got to go 100 km; how much gas do I need?" But Americans say, "I've got 10 gallons in the tank. How far can I drive?" In much the same way, re-expressions "think" about the data differently but don't change what they mean.

The direction of the association is positive now, since we're measuring gas consumption and heavier cars consume more gas per mile. The relationship is much straighter, as we can see from a scatterplot of the regression residuals.

This is more the kind of boring residuals plot (no direction, no particular shape, no outliers, no bends) that we hope to see, so we have reason to think that the Straight Enough Condition is now satisfied. Now here's the payoff: What does

---

[1] Multiplying by 100 to get gallons per 100 miles simply makes the numbers easier to think about: You might have a good idea of how many gallons your car needs to drive 100 miles, but probably a much poorer sense of how much gas you need to go just 1 mile.

FIGURE 10.3
The reciprocal (1/y) is measured in gallons per mile. Gallons per 100 miles gives more meaningful numbers. The reciprocal is more nearly linear against *Weight* than the original variable, but the re-expression changes the direction of the relationship. The residuals from the regression of *Fuel Consumption* (gal/100 mi) *on Weight* show less of a pattern than before.

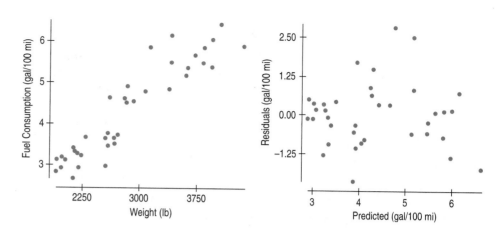

the reciprocal model say about the Hummer? The regression line fit to *Fuel Consumption vs. Weight* predicts somewhere near 9.7 for a car weighing 6400 pounds. What does this mean? It means the car is predicted to use 9.7 gallons for every 100 miles, or in other words,

$$\frac{100 \; miles}{9.7 \; gallons} = 10.3 \; mpg.$$

That's a much more reasonable prediction and very close to the reported value of 11.0 miles per gallon (of course, *your* mileage may vary . . .).

Goals of Re-expression

We re-express data for several reasons. Each of these goals helps make the data more suitable for analysis by our methods.

Goal 1

Make the distribution of a variable (as seen in its histogram, for example) more symmetric. It's easier to summarize the center of a symmetric distribution, and for nearly symmetric distributions, we can use the mean and standard deviation. If the distribution is unimodal, then the resulting distribution may be closer to the Normal model, allowing us to use the 68–95–99.7 Rule.

Here are a histogram, quite skewed, showing the *Assets* of 77 companies selected from the Forbes 500 list (in $100,000) and the more symmetric histogram after taking logs.

WHO	77 large companies
WHAT	Assets, sales, and market sector
UNITS	$100,000
HOW	Public records
WHEN	1986
WHY	By *Forbes* magazine in reporting on the Forbes 500 for that year

A S   *Simulation:* **Re-expression in Action.** Slide the re-expression power and watch the histogram change.

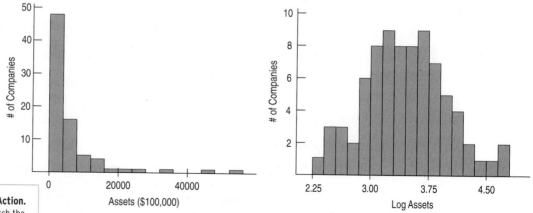

FIGURE 10.4
The distribution of the *Assets* of large companies is skewed to the right. Data on wealth often look like this. Taking logs makes the distribution more nearly symmetric.

Goal 2

Make the spread of several groups (as seen in side-by-side boxplots) more alike, even if their centers differ. Groups that share a common spread are easier to compare. We'll see methods later in the book that can be applied only to groups with a common standard deviation. We saw an example of re-expression for comparing groups with boxplots in Chapter 5.

Here are the *Assets* of these companies by *Market Sector:*

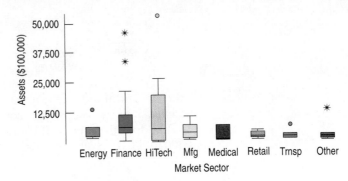

FIGURE 10.5

Assets of large companies by *Market Sector.*

It's hard to compare centers or spreads, and there seem to be a number of high outliers.

Taking logs makes the individual boxplots more symmetric and gives them spreads that are more nearly equal.

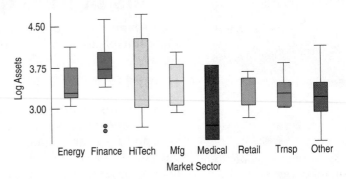

FIGURE 10.6

After re-expressing by logs, it's much easier to compare across market sectors. The boxplots are more nearly symmetric, most have similar spreads, and the companies that seemed to be outliers before are no longer extraordinary. Two new outliers have appeared in the finance sector. They are the only companies in that sector that are not banks. Perhaps they don't belong there.

Doing this makes it easier to compare assets across market sectors. It can also reveal problems in the data. Some companies that looked like outliers on the high end turned out to be more typical. But two companies in the finance sector now stick out. Unlike the rest of the companies in that sector, they are not banks. They may have been placed in the wrong sector, but we couldn't see that in the original data.

Goal 3

Make the form of a scatterplot more nearly linear. Linear scatterplots are easier to model. We saw an example of scatterplot straightening in Chapter 7. The greater value of re-expression to straighten a relationship is that we can fit a linear model once the relationship is straight.

Here are *Assets* of the companies plotted against the logarithm of *Sales*, clearly bent. Taking logs of *Assets* makes things much more linear.

FIGURE 10.7
Assets vs. log *Sales* shows a positive association (bigger sales go with bigger assets) but a bent shape. Note also that the points go from tightly bunched at the left to widely scattered at the right; the plot "thickens." In the second plot, log *Assets* vs. log *Sales* shows a clean, positive, linear association. And the variability at each value of *x* is about the same.

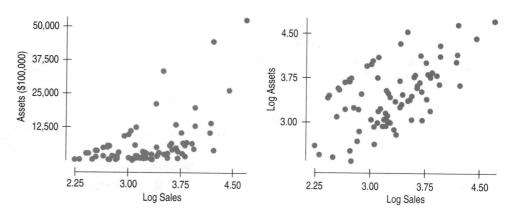

Goal 4

Make the scatter in a scatterplot spread out evenly rather than thickening at one end. Having an even scatter is a condition of many methods of Statistics, as we'll see in later chapters. This goal is closely related to Goal 2, but it often comes along with Goal 3. Indeed, a glance back at the scatterplot (Figure 10.7) shows that the plot for *Assets* is much more spread out on the right than on the left, while the plot for log *Assets* has roughly the same variation in log *Assets* for any *x*-value.

FOR EXAMPLE Recognizing When a Re-expression Can Help

In Chapter 9, we saw the awesome ability of emperor penguins to slow their heart rates while diving. Here are three displays relating to the diving heart rates:

(The boxplots show the diving heart rates for each of the 9 penguins whose dives were tracked. The names are those given by the researchers; EP = emperor penguin.)

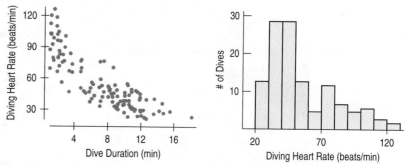

QUESTION: What features of each of these displays suggest that a re-expression might be helpful?

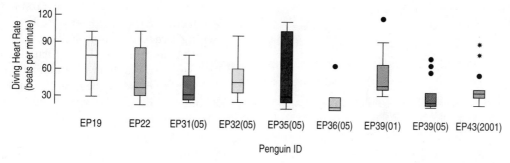

The scatterplot shows a curved relationship, concave upward, between the duration of the dives and penguins' heart rates. Re-expressing either variable may help to straighten the pattern.

The histogram of heart rates is skewed to the high end. Re-expression often helps to make skewed distributions more nearly symmetric.

The boxplots each show skewness to the high end as well. The medians are low in the boxes, and several show high outliers.

The Ladder of Powers

How can we pick a re-expression to use? Some kinds of data favor certain re-expressions. But even starting from a suggested one, it's always a good idea to look around a bit. Fortunately, the re-expressions line up in order, so it's easy to slide up and down to find the best one. The trick is to choose our re-expressions from a simple family that includes the most common ways to re-express data. More important, the members of the family line up in order, so that the farther you move away from the original data (the "1" position), the greater is the effect on the data. This fact lets you search systematically for a re-expression that works, stepping a bit farther from "1" or taking a step back toward "1" as you see the results.

Where to start? It turns out that certain kinds of data are more likely to be helped by particular re-expressions. Knowing that gives you a good place to start your search for a re-expression. We call this collection of re-expressions the **Ladder of Powers**.

Power	Name	Comment
2	The square of the data values, y^2.	Try this for unimodal distributions that are skewed to the left.
1	The raw data—no change at all. This is "home base." The farther you step from here up or down the ladder, the greater the effect.	Data that can take on both positive and negative values with no bounds are less likely to benefit from re-expression.
1/2	The square root of the data values, $\sqrt{y}$.	Counts often benefit from a square root re-expression. For counted data, start here.
"0"	Although mathematicians define the "0-th" power differently,[2] for us the place is held by the logarithm. You may feel uneasy about logarithms. Don't worry; the computer or calculator does the work.[3]	Measurements that cannot be negative, and especially values that grow by percentage increases such as salaries or populations, often benefit from a log re-expression. When in doubt, start here. If your data have zeros, try adding a small constant to all values before finding the logs.
$-1/2$	The (negative) reciprocal square root, $-1/\sqrt{y}$.	An uncommon re-expression, but sometimes useful. Changing the sign to take the *negative* of the reciprocal square root preserves the direction of relationships, making things a bit simpler.
-1	The (negative) reciprocal, $-1/y$.	Ratios of two quantities (miles per hour, for example) often benefit from a reciprocal. (You have about a 50–50 chance that the original ratio was taken in the "wrong" order for simple statistical analysis and would benefit from re-expression.) Often, the reciprocal will have simple units (hours per mile). Change the sign if you want to preserve the direction of relationships. If your data have zeros, try adding a small constant to all values before finding the reciprocal.

---

[2] You may remember that for any nonzero number y, $y^0 = 1$. This is not a very exciting transformation for data; every data value would be the same. We use the logarithm in its place.

[3] Your calculator or software package probably gives you a choice between "base 10" logarithms and "natural (base e)" logarithms. Don't worry about that. It doesn't matter at all which you use; they have exactly the same effect on the data. If you want to choose, base 10 logarithms can be a bit easier to interpret.

JUST CHECKING

1. You want to model the relationship between the number of birds counted at a nesting site and the temperature (in degrees Celsius). The scatterplot of counts vs. temperature shows an upwardly curving pattern, with more birds spotted at higher temperatures. What transformation (if any) of the bird counts might you start with?

2. You want to model the relationship between prices for various items in Paris and in Hong Kong. The scatterplot of Hong Kong prices vs. Parisian prices shows a generally straight pattern with a small amount of scatter. What transformation (if any) of the Hong Kong prices might you start with?

3. You want to model the population growth of the United States over the past 200 years. The scatterplot shows a strongly upwardly curved pattern. What transformation (if any) of the population might you start with?

Scientific laws often include simple re-expressions. For example, in Psychology, Fechner's Law states that sensation increases as the logarithm of stimulus intensity ($S = k \log R$).

The Ladder of Powers orders the effects that the re-expressions have on data. If you try, say, taking the square roots of all the values in a variable and it helps, but not enough, then move farther down the ladder to the logarithm or reciprocal root. Those re-expressions will have a similar, but even stronger, effect on your data. If you go too far, you can always back up. But don't forget—when you take a negative power, the *direction* of the relationship will change. That's OK. You can always change the sign of the response variable if you want to keep the same direction. With modern technology, finding a suitable re-expression is no harder than the push of a button.

FOR EXAMPLE   Trying a Re-expression

RECAP: We've seen curvature in the relationship between emperor penguins' diving heart rates and the duration of the dive. Let's start the process of finding a good re-expression. Heart rate is in beats per minute; maybe heart "speed" in minutes per beat would be a better choice. Here are the corresponding displays for this reciprocal re-expression (as we often do, we've changed the sign to preserve the order of the data values):

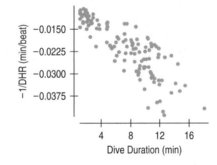 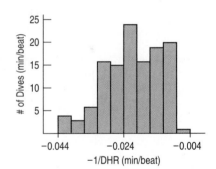

QUESTION: Were the re-expressions successful?

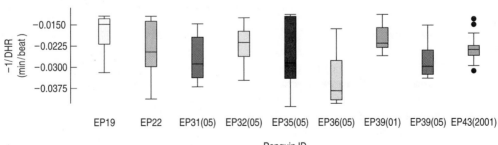

The scatterplot bends less than before, but now may be slightly concave downward. The histogram is now slightly skewed to the low end. Most of the boxplots have no outliers. These boxplots seem better than the ones for the raw heart rates.

Overall, it looks like I may have moved a bit "too far" on the ladder of powers. Halfway between "1" (the original data) and "−1" (the reciprocal) is "0", which represents the logarithm. I'd try that for comparison.

STEP-BY-STEP EXAMPLE | **Re-expressing to Straighten a Scatterplot**

Standard (monofilament) fishing line comes in a range of strengths, usually expressed as "test pounds." Five-pound test line, for example, can be expected to withstand a pull of up to five pounds without breaking. The convention in selling fishing line is that the price of a spool doesn't vary with strength. Instead, the length of line on the spool varies. Higher pound test line is thicker, though, so spools of fishing line hold about the same amount of material. Some spools hold line that is thinner and longer, some fatter and shorter. Let's look at the *Length* and *Strength* of spools of monofilament line manufactured by the same company and sold for the same price at one store.

Questions: How are the *Length* on the spool and the *Strength* related? And what re-expression will straighten the relationship?

 THINK

Plan State the problem.

Variables Identify the variables and report the W's.

I want to fit a linear model for the length and strength of monofilament fishing line.

I have the *length* and "pound test" strength of monofilament fishing line sold by a single vendor at a particular store. Each case is a different strength of line, but all spools of line sell for the same price.

Let *Length* = length (in yards) of fishing line on the spool

 Strength = the test strength (in pounds).

Plot Check that even if there is a curve, the overall pattern does not reach a minimum or maximum and then turn around and go back. An up-and-down curve can't be fixed by re-expression.

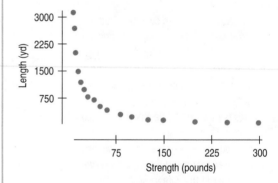

The plot shows a negative direction and an association that has little scatter but is not straight.

 SHOW

Mechanics Try a re-expression.

The lesson of the Ladder of Powers is that if we're moving in the right direction but have not had sufficient effect, we should go farther along the ladder. This example shows improvement, but is still not straight.

(Because *Length* is an amount of something and cannot be negative, we probably should have started with logs. This plot is here in part to illustrate how the Ladder of Powers works.)

Here's a plot of the square root of *Length* against *Strength*:

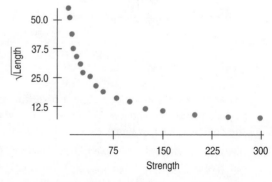

The plot is less bent, but still not straight.

Stepping from the 1/2 power to the "0" power, we try the logarithm of *Length* against *Strength*.

The scatterplot of the logarithm of *Length* against *Strength* is even less bent:

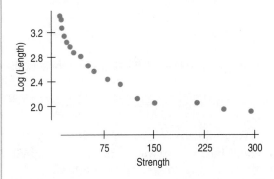

The straightness is improving, so we know we're moving in the right direction. But since the plot of the logarithms is not yet straight, we know we haven't gone far enough. To keep the direction consistent, change the sign and re-express to $-1/Length$.

This is much better, but still not straight, so I'll take another step to the "−1" power, or reciprocal.

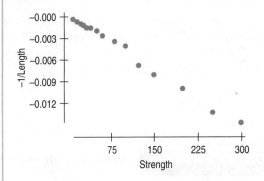

Maybe now I moved too far along the ladder.

A half-step back is the −1/2 power: the reciprocal square root.

We may have to choose between two adjacent re-expressions. For most data analyses, it really doesn't matter which we choose.

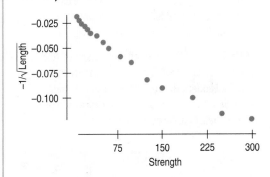

Conclusion Specify your choice of re-expression. If there's some natural interpretation (as for gallons per 100 miles), give that.

It's hard to choose between the last two alternatives. Either of the last two choices is good enough. I'll choose the −1/2 power.

Now that the re-expressed data satisfy the Straight Enough Condition, we can fit a linear model by least squares. We find that

$$\frac{-1}{\sqrt{\widehat{Length}}} = -0.023 - 0.000373 \; Strength.$$

We can use this model to predict the length of a spool of, say, 35-pound test line:

$$\frac{-1}{\sqrt{\widehat{Length}}} = -0.023 - 0.000373 \times 35 = -0.036$$

We could leave the result in these units ($-1/\sqrt{\textbf{yards}}$). Sometimes the new units may be as meaningful as the original, but here we want to transform the predicted value back into yards. Fortunately, each of the re-expressions in the Ladder of Powers can be reversed.

To reverse the process, we first take the reciprocal: $\sqrt{\widehat{Length}} = -1/(-0.036) = 27.778$. Then squaring gets us back to the original units:

$$\widehat{Length} = 27.778^2 = 771.6 \; yards.$$

This may be the most painful part of the re-expression. Getting back to the original units can sometimes be a little work. Nevertheless, it's worth the effort to always consider re-expression. Re-expressions extend the reach of all of your Statistics tools by helping more data to satisfy the conditions they require. Just think how much more useful this course just became!

FOR EXAMPLE   Comparing Re-expressions

RECAP: We've concluded that in trying to straighten the relationship between *Diving Heart Rate* and *Dive Duration* for emperor penguins, using the reciprocal re-expression goes a bit "too far" on the ladder of powers. Now we try the logarithm. Here are the resulting displays:

QUESTIONS: Comment on these displays. Now that we've looked at the original data (rung 1 on the Ladder), the reciprocal (rung −1), and the logarithm (rung 0), which re-expression of *Diving Heart Rate* would you choose?

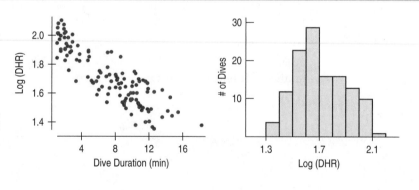

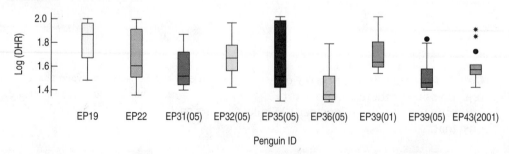

The scatterplot is now more linear and the histogram is symmetric. The boxplots are still a bit skewed to the high end, but less so than for the original Diving Heart Rate values. We don't expect real data to cooperate perfectly, and the logarithm seems like the best compromise re-expression, improving several different aspects of the data.

Plan B: Attack of the Logarithms

The Ladder of Powers is often successful at finding an effective re-expression. Sometimes, though, the curvature is more stubborn, and we're not satisfied with the residual plots. What then?

When none of the data values is zero or negative, logarithms can be a helpful ally in the search for a useful model. Try taking the logs of both the *x*- and *y*-variables. Then re-express the data using some combination of *x* or log(*x*) vs. *y* or log(*y*). You may find that one of these works pretty well.

Model Name	*x*-axis	*y*-axis	Comment
Exponential	*x*	log(*y*)	This model is the "0" power in the ladder approach, useful for values that grow by percentage increases.
Logarithmic	log(*x*)	*y*	A wide range of *x*-values, or a scatterplot descending rapidly at the left but leveling off toward the right, may benefit from trying this model.
Power	log(*x*)	log(*y*)	The Goldilocks model: When one of the ladder's powers is too big and the next is too small, this one may be just right.

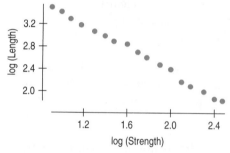

FIGURE 10.8
Plotting log (*Length*) against log (*Strength*) gives a straighter shape.

When we tried to model the relationship between the length of fishing line and its strength, we were torn between the "−1" power and the "−1/2" power. The first showed slight upward curvature, and the second downward.

The scatterplot shows what happens when we graph the logarithm of *Length* against the logarithm of *Strength*. Technology reveals that the equation of our log–log model is

$$\widehat{\log(Length)} = 4.49 - 1.08 \log(\text{Strength})$$

This approach comes out closer to the reciprocal (−1).

A warning, though! Don't expect to be able to straighten every curved scatterplot you find. It may be that there just isn't a very effective re-expression to be had. You'll certainly encounter situations when nothing seems to work the way you wish it would. Don't set your sights too high—you won't find a perfect model. Keep in mind: We seek a *useful* model, not perfection (or even "the best").

Multiple Benefits

We often choose a re-expression for one reason and then discover that it has helped other aspects of an analysis. For example, we might re-express a variable to make its histogram more nearly symmetric. It wouldn't be surprising to find that the same re-expression also straightens scatterplots or makes spreads more nearly equal. This phenomenon is one reason we encourage that data re-expression be used more often. Sometimes there's an obvious "best" or "right" re-expression for a variable. For example, it makes sense that things that tend to grow by a constant percentage, so that larger values increase faster (populations, bacteria counts, wealth), grow exponentially. Logarithms straighten out the exponential trend and pull in the long right tail in the histogram.

We saw just this phenomenon in the companies example earlier. Each of our goals was met by re-expressing *Assets* with logarithms. That single re-expression improved all four of the goals at the same time. Multiple successes like this turn out not to be all that unusual.

Measurement errors are often larger when measuring larger quantities than when measuring smaller ones. (The error in your height may be only a centimeter or two, but the error in the height of a tree could be ten times that much.) Again, logarithms are likely to help. Measurements of rates (e.g., time

OCCAM'S RAZOR

If you think that simpler explanations and simpler models are more likely to give a true picture of the way things work, then you should look for opportunities to re-express your data and simplify your analyses.

The general principle that simpler explanations are likely to be the better ones is known as Occam's Razor, after the English philosopher and theologian William of Occam (1284–1347).

to complete a task) are often plagued by infinities for those who never finish. The reciprocal of "minutes per task" is "tasks per minute"—a speed measure. And the unfinished tasks that once took "infinite" time now simply rate a zero speed. Here once more the re-expression seems natural.

In other cases, the only evidence we have to favor re-expression is that it seems to work well and that it leads to simpler models. Often we can find a re-expression for a variable that simplifies our analysis in several ways at once, making the distribution symmetric, making the relationship linear in terms of other variables of interest, or stabilizing its variance. If so, re-expressing the variable certainly simplifies our efforts to analyze and understand it.

Why Not Just Use a Curve?

When a clearly curved pattern shows up in the scatterplot, why not just fit a curve to the data? We saw earlier that the association between the *Weight* of a car and its *Fuel Efficiency* was not a straight line. Instead of trying to find a way to straighten the plot, why not find a curve that seems to describe the pattern well?

We can find "curves of best fit" using essentially the same approach that led us to linear models. You won't be surprised, though, to learn that the mathematics and the calculations are considerably more difficult for curved models. Many calculators and computer packages do have the ability to fit curves to data, but this approach has many drawbacks.

Straight lines are easy to understand. We know how to think about the slope and the *y*-intercept, for example. We often want some of the other benefits mentioned earlier, such as making the spread around the model more nearly the same everywhere. In later chapters you will learn more advanced statistical methods for analyzing linear associations.

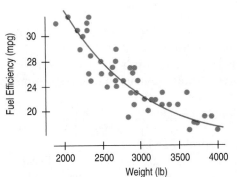

We give all of that up when we fit a model that is not linear. For many reasons, then, it is usually better to re-express the data to straighten the plot.

What Can Go Wrong?

- **Don't expect your model to be perfect.** In Chapter 6 we quoted statistician George Box: "All models are wrong, but some are useful." Be aware that the real world is a messy place and data can be uncooperative. Don't expect to find one elusive re-expression that magically irons out every kink in your scatterplot and produces perfect residuals. You aren't looking for the Right Model, because that mythical creature doesn't exist. Find a useful model and use it wisely.

- **Don't stray too far from the ladder.** It's wise not to stray too far from the powers that we suggest in the Ladder of Powers. Taking the *y*-values to an extremely high power may artificially inflate R^2, but it won't give a useful or meaningful model, so it doesn't really simplify anything. It's better to stick to powers between 2 and −2. Even in that range, you should prefer the simpler powers in the ladder to those in the cracks. A square root is easier to understand than the 0.413 power. That simplicity may compensate for a slightly less straight relationship.

- **Don't choose a model based on R^2 alone.** You've tried re-expressing your data to straighten a curved relationship and found a model with a high R^2. Beware: That doesn't mean the pattern is straight now. On the next page is a plot of a relationship with an R^2 of 98.3%.

The R^2 is about as high as we could ask for, but if you look closely, you'll see that there's a consistent bend. Plotting the residuals from the least squares line makes the bend much easier to see.

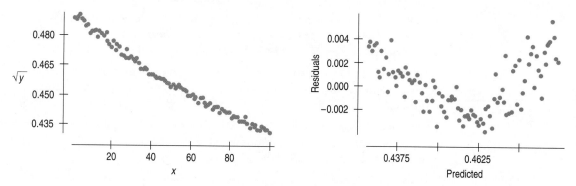

Remember the basic rule of data analysis: *Make a picture*. Before you fit a line, always look at the pattern in the scatterplot. After you fit the line, check for linearity again by plotting the residuals.

- **Beware of multiple modes.** Re-expression can often make a skewed unimodal histogram more nearly symmetric, but it cannot pull separate modes together. A suitable re-expression may, however, make the separation of the modes clearer, simplifying their interpretation and making it easier to separate them to analyze individually.

- **Watch out for scatterplots that turn around.** Re-expression can straighten many bent relationships but not those that go up and then down or down and then up. You should refuse to analyze such data with methods that require a linear form.

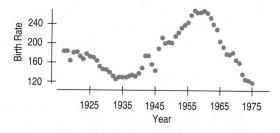

FIGURE 10.9
The shape of the scatterplot of *Birth Rates* (births per 100,000 women) in the United States shows an oscillation that cannot be straightened by re-expressing the data.

- **Watch out for negative data values.** It's impossible to re-express negative values by any power that is not a whole number on the Ladder of Powers or to re-express values that are zero for negative powers. Most statistics programs will just mark the result of trying to re-express such values "missing" if they can't be re-expressed. But that might mean that when you try a re-expression, you inadvertently lose a bunch of data values. The effect of that loss may be surprising and may substantially change your analysis. Because you are likely to be working with a computer package or calculator, take special care that you do not lose otherwise good data values when you choose a re-expression.
  One possible cure for zeros and small negative values is to add a constant ($\frac{1}{2}$ and $\frac{1}{6}$ are often used) to bring all the data values above zero.

- **Watch for data far from 1.** Data values that are all very far from 1 may not be much affected by re-expression unless the range is very large. Re-expressing numbers between 1 and 100 will have a much greater effect than re-expressing numbers between 100,001 and 100,100. When all your data values are large (for example, working with years), consider subtracting a constant to bring them back near 1. (For example, consider "years since 1950" as an alternative variable for re-expression. Unless your data start at 1950, then avoid creating a zero by using "years since 1949.")

We have seen several ways to model or summarize data. Each requires that the data have a particular simple structure. We seek symmetry for summaries of center and spread and to use a Normal model. We seek equal variation across groups when we compare groups with boxplots or want to compare their centers. We seek linear shape in a scatterplot so that we can use correlation to summarize the scatter and regression to fit a linear model.

Data do often satisfy the requirements to use Statistics methods. But often they do not. Our choice is to stop with just displays, to use much more complex methods, or to re-express the data so that we can use the simpler methods we have developed.

In this fundamental sense, this chapter connects to everything we have done thus far and to all of the methods we will introduce throughout the rest of the book. Re-expression greatly extends the reach and applicability of all of these methods.

WHAT HAVE WE LEARNED?

We've learned that when the conditions for regression are not met, a simple re-expression of the data may help. There are several reasons to consider a re-expression:

▶ To make the distribution of a variable more symmetric (as we saw in Chapter 5)

▶ To make the spread across different groups more similar

▶ To make the form of a scatterplot straighter

▶ To make the scatter around the line in a scatterplot more consistent

We've learned that when seeking a useful re-expression, taking logs is often a good, simple starting point. To search further, the Ladder of Powers or the log-log approach can help us find a good re-expression.

We've come to understand that our models won't be perfect, but that re-expression can lead us to a useful model.

Terms

Re-expression   We re-express data by taking the logarithm, the square root, the reciprocal, or some other mathematical operation of all values of a variable (p. 239).

Ladder of Powers   The Ladder of Powers places in order the effects that many re-expressions have on the data (p. 242).

Skills

THINK

▶ Recognize when a well-chosen re-expression may help you improve and simplify your analysis.

▶ Understand the value of re-expressing data to improve symmetry, to make the scatter around a line more constant, or to make a scatterplot more linear.

▶ Recognize when the pattern of the data indicates that no re-expression can improve the structure of the data.

SHOW

▶ Know how to re-express data with powers and how to find an effective re-expression for your data using your statistics software or calculator.

▶ Be able to reverse any of the common re-expressions to put a predicted value or residual back into the original units.

TELL

▶ Be able to describe a summary or display of a re-expressed variable, making clear how it was re-expressed and giving its re-expressed units.

▶ Be able to describe a regression model fit to re-expressed data in terms of the re-expressed variables.

RE-EXPRESSION ON THE COMPUTER

Computers and calculators make it easy to re-express data. Most statistics packages offer a way to re-express and compute with variables. Some packages permit you to specify the power of a re-expression with a slider or other moveable control, possibly while watching the consequences of the re-expression on a plot or analysis. This, of course, is a very effective way to find a good re-expression.

DATA DESK

To re-express a variable in Data Desk, select the variable and Choose the function to re-express it from the **Manip** > **Transform** menu. Square root, log, reciprocal, and reciprocal root are immediately available. For others, make a derived variable and type the function. Data Desk makes a new derived variable that holds the re-expressed values. Any value changed in the original variable will immediately be re-expressed in the derived variable.

COMMENTS

Or choose **Manip** > **Transform** > **Dynamic** > **Box-Cox** to generate a continuously changeable variable and a slider that specifies the power. Set plots to **Automatic Update** in their HyperView menus and watch them change dynamically as you drag the slider.

EXCEL

To re-express a variable in Excel, use Excel's built-in functions as you would for any calculation. Changing a value in the original column will change the re-expressed value.

JMP

To re-express a variable in JMP, double-click to the right of the last column of data to create a new column. Name the new column and select it. Choose **Formula** from the **Cols** menu. In the Formula dialog, choose the transformation and variable that you wish to assign to the new column. Click the **OK** button. JMP places the re-expressed data in the new column.

COMMENTS

The log and square root re-expressions are found in the **Transcendental** menu of functions in the formula dialog.

MINITAB

To re-express a variable in MINITAB, choose **Calculator** from the **Calc** menu. In the Calculator dialog, specify a name for the new re-expressed variable. Use the

Functions List, the calculator buttons, and the **Variables list** box to build the expression. Click **OK**.

SPSS

To re-express a variable in SPSS, choose **Compute** from the **Transform** menu. Enter a name in the Target Variable field. Use the calculator and Function List to build the expression.

Move a variable to be re-expressed from the source list to the Numeric Expression field. Click the **OK** button.

TI-83/84 PLUS

To re-express data stored in a list, perform the re-expression on the whole list and store it in another list. For example, to use the logarithms of the data in L1, enter the command **log(L1) STO L2**.

(continued)

TI-89

To re-express data stored in a list, perform the re-expression on the whole list and store it in another list. For example, to use the common (base 10) logarithms of the data in list1, on the home screen, enter the command **log(list1)** STO▶ **list2**.

COMMENTS

- To find the log command, press CATALOG then 4 (L) arrow to log, and press ENTER.
- Natural logs are **LN** (press 2nd ×).
- For square roots, press 2nd ×

EXERCISES

1. **Residuals.** Suppose you have fit a linear model to some data and now take a look at the residuals. For each of the following possible residuals plots, tell whether you would try a re-expression and, if so, why.

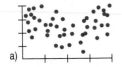  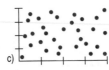

2. **Residuals.** Suppose you have fit a linear model to some data and now take a look at the residuals. For each of the following possible residuals plots, tell whether you would try a re-expression and, if so, why.

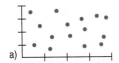 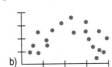 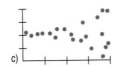

3. **Airline passengers revisited.** In Chapter 9, Exercise 9, we created a linear model describing the trend in the number of passengers departing from the Oakland (CA) airport each month since the start of 1997. Here's the residual plot, but with lines added to show the order of the values in time:

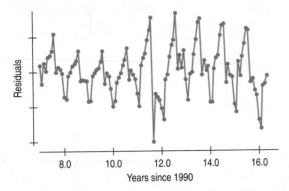

a) Can you account for the pattern shown here?
b) Would a re-expression help us deal with this pattern? Explain.

4. **Hopkins winds, revisited.** In Chapter 5, we examined the wind speeds in the Hopkins forest over the course of a year. Here's the scatterplot we saw then:

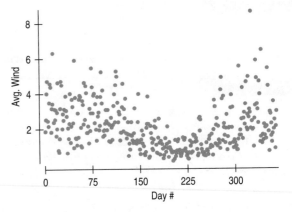

a) Describe the pattern you see here.
b) Should we try re-expressing either variable to make this plot straighter? Explain.

5. **Models.** For each of the models listed below, predict y when $x = 2$.
   a) $\ln \hat{y} = 1.2 + 0.8x$
   b) $\sqrt{\hat{y}} = 1.2 + 0.8x$
   c) $\dfrac{1}{\hat{y}} = 1.2 + 0.8x$
   d) $\hat{y} = 1.2 + 0.8 \ln x$
   e) $\log \hat{y} = 1.2 + 0.8 \log x$

6. **More models.** For each of the models listed below, predict y when $x = 2$.
   a) $\hat{y} = 1.2 + 0.8 \log x$
   b) $\log \hat{y} = 1.2 + 0.8x$
   c) $\ln \hat{y} = 1.2 + 0.8 \ln x$
   d) $\hat{y}^2 = 1.2 + 0.8x$
   e) $\dfrac{1}{\sqrt{\hat{y}}} = 1.2 + 0.8x$

7. **Gas mileage.** As the example in the chapter indicates, one of the important factors determining a car's *Fuel Efficiency* is its *Weight*. Let's examine this relationship again, for 11 cars.
   a) Describe the association between these variables shown in the scatterplot on the next page.

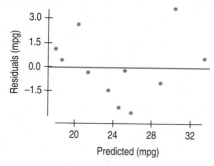

b) Here is the regression analysis for the linear model. What does the slope of the line say about this relationship?

Dependent variable is: Fuel Efficiency
R-squared = 85.9%

Variable	Coefficient
Intercept	47.9636
Weight	−7.65184

c) Do you think this linear model is appropriate? Use the residuals plot to explain your decision.

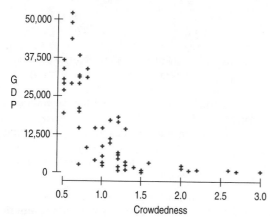

8. Crowdedness. In a *Chance* magazine article (Summer 2005), Danielle Vasilescu and Howard Wainer used data from the United Nations Center for Human Settlements to investigate aspects of living conditions for several countries. Among the variables they looked at were the country's per capita gross domestic product (*GDP*, in $) and *Crowdedness*, defined as the average number of persons per room living in homes there. This scatterplot displays these data for 56 countries:

a) Explain why you should re-express these data before trying to fit a model.

b) What re-expression of *GDP* would you try as a starting point?

9. Gas mileage, revisited. Let's try the re-expressed variable *Fuel Consumption* (gal/100 mi) to examine the fuel efficiency of the 11 cars in Exercise 7. Here are the revised regression analysis and residuals plot:

Dependent variable is: Fuel Consumption
R-squared = 89.2%

Variable	Coefficient
Intercept	0.624932
Weight	1.17791

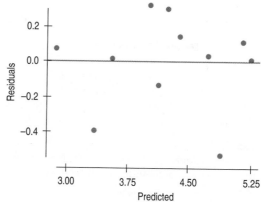

a) Explain why this model appears to be better than the linear model.
b) Using the regression analysis above, write an equation of this model.
c) Interpret the slope of this line.
d) Based on this model, how many miles per gallon would you expect a 3500-pound car to get?

10. Crowdedness again. In Exercise 8 we looked at United Nations data about a country's *GDP* and the average number of people per room (*Crowdedness*) in housing there. For a re-expression, a student tried the reciprocal −10000/*GDP*, representing the number of people per $10,000 of gross domestic product. Here are the results, plotted against *Crowdedness*:

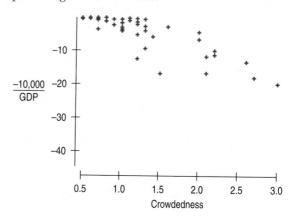

a) Is this a useful re-expression? Explain.
b) What re-expression would you suggest this student try next?

11. **GDP.** The scatterplot shows the gross domestic product (GDP) of the United States in billions of dollars plotted against years since 1950.

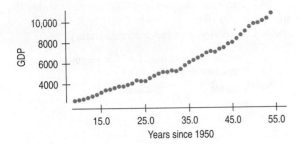

A linear model fit to the relationship looks like this:

Dependent variable is: GDP
R-squared = 97.2%   s = 406.6

Variable	Coefficient
Intercept	240.171
Year-1950	177.689

a) Does the value 97.2% suggest that this is a good model? Explain.
b) Here's a scatterplot of the residuals. Now do you think this is a good model for these data? Explain?

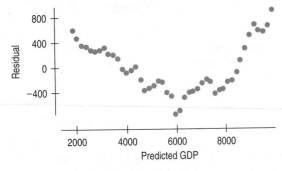

T 12. **Treasury bills.** The 3-month Treasury bill interest rate is watched by investors and economists. Here's a scatterplot of the 3-month Treasury bill rate since 1950:

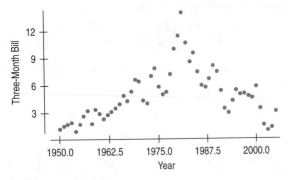

Clearly, the relationship is not linear. Can it be made nearly linear with a re-expression? If so, which one would you suggest? If not, why not?

13. **Better GDP model?** Consider again the post-1950 trend in U.S. GDP we examined in Exercise 11. Here are a regression and residual plot when we use the log of GDP in the model. Is this a better model for GDP? Explain.

Dependent variable is: LogGDP
R-squared = 99.4%   s = 0.0150

Variable	Coefficient
Intercept	3.29092
Year-1950	0.013881

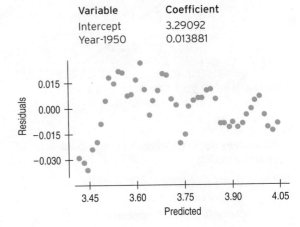

T 14. **Pressure.** Scientist Robert Boyle examined the relationship between the volume in which a gas is contained and the pressure in its container. He used a cylindrical container with a moveable top that could be raised or lowered to change the volume. He measured the *Height* in inches by counting equally spaced marks on the cylinder, and measured the *Pressure* in inches of mercury (as in a barometer). Some of his data are listed in the table. Create an appropriate model.

Height	48	44	40	36	32	28
Pressure	29.1	31.9	35.3	39.3	44.2	50.3
Height	24	20	18	16	14	12
Pressure	58.8	70.7	77.9	87.9	100.4	117.6

T 15. **Brakes.** The following table shows stopping distances in feet for a car tested 3 times at each of 5 speeds. We hope to create a model that predicts *Stopping Distance* from the *Speed* of the car.

Speed (mph)	Stopping Distances (ft)
20	64, 62, 59
30	114, 118, 105
40	153, 171, 165
50	231, 203, 238
60	317, 321, 276

a) Explain why a linear model is not appropriate.
b) Re-express the data to straighten the scatterplot.
c) Create an appropriate model.
d) Estimate the stopping distance for a car traveling 55 mph.
e) Estimate the stopping distance for a car traveling 70 mph.
f) How much confidence do you place in these predictions? Why?

T 16. **Pendulum.** A student experimenting with a pendulum counted the number of full swings the pendulum made in 20 seconds for various lengths of string. Her data are shown on the next page.

Length (in.)	6.5	9	11.5	14.5	18	21	24	27	30	37.5
Number of Swings	22	20	17	16	14	13	13	12	11	10

a) Explain why a linear model is not appropriate for using the *Length* of a pendulum to predict the *Number of Swings* in 20 seconds.

b) Re-express the data to straighten the scatterplot.

c) Create an appropriate model.

d) Estimate the number of swings for a pendulum with a 4-inch string.

e) Estimate the number of swings for a pendulum with a 48-inch string.

f) How much confidence do you place in these predictions? Why?

17. **Baseball salaries 2009.** Ballplayers have been signing ever larger contracts. The highest salaries (in millions of dollars per season) for some notable players are given in the following table.

Player	Year	Salary (million $)
Nolan Ryan	1980	1.0
George Foster	1982	2.0
Kirby Puckett	1990	3.0
Jose Canseco	1990	4.7
Roger Clemens	1991	5.3
Ken Griffey, Jr.	1986	8.5
Albert Belle	1997	11.0
Pedro Martinez	1998	12.5
Mike Piazza	1999	12.5
Mo Vaughn	1999	13.3
Kevin Brown	1999	15.0
Carlos Delgado	2001	17.0
Alex Rodriguez	2001	22.0
Manny Ramirez	2004	22.5
Alex Rodriguez	2005	26.0
Derek Jeter	2006	20.6
Jason Giambi	2007	23.4
Alex Rodriguez	2009	33.0

a) Examine a scatterplot of the data. Does it look straight?

b) Find the regression of *Salary* vs. *Year* and plot the residuals. Do they look straight?

c) Re-express the data, if necessary, to straighten the relationship.

d) What model would you report for the trend in salaries?

T 18. **Planet distances and years 2006.** At a meeting of the International Astronomical Union (IAU) in Prague in 2006, Pluto was determined not to be a planet, but rather the largest member of the Kuiper belt of icy objects. Let's examine some facts. Here is a table of the 9 sun-orbiting objects formerly known as planets:

Planet	Position Number	Distance from Sun (million miles)	Length of Year (Earth years)
Mercury	1	36	0.24
Venus	2	67	0.61
Earth	3	93	1.00
Mars	4	142	1.88
Jupiter	5	484	11.86
Saturn	6	887	29.46
Uranus	7	1784	84.07
Neptune	8	2796	164.82
Pluto	9	3707	247.68

a) Plot the *Length* of the year against the *Distance* from the sun. Describe the shape of your plot.

b) Re-express one or both variables to straighten the plot. Use the re-expressed data to create a model describing the length of a planet's year based on its distance from the sun.

c) Comment on how well your model fits the data.

T 19. **Planet distances and order 2006.** Let's look again at the pattern in the locations of the planets in our solar system seen in the table in Exercise 18.

a) Re-express the distances to create a model for the *Distance* from the sun based on the planet's *Position*.

b) Based on this model, would you agree with the International Astronomical Union that Pluto is not a planet? Explain.

T 20. **Planets 2006, part 3.** The asteroid belt between Mars and Jupiter may be the remnants of a failed planet. If so, then Jupiter is really in position 6, Saturn is in 7, and so on. Repeat Exercise 19, using this revised method of numbering the positions. Which method seems to work better?

T 21. **Eris: Planets 2006, part 4.** In July 2005, astronomers Mike Brown, Chad Trujillo, and David Rabinowitz announced the discovery of a sun-orbiting object, since named Eris,[5] that is 5% larger than Pluto. Eris orbits the sun once every 560 earth years at an average distance of about 6300 million miles from the sun. Based on its *Position*, how does Eris's *Distance* from the sun (re-expressed to logs) compare with the prediction made by your model of Exercise 19?

T 22. **Models and laws: Planets 2006, part 5.** The model you found in Exercise 18 is a relationship noted in the 17th century by Kepler as his Third Law of Planetary Motion. It was subsequently explained as a consequence of Newton's Law of Gravitation. The models for Exercises 19–21 relate to what is sometimes called the Titius-Bode "law," a

---

[5] Eris is the Greek goddess of warfare and strife who caused a quarrel among the other goddesses that led to the Trojan war. In the astronomical world, Eris stirred up trouble when the question of its proper designation led to the raucous meeting of the IAU in Prague where IAU members voted to demote Pluto and Eris to dwarf-planet status—http://www.gps.caltech.edu/~mbrown/planetlila/#paper.

pattern noticed in the 18th century but lacking any scientific explanation.

Compare how well the re-expressed data are described by their respective linear models. What aspect of the model of Exercise 18 suggests that we have found a physical law? In the future, we may learn enough about a planetary system around another star to tell whether the Titius-Bode pattern applies there. If you discovered that another planetary system followed the same pattern, how would it change your opinion about whether this is a real natural "law"? What would you think if the next system we find does not follow this pattern?

23. **Logs (not logarithms).** The value of a log is based on the number of board feet of lumber the log may contain. (A board foot is the equivalent of a piece of wood 1 inch thick, 12 inches wide, and 1 foot long. For example, a 2" × 4" piece that is 12 feet long contains 8 board feet.) To estimate the amount of lumber in a log, buyers measure the diameter inside the bark at the smaller end. Then they look in a table based on the Doyle Log Scale. The table below shows the estimates for logs 16 feet long.

Diameter of Log	8"	12"	16"	20"	24"	28"
Board Feet	16	64	144	256	400	576

a) What model does this scale use?
b) How much lumber would you estimate that a log 10 inches in diameter contains?
c) What does this model suggest about logs 36 inches in diameter?

24. **Weightlifting 2004.** Listed below are the gold medal-winning men's weight-lifting performances at the 2004 Olympics.

Weight Class (kg)	Winner (country)	Weight Lifted (kg)
56	Halil Mutlu (Turkey)	295.0
62	Zhiyong Shi (China)	325.0
69	Guozheng Zhang (China)	347.5
77	Taner Sagir (Turkey)	375.0
85	George Asanidze (Georgia)	382.5
94	Milen Dobrev (Bulgaria)	407.5
105	Dmitry Berestov (Russia)	425.0

a) Create a linear model for the *Weight Lifted* in each *Weight Class*.
b) Check the residuals plot. Is your linear model appropriate?
c) Create a better model.
d) Explain why you think your new model is better.
e) Based on your model, which of the medalists turned in the most surprising performance? Explain.

25. **Life expectancy.** The data in the next column list the *Life Expectancy* for white males in the United States every

decade during the last century (1 = 1900 to 1910, 2 = 1911 to 1920, etc.). Create a model to predict future increases in life expectancy. (National Vital Statistics Report)

Decade	1	2	3	4	5	6	7	8	9	10	11
Life exp.	46.6	48.6	54.4	59.7	62.1	66.5	67.4	68.0	70.7	72.7	74.9

26. **Lifting more weight 2004.** In Exercise 24 you examined the winning weight-lifting performances for the 2004 Olympics. One of the competitors turned in a performance that appears not to fit the model you created.
a) Consider that competitor to be an outlier. Eliminate that data point and re-create your model.
b) Using this revised model, how much would you have expected the outlier competitor to lift?
c) Explain the meaning of the residual from your new model for that competitor.

27. **Slower is cheaper?** Researchers studying how a car's *Fuel Efficiency* varies with its *Speed* drove a compact car 200 miles at various speeds on a test track. Their data are shown in the table.

Speed (mph)	35	40	45	50	55	60	65	70	75
Fuel Eff. (mpg)	25.9	27.7	28.5	29.5	29.2	27.4	26.4	24.2	22.8

Create a linear model for this relationship and report any concerns you may have about the model.

28. **Orange production.** The table below shows that as the number of oranges on a tree increases, the fruit tends to get smaller. Create a model for this relationship, and express any concerns you may have.

Number of Oranges/Tree	Average Weight/Fruit (lb)
50	0.60
100	0.58
150	0.56
200	0.55
250	0.53
300	0.52
350	0.50
400	0.49
450	0.48
500	0.46
600	0.44
700	0.42
800	0.40
900	0.38

29. **Years to live 2003.** Insurance companies and other organizations use actuarial tables to estimate the remaining lifespans of their customers. On the next page are the estimated additional years of life for black males in the

United States, according to a 2003 National Vital Statistics Report. (www.cdc.gov/nchs/deaths.htm)

Age	10	20	30	40	50	60	70	80	90	100
Years Left	60.3	50.7	41.8	32.9	24.8	17.9	12.1	7.9	5.0	3.0

a) Find a re-expression to create an appropriate model.
b) Predict the remaining lifespan of an 18-year-old black man.
c) Are you satisfied that your model has accounted for the relationship between *Years Left* and *Age*? Explain.

30. **Tree growth.** A 1996 study examined the growth of grapefruit trees in Texas, determining the average trunk *Diameter* (in inches) for trees of varying *Ages*:

Age (yr)	2	4	6	8	10	12	14	16	18	20
Diameter (in.)	2.1	3.9	5.2	6.2	6.9	7.6	8.3	9.1	10.0	11.4

a) Fit a linear model to these data. What concerns do you have about the model?
b) If data had been given for individual trees instead of averages, would you expect the fit to be stronger, less strong, or about the same? Explain.

JUST CHECKING

ANSWERS

1. Counts are often best transformed by using the square root.

2. None. The relationship is already straight.

3. Even though, technically, the population values are counts, you should probably try a stronger transformation like log(population) because populations grow in proportion to their size.

REVIEW OF PART II

Exploring Relationships Between Variables

Quick Review

You have now survived your second major unit of Statistics. Here's a brief summary of the key concepts and skills:

▶ We treat data two ways: as categorical and as quantitative.

▶ To explore relationships in categorical data, check out Chapter 3.

▶ To explore relationships in quantitative data:

- Make a picture. Use a scatterplot. Put the explanatory variable on the *x*-axis and the response variable on the *y*-axis.
- Describe the association between two quantitative variables in terms of direction, form, and strength.
- The amount of scatter determines the strength of the association.
- If, as one variable increases so does the other, the association is positive. If one increases as the other decreases, it's negative.

- If the form of the association is linear, calculate a correlation to measure its strength numerically, and do a regression analysis to model it.
- Correlations closer to −1 or +1 indicate stronger linear associations. Correlations near 0 indicate weak linear relationships, but other forms of association may still be present.
- The line of best fit is also called the least squares regression line because it minimizes the sum of the squared residuals.
- The regression line predicts values of the response variable from values of the explanatory variable.
- A residual is the difference between the true value of the response variable and the value predicted by the regression model.
- The slope of the line is a rate of change, best described in "*y*-units" per "*x*-unit."
- R^2 gives the fraction of the variation in the response variable that is accounted for by the model.
- The standard deviation of the residuals measures the amount of scatter around the line.

- Outliers and influential points can distort any of our models.
- If you see a pattern (a curve) in the residuals plot, your chosen model is not appropriate; use a different model. You may, for example, straighten the relationship by re-expressing one of the variables.
- To straighten bent relationships, re-express the data using logarithms or a power (squares, square roots, reciprocals, etc.).

- Always remember that an association is not necessarily an indication that one of the variables causes the other.

Need more help with some of this? Try rereading some sections of Chapters 7 through 10. And see below for more opportunities to review these concepts and skills.

"One must learn by doing the thing; though you think you know it, you have no certainty until you try."
—Sophocles (495–406 B.C.E.)

REVIEW EXERCISES

1. **College.** Every year *US News and World Report* publishes a special issue on many U.S. colleges and universities. The scatterplots below have *Student/Faculty Ratio* (number of students per faculty member) for the colleges and universities on the *y*-axes plotted against 4 other variables. The correct correlations for these scatterplots appear in this list. Match them.

−0.98   −0.71   −0.51   0.09   0.23   0.69

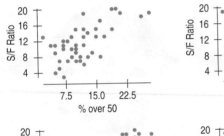

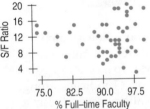

2. **Togetherness.** Are good grades in high school associated with family togetherness? A random sample of 142 high school students was asked how many meals per week their families ate together. Their responses produced a mean of 3.78 meals per week, with a standard deviation of 2.2. Researchers then matched these responses against the students' grade point averages (GPAs). The scatterplot appeared to be reasonably linear, so they created a line of regression. No apparent pattern emerged in the residuals plot. The equation of the line was $\widehat{GPA} = 2.73 + 0.11\,Meals$.
   a) Interpret the *y*-intercept in this context.
   b) Interpret the slope in this context.
   c) What was the mean GPA for these students?
   d) If a student in this study had a negative residual, what did that mean?
   e) Upon hearing of this study, a counselor recommended that parents who want to improve the grades their children get should get the family to eat together more often. Do you agree with this interpretation? Explain.

3. **Vineyards.** Here are the scatterplot and regression analysis for *Case Prices* of 36 wines from vineyards in the Finger Lakes region of New York State and the *Ages* of the vineyards.

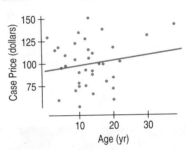

Dependent variable is: Case Price
R-squared = 2.7%

Variable	Coefficient
Constant	92.7650
Age	0.567284

   a) Does it appear that vineyards in business longer get higher prices for their wines? Explain.
   b) What does this analysis tell us about vineyards in the rest of the world?
   c) Write the regression equation.
   d) Explain why that equation is essentially useless.

4. **Vineyards again.** Instead of *Age*, perhaps the *Size* of the vineyard (in acres) is associated with the price of the wines. Look at the scatterplot:

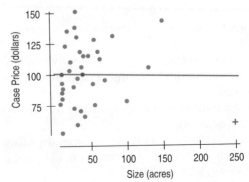

a. Do you see any evidence of an association?
b. What concern do you have about this scatterplot?
c. If the red "+" data point is removed, would the correlation become stronger or weaker? Explain.
d. If the red "+" data point is removed, would the slope of the line increase or decrease? Explain.

T 5. **More twins 2006?** As the table shows, the number of twins born in the United States has been increasing. (www.cdc.gov/nchs/births.htm)

Year	Twin Births	Year	Twin Births
1980	68,339	1994	97,064
1981	70,049	1995	96,736
1982	71,631	1996	100,750
1983	72,287	1997	104,137
1984	72,949	1998	110,670
1985	77,102	1999	114,307
1986	79,485	2000	118,916
1987	81,778	2001	121,246
1988	85,315	2002	125,134
1989	90,118	2003	128,665
1990	93,865	2004	132,219
1991	94,779	2005	133,122
1992	95,372	2006	137,085
1993	96,445		

a) Find the equation of the regression line for predicting the number of twin births.
b) Explain in this context what the slope means.
c) Predict the number of twin births in the United States for the year 2010. Comment on your faith in that prediction.
d) Comment on the residuals plot.

6. **Dow Jones 2006.** The Dow Jones stock index measures the performance of the stocks of America's largest companies (http://finance.yahoo.com). A regression of the Dow prices on years 1972–2006 looks like this:

Dependent variable is: Dow Index
R-squared = 83.5%   s = 1577

Variable	Coefficient
Intercept	−2294.01
Year since 1970	341.095

a) What is the correlation between *Dow Index* and *Year*?
b) Write the regression equation.
c) Explain in this context what the equation says.
d) Here's a scatterplot of the residuals. Which assumption(s) of the regression analysis appear to be violated?

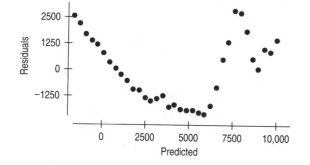

7. **Acid rain.** Biologists studying the effects of acid rain on wildlife collected data from 163 streams in the Adirondack Mountains. They recorded the *pH* (acidity) of the water and the *BCI*, a measure of biological diversity, and they calculated $R^2 = 27\%$. Here's a scatterplot of *BCI* against *pH*:

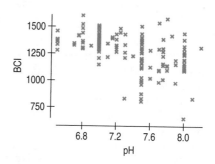

a) What is the correlation between *pH* and *BCI*?
b) Describe the association between these two variables.
c) If a stream has average *pH*, what would you predict about the *BCI*?
d) In a stream where the *pH* is 3 standard deviations above average, what would you predict about the *BCI*?

T 8. **Manatees 2005.** Marine biologists warn that the growing number of powerboats registered in Florida threatens the existence of manatees. The data below come from the Florida Fish and Wildlife Conservation Commission (www.floridamarine.org) and the National Marine Manufacturers Association (www.nmma.org/facts).

Year	Manatees Killed	Powerboat Registrations (in 1000s)
1982	13	447
1983	21	460
1984	24	481
1985	16	498
1986	24	513
1987	20	512
1988	15	527
1989	34	559
1990	33	585
1992	33	614
1993	39	646
1994	43	675
1995	50	711
1996	47	719
1997	53	716
1998	38	716
1999	35	716
2000	49	735
2001	81	860
2002	95	923
2003	73	940
2004	69	946
2005	79	974

a) In this context, which is the explanatory variable?
b) Make a scatterplot of these data and describe the association you see.
c) Find the correlation between *Boat Registrations* and *Manatee Deaths*.
d) Interpret the value of R^2.
e) Does your analysis prove that powerboats are killing manatees?

T 9. A manatee model 2005. Continue your analysis of the manatee situation from the previous exercise.

a) Create a linear model of the association between *Manatee Deaths* and *Powerboat Registrations*.
b) Interpret the slope of your model.
c) Interpret the *y*-intercept of your model.
d) How accurately did your model predict the high number of manatee deaths in 2005?
e) Which is better for the manatees, positive residuals or negative residuals? Explain.
f) What does your model suggest about the future for the manatee?

T 10. Grades. A Statistics instructor created a linear regression equation to predict students' final exam scores from their midterm exam scores. The regression equation was $\widehat{Fin} = 10 + 0.9\,Mid$.

a) If Susan scored a 70 on the midterm, what did the instructor predict for her score on the final?
b) Susan got an 80 on the final. How big is her residual?
c) If the standard deviation of the final was 12 points and the standard deviation of the midterm was 10 points, what is the correlation between the two tests?
d) How many points would someone need to score on the midterm to have a predicted final score of 100?
e) Suppose someone scored 100 on the final. Explain why you can't estimate this student's midterm score from the information given.
f) One of the students in the class scored 100 on the midterm but got overconfident, slacked off, and scored only 15 on the final exam. What is the residual for this student?
g) No other student in the class "achieved" such a dramatic turnaround. If the instructor decides not to include this student's scores when constructing a new regression model, will the R^2 value of the regression increase, decrease, or remain the same? Explain.
h) Will the slope of the new line increase or decrease?

T 11. Traffic. Highway planners investigated the relationship between traffic *Density* (number of automobiles per mile) and the average *Speed* of the traffic on a moderately large city thoroughfare. The data were collected at the same location at 10 different times over a span of 3 months. They found a mean traffic *Density* of 68.6 cars per mile (cpm) with standard deviation of 27.07 cpm. Overall, the cars' average *Speed* was 26.38 mph, with standard deviation of 9.68 mph. These researchers found the regression line for these data to be $\widehat{Speed} = 50.55 - 0.352\,Density$.

a) What is the value of the correlation coefficient between *Speed* and *Density*?
b) What percent of the variation in average *Speed* is explained by traffic *Density*?

c) Predict the average *Speed* of traffic on the thoroughfare when the traffic *Density* is 50 cpm.
d) What is the value of the residual for a traffic *Density* of 56 cpm with an observed *Speed* of 32.5 mph?
e) The data set initially included the point *Density* = 125 cpm, *Speed* = 55 mph. This point was considered an outlier and was not included in the analysis. Will the slope increase, decrease, or remain the same if we redo the analysis and include this point?
f) Will the correlation become stronger, weaker, or remain the same if we redo the analysis and include this point (125,55)?
g) A European member of the research team measured the *Speed* of the cars in kilometers per hour (1 km ≈ 0.62 miles) and the traffic *Density* in cars per kilometer. Find the value of his calculated correlation between speed and density.

T 12. Cramming. One Thursday, researchers gave students enrolled in a section of basic Spanish a set of 50 new vocabulary words to memorize. On Friday the students took a vocabulary test. When they returned to class the following Monday, they were retested—without advance warning. Here are the test scores for the 25 students.

Fri.	Mon.	Fri.	Mon.	Fri.	Mon.
42	36	48	37	39	41
44	44	43	41	46	32
45	46	45	32	37	36
48	38	47	44	40	31
44	40	50	47	41	32
43	38	34	34	48	39
41	37	38	31	37	31
35	31	43	40	36	41
43	32				

a) What is the correlation between *Friday* and *Monday* scores?
b) What does a scatterplot show about the association between the scores?
c) What does it mean for a student to have a positive residual?
d) What would you predict about a student whose *Friday* score was one standard deviation below average?
e) Write the equation of the regression line.
f) Predict the *Monday* score of a student who earned a 40 on Friday.

T 13. Car correlations. What factor most explains differences in *Fuel Efficiency* among cars? Below is a correlation matrix exploring that relationship for the car's *Weight*, *Horsepower*, engine *size* (*Displacement*), and number of *Cylinders*.

	MPG	Weight	Horse-power	Displacement	Cylinders
MPG	1.000				
Weight	−0.903	1.000			
Horsepower	−0.871	0.917	1.000		
Displacement	−0.786	0.951	0.872	1.000	
Cylinders	−0.806	0.917	0.864	0.940	1.000

a) Which factor seems most strongly associated with *Fuel Efficiency*?
b) What does the negative correlation indicate?
c) Explain the meaning of R^2 for that relationship.

T 14. Autos, revisited. Look again at the correlation table for cars in the previous exercise.
a) Which two variables in the table exhibit the strongest association?
b) Is that strong association necessarily cause-and-effect? Offer at least two explanations why that association might be so strong.
c) Engine displacements for U.S.-made cars are often measured in cubic inches. For many foreign cars, the units are either cubic centimeters or liters. How would changing from cubic inches to liters affect the calculated correlations involving *Displacement?*
d) What would you predict about the *Fuel Efficiency* of a car whose engine *Displacement* is one standard deviation above the mean?

T 15. Cars, one more time! Can we predict the *Horsepower* of the engine that manufacturers will put in a car by knowing the *Weight* of the car? Here are the regression analysis and residuals plot:

Dependent variable is: Horsepower
R-squared = 84.1%

Variable	Coefficient
Intercept	3.49834
Weight	34.3144

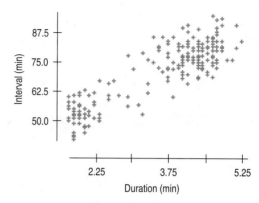

a) Write the equation of the regression line.
b) Do you think the car's *Weight* is measured in pounds or thousands of pounds? Explain.
c) Do you think this linear model is appropriate? Explain.
d) The highest point in the residuals plot, representing a residual of 22.5 horsepower, is for a Chevy weighing 2595 pounds. How much horsepower does this car have?

16. Colorblind. Although some women are colorblind, this condition is found primarily in men. Why is it wrong to say there's a strong correlation between *Sex* and *Colorblindness?*

T 17. Old Faithful. There is evidence that eruptions of Old Faithful can best be predicted by knowing the duration of the previous eruption.

a) Describe what you see in the scatterplot of *Intervals* between eruptions vs. *Duration* of the previous eruption.

b) Write the equation of the line of best fit. Here's the regression analysis:

Dependent variable is: Interval
R-squared = 77.0%

Variable	Coefficient
Intercept	33.9668
Duration	10.3582

c) Carefully explain what the slope of the line means in this context.
d) How accurate do you expect predictions based on this model to be? Cite statistical evidence.
e) If you just witnessed an eruption that lasted 4 minutes, how long do you predict you'll have to wait to see the next eruption?
f) So you waited, and the next eruption came in 79 minutes. Use this as an example to define a residual.

T 18. Crocodile lengths. The ranges inhabited by the Indian gharial crocodile and the Australian saltwater crocodile overlap in Bangladesh. Suppose a very large crocodile skeleton is found there, and we wish to determine the species of the animal. Wildlife scientists have measured the lengths of the heads and the complete bodies of several crocs (in centimeters) of each species, creating the regression analyses below:

Indian Crocodile
Dependent variable is: IBody
R-squared = 97.2%

Variable	Coefficient
Intercept	−69.3693
IHead	7.40004

Australian Crocodile
Dependent variable is: ABody
R-squared = 98.0%

Variable	Coefficient
Intercept	−20.2245
AHead	7.71726

a) Do the associations between the sizes of the heads and bodies of the two species appear to be strong? Explain.
b) In what ways are the two relationships similar? Explain.
c) What is different about the two models? What does that mean?
d) The crocodile skeleton found had a head length of 62 cm and a body length of 380 cm. Which species do you think it was? Explain why.

19. How old is that tree? One can determine how old a tree is by counting its rings, but that requires cutting the tree down. Can we estimate the tree's age simply from its diameter? A forester measured 27 trees of the same species that had been cut down, and counted the rings to determine the ages of the trees.

Diameter (in.)	Age (yr)	Diameter (in.)	Age (yr)
1.8	4	10.3	23
1.8	5	14.3	25
2.2	8	13.2	28
4.4	8	9.9	29
6.6	8	13.2	30
4.4	10	15.4	30
7.7	10	17.6	33
10.8	12	14.3	34
7.7	13	15.4	35
5.5	14	11.0	38
9.9	16	15.4	38
10.1	18	16.5	40
12.1	20	16.5	42
12.8	22		

a) Find the correlation between *Diameter* and *Age*. Does this suggest that a linear model may be appropriate? Explain.
b) Create a scatterplot and describe the association.
c) Create the linear model.
d) Check the residuals. Explain why a linear model is probably not appropriate.
e) If you used this model, would it generally overestimate or underestimate the ages of very large trees? Explain.

20. Improving trees. In the last exercise you saw that the linear model had some deficiencies. Let's create a better model.
a) Perhaps the cross-sectional area of a tree would be a better predictor of its age. Since area is measured in square units, try re-expressing the data by squaring the diameters. Does the scatterplot look better?
b) Create a model that predicts *Age* from the square of the *Diameter*.
c) Check the residuals plot for this new model. Is this model more appropriate? Why?
d) Estimate the age of a tree 18 inches in diameter.

21. New homes. A real estate agent collects data to develop a model that will use the *Size* of a new home (in square feet) to predict its *Sale Price* (in thousands of dollars). Which of these is most likely to be the slope of the regression line: 0.008, 0.08, 0.8, or 8? Explain.

22. Smoking and pregnancy 2003. The organization Kids Count monitors issues related to children. The table shows a 50-state average of the percent of expectant mothers who smoked cigarettes during their pregnancies.

Year	% Smoking While Pregnant	Year	% Smoking While Pregnant
1990	19.2	1997	14.9
1991	18.7	1998	14.8
1992	17.9	1999	14.1
1993	16.8	2000	14.0
1994	16.0	2001	13.8
1995	15.4	2002	13.3
1996	15.3	2003	12.7

a) Create a scatterplot and describe the trend you see.
b) Find the correlation.
c) How is the value of the correlation affected by the fact that the data are averages rather than percentages for each of the 50 states?
d) Write a linear model and interpret the slope in context.

23. No smoking? The downward trend in smoking you saw in the last exercise is good news for the health of babies, but will it ever stop?
a) Explain why you can't use the linear model you created in Exercise 22 to see when smoking during pregnancy will cease altogether.
b) Create a model that could estimate the year in which the level of smoking would be 0%.
c) Comment on the reliability of such a prediction.

24. Tips. It's commonly believed that people use tips to reward good service. A researcher for the hospitality industry examined tips and ratings of service quality from 2645 dining parties at 21 different restaurants. The correlation between ratings of service and tip percentages was 0.11. (M. Lynn and M. McCall, "Gratitude and Gratuity." *Journal of Socio-Economics* 29: 203–214)
a) Describe the relationship between *Quality of Service* and *Tip Size*.
b) Find and interpret the value of R^2 in this context.

25. U.S. cities. Data from 50 large U.S. cities show the mean *January Temperature* and the *Latitude*. Describe what you see in the scatterplot.

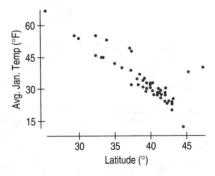

26. Correlations. The study of U.S. cities in Exercise 25 found the mean *January Temperature* (degrees Fahrenheit), *Altitude* (feet above sea level), and *Latitude* (degrees north of the equator) for 55 cities. Here's the correlation matrix:

	Jan. Temp	Latitude	Altitude
Jan. Temp	1.000		
Latitude	−0.848	1.000	
Altitude	−0.369	0.184	1.000

a) Which seems to be more useful in predicting *January Temperature—Altitude* or *Latitude*? Explain.

b) If the *Temperature* were measured in degrees Celsius, what would be the correlation between Temperature and Latitude?

c) If the *Temperature* were measured in degrees Celsius and the *Altitude* in meters, what would be the correlation? Explain.

d) What would you predict about the January *Temperature* in a city whose *Altitude* is two standard deviations higher than the average *Altitude*?

27. **Winter in the city.** Summary statistics for the data relating the latitude and average January temperature for 55 large U.S. cities are given below.

Variable	Mean	StdDev
Latitude	39.02	5.42
JanTemp	26.44	13.49
Correlation = −0.848		

a) What percent of the variation in January *Temperature* can be explained by variation in *Latitude*?

b) What is indicated by the fact that the correlation is negative?

c) Write the equation of the line of regression for predicting average January *Temperature* from *Latitude*.

d) Explain what the slope of the line means.

e) Do you think the *y*-intercept is meaningful? Explain.

f) The latitude of Denver is 40°N. Predict the mean January temperature there.

g) What does it mean if the residual for a city is positive?

28. **Depression.** The September 1998 issue of the *American Psychologist* published an article by Kraut et al. that reported on an experiment examining "the social and psychological impact of the Internet on 169 people in 73 households during their first 1 to 2 years online." In the experiment, 73 households were offered free Internet access for 1 or 2 years in return for allowing their time and activity online to be tracked. The members of the households who participated in the study were also given a battery of tests at the beginning and again at the end of the study. The conclusion of the study made news headlines: Those who spent more time online tended to be more depressed at the end of the experiment. Although the paper reports a more complex model, the basic result can be summarized in the following regression of *Depression* (at the end of the study, in "depression scale units") vs. *Internet Use* (in mean hours per week):

```
Dependent variable is: Depression
R-squared = 4.6%
s = 0.4563
```

Variable	Coefficient
Intercept	0.5655
Internet use	0.0199

The news reports about this study clearly concluded that using the Internet causes depression. Discuss whether such a conclusion can be drawn from this regression. If so, discuss the supporting evidence. If not, say why not.

29. **Jumps 2008.** How are Olympic performances in various events related? The plot shows winning long-jump

and high-jump distances, in meters, for the Summer Olympics from 1912 through 2008.

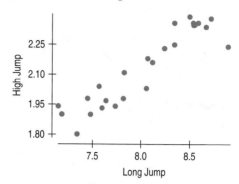

a) Describe the association.

b) Do long-jump performances somehow influence the high-jumpers? How do you account for the relationship you see?

c) The correlation for the given scatterplot is 0.920. If we converted the jump lengths to centimeters by multiplying by 100, would that make the actual correlation higher or lower?

d) What would you predict about the long jump in a year when the high-jumper jumped one standard deviation better than the average high jump?

30. **Modeling jumps 2008.** Here are the summary statistics for the Olympic long jumps and high jumps displayed in the previous exercise.

Event	Mean	StdDev
High Jump	2.13880	0.191884
Long Jump	8.03960	0.521380
Correlation = 0.920		

a) Write the equation of the line of regression for estimating *High Jump* from *Long Jump*.

b) Interpret the slope of the line.

c) In a year when the long jump is 8.9 m, what high jump would you predict?

d) Why can't you use this line to estimate the long jump for a year when you know the high jump was 2.25 m?

e) Write the equation of the line you need to make that prediction.

31. **French.** Consider the association between a student's score on a French vocabulary test and the weight of the student. What direction and strength of correlation would you expect in each of the following situations? Explain.

a) The students are all in third grade.

b) The students are in third through twelfth grades in the same school district.

c) The students are in tenth grade in France.

d) The students are in third through twelfth grades in France.

32. **Twins.** Twins are often born at less than 9 months' gestation. The graph from the *Journal of the American Medical Association (JAMA)* shows the rate of preterm twin births in the United States over the past 20 years. In this study, *JAMA* categorized mothers by the level of prenatal medical care they received: inadequate, adequate, or intensive.

a) Describe the overall trend in preterm twin births.
b) Describe any differences you see in this trend, depending on the level of prenatal medical care the mother received.
c) Should expectant mothers be advised to cut back on the level of medical care they seek in the hope of avoiding preterm births? Explain.

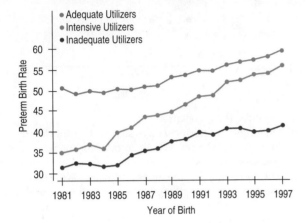

Preterm Birth Rate per 100 live twin births among U.S. twins by intensive, adequate, and less than adequate prenatal care utilization, 1981–1997. (*JAMA* 284[2000]: 335–341)

33. Lunchtime. Does how long toddlers sit at the lunch table help predict how much they eat? The table and graph show the number of minutes the kids stayed at the table and the number of calories they consumed. Create and interpret a model for these data.

Calories	Time	Calories	Time
472	21.4	450	42.4
498	30.8	410	43.1
465	37.7	504	29.2
456	33.5	437	31.3
423	32.8	489	28.6
437	39.5	436	32.9
508	22.8	480	30.6
431	34.1	439	35.1
479	33.9	444	33.0
454	43.8	408	43.7

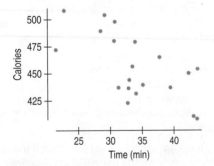

34. Gasoline. Since clean-air regulations have dictated the use of unleaded gasoline, the supply of leaded gas in New York state has diminished. The following table was given on the August 2001 New York State Math B exam, a statewide achievement test for high school students.

Year	1984	1988	1992	1996	2000
Gallons (1000's)	150	124	104	76	50

a) Create a linear model to predict the number of gallons that will be available in 2005.
b) The exam then asked students to estimate the year when leaded gasoline will first become unavailable, expecting them to use the model from part a to answer the question. Explain why that method is incorrect.
c) Create a model that *would* be appropriate for that task, and make the estimate.
d) The "wrong" answer from the other model is fairly accurate in this case. *Why?*

35. Tobacco and alcohol. Are people who use tobacco products more likely to consume alcohol? Here are data on household spending (in pounds) taken by the British government on 11 regions in Great Britain. Do tobacco and alcohol spending appear to be related? What questions do you have about these data? What conclusions can you draw?

Region	Alcohol	Tobacco
North	6.47	4.03
Yorkshire	6.13	3.76
Northeast	6.19	3.77
East Midlands	4.89	3.34
West Midlands	5.63	3.47
East Anglia	4.52	2.92
Southeast	5.89	3.20
Southwest	4.79	2.71
Wales	5.27	3.53
Scotland	6.08	4.51
Northern Ireland	4.02	4.56

36. Football weights. The Sears Cup was established in 1993 to honor institutions that maintain a broad-based athletic program, achieving success in many sports, both men's and women's. Since its Division III inception in 1995, the cup has been won by Williams College in every year except one. Their football team has a 85.3% winning record under their current coach. Why does the football team win so much? Is it because they're heavier than their opponents? The table shows the average team weights for selected years from 1973 to 1993.

Year	Weight (lb)	Year	Weight (lb)
1973	185.5	1983	192.0
1975	182.4	1987	196.9
1977	182.1	1989	202.9
1979	191.1	1991	206.0
1981	189.4	1993	198.7

a) Fit a straight line to the relationship between *Weight* and *Year*.

b) Does a straight line seem reasonable?

c) Predict the average weight of the team for the year 2015. Does this seem reasonable?

d) What about the prediction for the year 2103? Explain.

e) What about the prediction for the year 3003? Explain.

37. **Models.** Find the predicted value of *y*, using each model for $x = 10$.

a) $\hat{y} = 2 + 0.8 \ln x$ c) $\dfrac{1}{\sqrt{\hat{y}}} = 17.1 - 1.66x$

b) $\log \hat{y} = 5 - 0.23x$

T 38. **Williams vs. Texas.** Here are the average weights of the football team for the University of Texas for various years in the 20th century.

Year	1905	1919	1932	1945	1955	1965
Weight (lb)	164	163	181	192	195	199

a) Fit a straight line to the relationship of *Weight* by *Year* for Texas football players.

b) According to these models, in what year will the predicted weight of the Williams College team from Exercise 36 first be more than the weight of the University of Texas team?

c) Do you believe this? Explain.

39. **Vehicle weights.** The Minnesota Department of Transportation hoped that they could measure the weights of big trucks without actually stopping the vehicles by using a newly developed "weigh-in-motion" scale. After installation of the scale, a study was conducted to find out whether the scale's readings correspond to the true weights of the trucks being monitored. In Exercise 40 of Chapter 7, you examined the scatterplot for the data they collected, finding the association to be approximately linear with $R^2 = 93\%$. Their regression equation is $\widehat{Wt} = 10.85 + 0.64\, Scale$, where both the scale reading and the predicted weight of the truck are measured in thousands of pounds.

a) Estimate the weight of a truck if this scale read 31,200 pounds.

b) If that truck actually weighed 32,120 pounds, what was the residual?

c) If the scale reads 35,590 pounds, and the truck has a residual of −2440 pounds, how much does it actually weigh?

d) In general, do you expect estimates made using this equation to be reasonably accurate? Explain.

e) If the police plan to use this scale to issue tickets to trucks that appear to be overloaded, will negative or positive residuals be a greater problem? Explain.

40. **Profit.** How are a company's profits related to its sales? Let's examine data from 71 large U.S. corporations. All amounts are in millions of dollars.

a) Histograms of *Profits* and *Sales* and histograms of the logarithms of *Profits* and *Sales* are seen below. Why are the re-expressed data better for regression?

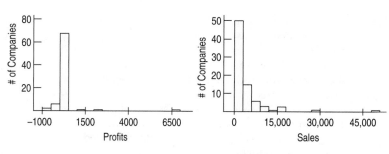

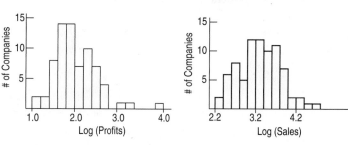

b) Here are the scatterplot and residuals plot for the regression of logarithm of *Profits* vs. log of *Sales*. Do you think this model is appropriate? Explain.

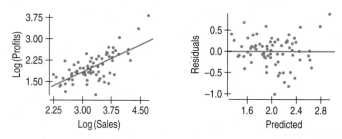

c) Here's the regression analysis. Write the equation.

Dependent variable is: Log Profit
R-squared = 48.1%

Variable	Coefficient
Intercept	−0.106259
LogSales	0.647798

d) Use your equation to estimate profits earned by a company with sales of 2.5 billion dollars. (That's 2500 million.)

T 41. **Down the drain.** Most water tanks have a drain plug so that the tank may be emptied when it's to be moved or repaired. How long it takes a certain size of tank to drain depends on the size of the plug, as shown in the table. Create a model.

Plug Dia (in.)	$\frac{3}{8}$	$\frac{1}{2}$	$\frac{3}{4}$	1	$1\frac{1}{4}$	$1\frac{1}{2}$	2
Drain Time (min.)	140	80	35	20	13	10	5

T 42. Chips. A start-up company has developed an improved electronic chip for use in laboratory equipment. The company needs to project the manufacturing cost, so it develops a spreadsheet model that takes into account the purchase of production equipment, overhead, raw materials, depreciation, maintenance, and other business costs. The spreadsheet estimates the cost of producing 10,000 to 200,000 chips per year, as seen in the table. Develop a regression model to predict *Costs* based on the *Level* of production.

Chips Produced (1000s)	Cost per Chip ($)	Chips Produced (1000s)	Cost per Chip ($)
10	146.10	90	47.22
20	105.80	100	44.31
30	85.75	120	42.88
40	77.02	140	39.05
50	66.10	160	37.47
60	63.92	180	35.09
70	58.80	200	34.04
80	50.91		

Understanding Randomness

Where are we going?

Few things in life are certain. But that doesn't mean that we all understand randomness. We'll be formal about it in a few chapters, but for now we'll simulate what we want to understand to see the patterns that emerge.

"The most decisive conceptual event of twentieth century physics has been the discovery that the world is not deterministic. . . . A space was cleared for chance."

—Ian Hocking,
The Taming of Chance

We all know what it means for something to be random. Or do we? Many children's games rely on chance outcomes. Rolling dice, spinning spinners, and shuffling cards all select at random. Adult games use randomness as well, from card games to lotteries to bingo. What's the most important aspect of the randomness in these games? It must be fair.

What is it about random selection that makes it seem fair? It's really two things. First, nobody can guess the outcome before it happens. Second, when we want things to be fair, usually some underlying set of outcomes will be equally likely (although in many games, some combinations of outcomes are more likely than others).

Randomness is not always what we might think of as "at random." Random outcomes have a lot of structure, especially when viewed in the long run. You can't predict how a fair coin will land on any single toss, but you're pretty confident that if you flipped it thousands of times you'd see about 50% heads. As we will see, randomness is an essential tool of Statistics. Statisticians don't think of randomness as the annoying tendency of things to be unpredictable or haphazard. Statisticians use randomness as a tool. In fact, without deliberately applying randomness, we couldn't do most of Statistics, and this book would stop right about here.[1]

But truly random values are surprisingly hard to get. Just to see how fair humans are at selecting, pick a number at random from the top of the next page. Go ahead. Turn the page, look at the numbers quickly, and pick a number at random.

Ready?

Go.

---

[1] Don't get your hopes up.

1 2 3 4

It's Not Easy Being Random

Did you pick 3? If so, you've got company. Almost 75% of all people pick the number 3. About 20% pick either 2 or 4. If you picked 1, well, consider yourself a little different. Only about 5% choose 1. Psychologists have proposed reasons for this phenomenon, but for us, it simply serves as a lesson that we've got to find a better way to choose things at random.

So how should we generate **random numbers**? It's surprisingly difficult to get random values even when they're equally likely. Computers have become a popular way to generate random numbers. Even though they often do much better than humans, computers can't generate truly random numbers either. Computers follow programs. Start a computer from the same place, and it will always follow exactly the same path. So numbers generated by a computer program are not truly random. Technically, "random" numbers generated this way are *pseudorandom* numbers. Pseudorandom values are generated in a fixed sequence, and because computers can represent only a finite number of distinct values, the sequence of pseudorandom numbers must eventually repeat itself. Fortunately, pseudorandom values are good enough for most purposes because they are virtually indistinguishable from truly random numbers.

A S
Activity: Random Behavior.
ActivStats' Random Experiment Tool lets you experiment with truly random outcomes. We'll use it a lot in the coming chapters.

A S
Activity: Truly Random Values on the Internet. This activity will take you to an Internet site (www.random.org) that generates all the truly random numbers you could want.

There *are* ways to generate random numbers so that they are both equally likely and truly random. In the past, entire books of carefully generated random numbers were published. The books never made the best-seller lists and probably didn't make for great reading, but they were quite valuable to those who needed truly random values. Today, we have a choice. We can use these books or find genuinely random digits from several Internet sites. The sites use methods like timing the decay of a radioactive element or

even the random changes of lava lamps to generate truly random digits.[2] In either case, a string of random digits might look like this:

2217726304387410092537086270581997622725849795907032825001108963
3217535822643800292254644943760642389043766557204107354186024508
8906427308645681412198226653885873285801699027843110380420067664
8740522639824530519902027044464984322000946238678577902639002954
8887003319933147508331265192321413908608674496383528968974910533
6944182713168919406022181281304751019321546303870481407676636740
6070204916508913632855351361361043794293428486909462881431793360
7706356513310563210508993624272872250535395513645991015328128202

You probably have more interesting things to download than a few million random digits, but we'll discuss ways to use such random digits to apply randomness to real situations soon. The best ways we know to generate data that give a fair and accurate picture of the world rely on randomness, and the ways in which we draw conclusions from those data depend on the randomness, too.

An ordinary deck of playing cards, like the ones used in bridge and many other card games, consists of 52 cards. There are numbered cards (2 through 10), and face cards (Jack, Queen, King, Ace) whose value depends on the game you are playing. Each card is also marked by one of four suits (clubs, diamonds, hearts, or spades) whose significance is also game-specific.

Aren't you done shuffling yet?
Even something as common as card shuffling may not be as random as you might think. If you shuffle cards by the usual method in which you split the deck in half and try to let cards fall roughly alternately from each half, you're doing a "riffle shuffle."

How many times should you shuffle cards to make the deck random? A surprising fact was discovered by statisticians Persi Diaconis, Ronald Graham, and W. M. Kantor. It takes seven riffle shuffles. Fewer than seven leaves order in the deck, but after that, more shuffling does little good. Most people, though, don't shuffle that many times.

When computers were first used to generate hands in bridge tournaments, some professional bridge players complained that the computer was making too many "weird" hands—hands with 10 cards of one suit, for example. Suddenly these hands were appearing more often than players were used to when cards were shuffled by hand. The players assumed that the computer was doing something wrong. But it turns out that it's humans who hadn't been shuffling enough to make the decks really random and have those "weird" hands appear as often as they should.

Practical Randomness

Suppose a cereal manufacturer puts pictures of famous athletes on cards in boxes of cereal in the hope of boosting sales. The manufacturer announces that 20% of the boxes contain a picture of LeBron James, 30% a picture of David Beckham, and the rest a picture of Serena Williams. You want all three pictures. How many boxes of cereal do you expect to have to buy in order to get the complete set?

How can we answer questions like this? Well, one way is to buy hundreds of boxes of cereal to see what might happen. But let's not. Instead, we'll consider using a random model. Why random? When we pick a box of cereal off the shelf, we don't know what picture is inside. We'll assume that the pictures

---

[2] For example, www.random.org or www.randomnumbers.info.

are randomly placed in the boxes and that the boxes are distributed randomly to stores around the country. Why a model? Because we won't actually buy the cereal boxes. We can't afford all those boxes and we don't want to waste food. So we need an imitation of the real process that we can manipulate and control. In short, we're going to **simulate** reality.

A Simulation

The question we've asked is how many boxes do you expect to buy to get a complete card collection. But we can't answer our question by completing a card collection just once. We want to understand the *typical* number of boxes to open, how that number varies, and, often, the shape of the distribution. So we'll have to do this over and over. We call each time we obtain a simulated answer to our question a **trial**.

For the sports cards, a trial's outcome is the number of boxes. We'll need at least 3 boxes to get one of each card, but with really bad luck, you could empty the shelves of several supermarkets before finding the card you need to get all 3. So, the possible outcomes of a trial are 3, 4, 5, or lots more. But we can't simply pick one of those numbers at random, because they're not equally likely. We'd be surprised if we only needed 3 boxes to get all the cards, but we'd probably be even more surprised to find that it took exactly 7,359 boxes. In fact, the reason we're doing the simulation is that it's hard to guess how many boxes we'd expect to open.

Building a Simulation

We know how to find equally likely random digits. How can we get from there to simulating the trial outcomes? We know the relative frequencies of the cards: 20% LeBron, 30% Beckham, and 50% Serena. So, we can interpret the digits 0 and 1 as finding LeBron; 2, 3, and 4 as finding Beckham; and 5 through 9 as finding Serena to simulate opening one box. Opening one box is the basic building block, called a **component** of our simulation. But the component's outcome isn't the result we want. We need to observe a sequence of components until our card collection is complete. The *trial's* outcome is called the **response variable**; for this simulation that's the *number* of components (boxes) in the sequence.

Let's look at the steps for making a simulation:

Specify how to model a component outcome using equally likely random digits:

1. **Identify the component to be repeated.** In this case, our component is the opening of a box of cereal.
2. **Explain how you will model the component's outcome.** The digits from 0 to 9 are equally likely to occur. Because 20% of the boxes contain LeBron's picture, we'll use 2 of the 10 digits to represent that outcome. Three of the 10 digits can model the 30% of boxes with David Beckham cards, and the remaining 5 digits can represent the 50% of boxes with Serena. One possible assignment of the digits, then, is

   0, 1 LeBron   2, 3, 4 Beckham   5, 6, 7, 8, 9 Serena.

Specify how to simulate trials:

3. **Explain how you will combine the components to model a trial.** We pretend to open boxes (repeat components) until our collection is complete.

> Modern physics has shown that randomness is not just a mathematical game; it is fundamentally the way the universe works.
>
> *Regardless of improvements in data collection or in computer power, the best we can ever do, according to quantum mechanics . . . is predict the probability that an electron, or a proton, or a neutron, or any other of nature's constituents, will be found here or there. Probability reigns supreme in the microcosmos.*
> —Brian Greene, *The Fabric of the Cosmos: Space, Time, and the Texture of Reality* (p. 91)

We do this by looking at each random digit and indicating what picture it represents. We continue until we've found all three.

4. **State clearly what the response variable is.** What are we interested in? We want to find out the number of boxes it might take to get all three pictures.

Put it all together to run the simulation:

5. **Run several trials.** For example, consider the third line of random digits shown earlier (p. 269):

89064273086456814121982266538858732858016990278431103804200676 64.

Let's see what happened.

The first random digit, 8, means you get Serena's picture. So the first component's outcome is Serena. The second digit, 9, means Serena's picture is also in the next box. Continuing to interpret the random digits, we get LeBron's picture (0) in the third, Serena's (6) again in the fourth, and finally Beckham (4) on the fifth box. Since we've now found all three pictures, we've finished one trial of our simulation. This trial's outcome is 5 boxes.

Now we keep going, running more trials by looking at the rest of our line of random digits:

89064 2730 8645681 41219 822665388587328580 169902 78431 1038 042006 7664.

It's best to create a chart to keep track of what happens:

Trial Number	Component Outcomes	Trial Outcomes: y = Number of Boxes
1	89064 = Serena, Serena, **LeBron**, Serena, **Beckham**	5
2	2730 = **Beckham**, Serena, Beckham, **LeBron**	4
3	8645681 = Serena, Serena, **Beckham**, . . . , **LeBron**	7
4	41219 = **Beckham**, **LeBron**, Beckham, LeBron, Serena	5
5	822665388587328580 = Serena, **Beckham**, . . . , **LeBron**	18
6	169902 = **LeBron**, Serena, Serena, Serena, LeBron, **Beckham**	6
7	78431 = Serena, Serena, **Beckham**, Beckham, **LeBron**	5
8	1038 = **LeBron**, LeBron, **Beckham**, Serena	4
9	042006 = **LeBron**, **Beckham**, Beckham, LeBron, LeBron, Serena	6
10	7664 . . . = Serena, Serena, Serena, **Beckham** . . .	?

Analyze the response variable:

6. **Collect and summarize the results of all the trials.** You know how to summarize and display a response variable. You'll certainly want to report the shape, center, and spread, and depending on the question asked, you may want to include more.

7. **State your conclusion,** as always, in the context of the question you wanted to answer. Based on this simulation, we estimate that customers hoping to complete their card collection will need to open a median of 5 boxes, but it could take a lot more.

If you fear that these may not be accurate estimates because we ran only nine trials, you are absolutely correct. The more trials the better, and nine is woefully inadequate. Twenty trials is probably a reasonable minimum if you are doing this by hand. Even better, use a computer and run a few hundred trials.

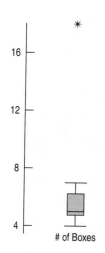

16 —

12 —

8 —

4 —

\# of Boxes

A S

Activity: **Bigger Samples Are Better.** The random simulation tool can generate lots of outcomes with a single click, so you can see more of the long run with less effort.

FOR EXAMPLE   Simulating a Dice Game

The game of 21 can be played with an ordinary 6-sided die. Competitors each roll the die repeatedly, trying to get the highest total less than or equal to 21. If your total exceeds 21, you lose.

Suppose your opponent has rolled an 18. Your task is to try to beat him by getting more than 18 points without going over 21. How many rolls do you expect to make, and what are your chances of winning?

QUESTION: How will you simulate the components?

A component is one roll of the die. I'll simulate each roll by looking at a random digit from a table or an Internet site. The digits 1 through 6 will represent the results on the die; I'll ignore digits 7–9 and 0.

QUESTION: How will you combine components to model a trial? What's the response variable?

I'll add components until my total is greater than 18, counting the number of rolls. If my total is greater than 21, it is a loss; if not, it is a win. There are two response variables. I'll count the number of times I roll the die, and I'll keep track of whether I win or lose.

QUESTION: How would you use these random digits to run trials? Show your method clearly for two trials.

91129 58757 69274 92380 82464 33089

I've marked the discarded digits in color.

Trial #1:	9	1	1	2	9	5	8	7	5	7	6	
Total:		1	2	4		9			14		20	Outcomes: 6 rolls, won
Trial #2:	9	2	7	4	9	2	3	8	0	8	2	4	6
Total:		2		6		8	11				13	17	23	Outcomes: 7 rolls, lost

QUESTION: Suppose you run 30 trials, getting the outcomes tallied here. What is your conclusion?

Based on my simulation, when competing against an opponent who has a score of 18, I expect my turn to usually last 5 or 6 rolls, and I should win about 70% of the time.

Number of rolls		Result	
4	///	Won	ℍℍ ℍℍ ℍℍ ℍℍ /
5	ℍℍ ℍℍ	Lost	ℍℍ ////
6	ℍℍ ℍℍ /		
7	ℍℍ		
8	/		

The baseball World Series consists of up to seven games. The first team to win four games wins the series. The first two are played at one team's home ballpark, the next three at the other team's park, and the final two (if needed) are played back at the first park. Records over the past century show that there is a home field advantage; the home team has about a 55% chance of winning. Does the current system of alternating ballparks even out the home field advantage? How often will the team that begins at home win the series?

Let's set up the simulation:

1. What is the component to be repeated?

2. How will you model each component from equally likely random digits?

3. How will you model a trial by combining components?

4. What is the response variable?

5. How will you analyze the response variable?

STEP-BY-STEP EXAMPLE   Simulation

Fifty-seven students participated in a lottery for a particularly desirable dorm room—a large triple close to the student center. Twenty of the participants were members of the same varsity team. When all three winners were members of the team, the other students cried foul.

Question: Could an all-team outcome reasonably be expected to happen if everyone had a fair shot at the room?

THINK

Plan State the problem. Identify the important parts of your simulation.

I'll use a simulation to investigate whether it's unlikely that three varsity athletes would get the great room in the dorm if the lottery were fair.

Components Identify the components.

A component is the selection of a student.

Outcomes State how you will model each component using equally likely random digits. You can't just use the digits from 0 to 9 because the outcomes you are simulating are not multiples of 10%.

I'll look at two-digit random numbers.

Let 00–19 represent the 20 varsity applicants.

Let 20–56 represent the other 37 applicants.

There are 20 and 37 students in the two groups. This time you must use *pairs* of random digits (and ignore some of them) to represent the 57 students.

Skip 57–99. If I get a number in this range, I'll throw it away and go back for another two-digit random number.

Trial Explain how you will combine the components to simulate a trial. In each of these trials, you can't choose the same student twice, so you'll need to ignore a random number if it comes up a second or third time. Be sure to mention this in describing your simulation.

Each trial consists of identifying pairs of digits as V (varsity) or N (nonvarsity) until 3 people are chosen, ignoring out-of-range or repeated numbers (X)—I can't put the same person in the room twice.

Response Variable Define your response variable.

The response variable is whether or not all three selected students are on the varsity team.

SHOW

Mechanics Run several trials. Carefully record the random numbers, indicating

1) the corresponding component outcomes (Varsity, Nonvarsity, or ignored number) and
2) the value of the response variable.

Trial Number	Component Outcomes	All Varsity?
1	74 02 94 39 02 77 55 X V X N X X N	No
2	18 63 33 25 V X N N	No
3	05 45 88 91 56 V N X X N	No
4	39 09 07 N V V	No
5	65 39 45 95 43 X N N X N	No
6	98 95 11 68 77 12 17 X X V X X V V	Yes

(continued)

7	26 19 89 93 77 27 N V X X X N	No
8	23 52 37 N N N	No
9	16 50 83 44 V N X N	No
10	74 17 46 85 09 X V N X V	No

Analyze Summarize the results across all trials to answer the initial question.

"All varsity" occurred once, or 10% of the time.

Conclusion Describe what the simulation shows, and interpret your results in the context of the real world.

In my simulation of "fair" room draws, the three people chosen were all varsity team members only 10% of the time. While this result could happen by chance, it is not particularly likely. I'm suspicious, but I'd need many more trials and a smaller frequency of the all-varsity outcome before I would make an accusation of unfairness.

What Can Go Wrong?

Activity: **Estimating Summaries from Random Outcomes.** See how well you can estimate something you can't know just by generating random outcomes.

- **Don't overstate your case.** Let's face it: In some sense, a simulation is *always* wrong. After all, it's not the real thing. We didn't buy any cereal or run a room draw. So beware of confusing what *really* happens with what a simulation suggests *might* happen. Never forget that future results will not match your simulated results exactly.

- **Model outcome chances accurately.** A common mistake in constructing a simulation is to adopt a strategy that may appear to produce the right kind of results, but that does not accurately model the situation. For example, in our room draw, we could have gotten 0, 1, 2, or 3 team members. Why not just see how often these digits occur in random digits from 0 to 9, ignoring the digits 4 and up?

  3 2 1 7 9 0 0 5 9 7 3 7 9 2 5 2 4 1 3 8
  3 2 1 x x 0 0 x x x 3 x x 2 x 2 x 1 3 x

  This "simulation" makes it seem fairly likely that three team members would be chosen. There's a big problem with this approach, though: The digits 0, 1, 2, and 3 occur with equal frequency among random digits, making each outcome appear to happen 25% of the time. In fact, the selection of 0, 1, 2, or all 3 team members are not all equally likely outcomes. In our correct simulation, we estimated that all 3 would be chosen only about 10% of the time. If your simulation overlooks important aspects of the real situation, your model will not be accurate.

- **Run enough trials.** Simulation is cheap and fairly easy to do. Don't try to draw conclusions based on 5 or 10 trials (even though we did for illustration purposes here). We'll see how many trials to use in later chapters. For now, err on the side of large numbers of trials.

CONNECTIONS

Simulations often generate many outcomes of a response variable, and we are often interested in the distribution of these responses. The tools we use to display and summarize the distribution of any real variable are appropriate for displaying and summarizing randomly generated responses as well.

Make histograms, boxplots, and Normal probability plots of the response variables from simulations, and summarize them with measures of center and spread. Be especially careful to report the variation of your response variable.

Don't forget to think about your analyses. Simulations can hide subtle errors. A careful analysis of the responses can save you from erroneous conclusions based on a faulty simulation.

You may be less likely to find an outlier in simulated responses, but if you find one, you should certainly determine how it happened.

WHAT HAVE WE LEARNED?

We've learned to harness the power of randomness. We've learned that a simulation model can help us investigate a question for which many outcomes are possible, we can't (or don't want to) collect data, and a mathematical answer is hard to calculate. We've learned how to base our simulation on random values generated by a computer, generated by a randomizing device such as a die or spinner, or found on the Internet. Like all models, simulations can provide us with useful insights about the real world.

Terms

Random
An outcome is random if we know the possible values it can have, but not which particular value it takes (p. 267).

Generating random numbers
Random numbers are hard to generate. Nevertheless, several Internet sites offer an unlimited supply of equally likely random values (p. 268).

Simulation
A simulation models a real-world situation by using random-digit outcomes to mimic the uncertainty of a response variable of interest (p. 270).

Trial
The sequence of several components representing events that we are pretending will take place (p. 270).

Simulation component
A component uses equally likely random digits to model simple random occurrences whose outcomes may not be equally likely (p. 270).

Response variable
Values of the response variable record the results of each trial with respect to what we were interested in (p. 270).

Skills

 ▶ Be able to recognize random outcomes in a real-world situation.

▶ Be able to recognize when a simulation might usefully model random behavior in the real world.

 ▶ Know how to perform a simulation either by generating random numbers on a computer or calculator, or by using some other source of random values, such as dice, a spinner, or a table of random numbers.

▶ Be able to describe a simulation so that others can repeat it.

▶ Be able to discuss the results of a simulation study and draw conclusions about the question being investigated.

SIMULATION ON THE COMPUTER

Simulations are best done with the help of technology simply because more trials make a better simulation, and computers are fast. There are special computer programs designed for simulation, but most statistics packages can generate random numbers and support a simulation.

All technology-generated random numbers are *pseudorandom*. The random numbers available on the Internet may technically be better, but the differences won't matter for any simulation of modest size. Pseudorandom numbers generate the next random value from the previous one by a specified algorithm. But they have to start somewhere. This starting point is called the "seed." Most programs let you set the seed. There's usually little reason to do this, but if you wish to, go ahead. If you reset the seed to the same value, the programs will generate the same sequence of "random" numbers.

A S *Activity:* **Creating Random Values.** Learn to use your statistics package to generate random outcomes.

DATA DESK

Generate random numbers in Data Desk with the **Generate Random Numbers . . .** command in the **Manip** menu. A dialog guides you in specifying the number of variables to fill, the number of cases, and details about the values. For most simulations, generate random uniform values.

COMMENTS

Bernoulli Trials generate random values that are 0 or 1, with a specified chance of a 1.

Binomial Experiments automatically generate a specified number of Bernoulli trials and count the number of 1's.

EXCEL

The **RAND** function generates a random value between 0 and 1. You can multiply to scale it up to any range you like and use the INT function to turn the result into an integer.

COMMENTS

Published tests of Excel's random-number generation have declared it to be inadequate. However, for simple simulations, it should be OK. Don't trust it for important large simulations.

JMP

In a new column, in the **Cols** menu choose **Column Info. . . .** In the dialog, click the **New Property** button, and choose **Formula** from the drop-down menu.

Click the **Edit Formula** button, and in the **Functions(grouped)** window click on **Random. Random Integer (10)**, for example, will generate a random integer between 1 and 10.

MINITAB

In the **Calc** menu, choose **Random Data. . . .** In the Random Data submenu, choose **Uniform. . . .**

A dialog guides you in specifying details of range and number of columns to generate.

SPSS

The **RV.UNIFORM(min, max)** function returns a random value that is equally likely between the min and max limits.

(continued)

TI-83/84 Plus	COMMENTS
To generate random numbers, use **5:RandInt** from the **Math PRB** menu. This command will produce any number of random integers in a specified range.	Some examples: **RandInt(0,1)** randomly chooses a 0 or a 1. This is an effective simulation of a coin toss. **RandInt(1,6,2)** randomly returns two integers between 1 and 6. This is a good way to simulate rolling two dice. To run several trials, just hit ENTER repeatedly. **RandInt(0,57,3)** produces three random integers between 0 and 57—a nice way to simulate the chapter's dorm room lottery.
TI-89	**COMMENTS**
To generate random numbers, move the cursor to highlight the name of a blank list. Use **5:RandInt** from the F4 (Calc) Probability menu. This command will produce any number of random integers in the specified range.	Some examples: **RandInt(0,10)** randomly chooses a 0 or a 1. This is an effective simulation of 10 coin tosses. **RandInt(1,6,2)** randomly returns two integers between 1 and 6. This is a good way to simulate rolling two dice. **RandInt(0,56,3)** produces three random integers between 0 and 56, a nice way to simulate the chapter's dorm room lottery.

EXERCISES

1. **Coin toss.** Is a coin flip random? Why or why not?

2. **Casino.** A casino claims that its electronic "video roulette" machine is truly random. What should that claim mean?

3. **The lottery.** Many states run lotteries, giving away millions of dollars if you match a certain set of winning numbers. How are those numbers determined? Do you think this method guarantees randomness? Explain.

4. **Games.** Many kinds of games people play rely on randomness. Cite three different methods commonly used in the attempt to achieve this randomness, and discuss the effectiveness of each.

5. **Birth defects.** The American College of Obstetricians and Gynecologists says that out of every 100 babies born in the United States, 3 have some kind of major birth defect. How would you assign random numbers to conduct a simulation based on this statistic?

6. **Colorblind.** By some estimates, about 10% of all males have some color perception defect, most commonly red–green colorblindness. How would you assign random numbers to conduct a simulation based on this statistic?

7. **Geography.** An elementary school teacher with 25 students plans to have each of them make a poster about two different states. The teacher first numbers the states (in alphabetical order, from 1-Alabama to 50-Wyoming),

then uses a random number table to decide which states each student gets. Here are the random digits:

45921 01710 22892 37076

a) Which two state numbers does the first student get?
b) Which two state numbers go to the second student?

8. **Get rich.** Your state's BigBucks Lottery prize has reached $100,000,000, and you decide to play. You have to pick five numbers between 1 and 60, and you'll win if your numbers match those drawn by the state. You decide to pick your "lucky" numbers using a random number table. Which numbers do you play, based on these random digits?

43680 98750 13092 76561 58712

9. **Play the lottery.** Some people play state-run lotteries by always playing the same favorite "lucky" number. Assuming that the lottery is truly random, is this strategy better, worse, or the same as choosing different numbers for each play? Explain.

10. **Play it again, Sam.** In Exercise 8 you imagined playing the lottery by using random digits to decide what numbers to play. Is this a particularly good or bad strategy? Explain.

11. **Bad simulations.** Explain why each of the following simulations fails to model the real situation properly:

a) Use a random integer from 0 through 9 to represent the number of heads when 9 coins are tossed.

b) A basketball player takes a foul shot. Look at a random digit, using an odd digit to represent a good shot and an even digit to represent a miss.

c) Use random digits from 1 through 13 to represent the denominations of the cards in a five-card poker hand.

12. **More bad simulations.** Explain why each of the following simulations fails to model the real situation:

a) Use random numbers 2 through 12 to represent the sum of the faces when two dice are rolled.

b) Use a random integer from 0 through 5 to represent the number of boys in a family of 5 children.

c) Simulate a baseball player's performance at bat by letting $0 = $ an out, $1 = $ a single, $2 = $ a double, $3 = $ a triple, and $4 = $ a home run.

13. **Wrong conclusion.** A Statistics student properly simulated the length of checkout lines in a grocery store and then reported, "The average length of the line will be 3.2 people." What's wrong with this conclusion?

14. **Another wrong conclusion.** After simulating the spread of a disease, a researcher wrote, "24% of the people contracted the disease." What should the correct conclusion be?

15. **Election.** You're pretty sure that your candidate for class president has about 55% of the votes in the entire school. But you're worried that only 100 students will show up to vote. How often will the underdog (the one with 45% support) win? To find out, you set up a simulation.

a) Describe how you will simulate a component.

b) Describe how you will simulate a trial.

c) Describe the response variable.

16. **Two pair or three of a kind?** When drawing five cards randomly from a deck, which is more likely, two pairs or three of a kind? A pair is exactly two of the same denomination. Three of a kind is exactly 3 of the same denomination. (Don't count three 8's as a pair—that's 3 of a kind. And don't count 4 of the same kind as two pair—that's 4 of a kind, a very special hand.) How could you simulate 5-card hands? Be careful; once you've picked the 8 of spades, you can't get it again in that hand.

a) Describe how you will simulate a component.

b) Describe how you will simulate a trial.

c) Describe the response variable.

17. **Cereal.** In the chapter's example, 20% of the cereal boxes contained a picture of LeBron James, 30% David Beckham, and the rest Serena Williams. Suppose you buy five boxes of cereal. Estimate the probability that you end up with a complete set of the pictures. Your simulation should have at least 20 runs.

18. **Cereal, again.** Suppose you really want the LeBron James picture. How many boxes of cereal do you need to buy to be pretty sure of getting at least one? Your simulation should use at least 10 trials.

19. **Multiple choice.** You take a quiz with 6 multiple choice questions. After you studied, you estimated that you would have about an 80% chance of getting any individual question right. What are your chances of getting them all right? Use at least 20 trials.

20. **Lucky guessing?** A friend of yours who took the multiple choice quiz in Exercise 19 got all 6 questions right, but now claims to have guessed blindly on every question. If each question offered 4 possible answers, do you believe her? Explain, basing your argument on a simulation involving at least 10 trials.

21. **Beat the lottery.** Many states run lotteries to raise money. A website advertises that it knows "how to increase YOUR chances of Winning the Lottery." They offer several systems and criticize others as foolish. One system is called *Lucky Numbers*. People who play the *Lucky Numbers* system just pick a "lucky" number to play, but maybe some numbers are luckier than others. Let's use a simulation to see how well this system works.

To make the situation manageable, simulate a simple lottery in which a single digit from 0 to 9 is selected as the winning number. Pick a single value to bet, such as 1, and keep playing it over and over. You'll want to run at least 100 trials. (If you can program the simulations on a computer, run several hundred. Or generalize the questions to a lottery that chooses two- or three-digit numbers—for which you'll need thousands of trials.)

a) What proportion of the time do you expect to win?

b) Would you expect better results if you picked a "luckier" number, such as 7? (Try it if you don't know.) Explain.

22. **Random is as random does.** The "beat the lottery" website discussed in Exercise 21 suggests that because lottery numbers are random, it is better to select your bet randomly. For the same simple lottery in Exercise 21 (random values from 0 to 9), generate each bet by choosing a separate random value between 0 and 9. Play many games. What proportion of the time do you win?

23. **It evens out in the end.** The "beat the lottery" website of Exercise 21 notes that in the long run we expect each value to turn up about the same number of times. That leads to their recommended strategy. First, watch the lottery for a while, recording the winners. Then bet the value that has turned up the least, because it will need to turn up more often to even things out. If there is more than one "rarest" value, just take the lowest one (since it doesn't matter). Simulating the simplified lottery described in Exercise 21, play many games with this system. What proportion of the time do you win?

24. **Play the winner?** Another strategy for beating the lottery is the reverse of the system described in Exercise 23. Simulate the simplified lottery described in Exercise 21. Each time, bet the number that just turned up. The website suggests that this method should do worse. Does it? Play many games and see.

25. **Driving test.** You are about to take the road test for your driver's license. You hear that only 34% of candidates pass the test the first time, but the percentage rises to 72% on subsequent retests. Estimate the average number of tests drivers take in order to get a license. Your simulation should use at least 20 runs.

26. **Still learning?** As in Exercise 25, assume that your chance of passing the driver's test is 34% the first time and 72% for subsequent retests. Estimate the percentage of those tested who still do not have a driver's license after two attempts.

27. **Basketball strategy.** Late in a basketball game, the team that is behind often fouls someone in an attempt to get the ball back. Usually the opposing player will get to shoot foul shots "one and one," meaning he gets a shot, and then a second shot only if he makes the first one. Suppose the opposing player has made 72% of his foul shots this season. Estimate the number of points he will score in a one-and-one situation.

28. **Blood donors.** A person with type O-positive blood can receive blood only from other type O donors. About 44% of the U.S. population has type O blood. At a blood drive, how many potential donors do you expect to examine in order to get three units of type O blood?

29. **Free groceries.** To attract shoppers, a supermarket runs a weekly contest that involves "scratch-off" cards. With each purchase, customers get a card with a black spot obscuring a message. When the spot is scratched away, most of the cards simply say, "Sorry—please try again." But during the week, 100 customers will get cards that make them eligible for a drawing for free groceries. Ten of the cards say they may be worth $200, 10 others say $100, 20 may be worth $50, and the rest could be worth $20. To register those cards, customers write their names on them and put them in a barrel at the front of the store. At the end of the week the store manager draws cards at random, awarding the lucky customers free groceries in the amount specified on their card. The drawings continue until the store has given away more than $500 of free groceries. Estimate the average number of winners each week.

30. **Find the ace.** A new electronics store holds a contest to attract shoppers. Once an hour someone in the store is chosen at random to play the Music Game. Here's how it works: An ace and four other cards are shuffled and placed face down on a table. The customer gets to turn cards over one at a time, looking for the ace. The person wins $100 worth of free CDs or DVDs if the ace is the first card, $50 if it is the second card, and $20, $10, or $5 if it is the third, fourth, or fifth card chosen. What is the average dollar amount of music the store will give away?

31. **The family.** Many couples want to have both a boy and a girl. If they decide to continue to have children until they have one child of each sex, what would the average family size be? Assume that boys and girls are equally likely.

32. **A bigger family.** Suppose a couple will continue having children until they have at least two children of each sex (two boys *and* two girls). How many children might they expect to have?

33. **Dice game.** You are playing a children's game in which the number of spaces you get to move is determined by the rolling of a die. You must land exactly on the final space in order to win. If you are 10 spaces away, how many turns might it take you to win?

34. **Parcheesi.** You are three spaces from a win in Parcheesi. On each turn, you will roll two dice. To win, you must roll a total of 3 or roll a 3 on one of the dice. How many turns might you expect this to take?

35. **The hot hand.** A basketball player with a 65% shooting percentage has just made 6 shots in a row. The announcer says this player "is hot tonight! She's in the zone!" Assume the player takes about 20 shots per game. Is it unusual for her to make 6 or more shots in a row during a game?

36. **The World Series.** The World Series ends when a team wins 4 games. Suppose that sports analysts consider one team a bit stronger, with a 55% chance to win any individual game. Estimate the likelihood that the underdog wins the series.

37. **Teammates.** Four couples at a dinner party play a board game after the meal. They decide to play as teams of two and to select the teams randomly. All eight people write their names on slips of paper. The slips are thoroughly mixed, then drawn two at a time. How likely is it that every person will be teamed with someone other than the person he or she came to the party with?

38. **Second team.** Suppose the couples in Exercise 37 choose the teams by having one member of each couple write their names on the cards and the other people each pick a card at random. How likely is it that every person will be teamed with someone other than the person he or she came with?

39. **Job discrimination?** A company with a large sales staff announces openings for three positions as regional managers. Twenty-two of the current salespersons apply, 12 men and 10 women. After the interviews, when the company announces the newly appointed managers, all three positions go to women. The men complain of job discrimination. Do they have a case? Simulate a random selection of three people from the applicant pool, and make a decision about the likelihood that a fair process would result in hiring all women.

40. **Cell phones.** A proud legislator claims that your state's new law against talking on a cell phone while driving has reduced cell phone use to less than 12% of all drivers. While waiting for your bus the next morning, you notice that 4 of the 10 people who drive by are using their cell phones. Does this cast doubt on the legislator's figure of 12%? Use a simulation to estimate the likelihood of seeing at least 4 of 10 randomly selected drivers talking on their cell phones if the actual rate of usage is 12%. Explain your conclusion clearly.

JUST CHECKING

ANSWERS

1. The component is one game.

2. I'll generate random numbers and assign numbers from 00 to 54 to the home team's winning and from 55 to 99 to the visitors' winning.

3. I'll generate components until one team wins 4 games. I'll record which team wins the series.

4. The response is who wins the series.

5. I'll calculate the proportion of wins by the team that starts at home.

Sample Surveys

Where are we going?

We see surveys all the time. How can asking just a thousand people tell us much about a national election? And how *do* they select the respondents? It turns out that there are many ways to select a good sample. But there are just three main ideas to understand.

I n 2007, Pew Research conducted a survey to assess Americans' knowledge of current events. They asked a random sample of 1,502 U.S. adults 23 factual questions about topics currently in the news.[1] Pew also asked respondents where they got their news. Those who frequented major newspaper websites or who are regular viewers of the *Daily Show* or *Colbert Report* scored best on knowledge of current events.[2] Even among those viewers, only 54% responded correctly to 15 or more of the questions. Pew claimed that this was close to the true percentage responding correctly that they would have found if they had asked all U.S. adults who got their news from those sources. That step from a small sample to the entire population is impossible without understanding Statistics. To make business decisions, to do science, to choose wise investments, or to understand what voters think they'll do the next election, we need to stretch beyond the data at hand to the world at large.

To make that stretch, we need three ideas. You'll find the first one natural. The second may be more surprising. The third is one of the strange but true facts that often confuse those who don't know Statistics.

---

[1] For example, two of the questions were "Who is the vice-president of the United States?" and "What party controls Congress?"

[2] The lowest scores came from those whose main source of news was network morning shows or *Fox News*.

THE W'S AND SAMPLING

The population we are interested in is usually determined by the *Why* of our study. The sample we draw will be the *Who*. *When* and *How* we draw the sample may depend on what is practical.

In 1936, a young pollster named George Gallup used a subsample of only 3000 of the 2.4 million responses that the *Literary Digest* received to reproduce the wrong prediction of Landon's victory over Roosevelt. He then used an entirely different sample of 50,000 and predicted that Roosevelt would get 56% of the vote to Landon's 44%. His sample was apparently much more *representative* of the actual voting populace. The Gallup Organization went on to become one of the leading polling companies.

Idea 1: Examine a Part of the Whole

The first idea is to draw a sample. We'd like to know about an entire **population** of individuals, but examining all of them is usually impractical, if not impossible. So we settle for examining a smaller group of individuals—a **sample**—selected from the population.

You do this every day. For example, suppose you wonder how the vegetable soup you're cooking for dinner tonight is going to go over with your friends. To decide whether it meets your standards, you only need to try a small amount. You might taste just a spoonful or two. You certainly don't have to consume the whole pot. You trust that the taste will *represent* the flavor of the entire pot. The idea behind your tasting is that a small sample, if selected properly, can represent the entire population.

It's hard to go a day without hearing about the latest opinion poll. These polls are examples of **sample surveys**, designed to ask questions of a small group of people in the hope of learning something about the entire population. Most likely, you've never been selected to be part of one of these national opinion polls. That's true of most people. So how can the pollsters claim that a sample is representative of the entire population? The answer is that professional pollsters work quite hard to ensure that the "taste"—the sample that they take—represents the population. If not, the sample can give misleading information about the population.

Bias

Selecting a sample to represent the population fairly is more difficult than it sounds. Polls or surveys most often fail because they use a sampling method that tends to over- or underrepresent parts of the population. The method may overlook subgroups that are harder to find (such as the homeless or those who use only cell phones) or favor others (such as Internet users who like to respond to online surveys). Sampling methods that, by their nature, tend to over- or underemphasize some characteristics of the population are said to be **biased**. Bias is the bane of sampling—the one thing above all to avoid. Conclusions based on samples drawn with biased methods are inherently flawed. There is usually no way to fix bias after the sample is drawn and no way to salvage useful information from it.

Here's a famous example of a really dismal failure. By the beginning of the 20th century, it was common for newspapers to ask readers to return "straw" ballots on a variety of topics. (Today's Internet surveys are the same idea, gone electronic.) The earliest known example of such a straw vote in the United States dates back to 1824.

During the period from 1916 to 1936, the magazine *Literary Digest* regularly surveyed public opinion and forecast election results correctly. During the 1936 presidential campaign between Alf Landon and Franklin Delano Roosevelt, it mailed more than 10 million ballots and got back an astonishing 2.4 million. (Polls were still a relatively novel idea, and many people thought it was important to send back their opinions.) The results were clear: Alf Landon would be the next president by a landslide, 57% to 43%. You remember President Landon? No? In fact, Landon carried only two states. Roosevelt won, 62% to 37%, and, perhaps coincidentally, the *Digest* went bankrupt soon afterward.

What went wrong? One problem was that the *Digest*'s sample wasn't representative. Where would *you* find 10 million names and addresses to sample?

The *Digest* used the phone book, as many surveys do.[3] But in 1936, at the height of the Great Depression, telephones were a real luxury, so they sampled more rich than poor voters. The campaign of 1936 focused on the economy, and those who were less well off were more likely to vote for the Democrat. So the *Digest*'s sample was hopelessly biased.

How do modern polls get their samples to *represent* the entire population? You might think that they'd handpick individuals to sample with care and precision. But in fact, they do something quite different: They select individuals to sample *at random*. The importance of deliberately using randomness is one of the great insights of Statistics.

Idea 2: Randomize

Think back to the soup sample. Suppose you add some salt to the pot. If you sample it from the top before stirring, you'll get the misleading idea that the whole pot is salty. If you sample from the bottom, you'll get an equally misleading idea that the whole pot is bland. By stirring, you *randomize* the amount of salt throughout the pot, making each taste more typical of the whole pot.

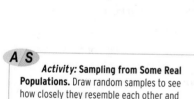

Not only does randomization protect you against factors that you know are in the data, it can also help protect against factors that you didn't even know were there. Suppose, while you weren't looking, a friend added a handful of peas to the soup. If they're down at the bottom of the pot, and you don't randomize the soup by stirring, your test spoonful won't have any peas. By stirring in the salt, you *also* randomize the peas throughout the pot, making your sample taste more typical of the overall pot *even though you didn't know the peas were there*. So randomizing protects us even in this case.

How do we "stir" people in a survey? We select them at random. **Randomizing** protects us from the influences of *all* the features of our population by making sure that, *on average*, the sample looks like the rest of the population.

> **Why not match the sample to the population?** Rather than randomizing, we could try to design our sample so that the people we choose are typical in terms of every characteristic we can think of. We might want the income levels of those we sample to match the population. How about age? Political affiliation? Marital status? Having children? Living in the suburbs? We can't possibly think of all the things that might be important. Even if we could, we wouldn't be able to match our sample to the population for all these characteristics.

FOR EXAMPLE   Is a Random Sample Representative?

Here are summary statistics comparing two samples of 8000 drawn at random from a company's database of 3.5 million customers:

Age (yr)	White (%)	Female (%)	# of Children	Income Bracket (1–7)	Wealth Bracket (1–9)	Homeowner? (% Yes)
61.4	85.12	56.2	1.54	3.91	5.29	71.36
61.2	84.44	56.4	1.51	3.88	5.33	72.30

QUESTION: Do you think these samples are representative of the population? Explain.

The two samples look very similar with respect to these seven variables. It appears that randomizing has automatically matched them pretty closely. We can reasonably assume that since the two samples don't differ too much from each other, they don't differ much from the rest of the population either.

---

[3] Today phone numbers are computer-generated to make sure that unlisted numbers are included. But even now, cell phones and VOIP Internet phones are often not included.

A friend who knows that you are taking Statistics asks your advice on her study. What can you possibly say that will be helpful? Just say, "If you could just get a larger sample, it would probably improve your study." Even though a larger sample might not be worth the cost, it will almost always make the results more precise.

Idea 3: It's the Sample Size

How large a random sample do we need for the sample to be reasonably representative of the population? Most people think that we need a large percentage, or *fraction*, of the population, but it turns out that what matters is the *number of individuals in the sample*, not the size of the population. A random sample of 100 students in a college represents the student body just about as well as a random sample of 100 voters represents the entire electorate of the United States. This is the *third* idea and probably the most surprising one in designing surveys.

How can it be that only the size of the sample, and not the population, matters? Well, let's return one last time to that pot of soup. If you're cooking for a banquet rather than just for a few people, your pot will be bigger, but do you need a bigger spoon to decide how the soup tastes? Of course not. The same-size spoonful is probably enough to make a decision about the entire pot, no matter how large the pot. The *fraction* of the population that you've sampled doesn't matter.[4] It's the **sample size** itself that's important.

How big a sample do you need? That depends on what you're estimating. To get an idea of what's really in the soup, you'll need a large enough taste to get a *representative* sample from the pot. For a survey that tries to find the proportion of the population falling into a category, you'll usually need several hundred respondents to say anything precise enough to be useful.[5]

> **What do the pollsters do?** How do professional polling agencies do their work? The most common polling method today is to contact respondents by telephone. Computers generate random telephone numbers, so pollsters can even call some people with unlisted phone numbers. The interviewer may then ask to speak with the adult who is home who had the most recent birthday or use some other essentially random way to select among those available. In phrasing questions, pollsters often list alternative responses (such as candidates' names) in different orders to avoid biases that might favor the first name on the list.
>
> Do these methods work? The Pew Research Center for the People and the Press, reporting on one survey, says that
>
> *Across five days of interviewing, surveys today are able to make some kind of contact with the vast majority of households (76%), and there is no decline in this contact rate over the past seven years. But because of busy schedules, skepticism and outright refusals, interviews were completed in just 38% of households that were reached using standard polling procedures.*
>
> Nevertheless, studies indicate that those actually sampled can give a good snapshot of larger populations from which the surveyed households were drawn.

Does a Census Make Sense?

Why bother determining the right sample size? Wouldn't it be better to just include everyone and "sample" the entire population? Such a special sample is called a **census**. Although a census would appear to provide the best possible information about the population, there are a number of reasons why it might not.

First, it can be difficult to complete a census. Some individuals in the population will be hard (and expensive) to locate. Or a census might just be impractical. If you were a taste tester for the Hostess™ Company, you probably

---

[4] Well, that's not exactly true. If the population is small enough and the sample is more than 10% of the whole population, it *can* matter. It doesn't matter whenever, as usual, our sample is a very small fraction of the population.

[5] Chapter 19 gives the details behind this statement and shows how to decide on a sample size for a survey.

wouldn't want to census *all* the Twinkies on the production line. Not only might this be life-endangering, but you wouldn't have any left to sell.

Second, populations rarely stand still. In populations of people, babies are born and folks die or leave the country. In opinion surveys, events may cause a shift in opinion during the survey. A census takes longer to complete and the population changes while you work. A sample surveyed in just a few days may give more accurate information.

Third, taking a census can be more complex than sampling. For example, the U.S. Census records too many college students. Many are counted once with their families and are then counted a second time in a report filed by their schools.

> **The undercount.** It's particularly difficult to compile a complete census of a population as large, complex, and spread out as the U.S. population. The U.S. Census is known to miss some residents. On occasion, the undercount has been striking. For example, there have been blocks in inner cities in which the number of residents recorded by the Census was smaller than the number of electric meters for which bills were being paid. What makes the problem particularly important is that some groups have a higher probability of being missed than others—undocumented immigrants, the homeless, the poor. The Census Bureau proposed the use of random sampling to estimate the number of residents missed by the ordinary census. Unfortunately, the resulting debate has become more political than statistical.

Populations and Parameters

A study found that teens were less likely to "buckle up." The National Center for Chronic Disease Prevention and Health Promotion reports that 21.7% of U.S. teens never or rarely wear seat belts. We're sure they didn't take a census, so what *does* the 21.7% mean? We can't know what percentage of teenagers wear seat belts. Reality is just too complex. But we can simplify the question by building a model.

Models use mathematics to represent reality. Parameters are the key numbers in those models. A parameter used in a model for a population is sometimes called (redundantly) a **population parameter**.

> Any quantity that we calculate from data could be called a "statistic." But in practice, we usually use a statistic to estimate a population parameter.

But let's not forget about the data. We use summaries of the data to estimate the population parameters. As we know, any summary found from the data is a **statistic**. Sometimes you'll see the (also redundant) term **sample statistic**.[6]

We've already met two parameters in Chapter 6: the mean, μ, and the standard deviation, σ. We'll try to keep denoting population model parameters with Greek letters and the corresponding statistics with Latin letters. Usually, but not always, the letter used for the statistic and the parameter correspond in a natural way. So the standard deviation of the data is s, and the corresponding parameter is σ (Greek for s). In Chapter 7, we used r to denote the sample correlation. The corresponding correlation in a model for the population would be called ρ (rho). In Chapter 8, b_1 represented the slope of a linear regression estimated from the data. But when we think about a (linear) *model* for the population, we denote the slope parameter β_1 (beta).

> Remember: Population model parameters are not just unknown—usually they are *unknowable*. We have to settle for sample statistics.

Get the pattern? Good. Now it breaks down. We denote the mean of a population model with μ (because μ is the Greek letter for m). It might make sense to denote the sample mean with m, but long-standing convention is to put a bar over anything when we average it, so we write $\bar{y}$. What about proportions? Suppose we want to talk about the proportion of teens who don't wear seat belts. If we use p to denote the proportion from the data, what is the corresponding model parameter? By all rights it should be π. But statements like $\pi = 0.25$ might be confusing because π has been equal to 3.1415926 . . . for so long, and it's worked so *well*. So, once again we violate the rule. We'll use p for

---

[6] Where else besides a sample *could* a statistic come from?

the population model parameter and $\hat{p}$ for the proportion from the data (since, like $\hat{y}$ in regression, it's an estimated value).

Here's a table summarizing the notation:

Name	Statistic	Parameter
Mean	$\bar{y}$	μ (mu, pronounced "meeoo," not "moo")
Standard deviation	s	σ (sigma)
Correlation	r	ρ (rho)
Regression coefficient	b	β (beta, pronounced "baytah"[7])
Proportion	$\hat{p}$	p (pronounced "pee"[8])

We draw samples because we can't work with the entire population, but we want the statistics we compute from a sample to reflect the corresponding parameters accurately. A sample that does this is said to be **representative**. A biased sampling methodology tends to over- or underestimate the parameter of interest.

JUST CHECKING

1. Various claims are often made for surveys. Why is each of the following claims not correct?

   a) It is always better to take a census than to draw a sample.
   b) Stopping students on their way out of the cafeteria is a good way to sample if we want to know about the quality of the food there.
   c) We drew a sample of 100 from the 3000 students in a school. To get the same level of precision for a town of 30,000 residents, we'll need a sample of 1000.
   d) A poll taken at a statistics support website garnered 12,357 responses. The majority said they enjoy doing statistics homework. With a sample size that large, we can be pretty sure that most Statistics students feel this way, too.
   e) The true percentage of all Statistics students who enjoy the homework is called a "population statistic."

Simple Random Samples

How would you select a representative sample? Most people would say that every individual in the population should have an equal chance to be selected, and certainly that seems fair. But it's not sufficient. There are many ways to give everyone an equal chance that still wouldn't give a representative sample. Consider, for example, a school that has equal numbers of males and females. We could sample like this: Flip a coin. If it comes up heads, select 100 female students at random. If it comes up tails, select 100 males at random. Everyone has an equal chance of selection, but every sample is of only a single sex—hardly representative.

We need to do better. Suppose we insist that every possible *sample* of the size we plan to draw has an equal chance to be selected. This ensures that situations like the one just described are not likely to occur and still guarantees that each person has an equal chance of being selected. What's different is that with this method, each *combination* of people has an equal chance of being selected as well. A sample drawn in this way is called a **Simple Random Sample**, usually abbreviated **SRS**. An SRS is the standard against which we measure other sampling methods, and the sampling method on which the theory of working with sampled data is based.

To select a sample at random, we first need to define where the sample will come from. The **sampling frame** is a list of individuals from which the sample

---

[7] If you're from the United States. If you're British or Canadian, it's "beetah."
[8] Just in case you weren't sure.

is drawn. For example, to draw a random sample of students at a college, we might obtain a list of all registered full-time students and sample from that list. In defining the sampling frame, we must deal with the details of defining the population. Are part-time students included? How about those who are attending school elsewhere and transferring credits back to the college?

Once we have a sampling frame, the easiest way to choose an SRS is to assign a random number to each individual in the sampling frame. We then select only those whose random numbers satisfy some rule.[9] Let's look at some ways to do this.

FOR EXAMPLE   Using Random Numbers to Get an SRS

There are 80 students enrolled in an introductory Statistics class; you are to select a sample of 5.

QUESTION: How can you select an SRS of 5 students using these random digits found on the Internet: 05166 29305 77482?

First I'll number the students from 00 to 79. Taking the random numbers two digits at a time gives me 05, 16, 62, 93, 05, 77, and 48. I'll ignore 93 because the students were numbered only up to 79. And, so as not to pick the same person twice, I'll skip the repeated number 05. My simple random sample consists of students with the numbers 05, 16, 62, 77, and 48.

> **ERROR OKAY, BIAS BAD!**
> Sampling variability is sometimes referred to as *sampling error*, making it sound like it's some kind of mistake. It's not. We understand that samples will vary, so "sampling error" is to be expected. It's *bias* we must strive to avoid. Bias means our sampling method distorts our view of the population, and that will surely lead to mistakes.

- We can be more efficient when we're choosing a larger sample from a sampling frame stored in a data file. First we assign a random number with several digits (say, from 0 to 10,000) to each individual. Then we arrange the random numbers in numerical order, keeping each name with its number. Choosing the first n names from this re-arranged list will give us a random sample of that size.
- Often the sampling frame is so large that it would be too tedious to number everyone consecutively. If our intended sample size is approximately 10% of the sampling frame, we can assign each individual a single random digit 0 to 9. Then we select only those with a specific random digit, say, 5.

Samples drawn at random generally differ one from another. Each draw of random numbers selects *different* people for our sample. These differences lead to different values for the variables we measure. We call these sample-to-sample differences **sampling variability**. Surprisingly, sampling variability isn't a problem; it's an opportunity. In future chapters we'll investigate what the variation in a sample can tell us about its population.

Stratified Sampling

Simple random sampling is not the only fair way to sample. More complicated designs may save time or money or help avoid sampling problems. All statistical sampling designs have in common the idea that chance, rather than human choice, is used to select the sample.

Designs that are used to sample from large populations—especially populations residing across large areas—are often more complicated than simple random samples. Sometimes the population is first sliced into homogeneous groups, called **strata**, before the sample is selected. Then simple random sampling is used within each stratum before the results are combined. This common sampling design is called **stratified random sampling**.

Why would we want to complicate things? Here's an example. Suppose we want to learn how students feel about funding for the football team at a large

---

[9] Chapter 11 presented ways of finding and working with random numbers.

university. The campus is 60% men and 40% women, and we suspect that men and women have different views on the funding. If we use simple random sampling to select 100 people for the survey, we could end up with 70 men and 30 women or 35 men and 65 women. Our resulting estimates of the level of support for the football funding could vary widely. To help reduce this sampling variability, we can decide to force a representative balance, selecting 60 men at random and 40 women at random. This would guarantee that the proportions of men and women within our sample match the proportions in the population, and that should make such samples more accurate in representing population opinion.

You can imagine the importance of stratifying by race, income, age, and other characteristics, depending on the questions in the survey. Samples taken within a stratum vary less, so our estimates can be more precise. This reduced sampling variability is the most important benefit of stratifying.

Stratified sampling can also help us notice important differences among groups. As we saw in Chapter 3, if we unthinkingly combine group data, we risk reaching the wrong conclusion, becoming victims of Simpson's paradox.

FOR EXAMPLE   Stratifying the Sample

RECAP: You're trying to find out what freshmen think of the food served on campus. Food Services believes that men and women typically have different opinions about the importance of the salad bar.

QUESTION: How should you adjust your sampling strategy to allow for this difference?

I will stratify my sample by drawing an SRS of men and a separate SRS of women—assuming that the data from the registrar include information about each person's sex.

Cluster and Multistage Sampling

Suppose we wanted to assess the reading level of this textbook based on the length of the sentences. Simple random sampling could be awkward; we'd have to number each sentence, then find, for example, the 576th sentence or the 2482nd sentence, and so on. Doesn't sound like much fun, does it?

It would be much easier to pick a few *pages* at random and count the lengths of the sentences on those pages. That works if we believe that each page is representative of the entire book in terms of reading level. Splitting the population into representative **clusters** can make sampling more practical. Then we could simply select one or a few clusters at random and perform a census within each of them. This sampling design is called **cluster sampling**. If each cluster represents the full population fairly, cluster sampling will be unbiased.

FOR EXAMPLE   Cluster Sampling

RECAP: In trying to find out what freshmen think about the food served on campus, you've considered both an SRS and a stratified sample. Now you have run into a problem: It's simply too difficult and time consuming to track down the individuals whose names were chosen for your sample. Fortunately, freshmen at your school are all housed in 10 freshman dorms.

QUESTIONS: How could you use this fact to draw a cluster sample? How might that alleviate the problem? What concerns do you have?

To draw a cluster sample, I would select one or two dorms at random and then try to contact everyone in each selected dorm. I could save time by simply knocking on doors on a given evening and interviewing people. I'd have to assume that freshmen were assigned to dorms pretty much at random and that the people I'm able to contact are representative of everyone in the dorm.

What's the difference between cluster sampling and stratified sampling? We stratify to ensure that our sample represents different groups in the population, and we sample randomly within each stratum. Strata are internally homogeneous, but differ from one another. By contrast, clusters are internally heterogeneous, each resembling the overall population. We select clusters to make sampling more practical or affordable.

Stratified vs. cluster sampling. Boston cream pie consists of a layer of yellow cake, a layer of pastry creme, another cake layer, and then a chocolate frosting. Suppose you are a professional taster (yes, there really are such people) whose job is to check your company's pies for quality. You'd need to eat small samples of randomly selected pies, tasting all three components: the cake, the creme, and the frosting.

One approach is to cut a thin vertical slice out of the pie. Such a slice will be a lot like the entire pie, so by eating that slice, you'll learn about the whole pie. This vertical slice containing all the different ingredients in the pie would be a *cluster* sample.

Another approach is to sample in *strata*: Select some tastes of the cake at random, some tastes of creme at random, and some bits of frosting at random. You'll end up with a reliable judgment of the pie's quality.

Many populations you might want to learn about are like this Boston cream pie. You can think of the subpopulations of interest as horizontal strata, like the layers of pie. Cluster samples slice vertically across the layers to obtain clusters, each of which is representative of the entire population. Stratified samples represent the population by drawing some from each layer, reducing variability in the results that could arise because of the differences among the layers.

STRATA OR CLUSTERS?

We may split a population into strata or clusters. What's the difference? We create strata by dividing the population into groups of similar individuals so that each stratum is different from the others. By contrast, since clusters each represent the entire population, they all look pretty much alike.

Sometimes we use a variety of sampling methods together. In trying to assess the reading level of this book, we might worry that it starts out easy and then gets harder as the concepts become more difficult. If so, we'd want to avoid samples that selected heavily from early or from late chapters. To guarantee a fair mix of chapters, we could randomly choose one chapter from each of the seven parts of the book and then randomly select a few pages from each of those chapters. If, altogether, that made too many sentences, we might select a few sentences at random from each of the chosen pages. So, what is our sampling strategy? First we stratify by the part of the book and randomly choose a chapter to represent each stratum. Within each selected chapter, we choose pages as clusters. Finally, we consider an SRS of sentences within each cluster. Sampling schemes that combine several methods are called **multistage samples**. Most surveys conducted by professional polling organizations use some combination of stratified and cluster sampling as well as simple random samples.

FOR EXAMPLE   Multistage Sampling

RECAP: Having learned that freshmen are housed in separate dorms allowed you to sample their attitudes about the campus food by going to dorms chosen at random, but you're still concerned about possible differences in opinions between men and women. It turns out that these freshman dorms house the sexes on alternate floors.

QUESTION: How can you design a sampling plan that uses this fact to your advantage?

Now I can stratify my sample by sex. I would first choose one or two dorms at random and then select the same number of dorm floors at random from among those that house men and, separately, from among those that house women. I could then draw an SRS of residents on the selected floors.

Systematic Samples

Some samples select individuals systematically. For example, you might survey every 10th person on an alphabetical list of students. To make it random, you still must start the systematic selection from a randomly selected individual. When the order of the list is not associated in any way with the responses sought, **systematic sampling** can give a representative sample. Systematic sampling can be much less expensive than true random sampling. When you use a systematic sample, you should justify the assumption that the systematic method is not associated with any of the measured variables.

Think about the reading-level sampling example again. Suppose we have chosen a chapter of the book at random, then three pages at random from that chapter, and now we want to select a sample of 10 sentences from the 73 sentences found on those pages. Instead of numbering each sentence so we can pick a simple random sample, it would be easier to sample systematically. A quick calculation shows $73/10 = 7.3$, so we can get our sample by just picking every seventh sentence on the page. But where should you start? At random, of course. We've accounted for $10 \times 7 = 70$ of the sentences, so we'll throw the extra 3 into the starting group and choose a sentence at random from the first 10. Then we pick every seventh sentence after that and record its length.

2. We need to survey a random sample of the 300 passengers on a flight from San Francisco to Tokyo. Name each sampling method described below.

a) Pick every 10th passenger as people board the plane.
b) From the boarding list, randomly choose 5 people flying first class and 25 of the other passengers.
c) Randomly generate 30 seat numbers and survey the passengers who sit there.
d) Randomly select a seat position (right window, right center, right aisle, etc.) and survey all the passengers sitting in those seats.

STEP-BY-STEP EXAMPLE   Sampling

The assignment says, "Conduct your own sample survey to find out how many hours per week students at your school spend watching TV during the school year." Let's see how we might do this step by step. (Remember, though—actually collecting the data from your sample can be difficult and time consuming.)

Question: How would you design this survey?

Plan State what you want to know.

Population and Parameter Identify the W's of the study. The *Why* determines the population and the associated sampling frame. The *What* identifies the parameter of interest and the variables measured. The *Who* is the sample we actually draw. The *How, When,* and *Where* are given by the sampling plan.

I wanted to design a study to find out how many hours of TV students at my school watch.

The population studied was students at our school. I obtained a list of all students currently enrolled and used it as the sampling frame. The parameter of interest was the number of TV hours watched per week during the school year, which I attempted to measure by asking students how much TV they watched during the previous week.

Often, thinking about the *Why* will help us see whether the sampling frame and plan are adequate to learn about the population.

Sampling Plan Specify the sampling method and the sample size, *n*. Specify how the sample was actually drawn. What is the sampling frame? How was the randomization performed?

A good description should be complete enough to allow someone to replicate the procedure, drawing another sample from the same population in the same manner.

I decided against stratifying by class or sex because I didn't think TV watching would differ much between males and females or across classes. I selected a simple random sample of students from the list. I obtained an alphabetical list of students, assigned each a random digit between 0 and 9, and then selected all students who were assigned a "4." This method generated a sample of 212 students from the population of 2133 students.

Sampling Practice Specify *When, Where,* and *How* the sampling was performed. Specify any other details of your survey, such as how respondents were contacted, what incentives were offered to encourage them to respond, how nonrespondents were treated, and so on.

The survey was taken over the period Oct. 15 to Oct. 25. Surveys were sent to selected students by e-mail, with the request that they respond by e-mail as well. Students who could not be reached by e-mail were handed the survey in person.

Summary and Conclusion This report should include a discussion of all the elements. In addition, it's good practice to discuss any special circumstances. Professional polling organizations report the *When* of their samples but will also note, for example, any important news that might have changed respondents' opinions during the sampling process. In this survey, perhaps, a major news story or sporting event might change students' TV viewing behavior.

The question you ask also matters. It's better to be specific ("How many hours did you watch TV last week?") than to ask a general question ("How many hours of TV do you usually watch in a week?").

During the period Oct. 15 to Oct. 25, 212 students were randomly selected, using a simple random sample from a list of all students currently enrolled. The survey they received asked the following question: "How many hours did you spend watching television last week?"

Of the 212 students surveyed, 110 responded. It's possible that the nonrespondents differ in the number of TV hours watched from those who responded, but I was unable to follow up on them due to limited time and funds. The 110 respondents reported an average 3.62 hours of TV watching per week. The median was only 2 hours per week. A histogram of the data shows that the distribution is highly right-skewed, indicating that the median might be a more appropriate summary of the typical TV watching of the students.

The report should show a display of the data, provide and interpret the statistics from the sample, and state the conclusions that you reached about the population.

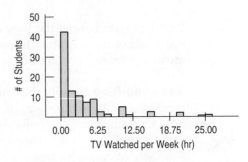

Most of the students (90%) watch between 0 and 10 hours per week, while 30% reported watching less than 1 hour per week. A few watch much more. About 3% reported watching more than 20 hours per week.

Defining the "Who": You Can't Always Get What You Want

Before you start a survey, think first about the population you want to study. You may find that it's not the well-defined group you thought it was. Who, exactly, is a student, for example? Even if the population seems well defined, it may not be a practical group from which to draw a sample. For example, election polls want to sample from all those who will vote in the next election—a population that is impossible to identify before Election Day.

Next, you must specify the sampling frame. (Do you have a list of students to sample from? How about a list of registered voters?) Usually, the sampling frame is not the group you *really* want to know about. (All those registered to vote are not equally likely to show up.) The sampling frame limits what your survey can find out.

Then there's your target sample. These are the individuals for whom you *intend* to measure responses. You're not likely to get responses from all of them. ("I know it's dinnertime, but I'm sure you wouldn't mind answering a few questions. It'll only take 20 minutes or so. Oh, you're busy?") Nonresponse is a problem in many surveys.

Finally, there's your sample—the actual respondents. These are the individuals about whom you *do* get data and can draw conclusions. Unfortunately, they might not be representative of the sampling frame or the population.

> The population is determined by the *Why* of the study. Unfortunately, the sample is just those we can reach to obtain responses—the *Who* of the study. This difference could undermine even a well-designed study.

CALVIN AND HOBBES © 1993 Watterson. Reprinted with permission of Universal Press Syndicate. All rights reserved.

At each step, the group we can study may be constrained further. The *Who* keeps changing, and each constraint can introduce biases. A careful study should address the question of how well each group matches the population of interest. One of the main benefits of simple random sampling is that it never loses its sense of who's *Who*. The *Who* in an SRS is the population of interest from which we've drawn a representative sample. That's not always true for other kinds of samples.

The Valid Survey

It isn't sufficient to just draw a sample and start asking questions. We'll want our survey to be *valid*. A valid survey yields the information we are seeking about the population we are interested in. Before setting out to survey, ask yourself:

- What do I want to know?
- Am I asking the right respondents?
- Am I asking the right questions?
- What would I do with the answers if I had them; would they address the things I want to know?

These questions may sound obvious, but there are a number of pitfalls to avoid.

Know what you want to know. Before considering a survey, understand what you hope to learn and about whom you hope to learn it. Far too often, people decide to perform a survey without any clear idea of what they hope to learn.

Use the right frame. A valid survey obtains responses from the appropriate respondents. Be sure you have a suitable *sampling frame.* Have you identified the population of interest and sampled from it appropriately? A company might survey customers who returned warranty registration cards, a readily available sampling frame. But if the company wants to know how to make their product more attractive, the most important population is the customers who rejected their product in favor of one from a competitor.

Tune your instrument. It is often tempting to ask questions you don't really need, but beware—longer questionnaires yield fewer responses and thus a greater chance of nonresponse bias.

Ask specific rather than general questions. People are not very good at estimating their typical behavior, so it is better to ask "How many hours did you sleep last night?" than "How much do you usually sleep?" Sure, some responses will include some unusual events (My dog was sick; I was up all night.), but overall you'll get better data.

Ask for quantitative results when possible. "How many magazines did you read last week?" is better than "How much do you read: A lot, A moderate amount, A little, or None at all?"

Be careful in phrasing questions. A respondent may not understand the question—or may understand the question differently than the researcher intended it. ("Does anyone in your family belong to a union?" Do you mean just me, my spouse, and my children? Or does "family" include my father, my siblings, and my second cousin once removed? What about my grandfather, who is staying with us? I think he once belonged to the Autoworkers Union.) Respondents are unlikely (or may not have the opportunity) to ask for clarification. A question like "Do you approve of the recent actions of the Secretary of Labor?" is likely not to measure what you want if many respondents don't know who the Secretary of Labor is or what actions he or she recently made.

Respondents may even lie or shade their responses if they feel embarrassed by the question ("Did you have too much to drink last night?"), are intimidated or insulted by the question ("Could you understand our new *Instructions for Dummies* manual, or was it too difficult for you?"), or if they want to avoid offending the interviewer ("Would you hire a man with a tattoo?" asked by a tattooed interviewer). Also, be careful to avoid phrases that have double or regional meanings. "How often do you go to town?" might be interpreted differently by different people and cultures.

Even subtle differences in phrasing can make a difference. In January 2006, the *New York Times* asked half of the 1229 U.S. adults in their sample the following question:

> *After 9/11, President Bush authorized government wiretaps on some phone calls in the U.S. without getting court warrants, saying this was necessary to reduce the threat of terrorism. Do you approve or disapprove of this?*

They found that 53% of respondents approved. But when they asked the other half of their sample a question with only slightly different phrasing,

> *After 9/11, George W. Bush authorized government wiretaps on some phone calls in the U.S. without getting court warrants. Do you approve or disapprove of this?*

only 46% approved.

Be careful in phrasing answers. It's often a good idea to offer choices rather than inviting a free response. Open-ended answers can be difficult to analyze. "How did you like the movie?" may start an interesting debate, but it may be better to give a range of possible responses. Be sure to phrase them in a neutral way. When asking "Do you support higher school taxes?" positive responses could be worded "Yes," "Yes, it is important for our children," or "Yes, our future depends on it." But those are not equivalent answers.

THE WIZARD OF ID — parker and hart

A researcher distributed a survey to an organization before some economizing changes were made. She asked how people felt about a proposed cutback in secretarial and administrative support on a seven-point scale from Very Happy to Very Unhappy.

But virtually all respondents were very unhappy about the cutbacks, so the results weren't particularly useful. If she had pretested the question, she might have chosen a scale that ran from unhappy to outraged.

The best way to protect a survey from such unanticipated measurement errors is to perform a pilot survey. A **pilot** is a trial run of the survey you eventually plan to give to a larger group, using a draft of your survey questions administered to a small sample drawn from the same sampling frame you intend to use. By analyzing the results from this smaller survey, you can often discover ways to improve your instrument.

Sampling Errors: How to Sample Badly

Bad sample designs yield worthless data. Many of the most convenient forms of sampling can be seriously biased. And there is no way to correct for the bias from a bad sample. So it's wise to pay attention to sample design—and to beware of reports based on poor samples.

Sample Badly with Volunteers

One of the most common dangerous sampling methods is a voluntary response sample. In a **voluntary response sample**, a large group of individuals is invited to respond, and all who do respond are counted. This method is used by call-in shows, 900 numbers, Internet polls, and letters written to members of Congress. Voluntary response samples are almost always biased, and so conclusions drawn from them are almost always wrong.

It's often hard to define the sampling frame of a voluntary response study. Practically, the frames are groups such as Internet users who frequent a particular website or those who happen to be watching a particular TV show at the moment. But those sampling frames don't correspond to interesting populations.

Even within the sampling frame, voluntary response samples are often biased toward those with strong opinions or those who are strongly motivated. People with very negative opinions tend to respond more often than those with equally strong positive opinions. The sample is not representative, even though every individual in the population may have been offered the chance to respond. The resulting **voluntary response bias** invalidates the survey.

A S

Activity: **Sources of Sampling Bias.** Here's a narrated exploration of sampling bias.

If you had it to do over again, would you have children? Ann Landers, the advice columnist, asked parents this question. The overwhelming majority–70% of the more than 10,000 people who wrote in–said no, kids weren't worth it. A more carefully designed survey later showed that about 90% of parents actually are happy with their decision to have children. What accounts for the striking difference in these two results? What parents do you think are most likely to respond to the original question?

FOR EXAMPLE   Bias in Sampling

RECAP: You're trying to find out what freshmen think of the food served on campus, and have thought of a variety of sampling methods, all time consuming. A friend suggests that you set up a "Tell Us What You Think" website and invite freshmen to visit the site to complete a questionnaire.

QUESTION: What's wrong with this idea?

Letting each freshman decide whether to participate makes this a voluntary response survey. Students who were dissatisfied might be more likely to go to the website to record their complaints, and this could give me a biased view of the opinions of all freshmen.

Sample Badly, but Conveniently

```
Do you use the Internet?
Click here O for yes
Click here O for no
```

Another sampling method that doesn't work is convenience sampling. As the name suggests, in **convenience sampling** we simply include the individuals who are convenient for us to sample. Unfortunately, this group may not be representative of the population. A recent survey of 437 potential home buyers in Orange County, California, found, among other things, that

All but 2 percent of the buyers have at least one computer at home, and 62 percent have two or more. Of those with a computer, 99 percent are connected to the Internet (Jennifer Hieger, "Portrait of Homebuyer Household: 2 Kids and a PC," Orange County Register, 27 July 2001).

Later in the article, we learn that the survey was conducted via the Internet! That was a convenient way to collect data and surely easier than drawing a simple random sample, but perhaps home builders shouldn't conclude from this study that *every* family has a computer and an Internet connection.

Many surveys conducted at shopping malls suffer from the same problem. People in shopping malls are not necessarily representative of the population of interest. Mall shoppers tend to be more affluent and include a larger percentage of teenagers and retirees than the population at large. To make matters worse, survey interviewers tend to select individuals who look "safe," or easy to interview.

> Internet convenience surveys are worthless. As voluntary response surveys, they have no well-defined sampling frame (all those who use the Internet and visit their site?) and thus report no useful information. Do not believe them.

FOR EXAMPLE   Bias in Sampling

RECAP: To try to gauge freshman opinion about the food served on campus, Food Services suggests that you just stand outside a school cafeteria at lunchtime and stop people to ask them questions.

QUESTIONS: What's wrong with this sampling strategy?

This would be a convenience sample, and it's likely to be biased. I would miss people who use the cafeteria for dinner, but not for lunch, and I'd never hear from anyone who hates the food so much that they have stopped coming to the school cafeteria.

Sample from a Bad Sampling Frame

An SRS from an incomplete sampling frame introduces bias because the individuals included may differ from the ones not in the frame. People in prison, homeless people, students, and long-term travelers are all likely to be missed. In telephone surveys, people who have only cell phones or who use VOIP Internet phones are often missing from the sampling frame.

Undercoverage

Many survey designs suffer from **undercoverage**, in which some portion of the population is not sampled at all or has a smaller representation in the sample than it has in the population. Undercoverage can arise for a number of reasons, but it's always a potential source of bias.

Telephone surveys are usually conducted when you are likely to be home, interrupting your dinner. If you eat out often, you may be less likely to be surveyed, a possible source of undercoverage.

What Can Go Wrong?

- **Watch out for nonrespondents.** A common and serious potential source of bias for most surveys is **nonresponse bias**. No survey succeeds in getting responses from everyone. The problem is that those who don't respond may differ from those who do. And they may differ on just the variables we care

(continued)

about. The lack of response will bias the results. Rather than sending out a large number of surveys for which the response rate will be low, it is often better to design a smaller randomized survey for which you have the resources to ensure a high response rate. One of the problems with nonresponse bias is that it's usually impossible to tell what the nonrespondents might have said.

A S

Video: **Biased Question Wording.** Watch a hapless interviewer make every mistake in the book.

> **Remember the *Literary Digest* survey?** It turns out that they were wrong on two counts. First, their list of 10 million people was not representative. There was a selection bias in their sampling frame. There was also a nonresponse bias. We know this because the *Digest* also surveyed a *systematic* sample in Chicago, sending the same question used in the larger survey to every third registered voter. They *still* got a result in favor of Landon, even though Chicago voted overwhelmingly for Roosevelt in the election. This suggests that the Roosevelt supporters were less likely to respond to the *Digest* survey. There's a modern version of this problem: It's been suggested that those who screen their calls with caller ID or an answering machine, and so might not talk to a pollster, may differ in wealth or political views from those who just answer the phone.

■ **Work hard to avoid influencing responses. Response bias**[10] refers to anything in the survey design that influences the responses. Response biases include the tendency of respondents to tailor their responses to try to please the interviewer, the natural unwillingness of respondents to reveal personal facts or admit to illegal or unapproved behavior, and the ways in which the wording of the questions can influence responses.

How to Think About Biases

■ **Look for biases in any survey you encounter.** If you design one of your own, ask someone else to help look for biases that may not be obvious to you. And do this *before* you collect your data. There's no way to recover from a biased sampling method or a survey that asks biased questions. Sorry, it just can't be done.

A bigger sample size for a biased study just gives you a bigger useless study. A really big sample gives you a really big useless study. (Think of the 2.4 million *Literary Digest* responses.)

■ **Spend your time and resources reducing biases.** No other use of resources is as worthwhile as reducing the biases.

■ **If you can, pilot-test your survey.** Administer the survey in the exact form that you intend to use it to a small sample drawn from the population you intend to sample. Look for misunderstandings, misinterpretation, confusion, or other possible biases. Then refine your survey instrument.

■ **Always report your sampling methods in detail.** Others may be able to detect biases where you did not expect to find them.

---

[10] Response bias is not the opposite of nonresponse bias. (We don't make these terms up; we just try to explain them.)

With this chapter, we take our first formal steps to relate our sample data to a larger population. Some of these ideas have been lurking in the background as we sought patterns and summaries for data. Even when we only worked with the data at hand, we often thought about implications for a larger population of individuals.

Notice the ongoing central importance of models. We've seen models in several ways in previous chapters. Here we recognize the value of a model for a population. The parameters of such a model are values we will often want to estimate using statistics such as those we've been calculating. The connections to summary statistics for center, spread, correlation, and slope are obvious.

We now have a specific application for random numbers. The idea of applying randomness deliberately showed up in Chapter 11 for simulation. Now we need randomization to get good-quality data from the real world.

WHAT HAVE WE LEARNED?

We've learned that a representative sample can offer us important insights about populations. It's the size of the sample—and not its fraction of the larger population—that determines the precision of the statistics it yields.

We've learned several ways to draw samples, all based on the power of randomness to make them representative of the population of interest:

▶ A Simple Random Sample (SRS) is our standard. Every possible group of n individuals has an equal chance of being our sample. That's what makes it *simple*.

▶ Stratified samples can reduce sampling variability by identifying homogeneous subgroups and then randomly sampling within each.

▶ Cluster samples randomly select among heterogeneous subgroups that each resemble the population at large, making our sampling tasks more manageable.

▶ Systematic samples can work in some situations and are often the least expensive method of sampling. But we still want to start them randomly.

▶ Multistage samples combine several random sampling methods.

We've learned that bias can destroy our ability to gain insights from our sample:

▶ Nonresponse bias can arise when sampled individuals will not or cannot respond.

▶ Response bias arises when respondents' answers might be affected by external influences, such as question wording or interviewer behavior.

We've learned that bias can also arise from poor sampling methods:

▶ Voluntary response samples are almost always biased and should be avoided and distrusted.

▶ Convenience samples are likely to be flawed for similar reasons.

▶ Even with a reasonable design, sample frames may not be representative. Undercoverage occurs when individuals from a subgroup of the population are selected less often than they should be.

Finally, we've learned to look for biases in any survey we find and to be sure to report our methods whenever we perform a survey so that others can evaluate the fairness and accuracy of our results.

Terms

Population	The entire group of individuals or instances about whom we hope to learn (p. 282).
Sample	A (representative) subset of a population, examined in the hope of learning about the population (p. 282).
Sample survey	A study that asks questions of a sample drawn from some population in the hope of learning something about the entire population. Polls taken to assess voter preferences are common sample surveys (p. 282).
Bias	Any systematic failure of a sampling method to represent its population is bias. Biased sampling methods tend to over- or underestimate parameters. It is almost impossible to recover from bias, so efforts to avoid it are well spent (p. 282). Common errors include:
	▶ relying on voluntary response.
	▶ undercoverage of the population.
	▶ nonresponse bias.
	▶ response bias.
Randomization	The best defense against bias is randomization, in which each individual is given a fair, random chance of selection (p. 283).
Sample size	The number of individuals in a sample. The sample size determines how well the sample represents the population, not the fraction of the population sampled (p. 284).
Census	A sample that consists of the entire population is called a census (p. 284).
Population parameter	A numerically valued attribute of a model for a population. We rarely expect to know the true value of a population parameter, but we do hope to estimate it from sampled data. For example, the mean income of all employed people in the country is a population parameter (p. 285).
Statistic, sample statistic	Statistics are values calculated for sampled data. Those that correspond to, and thus estimate, a population parameter, are of particular interest. For example, the mean income of all employed people in a representative sample can provide a good estimate of the corresponding population parameter. The term "sample statistic" is sometimes used, usually to parallel the corresponding term "population parameter" (p. 285).
Representative	A sample is said to be representative if the statistics computed from it accurately reflect the corresponding population parameters (p. 286).
Simple Random Sample (SRS)	A simple random sample of sample size n is a sample in which each set of n elements in the population has an equal chance of selection (p. 286).
Sampling frame	A list of individuals from whom the sample is drawn is called the sampling frame. Individuals who may be in the population of interest, but who are not in the sampling frame, cannot be included in any sample (p. 286).
Sampling variability	The natural tendency of randomly drawn samples to differ, one from another. Sometimes, unfortunately, called *sampling error*, sampling variability is no error at all, but just the natural result of random sampling (p. 287).
Stratified random sample	A sampling design in which the population is divided into several subpopulations, or **strata**, and random samples are then drawn from each stratum. If the strata are homogeneous, but are different from each other, a stratified sample may yield more consistent results than an SRS (p. 287).
Cluster sample	A sampling design in which entire groups, or **clusters**, are chosen at random. Cluster sampling is usually selected as a matter of convenience, practicality, or cost. Each cluster should be representative of the population, so all the clusters should be heterogeneous and similar to each other (p. 288).
Multistage sample	Sampling schemes that combine several sampling methods are called multistage samples. For example, a national polling service may stratify the country by geographical regions, select a random sample of cities from each region, and then interview a cluster of residents in each city (p. 289).
Systematic sample	A sample drawn by selecting individuals systematically from a sampling frame. When there is no relationship between the order of the sampling frame and the variables of interest, a systematic sample can be representative (p. 290).

Pilot	A small trial run of a survey to check whether questions are clear. A pilot study can reduce errors due to ambiguous questions (p. 294).
Voluntary response bias	Bias introduced to a sample when individuals can choose on their own whether to participate in the sample. Samples based on voluntary response are always invalid and cannot be recovered, no matter how large the sample size (p. 295).
Convenience sample	A convenience sample consists of the individuals who are conveniently available. Convenience samples often fail to be representative because every individual in the population is not equally convenient to sample (p. 295).
Undercoverage	A sampling scheme that biases the sample in a way that gives a part of the population less representation than it has in the population suffers from undercoverage (p. 296).
Nonresponse bias	Bias introduced when a large fraction of those sampled fails to respond. Those who do respond are likely to not represent the entire population. Voluntary response bias is a form of nonresponse bias, but nonresponse may occur for other reasons. For example, those who are at work during the day won't respond to a telephone survey conducted only during working hours (p. 296).
Response bias	Anything in a survey design that influences responses falls under the heading of response bias. One typical response bias arises from the wording of questions, which may suggest a favored response. Voters, for example, are more likely to express support of "the president" than support of the particular person holding that office at the moment (p. 297).

Skills

▶ Know the basic concepts and terminology of sampling (see the preceding list).

▶ Recognize population parameters in descriptions of populations and samples.

▶ Understand the value of randomization as a defense against bias.

▶ Understand the value of sampling to estimate population parameters from statistics calculated on representative samples drawn from the population.

▶ Understand that the size of the sample (not the fraction of the population) determines the precision of estimates.

▶ Know how to draw a simple random sample from a master list of a population, using a computer or a table of random numbers.

▶ Know what to report about a sample as part of your account of a statistical analysis.

▶ Report possible sources of bias in sampling methods. Recognize voluntary response and nonresponse as sources of bias in a sample survey.

SAMPLING ON THE COMPUTER

Computer-generated pseudorandom numbers are usually good enough for drawing random samples. But there is little reason not to use the truly random values available on the Internet.

Here's a convenient way to draw an SRS of a specified size using a computer-based sampling frame. The sampling frame can be a list of names or of identification numbers arrayed, for example, as a column in a spreadsheet, statistics program, or database:

1. Generate random numbers of enough digits so that each exceeds the size of the sampling frame list by several digits. This makes duplication unlikely.

2. Assign the random numbers arbitrarily to individuals in the sampling frame list. For example, put them in an adjacent column.

3. Sort the list of random numbers, carrying along the sampling frame list.

4. Now the first *n* values in the sorted sampling frame column are an SRS of *n* values from the entire sampling frame.

EXERCISES

1. **Roper.** Through their *Roper Reports Worldwide*, GfK Roper conducts a global consumer survey to help multinational companies understand different consumer attitudes throughout the world. Within 30 countries, the researchers interview 1000 people aged 13–65. Their samples are designed so that they get 500 males and 500 females in each country. (www.gfkamerica.com)
   a) Are they using a simple random sample? Explain.
   b) What kind of design do you think they are using?

2. **Student center survey.** For their class project, a group of Statistics students decide to survey the student body to assess opinions about the proposed new student center. Their sample of 200 contained 50 first-year students, 50 sophomores, 50 juniors, and 50 seniors.
   a) Do you think the group was using an SRS? Why?
   b) What sampling design do you think they used?

3. **Emoticons.** The website www.gamefaqs.com asked, as their question of the day to which visitors to the site were invited to respond, *"Do you ever use emoticons when you type online?"* Of the 87,262 respondents, 27% said that they did not use emoticons. ;-(
   a) What kind of sample was this?
   b) How much confidence would you place in using 27% as an estimate of the fraction of people who use emoticons?

4. **Drug tests.** Major League Baseball tests players to see whether they are using performance-enhancing drugs. Officials select a team at random, and a drug-testing crew shows up unannounced to test all 40 players on the team. Each testing day can be considered a study of drug use in Major League Baseball.
   a) What kind of sample is this?
   b) Is that choice appropriate?

5. **Gallup.** At its website (www.gallup.com) the Gallup Poll publishes results of a new survey each day. Scroll down to the end, and you'll find a statement that includes words such as these:

   Results are based on telephone interviews with 1,008 national adults, aged 18 and older, conducted April 2–5, 2007. . . . In addition to sampling error, question wording and practical difficulties in conducting surveys can introduce error or bias into the findings of public opinion polls.

   a) For this survey, identify the population of interest.
   b) Gallup performs its surveys by phoning numbers generated at random by a computer program. What is the sampling frame?
   c) What problems, if any, would you be concerned about in matching the sampling frame with the population?

6. **Gallup World.** At its website (www.gallupworldpoll.com) the Gallup World Poll describes their methods. After one report they explained:

   Results are based on face-to-face interviews with randomly selected national samples of approximately 1,000 adults, aged 15 and older, who live permanently in each of the 21 sub-Saharan African nations surveyed. Those countries include Angola (areas where land mines might be expected were excluded), Benin, Botswana, Burkina Faso, Cameroon, Ethiopia, Ghana, Kenya, Madagascar (areas where interviewers had to walk more than 20 kilometers from a road were excluded), Mali, Mozambique, Niger, Nigeria, Senegal, Sierra Leone, South Africa, Tanzania, Togo, Uganda (the area of activity of the Lord's Resistance Army was excluded from the survey), Zambia, and Zimbabwe. . . . In all countries except Angola, Madagascar, and Uganda, the sample is representative of the entire population.

   a) Gallup is interested in sub-Saharan Africa. What kind of survey design are they using?
   b) Some of the countries surveyed have large populations. (Nigeria is estimated to have about 130 million people.) Some are quite small. (Togo's population is estimated at 5.4 million.) Nonetheless, Gallup sampled 1000 adults in each country. How does this affect the precision of its estimates for these countries?

7–14. *What did they do?* For the following reports about statistical studies, identify the following items (if possible). If you can't tell, then say so—this often happens when we read about a survey.
   a) The population
   b) The population parameter of interest
   c) The sampling frame
   d) The sample
   e) The sampling method, including whether or not randomization was employed
   f) Any potential sources of bias you can detect and any problems you see in generalizing to the population of interest

7. Consumers Union asked all subscribers whether they had used alternative medical treatments and, if so, whether they had benefited from them. For almost all of the treatments, approximately 20% of those responding reported cures or substantial improvement in their condition.

8. A question posted on the Lycos website on 18 June 2000 asked visitors to the site to say whether they thought that marijuana should be legally available for medicinal purposes. (www.lycos.com)

9. Researchers waited outside a bar they had randomly selected from a list of such establishments. They stopped every 10th person who came out of the bar and asked whether he or she thought drinking and driving was a serious problem.

10. Hoping to learn what issues may resonate with voters in the coming election, the campaign director for a mayoral candidate selects one block from each of the city's election districts. Staff members go there and interview all the residents they can find.

11. The Environmental Protection Agency took soil samples at 16 locations near a former industrial waste dump and checked each for evidence of toxic chemicals. They found no elevated levels of any harmful substances.

12. State police set up a roadblock to estimate the percentage of cars with up-to-date registration, insurance, and safety inspection stickers. They usually find problems with about 10% of the cars they stop.

13. A company packaging snack foods maintains quality control by randomly selecting 10 cases from each day's production and weighing the bags. Then they open one bag from each case and inspect the contents.

14. Dairy inspectors visit farms unannounced and take samples of the milk to test for contamination. If the milk is found to contain dirt, antibiotics, or other foreign matter, the milk will be destroyed and the farm reinspected until purity is restored.

15. **Mistaken poll.** A local TV station conducted a "PulsePoll" about the upcoming mayoral election. Evening news viewers were invited to phone in their votes, with the results to be announced on the late-night news. Based on the phone calls, the station predicted that Amabo would win the election with 52% of the vote. They were wrong: Amabo lost, getting only 46% of the vote. Do you think the station's faulty prediction is more likely to be a result of bias or sampling error? Explain.

16. **Another mistaken poll.** Prior to the mayoral election discussed in Exercise 15, the newspaper also conducted a poll. The paper surveyed a random sample of registered voters stratified by political party, age, sex, and area of residence. This poll predicted that Amabo would win the election with 52% of the vote. The newspaper was wrong: Amabo lost, getting only 46% of the vote. Do you think the newspaper's faulty prediction is more likely to be a result of bias or sampling error? Explain.

17. **Parent opinion, part 1.** In a large city school system with 20 elementary schools, the school board is considering the adoption of a new policy that would require elementary students to pass a test in order to be promoted to the next grade. The PTA wants to find out whether parents agree with this plan. Listed below are some of the ideas proposed for gathering data. For each, indicate what kind of sampling strategy is involved and what (if any) biases might result.
    a) Put a big ad in the newspaper asking people to log their opinions on the PTA website.
    b) Randomly select one of the elementary schools and contact every parent by phone.
    c) Send a survey home with every student, and ask parents to fill it out and return it the next day.
    d) Randomly select 20 parents from each elementary school. Send them a survey, and follow up with a phone call if they do not return the survey within a week.

18. **Parent opinion, part 2.** Let's revisit the school system described in Exercise 17. Four new sampling strategies have been proposed to help the PTA determine whether parents favor requiring elementary students to pass a test in order to be promoted to the next grade. For each, indicate what kind of sampling strategy is involved and what (if any) biases might result.
    a) Run a poll on the local TV news, asking people to dial one of two phone numbers to indicate whether they favor or oppose the plan.
    b) Hold a PTA meeting at each of the 20 elementary schools, and tally the opinions expressed by those who attend the meetings.
    c) Randomly select one class at each elementary school and contact each of those parents.
    d) Go through the district's enrollment records, selecting every 40th parent. PTA volunteers will go to those homes to interview the people chosen.

19. **Churches.** For your political science class, you'd like to take a survey from a sample of all the Catholic Church members in your city. A list of churches shows 17 Catholic churches within the city limits. Rather than try to obtain a list of all members of all these churches, you decide to pick 3 churches at random. For those churches, you'll ask to get a list of all current members and contact 100 members at random.
    a) What kind of design have you used?
    b) What could go wrong with your design?

20. **Playground.** Some people have been complaining that the children's playground at a municipal park is too small and is in need of repair. Managers of the park decide to survey city residents to see if they believe the playground should be rebuilt. They hand out questionnaires to parents who bring children to the park. Describe possible biases in this sample.

21. **Roller coasters.** An amusement park has opened a new roller coaster. It is so popular that people are waiting for up to 3 hours for a 2-minute ride. Concerned about how patrons (who paid a large amount to enter the park and ride on the rides) feel about this, they survey every 10th person on the line for the roller coaster, starting from a randomly selected individual.
    a) What kind of sample is this?
    b) What is the sampling frame?
    c) Is it likely to be representative?

22. **Playground, act two.** The survey described in Exercise 20 asked,

    Many people believe this playground is too small and in need of repair. Do you think the playground should be repaired and expanded even if that means raising the entrance fee to the park?

    Describe two ways this question may lead to response bias.

23. **Wording the survey.** Two members of the PTA committee in Exercises 17 and 18 have proposed different questions to ask in seeking parents' opinions.

    Question 1: Should elementary school–age children have to pass high-stakes tests in order to remain with their classmates?
    Question 2: Should schools and students be held accountable for meeting yearly learning goals by testing students before they advance to the next grade?

    a) Do you think responses to these two questions might differ? How? What kind of bias is this?
    b) Propose a question with more neutral wording that might better assess parental opinion.

24. Banning ephedra. An online poll at a website asked:

A nationwide ban of the diet supplement ephedra went into effect recently. The herbal stimulant has been linked to 155 deaths and many more heart attacks and strokes. Ephedra manufacturer NVE Pharmaceuticals, claiming that the FDA lacked proof that ephedra is dangerous if used as directed, was denied a temporary restraining order on the ban yesterday by a federal judge. Do you think that ephedra should continue to be banned nationwide?

65% of 17,303 respondents said "yes." Comment on each of the following statements about this poll:

a) With a sample size that large, we can be pretty certain we know the true proportion of Americans who think ephedra should be banned.

b) The wording of the question is clearly very biased.

c) The sampling frame is all Internet users.

d) Results of this voluntary response survey can't be reliably generalized to any population of interest.

25. Survey questions. Examine each of the following questions for possible bias. If you think the question is biased, indicate how and propose a better question.

a) Should companies that pollute the environment be compelled to pay the costs of cleanup?

b) Given that 18-year-olds are old enough to vote and to serve in the military, is it fair to set the drinking age at 21?

26. More survey questions. Examine each of the following questions for possible bias. If you think the question is biased, indicate how and propose a better question.

a) Do you think high school students should be required to wear uniforms?

b) Given humanity's great tradition of exploration, do you favor continued funding for space flights?

27. Phone surveys. Any time we conduct a survey, we must take care to avoid undercoverage. Suppose we plan to select 500 names from the city phone book, call their homes between noon and 4 p.m., and interview whoever answers, anticipating contacts with at least 200 people.

a) Why is it difficult to use a simple random sample here?

b) Describe a more convenient, but still random, sampling strategy.

c) What kinds of households are likely to be included in the eventual sample of opinion? Excluded?

d) Suppose, instead, that we continue calling each number, perhaps in the morning or evening, until an adult is contacted and interviewed. How does this improve the sampling design?

e) Random-digit dialing machines can generate the phone calls for us. How would this improve our design? Is anyone still excluded?

28. Cell phone survey. What about drawing a random sample only from cell phone exchanges? Discuss the advantages and disadvantages of such a sampling method compared with surveying randomly generated telephone numbers from non–cell phone exchanges. Do you think these advantages and disadvantages have changed over time? How do you expect they'll change in the future?

29. Arm length. How long is your arm compared with your hand size? Put your right thumb at your left shoulder

bone, stretch your hand open wide, and extend your hand down your arm. Put your thumb at the place where your little finger is, and extend down the arm again. Repeat this a third time. Now your little finger will probably have reached the back of your left hand. If your arm is less than four hand widths, turn your hand sideways and count finger widths until you reach the end of your middle finger.

a) How many hand and finger widths is your arm?

b) Suppose you repeat your measurement 10 times and average your results. What parameter would this average estimate? What is the population?

c) Suppose you now collect arm lengths measured in this way from 9 friends and average these 10 measurements. What is the population now? What parameter would this average estimate?

d) Do you think these 10 arm lengths are likely to be representative of the population of arm lengths in your community? In the country? Why or why not?

30. Fuel economy. Occasionally, when I fill my car with gas, I figure out how many miles per gallon my car got. I wrote down those results after six fill-ups in the past few months. Overall, it appears my car gets 28.8 miles per gallon.

a) What statistic have I calculated?

b) What is the parameter I'm trying to estimate?

c) How might my results be biased?

d) When the Environmental Protection Agency (EPA) checks a car like mine to predict its fuel economy, what parameter is it trying to estimate?

31. Accounting. Between quarterly audits, a company likes to check on its accounting procedures to address any problems before they become serious. The accounting staff processes payments on about 120 orders each day. The next day, the supervisor rechecks 10 of the transactions to be sure they were processed properly.

a) Propose a sampling strategy for the supervisor.

b) How would you modify that strategy if the company makes both wholesale and retail sales, requiring different bookkeeping procedures?

32. Happy workers? A manufacturing company employs 14 project managers, 48 foremen, and 377 laborers. In an effort to keep informed about any possible sources of employee discontent, management wants to conduct job satisfaction interviews with a sample of employees every month.

a) Do you see any potential danger in the company's plan? Explain.

b) Propose a sampling strategy that uses a simple random sample.

c) Why do you think a simple random sample might not provide the representative opinion the company seeks?

d) Propose a better sampling strategy.

e) Listed below are the last names of the project managers. Use random numbers to select two people to be interviewed. Explain your method carefully.

Barrett	Bowman	Chen
DeLara	DeRoos	Grigorov
Maceli	Mulvaney	Pagliarulo
Rosica	Smithson	Tadros
Williams	Yamamoto	

33. Quality control. Sammy's Salsa, a small local company, produces 20 cases of salsa a day. Each case contains 12 jars and is imprinted with a code indicating the date and batch number. To help maintain consistency, at the end of each day, Sammy selects three jars of salsa, weighs the contents, and tastes the product. Help Sammy select the sample jars. Today's cases are coded 07N61 through 07N80.

a) Carefully explain your sampling strategy.

b) Show how to use random numbers to pick 3 jars.

c) Did you use a simple random sample? Explain.

34. A fish story. Concerned about reports of discolored scales on fish caught downstream from a newly sited chemical plant, scientists set up a field station in a shoreline public park. For one week they asked fishermen there to bring any fish they caught to the field station for a brief inspection. At the end of the week, the scientists said that 18% of the 234 fish that were submitted for inspection displayed the discoloration. From this information, can the researchers estimate what proportion of fish in the river have discolored scales? Explain.

35. Sampling methods. Consider each of these situations. Do you think the proposed sampling method is appropriate? Explain.

a) We want to know what percentage of local doctors accept Medicaid patients. We call the offices of 50 doctors randomly selected from local Yellow Pages listings.

b) We want to know what percentage of local businesses anticipate hiring additional employees in the upcoming month. We randomly select a page in the Yellow Pages and call every business listed there.

36. More sampling methods. Consider each of these situations. Do you think the proposed sampling method is appropriate? Explain.

a) We want to know if there is neighborhood support to turn a vacant lot into a playground. We spend a Saturday afternoon going door-to-door in the neighborhood, asking people to sign a petition.

b) We want to know if students at our college are satisfied with the selection of food available on campus. We go to the largest cafeteria and interview every 10th person in line.

JUST CHECKING

ANSWERS

1. a) It can be hard to reach all members of a population, and it can take so long that circumstances change, affecting the responses. A well-designed sample is often a better choice.

b) This sample is probably biased—students who didn't like the food at the cafeteria might not choose to eat there.

c) No, only the sample size matters, not the fraction of the overall population.

d) Students who frequent this website might be more enthusiastic about Statistics than the overall population of Statistics students. A large sample cannot compensate for bias.

e) It's the population "parameter." "Statistics" describe samples.

2. a) systematic

b) stratified

c) simple

d) cluster

Experiments and Observational Studies

Where are we going?

Experiments are the "Gold Standard" of data collection. No drug comes to market without at least one FDA-approved experiment to demonstrate its safety and effectiveness. Much of what we know in science and social science comes from carefully designed experiments.

The Four Principles of Experimental Design (Control what you can, Randomize for the rest, Replicate the trials, and, when appropriate, Block to remove identifiable variation) describe what makes a sound experiment and how to understand the results.

Who gets good grades? And, more important, why? Is there something schools and parents could do to help weaker students improve their grades? Some people think they have an answer: music! No, not your iPod, but an instrument. In a study conducted at Mission Viejo High School, in California, researchers compared the scholastic performance of music students with that of non-music students. Guess what? The music students had a much higher overall grade point average than the non-music students, 3.59 to 2.91. Not only that: A whopping 16% of the music students had all A's compared with only 5% of the non-music students.

As a result of this study and others, many parent groups and educators pressed for expanded music programs in the nation's schools. They argued that the work ethic, discipline, and feeling of accomplishment fostered by learning to play an instrument also enhance a person's ability to succeed in school. They thought that involving more students in music would raise academic performance. What do you think? Does this study provide solid evidence? Or are there other possible explanations for the difference in grades? Is there any way to really prove such a conjecture?

Observational Studies

This research tried to show an association between music education and grades. But it wasn't a survey. Nor did it assign students to get music education. Instead, it simply observed students "in the wild," recording the choices they made and the outcome. Such studies are called **observational studies**. In observational studies, researchers don't *assign* choices; they simply observe them.

In addition, this was a **retrospective study**, because researchers first identified subjects who studied music and then collected data on their past grades.

What's wrong with concluding that music education causes good grades? One high school during one academic year may not be representative of the whole United States. That's true, but the real problem is that the claim that music study *caused* higher grades depends on there being *no other differences* between the groups that could account for the differences in grades, and studying music was not the *only* difference between the two groups of students.

We can think of lots of lurking variables that might cause the groups to perform differently. Students who study music may have better work habits to start with, and this makes them successful in both music and course work. Music students may have more parental support (someone had to pay for all those lessons), and that support may have enhanced their academic performance, too. Maybe they came from wealthier homes and had other advantages. Or it could be that smarter kids just like to play musical instruments.

Observational studies are valuable for discovering trends and possible relationships. They are used widely in public health and marketing. Observational studies that try to discover variables related to rare outcomes, such as specific diseases, are often retrospective. They first identify people with the disease and then look into their history and heritage in search of things that may be related to their condition. But retrospective studies have a restricted view of the world because they are usually restricted to a small part of the entire population. And because retrospective records are based on historical data, they can have errors. (Do you recall *exactly* what you ate even yesterday? How about last Wednesday?)

A somewhat better approach is to observe individuals over time, recording the variables of interest and ultimately seeing how things turn out. For example, we might start by selecting young students who have not begun music lessons. We could then track their academic performance over several years, comparing those who later choose to study music with those who do not. Identifying subjects in advance and collecting data as events unfold would make this a **prospective study**.

Although an observational study may identify important variables related to the outcome we are interested in, there is no guarantee that we have found the right or the most important related variables. Students who choose to study an instrument might still differ from the others in some important way that we failed to observe. It may be this difference—whether we know what it is or not—rather than music itself that leads to better grades. It's just not possible for observational studies, whether prospective or retrospective, to demonstrate a causal relationship.

> For rare illnesses, it's not practical to draw a large enough sample to see many ill respondents, so the only option remaining is to develop retrospective data. For example, researchers can interview those who have become ill. The likely causes of both Legionnaires' disease and HIV were initially identified from such retrospective studies of the small populations who were initially infected. But to confirm the causes, researchers needed laboratory-based experiments.

FOR EXAMPLE   Designing an Observational Study

In early 2007, a larger-than-usual number of cats and dogs developed kidney failure; many died. Initially, researchers didn't know why, so they used an observational study to investigate.

QUESTION: Suppose you were called on to plan a study seeking the cause of this problem. Would your design be retrospective or prospective? Explain why.

I would use a retrospective observational study. Even though the incidence of disease was higher than usual, it was still rare. Surveying all pets would have been impractical. Instead, it makes sense to locate some who were sick and ask about their diets, exposure to toxins, and other possible causes.

Randomized, Comparative Experiments

Is it *ever* possible to get convincing evidence of a cause-and-effect relationship? Well, yes it is, but we would have to take a different approach. We could take a group of third graders, randomly assign half to take music lessons, and forbid the other half to do so. Then we could compare their grades several years later. This kind of study design is called an **experiment**.

An experiment requires a **random assignment** of subjects to treatments. Only an experiment can justify a claim like "Music lessons cause higher grades." Questions such as "Does taking vitamin C reduce the chance of getting a cold?" and "Does working with computers improve performance in Statistics class?" and "Is this drug a safe and effective treatment for that disease?" require a designed experiment to establish cause and effect.

Experiments study the relationship between two or more variables. An experimenter must identify at least one explanatory variable, called a **factor**, to manipulate and at least one **response variable** to measure. What distinguishes an experiment from other types of investigation is that the experimenter actively and deliberately manipulates the factors to control the details of the possible treatments, and assigns the subjects to those treatments *at random*. The experimenter then observes the response variable and *compares* responses for different groups of subjects who have been treated differently. For example, we might design an experiment to see whether the amount of sleep and exercise you get affects your performance.

The individuals on whom or which we experiment are known by a variety of terms. Humans who are experimented on are commonly called **subjects** or **participants**. Other individuals (rats, days, petri dishes of bacteria) are commonly referred to by the more generic term **experimental unit**. When we recruit subjects for our sleep deprivation experiment by advertising in Statistics class, we'll probably have better luck if we invite them to be participants than if we advertise that we need experimental units.

The specific values that the experimenter chooses for a factor are called the **levels** of the factor. We might assign our participants to sleep for 4, 6, or 8 hours. Often there are several factors at a variety of levels. (Our subjects will also be assigned to a treadmill for 0 or 30 minutes.) The combination of specific levels from all the factors that an experimental unit receives is known as its **treatment**. (Our subjects could have any one of six different treatments—three sleep levels, each at two exercise levels.)

How should we assign our participants to these treatments? Some students prefer 4 hours of sleep, while others need 8. Some exercise regularly; others are couch potatoes. Should we let the students choose the treatments they'd prefer? No. That would not be a good idea. To have any hope of drawing a fair conclusion, we must assign our participants to their treatments *at random*.

It may be obvious to you that we shouldn't let the students choose the treatment they'd prefer, but the need for random assignment is a lesson that was once hard for some to accept. For example, physicians might naturally prefer to assign patients to the therapy that they think best rather than have a random element such as a coin flip determine the treatment. But we've known for more than a century that for the results of an experiment to be valid, we must use deliberate randomization.

Experimental design was advanced in the 19th century by work in psychophysics by Gustav Fechner (1801-1887), the founder of experimental psychology. Fechner designed ingenious experiments that exhibited many of the features of modern designed experiments. Fechner was careful to control for the effects of factors that might affect his results. For example, in his 1860 book *Elemente der Psychophysik* he cautioned readers to group experiment trials together to minimize the possible effects of time of day and fatigue.

AN EXPERIMENT:

Manipulates the factor levels to create treatments. *Randomly assigns* subjects to these treatment levels. *Compares* the responses of the subject groups across treatment levels.

"He that leaves nothing to chance will do few things ill, but he will do very few things."

–Lord Halifax
(1633-1695)

The Women's Health Initiative is a major 15-year research program funded by the National Institutes of Health to address the most common causes of death, disability, and poor quality of life in older women. It consists of both an observational study with more than 93,000 participants and several randomized comparative experiments. The goals of this study include

(continued)

No drug can be sold in the United States without first showing, in a suitably designed experiment approved by the Food and Drug Administration (FDA), that it's safe and effective. The small print on the booklet that comes with many prescription drugs usually describes the outcomes of that experiment.

- giving reliable estimates of the extent to which known risk factors predict heart disease, cancers, and fractures;

- identifying "new" risk factors for these and other diseases in women;

- comparing risk factors, presence of disease at the start of the study, and new occurrences of disease during the study across all study components; and

- creating a future resource to identify biological indicators of disease, especially substances and factors found in blood.

That is, the study seeks to identify possible risk factors and assess how serious they might be. It seeks to build up data that might be checked retrospectively as the women in the study continue to be followed. There would be no way to find out these things with an experiment because the task includes identifying new risk factors. If we don't know those risk factors, we could never control them as factors in an experiment.

By contrast, one of the clinical trials (randomized experiments) that received much press attention randomly assigned postmenopausal women to take either hormone replacement therapy or an inactive pill. The results published in 2002 and 2004 concluded that hormone replacement with estrogen carried increased risks of stroke.

FOR EXAMPLE | Determining the Treatments and Response Variable

RECAP: In 2007, deaths of a large number of pet dogs and cats were ultimately traced to contamination of some brands of pet food. The manufacturer now claims that the food is safe, but before it can be released, it must be tested.

QUESTION: In an experiment to test whether the food is now safe for dogs to eat,[1] what would be the treatments and what would be the response variable?

The treatments would be ordinary-size portions of two dog foods: the new one from the company (the test food) and one that I was certain was safe (perhaps prepared in my kitchen or laboratory). The response would be a veterinarian's assessment of the health of the test animals.

The Four Principles of Experimental Design

A S
Video: An Industrial Experiment.
Manufacturers often use designed experiments to help them perfect new products. Watch this video about one such experiment.

1. **Control.** We control sources of variation other than the factors we are testing by making conditions as similar as possible for all treatment groups. For human subjects, we try to treat them alike. However, there is always a question of degree and practicality. Controlling extraneous sources of variation reduces the variability of the responses, making it easier to detect differences among the treatment groups.

   Making generalizations from the experiment to other levels of the controlled factor can be risky. For example, suppose we test two laundry detergents and carefully control the water temperature at 180°F. This would reduce the variation in our results due to water temperature, but what could we say about the detergents' performance in cold water? Not much. It would be hard to justify extrapolating the results to other temperatures.

   Although we control both experimental factors and other sources of variation, we think of them very differently. We control a factor by assigning subjects to different factor levels because we want to see how the

---

[1] It may disturb you (as it does us) to think of deliberately putting dogs at risk in this experiment, but in fact that is what is done. The risk is borne by a small number of dogs so that the far larger population of dogs can be kept safe.

The deep insight that experiments should use random assignment is quite an old one. It can be attributed to the American philosopher and scientist C. S. Peirce in his experiments with J. Jastrow, published in 1885.

A S
Activity: **The Three Rules of Experimental Design.** Watch an animated discussion of three rules of design.

A S
Activity: **Perform an Experiment.** How well can you read pie charts and bar charts? Find out as you serve as the subject in your own experiment.

response will change at those different levels. We control other sources of variation to *prevent* them from changing and affecting the response variable.

2. **Randomize.** As in sample surveys, **randomization** allows us to equalize the effects of unknown or uncontrollable sources of variation. It does not eliminate the effects of these sources, but it should spread them out across the treatment levels so that we can see past them. If experimental units were not assigned to treatments at random, we would not be able to use the powerful methods of Statistics to draw conclusions from an experiment. Assigning subjects to treatments at random reduces bias due to uncontrolled sources of variation. Randomization protects us even from effects we didn't know about. There's an adage that says "control what you can, and randomize the rest."

3. **Replicate.** Two kinds of replication show up in comparative experiments. First, we should apply each treatment to a number of subjects. Only with such replication can we estimate the variability of responses. If we have not assessed the variation, the experiment is not complete. The outcome of an experiment on a single subject is an anecdote, not data.

A second kind of replication shows up when the experimental units are not a representative sample from the population of interest. We may believe that what is true of the students in Psych 101 who volunteered for the sleep experiment is true of all humans, but we'll feel more confident if our results for the experiment are *replicated* in another part of the country, with people of different ages, and at different times of the year. **Replication** of an entire experiment with the controlled sources of variation at different levels is an essential step in science.

4. **Block.** The ability of randomizing to equalize variation across treatment groups works best in the long run. For example, if we're allocating players to two 6-player soccer teams from a pool of 12 children, we might do so at random to equalize the talent. But what if there were two 12-year-olds and ten 6-year-olds in the group? Randomizing may place both 12-year-olds on the same team. In the long run, if we did this over and over, it would all equalize. But wouldn't it be better to assign one 12-year-old to each group (at random) and five 6-year-olds to each team (at random)? By doing this, we would improve fairness in the short run. This approach makes the division more fair by recognizing the variation in *age* and allocating the players at random *within* each age level. When we do this, we call the variable *age* a **blocking variable.** The levels of *age* are called blocks.

Sometimes, attributes of the experimental units that we are not studying and that we can't control may nevertheless affect the outcomes of an experiment. If we group similar individuals together and then randomize within each of these **blocks**, we can remove much of the variability due to the difference among the blocks. Blocking is an important compromise between randomization and control. However, unlike the first three principles, blocking is not *required* in an experimental design.

Control, Randomize, and Replicate

RECAP: We're planning an experiment to see whether the new pet food is safe for dogs to eat. We'll feed some animals the new food and others a food known to be safe, comparing their health after a period of time.

QUESTION: In this experiment, how will you implement the principles of control, randomization, and replication?

I'd control the portion sizes eaten by the dogs. To reduce possible variability from factors other than the food, I'd standardize other aspects of their environments—housing the dogs in similar pens and ensuring that each got the same amount of water, exercise, play, and sleep time, for example. I might restrict the experiment to a single breed of dog and to adult dogs to further minimize variation.

To equalize traits, pre-existing conditions, and other unknown influences, I would assign dogs to the two feed treatments randomly.

I would replicate by assigning more than one dog to each treatment to allow for variability among individual dogs. If I had the time and funding, I might replicate the entire experiment using, for example, a different breed of dog.

Diagrams

An experiment is carried out over time with specific actions occurring in a specified order. A diagram of the procedure can help in thinking about experiments.[2]

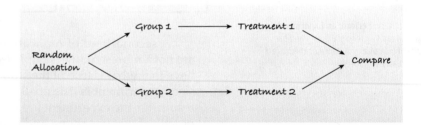

The diagram emphasizes the random allocation of subjects to treatment groups, the separate treatments applied to these groups, and the ultimate comparison of results. It's best to specify the responses that will be compared. A good way to start comparing results for the treatment groups is with boxplots.

STEP-BY-STEP EXAMPLE Designing an Experiment

An ad for OptiGro plant fertilizer claims that with this product you will grow "juicier, tastier" tomatoes. You'd like to test this claim, and wonder whether you might be able to get by with half the specified dose. How can you set up an experiment to check out the claim?

Of course, you'll have to get some tomatoes, try growing some plants with the product and some without, and see what happens. But you'll need a clearer plan than that. How should you design your experiment?

---

[2] Diagrams of this sort were introduced by David Moore in his textbooks and are widely used.

A **completely randomized experiment** is the ideal simple design, just as a *simple random sample* is the ideal simple sample—and for many of the same reasons.

Let's work through the design, step by step. We'll design the simplest kind of experiment, a **completely randomized experiment in one factor**. Since this is a *design* for an experiment, most of the steps are part of the *Think* stage. The statements in the right column are the kinds of things you would need to say in *proposing* an experiment. You'd need to include them in the "methods" section of a report once the experiment is run.

Question: How would you design an experiment to test OptiGro fertilizer?

Plan State what you want to know.

I want to know whether tomato plants grown with OptiGro yield juicier, tastier tomatoes than plants raised in otherwise similar circumstances but without the fertilizer.

Response Specify the response variable.

I'll evaluate the juiciness and taste of the tomatoes by asking a panel of judges to rate them on a scale from 1 to 7 in juiciness and in taste.

Treatments Specify the factor levels and the treatments.

The factor is fertilizer, specifically OptiGro fertilizer. I'll grow tomatoes at three different factor levels: some with no fertilizer, some with half the specified amount of OptiGro, and some with the full dose of OptiGro. These are the three treatments.

Experimental Units Specify the experimental units.

I'll obtain 24 tomato plants of the same variety from a local garden store.

Experimental Design Observe the principles of design:

> **Control** any sources of variability you know of and can control.

I'll locate the farm plots near each other so that the plants get similar amounts of sun and rain and experience similar temperatures. I will weed the plots equally and otherwise treat the plants alike.

> **Replicate** results by placing more than one plant in each treatment group.

I'll use 8 plants in each treatment group.

> **Randomly assign** experimental units to treatments, to equalize the effects of unknown or uncontrollable sources of variation.

To randomly divide the plants into three groups, first I'll label the plants with numbers 00–23. I'll look at pairs of digits across a random number table. The first 8 plants identified (ignoring numbers 24–99 and any repeats) will go in Group 1, the next 8 in Group 2, and the remaining plants in Group 3.

Describe how the randomization will be accomplished.

Make a Picture A diagram of your design can help you think about it clearly.

Specify any other experiment details. You must give enough details so that another experimenter could exactly replicate your experiment. It's generally better to include details that might seem irrelevant than to leave out matters that could turn out to make a difference.

Specify how to measure the response.

I will grow the plants until the tomatoes are mature, as judged by reaching a standard color.

I'll harvest the tomatoes when ripe and store them for evaluation.

I'll set up a numerical scale of juiciness and one of tastiness for the taste testers. Several people will taste slices of tomato and rate them.

Once you collect the data, you'll need to display them and compare the results for the three treatment groups.

I will display the results with side-by-side box-plots to compare the three treatment groups.

I will compare the means of the groups.

To answer the initial question, we ask whether the differences we observe in the means of the three groups are meaningful.

Because this is a randomized experiment, we can attribute significant differences to the treatments. To do this properly, we'll need methods from what is called "statistical inference," the subject of the rest of this book.

If the differences in taste and juiciness among the groups are greater than I would expect by knowing the usual variation among tomatoes, I may be able to conclude that these differences can be attributed to treatment with the fertilizer.

Does the Difference Make a Difference?

If the differences among the treatment groups are big enough, we'll attribute the differences to the treatments, but how can we decide whether the differences are big enough?

Would we expect the group means to be identical? Not really. Even if the treatment made no difference whatever, there would still be some variation. We assigned the tomato plants to treatments at random. But a different random assignment would have led to different results. Even a repeat of the *same* treatment on a different randomly assigned set of plants would lead to a different mean. The real question is whether the differences we observed are about as big as we might get just from the randomization alone, or whether they're bigger than that. If we decide that they're bigger, we'll attribute the differences to the treatments. In that case we say the differences are **statistically significant**.

A S

Activity: **Graph the Data.** Do you think there's a significant difference in your perception of pie charts and bar charts? Explore the data from your plot perception experiment.

How will we decide if something is different enough to be considered statistically significant? Later chapters will offer methods to help answer that question, but to get some intuition, think about deciding whether a coin is fair. If we flip a fair coin 100 times, we expect, *on average*, to get 50 heads. Suppose we get 54 heads out of 100. That doesn't seem very surprising. It's well within the bounds of ordinary random fluctuations. What if we'd seen 94 heads? That's clearly outside the bounds. We'd be pretty sure that the coin flips were not random. But what about 74 heads? Is that far enough from 50% to arouse our suspicions? That's the sort of question we need to ask of our experiment results.

In Statistics terminology, 94 heads would be a statistically significant difference from 50, and 54 heads would not. Whether 74 is *statistically significant* or not would depend on the chance of getting 74 heads in 100 flips of a fair coin and on our tolerance for believing that rare events can happen to us.

Back at the tomato patch, we ask whether the differences we see among the treatment groups are the kind of differences we'd expect from randomization. A good way to get a feeling for that is to look at how much our results vary among plants that get the *same* treatment. Boxplots of our results by treatment group can give us a general idea.

For example, Figure 13.1 shows two pairs of boxplots whose centers differ by exactly the same amount. In the upper set, that difference appears to be larger than we'd expect just by chance. Why? Because the variation is quite small *within* treatment groups, so the larger difference *between* the groups is unlikely to be just from the randomization. In the bottom pair, that same difference between the centers looks less impressive. There the variation *within* each group swamps the difference *between* the two medians. We'd say the difference is statistically significant in the upper pair and not statistically significant in the lower pair.

In later chapters we'll see statistical tests that quantify this intuition. For now, the important point is that a difference is statistically significant if we don't believe that it's likely to have occurred only by chance.

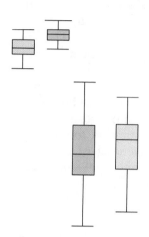

FIGURE 13.1
The boxplots in both pairs have centers the same distance apart, but when the spreads are large, the observed difference may be just from random fluctuation.

1. At one time, a method called "gastric freezing" was used to treat people with peptic ulcers. An inflatable bladder was inserted down the esophagus and into the stomach, and then a cold liquid was pumped into the bladder. Now you can find the following notice on the Internet site of a major insurance company:

[Our company] does not cover gastric freezing (intragastric hypothermia) for chronic peptic ulcer disease. . . .

Gastric freezing for chronic peptic ulcer disease is a non-surgical treatment which was popular about 20 years ago but now is seldom performed. It has been abandoned due to a high complication rate, only temporary improvement experienced by patients, and a lack of effectiveness when tested by double-blind, controlled clinical trials.

What did that "controlled clinical trial" (experiment) probably look like? (Don't worry about "double-blind"; we'll get to that soon.)

a) What was the factor in this experiment?
b) What was the response variable?
c) What were the treatments?

d) How did researchers decide which subjects received which treatment?
e) Were the results statistically significant?

Experiments and Samples

Both experiments and sample surveys use randomization to get unbiased data. But they do so in different ways and for different purposes. Sample surveys try to estimate population parameters, so the sample needs to be as representative of the population as possible. By contrast, experiments try to assess the effects of treatments. Experimental units are not always drawn randomly from the population. For example, a medical experiment may deal only with local patients who have the disease under study. The randomization is in the assignment of their therapy. We want a sample to exhibit the diversity and variability of the population, but for an experiment the more homogeneous the subjects the more easily we'll spot differences in the effects of the treatments.

Unless the experimental units are chosen from the population at random, you should be cautious about generalizing experiment results to larger populations until the experiment has been repeated under different circumstances. Results become more persuasive if they remain the same in completely different settings, such as in a different season, in a different country, or for a different species, to name a few.

Even without choosing experimental units from a population at random, experiments can draw stronger conclusions than surveys. By looking only at the differences across treatment groups, experiments cancel out many sources of bias. For example, the entire pool of subjects may be biased and not representative of the population. (College students may need more sleep, on average, than the general population.) When we assign subjects randomly to treatment groups, all the groups are still biased, but *in the same way*. When we consider the differences in their responses, these biases cancel out, allowing us to see the *differences* due to treatment effects more clearly.

Experiments are rarely performed on random samples from a population. Don't describe the subjects in an experiment as a random sample unless they really are. More likely, the randomization was in assigning subjects to treatments.

A S **Activity: Control Groups in Experiments.** Is a control group really necessary?

Control Treatments

Suppose you wanted to test a $300 piece of software designed to shorten download times. You could just try it on several files and record the download times, but you probably want to *compare* the speed with what would happen *without* the software installed. Such a baseline measurement is called a **control** treatment, and the experimental units to whom it is applied are called a **control group**.

This is a use of the word "control" in an entirely different context. Previously, we controlled extraneous sources of variation by keeping them constant. Here, we use a control treatment as another *level* of the factor in order to compare the treatment results to a situation in which "nothing happens." That's what we did in the tomato experiment when we used no fertilizer on the 8 tomato plants in Group 1.

Blinding

Humans are notoriously susceptible to errors in judgment.[3] All of us. When we know what treatment was assigned, it's difficult not to let that knowledge influence our assessment of the response, even when we try to be careful.

Suppose you were trying to advise your school on which brand of cola to stock in the school's vending machines. You set up an experiment to see which of the three competing brands students prefer (or whether they can tell the difference at all). But people have brand loyalties. You probably prefer one brand already. So if you knew which brand you were tasting, it might influence your rating. To avoid this problem, it would be better to disguise the brands as much as possible. This strategy is called **blinding** the participants to the treatment.[4]

But it isn't just the subjects who should be blind. Experimenters themselves often subconsciously behave in ways that favor what they believe. Even technicians may treat plants or test animals differently if, for example, they

---

[3] For example, here we are in Chapter 13 and you're still reading the footnotes.

[4] C. S. Peirce, in the same 1885 work in which he introduced randomization, also recommended blinding.

Social science experiments can sometimes blind subjects by misleading them about the purpose of a study. One of the authors participated as an undergraduate in a (now infamous) psychology experiment using such a blinding method. The subjects were told that the experiment was about three-dimensional spatial perception and were assigned to draw a horse. While they were drawing, they heard a loud noise and groaning from the next room. The *real* purpose of the experiment was to see whether the social pressure of being in groups made people react to disaster differently. Subjects had been randomly assigned to draw either in groups or alone; that was the treatment. The experimenter was not interested in the drawings, but the subjects were blinded to the treatment because they were misled.

expect them to die. An animal that starts doing a little better than others by showing an increased appetite may get fed a bit more than the experimental protocol specifies.

People are so good at picking up subtle cues about treatments that the best (in fact, the *only*) defense against such biases in experiments on human subjects is to keep *anyone* who could affect the outcome or the measurement of the response from knowing which subjects have been assigned to which treatments. So, not only should your cola-tasting subjects be blinded, but also *you,* as the experimenter, shouldn't know which drink is which, either—at least until you're ready to analyze the results.

There are two main classes of individuals who can affect the outcome of the experiment:

- those who could influence the results (the subjects, treatment administrators, or technicians)
- those who evaluate the results (judges, treating physicians, etc.)

When all the individuals in either one of these classes are blinded, an experiment is said to be **single-blind**. When everyone in *both* classes is blinded, we call the experiment **double-blind**. Even if several individuals in one class are blinded—for example, both the patients and the technicians who administer the treatment—the study would still be just single-blind. If only some of the individuals in a class are blind—for example, if subjects are not told of their treatment, but the administering technician is not blind—there is a substantial risk that subjects can discern their treatment from subtle cues in the technician's behavior or that the technician might inadvertently treat subjects differently. Such experiments cannot be considered truly blind.

In our tomato experiment, we certainly don't want the people judging the taste to know which tomatoes got the fertilizer. That makes the experiment single-blind. We might also not want the people caring for the tomatoes to know which ones were being fertilized, in case they might treat them differently in other ways, too. We can accomplish this double-blinding by having some fake fertilizer for them to put on the other plants. Read on.

FOR EXAMPLE   Blinding

RECAP: In our experiment to see if the new pet food is now safe, we're feeding one group of dogs the new food and another group a food we know to be safe. Our response variable is the health of the animals as assessed by a veterinarian.

QUESTIONS: Should the vet be blinded? Why or why not? How would you do this? (Extra credit: Can this experiment be double-blind? Would that mean that the test animals wouldn't know what they were eating?)

Whenever the response variable involves judgment, it is a good idea to blind the evaluator to the treatments. The veterinarian should not be told which dogs ate which foods.

Extra credit: There is a need for double-blinding. In this case, the workers who care for and feed the animals should not be aware of which dogs are receiving which food. We'll need to make the "safe" food look as much like the "test" food as possible.

Activity: Blinded Experiments. This narrated account of blinding isn't a placebo!

Placebos

Often, simply applying *any* treatment can induce an improvement. Every parent knows the medicinal value of a kiss to make a toddler's scrape or bump stop hurting. Some of the improvement seen with a treatment—even an effective treatment—can be due simply to the act of treating. To separate these two effects, we can use a control treatment that mimics the treatment itself.

The placebo effect is stronger when placebo treatments are administered with authority or by a figure who appears to be an authority. "Doctors" in white coats generate a stronger effect than salespeople in polyester suits. But the placebo effect is not reduced much even when subjects know that the effect exists. People often suspect that they've gotten the placebo if nothing at all happens. So, recently, drug manufacturers have gone so far in making placebos realistic that they cause the same side effects as the drug being tested! Such "active placebos" usually induce a stronger placebo effect. When those side effects include loss of appetite or hair, the practice may raise ethical questions.

A "fake" treatment that looks just like the treatments being tested is called a **placebo**. Placebos are the best way to blind subjects from knowing whether they are receiving the treatment or not. One common version of a placebo in drug testing is a "sugar pill." Especially when psychological attitude can affect the results, control group subjects treated with a placebo may show an improvement.

The fact is that subjects treated with a placebo sometimes improve. It's not unusual for 20% or more of subjects given a placebo to report reduction in pain, improved movement, or greater alertness, or even to demonstrate improved health or performance. This **placebo effect** highlights both the importance of effective blinding and the importance of comparing treatments with a control. Placebo controls are so effective that you should use them as an essential tool for blinding whenever possible.

The best experiments are usually

- randomized.
- comparative.
- double-blind.
- placebo-controlled.

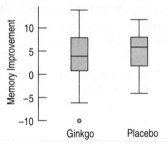

Does ginkgo biloba improve memory? Researchers investigated the purported memory-enhancing effect of ginkgo biloba tree extract (P. R. Solomon, F. Adams, A. Silver, J. Zimmer, R. De Veaux, "Ginkgo for Memory Enhancement. A Randomized Controlled Trial." *JAMA* 288 [2002]: 835–840). In a randomized, comparative, double-blind, placebo-controlled study, they administered treatments to 230 elderly community members. One group received Ginkoba™ according to the manufacturer's instructions. The other received a similar-looking placebo. Thirteen different tests of memory were administered before and after treatment. The placebo group showed greater improvement on 7 of the tests, the treatment group on the other 6. None showed any significant differences. In the margin are boxplots of one measure.

By permission of John L. Hart FLP and Creators Syndicate, Inc.

Blocking

We might want to use 18 tomato plants of the same variety for our experiment, but suppose the garden store had only 12 plants left. So we drove down to the nursery and bought 6 more plants of that variety. We worry that the tomato plants from the two stores are different somehow, and, in fact, they don't really look the same.

How can we design the experiment so that the differences between the stores don't mess up our attempts to see differences among fertilizer levels? We can't measure the effect of a store the same way as we can the fertilizer because we can't assign it as we would a factor in the experiment. You can't tell a tomato plant what store to come from.

Because stores may vary in the care they give plants or in the sources of their seeds, the plants from either store are likely to be more like each other than they are like the plants from the other store. When groups of experimental units are similar, it's often a good idea to gather them together into blocks. By blocking, we isolate the variability attributable to the differences between the blocks, so that we can see the differences caused by the treatments more clearly. Here, we would define the plants from each store to be a block. The randomization is introduced when we randomly assign treatments within each block.

In a completely randomized design, each of the 18 plants would have an equal chance to land in each of the three treatment groups. But we realize that the store may have an effect. To isolate the store effect, we block one store by assigning the plants from each store to treatments at random. So we now have six treatment groups, three for each block. Within each block, we'll randomly assign the same number of plants to each of the three treatments. The experiment is still fair because each treatment is still applied (at random) to the same number of plants and to the same proportion from each store: 4 from store A and 2 from store B. Because the randomization occurs only within the blocks (plants from one store cannot be assigned to treatment groups for the other), we call this a **randomized block design**.

In effect, we conduct two parallel experiments, one for tomato plants from each store, and then combine the results. The picture tells the story:

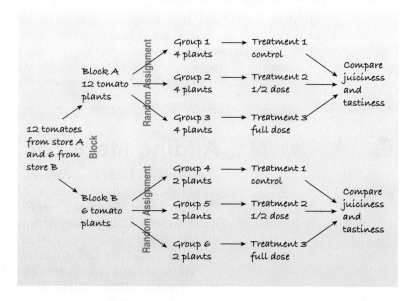

In a retrospective or prospective study, subjects are sometimes paired because they are similar in ways *not* under study. **Matching** subjects in this way can reduce variation in much the same way as blocking. For example, a retrospective study of music education and grades might match each student who studies an instrument with someone of the same sex who is similar in family income but didn't study an instrument. When we compare grades of music students with those of non-music students, the matching would reduce the variation due to income and sex differences.

Blocking is the same idea for experiments as stratifying is for sampling. Both methods group together subjects that are similar and randomize within those groups as a way to remove unwanted variation. (But be careful to keep the terms straight. Don't say that we "stratify" an experiment or "block" a sample.) We use blocks to reduce variability so we can see the effects of the factors; we're not usually interested in studying the effects of the blocks themselves.

RECAP: In 2007, pet food contamination put cats at risk, as well as dogs. Our experiment should probably test the safety of the new food on both animals.

QUESTIONS: Why shouldn't we randomly assign a mix of cats and dogs to the two treatment groups? What would you recommend instead?

Dogs and cats might respond differently to the foods, and that variability could obscure my results. Blocking by species can remove that superfluous variation. I'd randomize cats to the two treatments (test food and safe food) separately from the dogs. I'd measure their responses separately and look at the results afterward.

JUST CHECKING

2. Recall the experiment about gastric freezing, an old method for treating peptic ulcers that you read about in the first Just Checking. Doctors would insert an inflatable bladder down the patient's esophagus and into the stomach and then pump in a cold liquid. A major insurance company now states that it doesn't cover this treatment because "double-blind, controlled clinical trials" failed to demonstrate that gastric freezing was effective.

a) What does it mean that the experiment was double-blind?
b) Why would you recommend a placebo control?
c) Suppose that researchers suspected that the effectiveness of the gastric freezing treatment might depend on whether a patient had recently developed the peptic ulcer or had been suffering from the condition for a long time. How might the researchers have designed the experiment?

Adding More Factors

There are two kinds of gardeners. Some water frequently, making sure that the plants are never dry. Others let Mother Nature take her course and leave the watering to her. The makers of OptiGro want to ensure that their product will work under a wide variety of watering conditions. Maybe we should include the amount of watering as part of our experiment. Can we study a second factor at the same time and still learn as much about fertilizer?

We now have two factors (fertilizer at three levels and irrigation at two levels). We combine them in all possible ways to yield six treatments:

	No Fertilizer	Half Fertilizer	Full Fertilizer
No Added Water	1	2	3
Daily Watering	4	5	6

THINK LIKE A STATISTICIAN

With two factors, we can account for more of the variation. That lets us see the underlying patterns more clearly.

If we allocate the original 24 plants, 4 plants to each of these six treatments at random. This experiment is a **completely randomized two-factor experiment** because any plant could end up assigned at random to any of the six treatments (and we have two factors).

It's often important to include several factors in the same experiment in order to see what happens when the factor levels are applied in different *combinations*. A common misconception is that applying several factors at once makes it difficult to separate the effects of the individual factors. You may hear

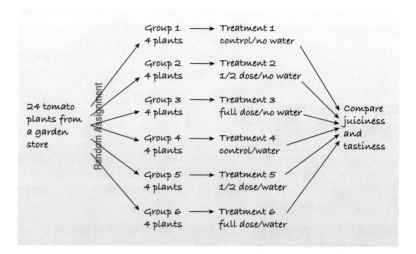

people say that experiments should always be run "one factor at a time." In fact, just the opposite is true: Experiments with more than one factor are both more efficient and provide more information than one-at-a-time experiments. There are many ways to design efficient multifactor experiments. You can take a whole course on the design and analysis of such experiments.

Confounding

Professor Stephen Ceci of Cornell University performed an experiment to investigate the effect of a teacher's classroom style on student evaluations. He taught a class in developmental psychology during two successive terms to a total of 472 students in two very similar classes. He kept everything about his teaching identical (same text, same syllabus, same office hours, etc.) and modified only his style in class. During the fall term, he maintained a subdued demeanor. During the spring term, he used expansive gestures and lectured with more enthusiasm, varying his vocal pitch and using more hand gestures. He administered a standard student evaluation form at the end of each term.

The students in the fall term class rated him only an average teacher. Those in the spring term class rated him an excellent teacher, praising his knowledge and accessibility, and even the quality of the textbook. On the question "How much did you learn in the course?" the average response changed from 2.93 to 4.05 on a 5-point scale.[5]

How much of the difference he observed was due to his difference in manner, and how much might have been due to the season of the year? Fall term in Ithaca, NY (home of Cornell University), starts out colorful and pleasantly warm but ends cold and bleak. Spring term starts out bitter and snowy and ends with blooming flowers and singing birds. Might students' overall happiness have been affected by the season and reflected in their evaluations?

Unfortunately, there's no way to tell. Nothing in the data enables us to tease apart these two effects, because all the students who experienced the subdued manner did so during the fall term and all who experienced the expansive manner did so during the spring. When the levels of one factor are associated with the levels of another factor, we say that these two factors are **confounded**.

In some experiments, such as this one, it's just not possible to avoid some confounding. Professor Ceci could have randomly assigned students to one of two classes during the same term, but then we might question whether

---

[5] But the two classes performed almost identically well on the final exam.

mornings or afternoons were better, or whether he really delivered the same class the second time (after practicing on the first class). Or he could have had another professor deliver the second class, but that would have raised more serious issues about differences in the two professors and concern over more serious confounding.

FOR EXAMPLE Confounding

RECAP: After many dogs and cats suffered health problems caused by contaminated foods, we're trying to find out whether a newly formulated pet food is safe. Our experiment will feed some animals the new food and others a food known to be safe, and a veterinarian will check the response.

QUESTION: Why would it be a bad design to feed the test food to some dogs and the safe food to cats?

This would create confounding. We would not be able to tell whether any differences in animals' health were attributable to the food they had eaten or to differences in how the two species responded.

A two-factor example Confounding can also arise from a badly designed multifactor experiment. Here's a classic. A credit card bank wanted to test the sensitivity of the market to two factors: the annual fee charged for a card and the annual percentage rate charged. Not wanting to scrimp on sample size, the bank selected 100,000 people at random from a mailing list. It sent out 50,000 offers with a low rate and no fee and 50,000 offers with a higher rate and a $50 annual fee. Guess what happened? That's right—people preferred the low-rate, no-fee card. No surprise. In fact, they signed up for that card at over twice the rate as the other offer. And because of the large sample size, the bank was able to estimate the difference precisely. But the question the bank really wanted to answer was "how much of the change was due to the rate, and how much was due to the fee?" Unfortunately, there's simply no way to separate out the two effects. If the bank had sent out all four possible different treatments—low rate with no fee, low rate with $50 fee, high rate with no fee, and high rate with $50 fee—each to 25,000 people, it could have learned about both factors and could have also seen what happens when the two factors occur in combination.

Lurking or Confounding?

Confounding may remind you of the problem of lurking variables we discussed back in Chapters 7 and 9. Confounding variables and lurking variables are alike in that they interfere with our ability to interpret our analyses simply. Each can mislead us, but there are important differences in both how and where the confusion may arise.

A lurking variable creates an association between two other variables that tempts us to think that one may cause the other. This can happen in a regression analysis or an observational study when a lurking variable influences both the explanatory and response variables. Recall that countries with more TV sets per capita tend to have longer life expectancies. We shouldn't conclude it's the TVs "causing" longer life. We suspect instead that a generally higher standard of living may mean that people can afford more TVs and get better health care, too. Our data revealed an association between TVs and life expectancy, but economic conditions were a likely lurking variable. A lurking variable, then, is usually thought of as a variable associated with both y and x that makes it appear that x may be causing y.

Confounding can arise in experiments when some other variable associated with a factor has an effect on the response variable. However, in a designed experiment, the experimenter *assigns* treatments (at random) to subjects rather than just observing them. A confounding variable can't be thought of as causing that assignment. Professor Ceci's choice of teaching styles was not caused by the weather, but because he used one style in the fall and the other in the spring, he was unable to tell how much of his students' reactions were attributable to his teaching and how much to the weather. A confounding variable, then, is associated in a noncausal way with a factor and affects the response. Because of the confounding, we find that we can't tell whether any effect we see was caused by our factor or by the confounding variable—or even by both working together.

Both confounding and lurking variables are outside influences that make it harder to understand the relationship we are modeling. However, the nature of the causation is different in the two situations. In regression and observational studies, we can only observe associations between variables. Although we can't demonstrate a causal relationship, we often imagine whether *x* could cause *y*. We can be misled by a lurking variable that influences both. In a designed experiment, we often hope to show that the factor causes a response. Here we can be misled by a confounding variable that's associated with the factor and causes or contributes to the differences we observe in the response.

It's worth noting that the role of blinding in an experiment is to combat a possible source of confounding. There's a risk that knowledge about the treatments could lead the subjects or those interacting with them to behave differently or could influence judgments made by the people evaluating the responses. That means we won't know whether the treatments really do produce different results or if we're being fooled by these confounding influences.

What Can Go Wrong?

- **Don't give up just because you can't run an experiment.** Sometimes we can't run an experiment because we can't identify or control the factors. Sometimes it would simply be unethical to run the experiment. (Consider randomly assigning students to take—and be graded in—a Statistics course deliberately taught to be boring and difficult or one that had an unlimited budget to use multimedia, real-world examples, and field trips to make the subject more interesting.) If we can't perform an experiment, often an observational study is a good choice.

- **Beware of confounding.** Use randomization whenever possible to ensure that the factors not in your experiment are not confounded with your treatment levels. Be alert to confounding that cannot be avoided, and report it along with your results.

- **Bad things can happen even to good experiments.** Protect yourself by recording additional information. An experiment in which the air conditioning failed for 2 weeks, affecting the results, was saved by recording the temperature (although that was not originally one of the factors) and estimating the effect the higher temperature had on the response.[6]

  It's generally good practice to collect as much information as possible about your experimental units and the circumstances of the experiment. For example, in the tomato experiment, it would be wise to record details of the weather (temperature, rainfall, sunlight) that might affect the plants

---

[6] R. D. DeVeaux and M. Szelewski, "Optimizing Automatic Splitless Injection Parameters for Gas Chromatographic Environmental Analysis." *Journal of Chromatographic Science* 27, no. 9 (1989): 513–518.

and any facts available about their growing situation. (Is one side of the field in shade sooner than the other as the day proceeds? Is one area lower and a bit wetter?) Sometimes we can use this extra information during the analysis to reduce biases.

■ **Don't spend your entire budget on the first run.** Just as it's a good idea to pretest a survey, it's always wise to try a small pilot experiment before running the full-scale experiment. You may learn, for example, how to choose factor levels more effectively, about effects you forgot to control, and about unanticipated confoundings.

CONNECTIONS

The fundamental role of randomization in experiments clearly points back to our discussions of randomization, to our experiments with simulations, and to our use of randomization in sampling. The similarities and differences between experiments and samples are important to keep in mind and can make each concept clearer.

If you think that blocking in an experiment resembles stratifying in a sample, you're quite right. Both are ways of removing variation we can identify to help us see past the variation in the data.

Experiments compare groups of subjects that have been treated differently. Graphics such as boxplots that help us compare groups are closely related to these ideas. Think about what we look for in a boxplot to tell whether two groups look really different, and you'll be thinking about the same issues as experiment designers.

Generally, we're going to consider how different the mean responses are for different treatment groups. And we're going to judge whether those differences are large by using standard deviations as rulers. (That's why we needed to replicate results for each treatment; we need to be able to estimate those standard deviations.) The discussion in Chapter 6 introduced this fundamental statistical thought, and it's going to keep coming back over and over again. Statistics is about variation.

We'll see a number of ways to analyze results from experiments in subsequent chapters.

WHAT HAVE WE LEARNED?

We've learned to recognize sample surveys, observational studies, and randomized comparative experiments. We know that these methods collect data in different ways and lead us to different conclusions.

We've learned to identify retrospective and prospective observational studies and understand the advantages and disadvantages of each.

We've learned that only well-designed experiments can allow us to reach cause-and-effect conclusions. We manipulate levels of treatments to see if the factor we have identified produces changes in our response variable.

We've learned the principles of experimental design:

▶ We want to be sure that variation in the response variable can be attributed to our factor, so we identify and control as many other sources of variability as possible.

▶ Because there are many possible sources of variability that we cannot identify, we try to equalize those by randomly assigning experimental units to treatments.

▶ We replicate the experiment on as many subjects as possible.

▶ We consider blocking to reduce variability from sources we recognize but cannot control.

We've learned the value of having a control group and of using blinding and placebo controls.

Finally, we've learned to recognize the problems posed by confounding variables in experiments and lurking variables in observational studies.

Terms

Observational study	A study based on data in which no manipulation of factors has been employed (p. 305).
Retrospective study	An observational study in which subjects are selected and then their previous conditions or behaviors are determined. Retrospective studies need not be based on random samples and they usually focus on estimating differences between groups or associations between variables (p. 306).
Prospective study	An observational study in which subjects are followed to observe future outcomes. Because no treatments are deliberately applied, a prospective study is not an experiment. Nevertheless, prospective studies typically focus on estimating differences among groups that might appear as the groups are followed during the course of the study (p. 306).
Experiment	An experiment *manipulates* factor levels to create treatments, *randomly assigns* subjects to these treatment levels, and then *compares* the responses of the subject groups across treatment levels (p. 307).
Random assignment	To be valid, an experiment must assign experimental units to treatment groups at random (p. 307).
Factor	A variable whose levels are manipulated by the experimenter. Experiments attempt to discover the effects that differences in factor levels may have on the responses of the experimental units (p. 307).
Response	A variable whose values are compared across different treatments. In a randomized experiment, large response differences can be attributed to the effect of differences in treatment level (p. 307).
Subjects or Participants	The individuals who participate in an experiment, especially when they are human. A more general term is experimental unit (p. 307).
Experimental units	Individuals on whom an experiment is performed. Usually called **subjects** or **participants** when they are human (p. 307).
Level	The specific values that the experimenter chooses for a factor are called the levels of the factor (p. 307).
Treatment	The process, intervention, or other controlled circumstance applied to randomly assigned experimental units. Treatments are the different levels of a single factor or are made up of combinations of levels of two or more factors (p. 307).
Principles of experimental design	▸ **Control** aspects of the experiment that we know may have an effect on the response, but that are not the factors being studied (p. 308).
	▸ **Randomize** subjects to treatments to even out effects that we cannot control (p. 309).
	▸ **Replicate** over as many subjects as possible. Results for a single subject are just anecdotes. If, as often happens, the subjects of the experiment are not a representative sample from the population of interest, replicate the entire study with a different group of subjects, preferably from a different part of the population (p. 309).
	▸ **Block** to reduce the effects of identifiable attributes of the subjects that cannot be controlled (p. 309).
Statistically significant	When an observed difference is too large for us to believe that it is likely to have occurred naturally, we consider the difference to be statistically significant. Subsequent chapters will show specific calculations and give rules, but the principle remains the same (p. 312).
Control group	The experimental units assigned to a baseline treatment level, typically either the default treatment, which is well understood, or a null, placebo treatment. Their responses provide a basis for comparison (p. 314).
Blinding	Any individual associated with an experiment who is not aware of how subjects have been allocated to treatment groups is said to be blinded (p. 314).
Single-blind Double-blind	There are two main classes of individuals who can affect the outcome of an experiment: ▸ those who could *influence the results* (the subjects, treatment administrators, or technicians). ▸ those who *evaluate the results* (judges, treating physicians, etc.). When every individual in *either* of these classes is blinded, an experiment is said to be single-blind. When everyone in *both* classes is blinded, we call the experiment double-blind (p. 315).

Placebo	A treatment known to have no effect, administered so that all groups experience the same conditions. Many subjects respond to such a treatment (a response known as a placebo effect). Only by comparing with a placebo can we be sure that the observed effect of a treatment is not due simply to the placebo effect (p. 316).
Placebo effect	The tendency of many human subjects (often 20% or more of experiment subjects) to show a response even when administered a placebo (p. 316).
Blocking	When groups of experimental units are similar, it is often a good idea to gather them together into blocks. By blocking, we isolate the variability attributable to the differences between the blocks so that we can see the differences caused by the treatments more clearly (p. 316).
Matching	In a retrospective or prospective study, subjects who are similar in ways not under study may be matched and then compared with each other on the variables of interest. Matching, like blocking, reduces unwanted variation (p. 317).
Designs	In **a completely randomized design,** all experimental units have an equal chance of receiving any treatment (p. 318). In **a randomized block design,** the randomization occurs only within blocks (p. 317).
Confounding	When the levels of one factor are associated with the levels of another factor in such a way that their effects cannot be separated, we say that these two factors are confounded (p. 319).

Skills

▶ Recognize when an observational study would be appropriate.

▶ Be able to identify observational studies as retrospective or prospective, and understand the strengths and weaknesses of each method.

▶ Know the four basic principles of sound experimental design—control, randomize, replicate, and block—and be able to explain each.

▶ Be able to recognize the factors, the treatments, and the response variable in a description of a designed experiment.

▶ Understand the essential importance of randomization in assigning treatments to experimental units.

▶ Understand the importance of replication to move from anecdotes to general conclusions.

▶ Understand the value of blocking so that variability due to differences in attributes of the subjects can be removed.

▶ Understand the importance of a control group and the need for a placebo treatment in some studies.

▶ Understand the importance of blinding and double-blinding in studies on human subjects, and be able to identify blinding and the need for blinding in experiments.

▶ Understand the value of a placebo in experiments with human participants.

SHOW

▶ Be able to design a completely randomized experiment to test the effect of a single factor.

▶ Be able to design an experiment in which blocking is used to reduce variation.

▶ Know how to use graphical displays to compare responses for different treatment groups. Understand that you should *never* proceed with any other analysis of a designed experiment without first looking at boxplots or other graphical displays.

TELL

▶ Know how to report the results of an observational study. Identify the subjects, how the data were gathered, and any potential biases or flaws you may be aware of. Identify the factors known and those that might have been revealed by the study.

▶ Know how to compare the responses in different treatment groups to assess whether the differences are larger than could be reasonably expected from ordinary sampling variability.

▶ Know how to report the results of an experiment. Tell who the subjects are and how their assignment to treatments was determined. Report how and in what measurement units the response variable was measured.

▶ Understand that your description of an experiment should be sufficient for another researcher to replicate the study with the same methods.

▶ Be able to report on the statistical significance of the result in terms of whether the observed group-to-group differences are larger than could be expected from ordinary sampling variation.

EXPERIMENTS ON THE COMPUTER

Most experiments are analyzed with a statistics package. You should almost always display the results of a comparative experiment with side-by-side boxplots. You may also want to display the means and standard deviations of the treatment groups in a table.

The analyses offered by statistics packages for comparative randomized experiments fall under the general heading of Analysis of Variance, usually abbreviated ANOVA. We'll see this method in Chapters 28 and 29.

EXERCISES

1. **Standardized test scores.** For his Statistics class experiment, researcher J. Gilbert decided to study how parents' income affects children's performance on standardized tests like the SAT. He proposed to collect information from a random sample of test takers and examine the relationship between parental income and SAT score.
   a) Is this an experiment? If not, what kind of study is it?
   b) If there is a relationship between parental income and SAT score, why can't we conclude that differences in score are caused by differences in parental income?

2. **Heart attacks and height.** Researchers who examined health records of thousands of males found that men who died of myocardial infarction (heart attack) tended to be shorter than men who did not.
   a) Is this an experiment? If not, what kind of study is it?
   b) Is it correct to conclude that shorter men are at higher risk for heart attack? Explain.

3. **MS and vitamin D.** Multiple sclerosis (MS) is an autoimmune disease that strikes more often the farther people live from the equator. Could vitamin D—which most people get from the sun's ultraviolet rays—be a factor? Researchers compared vitamin D levels in blood samples from 150 U.S. military personnel who have developed MS with blood samples of nearly 300 who have not. The samples were taken, on average, five years before the disease was diagnosed. Those with the highest blood vitamin D levels had a 62% lower risk of MS than those with the lowest levels. (The link was only in whites, not in blacks or Hispanics.)
   a) What kind of study was this?
   b) Is that an appropriate choice for investigating this problem? Explain.
   c) Who were the subjects?
   d) What were the variables?

4. **Super Bowl commercials.** When spending large amounts to purchase advertising time, companies want to know what audience they'll reach. In January 2007, a poll asked 1008 American adults whether they planned to watch the upcoming Super Bowl. Men and women were asked separately whether they were looking forward more to the football game or to watching the commercials. Among the men, 16% were planning to watch and were looking forward primarily to the commercials. Among women, 30% were looking forward primarily to the commercials.
   a) Was this a stratified sample or a blocked experiment? Explain.
   b) Was the design of the study appropriate for the advertisers' questions?

5. **Menopause.** Researchers studied the herb black cohosh as a treatment for hot flashes caused by menopause. They randomly assigned 351 women aged 45 to 55 who reported at least two hot flashes a day to one of five groups: (1) black cohosh, (2) a multiherb supplement with black cohosh, (3) the multiherb supplement plus advice to consume more soy foods, (4) estrogen replacement therapy, or (5) a placebo. After a year, only the women given estrogen replacement therapy had symptoms different from those of the placebo group. [*Annals of Internal Medicine* 145:12, 869–897]
   a) What kind of study was this?
   b) Is that an appropriate choice for this problem?
   c) Who were the subjects?
   d) Identify the treatment and response variables.

6. **Honesty.** Coffee stations in offices often just ask users to leave money in a tray to pay for their coffee, but many people cheat. Researchers at Newcastle University replaced the picture of flowers on the wall behind the

coffee station with a picture of staring eyes. They found that the average contribution increased significantly above the well-established standard when people felt they were being watched, even though the eyes were patently not real. (New York *Times* 12/10/06)

a) Was this a survey, an observational study, or an experiment? How can we tell?

b) Identify the variables.

c) What does "increased significantly" mean in a statistical sense?

7–20. *What's the design?* *Read each brief report of statistical research, and identify*

a) whether it was an observational study or an experiment.

If it was an observational study, identify (if possible)

b) whether it was retrospective or prospective.

c) the subjects studied and how they were selected.

d) the parameter of interest.

e) the nature and scope of the conclusion the study can reach.

If it was an experiment, identify (if possible)

b) the subjects studied.

c) the factor(s) in the experiment and the number of levels for each.

d) the number of treatments.

e) the response variable measured.

f) the design (completely randomized, blocked, or matched).

g) whether it was blind (or double-blind).

h) the nature and scope of the conclusion the experiment can reach.

7. Over a 4-month period, among 30 people with bipolar disorder, patients who were given a high dose (10 g/day) of omega-3 fats from fish oil improved more than those given a placebo. (*Archives of General Psychiatry* 56 [1999]: 407)

8. Among a group of disabled women aged 65 and older who were tracked for several years, those who had a vitamin B_{12} deficiency were twice as likely to suffer severe depression as those who did not. (*American Journal of Psychiatry* 157 [2000]: 715)

9. In a test of roughly 200 men and women, those with moderately high blood pressure (averaging 164/89 mm Hg) did worse on tests of memory and reaction time than those with normal blood pressure. (*Hypertension* 36 [2000]: 1079)

10. Is diet or exercise effective in combating insomnia? Some believe that cutting out desserts can help alleviate the problem, while others recommend exercise. Forty volunteers suffering from insomnia agreed to participate in a month-long test. Half were randomly assigned to a special no-desserts diet; the others continued desserts as usual. Half of the people in each of these groups were randomly assigned to an exercise program, while the others did not exercise. Those who ate no desserts and engaged in exercise showed the most improvement.

11. After menopause, some women take supplemental estrogen. There is some concern that if these women also drink alcohol, their estrogen levels will rise too high. Twelve volunteers who were receiving supplemental estrogen were randomly divided into two groups, as were 12 other volunteers not on estrogen. In each case, one group drank an alcoholic beverage, the other a non-alcoholic beverage. An hour later, everyone's estrogen level was checked. Only those on supplemental estrogen who drank alcohol showed a marked increase.

12. Researchers have linked an increase in the incidence of breast cancer in Italy to dioxin released by an industrial accident in 1976. The study identified 981 women who lived near the site of the accident and were under age 40 at the time. Fifteen of the women had developed breast cancer at an unusually young average age of 45. Medical records showed that they had heightened concentrations of dioxin in their blood and that each tenfold increase in dioxin level was associated with a doubling of the risk of breast cancer. (*Science News*, Aug. 3, 2002)

13. In 2002 the journal *Science* reported that a study of women in Finland indicated that having sons shortened the lifespans of mothers by about 34 weeks per son, but that daughters helped to lengthen the mothers' lives. The data came from church records from the period 1640 to 1870.

14. Scientists at a major pharmaceutical firm investigated the effectiveness of an herbal compound to treat the common cold. They exposed each subject to a cold virus, then gave him or her either the herbal compound or a sugar solution known to have no effect on colds. Several days later they assessed the patient's condition, using a cold severity scale ranging from 0 to 5. They found no evidence of benefits associated with the compound.

15. The May 4, 2000, issue of *Science News* reported that, contrary to popular belief, depressed individuals cry no more often in response to sad situations than nondepressed people. Researchers studied 23 men and 48 women with major depression and 9 men and 24 women with no depression. They showed the subjects a sad film about a boy whose father has died, noting whether or not the subjects cried. Women cried more often than men, but there were no significant differences between the depressed and nondepressed groups.

16. Some people who race greyhounds give the dogs large doses of vitamin C in the belief that the dogs will run faster. Investigators at the University of Florida tried three different diets in random order on each of five racing greyhounds. They were surprised to find that when the dogs ate high amounts of vitamin C they ran more slowly. (*Science News*, July 20, 2002)

17. Some people claim they can get relief from migraine headache pain by drinking a large glass of ice water. Researchers plan to enlist several people who suffer from migraines in a test. When a participant experiences a migraine headache, he or she will take a pill that may be a standard pain reliever or a placebo. Half of each group will also drink ice water. Participants will then report the level of pain relief they experience.

18. A dog food company wants to compare a new lower-calorie food with their standard dog food to see if it's effective in helping inactive dogs maintain a healthy weight. They have found several dog owners willing to participate in the trial. The dogs have been classified as small, medium, or large breeds, and the company will

supply some owners of each size of dog with one of the two foods. The owners have agreed not to feed their dogs anything else for a period of 6 months, after which the dogs' weights will be checked.

19. Athletes who had suffered hamstring injuries were randomly assigned to one of two exercise programs. Those who engaged in static stretching returned to sports activity in a mean of 15.2 days faster than those assigned to a program of agility and trunk stabilization exercises. (*Journal of Orthopaedic & Sports Physical Therapy* 34 [March 2004]: 3)

20. Pew Research compared respondents to an ordinary 5-day telephone survey with respondents to a 4-month-long rigorous survey designed to generate the highest possible response rate. They were especially interested in identifying any variables for which those who responded to the ordinary survey were different from those who could be reached only by the rigorous survey.

21. **Omega-3.** Exercise 7 describes an experiment that showed that high doses of omega-3 fats might be of benefit to people with bipolar disorder. The experiment involved a control group of subjects who received a placebo. Why didn't the experimenters just give everyone the omega-3 fats to see if they improved?

22. **Insomnia.** Exercise 10 describes an experiment showing that exercise helped people sleep better. The experiment involved other groups of subjects who didn't exercise. Why didn't the experimenters just have everyone exercise and see if their ability to sleep improved?

23. **Omega-3, revisited.** Exercises 7 and 21 describe an experiment investigating a dietary approach to treating bipolar disorder. Researchers randomly assigned 30 subjects to two treatment groups, one group taking a high dose of omega-3 fats and the other a placebo.
    a) Why was it important to randomize in assigning the subjects to the two groups?
    b) What would be the advantages and disadvantages of using 100 subjects instead of 30?

24. **Insomnia, again.** Exercises 10 and 22 describe an experiment investigating the effectiveness of exercise in combating insomnia. Researchers randomly assigned half of the 40 volunteers to an exercise program.
    a) Why was it important to randomize in deciding who would exercise?
    b) What would be the advantages and disadvantages of using 100 subjects instead of 40?

25. **Omega-3, finis.** Exercises 7, 21, and 23 describe an experiment investigating the effectiveness of omega-3 fats in treating bipolar disorder. Suppose some of the 30 subjects were very active people who walked a lot or got vigorous exercise several times a week, while others tended to be more sedentary, working office jobs and watching a lot of TV. Why might researchers choose to block the subjects by activity level before randomly assigning them to the omega-3 and placebo groups?

26. **Insomnia, at last.** Exercises 10, 22, and 24 describe an experiment investigating the effectiveness of exercise in combating insomnia. Suppose some of the 40 subjects had maintained a healthy weight, but others were quite overweight. Why might researchers choose to block the subjects by weight level before randomly assigning some of each group to the exercise program?

27. **Tomatoes.** Describe a strategy to randomly split the 24 tomato plants into the three groups for the chapter's completely randomized single factor test of OptiGro fertilizer.

28. **Tomatoes II.** The chapter also described a completely randomized two-factor experiment testing OptiGro fertilizer in conjunction with two different routines for watering the plants. Describe a strategy to randomly assign the 24 tomato plants to the six treatments.

29. **Shoes.** A running-shoe manufacturer wants to test the effect of its new sprinting shoe on 100-meter dash times. The company sponsors 5 athletes who are running the 100-meter dash in the 2004 Summer Olympic games. To test the shoe, it has all 5 runners run the 100-meter dash with a competitor's shoe and then again with their new shoe. The company uses the difference in times as the response variable.
    a) Suggest some improvements to the design.
    b) Why might the shoe manufacturer not be able to generalize the results they find to all runners?

30. **Swimsuits.** A swimsuit manufacturer wants to test the speed of its newly designed suit. The company designs an experiment by having 6 randomly selected Olympic swimmers swim as fast as they can with their old swimsuit first and then swim the same event again with the new, expensive swimsuit. The company will use the difference in times as the response variable. Criticize the experiment and point out some of the problems with generalizing the results.

31. **Hamstrings.** Exercise 19 discussed an experiment to see if the time it took athletes with hamstring injuries to be able to return to sports was different depending on which of two exercise programs they engaged in.
    a) Explain why it was important to assign the athletes to the two different treatments randomly.
    b) There was no control group consisting of athletes who did not participate in a special exercise program. Explain the advantage of including such a group.
    c) How might blinding have been used?
    d) One group returned to sports activity in a mean of 37.4 days (SD = 27.6 days) and the other in a mean of 22.2 days (SD = 8.3 days). Do you think this difference is statistically significant? Explain.

32. **Diet and blood pressure.** An experiment showed that subjects fed the DASH diet were able to lower their blood pressure by an average of 6.7 points compared to a group fed a "control diet." All meals were prepared by dieticians.
    a) Why were the subjects randomly assigned to the diets instead of letting people pick what they wanted to eat?
    b) Why were the meals prepared by dieticians?
    c) Why did the researchers need the control group? If the DASH diet group's blood pressure was lower at the end of the experiment than at the beginning, wouldn't that prove the effectiveness of that diet?
    d) What additional information would you want to know in order to decide whether an average reduction in blood pressure of 6.7 points was statistically significant?

33. **Mozart.** Will listening to a Mozart piano sonata make you smarter? In a 1995 study published in the journal *Psychological Science*, Rauscher, Shaw, and Ky reported that when students were given a spatial reasoning section of a standard IQ test, those who listened to Mozart for 10 minutes improved their scores more than those who simply sat quietly.

   a) These researchers said the differences were statistically significant. Explain what that means in context.

   b) Steele, Bass, and Crook tried to replicate the original study. In their study, also published in *Psychological Science* (1999), the subjects were 125 college students who participated in the experiment for course credit. Subjects first took the test. Then they were assigned to one of three groups: listening to a Mozart piano sonata, listening to music by Philip Glass, and sitting for 10 minutes in silence. Three days after the treatments, they were retested. Draw a diagram displaying the design of this experiment.

   c) These boxplots show the differences in score before and after treatment for the three groups. Did the Mozart group show improvement?

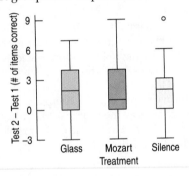

   d) Do you think the results prove that listening to Mozart is beneficial? Explain.

34. Contrast bath treatments use the immersion of an injured limb alternately in water of two contrasting temperatures. Those who use the method claim that it can reduce swelling. Researchers compared three treatments: (1) contrast baths and exercise, (2) contrast baths alone, and (3) exercise alone. (R. G. Janssen, D. A. Schwartz, and P.F. Velleman "A Randomized Controlled Study of Contrast Baths on Patients with Carpal Tunnel Syndrome." *Journal of Hand Therapy*, 2009). They report the following boxplots comparing the change in hand volume after treatment:

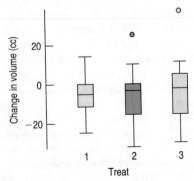

   a) The researchers conclude that the differences were not statistically significant. Explain what that means in context.

   b) The title says that the study was randomized and controlled. Explain what that probably means for this study.

   c) The study did not use a placebo treatment. What was done instead? Do you think that was an appropriate choice? Explain.

35. **Wine.** A 2001 Danish study published in the *Archives of Internal Medicine* casts significant doubt on suggestions that adults who drink wine have higher levels of "good" cholesterol and fewer heart attacks. These researchers followed a group of individuals born at a Copenhagen hospital between 1959 and 1961 for 40 years. Their study found that in this group the adults who drank wine were richer and better educated than those who did not.

   a) What kind of study was this?

   b) It is generally true that people with high levels of education and high socioeconomic status are healthier than others. How does this call into question the supposed health benefits of wine?

   c) Can studies such as these prove causation (that wine helps prevent heart attacks, that drinking wine makes one richer, that being rich helps prevent heart attacks, etc.)? Explain.

36. **Swimming.** Recently, a group of adults who swim regularly for exercise were evaluated for depression. It turned out that these swimmers were less likely to be depressed than the general population. The researchers said the difference was statistically significant.

   a) What does "statistically significant" mean in this context?

   b) Is this an experiment or an observational study? Explain.

   c) News reports claimed this study proved that swimming can prevent depression. Explain why this conclusion is not justified by the study. Include an example of a possible lurking variable.

   d) But perhaps it is true. We wonder if exercise can ward off depression, and whether anaerobic exercise (like weight training) is as effective as aerobic exercise (like swimming). We find 120 volunteers not currently engaged in a regular program of exercise. Design an appropriate experiment.

37. **Dowsing.** Before drilling for water, many rural homeowners hire a dowser (a person who claims to be able to sense the presence of underground water using a forked stick.) Suppose we wish to set up an experiment to test one dowser's ability. We get 20 identical containers, fill some with water, and ask him to tell which ones they are.

   a) How will we randomize this procedure?

   b) The dowser correctly identifies the contents of 12 out of 20 containers. Do you think this level of success is statistically significant? Explain.

   c) How many correct identifications (out of 20) would the dowser have to make to convince you that the forked-stick trick works? Explain.

38. **Healing.** A medical researcher suspects that giving post-surgical patients large doses of vitamin E will speed their recovery times by helping their incisions heal more quickly. Design an experiment to test this conjecture. Be sure to identify the factors, levels, treatments, response variable, and the role of randomization.

39. **Reading.** Some schools teach reading using phonics (the sounds made by letters) and others using whole language (word recognition). Suppose a school district wants to know which method works better. Suggest a design for an appropriate experiment.

40. **Gas mileage.** Do cars get better gas mileage with premium instead of regular unleaded gasoline? It might be possible to test some engines in a laboratory, but we'd rather use real cars and real drivers in real day-to-day driving, so we get 20 volunteers. Design the experiment.

41. **Weekend deaths.** A study published in the *New England Journal of Medicine* (Aug. 2001) suggests that it's dangerous to enter a hospital on a weekend. During a 10-year period, researchers tracked over 4 million emergency admissions to hospitals in Ontario, Canada. Their findings revealed that patients admitted on weekends had a much higher risk of death than those who went on weekdays.
    a) The researchers said the difference in death rates was "statistically significant." Explain in this context what that means.
    b) What kind of study was this? Explain.
    c) If you think you're quite ill on a Saturday, should you wait until Monday to seek medical help? Explain.
    d) Suggest some possible explanations for this troubling finding.

42. **Shingles.** A research doctor has discovered a new ointment that she believes will be more effective than the current medication in the treatment of shingles (a painful skin rash). Eight patients have volunteered to participate in the initial trials of this ointment. You are the statistician hired as a consultant to help design a completely randomized experiment.
    a) Describe how you will conduct this experiment.
    b) Suppose the eight patients' last names start with the letters A to H. Using the random numbers listed below, show which patients you will assign to each treatment. Explain your randomization procedure clearly.

    41098 18329 78458 31685 55259

    c) Can you make this experiment double-blind? How?
    d) The initial experiment revealed that males and females may respond differently to the ointment. Further testing of the drug's effectiveness is now planned, and many patients have volunteered. What changes in your first design, if any, would you make for this second stage of testing?

43. **Beetles.** Hoping to learn how to control crop damage by a certain species of beetle, a researcher plans to test two different pesticides in small plots of corn. A few days after application of the chemicals, he'll check the number of beetle larvae found on each plant. The researcher wants to know whether either pesticide works and whether there is a significant difference in effectiveness between them. Design an appropriate experiment.

44. **SAT prep.** Can special study courses actually help raise SAT scores? One organization says that the 30 students they tutored achieved an average gain of 60 points when they retook the test.
    a) Explain why this does not necessarily prove that the special course caused the scores to go up.
    b) Propose a design for an experiment that could test the effectiveness of the tutorial course.
    c) Suppose you suspect that the tutorial course might be more helpful for students whose initial scores were particularly low. How would this affect your proposed design?

45. **Safety switch.** An industrial machine requires an emergency shutoff switch that must be designed so that it can be easily operated with either hand. Design an experiment to find out whether workers will be able to deactivate the machine as quickly with their left hands as with their right hands. Be sure to explain the role of randomization in your design.

46. **Washing clothes.** A consumer group wants to test the effectiveness of a new "organic" laundry detergent and make recommendations to customers about how to best use the product. They intentionally get grass stains on 30 white T-shirts in order to see how well the detergent will clean them. They want to try the detergent in cold water and in hot water on both the "regular" and "delicates" wash cycles. Design an appropriate experiment, indicating the number of factors, levels, and treatments. Explain the role of randomization in your experiment.

47. **Skydiving, anyone?** A humor piece published in the *British Medical Journal* ("Parachute use to prevent death and major trauma related to gravitational challenge: Systematic review of randomized control trials," Gordon, Smith, and Pell, *BMJ*, 2003:327) notes that we can't tell for sure whether parachutes are safe and effective because there has never been a properly randomized, double-blind, placebo-controlled study of parachute effectiveness in skydiving. (Yes, this is the sort of thing statisticians find funny. . . .) Suppose you were designing such a study:
    a) What is the factor in this experiment?
    b) What experimental units would you propose?[7]
    c) What would serve as a placebo for this study?
    d) What would the treatments be?
    e) What would the response variable be?
    f) What sources of variability would you control?
    g) How would you randomize this "experiment"?
    h) How would you make the experiment double-blind?

---

[7] Don't include your Statistics instructor!

JUST CHECKING

ANSWERS

1. **a)** The factor was type of treatment for peptic ulcer.
   b) The response variable could be a measure of relief from gastric ulcer pain or an evaluation by a physician of the state of the disease.
   c) Treatments would be gastric freezing and some alternative control treatment.
   d) Treatments should be assigned randomly.
   e) No. The website reports "lack of effectiveness," indicating that no large differences in patient healing were noted.

2. **a)** Neither the patients who received the treatment nor the doctor who evaluated them afterward knew what treatment they had received.
   b) The placebo is needed to accomplish blinding. The best alternative would be using body-temperature liquid rather than the freezing liquid.
   c) The researchers should block the subjects by the length of time they had had the ulcer, then randomly assign subjects in each block to the freezing and placebo groups.

PART **III**

PART REVIEW

Gathering Data

Quick Review

Before you can make a boxplot, calculate a mean, describe a distribution, or fit a line, you must have meaningful data to work with. Getting good data is essential to any investigation. No amount of clever analysis can make up for badly collected data. Here's a brief summary of the key concepts and skills:

▶ The way you gather data depends both on what you want to discover and on what is practical.

▶ To get some insight into what might happen in a real situation, model it with a **simulation** using random numbers.

▶ To answer questions about a target population, collect information from a sample with a **survey** or poll.

- Choose the sample randomly. Random sampling designs include simple, stratified, systematic, cluster, and multistage.
- A simple random sample draws without restriction from the entire target population.
- When there are subgroups within the population that may respond differently, use a stratified sample.
- Avoid bias, a systematic distortion of the results. Sample designs that allow undercoverage or

response bias and designs such as voluntary response or convenience samples don't faithfully represent the population.

- Samples will naturally vary one from another. This sample-to-sample variation is called sampling error. Each sample only approximates the target population.

▶ **Observational studies** collect information from a sample drawn from a target population.

- Retrospective studies examine existing data. Prospective studies identify subjects in advance, then follow them to collect data as the data are created, perhaps over many years.
- Observational studies can spot associations between variables but cannot establish cause and effect. It's impossible to eliminate the possibility of lurking or confounding variables.

▶ To see how different treatments influence a response variable, design an **experiment**.

- Assign subjects to treatments randomly. If you don't assign treatments randomly, your experiment is not likely to yield valid results.

- Control known sources of variation as much as possible. Reduce variation that cannot be controlled by using blocking, if possible.
- Replicate the experiment, assigning several subjects to each treatment level.
- If possible, replicate the entire experiment with an entirely different collection of subjects.

- A well-designed experiment can provide evidence that changes in the factors cause changes in the response variable.

Now for more opportunities to review these concepts and skills . . .

REVIEW EXERCISES

1–18. *What design?* Analyze the design of each research example reported. Is it a sample survey, an observational study, or an experiment? If a sample, what are the population, the parameter of interest, and the sampling procedure? If an observational study, was it retrospective or prospective? If an experiment, describe the factors, treatments, randomization, response variable, and any blocking, matching, or blinding that may be present. In each, what kind of conclusions can be reached?

1. Researchers identified 242 children in the Cleveland area who had been born prematurely (at about 29 weeks). They examined these children at age 8 and again at age 20, comparing them to another group of 233 children not born prematurely. Their report, published in the *New England Journal of Medicine,* said the "preemies" engaged in significantly less risky behavior than the others. Differences showed up in the use of alcohol and marijuana, conviction of crimes, and teenage pregnancy.

2. The journal *Circulation* reported that among 1900 people who had heart attacks, those who drank an average of 19 cups of tea a week were 44% more likely than nondrinkers to survive at least 3 years after the attack.

3. Researchers at the Purina Pet Institute studied Labrador retrievers for evidence of a relationship between diet and longevity. At 8 weeks of age, 2 puppies of the same sex and weight were randomly assigned to one of two groups—a total of 48 dogs in all. One group was allowed to eat all they wanted, while the other group was fed a diet about 25% lower in calories. The median lifespan of dogs fed the restricted diet was 22 months longer than that of other dogs. (*Science News* 161, no. 19)

4. The radioactive gas radon, found in some homes, poses a health risk to residents. To assess the level of contamination in their area, a county health department wants to test a few homes. If the risk seems high, they will publicize the results to emphasize the need for home testing. Officials plan to use the local property tax list to randomly choose 25 homes from various areas of the county.

5. Almost 90,000 women participated in a 16-year study of the role of the vitamin folate in preventing colon cancer. Some of the women had family histories of colon cancer in close relatives. In this at-risk group, the incidence of colon cancer was cut in half among those who maintained a high folate intake. No such difference was observed in those with no family-based risk. (*Science News,* Feb. 9, 2002)

6. In the journal *Science,* a research team reported that plants in southern England are flowering earlier in the spring. Records of the first flowering dates for 385 species over a period of 47 years indicate that flowering has advanced an average of 15 days per decade, an indication of climate warming, according to the authors.

7. Fireworks manufacturers face a dilemma. They must be sure that the rockets work properly, but test-firing a rocket essentially destroys it. On the other hand, not testing the product leaves open the danger that they sell a bunch of duds, leading to unhappy customers and loss of future sales. The solution, of course, is to test a few of the rockets produced each day, assuming that if those tested work properly, the others are ready for sale.

8. Can makeup damage fetal development? Many cosmetics contain a class of chemicals called phthalates. Studies that exposed some laboratory animals to these chemicals found a heightened incidence of damage to male reproductive systems. Since traces of phthalates are found in the urine of women who use beauty products, there is growing concern that they may present a risk to male fetuses. (*Science News,* July 20, 2002)

9. Can long-term exposure to strong electromagnetic fields cause cancer? Researchers in Italy tracked down 13 years of medical records for people living near Vatican Radio's powerful broadcast antennas. A disproportionate share of the leukemia cases occurred among men and children who lived within 6 kilometers of the antennas. (*Science News,* July 20, 2002)

10. Some doctors have expressed concern that men who have vasectomies seemed more likely to develop prostate cancer. Medical researchers used a national cancer registry to identify 923 men who had had prostate cancer and 1224 men of similar ages who had not. Roughly one quarter of the men in each group had undergone a vasectomy, many more than 25 years before the study. The study's authors concluded that there is strong evidence that having the operation presents no long-term risk for developing prostate cancer. (*Science News,* July 20, 2002)

11. Researchers investigating appetite control as a means of losing weight found that female rats ate less and lost weight after injections of the hormone leptin, while male rats responded better to insulin. (*Science News,* July 20, 2002)

12. An artisan wants to create pottery that has the appearance of age. He prepares several samples of clay with four different glazes and test fires them in a kiln at three different temperature settings.

13. Tests of gene therapy on laboratory rats have raised hopes of stopping the degeneration of tissue that characterizes chronic heart failure. Researchers at the University of California, San Diego, used hamsters with cardiac disease, randomly assigning 30 to receive the gene therapy and leaving the other 28 untreated. Five weeks after treatment the gene therapy group's heart muscles stabilized, while those of the untreated hamsters continued to weaken. (*Science News*, July 27, 2002)

14. Researchers at the University of Bristol (England) investigated reasons why different species of birds begin to sing at different times in the morning. They captured and examined birds of 57 species at seven different sites. They measured the diameter of the birds' eyes and also recorded the time of day at which each species began to sing. These researchers reported a strong relationship between eye diameter and time of singing, saying that birds with bigger eyes tended to sing earlier. (*Science News*, 161, no. 16 [2002])

15. An orange-juice processing plant will accept a shipment of fruit only after several hundred oranges selected from various locations within the truck are carefully inspected. If too many show signs of unsuitability for juice (bruised, rotten, unripe, etc.), the whole truckload is rejected.

16. A soft-drink manufacturer must be sure the bottle caps on the soda are fully sealed and will not come off easily. Inspectors pull a few bottles off the production line at regular intervals and test the caps. If they detect any problems, they will stop the bottling process to adjust or repair the machine that caps the bottles.

17. Physically fit people seem less likely to die of cancer. A report in the May 2002 issue of *Medicine and Science in Sports and Exercise* followed 25,892 men aged 30 to 87 for 10 years. The most physically fit men had a 55% lower risk of death from cancer than the least fit group.

18. Does the use of computer software in Introductory Statistics classes lead to better understanding of the concepts? A professor teaching two sections of Statistics decides to investigate. She teaches both sections using the same lectures and assignments, but gives one class statistics software to help them with their homework. The classes take the same final exam, and graders do not know which students used computers during the semester. The professor is also concerned that students who have had calculus may perform differently from those who have not, so she plans to compare software vs. no-software scores separately for these two groups of students.

19. **Point spread.** When taking bets on sporting events, bookmakers often include a "point spread" that awards the weaker team extra points. In theory this makes the outcome of the bet a toss-up. Suppose a gambler places a $10 bet and picks the winners of five games. If he's right about fewer than three of the games, he loses. If he gets three, four, or all five correct, he's paid $10, $20, and $50,

respectively. Estimate the amount such a bettor might expect to lose over many weeks of gambling.

20. **The lottery.** Many people spend a lot of money trying to win huge jackpots in state lotteries. Let's play a simplified version using only the numbers from 1 to 20. You bet on three numbers. The state picks five winning numbers. If your three are all among the winners, you are rich!
    a) Simulate repeated plays. How long did it take you to win?
    b) In real lotteries, there are many more choices (often 54) and you must match all five winning numbers. Explain how these changes affect your chances of hitting the jackpot.

21. **Everyday randomness.** Aside from casinos, lotteries, and games, there are other situations you encounter in which something is described as "random" in some way. Give three different examples. Describe how randomness is (or is not) achieved in each.

22. **Cell phone risks.** Researchers at the Washington University School of Medicine randomly placed 480 rats into one of three chambers containing radio antennas. One group was exposed to digital cell phone radio waves, the second to analog cell phone waves, and the third group to no radio waves. Two years later the rats were examined for signs of brain tumors. In June 2002 the scientists said that differences among the three groups were not statistically significant.
    a) Is this a study or an experiment? Explain.
    b) Explain in this context what "not statistically significant" means.
    c) Comment on the fact that this research was funded by Motorola, a manufacturer of cell phones.

23. **Tips.** In restaurants, servers rely on tips as a major source of income. Does serving candy after the meal produce larger tips? To find out, two waiters determined randomly whether or not to give candy to 92 dining parties. They recorded the sizes of the tips and reported that guests getting candy tipped an average of 17.8% of the bill, compared with an average tip of only 15.1% from those who got no candy. ("Sweetening the Till: The Use of Candy to Increase Restaurant Tipping." *Journal of Applied Social Psychology* 32, no. 2 [2002]: 300–309)
    a) Was this an experiment or an observational study? Explain.
    b) Is it reasonable to conclude that the candy caused guests to tip more? Explain.
    c) The researchers said the difference was statistically significant. Explain in this context what that means.

24. **Tips, take 2.** In another experiment to see if getting candy after a meal would induce customers to leave a bigger tip, a waitress randomly decided what to do with 80 dining parties. Some parties received no candy, some just one piece, and some two pieces. Others initially got just one piece of candy, and then the waitress suggested that they take another piece. She recorded the tips received, finding that, in general, the more candy, the higher the tip, but the highest tips (23%) came from the parties who got one piece and then were offered more. ("Sweetening the Till: The Use of Candy to Increase

Restaurant Tipping." *Journal of Applied Social Psychology* 32, no. 2 [2002]: 300–309)

a) Diagram this experiment.
b) How many factors are there? How many levels?
c) How many treatments are there?
d) What is the response variable?
e) Did this experiment involve blinding? Double-blinding?
f) In what way might the waitress, perhaps unintentionally, have biased the results?

25. **Cloning.** In September 1998, *USA Weekend* magazine asked, "Should humans be cloned?" Readers were invited to vote "Yes" or "No" by calling one of two different 900 numbers. Based on 38,023 responses, the magazine reported that "9 out of 10 readers oppose cloning."

a) Explain why you think the conclusion is not justified. Describe the types of bias that may be present.
b) Reword the question in a way that you think might create a more positive response.

26. **Laundry.** An experiment to test a new laundry detergent, SparkleKleen, is being conducted by a consumer advocate group. They would like to compare its performance with that of a laboratory standard detergent they have used in previous experiments. They can stain 16 swatches of cloth with 2 tsp of a common staining compound and then use a well-calibrated optical scanner to detect the amount of the stain left after washing. To save time in the experiment, several suggestions have been made. Comment on the possible merits and drawbacks of each one.

a) Since data for the laboratory standard detergent are already available from previous experiments, for this experiment wash all 16 swatches with SparkleKleen, and compare the results with the previous data.
b) Use both detergents with eight separate runs each, but to save time, use only a 10-second wash time with very hot water.
c) To ease bookkeeping, first run all of the standard detergent washes on eight swatches, then run all of the SparkleKleen washes on the other eight swatches.
d) Rather than run the experiment, use data from the company that produced SparkleKleen, and compare them with past data from the standard detergent.

27. **When to stop?** You play a game that involves rolling a die. You can roll as many times as you want, and your score is the total for all the rolls. But . . . if you roll a 6 your score is 0 and your turn is over. What might be a good strategy for a game like this?

a) One of your opponents decides to roll 4 times, then stop (hoping not to get the dreaded 6 before then). Use a simulation to estimate his average score.
b) Another opponent decides to roll until she gets at least 12 points, then stop. Use a simulation to estimate her average score.
c) Propose another strategy that you would use to play this game. Using your strategy, simulate several turns. Do you think you would beat the two opponents?

28. **Rivets.** A company that manufactures rivets believes the shear strength of the rivets they manufacture follows a Normal model with a mean breaking strength of 950 pounds and a standard deviation of 40 pounds.

a) What percentage of rivets selected at random will break when tested under a 900-pound load?
b) You're trying to improve the rivets and want to examine some that fail. Use a simulation to estimate how many rivets you might need to test in order to find three that fail at 900 pounds (or below).

29. **Homecoming.** A college Statistics class conducted a survey concerning community attitudes about the college's large homecoming celebration. That survey drew its sample in the following manner: Telephone numbers were generated at random by selecting one of the local telephone exchanges (first three digits) at random and then generating a random four-digit number to follow the exchange. If a person answered the phone and the call was to a residence, then that person was taken to be the subject for interview. (Undergraduate students and those under voting age were excluded, as was anyone who could not speak English.) Calls were placed until a sample of 200 eligible respondents had been reached.

a) Did every telephone number that could occur in that community have an equal chance of being generated?
b) Did this method of generating telephone numbers result in a simple random sample (SRS) of local residences? Explain.
c) Did this method generate an SRS of local voters? Explain.
d) Is this method unbiased in generating samples of households? Explain.

30. **Youthful appearance.** *Readers' Digest* reported results of several surveys that asked graduate students to examine photographs of men and women and try to guess their ages. Researchers compared these guesses with the number of times the people in the pictures reported having sexual intercourse. It turned out that those who had been more sexually active were judged as looking younger, and that the difference was described as "statistically significant." Psychologist David Weeks, who compiled the research, speculated that lovemaking boosts hormones that "reduce fatty tissue and increase lean muscle, giving a more youthful appearance."

a) What does "statistically significant" mean in this context?
b) Explain in statistical terms why you might be skeptical about Dr. Weeks's conclusion. Propose an alternative explanation for these results.

31. **Smoking and Alzheimer's.** Medical studies indicate that smokers are less likely to develop Alzheimer's disease than people who never smoked.

a) Does this prove that smoking may offer some protection against Alzheimer's? Explain.
b) Offer an alternative explanation for this association.
c) How would you conduct a study to investigate this?

32. **Antacids.** A researcher wants to compare the performance of three types of antacid in volunteers suffering from acid reflux disease. Because men and women may react differently to this medication, the subjects are split into two groups, by sex. Subjects in each group are randomly assigned to take one of the antacids or to take

a sugar pill made to look the same. The subjects will rate their level of discomfort 30 minutes after eating.

a) What kind of design is this?

b) The experiment uses volunteers rather than a random sample of all people suffering from acid reflux disease. Does this make the results invalid? Explain.

c) How may the use of the placebo confound this experiment? Explain.

33. **Sex and violence.** Does the content of a television program affect viewers' memory of the products advertised in commercials? Design an experiment to compare the ability of viewers to recall brand names of items featured in commercials during programs with violent content, sexual content, or neutral content.

34. **Pubs.** In England, a Leeds University researcher said that the local watering hole's welcoming atmosphere helps men get rid of the stresses of modern life and is vital for their psychological well-being. Author of the report, Dr. Colin Gill, said rather than complain, women should encourage men to "pop out for a swift half." "Pub-time allows men to bond with friends and colleagues," he said. "Men need break-out time as much as women and are mentally healthier for it." Gill added that men might feel unfulfilled or empty if they had not been to the pub for a week. The report, commissioned by alcohol-free beer brand Kaliber, surveyed 900 men on their reasons for going to the pub. More than 40% said they went for the conversation, with relaxation and a friendly atmosphere being the other most common reasons. Only 1 in 10 listed alcohol as the overriding reason.

Let's examine this news story from a statistical perspective.

a) What are the W's: *Who, What, When, Where, Why, How?*

b) What population does the researcher think the study applies to?

c) What is the most important thing about the selection process that the article does *not* tell us?

d) How do *you* think the 900 respondents were selected? (Name a method of drawing a sample that is likely to have been used.)

e) Do you think the report that only 10% of respondents listed alcohol as an important reason for going to the pub might be a biased result? Why?

35. **Age and party.** The Gallup Poll conducted a representative telephone survey during the first quarter of 1999. Among its reported results was the following table concerning the preferred political party affiliation of respondents and their ages:

		Party		
		Republican	Democrat	Independent	Total
Age	18–29	241	351	409	1001
	30–49	299	330	370	999
	50–64	282	341	375	998
	65+	279	382	343	1004
	Total	1101	1404	1497	4002

a) What sampling strategy do you think the pollsters used? Explain.

b) What percentage of the people surveyed were Democrats?

c) Do you think this is a good estimate of the percentage of voters in the United States who are registered Democrats? Why or why not?

d) In creating this sample design, what question do you think the pollsters were trying to answer?

36. **Bias?** Political analyst Michael Barone has written that "conservatives are more likely than others to refuse to respond to polls, particularly those polls taken by media outlets that conservatives consider biased" (*The Weekly Standard*, March 10, 1997). The Pew Research Foundation tested this assertion by asking the same questions in a national survey run by standard methods and in a more rigorous survey that was a true SRS with careful follow-up to encourage participation. The response rate in the "standard survey" was 42%. The response rate in the "rigorous survey" was 71%.

a) What kind of bias does Barone claim may exist in polls?

b) What is the population for these surveys?

c) On the question of political position, the Pew researchers report the following table:

	Standard Survey	Rigorous Survey
Conservative	37%	35%
Moderate	40%	41%
Liberal	19%	20%

What makes you think these results are incomplete?

d) The Pew researchers report that differences between opinions expressed on the two surveys were not statistically significant. Explain what "not statistically significant" means in this context.

37. **Save the grapes.** Vineyard owners have problems with birds that like to eat the ripening grapes. Some vineyards use scarecrows to try to keep birds away. Others use netting that covers the plants. Owners really would like to know if either method works and, if so, which one is better. One owner has offered to let you use his vineyard this year for an experiment. Propose a design. Carefully indicate how you would set up the experiment, specifying the factor(s) and response variable.

38. **Bats.** It's generally believed that baseball players can hit the ball farther with aluminum bats than with the traditional wooden ones. Is that true? And, if so, how much farther? Players on your local high school baseball team have agreed to help you find out. Design an appropriate experiment.

39. **Knees.** Research reported in the spring of 2002 cast doubt on the effectiveness of arthroscopic knee surgery for patients with arthritis. Patients suffering from arthritis pain who volunteered to participate in the study were randomly divided into groups. One group received

arthroscopic knee surgery. The other group underwent "placebo surgery" during which incisions were made in their knees, but no surgery was actually performed. Follow-up evaluations over a period of 2 years found that differences in the amount of pain relief experienced by the two groups were not statistically significant. (*NEJM* 347:81–88, July 11, 2002)

a) Why did the researchers feel it was necessary to have some of the patients undergo "placebo surgery"?

b) Because patients had to consent to participate in this experiment, the subjects were essentially self-selected—a kind of voluntary response group. Explain why that does not invalidate the findings of the experiment.

c) What does "statistically significant" mean in this context?

40. **NBA draft lottery.** Professional basketball teams hold a "draft" each year in which they get to pick the best available college and high school players. In an effort to promote competition, teams with the worst records get to pick first, theoretically allowing them to add better players. To combat the fear that teams with no chance to make the playoffs might try to get better draft picks by intentionally losing late-season games, the NBA's Board of Governors adopted a weighted lottery system in 1990. Under this system, the 11 teams that did not make the playoffs were eligible for the lottery. The NBA prepared 66 cards, each naming one of the teams. The team with the worst win-loss record was named on 11 of the cards, the second-worst team on 10 cards, and so on, with the team having the best record among the nonplayoff clubs getting only one chance at having the first pick. The cards were mixed, then drawn randomly to determine the order in which the teams could draft players. (Since 1995, 13 teams have been involved in the lottery, using a complicated system with 14 numbered Ping-Pong balls drawn in groups of four.) Suppose there are two exceptional players available in this year's draft and your favorite team had the third-worst record. Use a simulation to find out how likely it is that your team gets to pick first or second. Describe your simulation carefully.

41. **Security.** There are 20 first-class passengers and 120 coach passengers scheduled on a flight. In addition to the usual security screening, 10% of the passengers will be subjected to a more complete search.

a) Describe a sampling strategy to randomly select those to be searched.

b) Here is the first-class passenger list and a set of random digits. Select two passengers to be searched, carefully demonstrating your process.

65436 71127 04879 41516 20451 02227 94769 23593

Bergman	Cox	Fontana	Perl
Bowman	DeLara	Forester	Rabkin
Burkhauser	Delli-Bovi	Frongillo	Roufaiel
Castillo	Dugan	Furnas	Swafford
Clancy	Febo	LePage	Testut

c) Explain how you would use a random number table to select the coach passengers to be searched.

42. **Profiling?** Among the 20 first-class passengers on the flight described in Exercise 41, there were four businessmen from the Middle East. Two of them were the two passengers selected to be searched. They complained of profiling, but the airline claims that the selection was random. What do you think? Support your conclusion with a simulation.

43. **Par 4.** In theory, a golfer playing a par-4 hole tees off, hitting the ball in the fairway, then hits an approach shot onto the green. The first putt (usually long) probably won't go in, but the second putt (usually much shorter) should. Sounds simple enough, but how many strokes might it really take? Use a simulation to estimate a pretty good golfer's score based on these assumptions:

- The tee shot hits the fairway 70% of the time.
- A first approach shot lands on the green 80% of the time from the fairway, but only 40% of the time otherwise.
- Subsequent approach shots land on the green 90% of the time.
- The first putt goes in 20% of the time, and subsequent putts go in 90% of the time.

44. **The back nine.** Use simulations to estimate more golf scores, similar to the procedure in Exercise 43.

a) On a par 3, the golfer hopes the tee shot lands on the green. Assume that the tee shot behaves like the first approach shot described in Exercise 43.

b) On a par 5, the second shot will reach the green 10% of the time and hit the fairway 60% of the time. If it does not hit the green, the golfer must play an approach shot as described in Exercise 43.

c) Create a list of assumptions that describe your golfing ability, and then simulate your score on a few holes. Explain your simulation clearly.

CHAPTER
14

From Randomness to Probability

Where are we going?

Flip a coin. Can you predict the outcome? It's hard to guess the outcome of just one flip because the outcome is random. If it's a fair coin, though, you can predict the *proportion* of heads you're likely to see in the long run.

It's this long-term predictability of randomness that we'll use throughout the rest of the book. To do that, we'll need to talk about the probability of different outcomes and learn some rules for dealing with them.

Early humans saw a world filled with random events. To help them make sense of the chaos around them, they sought out seers, consulted oracles, and read tea leaves. As science developed, we learned to recognize some events as predictable. We can now forecast the change of seasons, tell when eclipses will occur precisely, and even make a reasonably good guess at how warm it will be tomorrow. But many other events are still essentially random. Will the stock market go up or down today? When will the next car pass this corner? And we now know from quantum mechanics that the universe is in some sense random at the most fundamental levels of subatomic particles.

But we have also learned to understand randomness. The surprising fact is that in the long run, even truly random phenomena settle down in a way that's consistent and predictable. It's this property of random phenomena that makes the next steps we're about to take in Statistics possible.

Dealing with Random Phenomena

Every day you drive through the intersection at College and Main. Even though it may seem that the light is never green when you get there, you know this can't really be true. In fact, if you try really hard, you can recall just sailing through the green light once in a while.

What's random here? The light itself is governed by a timer. Its pattern isn't haphazard. In fact, the light may even be red at precisely the same times each day. It's the pattern of *your driving* that is random. No, we're certainly not insinuating that you can't keep the car on the road. At the precision level of the 30 seconds or so that the light spends being red or green, the time you arrive at the light *is random.* Even if you try to leave your house at exactly the same time every day, whether the light is red or green as *you* reach the intersection is a **random phenomenon.**[1]

Is the color of the light completely unpredictable? When you stop to think about it, it's clear that you do expect some kind of *regularity* in your long-run experience. Some *fraction* of the time, the light will be green as you get to the intersection. How can you figure out what that fraction is?

You might record what happens at the intersection each day and graph the *accumulated percentage* of green lights like this:

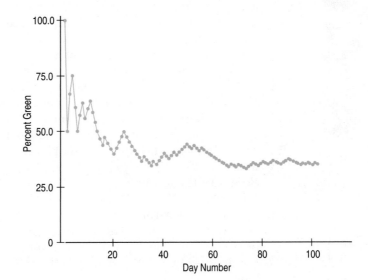

FIGURE 14.1
The overall percentage of times the light is green settles down as you see more outcomes.

Day	Light	% Green
1	Green	100
2	Red	50
3	Green	66.7
4	Green	75
5	Red	60
6	Red	50
⋮	⋮	⋮

The first day you recorded the light, it was green. Then on the next five days, it was red, then green again, then green, red, and red. If you plot the percentage of green lights against days, the graph would start at 100% (because the first time, the light was green, so 1 out of 1, for 100%). Then the next day it was red, so the accumulated percentage dropped to 50% (1 out of 2). The third day it was green again (2 out of 3, or 67% green), then green (3 out of 4, or 75%), then red twice in a row (3 out of 5, for 60% green, and then 3 out of 6, for 50%), and so on. As you collect a new data value for each day, each new outcome becomes a smaller and smaller fraction of the accumulated experience, so, in the long run, the graph settles down. As it settles down, you can see that, in fact, the light is green about 35% of the time.

When talking about random phenomena such as this, we should define our terms. You aren't interested in the traffic light *all* the time. You pull up to the intersection only once a day, so you care about the color of the light only at these particular times.[2] In general, each occasion upon which we observe a random phenomenon is called a **trial.** At each trial, we note the value of the random phenomenon, and call that the trial's **outcome.** (If this language reminds you of Chapter 11, that's *not* unintentional.)

For the traffic light, there are really three possible outcomes: red, yellow, or green. Often we're more interested in a combination of outcomes rather than in the individual ones. When you see the light turn yellow, what do *you* do? If you race through the intersection, then you treat the yellow more like a green light. If you step on the brakes, you treat it more like a red light. Either way, you

A phenomenon consists of trials. Each trial has an outcome. Outcomes combine to make events.

---

[1] If you somehow managed to leave your house at *precisely* the same time every day and there was *no* variation in the time it took you to get to the light, then there wouldn't be any randomness, but that's not very realistic.

[2] Even though the randomness here comes from the uncertainly in our arrival time, we can think of the light itself as showing a color at random.

might want to group the yellow with one or the other. When we combine outcomes like that, the resulting combination is an **event**.[3] We sometimes talk about the collection of *all possible outcomes* and call that event the **sample space**.[4] We'll denote the sample space **S**. (Some books are even fancier and use the Greek letter Ω.) For the traffic light, **S** = {red, green, yellow}.

The Law of Large Numbers

"For even the most stupid of men . . . is convinced that the more observations have been made, the less danger there is of wandering from one's goal."

–Jacob Bernoulli, 1713, discoverer of the LLN

EMPIRICAL PROBABILITY

For any event **A**,
$$P(\mathbf{A}) = \frac{\#\text{ times }\mathbf{A}\text{ occurs}}{\text{total }\#\text{ of trials}}$$
in the long run.

What's the *probability* of a green light at College and Main? Based on the graph, it looks like the relative frequency of green lights settles down to about 35%, so saying that the probability is about 0.35 seems like a reasonable answer. But do random phenomena always behave well enough for this to make sense? Perhaps the relative frequency of an event can bounce back and forth between two values forever, never settling on just one number.

Fortunately, a principle called the **Law of Large Numbers** (LLN) gives us the guarantee we need. It simplifies things if we assume that the events are **independent**.[5] Informally, this means that the outcome of one trial doesn't affect the outcomes of the others. (We'll see a formal definition of independent events in the next chapter.) The LLN says that as the number of independent trials increases, the long-run *relative frequency* of repeated events gets closer and closer to a single value.

Although the LLN wasn't proven until the 18th century, everyone expects the kind of long-run regularity that the Law describes from everyday experience. In fact, the first person to prove the LLN, Jacob Bernoulli, thought it was pretty obvious, too, as his remark quoted in the margin shows.[6]

Because the LLN guarantees that relative frequencies settle down in the long run, we can now officially give a name to the value that they approach. We call it the **probability** of the event. If the relative frequency of green lights at that intersection settles down to 35% in the long run, we say that the probability of encountering a green light is 0.35, and we write $P(\text{green}) = 0.35$. Because this definition is based on repeatedly observing the event's outcome, this definition of probability is often called **empirical probability**.

The Nonexistent Law of Averages

Don't let yourself think that there's a Law of Averages that promises short-term compensation for recent deviations from expected behavior. A belief in such a "Law" can lead to money lost in gambling and to poor business decisions.

Even though the LLN seems natural, it is often misunderstood because the idea of the *long run* is hard to grasp. Many people believe, for example, that an outcome of a random event that hasn't occurred in many trials is "due" to occur. Many gamblers bet on numbers that haven't been seen for a while, mistakenly believing that they're likely to come up sooner. A common term for this is the "Law of Averages." After all, we know that in the long run, the relative frequency will settle down to the probability of that outcome, so now we have some "catching up" to do, right?

Wrong. The Law of Large Numbers says nothing about short-run behavior. Relative frequencies even out *only in the long run*. And, according to the LLN, the long run is *really* long (*infinitely* long, in fact).

The so-called Law of Averages doesn't exist at all. But you'll hear people talk about it as if it does. Is a good hitter in baseball who has struck out the last

---

[3] Each individual outcome is also an event.

[4] Mathematicians like to use the term "space" as a fancy name for a set. Sort of like referring to that closet colleges call a dorm room as "living space." But remember that it's really just the set of all outcomes.

[5] There are stronger forms of the Law that don't require independence, but for our purposes, this form is general enough.

[6] Jacob's reputation was that he was every bit as nasty as this quotation suggests. He and his brother, who was also a mathematician, fought publicly over who had accomplished the most.

"Slump? I ain't in no slump. I just ain't hittin'."

–Yogi Berra

six times *due* for a hit his next time up? If you've been doing particularly well in weekly quizzes in Statistics class, are you *due* for a bad grade? No. This isn't the way random phenomena work. There is *no* Law of Averages for short runs.

The lesson of the LLN is that sequences of random events don't compensate in the *short* run and don't need to do so to get back to the right long-run probability. If the probability of an outcome doesn't change and the events are independent, the probability of any outcome in another trial is *always* what it was, no matter what has happened in other trials.

THE LAW OF AVERAGES IN EVERYDAY LIFE

"Dear Abby: My husband and I just had our eighth child. Another girl, and I am really one disappointed woman. I suppose I should thank God she was healthy, but, Abby, this one was supposed to have been a boy. Even the doctor told me that the law of averages was in our favor 100 to one." (Abigail Van Buren, 1974. Quoted in Karl Smith, *The Nature of Mathematics.* 6th ed. Pacific Grove, CA: Brooks/ Cole, 1991, p. 589)

Coins, Keno, and the Law of Averages You've just flipped a fair coin and seen six heads in a row. Does the coin "owe" you some tails? Suppose you spend that coin and your friend gets it in change. When she starts flipping the coin, should she expect a run of tails? Of course not. Each flip is a new event. The coin can't "remember" what it did in the past, so it can't "owe" any particular outcomes in the future.

Just to see how this works in practice, we ran a simulation of 100,000 flips of a fair coin. We collected 100,000 random numbers, letting the numbers 0 to 4 represent heads and the numbers 5 to 9 represent tails. In our 100,000 "flips," there were 2981 streaks of at least 5 heads. The "Law of Averages" suggests that the next flip after a run of 5 heads should be tails more often to even things out. Actually, the next flip was heads more often than tails: 1550 times to 1431 times. That's 51.9% heads. You can perform a similar simulation easily on a computer. Try it!

Of course, sometimes an apparent drift from what we expect means that the probabilities are, in fact, *not* what we thought. If you get 10 heads in a row, maybe the coin has heads on both sides!

Keno is a simple casino game in which numbers from 1 to 80 are chosen. The numbers, as in most lottery games, are supposed to be equally likely. Payoffs are made depending on how many of those numbers you match on your card. A group of graduate students from a Statistics department decided to take a field trip to Reno. They (*very* discreetly) wrote down the outcomes of the games for a couple of days, then drove back to test whether the numbers were, in fact, equally likely. It turned out that some numbers were *more likely* to come up than others. Rather than bet on the Law of Averages and put their money on the numbers that were "due," the students put their faith in the LLN—and all their (and their friends') money on the numbers that had come up before. After they pocketed more than $50,000, they were escorted off the premises and invited never to show their faces in that casino again.

JUST CHECKING

1. One common proposal for beating the lottery is to note which numbers have come up lately, eliminate those from consideration, and bet on numbers that have not come up for a long time. Proponents of this method argue that in the long run, every number should be selected equally often, so those that haven't come up are due. Explain why this is faulty reasoning.

Modeling Probability

Probability was first studied extensively by a group of French mathematicians who were interested in games of chance.[7] Rather than *experiment* with the games (and risk losing their money), they developed mathematical models of **theoretical probability**. To make things simple (as we usually do when we build models), they started by looking at games in which the different outcomes

---

[7] Ok, gambling.

We often use capital letters— and usually from the beginning of the alphabet—to denote events. We *always* use *P* to denote probability. So,

$$P(\mathbf{A}) = 0.35$$

means "the probability of the event **A** is 0.35."

When being formal, use decimals (or fractions) for the probability values, but sometimes, especially when talking more informally, it's easier to use percentages.

A S

Activity: **Multiple Discrete Outcomes.** The world isn't all heads or tails. Experiment with an event with 4 random alternative outcomes.

---

were equally likely. Fortunately, many games of chance are like that. Any of 52 cards is equally likely to be the next one dealt from a well-shuffled deck. Each face of a die is equally likely to land up (or at least it *should be*).

It's easy to find probabilities for events that are made up of several *equally likely* outcomes. We just count all the outcomes that the event contains. The probability of the event is the number of outcomes in the event divided by the total number of possible outcomes. We can write

$$P(\mathbf{A}) = \frac{\#\text{ outcomes in } \mathbf{A}}{\#\text{ of possible outcomes}}.$$

For example, the probability of drawing a face card (JQK) from a deck is

$$P(\text{face card}) = \frac{\#\text{ face cards}}{\#\text{ cards}} = \frac{12}{52} = \frac{3}{13}.$$

Is that all there is to it? Finding the probability of any event when the outcomes are equally likely is straightforward, but not necessarily easy. It gets hard when the number of outcomes in the event (and in the sample space) gets big. Think about flipping two coins. The sample space is **S** = {HH, HT, TH, TT} and each outcome is equally likely. So, what's the probability of getting *exactly* one head and one tail? Let's call that event **A**. Well, there are two outcomes in the event **A** = {HT, TH} out of the 4 possible equally likely ones in **S**, so $P(\mathbf{A}) = \frac{2}{4}$, or $\frac{1}{2}$.

OK, now flip 100 coins. What's the probability of exactly 67 heads? Well, first, how many outcomes are in the sample space? **S** = {HHHHHHHHHHH . . . H, HH . . . T, . . .} Hmm. A lot. In fact, there are 1,267,650,600,228,229,401,496,703,205,376 different outcomes possible when flipping 100 coins. To answer the question, we'd still have to figure out how many ways there are to get 67 heads. That's coming in Chapter 17; stay tuned!

Don't get trapped into thinking that random events are always equally likely. The chance of winning a lottery—especially lotteries with very large payoffs—is small. Regardless, people continue to buy tickets. In an attempt to understand why, an interviewer asked someone who had just purchased a lottery ticket, "What do you think your chances are of winning the lottery?" The reply was, "Oh, about 50–50." The shocked interviewer asked, "How do you get that?" to which the response was, "Well, the way I figure it, either I win or I don't!"

The moral of this story is that events are *not* always equally likely.

Personal Probability

What's the probability that your grade in this Statistics course will be an **A**? You may be able to come up with a number that seems reasonable. Of course, no matter how confident or depressed you feel about your chances for success, your probability should be between 0 and 1. How did you come up with this probability? Is it an empirical probability? Not unless you plan on taking the course over and over (and over . . .), calculating the proportion of times you get an **A**. And, unless you assume the outcomes are equally likely, it will be hard to find the theoretical probability. But people use probability in a third sense as well.

We use the language of probability in everyday speech to express a degree of uncertainty *without* basing it on long-run relative frequencies or mathematical models. Your personal assessment of your chances of getting an **A** expresses

The line between personal probability and the other two probabilities can be a fuzzy one. When a weather forecaster predicts a 40% probability of rain, is this a personal probability or a relative frequency probability? The claim may be that 40% of the time, when the map looks like this, it has rained (over some period of time). Or the forecaster may be stating a personal opinion that is based on years of experience and reflects a sense of what has happened in the past in similar situations. When you hear a probability stated, try to ascertain what kind of probability is intended.

your uncertainty about the outcome. That uncertainty may be based on how comfortable you're feeling in the course or on your midterm grade, but it can't be based on long-run behavior. We call this third kind of probability a subjective or **personal probability**.

Although personal probabilities may be based on experience, they're not based either on long-run relative frequencies or on equally likely events. So they don't display the kind of consistency that we'll need probabilities to have. For that reason, we'll stick to formally defined probabilities. You should be alert to the difference.

The First Three Rules for Working with Probability

1. Make a picture.
2. Make a picture.
3. Make a picture.

We're dealing with probabilities now, not data, but the three rules don't change. The most common kind of picture to make is called a Venn diagram. We'll use Venn diagrams throughout the rest of this chapter. Even experienced statisticians make Venn diagrams to help them think about probabilities of compound and overlapping events. You should, too.

John Venn (1834-1923) created the Venn diagram. His book on probability, *The Logic of Chance*, was "strikingly original and considerably influenced the development of the theory of Statistics," according to John Maynard Keynes, one of the luminaries of Economics.

Formal Probability

We've been careful to discuss probabilities only for situations in which the outcomes were finite, or even countably infinite. But if the outcomes can take on *any* numerical value at all (we say they are *continuous*), things can get surprising. For example, what is the probability that a randomly selected child will be *exactly* 3 feet tall? Well, if we mean 3.00000 . . . feet, the answer is zero. No randomly selected child—even one whose height would be recorded as 3 feet, will be *exactly* 3 feet tall (to an infinite number of decimal places). But, if you've grown taller than 3 feet, there must have been a time in your life when you actually *were* exactly 3 feet tall, even if only for a second. So this is an outcome with probability 0 that not only has happened—it has happened to *you*.

We've seen another example of this already in Chapter 6 when we worked with the Normal model. We said that the probability of any *specific* value—say, $z = 0.5$—is zero. The model gives a probability for any *interval* of values, such as $0.49 < z < 0.51$. The probability is smaller if we ask for $0.499 < z < 0.501$, and smaller still for $0.49999999 < z < 0.50000001$. Well, you get the idea. Continuous probabilities are useful for the mathematics behind much of what we'll do, but it's easier to deal with probabilities for countable outcomes.

For some people, the phrase "50/50" means something vague like "I don't know" or "whatever." But when we discuss probabilities of outcomes, it takes on the precise meaning of *equally likely*. Speaking vaguely about probabilities will get us into trouble, so whenever we talk about probabilities, we'll need to be precise.[8] And to do that, we'll need to develop some formal rules[9] about how probability works.

1. If the probability is 0, the event doesn't occur, and likewise if it has probability 1, it *always* occurs. Even if you think an event is very unlikely, its probability can't be negative, and even if you're sure it will happen, its probability can't be greater than 1. So we require that

   A probability is a number between 0 and 1.
   For any event A, $0 \leq P(A) \leq 1$.

---

[8] And to be precise, we will be talking only about sample spaces where we can enumerate all the outcomes. Mathematicians call this a countable number of outcomes.

[9] Actually, in mathematical terms, these are axioms—statements that we assume to be true of probability. We'll derive other rules from these in the next chapter.

2. If a random phenomenon has only one possible outcome, it's not very interesting (or very random). So we need to distribute the probabilities among all the outcomes a trial can have. How can we do that so that it makes sense? For example, consider what you're doing as you read this book. The possible outcomes might be

A: You read to the end of this chapter before stopping.
B: You finish this section but stop reading before the end of the chapter.
C: You bail out before the end of this section.

When we assign probabilities to these outcomes, the first thing to be sure of is that we distribute all of the available probability. Something always occurs, so the probability of the entire sample space is 1.

Making this more formal gives the **Probability Assignment Rule**.

The set of all possible outcomes of a trial must have probability 1.

$$P(S) = 1$$

3. Suppose the probability that you get to class on time is 0.8. What's the probability that you don't get to class on time? Yes, it's 0.2. The set of outcomes that are *not* in the event **A** is called the **complement** of **A**, and is denoted $\mathbf{A}^C$. This leads to the **Complement Rule**:

The probability of an event occurring is 1 minus the probability that it doesn't occur.

$$P(\mathbf{A}) = 1 - P(\mathbf{A}^C)$$

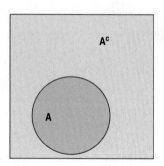

The set **A** and its complement $\mathbf{A}^C$. Together, they make up the entire sample space **S**.

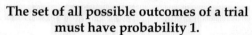
FOR EXAMPLE Applying the Complement Rule

RECAP: We opened the chapter by looking at the traffic light at the corner of College and Main, observing that when we arrive at that intersection, the light is green about 35% of the time.

QUESTION: If $P(\text{green}) = 0.35$, what's the probability the light isn't green when you get to College and Main?

"Not green" is the complement of "green," so $P(\text{not green}) = 1 - P(\text{green})$
$$= 1 - 0.35 = 0.65$$

There's a 65% chance I won't have a green light.

---

4. Suppose the probability that (**A**) a randomly selected student is a sophomore is 0.20, and the probability that (**B**) he or she is a junior is 0.30. What is the probability that the student is *either* a sophomore *or* a junior, written $P(\mathbf{A}$ *or* $\mathbf{B})$? If you guessed 0.50, you've deduced the **Addition Rule,** which says that you can add the probabilities of events that are disjoint.[10] To see whether two events are disjoint, we take them apart into their component outcomes and check whether they have any outcomes in common. **Disjoint**

---

[10] You may see P(**A** *or* **B**) written as $P(\mathbf{A} \cup \mathbf{B})$. The symbol $\cup$ means "union," representing outcomes that are in event **A** *or* event **B** (or both). The symbol $\cap$ means "intersection," representing outcomes that are in both event **A** *and* event **B**. You may sometimes see $P(\mathbf{A}$ *and* $\mathbf{B})$ written as $P(\mathbf{A} \cap \mathbf{B})$.

(or **mutually exclusive**) events have no outcomes in common. The **Addition Rule** states,

> **For two disjoint events A and B, the probability that one *or* the other occurs is the sum of the probabilities of the two events.**
>
> $P(A \ or \ B) = P(A) + P(B)$, **provided that A and B are disjoint.**

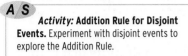

Activity: Addition Rule for Disjoint Events. Experiment with disjoint events to explore the Addition Rule.

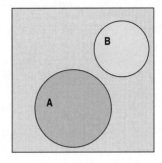

Two disjoint sets, **A** and **B**.

FOR EXAMPLE Applying the Addition Rule

RECAP: When you get to the light at College and Main, it's either red, green, or yellow. We know that $P(\text{green}) = 0.35$.

QUESTION: Suppose we find out that $P(\text{yellow})$ is about 0.04. What's the probability the light is red?

To find the probability that the light is green or yellow, I can use the Addition Rule because these are disjoint events: The light can't be both green and yellow at the same time.

$$P(\text{green OR yellow}) = 0.35 + 0.04 = 0.39$$

Red is the only remaining alternative, and the probabilities must add up to 1, so

$$
\begin{aligned}
P(\text{red}) &= P(\text{not (green OR yellow)}) \\
&= 1 - P(\text{green OR yellow}) \\
&= 1 - 0.39 = 0.61
\end{aligned}
$$

"Baseball is 90% mental. The other half is physical."

–Yogi Berra

Because sample space outcomes are disjoint, we have an easy way to check whether the probabilities we've assigned to the possible outcomes are **legitimate**. The Probability Assignment Rule tells us the sum of the probabilities of all possible outcomes must be exactly 1. No more, no less. For example, if we were told that the probabilities of selecting at random a freshman, sophomore, junior, or senior from all the undergraduates at a school were 0.25, 0.23, 0.22, and 0.20, respectively, we would know that something was wrong. These "probabilities" sum to only 0.90, so this is not a legitimate probability assignment. Either a value is wrong, or we just missed some possible outcomes, like "pre-freshman" or "postgraduate" categories that soak up the remaining 0.10. Similarly, a claim that the probabilities were 0.26, 0.27, 0.29, and 0.30 would be wrong because these "probabilities" sum to more than 1.

But be careful: The Addition Rule doesn't work for events that aren't disjoint. If the probability of owning a smart phone is 0.50 and the probability of owning a computer is 0.90, the probability of owning either a smart phone or a computer may be pretty high, but it is *not* 1.40! Why can't you add probabilities like this? Because these events are not disjoint. You *can* own both. In the next chapter, we'll see how to add probabilities for events like these, but we'll need another rule.

5. Suppose your job requires you to fly from Atlanta to Houston every Monday morning. The airline's website reports that this flight is on time 85% of the

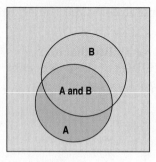

Two sets **A** and **B** that are not disjoint. The event (**A** and **B**) is their intersection.

A S *Activity:* **Multiplication Rule for Independent Events.** Experiment with independent random events to explore the Multiplication Rule.

A S *Activity:* **Probabilities of Compound Events.** The Random tool also lets you experiment with Compound random events to see if they are independent.

time. What's the chance that it will be on time two weeks in a row? That's the same as asking for the probability that your flight is on time this week *and* it's on time again next week. For independent events, the answer is very simple. Remember that independence means that the outcome of one event doesn't influence the outcome of the other. What happens with your flight this week doesn't influence whether it will be on time next week, so it's reasonable to assume that those events are independent. The **Multiplication Rule** says that for independent events, to find the probability that both events occur, we just multiply the probabilities together. Formally,

> **For two independent events A and B, the probability that both**
> **A *and* B occur is the product of the probabilities of the two events.**
> $$P(A \ and \ B) = P(A) \times P(B), \text{ provided that}$$
> **A and B are independent.**

This rule can be extended to more than two independent events. What's the chance of your flight being on time for a month—four Mondays in a row? We can multiply the probabilities of it happening each week:

$$0.85 \times 0.85 \times 0.85 \times 0.85 = 0.522$$

or just over 50–50. Of course, to calculate this probability, we have used the assumption that the four events are independent.

Many Statistics methods require an **Independence Assumption**, but *assuming* independence doesn't make it true. Always *Think* about whether that assumption is reasonable before using the Multiplication Rule.

FOR EXAMPLE   **Applying the Multiplication Rule (and others)**

RECAP: We've determined that the probability that we encounter a green light at the corner of College and Main is 0.35, a yellow light 0.04, and a red light 0.61. Let's think about how many times during your morning commute in the week ahead you might hit a red light there.

QUESTION: What's the probability you find the light red both Monday and Tuesday?

Because the color of the light I see on Monday doesn't influence the color I'll see on Tuesday, these are independent events; I can use the Multiplication Rule:

$$P(\text{red Monday AND red Tuesday}) = P(\text{Red}) \times P(\text{red})$$
$$= (0.61)(0.61)$$
$$= 0.3721$$

There's about a 37% chance I'll hit red lights both Monday and Tuesday mornings.

QUESTION: What's the probability you don't encounter a red light until Wednesday?

For that to happen, I'd have to see green or yellow on Monday, green or yellow on Tuesday, and then red on Wednesday. I can simplify this by thinking of it as not red on Monday and Tuesday and then red on Wednesday.

$$P(\text{not red}) = 1 - P(\text{red}) = 1 - 0.61 = 0.39, \text{ so}$$
$$P(\text{not red Monday AND not red Tuesday AND red Wednesday})$$
$$= P(\text{not red}) \times P(\text{not red}) \times P(\text{red})$$
$$= (0.39)(0.39)(0.61)$$
$$= 0.092781$$

There's about a 9% chance that this week I'll hit my first red light there on Wednesday morning.

QUESTION: What's the probability that you'll have to stop *at least once* during the week?

Having to stop at least once means that I have to stop for the light either 1, 2, 3, 4, or 5 times next week. It's easier to think about the complement: never having to stop at a red light. Having to stop at least once means that I didn't make it through the week with no red lights.

P(having to stop at the light at least once in 5 days)

$= 1 - P(\text{no red lights for 5 days in a row})$

$= 1 - P(\text{not red AND not red AND not red AND not red AND not red})$

$= 1 - (0.39)(0.39)(0.39)(0.39)(0.39)$

$= 1 - 0.0090$

$= 0.991$

I'm not likely to make it through the intersection of College and Main without having to stop sometime this week. There's over a 99% chance I'll hit at least one red light there.

Note that the phrase "at least" is often a tip-off to think about the complement. Something that happens *at least once* <u>does</u> happen. Happening at least once is the complement of not happening at all, and that's easier to find.

JUST CHECKING

2. Opinion polling organizations contact their respondents by telephone. Random telephone numbers are generated, and interviewers try to contact those households. In the 1990s this method could reach about 69% of U.S. households. According to the Pew Research Center for the People and the Press, by 2003 the contact rate had risen to 76%. We can reasonably assume each household's response to be independent of the others. What is the probability that:

In informal English, you may see "some" used to mean "at least one." "What's the probability that some of the eggs in that carton are broken?" means at least one.

a) the interviewer successfully contacts the next household on her list?
b) the interviewer successfully contacts both of the next two households on her list?
c) the interviewer's first successful contact is the third household on the list?
d) the interviewer makes at least one successful contact among the next five households on the list?

STEP-BY-STEP EXAMPLE   Probability

The five rules we've seen can be used in a number of different combinations to answer a surprising number of questions. Let's try one to see how we might go about it.

In 2001, Masterfoods, the manufacturers of M&M's® milk chocolate candies, decided to add another color to the standard color lineup of brown, yellow, red, orange, blue, and green. To decide which color to add, they surveyed kids in nearly every country of the world and asked them to vote among purple, pink, and teal. The global winner was purple! In the United States, 42% of those who voted said purple, 37% said teal, and only 19% said pink. But in Japan the percentages were 38% pink, 36% teal, and only 16% purple. Let's use Japan's percentages to ask some questions:

1. What's the probability that a Japanese M&M's survey respondent selected at random preferred either pink or teal?

2. If we pick two respondents at random, what's the probability that they both selected purple?

3. If we pick three respondents at random, what's the probability that *at least one* preferred purple?

The probability of an event is its long-term relative frequency. It can be determined in several ways: by looking at many replications of an event, by deducing it from equally likely events, or by using some other information. Here, we are told the relative frequencies of the three responses.

Make sure the probabilities are legitimate. Here, they're not. Either there was a mistake, or the other voters must have chosen a color other than the three given. A check of the reports from other countries shows a similar deficit, so probably we're seeing those who had no preference or who wrote in another color.

The M&M's website reports the proportions of Japanese votes by color. These give the probability of selecting a voter who preferred each of the colors:

$$P(\text{pink}) = 0.38$$
$$P(\text{teal}) = 0.36$$
$$P(\text{purple}) = 0.16$$

Each is between 0 and 1, but they don't all add up to 1. The remaining 10% of the voters must have not expressed a preference or written in another color. I'll put them together into "no preference" and add $P(\text{no preference}) = 0.10$.

With this addition, I have a legitimate assignment of probabilities.

Question 1. What's the probability that a Japanese M&M's survey respondent selected at random preferred either pink or teal?

 Plan Decide which rules to use and check the conditions they require.

The events "Pink" and "Teal" are individual outcomes (A respondent can't choose both colors), so they are disjoint. I can apply the Addition Rule.

 Mechanics Show your work.

$$P(\text{pink or teal}) = P(\text{pink}) + P(\text{teal})$$
$$= 0.38 + 0.36 = 0.74$$

 Conclusion Interpret your results in the proper context.

The probability that the respondent said pink or teal is 0.74.

Question 2. If we pick two respondents at random, what's the probability that they both said purple?

 Plan The word "both" suggests we want $P(\mathbf{A}$ *and* $\mathbf{B})$, which calls for the Multiplication Rule. Think about the assumption.

✔ **Independence Assumption:** It's unlikely that the choice made by one respondent affected the choice of the other, so the events seem to be independent. I can use the Multiplication Rule.

SHOW	**Mechanics** Show your work. For both respondents to say purple, each one has to say purple.	$P(\text{both purple})$ $= P(\text{first respondent says purple and}$ $\quad \text{second respondent says purple})$ $= P(\text{first respondent says purple}) \times$ $\quad P(\text{second respondent says purple})$ $= 0.16 \times 0.16 = 0.0256$
TELL	**Conclusion** Interpret your results in the proper context.	The probability that both respondents say purple is 0.0256.

Question 3. If we pick three respondents at random, what's the probability that at least one preferred purple?

THINK	**Plan** The phrase "at least . . ." often flags a question best answered by looking at the complement, and that's the best approach here. The complement of "At least one preferred purple" is "None of them preferred purple." Think about the assumption.	$P(\text{at least one preferred purple})$ $= P(\{\text{none preferred purple}\}^C)$ $= 1 - P(\text{none preferred purple}).$ $P(\text{none preferred purple}) = P(\text{not purple and}$ $\qquad\qquad\qquad\qquad\qquad\text{not purple and not}$ $\qquad\qquad\qquad\qquad\qquad\text{purple}).$ ✔ **Independence Assumption:** These are independent events because they are choices by three random respondents. I can use the Multiplication Rule.
SHOW	**Mechanics** We calculate $P(\text{none purple})$ by using the Multiplication Rule. Then we can use the Complement Rule to get the probability we want.	$P(\text{none preferred purple}) = P(\text{first not purple}) \times$ $\quad P(\text{second not purple}) \times P(\text{third not purple})$ $\qquad\qquad = [P(\text{not purple})]^3.$ $P(\text{not purple}) = 1 - P(\text{purple})$ $\qquad\qquad = 1 - 0.16 = 0.84.$ So $P(\text{none preferred purple}) = (0.84)^3 = 0.5927.$ $P(\text{at least 1 preferred purple})$ $\qquad\qquad = 1 - P(\text{none picked purple})$ $\qquad\qquad = 1 - 0.5927 = 0.4073.$
TELL	**Conclusion** Interpret your results in the proper context.	There's about a 40.7% chance that at least one of the respondents preferred purple.

What Can Go Wrong?

- **Beware of probabilities that don't add up to 1.** To be a legitimate probability assignment, the sum of the probabilities for all possible outcomes must total 1. If the sum is less than 1, you may need to add another category ("other") and assign the remaining probability to that outcome. If the sum is more than 1, check that the outcomes are disjoint. If they're not, then you can't assign probabilities by just counting relative frequencies.

- **Don't add probabilities of events if they're not disjoint.** Events must be disjoint to use the Addition Rule. The probability of being under 80 *or* a female is not the probability of being under 80 *plus* the probability of being female. That sum may be more than 1.

- **Don't multiply probabilities of events if they're not independent.** The probability of selecting a student at random who is over 6'10" tall *and* on the basketball team is *not* the probability the student is over 6'10" tall *times* the probability he's on the basketball team. Knowing that the student is over 6'10" changes the probability of his being on the basketball team. You can't multiply these probabilities. The multiplication of probabilities of events that are not independent is one of the most common errors people make in dealing with probabilities.

- **Don't confuse disjoint and independent.** Disjoint events *can't* be independent. If **A** = {you get an A in this class} and **B** = {you get a B in this class}, **A** and **B** are disjoint. Are they independent? If you find out that **A** is true, does that change the probability of **B**? You bet it does! So they can't be independent. We'll return to this issue in the next chapter.

CONNECTIONS

We saw in the previous three chapters that randomness plays a critical role in gathering data. That fact alone makes it important that we understand how random events behave. The rules and concepts of probability give us a language to talk and think about random phenomena. From here on, randomness will be fundamental to how we think about data, and probabilities will show up in every chapter.

We began thinking about independence back in Chapter 3 when we looked at contingency tables and asked whether the distribution of one variable was the same for each category of another. Then, in Chapter 12, we saw that independence was fundamental to drawing a Simple Random Sample. For computing compound probabilities, we again ask about independence. And we'll continue to think about independence throughout the rest of the book.

Our interest in probability extends back to the start of the book. We've talked about "relative frequencies" often. But—let's be honest—that's just a casual term for probability. For example, you can now rephrase the 68–95–99.7 Rule to talk about the *probability* that a random value selected from a Normal model will fall within 1, 2, or 3 standard deviations of the mean.

Why not just say "probability" from the start? Well, we didn't need any of the formal rules of this chapter (or the next one), so there was no point to weighing down the discussion with those rules. And "relative frequency" is the right intuitive way to think about probability in this course, so you've been thinking right all along.

Keep it up.

WHAT HAVE WE LEARNED?

We've learned that probability is based on long-run relative frequencies. We've thought about the Law of Large Numbers and noted that it speaks only of long-run behavior. Because the long run is a very long time, we need to be careful not to misinterpret the Law of Large Numbers. Even when we've observed a string of heads, we shouldn't expect extra tails in subsequent coin flips.

Also, we've learned some basic rules for combining probabilities of outcomes to find probabilities of more complex events. These include

▶ the Probability Assignment Rule,

▶ the Complement Rule,

▶ the Addition Rule for disjoint events, and

▶ the Multiplication Rule for independent events.

Terms

Random phenomenon	A phenomenon is random if we know what outcomes could happen, but not which particular values will happen (p. 336).
Trial	A single attempt or realization of a random phenomenon (p. 337).
Outcome	The outcome of a trial is the value measured, observed, or reported for an individual instance of that trial (p. 337).
Event	A collection of outcomes. Usually, we identify events so that we can attach probabilities to them. We denote events with bold capital letters such as **A**, **B**, or **C** (p. 338).
Sample Space	The collection of all possible outcome values. The sample space has a probability of 1 (p. 338).
Law of Large Numbers	The Law of Large Numbers states that the long-run *relative frequency* of repeated independent events gets closer and closer to the *true* relative frequency as the number of trials increases (p. 338).
Independence (informally)	Two events are *independent* if learning that one event occurs does not change the probability that the other event occurs (p. 338).
Probability	The probability of an event is a number between 0 and 1 that reports the likelihood of that event's occurrence. We write $P(\mathbf{A})$ for the probability of the event **A** (p. 338).
Empirical probability	When the probability comes from the long-run relative frequency of the event's occurrence, it is an **empirical probability** (p. 338).
Theoretical probability	When the probability comes from a model (such as equally likely outcomes), it is a **theoretical probability** (p. 339).
Personal probability	When the probability is subjective and represents your personal degree of belief, it is a **personal probability** (p. 340).
The Probability Assignment Rule	The probability of the entire sample space must be 1. $P(\mathbf{S}) = 1$ (p. 342).
Complement Rule	The probability of an event occurring is 1 minus the probability that it doesn't occur (p. 342).

$$P(\mathbf{A}) = 1 - P(\mathbf{A}^C)$$

Addition Rule	If **A** and **B** are disjoint events, then the probability of **A** *or* **B** is (p. 343)

$$P(\mathbf{A} \ or \ \mathbf{B}) = P(\mathbf{A}) + P(\mathbf{B}).$$

Disjoint (Mutually exclusive)	Two events are disjoint if they share no outcomes in common. If **A** and **B** are disjoint, then knowing that **A** occurs tells us that **B** cannot occur. Disjoint events are also called "mutually exclusive" (p. 343).
Legitimate probability assignment	An assignment of probabilities to outcomes is legitimate if (p. 343) ▶ each probability is between 0 and 1 (inclusive). ▶ the sum of the probabilities is 1.

Multiplication Rule	If **A** and **B** are independent events, then the probability of **A** *and* **B** is (p. 344)

$$P(A \text{ and } B) = P(A) \times P(B).$$

Independence Assumption	We often require events to be independent. (So you should think about whether this assumption is reasonable) (p. 344).

Skills

▶ Understand that random phenomena are unpredictable in the short term but show long-run regularity.

▶ Be able to recognize random outcomes in a real-world situation.

▶ Know that the relative frequency of a random event settles down to a value called the (empirical) probability. Know that this is guaranteed for independent events by the Law of Large Numbers.

▶ Know the basic definitions and rules of probability.

▶ Recognize when events are disjoint and when events are independent. Understand the difference and that disjoint events cannot be independent.

▶ Be able to use the facts about probability to determine whether an assignment of probabilities is legitimate. Each probability must be a number between 0 and 1, and the sum of the probabilities assigned to all possible outcomes must be 1.

▶ Know how and when to apply the Addition Rule. Know that events must be disjoint for the Addition Rule to apply.

▶ Know how and when to apply the Multiplication Rule. Know that events must be independent for the Multiplication Rule to apply. Be able to use the Multiplication Rule to find probabilities for combinations of independent events.

▶ Know how to use the Complement Rule to make calculating probabilities simpler. Recognize that probabilities of "at least . . ." are likely to be simplified in this way.

▶ Be able to use statements about probability in describing a random phenomenon. You will need this skill soon for making statements about statistical inference.

▶ Know and be able to use the terms "sample space," "disjoint events," and "independent events" correctly.

EXERCISES

1. **Sample spaces.** For each of the following, list the sample space and tell whether you think the events are equally likely:
   a) Toss 2 coins; record the order of heads and tails.
   b) A family has 3 children; record the number of boys.
   c) Flip a coin until you get a head or 3 consecutive tails; record each flip.
   d) Roll two dice; record the larger number.

2. **Sample spaces.** For each of the following, list the sample space and tell whether you think the events are equally likely:
   a) Roll two dice; record the sum of the numbers.
   b) A family has 3 children; record each child's sex in order of birth.
   c) Toss four coins; record the number of tails.
   d) Toss a coin 10 times; record the length of the longest run of heads.

3. **Roulette.** A casino claims that its roulette wheel is truly random. What should that claim mean?

4. **Rain.** The weather reporter on TV makes predictions such as a 25% chance of rain. What do you think is the meaning of such a phrase?

5. **Winter.** Comment on the following quotation:

   "What I think is our best determination is it will be a colder than normal winter," said Pamela Naber Knox, a Wisconsin state climatologist. "I'm basing that on a couple of different things. First, in looking at the past few winters, there has been a lack of really cold weather. Even though we are not supposed to use the law of averages, we are due." (Associated Press, fall 1992, quoted by Schaeffer et al.)

6. **Snow.** After an unusually dry autumn, a radio announcer is heard to say, "Watch out! We'll pay for these sunny days later on this winter." Explain what he's trying to say, and comment on the validity of his reasoning.

7. **Cold streak.** A batter who had failed to get a hit in seven consecutive times at bat then hits a game-winning home run. When talking to reporters afterward, he says he was very confident that last time at bat because he knew he was "due for a hit." Comment on his reasoning.

8. **Crash.** Commercial airplanes have an excellent safety record. Nevertheless, there are crashes occasionally, with the loss of many lives. In the weeks following a crash, airlines often report a drop in the number of passengers, probably because people are afraid to risk flying.
   a) A travel agent suggests that since the law of averages makes it highly unlikely to have two plane crashes within a few weeks of each other, flying soon after a crash is the safest time. What do you think?
   b) If the airline industry proudly announces that it has set a new record for the longest period of safe flights, would you be reluctant to fly? Are the airlines due to have a crash?

9. **Fire insurance.** Insurance companies collect annual payments from homeowners in exchange for paying to rebuild houses that burn down.
   a) Why should you be reluctant to accept a $300 payment from your neighbor to replace his house should it burn down during the coming year?
   b) Why can the insurance company make that offer?

10. **Jackpot.** On January 20, 2000, the International Gaming Technology company issued a press release:

    (*LAS VEGAS, Nev.*)—*Cynthia Jay was smiling ear to ear as she walked into the news conference at The Desert Inn Resort in Las Vegas today, and well she should. Last night, the 37-year-old cocktail waitress won the world's largest slot jackpot—$34,959,458—on a Megabucks machine. She said she had played $27 in the machine when the jackpot hit. Nevada Megabucks has produced 49 major winners in its 14-year history. The top jackpot builds from a base amount of $7 million and can be won with a 3-coin ($3) bet.*
    a) How can the Desert Inn afford to give away millions of dollars on a $3 bet?
    b) Why did the company issue a press release? Wouldn't most businesses want to keep such a huge loss quiet?

11. **Spinner.** The plastic arrow on a spinner for a child's game stops rotating to point at a color that will determine what happens next. Which of the following probability assignments are possible?

	Probabilities of . . .			
	Red	Yellow	Green	Blue
a)	0.25	0.25	0.25	0.25
b)	0.10	0.20	0.30	0.40
c)	0.20	0.30	0.40	0.50
d)	0	0	1.00	0
e)	0.10	0.20	1.20	−1.50

12. **Scratch off.** Many stores run "secret sales": Shoppers receive cards that determine how large a discount they get, but the percentage is revealed by scratching off that black stuff (what *is* that?) only after the purchase has been totaled at the cash register. The store is required to reveal (in the fine print) the distribution of discounts available. Which of these probability assignments are legitimate?

	Probabilities of . . .			
	10% off	20% off	30% off	50% off
a)	0.20	0.20	0.20	0.20
b)	0.50	0.30	0.20	0.10
c)	0.80	0.10	0.05	0.05
d)	0.75	0.25	0.25	−0.25
e)	1.00	0	0	0

13. **Vehicles.** Suppose that 46% of families living in a certain county own a car and 18% own an SUV. The Addition Rule might suggest, then, that 64% of families own either a car or an SUV. What's wrong with that reasoning?

14. **Homes.** Funding for many schools comes from taxes based on assessed values of local properties. People's homes are assessed higher if they have extra features such as garages and swimming pools. Assessment records in a certain school district indicate that 37% of the homes have garages and 3% have swimming pools. The Addition Rule might suggest, then, that 40% of residences have a garage or a pool. What's wrong with that reasoning?

15. **Speeders.** Traffic checks on a certain section of highway suggest that 60% of drivers are speeding there. Since $0.6 \times 0.6 = 0.36$, the Multiplication Rule might suggest that there's a 36% chance that two vehicles in a row are both speeding. What's wrong with that reasoning?

16. **Lefties.** Although it's hard to be definitive in classifying people as right- or left-handed, some studies suggest that about 14% of people are left-handed. Since $0.14 \times 0.14 = 0.0196$, the Multiplication Rule might suggest that there's about a 2% chance that a brother and a sister are both lefties. What's wrong with that reasoning?

17. **College admissions.** For high school students graduating in 2007, college admissions to the nation's most selective schools were the most competitive in memory (*The New York Times*, "A Great Year for Ivy League Schools, but Not So Good for Applicants to Them," April 4, 2007). Harvard accepted about 9% of its applicants, Stanford 10%, and Penn 16%. Jorge has applied to all three. Assuming that he's a typical applicant, he figures that his chances of getting into both Harvard and Stanford must be about 0.09%.
    a) How has he arrived at this conclusion?
    b) What additional assumption is he making?
    c) Do you agree with his conclusion?

18. **College admissions II.** In Exercise 17, we saw that in 2007 Harvard accepted about 9% of its applicants, Stanford 10%, and Penn 16%. Jorge has applied to all three. He figures that his chances of getting into at least one of the three must be about 35%.
    a) How has he arrived at this conclusion?
    b) What assumption is he making?
    c) Do you agree with his conclusion?

19. **Car repairs.** A consumer organization estimates that over a 1-year period 17% of cars will need to be repaired once, 7% will need repairs twice, and 4% will require three or more repairs. What is the probability that a car chosen at random will need
    a) no repairs?
    b) no more than one repair?
    c) some repairs?

20. **Stats projects.** In a large Introductory Statistics lecture hall, the professor reports that 55% of the students enrolled have never taken a Calculus course, 32% have taken only one semester of Calculus, and the rest have taken two or more semesters of Calculus. The professor randomly assigns students to groups of three to work on a project for the course. What is the probability that the first groupmate you meet has studied
    a) two or more semesters of Calculus?
    b) some Calculus?
    c) no more than one semester of Calculus?

21. **More repairs.** Consider again the auto repair rates described in Exercise 19. If you own two cars, what is the probability that
    a) neither will need repair?
    b) both will need repair?
    c) at least one car will need repair?

22. **Another project.** You are assigned to be part of a group of three students from the Intro Stats class described in Exercise 20. What is the probability that of your other two groupmates,
    a) neither has studied Calculus?
    b) both have studied at least one semester of Calculus?
    c) at least one has had more than one semester of Calculus?

23. **Repairs, again.** You used the Multiplication Rule to calculate repair probabilities for your cars in Exercise 21.
    a) What must be true about your cars in order to make that approach valid?
    b) Do you think this assumption is reasonable? Explain.

24. **Final project.** You used the Multiplication Rule to calculate probabilities about the Calculus background of your Statistics groupmates in Exercise 22.
    a) What must be true about the groups in order to make that approach valid?
    b) Do you think this assumption is reasonable? Explain.

25. **Energy 2007.** A Gallup Poll in March 2007 asked 1005 U.S. adults whether increasing domestic energy production or protecting the environment should be given a higher priority. Here are the results:

Response	Number
Increase production	342
Protect environment	583
Equally important	30
No opinion	50
Total	**1005**

If we select a person at random from this sample of 1005 adults,
    a) what is the probability that the person responded "Increase production"?
    b) what is the probability that the person responded "Equally important" or had no opinion?

26. **Failing fathers?** A Pew Research poll in 2007 asked 2020 U.S. adults whether fathers today were doing as good a job of fathering as fathers of 20–30 years ago. Here's how they responded:

Response	Number
Better	424
Same	566
Worse	950
No Opinion	80
Total	**2020**

If we select a respondent at random from this sample of 2020 adults,
    a) what is the probability that the selected person responded "Worse"?
    b) what is the probability that the person responded the "Same" or "Better"?

27. **More energy.** Exercise 25 shows the results of a Gallup Poll about energy. Suppose we select three people at random from this sample.
    a) What is the probability that all three responded "Protect the environment"?
    b) What is the probability that none responded "Equally important"?
    c) What assumption did you make in computing these probabilities?
    d) Explain why you think that assumption is reasonable.

28. **Fathers, revisited.** Consider again the results of the poll about fathering discussed in Exercise 26. If we select two people at random from this sample,
    a) what is the probability that both think fathers are better today?
    b) what is the probability that neither thinks fathers are better today?
    c) what is the probability that one person thinks fathers are better today and the other doesn't?
    d) What assumption did you make in computing these probabilities?
    e) Explain why you think that assumption is reasonable.

29. **Polling.** As mentioned in the chapter, opinion-polling organizations contact their respondents by sampling random telephone numbers. Although interviewers now can reach about 76% of U.S. households, the percentage of those contacted who agree to cooperate with the survey has fallen from 58% in 1997 to only 38% in 2003 (Pew Research Center for the People and the Press). Each household, of course, is independent of the others.
    a) What is the probability that the next household on the list will be contacted but will refuse to cooperate?

b) What is the probability (in 2003) of failing to contact a household or of contacting the household but not getting them to agree to the interview?

c) Show another way to calculate the probability in part b.

30. **Polling, part II.** According to Pew Research, the contact rate (probability of contacting a selected household) was 69% in 1997 and 76% in 2003. However, the cooperation rate (probability of someone at the contacted household agreeing to be interviewed) was 58% in 1997 and dropped to 38% in 2003.

a) What is the probability (in 2003) of obtaining an interview with the next household on the sample list? (To obtain an interview, an interviewer must both contact the household and then get agreement for the interview.)

b) Was it more likely to obtain an interview from a randomly selected household in 1997 or in 2003?

31. **M&M's.** The Masterfoods company says that before the introduction of purple, yellow candies made up 20% of their plain M&M's, red another 20%, and orange, blue, and green each made up 10%. The rest were brown.

a) If you pick an M&M at random, what is the probability that
 1. it is brown?
 2. it is yellow or orange?
 3. it is not green?
 4. it is striped?

b) If you pick three M&M's in a row, what is the probability that
 1. they are all brown?
 2. the third one is the first one that's red?
 3. none are yellow?
 4. at least one is green?

32. **Blood.** The American Red Cross says that about 45% of the U.S. population has Type O blood, 40% Type A, 11% Type B, and the rest Type AB.

a) Someone volunteers to give blood. What is the probability that this donor
 1. has Type AB blood?
 2. has Type A or Type B?
 3. is not Type O?

b) Among four potential donors, what is the probability that
 1. all are Type O?
 2. no one is Type AB?
 3. they are not all Type A?
 4. at least one person is Type B?

33. **Disjoint or independent?** In Exercise 31 you calculated probabilities of getting various M&M's. Some of your answers depended on the assumption that the outcomes described were *disjoint*; that is, they could not both happen at the same time. Other answers depended on the assumption that the events were *independent*; that is, the occurrence of one of them doesn't affect the probability of the other. Do you understand the difference between disjoint and independent?

a) If you draw one M&M, are the events of getting a red one and getting an orange one disjoint, independent, or neither?

b) If you draw two M&M's one after the other, are the events of getting a red on the first and a red on the second disjoint, independent, or neither?

c) Can disjoint events ever be independent? Explain.

34. **Disjoint or independent?** In Exercise 32 you calculated probabilities involving various blood types. Some of your answers depended on the assumption that the outcomes described were *disjoint*; that is, they could not both happen at the same time. Other answers depended on the assumption that the events were *independent*; that is, the occurrence of one of them doesn't affect the probability of the other. Do you understand the difference between disjoint and independent?

a) If you examine one person, are the events that the person is Type A and that the person is Type B disjoint, independent, or neither?

b) If you examine two people, are the events that the first is Type A and the second Type B disjoint, independent, or neither?

c) Can disjoint events ever be independent? Explain.

35. **Dice.** You roll a fair die three times. What is the probability that

a) you roll all 6's?
b) you roll all odd numbers?
c) none of your rolls gets a number divisible by 3?
d) you roll at least one 5?
e) the numbers you roll are not all 5's?

36. **Slot machine.** A slot machine has three wheels that spin independently. Each has 10 equally likely symbols: 4 bars, 3 lemons, 2 cherries, and a bell. If you play, what is the probability that

a) you get 3 lemons?
b) you get no fruit symbols?
c) you get 3 bells (the jackpot)?
d) you get no bells?
e) you get at least one bar (an automatic loser)?

37. **Champion bowler.** A certain bowler can bowl a strike 70% of the time. What's the probability that she

a) goes three consecutive frames without a strike?
b) makes her first strike in the third frame?
c) has at least one strike in the first three frames?
d) bowls a perfect game (12 consecutive strikes)?

38. **The train.** To get to work, a commuter must cross train tracks. The time the train arrives varies slightly from day to day, but the commuter estimates he'll get stopped on about 15% of work days. During a certain 5-day work week, what is the probability that he

a) gets stopped on Monday and again on Tuesday?
b) gets stopped for the first time on Thursday?
c) gets stopped every day?
d) gets stopped at least once during the week?

39. **Voters.** Suppose that in your city 37% of the voters are registered as Democrats, 29% as Republicans, and 11% as members of other parties (Liberal, Right to Life, Green, etc.). Voters not aligned with any official party are termed "Independent." You are conducting a poll

by calling registered voters at random. In your first three calls, what is the probability you talk to

a) all Republicans?
b) no Democrats?
c) at least one Independent?

40. **Religion.** Census reports for a city indicate that 62% of residents classify themselves as Christian, 12% as Jewish, and 16% as members of other religions (Muslims, Buddhists, etc.). The remaining residents classify themselves as nonreligious. A polling organization seeking information about public opinions wants to be sure to talk with people holding a variety of religious views, and makes random phone calls. Among the first four people they call, what is the probability they reach

a) all Christians?
b) no Jews?
c) at least one person who is nonreligious?

41. **Tires.** You bought a new set of four tires from a manufacturer who just announced a recall because 2% of those tires are defective. What is the probability that at least one of yours is defective?

42. **Pepsi.** For a sales promotion, the manufacturer places winning symbols under the caps of 10% of all Pepsi bottles. You buy a six-pack. What is the probability that you win something?

43. **9/11?** On September 11, 2002, the first anniversary of the terrorist attack on the World Trade Center, the New York State Lottery's daily number came up 9–1–1. An interesting coincidence or a cosmic sign?

a) What is the probability that the winning three numbers match the date on any given day?
b) What is the probability that a whole year passes without this happening?
c) What is the probability that the date and winning lottery number match at least once during any year?
d) If every one of the 50 states has a three-digit lottery, what is the probability that at least one of them will come up 9–1–1 on September 11?

44. **Red cards.** You shuffle a deck of cards and then start turning them over one at a time. The first one is red. So is the second. And the third. In fact, you are surprised to get 10 red cards in a row. You start thinking, "The next one is due to be black!"

a) Are you correct in thinking that there's a higher probability that the next card will be black than red? Explain.
b) Is this an example of the Law of Large Numbers? Explain.

JUST CHECKING

ANSWERS

1. The LLN works only in the long run, not in the short run. The random methods for selecting lottery numbers have no memory of previous picks, so there is no change in the probability that a certain number will come up.

2. **a)** 0.76
   b) $0.76(0.76) = 0.5776$
   c) $(1 - 0.76)^2(0.76) = 0.043776$
   d) $1 - (1 - 0.76)^5 = 0.9992$

Probability Rules!

Where are we going?

Is the probability of a car accident the same for all age groups? Insurance companies want to know, so they can set their rates. Is everyone at equal risk for getting the flu next winter? Medical researchers use factors such as age, sex, lifestyle, and family history to estimate the probability for each individual. These are examples of **conditional** probabilities. We'll see how the probability of an event can change when we learn more about the situation at hand. Sometimes the results are surprising.

Pull a bill from your wallet or pocket without looking at it. An outcome of this trial is the bill you select. The sample space is all the bills in circulation: **S** = {$1 bill, $2 bill, $5 bill, $10 bill, $20 bill, $50 bill, $100 bill}.[1] These are *all* the possible outcomes. (In spite of what you may have seen in bank robbery movies, there are no $500 or $1000 bills.)

We can combine the outcomes in different ways to make many different events. For example, the event **A** = {$1, $5, $10} represents selecting a $1, $5, or $10 bill. The event **B** = {a bill that does not have a president on it} is the collection of outcomes (Don't look! Can you name them?): {$10 (Hamilton), $100 (Franklin)}. The event **C** = {enough money to pay for a $12 meal with one bill} is the set of outcomes {$20, $50, $100}.

Notice that these outcomes are not equally likely. You'd no doubt be more surprised (and pleased) to pull out a $100 bill than a $1 bill—it's not very likely, though. You probably carry many more $1 than $100 bills, but without information about the probability of each outcome, we can't calculate the probability of an event.

The probability of the event **C** (getting a bill worth more than $12) is *not* 3/7. There are 7 possible outcomes, and 3 of them exceed $12, but they are *not equally* likely. (Remember the probability that your lottery ticket will win rather than lose still isn't 1/2.)

---

[1] Well, technically, the sample space is all the bills in your pocket. You may be quite sure there isn't a $100 bill in there, but *we* don't know that, so humor us that it's at least *possible* that any legal bill could be there.

The General Addition Rule

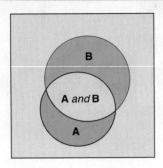

Events **A** and **B** and their intersection.

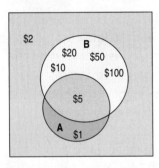

Denominations of bills that are odd (**A**) or that have a building on the reverse side (**B**). The two sets both include the $5 bill, and both exclude the $2 bill.

Now look at the bill in your hand. There are images of famous buildings in the center of the backs of all but two bills in circulation. The $1 bill has the word ONE in the center, and the $2 bill shows the signing of the Declaration of Independence.

What's the probability of randomly selecting **A** = {a bill with an odd-numbered value} *or* **B** = {a bill with a building on the reverse}? We know **A** = {$1, $5} and **B** = {$5, $10, $20, $50, $100}. But $P(\mathbf{A}\text{ or }\mathbf{B})$ is not simply the sum $P(\mathbf{A}) + P(\mathbf{B})$, because the events **A** and **B** are not disjoint. The $5 bill is in both sets. So what can we do? We'll need a new probability rule.

As the diagrams show, we can't use the Addition Rule and add the two probabilities because the events are not disjoint; they overlap. There's an outcome (the $5 bill) in the *intersection* of **A** and **B**. The Venn diagram represents the sample space. Notice that the $2 bill has neither a building nor an odd denomination, so it sits outside both circles.

The $5 bill plays a crucial role here because it is both odd *and* has a building on the reverse. It's in both **A** and **B**, which places it in the *intersection* of the two circles. The reason we can't simply add the probabilities of **A** and **B** is that we'd count the $5 bill twice.

If we did add the two probabilities, we could compensate by *subtracting* out the probability of that $5 bill. So,

P(odd number value or building)

$\quad = P$(odd number value) $+ P$(building) $- P$(odd number value *and* building)

$\quad = P(\$1, \$5) + P(\$5, \$10, \$20, \$50, \$100) - P(\$5).$

This method works in general. We add the probabilities of two events and then subtract out the probability of their intersection. This approach gives us the **General Addition Rule**, which does not require disjoint events:

$$P(\mathbf{A}\text{ or }\mathbf{B}) = P(\mathbf{A}) + P(\mathbf{B}) - P(\mathbf{A}\text{ and }\mathbf{B}).$$

FOR EXAMPLE   Using the General Addition Rule

A survey of college students found that 56% live in a campus residence hall, 62% participate in a campus meal program, and 42% do both.

QUESTION: What's the probability that a randomly selected student either lives or eats on campus?

Let **L** = {student lives on campus} and **M** = {student has a campus meal plan}.

$$P(\text{a student either lives or eats on campus}) = P(\mathbf{L}\text{ or }\mathbf{M})$$
$$= P(\mathbf{L}) + P(\mathbf{M}) - P(\mathbf{L}\text{ and }\mathbf{M})$$
$$= 0.56 + 0.62 - 0.42$$
$$= 0.76$$

There's a 76% chance that a randomly selected college student either lives or eats on campus.

Would you like dessert or coffee? Natural language can be ambiguous. In this question, is the answer one of the two alternatives, or simply "yes"? Must you decide between them, or may you have both? That kind of ambiguity can confuse our probabilities.

Suppose we had been asked a different question: What is the probability that the bill we draw has *either* an odd value *or* a building but *not both*? Which bills are we talking about now? The set we're interested in would be {$1, $10, $20, $50, $100}. We don't include the $5 bill in the set because it has both characteristics.

Why isn't this the same answer as before? The problem is that when we say the word "or," we usually mean *either* one *or* both. We don't usually mean the *exclusive* version of "or" as in, "Would you like the steak *or* the vegetarian entrée?" Ordinarily when we ask for the probability that **A** *or* **B** occurs, we mean **A** *or* **B** or both. And we know *that* probability is $P(A) + P(B) - P(A \text{ and } B)$. The General Addition Rule subtracts the probability of the outcomes in **A** *and* **B** because we've counted those outcomes *twice*. But they're still there.

If we really mean **A** or **B**, but NOT both, we have to get rid of the outcomes in {**A** *and* **B**}. So $P(A \text{ or } B, \text{ but } not \text{ both}) = P(A \text{ or } B) - P(A \text{ and } B) = P(A) + P(B) - 2 \times P(A \text{ and } B)$. Now we've subtracted $P(A \text{ and } B)$ twice–once because we don't want to double-count these events and a second time because we really didn't want to count them at all. Confused? *Make a picture.* It's almost always easier to think about such situations by looking at a Venn diagram.

FOR EXAMPLE   Using Venn Diagrams

RECAP: We return to our survey of college students: 56% live on campus, 62% have a campus meal program, and 42% do both.

QUESTIONS: Based on a Venn diagram, what is the probability that a randomly selected student

a) lives off campus and doesn't have a meal program?
b) lives in a residence hall but doesn't have a meal program?

Let **L** = {student lives on campus} and **M** = {student has a campus meal plan}. In the Venn diagram, the intersection of the circles is $P(L \text{ and } M) = 0.42$. Since $P(L) = 0.56$, $P(L \text{ and } M^C) = 0.56 - 0.42 = 0.14$. Also, $P(L^C \text{ and } M) = 0.62 - 0.42 = 0.20$. Now, $0.14 + 0.42 + 0.20 = 0.76$, leaving $1 - 0.76 = 0.24$ for the region outside both circles.

Now, $P(\text{off campus and no meal program}) = P(L^C \text{ and } M^C) = 0.24$

$P(\text{on campus and no meal program}) = P(L \text{ and } M^C) = 0.14$

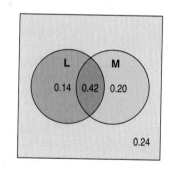

JUST CHECKING

1. Back in Chapter 1 we suggested that you sample some pages of this book at random to see whether they held a graph or other data display. We actually did just that. We drew a representative sample and found the following:

48% of pages had some kind of data display,

27% of pages had an equation, and

7% of pages had both a data display and an equation.

a) Display these results in a Venn diagram.
b) What is the probability that a randomly selected sample page had neither a data display nor an equation?
c) What is the probability that a randomly selected sample page had a data display but no equation?

STEP-BY-STEP EXAMPLE   **Using the General Addition Rule**

Police report that 78% of drivers stopped on suspicion of drunk driving are given a breath test, 36% a blood test, and 22% both tests.

Question: What is the probability that a randomly selected DUI (DWI) suspect is given
1. a test?
2. a blood test or a breath test, but not both?
3. neither test?

Plan Define the events we're interested in. There are no conditions to check; the General Addition Rule works for any events!	Let **A** = {suspect is given a breath test}. Let **B** = {suspect is given a blood test}.

Plot Make a picture, and use the given probabilities to find the probability for each region.

The blue region represents **A** but not **B**. The green intersection region represents **A** *and* **B**. Note that since $P(\mathbf{A}) = 0.78$ and $P(\mathbf{A}\ and\ \mathbf{B}) = 0.22$, the probability of **A** but not **B** must be $0.78 - 0.22 = 0.56$.

The yellow region is **B** but not **A**.

The gray region outside both circles represents the outcome neither **A** nor **B**. All the probabilities must total 1, so you can determine the probability of that region by subtraction.

Now, figure out what you want to know. The probabilities can come from the diagram or a formula. Sometimes translating the words to equations is the trickiest step.

I know that

$$P(\mathbf{A}) = 0.78$$
$$P(\mathbf{B}) = 0.36$$
$$P(\mathbf{A}\ and\ \mathbf{B}) = 0.22$$

So

$$P(\mathbf{A}\ and\ \mathbf{B}^C) = 0.78 - 0.22 = 0.56$$
$$P(\mathbf{B}\ and\ \mathbf{A}^C) = 0.36 - 0.22 = 0.14$$
$$P(\mathbf{A}^C\ and\ \mathbf{B}^C) = 1 - (0.56 + 0.22 + 0.14)$$
$$= 0.08$$

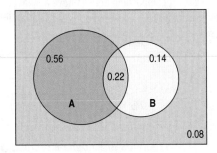

Question 1. What is the probability that the suspect is given a test?

 Mechanics The probability the suspect is given a test is $P(\mathbf{A}\ or\ \mathbf{B})$. We can use the General Addition Rule, or we can add the probabilities seen in the diagram.

$$P(\mathbf{A}\ or\ \mathbf{B}) = P(\mathbf{A}) + P(\mathbf{B}) - P(\mathbf{A}\ and\ \mathbf{B})$$
$$= 0.78 + 0.36 - 0.22$$
$$= 0.92$$

OR

$$P(\mathbf{A}\ or\ \mathbf{B}) = 0.56 + 0.22 + 0.14 = 0.92$$

 Conclusion Don't forget to interpret your result in context.

92% of all suspects are given a test.

Question 2. What is the probability that the suspect gets either a blood test or a breath test but NOT both?

 Mechanics We can use the rule, or just add the appropriate probabilities seen in the Venn diagram.

$P(\mathbf{A} \text{ or } \mathbf{B} \text{ but NOT both}) = P(\mathbf{A} \text{ or } \mathbf{B}) - P(\mathbf{A} \text{ and } \mathbf{B})$
$$= 0.92 - 0.22 = 0.70$$

OR

$P(\mathbf{A} \text{ or } \mathbf{B} \text{ but NOT both})$
$$= P(\mathbf{A} \text{ and } \mathbf{B}^c) + P(\mathbf{B} \text{ and } \mathbf{A}^c)$$
$$= 0.56 + 0.14 = 0.70$$

 Conclusion Interpret your result in context.

70% of the suspects get exactly one of the tests.

Question 3. What is the probability that the suspect gets neither test?

 Mechanics Getting neither test is the complement of getting one or the other. Use the Complement Rule or just notice that "neither test" is represented by the region outside both circles.

$P(\text{neither test}) = 1 - P(\text{either test})$
$$= 1 - P(\mathbf{A} \text{ or } \mathbf{B})$$
$$= 1 - 0.92 = 0.08$$

OR

$P(\mathbf{A}^c \text{ and } \mathbf{B}^c) = 0.08$

Conclusion Interpret your result in context.

Only 8% of the suspects get no test.

It Depends . . .

Two psychologists surveyed 478 children in grades 4, 5, and 6 in elementary schools in Michigan. They stratified their sample, drawing roughly 1/3 from rural, 1/3 from suburban, and 1/3 from urban schools. Among other questions, they asked the students whether their primary goal was to get good grades, to be popular, or to be good at sports. One question of interest was whether boys and girls at this age had similar goals.

Here's a *contingency table* giving counts of the students by their goals and sex:

		Goals			
		Grades	Popular	Sports	Total
Sex	Boy	117	50	60	227
	Girl	130	91	30	251
	Total	247	141	90	478

TABLE 15.1
The distribution of goals for boys and girls.

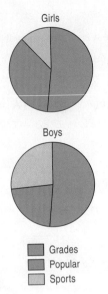

Girls

Boys

- ■ Grades
- ■ Popular
- ■ Sports

FIGURE 15.1
The distribution of goals for boys and girls.

NOTATION ALERT

$P(\mathbf{B}|\mathbf{A})$ is the conditional probability of **B** given **A**.

We looked at contingency tables and graphed *conditional distributions* back in Chapter 3. The pie charts show the *relative frequencies* with which boys and girls named the three goals. It's only a short step from these relative frequencies to probabilities.

Let's focus on this study and make the sample space just the set of these 478 students. If we select a student at random from this study, the probability we select a girl is just the corresponding relative frequency (since we're equally likely to select any of the 478 students). There are 251 girls in the data out of a total of 478, giving a probability of

$$P(\text{girl}) = 251/478 = 0.525$$

The same method works for more complicated events like intersections. For example, what's the probability of selecting a girl whose goal is to be popular? Well, 91 girls named popularity as their goal, so the probability is

$$P(\text{girl } and \text{ popular}) = 91/478 = 0.190$$

The probability of selecting a student whose goal is to excel at sports is

$$P(\text{sports}) = 90/478 = 0.188$$

What if we are given the information that the selected student is a girl? Would that change the probability that the selected student's goal is sports? You bet it would! The pie charts show that girls are much less likely to say their goal is to excel at sports than are boys. When we restrict our focus to girls, we look only at the girls' row of the table. Of the 251 girls, only 30 of them said their goal was to excel at sports.

We write the probability that a selected student wants to excel at sports *given that we have selected a girl* as

$$P(\text{sports}|\text{girl}) = 30/251 = 0.120$$

For boys, we look at the conditional distribution of goals given "boy" shown in the top row of the table. There, of the 227 boys, 60 said their goal was to excel at sports. So, $P(\text{sports}|\text{boy}) = 60/227 = 0.264$, more than twice the girls' probability.

In general, when we want the probability of an event from a *conditional* distribution, we write $P(\mathbf{B}|\mathbf{A})$ and pronounce it "the probability of **B** *given* **A**." A probability that takes into account a given *condition* such as this is called a **conditional probability**.

Let's look at what we did. We worked with the counts, but we could work with the probabilities just as well. There were 30 students who both were girls and had sports as their goal, and there are 251 girls. So we found the probability to be 30/251. To find the probability of the event **B** *given* the event **A**, we restrict our attention to the outcomes in **A**. We then find in what fraction of *those* outcomes **B** also occurred. Formally, we write:

$$P(\mathbf{B}|\mathbf{A}) = \frac{P(\mathbf{A} \text{ and } \mathbf{B})}{P(\mathbf{A})}.$$

Thinking this through, we can see that it's just what we've been doing, but now with probabilities rather than with counts. Look back at the girls for whom sports was the goal. How did we calculate $P(\text{sports}|\text{girl})$?

The rule says to use probabilities. It says to find $P(\mathbf{A} \text{ and } \mathbf{B})/P(\mathbf{A})$. The result is the same whether we use counts or probabilities because the total number in the sample cancels out:

$$\frac{P(\text{sports } and \text{ girl})}{P(\text{girl})} = \frac{30/478}{251/478} = \frac{30}{251}.$$

To use the formula for conditional probability, we're supposed to insist on one restriction. The formula doesn't work if $P(\mathbf{A})$ is 0. After all, we can't be "given" the fact that $\mathbf{A}$ was true if the probability of $\mathbf{A}$ is 0!

Let's take our rule out for a spin. What's the probability that we have selected a girl *given* that the selected student's goal is popularity? Applying the rule, we get

$$P(\text{girl}|\text{popular}) = \frac{P(\text{girl } and \text{ popular})}{P(\text{popular})}$$

$$= \frac{91/478}{141/478} = \frac{91}{141}.$$

FOR EXAMPLE | Finding a Conditional Probability

RECAP: Our survey found that 56% of college students live on campus, 62% have a campus meal program, and 42% do both.

QUESTION: While dining in a campus facility open only to students with meal plans, you meet someone interesting. What is the probability that your new acquaintance lives on campus?

Let **L** = {student lives on campus} and **M** = {student has a campus meal plan}.

P(student lives on campus given that the student has a meal plan) = $P(\mathbf{L}|\mathbf{M})$

$$= \frac{P(\mathbf{L} \text{ and } \mathbf{M})}{P(\mathbf{M})}$$

$$= \frac{0.42}{0.62}$$

$$\approx 0.677$$

There's a probability of about 0.677 that a student with a meal plan lives on campus.

The General Multiplication Rule

Remember the Multiplication Rule for the probability of **A** *and* **B**? It said

$$P(\mathbf{A} \text{ and } \mathbf{B}) = P(\mathbf{A}) \times P(\mathbf{B}) \text{ when } \mathbf{A} \text{ and } \mathbf{B} \text{ are independent.}$$

Now we can write a more general rule that doesn't require independence. In fact, we've *already* written it down. We just need to rearrange the equation a bit.

The equation in the definition for conditional probability contains the probability of **A** *and* **B**. Rewriting the equation gives

$$P(\mathbf{A} \text{ and } \mathbf{B}) = P(\mathbf{A}) \times P(\mathbf{B}|\mathbf{A}).$$

This is a **General Multiplication Rule** for compound events that does not require the events to be independent. Better than that, it even makes sense. The probability that two events, **A** and **B**, *both* occur is the probability that event **A** occurs multiplied by the probability that event **B** *also* occurs—that is, by the probability that event **B** occurs *given* that event **A** occurs.

Of course, there's nothing special about which set we call **A** and which one we call **B**. We should be able to state this the other way around. And indeed we can. It is equally true that

$$P(\mathbf{A} \text{ and } \mathbf{B}) = P(\mathbf{B}) \times P(\mathbf{A}|\mathbf{B}).$$

Independence

> If we had to pick one idea in this chapter that you should understand and remember, it's the definition and meaning of independence. We'll need this idea in every one of the chapters that follow.

Let's return to the question of just what it means for events to be independent. We've said informally that what we mean by independence is that the outcome of one event does not influence the probability of the other. With our new notation for conditional probabilities, we can write a formal definition: Events **A** and **B** are **independent** whenever

$$P(\mathbf{B}|\mathbf{A}) = P(\mathbf{B}).$$

Now we can see that the Multiplication Rule for independent events we saw in Chapter 14 is just a special case of the General Multiplication Rule. The general rule says

$$P(\mathbf{A} \text{ and } \mathbf{B}) = P(\mathbf{A}) \times P(\mathbf{B}|\mathbf{A}).$$

A S *Activity:* **Independence.** Are *Smoking* and *Low Birthweight* independent?

whether the events are independent or not. But when events **A** and **B** are independent, we can write $P(\mathbf{B})$ for $P(\mathbf{B}|\mathbf{A})$ and we get back our simple rule:

$$P(\mathbf{A} \text{ and } \mathbf{B}) = P(\mathbf{A}) \times P(\mathbf{B}).$$

Sometimes people use this statement as the definition of independent events, but we find the other definition more intuitive. Either way, the idea is that for independent events, the probability of one doesn't change when the other occurs.

> In earlier chapters we said informally that two events were independent if learning that one occurred didn't change what you thought about the other occurring. Now we can be more formal. Events **A** and **B** are independent if (and only if) the probability of **A** is the same when we are given that **B** has occurred. That is, $P(\mathbf{A}) = P(\mathbf{A}|\mathbf{B})$.
>
> Although sometimes your intuition is enough, now that we have the formal rule, use it whenever you can.

Is the probability of having good grades as a goal independent of the sex of the responding student? Looks like it might be. We need to check whether

$$P(\text{grades}|\text{girl}) = P(\text{grades})$$

$$\frac{130}{251} = 0.52 \overset{?}{=} \frac{247}{478} = 0.52$$

To two decimal place accuracy, it looks like we can consider choosing good grades as a goal to be independent of sex.

On the other hand, $P(\text{sports})$ is $90/478$, or about 18.8%, but $P(\text{sports}|\text{boy})$ is $60/227 = 26.4\%$. Because these probabilities aren't equal, we can be pretty sure that choosing success in sports as a goal is not independent of the student's sex.

FOR EXAMPLE Checking for Independence

RECAP: Our survey told us that 56% of college students live on campus, 62% have a campus meal program, and 42% do both.

QUESTION: Are living on campus and having a meal plan independent? Are they disjoint?

Let **L** = {student lives on campus} and **M** = {student has a campus meal plan}. If these events are independent, then knowing that a student lives on campus doesn't affect the probability that he or she has a meal plan. I'll check to see if $P(\mathbf{M}|\mathbf{L}) = P(\mathbf{M})$:

$$P(\mathbf{M}|\mathbf{L}) = \frac{P(\mathbf{L} \text{ and } \mathbf{M})}{P(\mathbf{L})}$$

$$= \frac{0.42}{0.56}$$

$$= 0.75, \quad \text{but } P(\mathbf{M}) = 0.62.$$

Because $0.75 \neq 0.62$, the events are not independent; students who live on campus are more likely to have meal plans. Living on campus and having a meal plan are not disjoint either; in fact, 42% of college students do both.

Independent ≠ Disjoint

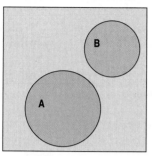

FIGURE 15.2
Because events **A** and **B** are mutually exclusive, learning that **A** happened tells us that **B** didn't. The probability of **B** has changed from whatever it was to zero. So disjoint events **A** and **B** are not independent.

Are disjoint events independent? Both disjoint and independent seem to imply separation and distinctness, but in fact disjoint events *cannot* be independent.[2] Let's see why. Consider the two disjoint events {you get an A in this course} and {you get a B in this course}. They're disjoint because they have no outcomes in common. Suppose you learn that you *did* get an A in the course. Now what is the probability that you got a B? You can't get both grades, so it must be 0.

Think about what that means. Knowing that the first event (getting an A) occurred changed your probability for the second event (down to 0). So these events aren't independent.

Mutually exclusive events can't be independent. They have no outcomes in common, so if one occurs, the other doesn't. A common error is to treat disjoint events as if they were independent, and apply the Multiplication Rule for independent events. Don't make that mistake.

JUST CHECKING

2. The American Association for Public Opinion Research (AAPOR) is an association of about 1600 individuals who share an interest in public opinion and survey research. They report that typically as few as 10% of random phone calls result in a completed interview. Reasons are varied, but some of the most common include no answer, refusal to cooperate, and failure to complete the call.

Which of the following events are independent, which are disjoint, and which are neither independent nor disjoint?

a) **A** = Your telephone number is randomly selected. **B** = You're not at home at dinnertime when they call.

b) **A** = As a selected subject, you complete the interview. **B** = As a selected subject, you refuse to cooperate.

c) **A** = You are not at home when they call at 11 a.m. **B** = You are employed full-time.

Depending on Independence

A S

Video: **Is There a Hot Hand in Basketball?** Most coaches and fans believe that basketball players sometimes get "hot" and make more of their shots. What do the conditional probabilities say?

It's much easier to think about independent events than to deal with conditional probabilities. It seems that most people's natural intuition for probabilities breaks down when it comes to conditional probabilities. Someone may estimate the probability of a compound event by multiplying the probabilities of its component events together without asking seriously whether those probabilities are independent.

For example, experts have assured us that the probability of a major commercial nuclear plant failure is so small that we should not expect such a failure to occur even in a span of hundreds of years. After only a few decades of commercial nuclear power, however, the world has seen two failures (Chernobyl and Three Mile Island). How could the estimates have been so wrong?

---

[2] Well, technically two disjoint events *can* be independent, but only if the probability of one of the events is 0. For practical purposes, though, we can ignore this case. After all, as statisticians we don't anticipate having data about things that never happen.

A S   *Activity:* **Hot Hand Simulation.** Can you tell the difference between real and simulated sequences of basketball shot hits and misses?

One simple part of the failure calculation is to test a particular valve and determine that valves such as this one fail only once in, say, 100 years of normal use. For a coolant failure to occur, several valves must fail. So we need the compound probability, *P*(valve 1 fails *and* valve 2 fails *and* . . .). A simple risk assessment might multiply the small probability of one valve failure together as many times as needed.

But if the valves all came from the same manufacturer, a flaw in one might be found in the others. And maybe when the first fails, it puts additional pressure on the next one in line. In either case, the events aren't independent and so we can't simply multiply the probabilities together.

Whenever you see probabilities multiplied together, stop and ask whether you think they are really independent.

Tables and Conditional Probability

One of the easiest ways to think about conditional probabilities is with contingency tables. We did that earlier in the chapter when we began our discussion. But sometimes we're given probabilities without a table. You can often construct a simple table to correspond to the probabilities.

In the drunk driving example, we were told that 78% of suspect drivers get a breath test, 36% a blood test, and 22% both. That's enough information. Translating percentages to probabilities, what we know looks like this:

		Breath Test		
		Yes	No	Total
Blood Test	Yes	0.22		0.36
	No			
	Total	0.78		1.00

Notice that the 0.78 and 0.36 are *marginal* probabilities and so they go into the *margins*. The 0.22 is the probability of getting both tests—a breath test *and* a blood test—so that's a *joint* probability. Those belong in the interior of the table.

Because the cells of the table show disjoint events, the probabilities always add to the marginal totals going across rows or down columns. So, filling in the rest of the table is quick:

		Breath Test		
		Yes	No	Total
Blood Test	Yes	0.22	0.14	0.36
	No	0.56	0.08	0.64
	Total	0.78	0.22	1.00

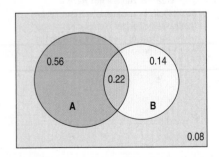

Compare this with the Venn diagram. Notice which entries in the table match up with the sets in this diagram. Whether a Venn diagram or a table is better to use will depend on what you are given and the questions you're being asked. Try both.

STEP-BY-STEP EXAMPLE   Are the events disjoint? Independent?

Let's take another look at the drunk driving situation. Police report that 78% of drivers are given a breath test, 36% a blood test, and 22% both tests.

Questions: 1. Are giving a DUI (DWI) suspect a blood test and a breath test mutually exclusive?
2. Are giving the two tests independent?

 THINK

Plan Define the events we're interested in.

State the given probabilities.

Let **A** = {suspect is given a breath test}.
Let **B** = {suspect is given a blood test}.

I know that $P(A) = 0.78$
$P(B) = 0.36$
$P(A \text{ and } B) = 0.22$

Question 1. Are giving a DUI (DWI) suspect a blood test and a breath test mutually exclusive?

 SHOW

Mechanics Disjoint events cannot *both* happen at the same time, so check to see if $P(A \text{ and } B) = 0$.

$P(A \text{ and } B) = 0.22$. Since some suspects are given both tests, $P(A \text{ and } B) \neq 0$. The events are not mutually exclusive.

 TELL

Conclusion State your conclusion in context.

22% of all suspects get both tests, so a breath test and a blood test are not disjoint events.

Question 2. Are the two tests independent?

 THINK

Plan Make a table.

		Breath Test		
		Yes	No	Total
Blood Test	Yes	0.22	0.14	0.36
	No	0.56	0.08	0.64
	Total	0.78	0.22	1.00

 SHOW

Mechanics Does getting a breath test change the probability of getting a blood test? That is, does $P(B|A) = P(B)$?

Because the two probabilities are *not* the same, the events are not independent.

$P(B|A) = \dfrac{P(A \text{ and } B)}{P(A)} = \dfrac{0.22}{0.78} \approx 0.28$
$P(B) = 0.36$
$P(B|A) \neq P(B)$

TELL

Conclusion Interpret your results in context.

Overall, 36% of the drivers get blood tests, but only 28% of those who get a breath test do. Since suspects who get a breath test are less likely to have a blood test, the two events are not independent.

JUST CHECKING

3. Remember our sample of pages in this book from the earlier Just Checking . . . ?

48% of pages had a data display.

27% of pages had an equation, and

7% of pages had both a data display and an equation.

a) Make a contingency table for the variables *display* and *equation*.
b) What is the probability that a randomly selected sample page with an equation also had a data display?
c) Are having an equation and having a data display disjoint events?
d) Are having an equation and having a data display independent events?

Drawing Without Replacement

Room draw is a process for assigning dormitory rooms to students who live on campus. Sometimes, when students have equal priority, they are randomly assigned to the currently available dorm rooms. When it's time for you and your friend to draw, there are 12 rooms left. Three are in Gold Hall, a very desirable dorm with spacious wood-paneled rooms. Four are in Silver Hall, centrally located, but not quite as desirable. And five are in Wood Hall, a new dorm with cramped rooms, located half a mile from the center of campus on the edge of the woods.

You get to draw first, and then your friend will draw. Naturally, you would both like to score rooms in Gold. What are your chances? In particular, what's the chance that you *both* can get rooms in Gold?

When you go first, the chance that *you* will draw one of the Gold rooms is 3/12. Suppose you do. Now, with you clutching your prized room assignment, what chance does your friend have? At this point there are only 11 rooms left and just 2 left in Gold, so your friend's chance is now 2/11.

Using our notation, we write

$$P(\text{friend draws Gold}|\text{you draw Gold}) = 2/11.$$

The reason the denominator changes is that we draw these rooms *without replacement*. That is, once one is drawn, it doesn't go back into the pool.

We often sample without replacement. When we draw from a very large population, the change in the denominator is too small to worry about. But when there's a small population to draw from, as in this case, we need to take note and adjust the probabilities.

What are the chances that *both* of you will luck out? Well, now we've calculated the two probabilities we need for the General Multiplication Rule, so we can write:

$$P(\text{you draw Gold } and \text{ friend draws Gold})$$
$$= P(\text{you draw Gold}) \times P(\text{friend draws Gold}|\text{you draw Gold})$$
$$= 3/12 \times 2/11 = 1/22 = 0.045$$

$$\xrightarrow{3/12} \text{Gold} \xrightarrow{2/11} \text{Gold}|\text{Gold}$$

In this instance, it doesn't matter who went first, or even if the rooms were drawn simultaneously. Even if the room draw was accomplished by shuffling cards containing the names of the dormitories and then dealing them out to 12 applicants (rather than by each student drawing a room in turn), we can still *think* of the calculation as having taken place in two steps:

That is, one of you has a probability of 3/12 of drawing a Gold room and the other then has a probability of 2/11 of also drawing a Gold room. It doesn't matter whose draw we think of first. The probability changes for the second person nonetheless. The diagram shows this ordering of our thoughts.

Diagramming conditional probabilities leads to a more general way of helping us think with pictures—one that works for calculating conditional probabilities even when they involve different variables.

Tree Diagrams

For men, binge drinking is defined as having five or more drinks in a row, and for women as having four or more drinks in a row. (The difference is because of the average difference in weight.) According to a study by the Harvard School of Public Health (H. Wechsler, G. W. Dowdall, A. Davenport, and W. DeJong, "Binge Drinking on Campus: Results of a National Study"), 44% of college students engage in binge drinking, 37% drink moderately, and 19% abstain entirely. Another study, published in the *American Journal of Health Behavior,* finds that among binge drinkers aged 21 to 34, 17% have been involved in an alcohol-related automobile accident, while among non-bingers of the same age, only 9% have been involved in such accidents.

What's the probability that a randomly selected college student will be a binge drinker who has had an alcohol-related car accident?

To start, we see that the probability of selecting a binge drinker is about 44%. To find the probability of selecting someone who is both a binge drinker and a driver with an alcohol-related accident, we would need to pull out the General Multiplication Rule and multiply the probability of one of the events by the conditional probability of the other given the first.

Or we *could* make a picture. Which would you prefer?

We thought so.

The kind of picture that helps us think through this kind of reasoning is called a **tree diagram**, because it shows sequences of events, like those we had in room draw, as paths that look like branches of a tree. It is a good idea to make a tree diagram almost any time you plan to use the General Multiplication Rule. The number of different paths we can take can get large, so we usually draw the tree starting from the left and growing vine-like across the page, although sometimes you'll see them drawn from the bottom up or top down.

The first branch of our tree separates students according to their drinking habits. We label each branch of the tree with a possible outcome and its corresponding probability.

"Why," said the Dodo, "the best way to explain it is to do it."

—Lewis Carroll

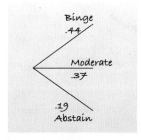

FIGURE 15.3
We can diagram the three outcomes of drinking and indicate their respective probabilities with a simple tree diagram.

Notice that we cover all possible outcomes with the branches. The probabilities add up to one. But we're also interested in car accidents. The probability of having an alcohol-related accident *depends* on one's drinking behavior.

Because the probabilities are *conditional,* we draw the alternatives separately on each branch of the tree:

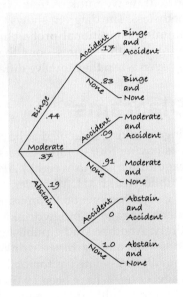

FIGURE 15.4
Extending the tree diagram, we can show both drinking and accident outcomes. The accident probabilities are conditional on the drinking outcomes, and they change depending on which branch we follow. Because we are concerned only with *alcohol-related* accidents, the conditional probability $P(\text{accident}\mid\text{abstinence})$ must be 0.

On each of the second set of branches, we write the possible outcomes associated with having an alcohol-related car accident (having an accident or not) and the associated probability. These probabilities are different because they are *conditional* depending on the student's drinking behavior. (It shouldn't be too surprising that those who binge drink have a higher probability of alcohol-related accidents.) The probabilities add up to one, because given the outcome on the first branch, these outcomes cover all the possibilities. Looking back at the General Multiplication Rule, we can see how the tree depicts the calculation. To find the probability that a randomly selected student will be a binge drinker who has had an alcohol-related car accident, we follow the top branches. The probability of selecting a binger is 0.44. The conditional probability of an accident *given* binge drinking is 0.17. The General Multiplication Rule tells us that to find the *joint* probability of being a binge drinker and having an accident, we multiply these two probabilities together:

$$P(\text{binge }and\text{ accident}) = P(\text{binge}) \times P(\text{accident}\mid\text{binge})$$
$$= 0.44 \times 0.17 = 0.075$$

And we can do the same for each combination of outcomes:

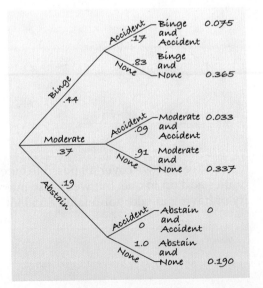

FIGURE 15.5
We can find the probabilities of compound events by multiplying the probabilities along the branch of the tree that leads to the event, just the way the General Multiplication Rule specifies.

The probability of abstaining and having an alcohol-related accident is, of course, zero.

REVERSING THE CONDITIONING

All the outcomes at the far right are disjoint because at each branch of the tree we chose between disjoint alternatives. And they are *all* the possibilities, so the probabilities on the far right must add up to one.

Because the final outcomes are disjoint, we can add up their probabilities to get probabilities for compound events. For example, what's the probability that a selected student has had an alcohol-related car accident? We simply find *all* the outcomes on the far right in which an accident has happened. There are three and we can add their probabilities: $0.075 + 0.033 + 0 = 0.108$—almost an 11% chance.

Reversing the Conditioning

If we know a student has had an alcohol-related accident, what's the probability that the student is a binge drinker? That's an interesting question, but we can't just read it from the tree. The tree gives us $P(\text{accident}|\text{binge})$, but we want $P(\text{binge}|\text{accident})$—conditioning in the other direction. The two probabilities are definitely *not* the same. We have reversed the conditioning.

We may not have the conditional probability we want, but we do know everything we need to know to find it. To find a conditional probability, we need the probability that both events happen divided by the probability that the given event occurs. We have already found the probability of an alcohol-related accident: $0.075 + 0.033 + 0 = 0.108$.

The joint probability that a student is both a binge drinker and someone who's had an alcohol-related accident is found at the top branch: 0.075. We've restricted the *Who* of the problem to the students with alcohol-related accidents, so we divide the two to find the conditional probability:

$$P(\text{binge}|\text{accident}) = \frac{P(\text{binge } and \text{ accident})}{P(\text{accident})}$$

$$= \frac{0.075}{0.108} = 0.694$$

The chance that a student who has an alcohol-related car accident is a binge drinker is more than 69%! As we said, reversing the conditioning is rarely intuitive, but tree diagrams help us keep track of the calculation when there aren't too many alternatives to consider.

STEP-BY-STEP EXAMPLE | **Reversing the Conditioning**

When the authors were in college, there were only three requirements for graduation that were the same for all students: You had to be able to tread water for 2 minutes, you had to learn a foreign language, and you had to be free of tuberculosis. For the last requirement, all freshmen had to take a TB screening test that consisted of a nurse jabbing what looked like a corncob holder into your forearm. You were then expected to report back in 48 hours to have it checked. If you were healthy and TB-free, your arm was supposed to look as though you'd never had the test.

Sometime during the 48 hours, one of us had a reaction. When he finally saw the nurse, his arm was about 50% bigger than normal and a very unhealthy red. Did he have TB? The nurse had said that the test was about 99% effective, so it seemed that the chances must be pretty high that he had TB. How high do you think the chances were? Go ahead and guess. Guess low.

We'll call **TB** the event of actually having TB and **+** the event of testing positive. To start a tree, we need to know $P(\textbf{TB})$, the probability of having TB.[3] We also need to know the conditional

---

[3] This isn't given, so we looked it up. Although TB is a matter of serious concern to public health officials, it is a fairly uncommon disease, with an incidence of about 5 cases per 100,000 in the United States (see *http://www.cdc.gov/tb/default.htm*).

probabilities $P(+|\mathbf{TB})$ and $P(+|\mathbf{TB}^C)$. Diagnostic tests can make two kinds of errors. They can give a positive result for a healthy person (a *false positive*) or a negative result for a sick person (a *false negative*). Being 99% accurate usually means a false-positive rate of 1%. That is, someone who doesn't have the disease has a 1% chance of testing positive anyway. We can write $P(+|\mathbf{TB}^C) = 0.01$.

Since a false negative is more serious (because a sick person might not get treatment), tests are usually constructed to have a lower false-negative rate. We don't know exactly, but let's assume a 0.1% false-negative rate. So only 0.1% of sick people test negative. We can write $P(-|\mathbf{TB}) = 0.001$.

	Plan Define the events we're interested in and their probabilities.	Let $\mathbf{TB} = \{$having TB$\}$ and $\mathbf{TB}^C = \{$no TB$\}$ $+ = \{$testing positive$\}$ and $- = \{$testing negative$\}$
	Figure out what you want to know in terms of the events. Use the notation of conditional probability to write the event whose probability you want to find.	I know that $P(+|\mathbf{TB}^C) = 0.01$ and $P(-|\mathbf{TB}) = 0.001$. I also know that $P(\mathbf{TB}) = 0.00005$. I'm interested in the probability that the author had TB given that he tested positive: $P(\mathbf{TB}|+)$.
	Plot Draw the tree diagram. When probabilities are very small like these are, be careful to keep all the significant digits. To finish the tree we need $P(\mathbf{TB}^C)$, $P(-|\mathbf{TB}^C)$, and $P(-|\mathbf{TB})$. We can find each of these from the Complement Rule: $$P(\mathbf{TB}^C) = 1 - P(\mathbf{TB}) = 0.99995$$ $$P(-|\mathbf{TB}^C) = 1 - P(+|\mathbf{TB}^C)$$ $$= 1 - 0.01 = 0.99 \text{ and}$$ $$P(+|\mathbf{TB}) = 1 - P(-|\mathbf{TB})$$ $$= 1 - 0.01 = 0.999$$	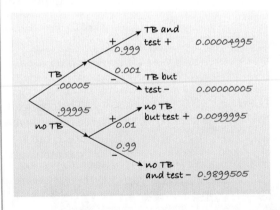
	Mechanics Multiply along the branches to find the probabilities of the four possible outcomes. Check your work by seeing if they total 1. Add up the probabilities corresponding to the condition of interest—in this case, testing positive. We can add the probabilities from the tree twigs that correspond to testing positive because the tree shows disjoint events.	(Check: $0.00004995 + 0.00000005 +$ $0.0099995 + 0.98995050 = 1$) $P(+) = P(\mathbf{TB} \text{ and } +) + P(\mathbf{TB}^C \text{ and } +)$ $P(+) = 0.00004995 + 0.0099995$ $= 0.01004945$

Divide the probability of both events occurring (here, having TB and a positive test) by the probability of satisfying the condition (testing positive).	$$P(TB\,|\,+) = \frac{P(TB\text{ and }+)}{P(+)}$$ $$= \frac{0.00004995}{0.01004945}$$ $$= 0.00497$$
Conclusion Interpret your result in context.	The chance of having TB after you test positive is less than 0.5%.

When we reverse the order of conditioning, we change the *Who* we are concerned with. With events of low probability, the result can be surprising. That's the reason patients who test positive for HIV, for example, are always told to seek medical counseling. They may have only a small chance of actually being infected. That's why global drug or disease testing can have unexpected consequences if people interpret *testing* positive as *being* positive.

Bayes' Rule

The Reverend Thomas Bayes is credited posthumously with the rule that is the foundation of Bayesian Statistics.

When we have $P(\mathbf{A}\,|\,\mathbf{B})$ but want the *reverse* probability $P(\mathbf{B}\,|\,\mathbf{A})$, we need to find $P(\mathbf{A}\text{ and }\mathbf{B})$ and $P(\mathbf{A})$. A tree is often a convenient way of finding these probabilities. It can work even when we have more than two possible events, as we saw in the binge-drinking example. Instead of using the tree, we *could* write the calculation algebraically, showing exactly how we found the quantities that we needed: $P(\mathbf{A}\text{ and }\mathbf{B})$ and $P(\mathbf{A})$. The result is a formula known as Bayes' Rule, after the Reverend Thomas Bayes (1702?–1761), who was credited with the rule after his death, when he could no longer defend himself. Bayes' Rule is quite important in Statistics and is the foundation of an approach to Statistical analysis known as Bayesian Statistics. Although the simple rule deals with two alternative outcomes, the rule can be extended to the situation in which there are more than two branches to the first split of the tree. The principle remains the same (although the math gets more difficult). Bayes' Rule is just a formula for reversing the probability from the conditional probability that you're originally given. Bayes' Rule for two events says that $P(\mathbf{B}\,|\,\mathbf{A}) = \dfrac{P(\mathbf{A}\,|\,\mathbf{B})P(\mathbf{B})}{P(\mathbf{A}\,|\,\mathbf{B})P(\mathbf{B}) + P(\mathbf{A}\,|\,\mathbf{B^{C}})P(\mathbf{B^{C}})}$. Masochists may wish to try it with the TB testing probabilities. (It's easier to just draw the tree, isn't it?)

FOR EXAMPLE   Reversing the Conditioning

A recent Maryland highway safety study found that in 77% of all accidents the driver was wearing a seatbelt. Accident reports indicated that 92% of those drivers escaped serious injury (defined as hospitalization or death), but only 63% of the non-belted drivers were so fortunate.

QUESTION: What's the probability that a driver who was seriously injured wasn't wearing a seatbelt?

Let B = the driver was wearing a seatbelt, and NB = no belt.

Let I = serious injury or death, and OK = not seriously injured.

I know P(B) = 0.77, so P(NB) = 1 − 0.77 = 0.23.

Also, P(OK|B) = 0.92, so P(I|B) = 0.08
and P(OK|B) = 0.63, so P(I|NB) = 0.37

(continued)

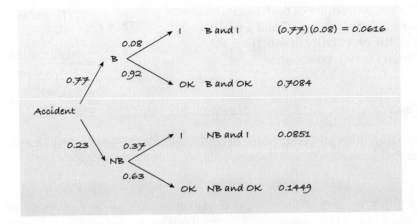

$$P(NB|I) = \frac{P(NB \text{ and } I)}{P(I)} = \frac{0.0851}{0.0616 + 0.0851} = 0.58$$

Even though only 23% of drivers weren't wearing seatbelts, they accounted for 58% of all the deaths and serious injuries.

Just some advice from your friends, the authors: *Please buckle up!* (We want you to finish this course.)

MATH BOX

How do we get from the tree to Bayes' Rule? It's just a matter of combining the things we already know.

Suppose we have two events **A** and **B**. If we want to draw a tree diagram for these two events, we'll need to know some probabilities. When we filled in the tree, we labeled the branch that went to **B** with $P(\mathbf{B})$ and the branch from that node that then went to **A** with $P(\mathbf{A}|\mathbf{B})$.

If we know $P(\mathbf{B})$, $P(\mathbf{A}|\mathbf{B})$, and $P(\mathbf{A}|\mathbf{B}^c)$, we can fill in the other branches as shown below:

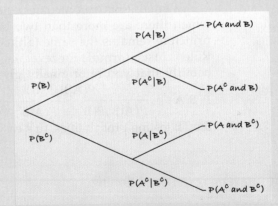

Notice that the probabilities at the end of the branches are joint probabilities where two events have occurred.

What we want is $P(\mathbf{B}|\mathbf{A})$, and we don't have that yet.

Let's start with the definition:

$$P(\mathbf{B}|\mathbf{A}) = \frac{P(\mathbf{A} \text{ and } \mathbf{B})}{P(\mathbf{A})}$$

We have $P(\mathbf{A}\ and\ \mathbf{B})$ from the top branch. We got that by using the General Multiplication Rule:

$$P(\mathbf{A}\ and\ \mathbf{B}) = P(\mathbf{B})P(\mathbf{A}|\mathbf{B}),$$

and we are given both of these probabilities.

What about $P(\mathbf{A})$?

That calls for a simple trick. Think about the set **A**. Whenever **A** occurs, either **B** also occurs, or it doesn't. (We said it was simple.) So we can write **A** as the combination of those times **A** occurs *with* **B** and those times it occurs *without* **B**:

$$\mathbf{A} = \{(\mathbf{A}\ and\ \mathbf{B})\ \text{or}\ (\mathbf{A}\ and\ \mathbf{B}^c)\}.$$

(Check the Venn diagram.)

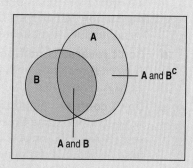

The two sets are disjoint because **B** and **B**^{c} are disjoint, so the Simple Addition Rule applies:
$P(\mathbf{A}) = P(\mathbf{A}\ and\ \mathbf{B}) + P(\mathbf{A}\ and\ \mathbf{B}^c)$

We already know $P(\mathbf{A}\ and\ \mathbf{B})$, and we can get $P(\mathbf{A}\ and\ \mathbf{B}^c)$ in just the same way from the tree: $P(\mathbf{A}\ and\ \mathbf{B}^c) = P(\mathbf{B}^c)P(\mathbf{A}|\mathbf{B}^c)$.

Adding them gives

$$P(\mathbf{A}) = P(\mathbf{A}\ and\ \mathbf{B}) + P(\mathbf{A}\ and\ \mathbf{B}^c) = P(\mathbf{B})P(\mathbf{A}|\mathbf{B}) + P(\mathbf{B}^c)P(\mathbf{A}|\mathbf{B}^c).$$

Putting it all together, we get Bayes' Rule:

$$P(\mathbf{B}|\mathbf{A}) = \frac{P(\mathbf{A}\ and\ \mathbf{B})}{P(\mathbf{A})} = \frac{P(\mathbf{B})P(\mathbf{A}|\mathbf{B})}{P(\mathbf{B})P(\mathbf{A}|\mathbf{B}) + P(\mathbf{B}^c)P(\mathbf{A}|\mathbf{B}^c)}.$$

Nicholas Saunderson (1682-1739) was a blind English mathematician who invented a tactile board to help other blind people do mathematics. And he may have been the true originator of "Bayes' Rule."

Who discovered Bayes' Rule? Stigler's "Law of Eponymy" states that discoveries named for someone (eponyms) are usually named for the wrong person. Steven Stigler, who admits he didn't originate the law, is an expert on the history of Statistics, and he suspected that the law might apply to Bayes' Rule. He looked at the possibility that another candidate—one Nicholas Saunderson—was the real discoverer, not the Reverend Bayes. He assembled historical evidence and compared probabilities that the historical events would have happened *given* that Bayes was the discoverer of the rule, with the corresponding probabilities *given* that Saunderson was the discoverer. Of course, what he really wanted to know were the probabilities that Bayes or Saunderson was the discoverer *given* the historical events. How did he *reverse* the conditional probabilities? He used Bayes' Rule and concluded that, actually, it's more likely that Saunderson is the real originator of the rule.

But that doesn't change our tradition of naming the rule for Bayes and calling the branch of Statistics arising from this approach Bayesian Statistics.

What Can Go Wrong?

- **Don't use a simple probability rule where a general rule is appropriate.** Don't assume independence without reason to believe it. Don't assume that outcomes are disjoint without checking that they are. Remember that the general rules always apply, even when outcomes are in fact independent or disjoint.

- **Don't find probabilities for samples drawn without replacement as if they had been drawn with replacement.** Remember to adjust the denominator of your probabilities. This warning applies only when we draw from small populations or draw a large fraction of a finite population. When the population is very large relative to the sample size, the adjustments make very little difference, and we ignore them.

- **Don't reverse conditioning naïvely.** As we have seen, the probability of **A** *given* **B** may not, and, in general does not, resemble the probability of **B** *given* **A**. The true probability may be counterintuitive.

- **Don't confuse "disjoint" with "independent."** Disjoint events *cannot* happen at the same time. When one happens, you know the other did not, so $P(\mathbf{B}|\mathbf{A}) = 0$. Independent events *must* be able to happen at the same time. When one happens, you know it has no effect on the other, so $P(\mathbf{B}|\mathbf{A}) = P(\mathbf{B})$.

CONNECTIONS

This chapter shows the unintuitive side of probability. If you've been thinking, "My mind doesn't work this way," you're probably right. Humans don't seem to find conditional and compound probabilities natural and often have trouble with them. Even statisticians make mistakes with conditional probability.

Our central connection is to the guiding principle that Statistics is about understanding the world. The events discussed in this chapter are close to the kinds of real-world situations in which understanding probabilities matters. The methods and concepts of this chapter are the tools you need to understand the part of the real world that deals with the outcomes of complex, uncertain events.

WHAT HAVE WE LEARNED?

The last chapter's basic rules of probability are important, but they work only in special cases—when events are disjoint or independent. Now we've learned the more versatile General Addition Rule and General Multiplication Rule. We've also learned about conditional probabilities, and seen that reversing the conditioning can give surprising results.

We've learned the value of Venn diagrams, tables, and tree diagrams to help organize our thinking about probabilities.

Most important, we've learned to think clearly about independence. We've seen how to use conditional probability to determine whether two events are independent and to work with events that are not independent. A sound understanding of independence will be important throughout the rest of this book.

Terms

General Addition Rule	For any two events, **A** and **B**, the probability of **A** *or* **B** is (p. 356).

$$P(\mathbf{A}\ or\ \mathbf{B}) = P(\mathbf{A}) + P(\mathbf{B}) - P(\mathbf{A}\ and\ \mathbf{B}).$$

Conditional probability	$P(\mathbf{B}\|\mathbf{A}) = \dfrac{P(\mathbf{A}\ and\ \mathbf{B})}{P(\mathbf{A})}$
	$P(\mathbf{B}\|\mathbf{A})$ is read "the probability of **B** *given* **A**" (p. 360).
General Multiplication Rule	For any two events, **A** and **B**, the probability of **A** and **B** is (p. 361)

$$P(\mathbf{A}\ and\ \mathbf{B}) = P(\mathbf{A}) \times P(\mathbf{B}\|\mathbf{A}).$$

Independence (used formally)	Events **A** and **B** are independent when $P(\mathbf{B}\|\mathbf{A}) = P(\mathbf{B})$ (p. 362).
Tree diagram	A display of conditional events or probabilities that is helpful in thinking through conditioning (p. 367).

Skills

▶ Understand the concept of conditional probability as redefining the *Who* of concern, according to the information about the event that is *given*.

▶ Understand the concept of independence.

▶ Know how and when to apply the General Addition Rule.

▶ Know how to find probabilities for compound events as fractions of counts of occurrences in a two-way table.

▶ Know how and when to apply the General Multiplication Rule.

▶ Know how to make and use a tree diagram to understand conditional probabilities and reverse conditioning.

▶ Be able to make a clear statement about a conditional probability that makes clear how the condition affects the probability.

▶ Avoid making statements that assume independence of events when there is no clear evidence that they are in fact independent.

EXERCISES

1. **Homes.** Real estate ads suggest that 64% of homes for sale have garages, 21% have swimming pools, and 17% have both features. What is the probability that a home for sale has
   a) a pool or a garage?
   b) neither a pool nor a garage?
   c) a pool but no garage?

2. **Travel.** Suppose the probability that a U.S. resident has traveled to Canada is 0.18, to Mexico is 0.09, and to both countries is 0.04. What's the probability that an American chosen at random has
   a) traveled to Canada but not Mexico?
   b) traveled to either Canada or Mexico?
   c) not traveled to either country?

3. **Amenities.** A check of dorm rooms on a large college campus revealed that 38% had refrigerators, 52% had TVs, and 21% had both a TV and a refrigerator. What's the probability that a randomly selected dorm room has
   a) a TV but no refrigerator?
   b) a TV or a refrigerator, but not both?
   c) neither a TV nor a refrigerator?

4. **Workers.** Employment data at a large company reveal that 72% of the workers are married, that 44% are college graduates, and that half of the college grads are married. What's the probability that a randomly chosen worker
   a) is neither married nor a college graduate?
   b) is married but not a college graduate?
   c) is married or a college graduate?

5. **Global survey.** The marketing research organization GfK Roper conducts a yearly survey on consumer attitudes worldwide. They collect demographic information on the roughly 1500 respondents from each country that they survey. Here is a table showing the number of people with various levels of education in five countries:

Educational Level by Country

	Post-graduate	College	Some high school	Primary or less	No answer	Total
China	7	315	671	506	3	1502
France	69	388	766	309	7	1539
India	161	514	622	227	11	1535
U.K.	58	207	1240	32	20	1557
USA	84	486	896	87	4	1557
Total	379	1910	4195	1161	45	7690

If we select someone at random from this survey,

a) what is the probability that the person is from the United States?

b) what is the probability that the person completed his or her education before college?

c) what is the probability that the person is from France *or* did some post-graduate study?

d) what is the probability that the person is from France *and* finished only primary school or less?

6. **Birth order.** A survey of students in a large Introductory Statistics class asked about their birth order (1 = oldest or only child) and which college of the university they were enrolled in. Here are the data:

Birth Order

		1 or only	2 or more	Total
College	Arts & Sciences	34	23	57
	Agriculture	52	41	93
	Human Ecology	15	28	43
	Other	12	18	30
	Total	113	110	223

Suppose we select a student at random from this class. What is the probability that the person is:

a) a Human Ecology student?

b) a firstborn student?

c) firstborn *and* a Human Ecology student?

d) firstborn *or* a Human Ecology student?

7. **Cards.** You draw a card at random from a standard deck of 52 cards. Find each of the following conditional probabilities:

a) The card is a heart, given that it is red.

b) The card is red, given that it is a heart.

c) The card is an ace, given that it is red.

d) The card is a queen, given that it is a face card.

8. **Pets.** In its monthly report, the local animal shelter states that it currently has 24 dogs and 18 cats available for adoption. Eight of the dogs and 6 of the cats are male.

Find each of the following conditional probabilities if an animal is selected at random:

a) The pet is male, given that it is a cat.

b) The pet is a cat, given that it is female.

c) The pet is female, given that it is a dog.

9. **Health.** The probabilities that an adult American man has high blood pressure and/or high cholesterol are shown in the table.

Blood Pressure

Cholesterol		High	OK
	High	0.11	0.21
	OK	0.16	0.52

What's the probability that:

a) a man has both conditions?

b) a man has high blood pressure?

c) a man with high blood pressure has high cholesterol?

d) a man has high blood pressure if it's known that he has high cholesterol?

10. **Death penalty.** The table shows the political affiliations of American voters and their positions on the death penalty.

Death Penalty

Party		Favor	Oppose
	Republican	0.26	0.04
	Democrat	0.12	0.24
	Other	0.24	0.10

a) What's the probability that

   i) a randomly chosen voter favors the death penalty?

   ii) a Republican favors the death penalty?

   iii) a voter who favors the death penalty is a Democrat?

b) A candidate thinks she has a good chance of gaining the votes of anyone who is a Republican or in favor of the death penalty. What portion of the voters is that?

11. **Global survey, take 2.** Look again at the table summarizing the Roper survey in Exercise 5.

a) If we select a respondent at random, what's the probability we choose a person from the United States who has done post-graduate study?

b) Among the respondents who have done post-graduate study, what's the probability the person is from the United States?

c) What's the probability that a respondent from the United States has done post-graduate study?

d) What's the probability that a respondent from China has only a primary-level education?

e) What's the probability that a respondent with only a primary-level education is from China?

12. **Birth order, take 2.** Look again at the data about birth order of Intro Stats students and their choices of colleges shown in Exercise 6.

a) If we select a student at random, what's the probability the person is an Arts and Sciences student who is a second child (or more)?

b) Among the Arts and Sciences students, what's the probability a student was a second child (or more)?

c) Among second children (or more), what's the probability the student is enrolled in Arts and Sciences?

d) What's the probability that a first or only child is enrolled in the Agriculture College?

e) What is the probability that an Agriculture student is a first or only child?

13. **Sick kids.** Seventy percent of kids who visit a doctor have a fever, and 30% of kids with a fever have sore throats. What's the probability that a kid who goes to the doctor has a fever and a sore throat?

14. **Sick cars.** Twenty percent of cars that are inspected have faulty pollution control systems. The cost of repairing a pollution control system exceeds $100 about 40% of the time. When a driver takes her car in for inspection, what's the probability that she will end up paying more than $100 to repair the pollution control system?

15. **Cards.** You are dealt a hand of three cards, one at a time. Find the probability of each of the following.

a) The first heart you get is the third card dealt.

b) Your cards are all red (that is, all diamonds or hearts).

c) You get no spades.

d) You have at least one ace.

16. **Another hand.** You pick three cards at random from a deck. Find the probability of each event described below.

a) You get no aces.

b) You get all hearts.

c) The third card is your first red card.

d) You have at least one diamond.

17. **Batteries.** A junk box in your room contains a dozen old batteries, five of which are totally dead. You start picking batteries one at a time and testing them. Find the probability of each outcome.

a) The first two you choose are both good.

b) At least one of the first three works.

c) The first four you pick all work.

d) You have to pick 5 batteries to find one that works.

18. **Shirts.** The soccer team's shirts have arrived in a big box, and people just start grabbing them, looking for the right size. The box contains 4 medium, 10 large, and 6 extra-large shirts. You want a medium for you and one for your sister. Find the probability of each event described.

a) The first two you grab are the wrong sizes.

b) The first medium shirt you find is the third one you check.

c) The first four shirts you pick are all extra-large.

d) At least one of the first four shirts you check is a medium.

19. **Eligibility.** A university requires its biology majors to take a course called BioResearch. The prerequisite for this course is that students must have taken either a Statistics course or a computer course. By the time they are juniors, 52% of the Biology majors have taken Statistics, 23% have had a computer course, and 7% have done both.

a) What percent of the junior Biology majors are ineligible for BioResearch?

b) What's the probability that a junior Biology major who has taken Statistics has also taken a computer course?

c) Are taking these two courses disjoint events? Explain.

d) Are taking these two courses independent events? Explain.

20. **Benefits.** Fifty-six percent of all American workers have a workplace retirement plan, 68% have health insurance, and 49% have both benefits. We select a worker at random.

a) What's the probability he has neither employer-sponsored health insurance nor a retirement plan?

b) What's the probability he has health insurance if he has a retirement plan?

c) Are having health insurance and a retirement plan independent events? Explain.

d) Are having these two benefits mutually exclusive? Explain.

21. **For sale.** In the real-estate ads described in Exercise 1, 64% of homes for sale have garages, 21% have swimming pools, and 17% have both features.

a) If a home for sale has a garage, what's the probability that it has a pool too?

b) Are having a garage and a pool independent events? Explain.

c) Are having a garage and a pool mutually exclusive? Explain.

22. **On the road again.** According to Exercise 2, the probability that a U.S. resident has traveled to Canada is 0.18, to Mexico is 0.09, and to both countries is 0.04.

a) What's the probability that someone who has traveled to Mexico has visited Canada too?

b) Are traveling to Mexico and to Canada disjoint events? Explain.

c) Are traveling to Mexico and to Canada independent events? Explain.

23. **Cards.** If you draw a card at random from a well-shuffled deck, is getting an ace independent of the suit? Explain.

24. **Pets, again.** The local animal shelter in Exercise 8 reported that it currently has 24 dogs and 18 cats available for adoption; 8 of the dogs and 6 of the cats are male. Are the species and sex of the animals independent? Explain.

25. **Unsafe food.** Early in 2007 *Consumer Reports* published the results of an extensive investigation of broiler chickens purchased from food stores in 23 states. Tests for bacteria in the meat showed that 81% of the chickens were contaminated with campylobacter, 15% with salmonella, and 13% with both.

a) What's the probability that a tested chicken was not contaminated with either kind of bacteria?

b) Are contamination with the two kinds of bacteria disjoint? Explain.

c) Are contamination with the two kinds of bacteria independent? Explain.

26. **Birth order, finis.** In Exercises 6 and 12 we looked at the birth orders and college choices of some Intro Stats students. For these students:
    a) Are enrolling in Agriculture and Human Ecology disjoint? Explain.
    b) Are enrolling in Agriculture and Human Ecology independent? Explain.
    c) Are being firstborn and enrolling in Human Ecology disjoint? Explain.
    d) Are being firstborn and enrolling in Human Ecology independent? Explain.

27. **Men's health, again.** Given the table of probabilities from Exercise 9, are high blood pressure and high cholesterol independent? Explain.

		Blood Pressure	
		High	OK
Cholesterol	High	0.11	0.21
	OK	0.16	0.52

28. **Politics.** Given the table of probabilities from Exercise 10, are party affiliation and position on the death penalty independent? Explain.

		Death Penalty	
		Favor	Oppose
Party	Republican	0.26	0.04
	Democrat	0.12	0.24
	Other	0.24	0.10

29. **Phone service.** According to estimates from the federal government's 2003 National Health Interview Survey, based on face-to-face interviews in 16,677 households, approximately 58.2% of U.S. adults have both a landline in their residence and a cell phone, 2.8% have only cell phone service but no landline, and 1.6% have no telephone service at all.
    a) Polling agencies won't phone cell phone numbers because customers object to paying for such calls. What proportion of U.S. households can be reached by a landline call?
    b) Are having a cell phone and having a landline independent? Explain.

30. **Snoring.** After surveying 995 adults, 81.5% of whom were over 30, the National Sleep Foundation reported that 36.8% of all the adults snored. 32% of the respondents were snorers over the age of 30.
    a) What percent of the respondents were under 30 and did not snore?
    b) Is snoring independent of age? Explain.

31. **Montana.** A 1992 poll conducted by the University of Montana classified respondents by sex and political party, as shown in the table. Is party affiliation independent of the respondents' sex? Explain.

	Democrat	Republican	Independent
Male	36	45	24
Female	48	33	16

32. **Cars.** A random survey of autos parked in student and staff lots at a large university classified the brands by country of origin, as seen in the table. Is country of origin independent of type of driver?

		Driver	
		Student	Staff
Origin	American	107	105
	European	33	12
	Asian	55	47

33. **Luggage.** Leah is flying from Boston to Denver with a connection in Chicago. The probability her first flight leaves on time is 0.15. If the flight is on time, the probability that her luggage will make the connecting flight in Chicago is 0.95, but if the first flight is delayed, the probability that the luggage will make it is only 0.65.
    a) Are the first flight leaving on time and the luggage making the connection independent events? Explain.
    b) What is the probability that her luggage arrives in Denver with her?

34. **Graduation.** A private college report contains these statistics:

    70% of incoming freshmen attended public schools.

    75% of public school students who enroll as freshmen eventually graduate.

    90% of other freshmen eventually graduate.

    a) Is there any evidence that a freshman's chances to graduate may depend upon what kind of high school the student attended? Explain.
    b) What percent of freshmen eventually graduate?

35. **Late luggage.** Remember Leah (Exercise 33)? Suppose you pick her up at the Denver airport, and her luggage is not there. What is the probability that Leah's first flight was delayed?

36. **Graduation, part II.** What percent of students who graduate from the college in Exercise 34 attended a public high school?

37. Absenteeism. A company's records indicate that on any given day about 1% of their day-shift employees and 2% of the night-shift employees will miss work. Sixty percent of the employees work the day shift.

a) Is absenteeism independent of shift worked? Explain.
b) What percent of employees are absent on any given day?

38. Lungs and smoke. Suppose that 23% of adults smoke cigarettes. It's known that 57% of smokers and 13% of nonsmokers develop a certain lung condition by age 60.

a) Explain how these statistics indicate that lung condition and smoking are not independent.
b) What's the probability that a randomly selected 60-year-old has this lung condition?

39. Absenteeism, part II. At the company described in Exercise 37, what percent of the absent employees are on the night shift?

40. Lungs and smoke again. Based on the statistics in Exercise 38, what's the probability that someone with the lung condition was a smoker?

41. Drunks. Police often set up sobriety checkpoints—roadblocks where drivers are asked a few brief questions to allow the officer to judge whether or not the person may have been drinking. If the officer does not suspect a problem, drivers are released to go on their way. Otherwise, drivers are detained for a Breathalyzer test that will determine whether or not they will be arrested. The police say that based on the brief initial stop, trained officers can make the right decision 80% of the time. Suppose the police operate a sobriety checkpoint after 9:00 p.m. on a Saturday night, a time when national traffic safety experts suspect that about 12% of drivers have been drinking.

a) You are stopped at the checkpoint and, of course, have not been drinking. What's the probability that you are detained for further testing?
b) What's the probability that any given driver will be detained?
c) What's the probability that a driver who is detained has actually been drinking?
d) What's the probability that a driver who was released had actually been drinking?

42. No-shows. An airline offers discounted "advance-purchase" fares to customers who buy tickets more than 30 days before travel and charges "regular" fares for tickets purchased during those last 30 days. The company has noticed that 60% of its customers take advantage of the advance-purchase fares. The "no-show" rate among people who paid regular fares is 30%, but only 5% of customers with advance-purchase tickets are no-shows.

a) What percent of all ticket holders are no-shows?
b) What's the probability that a customer who didn't show had an advance-purchase ticket?
c) Is being a no-show independent of the type of ticket a passenger holds? Explain.

43. Dishwashers. Dan's Diner employs three dishwashers. Al washes 40% of the dishes and breaks only 1% of those he handles. Betty and Chuck each wash 30% of the dishes, and Betty breaks only 1% of hers, but Chuck breaks 3% of the dishes he washes. (He, of course, will need a new job soon. . . .) You go to Dan's for supper one night and hear a dish break at the sink. What's the probability that Chuck is on the job?

44. Parts. A company manufacturing electronic components for home entertainment systems buys electrical connectors from three suppliers. The company prefers to use supplier A because only 1% of those connectors prove to be defective, but supplier A can deliver only 70% of the connectors needed. The company must also purchase connectors from two other suppliers, 20% from supplier B and the rest from supplier C. The rates of defective connectors from B and C are 2% and 4%, respectively. You buy one of these components, and when you try to use it you find that the connector is defective. What's the probability that your component came from supplier A?

45. HIV testing. In July 2005 the journal *Annals of Internal Medicine* published a report on the reliability of HIV testing. Results of a large study suggested that among people with HIV, 99.7% of tests conducted were (correctly) positive, while for people without HIV 98.5% of the tests were (correctly) negative. A clinic serving an at-risk population offers free HIV testing, believing that 15% of the patients may actually carry HIV. What's the probability that a patient testing negative is truly free of HIV?

46. Polygraphs. Lie detectors are controversial instruments, barred from use as evidence in many courts. Nonetheless, many employers use lie detector screening as part of their hiring process in the hope that they can avoid hiring people who might be dishonest. There has been some research, but no agreement, about the reliability of polygraph tests. Based on this research, suppose that a polygraph can detect 65% of lies, but incorrectly identifies 15% of true statements as lies.

A certain company believes that 95% of its job applicants are trustworthy. The company gives everyone a polygraph test, asking, "Have you ever stolen anything from your place of work?" Naturally, all the applicants answer "No," but the polygraph identifies some of those answers as lies, making the person ineligible for a job. What's the probability that a job applicant rejected under suspicion of dishonesty was actually trustworthy?

JUST CHECKING

1. a)

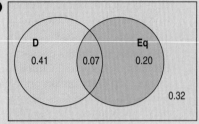

b) 0.32
c) 0.41

2. a) Independent
b) Disjoint
c) Neither

3. a)

		Equation		
		Yes	No	Total
Display	Yes	0.07	0.41	**0.48**
	No	0.20	0.32	**0.52**
	Total	**0.27**	**0.73**	**1.00**

b) $P(\mathbf{D}|\mathbf{Eq}) = P(\mathbf{D}\ and\ \mathbf{Eq})/P(\mathbf{Eq}) = 0.07/0.27 = 0.259$
c) No, pages can (and 7% do) have both.
d) To be independent, we'd need $P(\mathbf{D}|\mathbf{Eq}) = P(\mathbf{D})$. $P(\mathbf{D}|\mathbf{Eq}) = 0.259$, but $P(\mathbf{D}) = 0.48$. Overall, 48% of pages have data displays, but only about 26% of pages with equations do. They do not appear to be independent.

Random Variables

Where are we going?

How long do products last? Should you expect your computer to die just after the warranty runs out? How can you reduce your risk for developing hepatitis? Businesses, medical researchers, and other scientists all use probability to determine risk factors to help answer questions like these. To do that, they model the probability of outcomes using a special kind of variable—a *random* variable. Using random variables can help us talk about and predict random behavior.

WHAT IS AN ACTUARY?

Actuaries are the daring people who put a price on risk, estimating the likelihood and costs of rare events, so they can be insured. That takes financial, statistical, and business skills. It also makes them invaluable to many businesses. Actuaries are rather rare themselves; only about 19,000 work in North America. Perhaps because of this, they are well paid. If you're enjoying this course, you may want to look into a career as an actuary. Contact the Society of Actuaries or the Casualty Actuarial Society (who, despite what you may think, did not pay for this blurb).

Insurance companies make bets. They bet that you're going to live a long life. You bet that you're going to die sooner. Both you and the insurance company want the company to stay in business, so it's important to find a "fair price" for your bet. Of course, the right price for *you* depends on many factors, and nobody can predict exactly how long you'll live. But when the company averages over enough customers, it can make reasonably accurate estimates of the amount it can expect to collect on a policy before it has to pay its benefit.

Here's a simple example. An insurance company offers a "death and disability" policy that pays $10,000 when you die or $5000 if you are permanently disabled. It charges a premium of only $50 a year for this benefit. Is the company likely to make a profit selling such a plan? To answer this question, the company needs to know the *probability* that its clients will die or be disabled in any year. From actuarial information like this, the company can calculate the expected value of this policy.

Expected Value: Center

We'll want to build a probability model in order to answer the questions about the insurance company's risk. First we need to define a few terms. The amount the company pays out on an individual policy is called a **random variable**

NOTATION ALERT

The most common letters for random variables are X, Y, and Z. But be cautious. If you see any capital letter, it just might denote a random variable.

A S *Activity:* **Random Variables.** Learn more about random variables from this animated tour.

because its numeric value is based on the outcome of a random event. We use a capital letter, like X, to denote a random variable. We'll denote a particular value that it can have by the corresponding lowercase letter, in this case x. For the insurance company, x can be $10,000 (if you die that year), $5000 (if you are disabled), or $0 (if neither occurs). Because we can list all the outcomes, we might formally call this random variable a **discrete** random variable. Otherwise, we'd call it a **continuous** random variable. The collection of all the possible values and the probabilities that they occur is called the **probability model** for the random variable.

Suppose, for example, that the death rate in any year is 1 out of every 1000 people, and that another 2 out of 1000 suffer some kind of disability. Then we can display the probability model for this insurance policy in a table like this:

Policyholder Outcome	Payout x	Probability $P(X = x)$
Death	10,000	$\dfrac{1}{1000}$
Disability	5000	$\dfrac{2}{1000}$
Neither	0	$\dfrac{997}{1000}$

To see what the insurance company can expect, imagine that it insures exactly 1000 people. Further imagine that, in perfect accordance with the probabilities, 1 of the policyholders dies, 2 are disabled, and the remaining 997 survive the year unscathed. The company would pay $10,000 to one client and $5000 to each of 2 clients. That's a total of $20,000, or an average of 20000/1000 = $20 per policy. Since it is charging people $50 for the policy, the company expects to make a profit of $30 per customer. Not bad!

We can't predict what *will* happen during any given year, but we can say what we *expect* to happen. To do this, we (or, rather, the insurance company) need the probability model. The expected value of a policy is a parameter of this model. In fact, it's the mean. We'll signify this with the notation μ (for population mean) or $E(X)$ for expected value. This isn't an average of some data values, so we won't estimate it. Instead, we assume that the probabilities are known and simply calculate the expected value from them.

NOTATION ALERT

The expected value (or mean) of a random variable is written $E(X)$ or μ.

How did we come up with $20 as the expected value of a policy payout? Here's the calculation. As we've seen, it often simplifies probability calculations to think about some (convenient) number of outcomes. For example, we could imagine that we have exactly 1000 clients. Of those, exactly 1 died and 2 were disabled, corresponding to what the probabilities would say.

$$\mu = E(X) = \frac{10,000(1) + 5000(2) + 0(997)}{1000}$$

So our total payout comes to $20,000, or $20 per policy.

Instead of writing the expected value as one big fraction, we can rewrite it as separate terms with a common denominator of 1000.

$$\mu = E(X)$$
$$= \$10,000\left(\frac{1}{1000}\right) + \$5000\left(\frac{2}{1000}\right) + \$0\left(\frac{997}{1000}\right)$$
$$= \$20.$$

How convenient! See the probabilities? For each policy, there's a 1/1000 chance that we'll have to pay $10,000 for a death and a 2/1000 chance that we'll have to pay $5000 for a disability. Of course, there's a 997/1000 chance that we won't have to pay anything.

Take a good look at the expression now. It's easy to calculate the **expected value** of a (discrete) random variable—just multiply each possible value by the probability that it occurs, and find the sum:

$$\mu = E(X) = \sum xP(x).$$

Be sure that every possible outcome is included in the sum. And verify that you have a valid probability model to start with—the probabilities should each be between 0 and 1 and should sum to one.

FOR EXAMPLE | **Love and Expected Values**

On Valentine's Day the *Quiet Nook* restaurant offers a *Lucky Lovers Special* that could save couples money on their romantic dinners. When the waiter brings the check, he'll also bring the four aces from a deck of cards. He'll shuffle them and lay them out face down on the table. The couple will then get to turn one card over. If it's a black ace, they'll owe the full amount, but if it's the ace of hearts, the waiter will give them a $20 Lucky Lovers discount. If they first turn over the ace of diamonds (hey–at least it's red!), they'll then get to turn over one of the remaining cards, earning a $10 discount for finding the ace of hearts this time.

QUESTION: Based on a probability model for the size of the Lucky Lovers discounts the restaurant will award, what's the expected discount for a couple?

Let X = the Lucky Lovers discount. The probabilities of the three outcomes are:

$$P(X = 20) = P(A\heartsuit) = \frac{1}{4}$$

$$P(X = 10) = P(A\diamondsuit, \text{then } A\heartsuit) = P(A\diamondsuit) \times P(A\heartsuit|A\diamondsuit) = \frac{1}{4} \times \frac{1}{3} = \frac{1}{12}$$

$$P(X = 0) = P(X \neq 20 \text{ or } 10) = 1 - \left(\frac{1}{4} + \frac{1}{12}\right) = \frac{2}{3}.$$

My probability model is:

Outcome	A♥	A♦, then A♥	Black Ace
x	20	10	0
$P(X = x)$	$\frac{1}{4}$	$\frac{1}{12}$	$\frac{2}{3}$

$$E(X) = 20 \times \frac{1}{4} + 10 \times \frac{1}{12} + 0 \times \frac{2}{3} = \frac{70}{12} \approx 5.83$$

Couples dining at the Quiet Nook can expect an average discount of $5.83.

JUST CHECKING

1. One of the authors took his minivan in for repair recently because the air conditioner was cutting out intermittently. The mechanic identified the problem as dirt in a control unit. He said that in about 75% of such cases, drawing down and then recharging the coolant a couple of times cleans up the problem—and costs only $60. If that fails, then the control unit must be replaced at an additional cost of $100 for parts and $40 for labor.

(continued)

a) Define the random variable and construct the probability model.
b) What is the expected value of the cost of this repair?
c) What does that mean in this context?

Oh—in case you were wondering—the $60 fix worked!

First Center, Now Spread . . .

Of course, this expected value (or mean) is not what actually happens to any *particular* policyholder. No individual policy actually costs the company $20. We are dealing with random events, so some policyholders receive big payouts, others nothing. Because the insurance company must anticipate this variability, it needs to know the *standard deviation* of the random variable.

For data, we calculated the **standard deviation** by first computing the deviation from the mean and squaring it. We do that with (discrete) random variables as well. First, we find the deviation of each payout from the mean (expected value):

Policyholder Outcome	Payout x	Probability $P(X = x)$	Deviation $(x - \mu)$
Death	10,000	$\frac{1}{1000}$	$(10,000 - 20) = 9980$
Disability	5000	$\frac{2}{1000}$	$(5000 - 20) = 4980$
Neither	0	$\frac{997}{1000}$	$(0 - 20) = -20$

Next, we square each deviation. The **variance** is the expected value of those squared deviations, so we multiply each by the appropriate probability and sum those products. That gives us the variance of X. Here's what it looks like:

$$Var(X) = 9980^2 \left(\frac{1}{1000}\right) + 4980^2 \left(\frac{2}{1000}\right) + (-20)^2 \left(\frac{997}{1000}\right) = 149,600.$$

Finally, we take the square root to get the standard deviation:

$$SD(X) = \sqrt{149,600} \approx \$386.78.$$

The insurance company can expect an average payout of $20 per policy, with a standard deviation of $386.78.

Think about that. The company charges $50 for each policy and expects to pay out $20 per policy. Sounds like an easy way to make $30. In fact, most of the time (probability 997/1000) the company pockets the entire $50. But would you consider selling your roommate such a policy? The problem is that occasionally the company loses big. With probability 1/1000, it will pay out $10,000, and with probability 2/1000, it will pay out $5000. That may be more risk than you're willing to take on. The standard deviation of $386.78 gives an indication that it's no sure thing. That's a pretty big spread (and risk) for an average profit of $30.

Here are the formulas for what we just did. Because these are parameters of our probability model, the variance and standard deviation can also be written as σ^2 and σ. You should recognize both kinds of notation.

$$\sigma^2 = Var(X) = \sum (x - \mu)^2 P(x)$$
$$\sigma = SD(X) = \sqrt{Var(X)}$$

| FOR EXAMPLE | Finding the Standard Deviation |

RECAP: The probability model for the Lucky Lovers restaurant discount is

Outcome	A♥	A♦, then A♥	Black Ace
x	20	10	0
$P(X = x)$	$\dfrac{1}{4}$	$\dfrac{1}{12}$	$\dfrac{2}{3}$

We found that couples can expect an average discount of $\mu = \$5.83$.

QUESTION: What's the standard deviation of the discounts?

First find the variance: $Var(X) = \sum (x - \mu)^2 \times P(x)$

$$= (20 - 5.83)^2 \times \frac{1}{4} + (10 - 5.83)^2 \times \frac{1}{12} + (0 - 5.83)^2 \times \frac{2}{3}$$

$$\approx 74.306.$$

$$\text{So, } SD(X) = \sqrt{74.306} \approx \$8.62$$

Couples can expect the Lucky Lovers discounts to average $5.83, with a standard deviation of $8.62.

| STEP-BY-STEP EXAMPLE | **Expected Values and Standard Deviations for Discrete Random Variables** |

As the head of inventory for Knowway Computer Company, you were thrilled that you had managed to ship 2 computers to your biggest client the day the order arrived. You are horrified, though, to find out that someone had restocked refurbished computers in with the new computers in your storeroom. The shipped computers were selected randomly from the 15 computers in stock, but 4 of those were actually refurbished.

If your client gets 2 new computers, things are fine. If the client gets one refurbished computer, it will be sent back at your expense—$100—and you can replace it. However, if both computers are refurbished, the client will cancel the order this month and you'll lose a total of $1000.

Question: What's the expected value and the standard deviation of the company's loss?

| THINK ▶ | **Plan** State the problem. | I want to find the company's expected loss for shipping refurbished computers and the standard deviation. |
| | **Variable** Define the random variable. | Let X = amount of loss. |

Plot Make a picture. This is another job for tree diagrams.

If you prefer calculation to drawing, find $P(NN)$ and $P(RR)$, then use the Complement Rule to find $P(NR \ or \ RN)$.

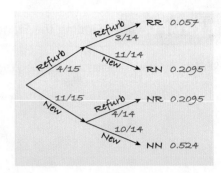

Model List the possible values of the random variable, and determine the probability model.

Outcome	x	P(X = x)
Two refurbs	1000	$P(RR) = 0.057$
One refurb	100	$P(NR \ or \ RN) = 0.2095$ $+ \ 0.2095 = 0.419$
New/new	0	$P(NN) = 0.524$

Mechanics Find the expected value.

$$E(X) = 0(0.524) + 100(0.419) + 1000(0.057)$$
$$= \$98.90$$

Find the variance.

$$Var(X) = (0 - 98.90)^2(0.524)$$
$$+ (100 - 98.90)^2(0.419)$$
$$+ (1000 - 98.90)^2(0.057)$$
$$= 51,408.79$$

Find the standard deviation.

$$SD(X) = \sqrt{51,408.79} = \$226.735$$

Conclusion Interpret your results in context.

REALITY CHECK Both numbers seem reasonable. The expected value of $98.90 is between the extremes of $0 and $1000, and there's great variability in the outcome values.

I expect this mistake to cost the firm $98.90, with a standard deviation of $226.74. The large standard deviation reflects the fact that there's a pretty large range of possible losses.

More About Means and Variances

Our insurance company expected to pay out an average of $20 per policy, with a standard deviation of about $387. If we take the $50 premium into account, we see the company makes a profit of $50 - 20 = \$30$ per policy. Suppose the company lowers the premium by $5 to $45. It's pretty clear that the expected profit also drops an average of $5 per policy, to $45 - 20 = \$25$.

What about the standard deviation? We know that adding or subtracting a constant from data shifts the mean but doesn't change the variance or standard deviation. The same is true of random variables.[1]

$$E(X \pm c) = E(X) \pm c \qquad Var(X \pm c) = Var(X).$$

---

[1] The rules in this section are true for both discrete *and* continuous random variables.

FOR EXAMPLE — Adding a Constant

RECAP: We've determined that couples dining at the *Quiet Nook* can expect Lucky Lovers discounts averaging $5.83 with a standard deviation of $8.62. Suppose that for several weeks the restaurant has also been distributing coupons worth $5 off any one meal (one discount per table).

QUESTION: If every couple dining there on Valentine's Day brings a coupon, what will be the mean and standard deviation of the total discounts they'll receive?

Let D = total discount (Lucky Lovers plus the coupon); then $D = X + 5$.

$$E(D) = E(X + 5) = E(X) + 5 = 5.83 + 5 = \$10.83$$
$$Var(D) = Var(X + 5) = Var(X) = 8.62^2$$
$$SD(D) = \sqrt{Var(X)} = \$8.62$$

Couples with the coupon can expect total discounts averaging $10.83. The standard deviation is still $8.62.

Back to insurance . . . What if the company decides to double all the payouts—that is, pay $20,000 for death and $10,000 for disability? This would double the average payout per policy and also increase the variability in payouts. We have seen that multiplying or dividing all data values by a constant changes both the mean and the standard deviation by the same factor. Variance, being the square of standard deviation, changes by the square of the constant. The same is true of random variables. In general, multiplying each value of a random variable by a constant multiplies the mean by that constant and the variance by the *square* of the constant.

$$E(aX) = aE(X) \qquad Var(aX) = a^2Var(X)$$

FOR EXAMPLE — Double the Love

RECAP: On Valentine's Day at the *Quiet Nook*, couples may get a Lucky Lovers discount averaging $5.83 with a standard deviation of $8.62. When two couples dine together on a single check, the restaurant doubles the discount offer—$40 for the ace of hearts on the first card and $20 on the second.

QUESTION: What are the mean and standard deviation of discounts for such foursomes?

$$E(2X) = 2E(X) = 2(5.83) = \$11.66$$
$$Var(2x) = 2^2Var(x) = 2^2 \cdot 8.62^2 = 297.2176$$
$$SD(2X) = \sqrt{297.2176} = \$17.24$$

If the restaurant doubles the discount offer, two couples dining together can expect to save an average of $11.66 with a standard deviation of $17.24.

This insurance company sells policies to more than just one person. How can we figure means and variances for a collection of customers? For example, how can the company find the total expected value (and standard deviation) of policies taken over all policyholders? Consider a simple case: just two customers, Mr. Ecks and Ms. Wye. With an expected payout of $20 on each policy, we might predict a total of $20 + $20 = $40 to be paid out on the two policies. Nothing surprising there. The expected value of the sum is the sum of the expected values.

$$E(X + Y) = E(X) + E(Y).$$

The variability is another matter. Is the risk of insuring two people the same as the risk of insuring one person for twice as much? We wouldn't expect both clients to die or become disabled in the same year. Because we've spread the risk, the standard deviation should be smaller. Indeed, this is the fundamental principle behind insurance. By spreading the risk among many policies, a company can keep the standard deviation quite small and predict costs more accurately.

But how much smaller is the standard deviation of the sum? It turns out that, if the random variables are independent, there is a simple Addition Rule for variances: *The variance of the sum of two independent random variables is the sum of their individual variances.*

For Mr. Ecks and Ms. Wye, the insurance company can expect their outcomes to be independent, so (using X for Mr. Ecks's payout and Y for Ms. Wye's)

$$Var(X + Y) = Var(X) + Var(Y)$$
$$= 149{,}600 + 149{,}600$$
$$= 299{,}200.$$

If they had insured only Mr. Ecks for twice as much, there would only be one outcome rather than two *independent* outcomes, so the variance would have been

$$Var(2X) = 2^2 Var(X) = 4 \times 149{,}600 = 598{,}400, \text{ or}$$

twice as big as with two independent policies.

Of course, variances are in squared units. The company would prefer to know standard deviations, which are in dollars. The standard deviation of the payout for two independent policies is $\sqrt{299{,}200} = \$546.99$. But the standard deviation of the payout for a single policy of twice the size is $\sqrt{598{,}400} = \$773.56$, or about 40% more.

If the company has two customers, then, it will have an expected annual total payout of $40 with a standard deviation of about $547.

FOR EXAMPLE   Adding the Discounts

RECAP: The Valentine's Day Lucky Lovers discount for couples averages $5.83 with a standard deviation of $8.62. We've seen that if the restaurant doubles the discount offer for two couples dining together on a single check, they can expect to save $11.66 with a standard deviation of $17.24. Some couples decide instead to get separate checks and pool their two discounts.

QUESTION: You and your amour go to this restaurant with another couple and agree to share any benefit from this promotion. Does it matter whether you pay separately or together?

Let X_1 and X_2 represent the two separate discounts, and T the total; then $T = X_1 + X_2$.

$$E(T) = E(X_1 + X_2) = E(X_1) + E(X_2) = 5.83 + 5.83 = \$11.66,$$

so the expected saving is the same either way.

The cards are reshuffled for each couple's turn, so the discounts couples receive are independent. It's okay to add the variances:

$$Var(T) = Var(X_1 + X_2) = Var(X_1) + Var(X_2) = 8.62^2 + 8.62^2 = 148.6088$$
$$SD(T) = \sqrt{148.6088} = \$12.19$$

When two couples get separate checks, there's less variation in their total discount. The standard deviation is $12.19, compared to $17.24 for couples who play for the double discount on a single check.

PYTHAGOREAN THEOREM OF STATISTICS

We often use the standard deviation to measure variability, but when we add independent random variables, we use their variances. Think of the Pythagorean Theorem. In a right triangle (only), the *square* of the length of the hypotenuse is the sum of the *squares* of the lengths of the other two sides:

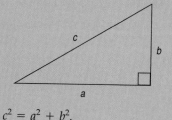

$$c^2 = a^2 + b^2.$$

For independent random variables (only), the *square* of the standard deviation of their sum is the sum of the *squares* of their standard deviations:

$$SD^2(X + Y) = SD^2(X) + SD^2(Y).$$

It's simpler to write this with *variances:*

For independent random variables, X and Y,
$$Var(X + Y) = Var(X) + Var(Y).$$

In general,

- *The mean of the sum of two random variables is the sum of the means.*
- *The mean of the difference of two random variables is the difference of the means.*
- *If the random variables are independent, the variance of their sum or difference is always the sum of the variances.*

$$E(X \pm Y) = E(X) \pm E(Y) \quad Var(X \pm Y) = Var(X) + Var(Y)$$

Wait a minute! Is that third part correct? Do we always *add* variances? Yes. Think about the two insurance policies. Suppose we want to know the mean and standard deviation of the *difference* in payouts to the two clients. Since each policy has an expected payout of $20, the expected difference is $20 - 20 = \$0$. If we also subtract variances, we get $0, too, and that surely doesn't make sense. Note that if the outcomes for the two clients are independent, the difference in payouts could range from $\$10{,}000 - \$0 = \$10{,}000$ to $\$0 - \$10{,}000 = -\$10{,}000$, a spread of $20,000. The variability in differences increases as much as the variability in sums. If the company has two customers, the difference in payouts has a mean of $0 and a standard deviation of about $547 (again).

FOR EXAMPLE | Working with Differences

RECAP: The Lucky Lovers discount at the *Quiet Nook* averages $5.83 with a standard deviation of $8.62. Just up the street, the *Wise Fool* restaurant has a competing Lottery of Love promotion. There a couple can select a specially prepared chocolate from a large bowl and unwrap it to learn the size of their discount. The restaurant's manager says the discounts vary with an average of $10.00 and a standard deviation of $15.00.

QUESTION: How much more can you expect to save at the *Wise Fool*? With what standard deviation?

Let W = discount at the Wise Fool, X = the discount at the Quiet Nook, and D = the difference: $D = W - X$. These are different promotions at separate restaurants, so the outcomes are independent.

$$E(W - X) = E(W) - E(X) = 10.00 - 5.83 = \$4.17$$
$$\begin{aligned} SD(W - X) &= \sqrt{Var(W - X)} \\ &= \sqrt{Var(W) + Var(X)} \\ &= \sqrt{15^2 + 8.62^2} \\ &\approx \$17.30 \end{aligned}$$

Discounts at the Wise Fool will average $4.17 more than at the Quiet Nook, with a standard deviation of $17.30.

For random variables, does $X + X + X = 3X$? Maybe, but be careful. As we've just seen, insuring one person for $30,000 is not the same risk as insuring three people for $10,000 each. When each instance represents a different outcome for the same random variable, it's easy to fall into the trap of writing all of them with the same symbol. Don't make this common mistake. Make sure you write each instance as a *different* random variable. Just because each random variable describes a similar situation doesn't mean that each random outcome will be the same.

These are *random* variables, not the variables you saw in Algebra. Being random, they take on different values each time they're evaluated. So what you really mean is $X_1 + X_2 + X_3$. Written this way, it's clear that the sum shouldn't necessarily equal 3 times *anything*.

FOR EXAMPLE   **Summing a Series of Outcomes**

RECAP: The *Quiet Nook*'s Lucky Lovers promotion offers couples discounts averaging $5.83 with a standard deviation of $8.62. The restaurant owner is planning on serving 40 couples on Valentine's Day.

QUESTION: What's the expected total of the discounts the owner will give? With what standard deviation?

Let $X_1, X_2, X_3, \ldots, X_{40}$ represent the discounts to the 40 couples, and T the total of all the discounts. Then:

$$T = X_1 + X_2 + X_3 + \cdots + X_{40}$$
$$E(T) = E(X_1 + X_2 + X_3 + \cdots + X_{40})$$
$$= E(X_1) + E(X_2) + E(X_3) + \cdots + E(X_{40})$$
$$= 5.83 + 5.83 + 5.83 + \cdots + 5.83$$
$$= \$233.20$$

Reshuffling cards between couples makes the discounts independent, so:

$$SD(T) = \sqrt{Var(X_1 + X_2 + X_3 + \cdots + X_{40})}$$
$$= \sqrt{Var(X_1) + Var(X_2) + Var(X_3) + \cdots + Var(X_{40})}$$
$$= \sqrt{8.62^2 + 8.62^2 + 8.62^2 + \cdots + 8.62^2}$$
$$\approx \$54.52$$

The restaurant owner can expect the 40 couples to win discounts totaling $233.20, with a standard deviation of $54.52.

JUST CHECKING

2. Suppose the time it takes a customer to get and pay for seats at the ticket window of a baseball park is a random variable with a mean of 100 seconds and a standard deviation of 50 seconds. When you get there, you find only two people in line in front of you.

    a) How long do you expect to wait for your turn to get tickets?
    b) What's the standard deviation of your wait time?
    c) What assumption did you make about the two customers in finding the standard deviation?

Combining Random Variables (The Bad News)

Although we know how to find the expected value and the variance for the sum of two independent random variables, that's not all we'd like to know. It would be nice if we could go directly from models of each random variable to a model for their sum. But the probability model for the sum of two random variables is *not* necessarily the same as the model we started with *even when the variables are independent.*

That's easy to see for discrete random variables because we can just think about their possible values. When we looked at insuring one person, either we paid out $0, $5000, or $10,000. Suppose we insure three people. The possibilities are not just $0, $15,000, or $30,000. We might have to pay out $5000, or $20,000, or other possible values. Even though the expected values add, the probability model is different. The probability model for $3X$ is not the same as the model for $X_1 + X_2 + X_3$. That's another reason to use the subscripted form rather than write $X + X + X$.

STEP-BY-STEP EXAMPLE   Hitting the Road: Means and Variances

You're planning to spend next year wandering through the mountains of Kyrgyzstan. You plan to sell your used SUV so you can purchase an off-road Honda motor scooter when you get there. Used SUVs of the year and mileage of yours are selling for a mean of $6940 with a standard deviation of $250. Your research shows that scooters in Kyrgyzstan are going for about 65,000 Kyrgyzstan som with a standard deviation of 500 som. One U.S. dollar is worth about 43.1 Kyrgyzstan som (43 som and 10 tylyn).

Question: How much cash can you expect to pocket after you sell your SUV and buy the scooter?

Plan State the problem.	I want to model how much money I'd have (in som) after selling my SUV and buying the scooter.
Variables Define the random variables.	Let A = sale price of my SUV (in dollars), B = price of a scooter (in som), and D = profit (in som)
Write an appropriate equation. Think about the assumptions.	$D = 43.1A - B$ ✔ **Independence Assumption:** The prices are independent.
SHOW **Mechanics** Find the expected value, using the appropriate rules.	$\begin{aligned} E(D) &= E(43.1A - B) \\ &= 43.1E(A) - E(B) \\ &= 43.1(6,940) - (65,000) \\ E(D) &= 234,114 \text{ som} \end{aligned}$
Find the variance, using the appropriate rules. Be sure to check the assumptions first!	Since sale and purchase prices are independent, $\begin{aligned} Var(D) &= Var(43.1A - B) \\ &= Var(43.1A) + Var(B) \\ &= (43.1)^2 Var(A) + Var(B) \\ &= 1857.61(250)^2 + (500)^2 \\ Var(D) &= 116,350,625 \end{aligned}$
Find the standard deviation.	$SD(D) = \sqrt{116,350,625} = 10,786.59 \text{ som}$

Conclusion Interpret your results in context. (Here that means talking about dollars.)

I can expect to clear about 234,114 som ($5432) with a standard deviation of 10,786.59 som ($250).

REALITY CHECK Given the initial cost estimates, the mean and standard deviation seem reasonable.

A S *Activity:* **Numeric Outcomes.** You've seen how to simulate discrete random outcomes. There's a tool for simulating continuous outcomes, too.

A S *Activity:* **Means of Random Variables.** Experiment with continuous random variables to learn how their expected values behave.

Continuous Random Variables

A company manufactures small stereo systems. At the end of the production line, the stereos are packaged and prepared for shipping. Stage 1 of this process is called "packing." Workers must collect all the system components (a main unit, two speakers, a power cord, an antenna, and some wires), put each in plastic bags, and then place everything inside a protective styrofoam form. The packed form then moves on to Stage 2, called "boxing." There, workers place the form and a packet of instructions in a cardboard box, close it, then seal and label the box for shipping.

The company says that times required for the packing stage can be described by a Normal model with a mean of 9 minutes and standard deviation of 1.5 minutes. The times for the boxing stage can also be modeled as Normal, with a mean of 6 minutes and standard deviation of 1 minute.

This is a common way to model events. Do our rules for random variables apply here? What's different? We no longer have a list of discrete outcomes, with their associated probabilities. Instead, we have **continuous random variables** that can take on any value. Now any single value won't have a probability. We saw this back in Chapter 6 when we first saw the Normal model (although we didn't talk then about "random variables" or "probability"). We know that the probability that $z = 1.5$ doesn't make sense, but we *can* talk about the probability that z lies *between* 0.5 and 1.5. For a Normal random variable, the probability that it falls within an interval is just the area under the Normal curve over that interval.

Some continuous random variables have Normal models; others may be skewed, uniform, or bimodal. Regardless of shape, all continuous random variables have means (which we also call *expected values*) and variances. In this book we won't worry about how to calculate them, but we can still work with models for continuous random variables when we're given these parameters.

Combining Random Variables (The Good News)

We know that when we add two random variables together, the sum generally has a different probability model than the one we started with. Amazingly, independent Normal random variables behave better. The probability model for the sum of any number of independent Normal random variables is still Normal. And we already know what the mean and variance of that sum will be because those both add.

This is just one of the things that is special about the Normal. However, if this mathematical good behavior alone doesn't impress you, here's a real-world example of how we can use this fact.

STEP-BY-STEP EXAMPLE   Packaging Stereos

Consider the company that manufactures and ships small stereo systems that we just discussed.

Recall that times required to pack the stereos can be described by a Normal model with a mean of 9 minutes and standard deviation of 1.5 minutes. The times for the boxing stage can also be modeled as Normal, with a mean of 6 minutes and standard deviation of 1 minute.

Questions:

1. What is the probability that packing two consecutive systems takes over 20 minutes?
2. What percentage of the stereo systems take longer to pack than to box?

Question 1: What is the probability that packing two consecutive systems takes over 20 minutes?

Plan State the problem.

I want to estimate the probability that packing two consecutive systems takes over 20 minutes.

Variables Define your random variables.

Let P_1 = time for packing the first system
P_2 = time for packing the second
T = total time to pack two systems

Write an appropriate equation.

$T = P_1 + P_2$

Think about the assumptions. Sums of independent Normal random variables follow a Normal model. Such simplicity isn't true in general.

✔ **Normal Model Assumption:** *We are told that both random variables follow Normal models.*

✔ **Independence Assumption:** *We can reasonably assume that the two packing times are independent.*

Mechanics Find the expected value.

$E(T) = E(P_1 + P_2)$
$= E(P_1) + E(P_2)$
$= 9 + 9 = 18$ minutes

For sums of independent random variables, variances add. (We don't need the variables to be Normal for this to be true—just independent.)

Since the times are independent,
$Var(T) = Var(P_1 + P_2)$
$= Var(P_1) + Var(P_2)$
$= 1.5^2 + 1.5^2$
$Var(T) = 4.50$

Find the standard deviation.

$SD(T) = \sqrt{4.50} \approx 2.12$ minutes

Now we use the fact that both random variables follow Normal models to say that their sum is also Normal.

I'll model T with $N(18, 2.12)$.

Sketch a picture of the Normal model for the total time, shading the region representing over 20 minutes.

Find the z-score for 20 minutes.

Use technology or Table Z to find the probability.

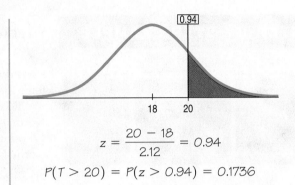

$$z = \frac{20 - 18}{2.12} = 0.94$$

$$P(T > 20) = P(z > 0.94) = 0.1736$$

Conclusion Interpret your result in context.

There's a little more than a 17% chance that it will take a total of over 20 minutes to pack two consecutive stereo systems.

Question 2: What percentage of the stereo systems take longer to pack than to box?

Plan State the question.

I want to estimate the percentage of the stereo systems that take longer to pack than to box.

Variables Define your random variables.

Let P = time for packing a system
   B = time for boxing a system
   D = difference in times to pack and box a system

Write an appropriate equation.

$D = P - B$

What are we trying to find? Notice that we can tell which of two quantities is greater by subtracting and asking whether the difference is positive or negative.

The probability that it takes longer to pack than to box a system is the probability that the difference $P - B$ is greater than zero.

Don't forget to think about the assumptions.

✔ **Normal Model Assumption:** We are told that both random variables follow Normal models.

✔ **Independence Assumption:** We can assume that the times it takes to pack and to box a system are independent.

Mechanics Find the expected value.

$$\begin{aligned} E(D) &= E(P - B) \\ &= E(P) - E(B) \\ &= 9 - 6 = 3 \text{ minutes} \end{aligned}$$

For the difference of independent random variables, variances add.

Since the times are independent,

$$\begin{aligned} Var(D) &= Var(P - B) \\ &= Var(P) + Var(B) \\ &= 1.5^2 + 1^2 \end{aligned}$$
$$Var(D) = 3.25$$

Find the standard deviation.

$SD(D) = \sqrt{3.25} \approx 1.80 \text{ minutes}$

State what model you will use.

I'll model D with $N(3, 1.80)$

Sketch a picture of the Normal model for the difference in times, and shade the region representing a difference greater than zero.

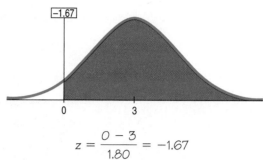

Find the z-score for 0 minutes, then use Table Z or technology to find the probability.

$$z = \frac{0 - 3}{1.80} = -1.67$$

$$P(D > 0) = P(z > -1.67) = 0.9525$$

Conclusion Interpret your result in context.

About 95% of all the stereo systems will require more time for packing than for boxing.

---

*Correlation and Covariance

In Chapter 7 we saw that the association of two variables could be measured with their correlation. What about random variables? We can talk about the correlation between random variables, too. But it's easier to start with a related concept called covariance.

If X is a random variable with expected value $E(X) = \mu$ and Y is a random variable with expected value $E(Y) = \nu$, then the **covariance** of X and Y is defined as

$$Cov(X, Y) = E((X - \mu)(Y - \nu)).$$

The covariance, like the correlation, measures how X and Y vary together (*co = together*). Think about a scatterplot of two highly correlated variables. When X is above its mean, Y tends to be, too. And when X is below its mean, Y tends to be below its mean. So the product in the covariance formula is positive.

Covariance has a few properties worth remembering:

- $Cov(X, Y) = Cov(Y, X)$
- $Cov(X, X) = Var(X)$
- $Cov(cX, dY) = cdCov(X, Y)$, for any constants c and d
- $Cov(X, Y) = E(XY) - \mu\nu$
- If X and Y are independent, then $Cov(X, Y) = 0$—but be careful not to assume that the converse is true; it is possible for two variables to have zero covariance yet not be independent.

The covariance gives us the extra information we need to find the variance of the sum or difference of two random variables when they are not independent:

$$Var(X \pm Y) = Var(X) + Var(Y) \pm 2Cov(X, Y).$$

When X and Y are independent, their covariance is zero, so we have the Pythagorean theorem of statistics, as we saw earlier.[2]

---

[2] If you want to follow the geometry one step further, the more general form for correlated variables is related to the Law of Cosines.

Covariance, unlike correlation, doesn't have to be between −1 and 1. If X and Y have large values, the covariance will be large as well. And that's its problem. If we know the covariance between X and Y is −64.5, it doesn't give us a real sense of how negatively related they are. To fix the "problem," we can divide the covariance by each of the standard deviations to get the **correlation**:

$$Corr(X, Y) = \frac{Cov(X, Y)}{\sigma_X \sigma_Y}.$$

Correlation of random variables is usually denoted as ρ. Correlation of random variables has many of the properties we saw in Chapter 7 for the correlation coefficient, r.

- The sign of a correlation coefficient gives the direction of the association between random variables.
- Correlation is always between −1 and +1.
- Correlation treats X and Y symmetrically. The correlation of X with Y is the same as the correlation of Y with X.
- Unlike covariance, correlation has no units.
- Correlation is not affected by changes in the center or scale of either variable.

What Can Go Wrong?

■ **Probability models are still just models.** Models can be useful, but they are not reality. Think about the assumptions behind your models. Are your dice really perfectly fair? (They are probably pretty close.) But when you hear that the probability of a nuclear accident is 1/10,000,000 per year, is that likely to be a precise value? Question probabilities as you would data.

■ **If the model is wrong, so is everything else.** Before you try to find the mean or standard deviation of a random variable, check to make sure the probability model is reasonable. As a start, the probabilities in your model should add up to 1. If not, you may have calculated a probability incorrectly or left out a value of the random variable. For instance, in the insurance example, the description mentions only death and disability. Good health is by far the most likely outcome, not to mention the best for both you and the insurance company (who gets to keep your money). Don't overlook that.

■ **Don't assume everything's Normal.** Just because a random variable is continuous or you happen to know a mean and standard deviation doesn't mean that a Normal model will be useful. You must *Think* about whether the **Normality Assumption** is justified. Using a Normal model when it really does not apply will lead to wrong answers and misleading conclusions.

To find the expected value of the sum or difference of random variables, we simply add or subtract means. Center is easy; spread is trickier. Watch out for some common traps.

■ **Watch out for variables that aren't independent.** You can add expected values of *any* two random variables, but you can only add variances of independent random variables. Suppose a survey includes questions about the number of hours of sleep people get each night and also the number of hours they are awake each day. From their answers, we find the mean and standard deviation of hours asleep and hours awake. The expected total must be 24 hours; after all, people are either asleep or awake.[3] The means still add just fine. Since all the totals are exactly 24 hours, however, the standard deviation of the total will be 0. We can't add variances here because the number of hours you're awake depends on the number of hours you're asleep. Be sure to check for independence before adding variances.

■ **Don't forget: Variances of independent random variables add. Standard deviations don't.**

■ **Don't forget: Variances of independent random variables add, even when you're looking at the difference between them.**

■ **Don't write independent instances of a random variable with notation that looks like they are the same variables.** Make sure you write each instance as a different random variable. Just because each random variable describes a similar situation doesn't mean that each random outcome will be the same. These are *random* variables, not the variables you saw in Algebra. Write $X_1 + X_2 + X_3$ rather than $X + X + X$.

CONNECTIONS

We've seen means, variances, and standard deviations of data. We know that they estimate parameters of models for these data. Now we're looking at the probability models directly. We have only parameters because there are no data to summarize.

It should be no surprise that expected values and standard deviations adjust to shifts and changes of units in the same way as the corresponding data summaries. The fact that we can add variances of independent random quantities is fundamental and will explain why a number of statistical methods work the way they do.

WHAT HAVE WE LEARNED?

We've learned to work with random variables. We can use the probability model for a discrete random variable to find its expected value and its standard deviation.

We've learned that the mean of the sum or difference of two random variables, discrete or continuous, is just the sum or difference of their means. And we've learned the Pythagorean Theorem of Statistics: *For independent random variables*, the variance of their sum or difference is always the *sum* of their variances.

Finally, we've learned that Normal models are once again special. Sums or differences of Normally distributed random variables also follow Normal models.

Terms

Random variable
: A random variable assumes any of several different numeric values as a result of some random event. Random variables are denoted by a capital letter such as X (p. 381).

Discrete random variable
: A random variable that can take one of a finite number[4] of distinct outcomes is called a discrete random variable (p. 382).

---

[3] Although some students do manage to attain a state of consciousness somewhere between sleeping and wakefulness during Statistics class.

[4] Technically, there could be an infinite number of outcomes, as long as they're *countable*. Essentially that means we can imagine listing them all in order, like the counting numbers 1, 2, 3, 4, 5, . . .

Continuous random variable	A random variable that can take any numeric value within a range of values is called a continuous random variable. The range may be infinite or bounded at either or both ends (pp. 382, 392).
Probability model	The probability model is a function that associates a probability P with each value of a discrete random variable X, denoted $P(X = x)$, or with any interval of values of a continuous random variable (p. 382).
Expected value	The expected value of a random variable is its theoretical long-run average value, the center of its model. Denoted μ or $E(X)$, it is found (if the random variable is discrete) by summing the products of variable values and probabilities (p. 382): $$\mu = E(X) = \sum xP(x).$$
Variance	The variance of a random variable is the expected value of the squared deviation from the mean. For discrete random variables, it can be calculated as (p. 384): $$\sigma^2 = Var(X) = \sum (x - \mu)^2 P(x).$$
Standard deviation	The standard deviation of a random variable describes the spread in the model, and is the square root of the variance (p. 384): $$\sigma = SD(X) = \sqrt{Var(X)}.$$
Changing a random variable by a constant:	$E(X \pm c) = E(X) \pm c \qquad Var(X \pm c) = Var(X)$ (p. 386). $E(aX) = aE(X) \qquad\qquad Var(aX) = a^2 Var(X)$ (p. 387).
Adding or subtracting random variables:	$E(X \pm Y) = E(X) \pm E(Y)$ (p. 387). $Var(X \pm Y) = Var(X) + Var(Y)$ If X and Y are independent, (p. 388). (The Pythagorean Theorem of Statistics) (p. 389).
***Covariance**	The covariance of random variables X and Y is $Cov(X, Y) = E((X - \mu)(Y - \nu))$ where $\mu = E(X)$ and $\nu = E(Y)$. In general (no need to assume independence) $Var(X \pm Y) = Var(X) + Var(Y) \pm 2Cov(X, Y)$ (p. 395).

Skills

▶ Be able to recognize random variables.

▶ Understand that random variables must be independent in order to determine the variability of their sum or difference by adding variances.

▶ Be able to find the probability model for a discrete random variable.

▶ Know how to find the mean (expected value) and the variance of a random variable.

▶ Always use the proper notation for these population parameters: μ or $E(X)$ for the mean, and σ, $SD(X)$, σ^2, or $Var(X)$ when discussing variability.

▶ Know how to determine the new mean and standard deviation after adding a constant, multiplying by a constant, or adding or subtracting two independent random variables.

▶ Be able to interpret the meaning of the expected value and standard deviation of a random variable in the proper context.

RANDOM VARIABLES ON THE COMPUTER

Statistics packages deal with data, not with random variables. Nevertheless, the calculations needed to find means and standard deviations of random variables are little more than weighted means. Most packages can manage that, but then they are just being overblown calculators. For technological assistance with these calculations, we recommend you pull out your calculator.

TI-83/84 Plus

To calculate the mean and standard deviation of a discrete random variable, enter the probability model in two lists:

- In one list (say, L1) enter the x-values of the variable.
- In a second list (say, L2) enter the associated probabilities $P(X = x)$.
- From the **STAT CALC** menu select **1-VarStats** and specify the two lists. You'll see the mean and standard deviation.

COMMENTS

You can enter the probabilities as fractions; the calculator will change them to decimals for you.

Notice that the calculator knows enough to call the standard deviation σ, but mistakenly uses $\bar{x}$ when it should say μ. Make sure you don't make that mistake!

TI-89

To calculate the mean and standard deviation of a discrete random variable, enter the probability model in two lists:

- In one list (say, list1) enter the x-values of the variable.
- In a second list (say, list2) enter the associated probabilities $P(X = x)$.
- From the **STAT CALC** ([F4]) menu select **1-VarStats**. Use VAR-LINK to enter the list name list1 in the List box and list2 in the Freq box.

COMMENTS

You can enter the probabilities as fractions; the calculator will change them to decimals for you.

Notice that the calculator knows enough to compute only the standard deviation σ, but mistakenly uses $\bar{x}$ when it should say μ. Make sure you don't make that mistake!

EXERCISES

1. **Expected value.** Find the expected value of each random variable:

   a)

x	10	20	30
$P(X = x)$	0.3	0.5	0.2

   b)

x	2	4	6	8
$P(X = x)$	0.3	0.4	0.2	0.1

2. **Expected value.** Find the expected value of each random variable:

   a)

x	0	1	2
$P(X = x)$	0.2	0.4	0.4

   b)

x	100	200	300	400
$P(X = x)$	0.1	0.2	0.5	0.2

3. **Pick a card, any card.** You draw a card from a deck. If you get a red card, you win nothing. If you get a spade, you win $5. For any club, you win $10 plus an extra $20 for the ace of clubs.

   a) Create a probability model for the amount you win.
   b) Find the expected amount you'll win.
   c) What would you be willing to pay to play this game?

4. **You bet!** You roll a die. If it comes up a 6, you win $100. If not, you get to roll again. If you get a 6 the second time, you win $50. If not, you lose.

   a) Create a probability model for the amount you win.
   b) Find the expected amount you'll win.
   c) What would you be willing to pay to play this game?

5. **Kids.** A couple plans to have children until they get a girl, but they agree that they will not have more than three children even if all are boys. (Assume boys and girls are equally likely.)

   a) Create a probability model for the number of children they might have.
   b) Find the expected number of children.
   c) Find the expected number of boys they'll have.

6. **Carnival.** A carnival game offers a $100 cash prize for anyone who can break a balloon by throwing a dart at it. It costs $5 to play, and you're willing to spend up to $20 trying to win. You estimate that you have about a 10% chance of hitting the balloon on any throw.

   a) Create a probability model for this carnival game.
   b) Find the expected number of darts you'll throw.
   c) Find your expected winnings.

7. **Software.** A small software company bids on two contracts and knows it can only get one of them. It anticipates a profit of $50,000 if it gets the larger contract and a profit of $20,000 on the smaller contract. The company estimates there's a 30% chance it will get the larger contract and a 60% chance it will get the smaller contract. Assuming the contracts will be awarded independently, what's the expected profit?

8. **Racehorse.** A man buys a racehorse for $20,000 and enters it in two races. He plans to sell the horse afterward, hoping to make a profit. If the horse wins both races, its value will jump to $100,000. If it wins one of the races, it will be worth $50,000. If it loses both races, it will be worth only $10,000. The man believes there's a 20% chance that the horse will win the first race and a 30% chance it will win the second one. Assuming that the two races are independent events, find the man's expected profit.

9. **Variation 1.** Find the standard deviations of the random variables in Exercise 1.

10. **Variation 2.** Find the standard deviations of the random variables in Exercise 2.

11. **Pick another card.** Find the standard deviation of the amount you might win drawing a card in Exercise 3.

12. **The die.** Find the standard deviation of the amount you might win rolling a die in Exercise 4.

13. **Kids, again.** Find the standard deviation of the number of children the couple in Exercise 5 may have.

14. **Darts.** Find the standard deviation of your winnings throwing darts in Exercise 6.

15. **Repairs.** The probability model below describes the number of repair calls that an appliance repair shop may receive during an hour.

Repair Calls	0	1	2	3
Probability	0.1	0.3	0.4	0.2

   a) How many calls should the shop expect per hour?
   b) What is the standard deviation?

16. **Red lights.** A commuter must pass through five traffic lights on her way to work and will have to stop at each one that is red. She estimates the probability model for the number of red lights she hits, as shown below.

X = # of red	0	1	2	3	4	5
P(X = x)	0.05	0.25	0.35	0.15	0.15	0.05

   a) How many red lights should she expect to hit each day?
   b) What's the standard deviation?

17. **Defects.** A consumer organization inspecting new cars found that many had appearance defects (dents, scratches, paint chips, etc.). While none had more than three of these defects, 7% had three, 11% two, and 21% one defect. Find the expected number of appearance defects in a new car and the standard deviation.

18. **Insurance.** An insurance policy costs $100 and will pay policyholders $10,000 if they suffer a major injury (resulting in hospitalization) or $3000 if they suffer a minor injury (resulting in lost time from work). The company estimates that each year 1 in every 2000 policyholders may have a major injury, and 1 in 500 a minor injury only.
   a) Create a probability model for the profit on a policy.
   b) What's the company's expected profit on this policy?
   c) What's the standard deviation?

19. **Cancelled flights.** Mary is deciding whether to book the cheaper flight home from college after her final exams, but she's unsure when her last exam will be. She thinks there is only a 20% chance that the exam will be scheduled after the last day she can get a seat on the cheaper flight. If it is and she has to cancel the flight, she will lose $150. If she can take the cheaper flight, she will save $100.
   a) If she books the cheaper flight, what can she expect to gain, on average?
   b) What is the standard deviation?

20. **Day trading.** An option to buy a stock is priced at $200. If the stock closes above 30 on May 15, the option will be worth $1000. If it closes below 20, the option will be worth nothing, and if it closes between 20 and 30 (inclusively), the option will be worth $200. A trader thinks there is a 50% chance that the stock will close in the 20–30 range, a 20% chance that it will close above 30, and a 30% chance that it will fall below 20 on May 15.
   a) Should she buy the stock option?
   b) How much does she expect to gain?
   c) What is the standard deviation of her gain?

21. **Contest.** You play two games against the same opponent. The probability you win the first game is 0.4. If you win the first game, the probability you also win the second is 0.2. If you lose the first game, the probability that you win the second is 0.3.
   a) Are the two games independent? Explain.
   b) What's the probability you lose both games?
   c) What's the probability you win both games?
   d) Let random variable X be the number of games you win. Find the probability model for X.
   e) What are the expected value and standard deviation?

22. **Contracts.** Your company bids for two contracts. You believe the probability you get contract #1 is 0.8. If you get contract #1, the probability you also get contract #2 will be 0.2, and if you do not get #1, the probability you get #2 will be 0.3.
   a) Are the two contracts independent? Explain.
   b) Find the probability you get both contracts.
   c) Find the probability you get no contract.
   d) Let X be the number of contracts you get. Find the probability model for X.
   e) Find the expected value and standard deviation.

23. **Batteries.** In a group of 10 batteries, 3 are dead. You choose 2 batteries at random.
   a) Create a probability model for the number of good batteries you get.
   b) What's the expected number of good ones you get?
   c) What's the standard deviation?

24. Kittens. In a litter of seven kittens, three are female. You pick two kittens at random.
   a) Create a probability model for the number of male kittens you get.
   b) What's the expected number of males?
   c) What's the standard deviation?

25. Random variables. Given independent random variables with means and standard deviations as shown, find the mean and standard deviation of:
   a) $3X$
   b) $Y + 6$
   c) $X + Y$
   d) $X - Y$
   e) $X_1 + X_2$

	Mean	SD
X	10	2
Y	20	5

26. Random variables. Given independent random variables with means and standard deviations as shown, find the mean and standard deviation of:
   a) $X - 20$
   b) $0.5Y$
   c) $X + Y$
   d) $X - Y$
   e) $Y_1 + Y_2$

	Mean	SD
X	80	12
Y	12	3

27. Random variables. Given independent random variables with means and standard deviations as shown, find the mean and standard deviation of:
   a) $0.8Y$
   b) $2X - 100$
   c) $X + 2Y$
   d) $3X - Y$
   e) $Y_1 + Y_2$

	Mean	SD
X	120	12
Y	300	16

28. Random variables. Given independent random variables with means and standard deviations as shown, find the mean and standard deviation of:
   a) $2Y + 20$
   b) $3X$
   c) $0.25X + Y$
   d) $X - 5Y$
   e) $X_1 + X_2 + X_3$

	Mean	SD
X	80	12
Y	12	3

29. Eggs. A grocery supplier believes that in a dozen eggs, the mean number of broken ones is 0.6 with a standard deviation of 0.5 eggs. You buy 3 dozen eggs without checking them.
   a) How many broken eggs do you expect to get?
   b) What's the standard deviation?
   c) What assumptions did you have to make about the eggs in order to answer this question?

30. Garden. A company selling vegetable seeds in packets of 20 estimates that the mean number of seeds that will actually grow is 18, with a standard deviation of 1.2 seeds. You buy 5 different seed packets.
   a) How many bad (non-growing) seeds do you expect to get?
   b) What's the standard deviation?
   c) What assumptions did you make about the seeds? Do you think that assumption is warranted? Explain.

31. Repair calls. Find the mean and standard deviation of the number of repair calls the appliance shop in Exercise 15 should expect during an 8-hour day.

32. Stop! Find the mean and standard deviation of the number of red lights the commuter in Exercise 16 should expect to hit on her way to work during a 5-day work week.

33. Tickets. A delivery company's trucks occasionally get parking tickets, and based on past experience, the company plans that the trucks will average 1.3 tickets a month, with a standard deviation of 0.7 tickets.
   a) If they have 18 trucks, what are the mean and standard deviation of the total number of parking tickets the company will have to pay this month?
   b) What assumption did you make in answering?

34. Donations. Organizers of a televised fundraiser know from past experience that most people donate small amounts ($10–$25), some donate larger amounts ($50–$100), and a few people make very generous donations of $250, $500, or more. Historically, pledges average about $32 with a standard deviation of $54.
   a) If 120 people call in pledges, what are the mean and standard deviation of the total amount raised?
   b) What assumption did you make in answering this question?

35. Fire! An insurance company estimates that it should make an annual profit of $150 on each homeowner's policy written, with a standard deviation of $6000.
   a) Why is the standard deviation so large?
   b) If it writes only two of these policies, what are the mean and standard deviation of the annual profit?
   c) If it writes 10,000 of these policies, what are the mean and standard deviation of the annual profit?
   d) Is the company likely to be profitable? Explain.
   e) What assumptions underlie your analysis? Can you think of circumstances under which those assumptions might be violated? Explain.

36. Casino. A casino knows that people play the slot machines in hopes of hitting the jackpot but that most of them lose their dollar. Suppose a certain machine pays out an average of $0.92, with a standard deviation of $120.
   a) Why is the standard deviation so large?
   b) If you play 5 times, what are the mean and standard deviation of the casino's profit?
   c) If gamblers play this machine 1000 times in a day, what are the mean and standard deviation of the casino's profit?
   d) Is the casino likely to be profitable? Explain.

37. Cereal. The amount of cereal that can be poured into a small bowl varies with a mean of 1.5 ounces and a standard deviation of 0.3 ounces. A large bowl holds a mean of 2.5 ounces with a standard deviation of 0.4 ounces. You open a new box of cereal and pour one large and one small bowl.
   a) How much more cereal do you expect to be in the large bowl?
   b) What's the standard deviation of this difference?
   c) If the difference follows a Normal model, what's the probability the small bowl contains more cereal than the large one?

d) What are the mean and standard deviation of the total amount of cereal in the two bowls?

e) If the total follows a Normal model, what's the probability you poured out more than 4.5 ounces of cereal in the two bowls together?

f) The amount of cereal the manufacturer puts in the boxes is a random variable with a mean of 16.3 ounces and a standard deviation of 0.2 ounces. Find the expected amount of cereal left in the box and the standard deviation.

38. **Pets.** The American Veterinary Association claims that the annual cost of medical care for dogs averages $100, with a standard deviation of $30, and for cats averages $120, with a standard deviation of $35.

a) What's the expected difference in the cost of medical care for dogs and cats?

b) What's the standard deviation of that difference?

c) If the costs can be described by Normal models, what's the probability that medical expenses are higher for someone's dog than for her cat?

d) What concerns do you have?

39. **More cereal.** In Exercise 37 we poured a large and a small bowl of cereal from a box. Suppose the amount of cereal that the manufacturer puts in the boxes is a random variable with mean 16.2 ounces and standard deviation 0.1 ounces.

a) Find the expected amount of cereal left in the box.

b) What's the standard deviation?

c) If the weight of the remaining cereal can be described by a Normal model, what's the probability that the box still contains more than 13 ounces?

40. **More pets.** You're thinking about getting two dogs and a cat. Assume that annual veterinary expenses are independent and have a Normal model with the means and standard deviations described in Exercise 38.

a) Define appropriate variables and express the total annual veterinary costs you may have.

b) Describe the model for this total cost. Be sure to specify its name, expected value, and standard deviation.

c) What's the probability that your total expenses will exceed $400?

41. **Medley.** In the 4 × 100 medley relay event, four swimmers swim 100 yards, each using a different stroke. A college team preparing for the conference championship looks at the times their swimmers have posted and creates a model based on the following assumptions:

- The swimmers' performances are independent.
- Each swimmer's times follow a Normal model.
- The means and standard deviations of the times (in seconds) are as shown:

Swimmer	Mean	SD
1 (backstroke)	50.72	0.24
2 (breaststroke)	55.51	0.22
3 (butterfly)	49.43	0.25
4 (freestyle)	44.91	0.21

a) What are the mean and standard deviation for the relay team's total time in this event?

b) The team's best time so far this season was 3:19.48. (That's 199.48 seconds.) Do you think the team is likely to swim faster than this at the conference championship? Explain.

42. **Bikes.** Bicycles arrive at a bike shop in boxes. Before they can be sold, they must be unpacked, assembled, and tuned (lubricated, adjusted, etc.). Based on past experience, the shop manager makes the following assumptions about how long this may take:

- The times for each setup phase are independent.
- The times for each phase follow a Normal model.
- The means and standard deviations of the times (in minutes) are as shown:

Phase	Mean	SD
Unpacking	3.5	0.7
Assembly	21.8	2.4
Tuning	12.3	2.7

a) What are the mean and standard deviation for the total bicycle setup time?

b) A customer decides to buy a bike like one of the display models but wants a different color. The shop has one, still in the box. The manager says they can have it ready in half an hour. Do you think the bike will be set up and ready to go as promised? Explain.

43. **Farmers' market.** A farmer has 100 lb of apples and 50 lb of potatoes for sale. The market price for apples (per pound) each day is a random variable with a mean of 0.5 dollars and a standard deviation of 0.2 dollars. Similarly, for a pound of potatoes, the mean price is 0.3 dollars and the standard deviation is 0.1 dollars. It also costs him 2 dollars to bring all the apples and potatoes to the market. The market is busy with eager shoppers, so we can assume that he'll be able to sell all of each type of produce at that day's price.

a) Define your random variables, and use them to express the farmer's net income.

b) Find the mean.

c) Find the standard deviation of the net income.

d) Do you need to make any assumptions in calculating the mean? How about the standard deviation?

44. **Bike sale.** The bicycle shop in Exercise 42 will be offering 2 specially priced children's models at a sidewalk sale. The basic model will sell for $120 and the deluxe model for $150. Past experience indicates that sales of the basic model will have a mean of 5.4 bikes with a standard deviation of 1.2, and sales of the deluxe model will have a mean of 3.2 bikes with a standard deviation of 0.8 bikes. The cost of setting up for the sidewalk sale is $200.

a) Define random variables and use them to express the bicycle shop's net income.

b) What's the mean of the net income?

c) What's the standard deviation of the net income?

d) Do you need to make any assumptions in calculating the mean? How about the standard deviation?

45. Coffee and doughnuts. At a certain coffee shop, all the customers buy a cup of coffee; some also buy a doughnut. The shop owner believes that the number of cups he sells each day is normally distributed with a mean of 320 cups and a standard deviation of 20 cups. He also believes that the number of doughnuts he sells each day is independent of the coffee sales and is normally distributed with a mean of 150 doughnuts and a standard deviation of 12.

a) The shop is open every day but Sunday. Assuming day-to-day sales are independent, what's the probability he'll sell more than 2000 cups of coffee in a week?

b) If he makes a profit of 50 cents on each cup of coffee and 40 cents on each doughnut, can he reasonably expect to have a day's profit of over $300? Explain.

c) What's the probability that on any given day he'll sell a doughnut to more than half of his coffee customers?

46. Weightlifting. The Atlas BodyBuilding Company (ABC) sells "starter sets" of barbells that consist of one bar, two 20-pound weights, and four 5-pound weights. The bars weigh an average of 10 pounds with a standard deviation of 0.25 pounds. The weights average the specified amounts, but the standard deviations are 0.2 pounds for the 20-pounders and 0.1 pounds for the 5-pounders. We can assume that all the weights are normally distributed.

a) ABC ships these starter sets to customers in two boxes: The bar goes in one box and the six weights go in another. What's the probability that the total weight in that second box exceeds 60.5 pounds? Define your variables clearly and state any assumptions you make.

b) It costs ABC $0.40 per pound to ship the box containing the weights. Because it's an odd-shaped package, though, shipping the bar costs $0.50 a pound plus a $6.00 surcharge. Find the mean and standard deviation of the company's total cost for shipping a starter set.

c) Suppose a customer puts a 20-pound weight at one end of the bar and the four 5-pound weights at the other end. Although he expects the two ends to weigh the same, they might differ slightly. What's the probability the difference is more than a quarter of a pound?

JUST CHECKING

ANSWERS

1. a)

Outcome	X = cost	Probability
Recharging works	$60	0.75
Replace control unit	$200	0.25

b) $60(0.75) + 200(0.25) = \$95$

c) Car owners with this problem will spend an average of $95 to get it fixed.

2. a) $100 + 100 = 200$ seconds

b) $\sqrt{50^2 + 50^2} = 70.7$ seconds

c) The times for the two customers are independent.

Probability Models

Where are we going?

How many frogs do you have to kiss before you meet Prince Charming? Is finding 10 leukemia cases in a neighborhood adjacent to an EPA superfund site grounds for a lawsuit? We could simulate to answer almost any question involving probabilities. But in some situations, we can do better with a simple probability model. We'll meet a few of the most common models in this chapter.

S uppose a cereal manufacturer puts pictures of famous athletes on cards in boxes of cereal, in the hope of increasing sales. The manufacturer announces that 20% of the boxes contain a picture of LeBron James, 30% a picture of David Beckham, and the rest a picture of Serena Williams.

Sound familiar? In Chapter 11 we simulated to find the number of boxes we'd need to open to get one of each card. That's a fairly complex question and one well suited for simulation. But many important questions can be answered more directly by using simple probability models.

Searching for LeBron

You're a huge LeBron James fan. You don't care about completing the whole sports card collection, but you've just *got* to have the LeBron James picture. How many boxes do you expect you'll have to open before you find him? This isn't the same question that we asked before, but this situation is simple enough for a probability model.

We'll keep the assumption that pictures are distributed at random and we'll trust the manufacturer's claim that 20% of the cards are LeBron. So, when you open the box, the probability that you succeed in finding LeBron is 0.20. Now we'll call the act of opening *each* box a trial, and note that:

- There are only two possible outcomes (called *success* and *failure*) on each trial. Either you get LeBron's picture (success), or you don't (failure).

- The probability of success, denoted p, is the same on every trial. Here $p = 0.20$.
- The trials are independent. Finding LeBron in the first box does not change what might happen when you reach for the next box.

Situations like this occur often, and are called **Bernoulli trials**. Common examples of Bernoulli trials include tossing a coin, looking for defective products rolling off an assembly line, or even shooting free throws in a basketball game. Just as we found equally likely random digits to be the building blocks for our simulation, we can use Bernoulli trials to build a wide variety of useful probability models.

Daniel Bernoulli (1700-1782) was the nephew of Jacob, whom you saw in Chapter 14. He was the first to work out the mathematics for what we now call Bernoulli trials.

A S *Activity:* **Bernoulli Trials.** Guess what! We've been generating Bernoulli trials all along. Look at the Random Simulation Tool in a new way.

CALVIN AND HOBBES © 1993 Watterson. Reprinted with permission of UNIVERSAL PRESS SYNDICATE. All rights reserved.

Back to LeBron. We want to know how many boxes we'll need to open to find his card. Let's call this random variable $Y = \#$ boxes, and build a probability model for it. What's the probability you find his picture in the first box of cereal? It's 20%, of course. We could write $P(Y = 1) = 0.20$.

How about the probability that you don't find LeBron until the second box? Well, that means you fail on the first trial and then succeed on the second. With the probability of success 20%, the probability of failure, denoted q, is $1 - 0.2 = 80\%$. Since the trials are independent, the probability of getting your first success on the second trial is $P(Y = 2) = (0.8)(0.2) = 0.16$.

Of course, you could have a run of bad luck. Maybe you won't find LeBron until the fifth box of cereal. What are the chances of that? You'd have to fail 4 straight times and then succeed, so $P(Y = 5) = (0.8)^4(0.2) = 0.08192$.

How many boxes might you expect to have to open? We could reason that since LeBron's picture is in 20% of the boxes, or 1 in 5, we expect to find his picture, on average, in the fifth box; that is, $E(Y) = \frac{1}{0.2} = 5$ boxes. That's correct, but not easy to prove.

The Geometric Model

We want to model how long it will take to achieve the first success in a series of Bernoulli trials. The model that tells us this probability is called the **Geometric probability model.** Geometric models are completely specified by one parameter, p, the probability of success, and are denoted Geom(p). Since achieving the first success on trial number x requires first experiencing $x - 1$ failures, the probabilities are easily expressed by a formula.

NOTATION ALERT

Now we have two more reserved letters. Whenever we deal with Bernoulli trials, p represents the probability of success, and q the probability of failure. (Of course, $q = 1 - p$.)

GEOMETRIC PROBABILITY MODEL FOR BERNOULLI TRIALS: Geom(p)

p = probability of success (and $q = 1 - p$ = probability of failure)
X = number of trials until the first success occurs

$$P(X = x) = q^{x-1}p$$

Expected value: $E(X) = \mu = \dfrac{1}{p}$

Standard deviation: $\sigma = \sqrt{\dfrac{q}{p^2}}$

FOR EXAMPLE   Spam and the Geometric Model

Postini is a global company specializing in communications security. The company monitors over 1 billion Internet messages per day and recently reported that 91% of e-mails are spam!

Let's assume that your e-mail is typical–91% spam. We'll also assume you aren't using a spam filter, so every message gets dumped in your inbox. And, since spam comes from many different sources, we'll consider your messages to be independent.

QUESTIONS: Overnight your inbox collects e-mail. When you first check your e-mail in the morning, about how many spam e-mails should you expect to have to wade through and discard before you find a real message? What's the probability that the fourth message in your inbox is the first one that isn't spam?

There are two outcomes: a real message (success) and spam (failure). Since 91% of e-mails are spam, the probability of success $p = 1 - 0.91 = 0.09$.

Let X = the number of e-mails I'll check until I find a real message. I assume that the messages arrive independently and in a random order. I can use the model Geom(0.09).

$$E(X) = \frac{1}{p} = \frac{1}{0.09} = 11.1$$

$$P(X = 4) = (0.91)^3(0.09) = 0.0678$$

On average, I expect to have to check just over 11 e-mails before I find a real message. There's slightly less than a 7% chance that my first real message will be the fourth one I check.

Note that the probability calculation isn't new. It's simply Chapter 14's Multiplication Rule used to find P(spam ∩ spam ∩ spam ∩ real).

MATH BOX

We want to find the mean (expected value) of random variable X, using a geometric model with probability of success p.

First, write the probabilities:

x	1	2	3	4	⋯
$P(X = x)$	p	qp	q^2p	q^3p	⋯

The expected value is:
$$E(X) = 1p + 2qp + 3q^2p + 4q^3p + \cdots$$

Let $p = 1 - q$:
$$= (1 - q) + 2q(1 - q) + 3q^2(1 - q) + 4q^3(1 - q) + \cdots$$

Simplify:
$$= 1 - q + 2q - 2q^2 + 3q^2 - 3q^3 + 4q^3 - 4q^4 + \cdots$$

That's an infinite geometric series, with first term 1 and common ratio q:

$$= 1 + q + q^2 + q^3 + \cdots$$

$$= \frac{1}{1 - q}$$

So, finally . . .

$$E(X) = \frac{1}{p}.$$

Independence

One of the important requirements for Bernoulli trials is that the trials be independent. Sometimes that's a reasonable assumption—when tossing a coin or rolling a die, for example. But that becomes a problem when (often!) we're looking at situations involving samples chosen without replacement. We said that whether we find a LeBron James card in one box has no effect on the probabilities in other boxes. This is *almost* true. Technically, if exactly 20% of the boxes have LeBron James cards, then when you find one, you've reduced the number of remaining LeBron James cards. With a few million boxes of cereal, though, the difference is hardly worth mentioning.

But if you knew there were 2 LeBron James cards hiding in the 10 boxes of cereal on the market shelf, then finding one in the first box you try would clearly change your chances of finding LeBron in the next box.

If we had an infinite number of boxes, there wouldn't be a problem. It's selecting from a finite population that causes the probabilities to change, making the trials not independent. Obviously, taking 2 out of 10 boxes changes the probability. Taking even a few hundred out of millions, though, makes very little difference. Fortunately, we have a rule of thumb for the in-between cases. It turns out that if we look at less than 10% of the population, we can pretend that the trials are independent and still calculate probabilities that are quite accurate.

The 10% Condition: Bernoulli trials must be independent. If that assumption is violated, it is still okay to proceed as long as the sample is smaller than 10% of the population.

STEP-BY-STEP EXAMPLE | **Working with a Geometric Model**

People with O-negative blood are called "universal donors" because O-negative blood can be given to anyone else, regardless of the recipient's blood type. Only about 6% of people have O-negative blood.

Questions:

1. If donors line up at random for a blood drive, how many do you expect to examine before you find someone who has O-negative blood?

2. What's the probability that the first O-negative donor found is one of the first four people in line?

THINK

Plan State the questions.

I want to estimate how many people I'll need to check to find an O-negative donor, and the probability that 1 of the first 4 people is O-negative.

Check to see that these are Bernoulli trials.	✔ There are two outcomes: success = O-negative failure = other blood types
	✔ The probability of success for each person is $p = 0.06$, because they lined up randomly.
	✔ **10% Condition:** Trials aren't independent because the population is finite, but the donors lined up are fewer than 10% of all possible donors.
Variable Define the random variable.	Let X = number of donors until one is O-negative.
Model Specify the model.	I can model X with Geom(0.06).

SHOW **Mechanics** Find the mean. Calculate the probability of success on one of the first four trials. That's the probability that $X = 1, 2, 3,$ *or* 4.	$E(X) = \dfrac{1}{0.06} \approx 16.7$ $\begin{aligned}P(X \le 4) = {} & P(X = 1) + P(X = 2) + \\ & P(X = 3) + P(X = 4) \\ = {} & (0.06) + (0.94)(0.06) + \\ & (0.94)^2(0.06) + (0.94)^3(0.06) \\ \approx {} & 0.2193\end{aligned}$

TELL **Conclusion** Interpret your results in context.	Blood drives such as this one expect to examine an average of 16.7 people to find a universal donor. About 22% of the time there will be one within the first 4 people in line.

The Binomial Model

A S
Activity: The Binomial Distribution.
It's more interesting to combine Bernoulli trials. Simulate this with the Random Tool to get a sense of how Binomial models behave.

We can use the Bernoulli trials to answer other questions. Suppose you buy 5 boxes of cereal. What's the probability you get *exactly* 2 pictures of LeBron James? Before, we asked how long it would take until our first success. Now we want to find the probability of getting 2 successes among the 5 trials. We are still talking about Bernoulli trials, but we're asking a different question.

This time we're interested in the *number of successes* in the 5 trials, so we'll call it X = number of successes. We want to find $P(X = 2)$. This is an example of a **Binomial probability.** It takes two parameters to define this **Binomial model:** the number of trials, n, and the probability of success, p. We denote this model Binom(n, p). Here, $n = 5$ trials, and $p = 0.2$, the probability of finding a LeBron James card in any trial.

Exactly 2 successes in 5 trials means 2 successes and 3 failures. It seems logical that the probability should be $(0.2)^2(0.8)^3$. Too bad! It's not that easy. That calculation would give you the probability of finding LeBron in the first 2 boxes and not in the next 3—*in that order.* But you could find LeBron in the third and fifth boxes and still have 2 successes. The probability of those outcomes in that particular order is $(0.8)(0.8)(0.2)(0.8)(0.2)$. That's also $(0.2)^2(0.8)^3$. In fact, the probability will always be the same, no matter what order the successes and failures occur in. Anytime we get 2 successes in 5 trials, no matter what the order, the probability will be $(0.2)^2(0.8)^3$. We just need to take account of all the possible orders in which the outcomes can occur.

Fortunately, these possible orders are *disjoint*. (For example, if your two successes came on the first two trials, they couldn't come on the last two.) So we can use the Addition Rule and add up the probabilities for all the possible orderings. Since the probabilities are all the same, we only need to know how many orders are possible. For small numbers, we can just make a tree diagram and count the branches. For larger numbers this isn't practical, so we let the computer or calculator do the work.

Each different order in which we can have k successes in n trials is called a "combination." The total number of ways that can happen is written $\binom{n}{k}$ or $_nC_k$ and pronounced "*n* choose *k*."

$$_nC_k = \frac{n!}{k!(n-k)!} \text{ where } n! \text{ (pronounced "n factorial")} = n \times (n-1) \times \cdots \times 1$$

For 2 successes in 5 trials,

$$_5C_2 = \frac{5!}{2!(5-2)!} = \frac{5 \times 4 \times 3 \times 2 \times 1}{2 \times 1 \times 3 \times 2 \times 1} = \frac{5 \times 4}{2 \times 1} = 10.$$

So there are 10 ways to get 2 LeBron pictures in 5 boxes, and the probability of each is $(0.2)^2(0.8)^3$. Now we can find what we wanted:

$$P(\#success = 2) = 10(0.2)^2(0.8)^3 = 0.2048$$

In general, the probability of exactly k successes in n trials is $_nC_k p^k q^{n-k}$.

Using this formula, we could find the expected value by adding up $xP(X = x)$ for all values, but it would be a long, hard way to get an answer that you already know intuitively. What's the expected value? If we have 5 boxes, and LeBron's picture is in 20% of them, then we would expect to have $5(0.2) = 1$ success. If we had 100 trials with probability of success 0.2, how many successes would you expect? Can you think of any reason not to say 20? It seems so simple that most people wouldn't even stop to think about it. You just multiply the probability of success by n. In other words, $E(X) = np$.

The standard deviation is less obvious; you can't just rely on your intuition. Fortunately, the formula for the standard deviation also boils down to something simple: $SD(X) = \sqrt{npq}$. (If you're curious about where that comes from, it's in the Math Box, too!) In 100 boxes of cereal, we expect to find 20 LeBron James cards, with a standard deviation of $\sqrt{100 \times 0.8 \times 0.2} = 4$ pictures.

Time to summarize. A Binomial probability model describes the number of successes in a specified number of trials. It takes two parameters to specify this model: the number of trials n and the probability of success p.

BINOMIAL PROBABILITY MODEL FOR BERNOULLI TRIALS: BINOM(*n*, *p*)

n = number of trials
p = probability of success (and $q = 1 - p$ = probability of failure)
X = number of successes in n trials

$$P(X = x) = {_nC_x}\, p^x q^{n-x}, \text{ where } {_nC_x} = \frac{n!}{x!(n-x)!}$$

Mean: $\mu = np$
Standard Deviation: $\sigma = \sqrt{npq}$

MATH BOX

To derive the formulas for the mean and standard deviation of a Binomial model we start with the most basic situation.

Consider a single Bernoulli trial with probability of success p. Let's find the mean and variance of the number of successes.

Here's the probability model for the number of successes:

x	0	1
$P(X = x)$	q	p

Find the expected value:

$$E(X) = 0q + 1p$$
$$E(X) = p$$

And now the variance:

$$Var(X) = (0 - p)^2 q + (1 - p)^2 p$$
$$= p^2 q + q^2 p$$
$$= pq(p + q)$$
$$= pq(1)$$
$$Var(X) = pq$$

What happens when there is more than one trial, though? A Binomial model simply counts the number of successes in a series of n independent Bernoulli trials. That makes it easy to find the mean and standard deviation of a binomial random variable, Y.

$$\text{Let } Y = X_1 + X_2 + X_3 + \cdots + X_n$$
$$E(Y) = E(X_1 + X_2 + X_3 + \cdots + X_n)$$
$$= E(X_1) + E(X_2) + E(X_3) + \cdots + E(X_n)$$
$$= p + p + p + \cdots + p \text{ (There are } n \text{ terms.)}$$

So, as we thought, the mean is $E(Y) = np$.

And since the trials are independent, the variances add:

$$Var(Y) = Var(X_1 + X_2 + X_3 + \cdots + X_n)$$
$$= Var(X_1) + Var(X_2) + Var(X_3) + \cdots + Var(X_n)$$
$$= pq + pq + pq + \cdots + pq \text{ (Again, } n \text{ terms.)}$$
$$Var(Y) = npq$$

Voilà! The standard deviation is $SD(Y) = \sqrt{npq}$.

FOR EXAMPLE   **Spam and the Binomial Model**

RECAP: The communications monitoring company *Postini* has reported that 91% of e-mail messages are spam. Suppose your inbox contains 25 messages.

QUESTIONS: What are the mean and standard deviation of the number of real messages you should expect to find in your inbox? What's the probability that you'll find only 1 or 2 real messages?

I assume that messages arrive independently and at random, with the probability of success (a real message) $p = 1 - 0.91 = 0.09$. Let $X =$ the number of real messages among 25. I can use the model Binom(25, 0.09).

$$E(X) = np = 25(0.09) = 2.25$$
$$SD(X) = \sqrt{npq} = \sqrt{25(0.09)(0.91)} = 1.43$$

$$P(X = 1 \text{ or } 2) = P(X = 1) + P(X = 2)$$
$$= {}_{25}C_1(0.09)^1(0.91)^{24} + {}_{25}C_2(0.09)^2(0.91)^{23}$$
$$= 0.2340 + 0.2777$$
$$= 0.5117$$

Among 25 e-mail messages, I expect to find an average of 2.25 that aren't spam, with a standard deviation of 1.43 messages. There's just over a 50% chance that 1 or 2 of my 25 e-mails will be real messages.

STEP-BY-STEP EXAMPLE Working with a Binomial Model

Suppose 20 donors come to a blood drive. Recall that 6% of people are "universal donors."

Questions:

1. What are the mean and standard deviation of the number of universal donors among them?
2. What is the probability that there are 2 or 3 universal donors?

	Plan State the question.	I want to know the mean and standard deviation of the number of universal donors among 20 people, and the probability that there are 2 or 3 of them.
	Check to see that these are Bernoulli trials.	✔ There are two outcomes: success = O-negative failure = other blood types ✔ $p = 0.06$, because people have lined up at random. ✔ **10% Condition:** Trials are not independent, because the population is finite, but fewer than 10% of all possible donors are lined up.
	Variable Define the random variable.	Let X = number of O-negative donors among $n = 20$ people.
	Model Specify the model.	I can model X with Binom(20, 0.06).
	Mechanics Find the expected value and standard deviation.	$E(X) = np = 20(0.06) = 1.2$ $SD(X) = \sqrt{npq} = \sqrt{20(0.06)(0.94)} \approx 1.06$ $P(X = 2 \text{ or } 3) = P(X = 2) + P(X = 3)$ $\qquad = {}_{20}C_2(0.06)^2(0.94)^{18}$ $\qquad \quad + {}_{20}C_3(0.06)^3(0.94)^{17}$ $\qquad \approx 0.2246 + 0.0860$ $\qquad = 0.3106$

Conclusion Interpret your results in context.

In groups of 20 randomly selected blood donors, I expect to find an average of 1.2 universal donors, with a standard deviation of 1.06. About 31% of the time, I'd find 2 or 3 universal donors among the 20 people.

The Normal Model to the Rescue!

Suppose the Tennessee Red Cross anticipates the need for at least 1850 units of O-negative blood this year. It estimates that it will collect blood from 32,000 donors. How great is the risk that the Tennessee Red Cross will fall short of meeting its need? We've just learned how to calculate such probabilities. We can use the Binomial model with $n = 32,000$ and $p = 0.06$. The probability of getting *exactly* 1850 units of O-negative blood from 32,000 donors is $_{32000}C_{1850} \times 0.06^{1850} \times 0.94^{30150}$. No calculator on earth can calculate that first term (it has more than 100,000 digits).[1] And that's just the beginning. The problem said *at least* 1850, so we have to do it again for 1851, for 1852, and all the way up to 32,000. No thanks.

When we're dealing with a large number of trials like this, making direct calculations of the probabilities becomes tedious (or outright impossible). Here an old friend—the Normal model—comes to the rescue.

The Binomial model has mean $np = 1920$ and standard deviation $\sqrt{npq} \approx 42.48$. We could try approximating its distribution with a Normal model, using the same mean and standard deviation. Remarkably enough, that turns out to be a very good approximation. (We'll see why in the next chapter.) With that approximation, we can find the *probability*:

$$P(X < 1850) = P\left(z < \frac{1850 - 1920}{42.48}\right) \approx P(z < -1.65) \approx 0.05$$

There seems to be about a 5% chance that this Red Cross chapter will run short of O-negative blood.

A S *Activity: Normal Approximation.* Binomial probabilities can be hard to calculate. With the Simulation Tool you'll see how well the Normal model can approximate the Binomial—a much easier method.

Can we always use a Normal model to make estimates of Binomial probabilities? No. Consider the LeBron James situation—pictures in 20% of the cereal boxes. If we buy five boxes, the actual Binomial probabilities that we get 0, 1, 2, 3, 4, or 5 pictures of LeBron are 33%, 41%, 20%, 5%, 1%, and 0.03%, respectively. The first histogram shows that this probability model is skewed. That makes it clear that we should not try to estimate these probabilities by using a Normal model.

Now suppose we open 50 boxes of this cereal and count the number of LeBron James pictures we find. The second histogram shows this probability model. It is centered at $np = 50(0.2) = 10$ pictures, as expected, and it appears to be fairly symmetric around that center. Let's have a closer look.

The third histogram again shows Binom(50, 0.2), this time magnified somewhat and centered at the expected value of 10 pictures of LeBron. It looks close to Normal, for sure. With this larger sample size, it appears that a Normal model might be a useful approximation.

A Normal model, then, is a close enough approximation only for a large enough number of trials. And what we mean by "large enough" depends on the probability of success. We'd need a larger sample if the probability of

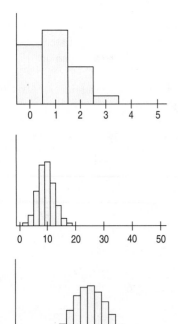

[1] If your calculator *can* find Binom(32000, 0.06), then it's smart enough to use an approximation. Read on to see how you can, too.

success were very low (or very high). It turns out that a Normal model works pretty well if we expect to see at least 10 successes and 10 failures. That is, we check the **Success/Failure Condition.**

The Success/Failure Condition: A Binomial model is approximately Normal if we expect at least 10 successes and 10 failures:

$$np \geq 10 \text{ and } nq \geq 10.$$

MATH BOX

It's easy to see where the magic number 10 comes from. You just need to remember how Normal models work. The problem is that a Normal model extends infinitely in both directions. But a Binomial model must have between 0 and n successes, so if we use a Normal to approximate a Binomial, we have to cut off its tails. That's not very important if the center of the Normal model is so far from 0 and n that the lost tails have only a negligible area. More than three standard deviations should do it, because a Normal model has little probability past that.

So the mean needs to be at least 3 standard deviations from 0 and at least 3 standard deviations from n. Let's look at the 0 end.

We require:	$\mu - 3\sigma > 0$
Or in other words:	$\mu > 3\sigma$
For a Binomial, that's:	$np > 3\sqrt{npq}$
Squaring yields:	$n^2p^2 > 9npq$
Now simplify:	$np > 9q$
Since $q \leq 1$, we can require:	$np > 9$

For simplicity, we usually require that np (and nq for the other tail) be at least 10 to use the Normal approximation, the Success/Failure Condition.[2]

FOR EXAMPLE   Spam and the Normal Approximation to the Binomial

RECAP: The communications monitoring company *Postini* has reported that 91% of e-mail messages are spam. Recently, you installed a spam filter. You observe that over the past week it okayed only 151 of 1422 e-mails you received, classifying the rest as junk. Should you worry that the filtering is too aggressive?

QUESTION: What's the probability that no more than 151 of 1422 e-mails is a real message?

I assume that messages arrive randomly and independently, with a probability of success (a real message) $p = 0.09$. The model Binom(1422, 0.09) applies, but will be hard to work with. Checking conditions for the Normal approximation, I see that:

✔ These messages represent less than 10% of all e-mail traffic.

✔ I expect $np = (1422)(0.09) = 127.98$ real messages and $nq = (1422)(0.91) = 1294.02$ spam messages, both far greater than 10.

(continued)

---

[2] Looking at the final step, we see that we need $np > 9$ in the worst case, when q (or p) is near 1, making the Binomial model quite skewed. When q and p are near 0.5—say between 0.4 and 0.6—the Binomial model is nearly symmetric and $np > 5$ ought to be safe enough. Although we'll always check for 10 expected successes and failures, keep in mind that for values of p near 0.5, we can be somewhat more forgiving.

It's okay to approximate this binomial probability by using a Normal model.

$\mu = np = 1422(0.09) = 127.98$

$\sigma = \sqrt{npq} = \sqrt{1422(0.09)(0.91)} \approx 10.79$

$P(x \le 151) = P\left(z \le \dfrac{151 - 127.98}{10.79}\right)$

$\qquad\qquad\quad = P(z \le 2.13)$

$\qquad\qquad\quad = 0.9834$

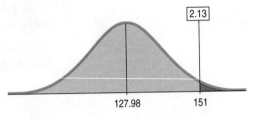

Among my 1422 e-mails, there's over a 98% chance that no more than 151 of them were real messages, so the filter may be working properly.

Continuous Random Variables

There's a problem with approximating a Binomial model with a Normal model. The Binomial is discrete, giving probabilities for specific counts, but the Normal models a **continuous random variable** that can take on *any value*. For continuous random variables, we can no longer list all the possible outcomes and their probabilities, as we could for discrete random variables.[3]

As we saw in the previous chapter, models for continuous random variables give probabilities for *intervals* of values. So, when we use the Normal model, we no longer calculate the probability that the random variable equals a *particular* value, but only that it lies *between* two values. We won't calculate the probability of getting exactly 1850 units of blood, but we have no problem approximating the probability of getting 1850 *or more*, which was, after all, what we really wanted.[4]

The Poisson Model

In the early 1990s, a leukemia cluster was identified in the Massachusetts town of Woburn. Many more cases of leukemia, a cancer that originates in a cell in the marrow of bone, appeared in this small town than would be predicted. Was it evidence of a problem in the town, or was it chance? That question led to a famous trial in which the families of eight leukemia victims sued, and became grist for the book and movie *A Civil Action*. Following an 80-day trial, the judge called for a retrial after dismissing the jury's contradictory and confusing findings. Shortly thereafter, the chemical companies and the families settled.

When rare events occur together or in clusters, people often want to know if that happened just by chance or whether something else is going on. If we assume that the events occur independently, we can use a Binomial model to find the probability that a cluster of events like this occurs. For rare events, p will be quite small, and when n is large, it may be difficult to compute the exact probability that a certain size cluster occurs. To see why, let's try to compute the probability that a cluster of cases of size x occurs in Woburn. We'll use the *national average* of leukemia incidence to get a value for p, and the population of Woburn as our value for n. In the United States in the early 1990s, there were about 30,800 new cases of leukemia each year and about 280,000,000 people, giving a value for p of about 0.00011. The population of Woburn was about $n = 35,000$. We'd expect $np = 3.85$ new cases of leukemia in Woburn. How unlikely would 8 or more

---

[3] In fact, some people use an adjustment called the "continuity correction" to help with this problem. It's related to the suggestion we make in the next footnote, and is discussed in more advanced textbooks.

[4] If we really had been interested in a single value, we might have approximated it by finding the probability of getting between 1849.5 and 1850.5 units of blood.

cases be? To answer that, we'll need to calculate the complement, adding the probabilities of no cases, exactly 1 case, etc., up to 7 cases. Each of those probabilities would have the form $_{35000}C_x p^x q^{35000-x}$. To find $_{35000}C_x$, we'll need 35,000!—a daunting number even for computers. In the 18th century, when people first were interested in computing such probabilities, it was impossible. We can easily go beyond the capabilities of today's computers by making n large enough and p small enough. We could use the Normal model to help us approximate the probabilities as long as np is at least 10, but for rare events such as leukemia in Woburn, when np falls much below 10, the Normal probabilities won't be accurate.

Simeon Denis Poisson[5] was a French mathematician interested in events with very small probability. He originally derived his model to approximate the Binomial model when the probability of a success, p, is very small and the number of trials, n, is very large. Poisson's contribution was providing a simple approximation to find that probability. When you see the formula, however, you won't necessarily see the connection to the Binomial. The Poisson's parameter is often denoted by λ, the mean of the distribution. To use the **Poisson model** to approximate the Binomial, we'll make their means match, so we set $\lambda = np$.

POISSON PROBABILITY MODEL FOR SUCCESSES: *Poisson* (λ)

λ = mean number of successes.
X = number of successes.

$$P(X = x) = \frac{e^{-\lambda}\lambda^x}{x!}$$

Expected value: $\quad E(X) = \lambda$

Standard deviation: $\quad SD(X) = \sqrt{\lambda}$

The Poisson model is a reasonably good approximation of the Binomial when $n \geq 20$ with $p \leq 0.05$ or $n \geq 100$ with $p \leq 0.10$.

Using Poisson's model, we can easily find the probabilities of a given size cluster of leukemia cases. Using $\lambda = np = 35000 \times 0.00011 = 3.85$ new cases a year, we find that the probability of seeing exactly x cases in a year is $P(X = x) = \frac{e^{-3.85}3.85^x}{x!}$. By adding up these probabilities for $x = 0, 1, \ldots 7$, we'd find that the probability of 8 or more cases with $\lambda = 3.85$ is about 0.043. That's small but not terribly unusual.

Where does *e* come from? You might know that $e = 2.7182818\ldots$ (to 7 decimal places), but does it have any real meaning? Yes! In fact, one of the places e originally turned up was in calculating how much more money you'd earn if you could get interest compounded more often. Suppose you receive an offer to earn 100% per year simple interest. (Take it!) At the end of the year, you'd have twice as much money as when you started. But suppose that you have the option of having the interest calculated and paid at the end of every month instead. Then each month you'll earn $(100/12)\%$ interest. At the end of the year, you'll wind up with $(1 + 1/12)^{12} = 2.613$ times as much instead of 2. If you could get the interest paid every day, you'd get $(1 + 1/365)^{365} = 2.715$ times as much. If you could convince the bank to compound and pay the interest every second, you'd get $(1 + 1/31536000)^{31536000} = 2.7182818$ times as much at year's end. This is where e shows up. In the limit, if you could get the interest compounded continually, you'd get e times as much. In other words, as n gets large, the limit of $(1 + 1/n)^n = e$. Who discovered this fact? Jacob Bernoulli, in 1683!

---

[5] *Poisson* is a French name (meaning "fish"), properly pronounced "pwa sohn" and not with an "oy" sound as in poison.

In spite of its origins as an approximation to the Binomial, the Poisson model is also used directly to model the probability of the occurrence of events for a variety of phenomena. It's a good model to consider whenever your data consist of counts of occurrences. It requires only that the events be independent and that the mean number of occurrences stay constant for the duration of the data collection (and beyond, if we hope to make predictions).

One nice feature of the Poisson model is that it scales according to the sample size.[6] For example, if we know that the average number of occurrences in a town the size of Woburn, with 35,000 people, is 3.85, we know that the average number of occurrences in a town of only 3500 residents is 0.385. We can use that new value of λ to calculate the probabilities for the smaller town. Using the Poisson formula, we'd find that the probability of 0 occurrences for a town of 3500 is about 0.68.

One of the consequences of the Poisson model is that, as long as the mean rate of occurrences stays constant, the occurrence of past events doesn't change the probability of future events. That is, even though events that occur according to the Poisson model *appear* to cluster, the probability of *another* event occurring is still the same. This is counterintuitive and may be one reason why so many people believe in the form of the "Law of Averages" that we argued against in Chapter 14. When the Poisson model was used to model bombs dropping over London in World War II, the model said that even though several bombs had hit a particular sector, the probability of *another* hit was still the same. So, there was no point in moving people from one sector to another, even after several hits. You can imagine how difficult it must have been trying to convince people of that! This same phenomenon leads many people to think they see patterns in which an athlete, slot machine, or financial market gets "hot." But the occurrence of several successes in a row is not unusual and is rarely evidence that the mean has changed. Careful studies have been made of "hot streaks," but none has ever demonstrated that they actually exist.

JUST CHECKING

As we noted a few chapters ago, the Pew Research Center (www.pewresearch.org) reports that they are actually able to contact only 76% of the randomly selected households drawn for a telephone survey.

1. Explain why these phone calls can be considered Bernoulli trials.

2. Which of the models of this chapter (Geometric, Binomial, Normal) would you use to model the number of successful contacts from a list of 1000 sampled households? Explain.

3. Pew further reports that even after they contacted a household, only 38% agree to be interviewed, so the probability of getting a completed interview for a randomly selected household is only 0.29. Which of the models of this chapter would you use to model the number of households Pew has to call before they get the first completed interview?

4. Suppose that in the course of a year, legitimate survey organizations (not folks pretending to take a survey but actually trying to sell you something or change your vote) sample 70,000 of the approximately 107,000,000 households in the United States. You wonder if people from your school are represented in these polls. Counting all the living alumni from your school, there are 10,000 households. How would you model the number of these alumni households you'd expect will be contacted next year by a legitimate poll?

---

[6] Because *Poisson* means "fish" in French, you can remember this fact by the bilingual pun "*Poisson* scales"!

*The Exponential Model

We saw that the Poisson model is a good model for the arrival, or occurrence, of events. We found, for example, the probability that x visits to our website will occur within the next minute. The exponential model with parameter λ can be used to model the time *between* those events. Its probability model[7] has the form:

$$f(x) = \lambda e^{-\lambda x} \quad for\, x \geq 0\, and\, \lambda > 0$$

The use of the parameter λ again is not coincidental. It highlights the relationship between the exponential and the Poisson.

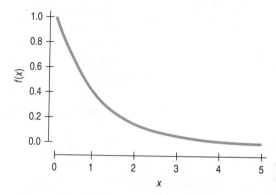

The exponential probability model (with $\lambda = 1$). The probability that x lies between any two values corresponds to the area under the curve between the two values.

If a discrete random variable can be modeled by a Poisson model with rate λ, then the times between events can be modeled by an exponential model with the same parameter λ. The mean of the exponential is $1/\lambda$. The inverse relationship between the two means makes intuitive sense. If λ increases and we expect *more* hits per minute, then the expected time between hits should go down. The standard deviation of an exponential random variable is $1/\lambda$.

Like any continuous random variable, probabilities of an exponential random variable can be found only through the probability model. Fortunately, the probability that x lies between any two values, s and t ($s \leq t$), has a particularly easy form:

$$P(s \leq X \leq t) = e^{-\lambda s} - e^{-\lambda t}.$$

In particular, by setting s to be 0, we can find the probability that the waiting time will be less than t from[8]

$$P(X \leq t) = P(0 \leq X \leq t) = e^{-\lambda 0} - e^{-\lambda t} = 1 - e^{-\lambda t}.$$

If arrivals of hits to our website can be well modeled by a Poisson with $\lambda = 4/$minute, then the probability that we'll have to wait less than 20 seconds (1/3 of a minute) is $F(1/3) = P(0 \leq X \leq 1/3) = 1 - e^{-4/3} = 0.736$. That seems about right. Arrivals are coming about every 15 seconds on average, so we shouldn't be surprised that nearly 75% of the time we won't have to wait more than 20 seconds for the next hit.

---

[7] Probability models for continuous random variables are also called density functions.
[8] The function $P(X \leq t) = F(t)$ is called the cumulative distribution function (cdf) of the random variable X. The probability model is formally called the density.

What Can Go Wrong?

- **Be sure you have Bernoulli trials.** Be sure to check the requirements first: two possible outcomes per trial ("success" and "failure"), a constant probability of success, and independence. Remember to check the 10% Condition when sampling without replacement.

- **Don't confuse Geometric and Binomial models.** Both involve Bernoulli trials, but the issues are different. If you are repeating trials until your first success, that's a Geometric probability. You don't know in advance how many trials you'll need—theoretically, it could take forever. If you are counting the number of successes in a specified number of trials, that's a Binomial probability.

- **Don't use the Normal approximation with small *n*.** To use a Normal approximation in place of a Binomial model, there must be at least 10 expected successes and 10 expected failures.

This chapter builds on what we know about random variables. We now have two more probability models to join the Normal model.

There are a number of "forward" connections from this chapter. We'll see the **10% Condition** and the **Success/Failure Condition** often. And the facts about the Binomial distribution can help explain how proportions behave, as we'll see in the next chapter.

WHAT HAVE WE LEARNED?

We've learned that Bernoulli trials show up in lots of places. Depending on the random variable of interest, we can use one of three models to estimate probabilities for Bernoulli trials:

- a Geometric model when we're interested in the number of Bernoulli trials until the next success;

- a Binomial model when we're interested in the number of successes in a certain number of Bernoulli trials;

- a Normal model to approximate a Binomial model when we expect at least 10 successes and 10 failures;

- We can use a Poisson model to approximate a Binomial model when the probability of success, *p*, is very small and the number of trials, *n*, is very large;

- an Exponential model can model the time between two random events.

Terms	
Bernoulli trials, if . . .	1. there are two possible outcomes (p. 405).
	2. the probability of success is constant.
	3. the trials are independent.
Geometric probability model	A Geometric model is appropriate for a random variable that counts the number of Bernoulli trials until the first success (p. 405).

Binomial probability model	A Binomial model is appropriate for a random variable that counts the number of successes in a fixed number of Bernoulli trials (p. 408).
Success/Failure Condition	For a Normal model to be a good approximation of a Binomial model, we must expect at least 10 successes and 10 failures. That is, $np \geq 10$ and $nq \geq 10$ (p. 413).
Continuous random variable	A continuous random variable is one that can take on any value in an interval of values (p. 414).
Poisson probability model	A Poisson model can be used to approximate a Binomial model when p is small and n is large. Use $\lambda = np$ as the mean for the Poisson. The Poisson can also be used to model counts of a wide variety of phenomena (p. 415).
Exponential probability model	An Exponential model can be used to estimate the time between two random events that occur following a Poisson probability model (p. 417).

Skills

▶ Know how to tell if a situation involves Bernoulli trials.

▶ Be able to choose whether to use a Geometric or a Binomial model for a random variable involving Bernoulli trials.

▶ Know the appropriate conditions for using a Geometric, Binomial, or Normal model.

▶ Know how to find the expected value of a Geometric model.

▶ Be able to calculate Geometric probabilities.

▶ Know how to find the mean and standard deviation of a Binomial model.

▶ Be able to calculate Binomial probabilities, perhaps approximating with a Normal or Poisson model.

▶ Be able to interpret means, standard deviations, and probabilities in the Bernoulli trial context.

THE BINOMIAL, THE GEOMETRIC, AND THE POISSON ON THE COMPUTER

Most statistics packages offer functions that compute Binomial probabilities, and many offer functions for Geometric and Poisson probabilities as well. Some technology solutions automatically use the Normal approximation for the Binomial when the exact calculations become unmanageable.

The only important differences among these functions are in what they are named and the order of their arguments. In these functions, pdf stands for "probability density function"—what we've been calling a probability model. The letters cdf stand for "cumulative distribution function," the technical term when we want to accumulate probabilities over a range of values. These technical terms show up in many of the function names. The term "cumulative" in a function name says that it corresponds to a cdf.

Generically, the four functions are as follows:

Geometric pdf(prob, x)	Finds the individual geometric probability of getting the first success on trial x when the probability of success is prob.	For example, the probability of finding the first LeBron James picture in the fifth cereal box is Geometric pdf(0.2, 5)
Geometric cdf(prob, x)	Finds the cumulative probability of getting the first success on or before trial x, when the probability of success is prob.	For example, the total probability of finding LeBron's picture in one of the first 4 boxes is Geometric cdf(0.2, 4)
Binomial pdf(n, prob, x)	Finds the probability of getting x successes in n trials when the probability of success is prob.	For example, Binomial pdf(5, 0.2, 2) is the probability of finding LeBron's picture exactly twice among 5 boxes of cereal.
Poisson pdf(λ, x)	Finds the probability of x successes when the mean number of successes is $\lambda = np$.	Best for large n and small probability of success.

DATA DESK

BinomDistr(*x, n, prob*) (pdf)
CumBinomDistr(*x, n, prob*) (cdf)
PoisDistr(*x, mean*) (pdf)
CumPoisDistr(*x, mean*) (cdf)

COMMENTS

Data Desk does not compute Geometric probabilities.
These functions work in derived variables or in scratchpads.

EXCEL

Binomdist(*x, n, prob, cumulative*)

COMMENTS

Set cumulative = *true* for cdf, *false* for pdf.
Excel's function fails when *x* or *n* is large.
Possibly, it does not use the Normal approximation.
Excel does not compute Geometric or Poisson probabilities.

JMP

Binomial Probability (*prob, n, x*) (pdf)
Binomial Distribution (*prob, n, x*) (cdf)
Poisson Probability (*mean, k*) (pdf)
Poisson Distribution (*mean, k*) (cdf)

COMMENTS

JMP does not compute Geometric probabilities.

MINITAB

Choose **Probability Distributions** from the **Calc** menu.
Choose **Binomial** from the Probability Distributions submenu.
To calculate the probability of getting *x* successes in *n* trials, choose **Probability.**

To calculate the probability of getting *x* or fewer successes among *n* trials, choose **Cumulative Probability.**
For Poisson, choose **Poisson** from the Probability Distribution submenu.
For Geometric, choose **Geometric** from the Probability Distribution submenu.

SPSS

PDF.GEOM(*x, prob*)
CDF.GEOM(*x, prob*)
PDF.Poisson(*x, mean*)

CDF.Poisson(*x, mean*)
PDF.BINOM(*x, n, prob*)
CDF.BINOM(*x, n, prob*)

TI-83/84 PLUS

geometpdf(*prob, x*)
geometcdf(*prob, x*)
binompdf(*n, prob, x*)
binomcdf(*n, prob, x*)
poissonpdf(*mean, x*)
poissoncdf(*mean, x*)

COMMENTS

Find these commands in the **2nd DISTR** menu (the calculator refers to models as "distributions").

TI-89

Find the commands under the [F5] (Distributions) menu.

- F: **Geometric Pdf** will ask for *p* and *x*. It returns the probability of the first success occurring on the *x*th trial.

- G: **Geometric Cdf** will ask for *p* and the upper and lower values of interest, say *a* and *b*. It returns $P(a \leq X \leq b)$, the probability the first success occurs between the *a*th and *b*th trials, inclusive.

- A: **Binomial Pdf** asks for *n, p,* and *x*.

- B: **Binomial Cdf** asks for *n, p,* and the lower and upper values of interest.

- D: **Poisson Pdf** asks for mean and *x*.

- E: **Poisson Cdf** asks for mean, *x*, and the lower and upper values of interest.

COMMENTS

For Geometric variables, when finding $P(X \geq a)$ specify an upper value of infinity, 1 [EE] 99, or a very large number.
For Binomial variables, when finding $P(X \geq a)$ the upper value is *n*.

EXERCISES

1. **Bernoulli.** Can we use probability models based on Bernoulli trials to investigate the following situations? Explain.

   a) We roll 50 dice to find the distribution of the number of spots on the faces.

   b) How likely is it that in a group of 120 the majority may have Type A blood, given that Type A is found in 43% of the population?

   c) We deal 7 cards from a deck and get all hearts. How likely is that?

   d) We wish to predict the outcome of a vote on the school budget, and poll 500 of the 3000 likely voters to see how many favor the proposed budget.

   e) A company realizes that about 10% of its packages are not being sealed properly. In a case of 24, is it likely that more than 3 are unsealed?

2. **Bernoulli 2.** Can we use probability models based on Bernoulli trials to investigate the following situations? Explain.

   a) You are rolling 5 dice and need to get at least two 6's to win the game.

   b) We record the distribution of eye colors found in a group of 500 people.

   c) A manufacturer recalls a doll because about 3% have buttons that are not properly attached. Customers return 37 of these dolls to the local toy store. Is the manufacturer likely to find any dangerous buttons?

   d) A city council of 11 Republicans and 8 Democrats picks a committee of 4 at random. What's the probability they choose all Democrats?

   e) A 2002 Rutgers University study found that 74% of high-school students have cheated on a test at least once. Your local high-school principal conducts a survey in homerooms and gets responses that admit to cheating from 322 of the 481 students.

3. **Simulating the model.** Think about the LeBron James picture search again. You are opening boxes of cereal one at a time looking for his picture, which is in 20% of the boxes. You want to know how many boxes you might have to open in order to find LeBron.

   a) Describe how you would simulate the search for LeBron using random numbers.

   b) Run at least 30 trials.

   c) Based on your simulation, estimate the probabilities that you might find your first picture of LeBron in the first box, the second, etc.

   d) Calculate the actual probability model.

   e) Compare the distribution of outcomes in your simulation to the probability model.

4. **Simulation II.** You are one space short of winning a child's board game and must roll a 1 on a die to claim victory. You want to know how many rolls it might take.

   a) Describe how you would simulate rolling the die until you get a 1.

   b) Run at least 30 trials.

   c) Based on your simulation, estimate the probabilities that you might win on the first roll, the second, the third, etc.

   d) Calculate the actual probability model.

   e) Compare the distribution of outcomes in your simulation to the probability model.

5. **LeBron, again.** Let's take one last look at the LeBron James picture search. You know his picture is in 20% of the cereal boxes. You buy five boxes to see how many pictures of LeBron you might get.

   a) Describe how you would simulate the number of pictures of LeBron you might find in five boxes of cereal.

   b) Run at least 30 trials.

   c) Based on your simulation, estimate the probabilities that you get no pictures of LeBron, 1 picture, 2 pictures, etc.

   d) Find the actual probability model.

   e) Compare the distribution of outcomes in your simulation to the probability model.

6. **Seatbelts.** Suppose 75% of all drivers always wear their seatbelts. Let's investigate how many of the drivers might be belted among five cars waiting at a traffic light.

   a) Describe how you would simulate the number of seatbelt-wearing drivers among the five cars.

   b) Run at least 30 trials.

   c) Based on your simulation, estimate the probabilities there are no belted drivers, exactly one, two, etc.

   d) Find the actual probability model.

   e) Compare the distribution of outcomes in your simulation to the probability model.

7. **On time.** A Department of Transportation report about air travel found that, nationwide, 76% of all flights are on time. Suppose you are at the airport and your flight is one of 50 scheduled to take off in the next two hours. Can you consider these departures to be Bernoulli trials? Explain.

8. **Lost luggage.** A Department of Transportation report about air travel found that airlines misplace about 5 bags per 1000 passengers. Suppose you are traveling with a group of people who have checked 22 pieces of luggage on your flight. Can you consider the fate of these bags to be Bernoulli trials? Explain.

9. **Hoops.** A basketball player has made 80% of his foul shots during the season. Assuming the shots are independent, find the probability that in tonight's game he

   a) misses for the first time on his fifth attempt.

   b) makes his first basket on his fourth shot.

   c) makes his first basket on one of his first 3 shots.

10. **Chips.** Suppose a computer chip manufacturer rejects 2% of the chips produced because they fail presale testing.
    a) What's the probability that the fifth chip you test is the first bad one you find?
    b) What's the probability you find a bad one within the first 10 you examine?

11. **More hoops.** For the basketball player in Exercise 9, what's the expected number of shots until he misses?

12. **Chips ahoy.** For the computer chips described in Exercise 10, how many do you expect to test before finding a bad one?

13. **Customer center operator.** Raaj works at the customer service call center of a major credit card bank. Cardholders call for a variety of reasons, but regardless of their reason for calling, if they hold a platinum card, Raaj is instructed to offer them a double-miles promotion. About 10% of all cardholders hold platinum cards, and about 50% of those will take the double-miles promotion. On average, how many calls will Raaj have to take before finding the first cardholder to take the double-miles promotion?

14. **Cold calls.** Justine works for an organization committed to raising money for Alzheimer's research. From past experience, the organization knows that about 20% of all potential donors will agree to give something if contacted by phone. They also know that of all people donating, about 5% will give $100 or more. On average, how many potential donors will she have to contact until she gets her first $100 donor?

15. **Blood.** Only 4% of people have Type AB blood.
    a) On average, how many donors must be checked to find someone with Type AB blood?
    b) What's the probability that there is a Type AB donor among the first 5 people checked?
    c) What's the probability that the first Type AB donor will be found among the first 6 people?
    d) What's the probability that we won't find a Type AB donor before the 10th person?

16. **Color blindness.** About 8% of males are color-blind. A researcher needs some color-blind subjects for an experiment and begins checking potential subjects.
    a) On average, how many men should the researcher expect to check to find one who is color-blind?
    b) What's the probability that she won't find anyone color-blind among the first 4 men she checks?
    c) What's the probability that the first color-blind man found will be the sixth person checked?
    d) What's the probability that she finds someone who is color-blind before checking the 10th man?

17. **Coins and intuition.** If you flip a fair coin 100 times,
    a) Intuitively, how many heads do you expect?
    b) Use the formula for expected value to verify your intuition.

18. **Roulette and intuition.** An American roulette wheel has 38 slots, of which 18 are red, 18 are black, and 2 are green (0 and 00). If you spin the wheel 38 times,
    a) Intuitively, how many times would you expect the ball to wind up in a green slot?
    b) Use the formula for expected value to verify your intuition.

19. **Lefties.** Assume that 13% of people are left-handed. If we select 5 people at random, find the probability of each outcome.
    a) The first lefty is the fifth person chosen.
    b) There are some lefties among the 5 people.
    c) The first lefty is the second or third person.
    d) There are exactly 3 lefties in the group.
    e) There are at least 3 lefties in the group.
    f) There are no more than 3 lefties in the group.

20. **Arrows.** An Olympic archer is able to hit the bull's-eye 80% of the time. Assume each shot is independent of the others. If she shoots 6 arrows, what's the probability of each of the following results?
    a) Her first bull's-eye comes on the third arrow.
    b) She misses the bull's-eye at least once.
    c) Her first bull's-eye comes on the fourth or fifth arrow.
    d) She gets exactly 4 bull's-eyes.
    e) She gets at least 4 bull's-eyes.
    f) She gets at most 4 bull's-eyes.

21. **Lefties, redux.** Consider our group of 5 people from Exercise 19.
    a) How many lefties do you expect?
    b) With what standard deviation?
    c) If we keep picking people until we find a lefty, how long do you expect it will take?

22. **More arrows.** Consider our archer from Exercise 20.
    a) How many bull's-eyes do you expect her to get?
    b) With what standard deviation?
    c) If she keeps shooting arrows until she hits the bull's-eye, how long do you expect it will take?

23. **Still more lefties.** Suppose we choose 12 people instead of the 5 chosen in Exercise 19.
    a) Find the mean and standard deviation of the number of right-handers in the group.
    b) What's the probability that they're not all right-handed?
    c) What's the probability that there are no more than 10 righties?
    d) What's the probability that there are exactly 6 of each?
    e) What's the probability that the majority is right-handed?

24. **Still more arrows.** Suppose the archer from Exercise 20 shoots 10 arrows.
    a) Find the mean and standard deviation of the number of bull's-eyes she may get.
    b) What's the probability that she never misses?
    c) What's the probability that there are no more than 8 bull's-eyes?
    d) What's the probability that there are exactly 8 bull's-eyes?
    e) What's the probability that she hits the bull's-eye more often than she misses?

25. **Vision.** It is generally believed that nearsightedness affects about 12% of all children. A school district tests the vision of 169 incoming kindergarten children. How many would you expect to be nearsighted? With what standard deviation?

26. **International students.** At a certain college, 6% of all students come from outside the United States. Incoming students there are assigned at random to freshman

dorms, where students live in residential clusters of 40 freshmen sharing a common lounge area. How many international students would you expect to find in a typical cluster? With what standard deviation?

27. **Tennis, anyone?** A certain tennis player makes a successful first serve 70% of the time. Assume that each serve is independent of the others. If she serves 6 times, what's the probability she gets

a) all 6 serves in?
b) exactly 4 serves in?
c) at least 4 serves in?
d) no more than 4 serves in?

28. **Frogs.** A wildlife biologist examines frogs for a genetic trait he suspects may be linked to sensitivity to industrial toxins in the environment. Previous research had established that this trait is usually found in 1 of every 8 frogs. He collects and examines a dozen frogs. If the frequency of the trait has not changed, what's the probability he finds the trait in

a) none of the 12 frogs?
b) at least 2 frogs?
c) 3 or 4 frogs?
d) no more than 4 frogs?

29. **And more tennis.** Suppose the tennis player in Exercise 27 serves 80 times in a match.

a) What are the mean and standard deviation of the number of good first serves expected?
b) Verify that you can use a Normal model to approximate the distribution of the number of good first serves.
c) Use the 68–95–99.7 Rule to describe this distribution.
d) What's the probability she makes at least 65 first serves?

30. **More arrows.** The archer in Exercise 20 will be shooting 200 arrows in a large competition.

a) What are the mean and standard deviation of the number of bull's-eyes she might get?
b) Is a Normal model appropriate here? Explain.
c) Use the 68–95–99.7 Rule to describe the distribution of the number of bull's-eyes she may get.
d) Would you be surprised if she made only 140 bull's-eyes? Explain.

31. **Apples.** An orchard owner knows that he'll have to use about 6% of the apples he harvests for cider because they will have bruises or blemishes. He expects a tree to produce about 300 apples.

a) Describe an appropriate model for the number of cider apples that may come from that tree. Justify your model.
b) Find the probability there will be no more than a dozen cider apples.
c) Is it likely there will be more than 50 cider apples? Explain.

32. **Frogs, part II.** Based on concerns raised by his preliminary research, the biologist in Exercise 28 decides to collect and examine 150 frogs.

a) Assuming the frequency of the trait is still 1 in 8, determine the mean and standard deviation of the number of frogs with the trait he should expect to find in his sample.

b) Verify that he can use a Normal model to approximate the distribution of the number of frogs with the trait.
c) He found the trait in 22 of his frogs. Do you think this proves that the trait has become more common? Explain.

33. **Lefties, again.** A lecture hall has 200 seats with folding arm tablets, 30 of which are designed for left-handers. The average size of classes that meet there is 188, and we can assume that about 13% of students are left-handed. What's the probability that a right-handed student in one of these classes is forced to use a lefty arm tablet?

34. **No-shows.** An airline, believing that 5% of passengers fail to show up for flights, overbooks (sells more tickets than there are seats). Suppose a plane will hold 265 passengers, and the airline sells 275 tickets. What's the probability the airline will not have enough seats, so someone gets bumped?

35. **Annoying phone calls.** A newly hired telemarketer is told he will probably make a sale on about 12% of his phone calls. The first week he called 200 people, but only made 10 sales. Should he suspect he was misled about the true success rate? Explain.

36. **The euro.** Shortly after the introduction of the euro coin in Belgium, newspapers around the world published articles claiming the coin is biased. The stories were based on reports that someone had spun the coin 250 times and gotten 140 heads—that's 56% heads. Do you think this is evidence that spinning a euro is unfair? Explain.

37. **Hurricanes, redux.** We first looked at the occurrences of hurricanes in Chapter 4 (Exercise 41) and found that they arrive with a mean of 2.45 per year. Suppose the number of hurricanes can be modeled by a Poisson distribution with this mean.

a) What's the probability of no hurricanes next year?
b) What's the probability that during the next two years, there's exactly 1 hurricane?

38. **Bank tellers.** I am the only bank teller on duty at my local bank. I need to run out for 10 minutes, but I don't want to miss any customers. Suppose the arrival of customers can be modeled by a Poisson distribution with mean 2 customers per hour.

a) What's the probability that no one will arrive in the next 10 minutes?
b) What's the probability that 2 or more people arrive in the next 10 minutes?
c) You've just served 2 customers who came in one after the other. Is this a better time to run out?

39. **TB, again.** In Chapter 15 we saw that the probability of contracting TB is small, with p about 0.0005 for a new case in a given year. In a town of 8000 people:

a) What's the expected number of new cases?
b) Use the Poisson model to approximate the probability that there will be at least one new case of TB next year.

40. **Earthquakes.** Suppose the probability of a major earthquake on a given day is 1 out of 10,000.

a) What's the expected number of major earthquakes in the next 1000 days?

b) Use the Poisson model to approximate the probability that there will be at least one major earthquake in the next 1000 days.

41. **Seatbelts II.** Police estimate that 80% of drivers now wear their seatbelts. They set up a safety roadblock, stopping cars to check for seatbelt use.
    a) How many cars do they expect to stop before finding a driver whose seatbelt is not buckled?
    b) What's the probability that the first unbelted driver is in the sixth car stopped?
    c) What's the probability that the first 10 drivers are all wearing their seatbelts?
    d) If they stop 30 cars during the first hour, find the mean and standard deviation of the number of drivers expected to be wearing seatbelts.
    e) If they stop 120 cars during this safety check, what's the probability they find at least 20 drivers not wearing their seatbelts?

42. **Rickets.** Vitamin D is essential for strong, healthy bones. Our bodies produce vitamin D naturally when sunlight falls upon the skin, or it can be taken as a dietary supplement. Although the bone disease rickets was largely eliminated in England during the 1950s, some people there are concerned that this generation of children is at increased risk because they are more likely to watch TV or play computer games than spend time outdoors. Recent research indicated that about 20% of British children are deficient in vitamin D. Suppose doctors test a group of elementary school children.
    a) What's the probability that the first vitamin D–deficient child is the eighth one tested?
    b) What's the probability that the first 10 children tested are all okay?
    c) How many kids do they expect to test before finding one who has this vitamin deficiency?
    d) They will test 50 students at the third-grade level. Find the mean and standard deviation of the number who may be deficient in vitamin D.
    e) If they test 320 children at this school, what's the probability that no more than 50 of them have the vitamin deficiency?

43. **ESP.** Scientists wish to test the mind-reading ability of a person who claims to "have ESP." They use five cards with different and distinctive symbols (square, circle, triangle, line, squiggle). Someone picks a card at random and thinks about the symbol. The "mind reader" must correctly identify which symbol was on the card. If the test consists of 100 trials, how many would this person need to get right in order to convince you that ESP may actually exist? Explain.

44. **True–false.** A true–false test consists of 50 questions. How many does a student have to get right to convince you that he is not merely guessing? Explain.

45. **Hot hand.** A basketball player who ordinarily makes about 55% of his free throw shots has made 4 in a row. Is this evidence that he has a "hot hand" tonight? That is, is this streak so unusual that it means the probability he makes a shot must have changed? Explain.

46. **New bow.** The archer in Exercise 20 purchases a new bow, hoping that it will improve her success rate to more than 80% bull's-eyes. She is delighted when she first tests her new bow and hits 6 consecutive bull's-eyes. Do you think this is compelling evidence that the new bow is better? In other words, is a streak like this unusual for her? Explain.

47. **Hotter hand.** The basketball player in Exercise 45 has new sneakers, which he thinks improve his game. Over his past 40 shots, he's made 32—much better than the 55% he usually shoots. Do you think his chances of making a shot really increased? In other words, is making at least 32 of 40 shots really unusual for him? (Do you think it's his sneakers?)

48. **New bow, again.** The archer in Exercise 46 continues shooting arrows, ending up with 45 bull's-eyes in 50 shots. Now are you convinced that the new bow is better? Explain.

49. **Web visitors.** A website manager has noticed that during the evening hours, about 3 people per minute check out from their shopping cart and make an online purchase. She believes that each purchase is independent of the others and wants to model the number of purchases per minute.
    a) What model might you suggest to model the number of purchases per minute?
    b) What is the probability that in any one minute at least one purchase is made?
    c) What is the probability that no one makes a purchase in the next two minutes?

50. **Quality control.** In an effort to improve the quality of their cell phones, a manufacturing manager records the number of faulty phones in each day's production run. The manager notices that the number of faulty cell phones in a production run of cell phones is usually small and that the quality of one day's run seems to have no bearing on the next day.
    a) What model might you use to model the number of faulty cell phones produced in one day?
    b) If the mean number of faulty cell phones is 2 per day, what is the probability that no faulty cell phones will be produced tomorrow?
    c) If the mean number of faulty cell phones is 2 per day, what is the probability that 3 or more faulty cell phones were produced in today's run?

*51. **Web visitors, part 2.** The website manager in Exercise 49 wants to model the time between purchases. Recall that the mean number of purchases in the evening is 3 per minute.
    a) What model would you use to model the time between events?
    b) What is the mean time between purchases?
    c) What is the probability that the time to the next purchase will be between 1 and 2 minutes?

*52. **Quality control, part 2.** The cell phone manufacturer in Exercise 50 wants to model the time between events. The mean number of defective cell phones is 2 per day.
    a) What model would you use to model the time between events?
    b) What would the probability be that the time to the next failure is 1 day or less?
    c) What is the mean time between failures?

ANSWERS

1. There are two outcomes (contact, no contact), the probability of contact is 0.76, and random calls should be independent.

2. Binomial, with $n = 1000$ and $p = 0.76$. For actual calculations, we could approximate using a Normal model with $\mu = np = 1000(0.76) = 760$ and $\sigma = \sqrt{npq} = \sqrt{1000(0.76)(0.24)} \approx 13.5$.

3. Geometric, with $p = 0.29$.

4. Poisson, with $\lambda = np = 10000\left(\dfrac{70000}{107000000}\right) = 6.54$.

PART REVIEW

Randomness and Probability

Quick Review

Here's a brief summary of the key concepts and skills in probability and probability modeling:

▶ The Law of Large Numbers says that the more times we try something, the closer the results will come to theoretical perfection.

- Don't mistakenly misinterpret the Law of Large Numbers as the "Law of Averages." There's no such thing.

▶ Basic rules of probability can handle most situations:

- To find the probability that an event OR another event happens, add their probabilities and subtract the probability that both happen.
- To find the probability that an event AND another independent event both happen, multiply probabilities.
- Conditional probabilities tell you how likely one event is to happen, knowing that another event has happened.
- Mutually exclusive events (also called "disjoint") cannot both happen at the same time.
- Two events are independent if the occurrence of one doesn't change the probability that the other happens.

▶ A probability model for a random variable describes the theoretical distribution of outcomes.

- The mean of a random variable is its expected value.
- For sums or differences of independent random variables, variances add.
- To estimate probabilities involving quantitative variables, you may be able to use a Normal model—but only if the distribution of the variable is unimodal and symmetric.
- To estimate the probability you'll get your first success on a certain trial, use a Geometric model.
- To estimate the probability you'll get a certain number of successes in a specified number of independent trials, use a Binomial model.
- To estimate the probability of the number of occurrences of a relatively rare phenomenon, consider using a Poisson model.
- To estimate the interval between two occurrences of a random phenomenon, use the Exponential model.

Ready? Here are some opportunities to check your understanding of these ideas.

REVIEW EXERCISES

1. **Quality control.** A consumer organization estimates that 29% of new cars have a cosmetic defect, such as a scratch or a dent, when they are delivered to car dealers. This same organization believes that 7% have a functional defect—something that does not work properly—and that 2% of new cars have both kinds of problems.

   a) If you buy a new car, what's the probability that it has some kind of defect?

   b) What's the probability it has a cosmetic defect but no functional defect?

   c) If you notice a dent on a new car, what's the probability it has a functional defect?

   d) Are the two kinds of defects disjoint events? Explain.

   e) Do you think the two kinds of defects are independent events? Explain.

2. **Workers.** A company's human resources officer reports a breakdown of employees by job type and sex shown in the table.

	Sex	
Job Type	Male	Female
Management	7	6
Supervision	8	12
Production	45	72

   a) What's the probability that a worker selected at random is
      i) female?
      ii) female or a production worker?
      iii) female, if the person works in production?
      iv) a production worker, if the person is female?

   b) Do these data suggest that job type is independent of being male or female? Explain.

3. **Airfares.** Each year a company must send 3 officials to a meeting in China and 5 officials to a meeting in France. Airline ticket prices vary from time to time, but the company purchases all tickets for a country at the same price. Past experience has shown that tickets to China have a mean price of $1000, with a standard deviation of $150, while the mean airfare to France is $500, with a standard deviation of $100.

   a) Define random variables and use them to express the total amount the company will have to spend to send these delegations to the two meetings.

   b) Find the mean and standard deviation of this total cost.

   c) Find the mean and standard deviation of the difference in price of a ticket to China and a ticket to France.

   d) Do you need to make any assumptions in calculating these means? How about the standard deviations?

4. **Bipolar.** Psychiatrists estimate that about 1 in 100 adults suffers from bipolar disorder. What's the probability that in a city of 10,000 there are more than 200 people with this condition? Be sure to verify that a Normal model can be used here.

5. **A game.** To play a game, you must pay $5 for each play. There is a 10% chance you will win $5, a 40% chance you will win $7, and a 50% chance you will win only $3.

   a) What are the mean and standard deviation of your net winnings?

   b) You play twice. Assuming the plays are independent events, what are the mean and standard deviation of your total winnings?

6. **Emergency switch.** Safety engineers must determine whether industrial workers can operate a machine's emergency shutoff device. Among a group of test subjects, 66% were successful with their left hands, 82% with their right hands, and 51% with both hands.

   a) What percent of these workers could not operate the switch with either hand?

   b) Are success with right and left hands independent events? Explain.

   c) Are success with right and left hands mutually exclusive? Explain.

7. **Twins.** In the United States, the probability of having twins (usually about 1 in 90 births) rises to about 1 in 10 for women who have been taking the fertility drug Clomid. Among a group of 10 pregnant women, what's the probability that

   a) at least one will have twins if none were taking a fertility drug?

   b) at least one will have twins if all were taking Clomid?

   c) at least one will have twins if half were taking Clomid?

8. **Deductible.** A car owner may buy insurance that will pay the full price of repairing the car after an at-fault accident, or save $12 a year by getting a policy with a $500 deductible. Her insurance company says that about 0.5% of drivers in her area have an at-fault auto accident during any given year. Based on this information, should she buy the policy with the deductible or not? How does the value of her car influence this decision?

9. **More twins.** A group of 5 women became pregnant while undergoing fertility treatments with the drug Clomid, discussed in Exercise 7. What's the probability that

   a) none will have twins?

   b) exactly 1 will have twins?

   c) at least 3 will have twins?

10. **At fault.** The car insurance company in Exercise 8 believes that about 0.5% of drivers have an at-fault accident during a given year. Suppose the company insures 1355 drivers in that city.

    a) What are the mean and standard deviation of the number who may have at-fault accidents?

    b) Can you describe the distribution of these accidents with a Normal model? Explain.

11. **Twins, part III.** At a large fertility clinic, 152 women became pregnant while taking Clomid. (See Exercise 7.)

    a) What are the mean and standard deviation of the number of twin births we might expect?

    b) Can we use a Normal model in this situation? Explain.

    c) What's the probability that no more than 10 of the women have twins?

12. **Child's play.** In a board game you determine the number of spaces you may move by spinning a spinner and rolling a die. The spinner has three regions: Half of the spinner is marked "5," and the other half is equally divided between "10" and "20." The six faces of the die show 0, 0, 1, 2, 3, and 4 spots. When it's your turn, you spin and roll, adding the numbers together to determine how far you may move.

    a) Create a probability model for the outcome on the spinner.

    b) Find the mean and standard deviation of the spinner results.

    c) Create a probability model for the outcome on the die.

    d) Find the mean and standard deviation of the die results.

    e) Find the mean and standard deviation of the number of spaces you get to move.

13. **Language.** Neurological research has shown that in about 80% of people, language abilities reside in the brain's left side. Another 10% display right-brain language centers, and the remaining 10% have two-sided language control. (The latter two groups are mainly left-handers; *Science News*, 161 no. 24 [2002].)

    a) Assume that a freshman composition class contains 25 randomly selected people. What's the probability that no more than 15 of them have left-brain language control?

    b) In a randomly assigned group of 5 of these students, what's the probability that no one has two-sided language control?

    c) In the entire freshman class of 1200 students, how many would you expect to find of each type?

    d) What are the mean and standard deviation of the number of these freshmen who might be right-brained in language abilities?

    e) If an assumption of Normality is justified, use the 68–95–99.7 Rule to describe how many students in the freshman class might have right-brain language control.

14. **Play, again.** If you land in a "penalty zone" on the game board described in Exercise 12, your move will be determined by subtracting the roll of the die from the result on the spinner. Now what are the mean and standard deviation of the number of spots you may move?

15. **Beanstalks.** In some cities tall people who want to meet and socialize with other tall people can join Beanstalk Clubs. To qualify, a man must be over 6′2″ tall, and a woman over 5′10″. According to the National Health Survey, heights of adults may have a Normal model with mean heights of 69.1″ for men and 64.0″ for women. The respective standard deviations are 2.8″ and 2.5″.

    a) You're probably not surprised to learn that men are generally taller than women, but what does the greater standard deviation for men's heights indicate?

    b) Are men or women more likely to qualify for Beanstalk membership?

    c) Beanstalk members believe that height is an important factor when people select their spouses. To investigate, we select at random a married man and, independently, a married woman. Define two random variables, and use them to express how many inches taller the man is than the woman.

    d) What's the mean of this difference?

    e) What's the standard deviation of this difference?

    f) What's the probability that the man is taller than the woman (that the difference in heights is greater than 0)?

    g) Suppose a survey of married couples reveals that 92% of the husbands were taller than their wives. Based on your answer to part f, do you believe that people's choice of spouses is independent of height? Explain.

16. **Stocks.** Since the stock market began in 1872, stock prices have risen in about 73% of the years. Assuming that market performance is independent from year to year, what's the probability that

    a) the market will rise for 3 consecutive years?

    b) the market will rise 3 years out of the next 5?

    c) the market will fall during at least 1 of the next 5 years?

    d) the market will rise during a majority of years over the next decade?

17. **Multiple choice.** A multiple choice test has 50 questions, with 4 answer choices each. You must get at least 30 correct to pass the test, and the questions are very difficult.

    a) Are you likely to be able to pass by guessing on every question? Explain.

    b) Suppose, after studying for a while, you believe you have raised your chances of getting each question right to 70%. How likely are you to pass now?

    c) Assuming you are operating at the 70% level and the instructor arranges questions randomly, what's the probability that the third question is the first one you get right?

18. **Stock strategy.** Many investment advisors argue that after stocks have declined in value for 2 consecutive years, people should invest heavily because the market rarely declines 3 years in a row.

    a) Since the stock market began in 1872, there have been two consecutive losing years eight times. In six of those cases, the market rose during the following year. Does this confirm the advice?

    b) Overall, stocks have risen in value during 95 of the 130 years since the market began in 1872. How is this fact relevant in assessing the statistical reasoning of the advisors?

19. **Insurance.** A 65-year-old woman takes out a $100,000 term life insurance policy. The company charges an annual premium of $520. Estimate the company's expected profit on such policies if mortality tables indicate that only 2.6% of women age 65 die within a year.

20. **Teen smoking.** The Centers for Disease Control and Prevention say that about 30% of high-school students smoke tobacco (down from a high of 38% in 1997). Suppose you randomly select high-school students to survey them on their attitudes toward scenes of smoking in the movies. What's the probability that
    a) none of the first 4 students you interview is a smoker?
    b) the first smoker is the sixth person you choose?
    c) there are no more than 2 smokers among 10 people you choose?

21. **Passing stats.** Molly's college offers two sections of Statistics 101. From what she has heard about the two professors listed, Molly estimates that her chances of passing the course are 0.80 if she gets Professor Scedastic and 0.60 if she gets Professor Kurtosis. The registrar uses a lottery to randomly assign the 120 enrolled students based on the number of available seats in each class. There are 70 seats in Professor Scedastic's class and 50 in Professor Kurtosis's class.
    a) What's the probability that Molly will pass Statistics?
    b) At the end of the semester, we find out that Molly failed. What's the probability that she got Professor Kurtosis?

22. **Teen smoking II.** Suppose that, as reported by the Centers for Disease Control, about 30% of high school students smoke tobacco. You randomly select 120 high school students to survey them on their attitudes toward scenes of smoking in the movies.
    a) What's the expected number of smokers?
    b) What's the standard deviation of the number of smokers?
    c) The number of smokers among 120 randomly selected students will vary from group to group. Explain why that number can be described with a Normal model.
    d) Using the 68–95–99.7 Rule, create and interpret a model for the number of smokers among your group of 120 students.

23. **Random variables.** Given independent random variables with means and standard deviations as shown, find the mean and standard deviation of each of these variables:
    a) $X + 50$
    b) $10Y$
    c) $X + 0.5Y$
    d) $X - Y$
    e) $X + Y$

	Mean	SD
X	50	8
Y	100	6

24. **Merger.** Explain why the facts you know about variances of independent random variables might encourage two small insurance companies to merge. (*Hint:* Think about the expected amount and potential variability in payouts for the separate and the merged companies.)

25. **Youth survey.** According to a recent Gallup survey, 93% of teens use the Internet, but there are differences in how teen boys and girls say they use computers. The telephone poll found that 77% of boys had played computer games in the past week, compared with 65% of girls. On the other hand, 76% of girls said they had e-mailed friends in the past week, compared with only 65% of boys.

    a) For boys, the cited percentages are 77% playing computer games and 65% using e-mail. That total is 142%, so there is obviously a mistake in the report. No? Explain.
    b) Based on these results, do you think playing games and using e-mail are mutually exclusive? Explain.
    c) Do you think whether a child e-mails friends is independent of being a boy or a girl? Explain.
    d) Suppose that in fact 93% of the teens in your area do use the Internet. You want to interview a few who do not, so you start contacting teenagers at random. What is the probability that it takes you 5 interviews until you find the first person who does not use the Internet?

26. **Meals.** A college student on a seven-day meal plan reports that the amount of money he spends daily on food varies with a mean of $13.50 and a standard deviation of $7.
    a) What are the mean and standard deviation of the amount he might spend in two consecutive days?
    b) What assumption did you make in order to find that standard deviation? Are there any reasons you might question that assumption?
    c) Estimate his average weekly food costs, and the standard deviation.
    d) Do you think it likely he might spend less than $50 in a week? Explain, including any assumptions you make in your analysis.

27. **Travel to Kyrgyzstan.** Your pocket copy of *Kyrgyzstan on* 4237 ± 360 *Som a Day* claims that you can expect to spend about 4237 som each day with a standard deviation of 360 som. How well can you estimate your expenses for the trip?
    a) Your budget allows you to spend 90,000 som. To the nearest day, how long can you afford to stay in Kyrgyzstan, on average?
    b) What's the standard deviation of your expenses for a trip of that duration?
    c) You doubt that your total expenses will exceed your expectations by more than two standard deviations. How much extra money should you bring? On average, how much of a "cushion" will you have per day?

28. **Picking melons.** Two stores sell watermelons. At the first store the melons weigh an average of 22 pounds, with a standard deviation of 2.5 pounds. At the second store the melons are smaller, with a mean of 18 pounds and a standard deviation of 2 pounds. You select a melon at random at each store.
    a) What's the mean difference in weights of the melons?
    b) What's the standard deviation of the difference in weights?
    c) If a Normal model can be used to describe the difference in weights, what's the probability that the melon you got at the first store is heavier?

29. **Home, sweet home.** According to the 2000 Census, 66% of U.S. households own the home they live in. A mayoral candidate conducts a survey of 820 randomly selected homes in your city and finds only 523 owned by the current residents. The candidate then attacks the incumbent mayor, saying that there is an unusually low level of homeownership in the city. Do you agree? Explain.

30. **Buying melons.** The first store in Exercise 28 sells watermelons for 32 cents a pound. The second store is having a sale on watermelons—only 25 cents a pound. Find the mean and standard deviation of the difference in the price you may pay for melons randomly selected at each store.

31. **Who's the boss?** The 2000 Census revealed that 26% of all firms in the United States are owned by women. You call some firms doing business locally, assuming that the national percentage is true in your area.
    a) What's the probability that the first 3 you call are all owned by women?
    b) What's the probability that none of your first 4 calls finds a firm that is owned by a woman?
    c) Suppose none of your first 5 calls found a firm owned by a woman. What's the probability that your next call does?

32. **Jerseys.** A Statistics professor comes home to find that all four of his children got white team shirts from soccer camp this year. He concludes that this year, unlike other years, the camp must not be using a variety of colors. But then he finds out that in each child's age group there are 4 teams, only 1 of which wears white shirts. Each child just happened to get on the white team at random.
    a) Why was he so surprised? If each age group uses the same 4 colors, what's the probability that all four kids would get the same-color shirt?
    b) What's the probability that all 4 would get white shirts?
    c) We lied. Actually, in the oldest child's group there are 6 teams instead of the 4 teams in each of the other three groups. How does this change the probability you calculated in part b?

33. **When to stop?** In Exercise 27 of the Review Exercises for Part III, we posed this question:

    You play a game that involves rolling a die. You can roll as many times as you want, and your score is the total for all the rolls. But . . . if you roll a 6, your score is 0 and your turn is over. What might be a good strategy for a game like this?

    You attempted to devise a good strategy by simulating several plays to see what might happen. Let's try calculating a strategy.
    a) On what roll would you expect to get a 6 for the first time?
    b) So, roll *one time less* than that. Assuming all those rolls were not 6's, what's your expected score?
    c) What's the probability that you can roll that many times without getting a 6?

34. **Plan B.** Here's another attempt at developing a good strategy for the dice game in Exercise 33. Instead of stopping after a certain number of rolls, you could decide to stop when your score reaches a certain number of points.
    a) How many points would you expect a roll to *add* to your score?
    b) In terms of your current score, how many points would you expect a roll to *subtract* from your score?
    c) Based on your answers in parts a and b, at what score will another roll "break even"?
    d) Describe the strategy this result suggests.

35. **Technology on campus.** Every 5 years the Conference Board of the Mathematical Sciences surveys college math departments. In 2000 the board reported that 51% of all undergraduates taking Calculus I were in classes that used graphing calculators and 31% were in classes that used computer assignments. Suppose that 16% used both calculators and computers.
    a) What percent used neither kind of technology?
    b) What percent used calculators but not computers?
    c) What percent of the calculator users had computer assignments?
    d) Based on this survey, do calculator and computer use appear to be independent events? Explain.

36. **Dogs.** A census by the county dog control officer found that 18% of homes kept one dog as a pet, 4% had two dogs, and 1% had three or more. If a salesman visits two homes selected at random, what's the probability he encounters
    a) no dogs?
    b) some dogs?
    c) dogs in each home?
    d) more than one dog in each home?

37. **O-rings.** Failures of O-rings on the space shuttle are fairly rare, but often disastrous, events. If we are testing O-rings, suppose that the probability of a failure of any one O-ring is 0.01. Let X be the number of failures in the next 10 O-rings tested.
    a) What model might you use to model X?
    b) What is the mean number of failures in the next 10 O-rings?
    c) What is the probability that there is exactly one failure in the next 10 O-rings?
    d) What is the probability that there is at least one failure in the next 10 O-rings?

38. **Volcanoes.** Almost every year, there is some incidence of volcanic activity on the island of Japan. In 2005 there were 5 volcanic episodes, defined as either eruptions or sizable seismic activity. Suppose the mean number of episodes is 2.4 per year. Let X be the number of episodes in the 2-year period 2010–2011.
    a) What model might you use to model X?
    b) What is the mean number of episodes in this period?
    c) What is the probability that there will be no episodes in this period?
    d) What is the probability that there are more than three episodes in this period?

39. **Socks.** In your sock drawer you have 4 blue socks, 5 gray socks, and 3 black ones. Half asleep one morning, you

grab 2 socks at random and put them on. Find the probability you end up wearing

a) 2 blue socks.
b) no gray socks.
c) at least 1 black sock.
d) a green sock.
e) matching socks.

40. **Coins.** A coin is to be tossed 36 times.

a) What are the mean and standard deviation of the number of heads?
b) Suppose the resulting number of heads is unusual, two standard deviations above the mean. How many "extra" heads were observed?
c) If the coin were tossed 100 times, would you still consider the same number of extra heads unusual? Explain.
d) In the 100 tosses, how many extra heads would you need to observe in order to say the results were unusual?
e) Explain how these results refute the "Law of Averages" but confirm the Law of Large Numbers.

41. **The Drake equation.** In 1961 astronomer Frank Drake developed an equation to try to estimate the number of extraterrestrial civilizations in our galaxy that might be able to communicate with us via radio transmissions. Now largely accepted by the scientific community, the Drake equation has helped spur efforts by radio astronomers to search for extraterrestrial intelligence.

Here is the equation:

$$N_C = N \cdot f_p \cdot n_e \cdot f_l \cdot f_i \cdot f_c \cdot f_L$$

OK, it looks a little messy, but here's what it means:

Factor	What It Represents	Possible Value
N	Number of stars in the Milky Way Galaxy	200–400 billion
f_p	Probability that a star has planets	20%–50%
n_e	Number of planets in a solar system capable of sustaining earth-type life	1? 2?
f_l	Probability that life develops on a planet with a suitable environment	1%–100%
f_i	Probability that life evolves intelligence	50%?
f_c	Probability that intelligent life develops radio communication	10%–20%
f_L	Fraction of the planet's life for which the civilization survives	$\frac{1}{1,000,000}$?
N_c	Number of extraterrestrial civilizations in our galaxy with which we could communicate	?

So, how many ETs are out there? That depends; values chosen for the many factors in the equation depend on ever-evolving scientific knowledge and one's personal guesses. But now, some questions.

a) What quantity is calculated by the first product, $N \cdot f_p$?
b) What quantity is calculated by the product, $N \cdot f_p \cdot n_e \cdot f_l$?
c) What probability is calculated by the product $f_l \cdot f_i$?
d) Which of the factors in the formula are conditional probabilities? Restate each in a way that makes the condition clear.

Note: A quick Internet search will find you a site where you can play with the Drake equation yourself.

42. **Recalls.** In a car rental company's fleet, 70% of the cars are American brands, 20% are Japanese, and the rest are German. The company notes that manufacturers' recalls seem to affect 2% of the American cars, but only 1% of the others.

a) What's the probability that a randomly chosen car is recalled?
b) What's the probability that a recalled car is American?

43. **Pregnant?** Suppose that 70% of the women who suspect they may be pregnant and purchase an in-home pregnancy test are actually pregnant. Further suppose that the test is 98% accurate. What's the probability that a woman whose test indicates that she is pregnant actually is?

44. **Door prize.** You are among 100 people attending a charity fundraiser at which a large-screen TV will be given away as a door prize. To determine who wins, 99 white balls and 1 red ball have been placed in a box and thoroughly mixed. The guests will line up and, one at a time, pick a ball from the box. Whoever gets the red ball wins the TV, but if the ball is white, it is returned to the box. If none of the 100 guests gets the red ball, the TV will be auctioned off for additional benefit of the charity.

a) What's the probability that the first person in line wins the TV?
b) You are the third person in line. What's the probability that you win the TV?
c) What's the probability that the charity gets to sell the TV because no one wins?
d) Suppose you get to pick your spot in line. Where would you want to be in order to maximize your chances of winning?
e) After hearing some protest about the plan, the organizers decide to award the prize by not returning the white balls to the box, thus ensuring that 1 of the 100 people will draw the red ball and win the TV. Now what position in line would you choose in order to maximize your chances?

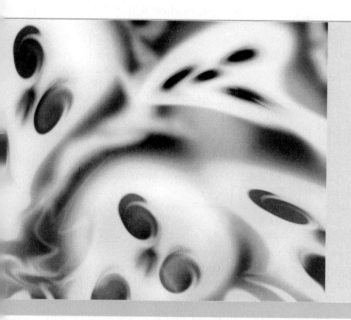

Sampling Distribution Models

Where are we going?

A poll based on a random sample of U.S. adults reports that 29% of those asked admit that they don't always drive as safely as they should because they are on their cell phone. How much can we trust this statistic? After all, they didn't ask *everyone*. Maybe the true proportion is 33% or 25%? How reliable *are* proportions based on random samples? We'll see that we can be surprisingly precise about how much we expect proportions from random samples to vary. This will enable us to start generalizing from samples we have at hand to the population at large.

WHO	U.S. adults
WHAT	Belief in ghosts
WHEN	November 2005
WHERE	United States
WHY	Public attitudes

In November 2005 the Harris Poll asked 889 U.S. adults, "Do you believe in ghosts?" 40% said they did. At almost the same time, CBS News polled 808 U.S. adults and asked the same question. 48% of their respondents professed a belief in ghosts. Why the difference? This seems like a simple enough question. Should we be surprised to find that we could get proportions this different from properly selected random samples drawn from the same population? You're probably used to seeing that observations vary, but how much variability among polls should we expect to see?

Why do sample proportions vary at all? How can surveys conducted at essentially the same time by organizations asking the same questions get different results? The answer is at the heart of Statistics. The proportions vary from sample to sample because the samples are composed of different people.

It's actually pretty easy to predict how much a proportion will vary under circumstances like this. Understanding the variability of our estimates will let us actually use that variability to better understand the world.

The Central Limit Theorem for Sample Proportions

We've talked about *Think*, *Show*, and *Tell*. Now we have to add *Imagine*. In order to understand the CBS poll, we want to imagine the results from all the random samples of size 808 that CBS News didn't take. What would the histogram of all the sample proportions look like?

A S *Activity:* **Sampling Distribution of a Proportion.** You don't have to imagine—you can simulate.

NOTATION ALERT

The letter p is our choice for the *parameter* of the model for proportions. It violates our "Greek letters for parameters" rule, but if we stuck to that, our natural choice would be π. We could use π to be perfectly consistent, but then we'd have to write statements like $\pi = 0.46$. That just seems a bit weird to us. After all, we've known that $\pi = 3.1415926 \ldots$ since the Greeks, and it's a hard habit to break.

So, we'll use p for the model parameter (the probability of a success) and $\hat{p}$ for the observed proportion in a sample. We'll also use q for the probability of a failure $(q = 1 - p)$ and $\hat{q}$ for its observed value.

But be careful. We've already used capital P for a general probability. And we'll soon see another use of P in the next chapter! There are a lot of p's in this course; you'll need to think clearly about the context to keep them straight.

For people's belief in ghosts, where do you expect the center of that histogram to be? Of course, we don't *know* the answer to that (and probably never will). But we know that it will be at the true proportion in the population, and we can call that p. (See the Notation Alert.) For the sake of discussion here, let's suppose that 45% of all American adults believe in ghosts, so we'll use $p = 0.45$.

How about the *shape* of the histogram? We don't have to just imagine. We can simulate a bunch of random samples that we didn't really draw. Here's a histogram of the proportions saying they believe in ghosts for 2000 simulated independent samples of 808 adults when the true proportion is $p = 0.45$.

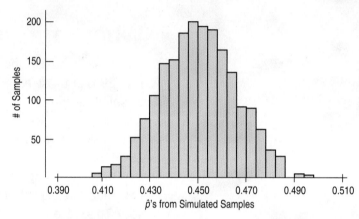

FIGURE 18.1

A histogram of sample proportions for 2000 simulated samples of 808 adults drawn from a population with $p = 0.45$. The sample proportions vary, but their distribution is centered at the true proportion, p.

It should be no surprise that we don't get the same proportion for each sample we draw, even though the underlying true value is the same for the population. Each $\hat{p}$ comes from a different simulated sample. The histogram above is a simulation of what we'd get if we could see *all the proportions from all possible samples.* That distribution has a special name. It is called the **sampling distribution** of the proportions.[1]

Does it surprise you that the histogram is unimodal? Symmetric? That it is centered at p? You probably don't find any of this shocking. Does the shape remind you of any model that we've discussed? It's an amazing and fortunate fact that a Normal model is just the right one for the histogram of sample proportions.

As we'll see in a few pages, this fact was proved in 1810 by the great French mathematician Pierre-Simon Laplace as part of a more general result. There is no reason you should guess that the Normal model would be the one we need here,[2] and, indeed, the importance of Laplace's result was not immediately understood by his contemporaries. But (unlike Laplace's contemporaries in 1810) we know how useful the Normal model can be.

Modeling how sample proportions vary from sample to sample is one of the most powerful ideas we'll see in this course. A **sampling distribution model** for how a sample proportion varies from sample to sample allows us to quantify that variation and to talk about how likely it is that we'd observe a sample proportion in any particular interval.

---

[1] A word of caution. Until now we've been plotting the *distribution of the sample,* a display of the actual data that were collected in that one sample. But now we've plotted the *sampling distribution;* a display of summary statistics ($\hat{p}$'s, for example) for many different samples. "Sample distribution" and "sampling distribution" sound a lot alike, but they refer to very different things. (Sorry about that—we didn't make up the terms. It's just the way it is.) And the distinction is critical. Whenever you read or write something about one of these, think very carefully about what the words signify.

[2] Well, the fact that we spent most of Chapter 6 on the Normal model might have been a hint.

Pierre-Simon Laplace, 1749-1827.

To use a Normal model, we need to specify two parameters: its mean and standard deviation. The center of the histogram is naturally at p, so we'll put μ, the mean of the Normal, at p.

What about the standard deviation? Usually the mean gives us no information about the standard deviation. Suppose we told you that a batch of bike helmets had a mean diameter of 26 centimeters and asked what the standard deviation was. If you said, "I have no idea," you'd be exactly right. There's no information about σ from knowing the value of μ.

But there's a special fact about proportions. With proportions we get something for free. Once we know the mean, p, we automatically also know the standard deviation. We saw in the last chapter that for a Binomial model the standard deviation of the *number* of successes is $\sqrt{npq}$. Now we want the standard deviation of the *proportion* of successes, $\hat{p}$. The sample proportion $\hat{p}$ is the number of successes divided by the number of trials, n, so the standard deviation is also divided by n:

$$\sigma(\hat{p}) = SD(\hat{p}) = \frac{\sqrt{npq}}{n} = \sqrt{\frac{pq}{n}}.$$

When we draw simple random samples of n individuals, the proportions we find will vary from sample to sample. As long as n is reasonably large,[3] we can model the distribution of these sample proportions with a probability model that is

$$N\left(p, \sqrt{\frac{pq}{n}}\right).$$

FIGURE 18.2
A Normal model centered at p with a standard deviation of $\sqrt{\dfrac{pq}{n}}$ is a good model for a collection of proportions found for many random samples of size n from a population with success probability p.

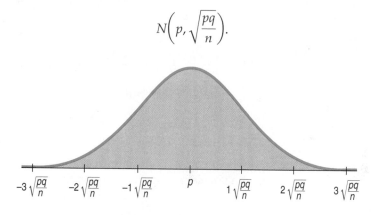

NOTATION ALERT

In Chapter 8 we introduced $\hat{y}$ as the predicted value for y. The "hat" here plays a similar role. It indicates that $\hat{p}$—the observed proportion in our data—is our *estimate* of the parameter p.

Although we'll never know the true proportion of adults who believe in ghosts, we're supposing it to be 45%. Once we put the center at $p = 0.45$, the standard deviation for the CBS poll is

$$SD(\hat{p}) = \sqrt{\frac{pq}{n}} = \sqrt{\frac{(0.45)(0.55)}{808}} = 0.0175, \text{ or } 1.75\%.$$

Here's a picture of the Normal model for our simulation histogram:

FIGURE 18.3
Using 0.45 for p gives this Normal model for Figure 18.1's histogram of the sample proportions of adults believing in ghosts ($n = 808$).

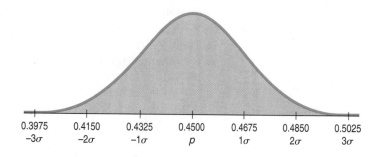

---

[3] For smaller n, we can just use a Binomial model.

Because we have a Normal model, we can use the 68–95–99.7 Rule or look up other probabilities using a table or technology. For example, we know that 95% of Normally distributed values are within two standard deviations of the mean, so we should not be surprised if 95% of various polls gave results that were near 45% but varied above and below that by no more than two standard deviations. Since $2 \times 1.75\% = 3.5\%$,[4] we see that the CBS poll estimating belief in ghosts at 48% is *consistent* with our guess of 45%. This is what we mean by **sampling error**. It's not really an *error* at all, but just *variability* you'd expect to see from one sample to another. A better term would be **sampling variability**.

MATH BOX

How do we know that proportions have a mean of p and a standard deviation of $\sqrt{\dfrac{pq}{n}}$?

Remember from Chapter 17 that if we have a sequence of independent trials where we record a success or failure on each trial (Bernoulli trials), that the total number of successes, Y, out of n trials has a mean of np and a *standard deviation* of $\sqrt{npq}$.

The sample proportion $\hat{p}$ is found by taking the total number of successes and dividing by the sample size:

$$\hat{p} = \frac{Y}{n}.$$

So,

$$E(\hat{p}) = E\left(\frac{Y}{n}\right) = \frac{E(Y)}{n} = \frac{np}{n} = p,$$

the true proportion of the population. And

$$SD(\hat{p}) = SD\left(\frac{Y}{n}\right) = \frac{SD(Y)}{n} = \frac{\sqrt{npq}}{\sqrt{n^2}} = \sqrt{\frac{npq}{n^2}} = \sqrt{\frac{pq}{n}}.$$

If we know the mean, p, of the sample proportion, then we know the standard deviation as well. This is a special property of proportions.

How Good Is the Normal Model?

Stop and think for a minute about what we've just said. It's a remarkable claim. We've said that if we draw repeated random samples of the same size, n, from some population and measure the proportion, $\hat{p}$, we see in each sample, then the collection of these proportions will pile up around the underlying population proportion, p, and that a histogram of the sample proportions can be modeled well by a Normal model.

There must be a catch. Suppose the samples were of size 2, for example. Then the only possible proportion values would be 0, 0.5, and 1. There's no way the histogram could ever look like a Normal model with only three possible values for the variable.

Well, there *is* a catch. The claim is only approximately true. (But, that's OK. After all, models are only supposed to be approximately true.) And the model

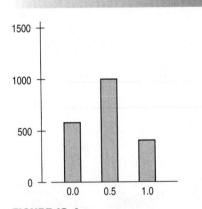

FIGURE 18.4
Proportions from samples of size 2 can take on only three possible values. A Normal model does not work well.

---

[4] The standard deviation is 1.75%. Remember that the standard deviation always has the same units as the data. Here our units are %. But that can be confusing, because the standard deviation is not 1.75% of anything. It is 1.75 percentage points. If that's confusing, try writing the units as "percentage points" instead of %.

becomes a better and better representation of the distribution of the sample proportions as the sample size gets bigger.[5] Samples of size 1 or 2 just aren't going to work very well. But the distributions of proportions of many larger samples do have histograms that are remarkably close to a Normal model.

Assumptions and Conditions

To use a model, we usually must make some assumptions. To use the sampling distribution model for sample proportions, we need two assumptions:

> **The Independence Assumption:** The sampled values must be independent of each other.
>
> **The Sample Size Assumption:** The sample size, n, must be large enough.

Of course, assumptions are hard—often impossible—to check. That's why we *assume* them. But, as we saw in Chapter 8, we should check to see whether the assumptions are reasonable. To think about the Independence Assumption, we often wonder whether there is any reason to think that the data values might affect each other. Fortunately, we can often check *conditions* that provide information about the assumptions. Check these conditions before using the Normal to model the distribution of sample proportions:

> **Randomization Condition:** If your data come from an experiment, subjects should have been randomly assigned to treatments. If you have a survey, your sample should be a simple random sample of the population. If some other sampling design was used, be sure the sampling method was not biased and that the data are representative of the population.
>
> **10% Condition:** The sample size, n, must be no larger than 10% of the population. For national polls, the total population is usually very large, so the sample is a small fraction of the population.
>
> **Success/Failure Condition:** The sample size has to be big enough so that we expect at least 10 successes and at least 10 failures. When np and nq are at least 10, we have enough data for sound conclusions. For the CBS survey, a "success" might be believing in ghosts. With $p = 0.45$, we expect $808 \times 0.45 = 364$ successes and $808 \times 0.55 = 444$ failures. Both are at least 10, so we certainly expect enough successes and enough failures for the condition to be satisfied.

The terms "success" and "failure" for the outcomes that have probability p and q are common in Statistics. But they are completely arbitrary labels. When we say that a disease occurs with probability p, we certainly don't mean that getting sick is a "success" in the ordinary sense of the word.

These last two conditions seem to conflict with each other. The **Success/ Failure Condition** wants sufficient data. How much depends on p. If p is near 0.5, we need a sample of only 20 or so. If p is only 0.01, however, we'd need 1000. But the **10% Condition** says that a sample should be no larger than 10% of the population. If you're thinking, "Wouldn't a larger sample be better?" you're right of course. It's just that if the sample were more than 10% of the population, we'd need to use different methods to analyze the data. Fortunately, this isn't usually a problem in practice. Often, as in polls that sample from all U.S. adults or industrial samples from a day's production, the populations are much larger than 10 times the sample size.

A Sampling Distribution Model for a Proportion

We've simulated repeated samples and looked at a histogram of the sample proportions. We modeled that histogram with a Normal model. Why do we bother to model it? Because this model will give us insight into how much the sample

---

[5] Formally, we say the claim is true in the limit as n grows.

proportion can vary from sample to sample. We've simulated many of the other random samples we might have gotten. The model is an attempt to show the distribution from *all* the random samples. But how do we know that a Normal model will really work? Is this just an observation based on some simulations that *might* be approximately true some of the time?

It turns out that this model can be justified theoretically and that the larger the sample size, the better the model works. That's the result Laplace proved. We won't bother you with the math because, in this instance, it really wouldn't help your understanding.[6] Nevertheless, the fact that we can think of the sample proportion as a random variable taking on a different value in each random sample, and then say something this specific about the distribution of those values, is a fundamental insight—one that we will use in each of the next four chapters.

We have changed our point of view in a very important way. No longer is a proportion something we just compute for a set of data. We now see it as a random variable quantity that has a probability distribution, and thanks to Laplace we have a model for that distribution. We call that the **sampling distribution model** for the proportion, and we'll make good use of it.

Simulation: Simulate the Sampling Distribution Model of a Proportion. You probably don't want to work through the formal mathematical proof; a simulation is far more convincing!

We have now answered the question raised at the start of the chapter. To know how variable a sample proportion is, we need to know the proportion and the size of the sample. That's all.

THE SAMPLING DISTRIBUTION MODEL FOR A PROPORTION

Provided that the sampled values are independent and the sample size is large enough, the sampling distribution of $\hat{p}$ is modeled by a Normal model with mean $\mu(\hat{p})$ and standard deviation $SD(\hat{p}) = \sqrt{\dfrac{pq}{n}}$.

Without the sampling distribution model, the rest of Statistics just wouldn't exist. Sampling models are what makes Statistics work. They inform us about the amount of variation we should expect when we sample. Suppose we flip a coin 100 times in order to decide whether it's fair or not. If we get 52 heads, we're probably not surprised. Although we'd expect 50 heads, 52 doesn't seem particularly unusual for a fair coin. But we would be surprised to see 90 heads; that might really make us doubt that the coin is fair. How about 64 heads? Harder to say. That's a case where we need the sampling distribution model. The sampling model quantifies the variability, telling us how surprising any sample proportion is. And it enables us to make informed decisions about how precise our estimate of the true proportion might be. That's exactly what we'll be doing for the rest of this book.

Sampling distribution models act as a bridge from the real world of data to the imaginary model of the statistic and enable us to say something about the population when all we have is data from the real world. This is the huge leap of Statistics. Rather than thinking about the sample proportion as a fixed quantity calculated from our data, we now think of it as a random variable—our value is just one of many we might have seen had we chosen a different random sample. By imagining what *might* happen if we were to draw many, many samples from the same population, we can learn a lot about how close the statistics computed from our one particular sample may be to the corresponding population parameters they estimate. That's the path to the *margin of error* you hear about in polls and surveys. We'll see how to determine that in the next chapter.

---

[6] The proof is pretty technical. We're not sure it helps *our* understanding all that much either.

FOR EXAMPLE   Using the Sampling Distribution Model for Proportions

The Centers for Disease Control and Prevention report that 22% of 18-year-old women in the United States have a body mass index (BMI)[7] of 25 or more—a value considered by the National Heart Lung and Blood Institute to be associated with increased health risk.

As part of a routine health check at a large college, the physical education department usually requires students to come in to be measured and weighed. This year, the department decided to try out a self-report system. It asked 200 randomly selected female students to report their heights and weights (from which their BMIs could be calculated). Only 31 of these students had BMIs greater than 25.

QUESTION: Is this proportion of high-BMI students unusually small?

First, check the conditions:

✔ Randomization Condition: The department drew a random sample, so the respondents should be independent and randomly selected from the population.

✔ 10% Condition: 200 respondents is less than 10% of all the female students at a "large college."

✔ Success/Failure Condition: The department expected $np = 200(0.22) = 44$ "successes" and $nq = 200(0.78) = 156$ "failures," both at least 10.

It's okay to use a Normal model to describe the sampling distribution of the proportion of respondents with BMIs above 25.

The phys ed department observed $\hat{p} = \dfrac{31}{200} = 0.155$.

The department expected $E(\hat{p}) = p = 0.22$, with $SD(\hat{p}) = \sqrt{\dfrac{pq}{n}} = \sqrt{\dfrac{(0.22)(0.78)}{200}} = 0.029$,

so $z = \dfrac{\hat{p} - p}{SD(\hat{p})} = \dfrac{0.155 - 0.22}{0.029} = -2.24$.

By the 68–95–99.7 Rule, I know that values more than 2 standard deviations below the mean of a Normal model show up less than 2.5% of the time. Perhaps women at this college differ from the general population, or self-reporting may not provide accurate heights and weights.

JUST CHECKING

1. You want to poll a random sample of 100 students on campus to see if they are in favor of the proposed location for the new student center. Of course, you'll get just one number, your sample proportion, $\hat{p}$. But if you imagined all the possible samples of 100 students you could draw and imagined the histogram of all the sample proportions from these samples, what shape would it have?

2. Where would the center of that histogram be?

3. If you think that about half the students are in favor of the plan, what would the standard deviation of the sample proportions be?

---

[7] BMI = weight in kg/(height in m)^2.

STEP-BY-STEP EXAMPLE   Working with Sampling Distribution Models for Proportions

Suppose that about 13% of the population is left-handed.[8] A 200-seat school auditorium has been built with 15 "lefty seats," seats that have the built-in desk on the left rather than the right arm of the chair. (For the right-handed readers among you, have you ever tried to take notes in a chair with the desk on the left side?)

Question: In a class of 90 students, what's the probability that there will not be enough seats for the left-handed students?

THINK

Plan State what we want to know.

I want to find the probability that in a group of 90 students, more than 15 will be left-handed. Since 15 out of 90 is 16.7%, I need the probability of finding more than 16.7% left-handed students out of a sample of 90 if the proportion of lefties is 13%.

Model Think about the assumptions and check the conditions.

You might be able to think of cases where the **Independence Assumption** is not plausible—for example, if the students are all related, or if they were selected for being left- or right-handed. But for a random sample, the assumption of independence seems reasonable.

✔ **Independence Assumption:** It is reasonable to assume that the probability that one student is left-handed is not changed by the fact that another student is right- or left-handed.

✔ **Randomization Condition:** The 90 students in the class can be thought of as a random sample of students.

✔ **10% Condition:** 90 is surely less than 10% of the population of all students. (Even if the school itself is small, I'm thinking of the population of all *possible* students who could have gone to the school.)

✔ **Success/Failure Condition:**

$$np = 90(0.13) = 11.7 \geq 10$$
$$nq = 90(0.87) = 78.3 \geq 10$$

State the parameters and the sampling distribution model.

The population proportion is $p = 0.13$. The conditions are satisfied, so I'll model the sampling distribution of $\hat{p}$ with a Normal model with mean 0.13 and a standard deviation of

$$SD(\hat{p}) = \sqrt{\frac{pq}{n}} = \sqrt{\frac{(0.13)(0.87)}{90}} \approx 0.035$$

My model for $\hat{p}$ is $N(0.13, 0.035)$.

---

[8] Actually, it's quite difficult to get an accurate estimate of the proportion of lefties in the population. Estimates range from 8% to 15%.

Plot Make a picture. Sketch the model and shade the area we're interested in, in this case the area to the right of 16.7%.

Mechanics Use the standard deviation as a ruler to find the z-score of the cutoff proportion. We see that 16.7% lefties would be just over one standard deviation above the mean.

Find the resulting probability from a table of Normal probabilities, a computer program, or a calculator.

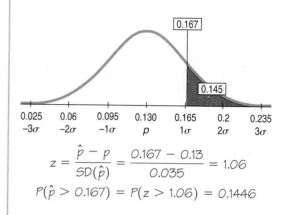

$$z = \frac{\hat{p} - p}{SD(\hat{p})} = \frac{0.167 - 0.13}{0.035} = 1.06$$

$$P(\hat{p} > 0.167) = P(z > 1.06) = 0.1446$$

Conclusion Interpret the probability in the context of the question.

There is about a 14.5% chance that there will not be enough seats for the left-handed students in the class.

What About Quantitative Data?

Proportions summarize categorical variables. And the Normal sampling distribution model looks like it is going to be very useful. But can we do something similar with quantitative data?

Of course we can (or we wouldn't have asked). Even more remarkable, not only can we use all of the same concepts, but almost the same model, too.

What are the concepts? We know that when we sample at random or randomize an experiment, the results we get will vary from sample-to-sample and from experiment-to-experiment. The Normal model seems an incredibly simple way to summarize all that variation. Could something that simple work for means? We won't keep you in suspense. It turns out that means also have a sampling distribution that we can model with a Normal model. And it turns out that Laplace's theoretical result applies to means, too. As we did with proportions, we can get some insight from a simulation.

Simulating the Sampling Distribution of a Mean

Here's a simple simulation. Let's start with one fair die. If we toss this die 10,000 times, what should the histogram of the numbers on the face of the die look like? Figure 18.5 shows the results of a simulated 10,000 tosses.

Now let's toss a *pair* of dice and record the average of the two. If we repeat this (or at least simulate repeating it) 10,000 times, recording the average of each pair, what will the histogram of these 10,000 averages look like? Before you look, think a minute. Is getting an average of 1 on *two* dice as likely as getting an average of 3 or 3.5?

Let's see.

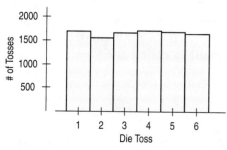

FIGURE 18.5

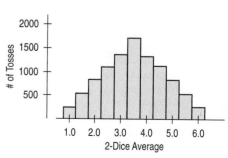

FIGURE 18.6

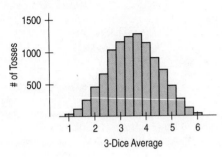

FIGURE 18.7

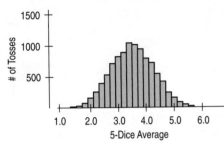

FIGURE 18.8

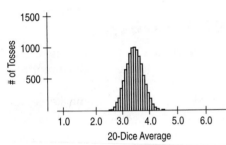

FIGURE 18.9

We're much more likely to get an average near 3.5 than we are to get one near 1 or 6. Without calculating those probabilities exactly, it's fairly easy to see that the *only* way to get an average of 1 is to get two 1's. To get a total of 7 (for an average of 3.5), though, there are many more possibilities. This distribution even has a name: the *triangular* distribution.

What if we average 3 dice? We'll simulate 10,000 tosses of 3 dice and take their average. Figure 18.7 shows the result.

What's happening? First notice that it's getting harder to have averages near the ends. Getting an average of 1 or 6 with 3 dice requires all three to come up 1 or 6, respectively. That's less likely than for 2 dice to come up both 1 or both 6. The distribution is being pushed toward the middle. But what's happening to the shape? (This distribution doesn't have a name, as far as we know.)

Let's continue this simulation to see what happens with larger samples. Figure 18.8 shows a histogram of the averages for 10,000 tosses of 5 dice.

The pattern is becoming clearer. Two things continue to happen. The first fact we knew already from the Law of Large Numbers. It says that as the sample size (number of dice) gets larger, each sample average is more likely to be closer to the population mean. So, we see the shape continuing to tighten around 3.5. But the shape of the distribution is the surprising part. It's becoming bell-shaped. And not just bell-shaped; it's approaching the Normal model.

Are you convinced? Let's skip ahead and try 20 dice. The histogram of averages for 10,000 throws of 20 dice is in Figure 18.9.

Now we see the Normal shape again (and notice how much smaller the spread is). But can we count on this happening for situations other than dice throws? What kinds of sample means have sampling distributions that we can model with a Normal model? It turns out that Normal models work well amazingly often.

"The theory of probabilities is at bottom nothing but common sense reduced to calculus."

–Laplace, in *Théorie analytique des probabilités*, 1812

The Central Limit Theorem: The Fundamental Theorem of Statistics

The dice simulation may look like a special situation, but it turns out that what we saw with dice is true for means of repeated samples for almost every situation. When we looked at the sampling distribution of a proportion, we had to check only a few conditions. For means, the result is even more remarkable. *There are almost no conditions at all.*

Let's say that again: The sampling distribution of *any* mean becomes more nearly Normal as the sample size grows. All we need is for the observations to be independent and collected with randomization. We don't even care about the shape of the population distribution![9] This surprising fact is the result Laplace proved in a fairly general form in 1810. At the time, Laplace's theorem caused quite a stir (at least in mathematics circles) because it is so unintuitive. Laplace's result is called the **Central Limit Theorem**[10] (CLT).

---

[9] OK, one technical condition. The data must come from a population with a finite variance. You probably can't imagine a population with an infinite variance, but statisticians can construct such things, so we have to discuss them in footnotes like this. It really makes no difference in how you think about the important stuff, so you can just forget we mentioned it.

[10] The word "central" in the name of the theorem means "fundamental." It doesn't refer to the center of a distribution.

Laplace was one of the greatest scientists and mathematicians of his time. In addition to his contributions to probability and statistics, he published many new results in mathematics, physics, and astronomy (where his nebular theory was one of the first to describe the formation of the solar system in much the way it is understood today). He also played a leading role in establishing the metric system of measurement.

His brilliance, though, sometimes got him into trouble. A visitor to the Académie des Sciences in Paris reported that Laplace let it be widely known that he considered himself the best mathematician in France. The effect of this on his colleagues was not eased by the fact that Laplace was right.

Why should the Normal model show up again for the sampling distribution of means as well as proportions? We're not going to try to persuade you that it is obvious, clear, simple, or straightforward. In fact, the CLT is surprising and a bit weird. Not only does the distribution of means of many random samples get closer and closer to a Normal model as the sample size grows, *this is true regardless of the shape of the population distribution!* Even if we sample from a skewed or bimodal population, the Central Limit Theorem tells us that means of repeated random samples will tend to follow a Normal model as the sample size grows. Of course, you won't be surprised to learn that it works better and faster the closer the population distribution is to a Normal model. And it works better for larger samples. If the data come from a population that's exactly Normal to start with, then the observations themselves are Normal. If we take samples of size 1, their "means" are just the observations—so, of course, they have Normal sampling distribution. But now suppose the population distribution is very skewed (like the CEO data from Chapter 5, for example). The CLT works, although it may take a sample size of dozens or even hundreds of observations for the Normal model to work well.

For example, think about a really bimodal population, one that consists of only 0's and 1's. The CLT says that even means of samples from this population will follow a Normal sampling distribution model. But wait. Suppose we have a categorical variable and we assign a 1 to each individual in the category and a 0 to each individual not in the category. And then we find the mean of these 0's and 1's. That's the same as counting the number of individuals who are in the category and dividing by n. That mean will be . . . the *sample proportion*, $\hat{p}$, of individuals who are in the category (a "success"). So maybe it wasn't so surprising after all that proportions, like means, have Normal sampling distribution models; they are actually just a special case of Laplace's remarkable theorem. Of course, for such an extremely bimodal population, we'll need a reasonably large sample size—and that's where the special conditions for proportions come in.

THE CENTRAL LIMIT THEOREM (CLT)

The mean of a random sample is a random variable whose sampling distribution can be approximated by a Normal model. The larger the sample, the better the approximation will be.

Assumptions and Conditions

The CLT requires essentially the same assumptions as we saw for modeling proportions:

> **Independence Assumption:** The sampled values must be independent of each other.
>
> **Sample Size Assumption:** The sample size must be sufficiently large.

We can't check these directly, but we can think about whether the **Independence Assumption** is plausible. We can also check some related conditions:

> **Randomization Condition:** The data values must be sampled randomly, or the concept of a sampling distribution makes no sense.
>
> **10% Condition:** When the sample is drawn without replacement (as is usually the case), the sample size, n, should be no more than 10% of the population.
>
> **Large Enough Sample Condition:** Although the CLT tells us that a Normal model is useful in thinking about the behavior of sample means when the sample size is large enough, it doesn't tell us how large a sample we need.

Activity: The Central Limit Theorem. Does it really work for samples from non-Normal populations?

The truth is, it depends; there's no one-size-fits-all rule. If the population is unimodal and symmetric, even a fairly small sample is okay. If the population is strongly skewed, like the compensation for CEOs we looked at in Chapter 5, it can take a pretty large sample to allow use of a Normal model to describe the distribution of sample means. For now you'll just need to think about your sample size in the context of what you know about the population, and then tell whether you believe the **Large Enough Sample Condition** has been met.

A S *Activity:* **The Standard Deviation of Means.** Experiment to see how the variability of the mean changes with the sample size.

But Which Normal?

The CLT says that the sampling distribution of any mean or proportion is approximately Normal. But which Normal model? We know that any Normal is specified by its mean and standard deviation. For proportions, the sampling distribution is centered at the population proportion. For means, it's centered at the population mean. What else would we expect?

What about the standard deviations, though? We noticed in our dice simulation that the histograms got narrower as we averaged more and more dice together. This shouldn't be surprising. Means vary less than the individual observations. Think about it for a minute. Which would be more surprising, having *one* person in your Statistics class who is over 6'9" tall or having the *mean* of 100 students taking the course be over 6'9"? The first event is fairly rare.[11] You may have seen somebody this tall in one of your classes sometime. But finding a class of 100 whose mean height is over 6'9" tall just won't happen. Why? Because *means have smaller standard deviations than individuals.*

How much smaller? Well, we have good news and bad news. The good news is that the standard deviation of y falls as the sample size grows. The bad news is that it doesn't drop as fast as we might like. It only goes down by the *square root* of the sample size. Why? The Math Box will show you that the Normal model for the sampling distribution of the mean has a standard deviation equal to

$$SD(\bar{y}) = \frac{\sigma}{\sqrt{n}}$$

where σ is the standard deviation of the population. To emphasize that this is a standard deviation *parameter* of the sampling distribution model for the sample mean, $\bar{y}$, we write $SD(\bar{y})$ or $\sigma(\bar{y})$.

A S *Activity:* **The Sampling Distribution of the Mean.** The CLT tells us what to expect. In this activity you can work with the CLT or simulate it if you prefer.

> ### THE SAMPLING DISTRIBUTION MODEL FOR A MEAN (CLT)
>
> When a random sample is drawn from any population with mean μ and standard deviation σ, its sample mean, $\bar{y}$, has a sampling distribution with the same *mean* μ but whose *standard deviation* is $\frac{\sigma}{\sqrt{n}}$ (and we write $\sigma(\bar{y}) = SD(\bar{y}) = \frac{\sigma}{\sqrt{n}}$). No matter what population the random sample comes from, the *shape* of the sampling distribution is approximately Normal as long as the sample size is large enough. The larger the sample used, the more closely the Normal approximates the sampling distribution for the mean.

---

[11] If students are a random sample of adults, fewer than 1 out of 10,000 should be taller than 6'9". Why might college students not really be a random sample with respect to height? Even if they're not a perfectly random sample, a college student over 6'9" tall is still rare.

MATH BOX

Why is $SD(\bar{y}) = \dfrac{\sigma}{\sqrt{n}}$? We know that $\bar{y}$ is a sum divided by n:

$$\bar{y} = \frac{y_1 + y_2 + y_3 + \cdots + y_n}{n}.$$

As we saw in Chapter 16, when a random variable is divided by a constant its variance is divided by the *square* of the constant:

$$Var(\bar{y}) = \frac{Var(y_1 + y_2 + y_3 + \cdots + y_n)}{n^2}.$$

To get our sample, we draw the y's randomly, ensuring they are independent. For independent random variables, variances add:

$$Var(\bar{y}) = \frac{Var(y_1) + Var(y_2) + Var(y_3) + \cdots + Var(y_n)}{n^2}.$$

All n of the y's were drawn from our population, so they all have the same variance, σ^2:

$$Var(\bar{y}) = \frac{\sigma^2 + \sigma^2 + \sigma^2 + \cdots + \sigma^2}{n^2} = \frac{n\sigma^2}{n^2} = \frac{\sigma^2}{n}.$$

The standard deviation of $\bar{y}$ is the square root of this variance:

$$SD(\bar{y}) = \sqrt{\frac{\sigma^2}{n}} = \frac{\sigma}{\sqrt{n}}.$$

We now have two closely related sampling distribution models that we can use when the appropriate assumptions and conditions are met. Which one we use depends on which kind of data we have:

- When we have categorical data, we calculate a sample proportion, $\hat{p}$; the sampling distribution of this random variable has a Normal model with a mean at the true proportion ("Greek letter") p and a standard deviation of $SD(\hat{p}) = \sqrt{\dfrac{pq}{n}} = \dfrac{\sqrt{pq}}{\sqrt{n}}$. We'll use this model in Chapters 19 through 22.
- When we have quantitative data, we calculate a sample mean, $\bar{y}$; the sampling distribution of this random variable has a Normal model with a mean at the true mean, μ, and a standard deviation of $SD(\bar{y}) = \dfrac{\sigma}{\sqrt{n}}$. We'll use this model in Chapters 23, 24, and 25.

The means of these models are easy to remember, so all you need to be careful about is the standard deviations. Remember that these are standard deviations of the *statistics* $\hat{p}$ and $\bar{y}$. They both have a square root of n in the denominator. That tells us that the larger the sample, the less either statistic will vary. The only difference is in the numerator. If you just start by writing $SD(\bar{y})$ for quantitative data and $SD(\hat{p})$ for categorical data, you'll be able to remember which formula to use.

FOR EXAMPLE | Using the CLT for Means

RECAP: A college physical education department asked a random sample of 200 female students to self-report their heights and weights, but the percentage of students with body mass indexes over 25 seemed suspiciously low. One possible explanation may be that the respondents "shaded" their weights down a bit. The CDC reports that the mean weight of 18-year-old women is 143.74 lb, with a standard deviation of 51.54 lb, but these 200 randomly selected women reported a mean weight of only 140 lb.

QUESTION: Based on the Central Limit Theorem and the 68-95-99.7 Rule, does the mean weight in this sample seem exceptionally low, or might this just be random sample-to-sample variation?

The conditions check out okay:

✔ Randomization Condition: The women were a random sample and their weights can be assumed to be independent.

✔ 10% Condition: They sampled fewer than 10% of all women at the college.

✔ Large Enough Sample Condition: The distribution of college women's weights is likely to be unimodal and reasonably symmetric, so the CLT applies to means of even small samples; 200 values is plenty.

The sampling model for sample means is approximately Normal with $E(\bar{y}) = 143.7$ and $SD(\bar{y}) = \dfrac{\sigma}{\sqrt{n}} = \dfrac{51.54}{\sqrt{200}} = 3.64$. The expected distribution of sample means is:

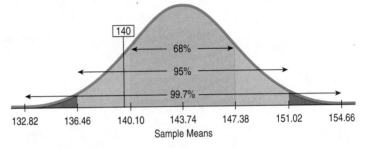

The 68–95–99.7 Rule suggests that although the reported mean weight of 140 pounds is somewhat lower than expected, it does not appear to be unusual. Such variability is not all that extraordinary for samples of this size.

STEP-BY-STEP EXAMPLE | Working with the Sampling Distribution Model for the Mean

The Centers for Disease Control and Prevention reports that the mean weight of adult men in the United States is 190 lb with a standard deviation of 59 lb.[12]

Question: An elevator in our building has a weight limit of 10 persons or 2500 lb. What's the probability that if 10 men get on the elevator, they will overload its weight limit?

THINK

Plan State what we want to know.

Asking the probability that the total weight of a sample of 10 men exceeds 2500 pounds is equivalent to asking the probability that their mean weight is greater than 250 pounds.

Model Think about the assumptions and check the conditions.

✔ **Independence Assumption:** It's reasonable to think that the weights of 10 randomly sampled men will be independent of each other. (But there could be exceptions—for example, if they were all from the same family or if the elevator were in a building with a diet clinic!)

---

[12] Cynthia L. Ogden, Cheryl D. Fryar, Margaret D. Carroll, and Katherine M. Flegal, *Mean Body Weight, Height, and Body Mass Index, United States 1960–2002, Advance Data from Vital and Health Statistics Number 347*, Oct. 27, 2004. www.cdc.gov/nchs

Note that if the sample were larger we'd be less concerned about the shape of the distribution of all weights.

✔ **Randomization Condition:** I'll assume that the 10 men getting on the elevator are a random sample from the population.

✔ **10% Condition:** 10 men is surely less than 10% of the population of possible elevator riders.

✔ **Large Enough Sample Condition:** I suspect the distribution of population weights is roughly unimodal and symmetric, so my sample of 10 men seems large enough.

State the parameters and the sampling model.

The mean for all weights is $\mu = 190$ and the standard deviation is $\sigma = 59$ pounds. Since the conditions are satisfied, the CLT says that the sampling distribution of $\bar{y}$ has a Normal model with mean 190 and standard deviation

$$SD(\bar{y}) = \frac{\sigma}{\sqrt{n}} = \frac{59}{\sqrt{10}} \approx 18.66$$

 Plot Make a picture. Sketch the model and shade the area we're interested in. Here the mean weight of 250 pounds appears to be far out on the right tail of the curve.

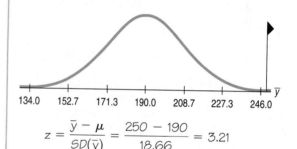

Mechanics Use the standard deviation as a ruler to find the z-score of the cutoff mean weight. We see that an average of 250 pounds is more than 3 standard deviations above the mean.

$$z = \frac{\bar{y} - \mu}{SD(\bar{y})} = \frac{250 - 190}{18.66} = 3.21$$

Find the resulting probability from a table of Normal probabilities such as Table Z, a computer program, or a calculator.

$$P(\bar{y} > 250) = P(z > 3.21) = 0.0007$$

 Conclusion Interpret your result in the proper context, being careful to relate it to the original question.

The chance that a random collection of 10 men will exceed the elevator's weight limit is only 0.0007. So, if they are a random sample, it is quite unlikely that 10 people will exceed the total weight allowed on the elevator.

About Variation

Means vary less than individual data values. That makes sense. If the same test is given to many sections of a large course and the class average is, say, 80%, some students may score 95% because individual scores vary a lot. But we'd be shocked (and pleased!) if the *average* score of the students in any section was 95%. Averages are much less variable. Not only do group averages vary less than individual values, but common sense suggests that averages should be more consistent for larger groups. The Central Limit Theorem confirms this

"The n's justify the means."

—Apocryphal statistical saying

hunch; the fact that $SD(\bar{y}) = \dfrac{\sigma}{\sqrt{n}}$ has n in the denominator shows that the

<table>
<tr><td>

A Billion Dollar Misunderstanding? In the late 1990s the Bill and Melinda Gates Foundation began funding an effort to encourage the breakup of large schools into smaller schools. Why? It had been noticed that smaller schools were more common among the best-performing schools than one would expect. In time, the Annenberg Foundation, the Carnegie Corporation, the Center for Collaborative Education, the Center for School Change, Harvard's Change Leadership Group, the Open Society Institute, Pew Charitable Trusts, and the U.S. Department of Education's Smaller Learning Communities Program all supported the effort. Well over a billion dollars was spent to make schools smaller.

But was it all based on a misunderstanding of sampling distributions? Statisticians Howard Wainer and Harris Zwerling[13] looked at the mean test scores of schools in Pennsylvania. They found that indeed 12% of the top-scoring 50 schools were from the smallest 3% of Pennsylvania schools—substantially more than the 3% we'd naively expect. But then they looked at the *bottom* 50. There they found that 18% were small schools! The explanation? Mean test scores are, well, means. We are looking at a rough real-world simulation in which each school is a trial. Even if all Pennsylvania schools were equivalent, we'd expect their mean scores to vary. How much? The CLT tells us that means of test scores vary according to $\frac{\sigma}{\sqrt{n}}$. Smaller schools have (by definition) smaller n's, so the sampling distributions of their mean scores naturally have larger standard deviations. It's natural, then, that small schools have both higher and lower mean scores.

On October 26, 2005, *The Seattle Times* reported:

> [T]he Gates Foundation announced last week it is moving away from its emphasis on converting large high schools into smaller ones and instead giving grants to specially selected school districts with a track record of academic improvement and effective leadership. Education leaders at the Foundation said they concluded that improving classroom instruction and mobilizing the resources of an entire district were more important first steps to improving high schools than breaking down the size.

</td></tr>
</table>

variability of sample means decreases as the sample size increases. There's a catch, though. The standard deviation of the sampling distribution declines only with the square root of the sample size and not, for example, with $1/n$.

The mean of a random sample of 4 has half $\left(\frac{1}{\sqrt{4}} = \frac{1}{2} \right)$ the standard deviation of an individual data value. To cut the standard deviation in half again, we'd need a sample of 16, and a sample of 64 to halve it once more.

If only we had a much larger sample, we could get the standard deviation of the sampling distribution *really* under control so that the sample mean could tell us still more about the unknown population mean, but larger samples cost more and take longer to survey. And while we're gathering all that extra data, the population itself may change, or a news story may alter opinions. There are practical limits to most sample sizes. As we shall see, that nasty square root limits how much we can make a sample tell about the population. This is an example of something that's known as the Law of Diminishing Returns.

The Real World and the Model World

Be careful. We have been slipping smoothly between the real world, in which we draw random samples of data, and a magical mathematical model world, in which we describe how the sample means and proportions we observe in the real world behave as random variables in all the random samples that we might have drawn. Now we have *two* distributions to deal with. The first is the real-world distribution of the sample, which we might display with a histogram (for quantitative data) or with a bar chart or table (for categorical data). The second is the math world *sampling distribution model* of the statistic, a Normal model based on the Central Limit Theorem. Don't confuse the two.

For example, don't mistakenly think the CLT says that the *data* are Normally distributed as long as the sample is large enough. In fact, as samples get larger, we expect the distribution of the data to look more and more like the population from which they are drawn—skewed, bimodal, whatever—but not necessarily Normal. You can collect a sample of CEO salaries for the next 1000 years,[14] but the histogram will never look Normal. It will be skewed to the right. The Central Limit Theorem doesn't talk about the distribution of the data from the sample. It talks about the sample *means* and sample *proportions* of many different random samples drawn from the same population. Of course, the CLT does require that the sample be big enough when the population shape is not unimodal and symmetric, but the fact that, even then, a Normal model is useful is still a very surprising and powerful result.

---

[13] Wainer, H. and Zwerling, H., "Legal and empirical evidence that smaller schools do not improve student achievement," *The Phi Delta Kappan* 2006 87:300–303. Discussed in Howard Wainer, "The Most Dangerous Equation," *American Scientist*, May–June 2007, pp. 249–256; also at www.Americanscientist.org.
[14] Don't forget to adjust for inflation.

4. Human gestation times have a mean of about 266 days, with a standard deviation of about 16 days. If we record the gestation times of a sample of 100 women, do we know that a histogram of the times will be well modeled by a Normal model?

5. Suppose we look at the *average* gestation times for a sample of 100 women. If we imagined all the possible random samples of 100 women we could take and looked at the histogram of all the sample means, what shape would it have?

6. Where would the center of that histogram be?

7. What would be the standard deviation of that histogram?

Sampling Distribution Models

Let's summarize what we've learned about sampling distributions. At the heart is the idea that *the statistic itself is a random variable*. We can't know what our statistic will be because it comes from a random sample. It's just one instance of something that happened for our particular random sample. A different random sample would have given a different result. This sample-to-sample variability is what generates the sampling distribution. The sampling distribution shows us the distribution of possible values that the statistic could have had.

We could simulate that distribution by pretending to take lots of samples. Fortunately, for the mean and the proportion, the CLT tells us that we can model their sampling distribution directly with a Normal model.

The two basic truths about sampling distributions are:

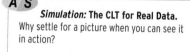

Simulation: The CLT for Real Data.
Why settle for a picture when you can see it in action?

1. Sampling distributions arise because samples vary. Each random sample will contain different cases and, so, a different value of the statistic.
2. Although we can always simulate a sampling distribution, the Central Limit Theorem saves us the trouble for means and proportions.

Here's a picture showing the process going into the sampling distribution model:

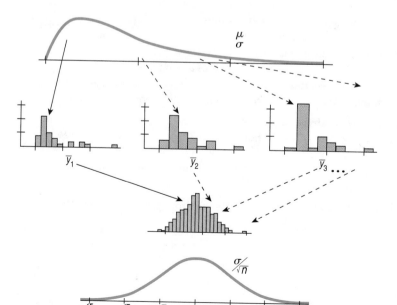

FIGURE 18.10

We start with a population model, which can have any shape. It can even be bimodal or skewed (as this one is). We label the mean of this model μ and its standard deviation, σ.

We draw one real sample (solid line) of size n and show its histogram and summary statistics. We imagine (or simulate) drawing many other samples (dotted lines), which have their own histograms and summary statistics.

We (imagine) gathering all the means into a histogram.

The CLT tells us we can model the shape of this histogram with a Normal model. The mean of this Normal is μ, and the standard deviation is $SD(\bar{y}) = \dfrac{\sigma}{\sqrt{n}}$.

What Can Go Wrong?

- **Don't confuse the sampling distribution with the distribution of the sample.** When you take a sample, you always look at the distribution of the values, usually with a histogram, and you may calculate summary statistics. Examining the distribution of the sample data is wise. But that's not the sampling distribution. The sampling distribution is an imaginary collection of all the values that a statistic *might* have taken for all possible random samples—the one you got and the ones that you didn't get. We use the sampling distribution model to make statements about how the statistic varies.

- **Beware of observations that are not independent.** The CLT depends crucially on the assumption of independence. If our elevator riders are related, are all from the same school (for example, an elementary school), or in some other way aren't a random sample, then the statements we try to make about the mean are going to be wrong. Unfortunately, this isn't something you can check in your data. You have to think about how the data were gathered. Good sampling practice and well-designed randomized experiments ensure independence.

- **Watch out for small samples from skewed populations.** The CLT assures us that the sampling distribution model is Normal if n is large enough. If the population is nearly Normal, even small samples (like our 10 elevator riders) work. If the population is very skewed, then n will have to be large before the Normal model will work well. If we sampled 15 or even 20 CEOs and used $\bar{y}$ to make a statement about the mean of all CEOs' compensation, we'd likely get into trouble because the underlying data distribution is so skewed. Unfortunately, there's no good rule of thumb.[15] It just depends on how skewed the data distribution is. Always plot the data to check.

CONNECTIONS

The concept of a sampling distribution connects to almost everything we have done. The fundamental connection is to the deliberate application of randomness in random sampling and randomized comparative experiments. If we didn't employ randomness to generate unbiased data, then repeating the data collection would just get the same data values again (with perhaps a few new measurement or recording errors). The distribution of statistic values arises directly because different random samples and randomized experiments would generate different statistic values.

The connection to the Normal distribution is obvious. We first introduced the Normal model before because it was "nice." As a unimodal, symmetric distribution with 99.7% of its area within three standard deviations of the mean, the Normal model is easy to work with. Now we see that the Normal holds a special place among distributions because we can use it to model the sampling distributions of the mean and the proportion.

We use simulation to understand sampling distributions. In fact, some important sampling distributions were discovered first by simulation.

---

[15] For proportions, of course, there is a rule: the **Success/Failure Condition**. That works for proportions because the standard deviation of a proportion is linked to its mean.

WHAT HAVE WE LEARNED?

Way back in Chapter 1 we said that Statistics is about variation. We know that no sample fully and exactly describes the population; sample proportions and means will vary from sample to sample. That's sampling error (or, better, sampling variability). We know it will always be present—indeed, the world would be a boring place if variability didn't exist. You might think that sampling variability would prevent us from learning anything reliable about a population by looking at a sample, but that's just not so. The fortunate fact is that sampling variability is not just unavoidable—it's predictable!

We've learned how the Central Limit Theorem describes the behavior of sample proportions—shape, center, and spread—as long as certain assumptions and conditions are met. The sample must be independent, random, and large enough that we expect at least 10 successes and failures. Then:

▶ The sampling distribution (the imagined histogram of the proportions from all possible samples) is shaped like a Normal model.

▶ The mean of the sampling model is the true proportion in the population.

▶ The standard deviation of the sample proportions is $\sqrt{\dfrac{pq}{n}}$.

And we've learned to describe the behavior of sample means as well, based on this amazing result known as the Central Limit Theorem—the Fundamental Theorem of Statistics. Again the sample must be independent and random—no surprise there—and needs to be larger if our data come from a population that's not roughly unimodal and symmetric. Then:

▶ Regardless of the shape of the original population, the shape of the distribution of the means of all possible samples can be described by a Normal model, provided the samples are large enough.

▶ The center of the sampling model will be the true mean of the population from which we took the sample.

▶ The standard deviation of the sample means is the population's standard deviation divided by the square root of the sample size, $\dfrac{\sigma}{\sqrt{n}}$.

Terms

Sampling distribution model
Different random samples give different values for a statistic. The sampling distribution model shows the behavior of the statistic over all the possible samples for the same size n (p. 432).

Sampling variability
Sampling error
The variability we expect to see from one random sample to another. It is sometimes called sampling error, but sampling variability is the better term (p. 434).

Sampling distribution model for a proportion
If assumptions of independence and random sampling are met, and we expect at least 10 successes and 10 failures, then the sampling distribution of a proportion is modeled by a Normal model with a mean equal to the true proportion value, p, and a standard deviation equal to $\sqrt{\dfrac{pq}{n}}$ (p. 436).

Central Limit Theorem
The Central Limit Theorem (CLT) states that the sampling distribution model of the sample mean (and proportion) from a random sample is approximately Normal for large n, *regardless of the distribution of the population, as long as the observations are independent* (p. 440).

Sampling distribution model for a mean
If assumptions of independence and random sampling are met, and the sample size is large enough, the sampling distribution of the sample mean is modeled by a Normal model with a mean equal to the population mean, μ, and a standard deviation equal to $\dfrac{\sigma}{\sqrt{n}}$ (p. 442).

Skills

THINK

▶ Understand that the variability of a statistic (as measured by the standard deviation of its sampling distribution) depends on the size of the sample. Statistics based on larger samples are less variable.

▶ Understand that the Central Limit Theorem gives the sampling distribution model of the mean for sufficiently large samples regardless of the underlying population.

SHOW

▶ Be able to demonstrate a sampling distribution by simulation.

▶ Be able to use a sampling distribution model to make simple statements about the distribution of a proportion or mean under repeated sampling.

TELL

▶ Be able to interpret a sampling distribution model as describing the values taken by a statistic in all possible realizations of a sample or randomized experiment under the same conditions.

EXERCISES

1. **Send money.** When they send out their fundraising letter, a philanthropic organization typically gets a return from about 5% of the people on their mailing list. To see what the response rate might be for future appeals, they did a simulation using samples of size 20, 50, 100, and 200. For each sample size, they simulated 1000 mailings with success rate $p = 0.05$ and constructed the histogram of the 1000 sample proportions, shown below. Explain how these histograms demonstrate what the Central Limit Theorem says about the sampling distribution model for sample proportions. Be sure to talk about shape, center, and spread.

2. **Character recognition.** An automatic character recognition device can successfully read about 85% of handwritten credit card applications. To estimate what might happen when this device reads a stack of applications, the company did a simulation using samples of size 20, 50, 75, and 100. For each sample size, they simulated 1000 samples with success rate $p = 0.85$ and constructed the histogram of the 1000 sample proportions, shown here. Explain how these histograms demonstrate what the Central Limit Theorem says about the sampling distribution model for sample proportions. Be sure to talk about shape, center, and spread.

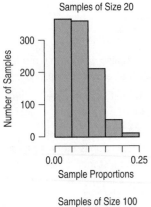

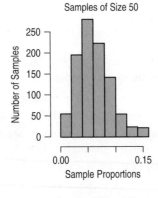

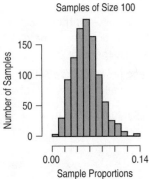

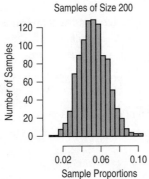

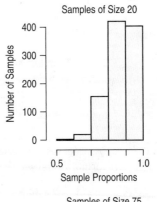

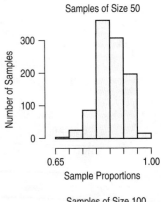

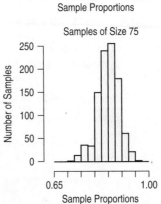

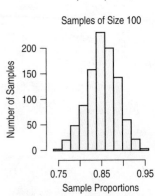

3. **Send money, again.** The philanthropic organization in Exercise 1 expects about a 5% success rate when they send fundraising letters to the people on their mailing list. In Exercise 1 you looked at the histograms showing distributions of sample proportions from 1000 simulated mailings for samples of size 20, 50, 100, and 200. The sample statistics from each simulation were as follows:

n	mean	st. dev.
20	0.0497	0.0479
50	0.0516	0.0309
100	0.0497	0.0215
200	0.0501	0.0152

a) According to the Central Limit Theorem, what should the theoretical mean and standard deviations be for these sample sizes?
b) How close are those theoretical values to what was observed in these simulations?
c) Looking at the histograms in Exercise 1, at what sample size would you be comfortable using the Normal model as an approximation for the sampling distribution?
d) What does the Success/Failure Condition say about the choice you made in part c?

4. **Character recognition, again.** The automatic character recognition device discussed in Exercise 2 successfully reads about 85% of handwritten credit card applications. In Exercise 2 you looked at the histograms showing distributions of sample proportions from 1000 simulated samples of size 20, 50, 75, and 100. The sample statistics from each simulation were as follows:

n	mean	st. dev.
20	0.8481	0.0803
50	0.8507	0.0509
75	0.8481	0.0406
100	0.8488	0.0354

a) According to the Central Limit Theorem, what should the theoretical mean and standard deviations be for these sample sizes?
b) How close are those theoretical values to what was observed in these simulations?
c) Looking at the histograms in Exercise 2, at what sample size would you be comfortable using the Normal model as an approximation for the sampling distribution?
d) What does the Success/Failure Condition say about the choice you made in part c?

5. **Coin tosses.** In a large class of introductory Statistics students, the professor has each person toss a coin 16 times and calculate the proportion of his or her tosses that were heads. The students then report their results, and the professor plots a histogram of these several proportions.
a) What shape would you expect this histogram to be? Why?
b) Where do you expect the histogram to be centered?

c) How much variability would you expect among these proportions?
d) Explain why a Normal model should not be used here.

6. **M&M's.** The candy company claims that 10% of the M&M's it produces are green. Suppose that the candies are packaged at random in small bags containing about 50 M&M's. A class of elementary school students learning about percents opens several bags, counts the various colors of the candies, and calculates the proportion that are green.
a) If we plot a histogram showing the proportions of green candies in the various bags, what shape would you expect it to have?
b) Can that histogram be approximated by a Normal model? Explain.
c) Where should the center of the histogram be?
d) What should the standard deviation of the proportion be?

7. **More coins.** Suppose the class in Exercise 5 repeats the coin-tossing experiment.
a) The students toss the coins 25 times each. Use the 68–95–99.7 Rule to describe the sampling distribution model.
b) Confirm that you can use a Normal model here.
c) They increase the number of tosses to 64 each. Draw and label the appropriate sampling distribution model. Check the appropriate conditions to justify your model.
d) Explain how the sampling distribution model changes as the number of tosses increases.

8. **Bigger bag.** Suppose the class in Exercise 6 buys bigger bags of candy, with 200 M&M's each. Again the students calculate the proportion of green candies they find.
a) Explain why it's appropriate to use a Normal model to describe the distribution of the proportion of green M&M's they might expect.
b) Use the 68–95–99.7 Rule to describe how this proportion might vary from bag to bag.
c) How would this model change if the bags contained even more candies?

9. **Just (un)lucky?** One of the students in the introductory Statistics class in Exercise 7 claims to have tossed her coin 200 times and found only 42% heads. What do you think of this claim? Explain.

10. **Too many green ones?** In a really large bag of M&M's, the students in Exercise 8 found 500 candies, and 12% of them were green. Is this an unusually large proportion of green M&M's? Explain.

11. **Speeding.** State police believe that 70% of the drivers traveling on a major interstate highway exceed the speed limit. They plan to set up a radar trap and check the speeds of 80 cars.
a) Using the 68–95–99.7 Rule, draw and label the distribution of the proportion of these cars the police will observe speeding.
b) Do you think the appropriate conditions necessary for your analysis are met? Explain.

12. **Smoking.** Public health statistics indicate that 26.4% of American adults smoke cigarettes. Using the 68–95–99.7 Rule, describe the sampling distribution model for the proportion of smokers among a randomly selected group of 50 adults. Be sure to discuss your assumptions and conditions.

13. **Vision.** It is generally believed that nearsightedness affects about 12% of all children. A school district has registered 170 incoming kindergarten children.
    a) Can you apply the Central Limit Theorem to describe the sampling distribution model for the sample proportion of children who are nearsighted? Check the conditions and discuss any assumptions you need to make.
    b) Sketch and clearly label the sampling model, based on the 68–95–99.7 Rule.
    c) How many of the incoming students might the school expect to be nearsighted? Explain.

14. **Mortgages.** In early 2007 the Mortgage Lenders Association reported that homeowners, hit hard by rising interest rates on adjustable-rate mortgages, were defaulting in record numbers. The foreclosure rate of 1.6% meant that millions of families were losing their homes. Suppose a large bank holds 1731 adjustable-rate mortgages.
    a) Can you apply the Central Limit Theorem to describe the sampling distribution model for the sample proportion of foreclosures? Check the conditions and discuss any assumptions you need to make.
    b) Sketch and clearly label the sampling model, based on the 68–95–99.7 Rule.
    c) How many of these homeowners might the bank expect will default on their mortgages? Explain.

15. **Loans.** Based on past experience, a bank believes that 7% of the people who receive loans will not make payments on time. The bank has recently approved 200 loans.
    a) What are the mean and standard deviation of the proportion of clients in this group who may not make timely payments?
    b) What assumptions underlie your model? Are the conditions met? Explain.
    c) What's the probability that over 10% of these clients will not make timely payments?

16. **Contacts.** Assume that 30% of students at a university wear contact lenses.
    a) We randomly pick 100 students. Let $\hat{p}$ represent the proportion of students in this sample who wear contacts. What's the appropriate model for the distribution of $\hat{p}$? Specify the name of the distribution, the mean, and the standard deviation. Be sure to verify that the conditions are met.
    b) What's the approximate probability that more than one third of this sample wear contacts?

17. **Back to school?** Best known for its testing program, ACT, Inc., also compiles data on a variety of issues in education. In 2004 the company reported that the national college freshman-to-sophomore retention rate held steady at 74% over the previous four years. Consider random samples of 400 freshmen who took the ACT. Use the 68–95–99.7 Rule to describe the sampling distribution model for the percentage of those students we expect to return to that school for their sophomore years. Do you think the appropriate conditions are met?

18. **Binge drinking.** As we learned in Chapter 15, a national study found that 44% of college students engage in binge drinking (5 drinks at a sitting for men, 4 for women). Use the 68–95–99.7 Rule to describe the sampling distribution model for the proportion of students in a randomly selected group of 200 college students who engage in binge drinking. Do you think the appropriate conditions are met?

19. **Back to school, again.** Based on the 74% national retention rate described in Exercise 17, does a college where 522 of the 603 freshman returned the next year as sophomores have a right to brag that it has an unusually high retention rate? Explain.

20. **Binge sample.** After hearing of the national result that 44% of students engage in binge drinking (5 drinks at a sitting for men, 4 for women), a professor surveyed a random sample of 244 students at his college and found that 96 of them admitted to binge drinking in the past week. Should he be surprised at this result? Explain.

21. **Polling.** Just before a referendum on a school budget, a local newspaper polls 400 voters in an attempt to predict whether the budget will pass. Suppose that the budget actually has the support of 52% of the voters. What's the probability the newspaper's sample will lead them to predict defeat? Be sure to verify that the assumptions and conditions necessary for your analysis are met.

22. **Seeds.** Information on a packet of seeds claims that the germination rate is 92%. What's the probability that more than 95% of the 160 seeds in the packet will germinate? Be sure to discuss your assumptions and check the conditions that support your model.

23. **Apples.** When a truckload of apples arrives at a packing plant, a random sample of 150 is selected and examined for bruises, discoloration, and other defects. The whole truckload will be rejected if more than 5% of the sample is unsatisfactory. Suppose that in fact 8% of the apples on the truck do not meet the desired standard. What's the probability that the shipment will be accepted anyway?

24. **Genetic defect.** It's believed that 4% of children have a gene that may be linked to juvenile diabetes. Researchers hoping to track 20 of these children for several years test 732 newborns for the presence of this gene. What's the probability that they find enough subjects for their study?

25. **Nonsmokers.** While some nonsmokers do not mind being seated in a smoking section of a restaurant, about 60% of the customers demand a smoke-free area. A new restaurant with 120 seats is being planned. How many seats should be in the nonsmoking area in order to be very sure of having enough seating there? Comment on the assumptions and conditions that support your model, and explain what "very sure" means to you.

26. **Meals.** A restauranteur anticipates serving about 180 people on a Friday evening, and believes that about 20% of the patrons will order the chef's steak

special. How many of those meals should he plan on serving in order to be pretty sure of having enough steaks on hand to meet customer demand? Justify your answer, including an explanation of what "pretty sure" means to you.

27. **Sampling.** A sample is chosen randomly from a population that can be described by a Normal model.
    a) What's the sampling distribution model for the sample mean? Describe shape, center, and spread.
    b) If we choose a larger sample, what's the effect on this sampling distribution model?

28. **Sampling, part II.** A sample is chosen randomly from a population that was strongly skewed to the left.
    a) Describe the sampling distribution model for the sample mean if the sample size is small.
    b) If we make the sample larger, what happens to the sampling distribution model's shape, center, and spread?
    c) As we make the sample larger, what happens to the expected distribution of the data in the sample?

29. **Waist size.** A study measured the *Waist Size* of 250 men, finding a mean of 36.33 inches and a standard deviation of 4.02 inches. Here is a histogram of these measurements.

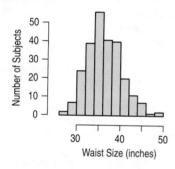

a) Describe the histogram of *Waist Size*.
b) To explore how the mean might vary from sample to sample, they simulated by drawing many samples of size 2, 5, 10, and 20, with replacement, from the 250 measurements. Here are histograms of the sample means for each simulation. Explain how these histograms demonstrate what the Central Limit Theorem says about the sampling distribution model for sample means.

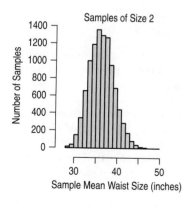

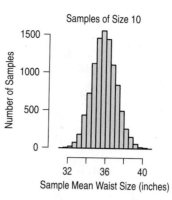

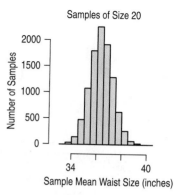

30. **CEO compensation.** In Chapter 5 we saw the distribution of the total compensation of the chief executive officers (CEOs) of the 800 largest U.S. companies (the Fortune 800). The average compensation (in thousands of dollars) is 10,307.31 and the standard deviation is 17,964.62. Here is a histogram of their annual compensations (in $1000):

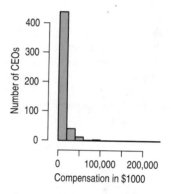

a) Describe the histogram of *Total Compensation*.
   A research organization simulated sample means by drawing samples of 30, 50, 100, and 200, with replacement, from the 800 CEOs. The histograms show the distributions of means for many samples of each size.

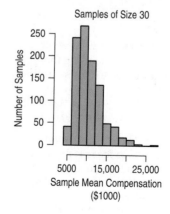

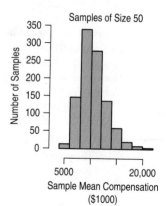

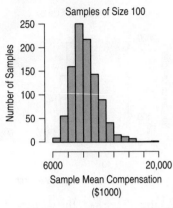

Samples of Size 100

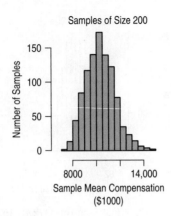

Samples of Size 200

a) According to the Central Limit Theorem, what should the theoretical mean and standard deviation be for each of these sample sizes?

b) How close are the theoretical values to what was observed from the simulation?

c) Looking at the histograms in Exercise 30, at what sample size would you be comfortable using the Normal model as an approximation for the sampling distribution?

d) What about the shape of the distribution of *Total Compensation* explains your answer in part c?

b) Explain how these histograms demonstrate what the Central Limit Theorem says about the sampling distribution model for sample means. Be sure to talk about shape, center, and spread.

c) Comment on the "rule of thumb" that "With a sample size of at least 30, the sampling distribution of the mean is Normal."

31. **Waist size, revisited.** Researchers measured the *Waist Sizes* of 250 men in a study on body fat. The true mean and standard deviation of the *Waist Sizes* for the 250 men are 36.33 in and 4.019 inches, respectively. In Exercise 29 you looked at the histograms of simulations that drew samples of sizes 2, 5, 10, and 20 (with replacement). The summary statistics for these simulations were as follows:

n	mean	st. dev.
2	36.314	2.855
5	36.314	1.805
10	36.341	1.276
20	36.339	0.895

a) According to the Central Limit Theorem, what should the theoretical mean and standard deviation be for each of these sample sizes?

b) How close are the theoretical values to what was observed in the simulation?

c) Looking at the histograms in Exercise 29, at what sample size would you be comfortable using the Normal model as an approximation for the sampling distribution?

d) What about the shape of the distribution of *Waist Size* explains your choice of sample size in part c?

32. **CEOs, revisited.** In Exercise 30 you looked at the annual compensation for 800 CEOs, for which the true mean and standard deviation were (in thousands of dollars) 10,307.31 and 17,964.62, respectively. A simulation drew samples of sizes 30, 50, 100, and 200 (with replacement) from the total annual compensations of the Fortune 800 CEOs. The summary statistics for these simulations were as follows:

n	mean	st. dev.
30	10,251.73	3359.64
50	10,343.93	2483.84
100	10,329.94	1779.18
200	10,340.37	1230.79

33. **GPAs.** A college's data about the incoming freshmen indicates that the mean of their high school GPAs was 3.4, with a standard deviation of 0.35; the distribution was roughly mound-shaped and only slightly skewed. The students are randomly assigned to freshman writing seminars in groups of 25. What might the mean GPA of one of these seminar groups be? Describe the appropriate sampling distribution model—shape, center, and spread—with attention to assumptions and conditions. Make a sketch using the 68–95–99.7 Rule.

34. **Home values.** Assessment records indicate that the value of homes in a small city is skewed right, with a mean of $140,000 and standard deviation of $60,000. To check the accuracy of the assessment data, officials plan to conduct a detailed appraisal of 100 homes selected at random. Using the 68–95–99.7 Rule, draw and label an appropriate sampling model for the mean value of the homes selected.

35. **Lucky spot?** A reporter working on a story about the New York lottery contacted one of the authors of this book, wanting help analyzing data to see if some ticket sales outlets were more likely to produce winners. His data for each of the 966 New York lottery outlets are graphed below; the scatterplot shows the ratio *TotalPaid/TotalSales* vs. *TotalSales* for the state's "instant winner" games for all of 2007.

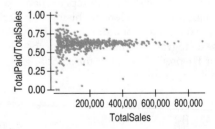

The reporter thinks that by identifying the outlets with the highest fraction of bets paid out, players might be able to increase their chances of winning. Typically—but not always—instant winners are paid immediately (instantly) at the store at which they are purchased. However, the fact that tickets may be scratched off and then cashed in at any outlet may account for some outlets paying out more than they take in. The few with very low payouts may be on interstate highways where players may purchase cards but then leave.

a) Explain why the plot has this funnel shape.

b) Explain why the reporter's idea wouldn't have worked anyway.

36. Safe cities. Allstate Insurance Company identified the 10 safest and 10 least-safe U.S. cities from among the 200 largest cities in the United States, based on the mean number of years drivers went between automobile accidents. The cities on both lists were all smaller than the 10 largest cities. Using facts about the sampling distribution model of the mean, explain why this is not surprising.

37. Pregnancy. Assume that the duration of human pregnancies can be described by a Normal model with mean 266 days and standard deviation 16 days.

a) What percentage of pregnancies should last between 270 and 280 days?

b) At least how many days should the longest 25% of all pregnancies last?

c) Suppose a certain obstetrician is currently providing prenatal care to 60 pregnant women. Let $\bar{y}$ represent the mean length of their pregnancies. According to the Central Limit Theorem, what's the distribution of this sample mean, $\bar{y}$? Specify the model, mean, and standard deviation.

d) What's the probability that the mean duration of these patients' pregnancies will be less than 260 days?

38. Rainfall. Statistics from Cornell's Northeast Regional Climate Center indicate that Ithaca, NY, gets an average of 35.4" of rain each year, with a standard deviation of 4.2". Assume that a Normal model applies.

a) During what percentage of years does Ithaca get more than 40" of rain?

b) Less than how much rain falls in the driest 20% of all years?

c) A Cornell University student is in Ithaca for 4 years. Let $\bar{y}$ represent the mean amount of rain for those 4 years. Describe the sampling distribution model of this sample mean, $\bar{y}$.

d) What's the probability that those 4 years average less than 30" of rain?

39. Pregnant again. The duration of human pregnancies may not actually follow the Normal model described in Exercise 37.

a) Explain why it may be somewhat skewed to the left.

b) If the correct model is in fact skewed, does that change your answers to parts a, b, and c of Exercise 37? Explain why or why not for each.

40. At work. Some business analysts estimate that the length of time people work at a job has a mean of 6.2 years and a standard deviation of 4.5 years.

a) Explain why you suspect this distribution may be skewed to the right.

b) Explain why you could estimate the probability that 100 people selected at random had worked for their employers an average of 10 years or more, but you could not estimate the probability that an individual had done so.

41. Dice and dollars. You roll a die, winning nothing if the number of spots is odd, $1 for a 2 or a 4, and $10 for a 6.

a) Find the expected value and standard deviation of your prospective winnings.

b) You play twice. Find the mean and standard deviation of your total winnings.

c) You play 40 times. What's the probability that you win at least $100?

42. New game. You pay $10 and roll a die. If you get a 6, you win $50. If not, you get to roll again. If you get a 6 this time, you get your $10 back.

a) Create a probability model for this game.

b) Find the expected value and standard deviation of your prospective winnings.

c) You play this game five times. Find the expected value and standard deviation of your average winnings.

d) 100 people play this game. What's the probability the person running the game makes a profit?

43. AP Stats 2006. The College Board reported the score distribution shown in the table for all students who took the 2006 AP Statistics exam.

Score	Percent of Students
5	12.6
4	22.2
3	25.3
2	18.3
1	21.6

a) Find the mean and standard deviation of the scores.

b) If we select a random sample of 40 AP Statistics students, would you expect their scores to follow a Normal model? Explain.

c) Consider the mean scores of random samples of 40 AP Statistics students. Describe the sampling model for these means (shape, center, and spread).

44. Museum membership. A museum offers several levels of membership, as shown in the table.

Member Category	Amount of Donation ($)	Percent of Members
Individual	50	41
Family	100	37
Sponsor	250	14
Patron	500	7
Benefactor	1000	1

a) Find the mean and standard deviation of the donations.

b) During their annual membership drive, they hope to sign up 50 new members each day. Would you expect the distribution of the donations for a day to follow a Normal model? Explain.

c) Consider the mean donation of the 50 new members each day. Describe the sampling model for these means (shape, center, and spread).

45. AP Stats 2006, again. An AP Statistics teacher had 63 students preparing to take the AP exam discussed in Exercise 43. Though they were obviously not a random sample, he considered his students to be "typical" of all the national students. What's the probability that his students will achieve an average score of at least 3?

46. **Joining the museum.** One of the museum's phone volunteers sets a personal goal of getting an average donation of at least $100 from the new members she enrolls during the membership drive. If she gets 80 new members and they can be considered a random sample of all the museum's members, what is the probability that she can achieve her goal?

47. **Pollution.** Carbon monoxide (CO) emissions for a certain kind of car vary with mean 2.9 g/mi and standard deviation 0.4 g/mi. A company has 80 of these cars in its fleet. Let $\bar{y}$ represent the mean CO level for the company's fleet.
    a) What's the approximate model for the distribution of $\bar{y}$? Explain.
    b) Estimate the probability that $\bar{y}$ is between 3.0 and 3.1 g/mi.
    c) There is only a 5% chance that the fleet's mean CO level is greater than what value?

48. **Potato chips.** The weight of potato chips in a medium-size bag is stated to be 10 ounces. The amount that the packaging machine puts in these bags is believed to have a Normal model with mean 10.2 ounces and standard deviation 0.12 ounces.
    a) What fraction of all bags sold are underweight?
    b) Some of the chips are sold in "bargain packs" of 3 bags. What's the probability that none of the 3 is underweight?
    c) What's the probability that the mean weight of the 3 bags is below the stated amount?
    d) What's the probability that the mean weight of a 24-bag case of potato chips is below 10 ounces?

49. **Tips.** A waiter believes the distribution of his tips has a model that is slightly skewed to the right, with a mean of $9.60 and a standard deviation of $5.40.
    a) Explain why you cannot determine the probability that a given party will tip him at least $20.
    b) Can you estimate the probability that the next 4 parties will tip an average of at least $15? Explain.
    c) Is it likely that his 10 parties today will tip an average of at least $15? Explain.

50. **Groceries.** A grocery store's receipts show that Sunday customer purchases have a skewed distribution with a mean of $32 and a standard deviation of $20.
    a) Explain why you cannot determine the probability that the next Sunday customer will spend at least $40.
    b) Can you estimate the probability that the next 10 Sunday customers will spend an average of at least $40? Explain.
    c) Is it likely that the next 50 Sunday customers will spend an average of at least $40? Explain.

51. **More tips.** The waiter in Exercise 49 usually waits on about 40 parties over a weekend of work.
    a) Estimate the probability that he will earn at least $500 in tips.
    b) How much does he earn on the best 10% of such weekends?

52. **More groceries.** Suppose the store in Exercise 50 had 312 customers this Sunday.

a) Estimate the probability that the store's revenues were at least $10,000.
b) If, on a typical Sunday, the store serves 312 customers, how much does the store take in on the worst 10% of such days?

53. **IQs.** Suppose that IQs of East State University's students can be described by a Normal model with mean 130 and standard deviation 8 points. Also suppose that IQs of students from West State University can be described by a Normal model with mean 120 and standard deviation 10.
    a) We select a student at random from East State. Find the probability that this student's IQ is at least 125 points.
    b) We select a student at random from each school. Find the probability that the East State student's IQ is at least 5 points higher than the West State student's IQ.
    c) We select 3 West State students at random. Find the probability that this group's average IQ is at least 125 points.
    d) We also select 3 East State students at random. What's the probability that their average IQ is at least 5 points higher than the average for the 3 West Staters?

54. **Milk.** Although most of us buy milk by the quart or gallon, farmers measure daily production in pounds. Ayrshire cows average 47 pounds of milk a day, with a standard deviation of 6 pounds. For Jersey cows, the mean daily production is 43 pounds, with a standard deviation of 5 pounds. Assume that Normal models describe milk production for these breeds.
    a) We select an Ayrshire at random. What's the probability that she averages more than 50 pounds of milk a day?
    b) What's the probability that a randomly selected Ayrshire gives more milk than a randomly selected Jersey?
    c) A farmer has 20 Jerseys. What's the probability that the average production for this small herd exceeds 45 pounds of milk a day?
    d) A neighboring farmer has 10 Ayrshires. What's the probability that his herd average is at least 5 pounds higher than the average for part c's Jersey herd?

JUST CHECKING

ANSWERS

1. A Normal model (approximately).

2. At the actual proportion of all students who are in favor.

3. $SD(\hat{p}) = \sqrt{\dfrac{(0.5)(0.5)}{100}} = 0.05$

4. No, this is a histogram of individuals. It may or may not be Normal, but we can't tell from the information provided.

5. A Normal model (approximately).

6. 266 days

7. $\dfrac{16}{\sqrt{100}} = 1.6$ days

Confidence Intervals for Proportions

Where are we going?

In March of 2006, the Pew Research Center surveyed 1286 cell phone users.[1] They found that, overall, 22% admitted that "When I'm on my cell phone I'm not always truthful about exactly where I am." Pew reports that there is a "3% margin of error" for this result. Among younger respondents, those 18 to 29 years old, the percentage increased to 39%, and Pew warns that the margin of error for this group is larger, too. What does that mean? Are 18- to 29-year-olds less reliable (as well as, apparently, less honest)? What exactly is a margin of error, how do we find one, and how should we interpret them? We'll see what they are all about in this chapter.

WHO	Sea fans
WHAT	Percent infected
WHEN	June 2000
WHERE	Las Redes Reef, Akumal, Mexico, 40 feet deep
WHY	Research

Coral reef communities are home to one quarter of all marine plants and animals worldwide. These reefs support large fisheries by providing breeding grounds and safe havens for young fish of many species. Coral reefs are seawalls that protect shorelines against tides, storm surges, and hurricanes, and are sand "factories" that produce the limestone and sand of which beaches are made. Beyond the beach, these reefs are major tourist attractions for snorkelers and divers, driving a tourist industry worth tens of billions of dollars.

But marine scientists say that 10% of the world's reef systems have been destroyed in recent times. At current rates of loss, 70% of the reefs could be gone in 40 years. Pollution, global warming, outright destruction of reefs, and increasing acidification of the oceans are all likely factors in this loss.

Dr. Drew Harvell's lab studies corals and the diseases that affect them. They sampled sea fans[2] at 19 randomly selected reefs along the Yucatan peninsula and diagnosed whether the animals were affected by the disease *aspergillosis*.[3] In specimens collected at a depth of 40 feet at the Las Redes Reef in Akumal, Mexico, these scientists found that 54 of 104 sea fans sampled were infected with that disease.

---

[1] www.pewinternet.org
[2] That's a sea fan in the picture. Although they look like trees, they are actually colonies of genetically identical animals.
[3] K. M. Mullen, C. D. Harvell, A. P. Alker, D. Dube, E. Jordán-Dahlgren, J. R. Ward, and L. E. Petes, "Host range and resistance to aspergillosis in three sea fan species from the Yucatan," *Marine Biology* (2006), Springer-Verlag.

Of course, we care about much more than these particular 104 sea fans. We care about the health of coral reef communities throughout the Caribbean. What can this study tell us about the prevalence of the disease among sea fans?

We have a sample proportion, which we write as $\hat{p}$, of 54/104, or 51.9%. Our first guess might be that this observed proportion is close to the population proportion, p. But we also know that because of natural sampling variability, if the researchers had drawn a second sample of 104 sea fans at roughly the same time, the proportion infected from that sample probably wouldn't have been exactly 51.9%.

What *can* we say about the population proportion, p? To start to answer this question, think about how different the sample proportion might have been if we'd taken another random sample from the same population. But wait. Remember—we aren't actually going to take more samples. We just want to *imagine* how the sample proportions might vary from sample to sample. In other words, we want to know about the *sampling distribution* of the sample proportion of infected sea fans.

A Confidence Interval

Let's look at our model for the sampling distribution. What do we know about it? We know it's approximately Normal (under certain assumptions, which we should be careful to check) and that its mean is the proportion of all infected sea fans on the Las Redes Reef. Is the infected proportion of *all* sea fans 51.9%? No, that's just $\hat{p}$, our estimate. We don't know the proportion, p, of all the infected sea fans; that's what we're trying to find out. We do know, though, that the sampling distribution model of $\hat{p}$ is centered at p, and we know that the standard deviation of the sampling distribution is $\sqrt{\dfrac{pq}{n}}$.

Now we have a problem: Since we don't know p, we can't find the true standard deviation of the sampling distribution model. We do know the observed proportion, $\hat{p}$, so, of course we just use what we know, and we estimate. That may not seem like a big deal, but it gets a special name. Whenever we estimate the standard deviation of a sampling distribution, we call it a **standard error**.[4] For a sample proportion, $\hat{p}$, the standard error is

$$SE(\hat{p}) = \sqrt{\frac{\hat{p}\hat{q}}{n}}.$$

For the sea fans, then:

$$SE(\hat{p}) = \sqrt{\frac{\hat{p}\hat{q}}{n}} = \sqrt{\frac{(0.519)(0.481)}{104}} = 0.049 = 4.9\%.$$

Now we know that the sampling model for $\hat{p}$ should look like this:

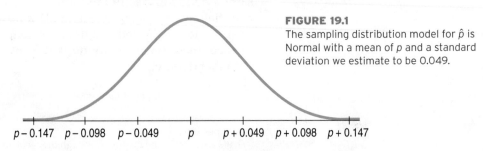

FIGURE 19.1
The sampling distribution model for $\hat{p}$ is Normal with a mean of p and a standard deviation we estimate to be 0.049.

$p - 0.147$ $p - 0.098$ $p - 0.049$ p $p + 0.049$ $p + 0.098$ $p + 0.147$

---

[4] This isn't such a great name because it isn't standard and nobody made an error. But it's much shorter and more convenient than saying, "the estimated standard deviation of the sampling distribution of the sample statistic."

A S   *Activity: Confidence Intervals and Sampling Distributions.* Simulate the sampling distribution, and see how it gives a confidence interval.

NOTATION ALERT

Remember that $\hat{p}$ is our sample-based estimate of the true proportion p. Recall also that q is just shorthand for $1 - p$, and $\hat{q} = 1 - \hat{p}$.

When we use $\hat{p}$ to estimate the standard deviation of the sampling distribution model, we call that the **standard error** and write $SE(\hat{p}) = \sqrt{\dfrac{\hat{p}\hat{q}}{n}}$.

Great. What does that tell us? Well, because it's Normal, it says that about 68% of all samples of 104 sea fans will have $\hat{p}$'s within 1 *SE*, 0.049, of *p*. And about 95% of all these samples will be within $p \pm 2$ *SE*s. But where is *our* sample proportion in this picture? And what value does *p* have? We still don't know!

We do know that for 95% of random samples, $\hat{p}$ will be no more than 2 *SE*s away from *p*. So let's look at this from $\hat{p}$'s point of view. If I'm $\hat{p}$, there's a 95% chance that *p* is no more than 2 *SE*s away from me. If I reach out 2 *SE*s, or 2×0.049, away from me on both sides, I'm 95% sure that *p* will be within my grasp. Now I've got him! Probably. Of course, even if my interval does catch *p*, I still don't know its true value. The best I can do is an interval, and even then I can't be positive it contains *p*.

So what can we really say about *p*? Here's a list of things we'd like to be able to say, in order of strongest to weakest and the reasons we can't say most of them:

1. **"51.9% of *all* sea fans on the Las Redes Reef are infected."** It would be nice to be able to make absolute statements about population values with certainty, but we just don't have enough information to do that. There's no way to be sure that the population proportion is the same as the sample proportion; in fact, it almost certainly isn't. Observations vary. Another sample would yield a different sample proportion.

2. **"It is *probably* true that 51.9% of all sea fans on the Las Redes Reef are infected."** No. In fact, we can be pretty sure that whatever the true proportion is, it's not exactly 51.900%. So the statement is not true.

3. **"We don't know exactly what proportion of sea fans on the Las Redes Reef is infected, but we *know* that it's within the interval 51.9% $\pm$ 2 $\times$ 4.9%. That is, it's between 42.1% and 61.7%."** This is getting closer, but we still can't be certain. We can't know *for sure* that the true proportion is in this interval—or in any particular interval.

4. **"We don't know exactly what proportion of sea fans on the Las Redes Reef is infected, but the interval from 42.1% to 61.7% *probably* contains the true proportion."** We've now fudged twice—first by giving an interval and second by admitting that we only think the interval "probably" contains the true value. And this statement is true.

That last statement may be true, but it's a bit wishy-washy. We can tighten it up a bit by quantifying what we mean by "probably." We saw that 95% of the time when we reach out 2 *SE*s from $\hat{p}$ we capture *p*, *so we can be 95% confident that this is one of those times.* After putting a number on the probability that this interval covers the true proportion, we've given our best guess of where the parameter is and how certain we are that it's within some range.

5. **"We are 95% confident that between 42.1% and 61.7% of Las Redes sea fans are infected."** Statements like these are called **confidence intervals**. They're the best we can do.

Each confidence interval discussed in the book has a name. You'll see many different kinds of confidence intervals in the following chapters. Some will be about more than *one* sample, some will be about statistics other than *proportions*, and some will use models other than the Normal. The interval calculated and interpreted here is sometimes called a **one-proportion z-interval**.[5]

FIGURE 19.2
Reaching out 2 *SE*s on either side of $\hat{p}$ makes us 95% confident that we'll trap the true proportion, *p*.

$\hat{p} - 2\,SE$ $\hat{p}$ $\hat{p} + 2\,SE$

ACME *p*-trap: Guaranteed* to capture *p*.

*with 95% confidence

A S

Activity: Can We Estimate a Parameter? Consider these four interpretations of a confidence interval by simulating to see whether they could be right.

"Far better an approximate answer to the right question, . . . than an exact answer to the wrong question."

—John W. Tukey

---

[5] In fact, this confidence interval is so standard for a single proportion that you may see it simply called a "confidence interval for the proportion."

JUST CHECKING

The Pew Research study regarding cell phones asked other questions about cell phone experience. One growing concern is unsolicited advertising in the form of text messages. Pew asked cell phone owners, "Have you ever received unsolicited text messages on your cell phone from advertisers?" and 17% reported that they had. Pew estimates a 95% confidence interval to be 0.17 ± 0.04, or between 13% and 21%.

Are the following statements about people who have cell phones correct? Explain.

1. In Pew's sample, somewhere between 13% and 21% of respondents reported that they had received unsolicited advertising text messages.

2. We can be 95% confident that 17% of U.S. cell phone owners have received unsolicited advertising text messages.

3. We are 95% confident that between 13% and 21% of all U.S. cell phone owners have received unsolicited advertising text messages.

4. We know that between 13% and 21% of all U.S. cell phone owners have received unsolicited advertising text messages.

5. 95% of all U.S. cell phone owners have received unsolicited advertising text messages.

What Does "95% Confidence" Really Mean?

A S *Activity:* **Confidence Intervals for Proportions.** This new interactive tool makes it easy to construct and experiment with confidence intervals. We'll use this tool for the rest of the course–sure beats calculating by hand!

What do we mean when we say we have 95% confidence that our interval contains the true proportion? Formally, what we mean is that "95% of samples of this size will produce confidence intervals that capture the true proportion." This is correct, but a little long winded, so we sometimes say, "we are 95% confident that the true proportion lies in our interval." Our uncertainty is about whether the particular sample we have at hand is one of the successful ones or one of the 5% that fail to produce an interval that captures the true value.

Back in Chapter 18 we saw that proportions vary from sample to sample. If other researchers select their own samples of sea fans, they'll also find some infected by the disease, but each person's sample proportion will almost certainly differ from ours. When they each try to estimate the true rate of infection in the entire population, they'll center *their* confidence intervals at the proportions they observed in their own samples. Each of us will end up with a different interval.

Our interval guessed the true proportion of infected sea fans to be between about 42% and 62%. Another researcher whose sample contained more infected fans than ours did might guess between 46% and 66%. Still another who happened to collect fewer infected fans might estimate the true proportion to be between 23% and 43%. And so on. Every possible sample would produce yet another confidence interval. Although wide intervals like these can't pin down the actual rate of infection very precisely, we expect that most of them should be winners, capturing the true value. Nonetheless, some will be duds, missing the population proportion entirely.

On the next page you'll see confidence intervals produced by simulating 20 different random samples. The red dots are the proportions of infected fans in each sample, and the blue segments show the confidence intervals found for each. The green line represents the true rate of infection in the population, so you can see that most of the intervals caught it—but a few missed. (And notice again that it is the *intervals* that vary from sample to sample; the green line doesn't move.)

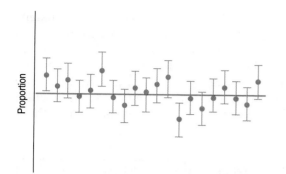

The horizontal green line shows the true percentage of all sea fans that are infected. Most of the 20 simulated samples produced confidence intervals that captured the true value, but a few missed.

Of course, there's a huge number of possible samples that *could* be drawn, each with its own sample proportion. These are just some of them. Each sample proportion can be used to make a confidence interval. That's a large pile of possible confidence intervals, and ours is just one of those in the pile. Did *our* confidence interval "work"? We can never be sure, because we'll never know the true proportion of all the sea fans that are infected. However, the Central Limit Theorem assures us that 95% of the intervals in the pile are winners, covering the true value, and only 5% are duds. *That's* why we're 95% confident that our interval is a winner!

FOR EXAMPLE | Polls and Margin of Error

On January 30-31, 2007, Fox News/Opinion Dynamics polled 900 registered voters nationwide.[6] When asked, "Do you believe global warming exists?" 82% said "Yes." Fox reported their margin of error to be ±3%.

QUESTION: It is standard among pollsters to use a 95% confidence level unless otherwise stated. Given that, what does Fox News mean by claiming a margin of error of ±3% in this context?

If this polling were done repeatedly, 95% of all random samples would yield estimates that come within ±3% of the true proportion of all registered voters who believe that global warming exists.

Margin of Error: Certainty vs. Precision

We've just claimed that with a certain confidence we've captured the true proportion of all infected sea fans. Our confidence interval had the form

$$\hat{p} \pm 2\,SE(\hat{p}).$$

The extent of the interval on either side of $\hat{p}$ is called the **margin of error** (*ME*). We'll want to use the same approach for many other situations besides estimating proportions. In general, confidence intervals look like this:

$$Estimate \pm ME.$$

The margin of error for our 95% confidence interval was 2 *SE*. What if we wanted to be more confident? To be more confident, we'll need to capture *p* more often, and to do that we'll need to make the interval wider. For example, if we want to be 99.7% confident, the margin of error will have to be 3 *SE*.

The more confident we want to be, the larger the margin of error must be. We can be 100% confident that the proportion of infected sea fans is between 0% and 100%, but this isn't likely to be very useful. On the other hand, we could give a confidence interval from 51.8% to 52.0%, but we can't be very confident about a precise statement like this. Every confidence interval is a balance between certainty and precision.

FIGURE 19.3
Reaching out 3 *SE*s on either side of $\hat{p}$ makes us 99.7% confident we'll trap the true proportion *p*. Compare with Figure 19.2.

$\hat{p} - 3\,SE$ $\hat{p}$ $\hat{p} + 3\,SE$

---

[6] www.foxnews.com, "Fox News Poll: Most Americans Believe in Global Warming," Feb. 7, 2007.

The tension between certainty and precision is always there. Fortunately, in most cases we can be both sufficiently certain and sufficiently precise to make useful statements. There is no simple answer to the conflict. You must choose a confidence level yourself. The data can't do it for you. The choice of confidence level is somewhat arbitrary. The most commonly chosen confidence levels are 90%, 95%, and 99%, but any percentage can be used. (In practice, though, using something like 92.9% or 97.2% is likely to make people think you're up to something.)

Garfield © 1999 Paws, Inc. Reprinted with permission of UNIVERSAL PRESS SYNDICATE. All rights reserved.

FOR EXAMPLE   **Finding the Margin of Error (Take 1)**

RECAP: A January 2007 Fox poll of 900 registered voters reported a margin of error of ±3%. It is a convention among pollsters to use a 95% confidence level and to report the "worst case" margin of error, based on $p = 0.5$.

QUESTION: How did Fox calculate their margin of error?

Assuming $p = 0.5$, for random samples of $n = 900$, $SD(\hat{p}) = \sqrt{\dfrac{pq}{n}} = \sqrt{\dfrac{(0.5)(0.5)}{900}} = 0.0167$

For a 95% confidence level, $ME = 2(0.0167) = 0.033$, so Fox's margin of error is just a bit over ±3%.

Critical Values

In our sea fans example we used $2SE$ to give us a 95% confidence interval. To change the confidence level, we'd need to change the *number* of SEs so that the size of the margin of error corresponds to the new level. This number of SEs is called the **critical value**. Here it's based on the Normal model, so we denote it z^*. For any confidence level, we can find the corresponding critical value from a computer, a calculator, or a Normal probability table, such as Table Z.

For a 95% confidence interval, you'll find the precise critical value is $z^* = 1.96$. That is, 95% of a Normal model is found within ±1.96 standard deviations of the mean. We've been using $z^* = 2$ from the 68–95–99.7 Rule because it's easy to remember.

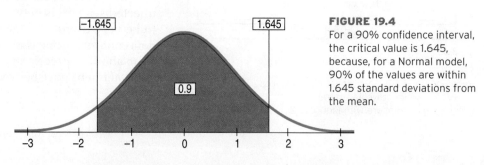

FIGURE 19.4
For a 90% confidence interval, the critical value is 1.645, because, for a Normal model, 90% of the values are within 1.645 standard deviations from the mean.

FOR EXAMPLE   Finding the Margin of Error (Take 2)

RECAP: In January 2007 a Fox News poll of 900 registered voters found that 82% of the respondents believed that global warming exists. Fox reported a 95% confidence interval with a margin of error of $\pm3\%$.

QUESTIONS: Using the critical value of z and the standard error based on the observed proportion, what would be the margin of error for a 90% confidence interval? What's good and bad about this change?

$$\text{With } n = 900 \text{ and } \hat{p} = 0.82, \quad SE(\hat{p}) = \sqrt{\frac{\hat{p}\hat{q}}{n}} = \sqrt{\frac{(0.82)(0.18)}{900}} = 0.0128$$

For a 90% confidence level, $z^* = 1.645$, so $ME = 1.645(0.0128) = 0.021$

Now the margin of error is only about $\pm2\%$, producing a narrower interval. That makes for a more precise estimate of voter belief, but provides less certainty that the interval actually contains the true proportion of voters believing in global warming.

JUST CHECKING

Think some more about the 95% confidence interval Fox News created for the proportion of registered voters who believe that global warming exists.

6. If Fox wanted to be 98% confident, would their confidence interval need to be wider or narrower?

7. Fox's margin of error was about $\pm3\%$. If they reduced it to $\pm2\%$, would their level of confidence be higher or lower?

8. If Fox News had polled more people, would the interval's margin of error have been larger or smaller?

Assumptions and Conditions

We've just made some pretty sweeping statements about sea fans. Those statements were possible because we used a Normal model for the sampling distribution. But is that model appropriate?

As we've seen, all statistical models make assumptions. Different models make different assumptions. If those assumptions are not true, the model might be inappropriate and our conclusions based on it may be wrong. Because the confidence interval is built on the Normal model for the sampling distribution, the assumptions and conditions are the same as those we discussed in Chapter 18. But, because they are so important, we'll go over them again.

We can never be certain that an assumption is true, but we can decide intelligently whether it is reasonable. When we have data, we can often decide whether an assumption is plausible by checking a related condition. However, we want to make a statement about the world at large, not just about the data we collected. So the assumptions we make are not just about how our data look, but about how representative they are.

Independence Assumption

Independence Assumption: We first need to *Think* about whether the independence assumption is plausible. We often look for reasons to suspect that it fails. We wonder whether there is any reason to believe that the data values somehow affect each other. (For example, might the disease in sea fans be contagious?) Whether you decide that the **Independence Assumption** is plausible depends on your knowledge of the situation. It's not one you can check by looking at the data.

A S   *Activity:* **Assumptions and Conditions.** Here's an animated review of the assumptions and conditions.

However, now that we have data, there are two conditions that we can check:

Randomization Condition: Were the data sampled at random or generated from a properly randomized experiment? Proper randomization can help ensure independence.

10% Condition: Samples are almost always drawn without replacement. Usually, of course, we'd like to have as large a sample as we can. But when the population itself is small we have another concern. When we sample from small populations, the probability of success may be different for the last few individuals we draw than it was for the first few. For example, if most of the women have already been sampled, the chance of drawing a woman from the remaining population is lower. If the sample exceeds 10% of the population, the probability of a success changes so much during the sampling that our Normal model may no longer be appropriate. But if less than 10% of the population is sampled, the effect on independence is negligible.

Sample Size Assumption

The model we use for inference is based on the Central Limit Theorem. The **Sample Size Assumption** addresses the question of whether the sample is large enough to make the sampling model for the sample proportions approximately Normal. It turns out that we need more data as the proportion gets closer and closer to either extreme (0 or 1). We can check this assumption with the:

Success/Failure Condition: We must expect at least 10 "successes" and at least 10 "failures." Recall that by tradition we arbitrarily label one alternative (usually the outcome being counted) as a "success" even if it's something bad (like a sick sea fan). The other alternative is, of course, then a "failure."

A S *Activity:* **A Confidence Interval for *p*.** View the video story of pollution in Chesapeake Bay, and make a confidence interval for the analysis with the interactive tool.

ONE-PROPORTION *z*-INTERVAL

When the conditions are met, we are ready to find the confidence interval for the population proportion, p. The confidence interval is $\hat{p} \pm z^* \times SE(\hat{p})$ where the standard deviation of the proportion is estimated by $SE(\hat{p}) = \sqrt{\dfrac{\hat{p}\hat{q}}{n}}$.

STEP-BY-STEP EXAMPLE   **A Confidence Interval for a Proportion**

WHO	Adults in the United States
WHAT	Response to a question about the death penalty
WHEN	October 2008
WHERE	United States
HOW	510 adults were randomly sampled and asked by the Gallup Poll
WHY	Public opinion research

In October 2008, the Gallup Poll[7] asked 510 randomly sampled adults the question "Generally speaking, do you believe the death penalty is applied fairly or unfairly in this country today?" Of these, 54% answered "Fairly," 38% said "Unfairly," and 8% said they didn't know.

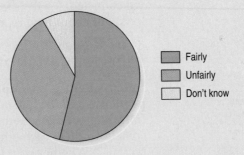

- Fairly
- Unfairly
- Don't know

---

[7]www.gallup.com/poll/1606/death-penalty.aspx

Question: From this survey, what can we conclude about the opinions of *all* adults?

To answer this question, we'll build a confidence interval for the proportion of all U.S. adults who believe the death penalty is applied fairly. There are four steps to building a confidence interval for proportions: Plan, Model, Mechanics, and Conclusion.

THINK ▶

Plan State the problem and the W's.

Identify the *parameter* you wish to estimate.

Identify the *population* about which you wish to make statements.

Choose and state a confidence level.

Model Think about the assumptions and check the conditions.

I want to find an interval that is likely, with 95% confidence, to contain the true proportion, p, of U.S. adults who think the death penalty is applied fairly. I have a random sample of 510 U.S. adults.

✓ **Independence Assumption:** Gallup phoned a random sample of U.S. adults. It is very unlikely that any of their respondents influenced each other.

✓ **Randomization Condition:** Gallup drew a random sample from all U.S. adults. I don't have details of their randomization but assume that I can trust it.

✓ **10% Condition:** Although sampling was necessarily without replacement, there are many more U.S. adults than were sampled. The sample is certainly less than 10% of the population.

✓ **Success/Failure Condition:**
$n\hat{p} = 510(54\%) = 275 \geq 10$ and
$n\hat{q} = 510(46\%) = 235 \geq 10$,
so the sample appears to be large enough to use the Normal model.

State the sampling distribution model for the statistic.

Choose your method.

The conditions are satisfied, so I can use a Normal model to find a **one-proportion z-interval.**

 SHOW ▶

Mechanics Construct the confidence interval.

First find the standard error. (Remember: It's called the "standard error" because we don't know p and have to use $\hat{p}$ instead.)

Next find the margin of error. We could informally use 2 for our critical value, but 1.96 is more accurate.

$n = 510, \hat{p} = 0.54$, so

$$SE(\hat{p}) = \sqrt{\frac{\hat{p}\hat{q}}{n}} = \sqrt{\frac{(0.54)(0.46)}{510}} = 0.022$$

Because the sampling model is Normal, for a 95% confidence interval, the critical value $z^* = 1.96$.

The margin of error is

$$ME = z^* \times SE(\hat{p}) = 1.96(0.022) = 0.043$$

Write the confidence interval (CI). The CI is centered at the sample proportion and about as wide as we might expect for a sample of 500.	*So the 95% confidence interval is* *0.54 ± 0.043 or (0.497, 0.583)*

TELL **Conclusion** Interpret the confidence interval in the proper context. We're 95% confident that our interval captured the true proportion.	*I am 95% confident that between 49.7% and 58.3% of all U.S. adults think that the death penalty is applied fairly.*

Choosing Your Sample Size

The question of how large a sample to take is an important step in planning any study. We weren't ready to make that calculation when we first looked at study design in Chapter 12, but now we can—and we always should.

Suppose a candidate is planning a poll and wants to estimate voter support within 3% with 95% confidence. How large a sample does she need?

Let's look at the margin of error:

$$ME = z^* \sqrt{\frac{\hat{p}\hat{q}}{n}}$$

$$0.03 = 1.96 \sqrt{\frac{\hat{p}\hat{q}}{n}}.$$

We want to find n, the sample size. To find n we need a value for $\hat{p}$. We don't know $\hat{p}$ because we don't have a sample yet, but we can probably guess a value. The worst case—the value that makes $\hat{p}\hat{q}$ (and therefore n) largest—is 0.50, so if we use that value for $\hat{p}$, we'll certainly be safe. Our candidate probably expects to be near 50% anyway.

Our equation, then, is

$$0.03 = 1.96 \sqrt{\frac{(0.5)(0.5)}{n}}.$$

WHAT DO I USE INSTEAD OF $\hat{p}$?

Often we have an estimate of the population proportion based on experience or perhaps a previous study. If so, use that value as $\hat{p}$ in calculating what size sample you need. If not, the cautious approach is to use $p = 0.5$ in the sample size calculation; that will determine the largest sample necessary regardless of the true proportion.

To solve for n, we first multiply both sides of the equation by $\sqrt{n}$ and then divide by 0.03:

$$0.03\sqrt{n} = 1.96\sqrt{(0.5)(0.5)}$$

$$\sqrt{n} = \frac{1.96\sqrt{(0.5)(0.5)}}{0.03} \approx 32.67$$

Notice that evaluating this expression tells us the *square root* of the sample size. We need to square that result to find n:

$$n \approx (32.67)^2 \approx 1067.1$$

To be safe, we round up and conclude that we need at least 1068 respondents to keep the margin of error as small as 3% with a confidence level of 95%.

FOR EXAMPLE Choosing a Sample Size

RECAP: The Fox News poll which estimated that 82% of all voters believed global warming exists had a margin of error of ±3%. Suppose an environmental group planning a follow-up survey of voters' opinions on global warming wants to determine a 95% confidence interval with a margin of error of no more than ±2%.

QUESTION: How large a sample do they need? Use the Fox News estimate as the basis for your calculation.

$$ME = z^* \sqrt{\frac{\hat{p}\hat{q}}{n}}$$

$$0.02 = 1.96 \sqrt{\frac{(0.82)(0.18)}{n}}$$

$$\sqrt{n} = \frac{1.96 \sqrt{(0.82)(0.18)}}{0.02} \approx 37.65$$

$$n = 37.65^2 = 1417.55$$

The environmental group's survey will need about 1418 respondents.

Public opinion polls often sample 1000 people, which gives an ME of 3% when $p = 0.5$. But businesses and nonprofit organizations typically use much larger samples to estimate the proportion who will accept a direct mail offer. Why? Because that proportion is very low—often far below 5%. An ME of 3% wouldn't be precise enough. An ME like 0.1% would be more useful, and that requires a very large sample size.

Unfortunately, bigger samples cost more money and more effort. Because the standard error declines only with the *square root* of the sample size, to cut the standard error (and thus the ME) in half, we must *quadruple* the sample size.

Generally a margin of error of 5% or less is acceptable, but different circumstances call for different standards. For a pilot study, a margin of error of 10% may be fine, so a sample of 100 will do quite well. In a close election, a polling organization might want to get the margin of error down to 2%. Drawing a large sample to get a smaller ME, however, can run into trouble. It takes time to survey 2400 people, and a survey that extends over a week or more may be trying to hit a target that moves during the time of the survey. An important event can change public opinion in the middle of the survey process.

Keep in mind that the sample size for a survey is the number of respondents, not the number of people to whom questionnaires were sent or whose phone numbers were dialed. And keep in mind that a low response rate turns any study essentially into a voluntary response study, which is of little value for inferring population values. It's almost always better to spend resources on increasing the response rate than on surveying a larger group. A full or nearly full response by a modest-size sample can yield useful results.

Surveys are not the only place where proportions pop up. Banks sample huge mailing lists to estimate what proportion of people will accept a credit card offer. Even pilot studies may mail offers to over 50,000 customers. Most don't respond; that doesn't make the sample smaller—they simply said "No thanks." Those who do respond want the card. To the bank, the response rate[8] is $\hat{p}$. With a typical success rate around 0.5%, the bank needs a very small margin of error—often as low as 0.1%—to make a sound business decision. That calls for a large sample, and the bank must take care in estimating the size needed. For our election poll calculation we used $p = 0.5$, both because it's safe and because we honestly believed p to be near 0.5. If the bank used 0.5, they'd get an absurd answer. Instead, they base their calculation on a proportion closer to the one they expect to find.

---

[8] In marketing studies every mailing yields a response—"yes" or "no"—and "response rate" means the proportion of customers who accept an offer. That's not the way we use the term for survey response.

Sample Size Revisited

A credit card company is about to send out a mailing to test the market for a new credit card. From that sample, they want to estimate the true proportion of people who will sign up for the card nationwide. A pilot study suggests that about 0.5% of the people receiving the offer will accept it.

QUESTION: To be within a tenth of a percentage point (0.001) of the true rate with 95% confidence, how big does the test mailing have to be?

Using the estimate $\hat{p} = 0.5\%$: $ME = 0.001 = z^* \sqrt{\dfrac{\hat{p}\hat{q}}{n}} = 1.96 \sqrt{\dfrac{(0.005)(0.995)}{n}}$

$$(0.001)^2 = 1.96^2 \frac{(0.005)(0.995)}{n} \Rightarrow n = \frac{1.96^2(0.005)(0.995)}{(0.001)^2}$$
$$= 19{,}111.96 \text{ or } 19{,}112$$

That's a lot, but it's actually a reasonable size for a trial mailing such as this. Note, however, that if they had assumed 0.50 for the value of p, they would have found

$$ME = 0.001 = z^* \sqrt{\frac{pq}{n}} = 1.96 \sqrt{\frac{(0.5)(0.5)}{n}}$$

$$(0.001)^2 = 1.96^2 \frac{(0.5)(0.5)}{n} \Rightarrow n = \frac{1.96^2(0.5)(0.5)}{(0.001)^2} = 960{,}400.$$

Quite a different (and unreasonable) result.

What Can Go Wrong?

Confidence intervals are powerful tools. Not only do they tell what we know about the parameter value, but—more important—they also tell what we *don't* know. In order to use confidence intervals effectively, you must be clear about what you say about them.

Don't Misstate What the Interval Means

- **Don't suggest that the parameter varies.** A statement like "There is a 95% chance that the true proportion is between 42.7% and 51.3%" sounds as though you think the population proportion wanders around and sometimes happens to fall between 42.7% and 51.3%. When you interpret a confidence interval, make it clear that *you* know that the population parameter is fixed and that it is the interval that varies from sample to sample.

- **Don't claim that other samples will agree with yours.** Keep in mind that the confidence interval makes a statement about the true population proportion. An interpretation such as "In 95% of samples of U.S. adults, the proportion who think marijuana should be decriminalized will be between 42.7% and 51.3%" is just wrong. The interval isn't about sample proportions but about the population proportion.

- **Don't be certain about the parameter.** Saying "Between 42.1% and 61.7% of sea fans are infected" asserts that the population proportion cannot be outside that interval. Of course, we can't be absolutely certain of that. (Just pretty sure.)

- **Don't forget: It's about the parameter.** Don't say, "I'm 95% confident that $\hat{p}$ is between 42.1% and 61.7%." Of course you are—in fact, we calculated that $\hat{p}$ = 51.9% of the fans in our sample were infected. So we already *know* the sample proportion. The confidence interval is about the (unknown) population parameter, p.

- **Don't claim to know too much.** Don't say, "I'm 95% confident that between 42.1% and 61.7% of all the sea fans in the world are infected." You didn't sample from all 500 species of sea fans found in coral reefs around the world. Just those of this type on the Las Redes Reef.

- **Do take responsibility.** Confidence intervals are about *uncertainty*. *You* are the one who is uncertain, not the parameter. You have to accept the responsibility and consequences of the fact that not all the intervals you compute will capture the true value. In fact, about 5% of the 95% confidence intervals you find will fail to capture the true value of the parameter. You *can* say, "I am 95% confident that between 42.1% and 61.7% of the sea fans on the Las Redes Reef are infected."[9]

- **Do treat the whole interval equally.** Although a confidence interval is a set of plausible values for the parameter, don't think that the values in the middle of a confidence interval are somehow "more plausible" than the values near the edges. Your interval provides no information about where in your current interval (if at all) the parameter value is most likely to be hiding.

What *Can* I Say?

Confidence intervals are based on random samples, so the interval is random, too. The CLT tells us that 95% of the random samples will yield intervals that capture the true value. That's what we mean by being 95% confident.

Technically, we should say, "I am 95% confident that the interval from 42.1% to 61.7% captures the true proportion of infected sea fans." That formal phrasing emphasizes that *our confidence (and our uncertainty) is about the interval, not the true proportion.* But you may choose a more casual phrasing like "I am 95% confident that between 42.1% and 61.7% of the Las Redes fans are infected." Because you've made it clear that the uncertainty is yours and you didn't suggest that the randomness is in the true proportion, this is OK. Keep in mind that it's the interval that's random and is the focus of both our confidence and doubt.

Margin of Error Too Large to Be Useful

We know we can't be exact, but how precise do we need to be? A confidence interval that says that the percentage of infected sea fans is between 10% and 90% wouldn't be of much use. Most likely, you have some sense of how large a margin of error you can tolerate. What can you do?

One way to make the margin of error smaller is to reduce your level of confidence. But that may not be a useful solution. It's a rare study that reports confidence levels lower than 80%. Levels of 95% or 99% are more common.

The time to think about whether your margin of error is small enough to be useful is when you design your study. Don't wait until you compute your confidence interval. To get a narrower interval without giving up confidence, you need to have less variability in your sample proportion. How can you do that? Choose a larger sample.

---

[9] When we are being very careful we say, "95% of samples of this size will produce confidence intervals that capture the true proportion of infected sea fans on the Las Redes Reef."

Violations of Assumptions

Confidence intervals and margins of error are often reported along with poll results and other analyses. But it's easy to misuse them and wise to be aware of the ways things can go wrong.

- **Watch out for biased sampling.** Don't forget about the potential sources of bias in surveys that we discussed in Chapter 12. Just because we have more statistical machinery now doesn't mean we can forget what we've already learned. A questionnaire that finds that 85% of people enjoy filling out surveys still suffers from nonresponse bias even though now we're able to put confidence intervals around this (biased) estimate.

- **Think about independence.** The assumption that the values in our sample are mutually independent is one that we usually cannot check. It always pays to think about it, though. For example, the disease affecting the sea fans might be contagious, so that fans growing near a diseased fan are more likely themselves to be diseased. Such contagion would violate the Independence Assumption and could severely affect our sample proportion. It could be that the proportion of infected sea fans on the entire reef is actually quite small, and the researchers just happened to find an infected area. To avoid this, the researchers should be careful to sample sites far enough apart to make contagion unlikely.

CONNECTIONS

Now we can see a practical application of sampling distributions. To find a confidence interval, we lay out an interval measured in standard deviations. We're using the standard deviation as a ruler again. But now the standard deviation we need is the standard deviation of the sampling distribution. That's the one that tells how much the proportion varies. (And when we estimate it from the data, we call it a standard error.)

WHAT HAVE WE LEARNED?

The first 10 chapters of the book explored graphical and numerical ways of summarizing and presenting sample data. We've learned (at last!) to use the sample we have at hand to say something about the *world at large*. This process, called statistical inference, is based on our understanding of sampling models and will be our focus for the rest of the book.

As our first step in statistical inference, we've learned to use our sample to make a *confidence interval* that estimates what proportion of a population has a certain characteristic.

We've learned that:

▶ Our best estimate of the true population proportion is the proportion we observed in the sample, so we center our confidence interval there.

▶ Samples don't represent the population perfectly, so we create our interval with a *margin of error*.

▶ This method successfully captures the true population proportion most of the time, providing us with a level of confidence in our interval.

▶ The higher the level of confidence we want, the *wider* our confidence interval becomes.

▶ The larger the sample size we have, the *narrower* our confidence interval can be.

▶ When designing a study, we can calculate the sample size we'll need to be able to reach conclusions that have a desired degree of precision and level of confidence.

▶ There are important assumptions and conditions we must check before using this (or any) statistical inference procedure.

We've learned to interpret a confidence interval by *Telling* what we believe is true in the entire population from which we took our random sample. Of course, we can't be *certain*. We've learned not to overstate or misinterpret what the confidence interval says.

Terms

Standard error

When we estimate the standard deviation of a sampling distribution using statistics found from the data, the estimate is called a standard error (p. 458).

$$SE(\hat{p}) = \sqrt{\frac{\hat{p}\hat{q}}{n}}$$

Confidence interval

A level C confidence interval for a model parameter is an interval of values usually of the form (p. 459)

$$estimate \pm margin\ of\ error$$

found from data in such a way that C% of all random samples will yield intervals that capture the true parameter value.

One-proportion z-interval

A confidence interval for the true value of a proportion. The confidence interval is (p. 459)

$$\hat{p} \pm z^{*}SE(\hat{p}),$$

where z* is a critical value from the Standard Normal model corresponding to the specified confidence level.

Margin of error

In a confidence interval, the extent of the interval on either side of the observed statistic value is called the margin of error. A margin of error is typically the product of a critical value from the sampling distribution and a standard error from the data. A small margin of error corresponds to a confidence interval that pins down the parameter precisely. A large margin of error corresponds to a confidence interval that gives relatively little information about the estimated parameter. For a proportion (p. 461).

$$ME = z^{*}\sqrt{\frac{\hat{p}\hat{q}}{n}}$$

Critical value

The number of standard errors to move away from the mean of the sampling distribution to correspond to the specified level of confidence. The critical value, denoted z*, is usually found from a table or with technology (p. 462).

Skills

 THINK

▶ Understand confidence intervals as a balance between the precision and the certainty of a statement about a model parameter

▶ Understand that the margin of error of a confidence interval for a proportion changes with the sample size and the level of confidence.

▶ Know how to examine your data for violations of conditions that would make inference about a population proportion unwise or invalid.

SHOW

▶ Be able to construct a one-proportion z-interval.

TELL

▶ Be able to interpret a one-proportion z-interval in a simple sentence or two. Write such an interpretation so that it does not state or suggest that the parameter of interest is itself random, but rather that the bounds of the confidence interval are the random quantities about which we state our degree of confidence.

CONFIDENCE INTERVALS FOR PROPORTIONS ON THE COMPUTER

Confidence intervals for proportions are so easy and natural that many statistics packages don't offer special commands for them. Most statistics programs want the "raw data" for computations. For proportions, the raw data are the "success" and "failure" status for each case. Usually, these are given as 1 or 0, but they might be category names like "yes" and "no." Often we just know the proportion of successes, $\hat{p}$, and the total count, n. Computer packages don't usually deal with summary data like this easily, but the statistics routines found on many graphing calculators allow you to create confidence intervals from summaries of the data—usually all you need to enter are the number of successes and the sample size.

In some programs you can reconstruct variables of 0's and 1's with the given proportions. But even when you have (or can reconstruct) the raw data values, you may not get exactly the same margin of error from a computer package as you would find working by hand. The reason is that some packages make approximations or use other methods. The result is very close but not exactly the same. Fortunately, Statistics means never having to say you're certain, so the approximate result is good enough.

DATA DESK

Data Desk does not offer built-in methods for inference with proportions.

COMMENTS

For summarized data, open a Scratchpad to compute the standard deviation and margin of error by typing the calculation. Then use **z-interval for individual μs**.

EXCEL

Inference methods for proportions are not part of the standard Excel tool set.

COMMENTS

For summarized data, type the calculation into any cell and evaluate it.

JMP

For a **categorical** variable that holds category labels, the **Distribution** platform includes tests and intervals for proportions. For summarized data, put the category names in one variable and the frequencies in an adjacent variable. Designate the frequency column to have the **role** of **frequency.** Then use the **Distribution** platform.

COMMENTS

JMP uses slightly different methods for proportion inferences than those discussed in this text. Your answers are likely to be slightly different, especially for small samples.

MINITAB

Choose **Basic Statistics** from the **Stat** menu.

- Choose **1Proportion** from the Basic Statistics submenu.

- If the data are category names in a variable, assign the variable from the variable list box to the **Samples in columns** box. If you have summarized data, click the **Summarized Data** button and fill in the number of trials and the number of successes.

- Click the **Options** button and specify the remaining details.

- If you have a large sample, check **Use test and interval based on normal distribution.** Click the **OK** button.

COMMENTS

When working from a variable that names categories, MINITAB treats the last category as the "success" category. You can specify how the categories should be ordered.

SPSS

SPSS does not find confidence intervals for proportions.

TI-83/84 PLUS	**COMMENTS**
To calculate a confidence interval for a population proportion: • Go to the **STATS TESTS** menu and select **A:1-PropZInt.** • Enter the number of successes observed and the sample size. • Specify a confidence level. • Calculate the interval.	*Beware:* When you enter the value of *x*, you need the count, not the percentage. The count must be a whole number. If the number of successes are given as a percentage, you must first multiply *np* and round the result.
TI-89	**COMMENTS**
To calculate a confidence interval for a population proportion: • Go to the **Ints** menu (2nd F2) and select **5:1-PropZInt.** • Enter the number of successes observed and the sample size. • Specify a confidence level. • Calculate the interval.	*Beware:* When you enter the value of *x*, you need the count, not the percentage. The count must be a whole number. If the number of successes are given as a percentage, you must first multiply *np* and round the result.

EXERCISES

1. **Margin of error.** A TV newscaster reports the results of a poll of voters, and then says, "The margin of error is plus or minus 4%." Explain carefully what that means.

2. **Margin of error.** A medical researcher estimates the percentage of children exposed to lead-base paint, adding that he believes his estimate has a margin of error of about 3%. Explain what the margin of error means.

3. **Conditions.** For each situation described below, identify the population and the sample, explain what p and $\hat{p}$ represent, and tell whether the methods of this chapter can be used to create a confidence interval.

   a) Police set up an auto checkpoint at which drivers are stopped and their cars inspected for safety problems. They find that 14 of the 134 cars stopped have at least one safety violation. They want to estimate the percentage of all cars that may be unsafe.

   b) A TV talk show asks viewers to register their opinions on prayer in schools by logging on to a website. Of the 602 people who voted, 488 favored prayer in schools. We want to estimate the level of support among the general public.

   c) A school is considering requiring students to wear uniforms. The PTA surveys parent opinion by sending a questionnaire home with all 1245 students; 380 surveys are returned, with 228 families in favor of the change.

   d) A college admits 1632 freshmen one year, and four years later 1388 of them graduate on time. The college wants to estimate the percentage of all their freshman enrollees who graduate on time.

4. **More conditions.** Consider each situation described. Identify the population and the sample, explain what p and $\hat{p}$ represent, and tell whether the methods of this chapter can be used to create a confidence interval.

   a) A consumer group hoping to assess customer experiences with auto dealers surveys 167 people who recently bought new cars; 3% of them expressed dissatisfaction with the salesperson.

   b) What percent of college students have cell phones? 2883 students were asked as they entered a football stadium, and 2430 said they had phones with them.

   c) 240 potato plants in a field in Maine are randomly checked, and only 7 show signs of blight. How severe is the blight problem for the U.S. potato industry?

   d) 12 of the 309 employees of a small company suffered an injury on the job last year. What can the company expect in future years?

5. **Conclusions.** A catalog sales company promises to deliver orders placed on the Internet within 3 days. Follow-up calls to a few randomly selected customers show that a 95% confidence interval for the proportion of all orders that arrive on time is 88% ± 6%. What does this mean? Are these conclusions correct? Explain.

   a) Between 82% and 94% of all orders arrive on time.

   b) 95% of all random samples of customers will show that 88% of orders arrive on time.

   c) 95% of all random samples of customers will show that 82% to 94% of orders arrive on time.

   d) We are 95% sure that between 82% and 94% of the orders placed by the sampled customers arrived on time.

   e) On 95% of the days, between 82% and 94% of the orders will arrive on time.

6. **More conclusions.** In January 2002, two students made worldwide headlines by spinning a Belgian euro 250

times and getting 140 heads—that's 56%. That makes the 90% confidence interval (51%, 61%). What does this mean? Are these conclusions correct? Explain.

a) Between 51% and 61% of all euros are unfair.

b) We are 90% sure that in this experiment this euro landed heads on between 51% and 61% of the spins.

c) We are 90% sure that spun euros will land heads between 51% and 61% of the time.

d) If you spin a euro many times, you can be 90% sure of getting between 51% and 61% heads.

e) 90% of all spun euros will land heads between 51% and 61% of the time.

7. **Confidence intervals.** Several factors are involved in the creation of a confidence interval. Among them are the sample size, the level of confidence, and the margin of error. Which statements are true?

a) For a given sample size, higher confidence means a smaller margin of error.

b) For a specified confidence level, larger samples provide smaller margins of error.

c) For a fixed margin of error, larger samples provide greater confidence.

d) For a given confidence level, halving the margin of error requires a sample twice as large.

8. **Confidence intervals, again.** Several factors are involved in the creation of a confidence interval. Among them are the sample size, the level of confidence, and the margin of error. Which statements are true?

a) For a given sample size, reducing the margin of error will mean lower confidence.

b) For a certain confidence level, you can get a smaller margin of error by selecting a bigger sample.

c) For a fixed margin of error, smaller samples will mean lower confidence.

d) For a given confidence level, a sample 9 times as large will make a margin of error one third as big.

9. **Cars.** What fraction of cars is made in Japan? The computer output below summarizes the results of a random sample of 50 autos. Explain carefully what it tells you.

> z-Interval for proportion
> With 90.00% confidence,
> 0.29938661 < p(japan) < 0.46984416

10. **Parole.** A study of 902 decisions made by the Nebraska Board of Parole produced the following computer output. Assuming these cases are representative of all cases that may come before the Board, what can you conclude?

> z-Interval for proportion
> With 95.00% confidence,
> 0.56100658 < p(parole) < 0.62524619

11. **Contaminated chicken.** In January 2007 *Consumer Reports* published their study of bacterial contamination of chicken sold in the United States. They purchased 525 broiler chickens from various kinds of food stores in 23 states and tested them for types of bacteria that cause food-borne illnesses. Laboratory results indicated that 83% of these chickens were infected with *Campylobacter*.

a) Construct a 95% confidence interval.

b) Explain what your confidence interval says about chicken sold in the United States.

c) A spokesperson for the U.S. Department of Agriculture dismissed the *Consumer Reports* finding, saying, "That's 500 samples out of 9 billion chickens slaughtered a year. . . . With the small numbers they [tested], I don't know that one would want to change one's buying habits." Is this criticism valid? Explain.

12. **Contaminated chicken, second course.** The January 2007 *Consumer Reports* study described in Exercise 11 also found that 15% of the 525 broiler chickens tested were infected with *Salmonella*.

a) Are the conditions for creating a confidence interval satisfied? Explain.

b) Construct a 95% confidence interval.

c) Explain what your confidence interval says about chicken sold in the United States.

13. **Baseball fans.** In a poll taken in March of 2007, Gallup asked 1006 national adults whether they were baseball fans. 36% said they were. A year previously, 37% of a similar-size sample had reported being baseball fans.

a) Find the margin of error for the 2007 poll if we want 90% confidence in our estimate of the percent of national adults who are baseball fans.

b) Explain what that margin of error means.

c) If we wanted to be 99% confident, would the margin of error be larger or smaller? Explain.

d) Find that margin of error.

e) In general, if all other aspects of the situation remain the same, will smaller margins of error produce greater or less confidence in the interval?

f) Do you think there's been a change from 2006 to 2007 in the real proportion of national adults who are baseball fans? Explain.

14. **Cloning 2007.** A May 2007 Gallup Poll found that only 11% of a random sample of 1003 adults approved of attempts to clone a human.

a) Find the margin of error for this poll if we want 95% confidence in our estimate of the percent of American adults who approve of cloning humans.

b) Explain what that margin of error means.

c) If we only need to be 90% confident, will the margin of error be larger or smaller? Explain.

d) Find that margin of error.

e) In general, if all other aspects of the situation remain the same, would smaller samples produce smaller or larger margins of error?

15. **Contributions, please.** The Paralyzed Veterans of America is a philanthropic organization that relies on contributions. They send free mailing labels and greeting cards to potential donors on their list and ask for a voluntary contribution. To test a new campaign, they recently sent letters to a random sample of 100,000 potential donors and received 4781 donations.

a) Give a 95% confidence interval for the true proportion of their entire mailing list who may donate.

b) A staff member thinks that the true rate is 5%. Given the confidence interval you found, do you find that percentage plausible?

16. **Take the offer.** First USA, a major credit card company, is planning a new offer for their current cardholders. The

offer will give double airline miles on purchases for the next 6 months if the cardholder goes online and registers for the offer. To test the effectiveness of the campaign, First USA recently sent out offers to a random sample of 50,000 cardholders. Of those, 1184 registered.

a) Give a 95% confidence interval for the true proportion of those cardholders who will register for the offer.

b) If the acceptance rate is only 2% or less, the campaign won't be worth the expense. Given the confidence interval you found, what would you say?

17. **Teenage drivers.** An insurance company checks police records on 582 accidents selected at random and notes that teenagers were at the wheel in 91 of them.

a) Create a 95% confidence interval for the percentage of all auto accidents that involve teenage drivers.

b) Explain what your interval means.

c) Explain what "95% confidence" means.

d) A politician urging tighter restrictions on drivers' licenses issued to teens says, "In one of every five auto accidents, a teenager is behind the wheel." Does your confidence interval support or contradict this statement? Explain.

18. **Junk mail.** Direct mail advertisers send solicitations (a.k.a. "junk mail") to thousands of potential customers in the hope that some will buy the company's product. The acceptance rate is usually quite low. Suppose a company wants to test the response to a new flyer, and sends it to 1000 people randomly selected from their mailing list of over 200,000 people. They get orders from 123 of the recipients.

a) Create a 90% confidence interval for the percentage of people the company contacts who may buy something.

b) Explain what this interval means.

c) Explain what "90% confidence" means.

d) The company must decide whether to now do a mass mailing. The mailing won't be cost-effective unless it produces at least a 5% return. What does your confidence interval suggest? Explain.

19. **Safe food.** Some food retailers propose subjecting food to a low level of radiation in order to improve safety, but sale of such "irradiated" food is opposed by many people. Suppose a grocer wants to find out what his customers think. He has cashiers distribute surveys at checkout and ask customers to fill them out and drop them in a box near the front door. He gets responses from 122 customers, of whom 78 oppose the radiation treatments. What can the grocer conclude about the opinions of all his customers?

20. **Local news.** The mayor of a small city has suggested that the state locate a new prison there, arguing that the construction project and resulting jobs will be good for the local economy. A total of 183 residents show up for a public hearing on the proposal, and a show of hands finds only 31 in favor of the prison project. What can the city council conclude about public support for the mayor's initiative?

21. **Death penalty, again.** In the survey on the death penalty you read about in the chapter, the Gallup Poll actually split the sample at random, asking 510 respondents the question quoted earlier, "Generally speaking, do you believe the death penalty is applied fairly or unfairly in this country today?" The other 510 were asked "Generally speaking, do you believe the death penalty is applied unfairly or fairly in this country today?" Seems like the same question, but sometimes the order of the choices matters. Suppose that for the second way of phrasing it, 60% said they thought the death penalty was fairly applied.

a) What kind of bias may be present here?

b) If we combine them, considering the overall group to be one larger random sample of 1020 respondents, what is a 95% confidence interval for the proportion of the general public that thinks the death penalty is being fairly applied?

c) How does the margin of error based on this pooled sample compare with the margins of error from the separate groups? Why?

22. **Gambling.** A city ballot includes a local initiative that would legalize gambling. The issue is hotly contested, and two groups decide to conduct polls to predict the outcome. The local newspaper finds that 53% of 1200 randomly selected voters plan to vote "yes," while a college Statistics class finds 54% of 450 randomly selected voters in support. Both groups will create 95% confidence intervals.

a) Without finding the confidence intervals, explain which one will have the larger margin of error.

b) Find both confidence intervals.

c) Which group concludes that the outcome is too close to call? Why?

23. **Rickets.** Vitamin D, whether ingested as a dietary supplement or produced naturally when sunlight falls on the skin, is essential for strong, healthy bones. The bone disease rickets was largely eliminated in England during the 1950s, but now there is concern that a generation of children more likely to watch TV or play computer games than spend time outdoors is at increased risk. A recent study of 2700 children randomly selected from all parts of England found 20% of them deficient in vitamin D.

a) Find a 98% confidence interval.

b) Explain carefully what your interval means.

c) Explain what "98% confidence" means.

24. **Pregnancy.** In 1998 a San Diego reproductive clinic reported 49 live births to 207 women under the age of 40 who had previously been unable to conceive.

a) Find a 90% confidence interval for the success rate at this clinic.

b) Interpret your interval in this context.

c) Explain what "90% confidence" means.

d) Do these data refute the clinic's claim of a 25% success rate? Explain.

25. **Payments.** In a May 2007 Experian/Gallup Personal Credit Index poll of 1008 U.S. adults aged 18 and over, 8% of respondents said they were very uncomfortable with their ability to make their monthly payments on their current debt during the next three months. A more detailed poll surveyed 1288 adults, reporting similar overall results and also noting differences among four age groups: 18–29, 30–49, 50–64, and 65+.

a) Do you expect the 95% confidence interval for the true proportion of all 18- to 29-year-olds who are worried to be wider or narrower than the 95% confidence interval for the true proportion of all U.S. consumers? Explain.

b) Do you expect this second poll's overall margin of error to be larger or smaller than the Experian/Gallup poll's? Explain.

26. **Back to campus, again.** In 2004 ACT, Inc., reported that 74% of 1644 randomly selected college freshmen returned to college the next year. The study was stratified by type of college—public or private. The retention rates were 71.9% among 505 students enrolled in public colleges and 74.9% among 1139 students enrolled in private colleges.

a) Will the 95% confidence interval for the true national retention rate in private colleges be wider or narrower than the 95% confidence interval for the retention rate in public colleges? Explain.

b) Do you expect the margin of error for the overall retention rate to be larger or smaller? Explain.

27. **Deer ticks.** Wildlife biologists inspect 153 deer taken by hunters and find 32 of them carrying ticks that test positive for Lyme disease.

a) Create a 90% confidence interval for the percentage of deer that may carry such ticks.

b) If the scientists want to cut the margin of error in half, how many deer must they inspect?

c) What concerns do you have about this sample?

28. **Pregnancy, II.** The San Diego reproductive clinic in Exercise 24 wants to publish updated information on its success rate.

a) The clinic wants to cut the stated margin of error in half. How many patients' results must be used?

b) Do you have any concerns about this sample? Explain.

29. **Graduation.** It's believed that as many as 25% of adults over 50 never graduated from high school. We wish to see if this percentage is the same among the 25 to 30 age group.

a) How many of this younger age group must we survey in order to estimate the proportion of non-grads to within 6% with 90% confidence?

b) Suppose we want to cut the margin of error to 4%. What's the necessary sample size?

c) What sample size would produce a margin of error of 3%?

30. **Hiring.** In preparing a report on the economy, we need to estimate the percentage of businesses that plan to hire additional employees in the next 60 days.

a) How many randomly selected employers must we contact in order to create an estimate in which we are 98% confident with a margin of error of 5%?

b) Suppose we want to reduce the margin of error to 3%. What sample size will suffice?

c) Why might it not be worth the effort to try to get an interval with a margin of error of only 1%?

31. **Graduation, again.** As in Exercise 29, we hope to estimate the percentage of adults aged 25 to 30 who never graduated from high school. What sample size would allow us to increase our confidence level to 95% while reducing the margin of error to only 2%?

32. **Better hiring info.** Editors of the business report in Exercise 30 are willing to accept a margin of error of 4% but want 99% confidence. How many randomly selected employers will they need to contact?

33. **Pilot study.** A state's environmental agency worries that many cars may be violating clean air emissions standards. The agency hopes to check a sample of vehicles in order to estimate that percentage with a margin of error of 3% and 90% confidence. To gauge the size of the problem, the agency first picks 60 cars and finds 9 with faulty emissions systems. How many should be sampled for a full investigation?

34. **Another pilot study.** During routine screening, a doctor notices that 22% of her adult patients show higher than normal levels of glucose in their blood—a possible warning signal for diabetes. Hearing this, some medical researchers decide to conduct a large-scale study, hoping to estimate the proportion to within 4% with 98% confidence. How many randomly selected adults must they test?

35. **Approval rating.** A newspaper reports that the governor's approval rating stands at 65%. The article adds that the poll is based on a random sample of 972 adults and has a margin of error of 2.5%. What level of confidence did the pollsters use?

36. **Amendment.** A TV news reporter says that a proposed constitutional amendment is likely to win approval in the upcoming election because a poll of 1505 likely voters indicated that 52% would vote in favor. The reporter goes on to say that the margin of error for this poll was 3%.

a) Explain why the poll is actually inconclusive.

b) What confidence level did the pollsters use?

JUST CHECKING

ANSWERS

1. No. We know that in the sample 17% said "yes"; there's no need for a margin of error.

2. No, we are 95% confident that the percentage falls in some interval, not exactly on a particular value.

3. Yes. That's what the confidence interval means.

4. No. We don't know for sure that's true; we are only 95% confident.

5. No. That's our level of confidence, not the proportion of people receiving unsolicited text messages. The sample suggests the proportion is much lower.

6. Wider.

7. Lower.

8. Smaller.

Testing Hypotheses About Proportions

Where are we going?

Do people ages 18-24 really prefer Pepsi to Coke? Does this new allergy medication really reduce symptoms more than a placebo? There are times when we want to make a *decision*. To do that, we'll propose a model for the situation at hand and test a hypothesis about that model. The result will help us answer the real-world question.

A S   *Activity:* **Testing a Claim.** Can we really draw a reasonable conclusion from a random sample? Run this simulation before you read the chapter, and you'll gain a solid sense of what we're doing here.

Ingots are huge pieces of metal, often weighing more than 20,000 pounds, made in a giant mold. They must be cast in one large piece for use in fabricating large structural parts for cars and planes. If they crack while being made, the crack can propagate into the zone required for the part, compromising its integrity. Airplane manufacturers insist that metal for their planes be defect-free, so the ingot must be made over if any cracking is detected.

Even though the metal from the cracked ingot is recycled, the scrap cost runs into the tens of thousands of dollars. Metal manufacturers would like to avoid cracking if at all possible. But the casting process is complicated and not everything is completely under control. In one plant, only about 80% of the ingots have been free of cracks. In an attempt to reduce the cracking proportion, the plant engineers and chemists recently tried out some changes in the casting process. Since then, 400 ingots have been cast and only 17% of them have cracked. Should management declare victory? Has the cracking rate really decreased, or was 17% just due to luck?

We can treat the 400 ingots cast with the new method as a random sample. We know that each random sample will have a somewhat different proportion of cracked ingots. Is the 17% we observe merely a result of natural sampling variability, or is this lower cracking rate strong enough evidence to assure management that the true cracking rate now is really below 20%?

People want answers to questions like these all the time. Has the president's approval rating changed since last month? Has teenage smoking decreased in the past five years? Is the global temperature increasing? Did the Super Bowl ad we bought actually increase sales? To answer such questions, we test *hypotheses* about models.

"Half the money I spend on advertising is wasted; the trouble is I don't know which half."

–John Wanamaker
(attributed)

Hypothesis n.;
pl. {Hypotheses}.
   A supposition; a proposition or principle which is supposed or taken for granted, in order to draw a conclusion or inference for proof of the point in question; something not proved, but assumed for the purpose of argument.
—*Webster's Unabridged Dictionary, 1913*

Hypotheses

How can we state and test a hypothesis about ingot cracking? Hypotheses are working models that we adopt temporarily. To test whether the changes made by the engineers have *improved* the cracking rate, we assume that they have in fact made no difference and that any apparent improvement is just random fluctuation (sampling error). So, our starting hypothesis, called the **null hypothesis**, is that the proportion of cracks is still 20%.

The **null hypothesis**, which we denote H_0, specifies a population model parameter of interest and proposes a value for that parameter. We usually write down the null hypothesis in the form H_0: *parameter = hypothesized value*. This is a concise way to specify the two things we need most: the identity of the parameter we hope to learn about and a specific hypothesized value for that parameter. (We need a hypothesized value so we can compare our observed statistic value to it.)

Which value to use is often obvious from the *Who* and *What* of the data. But sometimes it takes a bit of thinking to translate the question we hope to answer into a hypothesis about a parameter. For the ingots we can write H_0: $p = 0.20$.

The **alternative hypothesis**, which we denote H_A, contains the values of the parameter that we consider plausible if we reject the null hypothesis. In the ingots example, our null hypothesis is that $p = 0.20$. What's the alternative? Management is interested in *reducing* the cracking rate, so their alternative is H_A: $p < 0.20$.

What would convince you that the cracking rate had actually gone down? If you observed a cracking rate *much lower* than 20% in your sample, you'd likely be convinced. If only 3 out of the next 400 ingots crack (for a rate of 0.75%), most folks would conclude that the changes helped. But if the sample cracking rate is only slightly lower than 20%, you should be skeptical. After all, observed proportions do vary, so we wouldn't be surprised to see some difference. How much smaller must the cracking rate be before we *are* convinced that it has changed? Whenever we ask about the size of a statistical difference, we naturally think of using the standard deviation as a ruler. So let's start by finding the standard deviation of the sample cracking rate.

Since the company changed the process, 400 new ingots have been cast. The sample size of 400 is big enough to satisfy the **Success/Failure Condition**. (We expect $0.20 \times 400 = 80$ ingots to crack.) We have no reason to think the ingots are not independent, so the Normal sampling distribution model should work well. The standard deviation of the sampling model is

To remind us that the parameter value comes from the null hypothesis, it is sometimes written as p_0 and the standard deviation as

$$SD(\hat{p}) = \sqrt{\frac{p_0 q_0}{n}}.$$

$$SD(\hat{p}) = \sqrt{\frac{pq}{n}} = \sqrt{\frac{(0.20)(0.80)}{400}} = 0.02$$

Why is this a standard deviation and not a standard error?
Because we haven't estimated anything. When we assume that the null hypothesis is true, it gives us a value for the model parameter p. With proportions, if we know p, then we also automatically know its standard deviation. And because we find the standard deviation from the model parameter, this is a standard deviation and not a standard error. When we found a confidence interval for p, we could not assume that we knew its value, so we estimated the standard deviation from the sample value $\hat{p}$.

Now we know both parameters of the Normal sampling distribution model: $p = 0.20$ and $SD(\hat{p}) = 0.02$, so we can find out how likely it would be to see the observed value of $\hat{p} = 17\%$. Since we are using a Normal model, we find the z-score:

$$z = \frac{0.17 - 0.20}{0.02} = -1.5$$

Then we ask, "How likely is it to observe a value at least 1.5 standard deviations below the mean of a Normal model?" The answer (from a calculator, computer program, or the Normal table) is about 0.067. This is the probability of observing a cracking rate of 17% or less in a sample of 400 if the null hypothesis is true. Management now must decide whether an event that would happen 6.7% of the time by chance is strong enough evidence to conclude that the true cracking proportion has decreased.

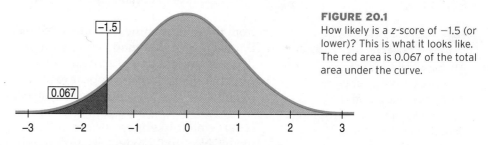

FIGURE 20.1
How likely is a *z*-score of −1.5 (or lower)? This is what it looks like. The red area is 0.067 of the total area under the curve.

A Trial as a Hypothesis Test

Does the reasoning of hypothesis tests seem backward? That could be because we usually prefer to think about getting things right rather than getting them wrong. You have seen this reasoning before in a different context. This is the logic of jury trials.

Let's suppose a defendant has been accused of robbery. In British common law and those systems derived from it (including U.S. law), the null hypothesis is that the defendant is innocent. Instructions to juries are quite explicit about this.

The evidence takes the form of facts that seem to contradict the presumption of innocence. For us, this means collecting data. In the trial, the prosecutor presents evidence. ("If the defendant were innocent, wouldn't it be remarkable that the police found him at the scene of the crime with a bag full of money in his hand, a mask on his face, and a getaway car parked outside?")

The next step is to judge the evidence. Evaluating the evidence is the responsibility of the jury in a trial, but it falls on your shoulders in hypothesis testing. The jury considers the evidence in light of the *presumption* of innocence and judges whether the evidence against the defendant would be plausible *if the defendant were in fact innocent.*

Like the jury, you ask, "Could these data plausibly have happened by chance if the null hypothesis were true?" If they are very unlikely to have occurred, then the evidence raises a reasonable doubt about the null hypothesis.

Ultimately, you must make a decision. The standard of "beyond a reasonable doubt" is wonderfully ambiguous because it leaves the jury to decide the degree to which the evidence contradicts the hypothesis of innocence. Juries don't explicitly use probability to help them decide whether to reject that hypothesis. But when you ask the same question of your null hypothesis, you have the advantage of being able to quantify exactly how surprising the evidence would be were the null hypothesis true.

How unlikely is unlikely? Some people set rigid standards, like 1 time out of 20 (0.05) or 1 time out of 100 (0.01). But if *you* have to make the decision, you must judge for yourself in each situation whether the probability of observing your data is small enough to constitute "reasonable doubt."

A S
Activity: **The Reasoning of Hypothesis Testing.** Our reasoning is based on a rule of logic that dates back to ancient scholars. Here's a modern discussion of it.

P-Values

The fundamental step in our reasoning is the question "Are the data surprising, given the null hypothesis?" And the key calculation is to determine exactly how likely the data we observed would be were the null hypothesis a true

"If the People fail to satisfy their burden of proof, you must find the defendant not guilty."

–NY state jury instructions

model of the world. So we need a *probability*. Specifically, we want to find the probability of seeing data like these (or something even less likely) *given* that the null hypothesis is true. Statisticians are so thrilled with their ability to measure precisely how surprised they are that they give this probability a special name. It's called a **P-value**.[1]

When the P-value is high, we haven't seen anything unlikely or surprising at all. Events that have a high probability of happening happen often. The data are thus consistent with the model from the null hypothesis, and we have no reason to reject the null hypothesis. But we realize that many other similar hypotheses could also account for the data we've seen, so *we haven't proven that the null hypothesis is true*. The most we can say is that it doesn't appear to be false. Formally, we "fail to reject" the null hypothesis. That's a pretty weak conclusion, but it's all we're entitled to.

When the P-value *is* low enough, it says that it's very unlikely we'd observe data like these if our null hypothesis were true. We started with a model. Now that model tells us that the data we have are unlikely to have happened. The model and data are at odds with each other, so we have to make a choice. Either the null hypothesis is correct and we've just seen something remarkable, or the null hypothesis is wrong, and we were wrong to use it as the basis for computing our P-value. Perhaps another model is correct, and the data really aren't that remarkable after all. If you believe in data more than in assumptions, then, given that choice, you should reject the null hypothesis.

What to Do with an "Innocent" Defendant

If the evidence is not strong enough to reject the defendant's presumption of innocence, what verdict does the jury return? They say "not guilty." Notice that they do not say that the defendant is innocent. All they say is that they have not seen sufficient evidence to convict, to reject innocence. The defendant may, in fact, be innocent, but the jury has no way to be sure.

Said statistically, the jury's null hypothesis is H_0: innocent defendant. If the evidence is too unlikely given this assumption, the jury rejects the null hypothesis and finds the defendant guilty. But—and this is an important distinction—if there is *insufficient evidence* to convict the defendant, the jury does not decide that H_0 is true and declare the defendant innocent. Juries can only *fail to reject* the null hypothesis and declare the defendant "not guilty."

In the same way, if the data are not particularly unlikely under the assumption that the null hypothesis is true, then the most we can do is to "fail to reject" our null hypothesis. We never declare the null hypothesis to be true (or "accept" the null), because we simply do not know whether it's true or not. (After all, more evidence may come along later.)

In the trial, the burden of proof is on the prosecution. In a hypothesis test, the burden of proof is on the unusual claim. The null hypothesis is the ordinary state of affairs, so it's the alternative to the null hypothesis that we consider unusual and for which we must marshal evidence.

Imagine a clinical trial testing the effectiveness of a new headache remedy. In Chapter 13 we saw the value of comparing such treatments to a placebo. The null hypothesis, then, is that the new treatment is no more effective than the placebo. This is important, because some patients will improve even when

---

[1] You'd think if they were so excited, they'd give it a better name, but "P-value" is about as excited as statisticians get.

administered the placebo treatment. If we use only six people to test the drug, the results are likely *not to be clear* and we'll be unable to reject the hypothesis. Does this mean the drug doesn't work? Of course not. It simply means that we don't have enough evidence to reject our assumption. That's why we don't start by assuming that the drug *is more effective*. If we were to do that, then we could test just a few people, find that the results aren't clear, and claim that since we've been unable to reject our original assumption the drug must be effective. The FDA is unlikely to be impressed by that argument.

JUST CHECKING

1. A research team wants to know if aspirin helps to thin blood. The null hypothesis says that it doesn't. They test 12 patients, observe the proportion with thinner blood, and get a P-value of 0.32. They proclaim that aspirin doesn't work. What would you say?

2. An allergy drug has been tested and found to give relief to 75% of the patients in a large clinical trial. Now the scientists want to see if the new, improved version works even better. What would the null hypothesis be?

3. The new drug is tested and the P-value is 0.0001. What would you conclude about the new drug?

The Reasoning of Hypothesis Testing

Hypothesis tests follow a carefully structured path. To avoid getting lost as we navigate down it, we divide that path into four distinct sections.

1. Hypotheses

> "The null hypothesis is never proved or established, but is possibly disproved, in the course of experimentation. Every experiment may be said to exist only in order to give the facts a chance of disproving the null hypothesis."
>
> –Sir Ronald Fisher,
> *The Design of Experiments*

Some folks pronounce the hypothesis labels "Ho!" and "Ha!" (but it makes them seem overexcitable). We prefer to pronounce H_0 "H naught" (as in "all is for naught").

First we state the null hypothesis. That's usually the skeptical claim that nothing's different. Are we considering a (New! Improved!) possibly better method? The null hypothesis says, "Oh yeah? Convince me!" To convert a skeptic, we must pile up enough evidence against the null hypothesis that we can reasonably reject it.

In statistical hypothesis testing, hypotheses are almost always about model parameters. To assess how unlikely our data may be, we need a null model. The null hypothesis specifies a particular parameter value to use in our model. In the usual shorthand, we write H_0: *parameter = hypothesized value*. The alternative hypothesis, H_A, contains the values of the parameter we consider plausible when we reject the null.

FOR EXAMPLE | Writing Hypotheses

A large city's Department of Motor Vehicles claimed that 80% of candidates pass driving tests, but a newspaper reporter's survey of 90 randomly selected local teens who had taken the test found only 61 who passed.

QUESTION: Does this finding suggest that the passing rate for teenagers is lower than the DMV reported? Write appropriate hypotheses.

I'll assume that the passing rate for teenagers is the same as the DMV's overall rate of 80%, unless there's strong evidence that it's lower.

$$H_0: p = 0.80$$
$$H_A: p < 0.80$$

2. Model

To plan a statistical hypothesis test, specify the *model* you will use to test the null hypothesis and the parameter of interest. Of course, all models require assumptions, so you will need to state them and check any corresponding conditions.

Your Model step should end with a statement such as

Because the conditions are satisfied, I can model the sampling distribution of the proportion with a Normal model.

Watch out, though. Your Model step could end with

Because the conditions are not satisfied, I can't proceed with the test.

If that's the case, stop and reconsider.

Each test in the book has a name that you should include in your report. We'll see many tests in the chapters that follow. Some will be about more than one sample, some will involve statistics other than proportions, and some will use models other than the Normal (and so will not use z-scores). The test about proportions is called a **one-proportion z-test**.[2]

WHEN THE CONDITIONS FAIL . . .

You might proceed with caution, explicitly stating your concerns. Or you may need to do the analysis with and without an outlier, or on different subgroups, or after re-expressing the response variable. Or you may not be able to proceed at all.

A S *Activity:* **Was the Observed Outcome Unlikely?** Complete the test you started in the first activity for this chapter. The narration explains the steps of the hypothesis test.

ONE-PROPORTION z-TEST

The conditions for the one-proportion z-test are the same as for the one-proportion z-interval. We test the hypothesis $H_0: p = p_0$ using the statistic $z = \dfrac{(\hat{p} - p_0)}{SD(\hat{p})}$. We use the hypothesized proportion to find the standard deviation, $SD(\hat{p}) = \sqrt{\dfrac{p_0 q_0}{n}}$.

When the conditions are met and the null hypothesis is true, this statistic follows the standard Normal model, so we can use that model to obtain a P-value.

FOR EXAMPLE   Checking the Conditions

RECAP: A large city's DMV claimed that 80% of candidates pass driving tests. A reporter has results from a survey of 90 randomly selected local teens who had taken the test.

QUESTION: Are the conditions for inference satisfied?

✔ The 90 teens surveyed were a random sample of local teenage driving candidates.

✔ 90 is fewer than 10% of the teenagers who take driving tests in a large city.

✔ We expect $np_0 = 90(0.80) = 72$ successes and $nq_0 = 90(0.20) = 18$ failures. Both are at least 10.

The conditions are satisfied, so it's okay to use a Normal model and perform a one-proportion z-test.

---

3. Mechanics

CONDITIONAL PROBABILITY

Did you notice that a P-value is a conditional probability? It's the probability that the observed results could have happened *if the null hypothesis is true*.

Under "Mechanics," we place the actual calculation of our test statistic from the data. Different tests we encounter will have different formulas and different test statistics. Usually, the mechanics are handled by a statistics program or calculator, but it's good to have the formulas recorded for reference and to

---

[2] It's also called the "one-sample test for a proportion."

know what's being computed. The ultimate goal of the calculation is to obtain a P-value—the probability that the observed statistic value (or an even more extreme value) occurs if the null model is correct. If the P-value is small enough, we'll reject the null hypothesis.

FOR EXAMPLE   Finding a P-Value

RECAP: A large city's DMV claimed that 80% of candidates pass driving tests, but a survey of 90 randomly selected local teens who had taken the test found only 61 who passed.

QUESTION: What's the P-value for the one-proportion z-test?

I have $n = 90$, $x = 61$, and a hypothesized $p = 0.80$.

$$\hat{p} = \frac{61}{90} \approx 0.678$$

$$SD(\hat{p}) = \sqrt{\frac{p_0 q_0}{n}} = \sqrt{\frac{(0.8)(0.2)}{90}} \approx 0.042$$

$$z = \frac{\hat{p} - p_0}{SD(\hat{p})} = \frac{0.678 - 0.800}{0.042} \approx -2.90$$

$$\text{P-value} = P(z < -2.90) = 0.002$$

---

4. Conclusion

The conclusion in a hypothesis test is always a statement about the null hypothesis. The conclusion must state either that we reject or that we fail to reject the null hypothesis. And, as always, the conclusion should be stated in context.

FOR EXAMPLE   Stating the Conclusion

RECAP: A large city's DMV claimed that 80% of candidates pass driving tests. Data from a reporter's survey of randomly selected local teens who had taken the test produced a P-value of 0.002.

QUESTION: What can the reporter conclude? And how might the reporter explain what the P-value means for the newspaper story?

Because the P-value of 0.002 is very low, I reject the null hypothesis. These survey data provide strong evidence that the passing rate for teenagers taking the driving test is lower than 80%.

If the passing rate for teenage driving candidates were actually 80%, we'd expect to see success rates this low in only about 1 in 500 samples (0.2%). This seems quite unlikely, casting doubt that the DMV's stated success rate applies to teens.

". . . They make things admirably plain,
But one hard question will remain:
If one hypothesis you lose,
Another in its place you choose . . ."

–James Russell Lowell,
Credidimus Jovem
Regnare

Your conclusion about the null hypothesis should never be the end of a testing procedure. Often there are actions to take or policies to change. In our ingot example, management must decide whether to continue the changes proposed by the engineers. The decision always includes the practical consideration of whether the new method is worth the cost. Suppose management decides to reject the null hypothesis of 20% cracking in favor of the alternative that the percentage has been reduced. They must still evaluate how much the cracking rate has been reduced and how much it cost to accomplish the reduction. The *size of the effect* is always a concern when we test hypotheses. A good way to look at the **effect size** is to examine a confidence interval.

> **How much does it cost?** Formal tests of a null hypothesis base the decision of whether to reject the null hypothesis solely on the size of the P-value. But in real life, we want to evaluate the costs of our decisions as well. How much would you be willing to pay for a faster computer? Shouldn't your decision depend on how much faster? And on how much more it costs? Costs are not just monetary either. Would you use the same standard of proof for testing the safety of an airplane as for the speed of your new computer?

Alternative Alternatives

Tests on the ingot data can be viewed in two different ways. We know the old cracking rate is 20%, so the null hypothesis is

$$H_0: p = 0.20$$

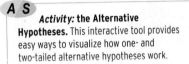

Activity: the Alternative Hypotheses. This interactive tool provides easy ways to visualize how one- and two-tailed alternative hypotheses work.

But we have a choice of alternative hypotheses. A metallurgist working for the company might be interested in *any* change in the cracking rate due to the new process. Even if the rate got worse, she might learn something useful from it. She's interested in possible changes on both sides of the null hypothesis. So she would write her alternative hypothesis as

$$H_A: p \neq 0.20$$

An alternative hypothesis such as this is known as a **two-sided alternative**,[3] because we are equally interested in deviations on either side of the null hypothesis value. For two-sided alternatives, the P-value is the probability of deviating in *either* direction from the null hypothesis value.

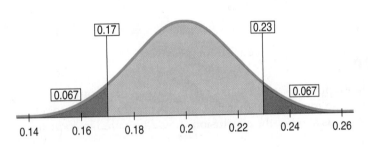

But management is really interested only in *lowering* the cracking rate below 20%. The scientific value of knowing how to *increase* the cracking rate may not appeal to them. The only alternative of interest to them is that the cracking rate *decreases*. They would write their alternative hypothesis as

$$H_A: p < 0.20$$

An alternative hypothesis that focuses on deviations from the null hypothesis value in only one direction is called a **one-sided alternative**.

For a hypothesis test with a one-sided alternative, the P-value is the probability of deviating *only in the direction of the alternative* away from the null hypothesis value. For the same data, the one-sided P-value is half the two-sided P-value. So, a one-sided test will reject the null hypothesis more often. If you aren't sure which to use, a two-sided test is always more conservative. Be sure you can justify the choice of a one-sided test from the *Why* of the situation.

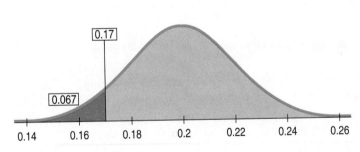

---

[3] It is also called a **two-tailed alternative,** because the probabilities we care about are found in both tails of the sampling distribution.

STEP-BY-STEP EXAMPLE | Testing a Hypothesis

Anyone who plays or watches sports has heard of the "home field advantage." Teams tend to win more often when they play at home. Or do they?

If there were no home field advantage, the home teams would win about half of all games played. In the 2009 Major League Baseball season, there were 2430 regular-season games. (Tied at the end of the regular season, the Colorado Rockies and San Diego Padres played an extra game to determine who won the Wild Card playoff spot.) It turns out that the home team won 1333 of the 2430 games, or 54.81% of the time.

Question: Could this deviation from 50% be explained just from natural sampling variability, or is it evidence to suggest that there really is a home field advantage, at least in professional baseball?

| THINK | **Plan** State what we want to know. | I want to know whether the home team in professional baseball is more likely to win. The data are all 2430-games from the 2009 Major League Baseball season. The variable is whether or not the home team won. The parameter of interest is the proportion of home team wins. If there's no advantage, I'd expect that proportion to be 0.50. |

Define the variables and discuss the W's.

$$H_0: p = 0.50$$
$$H_A: p > 0.50$$

Hypotheses The null hypothesis makes the claim of no difference from the baseline. Here, that means no home field advantage.

We are interested only in a home field *advantage*, so the alternative hypothesis is one-sided.

Model Think about the assumptions and check the appropriate conditions.

✔ **Independence Assumption:** Generally, the outcome of one game has no effect on the outcome of another game. But this may not be strictly true. For example, if a key player is injured, the probability that the team will win in the next couple of games may decrease slightly, but independence is still roughly true. The data come from one entire season, but I expect other seasons to be similar.

✔ **Randomization Condition:** I have results for all 2430 games of the 2009 season. But I'm not just interested in 2009, and those games, while not randomly selected, should be a reasonable representative sample of all Major League Baseball games in the recent past and near future.

✔ **10% Condition:** We are interested in home field advantage for Major League Baseball for all seasons. While not a random sample, these 2430 games are fewer than 10% of all games played over the years.

✔ **Success/Failure Condition:** Both $np_0 = 2430(0.50) = 1215.0$ and $nq_0 = 2431(0.50) = 1215.0$ are at least 10.

A S *Activity:* **Practice with Testing Hypotheses About Proportions.** Here's an interactive tool that makes it easy to see what's going on in a hypothesis test.

Specify the sampling distribution model. State what test you plan to use.	Because the conditions are satisfied, I'll use a Normal model for the sampling distribution of the proportion and do a **one-proportion z-test.**

 Mechanics The null model gives us the mean, and (because we are working with proportions) the mean gives us the standard deviation.

The null model is a Normal distribution with a mean of 0.50 and a standard deviation of

$$SD(\hat{p}) = \sqrt{\frac{p_0 q_0}{n}} = \sqrt{\frac{(0.5)(1 - 0.5)}{2430}}$$

$$= 0.01014$$

Next, we find the z-score for the observed proportion, to find out how many standard deviations it is from the hypothesized proportion.

The observed proportion, $\hat{p}$, is 0.5481.

So the z-value is

$$z = \frac{0.5481 - 0.5}{0.01014} = 4.74$$

The sample proportion lies 4.74 standard deviations above the mean.

From the z-score, we can find the P-value, which tells us the probability of observing a value that extreme (or more).

The probability of observing a value 4.20 or more standard deviations above the mean of a Normal model can be found by computer, calculator, or table to be <0.001.

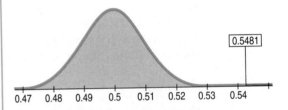

The corresponding P-value is < 0.001.

 Conclusion State your conclusion about the parameter—in context, of course!

The P-value of < 0.001 says that if the true proportion of home team wins were 0.50, then an observed value of 0.5481 (or larger) would occur less than 1 time in 1000. With a P-value so small, I reject H_0. I have evidence that the true proportion of home team wins is greater than 50%. It appears there is a home field advantage.

Ok, but how *big* is the home field advantage? Measuring the size of the effect involves a confidence interval. (Use your calculator.)

P-Values and Decisions: What to Tell About a Hypothesis Test

Hypothesis tests are particularly useful when we must make a decision. Is the defendant guilty or not? Should we choose print advertising or television? Questions like these cannot always be answered with the margins of error of confidence intervals. The absolute nature of the hypothesis test decision, however, makes some people (including the authors) uneasy. If possible, it's often a good idea to report a confidence interval for the parameter of interest as well.

How small should the P-value be in order for you to reject the null hypothesis? A jury needs enough evidence to show the defendant guilty "beyond a reasonable doubt." How does that translate to P-values? The answer is that it's highly context-dependent. When we're screening for a disease and want to be sure we treat all those who are sick, we may be willing to reject the null hypothesis of no disease with a P-value as large as 0.10. We would rather treat the occasional healthy person than fail to treat someone who was really sick. But a long-standing hypothesis, believed by many to be true, needs stronger evidence (and a correspondingly small P-value) to reject it.

See if you require the same P-value to reject each of the following null hypotheses:

- A renowned musicologist claims that she can distinguish between the works of Mozart and Haydn simply by hearing a randomly selected 20 seconds of music from any work by either composer. What's the null hypothesis? If she's just guessing, she'll get 50% of the pieces correct, on average. So our null hypothesis is that *p* equals 50%. If she's for real, she'll get more than 50% correct. Now, we present her with 10 pieces of Mozart or Haydn chosen at random. She gets 9 out of 10 correct. It turns out that the P-value associated with that result is 0.011. (In other words, if you tried to just guess, you'd get at least 9 out of 10 correct only about 1% of the time.) What would *you* conclude? Most people would probably reject the null hypothesis and be convinced that she has some ability to do as she claims. Why? Because the P-value is small and we don't have any particular reason to doubt the alternative.

- On the other hand, imagine a student who bets that he can make a flipped coin land the way he wants just by thinking hard. To test him, we flip a fair coin 10 times. Suppose he gets 9 out of 10 right. This also has a P-value of 0.011. Are you willing now to reject this null hypothesis? Are you convinced that he's not just lucky? What amount of evidence *would* convince you? We require more evidence if rejecting the null hypothesis would contradict long-standing beliefs or other scientific results. Of course, with sufficient evidence we would revise our opinions (and scientific theories). That's how science makes progress.

Another factor in choosing a P-value is the importance of the issue being tested. Consider the following two tests:

- A researcher claims that the proportion of college students who hold part-time jobs now is higher than the proportion known to hold such jobs a decade ago. You might be willing to believe the claim (and reject the null hypothesis of no change) with a P-value of 0.10.
- An engineer claims that the proportion of rivets holding the wing on an airplane that are likely to fail is below the proportion at which the wing would fall off. What P-value would be small enough to get you to fly on that plane?

Your conclusion about any null hypothesis should be accompanied by the P-value of the test. Don't just declare the null hypothesis rejected or not rejected. Report the P-value to show the strength of the evidence against the hypothesis and the effect size. This will let each reader decide whether or not to reject the null hypothesis and whether or not to consider the result important if it is statistically significant.

To complete your analysis, follow your test with a confidence interval for the parameter of interest, to report the size of the effect.

"Extraordinary claims require extraordinary proof."

–Carl Sagan

A S
Activity: Hypothesis Tests for Proportions. You've probably noticed that the tools for confidence intervals and for hypothesis tests are similar. See how tests and intervals for proportions are related—and an important way in which they differ.

4. A bank is testing a new method for getting delinquent customers to pay their past-due credit card bills. The standard way was to send a letter (costing about $0.40) asking the customer to pay. That worked 30% of the time. They want to test a new method that involves sending a DVD to customers encouraging them to contact the bank and set up a payment plan. Developing and sending the video costs about $10.00 per customer. What is the parameter of interest? What are the null and alternative hypotheses?

5. The bank sets up an experiment to test the effectiveness of the DVD. They mail it out to several randomly selected delinquent customers and keep track of how many actually do contact the bank to arrange payments. The bank's statistician calculates a P-value of 0.003. What does this P-value suggest about the DVD?

6. The statistician tells the bank's management that the results are clear and that they should switch to the DVD method. Do you agree? What else might you want to know?

STEP-BY-STEP EXAMPLE   Tests and Intervals

Advances in medical care such as prenatal ultrasound examination now make it possible to determine a child's sex early in a pregnancy. There is a fear that in some cultures some parents may use this technology to select the sex of their children. A study from Punjab, India (E. E. Booth, M. Verma, and R. S. Beri, "Fetal Sex Determination in Infants in Punjab, India: Correlations and Implications," *BMJ* 309 [12 November 1994]: 1259–1261), reports that, in 1993, in one hospital, 56.9% of the 550 live births that year were boys. It's a medical fact that male babies are slightly more common than female babies. The study's authors report a baseline for this region of 51.7% male live births.

Question: Is there evidence that the proportion of male births has changed?

 THINK

Plan State what we want to know.

Define the variables and discuss the W's.

Hypotheses The null hypothesis makes the claim of no difference from the baseline.

Before seeing the data, we were interested in any change in male births, so the alternative hypothesis is two-sided.

Model Think about the assumptions and check the appropriate conditions.

For testing proportions, the conditions are the same ones we had for making confidence intervals, except that we check the **Success/Failure Condition** with the *hypothesized* proportions rather than with the *observed* proportions.

I want to know whether the proportion of male births has changed from the established baseline of 51.7%. The data are the recorded sexes of the 550 live births from a hospital in Punjab, India, in 1993, collected for a study on fetal sex determination. The parameter of interest, p, is the proportion of male births:

$$H_O: p = 0.517$$
$$H_A: p \neq 0.517$$

✔ **Independence Assumption:** There is no reason to think that the sex of one baby can affect the sex of other babies, so births can reasonably be assumed to be independent with regard to the sex of the child.

✔ **Randomization Condition:** The 550 live births are not a random sample, so I must be cautious about any general conclusions. I hope that this is a representative year, and I think that the births at this hospital may be typical of this area of India.

✔ **10% Condition:** I would like to be able to make statements about births at similar hospitals in India. These 550 births are fewer than 10% of all of those births.

✔ **Success/Failure Condition:** Both $np_0 = 550(0.517) = 284.35$ and $nq_0 = 550(0.483) = 265.65$ are greater than 10; I expect the births of at least 10 boys and at least 10 girls, so the sample is large enough.

Specify the sampling distribution model.

Tell what test you plan to use.

The conditions are satisfied, so I can use a Normal model and perform a **one-proportion z-test.**

 Mechanics The null model gives us the mean, and (because we are working with proportions) the mean gives us the standard deviation.

We find the z-score for the observed proportion to find out how many standard deviations it is from the hypothesized proportion.

Make a picture. Sketch a Normal model centered at $p_0 = 0.517$. Shade the region to the right of the observed proportion, and because this is a two-tail test, also shade the corresponding region in the other tail.

From the z-score, we can find the P-value, which tells us the probability of observing a value that extreme (or more). Use technology or a table (see page 490).

Because this is a two-tail test, the P-value is the probability of observing an outcome more than 2.44 standard deviations from the mean of a Normal model *in either direction.* We must therefore *double* the probability we find in the upper tail.

The null model is a Normal distribution with a mean of 0.517 and a standard deviation of

$$SD(\hat{p}) = \sqrt{\frac{p_0q_0}{n}} = \sqrt{\frac{(0.517)(1 - 0.517)}{550}}$$
$$= 0.0213$$

The observed proportion, $\hat{p}$, is 0.569, so

$$z = \frac{\hat{p} - p_0}{SD(\hat{p})} = \frac{0.569 - 0.517}{0.0213} = 2.44$$

The sample proportion lies 2.44 standard deviations above the mean.

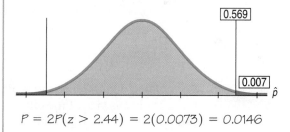

$$P = 2P(z > 2.44) = 2(0.0073) = 0.0146$$

 Conclusion State your conclusion in context.

This P-value is roughly 1 time in 70. That's clearly significant, but don't jump to other conclusions. We can't be sure how this deviation came about. For instance, we don't know whether this hospital is typical, or whether the time period studied was selected at random.

The P-value of 0.0146 says that if the true proportion of male babies were still at 51.7%, then an observed proportion as different as 56.9% male babies would occur at random only about 15 times in 1000. With a P-value this small, I reject H_0. This is strong evidence that the birth ratio of boys to girls is not equal to its natural level. It appears that the proportion of boys may have increased.

How big an increase are we talking about? Let's find a confidence interval for the proportion of male births.

Model Check the conditions.

The conditions are identical to those for the hypothesis test, with one difference. Now we are not given a hypothesized proportion, p_0, so we must instead work with the observed proportion $\hat{p}$.

Specify the sampling distribution model.

Tell what method you plan to use.

✓ **Success/Failure Condition:** Both $n\hat{p} = 550(0.569) = 313$ and $n\hat{q} = 237$ are at least 10.

The conditions are satisfied, so I can model the sampling distribution of the proportion with a Normal model and find a **one-proportion z-interval.**

Mechanics We can't find the sampling model standard deviation from the null model proportion. (In fact, we've just rejected it.) Instead, we find the standard error of $\hat{p}$ from the *observed* proportions. Other than that substitution, the calculation looks the same as for the hypothesis test.

With this large a sample size, the difference is negligible, but in smaller samples, it could make a bigger difference.

$$SE(\hat{p}) = \sqrt{\frac{\hat{p}\hat{q}}{n}} = \sqrt{\frac{(0.569)(1 - 0.569)}{550}}$$
$$= 0.0211$$

The sampling model is Normal, so for a 95% confidence interval, the critical value $z^* = 1.96$.

The margin of error is

$$ME = z^* \times SE(\hat{p}) = 1.96(0.0211) = 0.041$$

So the 95% confidence interval is

$$0.569 \pm 0.041 \text{ or } (0.528, 0.610).$$

Conclusion Confidence intervals help us think about the size of the effect. Here we can see that the change from the baseline of 51.7% male births might be quite substantial.

We are 95% confident that the true proportion of male births is between 52.8% and 61.0%.

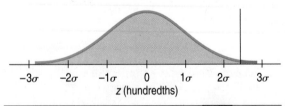

Here's a portion of a Normal table that gives the probability we needed for the hypothesis test. At $z = 2.44$, the table gives the percentile as 0.9927. The upper-tail probability (shaded red) is, therefore, $1 - 0.9927 = 0.0073$; so, for our two-sided test, the P-value is $2(0.0073) = 0.0146$.

z	0.00	0.01	0.02	0.03	0.04	0.05
1.9	0.9713	0.9719	0.9726	0.9732	0.9738	0.9744
2.0	0.9772	0.9778	0.9783	0.9788	0.9793	0.9798
2.1	0.9821	0.9826	0.9830	0.9834	0.9838	0.9842
2.2	0.9861	0.9864	0.9868	0.9871	0.9875	0.9878
2.3	0.9893	0.9896	0.9898	0.9901	0.9904	0.9906
2.4	0.9918	0.9920	0.9922	0.9925	0.9927	0.9929
2.5	0.9938	0.9940	0.9941	0.9943	0.9945	0.9946
2.6	0.9953	0.9955	0.9956	0.9957	0.9959	0.9960

What Can Go Wrong?

Hypothesis tests are so widely used—and so widely misused—that we've devoted all of the next chapter to discussing the pitfalls involved, but there are a few issues that we can talk about already.

- **Don't base your null hypothesis on what you see in the data.** You are not allowed to look at the data first and then adjust your null hypothesis so that it will be rejected. When your sample value turns out to be $\hat{p} = 51.8\%$, with a standard deviation of 1%, don't form a null hypothesis like $H_0: p = 49.8\%$, knowing that you can reject it. You should always *Think* about the situation you are investigating and make your null hypothesis describe the "nothing interesting" or "nothing has changed" scenario. No peeking at the data!

- **Don't base your alternative hypothesis on the data, either.** Again, you need to *Think* about the situation. Are you interested only in knowing whether something has *increased*? Then write a one-sided (upper-tail) alternative. Or would you be equally interested in a change in either direction? Then you want a two-sided alternative. You should decide whether to do a one- or two-sided test based on what results would be of interest to you, not what you see in the data.

- **Don't make your null hypothesis what you want to show to be true.** Remember, the null hypothesis is the status quo, the nothing-is-strange-here position a skeptic would take. You wonder whether the data cast doubt on that. You can reject the null hypothesis, but you can never "accept" or "prove" the null.

- **Don't forget to check the conditions.** The reasoning of inference depends on randomization. No amount of care in calculating a test result can recover from biased sampling. The probabilities we compute depend on the independence assumption. And the sample must be large enough to justify the use of a Normal model.

- **Don't accept the null hypothesis.** You may not have found enough evidence to reject it, but you surely have *not* proven it's true!

- **If you fail to reject the null hypothesis, don't think that a bigger sample would be more likely to lead to rejection.** If the results you looked at were "almost" significant, it's enticing to think that because you would have rejected the null had these same observations come from a larger sample, then a larger sample would surely lead to rejection. Don't be misled. Remember, each sample is different, and a larger sample won't necessarily duplicate your current observations. Indeed, the Central Limit Theorem tells us that statistics will vary *less* in larger samples. We should therefore expect such results to be less extreme. Maybe they'd be statistically significant but maybe (perhaps even probably) not. Even if you fail to reject the null hypothesis, it's a good idea to examine a confidence interval. If none of the plausible parameter values in the interval would matter to you (for example, because none would be *practically* significant), then even a larger study with a correspondingly smaller standard error is unlikely to be worthwhile.

CONNECTIONS

Hypothesis tests and confidence intervals share many of the same concepts. Both rely on sampling distribution models, and because the models are the same and require the same assumptions, both check the same conditions. They also calculate many of the same statistics. Like confidence intervals, hypothesis tests use the standard deviation of the sampling distribution as a ruler, as we first saw in Chapter 6.

For testing, we find ourselves looking once again at z-scores, and we compute the P-value by finding the distance of our test statistic from the center of the null model. P-values are conditional probabilities. They give the probability of observing the result we have seen (or one even more extreme) *given* that the null hypothesis is true.

The Standard Normal model is here again as our connection between z-score values and probabilities.

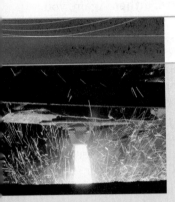

WHAT HAVE WE LEARNED?

We've learned to use what we see in a random sample to test a particular hypothesis about the world. This is our second step in statistical inference, complementing our use of confidence intervals.

We've learned that testing a hypothesis involves proposing a model, then seeing whether the data we observe are consistent with that model or are so unusual that we must reject it. We do this by finding a P-value—the probability that data like ours could have occurred if the model is correct.

We've learned that:

▶ We start with a null hypothesis specifying the parameter of a model we'll test using our data.

▶ Our alternative hypothesis can be one- or two-sided, depending on what we want to learn.

▶ We must check the appropriate assumptions and conditions before proceeding with our test.

▶ If the data are out of line with the null hypothesis model, the P-value will be small and we will reject the null hypothesis.

▶ If the data are consistent with the null hypothesis model, the P-value will be large and we will not reject the null hypothesis.

▶ We must always state our conclusion in the context of the original question.

And we've learned that confidence intervals and hypothesis tests go hand in hand in helping us think about models. A hypothesis test makes a yes/no decision about the plausibility of a parameter value. The confidence interval shows us the range of plausible values for the parameter.

Terms

Null hypothesis The claim being assessed in a hypothesis test is called the null hypothesis. Usually, the null hypothesis is a statement of "no change from the traditional value," "no effect," "no difference," or "no relationship." For a claim to be a testable null hypothesis, it must specify a value for some population parameter that can form the basis for assuming a sampling distribution for a test statistic (p. 478).

Alternative hypothesis The alternative hypothesis proposes what we should conclude if we find the null hypothesis to be unlikely (p. 478).

P-value

The probability of observing a value for a test statistic at least as far from the hypothesized value as the statistic value actually observed if the null hypothesis is true. A small P-value indicates either that the observation is improbable or that the probability calculation was based on incorrect assumptions. The assumed truth of the null hypothesis is the assumption under suspicion (p. 480).

One-proportion z-test

A test of the null hypothesis that the proportion of a single sample equals a specified value ($H_0: p = p_0$) by referring the statistic $z = \dfrac{\hat{p} - p_0}{SD(\hat{p})}$ to a Standard Normal model (p. 482).

Effect size

The difference between the null hypothesis value and the true value of a model parameter (p. 483).

Two-sided alternative
(Two-tailed alternative)

An alternative hypothesis is two-sided ($H_A: p \neq p_0$) when we are interested in deviations in *either* direction away from the hypothesized parameter value (p. 484).

One-sided alternative
(One-tailed alternative)

An alternative hypothesis is one-sided (e.g., $H_A: p > p_0$ or $H_A: p < p_0$) when we are interested in deviations in *only one* direction away from the hypothesized parameter value (p. 484).

Skills

▶ Be able to state the null and alternative hypotheses for a one-proportion z-test.

▶ Know the conditions that must be true for a one-proportion z-test to be appropriate, and know how to examine your data for violations of those conditions.

▶ Be able to identify and use the alternative hypothesis when testing hypotheses. Understand how to choose between a one-sided and two-sided alternative hypothesis, and be able to explain your choice.

▶ Be able to perform a one-proportion z-test.

▶ Be able to write a sentence interpreting the results of a one-proportion z-test.

▶ Know how to interpret the meaning of a P-value in nontechnical language, making clear that the probability claim is made about computed values under the assumption that the null model is true and not about the population parameter of interest.

HYPOTHESIS TESTS FOR PROPORTIONS ON THE COMPUTER

Hypothesis tests for proportions are so easy and natural that many statistics packages don't offer special commands for them. Most statistics programs want to know the "success" and "failure" status for each case. Usually these are given as 1 or 0, but they might be category names like "yes" and "no." Often you just know the proportion of successes, $\hat{p}$, and the total count, n. Computer packages don't usually deal naturally with summary data like this, but the statistics routines found on many graphing calculators do. These calculators allow you to test hypotheses from summaries of the data—usually, all you need to enter are the number of successes and the sample size.

In some programs you can reconstruct the original values. But even when you have reconstructed (or can reconstruct) the raw data values, often you won't get exactly the same test statistic from a computer package as you would find working by hand. The reason is that when the packages treat the proportion as a mean, they make some approximations. The result is very close, but not exactly the same.

DATA DESK

Data Desk does not offer built-in methods for inference with proportions. The **Replicate Y by X** command in the **Manip** menu will "reconstruct" summarized count data so that you can display it.

COMMENTS

For summarized data, open a Scratchpad to compute the standard deviation and margin of error by typing the calculation. Then perform the test with the **z-test for individual μs** found in the Test command.

EXCEL

Inference methods for proportions are not part of the standard Excel tool set.

COMMENTS

For summarized data, type the calculation into any cell and evaluate it.

JMP

For a **categorical** variable that holds category labels, the **Distribution** platform includes tests and intervals of proportions. For summarized data, put the category names in one variable and the frequencies in an adjacent variable. Designate the frequency column to have the **role** of **frequency.** Then use the **Distribution** platform.

COMMENTS

JMP uses slightly different methods for proportion inferences than those discussed in this text. Your answers are likely to be slightly different.

MINITAB

Choose **Basic Statistics** from the **Stat** menu.

- Choose **1Proportion** from the Basic Statistics submenu.
- If the data are category names in a variable, assign the variable from the variable list box to the **Samples in columns** box.
- If you have summarized data, click the **Summarized Data** button and fill in the number of trials and the number of successes.
- Click the **Options** button and specify the remaining details.
- If you have a large sample, check **Use test and interval based on Normal distribution.**
- Click the **OK** button.

COMMENTS

When working from a variable that names categories, Minitab treats the last category as the "success" category. You can specify how the categories should be ordered.

SPSS

SPSS does not offer hypothesis tests for proportions.

TI-83/84 PLUS

To do the mechanics of a hypothesis test for a proportion,

- Select **5:1-PropZTest** from the **STAT TESTS** menu.
- Specify the hypothesized proportion.
- Enter the observed value of x.
- Specify the sample size.
- Indicate what kind of test you want: one-tail lower tail, two-tail, or one-tail upper tail.
- Calculate the result.

COMMENTS

Beware: When you enter the value of *x*, you need the *count*, not the percentage. The count must be a whole number. If the number of successes is given as a percent, you must first multiply *np* and round the result to obtain *x.*

TI-89

To do the mechanics of a hypothesis test for a proportion,

- Select **5:1-PropZTest** from the **STAT TESTS** [2nd][F1] menu.
- Specify the hypothesized proportion.
- Enter the observed value of x.
- Specify the sample size.
- Indicate what kind of test you want: one-tail lower tail, two-tail, or one-tail upper tail.
- Specify whether to calculate the result or draw the result (a normal curve with *p*-value area shaded).

COMMENTS

Beware: When you enter the value of *x*, you need the *count*, not the percentage. The count must be a whole number. If the number of successes is given as a percent, you must first multiply *np* and round the result to obtain *x.*

EXERCISES

1. **Hypotheses.** Write the null and alternative hypotheses you would use to test each of the following situations:

    a) A governor is concerned about his "negatives"—the percentage of state residents who express disapproval of his job performance. His political committee pays for a series of TV ads, hoping that they can keep the negatives below 30%. They will use follow-up polling to assess the ads' effectiveness.

    b) Is a coin fair?

    c) Only about 20% of people who try to quit smoking succeed. Sellers of a motivational tape claim that listening to the recorded messages can help people quit.

2. **More hypotheses.** Write the null and alternative hypotheses you would use to test each situation.

    a) In the 1950s only about 40% of high school graduates went on to college. Has the percentage changed?

    b) 20% of cars of a certain model have needed costly transmission work after being driven between 50,000 and 100,000 miles. The manufacturer hopes that a redesign of a transmission component has solved this problem.

    c) We field-test a new-flavor soft drink, planning to market it only if we are sure that over 60% of the people like the flavor.

3. **Negatives.** After the political ad campaign described in Exercise 1a, pollsters check the governor's negatives. They test the hypothesis that the ads produced no change against the alternative that the negatives are now below 30% and find a P-value of 0.22. Which conclusion is appropriate? Explain.

    a) There's a 22% chance that the ads worked.

    b) There's a 78% chance that the ads worked.

    c) There's a 22% chance that their poll is correct.

    d) There's a 22% chance that natural sampling variation could produce poll results like these if there's really no change in public opinion.

4. **Dice.** The seller of a loaded die claims that it will favor the outcome 6. We don't believe that claim, and roll the die 200 times to test an appropriate hypothesis. Our P-value turns out to be 0.03. Which conclusion is appropriate? Explain.

    a) There's a 3% chance that the die is fair.

    b) There's a 97% chance that the die is fair.

    c) There's a 3% chance that a loaded die could randomly produce the results we observed, so it's reasonable to conclude that the die is fair.

    d) There's a 3% chance that a fair die could randomly produce the results we observed, so it's reasonable to conclude that the die is loaded.

5. **Relief.** A company's old antacid formula provided relief for 70% of the people who used it. The company tests a new formula to see if it is better and gets a P-value of 0.27. Is it reasonable to conclude that the new formula and the old one are equally effective? Explain.

6. **Cars.** A survey investigating whether the proportion of today's high school seniors who own their own cars is higher than it was a decade ago finds a P-value of 0.017. Is it reasonable to conclude that more high schoolers have cars? Explain.

7. **He cheats!** A friend of yours claims that when he tosses a coin he can control the outcome. You are skeptical and want him to prove it. He tosses the coin, and you call heads; it's tails. You try again and lose again.

    a) Do two losses in a row convince you that he really can control the toss? Explain.

    b) You try a third time, and again you lose. What's the probability of losing three tosses in a row if the process is fair?

    c) Would three losses in a row convince you that your friend cheats? Explain.

    d) How many times in a row would you have to lose in order to be pretty sure that this friend really can control the toss? Justify your answer by calculating a probability and explaining what it means.

8. **Candy.** Someone hands you a box of a dozen chocolate-covered candies, telling you that half are vanilla creams and the other half peanut butter. You pick candies at random and discover the first three you eat are all vanilla.

    a) If there really were 6 vanilla and 6 peanut butter candies in the box, what is the probability that you would have picked three vanillas in a row?

    b) Do you think there really might have been 6 of each? Explain.

    c) Would you continue to believe that half are vanilla if the fourth one you try is also vanilla? Explain.

9. **Cell phones.** Many people have trouble setting up all the features of their cell phones, so a company has developed what it hopes will be easier instructions. The goal is to have at least 96% of customers succeed. The company tests the new system on 200 people, of whom 188 were successful. Is this strong evidence that the new system fails to meet the company's goal? A student's test of this hypothesis is shown. How many mistakes can you find?

    $H_0: \hat{p} = 0.96$

    $H_A: \hat{p} \neq 0.96$

    SRS, $0.96(200) > 10$

    $\dfrac{188}{200} = 0.94; \quad SD(\hat{p}) = \sqrt{\dfrac{(0.94)(0.06)}{200}} = 0.017$

    $z = \dfrac{0.96 - 0.94}{0.017} = 1.18$

    $P = P(z > 1.18) = 0.12$

    There is strong evidence the new instructions don't work.

10. **Got milk?** In November 2001, the *Ag Globe Trotter* newsletter reported that 90% of adults drink milk. A regional farmers' organization planning a new marketing campaign across its multicounty area polls a random sample of 750 adults living there. In this sample, 657 people said that they drink milk. Do these responses provide strong evidence that the 90% figure is not accurate for this region? Correct the mistakes you find in a student's attempt to test an appropriate hypothesis.

$H_0: \hat{p} = 0.9$

$H_A: \hat{p} < 0.9$

SRS, $750 \geq 10$

$\dfrac{657}{750} = 0.876; \quad SD(\hat{p}) = \sqrt{\dfrac{(0.88)(0.12)}{750}} = 0.012$

$z = \dfrac{0.876 - 0.90}{0.012} = -2$

$P = P(z > -2) = 0.977$

There is more than a 97% chance that the stated percentage is correct for this region.

11. **Dowsing.** In a rural area, only about 30% of the wells that are drilled find adequate water at a depth of 100 feet or less. A local man claims to be able to find water by "dowsing"—using a forked stick to indicate where the well should be drilled. You check with 80 of his customers and find that 27 have wells less than 100 feet deep. What do you conclude about his claim?
   a) Write appropriate hypotheses.
   b) Check the necessary assumptions and conditions.
   c) Perform the mechanics of the test. What is the P-value?
   d) Explain carefully what the P-value means in context.
   e) What's your conclusion?

12. **Abnormalities.** In the 1980s it was generally believed that congenital abnormalities affected about 5% of the nation's children. Some people believe that the increase in the number of chemicals in the environment has led to an increase in the incidence of abnormalities. A recent study examined 384 children and found that 46 of them showed signs of an abnormality. Is this strong evidence that the risk has increased?
   a) Write appropriate hypotheses.
   b) Check the necessary assumptions and conditions.
   c) Perform the mechanics of the test. What is the P-value?
   d) Explain carefully what the P-value means in context.
   e) What's your conclusion?
   f) Do environmental chemicals cause congenital abnormalities?

13. **Absentees.** The National Center for Education Statistics monitors many aspects of elementary and secondary education nationwide. Their 1996 numbers are often used as a baseline to assess changes. In 1996, 34% of students had not been absent from school even once during the previous month. In the 2000 survey, responses from 8302 students showed that this figure had slipped to 33%. Officials would, of course, be concerned if student attendance were declining. Do these figures give evidence of a change in student attendance?

   a) Write appropriate hypotheses.
   b) Check the assumptions and conditions.
   c) Perform the test and find the P-value.
   d) State your conclusion.
   e) Do you think this difference is meaningful? Explain.

14. **Educated mothers.** The National Center for Education Statistics monitors many aspects of elementary and secondary education nationwide. Their 1996 numbers are often used as a baseline to assess changes. In 1996, 31% of students reported that their mothers had graduated from college. In 2000, responses from 8368 students found that this figure had grown to 32%. Is this evidence of a change in education level among mothers?
   a) Write appropriate hypotheses.
   b) Check the assumptions and conditions.
   c) Perform the test and find the P-value.
   d) State your conclusion.
   e) Do you think this difference is meaningful? Explain.

15. **Contributions, please, part II.** In Exercise 15 of Chapter 19 you learned that the Paralyzed Veterans of America is a philanthropic organization that relies on contributions. They send free mailing labels and greeting cards to potential donors on their list and ask for a voluntary contribution. To test a new campaign, the organization recently sent letters to a random sample of 100,000 potential donors and received 4781 donations. They've had a contribution rate of 5% in past campaigns, but a staff member worries that the rate will be lower if they run this campaign as currently designed.
   a) What are the hypotheses?
   b) Are the assumptions and conditions for inference met?
   c) Do you think the rate would drop? Explain.

16. **Take the offer, part II.** In Exercise 16 in Chapter 19 you learned that First USA, a major credit card company, is planning a new offer for their current cardholders. First USA will give double airline miles on purchases for the next 6 months if the cardholder goes online and registers for this offer. To test the effectiveness of this campaign, the company recently sent out offers to a random sample of 50,000 cardholders. Of those, 1184 registered. A staff member suspects that the success rate for the full campaign will be comparable to the standard 2% rate that they are used to seeing in similar campaigns. What do you predict?
   a) What are the hypotheses?
   b) Are the assumptions and conditions for inference met?
   c) Do you think the rate would change if they use this fundraising campaign? Explain.

17. **Law School.** According to the Law School Admission Council, in the fall of 2006, 63% of law school applicants were accepted to some law school.[4] The training program *LSATisfaction* claims that 163 of the 240 students trained in 2006 were admitted to law school. You can safely consider these trainees to be representative of the population of law school applicants. Has *LSATisfaction* demonstrated a real improvement over the national average?

---

[4] As reported by the Cornell office of career services in their *Class of 2006 Postgraduate Report.*

a) What are the hypotheses?

b) Check the conditions and find the P-value.

c) Would you recommend this program based on what you see here? Explain.

18. **Med School.** According to the Association of American Medical Colleges, only 46% of medical school applicants were admitted to a medical school in the fall of 2006.[5] Upon hearing this, the trustees of Striving College expressed concern that only 77 of the 180 students in their class of 2006 who applied to medical school were admitted. The college president assured the trustees that this was just the kind of year-to-year fluctuation in fortunes that is to be expected and that, in fact, the school's success rate was consistent with the national average. Who is right?

a) What are the hypotheses?

b) Check the conditions and find the P-value.

c) Are the trustees right to be concerned, or is the president correct? Explain.

19. **Pollution.** A company with a fleet of 150 cars found that the emissions systems of 7 out of the 22 they tested failed to meet pollution control guidelines. Is this strong evidence that more than 20% of the fleet might be out of compliance? Test an appropriate hypothesis and state your conclusion. Be sure the appropriate assumptions and conditions are satisfied before you proceed.

20. **Scratch and dent.** An appliance manufacturer stockpiles washers and dryers in a large warehouse for shipment to retail stores. Sometimes in handling them the appliances get damaged. Even though the damage may be minor, the company must sell those machines at drastically reduced prices. The company goal is to keep the level of damaged machines below 2%. One day an inspector randomly checks 60 washers and finds that 5 of them have scratches or dents. Is this strong evidence that the warehouse is failing to meet the company goal? Test an appropriate hypothesis and state your conclusion. Be sure the appropriate assumptions and conditions are satisfied before you proceed.

21. **Twins.** In 2001 a national vital statistics report indicated that about 3% of all births produced twins. Is the rate of twin births the same among very young mothers? Data from a large city hospital found that only 7 sets of twins were born to 469 teenage girls. Test an appropriate hypothesis and state your conclusion. Be sure the appropriate assumptions and conditions are satisfied before you proceed.

22. **Football 2006.** During the 2006 season, the home team won 136 of the 240 regular-season National Football League games. Is this strong evidence of a home field advantage in professional football? Test an appropriate hypothesis and state your conclusion. Be sure the appropriate assumptions and conditions are satisfied before you proceed.

23. **WebZine.** A magazine is considering the launch of an online edition. The magazine plans to go ahead only if it's convinced that more than 25% of current readers would subscribe. The magazine contacted a simple random sample of 500 current subscribers, and 137 of those surveyed expressed interest. What should the company do? Test an appropriate hypothesis and state your conclusion. Be sure the appropriate assumptions and conditions are satisfied before you proceed.

24. **Seeds.** A garden center wants to store leftover packets of vegetable seeds for sale the following spring, but the center is concerned that the seeds may not germinate at the same rate a year later. The manager finds a packet of last year's green bean seeds and plants them as a test. Although the packet claims a germination rate of 92%, only 171 of 200 test seeds sprout. Is this evidence that the seeds have lost viability during a year in storage? Test an appropriate hypothesis and state your conclusion. Be sure the appropriate assumptions and conditions are satisfied before you proceed.

25. **Women executives.** A company is criticized because only 13 of 43 people in executive-level positions are women. The company explains that although this proportion is lower than it might wish, it's not surprising given that only 40% of all its employees are women. What do you think? Test an appropriate hypothesis and state your conclusion. Be sure the appropriate assumptions and conditions are satisfied before you proceed.

26. **Jury.** Census data for a certain county show that 19% of the adult residents are Hispanic. Suppose 72 people are called for jury duty and only 9 of them are Hispanic. Does this apparent underrepresentation of Hispanics call into question the fairness of the jury selection system? Explain.

27. **Dropouts.** Some people are concerned that new tougher standards and high-stakes tests adopted in many states have driven up the high school dropout rate. The National Center for Education Statistics reported that the high school dropout rate for the year 2004 was 10.3%. One school district whose dropout rate has always been very close to the national average reports that 210 of their 1782 high school students dropped out last year. Is this evidence that their dropout rate may be increasing? Explain.

28. **Acid rain.** A study of the effects of acid rain on trees in the Hopkins Forest shows that 25 of 100 trees sampled exhibited some sort of damage from acid rain. This rate seemed to be higher than the 15% quoted in a recent *Environmetrics* article on the average proportion of damaged trees in the Northeast. Does the sample suggest that trees in the Hopkins Forest are more susceptible than trees from the rest of the region? Comment, and write up your own conclusions based on an appropriate confidence interval as well as a hypothesis test. Include any assumptions you made about the data.

29. **Lost luggage.** An airline's public relations department says that the airline rarely loses passengers' luggage. It further claims that on those occasions when luggage is lost, 90% is recovered and delivered to its owner within 24 hours. A consumer group that surveyed a large number of air travelers found that only 103 of 122 people who lost luggage on that airline were reunited with the missing items by the next day. Does this cast doubt on the airline's claim? Explain.

[5] *Ibid.*

30. **TV ads.** A start-up company is about to market a new computer printer. It decides to gamble by running commercials during the Super Bowl. The company hopes that name recognition will be worth the high cost of the ads. The goal of the company is that over 40% of the public recognize its brand name and associate it with computer equipment. The day after the game, a pollster contacts 420 randomly chosen adults and finds that 181 of them know that this company manufactures printers. Would you recommend that the company continue to advertise during Super Bowls? Explain.

31. **John Wayne.** Like a lot of other Americans, John Wayne died of cancer. But is there more to this story? In 1955 Wayne was in Utah shooting the film *The Conqueror*. Across the state line, in Nevada, the United States military was testing atomic bombs. Radioactive fallout from those tests drifted across the filming location. A total of 46 of the 220 people working on the film eventually died of cancer. Cancer experts estimate that one would expect only about 30 cancer deaths in a group this size.
    a) Is the death rate among the movie crew unusually high?
    b) Does this prove that exposure to radiation increases the risk of cancer?

32. **AP Stats.** The College Board reported that 60% of all students who took the 2006 AP Statistics exam earned scores of 3 or higher. One teacher wondered if the performance of her school was different. She believed that year's students to be typical of those who will take AP Stats at that school and was pleased when 65% of her 54 students achieved scores of 3 or better. Can she claim that her school is different? Explain.

JUST CHECKING

ANSWERS

1. You can't conclude that the null hypothesis is true. You can conclude only that the experiment was unable to reject the null hypothesis. They were unable, on the basis of 12 patients, to show that aspirin was effective.

2. The null hypothesis is $H_0: p = 0.75$.

3. With a P-value of 0.0001, this is very strong evidence against the null hypothesis. We can reject H_0 and conclude that the improved version of the drug gives relief to a higher proportion of patients.

4. The parameter of interest is the proportion, p, of all delinquent customers who will pay their bills. $H_0: p = 0.30$ and $H_A: p > 0.30$.

5. The very low P-value leads us to reject the null hypothesis. There is strong evidence that the DVD is more effective in getting people to start paying their debts than just sending a letter had been.

6. All we know is that there is strong evidence to suggest that $p > 0.30$. We don't know how much higher than 30% the new proportion is. We'd like to see a confidence interval to see if the new method is worth the cost.

More About Tests and Intervals

Where are we going?

A news headline reports a "statistically significant" increase in global temperatures. Another says that studies have found no "statistically significant" benefits of taking vitamin C to prevent colds. What does significance really mean? Can failing to reject the null hypothesis sometimes be as important as rejecting it? Knowing what the hypotheses were and how the researchers arrived at their conclusion can help you decide whether you agree with the findings of their studies. We'll look at hypothesis testing in more depth in this chapter.

WHO	Florida motorcycle riders aged 20 and younger involved in motorcycle accidents
WHAT	% wearing helmets
WHEN	2001–2003
WHERE	Florida
WHY	Assessment of injury rates commissioned by the National Highway Traffic Safety Administration (NHTSA)

In 2000 Florida changed its motorcycle helmet law. No longer are riders 21 and older required to wear helmets. Under the new law, those under 21 still must wear helmets, but a report by the Preusser Group (www.preussergroup.com) suggests that helmet use may have declined in this group, too.

It isn't practical to survey young motorcycle riders. (For example, how can you construct a sampling frame? If you contacted licensed riders, would they admit to riding illegally without a helmet?) The researchers adopted a different strategy. Police reports of motorcycle accidents record whether the rider wore a helmet and give the rider's age. Before the change in the helmet law, 60% of youths involved in a motorcycle accident had been wearing their helmets. The Preusser study looked at accident reports during 2001–2003, the three years following the law change, considering these riders to be a representative sample of the larger population. They observed 781 young riders who were involved in accidents. Of these, 396 (or 50.7%) were wearing helmets. Is this evidence of a decline in helmet-wearing, or just the natural fluctuation of such statistics?

Zero In on the Null

Null hypotheses have special requirements. In order to perform a statistical test of the hypothesis, the null must be a statement about the value of a parameter for a model. We use this value to compute the probability that the observed sample statistic—or something even farther from the null value—might occur.

How do we choose the null hypothesis? The appropriate null arises directly from the context of the problem. It is dictated, not by the data, but by the situation. One good way to identify both the null and alternative hypotheses is to think about the *Why* of the situation. Typical null hypotheses might be that the proportion of patients recovering after receiving a new drug is the same as we would expect of patients receiving a placebo or that the mean strength attained by athletes training with new equipment is the same as with the old equipment. The alternative hypotheses would be that the new drug cures a higher proportion of patients or that the new equipment results in a greater mean strength.

To write a null hypothesis, you can't just choose any parameter value you like. The null must relate to the question at hand. Even though the null usually means no difference or no change, you can't automatically interpret "null" to mean zero. A claim that "nobody" wears a motorcycle helmet would be absurd. The null hypothesis for the Florida study could be that the true rate of helmet use remained the same among young riders after the law changed. You need to find the value for the parameter in the null hypothesis from the context of the problem.

There is a temptation to state your *claim* as the null hypothesis. As we have seen, however, you cannot prove a null hypothesis true any more than you can prove a defendant innocent. So, it makes more sense to use what you want to show as the *alternative*. This way, if you reject the null, you are left with what you want to show.

FOR EXAMPLE   Writing Hypotheses

The diabetes drug Avandia® was approved to treat Type 2 diabetes in 1999. But in 2007 an article in the *New England Journal of Medicine* (*NEJM*)[1] raised concerns that the drug might carry an increased risk of heart attack. This study combined results from a number of other separate studies to obtain an overall sample of 4485 diabetes patients taking Avandia. People with Type 2 diabetes are known to have about a 20.2% chance of suffering a heart attack within a seven-year period. According to the article's author, Dr. Steven E. Nissen,[2] the risk found in the *NEJM* study was equivalent to a 28.9% chance of heart attack over seven years. The FDA is the government agency responsible for relabeling Avandia to warn of the risk if it is judged to be unsafe. Although the statistical methods they use are more sophisticated, we can get an idea of their reasoning with the tools we have learned.

QUESTION: What null hypothesis and alternative hypothesis about seven-year heart attack risk would you test? Explain.

$$H_0: p = 0.202$$
$$H_A: p > 0.202$$

The parameter of interest is the proportion of diabetes patients suffering a heart attack in seven years. The FDA is concerned only with whether Avandia increases the seven-year risk of heart attacks above the baseline value of 20.2%, so a one-sided upper-tail test is appropriate.

One-sided or two? In the 1930s, a series of experiments was performed at Duke University in an attempt to see whether humans were capable of extrasensory perception, or ESP. Psychologist Karl Zener designed a set of cards with 5 symbols later made infamous in the movie *Ghostbusters*:

(continued)

---

[1] Steven E. Nissen, M.D., and Kathy Wolski, M.P.H., "Effect of Rosiglitazone on the Risk of Myocardial Infarction and Death from Cardiovascular Causes," *NEJM* 2007; 356.
[2] Interview reported in the *New York Times* [May 26, 2007].

In the experiment, the "sender" selects one of the 5 cards at random from a deck and then concentrates on it. The "receiver" tries to determine which card it is. If we let p be the proportion of correct responses, what's the null hypothesis? The null hypothesis is that ESP makes no difference. Without ESP, the receiver would just be guessing, and since there are 5 possible responses, there would be a 20% chance of guessing each card correctly. So, H_0 is $p = 0.20$. What's the alternative? It seems that it should be $p > 0.20$, a one-sided alternative. But some ESP researchers have expressed the claim that if the proportion guessed were much *lower* than expected, that would show an "interference" and should be considered evidence for ESP as well. So they argue for a two-sided alternative.

STEP-BY-STEP EXAMPLE | Another One-Proportion z-Test

Let's try to answer the question raised at the start of the chapter.

Question: Has helmet use in Florida declined among riders under the age of 21 subsequent to the change in the helmet laws?

THINK

Plan State the problem and discuss the variables and the W's.

Hypotheses The null hypothesis is established by the rate set before the change in the law. The study was concerned with safety, so they'll want to know of any decline in helmet use, making this a lower-tail test.

I want to know whether the rate of helmet wearing among Florida's motorcycle riders under the age of 21 remained at 60% after the law changed to allow older riders to go without helmets. I have data from accident records showing 396 of 781 young riders were wearing helmets.

$$H_0: p = 0.60$$
$$H_A: p < 0.60$$

SHOW

Model Check the conditions.

The Risky Behavior Surveillance survey is in fact a complex, multistage sample, but it is randomized and great effort is taken to make it representative. It is safe to treat it as though it were a random sample.

✔ **Independence Assumption:** The data are for riders involved in accidents during a three-year period. Individuals are independent of one another.

✘ **Randomization Condition:** No randomization was applied, but we are considering these riders involved in accidents to be a representative sample of all riders. We should take care in generalizing our conclusions.

✔ **10% Condition:** These 781 riders are a small sample of a larger population of all young motorcycle riders.

✔ **Success/Failure Condition:** We'd expect $np = 781(0.6) = 468.6$ helmeted riders and $nq = 781(0.4) = 312.4$ non-helmeted. Both are at least 10.

Specify the sampling distribution model and name the test.

The conditions are satisfied, so I can use a Normal model and perform a **one-proportion z-test.**

Mechanics Find the standard deviation of the sampling model using the hypothesized proportion.

Find the z-score for the observed proportion.

There were 396 helmet wearers among the 781 accident victims.

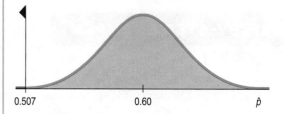

$$\hat{p} = \frac{396}{781} = 0.507$$

$$SD(\hat{p}) = \sqrt{\frac{p_0 q_0}{n}} = \sqrt{\frac{(0.60)(0.40)}{781}} = 0.0175$$

$$z = \frac{\hat{p} - p_0}{SD(\hat{p})} = \frac{0.507 - 0.60}{0.0175} = -5.31$$

Make a picture. Sketch a Normal model centered at the hypothesized helmet rate of 60%. This is a lower-tail test, so shade the region to the left of the observed rate.

Given this z-score, the P-value is obviously very low.

The observed helmet rate is 5.31 standard deviations below the former rate. The corresponding P-value is less than 0.001.

Conclusion Link the P-value to your decision about the null hypothesis, and then state your conclusion in context.

The very small P-value says that if the true rate of helmet-wearing among riders under 21 were still 60%, the probability of observing a rate no higher than 50.7% in a sample like this is less than 1 chance in 1000, so I reject the null hypothesis. There is strong evidence that there has been a decline in helmet use among riders under 21.

How to Think About P-Values

Suppose that as a political science major you are offered the chance to be a White House intern. There would be a very high probability that next summer you'd be in Washington, D.C. That is, $P(\text{Washington} \mid \text{Intern})$ would be high. But if we find a student in Washington, D.C., is it likely that he's a White House intern? Almost surely not; $P(\text{Intern} \mid \text{Washington})$ is low. You can't switch around conditional probabilities. The P-value is $P(\text{data} \mid H_0)$. We might wish we could report $P(H_0 \mid \text{data})$, but these two quantities are NOT the same.

A P-value actually is a conditional probability. It tells us the probability of getting results at least as unusual as the observed statistic, *given* that the null hypothesis is true. We can write P-value = $P(\text{observed statistic value [or even more extreme]} \mid H_0)$.

Writing the P-value this way helps to make clear that the P-value is *not* the probability that the null hypothesis is true. It is a probability about the data. Let's say that again:

The P-value is not the probability that the null hypothesis is true.

The P-value is not even the conditional probability that the null hypothesis is true given the data. We would write that probability as $P(H_0 \mid \text{observed statistic value})$. This is a conditional probability but in reverse. It would be nice to know this, but it's impossible to calculate without making additional assumptions. As we saw in Chapter 15, reversing the order in a conditional probability is difficult, and the results can be counterintuitive.

We can find the P-value, $P(\text{observed statistic value} \mid H_0)$, because H_0 gives the parameter values that we need to find the required probability. But there's no direct way to find $P(H_0 \mid \text{observed statistic value})$.[3] As tempting as it may

---

[3] The approach to statistical inference known as Bayesian Statistics addresses the question in just this way, but it requires more advanced mathematics and more assumptions. See p. 371 for more about the founding father of this approach.

be to say that a P-value of 0.03 means there's a 3% chance that the null hypothesis is true, that just isn't right. All we can say is that, given the null hypothesis, there's a 3% chance of observing the statistic value that we have actually observed (or one more unlike the null value).

"The wise man proportions his belief to the evidence."

–David Hume,
"Enquiry Concerning Human Understanding," 1748

"You're so guilty now."

–Rearview Mirror

> **How guilty is the suspect?** We might like to know $P(H_0 \mid \text{data})$, but when you think about it, we can't talk about the probability that the null hypothesis is true. The null is not a random event, so either it is true or it isn't. The data, however, are random in the sense that if we were to repeat a randomized experiment or draw another random sample, we'd get different data and expect to find a different statistic value. So we can talk about the probability of the data given the null hypothesis, and that's the P-value.
>
> But it does make sense that the smaller the P-value, the more confident we can be in declaring that we doubt the null hypothesis. Think again about the jury trial. Our null hypothesis is that the defendant is innocent. Then the evidence starts rolling in. A car the same color as his was parked in front of the bank. Well, there are lots of cars that color. The probability of that happening (given his innocence) is pretty high, so we're not persuaded that he's guilty. The bank's security camera showed the robber was male and about the defendant's height and weight. Hmmm. Could that be a coincidence? If he's innocent, then it's a little less likely that the car and description would *both* match, so our P-value goes down. We're starting to question his innocence a little. Witnesses said the robber wore a blue jacket just like the one the police found in a garbage can behind the defendant's house. Well, if he's innocent, then that doesn't seem very likely, does it? If he's really innocent, the probability that all of these could have happened is getting pretty low. Now our P-value may be small enough to be called "beyond a reasonable doubt" and lead to a conviction. Each new piece of evidence strains our skepticism a bit more. The more compelling the evidence–the more *unlikely* it would be were he innocent–the more convinced we become that he's guilty.
>
> But even though it may make *us* more confident in declaring him guilty, additional evidence does not make *him* any guiltier. Either he robbed the bank or he didn't. Additional evidence (like the teller picking him out of a police lineup) just makes us more confident that we did the right thing when we convicted him. The lower the P-value, the more comfortable we feel about our decision to reject the null hypothesis, but the null hypothesis doesn't get any more false.

FOR EXAMPLE   Thinking About the P-Value

RECAP: A *New England Journal of Medicine* paper reported that the seven-year risk of heart attack in diabetes patients taking the drug Avandia was increased from the baseline of 20.2% to an estimated risk of 28.9% and said the P-value was 0.03.

QUESTION: How should the P-value be interpreted in this context?

The P-value = $P(\hat{p} \geq 28.9\% \mid p = 20.2\%)$. That is, it's the probability of seeing such a high heart attack rate among the people studied if, in fact, taking Avandia really didn't increase the risk at all.

What to Do with a High P-Value

A S   *Video:* Is There Evidence for Therapeutic Touch? This video shows the experiment and tells the story.

A S   *Activity:* Testing Therapeutic Touch. Perform the one-proportion z-test using *ActivStats* technology. The test in *ActivStats* is two-sided. Do you think this is the appropriate choice?

Therapeutic touch (TT), taught in many schools of nursing, is a therapy in which the practitioner moves her hands near, but does not touch, a patient in an attempt to manipulate a "human energy field." Therapeutic touch practitioners believe that by adjusting this field they can promote healing. However, no instrument has ever detected a human energy field, and no experiment has ever shown that TT practitioners can detect such a field.

In 1998, the *Journal of the American Medical Association* published a paper reporting work by a then nine-year-old girl.[4] She had performed a simple

_____

[4] L. Rosa, E. Rosa, L. Sarner, and S. Barrett, "A Close Look at Therapeutic Touch," *JAMA* 279(13) [1 April 1998]: 1005–1010.

experiment in which she challenged 15 TT practitioners to detect whether her unseen hand was hovering over their left or right hand (selected by the flip of a coin).

The practitioners "warmed up" with a period during which they could see the experimenter's hand, and each said that they could detect the girl's human energy field. Then a screen was placed so that the practitioners could not see the girl's hand, and they attempted 10 trials each. Overall, of 150 trials, the TT practitioners were successful 70 times, for a success proportion of 46.7%. Is there evidence from this experiment that TT practitioners can successfully detect a "human energy field"?

When we see a small P-value, we could continue to believe the null hypothesis and conclude that we just witnessed a rare event. But instead, we trust the data and use it as evidence to reject the null hypothesis.

In the therapeutic touch example, the null hypothesis is that the practitioners are guessing, so we expect them to be right about half the time by chance. That's why we say $H_0: p = 0.5$. They claim that they can detect a "human energy field" and that their success rate should be well above chance, so our alternative is that they would do *better* than guessing. That's a one-sided alternative hypothesis: $H_A: p > 0.5$. With a one-sided hypothesis, our P-value is the probability the practitioners could achieve the observed number of successes or *more* even if they were just guessing.

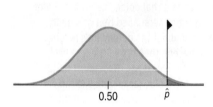

If the practitioners had been highly successful, that would have been unusually lucky for guessing, so we would have seen a correspondingly low P-value. Since we don't believe in rare events, we would then have concluded that they weren't guessing.

But that's not what happened. What we actually observed was that they did slightly *worse* than 50%, with a $\hat{p} = 0.467$ success rate.

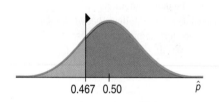

As the figure shows, the probability of a success rate of 0.467 *or more* is even bigger than 0.5. In this case, it turns out to be 0.793. Obviously, we won't be rejecting the null hypothesis; for us to reject it, the P-value would have to be quite small. But a P-value of 0.788 seems so big it is almost awkward. With a success rate even lower than chance, we could have concluded right away that we have no evidence for rejecting H_0.

Big P-values just mean that what we've observed isn't surprising. That is, the results are in line with our assumption that the null hypothesis models the world, so we have no reason to reject it. A big P-value doesn't prove that the null hypothesis is true, but it certainly offers no evidence that it's *not* true. When we see a large P-value, all we can say is that we "don't reject the null hypothesis."

FOR EXAMPLE   More About P-Values

RECAP: The question of whether the diabetes drug Avandia increased the risk of heart attack was raised by a study in the *New England Journal of Medicine*. This study estimated the seven-year risk of heart attack to be 28.9% and reported a P-value of 0.03 for a test of whether this risk was higher than the baseline seven-year risk of 20.2%. An earlier study (the ADOPT study) had estimated the seven-year risk to be 26.9% and reported a P-value of 0.27.

QUESTION: Why did the researchers in the ADOPT study not express alarm about the increased risk they had seen?

A P-value of 0.27 means that a heart attack rate at least as high as the one they observed could be expected in 27% of similar experiments even if, in fact, there were no increased risk from taking Avandia. That's not remarkable enough to reject the null hypothesis. In other words, the ADOPT study wasn't convincing.

Alpha Levels

Sir Ronald Fisher (1890–1962) was one of the founders of modern Statistics.

Sometimes we need to make a firm decision about whether or not to reject the null hypothesis. A jury must *decide* whether the evidence reaches the level of "beyond a reasonable doubt." A business must *select* a Web design. You need to decide which section of Statistics to enroll in.

When the P-value is small, it tells us that our data are rare, *given the null hypothesis*. As humans, we are suspicious of rare events. If the data are "rare enough," we just don't think that could have happened due to chance. Since the data *did* happen, something must be wrong. All we can do now is reject the null hypothesis.

But how rare is "rare"?

We can define "rare event" arbitrarily by setting a threshold for our P-value. If our P-value falls below that point, we'll reject the null hypothesis. We call such results **statistically significant**. The threshold is called an **alpha level**. Not surprisingly, it's labeled with the Greek letter α. Common α levels are 0.10, 0.05, 0.01, and 0.001. You have the option—almost the *obligation*—to consider your alpha level carefully and choose an appropriate one for the situation. If you're assessing the safety of air bags, you'll want a low alpha level; even 0.01 might not be low enough. If you're just wondering whether folks prefer their pizza with or without pepperoni, you might be happy with $\alpha = 0.10$. It can be hard to justify your choice of α, though, so often we arbitrarily choose 0.05. Note, however: You must select the alpha level *before* you look at the data. Otherwise you can be accused of cheating by tuning your alpha level to suit the data.

> **Where did the value 0.05 come from?** In 1931, in a famous book called *The Design of Experiments*, Sir Ronald Fisher discussed the amount of evidence needed to reject a null hypothesis. He said that it was *situation dependent*, but remarked, somewhat casually, that for many scientific applications, 1 out of 20 *might be* a reasonable value. Since then, some people—indeed some entire disciplines—have treated the number 0.05 as sacrosanct.

The alpha level is also called the **significance level**. When we reject the null hypothesis, we say that the test is "significant at that level." For example, we might say that we reject the null hypothesis "at the 5% level of significance."

What can you say if the P-value does not fall below α?

When you have not found sufficient evidence to reject the null according to the standard you have established, you should say that "The data have failed to provide sufficient evidence to reject the null hypothesis." Don't say that you "accept the null hypothesis." You certainly haven't proven or established it; it was merely assumed to begin with. Say that you've failed to reject it.

Think again about the therapeutic touch example. The P-value was 0.788. This is so much larger than any reasonable alpha level that we can't reject H_0. For this test, we'd conclude, "We fail to reject the null hypothesis. There is insufficient evidence to conclude that the practitioners are performing better than they would if they were just guessing."

The automatic nature of the reject/fail-to-reject decision when we use an alpha level may make you uncomfortable. If your P-value falls just slightly above your alpha level, you're not allowed to reject the null. Yet a P-value just barely below the alpha level leads to rejection. If this bothers you, you're in good company. Many statisticians think it better to report the P-value than to base a decision on an arbitrary alpha level.

> **It's in the stars** Some disciplines carry the idea further and code P-values by their size. In this scheme, a P-value between 0.05 and 0.01 gets highlighted by *. A P-value between 0.01 and 0.001 gets **, and a P-value less than 0.001 gets ***. This can be a convenient summary of the weight of evidence against the null hypothesis if it's not taken too literally. But we warn you against taking the distinctions too seriously and against making a black-and-white decision near the boundaries. The boundaries are a matter of tradition, not science; there is nothing special about 0.05. A P-value of 0.051 should be looked at very seriously and not casually thrown away just because it's larger than 0.05, and one that's 0.009 is not very different from one that's 0.011.

When you decide to declare a verdict, it's always a good idea to report the P-value as an indication of the strength of the evidence. Sometimes it's best to report that the conclusion is not yet clear and to suggest that more data be gathered. (In a trial, a jury may "hang" and be unable to return a verdict.) In these cases, the P-value is the best summary we have of what the data say or fail to say about the null hypothesis.

Significant vs. Important

PRACTICAL VS. STATISTICAL SIGNIFICANCE

A large insurance company mined its data and found a statistically significant ($P = 0.04$) difference between the mean value of policies sold in 2001 and 2002. The difference in the mean values was $9.83. Even though it was statistically significant, management did not see this as an important difference when a typical policy sold for more than $1000. On the other hand, even a clinically important improvement of 10% in cure rate with a new treatment is not likely to be statistically significant in a study of fewer than 225 patients. A small clinical trial would probably not be conclusive.

What do we mean when we say that a test is statistically significant? All we mean is that the test statistic had a P-value lower than our alpha level. Don't be lulled into thinking that statistical significance carries with it any sense of practical importance or impact.

For large samples, even small, unimportant ("insignificant") deviations from the null hypothesis can be statistically significant. On the other hand, if the sample is not large enough, even large financially or scientifically "significant" differences may not be statistically significant.

It's good practice to report the magnitude of the difference between the observed statistic value and the null hypothesis value (in the data units) along with the P-value on which we base statistical significance.

Critical Values Again

When making a confidence interval, we've found a **critical value**, z^*, to correspond to our selected confidence level. Critical values can also be used as a shortcut for hypothesis tests. Before computers and calculators were common, P-values were hard to find. It was easier to select a few common alpha levels (0.05, 0.01, 0.001, for example) and learn the corresponding critical values for the Normal model (that is, the critical values corresponding to confidence levels 0.95, 0.99, and 0.999, respectively). Rather than looking up the probability corresponding to your z-score in the table, you'd just check your z-score directly against these z^* values. Any z-score larger in magnitude (that is, more extreme) than a particular critical value has to be less likely, so it will have a P-value smaller than the corresponding alpha. If we are willing to settle for a flat reject/fail-to-reject decision, comparing an observed z-score with the critical value for a specified alpha level gives a shortcut path to that decision. For the motorcycle helmet example, if we choose $\alpha = 0.05$, then, in order to reject H_0, our z-score has to be less than the one-sided critical value of -1.645. Our calculated z-score is -5.31, so clearly we can reject the null hypothesis. This is perfectly correct and does give us a yes/no decision, but it gives us less information about the hypothesis because we don't have the P-value to think about.

With technology, P-values are easy to find. And since they give more information about the strength of the evidence, you should always report them.

Here are the traditional critical values from the Normal model:[5]

α	1-sided	2-sided
0.05	1.645	1.96
0.01	2.33	2.576
0.001	3.09	3.29

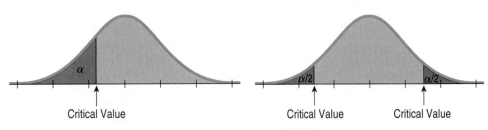

FIGURE 21.1
When the alternative is one-sided, the critical value puts all of α on one side.

FIGURE 21.2
When the alternative is two-sided, the critical value splits α equally into two tails.

Confidence Intervals and Hypothesis Tests

For the motorcycle helmet example, a 95% confidence interval would give $0.507 \pm 1.96 \times 0.0179 = (0.472, 0.542)$, or 47.2% to 54.2%. If the previous rate of helmet compliance had been, say, 50%, we would not have been able to reject the null hypothesis because 50% is in the interval, so it's a plausible value. Indeed, *any* hypothesized value for the true proportion of helmet wearers in this interval is consistent with the data. Any value outside the confidence interval would make a null hypothesis that we would reject, but we'd feel more strongly about values far outside the interval.

Confidence intervals and hypothesis tests are built from the same calculations.[6] They have the same assumptions and conditions. As we have just seen, you can approximate a hypothesis test by examining the confidence interval. Just ask whether the null hypothesis value is consistent with a confidence interval for the parameter at the corresponding confidence level. Because confidence intervals are naturally two-sided, they correspond to two-sided tests. For example, a 95% confidence interval corresponds to a two-sided hypothesis test at $\alpha = 5\%$. In general, a confidence interval with a confidence level of C% corresponds to a two-sided hypothesis test with an α level of $100 - C\%$.

---

[5] In a sense, these are the flip side of the 68–95–99.7 Rule. When we studied that rule, we chose simple statistical distances from the mean and recalled the areas of the tails. Now we select convenient tail areas (0.05, 0.01, and 0.001, either on one side or adding both together) and record the corresponding statistical distances.

[6] As we saw in Chapter 20, this is not *exactly* true for proportions. For a confidence interval, we estimate the standard deviation of $\hat{p}$ from $\hat{p}$ itself. Because we estimate it from the data, we have a *standard error*. For the corresponding hypothesis test, we use the model's standard deviation for $\hat{p}$, based on the null hypothesis value p_0. When $\hat{p}$ and p_0 are close, these calculations give similar results. When they differ, you're likely to reject H_0 (because the observed proportion is far from your hypothesized value). In that case, you're better off building your confidence interval with a standard error estimated from the data.

The relationship between confidence intervals and one-sided hypothesis tests is a little more complicated. For a one-sided test with $\alpha = 5\%$, the corresponding confidence interval has a confidence level of 90%—that's 5% in each tail. In general, a confidence interval with a confidence level of $C\%$ corresponds to a one-sided hypothesis test with an α level of $\frac{1}{2}(100 - C)\%$.

FOR EXAMPLE | Making a Decision Based on a Confidence Interval

RECAP: The baseline seven-year risk of heart attacks for diabetics is 20.2%. In 2007 a *NEJM* study reported a 95% confidence interval equivalent to 20.8% to 40.0% for the risk among patients taking the diabetes drug Avandia.

QUESTION: What did this confidence interval suggest to the FDA about the safety of the drug?

The FDA could be 95% confident that the interval from 20.8% to 40.0% included the true risk of heart attack for diabetes patients taking Avandia. Because the lower limit of this interval was higher than the baseline risk of 20.2%, there was evidence of an increased risk.

MATH BOX

Suppose we are testing a null hypothesis against a two-sided alternative hypothesis:

$$H_0: p = p_0 \text{ vs. } H_A: p \neq p_0$$

and we use $\alpha = 0.05$.

Our test statistic would be $z = \dfrac{\hat{p} - p_0}{SD(\hat{p})}$, and we would perform the test by checking

whether $\dfrac{\hat{p} - p_0}{SD(\hat{p})} < -1.96$ or $\dfrac{\hat{p} - p_0}{SD(\hat{p})} > 1.96$. If either were true, we'd reject H_0.

We can rearrange the inequalities to solve for p_0:

$$p_0 > \hat{p} + 1.96 SD(\hat{p}) \quad \text{or} \quad p_0 < \hat{p} - 1.96 SD(\hat{p}).$$

These inequalities are symmetric, so we can combine them and say that we reject H_0 if the null value p_0 falls outside $\hat{p} \pm 1.96 SD(\hat{p})$.

But that looks remarkably like a 95% confidence interval for p:

$$\hat{p} \pm 1.96 SE(\hat{p}).$$

Except for the fact that in the confidence interval we use $SE(\hat{p})$ instead of $SD(\hat{p})$, they are same statement. That is, we reject H_0 *when the confidence interval fails to cover the hypothesized value p_0*. In most cases, the difference between SE and SD is small. We said that a confidence interval holds the *plausible* values of p, so we shouldn't be surprised to find that we would reject a hypothesized p_0 if it isn't a plausible value and fail to reject it if it is plausible.

For a one-sided test, all of the α probability is in one tail. But confidence intervals are almost always two-sided, with a symmetric region in the other tail. So to test a one-sided test at $\alpha = 0.05$ with a confidence interval, just construct a 90% confidence interval and see if p_0 falls outside the interval on the appropriate side.

1. An experiment to test the fairness of a roulette wheel gives a z-score of 0.62. What would you conclude?

2. In the last chapter we encountered a bank that wondered if it could get more customers to make payments on delinquent balances by sending them a DVD urging them to set up a payment plan. Well, the bank just got back the results on their test of this strategy. A 90% confidence interval for the success rate is (0.29, 0.45). Their old send-a-letter method had worked 30% of the time. Can you reject the null hypothesis that the proportion is still 30% at $\alpha = 0.05$? Explain.

3. Given the confidence interval the bank found in their trial of DVDs, what would you recommend that they do? Should they scrap the DVD strategy?

STEP-BY-STEP EXAMPLE   Wear that Seat Belt!

Massachusetts is Serious About Saving Lives

Teens are at the greatest risk of being killed or injured in traffic crashes. According to the National Highway Traffic Safety Administration, 65% of young people killed were not wearing a safety belt. In 2001, a total of 3322 teens were killed in motor vehicle crashes, an average of 9 teenagers a day. Because many of these deaths could easily be prevented by the use of safety belts, several states have begun "Click It or Ticket" campaigns in which increased enforcement and publicity have resulted in significantly higher seat-belt use. Overall use in Massachusetts quickly increased from 51% in 2002 to 64.8% in 2006, with a goal of surpassing the national average of 82%. Recently, a local newspaper reported that a roadblock resulted in 23 tickets to drivers who were unbelted out of 134 stopped for inspection.

Question: Does this provide evidence that the goal of over 82% compliance was met?

Let's use a confidence interval to test this hypothesis.

 THINK

Plan State the problem and discuss the variables and the W's.

Hypotheses The null hypothesis is that the compliance rate is only 82%. The alternative is that it is now higher. It's clearly a one-sided test, so if we use a confidence interval, we'll have to be careful about what level we use.

Model Think about the assumptions and check the conditions.

We are finding a confidence interval, so we work from the data rather than the null model.

State your method.

The data come from a local newspaper report that tells the number of tickets issued and number of drivers stopped at a recent roadblock. I want to know whether the rate of compliance with the seat-belt law is greater than 82%.

$$H_0: p = 0.82$$
$$H_A: p > 0.82$$

✓ **Independence Assumption:** Drivers are not likely to influence one another when it comes to wearing a seat belt.

✓ **Randomization Condition:** This wasn't a random sample, but I assume these drivers are representative of the driving public.

✓ **10% Condition:** The police stopped fewer than 10% of all drivers.

✓ **Success/Failure Condition:** There were 111 successes and 23 failures, both at least 10. The sample is large enough.

Under these conditions, the sampling model is Normal. I'll create a **one-proportion z-interval**.

Mechanics Write down the given information, and determine the sample proportion.

To use a confidence interval, we need a confidence level that corresponds to the alpha level of the test. If we use $\alpha = 0.05$, we should construct a 90% confidence interval, because this is a one-sided test.

That will leave 5% on *each* side of the observed proportion. Determine the standard error of the sample proportion and the margin of error. The critical value is $z^* = 1.645$.

The confidence interval is

estimate $\pm$ margin of error.

$n = 134$, so

$$\hat{p} = \frac{111}{134} = 0.828 \text{ and}$$

$$SE(\hat{p}) = \sqrt{\frac{\hat{p}\hat{q}}{n}} = \sqrt{\frac{(0.828)(0.172)}{134}} = 0.033$$

$$ME = z^* \times SE(\hat{p})$$
$$= 1.645(0.033) = 0.054$$

The 90% confidence interval is

$$0.828 \pm 0.054 \text{ or}$$
$$(0.774, 0.882).$$

Conclusion Link the confidence interval to your decision about the null hypothesis, and then state your conclusion in context.

I am 90% confident that between 77.4% and 88.2% of all drivers wear their seat belts. Because the hypothesized rate of 82% is within this interval, I do not reject the null hypothesis. There is insufficient evidence to conclude that the campaign was truly effective and now more than 82% of all drivers are wearing seat belts.

The upper limit of the confidence interval shows it's possible that the campaign is quite successful, but the small sample size makes the interval too wide to be very specific.

A Confidence Interval for Small Samples

When the **Success/Failure Condition** fails, all is not lost. A simple adjustment to the calculation lets us make a 95% confidence interval anyway.

All we do is add four *phony* observations—two to the successes, two to the failures. So instead of the proportion $\hat{p} = \dfrac{y}{n}$, we use the adjusted proportion $\tilde{p} = \dfrac{y + 2}{n + 4}$ and, for convenience, we write $\tilde{n} = n + 4$. We modify the interval by using these adjusted values for both the center of the interval *and* the margin of error. Now the adjusted interval is

$$\tilde{p} \pm z^* \sqrt{\frac{\tilde{p}(1 - \tilde{p})}{\tilde{n}}}.$$

This adjusted form gives better performance overall[7] and works much better for proportions near 0 or 1. It has the additional advantage that we no longer need to check the **Success/Failure Condition** that $n\hat{p}$ and $n\hat{q}$ are greater than 10. This interval is called the Agresti-Coull interval or the "plus-four" interval. The plus-four method is not the most common way to find a confidence interval for a proportion, but you can use it safely for any sample size—as long as you explain clearly what you have done.

FOR EXAMPLE An Agresti-Coull "Plus-Four" Interval

Surgeons examined their results to compare two methods for a surgical procedure used to alleviate pain on the outside of the wrist. A new method was compared with the traditional "freehand" method for the procedure. Of 45 operations using the "freehand" method, three were unsuccessful, for a failure rate of 6.7%. With only 3 failures, the data don't satisfy the **Success/Failure Condition,** so we can't use a standard confidence interval.

QUESTION: What's the confidence interval using the "plus-four" method?

There were 42 successes and 3 failures. Adding 2 "pseudo-successes" and 2 "pseudo-failures," we find

$$\tilde{p} = \frac{3 + 2}{45 + 4} = 0.102$$

A 95% confidence interval (for example) is then

$$0.102 \pm 1.96\sqrt{\frac{0.102(1 - 0.102)}{49}} = 0.102 \pm 0.085 \; or \; (0.017, 0.187).$$

Notice that although the observed failure rate of 0.067 is contained in the interval, it is not at the center of the interval—something we haven't seen with any of the other confidence intervals we've considered.

A S **Activity: Type I and Type II Errors.** View an animated exploration of Type I and Type II errors–a good backup for the reading in this section.

FALSE POSITIVE CONSEQUENCES

Some false-positive results mean no more than an unnecessary chest X-ray. But for a drug test or a disease like AIDS, a false-positive result that is not kept confidential could have serious consequences.

Making Errors

Nobody's perfect. Even with lots of evidence, we can still make the wrong decision. In fact, when we perform a hypothesis test, we can make mistakes in *two* ways:

I. The null hypothesis is true, but we mistakenly reject it.
II. The null hypothesis is false, but we fail to reject it.

These two types of errors are known as **Type I** and **Type II errors**. One way to keep the names straight is to remember that we start by assuming the null hypothesis is true, so a Type I error is the first kind of error we could make.

In medical disease testing, the null hypothesis is usually the assumption that a person is healthy. The alternative is that he or she has the disease we're testing for. So a Type I error is a *false positive:* A healthy person is diagnosed with the disease. A Type II error, in which an infected person is diagnosed as disease free, is a *false negative.* These errors have other names, depending on the particular discipline and context.

Which type of error is more serious depends on the situation. In the jury trial, a Type I error occurs if the jury convicts an innocent person. A Type II error occurs if the jury fails to convict a guilty person. Which seems more

---

[7] By "better performance," we mean that a 95% confidence interval has more nearly a 95% chance of covering the true population proportion. Simulation studies have shown that our original, simpler confidence interval in fact is less likely than 95% to cover the true population proportion when the sample size is small or the proportion very close to 0 or 1. The original idea for this method can be attributed to E. B. Wilson. The simpler approach discussed here was proposed by Agresti and Coull (A. Agresti and B. A. Coull, "Approximate Is Better Than 'Exact' for Interval Estimation of Binomial Proportions," *The American Statistician,* 52[1998]: 119–129).

Why did Sir Ronald Fisher suggest 0.05 as a criterion for testing hypotheses? It turns out that he had in mind small initial studies. Small studies have relatively little power. Fisher was concerned that they might make too many Type II errors—failing to discover an important effect—if too strict a criterion were used. Once a test failed to reject a null hypothesis, it was unlikely that researchers would return to that hypothesis to try again.

On the other hand, the increased risk of Type I errors arising from a generous criterion didn't concern him as much for exploratory studies because these are ordinarily followed by a replication or a larger study. The probability of a Type I error is α—in this case, 0.05. The probability that two independent studies would both make Type I errors is $0.05 \times 0.05 = 0.0025$, so Fisher was confident that Type I errors in initial studies were not a major concern.

The widespread use of the relatively generous 0.05 criterion even in large studies is most likely not what Fisher had in mind.

NOTATION ALERT

In Statistics, α is almost always saved for the alpha level. But β has already been used for the parameters of a linear model. Fortunately, it's usually clear whether we're talking about a Type II error probability or the slope or intercept of a regression model.

The null hypothesis specifies a single value for the parameter. So it's easy to calculate the probability of a Type I error. But the alternative gives a whole range of possible values, and we may want to find a β for several of them.

We have seen ways to find a sample size by specifying the margin of error. Choosing the sample size to achieve a specified β (for a particular alternative value) is sometimes more appropriate, but the calculation is more complex and lies beyond the scope of this book.

---

serious? In medical diagnosis, a false negative could mean that a sick patient goes untreated. A false positive might mean that the person must undergo further tests. In a Statistics final exam (with H_0: the student has learned only 60% of the material), a Type I error would be passing a student who in fact learned less than 60% of the material, while a Type II error would be failing a student who knew enough to pass. Which of these errors seems more serious? It depends on the situation, the cost, and your point of view.

Here's an illustration of the situations:

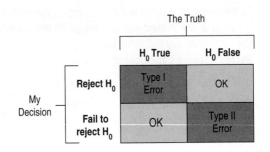

How often will a Type I error occur? It happens when the null hypothesis is true but we've had the bad luck to draw an unusual sample. To reject H_0, the P-value must fall below α. When H_0 is true, that happens *exactly* with probability α. So when you choose level α, you're setting the probability of a Type I error to α.

What if H_0 is not true? Then we can't possibly make a Type I error. You can't get a false positive from a sick person. A Type I error can happen only when H_0 is true.

When H_0 is false but we *fail* to reject it, we have made a Type II error. We assign the letter β to the probability of this mistake. What's the value of β? That's harder to assess than α because we don't know what the value of the parameter really is. When H_0 is true, it specifies a single parameter value. But when H_0 is false, we don't know the parameter value and there are many possible values. We can compute the probability β for any parameter value in H_A. But the one we should choose depends on the situation.

We could reduce β for *all* alternative parameter values by increasing α. By making it easier to reject the null, we'd be more likely to reject it whether it's true or not. So we'd reduce β, the chance that we fail to reject a false null—but we'd make more Type I errors. This tension between Type I and Type II errors is inevitable. In the political arena, think of the ongoing debate between those who favor provisions to reduce Type I errors in the courts (supporting defendants' rights, requiring warrants for wiretaps, providing legal representation for those who can't afford it) and those who advocate changes to reduce Type II errors (admitting into evidence confessions made when no lawyer is present, eavesdropping on conferences with lawyers, restricting paths of appeal, etc.).

The only way to reduce *both* types of error is to collect more evidence or, in statistical terms, to collect more data. Too often, studies fail because their sample sizes are too small to detect the change they are looking for.

Of course, what we really want to do is to detect a false null hypothesis. When H_0 is false and we reject it, we have done the right thing. A test's ability to detect a false null hypothesis is called the power of the test. In a jury trial, power is the ability of the criminal justice system to convict people who are guilty—a good thing!

FOR EXAMPLE | **Thinking About Errors**

RECAP: A published study found the risk of heart attack to be increased in patients taking the diabetes drug Avandia. The issue of the *New England Journal of Medicine* (*NEJM*) in which that study appeared also included an editorial that said, in part, "A few events either way might have changed the findings for myocardial infarction[8] or for death from cardiovascular causes. In this setting, the possibility that the findings were due to chance cannot be excluded."

QUESTION: What kind of error would the researchers have made if, in fact, their findings were due to chance? What could be the consequences of this error?

The null hypothesis said the risk didn't change, but the researchers rejected that model and claimed evidence of a higher risk. If these findings were just due to chance, they rejected a true null hypothesis—a Type I error.

If, in fact, Avandia carried no extra risk, then patients might be deprived of its benefits for no good reason.

Power

When we failed to reject the null hypothesis about TT practitioners, did we prove that they were just guessing? No, it could be that they actually *can* discern a human energy field but we just couldn't tell. For example, suppose they really have the ability to get 53% of the trials right but just happened to get only 47% in our experiment. Our confidence interval shows that with these data we wouldn't have rejected the null. And if we retained the null even though the true proportion was actually greater than 50%, we would have made a Type II error because we failed to detect their ability.

Remember, we can never prove a null hypothesis true. We can only fail to reject it. But when we fail to reject a null hypothesis, it's natural to wonder whether we looked hard enough. Might the null hypothesis actually be false and our test too weak to tell?

When the null hypothesis actually *is* false, we hope our test is strong enough to reject it. We'd like to know how likely we are to succeed. The power of the test gives us a way to think about that. The **power** of a test is the probability that it correctly rejects a false null hypothesis. When the power is high, we can be confident that we've looked hard enough. We know that β is the probability that a test *fails* to reject a false null hypothesis, so the power of the test is the probability that it *does* reject: $1 - \beta$.

Whenever a study fails to reject its null hypothesis, the test's power comes into question. Was the sample size big enough to detect an effect had there been one? Might we have missed an effect large enough to be interesting just because we failed to gather sufficient data or because there was too much variability in the data we could gather? The therapeutic touch experiment failed to reject the null hypothesis that the TT practitioners were just guessing. Might the problem be that the experiment simply lacked adequate power to detect their ability?

---

[8] Medical jargon for "heart attack."

RECAP: The study of Avandia published in the *NEJM* combined results from 47 different trials—a method called *meta-analysis*. The drug's manufacturer, GlaxoSmithKline (GSK), issued a statement that pointed out, "Each study is designed differently and looks at unique questions: For example, individual studies vary in size and length, in the type of patients who participated, and in the outcomes they investigate." Nevertheless, by combining data from many studies, meta-analyses can achieve a much larger sample size.

QUESTION: How could this larger sample size help?

If Avandia really did increase the seven-year heart attack rate, doctors needed to know. To overlook that would have been a Type II error (failing to detect a false null hypothesis), resulting in patients being put at greater risk. Increasing the sample size could increase the power of the analysis, making it more likely that researchers will detect the danger if there is one.

Effect Size

A S *Activity:* **The Power of a Test.** Power is a concept that's much easier to understand when you can visualize what's happening.

When we think about power, we imagine that the null hypothesis is false. The value of the power depends on how far the truth lies from the value we hypothesize. We call the distance between the null hypothesis value (for example), p_0, and the truth, p, the **effect size**. Not knowing the true value, we estimate the effect size as the difference between the null and the observed value.

The effect size is central to how we think about the power of a hypothesis test. A larger effect is easier to see and results in larger power. Small effects are naturally more difficult to detect. They'll result in more Type II errors and therefore lower power. Knowing the effect size and the sample size helps us determine the power. But when we design a study we won't know the effect size, so we can only imagine possible effect sizes and look at their consequences. How can we decide what effect sizes to look at?

One way to think about the effect size is to ask "*How big a difference would matter?*" The answer to this question depends on who is asking it and why. For example, if therapeutic touch practitioners could detect a human energy field, but only 53% of the time, that might not be a sufficient improvement over chance for health insurers to conclude it was worth covering. But *any* real ability to detect a previously unknown human energy field would be of great interest to scientists.[9]

The power of a test depends on the size of the effect and the standard deviation. For proportions, our test statistic is $\hat{p}$, and its standard deviation is $\sqrt{\dfrac{pq}{n}}$. When researchers design a study they often know what effect size would matter to their conclusions and have some idea of the proportion they are estimating. Once they know both of those, a bit of algebra gives them an estimate of n, the sample size they'll need. This is a common calculation to perform when designing a study so you can be sure to gather a large enough sample (and, speaking practically, to gather enough funds to pay for that large a sample). In Chapter 19 we made a similar calculation of a sample size based on how large a margin of error we could accept. This is essentially the same idea.

Effect size and power are also important at the *Tell* stage of an analysis whenever we fail to reject the null hypothesis. The natural question to wonder about in that event is whether we tried hard enough—whether we had sufficient power to discern a difference. After all, if we based our test on only three observations, we might easily have missed even a large effect.

---

[9]And would probably lead to a Nobel prize.

Whenever a hypothesis test fails to reject the null, the question of power can arise. For example, in the therapeutic touch experiment, with a sample size of 150, if the researchers took 75% as a reasonably interesting effect size (keeping in mind that 50% is the level of guessing), they could determine that the TT experiment would have been able to detect such an ability with a power of 99.99%. So there is only a very small chance that their study would have failed to detect a practitioner's ability at that level, had it existed.

Effect size is a concern whenever we use statistical inference. The details of the calculations depend on the situation and the appropriate statistic, but as we will see in the following chapters, the concept is the same.

JUST CHECKING

4. Remember the bank that's sending out DVDs to try to get customers to make payments on delinquent loans? It is looking for evidence that the costlier DVD strategy produces a higher success rate than the letters it has been sending. Explain what a Type I error is in this context and what the consequences would be to the bank.

5. What's a Type II error in the bank experiment context, and what would the consequences be?

6. For the bank, which situation has higher power: a strategy that works really well, actually getting 60% of people to pay off their balances, or a strategy that barely increases the payoff rate to 32%? Explain briefly.

A Picture Worth $\dfrac{1}{P(z > 3.09)}$ Words

It makes intuitive sense that the larger the effect size, the easier it should be to see it. Obtaining a larger sample size decreases the probability of a Type II error, so it increases the power. It also makes sense that the more we're willing to accept a Type I error, the less likely we will be to make a Type II error.

FIGURE 21.3
The power of a test is the probability that it rejects a false null hypothesis. The upper figure shows the null hypothesis model. We'd reject the null in a one-sided test if we observed a value of $\hat{p}$ in the red region to the right of the critical value, p^*. The lower figure shows the true model. If the true value of p is greater than p_0, then we're more likely to observe a value that exceeds the critical value and make the correct decision to reject the null hypothesis. The power of the test is the purple region on the right of the lower figure. Of course, even drawing samples whose observed proportions are distributed around p, we'll sometimes get a value in the red region on the left and make a Type II error of failing to reject the null.

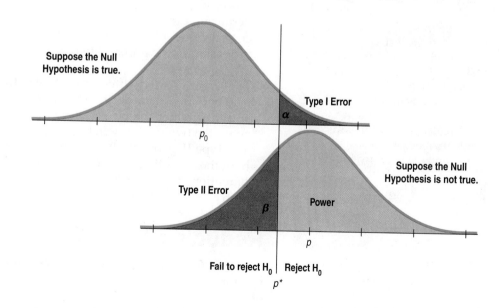

SENSITIVITY AND SPECIFICITY

The terms *sensitivity* and *specificity* often appear in medical studies. Both are conditional probabilities that a diagnostic test works the way we want it to.

In testing for a disease, the usual null hypothesis is that you are not sick. The *sensitivity* of a test is P(Reject | H_0 is false)—it tells how likely the test is to (correctly) detect that you are sick when in fact you are.

The *specificity* is P(Fail to Reject | H_0 is true). This is the flip-side—the probability the test will *not* claim you are sick when you are healthy.

So these are the probabilities that the test diagnoses your condition correctly—sensitivity if you are sick, specificity if you are healthy. Because it is important to diagnose an illness so it can be treated, diagnostic tests are often designed to maximize sensitivity, but the best tests have both high sensitivity and high specificity.

A S *Activity:* **Power and Sample Size.** Investigate how the power of a test changes with the sample size. The interactive tool is really the only way you can see this easily.

Figure 21.3 shows a good way to visualize the relationships among these concepts. Suppose we are testing $H_0: p = p_0$ against the alternative $H_A: p > p_0$. We'll reject the null if the observed proportion, $\hat{p}$, is big enough. By big enough, we mean $\hat{p} > p^*$ for some critical value, p^* (shown as the red region in the right tail of the upper curve). For example, we might be willing to believe the ability of therapeutic touch practitioners if they were successful in 65% of our trials. This is what the upper model shows. It's a picture of the sampling distribution model for the proportion if the null hypothesis were true. We'd make a Type I error whenever the sample gave us $\hat{p} > p^*$, because we would reject the (true) null hypothesis. And unusual samples like that would happen only with probability α.

In reality, though, the null hypothesis is rarely *exactly* true. The lower probability model supposes that H_0 is not true. In particular, it supposes that the true value is p, not p_0. (Perhaps the TT practitioner really can detect the human energy field 72% of the time.) It shows a distribution of possible observed $\hat{p}$ values around this true value. Because of sampling variability, sometimes $\hat{p} < p^*$ and we fail to reject the (false) null hypothesis. Suppose a TT practitioner with a true ability level of 72% is actually successful on fewer than 65% of our tests. Then we'd make a Type II error. The area under the curve to the left of p^* in the bottom model represents how often this happens. The probability is β. In this picture, β is less than half, so most of the time we *do* make the right decision. The *power* of the test—the probability that we make the right decision—is shown as the region to the right of p^*. It's $1 - \beta$.

We calculate p^* based on the upper model because p^* depends only on the null model and the alpha level. No matter what the true proportion, no matter whether the practitioners can detect a human energy field 90%, 53%, or 2% of the time, p^* doesn't change. After all, we don't *know* the truth, so we can't use it to determine the critical value. But we always reject H_0 when $\hat{p} > p^*$.

How often we correctly reject H_0 when it's *false* depends on the effect size. We can see from the picture that if the effect size were larger (the true proportion were farther above the hypothesized value), the bottom curve would shift to the right, making the power greater.

We can see several important relationships from this figure:

- Power = $1 - \beta$.
- Reducing α to lower the chance of committing a Type I error will move the critical value, p^*, to the right (in this example). This will have the effect of increasing β, the probability of a Type II error, and correspondingly reducing the power.
- The larger the real difference between the hypothesized value, p_0, and the true population value, p, the smaller the chance of making a Type II error and the greater the power of the test. If the two proportions are very far apart, the two models will barely overlap, and we will not be likely to make any Type II errors at all—but then, we are unlikely to really need a formal hypothesis-testing procedure to see such an obvious difference. If the TT practitioners were successful almost all the time, we'd be able to see that with even a small experiment.

Reducing Both Type I and Type II Errors

Figure 21.3 seems to show that if we reduce Type I error, we automatically must increase Type II error. But there is a way to reduce both. Can you think of it?

If we can make both curves narrower, as shown in Figure 21.4, then both the probability of Type I errors and the probability of Type II errors will decrease, and the power of the test will increase.

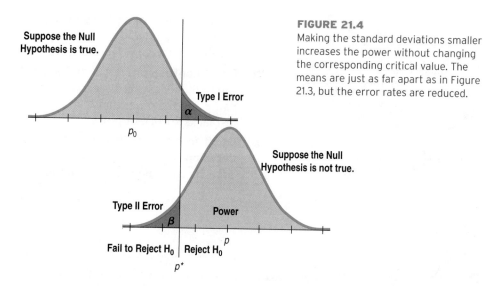

FIGURE 21.4
Making the standard deviations smaller increases the power without changing the corresponding critical value. The means are just as far apart as in Figure 21.3, but the error rates are reduced.

How can we accomplish that? The only way is to reduce the standard deviations by increasing the sample size. (Remember, these are pictures of sampling distribution models, not of data.) Increasing the sample size works regardless of the true population parameters. But recall the curse of diminishing returns. The standard deviation of the sampling distribution model decreases only as the *square root* of the sample size, so to halve the standard deviations we must *quadruple* the sample size.

FOR EXAMPLE Sample Size, Errors, and Power

RECAP: The meta-analysis of the risks of heart attacks in patients taking the diabetes drug Avandia combined results from 47 smaller studies. As GlaxoSmithKline (GSK), the drug's manufacturer, pointed out in their rebuttal, "Data from the ADOPT clinical trial did show a small increase in reports of myocardial infarction among the *Avandia*-treated group . . . however, the number of events is too small to reach a reliable conclusion about the role any of the medicines may have played in this finding."

QUESTION: Why would this smaller study have been less likely to detect the difference in risk? What are the appropriate statistical concepts for comparing the smaller studies?

Smaller studies are subject to greater sampling variability; that is, the sampling distributions they estimate have a larger standard deviation for the sample proportion. That gives small studies less power: They'd be less able to discern whether an apparently higher risk was merely the result of chance variation or evidence of real danger. The FDA doesn't want to restrict the use of a drug that's safe and effective (Type I error), nor do they want patients to continue taking a medication that puts them at risk (Type II error). Larger sample sizes can reduce the risk of both kinds of error. Greater power (the probability of rejecting a false null hypothesis) means a better chance of spotting a genuinely higher risk of heart attacks.

What Can Go Wrong?

- **Don't interpret the P-value as the probability that H_0 is true.** The P-value is about the data, not the hypothesis. It's the probability of observing data this unusual, *given* that H_0 is true, not the other way around.

- **Don't believe too strongly in arbitrary alpha levels.** There's not really much difference between a P-value of 0.051 and a P-value of 0.049, but sometimes it's regarded as the difference between night (having to refrain from rejecting H_0) and day (being able to shout to the world that your results are "statistically significant"). It may just be better to report the P-value and a confidence interval and let the world decide along with you.

- **Don't confuse practical and statistical significance.** A large sample size can make it easy to discern even a trivial change from the null hypothesis value. On the other hand, an important difference can be missed if your test lacks sufficient power.

- **Don't forget that in spite of all your care, you might make a wrong decision.** We can never reduce the probability of a Type I error (α) or of a Type II error (β) to zero (but increasing the sample size helps).

CONNECTIONS

All of the hypothesis tests we'll see boil down to the same question: "Is the difference between two quantities large?" We always measure "how large" by finding a ratio of this difference to the standard deviation of the sampling distribution of the statistic. Using the standard deviation as our ruler for inference is one of the core ideas of statistical thinking.

We've discussed the close relationship between hypothesis tests and confidence intervals. They are two sides of the same coin.

This chapter also has natural links to the discussion of probability, to the Normal model, and to the two previous chapters on inference.

WHAT HAVE WE LEARNED?

We've learned that there's a lot more to hypothesis testing than a simple yes/no decision.

▶ We've learned that the P-value can indicate evidence against the null hypothesis when it's small, but it does not tell us the probability that the null hypothesis is true.

▶ We've learned that the alpha level of the test establishes the level of proof we'll require. That determines the critical value of z that will lead us to reject the null hypothesis.

▶ We've also learned more about the connection between hypothesis tests and confidence intervals; they're really two ways of looking at the same question. The hypothesis test gives us the answer to a decision about a parameter; the confidence interval tells us the plausible values of that parameter.

We've learned about the two kinds of errors we might make, and we've seen why in the end we're never sure we've made the right decision.

▶ If the null hypothesis is really true and we reject it, that's a Type I error; the alpha level of the test is the probability that this could happen.

▶ If the null hypothesis is really false but we fail to reject it, that's a Type II error.

▶ The power of the test is the probability that we reject the null hypothesis when it's false. The larger the size of the effect we're testing for, the greater the power of the test to detect it.

▶ We've seen that tests with a greater likelihood of Type I error have more power and less chance of a Type II error. We can increase power while reducing the chances of both kinds of error by increasing the sample size.

Terms

Alpha level
The threshold P-value that determines when we reject a null hypothesis. If we observe a statistic whose P-value based on the null hypothesis is less than α, we reject that null hypothesis (p. 505).

Statistically significant
When the P-value falls below the alpha level, we say that the test is "statistically significant" at that alpha level (p. 505).

Significance level
The alpha level is also called the significance level, most often in a phrase such as a conclusion that a particular test is "significant at the 5% significance level" (p. 505).

Type I error
The error of rejecting a null hypothesis when in fact it is true (also called a "false positive"). The probability of a Type I error is α (p. 511).

Type II error
The error of failing to reject a null hypothesis when in fact it is false (also called a "false negative"). The probability of a Type II error is commonly denoted β and depends on the effect size (p. 511).

Power
The probability that a hypothesis test will correctly reject a false null hypothesis is the power of the test. To find power, we must specify a particular alternative parameter value as the "true" value. For any specific value in the alternative, the power is $1 - \beta$ (p. 513).

Effect size
The difference between the null hypothesis value and true value of a model parameter is called the effect size (p. 514).

Skills

▶ Understand that statistical significance does not measure the importance or magnitude of an effect. Recognize when others misinterpret statistical significance as proof of practical importance.

▶ Understand the close relationship between hypothesis tests and confidence intervals.

▶ Be able to identify and use the alternative hypothesis when testing hypotheses. Understand how to choose between a one-sided and two-sided alternative hypothesis, and know how to defend the choice of a one-sided alternative.

▶ Understand how the critical value for a test is related to the specified alpha level.

▶ Understand that the power of a test gives the probability that it correctly rejects a false null hypothesis when a specified alternative is true.

▶ Understand that the power of a test depends in part on the sample size. Larger sample sizes lead to greater power (and thus fewer Type II errors).

▶ Know how to complete a hypothesis test for a population proportion.

▶ Be able to interpret the meaning of a P-value in nontechnical language.

▶ Understand that the P-value of a test does not give the probability that the null hypothesis is correct.

▶ Know that we do not "accept" a null hypothesis if we cannot reject it but, rather, that we can only "fail to reject" the hypothesis for lack of evidence against it.

HYPOTHESIS TESTS ON THE COMPUTER

Reports about hypothesis tests generated by technologies don't follow a standard form. Most will name the test and provide the test statistic value, its standard deviation, and the P-value. But these elements may not be labeled clearly. For example, the expression "Prob > |z|" means the probability (the "Prob") of observing a test statistic whose magnitude (the absolute value tells us this) is larger than that of the one (the "z") found in the data (which, because it is written as "z," we know follows a Normal model). That is a fancy (and not very clear) way of saying P-value. In some packages, you can specify that the test be one-sided. Others might report three P-values, covering the ground for both one-sided tests and the two-sided test.

Sometimes a confidence interval and hypothesis test are automatically given together. The CI ought to be for the corresponding confidence level: $1 - \alpha$ for 2-tailed tests, $1 - 2\alpha$ for 1-tailed tests.

Often, the standard deviation of the statistic is called the "standard error," and usually that's appropriate because we've had to estimate its value from the data. That's not the case for proportions, however: We get the standard deviation for a proportion from the null hypothesis value. Nevertheless, you may see the standard deviation called a "standard error" even for tests with proportions.

It's common for statistics packages and calculators to report more digits of "precision" than could possibly have been found from the data. You can safely ignore them. Round values such as the standard deviation to one digit more than the number of digits reported in your data.

Here are the kind of results you might see. This is not from any program or calculator we know of, but it shows some of the things you might see in typical computer output.

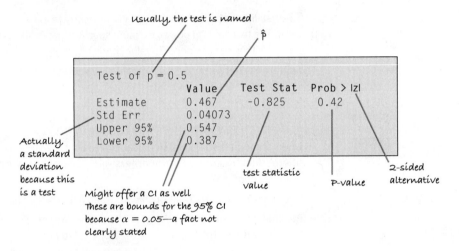

For information on hypothesis testing with particular statistics packages, see the Hypothesis Tests for Proportions on the Computer section in Chapter 20 on pages 493–494.

EXERCISES

1. **One sided or two?** In each of the following situations, is the alternative hypothesis one-sided or two-sided? What are the hypotheses?
   a) A business student conducts a taste test to see whether students prefer Diet Coke or Diet Pepsi.
   b) PepsiCo recently reformulated Diet Pepsi in an attempt to appeal to teenagers. They run a taste test to see if the new formula appeals to more teenagers than the standard formula.
   c) A budget override in a small town requires a two-thirds majority to pass. A local newspaper conducts a poll to see if there's evidence it will pass.
   d) One financial theory states that the stock market will go up or down with equal probability. A student collects data over several years to test the theory.

2. **Which alternative?** In each of the following situations, is the alternative hypothesis one-sided or two-sided? What are the hypotheses?
   a) A college dining service conducts a survey to see if students prefer plastic or metal cutlery.
   b) In recent years, 10% of college juniors have applied for study abroad. The dean's office conducts a survey to see if that's changed this year.
   c) A pharmaceutical company conducts a clinical trial to see if more patients who take a new drug experience headache relief than the 22% who claimed relief after taking the placebo.
   d) At a small computer peripherals company, only 60% of the hard drives produced passed all their performance tests the first time. Management recently invested a lot of resources into the production system and now conducts a test to see if it helped.

3. **P-value.** A medical researcher tested a new treatment for poison ivy against the traditional ointment. He concluded that the new treatment is more effective. Explain what the P-value of 0.047 means in this context.

4. **Another P-value.** Have harsher penalties and ad campaigns increased seat-belt use among drivers and passengers? Observations of commuter traffic failed to find evidence of a significant change compared with three years ago. Explain what the study's P-value of 0.17 means in this context.

5. **Alpha.** A researcher developing scanners to search for hidden weapons at airports has concluded that a new device is significantly better than the current scanner. He made this decision based on a test using $\alpha = 0.05$. Would he have made the same decision at $\alpha = 0.10$? How about $\alpha = 0.01$? Explain.

6. **Alpha, again.** Environmentalists concerned about the impact of high-frequency radio transmissions on birds found that there was no evidence of a higher mortality rate among hatchlings in nests near cell towers. They based this conclusion on a test using $\alpha = 0.05$. Would they have made the same decision at $\alpha = 0.10$? How about $\alpha = 0.01$? Explain.

7. **Significant?** Public health officials believe that 90% of children have been vaccinated against measles. A random survey of medical records at many schools across the country found that, among more than 13,000 children, only 89.4% had been vaccinated. A statistician would reject the 90% hypothesis with a P-value of $P = 0.011$.

   a) Explain what the P-value means in this context.
   b) The result is statistically significant, but is it important? Comment.

8. **Significant again?** A new reading program may reduce the number of elementary school students who read below grade level. The company that developed this program supplied materials and teacher training for a large-scale test involving nearly 8500 children in several different school districts. Statistical analysis of the results showed that the percentage of students who did not meet the grade-level goal was reduced from 15.9% to 15.1%. The hypothesis that the new reading program produced no improvement was rejected with a P-value of 0.023.

   a) Explain what the P-value means in this context.
   b) Even though this reading method has been shown to be significantly better, why might you not recommend that your local school adopt it?

9. **Success.** In August 2004, *Time* magazine reported the results of a random telephone poll commissioned by the Spike network. Of the 1302 men who responded, only 39 said that their most important measure of success was their work.

   a) Estimate the percentage of all American males who measure success primarily from their work. Use a 98% confidence interval. Check the conditions first.
   b) Some believe that few contemporary men judge their success primarily by their work. Suppose we wished to conduct a hypothesis test to see if the fraction has

fallen below the 5% mark. What does your confidence interval indicate? Explain.

   c) What is the level of significance of this test? Explain.

10. **Is the Euro fair?** Soon after the Euro was introduced as currency in Europe, it was widely reported that someone had spun a Euro coin 250 times and gotten heads 140 times. We wish to test a hypothesis about the fairness of spinning the coin.

    a) Estimate the true proportion of heads. Use a 95% confidence interval. Don't forget to check the conditions.
    b) Does your confidence interval provide evidence that the coin is unfair when spun? Explain.
    c) What is the significance level of this test? Explain.

11. **Approval 2007.** In May 2007, George W. Bush's approval rating stood at 30% according to a CBS News/*New York Times* national survey of 1125 randomly selected adults.

    a) Make a 95% confidence interval for his approval rating by all U.S. adults.
    b) Based on the confidence interval, test the null hypothesis that Bush's approval rating was no better than the 27% level established by Richard Nixon during the Watergate scandal.

12. **Superdads.** The Spike network commissioned a telephone poll of randomly sampled U.S. men. Of the 712 respondents who had children, 22% said "yes" to the question "Are you a stay-at-home dad?" (*Time*, August 23, 2004).

    a) To help market commercial time, Spike wants an accurate estimate of the true percentage of stay-at-home dads. Construct a 95% confidence interval.
    b) An advertiser of baby-carrying slings for dads will buy commercial time if at least 25% of men are stay-at-home dads. Use your confidence interval to test an appropriate hypothesis, and make a recommendation to the advertiser.
    c) Could Spike claim to the advertiser that it is possible that 25% of men with young children are stay-at-home dads? What is wrong with the reasoning?

13. **Dogs.** Canine hip dysplasia is a degenerative disease that causes pain in many dogs. Sometimes advanced warning signs appear in puppies as young as 6 months. A veterinarian checked 42 puppies whose owners brought them to a vaccination clinic, and she found 5 with early hip dysplasia. She considers this group to be a random sample of all puppies.

    a) Explain why we cannot use this information to construct a confidence interval for the rate of occurrence of early hip dysplasia among all 6-month-old puppies.
    b) Construct a "plus-four" confidence interval and interpret it in this context.

14. **Fans.** A survey of 81 randomly selected people standing in line to enter a football game found that 73 of them were home team fans.

    a) Explain why we cannot use this information to construct a confidence interval for the proportion of all people at the game who are fans of the home team.
    b) Construct a "plus-four" confidence interval and interpret it in this context.

15. **Loans.** Before lending someone money, banks must decide whether they believe the applicant will repay the loan. One strategy used is a point system. Loan officers assess information about the applicant, totaling points they award for the person's income level, credit history, current debt burden, and so on. The higher the point total, the more convinced the bank is that it's safe to make the loan. Any applicant with a lower point total than a certain cutoff score is denied a loan.

    We can think of this decision as a hypothesis test. Since the bank makes its profit from the interest collected on repaid loans, their null hypothesis is that the applicant will repay the loan and therefore should get the money. Only if the person's score falls below the minimum cutoff will the bank reject the null and deny the loan. This system is reasonably reliable, but, of course, sometimes there are mistakes.

    a) When a person defaults on a loan, which type of error did the bank make?
    b) Which kind of error is it when the bank misses an opportunity to make a loan to someone who would have repaid it?
    c) Suppose the bank decides to lower the cutoff score from 250 points to 200. Is that analogous to choosing a higher or lower value of α for a hypothesis test? Explain.
    d) What impact does this change in the cutoff value have on the chance of each type of error?

16. **Spam.** Spam filters try to sort your e-mails, deciding which are real messages and which are unwanted. One method used is a point system. The filter reads each incoming e-mail and assigns points to the sender, the subject, key words in the message, and so on. The higher the point total, the more likely it is that the message is unwanted. The filter has a cutoff value for the point total; any message rated lower than that cutoff passes through to your inbox, and the rest, suspected to be spam, are diverted to the junk mailbox.

    We can think of the filter's decision as a hypothesis test. The null hypothesis is that the e-mail is a real message and should go to your inbox. A higher point total provides evidence that the message may be spam; when there's sufficient evidence, the filter rejects the null, classifying the message as junk. This usually works pretty well, but, of course, sometimes the filter makes a mistake.

    a) When the filter allows spam to slip through into your inbox, which kind of error is that?
    b) Which kind of error is it when a real message gets classified as junk?
    c) Some filters allow the user (that's you) to adjust the cutoff. Suppose your filter has a default cutoff of 50 points, but you reset it to 60. Is that analogous to choosing a higher or lower value of α for a hypothesis test? Explain.
    d) What impact does this change in the cutoff value have on the chance of each type of error?

17. **Second loan.** Exercise 15 describes the loan score method a bank uses to decide which applicants it will lend money. Only if the total points awarded for various aspects of an applicant's financial condition fail to add up to a minimum cutoff score set by the bank will the loan be denied.

    a) In this context, what is meant by the power of the test?
    b) What could the bank do to increase the power?
    c) What's the disadvantage of doing that?

18. **More spam.** Consider again the points-based spam filter described in Exercise 16. When the points assigned to various components of an e-mail exceed the cutoff value you've set, the filter rejects its null hypothesis (that the message is real) and diverts that e-mail to a junk mailbox.

    a) In this context, what is meant by the power of the test?
    b) What could you do to increase the filter's power?
    c) What's the disadvantage of doing that?

19. **Homeowners 2005.** In 2005 the U.S. Census Bureau reported that 68.9% of American families owned their homes. Census data reveal that the ownership rate in one small city is much lower. The city council is debating a plan to offer tax breaks to first-time home buyers in order to encourage people to become homeowners. They decide to adopt the plan on a 2-year trial basis and use the data they collect to make a decision about continuing the tax breaks. Since this plan costs the city tax revenues, they will continue to use it only if there is strong evidence that the rate of home ownership is increasing.

    a) In words, what will their hypotheses be?
    b) What would a Type I error be?
    c) What would a Type II error be?
    d) For each type of error, tell who would be harmed.
    e) What would the power of the test represent in this context?

20. **Alzheimer's.** Testing for Alzheimer's disease can be a long and expensive process, consisting of lengthy tests and medical diagnosis. A group of researchers (Solomon et al., 1998) devised a 7-minute test to serve as a quick screen for the disease for use in the general population of senior citizens. A patient who tested positive would then go through the more expensive battery of tests and medical diagnosis. The authors reported a false positive rate of 4% and a false negative rate of 8%.

    a) Put this in the context of a hypothesis test. What are the null and alternative hypotheses?
    b) What would a Type I error mean?
    c) What would a Type II error mean?
    d) Which is worse here, a Type I or Type II error? Explain.
    e) What is the power of this test?

21. **Testing cars.** A clean air standard requires that vehicle exhaust emissions not exceed specified limits for various pollutants. Many states require that cars be tested annually to be sure they meet these standards. Suppose state regulators double-check a random sample of cars that a suspect repair shop has certified as okay. They will revoke the shop's license if they find significant evidence that the shop is certifying vehicles that do not meet standards.

    a) In this context, what is a Type I error?
    b) In this context, what is a Type II error?
    c) Which type of error would the shop's owner consider more serious?
    d) Which type of error might environmentalists consider more serious?

22. Quality control. Production managers on an assembly line must monitor the output to be sure that the level of defective products remains small. They periodically inspect a random sample of the items produced. If they find a significant increase in the proportion of items that must be rejected, they will halt the assembly process until the problem can be identified and repaired.

a) In this context, what is a Type I error?
b) In this context, what is a Type II error?
c) Which type of error would the factory owner consider more serious?
d) Which type of error might customers consider more serious?

23. Cars, again. As in Exercise 21, state regulators are checking up on repair shops to see if they are certifying vehicles that do not meet pollution standards.

a) In this context, what is meant by the power of the test the regulators are conducting?
b) Will the power be greater if they test 20 or 40 cars? Why?
c) Will the power be greater if they use a 5% or a 10% level of significance? Why?
d) Will the power be greater if the repair shop's inspectors are only a little out of compliance or a lot? Why?

24. Production. Consider again the task of the quality control inspectors in Exercise 22.

a) In this context, what is meant by the power of the test the inspectors conduct?
b) They are currently testing 5 items each hour. Someone has proposed that they test 10 instead. What are the advantages and disadvantages of such a change?
c) Their test currently uses a 5% level of significance. What are the advantages and disadvantages of changing to an alpha level of 1%?
d) Suppose that, as a day passes, one of the machines on the assembly line produces more and more items that are defective. How will this affect the power of the test?

25. Equal opportunity? A company is sued for job discrimination because only 19% of the newly hired candidates were minorities when 27% of all applicants were minorities. Is this strong evidence that the company's hiring practices are discriminatory?

a) Is this a one-tailed or a two-tailed test? Why?
b) In this context, what would a Type I error be?
c) In this context, what would a Type II error be?
d) In this context, what is meant by the power of the test?
e) If the hypothesis is tested at the 5% level of significance instead of 1%, how will this affect the power of the test?
f) The lawsuit is based on the hiring of 37 employees. Is the power of the test higher than, lower than, or the same as it would be if it were based on 87 hires?

26. Stop signs. Highway safety engineers test new road signs, hoping that increased reflectivity will make them more visible to drivers. Volunteers drive through a test course with several of the new- and old-style signs and rate which kind shows up the best.

a) Is this a one-tailed or a two-tailed test? Why?
b) In this context, what would a Type I error be?
c) In this context, what would a Type II error be?
d) In this context, what is meant by the power of the test?
e) If the hypothesis is tested at the 1% level of significance instead of 5%, how will this affect the power of the test?
f) The engineers hoped to base their decision on the reactions of 50 drivers, but time and budget constraints may force them to cut back to 20. How would this affect the power of the test? Explain.

27. Dropouts. A Statistics professor has observed that for several years about 13% of the students who initially enroll in his Introductory Statistics course withdraw before the end of the semester. A salesman suggests that he try a statistics software package that gets students more involved with computers, predicting that it will cut the dropout rate. The software is expensive, and the salesman offers to let the professor use it for a semester to see if the dropout rate goes down significantly. The professor will have to pay for the software only if he chooses to continue using it.

a) Is this a one-tailed or two-tailed test? Explain.
b) Write the null and alternative hypotheses.
c) In this context, explain what would happen if the professor makes a Type I error.
d) In this context, explain what would happen if the professor makes a Type II error.
e) What is meant by the power of this test?

28. Ads. A company is willing to renew its advertising contract with a local radio station only if the station can prove that more than 20% of the residents of the city have heard the ad and recognize the company's product. The radio station conducts a random phone survey of 400 people.

a) What are the hypotheses?
b) The station plans to conduct this test using a 10% level of significance, but the company wants the significance level lowered to 5%. Why?
c) What is meant by the power of this test?
d) For which level of significance will the power of this test be higher? Why?
e) They finally agree to use $\alpha = 0.05$, but the company proposes that the station call 600 people instead of the 400 initially proposed. Will that make the risk of Type II error higher or lower? Explain.

29. Dropouts, part II. Initially, 203 students signed up for the Stats course in Exercise 27. They used the software suggested by the salesman, and only 11 dropped out of the course.

a) Should the professor spend the money for this software? Support your recommendation with an appropriate test.
b) Explain what your P-value means in this context.

30. Testing the ads. The company in Exercise 28 contacts 600 people selected at random, and only 133 remember the ad.

a) Should the company renew the contract? Support your recommendation with an appropriate test.

b) Explain what your P-value means in this context.

31. **Two coins.** In a drawer are two coins. They look the same, but one coin produces heads 90% of the time when spun while the other one produces heads only 30% of the time. You select one of the coins. You are allowed to spin it *once* and then must decide whether the coin is the 90%- or the 30%-head coin. Your null hypothesis is that your coin produces 90% heads.

a) What is the alternative hypothesis?

b) Given that the outcome of your spin is tails, what would you decide? What if it were heads?

c) How large is α in this case?

d) How large is the power of this test? (*Hint:* How many possibilities are in the alternative hypothesis?)

e) How could you lower the probability of a Type I error and increase the power of the test at the same time?

32. **Faulty or not?** You are in charge of shipping computers to customers. You learn that a faulty disk drive was put into some of the machines. There's a simple test you can perform, but it's not perfect. All but 4% of the time, a good disk drive passes the test, but unfortunately, 35% of the bad disk drives pass the test, too. You have to decide on the basis of one test whether the disk drive is good or bad. Make this a hypothesis test.

a) What are the null and alternative hypotheses?

b) Given that a computer fails the test, what would you decide? What if it passes the test?

c) How large is α for this test?

d) What is the power of this test? (*Hint:* How many possibilities are in the alternative hypothesis?)

33. **Hoops.** A basketball player with a poor foul-shot record practices intensively during the off-season. He tells the coach that he has raised his proficiency from 60% to 80%. Dubious, the coach asks him to take 10 shots, and is surprised when the player hits 9 out of 10. Did the player prove that he has improved?

a) Suppose the player really is no better than before—still a 60% shooter. What's the probability he can hit at least 9 of 10 shots anyway? (*Hint:* Use a Binomial model.)

b) If that is what happened, now the coach thinks the player has improved when he has not. Which type of error is that?

c) If the player really can hit 80% now, and it takes at least 9 out of 10 successful shots to convince the coach, what's the power of the test?

d) List two ways the coach and player could increase the power to detect any improvement.

34. **Pottery.** An artist experimenting with clay to create pottery with a special texture has been experiencing difficulty with these special pieces. About 40% break in the kiln during firing. Hoping to solve this problem, she buys some more expensive clay from another supplier. She plans to make and fire 10 pieces and will decide to use the new clay if at most one of them breaks.

a) Suppose the new, expensive clay really is no better than her usual clay. What's the probability that this test convinces her to use it anyway? (*Hint:* Use a Binomial model.)

b) If she decides to switch to the new clay and it is no better, what kind of error did she commit?

c) If the new clay really can reduce breakage to only 20%, what's the probability that her test will not detect the improvement?

d) How can she improve the power of her test? Offer at least two suggestions.

JUST CHECKING

ANSWERS

1. With a z-score of 0.62, you can't reject the null hypothesis. The experiment shows no evidence that the wheel is not fair.

2. At $\alpha = 0.05$, you can't reject the null hypothesis because 0.30 is contained in the 90% confidence interval—it's plausible that sending the DVDs is no more effective than just sending letters.

3. The confidence interval is from 29% to 45%. The DVD strategy is more expensive and may not be worth it. We can't distinguish the success rate from 30% given the results of this experiment, but 45% would represent a large improvement. The bank should consider another trial, increasing their sample size to get a narrower confidence interval.

4. A Type I error would mean deciding that the DVD success rate is higher than 30% when it really isn't. They would adopt a more expensive method for collecting payments that's no better than the less expensive strategy.

5. A Type II error would mean deciding that there's not enough evidence to say that the DVD strategy works when in fact it does. The bank would fail to discover an effective method for increasing their revenue from delinquent accounts.

6. 60%; the larger the effect size, the greater the power. It's easier to detect an improvement to a 60% success rate than to a 32% rate.

Comparing Two Proportions

Where are we going?

Is the proportion of men who like our new Web page the same as the proportion of women who like it? Do people really feel better about the economy this month compared to last month, or was that increase just sampling variation? In practice, it's much more common to compare two proportions than to test whether one is equal to a given number. That's how we'll see whether two groups are different or whether there's been a change over time. Comparing two proportions is very much like testing one. The standard error is different, but the concepts are really the same.

WHO	6971 male drivers
WHAT	Seat-belt use
WHY	Highway safety
WHEN	2007
WHERE	Massachusetts

Do men take more risks than women? Psychologists have documented that in many situations, men choose riskier behavior than women do. But what is the effect of having a woman by their side? A recent seat-belt observation study in Massachusetts[1] found that, not surprisingly, male drivers wear seat belts less often than women do. The study also noted that men's belt-wearing jumped more than 16 percentage points when they had a female passenger. Seat-belt use was recorded at 161 locations in Massachusetts, using random-sampling methods developed by the National Highway Traffic Safety Administration (NHTSA). Female drivers wore belts more than 70% of the time, regardless of the sex of their passengers. Of 4208 male drivers with female passengers, 2777 (66.0%) were belted. But among 2763 male drivers with male passengers only, 1363 (49.3%) wore seat belts. This was only a random sample, but it suggests there may be a shift in men's risk-taking behavior when women are present. What would we estimate the true size of that gap to be?

Comparisons between two percentages are much more common than questions about isolated percentages. And they are more interesting. We often want to know how two groups differ, whether a treatment is better than a placebo control, or whether this year's results are better than last year's.

Another Ruler

We know the difference between the proportions of men wearing seat belts seen in the *sample*. It's 16.7%. But what's the *true* difference for all men? We know that our estimate probably isn't exactly right. To say more, we need a

---

[1] Massachusetts Traffic Safety Research Program [June 2007].

new ruler—the standard deviation of the sampling distribution model for the difference in the proportions. Now we have two proportions, and each will vary from sample to sample. We are interested in the difference between them. So what is the correct standard deviation?

The answer comes from Chapter 16. Remember the Pythagorean Theorem of Statistics?

> *The variance of the sum or difference of two independent random variables is the sum of their variances.*

This is such an important (and powerful) idea in Statistics that it's worth pausing a moment to review the reasoning. Here's some intuition about why variation increases even when we subtract two random quantities.

Grab a full box of cereal. The box claims to contain 16 ounces of cereal. We know that's not exact: There's some small variation from box to box. Now pour a bowl of cereal. Of course, your 2-ounce serving will not be exactly 2 ounces. There'll be some variation there, too. How much cereal would you guess was left in the box? Do you think your guess will be as close as your guess for the full box? *After* you pour your bowl, the amount of cereal in the box is still a random quantity (with a smaller mean than before), but it is even *more variable* because of the additional variation in the amount you poured.

According to the rule, the variance of the amount of cereal left in the box would now be the *sum* of the two *variances*.

We want a standard deviation, not a variance, but that's just a square root away. We can write symbolically what we've just said:

For independent random variables, **variances add.**

$$Var(X - Y) = Var(X) + Var(Y), \text{ so}$$
$$SD(X - Y) = \sqrt{SD^2(X) + SD^2(Y)} = \sqrt{Var(X) + Var(Y)}.$$

Be careful, though—this simple formula applies only when X and Y are independent. Just as the Pythagorean Theorem[2] works only for right triangles, the formula works only for independent random variables. Always check for independence before using it.

The Standard Deviation of the Difference Between Two Proportions

Combining independent random quantities always *increases* the overall variation, so even for *differences* of independent random variables, **variances add.**

Fortunately, proportions observed in independent random samples *are* independent, so we can put the two proportions in for X and Y and add their variances. We just need to use careful notation to keep things straight.

When we have two samples, each can have a different size and proportion value, so we keep them straight with subscripts. Often we choose subscripts that remind us of the groups. For our example, we might use "$_M$" and "$_F$", but generically we'll just use "$_1$" and "$_2$". We will represent the two sample proportions as $\hat{p}_1$ and $\hat{p}_2$, and the two sample sizes as n_1 and n_2.

The standard deviations of the sample proportions are $SD(\hat{p}_1) = \sqrt{\dfrac{p_1 q_1}{n_1}}$ and $SD(\hat{p}_2) = \sqrt{\dfrac{p_2 q_2}{n_2}}$, so the variance of the difference in the proportions is

$$Var(\hat{p}_1 - \hat{p}_2) = \left(\sqrt{\frac{p_1 q_1}{n_1}}\right)^2 + \left(\sqrt{\frac{p_2 q_2}{n_2}}\right)^2 = \frac{p_1 q_1}{n_1} + \frac{p_2 q_2}{n_2}.$$

---

[2] If you don't remember the formula, don't rely on the Scarecrow's version from *The Wizard of Oz*. He may have a brain and have been awarded his Th.D. (Doctor of Thinkology), but he gets the formula wrong.

The standard deviation is the square root of that variance:

$$SD(\hat{p}_1 - \hat{p}_2) = \sqrt{\frac{p_1 q_1}{n_1} + \frac{p_2 q_2}{n_2}}.$$

We usually don't know the true values of p_1 and p_2. When we have the sample proportions in hand from the data, we use them to estimate the variances. So the standard error is

$$SE(\hat{p}_1 - \hat{p}_2) = \sqrt{\frac{\hat{p}_1 \hat{q}_1}{n_1} + \frac{\hat{p}_2 \hat{q}_2}{n_2}}.$$

FOR EXAMPLE | Finding the Standard Error of a Difference in Proportions

A recent survey of 886 randomly selected teenagers (aged 12-17) found that more than half of them had online profiles.[3] Some researchers and privacy advocates are concerned about the possible access to personal information about teens in public places on the Internet. There appear to be differences between boys and girls in their online behavior. Among teens aged 15-17, 57% of the 248 boys had posted profiles, compared to 70% of the 256 girls. Let's start the process of estimating how large the true gender gap might be.

QUESTION: What's the standard error of the difference in sample proportions?

Because the boys and girls were selected at random, it's reasonable to assume their behaviors are independent, so it's okay to use the Pythagorean Theorem of Statistics and add the variances:

$$SE(\hat{p}_{boys}) = \sqrt{\frac{0.57 \times 0.43}{248}} = 0.0314 \qquad SE(\hat{p}_{girls}) = \sqrt{\frac{0.70 \times 0.30}{256}} = 0.0286$$

$$SE(\hat{p}_{girls} - \hat{p}_{boys}) = \sqrt{0.0314^2 + 0.0286^2} = 0.0425$$

Assumptions and Conditions

Before we look at our example, we need to check assumptions and conditions.

Independence Assumption

Independence Assumption: Within each group, the data should be based on results for independent individuals. We can't check that for certain, but we *can* check the following:

Randomization Condition: The data in each group should be drawn independently and at random from a homogeneous population or generated by a randomized comparative experiment.

The 10% Condition: If the data are sampled without replacement, the sample should not exceed 10% of the population.

Because we are comparing two groups in this way, we need an additional Independence Assumption. In fact, this is the most important of these assumptions. If it is violated, these methods just won't work.

Independent Groups Assumption: The two groups we're comparing must also be independent *of each other*. Usually, the independence of the groups from each other is evident from the way the data were collected.

Why is the Independent Groups Assumption so important? If we compare husbands with their wives, or a group of subjects before and after some treatment, we can't just add the variances. Subjects' performance before a treatment might very well be related to their performance after the treatment. So the proportions are not independent and the Pythagorean-style variance formula does

---

[3] Princeton Survey Research Associates International for the Pew Internet & American Life Project.

not hold. We'll see a way to compare a common kind of nonindependent sample in a later chapter.

Sample Size Assumption

Each of the groups must be big enough. As with individual proportions, we need larger groups to estimate proportions that are near 0% or 100%. We usually check the Success/Failure Condition for each group.

Success/Failure Condition: Both groups are big enough that at least 10 successes and at least 10 failures have been observed in each.

FOR EXAMPLE **Checking Assumptions and Conditions**

RECAP: Among randomly sampled teens aged 15-17, 57% of the 248 boys had posted online profiles, compared to 70% of the 256 girls.

QUESTION: Can we use these results to make inferences about all 15-17-year-olds?

✔ **Randomization Condition:** The sample of boys and the sample of girls were both chosen randomly.

✔ **10% Condition:** 248 boys and 256 girls are each less than 10% of all teenage boys and girls.

✔ **Independent Groups Assumption:** Because the samples were selected at random, it's reasonable to believe the boys' online behaviors are independent of the girls' online behaviors.

✔ **Success/Failure Condition:** Among the boys, 248(0.57) = 141 had online profiles and the other 248(0.43) = 107 did not. For the girls, 256(0.70) = 179 successes and 256(0.30) = 77 failures. All counts are at least 10.

Because all the assumptions and conditions are satisfied, it's okay to proceed with inference for the difference in proportions.

(Note that when we find the *observed* counts of successes and failures, we round off to whole numbers. We're using the reported percentages to recover the actual counts.)

The Sampling Distribution

We're almost there. We just need one more fact about proportions. We already know that for large enough samples, each of our proportions has an approximately Normal sampling distribution. The same is true of their difference.

WHY NORMAL?

In Chapter 16 we learned that sums and differences of independent Normal random variables also follow a Normal model. That's the reason we use a Normal model for the difference of two independent proportions.

THE SAMPLING DISTRIBUTION MODEL FOR A DIFFERENCE BETWEEN TWO INDEPENDENT PROPORTIONS

Provided that the sampled values are independent, the samples are independent, and the sample sizes are large enough, the sampling distribution of $\hat{p}_1 - \hat{p}_2$ is modeled by a Normal model with mean $\mu = p_1 - p_2$ and standard deviation

$$SD(\hat{p}_1 - \hat{p}_2) = \sqrt{\frac{p_1 q_1}{n_1} + \frac{p_2 q_2}{n_2}}.$$

The sampling distribution model and the standard deviation give us all we need to find a margin of error for the difference in proportions—or at least they would if we knew the true proportions, p_1 and p_2. However, we don't know the true values, so we'll work with the observed proportions, $\hat{p}_1$ and $\hat{p}_2$, and use $SE(\hat{p}_1 - \hat{p}_2)$ to estimate the standard deviation. The rest is just like a one-proportion z-interval.

A S   *Activity:* **Compare Two Proportions.**
Does a preschool program help
disadvantaged children later in life?

A TWO-PROPORTION z-INTERVAL

When the conditions are met, we are ready to find the confidence interval for the difference of two proportions, $p_1 - p_2$. The confidence interval is

$$(\hat{p}_1 - \hat{p}_2) \pm z^* \times SE(\hat{p}_1 - \hat{p}_2)$$

where we find the standard error of the difference,

$$SE(\hat{p}_1 - \hat{p}_2) = \sqrt{\frac{\hat{p}_1 \hat{q}_1}{n_1} + \frac{\hat{p}_2 \hat{q}_2}{n_2}},$$

from the observed proportions.

The critical value z^* depends on the particular confidence level, C, that we specify.

FOR EXAMPLE   Finding a Two-Proportion z-Interval

RECAP: Among randomly sampled teens aged 15–17, 57% of the 248 boys had posted online profiles, compared to 70% of the 256 girls. We calculated the standard error for the difference in sample proportions to be $SE(\hat{p}_{girls} - \hat{p}_{boys}) = 0.0425$ and found that the assumptions and conditions required for inference checked out okay.

QUESTION: What does a confidence interval say about the difference in online behavior?

A 95% confidence interval for $p_{girls} - p_{boys}$ is $(\hat{p}_{girls} - \hat{p}_{boys}) \pm z^* SE(\hat{p}_{girls} - \hat{p}_{boys})$

$$(0.70 - 0.57) \pm 1.96(0.0425)$$
$$0.13 \pm 0.083$$
$$(4.7\%, 21.3\%)$$

We can be 95% confident that among teens aged 15–17, the proportion of girls who post online profiles is between 4.7 and 21.3 percentage points higher than the proportion of boys who do. It seems clear that teen girls are more likely to post profiles than are boys the same age.

STEP-BY-STEP EXAMPLE   A Two-Proportion z-Interval

Now we are ready to be more precise about the passenger-based gap in male drivers' seat-belt use. We'll estimate the difference with a confidence interval using a method called the **two-proportion z-interval** and follow the four confidence interval steps.

Question: How much difference is there in the proportion of male drivers who wear seat belts when sitting next to a male passenger and the proportion who wear seat belts when sitting next to a female passenger?

| THINK | **Plan** State what you want to know. Discuss the variables and the W's.

Identify the parameter you wish to estimate. (It usually doesn't matter in which direction we subtract, so, for convenience, we usually choose the direction with a positive difference.) | I want to know the true difference in the population proportion, p_M, of male drivers who wear seat belts when sitting next to a man and p_F, the proportion who wear seat belts when sitting next to a woman. The data are from a random sample of drivers in Massachusetts in 2007, observed according to procedures developed by the NHTSA. The parameter of interest is the difference $p_F - p_M$. |

Choose and state a confidence level.	I will find a 95% confidence interval for this parameter.
Model Think about the assumptions and check the conditions.	✔ **Independence Assumption:** Driver behavior was independent from car to car. ✔ **Randomization Condition:** The NHTSA methods are more complex than an SRS, but they result in a suitable random sample. ✔ **10% Condition:** The samples include far fewer than 10% of all male drivers accompanied by male or by female passengers. ✔ **Independent Groups Assumption:** There's no reason to believe that seat-belt use among drivers with male passengers and those with female passengers are not independent.
The Success/Failure Condition must hold for each group.	✔ **Success/Failure Condition:** Among male drivers with female passengers, 2777 wore seat belts and 1431 did not; of those driving with male passengers, 1363 wore seat belts and 1400 did not. Each group contained far more than 10 successes and 10 failures.
State the sampling distribution model for the statistic. Choose your method.	Under these conditions, the sampling distribution of the difference between the sample proportions is approximately Normal, so I'll find a **two-proportion z-interval.**

 Mechanics Construct the confidence interval.

I know

$$n_F = 4208, n_M = 2763.$$

The observed sample proportions are

$$\hat{p}_F = \frac{2777}{4208} = 0.660, \hat{p}_M = \frac{1363}{2763} = 0.493$$

As often happens, the key step in finding the confidence interval is estimating the standard deviation of the sampling distribution model of the statistic. Here the statistic is the difference in the proportions of men who wear seat belts when they have a female passenger and the proportion who do so with a male passenger. Substitute the data values into the formula.

I'll estimate the SD of the difference with

$$SE(\hat{p}_F - \hat{p}_M) = \sqrt{\frac{\hat{p}_F \hat{q}_F}{n_F} + \frac{\hat{p}_M \hat{q}_M}{n_M}}$$

$$= \sqrt{\frac{(0.660)(0.340)}{4208} + \frac{(0.493)(0.507)}{2763}}$$

$$= 0.012$$

The sampling distribution is Normal, so the critical value for a 95% confidence interval, z^*, is 1.96. The margin of error is the critical value times the SE.

$$ME = z^* \times SE(\hat{p}_F - \hat{p}_M)$$
$$= 1.96(0.012) = 0.024$$

The confidence interval is the statistic $\pm$ME.

The observed difference in proportions is $\hat{p}_F - \hat{p}_M = 0.660 - 0.493 = 0.167$, so the 95% confidence interval is

$$0.167 \pm 0.024$$

or 14.3% to 19.1%

Conclusion Interpret your confidence interval in the proper context. (Remember: We're 95% confident that our interval captured the true difference.)

> I am 95% confident that the proportion of male drivers who wear seat belts when driving next to a female passenger is between 14.3 and 19.1 percentage points higher than the proportion who wear seat belts when driving next to a male passenger.

This is an interesting result—but be careful not to try to say too much! In Massachusetts, overall seat-belt use is lower than the national average, so we can't be certain that these results generalize to other states. And these were two different groups of men, so we can't say that, individually, men are more likely to buckle up when they have a woman passenger. You can probably think of several alternative explanations; we'll suggest just a couple. Perhaps age is a lurking variable: Maybe older men are more likely to wear seat belts and also more likely to be driving with their wives. Or maybe men who don't wear seat belts have trouble attracting women!

 A public broadcasting station plans to launch a special appeal for additional contributions from current members. Unsure of the most effective way to contact people, they run an experiment. They randomly select two groups of current members. They send the same request for donations to everyone, but it goes to one group by e-mail and to the other group by regular mail. The station was successful in getting contributions from 26% of the members they e-mailed but only from 15% of those who received the request by regular mail. A 90% confidence interval estimated the difference in donation rates to be 11% ± 7%.

1. Interpret the confidence interval in this context.

2. Based on this confidence interval, what conclusion would we reach if we tested the hypothesis that there's no difference in the response rates to the two methods of fund-raising? Explain.

Will I Snore When I'm 64?

WHO	Randomly selected U.S. adults over age 18
WHAT	Proportion who snore, categorized by age (less than 30, 30 or older)
WHEN	2001
WHERE	United States
WHY	To study sleep behaviors of U.S. adults

The National Sleep Foundation asked a random sample of 1010 U.S. adults questions about their sleep habits. The sample was selected in the fall of 2001 from random telephone numbers, stratified by region and sex, guaranteeing that an equal number of men and women were interviewed (2002 Sleep in America Poll, National Sleep Foundation, Washington, D.C.).

One of the questions asked about snoring. Of the 995 respondents, 37% of adults reported that they snored at least a few nights a week during the past year. Would you expect that percentage to be the same for all age groups? Split into two age categories, 26% of the 184 people under 30 snored, compared with 39% of the 811 in the older group. Is this difference of 13% real, or due only to natural fluctuations in the sample we've chosen?

The question calls for a hypothesis test. Now the parameter of interest is the true *difference* between the (reported) snoring rates of the two age groups.

What's the appropriate null hypothesis? That's easy here. We hypothesize that there is no difference in the proportions. This is such a natural null hypothesis that we rarely consider any other. But instead of writing $H_0: p_1 = p_2$, we usually express it in a slightly different way. To make it relate directly to the *difference*, we hypothesize that the difference in proportions is zero:

$$H_0: p_1 - p_2 = 0.$$

Everyone into the Pool

Our hypothesis is about a new parameter: the *difference* in proportions. We'll need a standard error for that. Wait—don't we know that already? Yes and no. We know that the standard error of the difference in proportions is

$$SE(\hat{p}_1 - \hat{p}_2) = \sqrt{\frac{\hat{p}_1\hat{q}_1}{n_1} + \frac{\hat{p}_2\hat{q}_2}{n_2}},$$

and we could just plug in the numbers, but we can do even better. The secret is that proportions and their standard deviations are linked. There are two proportions in the standard error formula—but look at the null hypothesis. It says that these proportions are equal. To do a hypothesis test, we *assume* that the null hypothesis is true. So there should be just a single value of $\hat{p}$ in the SE formula (and, of course, $\hat{q}$ is just $1 - \hat{p}$).

How would we do this for the snoring example? If the null hypothesis is true, then, among all adults, the two groups have the same proportion. Overall, we saw $48 + 318 = 366$ snorers out of a total of $184 + 811 = 995$ adults who responded to this question. The overall proportion of snorers was $366/995 = 0.3678$.

Combining the counts like this to get an overall proportion is called **pooling**. Whenever we have data from different sources or different groups but we believe that they really came from the same underlying population, we pool them to get better estimates.

When we have counts for each group, we can find the pooled proportion as

$$\hat{p}_{\text{pooled}} = \frac{Success_1 + Success_2}{n_1 + n_2},$$

where $Success_1$ is the number of successes in group 1 and $Success_2$ is the number of successes in group 2. That's the overall proportion of success.

When we have only proportions and not the counts, as in the snoring example, we have to reconstruct the number of successes by multiplying the sample sizes by the proportions:

$$Success_1 = n_1\hat{p}_1 \quad \text{and} \quad Success_2 = n_2\hat{p}_2.$$

If these calculations don't come out to whole numbers, round them first. There must have been a whole number of successes, after all. (This is the *only* time you should round values in the middle of a calculation.)

We then put this pooled value into the formula, substituting it for *both* sample proportions in the standard error formula:

$$SE_{pooled}(\hat{p}_1 - \hat{p}_2) = \sqrt{\frac{\hat{p}_{pooled}\,\hat{q}_{pooled}}{n_1} + \frac{\hat{p}_{pooled}\,\hat{q}_{pooled}}{n_2}}$$

$$= \sqrt{\frac{0.3678 \times (1 - 0.3678)}{184} + \frac{0.3678 \times (1 - 0.3678)}{811}}.$$

This comes out to 0.039.

> When finding the number of successes, round the values to integers. For example, the 48 snorers among the 184 under-30 respondents are actually 26.1% of 184. We round back to the nearest whole number to find the count that could have yielded the rounded percent we were given.

Improving the Success/Failure Condition

The vaccine Gardasil® was introduced to prevent the strains of human papillomavirus (HPV) that are responsible for almost all cases of cervical cancer. In randomized placebo-controlled clinical trials,[4] only 1 case of HPV was diagnosed

---

[4] *Quadrivalent Human Papillomavirus Vaccine: Recommendations of the Advisory Committee on Immunization Practices (ACIP)*, National Center for HIV/AIDS, Viral Hepatitis, STD and TB Prevention [May 2007].

among 7897 women who received the vaccine, compared with 91 cases diagnosed among 7899 who received a placebo. The one observed HPV case ("success") doesn't meet the at-least-10-successes criterion. Surely, though, we should not refuse to test the effectiveness of the vaccine just because it failed so rarely; that would be absurd.

For that reason, in a two-proportion z-test, the proper Success/Failure test uses the *expected* frequencies, which we can find from the pooled proportion. In this case,

$$\hat{p}_{pooled} = \frac{91 + 1}{7899 + 7897} = 0.0058$$

$$n_1 \hat{p}_{pooled} = 7899(0.0058) = 46$$

$$n_2 \hat{p}_{pooled} = 7897(0.0058) = 46,$$

so we can proceed with the hypothesis test.

Often it is easier just to check the observed numbers of successes and failures. If they are both greater than 10, you don't need to look further. But keep in mind that the correct test uses the *expected* frequencies rather than the *observed* ones.

Compared to What?

Naturally, we'll reject our null hypothesis if we see a large enough difference in the two proportions. How can we decide whether the difference we see, $\hat{p}_1 - \hat{p}_2$, is large? The answer is the same as always: We just compare it to its standard deviation.

Unlike previous hypothesis-testing situations, the null hypothesis doesn't provide a standard deviation, so we'll use a standard error (here, pooled). Since the sampling distribution is Normal, we can divide the observed difference by its standard error to get a z-score. The z-score will tell us how many standard errors the observed difference is away from 0. We can then use the 68–95–99.7 Rule to decide whether this is large, or some technology to get an exact P-value. The result is a **two-proportion z-test**.

A S

Activity: Test for a Difference Between Two Proportions. Is premium-brand chicken less likely to be contaminated than store-brand chicken?

TWO-PROPORTION z-TEST

The conditions for the two-proportion z-test are the same as for the two-proportion z-interval. We are testing the hypothesis

$$H_0: p_1 - p_2 = 0.$$

Because we hypothesize that the proportions are equal, we pool the groups to find

$$\hat{p}_{pooled} = \frac{Success_1 + Success_2}{n_1 + n_2}$$

and use that pooled value to estimate the standard error:

$$SE_{pooled}(\hat{p}_1 - \hat{p}_2) = \sqrt{\frac{\hat{p}_{pooled}\hat{q}_{pooled}}{n_1} + \frac{\hat{p}_{pooled}\hat{q}_{pooled}}{n_2}}.$$

Now we find the test statistic,

$$z = \frac{(\hat{p}_1 - \hat{p}_2) - 0}{SE_{pooled}(\hat{p}_1 - \hat{p}_2)}.$$

When the conditions are met and the null hypothesis is true, this statistic follows the standard Normal model, so we can use that model to obtain a P-value.

STEP-BY-STEP EXAMPLE | **A Two-Proportion z-Test**

Question: Are the snoring rates of the two age groups really different?

THINK

Plan State what you want to know. Discuss the variables and the W's.

I want to know whether snoring rates differ for those under and over 30 years old. The data are from a random sample of 1010 U.S. adults surveyed in the 2002 Sleep in America Poll. Of these, 995 responded to the question about snoring, indicating whether or not they had snored at least a few nights a week in the past year.

Hypotheses The study simply broke down the responses by age, so there is no sense that either alternative was preferred. A two-sided alternative hypothesis is appropriate.

H_0: There is no difference in snoring rates in the two age groups:

$$p_{old} - p_{young} = 0.$$

H_A: The rates are different: $p_{old} - p_{young} \neq 0$.

Model Think about the assumptions and check the conditions.

✔ **Independence Assumption:** The National Sleep Foundation selected respondents at random, so they should be independent.

✔ **Randomization Condition:** The respondents were randomly selected by telephone number and stratified by sex and region.

✔ **10% Condition:** The number of adults surveyed in each age group is certainly far less than 10% of that population.

✔ **Independent Groups Assumption:** The two groups are independent of each other because the sample was selected at random.

✔ **Success/Failure Condition:** In the younger age group, 48 snored and 136 didn't. In the older group, 318 snored and 493 didn't. The observed numbers of both successes and failures are much more than 10 for both groups.[5]

State the null model.

Choose your method.

Because the conditions are satisfied, I'll use a Normal model and perform a **two-proportion z-test.**

---

[5] This is one of those situations in which the traditional term "success" seems a bit weird. A success here could be that a person snores. "Success" and "failure" are arbitrary labels left over from studies of gambling games.

SHOW

Mechanics

$n_{young} = 184, \quad y_{young} = 48, \quad \hat{p}_{young} = 0.261$
$n_{old} = 811, \quad y_{old} = 318, \quad \hat{p}_{old} = 0.392$

The hypothesis is that the proportions are equal, so pool the sample data.

$$\hat{p}_{pooled} = \frac{y_{old} + y_{young}}{n_{old} + n_{young}} = \frac{318 + 48}{811 + 184} = 0.3678$$

Use the pooled SE to estimate $SD(p_{old} - p_{young})$.

$$SE_{pooled}(\hat{p}_{old} - \hat{p}_{young})$$
$$= \sqrt{\frac{\hat{p}_{pooled}\,\hat{q}_{pooled}}{n_{old}} + \frac{\hat{p}_{pooled}\,\hat{q}_{pooled}}{n_{young}}}$$
$$= \sqrt{\frac{(0.3678)(0.6322)}{811} + \frac{(0.3678)(0.6322)}{184}}$$
$$\approx 0.039375$$

The observed difference in sample proportions is
$\hat{p}_{old} - \hat{p}_{young} = 0.392 - 0.261 = 0.131$

Make a picture. Sketch a Normal model centered at the hypothesized difference of 0. Shade the region to the right of the observed difference, and because this is a two-tailed test, also shade the corresponding region in the other tail.

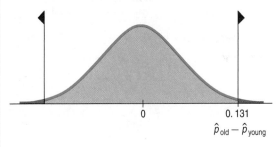

Find the z-score for the observed difference in proportions, 0.131.

Find the P-value using Table Z or technology. Because this is a two-tailed test, we must *double* the probability we find in the upper tail.

$$z = \frac{(\hat{p}_{old} - \hat{p}_{young}) - 0}{SE_{pooled}(\hat{p}_{old} - \hat{p}_{young})} = \frac{0.131 - 0}{0.039375} = 3.33$$
$$P = 2P(z \geq 3.33) = 0.0008$$

TELL

Conclusion Link the P-value to your decision about the null hypothesis, and state your conclusion in context.

The P-value of 0.0008 says that if there really were no difference in (reported) snoring rates between the two age groups, then the difference observed in this study would happen only 8 times in 10,000. This is so small that I reject the null hypothesis of no difference and conclude that there is a difference in the rate of snoring between older adults and younger adults. It appears that older adults are more likely to snore.

JUST CHECKING

3. A June 2004 public opinion poll asked 1000 randomly selected adults whether the United States should decrease the amount of immigration allowed; 49% of those responding said "yes." In June of 1995, a random sample of 1000 had found that 65% of adults thought immigration should be curtailed. To see if that percentage has decreased, why can't we just use a one-proportion z-test of $H_0: p = 0.65$ and see what the P-value for $\hat{p} = 0.49$ is?

4. For opinion polls like this, which has more variability: the percentage of respondents answering "yes" in either year or the difference in the percentages between the two years?

FOR EXAMPLE | Another Two-Proportion z-Test

RECAP: One concern of the study on teens' online profiles was safety and privacy. In the random sample, girls were less likely than boys to say that they are easy to find online from their profiles. Only 19% (62 girls) of 325 teen girls with profiles say that they are easy to find, while 28% (75 boys) of the 268 boys with profiles say the same.

QUESTION: Are these results evidence of a real difference between boys and girls? Perform a two-proportion z-test and discuss what you find.

$$H_O: p_{boys} - p_{girls} = 0$$
$$H_A: p_{boys} - p_{girls} \neq 0$$

✔ *Randomization Condition:* The sample of boys and the sample of girls were both chosen randomly.

✔ *10% Condition:* 268 boys and 325 girls are each less than 10% of all teenage boys and girls with online profiles.

✔ *Independent Groups Assumption:* Because the samples were selected at random, it's reasonable to believe the boys' perceptions are independent of the girls'.

✔ *Success/Failure Condition:* Among the girls, there were 62 "successes" and 263 failures, and among boys, 75 successes and 193 failures. These counts are at least 10 for each group.

Because all the assumptions and conditions are satisfied, it's okay to do a **two-proportion z-test**:

$$\hat{p}_{pooled} = \frac{75 + 62}{268 + 325} = 0.231$$

$$SE_{pooled}(\hat{p}_{boys} - \hat{p}_{girls}) = \sqrt{\frac{0.231 \times 0.769}{268} + \frac{0.231 \times 0.769}{325}} = 0.0348$$

$$z = \frac{(0.28 - 0.19) - 0}{0.0348} = 2.59$$

$$P(z > 2.59) = 0.0048$$

This is a two-tailed test, so the P-value = 2(0.0048) = 0.0096. Because this P-value is very small, I reject the null hypothesis. This study provides strong evidence that there really is a difference in the proportions of teen girls and boys who say they are easy to find online.

What Can Go Wrong?

- **Don't use two-sample proportion methods when the samples aren't independent.** These methods give wrong answers when this assumption of independence is violated. Good random sampling is usually the best insurance of independent groups. Make sure there is no relationship between the two groups. For example, you can't compare the proportion of respondents who own SUVs with the proportion of those same respondents who think the tax on gas should be eliminated. The responses are not independent because you've asked the same people. To use these methods to estimate or test the difference, you'd need to survey two different groups of people.

  Alternatively, if you have a random sample, you can split your respondents according to their answers to one question and treat the two resulting groups as independent samples. So, you could test whether the proportion of SUV owners who favored eliminating the gas tax was the same as the corresponding proportion among non-SUV owners.

- **Don't apply inference methods where there was no randomization.** If the data do not come from representative random samples or from a properly randomized experiment, then the inference about the differences in proportions may be wrong.

- **Don't interpret a significant difference in proportions causally.** It turns out that people with higher incomes are more likely to snore. Does that mean money affects sleep patterns? Probably not. We have seen that older people are more likely to snore, and they are also likely to earn more. In a prospective or retrospective study, there is always the danger that other lurking variables not accounted for are the real reason for an observed difference. Be careful not to jump to conclusions about causality.

CONNECTIONS

In Chapter 3 we looked at contingency tables for two categorical variables. Differences in proportions are just 2 × 2 contingency tables. You'll often see data presented in this way. For example, the snoring data could be shown as

	18–29	30 and over	Total
Snore	48	318	**366**
Don't snore	136	493	**629**
Total	**184**	**811**	**995**

We tested whether the column percentages of snorers were the same for the two age groups.

This chapter gives the first examples we've seen of inference methods for a parameter other than a simple proportion. Although we have a different standard error, the step-by-step procedures are almost identical. In particular, once again we divide the statistic (the difference in proportions) by its standard error and get a z-score. You should feel right at home.

WHAT HAVE WE LEARNED?

In the last few chapters we began our exploration of statistical inference; we learned how to create confidence intervals and test hypotheses about a proportion. Now we've looked at inference for the difference in two proportions. In doing so, perhaps the most important thing we've learned is that the concepts and interpretations are essentially the same—only the mechanics have changed slightly.

We've learned that hypothesis tests and confidence intervals for the difference in two proportions are based on Normal models. Both require us to find the standard error of the difference in two proportions. We do that by adding the variances of the two sample proportions, assuming our two groups are independent. When we test a hypothesis that the two proportions are equal, we pool the sample data; for confidence intervals, we don't pool.

Terms

Sampling distribution of the difference between two proportions

The sampling distribution of $\hat{p}_1 - \hat{p}_2$ is, under appropriate assumptions, modeled by a Normal model with mean $\mu = p_1 - p_2$ and standard deviation $SD(\hat{p}_1 - \hat{p}_2) = \sqrt{\dfrac{p_1 q_1}{n_1} + \dfrac{p_2 q_2}{n_2}}$ (p. 528).

Two-proportion z-interval	A two-proportion z-interval gives a confidence interval for the true difference in proportions, $p_1 - p_2$, in two independent groups (p. 529).

The confidence interval is $(\hat{p}_1 - \hat{p}_2) \pm z^* \times SE(\hat{p}_1 - \hat{p}_2)$, where z^* is a critical value from the standard Normal model corresponding to the specified confidence level. |
| Pooling | A better estimate of common proportion and its standard deviation are possible when data from different sources are believed to be homogeneous. The data are combined, or pooled, into a single group for the purpose of estimating the common proportion. The resulting pooled standard error is based on more data and is thus more reliable (if the null hypothesis is true and the groups are truly homogeneous) (p. 532). |
| Two-proportion z-test | Test the null hypothesis $H_0: p_1 - p_2 = 0$ by referring the statistic

$$z = \frac{\hat{p}_1 - \hat{p}_2}{SE_{pooled}(\hat{p}_1 - \hat{p}_2)}$$

to a standard Normal model (p. 533). |

Skills

▶ Be able to state the null and alternative hypotheses for testing the difference between two population proportions.

▶ Know how to examine your data for violations of conditions that would make inference about the difference between two population proportions unwise or invalid.

▶ Understand that the formula for the standard error of the difference between two independent sample proportions is based on the principle that when finding the sum or difference of two independent random variables, their variances add.

▶ Know how to find a confidence interval for the difference between two proportions.

▶ Be able to perform a significance test of the natural null hypothesis that two population proportions are equal.

▶ Know how to write a sentence describing what is said about the difference between two population proportions by a confidence interval.

▶ Know how to write a sentence interpreting the results of a significance test of the null hypothesis that two population proportions are equal.

▶ Be able to interpret the meaning of a P-value in nontechnical language, making clear that the probability claim is made about computed values and not about the population parameter of interest.

▶ Know that failure to reject a null hypothesis does not mean it is "accepted."

INFERENCES FOR THE DIFFERENCE BETWEEN TWO PROPORTIONS ON THE COMPUTER

It is so common to test against the null hypothesis of no difference between the two true proportions that most statistics programs simply assume this null hypothesis. And most will automatically use the pooled standard deviation. If you wish to test a different null (say, that the true difference is 0.3), you may have to search for a way to do it.

Many statistics packages don't offer special commands for inference for differences between proportions. As with inference for single proportions, most statistics programs want the "success" and "failure" status for each case. Usually these are given as 1 or 0, but they might be category names like "yes" and "no." Often you just know the proportions of successes, $\hat{p}_1$ and $\hat{p}_2$, and the counts, n_1 and n_2. Computer packages don't usually deal with summary data like these easily. Calculators typically do a better job.

In some programs, you can reconstruct the original values. But even when you have (or can reconstruct) the raw data values, often you won't get exactly the same test statistic from a computer package as you would find working by hand. The reason is that when the packages treat the proportion as a mean, they make some approximations. The result is very close, but not exactly the same.

DATA DESK	COMMENTS
Data Desk does not offer built-in methods for inference with proportions. Use **Replicate Y by X** to construct data corresponding to given proportions and totals.	For summarized data, open a Scratchpad to compute the standard deviations and margin of error by typing the calculation.

EXCEL	COMMENTS
Inference methods for proportions are not part of the standard Excel tool set.	For summarized data, type the calculation into any cell and evaluate it.

JMP	COMMENTS
For a **categorical** variable that holds category labels, the **Distribution** platform includes tests and intervals of proportions. For summarized data, put the category names in one variable and the frequencies in an adjacent variable. Designate the frequency column to have the **role** of **frequency**. Then use the **Distribution** platform.	JMP uses slightly different methods for proportion inferences than those discussed in this text. Your answers are likely to be slightly different.

MINITAB	COMMENTS
To find a hypothesis test for a proportion, Choose **Basic Statistics** from the **Stat** menu. Choose **2Proportions . . .** from the Basic Statistics submenu. If the data are organized as category names in one column and case IDs in another, assign the variables from the variable list box to the **Samples in one column** box. If the data are organized as two separate columns of responses, click on **Samples in different columns:** and assign the variables from the variable list box. If you have summarized data, click the **Summarized Data** button and fill in the number of trials and the number of successes for each group. Click the **Options** button and specify the remaining details. Remember to click the **Use pooled estimate of p for test** box when testing the null hypothesis of no difference between proportions. Click the **OK** button.	When working from a variable that names categories, Minitab treats the last category as the "success" category. You can specify how the categories should be ordered.

SPSS
SPSS does not perform hypothesis tests for proportions.

TI-83/84 PLUS	COMMENTS

To calculate a confidence interval for the difference between two population proportions,

- Select **B:2-PropZInt** from the **STAT TESTS** menu.
- Enter the observed counts and the sample sizes for both samples.
- Specify a confidence level.
- Calculate the interval.

To do the mechanics of a hypothesis test for equality of population proportions,

- Select **6:2-PropZTest** from the **STAT TESTS** menu.
- Enter the observed counts and sample sizes.
- Indicate what kind of test you want: one-tail upper tail, lower tail, or two-tail.
- Calculate the result.

Beware: When you enter the value of x, you need the *count*, not the percentage. The count must be a whole number. If the number of successes is given as a percent, you must first multiply np and round the result to obtain x.

TI-89	COMMENTS

To calculate a confidence interval for the difference between two population proportions,

- Select **6:2-PropZInt** from the **STAT Ints** menu.
- Enter the observed counts and the sample sizes for both samples.
- Specify a confidence level.
- Calculate the interval.

To do the mechanics of a hypothesis test for equality of population proportions,

- Select **6:2-PropZTest** from the **STAT Tests** menu.
- Enter the observed counts and sample sizes.
- Indicate what kind of test you want: one-tail upper tail, lower tail, or two-tail.
- Specify whether results should simply be calculated or displayed with the area corresponding to the P-value of the test shaded.

Beware: When you enter the value of x, you need the *count*, not the percentage. The count must be a whole number. If the number of successes is given as a percent, you must first multiply np and round the result to obtain x.

EXERCISES

1. **Online social networking.** The Parents & Teens 2006 Survey of 935 12- to 17-year-olds found that, among teens aged 15–17, girls were significantly more likely to have used social networking sites and online profiles. 70% of the girls surveyed had used an online social network, compared to 54% of the boys. What does it mean to say that the difference in proportions is "significant"?

2. **Science news.** In 2007 a Pew survey asked 1447 Internet users about their sources of news and information about science. Among those who had broadband access at home, 34% said they would turn to the Internet for most of their science news. The report on this survey claims that this is not significantly different from the percentage (33%) who said they ordinarily get their science news from television. What does it mean to say that the difference is not significant?

3. **Name recognition.** A political candidate runs a weeklong series of TV ads designed to attract public attention to his campaign. Polls taken before and after the ad campaign show some increase in the proportion of voters who now recognize this candidate's name, with a P-value of 0.033. Is it reasonable to believe the ads may be effective?

4. **Origins.** In a 1993 Gallup poll, 47% of the respondents agreed with the statement *"God created human beings pretty much in their present form at one time within the last 10,000 years or so."* When Gallup asked the same question in 2001, only 45% of those respondents agreed. Is it reasonable to conclude that there was a change in public opinion given that the P-value is 0.37? Explain.

5. **Revealing information.** 886 randomly sampled teens were asked which of several personal items of information they thought it okay to share with someone they had just met. 44% said it was okay to share their e-mail addresses, but only 29% said they would give out their cell phone numbers. A researcher claims that a two-proportion z-test could tell whether there was a real difference among all teens. Explain why that test would not be appropriate for these data.

6. **Regulating access.** When a random sample of 935 parents were asked about rules in their homes, 77% said they had rules about the kinds of TV shows their children could watch. Among the 790 of those parents whose teenage children had Internet access, 85% had rules about the kinds of Internet sites their teens could visit. That looks like a difference, but can we tell? Explain why a two-proportion z-test would not be appropriate here.

7. **Gender gap.** A presidential candidate fears he has a problem with women voters. His campaign staff plans to run a poll to assess the situation. They'll randomly sample 300 men and 300 women, asking if they have a favorable impression of the candidate. Obviously, the staff can't know this, but suppose the candidate has a positive image with 59% of males but with only 53% of females.
   a) What sampling design is his staff planning to use?
   b) What difference would you expect the poll to show?
   c) Of course, sampling error means the poll won't reflect the difference perfectly. What's the standard deviation for the difference in the proportions?
   d) Sketch a sampling model for the size difference in proportions of men and women with favorable impressions of this candidate that might appear in a poll like this.
   e) Could the campaign be misled by the poll, concluding that there really is no gender gap? Explain.

8. **Buy it again?** A consumer magazine plans to poll car owners to see if they are happy enough with their vehicles that they would purchase the same model again. They'll randomly select 450 owners of American-made cars and 450 owners of Japanese models. Obviously, the actual opinions of the entire population couldn't be known, but suppose 76% of owners of American cars and 78% of owners of Japanese cars would purchase another.
   a) What sampling design is the magazine planning to use?
   b) What difference would you expect their poll to show?
   c) Of course, sampling error means the poll won't reflect the difference perfectly. What's the standard deviation for the difference in the proportions?
   d) Sketch a sampling model for the difference in proportions that might appear in a poll like this.

   e) Could the magazine be misled by the poll, concluding that owners of American cars are much happier with their vehicles than owners of Japanese cars? Explain.

9. **Arthritis.** The Centers for Disease Control and Prevention reported a survey of randomly selected Americans age 65 and older, which found that 411 of 1012 men and 535 of 1062 women suffered from some form of arthritis.
   a) Are the assumptions and conditions necessary for inference satisfied? Explain.
   b) Create a 95% confidence interval for the difference in the proportions of senior men and women who have this disease.
   c) Interpret your interval in this context.
   d) Does this confidence interval suggest that arthritis is more likely to afflict women than men? Explain.

10. **Graduation.** In October 2000 the U.S. Department of Commerce reported the results of a large-scale survey on high school graduation. Researchers contacted more than 25,000 Americans aged 24 years to see if they had finished high school; 84.9% of the 12,460 males and 88.1% of the 12,678 females indicated that they had high school diplomas.
   a) Are the assumptions and conditions necessary for inference satisfied? Explain.
   b) Create a 95% confidence interval for the difference in graduation rates between males and females.
   c) Interpret your confidence interval.
   d) Does this provide strong evidence that girls are more likely than boys to complete high school? Explain.

11. **Pets.** Researchers at the National Cancer Institute released the results of a study that investigated the effect of weed-killing herbicides on house pets. They examined 827 dogs from homes where an herbicide was used on a regular basis, diagnosing malignant lymphoma in 473 of them. Of the 130 dogs from homes where no herbicides were used, only 19 were found to have lymphoma.
   a) What's the standard error of the difference in the two proportions?
   b) Construct a 95% confidence interval for this difference.
   c) State an appropriate conclusion.

12. **Carpal tunnel.** The painful wrist condition called carpal tunnel syndrome can be treated with surgery or less invasive wrist splints. In September 2002, *Time* magazine reported on a study of 176 patients. Among the half that had surgery, 80% showed improvement after three months, but only 48% of those who used the wrist splints improved.
   a) What's the standard error of the difference in the two proportions?
   b) Construct a 95% confidence interval for this difference.
   c) State an appropriate conclusion.

13. **Ear infections.** A new vaccine was recently tested to see if it could prevent the painful and recurrent ear infections that many infants suffer from. *The Lancet*, a medical journal, reported a study in which babies about a year

old were randomly divided into two groups. One group received vaccinations; the other did not. During the following year, only 333 of 2455 vaccinated children had ear infections, compared to 499 of 2452 unvaccinated children in the control group.

a) Are the conditions for inference satisfied?

b) Find a 95% confidence interval for the difference in rates of ear infection.

c) Use your confidence interval to explain whether you think the vaccine is effective.

14. **Anorexia.** The *Journal of the American Medical Association* reported on an experiment intended to see if the drug Prozac® could be used as a treatment for the eating disorder anorexia nervosa. The subjects, women being treated for anorexia, were randomly divided into two groups. Of the 49 who received Prozac, 35 were deemed healthy a year later, compared to 32 of the 44 who got the placebo.

a) Are the conditions for inference satisfied?

b) Find a 95% confidence interval for the difference in outcomes.

c) Use your confidence interval to explain whether you think Prozac is effective.

15. **Another ear infection.** In Exercise 13 you used a confidence interval to examine the effectiveness of a vaccine against ear infections in babies. Suppose that instead you had conducted a hypothesis test. (Answer these questions *without* actually doing the test.)

a) What hypotheses would you test?

b) State a conclusion based on your confidence interval.

c) What alpha level did your test use?

d) If that conclusion is wrong, which type of error did you make?

e) What would be the consequences of such an error?

16. **Anorexia, again.** In Exercise 14 you used a confidence interval to examine the effectiveness of Prozac in treating anorexia nervosa. Suppose that instead you had conducted a hypothesis test. (Answer these questions *without* actually doing the test.)

a) What hypotheses would you test?

b) State a conclusion based on your confidence interval.

c) What alpha level did your test use?

d) If that conclusion is wrong, which type of error did you make?

e) What would be the consequences of such an error?

17. **Teen smoking, part I.** A Vermont study published in December 2001 by the American Academy of Pediatrics examined parental influence on teenagers' decisions to smoke. A group of students who had never smoked were questioned about their parents' attitudes toward smoking. These students were questioned again two years later to see if they had started smoking. The researchers found that, among the 284 students who indicated that their parents disapproved of kids smoking, 54 had become established smokers. Among the 41 students who initially said their parents were lenient about smoking, 11 became smokers. Do these data provide strong evidence that parental attitude influences teenagers' decisions about smoking?

a) What kind of design did the researchers use?

b) Write appropriate hypotheses.

c) Are the assumptions and conditions necessary for inference satisfied?

d) Test the hypothesis and state your conclusion.

e) Explain in this context what your P-value means.

f) If that conclusion is actually wrong, which type of error did you commit?

18. **Depression.** A study published in the *Archives of General Psychiatry* in March 2001 examined the impact of depression on a patient's ability to survive cardiac disease. Researchers identified 450 people with cardiac disease, evaluated them for depression, and followed the group for 4 years. Of the 361 patients with no depression, 67 died. Of the 89 patients with minor or major depression, 26 died. Among people who suffer from cardiac disease, are depressed patients more likely to die than non-depressed ones?

a) What kind of design was used to collect these data?

b) Write appropriate hypotheses.

c) Are the assumptions and conditions necessary for inference satisfied?

d) Test the hypothesis and state your conclusion.

e) Explain in this context what your P-value means.

f) If your conclusion is actually incorrect, which type of error did you commit?

19. **Teen smoking, part II.** Consider again the Vermont study discussed in Exercise 17.

a) Create a 95% confidence interval for the difference in the proportion of children who may smoke and have approving parents and those who may smoke and have disapproving parents.

b) Interpret your interval in this context.

c) Carefully explain what "95% confidence" means.

20. **Depression, revisited.** Consider again the study of the association between depression and cardiac disease survivability in Exercise 18.

a) Create a 95% confidence interval for the difference in survival rates.

b) Interpret your interval in this context.

c) Carefully explain what "95% confidence" means.

21. **Pregnancy.** In 1998, a San Diego reproductive clinic reported 42 live births to 157 women under the age of 38, but only 7 live births for 89 clients aged 38 and older. Is this strong evidence of a difference in the effectiveness of the clinic's methods for older women?

a) Was this an experiment? Explain.

b) Test an appropriate hypothesis and state your conclusion in context.

c) If you concluded there was a difference, estimate that difference with a confidence interval and interpret your interval in context.

22. **Birth weight.** In 2003 the *Journal of the American Medical Association* reported a study examining the possible impact of air pollution caused by the 9/11 attack on New York's World Trade Center on the weight of babies. Researchers found that 8% of 182 babies born to mothers who were exposed to heavy doses of soot and ash on

September 11 were classified as having low birth weight. Only 4% of 2300 babies born in another New York City hospital whose mothers had not been near the site of the disaster were similarly classified. Does this indicate a possibility that air pollution might be linked to a significantly higher proportion of low birth-weight babies?

a) Was this an experiment? Explain.

b) Test an appropriate hypothesis and state your conclusion in context.

c) If you concluded there is a difference, estimate that difference with a confidence interval and interpret that interval in context.

23. **Politics and sex.** One month before the election, a poll of 630 randomly selected voters showed 54% planning to vote for a certain candidate. A week later it became known that he had had an extramarital affair, and a new poll showed only 51% of 1010 voters supporting him. Do these results indicate a decrease in voter support for his candidacy?

a) Test an appropriate hypothesis and state your conclusion.

b) If your conclusion turns out to be wrong, did you make a Type I or Type II error?

c) If you concluded there was a difference, estimate that difference with a confidence interval and interpret your interval in context.

24. **Shopping.** A survey of 430 randomly chosen adults found that 21% of the 222 men and 18% of the 208 women had purchased books online.

a) Is there evidence that men are more likely than women to make online purchases of books? Test an appropriate hypothesis and state your conclusion in context.

b) If your conclusion in fact proves to be wrong, did you make a Type I or Type II error?

c) Estimate this difference with a confidence interval.

d) Interpret your interval in context.

25. **Twins.** In 2001, one county reported that, among 3132 white women who had babies, 94 were multiple births. There were also 20 multiple births to 606 black women. Does this indicate any racial difference in the likelihood of multiple births?

a) Test an appropriate hypothesis and state your conclusion in context.

b) If your conclusion is incorrect, which type of error did you commit?

26. **Mammograms.** A 9-year study in Sweden compared 21,088 women who had mammograms with 21,195 who did not. Of the women who underwent screening, 63 died of breast cancer, compared to 66 deaths among the control group. (*The New York Times*, Dec 9, 2001)

a) Do these results support the effectiveness of regular mammograms in preventing deaths from breast cancer?

b) If your conclusion is incorrect, what kind of error have you committed?

27. **Pain.** Researchers comparing the effectiveness of two pain medications randomly selected a group of patients who had been complaining of a certain kind of joint pain. They randomly divided these people into two groups, then administered the pain killers. Of the 112 people in the group who received medication A, 84 said this pain reliever was effective. Of the 108 people in the other group, 66 reported that pain reliever B was effective.

a) Write a 95% confidence interval for the percent of people who may get relief from this kind of joint pain by using medication A. Interpret your interval.

b) Write a 95% confidence interval for the percent of people who may get relief by using medication B. Interpret your interval.

c) Do the intervals for A and B overlap? What do you think this means about the comparative effectiveness of these medications?

d) Find a 95% confidence interval for the difference in the proportions of people who may find these medications effective. Interpret your interval.

e) Does this interval contain zero? What does that mean?

f) Why do the results in parts c and e seem contradictory? If we want to compare the effectiveness of these two pain relievers, which is the correct approach? Why?

28. **Gender gap.** Candidates for political office realize that different levels of support among men and women may be a crucial factor in determining the outcome of an election. One candidate finds that 52% of 473 men polled say they will vote for him, but only 45% of the 522 women in the poll express support.

a) Write a 95% confidence interval for the percent of male voters who may vote for this candidate. Interpret your interval.

b) Write and interpret a 95% confidence interval for the percent of female voters who may vote for him.

c) Do the intervals for males and females overlap? What do you think this means about the gender gap?

d) Find a 95% confidence interval for the difference in the proportions of males and females who will vote for this candidate. Interpret your interval.

e) Does this interval contain zero? What does that mean?

f) Why do the results in parts c and e seem contradictory? If we want to see if there is a gender gap among voters with respect to this candidate, which is the correct approach? Why?

29. **Sensitive men.** In August 2004, *Time* magazine, reporting on a survey of men's attitudes, noted that "Young men are more comfortable than older men talking about their problems." The survey reported that 80 of 129 surveyed 18- to 24-year-old men and 98 of 184 25- to 34-year-old men said they were comfortable. What do you think? Is *Time*'s interpretation justified by these numbers?

30. **Retention rates.** In 2004 the testing company ACT, Inc., reported on the percentage of first-year students at four-year colleges who return for a second year. Their sample of 1139 students in private colleges showed a 74.9% retention rate, while the rate was 71.9% for the sample of 505 students at public colleges. Does this provide evidence that there's a difference in retention rates of first-year students at public and private colleges?

31. **Online activity checks.** Are more parents checking up on their teen's online activities? A Pew survey in 2004 found that 33% of 868 randomly sampled teens said that their parents checked to see what websites they visited. In 2006 the same question posed to 811 teens found 41% reporting such checks. Do these results provide evidence that more parents are checking?

32. **Computer gaming.** Who plays online or electronic games? A survey in 2006 found that 69% of 223 boys aged 12–14 said they "played computer or console games like Xbox or PlayStation . . . or games online." Of 248 boys aged 15–17, only 62% played these games. Is this evidence of a real age-based difference?

JUST CHECKING

ANSWERS

1. We're 90% confident that if members are contacted by e-mail, the donation rate will be between 4 and 18 percentage points higher than if they received regular mail.

2. Since a difference of 0 is not in the confidence interval, we'd reject the null hypothesis. There is evidence that more members will donate if contacted by e-mail.

3. The proportion from the sample in 1995 has variability, too. If we do a one-proportion z-test, we won't take that variability into account and our P-value will be incorrect.

4. The difference in the proportions between the two years has more variability than either individual proportion. The variance of the difference is the sum of the two variances.

PART
V

PART REVIEW

From the Data at Hand to the World at Large

Quick Review

What do samples really tell us about the populations from which they are drawn? Are the results of an experiment meaningful, or are they just sampling error? Statistical inference based on our understanding of sampling models can help answer these questions. Here's a brief summary of the key concepts and skills:

▶ Sampling models describe the variability of sample statistics using a remarkable result called the Central Limit Theorem.

- When the number of trials is sufficiently large, proportions found in different samples vary according to an approximately Normal model.
- When samples are sufficiently large, the means of different samples vary, with an approximately Normal model.
- The variability of sample statistics decreases as sample size increases.
- Statistical inference procedures are based on the Central Limit Theorem.
- No inference procedure is valid unless the underlying assumptions are true. Always check the conditions before proceeding.

▶ A confidence interval uses a sample statistic (such as a proportion) to estimate a range of plausible values for the parameter of a population model.

- All confidence intervals involve an estimate of the parameter, a margin of error, and a level of confidence.
- For confidence intervals based on a given sample, the greater the margin of error, the higher the confidence.
- At a given level of confidence, the larger the sample, the smaller the margin of error.

▶ A hypothesis test proposes a model for the population, then examines the observed statistics to see if that model is plausible.

- A null hypothesis suggests a parameter value for the population model. Usually, we assume there is nothing interesting, unusual, or different about the sample results.
- The alternative hypothesis states what we will believe if the sample results turn out to be inconsistent with our null model.
- We compare the difference between the statistic and the hypothesized value with the standard

deviation of the statistic. It's the sampling distribution of this ratio that gives us a P-value.

- The P-value of the test is the conditional probability that the null model could produce results at least as extreme as those observed in the sample or the experiment just as a result of sampling error.
- A low P-value indicates evidence against the null model. If it is sufficiently low, we reject the null model.
- A high P-value indicates that the sample results are not inconsistent with the null model, so we cannot reject it. However, this does not prove the null model is true.
- Sometimes we will mistakenly reject the null hypothesis even though it's actually true—that's

called a Type I error. If we fail to reject a false null hypothesis, we commit a Type II error.
- The power of a test measures its ability to detect a false null hypothesis.
- You can lower the risk of a Type I error by requiring a higher standard of proof (lower P-value) before rejecting the null hypothesis. But this will raise the risk of a Type II error and decrease the power of the test.
- The only way to increase the power of a test while decreasing the chance of committing either error is to design a study based on a larger sample.

And now for some opportunities to review these concepts and skills . . .

REVIEW EXERCISES

1. **Herbal cancer.** A report in the *New England Journal of Medicine* (June 6, 2000) notes growing evidence that the herb *Aristolochia fangchi* can cause urinary tract cancer in those who take it. Suppose you are asked to design an experiment to study this claim. Imagine that you have data on urinary tract cancers in subjects who have used this herb and similar subjects who have not used it and that you can measure incidences of cancer and precancerous lesions in these subjects. State the null and alternative hypotheses you would use in your study.

2. **Color-blind.** Medical literature says that about 8% of males are color-blind. A university's introductory psychology course is taught in a large lecture hall. Among the students, there are 325 males. Each semester when the professor discusses visual perception, he shows the class a test for color blindness. The percentage of males who are color-blind varies from semester to semester.
   a) Is the sampling distribution model for the sample proportion likely to be Normal? Explain.
   b) What are the mean and standard deviation of this sampling distribution model?
   c) Sketch the sampling model, using the 68–95–99.7 Rule.
   d) Write a few sentences explaining what the model says about this professor's class.

3. **Birth days.** During a 2-month period in 2002, 72 babies were born at the Tompkins Community Hospital in upstate New York. The table shows how many babies were born on each day of the week.
   a) If births are uniformly distributed across all days of the week, how many would you expect on each day?
   b) Only 7 births occurred on a Monday. Does this indicate that women might be less likely to give birth on a Monday? Explain.

Day	Births
Mon.	7
Tues.	17
Wed.	8
Thurs.	12
Fri.	9
Sat.	10
Sun.	9

   c) Are the 17 births on Tuesdays unusually high? Explain.
   d) Can you think of any reasons why births may not occur completely at random?

4. **Polling 2004.** In the 2004 U.S. presidential election, the official results showed that George W. Bush received 50.7% of the vote and John Kerry received 48.3%. Ralph Nader, running as a third-party candidate, picked up only 0.4%. After the election, there was much discussion about exit polls, which had initially indicated a different result. Suppose you had taken a random sample of 1000 voters in an exit poll and asked them for whom they had voted.
   a) Would you always get 507 votes for Bush and 483 for Kerry?
   b) In 95% of such polls, your sample proportion of voters for Bush should be between what two values?
   c) In 95% of such polls, your sample proportion of voters for Nader should be between what two numbers?
   d) Would you expect the sample proportion of Nader votes to vary more, less, or about the same as the sample proportion of Bush votes? Why?

5. **Leaky gas tanks.** Nationwide, it is estimated that 40% of service stations have gas tanks that leak to some extent. A new program in California is designed to lessen the prevalence of these leaks. We want to assess the effectiveness of the program by seeing if the percentage of service stations whose tanks leak has decreased. To do this, we randomly sample 27 service stations in California and determine whether there is any evidence of leakage. In our sample, only 7 of the stations exhibit any leakage. Is there evidence that the new program is effective?
   a) What are the null and alternative hypotheses?
   b) Check the assumptions necessary for inference.
   c) Test the null hypothesis.
   d) What do you conclude (in plain English)?
   e) If the program actually works, have you made an error? What kind?

f) What two things could you do to decrease the probability of making this kind of error?

g) What are the advantages and disadvantages of taking those two courses of action?

6. **Surgery and germs.** Joseph Lister (for whom Listerine is named!) was a British physician who was interested in the role of bacteria in human infections. He suspected that germs were involved in transmitting infection, so he tried using carbolic acid as an operating room disinfectant. In 75 amputations, he used carbolic acid 40 times. Of the 40 amputations using carbolic acid, 34 of the patients lived. Of the 35 amputations without carbolic acid, 19 patients lived. The question of interest is whether carbolic acid is effective in increasing the chances of surviving an amputation.

a) What kind of a study is this?

b) What do you conclude? Support your conclusion by testing an appropriate hypothesis.

c) What reservations do you have about the design of the study?

7. **Scrabble.** Using a computer to play many simulated games of Scrabble, researcher Charles Robinove found that the letter "A" occurred in 54% of the hands. This study had a margin of error of ±10%. (*Chance*, 15, no. 1 [2002])

a) Explain what the margin of error means in this context.

b) Why might the margin of error be so large?

c) Probability theory predicts that the letter "A" should appear in 63% of the hands. Does this make you concerned that the simulation might be faulty? Explain.

8. **Dice.** When one die is rolled, the number of spots showing has a mean of 3.5 and a standard deviation of 1.7. Suppose you roll 10 dice. What's the approximate probability that your total is between 30 and 40 (that is, the average for the 10 dice is between 3 and 4)? Specify the model you use and the assumptions and conditions that justify your approach.

9. **Net-Newsers** In June of 2008, the Pew Research Foundation sampled 3615 U.S. adults and asked about their choice of news sources. They identified 13% as "Net-Newsers" who regularly get their news from online sources rather than TV or newspapers.

a) Pew reports a margin of error of ±2% for this result. Explain what the margin of error means.

b) Pew's survey included 2802 respondents contacted by landline and 813 contacted by cell phone. If the percentage of Net-Newsers is the same in both groups and Pew estimated those percentages separately, which group would have the larger margin of error? Explain.

c) Pew reports that 82% of the 470 Net-Newsers in their survey get news during the course of the day, far more than other respondents. Find a 95% confidence interval for this statistic.

d) How does the margin of error for your confidence interval compare with the values in parts a and b? Explain why.

10. **Death penalty 2006.** In May of 2006, the Gallup Organization asked a random sample of 537 American adults this question:

If you could choose between the following two approaches, which do you think is the better penalty for murder, the death penalty or life imprisonment, with absolutely no possibility of parole?

Of those polled, 47% chose the death penalty, the lowest percentage in the 21 years that Gallup has asked this question.

a) Create a 95% confidence interval for the percentage of all American adults who favor the death penalty.

b) Based on your confidence interval, is it clear that the death penalty no longer has majority support? Explain.

c) If pollsters wanted to follow up on this poll with another survey that could determine the level of support for the death penalty to within 2% with 98% confidence, how many people should they poll?

11. **Bimodal.** We are sampling randomly from a distribution known to be bimodal.

a) As our sample size increases, what's the expected shape of the sample's distribution?

b) What's the expected value of our sample's mean? Does the size of the sample matter?

c) How is the variability of sample means related to the standard deviation of the population? Does the size of the sample matter?

d) How is the shape of the sampling distribution model affected by the sample size?

12. **Vitamin D.** In July 2002 the *American Journal of Clinical Nutrition* reported that 42% of 1546 African-American women studied had vitamin D deficiency. The data came from a national nutrition study conducted by the Centers for Disease Control and Prevention in Atlanta.

a) Do these data meet the assumptions necessary for inference? What would you like to know that you don't?

b) Create a 95% confidence interval.

c) Interpret the interval in this context.

d) Explain in this context what "95% confidence" means.

13. **Archery.** A champion archer can generally hit the bull's-eye 80% of the time. Suppose she shoots 200 arrows during competition. Let $\hat{p}$ represent the percentage of bull's-eyes she gets (the sample proportion).

a) What are the mean and standard deviation of the sampling distribution model for $\hat{p}$?

b) Is a Normal model appropriate here? Explain.

c) Sketch the sampling model, using the 68–95–99.7 Rule.

d) What's the probability that she gets at least 85% bull's-eyes?

14. **Free throws 2007.** During the 2006–2007 NBA season, Kyle Korver led the league by making 191 of 209 free throws, for a success rate of 91.39%. But Matt Carroll was close behind, with 188 of 208 (90.39%).

a) Find a 95% confidence interval for the difference in their free throw percentages.

b) Based on your confidence interval, is it certain that Korver is better than Carroll at making free throws?

15. **Twins.** There is some indication in medical literature that doctors may have become more aggressive in inducing labor or doing preterm cesarean sections when a woman is carrying twins. Records at a large hospital show that, of the 43 sets of twins born in 1990, 20 were delivered before the 37th week of pregnancy. In 2000, 26 of 48 sets of twins

were born preterm. Does this indicate an increase in the incidence of early births of twins? Test an appropriate hypothesis and state your conclusion.

16. **Eclampsia.** It's estimated that 50,000 pregnant women worldwide die each year of eclampsia, a condition involving elevated blood pressure and seizures. A research team from 175 hospitals in 33 countries investigated the effectiveness of magnesium sulfate in preventing the occurrence of eclampsia in at-risk patients. Results are summarized below. (*Lancet*, June 1, 2002)

	Total Subjects	Reported side effects	Developed eclampsia	Deaths
Magnesium sulfate	4999	1201	40	11
Placebo	4993	228	96	20

(Treatment)

a) Write a 95% confidence interval for the increase in the proportion of women who may develop side effects from this treatment. Interpret your interval.
b) Is there evidence that the treatment may be effective in preventing the development of eclampsia? Test an appropriate hypothesis and state your conclusion.

17. **Eclampsia.** Refer again to the research summarized in Exercise 16. Is there any evidence that when eclampsia does occur, the magnesium sulfate treatment may help prevent the woman's death?

a) Write an appropriate hypothesis.
b) Check the assumptions and conditions.
c) Find the P-value of the test.
d) What do you conclude about the magnesium sulfate treatment?
e) If your conclusion is wrong, which type of error have you made?
f) Name two things you could do to increase the power of this test.
g) What are the advantages and disadvantages of those two options?

18. **Eggs.** The ISA Babcock Company supplies poultry farmers with hens, advertising that a mature B300 Layer produces eggs with a mean weight of 60.7 grams. Suppose that egg weights follow a Normal model with standard deviation 3.1 grams.

a) What fraction of the eggs produced by these hens weigh more than 62 grams?
b) What's the probability that a dozen randomly selected eggs average more than 62 grams?
c) Using the 68–95–99.7 Rule, sketch a model of the total weights of a dozen eggs.

19. **Polling disclaimer.** A newspaper article that reported the results of an election poll included the following explanation:

The Associated Press poll on the 2000 presidential campaign is based on telephone interviews with 798 randomly selected registered voters from all states except Alaska and Hawaii. The interviews were conducted June 21–25 by ICR of Media, Pa.

The results were weighted to represent the population by demographic factors such as age, sex, region, and education.

No more than 1 time in 20 should chance variations in the sample cause the results to vary by more than 4 percentage points from the answers that would be obtained if all Americans were polled.

The margin of sampling error is larger for responses of subgroups, such as income categories or those in political parties. There are other sources of potential error in polls, including the wording and order of questions.

a) Did they describe the 5 W's well?
b) What kind of sampling design could take into account the several demographic factors listed?
c) What was the margin of error of this poll?
d) What was the confidence level?
e) Why is the margin of error larger for subgroups?
f) Which kinds of potential bias did they caution readers about?

20. **Enough eggs?** One of the important issues for poultry farmers is the production rate—the percentage of days on which a given hen actually lays an egg. Ideally, that would be 100% (an egg every day), but realistically, hens tend to lay eggs on about 3 of every 4 days. ISA Babcock wants to advertise the production rate for the B300 Layer (see Exercise 18) as a 95% confidence interval with a margin of error of ±2%. How many hens must they collect data on?

21. **Teen deaths.** Traffic accidents are the leading cause of death among people aged 15 to 20. In May 2002, the National Highway Traffic Safety Administration reported that even though only 6.8% of licensed drivers are between 15 and 20 years old, they were involved in 14.3% of all fatal crashes. Insurance companies have long known that teenage boys were high risks, but what about teenage girls? One insurance company found that the driver was a teenage girl in 44 of the 388 fatal accidents they investigated. Is this strong evidence that the accident rate is lower for girls than for teens in general?

a) Test an appropriate hypothesis and state your conclusion.
b) Explain what your P-value means in this context.

22. **Perfect pitch.** A recent study of perfect pitch tested students in American music conservatories. It found that 7% of 1700 non-Asian and 32% of 1000 Asian students have perfect pitch. A test of the difference in proportions resulted in a P-value of < 0.0001.

a) What are the researchers' null and alternative hypotheses?
b) State your conclusion.
c) Explain in this context what the P-value means.
d) The researchers claimed that the data prove that genetic differences between the two populations cause a difference in the frequency of occurrence of perfect pitch. Do you agree? Why or why not?

23. **Largemouth bass.** Organizers of a fishing tournament believe that the lake holds a sizable population of largemouth bass. They assume that the weights of these fish have a model that is skewed to the right with a mean of 3.5 pounds and a standard deviation of 2.2 pounds.

a) Explain why a skewed model makes sense here.

b) Explain why you cannot determine the probability that a largemouth bass randomly selected ("caught") from the lake weighs over 3 pounds.

c) Each fisherman in the contest catches 5 fish each day. Can you determine the probability that someone's catch averages over 3 pounds? Explain.

d) The 12 fishermen competing each caught the limit of 5 fish. What's the probability that the total catch of 60 fish averaged more than 3 pounds?

24. **Cheating.** A Rutgers University study released in 2002 found that many high school students cheat on tests. The researchers surveyed a random sample of 4500 high school students nationwide; 74% of them said they had cheated at least once.

a) Create a 90% confidence interval for the level of cheating among high school students. Don't forget to check the appropriate conditions.

b) Interpret your interval.

c) Explain what "90% confidence" means.

d) Would a 95% confidence interval be wider or narrower? Explain without actually calculating the interval.

25. **Language.** Neurological research has shown that in about 80% of people language abilities reside in the brain's left side. Another 10% display right-brain language centers, and the remaining 10% have two-sided language control. (The latter two groups are mainly left-handers.) (*Science News*, 161, no. 24 [2002])

a) We select 60 people at random. Is it reasonable to use a Normal model to describe the possible distribution of the proportion of the group that has left-brain language control? Explain.

b) What's the probability that our group has at least 75% left-brainers?

c) If the group had consisted of 100 people, would that probability be higher, lower, or about the same? Explain why, without actually calculating the probability.

d) How large a group would almost certainly guarantee at least 75% left-brainers? Explain.

26. **Cigarettes 2006.** In 1999 the Centers for Disease Control and Prevention estimated that about 34.8% of high school students smoked cigarettes. They established a national health goal of reducing that figure to 16% by the year 2010. To that end, they hoped to achieve a reduction to 20% by 2006. In 2006 they released a research study in which 23% of a random sample of 1815 high school students said they were current smokers. Is this evidence that progress toward the goal is off track?

a) Write appropriate hypotheses.

b) Verify that the appropriate assumptions are satisfied.

c) Find the P-value of this test.

d) Explain what the P-value means in this context.

e) State an appropriate conclusion.

f) Of course, your conclusion may be incorrect. If so, which kind of error did you commit?

27. **Crohn's disease.** In 2002 the medical journal *The Lancet* reported that 335 of 573 patients suffering from Crohn's disease responded positively to injections of the arthritis-fighting drug infliximab.

a) Create a 95% confidence interval for the effectiveness of this drug.

b) Interpret your interval in context.

c) Explain carefully what "95% confidence" means in this context.

28. **Teen smoking 2006.** The Centers for Disease Control and Prevention say that about 23% of teenagers smoke tobacco (down from a high of 38% in 1997). A college has 522 students in its freshman class. Is it likely that more than 30% of them are smokers? Explain.

29. **Alcohol abuse.** Growing concern about binge drinking among college students has prompted one large state university to conduct a survey to assess the size of the problem on its campus. The university plans to randomly select students and ask how many have been drunk during the past week. If the school hopes to estimate the true proportion among all its students with 90% confidence and a margin of error of ±4%, how many students must be surveyed?

30. **Errors.** An auto parts company advertises that its special oil additive will make the engine "run smoother, cleaner, longer, with fewer repairs." An independent laboratory decides to test part of this claim. It arranges to use a taxicab company's fleet of cars. The cars are randomly divided into two groups. The company's mechanics will use the additive in one group of cars but not in the other. At the end of a year the laboratory will compare the percentage of cars in each group that required engine repairs.

a) What kind of a study is this?

b) Will they do a one-tailed or a two-tailed test?

c) Explain in this context what a Type I error would be.

d) Explain in this context what a Type II error would be.

e) Which type of error would the additive manufacturer consider more serious?

f) If the cabs with the additive do indeed run significantly better, can the company conclude it is an effect of the additive? Can they generalize this result and recommend the additive for all cars? Explain.

31. **Preemies.** Among 242 Cleveland-area children born prematurely at low birth weights between 1977 and 1979, only 74% graduated from high school. Among a comparison group of 233 children of normal birth weight, 83% were high school graduates. ("Outcomes in Young Adulthood for Very-Low-Birth-Weight Infants," *New England Journal of Medicine*, 346, no. 3 [2002])

a) Create a 95% confidence interval for the difference in graduation rates between children of normal and children of very low birth weights. Be sure to check the appropriate assumptions and conditions.

b) Does this provide evidence that premature birth may be a risk factor for not finishing high school? Use your confidence interval to test an appropriate hypothesis.

c) Suppose your conclusion is incorrect. Which type of error did you make?

32. **Safety.** Observers in Texas watched children at play in eight communities. Of the 814 children seen biking, roller skating, or skateboarding, only 14% wore a helmet.

a) Create and interpret a 95% confidence interval.
b) What concerns do you have about this study that might make your confidence interval unreliable?
c) Suppose we want to do this study again, picking various communities and locations at random, and hope to end up with a 98% confidence interval having a margin of error of ±4%. How many children must we observe?

33. **Fried PCs.** A computer company recently experienced a disastrous fire that ruined some of its inventory. Unfortunately, during the panic of the fire, some of the damaged computers were sent to another warehouse, where they were mixed with undamaged computers. The engineer responsible for quality control would like to check out each computer in order to decide whether it's undamaged or damaged. Each computer undergoes a series of 100 tests. The number of tests it fails will be used to make the decision. If it fails more than a certain number, it will be classified as damaged and then scrapped. From past history, the distribution of the number of tests failed is known for both undamaged and damaged computers. The probabilities associated with each outcome are listed in the table below:

Number of tests failed	0	1	2	3	4	5	>5
Undamaged (%)	80	13	2	4	1	0	0
Damaged (%)	0	10	70	5	4	1	10

The table indicates, for example, that 80% of the undamaged computers have no failures, while 70% of the damaged computers have 2 failures.

a) To the engineers, this is a hypothesis-testing situation. State the null and alternative hypotheses.
b) Someone suggests classifying a computer as damaged if it fails any of the tests. Discuss the advantages and disadvantages of this test plan.
c) What number of tests would a computer have to fail in order to be classified as damaged if the engineers want to have the probability of a Type I error equal to 5%?
d) What's the power of the test plan in part c?
e) A colleague points out that by increasing α just 2%, the power can be increased substantially. Explain.

34. **Power.** We are replicating an experiment. How will each of the following changes affect the power of our test? Indicate whether it will increase, decrease, or remain the same, assuming that all other aspects of the situation remain unchanged.

a) We increase the number of subjects from 40 to 100.
b) We require a higher standard of proof, changing from $\alpha = 0.05$ to $\alpha = 0.01$.

35. **Approval 2007.** Of all the post–World War II presidents, Richard Nixon had the highest *disapproval* rating near the end of his presidency. His disapproval rating peaked at 66% in July 1974, just before he resigned. In May 2007, George W. Bush's disapproval rating was 63%, according to a Gallup poll of 1000 voters. Pundits started discussing whether his rating was still discernibly better than Nixon's. What do you think?

36. **Grade inflation.** In 1996, 20% of the students at a major university had an overall grade point average of 3.5 or higher (on a scale of 4.0). In 2000, a random sample of

1100 student records found that 25% had a GPA of 3.5 or higher. Is this evidence of grade inflation?

37. **Name recognition.** An advertising agency won't sign an athlete to do product endorsements unless it is sure the person is known to more than 25% of its target audience. The agency always conducts a poll of 500 people to investigate the athlete's name recognition before offering a contract. Then it tests $H_0: p = 0.25$ against $H_A: p > 0.25$ at a 5% level of significance.

a) Why does the company use upper tail tests in this situation?
b) Explain what Type I and Type II errors would represent in this context, and describe the risk that each error poses to the company.
c) The company is thinking of changing its test to use a 10% level of significance. How would this change the company's exposure to each type of risk?

38. **Name recognition, part II.** The advertising company described in Exercise 37 is thinking about signing a WNBA star to an endorsement deal. In its poll, 27% of the respondents could identify her.

a) Fans who never took Statistics can't understand why the company did not offer this WNBA player an endorsement contract even though the 27% recognition rate in the poll is above the 25% threshold. Explain it to them.
b) Suppose that further polling reveals that this WNBA star really is known to about 30% of the target audience. Did the company initially commit a Type I or Type II error in not signing her?
c) Would the power of the company's test have been higher or lower if the player were more famous? Explain.

39. **NIMBY.** In March 2007, the Gallup Poll split a sample of 1003 randomly selected U.S. adults into two groups at random. Half ($n = 502$) of the respondents were asked,

"Overall, do you strongly favor, somewhat favor, somewhat oppose, or strongly oppose the use of nuclear energy as one of the ways to provide electricity for the U.S.?"

They found that 53% were either "somewhat" or "strongly" in favor. The other half ($n = 501$) were asked,

"Overall, would you strongly favor, somewhat favor, somewhat oppose, or strongly oppose the construction of a nuclear energy plant in your area as one of the ways to provide electricity for the U.S.?"

Only 40% were somewhat or strongly in favor. This difference is an example of the *NIMBY* (Not In My BackYard) phenomenon and is a serious concern to policy makers and planners. How large is the difference between the proportion of American adults who think nuclear energy is a good idea and the proportion who would be willing to have a nuclear plant in their area? Construct and interpret an appropriate confidence interval.

40. **Dropouts.** One study comparing various treatments for the eating disorder anorexia nervosa initially enlisted 198 subjects, but found overall that 105 failed to complete their assigned treatment programs. Construct and interpret an appropriate confidence interval. Discuss any reservations you have about this inference.

CHAPTER 23

Inferences About Means

Where are we going?

We've learned how to generalize from the data at hand to the world at large for proportions. But not all data are as simple as Yes or No. In this chapter we'll learn how to make confidence intervals and test hypotheses for the mean of a quantitative variable.

WHO	Vehicles on Triphammer Road
WHAT	Speed
UNITS	Miles per hour
WHEN	April 11, 2000, 1 p.m.
WHERE	A small town in the northeastern United States
WHY	Concern over impact on residential neighborhood

Motor vehicle crashes are the leading cause of death for people between 4 and 33 years old. In the year 2006, motor vehicle accidents claimed the lives of 43,300 people in the United States. This means that, on average, motor vehicle crashes resulted in 119 deaths each day, or 1 death every 12 minutes. Speeding is a contributing factor in 31% of all fatal accidents, according to the National Highway Traffic Safety Administration.

Triphammer Road is a busy street that passes through a residential neighborhood. Residents there are concerned that vehicles traveling on Triphammer often exceed the posted speed limit of 30 miles per hour. The local police sometimes place a radar speed detector by the side of the road; as a vehicle approaches, this detector displays the vehicle's speed to its driver.

The local residents are not convinced that such a passive method is helping the problem. They wish to persuade the village to add extra police patrols to encourage drivers to observe the speed limit. To help their case, a resident stood where he could see the detector and recorded the speed of vehicles passing it during a 15-minute period one day. When clusters of vehicles went by, he noted only the speed of the front vehicle. Here are his data and the histogram.

FIGURE 23.1

The speeds of cars on Triphammer Road seem to be unimodal and symmetric, at least at this scale.

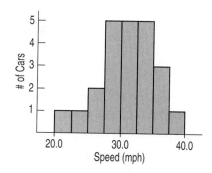

Speed		
29	29	24
34	34	34
34	32	36
28	31	31
30	27	34
29	37	36
38	29	21
31	26	

We're interested both in estimating the true mean speed and in testing whether it exceeds the posted speed limit. Although the sample of vehicles is a convenience sample, not a truly random sample, there's no compelling reason to believe that vehicles at one time of day are driving faster or slower than vehicles at another time of day,[1] so we can take the sample to be representative.

These data differ from data on proportions in one important way. Proportions are usually reported as summaries. After all, individual responses are just "success" and "failure" or "1" and "0." Quantitative data, though, usually report a value for each individual. When you have a value for each individual, you should remember the three rules of data analysis and plot the data, as we have done here.

We have quantitative data, so we summarize with means and standard deviations. Because we want to make inferences, we'll think about sampling distributions, too, and we already know most of the facts we need.

Getting Started

You've learned how to create confidence intervals and test hypotheses about proportions. We always center confidence intervals at our best guess of the unknown parameter. Then we add and subtract a margin of error. For proportions, that means $\hat{p} \pm ME$.

We found the margin of error as the product of the standard error, $SE(\hat{p})$, and a critical value, z^*, from the Normal table. So we had $\hat{p} \pm z^*SE(\hat{p})$.

We knew we could use z because the Central Limit Theorem told us (back in Chapter 18) that the sampling distribution model for proportions is Normal.

Now we want to do exactly the same thing for means, and fortunately, the Central Limit Theorem (still in Chapter 18) told us that the same Normal model works as the sampling distribution for means.

THE CENTRAL LIMIT THEOREM

When a random sample is drawn from any population with mean μ and standard deviation σ, its sample mean, $\bar{y}$, has a sampling distribution with the same *mean μ* but whose *standard deviation* is $\dfrac{\sigma}{\sqrt{n}}$ (and we write

$$\sigma(\bar{y}) = SD(\bar{y}) = \frac{\sigma}{\sqrt{n}}).$$

No matter what population the random sample comes from, the *shape* of the sampling distribution is approximately Normal as long as the sample size is large enough. The larger the sample used, the more closely the Normal approximates the sampling distribution for the mean.

---

[1] Except, perhaps, at rush hour. But at that time, traffic is slowed. Our concern is with ordinary traffic during the day.

FOR EXAMPLE Using the CLT (as if we knew σ)

Based on weighing thousands of animals, the American Angus Association reports that mature Angus cows have a mean weight of 1309 pounds with a standard deviation of 157 pounds. This result was based on a very large sample of animals from many herds over a period of 15 years, so let's assume that these summaries are the population parameters and that the distribution of the weights was unimodal and reasonably symmetric.

QUESTION: What does the CLT predict about the mean weight seen in random samples of 100 mature Angus cows?

It's given that weights of all mature Angus cows have $\mu = 1309$ and $\sigma = 157$ pounds. Because $n = 100$ animals is a fairly large sample, I can apply the Central Limit Theorem. I expect the resulting sample means $\bar{y}$ will average 1309 pounds and have a standard deviation of

$$SD(\bar{y}) = \frac{\sigma}{\sqrt{n}} = \frac{157}{\sqrt{100}} = 15.7 \text{ pounds.}$$

The CLT also says that the distribution of sample means follows a Normal model, so the 68–95–99.7 Rule applies. I'd expect that

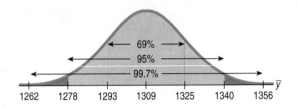

▶ in 68% of random samples of 100 mature Angus cows, the mean weight will be between $1309 - 15.7 = 1293.3$ and $1309 + 15.7 = 1324.7$ pounds;

▶ in 95% of such samples, $1277.6 \le \bar{y} \le 1340.4$ pounds;

▶ in 99.7% of such samples, $1261.9 \le \bar{y} \le 1356.1$ pounds.

The CLT says that all we need to model the sampling distribution of $\bar{y}$ is a random sample of quantitative data.

And the true population standard deviation, σ.

Uh oh. That could be a problem. How are we supposed to know σ? With proportions, we had a link between the proportion value and the standard deviation of the sample proportion: $SD(\hat{p}) = \sqrt{\frac{pq}{n}}$. And there was an obvious way to estimate the standard deviation from the data: $SE(\hat{p}) = \sqrt{\frac{\hat{p}\hat{q}}{n}}$. But for means, $SD(\bar{y}) = \frac{\sigma}{\sqrt{n}}$, so knowing $\bar{y}$ doesn't tell us anything about $SD(\bar{y})$. We know n, the sample size, but the population standard deviation, σ, could be *anything*. So what should we do? We do what any sensible person would do: We estimate the population parameter σ with s, the sample standard deviation based on the data. The resulting standard error is $SE(\bar{y}) = \frac{s}{\sqrt{n}}$.

A century ago, people used this standard error with the Normal model, assuming it would work. And for large sample sizes it *did* work pretty well. But they began to notice problems with smaller samples. The sample standard deviation, s, like any other statistic, varies from sample to sample. And this extra variation in the standard error was messing up the P-values and margins of error.

William S. Gosset is the man who first investigated this fact. He realized that not only do we need to allow for the extra variation with larger margins of error and P-values, but we even need a new sampling distribution model. In fact, we need a whole *family* of models, depending on the sample size, n. These models are unimodal, symmetric, bell-shaped models, but the smaller our sample, the more we must stretch out the tails. Gosset's work transformed Statistics, but most people who use his work don't even know his name.

> Because we estimate the standard deviation of the sampling distribution model from the data, it's a *standard error*. So we use the $SE(\bar{y})$ notation. Remember, though, that it's just the estimated standard deviation of the sampling distribution model for means.

A S *Activity: Estimating the Standard Error.* What's the average age at which people have heart attacks? A confidence interval gives a good answer, but we must estimate the standard deviation from the data to construct the interval.

Gosset's *t*

To find the sampling distribution of $\dfrac{\bar{y}}{s/\sqrt{n}}$, Gosset simulated it *by hand*. He drew paper slips of small samples from a hat *hundreds of times* and computed the means and standard deviations with a mechanically cranked calculator. Today you could repeat in seconds on a computer the experiment that took him over a year. Gosset's work was so meticulous that not only did he get the shape of the new histogram approximately right, but he even figured out the exact *formula* for it from his sample. The formula was not confirmed mathematically until years later by Sir R. A. Fisher.

Gosset had a job that made him the envy of many. He was the quality control engineer for the Guinness Brewery in Dublin, Ireland. His job was to make sure that the stout (a thick, dark beer) leaving the brewery was of high enough quality to meet the demands of the brewery's many discerning customers. It's easy to imagine why a large sample with many observations might be undesirable when testing stout, not to mention dangerous to one's health. So Gosset often used small samples of 3 or 4. But he noticed that with samples of this size, his tests for quality weren't quite right. He knew this because when the batches that he rejected were sent back to the laboratory for more extensive testing, too often they turned out to be OK.

Gosset checked the stout's quality by performing hypothesis tests. He knew that the test would make some Type I errors and reject about 5% of the *good* batches of stout. However, the lab told him that he was in fact rejecting about 15% of the good batches. Gosset knew something was wrong, and it bugged him.

Gosset took time off to study the problem (and earn a graduate degree in the emerging field of Statistics). He figured out that when he used the standard error, $\dfrac{s}{\sqrt{n}}$, as an estimate of the standard deviation, the shape of the sampling model changed. He even figured out what the new model should be and called it a *t*-distribution.

The Guinness Company didn't give Gosset a lot of support for his work. In fact, it had a policy against publishing results. Gosset had to convince the company that he was not publishing an industrial secret, and (as part of getting permission to publish) he had to use a pseudonym. The pseudonym he chose was "Student," and ever since, the model he found has been known as **Student's *t*.**

Gosset's model is always bell-shaped, but the details change with different sample sizes. So the Student's *t*-models form a whole *family* of related distributions that depend on a parameter known as **degrees of freedom.** We often denote degrees of freedom as *df* and the model as t_{df}, with the degrees of freedom as a subscript.

A Confidence Interval for Means

To make confidence intervals or test hypotheses for means, we need to use Gosset's model. Which one? Well, for means, it turns out the right value for degrees of freedom is $df = n - 1$.

A PRACTICAL SAMPLING DISTRIBUTION MODEL FOR MEANS

When certain assumptions and conditions[2] are met, the standardized sample mean,

$$t = \frac{\bar{y} - \mu}{SE(\bar{y})},$$

follows a Student's *t*-model with $n - 1$ degrees of freedom. We estimate the standard deviation with

$$SE(\bar{y}) = \frac{s}{\sqrt{n}}$$

---

[2] You can probably guess what they are. We'll see them in the next section.

When Gosset corrected the model for the extra uncertainty, the margin of error got bigger, as you might have guessed. When you use Gosset's model instead of the Normal model, your confidence intervals will be just a bit wider and your P-values just a bit larger. That's the correction you need. By using the *t*-model, you've compensated for the extra variability in precisely the right way.

A S

Activity: **Student's *t* in Practice.** Use a statistics package to find a *t*-based confidence interval; that's how it's almost always done.

ONE-SAMPLE *t*-INTERVAL FOR THE MEAN

When the assumptions and conditions[3] are met, we are ready to find the confidence interval for the population mean, μ. The confidence interval is

$$\bar{y} \pm t^*_{n-1} \times SE(\bar{y}),$$

where the standard error of the mean is $SE(\bar{y}) = \dfrac{s}{\sqrt{n}}$.

The critical value t^*_{n-1} depends on the particular confidence level, C, that you specify and on the number of degrees of freedom, $n - 1$, which we get from the sample size.

FOR EXAMPLE   **A One-Sample *t*-Interval for the Mean**

In 2004, a team of researchers published a study of contaminants in farmed salmon.[4] Fish from many sources were analyzed for 14 organic contaminants. The study expressed concerns about the level of contaminants found. One of those was the insecticide mirex, which has been shown to be carcinogenic and is suspected to be toxic to the liver, kidneys, and endocrine system. One farm in particular produced salmon with very high levels of mirex. After those outliers are removed, summaries for the mirex concentrations (in parts per million) in the rest of the farmed salmon are:

$$n = 150 \qquad \bar{y} = 0.0913 \text{ ppm} \qquad s = 0.0495 \text{ ppm}.$$

QUESTION: What does a 95% confidence interval say about mirex?

$$df = 150 - 1 = 149$$

$$SE(\bar{y}) = \frac{s}{\sqrt{n}} = \frac{0.0495}{\sqrt{150}} = 0.0040$$

$t^*_{149} \approx 1.977$ (from Table T, using 140 *df*)
(actually, $t^*_{149} \approx 1.976$ from technology)

So the confidence interval for μ is $\bar{y} \pm t^*_{149} \times SE(\bar{y}) = 0.0913 \pm 1.977(0.0040)$

$$= 0.0913 \pm 0.0079$$

$$= (0.0834, 0.0992)$$

I'm 95% confident that the mean level of mirex concentration in farm-raised salmon is between 0.0834 and 0.0992 parts per million.

A S

Activity: **Student's Distributions.** Interact with Gosset's family of *t*-models. Watch the shape of the model change as you slide the degrees of freedom up and down.

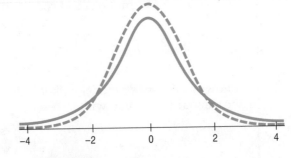

FIGURE 23.2
The *t*-model (solid curve) on 2 degrees of freedom has fatter tails than the Normal model (dashed curve). So the 68-95-99.7 Rule doesn't work for *t*-models with only a few degrees of freedom.

---

[3] Yes, the same ones, and they're still coming in the next section.
[4] Ronald A. Hites, Jeffery A. Foran, David O. Carpenter, M. Coreen Hamilton, Barbara A. Knuth, and Steven J. Schwager, "Global Assessment of Organic Contaminants in Farmed Salmon," *Science* 9 January 2004: Vol. 303, no. 5655, pp. 226–229.

Student's t-models are unimodal, symmetric, and bell-shaped, just like the Normal. But t-models with only a few degrees of freedom have much fatter tails than the Normal. (That's what makes the margin of error bigger.) As the degrees of freedom increase, the t-models look more and more like the Normal. In fact, the t-model with infinite degrees of freedom is exactly Normal.[5] This is great news if you happen to have an infinite number of data values. Unfortunately, that's not practical. Fortunately, above a few hundred degrees of freedom it's very hard to tell the difference. Of course, in the rare situation that we *know* σ, it would be foolish not to use that information. And if we don't have to estimate σ, we can use the Normal model.

> **z OR t?**
>
> If you know σ, use z. (That's rare!)
> Whenever you use s to estimate σ, use t.

> **When σ is known** Administrators of a hospital were concerned about the prenatal care given to mothers in their part of the city. To study this, they examined the gestation times of babies born there. They drew a sample of 25 babies born in their hospital in the previous 6 months. Human gestation times for healthy pregnancies are thought to be well-modeled by a Normal with a mean of 280 days and a standard deviation of 14 days. The hospital administrators wanted to test the mean gestation time of their sample of babies against the known standard. For this test, they should use the established value for the standard deviation, 14 days, rather than estimating the standard deviation from their sample. Because they use the model parameter value for σ, they should base their test on the Normal model rather than Student's t.

Assumptions and Conditions

Gosset found the t-model by simulation. Years later, when Sir Ronald A. Fisher[6] showed mathematically that Gosset was right, he needed to make some assumptions to make it work. These are the assumptions we need to use the Student's t-models.

Independence Assumption

Independence Assumption: The data values should be independent. There's really no way to check independence of the data by looking at the sample, but we should think about whether the assumption is reasonable.

Randomization Condition: The data arise from a random sample or suitably randomized experiment. Randomly sampled data—and especially data from a Simple Random Sample—are ideal.

When a sample is drawn without replacement, technically we ought to confirm that we haven't sampled a large fraction of the population, which would threaten the independence of our selections. We check the

10% Condition: The sample is no more than 10% of the population.

In practice, though, we often don't mention the 10% Condition for means. Why not? When we made inferences about proportions, this condition was crucial because we usually had large samples. But for means our samples are generally smaller, so the independence problem arises only if we're sampling from a small population. There's a correction formula, but it isn't often used so we won't bother with it here. And sometimes we're dealing with a randomized experiment; then there's no sampling at all.

---

[5] Formally, in the limit as n goes to infinity.
[6] We met Fisher back in Chapter 21. You can see his picture on page 514.

Normal Population Assumption

Student's *t*-models won't work for data that are badly skewed. How skewed is too skewed? Well, formally, we assume that the data are from a population that follows a Normal model. Practically speaking, there's no way to be certain this is true.

And it's almost certainly *not* true. Models are idealized; real data are, well, real—*never* Normal. The good news, however, is that even for small samples, it's sufficient to check the . . .

Nearly Normal Condition: The data come from a distribution that is unimodal and symmetric.

Check this condition by making a histogram or Normal probability plot. The importance of Normality for Student's *t* depends on the sample size. Just our luck: It matters most when it's hardest to check.[7]

For very small samples ($n < 15$ or so), the data should follow a Normal model pretty closely. Of course, with so little data, it's rather hard to tell. But if you do find outliers or strong skewness, don't use these methods.

For moderate sample sizes (n between 15 and 40 or so), the *t* methods will work well as long as the data are unimodal and reasonably symmetric. Make a histogram.

When the sample size is larger than 40 or 50, the *t* methods are safe to use unless the data are extremely skewed. Be sure to make a histogram. If you find outliers in the data, it's always a good idea to perform the analysis twice, once with and once without the outliers, even for large samples. They may well hold additional information about the data that deserves special attention. If you find multiple modes, you may well have different groups that should be analyzed and understood separately.

FOR EXAMPLE | Checking Assumptions and Conditions for Student's *t*

RECAP: Researchers purchased whole farmed salmon from 51 farms in eight regions in six countries. The histogram shows the concentrations of the insecticide mirex in 150 farmed salmon.

QUESTION: Are the assumptions and conditions for inference satisfied?

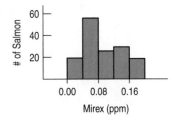

✓ **Independence Assumption:** The fish were raised in many different places, and samples were purchased independently from several sources.

✓ **Randomization Condition:** The fish were selected randomly from those available for sale.

✓ **10% Condition:** There's lots of fish in the sea (and at the fish farms); 150 is certainly far fewer than 10% of the population.

✓ **Nearly Normal Condition:** The histogram of the data is unimodal. Although it may be somewhat skewed to the right, this is not a concern with a sample size of 150.

It's okay to use these data for inference about farm-raised salmon.

---

[7] There are formal tests of Normality, but they don't really help. When we have a small sample—just when we really care about checking Normality—these tests have very little power. So it doesn't make much sense to use them in deciding whether to perform a *t*-test. We don't recommend that you use them.

JUST CHECKING

Every 10 years, the United States takes a census. The census tries to count every resident. There have been two forms, known as the "short form," answered by most people, and the "long form," slogged through by about one in six or seven households chosen at random. (For the 2010 census, the long form was replaced by the American Community Survey.) According to the Census Bureau (www.census.gov), ". . . each estimate based on the long form responses has an associated confidence interval."

1. Why does the Census Bureau need a confidence interval for long-form information but not for the questions that appear on both the long and short forms?

2. Why must the Census Bureau base these confidence intervals on *t*-models?

The Census Bureau goes on to say, "These confidence intervals are wider . . . for geographic areas with smaller populations and for characteristics that occur less frequently in the area being examined (such as the proportion of people in poverty in a middle-income neighborhood)."

3. Why is this so? For example, why should a confidence interval for the mean amount families spend monthly on housing be wider for a sparsely populated area of farms in the Midwest than for a densely populated area of an urban center? How does the formula show this will happen?

To deal with this problem, the Census Bureau reports long-form data only for ". . . geographic areas from which about two hundred or more long forms were completed—which are large enough to produce good quality estimates. If smaller weighting areas had been used, the confidence intervals around the estimates would have been significantly wider, rendering many estimates less useful. . . ."

4. Suppose the Census Bureau decided to report on areas from which only 50 long forms were completed. What effect would that have on a 95% confidence interval for, say, the mean cost of housing? Specifically, which values used in the formula for the margin of error would change? Which would change a lot and which would change only slightly?

5. Approximately how much wider would that confidence interval based on 50 forms be than the one based on 200 forms?

STEP-BY-STEP EXAMPLE   A One-Sample *t*-Interval for the Mean

Let's build a 90% confidence interval for the mean speed of all vehicles traveling on Triphammer Road. The interval that we'll make is called the **one-sample *t*-interval.**

Question: What can we say about the mean speed of all cars on Triphammer Road?

THINK **Plan** State what we want to know. Identify the parameter of interest. Identify the variables and review the W's. Make a picture. Check the distribution shape and look for skewness, multiple modes, and outliers.	I want to find a 90% confidence interval for the mean speed, μ, of vehicles driving on Triphammer Road. I have data on the speeds of 23 cars there, sampled on April 11, 2000. Here's a histogram of the 23 observed speeds. 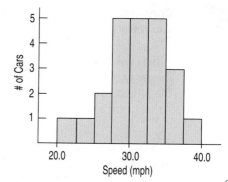

REALITY CHECK The histogram centers around 30 mph, and the data lie between 20 and 40 mph. We'd expect a confidence interval to place the population mean within a few mph of 30.

(continued)

Model Think about the assumptions and check the conditions.

Note that with this small sample we probably didn't need to check the 10% Condition.

On the other hand, doing so gives us a chance to think about what the population is.

✔ **Independence Assumption:** This is a convenience sample, but care was taken to select cars that were not driving near each other, so their speeds are plausibly independent.

✔ **Randomization Condition:** Not really met. This is a convenience sample, but I have reason to believe that it is representative.

✔ **10% Condition:** The cars I observed were fewer than 10% of all cars that travel Triphammer Road.

✔ **Nearly Normal Condition:** The histogram of the speeds is unimodal and symmetric.

State the sampling distribution model for the statistic.

Choose your method.

The conditions are satisfied, so I will use a Student's t-model with

$$(n - 1) = 22 \text{ degrees of freedom}$$

and find a **one-sample t-interval for the mean.**

Mechanics Construct the confidence interval.

Be sure to include the units along with the statistics.

Calculating from the data (see page 550):

$$n = 23 \text{ cars}$$
$$\bar{y} = 31.0 \text{ mph}$$
$$s = 4.25 \text{ mph.}$$

The standard error of $\bar{y}$ is

$$SE(\bar{y}) = \frac{s}{\sqrt{n}} = \frac{4.25}{\sqrt{23}} = 0.886 \text{ mph.}$$

The critical value we need to make a 90% interval comes from a Student's t table, a computer program, or a calculator. We have 23 − 1 = 22 degrees of freedom. The selected confidence level says that we want 90% of the probability to be caught in the middle, so we exclude 5% in *each* tail, for a total of 10%. The degrees of freedom and 5% tail probability are all we need to know to find the critical value.

The 90% critical value is $t^*_{22} = 1.717$, so the margin of error is

$$ME = t^*_{22} \times SE(\bar{y})$$
$$= 1.717(0.886)$$
$$= 1.521 \text{ mph.}$$

The 90% confidence interval for the mean speed is 31.0 ± 1.5 mph.

 The result looks plausible and in line with what we thought.

Conclusion Interpret the confidence interval in the proper context.

When we construct confidence intervals in this way, we expect 90% of them to cover the true mean and 10% to miss the true value. That's what "90% confident" means.

I am 90% confident that the interval from 29.5 mph to 32.5 mph contains the true mean speed of all vehicles on Triphammer Road.

Caveat: This was not a random sample of vehicles. It was a convenience sample taken at one time on one day. And the participants were not blinded. Drivers could see the police device, and some may have slowed down. I'm reluctant to extend this inference to other situations.

Here's the part of the Student's *t* table that gives the critical value we needed for the Step-by-Step confidence interval. (See Table T in the back of the book.) To find a critical value, locate the row of the table corresponding to the degrees of freedom and the column corresponding to the probability you want. Our 90% confidence interval leaves 5% of the values on either side, so look for 0.05 at the top of the column or 90% at the bottom. The value in the table at that intersection is the critical value we need: 1.717.

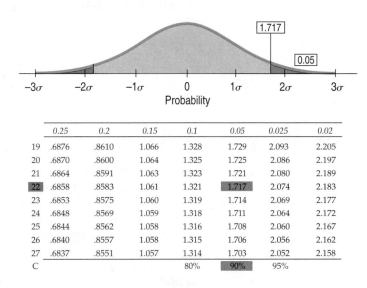

	0.25	0.2	0.15	0.1	0.05	0.025	0.02
19	.6876	.8610	1.066	1.328	1.729	2.093	2.205
20	.6870	.8600	1.064	1.325	1.725	2.086	2.197
21	.6864	.8591	1.063	1.323	1.721	2.080	2.189
22	.6858	.8583	1.061	1.321	1.717	2.074	2.183
23	.6853	.8575	1.060	1.319	1.714	2.069	2.177
24	.6848	.8569	1.059	1.318	1.711	2.064	2.172
25	.6844	.8562	1.058	1.316	1.708	2.060	2.167
26	.6840	.8557	1.058	1.315	1.706	2.056	2.162
27	.6837	.8551	1.057	1.314	1.703	2.052	2.158
C				80%	90%	95%	

A S *Activity:* **Building *t*-Intervals with the *t*-Table.** Interact with an animated version of Table T.

Of course, you can also create the confidence interval with computer software or a calculator.

More Cautions About Interpreting Confidence Intervals

Confidence intervals for means offer new, tempting, wrong interpretations. Here are some things you *shouldn't* say:

A S *Activity:* **Intuition for *t*-based Intervals.** A narrated review of Student's *t*.

- *Don't say,* "90% *of all vehicles* on Triphammer Road drive at a speed between 29.5 and 32.5 mph." The confidence interval is about the *mean* speed, not about the speeds of *individual* vehicles.
- *Don't say,* "We are 90% confident that *a randomly selected vehicle* will have a speed between 29.5 and 32.5 mph." This false interpretation is also about individual vehicles rather than about the *mean* of the speeds. We are 90% confident that the *mean* speed of all vehicles on Triphammer Road is between 29.5 and 32.5 mph.
- *Don't say,* "The mean speed of the vehicles is 31.0 mph 90% *of the time*." That's about means, but still wrong. It implies that the true mean varies, when in fact it is the confidence interval that would have been different had we gotten a different sample.
- Finally, *don't say,* "90% *of all samples* will have mean speeds between 29.5 and 32.5 mph." That statement suggests that *this* interval somehow sets a standard for every other interval. In fact, this interval is no more (or less) likely to be correct than any other. You could say that 90% of all possible samples will produce intervals that actually do contain the true mean speed. (The problem is that, because we'll never know where the true mean speed really is, we can't know if our sample was one of those 90%.)
- *Do say,* "90% of intervals that could be found in this way would cover the true value." Or make it more personal and say, "I am 90% confident that the true mean speed is between 29.5 and 32.5 mph."

SO WHAT *SHOULD* WE SAY?

Since 90% of random samples yield an interval that captures the true mean, we *should* say, "I am 90% confident that the interval from 29.5 to 32.5 mph contains the mean speed of all the vehicles on Triphammer Road." It's also okay to say something less formal: "I am 90% confident that the average speed of all vehicles on Triphammer Road is between 29.5 and 32.5 mph." Remember: *Our uncertainty is about the interval, not the true mean.* The interval varies randomly. The true mean speed is neither variable nor random—just unknown.

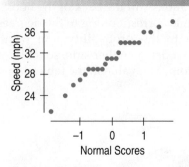

FIGURE 23.3
A Normal probability plot of speeds looks reasonably straight.

Make a Picture, Make a Picture, Make a Picture

The only reasonable way to check the Nearly Normal Condition is with graphs of the data. Make a histogram of the data and verify that its distribution is unimodal and symmetric and that it has no outliers. It's also a good idea to make a Normal probability plot to see that it's reasonably straight. You'll be able to spot deviations from the Normal model more easily with a Normal probability plot, but it's easier to understand the particular nature of the deviations from a histogram.

If you use a computer or graphing calculator to do the work, there's no excuse not to look at *both* displays as part of checking the Nearly Normal Condition.

A Test for the Mean

The residents along Triphammer Road have a more specific concern. It appears that the mean speed along the road is higher than it ought to be. To get the police to patrol more frequently, though, they'll need to show that the true mean speed is *in fact greater* than the 30 mph speed limit. This calls for a hypothesis test called the **one-sample *t*-test for the mean.**

You already know enough to construct this test. The test statistic looks just like the others we've seen. It compares the difference between the observed statistic and a hypothesized value to the standard error of the observed statistic. We've seen that, for means, the appropriate probability model to use for P-values is Student's *t* with $n - 1$ degrees of freedom.

A S

Activity: A *t*-Test for Wind Speed.
Watch the video in the preceding activity, and then use the interactive tool to test whether there's enough wind for electricity generation at a site under investigation.

> **ONE-SAMPLE *t*-TEST FOR THE MEAN**
>
> The assumptions and conditions for the one-sample *t*-test for the mean are the same as for the one-sample *t*-interval. We test the hypothesis $H_0: \mu = \mu_0$ using the statistic
>
> $$t_{n-1} = \frac{\bar{y} - \mu_0}{SE(\bar{y})}.$$
>
> The standard error of $\bar{y}$ is $SE(\bar{y}) = \dfrac{s}{\sqrt{n}}$.
>
> When the conditions are met and the null hypothesis is true, this statistic follows a Student's *t*-model with $n - 1$ degrees of freedom. We use that model to obtain a P-value.

FOR EXAMPLE   A One-Sample *t*-Test for the Mean

RECAP: Researchers tested 150 farm-raised salmon for organic contaminants. They found the mean concentration of the carcinogenic insecticide mirex to be 0.0913 parts per million, with standard deviation 0.0495 ppm. As a safety recommendation to recreational fishers, the Environmental Protection Agency's (EPA) recommended "screening value" for mirex is 0.08 ppm.

QUESTION: Are farmed salmon contaminated beyond the level permitted by the EPA? (We've already checked the conditions; see page 556.)

$$H_0: \mu = 0.08$$
$$H_A: \mu > 0.08$$

These data satisfy the conditions for inference; I'll do a one-sample t-test for the mean:

$n = 150, df = 149$

$\bar{y} = 0.0913, s = 0.0495$

$SE(\bar{y}) = \dfrac{0.0495}{\sqrt{150}} = 0.0040$

$t_{149} = \dfrac{0.0913 - 0.08}{0.0040} = 2.825$

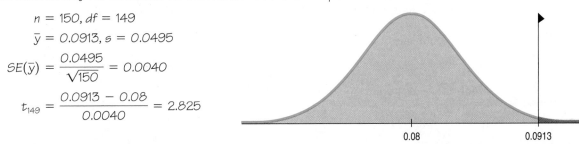

$P(t_{149} > 2.825) = 0.0027$ (from technology).

With a P-value that low, I reject the null hypothesis and conclude that, in farm-raised salmon, the mirex contamination level does exceed the EPA screening value.

STEP-BY-STEP EXAMPLE A One-Sample *t*-Test for the Mean

Let's apply the one-sample *t*-test to the Triphammer Road car speeds. The speed limit is 30 mph, so we'll use that as the null hypothesis value.

Question: Does the mean speed of all cars exceed the posted speed limit?

THINK **Plan** State what we want to know. Make clear what the population and parameter are.	I want to know whether the mean speed of vehicles on Triphammer Road exceeds the posted speed limit of 30 mph. I have a sample of 23 car speeds on April 11, 2000.

Identify the variables and review the W's.

Hypotheses The null hypothesis is that the true mean speed is equal to the limit. Because we're interested in whether the vehicles are speeding, the alternative is one-sided.

H_0: Mean speed, $\mu = 30$ mph
H_A: Mean speed, $\mu > 30$ mph

Make a picture. Check the distribution for skewness, multiple modes, and outliers.

REALITY CHECK The histogram of the observed speeds is clustered around 30, so we'd be surprised to find that the mean was much higher than that. (The fact that 30 is within the confidence interval that we've just found confirms this suspicion.)

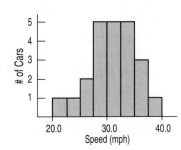

Model Think about the assumptions and check the conditions.

(We won't worry about the 10% Condition—it's a small sample.)

✔ **Independence Assumption:** These cars are a convenience sample, but they were selected so no two cars were driving near each other, so I am justified in believing that their speeds are independent.

✔ **Randomization Condition:** Although I have a convenience sample, I have reason to believe that it is a representative sample.

✔ **Nearly Normal Condition:** The histogram of the speeds is unimodal and reasonably symmetric.

(continued)

State the sampling distribution model. (Be sure to include the degrees of freedom.) Choose your method.	The conditions are satisfied, so I'll use a Student's *t*-model with $(n-1) = 22$ degrees of freedom to do a **one-sample t-test for the mean.**

Mechanics Be sure to include the units when you write down what you know from the data.

We use the null model to find the P-value. Make a picture of the *t*-model centered at $\mu = 30$. Since this is an upper-tail test, shade the region to the right of the observed mean speed.

The *t*-statistic calculation is just a standardized value, like *z*. We subtract the hypothesized mean and divide by the standard error.

The P-value is the probability of observing a sample mean as large as 31.0 (or larger) *if* the true mean were 30.0, as the null hypothesis states. We can find this P-value from a table, calculator, or computer program.

REALITY CHECK We're not surprised that the difference isn't statistically significant.

From the data,

$$n = 23 \text{ cars}$$
$$\bar{y} = 31.0 \text{ mph}$$
$$s = 4.25 \text{ mph}$$

$$SE(\bar{y}) = \frac{s}{\sqrt{n}} = \frac{4.25}{\sqrt{23}} = 0.886 \text{ mph}.$$

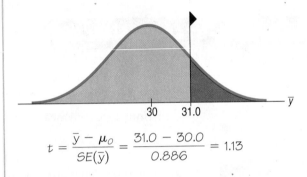

$$t = \frac{\bar{y} - \mu_0}{SE(\bar{y})} = \frac{31.0 - 30.0}{0.886} = 1.13$$

(The observed mean is 1.13 standard errors above the hypothesized value.)

$$\text{P-value} = P(t_{22} > 1.13) = 0.136$$

TELL

Conclusion Link the P-value to your decision about H_0, and state your conclusion in context.

Unfortunately for the residents, there is no course of action associated with failing to reject this particular null hypothesis.

The P-value of 0.136 says that if the true mean speed of vehicles on Triphammer Road were 30 mph, samples of 23 vehicles can be expected to have an observed mean of at least 31.0 mph 13.6% of the time. That P-value is not small enough for me to reject the hypothesis that the true mean is 30 mph at any reasonable alpha level. I conclude that there is not enough evidence to say the average speed is too high.

Finding *t*-Values by Hand

The Student's *t*-model is different for each value of degrees of freedom. Usually we find critical values and margins of error for Student's *t*-based intervals with technology. Calculators or statistics programs can give critical values for a *t*-model for any number of degrees of freedom and for any confidence level you please.

But you can also use tables such as Table T at the back of this book. The tables run down the page for as many degrees of freedom as can fit. As the degrees of freedom increase, the *t*-model gets closer and closer to the Normal, so the tables give a final row with the critical values from the Normal model and label it "∞ df."

Part of Table T.

As degrees of freedom increase, the shape of Student's *t*-models changes more gradually. Table T at the back of the book includes degrees of freedom between 100 and 1000 selected so that you can pin down the P-value for just about any df. If your df's aren't listed, take the cautious approach by using the next lower value, or use technology.

These tables are only a portion of the full tables, such as the one we used for the Normal model. We could have printed a table like Table Z for every df, but that's a lot of pages, and not likely to be a best seller. One way to shorten the book is to limit ourselves to only a few values. Although it might be nice to be able to get a critical value for a 93.4% confidence interval with 179 df, in practice we usually limit ourselves to 90%, 95%, 99%, and 99.9% and selected degrees of freedom. So, Table T fits on a single page with columns for selected confidence levels and rows for selected df's.[8]

For confidence intervals, the values in the table are usually enough to cover most cases of interest. If you can't find a row for the df you need, just use the next smaller df in the table.

For hypothesis tests, the computed *t*-statistic can take on any value, so the value you get is not likely to be one found in the table. The best we can do is to trap a calculated *t*-value between two columns. Just look across the row with the appropriate degrees of freedom to find where the *t*-statistic falls. The P-value will be between the two values at the heads of the columns.[9] Report that the P-value falls between these two values. Usually that's good enough.

Significance and Importance

Recall that "statistically significant" does not mean "actually important" or "meaningful," even though it sounds that way. In this example, it does seem that speeds may be a bit above 30 miles per hour. If so, it's possible that a larger sample would show statistical significance.

But would that be the right decision? The difference between 31 miles per hour and 30 miles per hour doesn't seem meaningful, and rejecting the null hypothesis wouldn't change that. Even with a statistically significant result, it would be hard to convince the police that vehicles on Triphammer Road were driving at dangerously fast speeds. It would probably also be difficult to persuade the town that spending more money to lower the average speed on Triphammer Road would be a good use of the town's resources. Looking at the confidence interval, we can say with 90% confidence that the mean speed is somewhere between 29.5 and 32.5 mph. Even in the worst case, if the mean speed is 32.5 mph, would this be a bad enough situation to convince the town to spend more money? Probably not. It's always a good idea when we test a hypothesis to also check the confidence interval and think about the likely values for the mean.

JUST CHECKING

In discussing estimates based on the long-form samples, the Census Bureau notes, "The disadvantage . . . is that . . . estimates of characteristics that are also reported on the short form will not match the [long-form estimates]."

The short-form estimates are values from a complete census, so they are the "true" values—something we don't usually have when we do inference.

6. Suppose we use long-form data to make 95% confidence intervals for the mean age of residents for each of 100 of the Census-defined areas. How many of these 100 intervals should we expect will fail to include the true mean age (as determined from the complete short-form Census data)?

7. Based only on the long-form sample, we might test the null hypothesis about the mean household income in a region. Would the power of the test increase or decrease if we used an area with more long forms?

---

[8] You can also find tables on the Internet. Search for terms like "statistical tables z t."
[9] Don't be confused that the *t*-values in the table *increase* from left to right while the P-values labeling the columns *decrease* from left to right. Think about it, and you'll see that it makes sense.

Intervals and Tests

The 90% confidence interval for the mean speed was 31.0 mph $\pm$ 1.5, or (29.5 mph, 32.5 mph). If someone hypothesized that the mean speed was really 30 mph, how would you feel about it? How about 35 mph?

Because the confidence interval included the speed limit of 30 mph, it certainly looked like 30 mph might be a plausible value for the true mean speed of the vehicles on Triphammer Road. In fact, 30 mph gave a P-value of 0.136—too large to reject the null hypothesis. We should have seen this coming. The hypothesized mean of 30 mph lies *within the confidence interval*. It's one of the reasonable values for the mean.

Confidence intervals and significance tests are built from the same calculations. In fact, they are really complementary ways of looking at the same question. Here's the connection: The confidence interval contains all the null hypothesis values we can't reject with these data.

More precisely, a level C confidence interval contains *all* of the plausible null hypothesis values that would *not* be rejected by a two-sided hypothesis test at alpha level $1 - C$. So a 95% confidence interval matches a $1 - 0.95 = 0.05$ level two-sided test for these data.

Confidence intervals are naturally two-sided, so they match exactly with two-sided hypothesis tests. When, as in our example, the hypothesis is one-sided, the corresponding alpha level is $(1 - C)/2$.

> **Fail to reject** Our 90% confidence interval was 29.5 to 32.5 mph. If any of these values had been the null hypothesis for the mean, then the corresponding hypothesis test at $\alpha = 0.05$ (because $\dfrac{1 - 0.90}{2} = 0.05$) would not have been able to reject the null. That is, the corresponding one-sided P-value for our observed mean of 31 mph would be greater than 0.05. So, we would not reject any hypothesized value between 29.5 and 32.5 mph.

Sample Size

A S *Activity:* **The Real Effect of Small Sample Size.** We know that smaller sample sizes lead to wider confidence intervals, but is that just because they have fewer degrees of freedom?

How large a sample do we need? The simple answer is "more." But more data cost money, effort, and time, so how much is enough? Suppose your computer just took an hour to download a movie you wanted to watch. You're not happy. You hear about a program that claims to download movies in under a half hour. You're interested enough to spend $29.95 for it, but only if it really delivers. So you get the free evaluation copy and test it by downloading that movie 5 different times. Of course, the mean download time is not exactly 30 minutes as claimed. Observations vary. If the margin of error were 8 minutes, though, you'd probably be able to decide whether the software is worth the money. Doubling the sample size would require another 5 hours of testing and would reduce your margin of error to a bit under 6 minutes. You'll need to decide whether that's worth the effort.

As we make plans to collect data, we should have some idea of how small a margin of error we need to be able to draw a conclusion or detect a difference we want to see. If the size of the effect we're studying is large, then we may be able to tolerate a larger *ME*. If we need great precision, however, we'll want a smaller *ME*, and, of course, that means a larger sample size.

Armed with the *ME* and confidence level, we can find the sample size we'll need. Almost.

We know that for a mean, $ME = t_{n-1}^* \times SE(\bar{y})$ and that $SE(\bar{y}) = \dfrac{s}{\sqrt{n}}$, so we can determine the sample size by solving this equation for n:

$$ME = t_{n-1}^* \frac{s}{\sqrt{n}}.$$

The good news is that we have an equation; the bad news is that we won't know most of the values we need to solve it. When we thought about sample size for proportions back in Chapter 19, we ran into a similar problem. There we had to guess a working value for p to compute a sample size. Here, we need to know s. We don't know s until we get some data, but we want to calculate the sample size *before* collecting the data. We might be able to make a good guess, and that is often good enough for this purpose. If we have no idea what the standard deviation might be, or if the sample size really matters (for example, because each additional individual is very expensive to sample or experiment on), it might be a good idea to run a small *pilot study* to get some feeling for the standard deviation.

That's not all. Without knowing n, we don't know the degrees of freedom and we can't find the critical value, t_{n-1}^*. One common approach is to use the corresponding z^* value from the Normal model. If you've chosen a 95% confidence level, then just use 2, following the 68–95–99.7 Rule. If your estimated sample size is, say, 60 or more, it's probably okay—z^* was a good guess. If it's smaller than that, you may want to add a step, using z^* at first, finding n, and then replacing z^* with the corresponding t_{n-1}^* and calculating the sample size once more.

Sample size calculations are *never* exact. The margin of error you find *after* collecting the data won't match exactly the one you used to find n. The sample size formula depends on quantities that you won't have until you collect the data, but using it is an important first step. Before you collect data, it's always a good idea to know whether the sample size is large enough to give you a good chance of being able to tell you what you want to know.

// FOR EXAMPLE   Finding Sample Size

A company claims its program will allow your computer to download movies quickly. We'll test the free evaluation copy by downloading a movie several times, hoping to estimate the mean download time with a margin of error of only 8 minutes. We think the standard deviation of download times is about 10 minutes.

QUESTION: How many trial downloads must we run if we want 95% confidence in our estimate with a margin of error of only 8 minutes?

Using $z^* = 1.96$, solve

$$8 = 1.96 \frac{10}{\sqrt{n}}$$

$$\sqrt{n} = \frac{1.96 \times 10}{8} = 2.45$$

$$n = (2.45)^2 = 6.0025$$

That's a small sample size, so I'll use $(6 - 1) = 5$ degrees of freedom[10] to substitute an appropriate t^* value. At 95%, $t_5^* = 2.571$. Solving the equation one more time:

$$8 = 2.571 \frac{10}{\sqrt{n}}$$

(continued)

---

[10] Ordinarily we'd round the sample size *up*. But at this stage of the calculation, rounding *down* is the safer choice. Can you see why?

$$\sqrt{n} = \frac{2.571 \times 10}{8} \approx 3.214$$

$$n = (3.214)^2 \approx 10.33$$

To make sure the ME is no larger, I'll round up, which gives $n = 11$ runs. So, to get an ME of 8 minutes, I'll find the downloading times for 11 movies.

Degrees of Freedom

Some calculators offer an alternative button for standard deviation that divides by n instead of $n - 1$. Why don't you stick a wad of gum over the "n" button so you won't be tempted to use it? Use $n - 1$.

The number of degrees of freedom, $(n - 1)$, might have reminded you of the value we divide by to find the standard deviation of the data (since, in fact, it's the same number). When we introduced that formula, we promised to say a bit more about why we divide by $n - 1$ rather than by n. The reason is closely tied to the reasoning behind the t-distribution.

If only we knew the true population mean, μ, we would find the sample standard deviation as

$$s = \sqrt{\frac{\sum (y - \mu)^2}{n}} \qquad \text{(Equation 23.1)}[11]$$

We use $\bar{y}$ instead of μ, though, and that causes a problem. For any sample, the data values will generally be closer to their own sample mean than to the true population mean, μ. Why is that? Imagine that we take a random sample of 10 high school seniors. The mean SAT verbal score is 500 in the United States. But the sample mean, $\bar{y}$, for *these* 10 seniors won't be exactly 500. Are the 10 seniors' scores closer to 500 or $\bar{y}$? They'll always be closer to their own average $\bar{y}$. If we used $\sum (y - \bar{y})^2$ instead of $\sum (y - \mu)^2$ in Equation 23.1 to calculate s, our standard deviation estimate would be too small. How can we fix it? The amazing mathematical fact is that we an compensate for the smaller sum exactly by dividing by $n - 1$ instead of by n. So that's all the $n - 1$ is doing in the denominator of s. And we call $n - 1$ the degrees of freedom.

*The Sign Test—Back to Yes and No

Another, and perhaps simpler, way to look at the Triphammer Road data would be to ignore the actual speed and just ask, "Is the car speeding?" Rather than record the speed, we might have recorded a "yes" (or "1") for cars going over 30 mph and a "no" (or "0") for the cars whose speed is below the posted limit (we'll ignore cars going exactly 30 mph).

What null hypothesis can we use? Well, if drivers were really trying to maintain a 30-mph speed, they might miss randomly above and below that target. We'd expect the number of cars driving faster than 30 mph to be about the same as the number driving slower than 30. In that case 30 mph would be the *median* speed, and so our null hypothesis says that the median is 30. If that null hypothesis were true, we'd expect the proportion of cars driving faster than 30 mph to be 0.50. On the other hand, if the median speed were greater than 30 mph, we'd expect to see more cars driving faster than 30.

What we've done is to turn the quantitative data about car speeds into a set of yes/no values (Bernoulli trials from Chapter 17). And we've turned a question about the median car speed into a test of a proportion (Is the proportion of cars that are going faster than the speed limit 0.50?). We already know how to do a test of proportions, so this isn't a new situation at all.

---

[11] Statistics textbooks usually have equation numbers so they can talk about equations by name. We haven't needed equation numbers yet, but we admit it's useful here, so this is our first.

When we test a median by counting the number of values above and below that value, it's called a **sign test**. The sign test is a *distribution-free* method, so named because there are no *distributional* assumptions or conditions on the data. Specifically, because we no longer have quantitative data, we're not requiring the Nearly Normal Condition.

We already know all we need to do the test Step-By-Step:

STEP-BY-STEP   *A Sign Test

THINK	**Plan** State what we want to know.	I want to know whether the median speed of cars on Triphammer Road is 30 mph.

Plan State what we want to know.

Identify the parameter of interest. Here it's the population median.

Identify the variables and review the W's.

Hypotheses Write the null and alternative hypotheses.

There is not a great need to plot the data. Medians are resistant to the effects of skewness or outliers.

Model Think about the assumptions and check the conditions. The sign test doesn't require the Nearly Normal Condition.

Choose your method. (The sign test is just a one-proportion z-test for $p_0 = 0.5$.)

I want to know whether the median speed of cars on Triphammer Road is 30 mph.

I have 22 car speeds (one car measured at exactly 30 mph was omitted) and have recorded whether their speed exceeded the 30 mph limit or not.

H_0: The median speed of cars on Triphammer Road is 30 mph. Equivalently, the proportion of cars exceeding 30 mph is 50%.

H_0: $p_0 = 0.50$.

H_A: The true proportion of speeders is more than 0.50. $p_0 > 0.50$.

✔ **Independence Assumption:** The cars were far apart so their speeds were probably independent.

✔ **Randomization Condition:** The data are a convenience sample, but they are likely to be representative.

✔ **10% Condition:** The data are from what could be a very large number of cars.

✔ **Success/Failure Condition:** Both $np_0 = 22(0.5) = 11$ and $nq_0 = 22(0.5) = 11$ are greater than 10, showing that I expect at least 10 successes and at least 10 failures.

Because the conditions are satisfied, I'll do a **sign test.**

SHOW **Mechanics** We use the null model to find the P-value, the probability of observing a proportion as far from the hypothesized proportion as the one we did observe, or even farther.

$$SD(\hat{p}) = \sqrt{\frac{0.5 \times 0.5}{22}} = 0.107.$$

Of the 22 cars, 13 had speeds over 30 mph (one of the original 23 was going 30 mph), so the observed proportion, $\hat{p}$, is 0.591.

The P-value is the conditional probability of observing a sample proportion as large as 0.591 (or larger) if the null hypothesis is true:

$$P = P(\hat{p} \geq 0.591 \,|\, p_0 = 0.50).$$

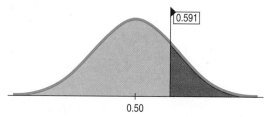

(continued)

The probability of observing a value 0.850 standard deviations or more above the mean of a Normal model can be found by computer, calculator, or table to be 0.197.

The observed value has a z-score of

$$z = \frac{0.591 - 0.5}{0.107} = 0.850,$$ so it is 0.85 standard deviations above the hypothesized proportion.

The P-value is 0.197.

Conclusion Link the P-value to your decision, then state your conclusion in the proper context.

The P-value of 0.197 is not very small, so I fail to reject the null hypothesis. There is insufficient evidence to suggest that cars are speeding on Triphammer Road.

The sign test *is* simpler than the *t*-test, and it requires fewer assumptions. We need only yes/no data. We still should check the Randomization Condition, but we no longer need the Nearly Normal Condition. When the data satisfy all the assumptions and conditions for a *t*-test on the mean, we usually prefer the *t*-test because it is more powerful than the sign test; for the same data, the P-value from the *t*-test would be smaller than the P-value from the sign test. (In fact, the P-value for the *t*-test was 0.136.) That's because the *t*-test uses the actual *values*, which contain much more information than just knowing whether those values are over 30. The more information we use, the more potential our conclusions have to be significant.

On the other hand, the sign test works even when the data have outliers or a skewed distribution—problems that can distort the results of the *t*-test and reduce its power. When we have doubts whether the conditions for the *t*-test are satisfied, it's a good idea to perform a sign test.[12]

What Can Go Wrong?

The most fundamental issue you face is knowing when to use Student's *t* methods.

- **Don't confuse proportions and means.** When you treat your data as categorical, counting successes and summarizing with a sample proportion, make inferences using the Normal model methods you learned about in Chapters 19 through 22. When you treat your data as quantitative, summarizing with a sample mean, make your inferences using Student's *t* methods.

Student's *t* methods work only when the Normality Assumption is true. Naturally, many of the ways things can go wrong turn out to be different ways that the Normality Assumption can fail. It's always a good idea to look for the most common kinds of failure. It turns out that you can even fix some of them.

- **Beware of multimodality.** The Nearly Normal Condition clearly fails if a histogram of the data has two or more modes. When you see this, look for the possibility that your data come from two groups. If so, your best bet is to try to separate the data into different groups. (Use the variables to help distinguish the modes, if possible. For example, if the modes seem to be composed mostly of men in one and women in the other, split the data according to sex.) Then you could analyze each group separately.

---

[12] It's probably a good idea to routinely compute both. If they agree, then the inference is clear. If they differ, it may be interesting and important to see why.

■ **Beware of skewed data.** Make a Normal probability plot and a histogram of the data. If the data are very skewed, you might try re-expressing the variable. Re-expressing may yield a distribution that is unimodal and symmetric, more appropriate for Student's *t* inference methods for means. Re-expression cannot help if the sample distribution is not unimodal. Some people may object to re-expressing the data, but unless your sample is very large, you just can't use the methods of this chapter on skewed data.

■ **Set outliers aside.** Student's *t* methods are built on the mean and standard deviation, so we should beware of outliers when using them. When you make a histogram to check the Nearly Normal Condition, be sure to check for outliers as well. If you find some, consider doing the analysis twice, both with the outliers excluded and with them included in the data, to get a sense of how much they affect the results.

> As tempting as it is to get rid of annoying values, you can't just throw away outliers and not discuss them. It isn't appropriate to lop off the highest or lowest values just to improve your results.

The suggestion that you can perform an analysis with outliers removed may be controversial in some disciplines. Setting aside outliers is seen by some as "cheating." But an analysis of data with outliers left in place is *always* wrong. The outliers violate the Nearly Normal Condition and also the implicit assumption of a homogeneous population, so they invalidate inference procedures. An analysis of the nonoutlying points, along with a separate discussion of the outliers, is often much more informative and can reveal important aspects of the data.

How can you tell whether there are outliers in your data? The "outlier nomination rule" of boxplots can offer some guidance, but it's just a rule of thumb and not an absolute definition. The best practical definition is that a value is an outlier if removing it substantially changes your conclusions about the data. You won't want a single value to determine your understanding of the world unless you are very, very sure that it is absolutely correct. Of course, when the outliers affect your conclusion, this can lead to the uncomfortable state of not really knowing what to conclude. Such situations call for you to use your knowledge of the real world and your understanding of the data you are working with.[13]

Of course, Normality issues aren't the only risks you face when doing inferences about means. Remember to *Think* about the usual suspects.

■ **Watch out for bias.** Measurements of all kinds can be biased. If your observations differ from the true mean in a systematic way, your confidence interval may not capture the true mean. And there is no sample size that will save you. A bathroom scale that's 5 pounds off will be 5 pounds off even if you weigh yourself 100 times and take the average. We've seen several sources of bias in surveys, and measurements can be biased, too. Be sure to think about possible sources of bias in your measurements.

■ **Make sure cases are independent.** Student's *t* methods also require the sampled values to be mutually independent. We check for random sampling and the 10% Condition. You should also think hard about whether there are likely violations of independence in the data collection method. If there are, be very cautious about using these methods.

■ **Make sure that data are from an appropriately randomized sample.** Ideally, all data that we analyze are drawn from a simple random sample or generated by a randomized experiment. When they're not, be careful about making inferences from them. You may still compute a confidence interval correctly, or get the mechanics of the P-value right, but this might not save you from making a serious mistake in inference.

---

[13] An important reason for you to know Statistics rather than let someone else analyze your data.

- **Interpret your confidence interval correctly.** Many statements that sound tempting are, in fact, misinterpretations of a confidence interval for a mean. You might want to have another look at some of the common mistakes, explained on page 559. Keep in mind that a confidence interval is about the mean of the population, not about the means of samples, individuals in samples, or individuals in the population.

CONNECTIONS

The steps for finding a confidence interval or hypothesis test for means are just like the corresponding steps for proportions. Even the form of the calculations is similar. As the z-statistic did for proportions, the t-statistic tells us how many standard errors our sample mean is from the hypothesized mean. For means, though, we have to estimate the standard error separately. This added uncertainty changes the model for the sampling distribution from z to t.

As with all of our inference methods, the randomization applied in drawing a random sample or in randomizing a comparative experiment is what generates the sampling distribution. Randomization is what makes inference in this way possible at all.

The new concept of degrees of freedom connects back to the denominator of the sample standard deviation calculation, as shown earlier.

There's just no escaping histograms and Normal probability plots. The Nearly Normal Condition required to use Student's t can be checked best by making appropriate displays of the data. Back when we first used histograms, we looked at their shape and, in particular, checked whether they were unimodal and symmetric, and whether they showed any outliers. Those are just the features we check for here. The Normal probability plot zeros in on the Normal model a little more precisely.

WHAT HAVE WE LEARNED?

We first learned to create confidence intervals and test hypotheses about proportions. Now we've turned our attention to means, and learned that statistical inference for means relies on the same concepts; only the mechanics and our model have changed.

▶ We've learned that what we can say about a population mean is inferred from data, using the mean of a representative random sample.

▶ We've learned to describe the sampling distribution of sample means using a new model we select from the Student's t family based on our degrees of freedom.

▶ We've learned that our ruler for measuring the variability in sample means is the standard error $SE(\bar{y}) = \dfrac{s}{\sqrt{n}}$.

▶ We've learned to find the margin of error for a confidence interval using that ruler and critical values based on a Student's t-model.

▶ And we've also learned to use that ruler to test hypotheses about the population mean.

Above all, we've learned that the reasoning of inference, the need to verify that the appropriate assumptions are met, and the proper interpretation of confidence intervals and P-values all remain the same regardless of whether we are investigating means or proportions.

Terms

Student's t Degrees of freedom (df) A family of distributions indexed by its degrees of freedom. The t-models are unimodal symmetric, and bell shaped, but have fatter tails and a narrower center than the Normal model. As the degrees of freedom increase, t-distributions approach the Normal (p. 553).

One-sample t-interval for the mean A one-sample t-interval for the population mean is (p. 554)

$$\bar{y} \pm t^*_{n-1} \times SE(\bar{y}), \text{ where } SE(\bar{y}) = \frac{s}{\sqrt{n}}.$$

The critical value t^*_{n-1} depends on the particular confidence level, C, that you specify and on the number of degrees of freedom, $n - 1$.

One-sample t-test for the mean The one-sample t-test for the mean tests the hypothesis $H_0: \mu = \mu_0$ using the statistic (p. 560)

$$t_{n-1} = \frac{\bar{y} - \mu_0}{SE(\bar{y})}.$$

The standard error of $\bar{y}$ is

$$SE(\bar{y}) = \frac{s}{\sqrt{n}}.$$

Skills

THINK
- ▶ Know the assumptions required for t-tests and t-based confidence intervals.
- ▶ Know how to examine your data for violations of conditions that would make inference about the population mean unwise or invalid.
- ▶ Understand that a confidence interval and a hypothesis test are essentially equivalent. You can do a two-tailed hypothesis test at level of significance α with a $1 - \alpha$ confidence interval, or a one-tailed test with a $1 - 2\alpha$ confidence interval.

SHOW
- ▶ Be able to compute and interpret a t-test for the population mean using a statistics package or working from summary statistics for a sample.
- ▶ Be able to compute and interpret a t-based confidence interval for the population mean using a statistics package or working from summary statistics for a sample.

TELL
- ▶ Be able to explain the meaning of a confidence interval for a population mean. Make clear that the randomness associated with the confidence level is a statement about the interval bounds and not about the population parameter value.
- ▶ Understand that a 95% confidence interval does not trap 95% of the sample values.
- ▶ Be able to interpret the result of a test of a hypothesis about a population mean.
- ▶ Do not "accept" a null hypothesis if it can't be rejected. Instead "fail to reject it."
- ▶ Understand that the P-value of a test does not give the probability that the null hypothesis is correct.

INFERENCE FOR MEANS ON THE COMPUTER

Statistics packages offer convenient ways to make histograms of the data. Even better for assessing near-Normality is a Normal probability plot. When you work on a computer, there is simply no excuse for skipping the step of plotting the data to check that it is nearly Normal. *Beware:* Statistics packages don't agree on whether to place the Normal scores on the x-axis (as we have done) or the y-axis. Read the axis labels.

Any standard statistics package can compute a hypothesis test. Here's what the package output might look like in general (although no package we know gives the results in exactly this form):[14]

> **A S** **Activity: Student's *t* in Practice.** We almost always use technology to do inference with Student's *t*. Here's a chance to do that as you investigate several questions.

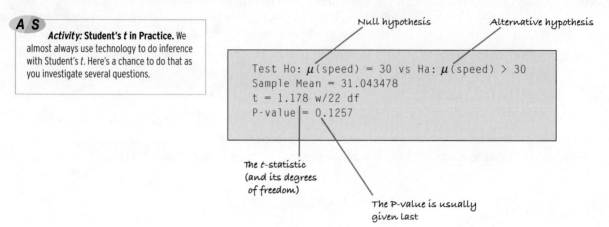

The package computes the sample mean and sample standard deviation of the variable and finds the P-value from the t-distribution based on the appropriate number of degrees of freedom. All modern statistics packages report P-values. The package may also provide additional information such as the sample mean, sample standard deviation, t-statistic value, and degrees of freedom. These are useful for interpreting the resulting P-value and telling the difference between a meaningful result and one that is merely statistically significant. Statistics packages that report the estimated standard deviation of the sampling distribution usually label it "standard error" or "SE."

Inference results are also sometimes reported in a table. You may have to read carefully to find the values you need. Often, test results and the corresponding confidence interval bounds are given together. And often you must read carefully to find the alternative hypotheses. Here's an example of that kind of output:

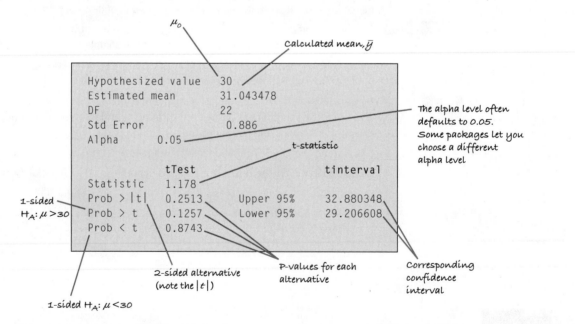

The commands to do inference for means on common statistics programs and calculators are not always obvious. (By contrast, the resulting output is usually clearly labeled and easy to read.) The guides for each program can help you start navigating.

---

[14] Many statistics packages keep as many as 16 digits for all intermediate calculations. If we had kept as many, our results in the Step-By-Step section would have been closer to these.

DATA DESK

Select variables.
From the **Calc** menu, choose **Estimate** for confidence intervals or **Test** for hypothesis tests. Select the interval or test from the drop-down menu and make other choices in the dialog.

EXCEL

Specify formulas. Find t^* with the TINV(alpha, df) function.

COMMENTS

Not really automatic. There's no easy way to find P-values in Excel.

JMP

From the **Analyze** menu, select **Distribution.** For a confidence interval, scroll down to the "Moments" section to find the interval limits. For a hypothesis test, click the red triangle next to the variable's name and choose **Test Mean** from the menu. Then fill in the resulting dialog.

COMMENTS

"Moment" is a fancy statistical term for means, standard deviations, and other related statistics.

MINITAB

From the **Stat** menu, choose the **Basic Statistics** submenu. From that menu, choose **1-sample t. . . .** Then fill in the dialog.

COMMENTS

The dialog offers a clear choice between confidence interval and test.

SPSS

From the **Analyze** menu, choose the **Compare Means** submenu. From that, choose the **One-Sample t-test** command.

COMMENTS

The commands suggest neither a single mean nor an interval. But the results provide both a test and an interval.

TI-83/84 PLUS

Finding a confidence interval:
In the **STAT TESTS** menu, choose **8:TInterval.** You may specify that you are using data stored in a list, or you may enter the mean, standard deviation, and sample size. You must also specify the desired level of confidence.

Testing a hypothesis:
In the **STAT TESTS** menu, choose **2:T-Test.** You may specify that you are using data stored in a list, or you may enter the mean, standard deviation, and size of your sample. You must also specify the hypothesized model mean and whether the test is to be two-tail, lower-tail, or upper-tail.

TI-89

Finding a confidence interval:
In the **STAT Ints** menu, choose **2:TInterval.** Specify whether you are using data stored in a list or whether you will enter the mean, standard deviation, and sample size. You must also specify the desired level of confidence.

Testing a hypothesis:
In the **STAT Tests** menu, choose **2:T-Test.** You must specify whether you are using data stored in a list or whether you will enter the mean, standard deviation, and size of your sample. You must also specify the hypothesized model mean and whether the test is to be two-tail, lower-tail, or upper-tail. Select whether the test is to be simply computed or whether to display the distribution curve and highlight the area corresponding to the P-value of the test.

EXERCISES

1. **t-models, part I.** Using the *t* tables, software, or a calculator, estimate
   a) the critical value of *t* for a 90% confidence interval with df = 17.
   b) the critical value of *t* for a 98% confidence interval with df = 88.
   c) the P-value for $t \geq 2.09$ with 4 degrees of freedom.
   d) the P-value for $|t| > 1.78$ with 22 degrees of freedom.

2. **t-models, part II.** Using the *t* tables, software, or a calculator, estimate
   a) the critical value of *t* for a 95% confidence interval with df = 7.
   b) the critical value of *t* for a 99% confidence interval with df = 102.
   c) the P-value for $t \leq 2.19$ with 41 degrees of freedom.
   d) the P-value for $|t| > 2.33$ with 12 degrees of freedom.

3. **t-models, part III.** Describe how the shape, center, and spread of *t*-models change as the number of degrees of freedom increases.

4. **t-models, part IV (last one!).** Describe how the critical value of *t* for a 95% confidence interval changes as the number of degrees of freedom increases.

5. **Cattle.** Livestock are given a special feed supplement to see if it will promote weight gain. Researchers report that the 77 cows studied gained an average of 56 pounds, and that a 95% confidence interval for the mean weight gain this supplement produces has a margin of error of ±11 pounds. Some students wrote the following conclusions. Did anyone interpret the interval correctly? Explain any misinterpretations.
   a) 95% of the cows studied gained between 45 and 67 pounds.
   b) We're 95% sure that a cow fed this supplement will gain between 45 and 67 pounds.
   c) We're 95% sure that the average weight gain among the cows in this study was between 45 and 67 pounds.
   d) The average weight gain of cows fed this supplement will be between 45 and 67 pounds 95% of the time.
   e) If this supplement is tested on another sample of cows, there is a 95% chance that their average weight gain will be between 45 and 67 pounds.

6. **Teachers.** Software analysis of the salaries of a random sample of 288 Nevada teachers produced the confidence interval shown below. Which conclusion is correct? What's wrong with the others?

   t-Interval for μ: with 90.00% Confidence,
   $38944 < \mu(\text{TchPay}) < 42893$

   a) If we took many random samples of 288 Nevada teachers, about 9 out of 10 of them would produce this confidence interval.

   b) If we took many random samples of Nevada teachers, about 9 out of 10 of them would produce a confidence interval that contained the mean salary of all Nevada teachers.
   c) About 9 out of 10 Nevada teachers earn between $38,944 and $42,893.
   d) About 9 out of 10 of the teachers surveyed earn between $38,944 and $42,893.
   e) We are 90% confident that the average teacher salary in the United States is between $38,944 and $42,893.

7. **Meal plan.** After surveying students at Dartmouth College, a campus organization calculated that a 95% confidence interval for the mean cost of food for one term (of three in the Dartmouth trimester calendar) is ($1102, $1290). Now the organization is trying to write its report and is considering the following interpretations. Comment on each.
   a) 95% of all students pay between $1102 and $1290 for food.
   b) 95% of the sampled students paid between $1102 and $1290.
   c) We're 95% sure that students in this sample averaged between $1102 and $1290 for food.
   d) 95% of all samples of students will have average food costs between $1102 and $1290.
   e) We're 95% sure that the average amount all students pay is between $1102 and $1290.

8. **Snow.** Based on meteorological data for the past century, a local TV weather forecaster estimates that the region's average winter snowfall is 23", with a margin of error of ±2 inches. Assuming he used a 95% confidence interval, how should viewers interpret this news? Comment on each of these statements:
   a) During 95 of the last 100 winters, the region got between 21" and 25" of snow.
   b) There's a 95% chance the region will get between 21" and 25" of snow this winter.
   c) There will be between 21" and 25" of snow on the ground for 95% of the winter days.
   d) Residents can be 95% sure that the area's average snowfall is between 21" and 25".
   e) Residents can be 95% confident that the average snowfall during the last century was between 21" and 25" per winter.

9. **Pulse rates.** A medical researcher measured the pulse rates (beats per minute) of a sample of randomly selected adults and found the following Student's *t*-based confidence interval:

   With 95.00% Confidence,
   $70.887604 < \mu(\text{Pulse}) < 74.497011$

   a) Explain carefully what the software output means.
   b) What's the margin of error for this interval?
   c) If the researcher had calculated a 99% confidence interval, would the margin of error be larger or smaller? Explain.

10. **Crawling.** Data collected by child development scientists produced this confidence interval for the average age (in weeks) at which babies begin to crawl:

t-Interval for μ
(95.00% Confidence): $29.202 < \mu(\text{age}) < 31.844$

a) Explain carefully what the software output means.
b) What is the margin of error for this interval?
c) If the researcher had calculated a 90% confidence interval, would the margin of error be larger or smaller? Explain.

11. **CEO compensation.** A sample of 20 CEOs from the Forbes 500 shows total annual compensations ranging from a minimum of $0.1 to $62.24 million. The average for these 20 CEOs is $7.946 million. Here's a histogram:

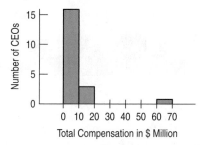

Based on these data, a computer program found that a 95% confidence interval for the mean annual compensation of all Forbes 500 CEOs is (1.69, 14.20) $ million. Why should you be hesitant to trust this confidence interval?

12. **Credit card charges.** A credit card company takes a random sample of 100 cardholders to see how much they charged on their card last month. Here's a histogram.

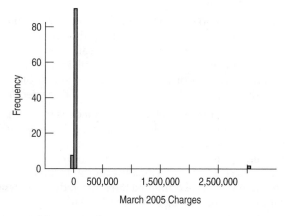

A computer program found that the resulting 95% confidence interval for the mean amount spent in March 2005 is (−$28366.84, $90691.49). Explain why the analysts didn't find the confidence interval useful, and explain what went wrong.

T 13. **Normal temperature.** The researcher described in Exercise 9 also measured the body temperatures of that randomly selected group of adults. Here are summaries of the data he collected. We wish to estimate the average (or "normal") temperature among the adult population.

Summary	Temperature
Count	52
Mean	98.285
Median	98.200
MidRange	98.600
StdDev	0.6824
Range	2.800
IntQRange	1.050

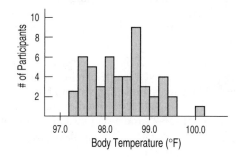

a) Check the conditions for creating a t-interval.
b) Find a 98% confidence interval for mean body temperature.
c) Explain the meaning of that interval.
d) Explain what "98% confidence" means in this context.
e) 98.6°F is commonly assumed to be "normal." Do these data suggest otherwise? Explain.

14. **Parking.** Hoping to lure more shoppers downtown, a city builds a new public parking garage in the central business district. The city plans to pay for the structure through parking fees. During a two-month period (44 weekdays), daily fees collected averaged $126, with a standard deviation of $15.

a) What assumptions must you make in order to use these statistics for inference?
b) Write a 90% confidence interval for the mean daily income this parking garage will generate.
c) Interpret this confidence interval in context.
d) Explain what "90% confidence" means in this context.
e) The consultant who advised the city on this project predicted that parking revenues would average $130 per day. Based on your confidence interval, do you think the consultant was correct? Why?

T 15. **Normal temperatures, part II.** Consider again the statistics about human body temperature in Exercise 13.

a) Would a 90% confidence interval be wider or narrower than the 98% confidence interval you calculated before? Explain. (Don't compute the new interval.)
b) What are the advantages and disadvantages of the 98% confidence interval?
c) If we conduct further research, this time using a sample of 500 adults, how would you expect the 98% confidence interval to change? Explain.
d) How large a sample might allow you to estimate the mean body temperature to within 0.1 degrees with 98% confidence?

16. **Parking II.** Suppose that, for budget planning purposes, the city in Exercise 14 needs a better estimate of the mean daily income from parking fees.
    a) Someone suggests that the city use its data to create a 95% confidence interval instead of the 90% interval first created. How would this interval be better for the city? (You need not actually create the new interval.)
    b) How would the 95% interval be worse for the planners?
    c) How could they achieve an interval estimate that would better serve their planning needs?
    d) How many days' worth of data should they collect to have 95% confidence of estimating the true mean to within $3?

17. **Speed of light.** In 1882 Michelson measured the speed of light (usually denoted c as in Einstein's famous equation $E = mc^2$). His values are in km/sec and have 299,000 subtracted from them. He reported the results of 23 trials with a mean of 756.22 and a standard deviation of 107.12.
    a) Find a 95% confidence interval for the true speed of light from these statistics.
    b) State in words what this interval means. Keep in mind that the speed of light is a physical constant that, as far as we know, has a value that is true throughout the universe.
    c) What assumptions must you make in order to use your method?

18. **Better light.** After his first attempt to determine the speed of light (described in Exercise 17), Michelson conducted an "improved" experiment. In 1897 he reported results of 100 trials with a mean of 852.4 and a standard deviation of 79.0.
    a) What is the standard error of the mean for these data?
    b) Without computing it, how would you expect a 95% confidence interval for the second experiment to differ from the confidence interval for the first? Note at least three specific reasons why they might differ, and indicate the ways in which these differences would change the interval.
    c) According to Stigler (who reports these values), the true speed of light is 299,710.5 km/sec, corresponding to a value of 710.5 for Michelson's 1897 measurements. What does this indicate about Michelson's two experiments? Find a new confidence interval and explain using your confidence interval.

19. **Departures 2009.** What are the chances your flight will leave on time? The U.S. Bureau of Transportation Statistics of the Department of Transportation publishes information about airline performance. Here are a histogram and summary statistics for the percentage of flights departing on time each month from 1995 thru March 2009. (http://www.transtats.bts.gov/HomeDrillChart.asp)

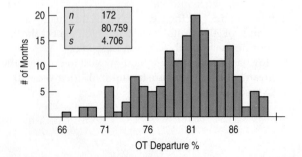

There is no evidence of a trend over time.
    a) Check the assumptions and conditions for inference.
    b) Find a 90% confidence interval for the true percentage of flights that depart on time.
    c) Interpret this interval for a traveler planning to fly.

20. **Arrivals 2009.** Will your flight get you to your destination on time? The U.S. Bureau of Transportation Statistics reported the percentage of flights that were late each month from 1995 through March of 2009. Here's a histogram, along with some summary statistics:

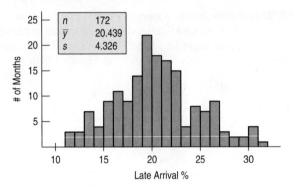

We can consider these data to be a representative sample of all months. There is no evidence of a time trend ($r = 0.07$).
    a) Check the assumptions and conditions for inference about the mean.
    b) Find a 99% confidence interval for the true percentage of flights that arrive late.
    c) Interpret this interval for a traveler planning to fly.

21. **Salmon, second look.** This chapter's For Examples looked at mirex contamination in farmed salmon. We first found a 95% confidence interval for the mean concentration to be 0.0834 to 0.0992 parts per million. Later we rejected the null hypothesis that the mean did not exceed the EPA's recommended safe level of 0.08 ppm based on a P-value of 0.0027. Explain how these two results are consistent. Your explanation should discuss the confidence level, the P-value, and the decision.

22. **Hot dogs.** A nutrition lab tested 40 hot dogs to see if their mean sodium content was less than the 325-mg upper limit set by regulations for "reduced sodium" franks. The lab failed to reject the hypothesis that the hot dogs did not meet this requirement, with a P-value of 0.142. A 90% confidence interval estimated the mean sodium content for this kind of hot dog at 317.2 to 326.8 mg. Explain how these two results are consistent. Your explanation should discuss the confidence level, the P-value, and the decision.

23. **Pizza.** A researcher tests whether the mean cholesterol level among those who eat frozen pizza exceeds the value considered to indicate a health risk. She gets a P-value of 0.07. Explain in this context what the "7%" represents.

24. **Golf balls.** The United States Golf Association (USGA) sets performance standards for golf balls. For example, the initial velocity of the ball may not exceed 250 feet per second when measured by an apparatus approved by the USGA. Suppose a manufacturer introduces a new kind of ball and provides a sample for testing. Based on the mean speed in the test, the USGA comes up with a P-value of 0.34. Explain in this context what the "34%" represents.

25. **TV safety.** The manufacturer of a metal stand for home TV sets must be sure that its product will not fail under the weight of the TV. Since some larger sets weigh nearly 300 pounds, the company's safety inspectors have set a standard of ensuring that the stands can support an average of over 500 pounds. Their inspectors regularly subject a random sample of the stands to increasing weight until they fail. They test the hypothesis $H_0: \mu = 500$ against $H_A: \mu > 500$, using the level of significance $\alpha = 0.01$. If the sample of stands fail to pass this safety test, the inspectors will not certify the product for sale to the general public.
    a) Is this an upper-tail or lower-tail test? In the context of the problem, why do you think this is important?
    b) Explain what will happen if the inspectors commit a Type I error.
    c) Explain what will happen if the inspectors commit a Type II error.

26. **Catheters.** During an angiogram, heart problems can be examined via a small tube (a catheter) threaded into the heart from a vein in the patient's leg. It's important that the company that manufactures the catheter maintain a diameter of 2.00 mm. (The standard deviation is quite small.) Each day, quality control personnel make several measurements to test $H_0: \mu = 2.00$ against $H_A: \mu \neq 2.00$ at a significance level of $\alpha = 0.05$. If they discover a problem, they will stop the manufacturing process until it is corrected.
    a) Is this a one-sided or two-sided test? In the context of the problem, why do you think this is important?
    b) Explain in this context what happens if the quality control people commit a Type I error.
    c) Explain in this context what happens if the quality control people commit a Type II error.

27. **TV safety, revisited.** The manufacturer of the metal TV stands in Exercise 25 is thinking of revising its safety test.
    a) If the company's lawyers are worried about being sued for selling an unsafe product, should they increase or decrease the value of α? Explain.
    b) In this context, what is meant by the power of the test?
    c) If the company wants to increase the power of the test, what options does it have? Explain the advantages and disadvantages of each option.

28. **Catheters, again.** The catheter company in Exercise 26 is reviewing its testing procedure.
    a) Suppose the significance level is changed to $\alpha = 0.01$. Will the probability of a Type II error increase, decrease, or remain the same?
    b) What is meant by the power of the test the company conducts?
    c) Suppose the manufacturing process is slipping out of proper adjustment. As the actual mean diameter of the catheters produced gets farther and farther above the desired 2.00 mm, will the power of the quality control test increase, decrease, or remain the same?
    d) What could they do to improve the power of the test?

29. **Marriage.** In 1960, census results indicated that the age at which American men first married had a mean of 23.3 years. It is widely suspected that young people today are waiting longer to get married. We want to find out if the mean age of first marriage has increased during the past 40 years.
    a) Write appropriate hypotheses.
    b) We plan to test our hypothesis by selecting a random sample of 40 men who married for the first time last year. Do you think the necessary assumptions for inference are satisfied? Explain.
    c) Describe the approximate sampling distribution model for the mean age in such samples.
    d) The men in our sample married at an average age of 24.2 years, with a standard deviation of 5.3 years. What's the P-value for this result?
    e) Explain (in context) what this P-value means.
    f) What's your conclusion?

30. **Fuel economy.** A company with a large fleet of cars hopes to keep gasoline costs down and sets a goal of attaining a fleet average of at least 26 miles per gallon. To see if the goal is being met, they check the gasoline usage for 50 company trips chosen at random, finding a mean of 25.02 mpg and a standard deviation of 4.83 mpg. Is this strong evidence that they have failed to attain their fuel economy goal?
    a) Write appropriate hypotheses.
    b) Are the necessary assumptions to make inferences satisfied?
    c) Describe the sampling distribution model of mean fuel economy for samples like this.
    d) Find the P-value.
    e) Explain what the P-value means in this context.
    f) State an appropriate conclusion.

31. **Ruffles.** Students investigating the packaging of potato chips purchased 6 bags of Lay's Ruffles marked with a net weight of 28.3 grams. They carefully weighed the contents of each bag, recording the following weights (in grams): 29.3, 28.2, 29.1, 28.7, 28.9, 28.5.
    a) Do these data satisfy the assumptions for inference? Explain.
    b) Find the mean and standard deviation of the weights.
    c) Create a 95% confidence interval for the mean weight of such bags of chips.
    d) Explain in context what your interval means.
    e) Comment on the company's stated net weight of 28.3 grams.

32. **Doritos.** Some students checked 6 bags of Doritos marked with a net weight of 28.3 grams. They carefully weighed the contents of each bag, recording the following weights (in grams): 29.2, 28.5, 28.7, 28.9, 29.1, 29.5.
    a) Do these data satisfy the assumptions for inference? Explain.
    b) Find the mean and standard deviation of the weights.
    c) Create a 95% confidence interval for the mean weight of such bags of chips.
    d) Explain in context what your interval means.
    e) Comment on the company's stated net weight of 28.3 grams.

33. **Popcorn.** Yvon Hopps ran an experiment to test optimum power and time settings for microwave popcorn. His goal was to find a combination of power and time that would deliver high-quality popcorn with

less than 10% of the kernels left unpopped, on average. After experimenting with several bags, he determined that power 9 at 4 minutes was the best combination.

a) He concluded that this popping method achieved the 10% goal. If it really does not work that well, what kind of error did Hopps make?
b) To be sure that the method was successful, he popped 8 more bags of popcorn (selected at random) at this setting. All were of high quality, with the following percentages of uncooked popcorn: 7, 13.2, 10, 6, 7.8, 2.8, 2.2, 5.2. Does this provide evidence that he met his goal of an average of no more than 10% uncooked kernels? Explain.

34. Ski wax. Bjork Larsen was trying to decide whether to use a new racing wax for cross-country skis. He decided that the wax would be worth the price if he could average less than 55 seconds on a course he knew well, so he planned to test the wax by racing on the course 8 times.

a) Suppose that he eventually decides not to buy the wax, but it really would lower his average time to below 55 seconds. What kind of error would he have made?
b) His 8 race times were 56.3, 65.9, 50.5, 52.4, 46.5, 57.8, 52.2, and 43.2 seconds. Should he buy the wax? Explain.

35. Chips Ahoy! In 1998, as an advertising campaign, the Nabisco Company announced a "1000 Chips Challenge," claiming that every 18-ounce bag of their Chips Ahoy! cookies contained at least 1000 chocolate chips. Dedicated Statistics students at the Air Force Academy (no kidding) purchased some randomly selected bags of cookies, and counted the chocolate chips. Some of their data are given below. (*Chance*, 12, no. 1[1999])

| 1219 | 1214 | 1087 | 1200 | 1419 | 1121 | 1325 | 1345 |
| 1244 | 1258 | 1356 | 1132 | 1191 | 1270 | 1295 | 1135 |

a) Check the assumptions and conditions for inference. Comment on any concerns you have.
b) Create a 95% confidence interval for the average number of chips in bags of Chips Ahoy! cookies.
c) What does this evidence say about Nabisco's claim? Use your confidence interval to test an appropriate hypothesis and state your conclusion.

36. Yogurt. *Consumer Reports* tested 14 brands of vanilla yogurt and found these numbers of calories per serving:

| 160 | 200 | 220 | 230 | 120 | 180 | 140 |
| 130 | 170 | 190 | 80 | 120 | 100 | 170 |

a) Check the assumptions and conditions for inference.
b) Create a 95% confidence interval for the average calorie content of vanilla yogurt.
c) A diet guide claims that you will get an average of 120 calories from a serving of vanilla yogurt. What does this evidence indicate? Use your confidence interval to test an appropriate hypothesis and state your conclusion.

37. Maze. Psychology experiments sometimes involve testing the ability of rats to navigate mazes. The mazes are classified according to difficulty, as measured by the mean length of time it takes rats to find the food at the

end. One researcher needs a maze that will take rats an average of about one minute to solve. He tests one maze on several rats, collecting the data shown.

Time (sec)	
38.4	57.6
46.2	55.5
62.5	49.5
38.0	40.9
62.8	44.3
33.9	93.8
50.4	47.9
35.0	69.2
52.8	46.2
60.1	56.3
55.1	

a) Plot the data. Do you think the conditions for inference are satisfied? Explain.
b) Test the hypothesis that the mean completion time for this maze is 60 seconds. What is your conclusion?
c) Eliminate the outlier, and test the hypothesis again. What is your conclusion?
d) Do you think this maze meets the "one-minute average" requirement? Explain.

38. Braking. A tire manufacturer is considering a newly designed tread pattern for its all-weather tires. Tests have indicated that these tires will provide better gas mileage and longer tread life. The last remaining test is for braking effectiveness. The company hopes the tire will allow a car traveling at 60 mph to come to a complete stop within an average of 125 feet after the brakes are applied. They will adopt the new tread pattern unless there is strong evidence that the tires do not meet this objective. The distances (in feet) for 10 stops on a test track were 129, 128, 130, 132, 135, 123, 102, 125, 128, and 130. Should the company adopt the new tread pattern? Test an appropriate hypothesis and state your conclusion. Explain how you dealt with the outlier and why you made the recommendation you did.

39. Driving distance 2009. How far do professional golfers drive a ball? (For non-golfers, the drive is the shot hit from a tee at the start of a hole and is typically the longest shot.) Here's a histogram of the average driving distances of the 192 leading professional golfers by end of June 2009 along with summary statistics.

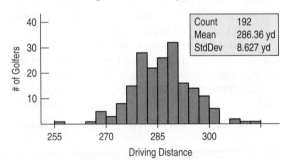

Count	192
Mean	286.36 yd
StdDev	8.627 yd

a) Find a 95% confidence interval for the mean drive distance.
b) Interpreting this interval raises some problems. Discuss.
c) The data are the mean driving distance for each golfer. Is that a concern in interpreting the interval? (*Hint:* Review the What Can Go Wrong warnings of Chapter 9. Chapter 9?! Yes, Chapter 9.)

40. Wind power. Should you generate electricity with your own personal wind turbine? That depends on whether you have enough wind on your site. To produce enough energy, your site should have an annual average wind speed above 8 miles per hour, according to the Wind Energy Association. One candidate site was monitored for a year, with wind speeds recorded every 6 hours. A total of 1114 readings of wind speed averaged 8.019 mph with a standard deviation of 3.813 mph. You've been asked to make a statistical report to help the landowner decide whether to place a wind turbine at this site.

a) Discuss the assumptions and conditions for using Student's *t* inference methods with these data. Here are some plots that may help you decide whether the methods can be used:

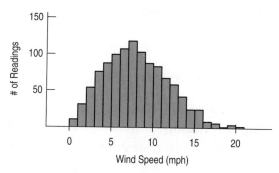

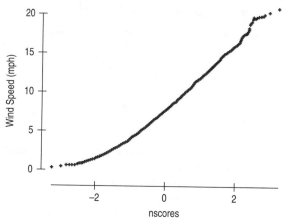

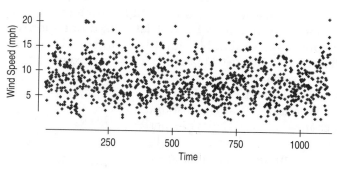

b) What would you tell the landowner about whether this site is suitable for a small wind turbine? Explain.

JUST CHECKING

ANSWERS

1. Questions on the short form are answered by everyone in the population. This is a census, so means or proportions *are* the true population values. The long forms are given just to a sample of the population. When we estimate parameters from a sample, we use a confidence interval to take sample-to-sample variability into account.

2. They don't know the population standard deviation, so they must use the sample SD as an estimate. The additional uncertainty is taken into account by *t*-models.

3. The margin of error for a confidence interval for a mean depends, in part, on the standard error,

$$SE(\bar{y}) = \frac{s}{\sqrt{n}}.$$

Since *n* is in the denominator, smaller sample sizes lead to larger SEs and correspondingly wider intervals. Long forms returned by one in every six or seven households in a less populous area will be a smaller sample.

4. The critical values for *t* with fewer degrees of freedom would be slightly larger. The $\sqrt{n}$ part of the standard error changes a lot, making the SE much larger. Both would increase the margin of error.

5. The smaller sample is one fourth as large, so the confidence interval would be roughly twice as wide.

6. We expect 95% of such intervals to cover the true value, so we would expect about 5 of the 100 intervals to miss.

7. The power would increase if we have a larger sample size.

CHAPTER
24

Comparing Means

Where are we going?

Does taking echinacea help you get over a cold faster? Does playing Mozart to babies in the womb make them smarter? Which last longer, generic or brand-name batteries? Many of the decisions made in business, medicine, and science compare the mean of two groups. In this chapter, we'll learn how.

WHO	AA alkaline batteries
WHAT	Length of battery life while playing a CD continuously
UNITS	Minutes
WHY	Class project
WHEN	1998

A S *Video: Can Diet Prolong Life?* Watch a video that tells the story of an experiment. We'll analyze the data later in this chapter.

Should you buy generic rather than brand-name batteries? A Statistics student designed a study to test battery life. He wanted to know whether there was any real difference between brand-name batteries and a generic brand. To estimate the difference in mean lifetimes, he kept a battery-powered CD player[1] continuously playing the same CD, with the volume control fixed at 5, and measured the time until no more music was heard through the headphones. (He ran an initial trial to find out approximately how long that would take so that he didn't have to spend the first 3 hours of each run listening to the same CD.) For his trials he used six sets of AA alkaline batteries from two major battery manufacturers: a well-known brand name and a generic brand. He measured the time in minutes until the sound stopped. To account for changes in the CD player's performance over time, he randomized the run order by choosing sets of batteries at random. The table to the right shows his data (times in minutes).

Studies that compare two groups are common throughout both science and industry. We might want to compare the effects of a new drug with the tradi-

Brand Name	Generic
194.0	190.7
205.5	203.5
199.2	203.5
172.4	206.5
184.0	222.5
169.5	209.4

tional therapy, the fuel efficiency of two car engine designs, or the sales of new products in two different test cities. In fact, battery manufacturers do research like this on their products and competitors' products themselves.

---

[1] Once upon a time, not so very long ago, there were no iPods. At the turn of the century, people actually carried CDs around—and devices to play them. We bet you can find one on eBay.

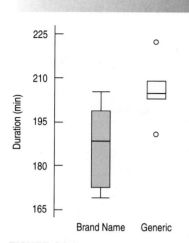

FIGURE 24.1
Boxplots comparing the brand-name and generic batteries suggest a difference in duration.

Plot the Data

The natural display for comparing two groups is boxplots of the data for the two groups, placed side by side. Although we can't make a confidence interval or test a hypothesis from the boxplots themselves, you should always start with boxplots when comparing groups. Let's look at the boxplots of the battery test data.

It sure looks like the generic batteries lasted longer. And we can see that they were also more consistent. But is the difference large enough to change our battery-buying behavior? Can we be confident that the difference is more than just random fluctuation? That's why we need statistical inference.

The boxplot for the generic data identifies two possible outliers. That's interesting, but with only six measurements in each group, the outlier nomination rule is not very reliable. Both of the extreme values are plausible results, and the range of the generic values is smaller than the range of the brand-name values, even with the outliers. So we're probably better off just leaving these values in the data.

Comparing Two Means

Comparing two means is not very different from comparing two proportions. In fact, it's not different in concept from any of the methods we've seen. Now, the population model parameter of interest is the difference between the *mean* battery lifetimes of the two brands, $\mu_1 - \mu_2$.

The rest is the same as before. The statistic of interest is the difference in the two observed means, $\bar{y}_1 - \bar{y}_2$. We'll start with this statistic to build our confidence interval, but we'll need to know its standard deviation and its sampling model. Then we can build confidence intervals and find P-values for hypothesis tests.

We know that, for independent random variables, the variance of their *difference* is the *sum* of their individual variances, $Var(Y - X) = Var(Y) + Var(X)$. To find the standard deviation of the difference between the two independent sample means, we add their variances and then take a square root:

$$SD(\bar{y}_1 - \bar{y}_2) = \sqrt{Var(\bar{y}_1) + Var(\bar{y}_2)}$$
$$= \sqrt{\left(\frac{\sigma_1}{\sqrt{n_1}}\right)^2 + \left(\frac{\sigma_2}{\sqrt{n_2}}\right)^2}$$
$$= \sqrt{\frac{\sigma_1^2}{n_1} + \frac{\sigma_2^2}{n_2}}.$$

Of course, we still don't know the true standard deviations of the two groups, σ_1 and σ_2, so as usual, we'll use the estimates, s_1 and s_2. Using the estimates gives us the *standard error*:

$$SE(\bar{y}_1 - \bar{y}_2) = \sqrt{\frac{s_1^2}{n_1} + \frac{s_2^2}{n_2}}.$$

We'll use the standard error to see how big the difference really is. Because we are working with means and estimating the standard error of their difference using the data, we shouldn't be surprised that the sampling model is a Student's *t*.

//// **FOR EXAMPLE** | **Finding the Standard Error of the Difference in Independent Sample Means**

Can you tell how much you are eating from how full you are? Or do you need visual cues? Researchers[2] constructed a table with two ordinary 18 oz soup bowls and two identical-looking bowls that had been modified to slowly, imperceptibly, refill as they were emptied. They assigned experiment participants to the bowls randomly and served them tomato soup. Those eating from the ordinary bowls had their bowls refilled by ladle whenever they were one-quarter full. If people judge their portions by internal cues, they should eat about the same amount. How big a difference was there in the amount of soup consumed? The table summarizes their results.

	Ordinary bowl	Refilling bowl
n	27	27
$\bar{y}$	8.5 oz	14.7 oz
s	6.1 oz	8.4 oz

QUESTION: How much variability do we expect in the difference between the two means? Find the standard error.

Participants were randomly assigned to bowls, so the two groups should be independent. It's okay to add variances.

$$SE(\bar{y}_{refill} - \bar{y}_{ordinary}) = \sqrt{\frac{s_r^2}{n_r} + \frac{s_o^2}{n_o}} = \sqrt{\frac{8.4^2}{27} + \frac{6.1^2}{27}} = 2.0 \text{ oz.}$$

The confidence interval we build is called a **two-sample *t*-interval** (for the difference in means). The corresponding hypothesis test is called a **two-sample *t*-test**. The interval looks just like all the others we've seen—the statistic plus or minus an estimated margin of error:

$$(\bar{y}_1 - \bar{y}_2) \pm ME$$
$$\text{where } ME = t^* \times SE(\bar{y}_1 - \bar{y}_2).$$

z OR *t*?

If you know σ, use z. (That's rare!) Whenever you use s to estimate σ, use t.

Compare this formula with the one for the confidence interval for the difference of two proportions we saw in Chapter 22 (page 529). The formulas are almost the same. It's just that here we use a Student's *t*-model instead of a Normal model to find the appropriate critical *t**-value corresponding to our chosen confidence level.

What are we missing? Only the degrees of freedom for the Student's *t*-model. Unfortunately, *that* formula is strange.

The deep, dark secret is that the sampling model isn't *really* Student's *t*, but only something close. The trick is that by using a special, adjusted degrees-of-freedom value, we can make it so close to a Student's *t*-model that nobody can tell the difference. The adjustment formula is straightforward but doesn't help our understanding much, so we leave it to the computer or calculator. (If you are curious and really want to see the formula, look in the footnote.[3])

---

[2] Brian Wansink, James E. Painter, and Jill North, "Bottomless Bowls: Why Visual Cues of Portion Size May Influence Intake," *Obesity Research*, Vol. 13, No. 1, January 2005.

[3]
$$df = \frac{\left(\frac{s_1^2}{n_1} + \frac{s_2^2}{n_2}\right)^2}{\frac{1}{n_1 - 1}\left(\frac{s_1^2}{n_1}\right)^2 + \frac{1}{n_2 - 1}\left(\frac{s_2^2}{n_2}\right)^2}$$

Are you sorry you looked? This formula usually doesn't even give a whole number. If you are using a table, you'll need a whole number, so round down to be safe. If you are using technology, it's even easier. The approximation formulas that computers and calculators use for the Student's *t*-distribution deal with fractional degrees of freedom automatically.

> ### A SAMPLING DISTRIBUTION FOR THE DIFFERENCE BETWEEN TWO MEANS
>
> When the conditions are met, the sampling distribution of the standardized sample difference between the means of two independent groups,
>
> $$t = \frac{(\bar{y}_1 - \bar{y}_2) - (\mu_1 - \mu_2)}{SE(\bar{y}_1 - \bar{y}_2)},$$
>
> can be modeled by a Student's t-model with a number of degrees of freedom found with a special formula. We estimate the standard error with
>
> $$SE(\bar{y}_1 - \bar{y}_2) = \sqrt{\frac{s_1^2}{n_1} + \frac{s_2^2}{n_2}}.$$

Assumptions and Conditions

Now we've got everything we need. Before we can make a two-sample t-interval or perform a two-sample t-test, though, we have to check the assumptions and conditions.

Independence Assumption

Independence Assumption: The data in each group must be drawn independently and at random from a homogeneous population, or generated by a randomized comparative experiment. We can't expect that the data, taken as one big group, come from a homogeneous population, because that's what we're trying to test. But without randomization of some sort, there are no sampling distribution models and no inference. We can check two conditions:

Randomization Condition: Were the data collected with suitable randomization? For surveys, are they a representative random sample? For experiments, was the experiment randomized?

10% Condition: We usually don't check this condition for differences of means. We'll check it only if we have a very small population or an extremely large sample. We needn't worry about it at all for randomized experiments.

Normal Population Assumption

As we did before with Student's t-models, we should check the assumption that the underlying populations are *each* Normally distributed. We check the . . .

Nearly Normal Condition: We must check this for *both* groups; a violation by either one violates the condition. As we saw for single sample means, the Normality Assumption matters most when sample sizes are small. For samples of $n < 15$ in either group, you should not use these methods if the histogram or Normal probability plot shows severe skewness. For n's closer to 40, a mildly skewed histogram is OK, but you should remark on any outliers you find and not work with severely skewed data. When both groups are bigger than 40, the Central Limit Theorem starts to kick in no matter how the data are distributed, so the Nearly Normal Condition for the data matters less. Even in large samples, however, you should still be on the lookout for outliers, extreme skewness, and multiple modes.

Independent Groups Assumption

Independent Groups Assumption: To use the two-sample t methods, the two groups we are comparing must be independent of each other. In fact, this test is sometimes called the two *independent samples t-test*. No statistical test can verify this assumption. You have to think about how the data were collected. The assumption would be violated, for example, if one group consisted of husbands and the other group their wives. Whatever we measure on couples might naturally be related. Similarly, if we compared subjects' performances before some treatment with their performances afterward, we'd expect a relationship of each "before" measurement with its corresponding "after" measurement. In cases such as these, where the observational units in the two groups are related or matched, *the two-sample methods of this chapter can't be applied.* When this happens, we need a different procedure that we'll see in the next chapter.

FOR EXAMPLE   Checking Assumptions and Conditions

RECAP: Researchers randomly assigned people to eat soup from one of two bowls: 27 got ordinary bowls that were refilled by ladle, and 27 others bowls that secretly refilled slowly as the people ate.

QUESTION: Can the researchers use their data to make inferences about the role of visual cues in determining how much people eat?

✔ **Independence Assumption:** The amount consumed by one person should be independent of the amount consumed by others.

✔ **Randomization Condition:** Subjects were randomly assigned to the treatments.

✔ **Nearly Normal Condition:** The histograms for both groups look unimodal but somewhat skewed to the right. I believe both groups are large enough (27) to allow use of t-methods.

✔ **Independent Groups Assumption:** Randomization to treatment groups guarantees this.

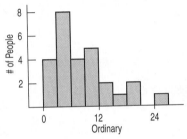

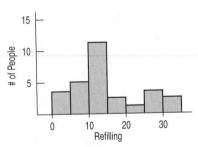

It's okay to construct a two-sample t-interval for the difference in means.

Note: When you check the Nearly Normal Condition it's important that you include the graphs you looked at (histograms or Normal probability plots).

AN EASIER RULE?

The formula for the degrees of freedom of the sampling distribution of the difference between two means is long, but the number of degrees of freedom is always at *least* the smaller of the two *n*'s, minus 1. Wouldn't it be easier to just use that value? You could, but *that* approximation can be a poor choice because it can give fewer than *half* the degrees of freedom you're entitled to from the correct formula.

TWO-SAMPLE *t*-INTERVAL FOR THE DIFFERENCE BETWEEN MEANS

When the conditions are met, we are ready to find the confidence interval for the difference between means of two independent groups, $\mu_1 - \mu_2$. The confidence interval is

$$(\bar{y}_1 - \bar{y}_2) \pm t^*_{df} \times SE(\bar{y}_1 - \bar{y}_2),$$

where the standard error of the difference of the means

$$SE(\bar{y}_1 - \bar{y}_2) = \sqrt{\frac{s_1^2}{n_1} + \frac{s_2^2}{n_2}}.$$

The critical value t^*_{df} depends on the particular confidence level, C, that you specify and on the number of degrees of freedom, which we get from the sample sizes and a special formula.

FOR EXAMPLE

Finding a Confidence Interval for the Difference in Sample Means

RECAP: Researchers studying the role of internal and visual cues in determining how much people eat conducted an experiment in which some people ate soup from bowls that secretly refilled. The results are summarized in the table.

We've already checked the assumptions and conditions, and have found the standard error for the difference in means to be $SE(\bar{y}_{refill} - \bar{y}_{ordinary}) = 2.0$ oz.

	Ordinary bowl	Refilling bowl
n	27	27
$\bar{y}$	8.5 oz	14.7 oz
s	6.1 oz	8.4 oz

QUESTION: What does a 95% confidence interval say about the difference in mean amounts eaten?

The observed difference in means is $\bar{y}_{refill} - \bar{y}_{ordinary} = (14.7 - 8.5) = 6.2$ oz

$$df = 47.46 \quad t^*_{47.46} = 2.011 \ (\text{Table gives } t^*_{45} = 2.014.)$$
$$ME = t^* \times SE(\bar{y}_{refill} - \bar{y}_{ordinary}) = 2.011(2.0) = 4.02 \text{ oz}$$

The 95% confidence interval for $\mu_{refill} - \mu_{ordinary}$ is 6.2 ± 4.02, or $(2.18, 10.22)$ oz.

I am 95% confident that people eating from a subtly refilling bowl will eat an average of between 2.18 and 10.22 more ounces of soup than those eating from an ordinary bowl.

STEP-BY-STEP EXAMPLE

A Two-Sample *t*-Interval

Judging from the boxplot, the generic batteries seem to have lasted about 20 minutes longer than the brand-name batteries. Before we change our buying habits, what should we expect to happen with the next batteries we buy?

Question: How much longer might the generic batteries last?

THINK

Plan State what we want to know.

Identify the *parameter* you wish to estimate. Here our parameter is the difference in the means, not the individual group means.

Identify the *population(s)* about which you wish to make statements. We hope to make decisions about purchasing batteries, so we're interested in all the AA batteries of these two brands.

Identify the variables and review the W's.

REALITY CHECK From the boxplots, it appears our confidence interval should be centered near a difference of 20 minutes. We don't have a lot of intuition about how far the interval should extend on either side of 20.

I have measurements of the lifetimes (in minutes) of 6 sets of generic and 6 sets of brand-name AA batteries from a randomized experiment. I want to find an interval that is likely, with 95% confidence, to contain the true difference $\mu_G - \mu_B$ between the mean lifetime of the generic AA batteries and the mean lifetime of the brand-name batteries.

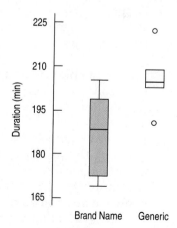

Model Think about the appropriate assumptions and check the conditions to be sure that a Student's *t*-model for the sampling distribution is appropriate.

For very small samples like these, we often don't worry about the 10% Condition.

✓ **Randomization Condition:** The batteries were selected at random from those available for sale. Not exactly an SRS, but a reasonably representative random sample.

✓ **Independence Assumption:** The batteries were packaged together, so they may not be independent. For example, a storage problem might affect all the batteries in the same pack. Repeating the study for several different packs of batteries would make the conclusions stronger.

✓ **Independent Groups Assumption:** Batteries manufactured by two different companies and purchased in separate packages should be independent.

✓ **Nearly Normal Condition:** The samples are small, but the histograms look unimodal and symmetric:

Make a picture. Boxplots are the display of choice for comparing groups, but now we want to check the *shape* of distribution of each group. Histograms or Normal probability plots do a better job there.

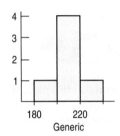
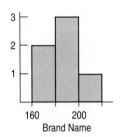

State the sampling distribution model for the statistic. Here the degrees of freedom will come from that messy approximation formula.

Under these conditions, it's okay to use a Student's t-model.

Specify your method.

I'll use a **two-sample t-interval.**

Mechanics Construct the confidence interval.

Be sure to include the units along with the statistics. Use meaningful subscripts to identify the groups.

Use the sample standard deviations to find the standard error of the sampling distribution.

We have three choices for degrees of freedom. The best alternative is to let the computer or calculator use the approximation formula for df. This gives a fractional degree of freedom (here df = 8.98), and technology can find a corresponding critical value. In this case, it is $t^* = 2.263$.

I know $\quad n_G = 6 \qquad\qquad n_B = 6$
$\qquad\qquad \bar{y}_G = 206.0 \text{ min} \quad \bar{y}_B = 187.4 \text{ min}$
$\qquad\qquad s_G = 10.3 \text{ min} \qquad s_B = 14.6 \text{ min}$

The groups are independent, so

$$SE(\bar{y}_G - \bar{y}_B) = \sqrt{SE^2(\bar{y}_G) + SE^2(\bar{y}_B)}$$

$$= \sqrt{\frac{s_G^2}{n_G} + \frac{s_B^2}{n_B}}$$

$$= \sqrt{\frac{10.3^2}{6} + \frac{14.6^2}{6}}$$

$$= \sqrt{\frac{106.09}{6} + \frac{213.16}{6}}$$

$$= \sqrt{53.208}$$

$$= 7.29 \text{ min.}$$

Or we could round the approximation formula's df value down to an integer so we can use a *t* table. That gives 8 df and a critical value *t** = 2.306.

The easy rule says to use only 6 − 1 = 5 df. That gives a critical value *t** = 2.571. The corresponding confidence interval is about 14% wider—a high price to pay for a small savings in effort.

The approximation formula calls for 8.98 degrees of freedom.[4]

The corresponding critical value for a 95% confidence level from a Student's *t*-model with 8.98 df is *t** = 2.263.

So the margin of error is

$$ME = t^* \times SE(\bar{y}_G - \bar{y}_B)$$
$$= 2.263(7.29)$$
$$= 16.50 \text{ min.}$$

The 95% confidence interval is

$$(206.0 - 187.4) \pm 16.5 \text{ min.}$$
$$\text{or } 18.6 \pm 16.5 \text{ min.}$$
$$= (2.1, 35.1) \text{ min.}$$

 TELL

Conclusion Interpret the confidence interval in the proper context.

Less formally, you could say, "I'm 95% confident that generic batteries last an average of 2.1 to 35.1 minutes longer than brand-name batteries."

I am 95% confident that the interval from 2.1 minutes to 35.1 minutes captures the mean amount of time by which generic batteries outlast brand-name batteries for this task. If generic batteries are cheaper, there seems little reason not to use them. If it is more trouble or costs more to buy them, then I'd consider whether the additional performance is worth it.

Another One Just Like the Other Ones?

Yes. That's been our point all along. Once again we see a statistic plus or minus the margin of error. And the ME is just a critical value times the standard error. Just look out for that crazy degrees of freedom formula.

A S

Activity: Find Two-Sample *t*-Intervals. Who wants to deal with that ugly df formula? We usually find these intervals with a statistics package. Learn how here.

 JUST CHECKING

Carpal tunnel syndrome (CTS) causes pain and tingling in the hand. It can be bad enough to keep sufferers awake at night and restrict their daily activities. Researchers studied the effectiveness of two alternative surgical treatments for CTS (Mackenzie, Hainer, and Wheatley, *Annals of Plastic Surgery*, 2000). Patients were randomly assigned to have endoscopic or open-incision surgery. Four weeks later the endoscopic surgery patients demonstrated a mean pinch strength of 9.1 kg compared to 7.6 kg for the open-incision patients.

1. Why is the randomization of the patients into the two treatments important?

2. A 95% confidence interval for the difference in mean strength is about (0.04 kg, 2.96 kg). Explain what this interval means.

3. Why might we want to examine such a confidence interval in deciding between these two surgical procedures?

4. Why might you want to see the data before trusting the confidence interval?

---

[4] If you try to find the degrees of freedom with that messy approximation formula (it's in the footnote on page 582) using the values above, you'll get 8.99. The minor discrepancy is because we rounded the standard deviations to make the exposition clearer.

Testing the Difference Between Two Means

If you bought a used camera in good condition from a friend, would you pay the same as you would if you bought the same item from a stranger? A researcher at Cornell University (J. J. Halpern, "The Transaction Index: A Method for Standardizing Comparisons of Transaction Characteristics Across Different Contexts," *Group Decision and Negotiation, 6*: 557–572) wanted to know how friendship might affect simple sales such as this. She randomly divided subjects into two groups and gave each group descriptions of items they might want to buy. One group was told to imagine buying from a friend whom they expected to see again. The other group was told to imagine buying from a stranger.

Here are the prices they offered for a used camera in good condition:

WHO	University students
WHAT	Prices offered for a used camera
UNITS	$
WHY	Study of the effects of friendship on transactions
WHEN	1990s
WHERE	U.C. Berkeley

PRICE OFFERED FOR A USED CAMERA ($)	
Buying from a Friend	Buying from a Stranger
275	260
300	250
260	175
300	130
255	200
275	225
290	240
300	

The researcher who designed this study had a specific concern. Previous theories had doubted that friendship had a measurable effect on pricing. She hoped to find an effect on friendship. This calls for a hypothesis test—in this case a **two-sample *t*-test for the difference between means.**[5]

A Test for the Difference Between Two Means

You already know enough to construct this test. The test statistic looks just like the others we've seen. It finds the difference between the observed group means and compares this with a hypothesized value for that difference. We'll call that hypothesized difference Δ_0 ("delta naught"). It's so common for that hypothesized difference to be zero that we often just assume $\Delta_0 = 0$. We then compare the difference in the means with the standard error of that difference. We already know that, for a difference between independent means, we can find P-values from a Student's *t*-model on that same special number of degrees of freedom.

NOTATION ALERT

Δ_0—delta naught—isn't so standard that you can assume everyone will understand it. We use it because it's the Greek letter (good for a parameter) "D" for "difference." You should say "delta naught" rather than "delta zero"—that's standard for parameters associated with null hypotheses.

TWO-SAMPLE *t*-TEST FOR THE DIFFERENCE BETWEEN MEANS

The conditions for the two-sample *t*-test for the difference between the means of two independent groups are the same as for the two-sample *t*-interval. We test the hypothesis

$$H_0: \mu_1 - \mu_2 = \Delta_0$$

---

[5] Because it is performed so often, this test is usually just called a "two-sample *t*-test."

where the hypothesized difference is almost always 0, using the statistic

$$t = \frac{(\bar{y}_1 - \bar{y}_2) - \Delta_0}{SE(\bar{y}_1 - \bar{y}_2)}.$$

The standard error of $\bar{y}_1 - \bar{y}_2$ is

$$SE(\bar{y}_1 - \bar{y}_2) = \sqrt{\frac{s_1^2}{n_1} + \frac{s_2^2}{n_2}}.$$

When the conditions are met and the null hypothesis is true, this statistic can be closely modeled by a Student's t-model with a number of degrees of freedom given by a special formula. We use that model to obtain a P-value.

STEP-BY-STEP EXAMPLE — A Two-Sample t-Test for the Difference Between Two Means

The usual null hypothesis is that there's no difference in means. That's just the right null hypothesis for the camera purchase prices.

Question: Is there a difference in the price people would offer a friend rather than a stranger?

THINK

Plan State what we want to know.

Identify the *parameter* you wish to estimate. Here our parameter is the difference in the means, not the individual group means.

Identify the variables and check the W's.

Hypotheses State the null and alternative hypotheses.

The research claim is that friendship changes what people are willing to pay.[6] The natural null hypothesis is that friendship makes no difference.

We didn't start with any knowledge of whether friendship might increase or decrease the price, so we choose a two-sided alternative.

Make a picture. Boxplots are the display of choice for comparing groups. We'll also want to check the distribution of each group. Histograms or Normal probability plots do a better job there.

I want to know whether people are likely to offer a different amount for a used camera when buying from a friend than when buying from a stranger. I wonder whether the difference between mean amounts is zero. I have bid prices from 8 subjects buying from a friend and 7 buying from a stranger, found in a randomized experiment.

H_0: The difference in mean price offered to friends and the mean price offered to strangers is zero:

$$\mu_F - \mu_S = 0.$$

H_A: The difference in mean prices is not zero:

$$\mu_F - \mu_S \neq 0.$$

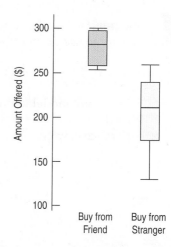

---

[6] This claim is a good example of what is called a "research hypothesis" in many social sciences. The only way to check it is to deny that it's true and see where the resulting null hypothesis leads us.

REALITY CHECK Looks like the prices are higher if you buy from a friend, but we can't be sure. You can't tell from looking at the boxplots whether the difference is statistically significant—the plot shows spreads on the same scale as the data, and we know those don't add. You'll need to add the *variances* to get a suitable ruler for comparing the difference.

Model Think about the assumptions and check the conditions. (Note that, because this is a randomized experiment, we haven't sampled at all, so the 10% Condition does not apply.)

✓ **Randomization Condition:** The experiment was randomized. Subjects were assigned to treatment groups at random.

✓ **Independence Assumption:** This is an experiment, so there is no need for the subjects to be randomly selected from any particular population. All we need to check is whether they were assigned randomly to treatment groups.

✓ **Independent Groups Assumption:** Randomizing the experiment gives independent groups.

✓ **Nearly Normal Condition:** Histograms of the two sets of prices are unimodal and symmetric:

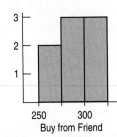
Buy from Friend

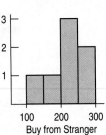
Buy from Stranger

State the sampling distribution model.

Specify your method.

The assumptions are reasonable and the conditions are okay, so I'll use a Student's t-model to perform a **two-sample t-test.**

SHOW **Mechanics** List the summary statistics. Be sure to use proper notation.

From the data:

$$n_F = 8 \qquad n_S = 7$$
$$\bar{y}_F = \$281.88 \qquad \bar{y}_S = \$211.43$$
$$s_F = \$18.31 \qquad s_S = \$46.43$$

Use the null model to find the P-value. First determine the standard error of the difference between sample means.

For independent groups,

$$SE(\bar{y}_F - \bar{y}_S) = \sqrt{SE^2(\bar{y}_F) + SE^2(\bar{y}_S)}$$
$$= \sqrt{\frac{s_F^2}{n_F} + \frac{s_S^2}{n_S}}$$
$$= \sqrt{\frac{18.31^2}{8} + \frac{46.43^2}{7}}$$
$$= 18.70$$

Make a picture. Sketch the *t*-model centered at the hypothesized difference of zero. Because this is a two-tailed test, shade the region to the right of the observed difference and the corresponding region in the other tail.

Find the *t*-value.

A statistics program or graphing calculator can find the P-value using the fractional degrees of freedom from the approximation formula. If you are doing a test like this without technology, you could use the smaller sample size to determine degrees of freedom. In this case, $n_2 - 1 = 6$.

The observed difference is

$$(\bar{y}_F - \bar{y}_S) = 281.88 - 211.43 = \$70.45$$

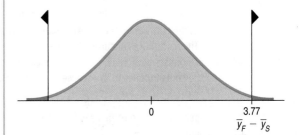

$$t = \frac{(\bar{y}_F - \bar{y}_S) - (0)}{SE(\bar{y}_F - \bar{y}_S)} = \frac{70.45}{18.70} = 3.77$$

The approximation formula gives 7.62 degrees of freedom.[7]

$$P\text{-value} = 2P(t_{7.62} > 3.77) = 0.006$$

 Conclusion Link the P-value to your decision about the null hypothesis, and state the conclusion in context.

Be cautious about generalizing to items whose prices are outside the range of those in this study.

If there were no difference in the mean prices, a difference this large would occur only 6 times in 1000. That's too rare to believe, so I reject the null hypothesis and conclude that people are likely to offer a friend more than they'd offer a stranger for a used camera (and possibly for other, similar items).

Recall the experiment comparing patients 4 weeks after surgery for carpal tunnel syndrome. The patients who had endoscopic surgery demonstrated a mean pinch strength of 9.1 kg compared to 7.6 kg for the open-incision patients.

5. What hypotheses would you test?

6. The P-value of the test was less than 0.05. State a brief conclusion.

7. The study reports work on 36 "hands," but there were only 26 patients. In fact, 7 of the endoscopic surgery patients had both hands operated on, as did 3 of the open-incision group. Does this alter your thinking about any of the assumptions? Explain.

---

[7] If you were daring enough to calculate that messy degrees of freedom formula by hand with the values given here, you'd get about 7.74. Computers work with more precision for the standard deviations than we gave in our example. Many computer programs will round the final result down to 7 degrees of freedom. All give about the same result for the P-value, so it doesn't really matter—the conclusion would be the same.

FOR EXAMPLE   A Two-Sample *t*-Test

Many office "coffee stations" collect voluntary payments for the food consumed. Researchers at the University of Newcastle upon Tyne performed an experiment to see whether the image of eyes watching would change employee behavior.[8] They alternated pictures of eyes looking at the viewer with pictures of flowers each week on the cupboard behind the "honesty box." They measured the consumption of milk to approximate the amount of food consumed and recorded the contributions (in £) each week per liter of milk. The table summarizes their results.

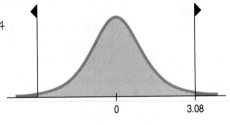

QUESTION: Do these results provide evidence that there really is a difference in honesty even when it's only photographs of eyes that are "watching"?

	Eyes	Flowers
n (# weeks)	5	5
$\bar{y}$	0.417 £/l	0.151 £/l
s	0.1811 £/l	0.067 £/l

$$H_O: \mu_{eyes} - \mu_{flowers} = 0$$
$$H_A: \mu_{eyes} - \mu_{flowers} \neq 0$$

✓ **Independence Assumption:** The amount paid by one person should be independent of the amount paid by others.

✓ **Randomization Condition:** This study was observational. Treatments alternated a week at a time and were applied to the same group of office workers.

✓ **Nearly Normal Condition:** I don't have the data to check, but it seems unlikely there would be outliers in either group. I could be more certain if I could see histograms for both groups.

✓ **Independent Groups Assumption:** The same workers were recorded each week, but week-to-week independence is plausible.

It's okay to do a two sample t-test for the difference in means:

$$SE(\bar{y}_{eyes} - \bar{y}_{flowers}) = \sqrt{\frac{s^2_{eyes}}{n_{eyes}} + \frac{s^2_{flowers}}{n_{flowers}}} = \sqrt{\frac{0.1811^2}{5} + \frac{0.067^2}{5}} = 0.0864$$

$$df = 5.07$$

$$t_5 = \frac{(\bar{y}_{eyes} - \bar{y}_{flowers}) - 0}{SE(\bar{y}_{eyes} - \bar{y}_{flowers})} = \frac{0.417 - 0.151}{0.0864} = 3.08$$

$$P(|t_5| > 3.08) = 0.027$$

Assuming the data were free of outliers, the very low P-value leads me to reject the null hypothesis. This study provides evidence that people will leave higher average voluntary payments for food if pictures of eyes are "watching."

(Note: In Table T we can see that at 5 df, $t = 3.08$ lies between the critical values for $P = 0.02$ and $P = 0.05$, so we could report $P < 0.05$.)

Back into the Pool

Remember that when we know a proportion, we know its standard deviation. When we tested the null hypothesis that two proportions were equal, that link meant we could assume their variances were equal as well. This led us to pool our data to estimate a standard error for the hypothesis test.

For means, there is also a pooled *t*-test. Like the two-proportions *z*-test, this test assumes that the variances in the two groups are equal. But be careful: Knowing the mean of some data doesn't tell you anything about their variance. And knowing that two means are equal doesn't say anything about whether their variances are equal. If we were willing to *assume* that their variances are

[8] Melissa Bateson, Daniel Nettle, and Gilbert Roberts, "Cues of Being Watched Enhance Cooperation in a Real-World Setting," *Biol. Lett. doi*:10.1098/rsbl.2006.0509.

equal, we could pool the data from two groups to estimate the common variance. We'd estimate this pooled variance from the data, so we'd still use a Student's *t*-model. This test is called a **pooled *t*-test (for the difference between means).**

Pooled *t*-tests have a couple of advantages. They often have a few more degrees of freedom than the corresponding two-sample test and a much simpler degrees of freedom formula. But these advantages come at a price: You have to pool the variances and think about another assumption. The assumption of equal variances is a strong one, is often not true, and is difficult to check. For these reasons, we recommend that you use a two-sample *t*-test instead.

The pooled *t*-test is the theoretically correct method only when we have a good reason to believe that the variances are equal. And (as we will see shortly) there are times when this makes sense. Keep in mind, however, that it's never wrong *not* to pool.

The Pooled *t*-Test

Termites cause billions of dollars of damage each year to homes and other buildings, but some tropical trees seem to be able to resist termite attack. A researcher extracted a compound from the sap of one such tree and tested it by feeding it at two different concentrations to randomly assigned groups of 25 termites.[9] After 5 days, 8 groups fed the lower dose had an average of 20.875 termites alive, with a standard deviation of 2.23. But 6 groups fed the higher dose had an average of only 6.667 termites alive, with a standard deviation of 3.14. Is this a large enough difference to declare the sap compound effective in killing termites? In order to use the pooled *t*-test, we must make the **Equal Variance Assumption** that the variances of the two populations from which the samples have been drawn are equal. That is, $\sigma_1^2 = \sigma_2^2$. (Of course, we could think about the standard deviations being equal instead.) The corresponding **Similar Spreads Condition** really just consists of looking at the boxplots to check that the spreads are not wildly different. We were going to make boxplots anyway, so there's really nothing new here.

Once we decide to pool, we estimate the common variance by combining numbers we already have:

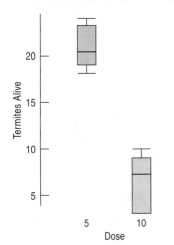

$$s_{\text{pooled}}^2 = \frac{(n_1 - 1)s_1^2 + (n_2 - 1)s_2^2}{(n_1 - 1) + (n_2 - 1)}.$$

(If the two sample sizes are equal, this is just the average of the two variances.)

Now we just substitute this pooled variance in place of each of the variances in the standard error formula.

$$SE_{\text{pooled}}(\bar{y}_1 - \bar{y}_2) = \sqrt{\frac{s_{\text{pooled}}^2}{n_1} + \frac{s_{\text{pooled}}^2}{n_2}} = s_{\text{pooled}}\sqrt{\frac{1}{n_1} + \frac{1}{n_2}}.$$

The formula for degrees of freedom for the Student's *t*-model is simpler, too. It was so complicated for the two-sample *t* that we stuck it in a footnote.[10] Now it's just df $= n_1 + n_2 - 2$.

Substitute the pooled-*t* estimate of the standard error and its degrees of freedom into the steps of the confidence interval or hypothesis test, and you'll be using the pooled-*t* method. For the termites, $\bar{y}_1 - \bar{y}_2 = 14.208$, giving a *t*-value $= 9.935$ with 12 df and a P-value ≤ 0.0001.

Of course, if you decide to use a pooled-*t* method, you must defend your assumption that the variances of the two groups are equal.

$$s_{\text{pooled}}^2 = \frac{(8 - 1)2.23^2 + (6 - 1)3.14^2}{(8 - 1) + (6 - 1)} = 7.01$$

$$SE_{\text{pooled}}(\bar{y}_1 - \bar{y}_2) = \sqrt{\frac{7.01}{8} + \frac{7.01}{6}} = 1.43$$

$$t = \frac{20.875 - 6.667}{1.43} = 9.935$$

[9] Adam Messer, Kevin McCormick, Sunjaya, H. H. Hagedorm, Ferny Tumbel, and J. Meinwald, "Defensive role of tropical tree resins: Antitermitic sesquiterpenes from Southeast Asian Diptero-carpaceae," *J Chem Ecology*, 16:122, pp. 3333–3352.
[10] But not this one. See page 582.

> ### POOLED *t*-TEST AND CONFIDENCE INTERVAL FOR MEANS
>
> The conditions for the pooled *t*-test for the difference between the means of two independent groups (commonly called a "pooled *t*-test") are the same as for the two-sample *t*-test with the additional assumption that the variances of the two groups are the same. We test the hypothesis
>
> $$H_0: \mu_1 - \mu_2 = \Delta_0$$
>
> where the hypothesized difference, Δ_0, is almost always 0, using the statistic
>
> $$t = \frac{(\bar{y}_1 - \bar{y}_2) - \Delta_0}{SE_{pooled}(\bar{y}_1 - \bar{y}_2)}.$$
>
> The standard error of $\bar{y}_1 - \bar{y}_2$ is
>
> $$SE_{pooled}(\bar{y}_1 - \bar{y}_2) = \sqrt{\frac{s_{pooled}^2}{n_1} + \frac{s_{pooled}^2}{n_2}} = s_{pooled}\sqrt{\frac{1}{n_1} + \frac{1}{n_2}},$$
>
> where the pooled variance is
>
> $$s_{pooled}^2 = \frac{(n_1 - 1)s_1^2 + (n_2 - 1)s_2^2}{(n_1 - 1) + (n_2 - 1)}.$$
>
> When the conditions are met and the null hypothesis is true, we can model this statistic's sampling distribution with a Student's *t*-model with $(n_1 - 1) + (n_2 - 1)$ degrees of freedom. We use that model to obtain a P-value for a test or a margin of error for a confidence interval.
>
> The corresponding confidence interval is
>
> $$(\bar{y}_1 - \bar{y}_2) \pm t_{df}^* \times SE_{pooled}(\bar{y}_1 - \bar{y}_2),$$
>
> where the critical value *t** depends on the confidence level and is found with $(n_1 - 1) + (n_2 - 1)$ degrees of freedom.

A S   *Activity: The Pooled t-Test.* It's those hot dogs again. The same interactive tool can handle a pooled *t*-test, too. Take it for a spin here.

Is the Pool All Wet?

We're testing whether the means are equal, so we admit that we don't *know* whether they are equal. Doesn't it seem a bit much to just *assume* that the variances are equal? Well, yes—but there are some special cases to consider. So when *should* you use pooled-*t* methods rather than two-sample *t* methods?

Never.

What, never?

Well, hardly ever.

You see, when the variances of the two groups are in fact equal, the two methods give pretty much the same result. (For the termites, the two-sample *t* statistic is barely different—9.436 with 8 df—and the P-value is still < 0.001.) Pooled methods have a small advantage (slightly narrower confidence intervals, slightly more powerful tests) mostly because they usually have a few more degrees of freedom, but the advantage is slight.

When the variances are *not* equal, the pooled methods are just not valid and can give poor results. You have to use the two-sample methods instead.

As the sample sizes get bigger, the advantages that come from a few more degrees of freedom make less and less difference. So the advantage (such as it is) of the pooled method is greatest when the samples are small—just when it's hardest to check the conditions. And the difference in the degrees of freedom is greatest when the variances are not equal—just when you can't use the pooled method anyway. Our advice is to use the two-sample *t* methods to compare means.

> Because the advantages of pooling are small, and you are allowed to pool only rarely (when the Equal Variances Assumption is met), *don't*.
> **It's never wrong *not* to pool.**

Why did we devote a whole section to a method that we don't recommend using? Good question. The answer is that pooled methods are actually very important in Statistics. It's just that the simplest of the pooled methods—those for comparing two means—have good alternatives in the two-sample methods that don't require the extra assumption. Lacking the burden of the Equal Variances Assumption, the two-sample methods apply to more situations and are safer to use.

Why Not Test the Assumption That the Variances Are Equal?

There is a hypothesis test that would do this. However, it is very sensitive to failures of the assumptions and works poorly for small sample sizes—just the situation in which we might care about a difference in the methods. When the choice between two-sample t and pooled-t methods makes a difference (that is, when the sample sizes are small), the test for whether the variances are equal hardly works at all.

Is There Ever a Time When Assuming Equal Variances Makes Sense?

Pooling may make sense in a randomized comparative experiment, where we start by assigning our experimental units to treatments at random, as the experimenter did with the termites. We know that at the start of the experiment each treatment group is a random sample from the same population,[11] so each treatment group begins with the same population variance. In this case, assuming that the variances are equal after we apply the treatment is the same as assuming that the treatment doesn't change the variance. When we test whether the true means are equal, we may be willing to go a bit farther and say that the treatments made no difference *at all*. For example, we might suspect that the treatment is no different from the placebo offered as a control. Then it's not much of a stretch to assume that the variances have remained equal. It's still an assumption, and there are conditions that need to be checked (make the boxplots, make the boxplots, make the boxplots), but at least it's a plausible assumption.

This line of reasoning is important. The methods used to analyze comparative experiments *do* pool variances in exactly this way and defend the pooling with a version of this argument. Chapters 28 and 29 introduce these methods.

FOR EXAMPLE   A Pooled *t*-Test

RECAP: Remember the experiment showing that people would consume more soup if their bowls were secretly being refilled as they ate? Could it be that those subjects with the refilling bowl just enjoyed eating soup more? The experimenters asked the participants to estimate how many ounces of soup they had eaten. Their responses are summarized in the table.

	Ordinary bowl	Refilling bowl
n	27	27
$\bar{y}$	8.2	9.8
s	6.9	9.2

---

[11] That is, the population of experimental subjects. Remember that to be valid, experiments do not need a representative sample drawn from a population because we are not trying to estimate a population model parameter.

QUESTION: Is there evidence that those who ate more knew what they had done?

$$H_O: \mu_{refill} - \mu_{ordinary} = 0$$
$$H_A: \mu_{refill} - \mu_{ordinary} \neq 0$$

The independence assumptions and randomization condition checked out earlier. Also,

? **Nearly Normal Condition:** The histograms for both groups look unimodal and skewed to the right. The presence of skewness and outliers in groups this size (27) suggests that using t-methods may be risky.

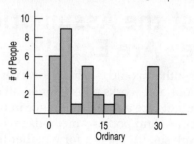
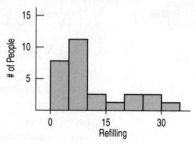

✔ **Equal Variances Assumption:** Subjects were assigned randomly to the two treatments, so I can assume that before they ate any soup, both groups' volume-estimating abilities had equal variances.

✔ **Similar Spreads Condition:** The boxplots show comparable spreads for the two groups.

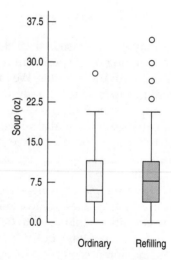

I'll perform a pooled t-test, with caution.

$$s_{pooled}^2 = \frac{(n_r - 1)s_r^2 + (n_o - 1)s_o^2}{(n_r - 1) + (n_o - 1)} = \frac{26 \times 9.2^2 + 26 \times 6.9^2}{(26) + (26)} = 8.13$$

$$SE_{pooled}(\bar{y}_o - \bar{y}_r) = \sqrt{\frac{s_{pooled}^2}{n_p} + \frac{s_{pooled}^2}{n_o}} = s_{pooled}\sqrt{\frac{1}{n_p} + \frac{1}{n_o}} = 5.09\sqrt{\frac{1}{27} + \frac{1}{27}} = 2.213$$

$$df = (27 - 1) + (27 - 1) = 52$$

$$t_{52} = \frac{(9.8 - 8.2) - 0}{2.213} = 0.723$$

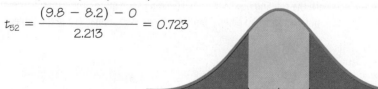

P-value $= P(|t_{52}| > 0.723) = 0.473$

An event that happens nearly half of the time by chance isn't very extraordinary, so I would not reject the null hypothesis. There's no evidence that the subjects who ate more soup from the refilling bowl could tell that they did so.

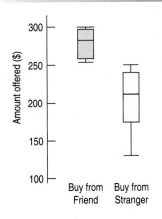

Buying from a Friend	Buying from a Stranger
$275	$260
300	250
260	175
300	130
255	200
275	225
290	240
300	

*Tukey's Quick Test

The famous statistician John Tukey[12] was once challenged to come up with a simpler alternative to the two-sample *t*-test that, like the 68-95-99.7 Rule, had critical values that could be remembered easily. The test he came up with asks you only to count and to remember three numbers: 7, 10, and 13.

When you first looked at the boxplots of the price data, you might have noticed that they didn't overlap very much. That's the basis for Tukey's test.

To use Tukey's test, one group must have the highest value and the other, the lowest. We just count how many values in the high group are higher than *all* the values of the lower group. Add to this the number of values in the low group that are lower than *all* the values of the higher group. (Count ties as 1/2.) Now if this total is 7 or more, we can reject the null hypothesis of equal means at $\alpha = 0.05$. The "critical values" of 10 and 13 give us α's of 0.01 and 0.001.

Let's try it. The "Friend" group has the highest value ($300) and the "Stranger" group has the lowest ($130). Six of the values in the Friend group (shown in red) are higher than the highest value of the Stranger group ($260) and one is a tie. Six of the Stranger values (shown in green) are lower than the lowest value for Friends. That's a total of 12 1/2 values that leak over. That's more than 10, but less than 13. So the P-value is between 0.01 and 0.001—just what we found with the two-sample *t*.

This is a remarkably good test. The only assumption it requires is that the two samples be independent. It's so simple to do that there's no reason not to do one to check your two-sample *t* results. If they disagree, check the assumptions. Tukey's quick test, however, is not as widely known or accepted as the two-sample *t*-test, so you still need to know and use the two-sample *t*.

*A Rank Sum Test

Another distribution-free test of whether two means are equal uses the *ranks* of the data in the two samples. Tukey's test ranks all the data and then counts the excesses on either side. Like Tukey's test, the **Wilcoxon rank sum (or Mann-Whitney) test** first ranks the combined sample from the two groups together from smallest to largest, assigning each observation a rank from 1 to N, where N is $n_1 + n_2$, the total number of observations in both groups. (If there are ties, we use the average rank.) The test statistic, W, is the sum of the ranks of the first group. If W is too large or too small, it is evidence that the groups do not have equal means (actually, the test is more general and tests whether the two distributions are the same).

The hypothesis test is based on a couple of facts that can be derived mathematically. When the null hypothesis is true, the test statistic, W, has mean $\mu_W = \dfrac{n_1(N + 1)}{2}$ and variance $Var(W) = \dfrac{n_1 n_2(N + 1)}{12}$. Now, for any but very small sample sizes (both larger than about 7 should do), we can use a z-test with $z = \dfrac{W - \mu_W}{SD(W)}$ to test the hypothesis.

Let's try it on the buying from friends data. Here are the data ranked from 1 to 15. There are two ties at ranks 8 and 9, two ties at ranks 10 and 11, and three ties at ranks 13, 14, and 15:

Data	130	175	200	225	240	250	255	260	260	275	275	290	300	300	300
Rank	1	2	3	4	5	6	7	8.5	8.5	10.5	10.5	12	14	14	14
Group	S	S	S	S	S	S	F	S	F	F	F	F	F	F	F

---

[12] You know some of his other work. Tukey originated the stem-and-leaf display and the boxplot. He also developed the Fast Fourier Transform, which is important in engineering, and coined the term "bit" for binary digit.

Now we'll add up all the ranks of the Friend's group:

$$W = 7 + 8.5 + 10.5 + 10.5 + 12 + 14 + 14 + 14 = 90.5$$

The mean of W when the null hypothesis is true is $\mu_W = \dfrac{8(15 + 1)}{2} = 64$ and the

SD is $SD(W) = \sqrt{Var(W)} = \sqrt{\dfrac{8 \times 7(15 + 1)}{12}} = 8.64$, so $z = \dfrac{90.5 - 64}{8.64} = 3.07$

with a two-sided P-value of 0.0021. As with both other tests we looked at, this P-value is between 0.01 and 0.001. We reach the same conclusion as we did with Tukey's test and with the two-sample t-test. Of course, like Tukey's test, the rank sum test has the advantage that it doesn't depend on the Nearly Normal Condition. And, like Tukey's test, it will be less powerful than the two-sample t-test when that condition *is* satisfied because it doesn't use all the information in the data.

What Can Go Wrong?

- **Watch out for paired data.** The Independent Groups Assumption deserves special attention. If the samples are not independent, you can't use these two-sample methods. This is probably the main thing that can go wrong when using these two-sample methods. The methods of this chapter can be used *only* if the observations in the two groups are *independent*. Matched-pairs designs in which the observations are deliberately related arise often and are important. The next chapter deals with them.

- **Look at the plots.** The usual cautions about checking for outliers and non-Normal distributions apply, of course. The simple defense is to make and examine boxplots. You may be surprised how often this simple step saves you from the wrong or even absurd conclusions that can be generated by a single undetected outlier. You don't want to conclude that two methods have very different means just because one observation is atypical.

> **Do what we say, not what we do . . .** Precision machines used in industry often have a bewildering number of parameters that have to be set, so experiments are performed in an attempt to try to find the best settings. Such was the case for a hole-punching machine used by a well-known computer manufacturer to make printed circuit boards. The data were analyzed by one of the authors, but because he was in a hurry, he didn't look at the boxplots first and just performed t-tests on the experimental factors. When he found extremely small P-values even for factors that made no sense, he plotted the data. Sure enough, there was one observation 1,000,000 times bigger than the others. It turns out that it had been recorded in microns (millionths of an inch), while all the rest were in inches.

CONNECTIONS

The structure and reasoning of inference methods for comparing two means are very similar to what we used for comparing two proportions. Here we must estimate the standard errors independent of the means, so we use Student's *t*-models rather than the Normal.

We first learned about side-by-side boxplots in Chapter 5. There we made general statements about the shape, center, and spread of each group. When we compared groups, we asked whether their centers looked different compared to how spread out the distributions were. Here we've made that kind of thinking precise, with confidence intervals for the difference and tests of whether the means are the same.

We use Student's *t* as we did for single sample means, and for the same reasons: We are using standard errors from the data to estimate the standard deviation of the sample statistic. As before, to work with Student's *t*-models, we need to check the Nearly Normal Condition. Histograms and Normal probability plots are the best methods for such checks.

As always, we've decided whether a statistic is large by comparing it with its standard error. In this case, our statistic is the difference in means.

We pooled data to find a standard deviation when we tested the hypothesis of equal proportions. For that test, the assumption of equal variances was a consequence of the null hypothesis that the proportions were equal, so it didn't require an extra assumption. When two proportions are equal, so are their variances. But means don't have a linkage with their corresponding variances; so to use pooled-*t* methods, we must make the additional assumption of equal variances. When we can make this assumption, the pooled variance calculations are very similar to those for proportions, combining the squared deviations of each group from its own mean to find a common variance.

WHAT HAVE WE LEARNED?

Are the means of two groups the same? If not, how different are they? We've learned to use statistical inference to compare the means of two independent groups.

▶ We've seen that confidence intervals and hypothesis tests about the difference between two means, like those for an individual mean, use *t*-models.

▶ Once again we've seen the importance of checking assumptions that tell us whether our method will work.

▶ We've seen that, as when comparing proportions, finding the standard error for the difference in sample means depends on believing that our data come from independent groups. Unlike proportions, however, pooling is usually not the best choice here.

▶ And we've seen once again that we can add variances of independent random variables to find the standard deviation of the difference in two independent means.

▶ Finally, we've learned that the reasoning of statistical inference remains the same; only the mechanics change.

Terms

Two-sample *t* methods — Two-sample *t* methods allow us to draw conclusions about the difference between the means of two independent groups. The two-sample methods make relatively few assumptions about the underlying populations, so they are usually the method of choice for comparing two sample means. However, the Student's *t*-models are only approximations for their true sampling distribution. To make that approximation work well, the two-sample *t* methods have a special rule for estimating degrees of freedom (p. 582).

Two-sample *t*-interval for the difference between means	A confidence interval for the difference between the means of two independent groups found as (p. 584) $$(\bar{y}_1 - \bar{y}_2) \pm t^*_{df} \times SE(\bar{y}_1 - \bar{y}_2)$$ where $$SE(\bar{y}_1 - \bar{y}_2) = \sqrt{\frac{s_1^2}{n_1} + \frac{s_2^2}{n_2}}$$ and the number of degrees of freedom is given by a special formula (see footnote 3 on page 582).
Two-sample *t*-test for the difference between means	A hypothesis test for the difference between the means of two independent groups. It tests the null hypothesis (p. 588) $$H_0: \mu_1 - \mu_2 = \Delta_0,$$ where the hypothesized difference, Δ_0, is almost always 0, using the statistic $$t_{df} = \frac{(\bar{y}_1 - \bar{y}_2) - \Delta_0}{SE(\bar{y}_1 - \bar{y}_2)},$$ with the number of degrees of freedom given by the special formula.
Pooling	Data from two or more populations may sometimes be combined, or *pooled*, to estimate a statistic (typically a pooled variance) when the estimated value is assumed to be the same in both populations. The resulting larger sample size may lead to an estimate with lower sample variance. However, pooled estimates are appropriate only when the required assumptions are true (p. 592).
*Pooled *t*-intervals	A confidence interval for the difference between the means of two independent groups used when we are willing and able to make the additional assumption that the variances of the groups are equal. It is found as $$(\bar{y}_1 - \bar{y}_2) \pm t^*_{df} \times SE_{pooled}(\bar{y}_1 - \bar{y}_2),$$ where $$SE_{pooled}(\bar{y}_1 - \bar{y}_2) = \sqrt{\frac{s_{pooled}^2}{n_1} + \frac{s_{pooled}^2}{n_2}} = s_{pooled}\sqrt{\frac{1}{n_1} + \frac{1}{n_2}},$$ the pooled variance is $$s_{pooled}^2 = \frac{(n_1 - 1)s_1^2 + (n_2 - 1)s_2^2}{(n_1 - 1) + (n_2 - 1)},$$ and the number of degrees of freedom is $(n_1 - 1) + (n_2 - 1)$ (p. 593).
Pooled *t*-test	A hypothesis test for the difference in the means of two independent groups when we are willing and able to assume that the variances of the groups are equal. It tests the null hypothesis $$H_0: \mu_1 - \mu_2 = \Delta_0,$$ where the hypothesized difference, Δ_0, is almost always 0, using the statistic $$t_{df} = \frac{(\bar{y}_1 - \bar{y}_2) - \Delta_0}{SE_{pooled}(\bar{y}_1 - \bar{y}_2)},$$ where the pooled standard error is defined as for the pooled interval and the degrees of freedom is $(n_1 - 1) + (n_2 - 1)$ (p. 594).
Tukey's quick test	A nonparametric test of whether two groups have different means (p. 597).
Rank sum test	A nonparametric test that uses ranks to test whether the means of two groups are the same (p. 597).

Skills

THINK

▶ Be able to recognize situations in which we want to do inference on the difference between the means of two independent groups.

► Know how to examine your data for violations of conditions that would make inference about the difference between two population means unwise or invalid.

► Be able to recognize when a pooled-*t* procedure might be appropriate and be able to explain why you decided to use a two-sample method anyway.

► Be able to perform a two-sample *t*-test using a statistics package or calculator (at least for finding the degrees of freedom).

► Be able to interpret a test of the null hypothesis that the means of two independent groups are equal. (If the test is a pooled *t*-test, your interpretation should include a defense of your assumption of equal variances.)

TWO-SAMPLE METHODS ON THE COMPUTER

Here's some typical computer package output with comments:

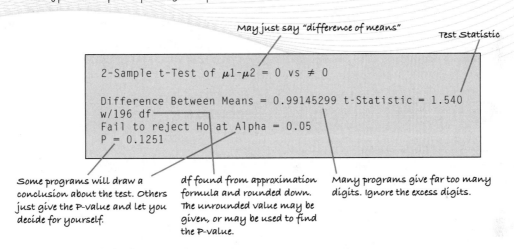

Most statistics packages compute the test statistic and report a P-value corresponding to that statistic. And, of course, statistics packages make it easy to examine the boxplots and histograms of the two groups, so there's no excuse for skipping this important check.

Some statistics software automatically tries to test whether the variances of the two groups are equal. Some automatically offer both the two-sample-*t* and pooled-*t* results. Ignore the test for the variances; it has little power in any situation in which its results could matter. If the pooled and two-sample methods differ in any important way, you should stick with the two-sample method. Most likely, the Equal Variance Assumption needed for the pooled method has failed.

The degrees of freedom approximation usually gives a fractional value. Most packages seem to round the approximate value down to the next smallest integer (although they may actually compute the P-value with the fractional value, gaining a tiny amount of power).

There are two ways to organize data when we want to compare two independent groups. The data can be in two lists, as in the table at the start of this chapter. Each list can be thought of as a variable. In this method, the variables in the batteries example would be *Brand Name* and *Generic*. Graphing calculators usually prefer this form, and some computer programs can use it as well.

There's another way to think about the data. What is the response variable for the battery life experiment? It's the *Time* until the music stopped. But the values of this variable are in both columns, and actually there's an experiment factor here, too—namely, the *Brand* of the battery. So, we could put the data into two different columns, one with the *Times* in it and one with the *Brand*. Then the data would look like this:

Time	Brand
194.0	Brand name
205.5	Brand name
199.2	Brand name
172.4	Brand name
184.0	Brand name
169.5	Brand name
190.7	Generic
203.5	Generic
203.5	Generic
206.5	Generic
222.5	Generic
209.4	Generic

This way of organizing the data makes sense as well. Now the factor and the response variables are clearly visible. You'll have to see which method your program requires. Some packages even allow you to structure the data either way.

The commands to do inference for two independent groups on common statistics technology are not always found in obvious places. Here are some starting guidelines.

DATA DESK

Select variables.
From the **Calc** menu, choose **Estimate** for confidence intervals or **Test** for hypothesis tests. Select the interval or test from the drop-down menu and make other choices in the dialog.

COMMENTS

Data Desk expects the two groups to be in separate variables.

EXCEL

From the Data Tab, Analysis Group, choose **Data Analysis.** Alternatively (if the Data Analysis Tool Pack is not installed), in the Formulas Tab, choose More functions > Statistical > TTEST, and specify Type = 3 in the resulting dialog.
Fill in the cell ranges for the two groups, the hypothesized difference, and the alpha level.

COMMENTS

Excel expects the two groups to be in separate cell ranges. Notice that, contrary to Excel's wording, we do not need to assume that the variances are *not* equal; we simply choose not to assume that they *are* equal.

JMP

From the **Analyze** menu, select **Fit y by x.** Select variables: a **Y, Response** variable that holds the data and an **X, Factor** variable that holds the group names. JMP will make a dotplot. Click the **red triangle** in the dotplot title, and choose **Unequal variances.** The *t*-test is at the bottom of the resulting table. Find the P-value from the Prob > F section of the table (they are the same).

Comments

JMP expects data in one variable and category names in the other. Don't be misled: There is no need for the variances to be unequal to use two-sample *t* methods.

MINITAB

From the **Stat** menu, choose the **Basic Statistics** submenu. From that menu, choose **2-sample t. . . .** Then fill in the dialog.

COMMENTS

The dialog offers a choice of data in two variables, or data in one variable and category names in the other.

SPSS

From the **Analyze** menu, choose the **Compare Means** submenu. From that, choose the **Independent-Samples t-test** command. Specify the data variable and "group variable." Then type in the labels used in the group variable. SPSS offers both the two-sample and pooled-t results in the same table.

COMMENTS

SPSS expects the data in one variable and group names in the other. If there are more than two group names in the group variable, only the two that are named in the dialog box will be compared.

TI-83/84 PLUS

For a confidence interval:
In the **STAT TESTS** menu, choose **0:2-SampTInt.** You may specify that you are using data stored in two lists, or you may enter the means, standard deviations, and sizes of both samples. You must also indicate whether to pool the variances (when in doubt, say no) and specify the desired level of confidence.

To test a hypothesis:
In the **STAT TESTS** menu, choose **4:2-SampTTest.** You may specify if you are using data stored in two lists, or you may enter the means, standard deviations, and sizes of both samples. You must also indicate whether to pool the variances (when in doubt, say no) and specify whether the test is to be two-tail, lower-tail, or upper-tail.

TI-89

For a confidence interval:
In the **STAT Ints** menu, choose **4:2-SampTInt.** You must specify if you are using data stored in two lists or if you will enter the means, standard deviations, and sizes of both samples. You must also indicate whether to pool the variances (when in doubt, say no) and specify the desired level of confidence.

To test a hypothesis:
In the **STAT TESTS** menu, choose **4:2-SampTTest.** You must specify if you are using data stored in two lists or if you will enter the means, standard deviations, and sizes of both samples. You must also indicate whether to pool the variances (when in doubt, say no) and specify whether the test is to be two-tail, lower-tail, or upper-tail.

EXERCISES

1. **Dogs and calories.** In the July 2007 issue, *Consumer Reports* examined the calorie content of two kinds of hot dogs: meat (usually a mixture of pork, turkey, and chicken) and all beef. The researchers purchased samples of several different brands. The meat hot dogs averaged 111.7 calories, compared to 135.4 for the beef hot dogs. A test of the null hypothesis that there's no difference in mean calorie content yields a P-value of 0.124. Would a 95% confidence interval for $\mu_{Meat} - \mu_{Beef}$ include 0? Explain.

2. **Dogs and sodium.** The *Consumer Reports* article described in Exercise 1 also listed the sodium content (in mg) for the various hot dogs tested. A test of the null hypothesis that beef hot dogs and meat hot dogs don't differ in the mean amounts of sodium yields a P-value of 0.11. Would a 95% confidence interval for $\mu_{Meat} - \mu_{Beef}$ include 0? Explain.

3. **Dogs and fat.** The *Consumer Reports* article described in Exercise 1 also listed the fat content (in grams) for samples of beef and meat hot dogs. The resulting 90% confidence interval for $\mu_{Meat} - \mu_{Beef}$ is $(-6.5, -1.4)$.
   a) The endpoints of this confidence interval are negative numbers. What does that indicate?

   b) What does the fact that the confidence interval does not contain 0 indicate?
   c) If we use this confidence interval to test the hypothesis that $\mu_{Meat} - \mu_{Beef} = 0$, what's the corresponding alpha level?

4. **Washers.** In the June 2007 issue, *Consumer Reports* examined top-loading and front-loading washing machines, testing samples of several different brands of each type. One of the variables the article reported was "cycle time," the number of minutes it took each machine to wash a load of clothes. Among the machines rated good to excellent, the 98% confidence interval for the difference in mean cycle time ($\mu_{Top} - \mu_{Front}$) is $(-40, -22)$.
   a) The endpoints of this confidence interval are negative numbers. What does that indicate?
   b) What does the fact that the confidence interval does not contain 0 indicate?
   c) If we use this confidence interval to test the hypothesis that $\mu_{Top} - \mu_{Front} = 0$, what's the corresponding alpha level?

5. **Dogs and fat, second helping.** In Exercise 3, we saw a 90% confidence interval of $(-6.5, -1.4)$ grams for $\mu_{Meat} - \mu_{Beef}$, the difference in mean fat content for meat vs. all-beef hot dogs. Explain why you think each of the following statements is true or false:
   a) If I eat a meat hot dog instead of a beef dog, there's a 90% chance I'll consume less fat.
   b) 90% of meat hot dogs have between 1.4 and 6.5 grams less fat than a beef hot dog.
   c) I'm 90% confident that meat hot dogs average 1.4–6.5 grams less fat than the beef hot dogs.
   d) If I were to get more samples of both kinds of hot dogs, 90% of the time the meat hot dogs would average 1.4–6.5 grams less fat than the beef hot dogs.
   e) If I tested more samples, I'd expect about 90% of the resulting confidence intervals to include the true difference in mean fat content between the two kinds of hot dogs.

6. **Second load of wash.** In Exercise 4, we saw a 98% confidence interval of $(-40, -22)$ minutes for $\mu_{Top} - \mu_{Front}$, the difference in time it takes top-loading and front-loading washers to do a load of clothes. Explain why you think each of the following statements is true or false:
   a) 98% of top loaders are 22 to 40 minutes faster than front loaders.
   b) If I choose the laundromat's top loader, there's a 98% chance that my clothes will be done faster than if I had chosen the front loader.
   c) If I tried more samples of both kinds of washing machines, in about 98% of these samples I'd expect the top loaders to be an average of 22 to 40 minutes faster.
   d) If I tried more samples, I'd expect about 98% of the resulting confidence intervals to include the true difference in mean cycle time for the two types of washing machines.
   e) I'm 98% confident that top loaders wash clothes an average of 22 to 40 minutes faster than front-loading machines.

7. **Learning math.** The Core Plus Mathematics Project (CPMP) is an innovative approach to teaching Mathematics that engages students in group investigations and mathematical modeling. After field tests in 36 high schools over a three-year period, researchers compared the performances of CPMP students with those taught using a traditional curriculum. In one test, students had to solve applied Algebra problems using calculators. Scores for 320 CPMP students were compared to those of a control group of 273 students in a traditional Math program. Computer software was used to create a confidence interval for the difference in mean scores. (*Journal for Research in Mathematics Education*, 31, no. 3[2000])

   Conf level: 95%   Variable: Mu(CPMP) − Mu(Ctrl)
   Interval: (5.573, 11.427)

   a) What's the margin of error for this confidence interval?
   b) If we had created a 98% CI, would the margin of error be larger or smaller?
   c) Explain what the calculated interval means in this context.

   d) Does this result suggest that students who learn Mathematics with CPMP will have significantly higher mean scores in Algebra than those in traditional programs? Explain.

8. **Stereograms.** Stereograms appear to be composed entirely of random dots. However, they contain separate images that a viewer can "fuse" into a three-dimensional (3D) image by staring at the dots while defocusing the eyes. An experiment was performed to determine whether knowledge of the form of the embedded image affected the time required for subjects to fuse the images. One group of subjects (group NV) received no information or just verbal information about the shape of the embedded object. A second group (group VV) received both verbal information and visual information (specifically, a drawing of the object). The experimenters measured how many seconds it took for the subject to report that he or she saw the 3D image.

   2-Sample t-Interval for $\mu 1 - \mu 2$
   Conf level − 90%   df = 70
   μ(NV) − μ(VV) interval: (0.55, 5.47)

   a) Interpret your interval in context.
   b) Does it appear that viewing a picture of the image helps people "see" the 3D image in a stereogram?
   c) What's the margin of error for this interval?
   d) Explain carefully what the 90% confidence level means.
   e) Would you expect a 99% confidence level to be wider or narrower? Explain.
   f) Might that change your conclusion in part b? Explain.

9. **CPMP, again.** During the study described in Exercise 7, students in both CPMP and traditional classes took another Algebra test that did not allow them to use calculators. The table below shows the results. Are the mean scores of the two groups significantly different?

Math Program	n	Mean	SD
CPMP	312	29.0	18.8
Traditional	265	38.4	16.2

   Performance on Algebraic Symbolic Manipulation Without Use of Calculators

   a) Write an appropriate hypothesis.
   b) Do you think the assumptions for inference are satisfied? Explain.
   c) Here is computer output for this hypothesis test. Explain what the P-value means in this context.

   2-Sample t-Test of $\mu 1 - \mu 2 \neq 0$
   t-Statistic = −6.451 w/574.8761 df
   P < 0.0001

   d) State a conclusion about the CPMP program.

10. **CPMP and word problems.** The study of the new CPMP Mathematics methodology described in Exercise 7 also tested students' abilities to solve word problems. This table shows how the CPMP and traditional groups performed. What do you conclude?

Math Program	n	Mean	SD
CPMP	320	57.4	32.1
Traditional	273	53.9	28.5

11. **Commuting.** A man who moves to a new city sees that there are two routes he could take to work. A neighbor who has lived there a long time tells him Route A will average 5 minutes faster than Route B. The man decides to experiment. Each day he flips a coin to determine which way to go, driving each route 20 days. He finds that Route A takes an average of 40 minutes, with standard deviation 3 minutes, and Route B takes an average of 43 minutes, with standard deviation 2 minutes. Histograms of travel times for the routes are roughly symmetric and show no outliers.

    a) Find a 95% confidence interval for the difference in average commuting time for the two routes.
    b) Should the man believe the old-timer's claim that he can save an average of 5 minutes a day by always driving Route A? Explain.

12. **Pulse rates.** A researcher wanted to see whether there is a significant difference in resting pulse rates for men and women. The data she collected are summarized and displayed in the boxplots below.

	Sex	
	Male	Female
Count	28	24
Mean	72.75	72.625
Median	73	73
StdDev	5.37225	7.69987
Range	20	29
IQR	9	12.5

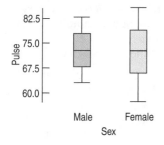

a) What do the boxplots suggest about differences between male and female pulse rates?
b) Is it appropriate to analyze these data using the methods of inference discussed in this chapter? Explain.
c) Create a 90% confidence interval for the difference in mean pulse rates.
d) Does the confidence interval confirm your answer to part a? Explain.

13. **Cereal.** The following data show the sugar content (as a percentage of weight) of several national brands of children's and adults' cereals. Create and interpret a 95% confidence interval for the difference in mean sugar content. Be sure to check the necessary assumptions and conditions.

Children's cereals: 40.3, 55, 45.7, 43.3, 50.3, 45.9, 53.5, 43, 44.2, 44, 47.4, 44, 33.6, 55.1, 48.8, 50.4, 37.8, 60.3, 46.6

Adults' cereals: 20, 30.2, 2.2, 7.5, 4.4, 22.2, 16.6, 14.5, 21.4, 3.3, 6.6, 7.8, 10.6, 16.2, 14.5, 4.1, 15.8, 4.1, 2.4, 3.5, 8.5, 10, 1, 4.4, 1.3, 8.1, 4.7, 18.4

14. **Egyptians.** Some archaeologists theorize that ancient Egyptians interbred with several different immigrant populations over thousands of years. To see if there is any indication of changes in body structure that might have resulted, they measured 30 skulls of male Egyptians dated from 4000 B.C.E. and 30 others dated from 200 B.C.E. (A. Thomson and R. Randall-Maciver, *Ancient Races of the Thebaid*, Oxford: Oxford University Press, 1905)

    a) Are these data appropriate for inference? Explain.
    b) Create a 95% confidence interval for the difference in mean skull breadth between these two eras.
    c) Do these data provide evidence that the mean breadth of males' skulls changed over this period? Explain.
    *d) Perform Tukey's test for the difference. Do your conclusions of part c change?
    *e) Perform a rank sum test for the difference. Do your conclusions of part c change?

Maximum Skull Breadth (mm)			
4000 B.C.E.		**200 B.C.E.**	
131	131	141	131
125	135	141	129
131	132	135	136
119	139	133	131
136	132	131	139
138	126	140	144
139	135	139	141
125	134	140	130
131	128	138	133
134	130	132	138
129	138	134	131
134	128	135	136
126	127	133	132
132	131	136	135
141	124	134	141

15. **Reading.** An educator believes that new reading activities for elementary school children will improve reading comprehension scores. She randomly assigns third graders to an eight-week program in which some will use these activities and others will experience traditional teaching methods. At the end of the experiment, both groups take a reading comprehension exam. Their scores are shown in the back-to-back stem-and-leaf display. Do these results suggest that the new activities are better? Test an appropriate hypothesis and state your conclusion. (*Would Tukey's test be appropriate here? Explain. *Use a rank sum test to test an appropriate hypothesis as well. Compare your conclusions with your previous one.)

New Activities		Control
	1	07
4	2	068
3	3	377
96333	4	12222238
9876432	5	355
721	6	02
1	7	
	8	5

T 16. **Streams.** Researchers collected samples of water from streams in the Adirondack Mountains to investigate the effects of acid rain. They measured the pH (acidity) of the water and classified the streams with respect to the kind of substrate (type of rock over which they flow). A lower pH means the water is more acidic. Here is a plot of the pH of the streams by substrate (limestone, mixed, or shale):

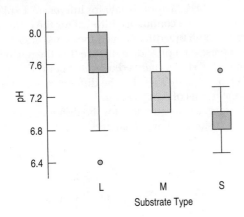

Here are selected parts of a software analysis comparing the pH of streams with limestone and shale substrates:

2-Sample t-Test of $\mu 1 - \mu 2$
Difference Between Means = 0.735
t-Statistic = 16.30 w/133 df
p ≤ 0.0001

a) State the null and alternative hypotheses for this test.
b) From the information you have, do the assumptions and conditions appear to be met?
c) What conclusion would you draw?

T 17. **Baseball 2006.** American League baseball teams play their games with the designated hitter rule, meaning that pitchers do not bat. The league believes that replacing the pitcher, traditionally a weak hitter, with another player in the batting order produces more runs and generates more interest among fans. Below are the average numbers of runs scored in American League and National League stadiums for the 2006 season.

American		National	
11.4	9.9	10.5	9.5
10.5	9.7	10.3	9.4
10.5	9.1	10.0	9.1
10.3	9.0	10.0	9.0
10.2	9.0	9.7	9.0
10.0	8.9	9.7	8.9
9.9	8.8	9.6	8.9
		9.5	7.9

a) Create an appropriate display of these data. What do you see?
b) With a 95% confidence interval, estimate the mean number of runs scored in American League games.
c) Coors Field, in Denver, stands a mile above sea level, an altitude far greater than that of any other National League ballpark. Some believe that the thinner air makes it harder for pitchers to throw curve balls and easier for batters to hit the ball a long way. Do you think the 10.5 runs scored per game at Coors is unusual? Explain.
d) Explain why you should not use two separate confidence intervals to decide whether the two leagues differ in average number of runs scored.

18. **Handy.** A factory hiring people to work on an assembly line gives job applicants a test of manual agility. This test counts how many strangely shaped pegs the applicant can fit into matching holes in a one-minute period. The table below summarizes the data by sex of the job applicant. Assume that all conditions necessary for inference are met.

	Male	Female
Number of subjects	50	50
Pegs placed:		
Mean	19.39	17.91
SD	2.52	3.39

a) Find 95% confidence intervals for the average number of pegs that males and females can each place.
b) Those intervals overlap. What does this suggest about any sex-based difference in manual agility?
c) Find a 95% confidence interval for the difference in the mean number of pegs that could be placed by men and women.
d) What does this interval suggest about any difference in manual agility between men and women?
e) The two results seem contradictory. Which method is correct: doing two-sample inference or doing one-sample inference twice?
f) Why don't the results agree?

T 19. **Baseball 2006, part 2.** Do the data in Exercise 17 suggest that the American League's designated hitter rule may lead to more runs?

a) Using a 95% confidence interval, estimate the difference between the mean number of runs scored in American and National League games.
b) Interpret your interval.
c) Does that interval suggest that the two leagues may differ in average number of runs scored per game?

T 20. **Derby hard water.** In an investigation of environmental causes of disease, data were collected on the annual mortality rate (deaths per 100,000) for males in 61 large towns in England and Wales. In addition, the water hardness was recorded as the calcium concentration (parts per million, ppm) in the drinking water. The data set also notes, for each town, whether it was south or north of Derby. Is there a significant difference in mortality rates in the two regions? Here are the summary statistics.

Summary of: mortality
For categories in: Derby

Group	Count	Mean	Median	StdDev
North	34	1631.59	1631	138.470
South	27	1388.85	1369	151.114

a) Test appropriate hypotheses and state your conclusion.

b) In the following figure, the boxplots of the two distributions show an outlier among the data north of Derby. What effect might that have had on your test?

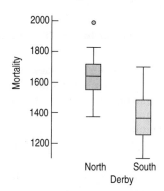

T 21. Job satisfaction. A company institutes an exercise break for its workers to see if this will improve job satisfaction, as measured by a questionnaire that assesses workers' satisfaction. Scores for 10 randomly selected workers before and after implementation of the exercise program are shown. The company wants to assess the effectiveness of the exercise program. Explain why you can't use the methods discussed in this chapter to do that. (Don't worry, we'll give you another chance to do this the right way.)

Worker Number	Job Satisfaction Index	
	Before	After
1	34	33
2	28	36
3	29	50
4	45	41
5	26	37
6	27	41
7	24	39
8	15	21
9	15	20
10	27	37

22. Summer school. Having done poorly on their math final exams in June, six students repeat the course in summer school, then take another exam in August. If we consider these students representative of all students who might attend this summer school in other years, do these results provide evidence that the program is worthwhile?

June	54	49	68	66	62	62
Aug.	50	65	74	64	68	72

23. Sex and violence. In June 2002, the *Journal of Applied Psychology* reported on a study that examined whether the content of TV shows influenced the ability of viewers to recall brand names of items featured in the commercials. The researchers randomly assigned volunteers to watch one of three programs, each containing the same nine commercials. One of the programs had violent content, another sexual content, and the third neutral content. After the shows ended, the subjects were asked to recall the brands of products that were advertised. Here are summaries of the results:

	Program Type		
	Violent	Sexual	Neutral
No. of subjects	108	108	108
Brands recalled			
Mean	2.08	1.71	3.17
SD	1.87	1.76	1.77

a) Do these results indicate that viewer memory for ads may differ depending on program content? A test of the hypothesis that there is no difference in ad memory between programs with sexual content and those with violent content has a P-value of 0.136. State your conclusion.

b) Is there evidence that viewer memory for ads may differ between programs with sexual content and those with neutral content? Test an appropriate hypothesis and state your conclusion.

24. Ad campaign. You are a consultant to the marketing department of a business preparing to launch an ad campaign for a new product. The company can afford to run ads during one TV show, and has decided not to sponsor a show with sexual content. You read the study described in Exercise 23, then use a computer to create a confidence interval for the difference in mean number of brand names remembered between the groups watching violent shows and those watching neutral shows.

> TWO-SAMPLE T
> 95% CI FOR MUviol − MUneut: (−1.578, −0.602)

a) At the meeting of the marketing staff, you have to explain what this output means. What will you say?

b) What advice would you give the company about the upcoming ad campaign?

25. Sex and violence II. In the study described in Exercise 23, the researchers also contacted the subjects again, 24 hours later, and asked them to recall the brands advertised. Results are summarized below.

	Program Type		
	Violent	Sexual	Neutral
No. of subjects	101	106	103
Brands recalled			
Mean	3.02	2.72	4.65
SD	1.61	1.85	1.62

a) Is there a significant difference in viewers' abilities to remember brands advertised in shows with violent vs. neutral content?

b) Find a 95% confidence interval for the difference in mean number of brand names remembered between the groups watching shows with sexual content and those watching neutral shows. Interpret your interval in this context.

26. **Ad recall.** In Exercises 23 and 25, we see the number of advertised brand names people recalled immediately after watching TV shows and 24 hours later. Strangely enough, it appears that they remembered more about the ads the next day. Should we conclude this is true in general about people's memory of TV ads?

    a) Suppose one analyst conducts a two-sample hypothesis test to see if memory of brands advertised during violent TV shows is higher 24 hours later. If his P-value is 0.00013, what might he conclude?

    b) Explain why his procedure was inappropriate. Which of the assumptions for inference was violated?

    c) How might the design of this experiment have tainted the results?

    d) Suggest a design that could compare immediate brand-name recall with recall one day later.

27. **Hungry?** Researchers investigated how the size of a bowl affects how much ice cream people tend to scoop when serving themselves.[13] At an "ice cream social," people were randomly given either a 17 oz or a 34 oz bowl (both large enough that they would not be filled to capacity). They were then invited to scoop as much ice cream as they liked. Did the bowl size change the selected portion size? Here are the summaries:

	Small Bowl		Large Bowl
n	26	n	22
$\bar{y}$	5.07 oz	$\bar{y}$	6.58 oz
s	1.84 oz	s	2.91 oz

Test an appropriate hypothesis and state your conclusions. Assume any assumptions and conditions that you cannot test are sufficiently satisfied to proceed.

28. **Thirsty?** Researchers randomly assigned participants either a tall, thin "highball" glass or a short, wide "tumbler," each of which held 355 ml. Participants were asked to pour a shot (1.5 oz = 44.3 ml) into their glass. Did the shape of the glass make a difference in how much liquid they poured?[14] Here are the summaries:

	highball		tumbler
n	99	n	99
$\bar{y}$	42.2 ml	$\bar{y}$	60.9 ml
s	16.2 ml	s	17.9 ml

Test an appropriate hypothesis and state your conclusions. Assume any assumptions and conditions that you cannot test are sufficiently satisfied to proceed.

29. **Lower scores?** Newspaper headlines recently announced a decline in science scores among high school seniors. In 2000, a total of 15,109 seniors tested by The National Assessment in Education Program (NAEP) scored a mean

of 147 points. Four years earlier, 7537 seniors had averaged 150 points. The standard error of the difference in the mean scores for the two groups was 1.22.

    a) Have the science scores declined significantly? Cite appropriate statistical evidence to support your conclusion.

    b) The sample size in 2000 was almost double that in 1996. Does this make the results more convincing or less? Explain.

30. **The Internet.** The NAEP report described in Exercise 29 compared science scores for students who had home Internet access to the scores of those who did not, as shown in the graph. They report that the differences are statistically significant.

    a) Explain what "statistically significant" means in this context.

    b) If their conclusion is incorrect, which type of error did the researchers commit?

    c) Does this prove that using the Internet at home can improve a student's performance in science?

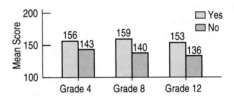

31. **Running heats.** In Olympic running events, preliminary heats are determined by random draw, so we should expect that the abilities of runners in the various heats to be about the same, on average. Here are the times (in seconds) for the 400-m women's run in the 2004 Olympics in Athens for preliminary heats 2 and 5. Is there any evidence that the mean time to finish is different for randomized heats? Explain. Be sure to include a discussion of assumptions and conditions for your analysis.

Country	Name	Heat	Time
USA	HENNAGAN Monique	2	51.02
BUL	DIMITROVA Mariyana	2	51.29
CHA	NADJINA Kaltouma	2	51.50
JAM	DAVY Nadia	2	52.04
BRA	ALMIRAO Maria Laura	2	52.10
FIN	MYKKANEN Kirsi	2	52.53
CHN	BO Fanfang	2	56.01
BAH	WILLIAMS-DARLING Tonique	5	51.20
BLR	USOVICH Svetlana	5	51.37
UKR	YEFREMOVA Antonina	5	51.53
CMR	NGUIMGO Mireille	5	51.90
JAM	BECKFORD Allison	5	52.85
TOG	THIEBAUD-KANGNI Sandrine	5	52.87
SRI	DHARSHA K V Damayanthi	5	54.58

32. **Swimming heats.** In Exercise 31 we looked at the times in two different heats for the 400-m women's run from the 2004 Olympics. Unlike track events, swimming heats are *not* determined at random. Instead, swimmers are seeded so that better swimmers are placed in later heats. Here are the times (in seconds)

---

[13] Brian Wansink, Koert van Ittersum, and James E. Painter, "Ice Cream Illusions: Bowls, Spoons, and Self-Served Portion Sizes," *Am J Prev Med* 2006.

[14] Brian Wansink and Koert van Ittersum, "Shape of Glass and Amount of Alcohol Poured: Comparative Study of Effect of Practice and Concentration," *BMJ* 2005;331;1512–1514.

for the women's 400-m freestyle from heats 2 and 5. Do these results suggest that the mean times of seeded heats are not equal? Explain. Include a discussion of assumptions and conditions for your analysis.

Country	Name	Heat	Time
ARG	BIAGIOLI Cecilia Elizabeth	2	256.42
SLO	CARMAN Anja	2	257.79
CHI	KOBRICH Kristel	2	258.68
MKD	STOJANOVSKA Vesna	2	259.39
JAM	ATKINSON Janelle	2	260.00
NZL	LINTON Rebecca	2	261.58
KOR	HA Eun-Ju	2	261.65
UKR	BERESNYEVA Olga	2	266.30
FRA	MANAUDOU Laure	5	246.76
JPN	YAMADA Sachiko	5	249.10
ROM	PADURARU Simona	5	250.39
GER	STOCKBAUER Hannah	5	250.46
AUS	GRAHAM Elka	5	251.67
CHN	PANG Jiaying	5	251.81
CAN	REIMER Brittany	5	252.33
BRA	FERREIRA Monique	5	253.75

33. **Tees.** Does it matter what kind of tee a golfer places the ball on? The company that manufactures "Stinger" tees claims that the thinner shaft and smaller head will lessen drag, reducing spin and allowing the ball to travel farther. In August 2003, Golf Laboratories, Inc., compared the distance traveled by golf balls hit off regular wooden tees to those hit off Stinger tees. All the balls were struck by the same golf club using a robotic device set to swing the club head at approximately 95 miles per hour. Summary statistics from the test are shown in the table. Assume that 6 balls were hit off each tee and that the data were suitable for inference.

		Total Distance (yards)	Ball Velocity (mph)	Club Velocity (mph)
Regular tee	Avg.	227.17	127.00	96.17
	SD	2.14	0.89	0.41
Stinger tee	Avg.	241.00	128.83	96.17
	SD	2.76	0.41	0.52

Is there evidence that balls hit off the Stinger tees would have a higher initial velocity?

34. **Golf again.** Given the test results on golf tees described in Exercise 33, is there evidence that balls hit off Stinger tees would travel farther? Again, assume that 6 balls were hit off each tee and that the data were suitable for inference.

35. **Crossing Ontario.** Between 1954 and 2003, swimmers have crossed Lake Ontario 43 times. Both women and men have made the crossing. Here are some plots (we've omitted a crossing by Vikki Keith, who swam a round trip—North to South to North—in 3390 minutes):

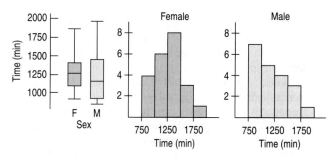

The summary statistics are:

	Summary of Time (min)		
Group	Count	Mean	StdDev
F	22	1271.59	261.111
M	20	1196.75	304.369

How much difference is there between the mean amount of time (in minutes) it would take female and male swimmers to swim the lake?

a) Construct and interpret a 95% confidence interval for the difference between female and male times.

b) Comment on the assumptions and conditions.

36. **Music and memory.** Is it a good idea to listen to music when studying for a big test? In a study conducted by some Statistics students, 62 people were randomly assigned to listen to rap music, music by Mozart, or no music while attempting to memorize objects pictured on a page. They were then asked to list all the objects they could remember. Here are summary statistics:

	Rap	Mozart	No Music
Count	29	20	13
Mean	10.72	10.00	12.77
SD	3.99	3.19	4.73

a) Does it appear that it is better to study while listening to Mozart than to rap music? Test an appropriate hypothesis and state your conclusion.

b) Create a 90% confidence interval for the mean difference in memory score between students who study to Mozart and those who listen to no music at all. Interpret your interval.

37. **Rap.** Using the results of the experiment described in Exercise 36, does it matter whether one listens to rap music while studying, or is it better to study without music at all?

a) Test an appropriate hypothesis and state your conclusion.

b) If you concluded there is a difference, estimate the size of that difference with a confidence interval and explain what your interval means.

T 38. Cuckoos. Cuckoos lay their eggs in the nests of other (host) birds. The eggs are then adopted and hatched by the host birds. But the potential host birds lay eggs of different sizes. Does the cuckoo change the size of her eggs for different foster species? The numbers in the table are lengths (in mm) of cuckoo eggs found in nests of three different species of other birds. The data are drawn from the work of O. M. Latter in 1902 and were used in a fundamental textbook on statistical quality control by L. H. C. Tippett (1902–1985), one of the pioneers in that field. (*Use Tukey's test to compare the mean egg length of each pair of species. Do your conclusions change? *Use a rank sum test to compare the mean egg length of each pair of species. Do your conclusions change?)

CUCKOO EGG LENGTH (MM)

Sparrow	Robin	Wagtail
20.85	21.05	21.05
21.65	21.85	21.85
22.05	22.05	21.85
22.85	22.05	21.85
23.05	22.05	22.05
23.05	22.25	22.45
23.05	22.45	22.65
23.05	22.45	23.05
23.45	22.65	23.05
23.85	23.05	23.25
23.85	23.05	23.45
23.85	23.05	24.05
24.05	23.05	24.05
25.05	23.05	24.05
	23.25	24.85
	23.85	

Foster Parent Species

Investigate the question of whether the mean length of cuckoo eggs is the same for different species, and state your conclusion.

JUST CHECKING

ANSWERS

1. Randomization should balance unknown sources of variability in the two groups of patients and helps us believe the two groups are independent.

2. We can be 95% confident that after 4 weeks endoscopic surgery patients will have a mean pinch strength between 0.04 kg and 2.96 kg higher than open-incision patients.

3. The lower bound of this interval is close to 0, so the difference may not be great enough that patients could actually notice the difference. We may want to consider other issues such as cost or risk in making a recommendation about the two surgical procedures.

4. Without data, we can't check the Nearly Normal Condition.

5. H_0: Mean pinch strength is the same after both surgeries. $(\mu_E - \mu_O = 0)$

H_A: Mean pinch strength is different after the two surgeries. $(\mu_E - \mu_O \neq 0)$

6. With a P-value this low, we reject the null hypothesis. We can conclude that mean pinch strength differs after 4 weeks in patients who undergo endoscopic surgery vs. patients who have open-incision surgery. Results suggest that the endoscopic surgery patients may be stronger, on average.

7. If some patients contributed two hands to the study, then the groups may not be internally independent. It is reasonable to assume that two hands from the same patient might respond in similar ways to similar treatments.

Paired Samples and Blocks

Where are we going?

How much will an LSAT prep course raise scores? Are boys better at computer games than their sisters? Questions like these look at paired variables. When pairs of observations go together naturally, they can't be independent, so the methods we used in the last chapter won't work. In this chapter you'll see what to do with paired data.

WHO	Olympic speed-skaters
WHAT	Time for women's 1500 m
UNITS	Seconds
WHEN	2006
WHERE	Torino, Italy
WHY	To see whether one lane is faster than the other

Speed-skating races are run in pairs. Two skaters start at the same time, one on the inner lane and one on the outer lane. Halfway through the race, they cross over, switching lanes so that each will skate the same distance in each lane. Even though this seems fair, at the 2006 Olympics some fans thought there might have been an advantage to starting on the outside. After all, the winner, Cindy Klassen, started on the outside and skated a remarkable 1.47 seconds faster than the silver medalist.

Here are the data for the women's 1500-m race:

Inner Lane		Outer Lane	
Name	Time	Name	Time
OLTEAN Daniela	129.24	(no competitor)	
ZHANG Xiaolei	125.75	NEMOTO Nami	122.34
ABRAMOVA Yekaterina	121.63	LAMB Maria	122.12
REMPEL Shannon	122.24	NOH Seon Yeong	123.35
LEE Ju-Youn	120.85	TIMMER Marianne	120.45
ROKITA Anna Natalia	122.19	MARRA Adelia	123.07
YAKSHINA Valentina	122.15	OPITZ Lucille	122.75
BJELKEVIK Hedvig	122.16	HAUGLI Maren	121.22
ISHINO Eriko	121.85	WOJCICKA Katarzyna	119.96
RANEY Catherine	121.17	BJELKEVIK Annette	121.03
OTSU Hiromi	124.77	LOBYSHEVA Yekaterina	118.87
SIMIONATO Chiara	118.76	JI Jia	121.85
ANSCHUETZ THOMS Daniela	119.74	WANG Fei	120.13
BARYSHEVA Varvara	121.60	van DEUTEKOM Paulien	120.15
GROENEWOLD Renate	119.33	GROVES Kristina	116.74
RODRIGUEZ Jennifer	119.30	NESBITT Christine	119.15
FRIESINGER Anni	117.31	KLASSEN Cindy	115.27
WUST Ireen	116.90	TABATA Maki	120.77

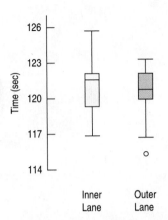

FIGURE 25.1
Using boxplots to compare times in the inner and outer lanes shows little because it ignores the fact that the skaters raced in pairs.

We can view this skating event as an experiment testing whether the lanes were equally fast. Skaters were assigned to lanes randomly. The boxplots of times recorded in the inner and outer lanes don't show much difference. But that's not the right way to compare these times. Conditions can change during the day. The data are recorded for races run two at a time, so the two groups are not independent.

Paired Data

Data such as these are called **paired**. We have the times for skaters in each lane for each race. The races are run in pairs, so they can't be independent. And since they're not independent, we can't use the two-sample *t* methods. Instead, we can focus on the *differences* in times for each racing pair.

Paired data arise in a number of ways. Perhaps the most common way is to compare subjects with themselves before and after a treatment. When pairs arise from an experiment, the pairing is a type of *blocking*. When they arise from an observational study, it is a form of *matching*.

FOR EXAMPLE   Identifying Paired Data

Do flexible schedules reduce the demand for resources? The Lake County, Illinois, Health Department experimented with a flexible four-day workweek. For a year, the department recorded the mileage driven by 11 field workers on an ordinary five-day workweek. Then it changed to a flexible four-day workweek and recorded mileage for another year.[1] The data are shown.

QUESTION: Why are these data paired?

The mileage data are paired because each driver's mileage is measured before and after the change in schedule. I'd expect drivers who drove more than others before the schedule change to continue to drive more afterwards, so the two sets of mileages can't be considered independent.

Name	5-Day mileage	4-Day mileage
Jeff	2798	2914
Betty	7724	6112
Roger	7505	6177
Tom	838	1102
Aimee	4592	3281
Greg	8107	4997
Larry G.	1228	1695
Tad	8718	6606
Larry M.	1097	1063
Leslie	8089	6392
Lee	3807	3362

Pairing isn't a problem; it's an opportunity. If you know the data are paired, you can take advantage of that fact—in fact, you *must* take advantage of it. You *may not* use the two-sample and pooled methods of the previous chapter when the data are paired. Remember: Those methods rely on the

---

[1] Charles S. Catlin, "Four-day Work Week Improves Environment," *Journal of Environmental Health*, Denver, 59:7.

Pythagorean Theorem of Statistics, and that requires the two samples be independent. Paired data aren't. There is no test to determine whether the data are paired. You must determine that from understanding how they were collected and what they mean (check the W's).

Once we recognize that the speed-skating data are matched pairs, it makes sense to consider the difference in times for each two-skater race. So we look at the *pairwise* differences:

<table>
<tr><th>Skating Pair</th><th>Inner Time</th><th>Outer Time</th><th>Inner − Outer</th></tr>
<tr><td>1</td><td>129.24</td><td></td><td>·</td></tr>
<tr><td>2</td><td>125.75</td><td>122.34</td><td>3.41</td></tr>
<tr><td>3</td><td>121.63</td><td>122.12</td><td>−0.49</td></tr>
<tr><td>4</td><td>122.24</td><td>123.35</td><td>−1.11</td></tr>
<tr><td>5</td><td>120.85</td><td>120.45</td><td>0.40</td></tr>
<tr><td>6</td><td>122.19</td><td>123.07</td><td>−0.88</td></tr>
<tr><td>7</td><td>122.15</td><td>122.75</td><td>−0.60</td></tr>
<tr><td>8</td><td>122.16</td><td>121.22</td><td>0.94</td></tr>
<tr><td>9</td><td>121.85</td><td>119.96</td><td>1.89</td></tr>
<tr><td>10</td><td>121.17</td><td>121.03</td><td>0.14</td></tr>
<tr><td>11</td><td>124.77</td><td>118.87</td><td>5.90</td></tr>
<tr><td>12</td><td>118.76</td><td>121.85</td><td>−3.09</td></tr>
<tr><td>13</td><td>119.74</td><td>120.13</td><td>−0.39</td></tr>
<tr><td>14</td><td>121.60</td><td>120.15</td><td>1.45</td></tr>
<tr><td>15</td><td>119.33</td><td>116.74</td><td>2.59</td></tr>
<tr><td>16</td><td>119.30</td><td>119.15</td><td>0.15</td></tr>
<tr><td>17</td><td>117.31</td><td>115.27</td><td>2.04</td></tr>
<tr><td>18</td><td>116.90</td><td>120.77</td><td>−3.87</td></tr>
</table>

The first skater raced alone, so we'll omit that race. Because it is the *differences* we care about, we'll treat them as if *they* were the data, ignoring the original two columns. Now that we have only one column of values to consider, we can use a simple one-sample *t*-test. Mechanically, a **paired *t*-test** is just a one-sample *t*-test for the means of these pairwise differences. The sample size is the number of pairs.

So you've already seen the *Show.*

Assumptions and Conditions

Paired Data Assumption

Paired Data Assumption: The data must be paired. You can't just decide to pair data when in fact the samples are independent. When you have two groups with the same number of observations, it may be tempting to match them up.

Don't, unless you are prepared to justify your claim that the data are paired.

On the other hand, be sure to recognize paired data when you have them. Remember, two-sample *t* methods aren't valid without independent groups, and paired groups aren't independent. Although this is a strictly required assumption, it is one that can be easy to check if you understand how the data were collected.

Independence Assumption

Independence Assumption: If the data are paired, the *groups* are not independent. For these methods, it's the *differences* that must be independent of each other. There's no reason to believe that the difference in speeds of one pair of races could affect the difference in speeds for another pair.

A S

Activity: Differences in Means of Paired Groups. Are married couples typically the same age, or do wives tend to be younger than their husbands, on average?

Randomization Condition: Randomness can arise in many ways. The pairs may be a random sample. In an experiment, the order of the two treatments may be randomly assigned, or the treatments may be randomly assigned to one member of each pair. In a before-and-after study, we may believe that the observed differences are a representative sample from a population of interest. If we have any doubts, we'll need to include a control group to be able to draw conclusions. What we want to know usually focuses our attention on where the randomness should be.

In our example, skaters were assigned to the lanes at random.

10% Condition: We're thinking of the speed-skating data as an experiment testing the difference between lanes. The 10% Condition doesn't apply to randomized experiments, where no sampling takes place.

Normal Population Assumption

We need to assume that the population of *differences* follows a Normal model. We don't need to check the individual groups.

Nearly Normal Condition: This condition can be checked with a histogram or Normal probability plot of the *differences*—but not of the individual groups. As with the one-sample *t*-methods, this assumption matters less the more pairs we have to consider. You may be pleasantly surprised when you check this condition. Even if your original measurements are skewed or bimodal, the *differences* may be nearly Normal. After all, the individual who was way out in the tail on an initial measurement is likely to still be out there on the second one, giving a perfectly ordinary difference.

FOR EXAMPLE   Checking Assumptions and Conditions

RECAP: Field workers for a health department compared driving mileage on a five-day work schedule with mileage on a new four-day schedule. To see if the new schedule changed the amount of driving they did, we'll look at paired differences in mileages before and after.

QUESTION: Is it okay to use these data to test whether the new schedule changed the amount of driving?

✔ **Paired Data Assumption:** The data are paired because each value is the mileage driven by the same person before and after a change in work schedule.

✔ **Independence Assumption:** The driving behavior of any individual worker is independent of the others, so the differences are mutually independent.

✔ **Randomization Condition:** The mileages are the sums of many individual trips, each of which experienced random events that arose while driving. Repeating the experiment in two new years would give randomly different values.

✔ **Nearly Normal Condition:** The histogram of the mileage differences is unimodal and symmetric:

Since the assumptions and conditions are satisfied, it's okay to use paired-*t* methods for these data.

Name	5-Day mileage	4-Day mileage	Difference
Jeff	2798	2914	−116
Betty	7724	6112	1612
Roger	7505	6177	1328
Tom	838	1102	−264
Aimee	4592	3281	1311
Greg	8107	4997	3110
Larry G.	1228	1695	−467
Tad	8718	6606	2112
Larry M.	1097	1063	34
Leslie	8089	6392	1697
Lee	3807	3362	445

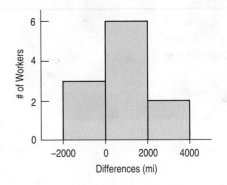

The steps in testing a hypothesis for paired differences are very much like the steps for a one-sample t-test for a mean.

THE PAIRED t-TEST

When the conditions are met, we are ready to test whether the mean of paired differences is significantly different from zero. We test the hypothesis

$$H_0: \mu_d = \Delta_0,$$

where the d's are the pairwise differences and Δ_0 is almost always 0.

We use the statistic

$$t_{n-1} = \frac{\bar{d} - \Delta_0}{SE((\bar{d})}'$$

where $\bar{d}$ is the mean of the pairwise differences, n is the number of *pairs*, and

$$SE(\bar{d}) = \frac{s_d}{\sqrt{n}}.$$

$SE(\bar{d})$ is the ordinary standard error for the mean, applied to the differences.

When the conditions are met and the null hypothesis is true, we can model the sampling distribution of this statistic with a Student's t-model with $n - 1$ degrees of freedom, and use that model to obtain a P-value.

STEP-BY-STEP EXAMPLE   A Paired t-Test

Question: Was there a difference in speeds between the inner and outer speed-skating lanes at the 2006 Winter Olympics?

THINK

Plan State what we want to know.

Identify the *parameter* we wish to estimate. Here our parameter is the mean difference in race times.

Identify the variables and check the W's.

Hypotheses State the null and alternative hypotheses.

Although fans suspected one lane was faster, we can't use the data we have to specify the direction of a test. We (and Olympic officials) would be interested in a difference in either direction, so we'd better test a two-sided alternative.

REALITY CHECK The individual differences are all in seconds. We should expect the mean difference to be comparable in magnitude.

I want to know whether there really was a difference in the speeds of the two lanes for speed skating at the 2006 Olympics. I have data for 17 pairs of racers at the women's 1500-m race.

H_0: Neither lane offered an advantage:

$$\mu_d = 0.$$

H_A: The mean difference is different from zero:

$$\mu_d \neq 0.$$

Model Think about the assumptions and check the conditions.

State why you think the data are paired. Simply having the same number of individuals in each group and displaying them in side-by-side columns doesn't make them paired.

Think about what we hope to learn and where the randomization comes from. Here, the randomization comes from the racer pairings and lane assignments.

Make a picture—just one. Don't plot separate distributions of the two groups—that entirely misses the pairing. For paired data, it's the Normality of the *differences* that we care about. Treat those paired differences as you would a single variable, and check the Nearly Normal Condition with a histogram or a Normal probability plot.

Specify the sampling distribution model.

Choose the method.

✔ **Independence Assumption:** Each race is independent of the others, so the differences are mutually independent.

✔ **Paired Data Assumption:** The data are paired because racers compete in pairs.

✔ **Randomization Condition:** Skaters are assigned to lanes at random. Repeating the experiment with different pairings and lane assignments would give randomly different values.

✔ **Nearly Normal Condition:** The histogram of the differences is unimodal and symmetric:

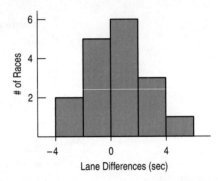

The conditions are met, so I'll use a Student's t-model with $(n - 1) = 16$ degrees of freedom, and perform a **paired t-test.**

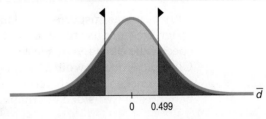 SHOW | **Mechanics**

n is the number of *pairs*—in this case, the number of races.

$\bar{d}$ is the mean difference.

s_d is the standard deviation of the differences.

Find the standard error and the *t*-score of the observed mean difference. There is nothing new in the mechanics of the paired-*t* methods. These are the mechanics of the *t*-test for a mean applied to the differences.

Make a picture. Sketch a *t*-model centered at the hypothesized mean of 0. Because this is a two-tail test, shade both the region to the right of the observed mean difference of 0.499 seconds and the corresponding region in the lower tail.

Find the P-value, using technology.

The data give

$$n = 17 \text{ pairs}$$
$$\bar{d} = 0.499 \text{ seconds}$$
$$s_d = 2.333 \text{ seconds.}$$

I estimate the standard deviation of $\bar{d}$ using

$$SE(\bar{d}) = \frac{s_d}{\sqrt{n}} = \frac{2.333}{\sqrt{17}} = 0.5658$$

$$\text{So } t_{16} = \frac{\bar{d} - 0}{SE(\bar{d})} = \frac{0.499}{0.5658} = 0.882$$

P-value $= 2P(t_{16} > 0.882) = 0.39$

REALITY CHECK The mean difference is 0.499 seconds. That may not seem like much, but a smaller difference determined the Silver and Bronze medals. The standard error is about this big, so a *t*-value less than 1.0 isn't surprising. Nor is a large P-value.

TELL **Conclusion** Link the P-value to your decision about H_0, and state your conclusion in context.

The P-value is large. Events that happen more than a third of the time are not remarkable. So, even though there is an observed difference between the lanes, I can't conclude that it isn't due simply to random chance. It appears the fans may have interpreted a random fluctuation in the data as favoring one lane. There's insufficient evidence to declare any lack of fairness.

FOR EXAMPLE   Doing a Paired *t*-Test

RECAP: We want to test whether a change from a five-day workweek to a four-day workweek could change the amount driven by field workers of a health department. We've already confirmed that the assumptions and conditions for a paired *t*-test are met.

QUESTION: Is there evidence that a four-day workweek would change how many miles workers drive?

H_0: The change in the health department workers' schedules didn't change the mean mileage driven; the mean difference is zero:

$$\mu_d = 0.$$

H_A: The mean difference is different from zero:

$$\mu_d \neq 0.$$

The conditions are met, so I'll use a Student's t-model with $(n - 1) = 10$ degrees of freedom and perform a **paired t-test.**

The data give

$$n = 11 \text{ pairs}$$
$$\overline{d} = 982 \text{ miles}$$
$$s_d = 1139.6 \text{ miles}.$$

$$SE(\overline{d}) = \frac{s_d}{\sqrt{n}} = \frac{1139.6}{\sqrt{11}} = 343.6$$

So $t_{10} = \dfrac{\overline{d} - 0}{SE(\overline{d})} = \dfrac{982.0}{343.6} = 2.86$

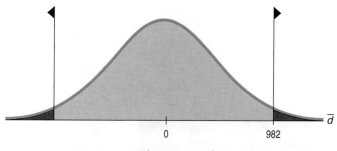

P-value $= 2P(t_{10} > 2.86) = 0.017$

The P-value is small, so I reject the null hypothesis and conclude that the change in workweek did lead to a change in average driving mileage. It appears that changing the work schedule may reduce the mileage driven by workers.

Note: We should propose a course of action, but it's hard to tell from the hypothesis test whether the reduction matters. Is the difference in mileage important in the sense of reducing air pollution or costs, or is it merely statistically significant? To help make that decision, we should look at a confidence interval. If the difference in mileage proves to be large in a practical sense, then we might recommend a change in schedule for the rest of the department.

Confidence Intervals for Matched Pairs

In developed countries, the average age of women is generally higher than that of men. After all, women tend to live longer. But if we look at *married couples*, husbands tend to be slightly older than wives. How much older, on average, are husbands? We have data from a random sample of 200 British couples, the first 7 of which are shown below. Only 170 couples provided ages for both husband and wife, so we can work only with that many pairs. Let's form a confidence interval for the mean difference of husband's and wife's ages for these 170 couples. Here are the first 7 pairs:

Wife's Age	Husband's Age	Difference (husband − wife)
43	49	6
28	25	−3
30	40	10
57	52	−5
52	58	6
27	32	5
52	43	−9
⋮	⋮	⋮

WHO	170 randomly sampled couples
WHAT	Ages
UNITS	Years
WHEN	Recently
WHERE	Britain

Clearly, these data are paired. The survey selected *couples* at random, not individuals. We're interested in the mean age difference within couples. How would we construct a confidence interval for the true mean difference in ages?

PAIRED *t*-INTERVAL

When the conditions are met, we are ready to find the confidence interval for the mean of the paired differences. The confidence interval is

$$\bar{d} \pm t^*_{n-1} \times SE(\bar{d}),$$

where the standard error of the mean difference is $SE(\bar{d}) = \dfrac{s_d}{\sqrt{n}}$.

The critical value t^* from the Student's t-model depends on the particular confidence level, C, that you specify and on the degrees of freedom, $n - 1$, which is based on the number of pairs, n.

Making confidence intervals for matched pairs follows exactly the steps for a one-sample t-interval.

STEP-BY-STEP EXAMPLE   A Paired *t*-Interval

Question: How big a difference is there, on average, between the ages of husbands and wives?

 THINK

Plan State what we want to know.

Identify the variables and check the W's.

I want to estimate the mean difference in age between husbands and wives. I have a random sample of 200 British couples, 170 of whom provided both ages.

Identify the parameter you wish to estimate. For a paired analysis, the parameter of interest is the mean of the differences. The population of interest is the population of differences.

Model Think about the assumptions and check the conditions.

✔ **Paired Data Assumption:** The data are paired because they are on members of married couples.

✔ **Independence Assumption:** The data are from a randomized survey, so couples should be independent of each other.

✔ **Randomization Condition:** These couples were randomly sampled.

✔ **10% Condition:** The sample is less than 10% of the population of married couples in Britain.

Make a picture. We focus on the differences, so a histogram or Normal probability plot is best here.

✔ **Nearly Normal Condition:** The histogram of the husband − wife differences is unimodal and symmetric:

REALITY CHECK ▶ The histogram shows husbands are often older than wives (because most of the differences are greater than 0). The mean difference seen here of about 2 years is reasonable.

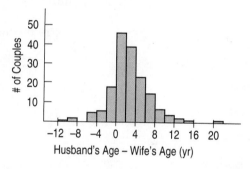

Husband's Age − Wife's Age (yr)

State the sampling distribution model.

Choose your method.

The conditions are met, so I can use a Student's t-model with $(n - 1) = 169$ degrees of freedom and find a **paired t-interval.**

SHOW ▶ **Mechanics**

n is the number of *pairs*, here, the number of couples.

$\bar{d}$ is the mean difference.

s_d is the standard deviation of the differences.

Be sure to include the units along with the statistics.

$n = 170$ couples
$\bar{d} = 2.2$ years
$s_d = 4.1$ years

I estimate the standard error of $\bar{d}$ as

$$SE(\bar{d}) = \frac{s_d}{\sqrt{n}} = \frac{4.1}{\sqrt{170}} = 0.31 \text{ years.}$$

The df for the t-model is $n - 1 = 169$.

The critical value we need to make a 95% interval comes from a Student's t table, a computer program, or a calculator.

The 95% critical value for t_{169} (from the table) is 1.97.

The margin of error is

$$ME = t^*_{169} \times SE(\bar{d}) = 1.97(0.31) = 0.61$$

REALITY CHECK	This result makes sense. Our everyday experience confirms that an average age difference of about 2 years is reasonable.	So the 95% confidence interval is 2.2 ± 0.6 years, or an interval of (1.6, 2.8) years.
TELL	**Conclusion** Interpret the confidence interval in context.	I am 95% confident that British husbands are, on average, 1.6 to 2.8 years older than their wives.

Effect Size

When we examined the speed-skating times, we failed to reject the null hypothesis, so we couldn't be certain whether there really was a difference between the lanes. Maybe there wasn't any difference, or maybe whatever difference there might have been was just too small to matter at all. Were the fans right to be concerned?

We can't tell from the hypothesis test, but using the same summary statistics, we can find that the corresponding 95% confidence interval for the mean difference is $(-0.70 < \mu_d < 1.70)$ seconds.

A confidence interval is a good way to get a sense for the size of the effect we're trying to understand. That gives us a plausible range of values for the true mean difference in lane times. If differences of 1.7 seconds were too small to matter in 1500-m Olympic speed skating, we'd be pretty sure there was no need for concern.

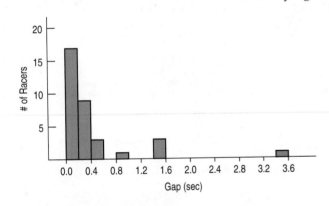

But in fact, except for the Gold − Silver gap, the successive gaps between each skater and the next-faster one were *all* less than the high end of this interval, and most were right around the middle of the interval.

So even though we were unable to discern a real difference, the confidence interval shows that the effects we're considering may be big enough to be important. We may want to continue this investigation by checking out other races on this ice and being alert for possible differences at other venues.

FOR EXAMPLE   Looking at Effect Size with a Paired-*t* Confidence Interval

RECAP: We know that, on average, the switch from a five-day workweek to a four-day workweek reduced the amount driven by field workers in that Illinois health department. However, finding that there is a significant difference doesn't necessarily mean that difference is meaningful or worthwhile. To assess the size of the effect, we need a confidence interval. We already know the assumptions and conditions are met.

QUESTION: By how much, on average, might a change in workweek schedule reduce the amount driven by workers?

$$\bar{d} = 982 \text{ mi} \quad SE(\bar{d}) = 343.6 \quad t^*_{10} = 2.228 \text{ (for 95%)}$$
$$ME = t^*_{10} \times SE(\bar{d}) = 2.228(343.6) = 765.54$$

So the 95% confidence interval for μ_d is 982 ± 765.54 or (216.46, 1747.54) fewer miles.

With 95% confidence, I estimate that by switching to a four-day workweek employees would drive an average of between 216 and 1748 fewer miles per year. With high gas prices, this could save a lot of money.

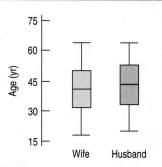

FIGURE 25.2
This display is worthless. It does no good to compare all the wives as a group with all the husbands. We care about the paired differences.

Blocking

Because the sample of British husbands and wives includes both older and younger couples, there's a lot of variation in the ages of the men and in the ages of the women. In fact, that variation is so great that a boxplot of the two groups would show little difference. But that would be the wrong plot. It's the *difference* we care about. Pairing isolates the extra variation and allows us to focus on the individual differences. In Chapter 13 we saw how we could design an experiment with blocking to isolate the variability between identifiable groups of subjects, allowing us to better see variability among treatment groups due to their response to the treatment. A paired design is an example of blocking.

When we pair, we have roughly half the degrees of freedom of a two-sample test. You may see discussions that suggest that in "choosing" a paired analysis we "give up" these degrees of freedom. This isn't really true, though. If the data are paired, then there never were additional degrees of freedom, and we have no "choice." The fact of the pairing determines how many degrees of freedom are available.

Matching pairs generally removes so much extra variation that it more than compensates for having only half the degrees of freedom. Of course, inappropriate matching when the groups are in fact independent (say, by matching on the first letter of the last name of subjects) would cost degrees of freedom without the benefit of reducing the variance. When you design a study or experiment, you should consider using a paired design if possible.

Think about each of the situations described below.

▸ Would you use a two-sample *t* or paired-*t* method (or neither)? Why?

▸ Would you perform a hypothesis test or find a confidence interval?

1. Random samples of 50 men and 50 women are asked to imagine buying a birthday present for their best friend. We want to estimate the difference in how much they are willing to spend.

2. Mothers of twins were surveyed and asked how often in the past month strangers had asked whether the twins were identical.

3. Are parents equally strict with boys and girls? In a random sample of families, researchers asked a brother and sister from each family to rate how strict their parents were.

4. Forty-eight overweight subjects are randomly assigned to either aerobic or stretching exercise programs. They are weighed at the beginning and at the end of the experiment to see how much weight they lost.

    a) We want to estimate the mean amount of weight lost by those doing aerobic exercise.
    b) We want to know which program is more effective at reducing weight.

5. Couples at a dance club were separated and each person was asked to rate the band. Do men or women like this band more?

*The Sign Test Again

Because we have paired data, we've been using a simple *t*-test for the paired differences. This suggests that if we want a distribution-free method, it would be natural to compute a sign test on the paired differences and test whether the *median* of the differences is 0. That's exactly what we do. The test is very simple. We record a 0 for every paired difference that's negative and a 1 for each positive difference, ignoring pairs for which the difference is exactly 0. We test the associated proportion $p = 0.5$ using a *z*-test if the number of pairs is at least

20 (so that we expect at least 10 successes and failures), or compute the exact Binomial probabilities if $n < 20$ (as discussed in Chapter 17).

Let's try it on the married couples data. Of the 170 couples, there are 119 with the husband older, 32 with the wife older, and 19 that have the same age. So we test $p = 0.5$, with a sample proportion of $119/151 = 0.788$. The Success/Failure Condition is easily satisfied. Applying the one-proportion z-test to the differences gives a z-score of 7.08, with a P-value < 0.00001. We can be pretty confident that the median is not 0. (Just for comparison, the t-statistic for testing $\mu_d = 0$ is 7.152 with 169 df—almost the identical result.)

As with other distribution-free tests, the advantage of the ***sign test for matched pairs*** is that we don't require the Nearly Normal Condition for the paired differences. Because it looks only at the direction of the difference, the sign test isn't affected by outliers—extraordinarily large differences—which can be an advantage. On the other hand, when the assumptions of the *paired t-test* are met, the paired t-test is more powerful than the sign test.

What Can Go Wrong?

- **Don't use a two-sample *t*-test when you have paired data.** See the What Can Go Wrong? discussion in Chapter 24.

- **Don't use a paired-*t* method when the samples aren't paired.** Just because two groups have the same number of observations doesn't mean they can be paired, even if they are shown side by side in a table. We might have 25 men and 25 women in our study, but they might be completely independent of one another. If they were siblings or spouses, we might consider them paired. Remember that you cannot *choose* which method to use based on your preferences. If the data are from two independent samples, use two-sample t methods. If the data are from an experiment in which observations were paired, you must use a paired method. If the data are from an observational study, you must be able to defend your decision to use matched pairs or independent groups.

- **Don't forget outliers.** The outliers we care about now are in the differences. A subject who is extraordinary both before and after a treatment may still have a perfectly typical difference. But one outlying difference can completely distort your conclusions. Be sure to plot the differences (even if you also plot the data).

- **Don't look for the difference between the means of paired groups with side-by-side boxplots.** The point of the paired analysis is to remove extra variation. The boxplots of each group still contain that variation. Comparing them is likely to be misleading.

CONNECTIONS

The most important connection is to the concept of blocking that we first discussed when we considered designed experiments in Chapter 13. Pairing is a basic and very effective form of blocking.

Of course, the details of the mechanics for paired t-tests and intervals are identical to those for the one-sample t-methods. Everything we know about those methods applies here.

The connection to the two-sample and pooled methods of the previous chapter is that when the data are naturally paired, those methods are not appropriate because paired data fail the required condition of independence.

WHAT HAVE WE LEARNED?

When we looked at various ways to design experiments, back in Chapter 13, we saw that pairing can be a very effective strategy. Because pairing can help control variability between individual subjects, paired methods are usually more powerful than methods that compare independent groups. Now we've learned that analyzing data from matched pairs requires different inference procedures.

▶ We've learned that paired *t*-methods look at pairwise differences. Based on these differences, we test hypotheses and generate confidence intervals. These procedures are mechanically identical to the one-sample *t*-methods we saw in Chapter 23.

▶ We've also learned to *Think* about the design of the study that collected the data before we proceed with inference. We must be careful to recognize pairing when it is present but not assume it when it is not. Making the correct decision about whether to use independent *t*-procedures or paired *t*-methods is the first critical step in analyzing the data.

Terms

Paired data

Data are paired when the observations are collected in pairs or the observations in one group are naturally related to observations in the other. The simplest form of pairing is to measure each subject twice—often before and after a treatment is applied. More sophisticated forms of pairing in experiments are a form of blocking and arise in other contexts. Pairing in observational and survey data is a form of matching (p. 612).

Paired *t*-test

A hypothesis test for the mean of the pairwise differences of two groups. It tests the null hypothesis

$$H_0: \mu_d = \Delta_0,$$

where the hypothesized difference is almost always 0, using the statistic

$$t = \frac{\bar{d} - \Delta_0}{SE(\bar{d})}$$

with $n - 1$ degrees of freedom, where $SE(\bar{d}) = \frac{s_d}{\sqrt{n}}$, and n is the number of pairs (p. 615).

Paired-*t* confidence interval

A confidence interval for the mean of the pairwise differences between paired groups found as

$$\bar{d} \pm t^*_{n-1} \times SE(\bar{d}), \text{ where } SE(\bar{d}) = \frac{s_d}{\sqrt{n}} \text{ and } n \text{ is the number of pairs (p. 618).}$$

*Sign Test

One application of the sign test is to test for differences between paired groups. We count the signs of pairwise differences to test whether they are equally likely to be positive or negative (p. 621).

Skills

▶ Be able to recognize whether a design that compares two groups is paired.

▶ Be able to find a paired confidence interval, recognizing that it is mechanically equivalent to doing a one-sample *t*-interval applied to the differences.

▶ Be able to perform a paired *t*-test, recognizing that it is mechanically equivalent to a one-sample *t*-test applied to the differences.

▶ Be able to interpret a paired *t*-test, recognizing that the hypothesis tested is about the mean of the differences between paired values rather than about the differences between the means of two independent groups.

▶ Be able to interpret a paired *t*-interval, recognizing that it gives an interval for the mean difference in the pairs.

PAIRED *t* ON THE COMPUTER

Most statistics programs can compute paired-*t* analyses. Some may want you to find the differences yourself and use the one-sample *t* methods. Those that perform the entire procedure will need to know the two variables to compare. The computer, of course, cannot verify that the variables are naturally paired. Most programs will check whether the two variables have the same number of observations, but some stop there, and that can cause trouble. Most programs will automatically omit any pair that is missing a value for either variable (as we did with the British couples). You must look carefully to see whether that has happened.

As we've seen with other inference results, some packages pack a lot of information into a simple table, but you must locate what you want for yourself. Here's a generic example with comments:

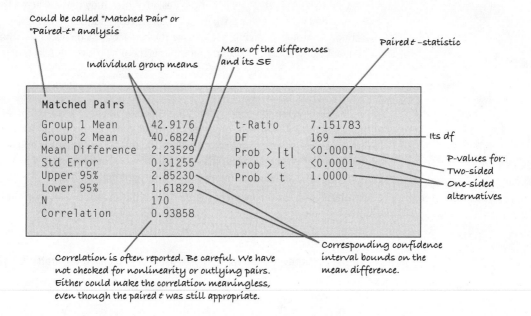

Other packages try to be more descriptive. It may be easier to find the results, but you may get less information from the output table.

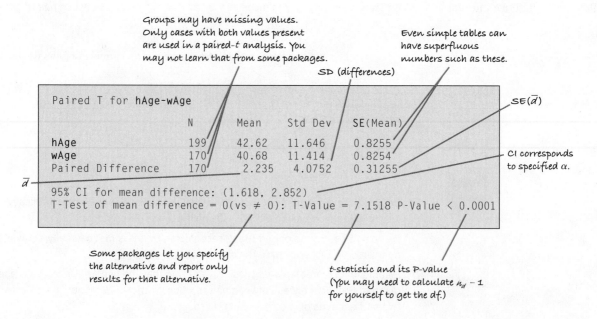

Computers make it easy to examine the boxplots of the two groups and the histogram of the differences—both important steps. Some programs offer a scatterplot of the two variables. That can be helpful. In terms of the scatterplot, a paired *t*-test is about whether the points tend to be above or below the line y = x. (Note, though, that pairing says nothing about whether the scatterplot should be straight. That doesn't matter for our *t*-methods.)

DATA DESK

Select variables.
From the **Calc** menu, choose **Estimate** for confidence intervals or **Test** for hypothesis tests. Select the interval or test from the drop-down menu, and make other choices in the dialog.

COMMENTS

Data Desk expects the two groups to be in separate variables and in the same "Relation"–that is, about the same cases.

EXCEL

In Excel 2003 and earlier, select **Data Analysis** from the **Tools** menu.
In Excel 2007, select **Data Analysis** from the **Analysis Group** on the **Data** Tab.
From the **Data Analysis** menu, choose **t-test: paired two-sample for Means.** Fill in the cell ranges for the two groups, the hypothesized difference, and the alpha level.

COMMENTS

Excel expects the two groups to be in separate cell ranges.
Warning: Do not compute this test in Excel without checking for missing values. If there are any missing values (empty cells), Excel will usually give a wrong answer. Excel compacts each list, pushing values up to cover the missing cells, and then checks only that it has the same number of values in each list. The result is mismatched pairs and an entirely wrong analysis.

JMP

From the **Analyze** menu, select **Matched Pairs.** Specify the columns holding the two groups in the **Y Paired Response** dialog. Click **OK.**

MINITAB

From the **Stat** menu, choose the **Basic Statistics** submenu. From that menu, choose **Paired t. . .** Then fill in the dialog.

COMMENTS

Minitab takes "First sample" minus "Second sample."

SPSS

From the **Analyze** menu, choose the **Compare Means** submenu. From that, choose the **Paired-Samples t-test** command. Select pairs of variables to compare, and click the arrow to add them to the selection box.

COMMENTS

You can compare several pairs of variables at once. Options include the choice to exclude cases missing in any pair from all tests.

*TI-83/84 PLUS

If the data are stored in two lists, say, L1 and L2, create a list of the differences:
L1 − L2 → L3. (The arrow is the STO button.) Since inference for paired differences uses one-sample *t*-procedures, select **2:T-Test** or **8:TInterval** from the **STAT TESTS** menu. Specify as the data the list of differences you just created in L3, and apply the procedure.

TI-89

If the data are stored in two lists, say, list1 and list2, create a list of the differences: Move the cursor to the name of an empty list, and then use VAR-LINK to enter the command list1-list2. Press ⏎ENTER⏎ to perform the subtraction.

Since inference for paired differences uses one-sample *t*-procedures, select **2:T-Test** or **2:TInterval** from the **STAT TESTS** or **Ints** menu. Specify as your data the list of differences you just created, and apply the procedure.

EXERCISES

1. **More eggs?** Can a food additive increase egg production? Agricultural researchers want to design an experiment to find out. They have 100 hens available. They have two kinds of feed: the regular feed and the new feed with the additive. They plan to run their experiment for a month, recording the number of eggs each hen produces.
   a) Design an experiment that will require a two-sample t procedure to analyze the results.
   b) Design an experiment that will require a matched-pairs t procedure to analyze the results.
   c) Which experiment would you consider the stronger design? Why?

2. **MTV.** Some students do homework with the TV on. (Anyone come to mind?) Some researchers want to see if people can work as effectively with as without distraction. The researchers will time some volunteers to see how long it takes them to complete some relatively easy crossword puzzles. During some of the trials, the room will be quiet; during other trials in the same room, a TV will be on, tuned to MTV.
   a) Design an experiment that will require a two-sample t procedure to analyze the results.
   b) Design an experiment that will require a matched-pairs t procedure to analyze the results.
   c) Which experiment would you consider the stronger design? Why?

3. **Sex sells?** Ads for many products use sexual images to try to attract attention to the product. But do these ads bring people's attention to the item that was being advertised? We want to design an experiment to see if the presence of sexual images in an advertisement affects people's ability to remember the product.
   a) Describe an experimental design requiring a matched-pairs t procedure to analyze the results.
   b) Describe an experimental design requiring an independent sample procedure to analyze the results.

4. **Freshman 15?** Many people believe that students gain weight as freshmen. Suppose we plan to conduct a study to see if this is true.
   a) Describe a study design that would require a matched-pairs t procedure to analyze the results.
   b) Describe a study design that would require a two-sample t procedure to analyze the results.

5. **Women.** Values for the labor force participation rate of women (LFPR) are published by the U.S. Bureau of Labor Statistics. We are interested in whether there was a difference between female participation in 1968 and 1972, a time of rapid change for women. We check LFPR values for 19 randomly selected cities for 1968 and 1972. Shown below is software output for two possible tests:

Paired t-Test of $\mu(1 - 2)$
Test Ho: $\mu(1972-1968) = 0$ vs Ha: $\mu(1972-1968) \neq 0$
Mean of Paired Differences = 0.0337
t-Statistic = 2.458 w/18 df
p = 0.0244

2-Sample t-Test of $\mu1 - \mu2$
Ho: $\mu1 - \mu2 = 0$ Ha: $\mu1 - \mu2 \neq 0$
Test Ho: $\mu(1972) - \mu(1968) = 0$ vs
Ha: $\mu(1972) - \mu(1968) \neq 0$
Difference Between Means = 0.0337
t-Statistic = 1.496 w/35 df
p = 0.1434

   a) Which of these tests is appropriate for these data? Explain.
   b) Using the test you selected, state your conclusion.

6. **Could seeding.** Simpson, Alsen, and Eden (*Technometrics* 1975) report the results of trials in which clouds were seeded and the amount of rainfall recorded. The authors report on 26 seeded and 26 unseeded clouds in order of the amount of rainfall, largest amount first. Here are two possible tests to study the question of whether cloud seeding works. Which test is appropriate for these data? Explain your choice. Using the test you select, state your conclusion.

Paired t-Test of $\mu(1 - 2)$
Mean of Paired Differences = −277.39615
t-Statistic = −3.641 w/25 df
p = 0.0012

2-Sample t-Test of $\mu1 - \mu2$
Difference Between Means = −277.4
t-Statistic = −1.998 w/33 df
p = 0.0538

   a) Which of these tests is appropriate for these data? Explain.
   b) Using the test you selected, state your conclusion.

7. **Friday the 13th, traffic.** In 1993 the *British Medical Journal* published an article titled, "Is Friday the 13th Bad for Your Health?" Researchers in Britain examined how Friday the 13th affects human behavior. One question was whether people tend to stay at home more on Friday the 13th. The data below are the number of cars passing Junctions 9 and 10 on the M25 motorway for consecutive Fridays (the 6th and 13th) for five different periods.

Year	Month	6th	13th
1990	July	134,012	132,908
1991	September	133,732	131,843
1991	December	121,139	118,723
1992	March	124,631	120,249
1992	November	117,584	117,263

Here are summaries of two possible analyses:

Paired t-Test of mu(1 − 2) = 0 vs. mu(1 − 2) > 0
Mean of Paired Differences: 2022.4
t-Statistic = 2.9377 w/4 df
P = 0.0212

2-Sample t-Test of mu1 = mu2 vs. mu1 > mu2
Difference Between Means: 2022.4
t-Statistic = 0.4273 w/7.998 df
P = 0.3402

a) Which of the tests is appropriate for these data? Explain.
b) Using the test you selected, state your conclusion.
c) Are the assumptions and conditions for inference met?

T 8. **Friday the 13th, accidents.** The researchers in Exercise 7 also examined the number of people admitted to emergency rooms for vehicular accidents on 12 Friday evenings (6 each on the 6th and 13th).

Year	Month	6th	13th
1989	October	9	13
1990	July	6	12
1991	September	11	14
1991	December	11	10
1992	March	3	4
1992	November	5	12

Based on these data, is there evidence that more people are admitted, on average, on Friday the 13th? Here are two possible analyses of the data:

Paired t-Test of mu(1 − 2) = 0 vs. mu(1 − 2) < 0
Mean of Paired Differences = 3.333
t-Statistic = 2.7116 w/5 df
P = 0.0211

2-Sample t-Test of mu1 = mu2 vs. mu1 < mu2
Difference Between Means = 3.333
t-Statistic = 1.6644 w/9.940 df
P = 0.0636

a) Which of these tests is appropriate for these data? Explain.
b) Using the test you selected, state your conclusion.
c) Are the assumptions and conditions for inference met?

T 9. **Online insurance I.** After seeing countless commercials claiming one can get cheaper car insurance from an online company, a local insurance agent was concerned that he might lose some customers. To investigate, he randomly selected profiles (type of car, coverage, driving record, etc.) for 10 of his clients and checked online price quotes for their policies. The comparisons are shown in the table below. His statistical software produced the following summaries (where *PriceDiff* = *Local* − *Online*):

Variable	Count	Mean	StdDev
Local	10	799.200	229.281
Online	10	753.300	256.267
PriceDiff	10	45.9000	175.663

Local	Online	PriceDiff
568	391	177
872	602	270
451	488	−37
1229	903	326
605	677	−72
1021	1270	−249
783	703	80
844	789	55
907	1008	−101
712	702	10

At first, the insurance agent wondered whether there was some kind of mistake in this output. He thought the Pythagorean Theorem of Statistics should work for finding the standard deviation of the price differences—in other words, that $SD(Local - Online) = \sqrt{SD^2(Local) + SD^2(Online)}$. But when he checked, he found that $\sqrt{(229.281)^2 + (256.267)^2} = 343.864$, not 175.663 as given by the software. Tell him where his mistake is.

T 10. **Wind speed, part I.** To select the site for an electricity-generating wind turbine, wind speeds were recorded at several potential sites every 6 hours for a year. Two sites not far from each other looked good. Each had a mean wind speed high enough to qualify, but we should choose the site with a higher average daily wind speed. Because the sites are near each other and the wind speeds were recorded at the same times, we should view the speeds as paired. Here are the summaries of the speeds (in miles per hour):

Variable	Count	Mean	StdDev
site2	1114	7.452	3.586
site4	1114	7.248	3.421
site2 − site4	1114	0.204	2.551

Is there a mistake in this output? Why doesn't the Pythagorean Theorem of Statistics work here? In other words, shouldn't

$$SD(site2 - site4) = \sqrt{SD^2(site2) + SD^2(site4)}?$$

But $\sqrt{(3.586)^2 + (3.421)^2} = 4.956$, not 2.551 as given by the software. Explain why this happened.

T 11. **Online insurance II.** In Exercise 9, we saw summary statistics for 10 drivers' car insurance premiums quoted by a local agent and an online company. Here are displays for each company's quotes and for the difference (*Local* − *Online*):

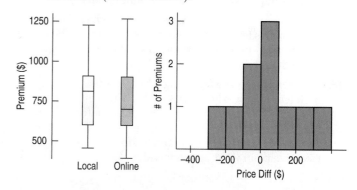

a) Which of the summaries would help you decide whether the online company offers cheaper insurance? Why?

b) The standard deviation of *PriceDiff* is quite a bit smaller than the standard deviation of prices quoted by either the local or online companies. Discuss why.

c) Using the information you have, discuss the assumptions and conditions for inference with these data.

12. Wind speed, part II. In Exercise 10, we saw summary statistics for wind speeds at two sites near each other, both being considered as locations for an electricity-generating wind turbine. The data, recorded every 6 hours for a year, showed each of the sites had a mean wind speed high enough to qualify, but how can we tell which site is best? Here are some displays:

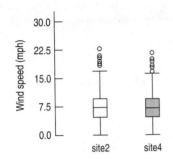

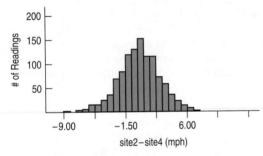

a) The boxplots show outliers for each site, yet the histogram shows none. Discuss why.

b) Which of the summaries would you use to select between these sites? Why?

c) Using the information you have, discuss the assumptions and conditions for paired *t* inference for these data. (*Hint:* Think hard about the independence assumption in particular.)

13. Online insurance III. Exercises 9 and 11 give summaries and displays for car insurance premiums quoted by a local agent and an online company. Test an appropriate hypothesis to see if there is evidence that drivers might save money by switching to the online company.

14. Wind speed, part III. Exercises 10 and 12 give summaries and displays for two potential sites for a wind turbine. Test an appropriate hypothesis to see if there is evidence that either of these sites has a higher average wind speed.

15. Temperatures. The following table gives the average high temperatures in January and July for several European cities. Write a 90% confidence interval for the mean temperature difference between summer and winter in Europe. Be sure to check conditions for inference, and clearly explain what your interval means.

City	Mean High Temperatures (°F)	
	Jan.	July
Vienna	34	75
Copenhagen	36	72
Paris	42	76
Berlin	35	74
Athens	54	90
Rome	54	88
Amsterdam	40	69
Madrid	47	87
London	44	73
Edinburgh	43	65
Moscow	21	76
Belgrade	37	84

16. NY Marathon 2008. The table below shows the winning times (in minutes) for men and women in the New York City Marathon between 1978 and 2008. Assuming that performances in the Big Apple resemble performances elsewhere, we can think of these data as a sample of performance in marathon competitions. Create a 90% confidence interval for the mean difference in winning times for male and female marathon competitors. (www.nycmarathon.org)

Year	Men	Women	Year	Men	Women
1978	132.2	152.5	1994	131.4	147.6
1979	131.7	147.6	1995	131.0	148.1
1980	129.7	145.7	1996	129.9	148.3
1981	128.2	145.5	1997	128.2	148.7
1982	129.5	147.2	1998	128.8	145.3
1983	129.0	147.0	1999	129.2	145.1
1984	134.9	149.5	2000	130.2	145.8
1985	131.6	148.6	2001	127.7	144.4
1986	131.1	148.1	2002	128.1	145.9
1987	131.0	150.3	2003	130.5	142.5
1988	128.3	148.1	2004	129.5	143.2
1989	128.0	145.5	2005	129.5	144.7
1990	132.7	150.8	2006	130.0	145.1
1991	129.5	147.5	2007	129.1	143.2
1992	129.5	144.7	2008	128.7	143.9
1993	130.1	146.4			

17. Push-ups. Every year the students at Gossett High School take a physical fitness test during their gym classes. One component of the test asks them to do as many push-ups as they can. Results for one class are shown below, separately for boys and girls. Assuming that students at Gossett are assigned to gym classes at random, create a 90% confidence interval for how many more push-ups boys can do than girls, on average, at that high school.

Boys	17	27	31	17	25	32	28	23	25	16	11	34
Girls	24	7	14	16	2	15	19	25	10	27	31	8

18. Brain waves. An experiment was performed to see whether sensory deprivation over an extended period of time has any effect on the alpha-wave patterns produced by the brain. To determine this, 20 subjects, inmates in a

Canadian prison, were randomly split into two groups. Members of one group were placed in solitary confinement. Those in the other group were allowed to remain in their own cells. Seven days later, alpha-wave frequencies were measured for all subjects, as shown in the following table. (P. Gendreau et al., "Changes in EEG Alpha Frequency and Evoked Response Latency During Solitary Confinement," *Journal of Abnormal Psychology* 79 [1972]: 54–59)

Nonconfined	Confined
10.7	9.6
10.7	10.4
10.4	9.7
10.9	10.3
10.5	9.2
10.3	9.3
9.6	9.9
11.1	9.5
11.2	9.0
10.4	10.9

a) What are the null and alternative hypotheses? Be sure to define all the terms and symbols you use.
b) Are the assumptions necessary for inference met?
c) Perform the appropriate test, indicating the formula you used, the calculated value of the test statistic, the df, and the P-value.
d) State your conclusion.

19. Job satisfaction. (When you first read about this exercise break plan in Chapter 24, you did not have an inference method that would work. Try again now.) A company institutes an exercise break for its workers to see if it will improve job satisfaction, as measured by a questionnaire that assesses workers' satisfaction. Scores for 10 randomly selected workers before and after the implementation of the exercise program are shown in the table below.

a) Identify the procedure you would use to assess the effectiveness of the exercise program, and check to see if the conditions allow the use of that procedure.
b) Test an appropriate hypothesis and state your conclusion.
c) If your conclusion turns out to be incorrect, what kind of error did you commit?
*d) Use a matched-pairs sign test to test the appropriate hypothesis. Do your conclusions change from those in part b?

Worker Number	Job Satisfaction Index	
	Before	After
1	34	33
2	28	36
3	29	50
4	45	41
5	26	37
6	27	41
7	24	39
8	15	21
9	15	20
10	27	37

20. Summer school. (When you first read about the summer school issue in Chapter 24 you did not have an inference method that would work. Try again now.) Having done poorly on their Math final exams in June, six students repeat the course in summer school and take another exam in August.

June	54	49	68	66	62	62
Aug.	50	65	74	64	68	72

a) If we consider these students to be representative of all students who might attend this summer school in other years, do these results provide evidence that the program is worthwhile?
b) This conclusion, of course, may be incorrect. If so, which type of error was made?

21. Yogurt. Is there a significant difference in calories between servings of strawberry and vanilla yogurt? Based on the data shown in the table, test an appropriate hypothesis and state your conclusion. Don't forget to check assumptions and conditions!

Brand	Calories per Serving	
	Strawberry	Vanilla
America's Choice	210	200
Breyer's Lowfat	220	220
Columbo	220	180
Dannon Light 'n Fit	120	120
Dannon Lowfat	210	230
Dannon la Crème	140	140
Great Value	180	80
La Yogurt	170	160
Mountain High	200	170
Stonyfield Farm	100	120
Yoplait Custard	190	190
Yoplait Light	100	100

22. Gasoline. Many drivers of cars that can run on regular gas actually buy premium in the belief that they will get better gas mileage. To test that belief, we use 10 cars from a company fleet in which all the cars run on regular gas. Each car is filled first with either regular or premium gasoline, decided by a coin toss, and the mileage for that tankful is recorded. Then the mileage is recorded again for the same cars for a tankful of the other kind of gasoline. We don't let the drivers know about this experiment.

Here are the results (miles per gallon):

Car #	1	2	3	4	5	6	7	8	9	10
Regular	16	20	21	22	23	22	27	25	27	28
Premium	19	22	24	24	25	25	26	26	28	32

a) Is there evidence that cars get significantly better fuel economy with premium gasoline?
b) How big might that difference be? Check a 90% confidence interval.
c) Even if the difference is significant, why might the company choose to stick with regular gasoline?
d) Suppose you had done a "bad thing." (We're sure you didn't.) Suppose you had mistakenly treated these data as two independent samples instead of matched pairs. What would the significance test have found? Carefully explain why the results are so different.
*e) Use a matched-pairs sign test to test the appropriate hypothesis. Do your conclusions change from those in part a?

T 23. Braking test. A tire manufacturer tested the braking performance of one of its tire models on a test track. The company tried the tires on 10 different cars, recording the stopping distance for each car on both wet and dry pavement. Results are shown in the table.

Car #	Stopping Distance (ft)	
	Dry Pavement	Wet Pavement
1	150	201
2	147	220
3	136	192
4	134	146
5	130	182
6	134	173
7	134	202
8	128	180
9	136	192
10	158	206

a) Write a 95% confidence interval for the mean dry pavement stopping distance. Be sure to check the appropriate assumptions and conditions, and explain what your interval means.
b) Write a 95% confidence interval for the mean increase in stopping distance on wet pavement. Be sure to check the appropriate assumptions and conditions, and explain what your interval means.

T 24. Braking test 2. For another test of the tires in Exercise 23, a car made repeated stops from 60 miles per hour. The test was run on both dry and wet pavement, with results as shown in the table. (Note that actual *braking distance*, which takes into account the driver's reaction time, is much longer, typically nearly 300 feet at 60 mph!)

a) Write a 95% confidence interval for the mean dry pavement stopping distance. Be sure to check the appropriate assumptions and conditions, and explain what your interval means.
b) Write a 95% confidence interval for the mean increase in stopping distance on wet pavement. Be sure to check the appropriate assumptions and conditions, and explain what your interval means.

Stopping Distance (ft)	
Dry Pavement	Wet Pavement
145	211
152	191
141	220
143	207
131	198
148	208
126	206
140	177
135	183
133	223

T 25. Tuition 2006. How much more do public colleges and universities charge out-of-state students for tuition per semester? A random sample of 19 public colleges and universities listed at www.collegeboard.com yielded the following data. Tuition figures per semester are rounded to the nearest hundred dollars.

Institution	Resident	Nonresident
Univ of Akron (OH)	4200	8800
Athens State (AL)	1900	3600
Ball State (IN)	3400	8600
Bloomsburg U (PA)	3200	7000
UC Irvine (CA)	3400	12700
Central State (OH)	2600	5700
Clarion U (PA)	3300	5900
Dakota State	2900	3400
Fairmont State (WV)	2200	4600
Johnson State (VT)	3400	7300
Lock Haven U (PA)	3200	6000
New College of Florida	1600	8300
Oakland U (MI)	3300	7700
U Pittsburgh	6100	10700
Savannah State (GA)	1600	5400
SE Louisiana	1700	4400
W Liberty State (WV)	2000	4800
W Texas College	800	1000
Worcester State (MA)	2800	5800

a) Create a 90% confidence interval for the mean difference in cost. Be sure to justify your procedure.
b) Interpret your interval in context.
c) A national magazine claims that public institutions charge state residents an average of $3500 less than out-of-staters for tuition each semester. What does your confidence interval indicate about this assertion?

T 26. Sex sells, part II. In Exercise 3 you considered the question of whether sexual images in ads affected people's abilities to remember the item being advertised. To investigate, a group of Statistics students cut ads out of magazines. They were careful to find two ads for each of 10 similar items, one with a sexual image and one without. They arranged the ads in random order and had 39 subjects look at them for one minute. Then they asked the subjects to list as many of the products as they could remember. Their data are shown in the table. Is there evidence that the sexual images mattered?

Subject Number	Ads Remembered		Subject Number	Ads Remembered	
	Sexual Image	No Sex		Sexual Image	No Sex
1	2	2	21	2	3
2	6	7	22	4	2
3	3	1	23	3	3
4	6	5	24	5	3
5	1	0	25	4	5
6	3	3	26	2	4
7	3	5	27	2	2
8	7	4	28	2	4
9	3	7	29	7	6
10	5	4	30	6	7
11	1	3	31	4	3
12	3	2	32	4	5
13	6	3	33	3	0
14	7	4	34	4	3
15	3	2	35	2	3
16	7	4	36	3	3
17	4	4	37	5	5
18	1	3	38	3	4
19	5	5	39	4	3
20	2	2			

27. **Strikes.** Advertisements for an instructional video claim that the techniques will improve the ability of Little League pitchers to throw strikes and that, after undergoing the training, players will be able to throw strikes on at least 60% of their pitches. To test this claim, we have 20 Little Leaguers throw 50 pitches each, and we record the number of strikes. After the players participate in the training program, we repeat the test. The table shows the number of strikes each player threw before and after the training.

a) Is there evidence that after training players can throw strikes more than 60% of the time?

b) Is there evidence that the training is effective in improving a player's ability to throw strikes?

*c) Use a matched-pairs sign test to test the appropriate hypothesis. Do your conclusions change from those in part a?

Number of Strikes (out of 50)		Number of Strikes (out of 50)	
Before	After	Before	After
28	35	33	33
29	36	33	35
30	32	34	32
32	28	34	30
32	30	34	33
32	31	35	34
32	32	36	37
32	34	36	33
32	35	37	35
33	36	37	32

28. **Freshman 15, revisited.** In Exercise 4 you thought about how to design a study to see if it's true that students tend to gain weight during their first year in college. Well, Cornell Professor of Nutrition David Levitsky did just that. He recruited students from two large sections of an introductory health course. Although they were volunteers, they appeared to match the rest of the freshman class in terms of demographic variables such as sex and ethnicity. The students were weighed during the first week of the semester, then again 12 weeks later. Based on Professor Levitsky's data, estimate the mean weight gain in first-semester freshmen and comment on the "freshman 15." (Weights are in pounds.)

Subject Number	Initial Weight	Terminal Weight	Subject Number	Initial Weight	Terminal Weight
1	171	168	35	148	150
2	110	111	36	164	165
3	134	136	37	137	138
4	115	119	38	198	201
5	150	155	39	122	124
6	104	106	40	146	146
7	142	148	41	150	151
8	120	124	42	187	192
9	144	148	43	94	96
10	156	154	44	105	105
11	114	114	45	127	130
12	121	123	46	142	144
13	122	126	47	140	143
14	120	115	48	107	107
15	115	118	49	104	105
16	110	113	50	111	112
17	142	146	51	160	162
18	127	127	52	134	134
19	102	105	53	151	151
20	125	125	54	127	130
21	157	158	55	106	108
22	119	126	56	185	188
23	113	114	57	125	128
24	120	128	58	125	126
25	135	139	59	155	158
26	148	150	60	118	120
27	110	112	61	149	150
28	160	163	62	149	149
29	220	224	63	122	121
30	132	133	64	155	158
31	145	147	65	160	161
32	141	141	66	115	119
33	158	160	67	167	170
34	135	134	68	131	131

29. **Wheelchair marathon 2009.** The Boston Marathon has had a wheelchair division since 1977. Who do you think is typically faster, the men's marathon winner on foot or the women's wheelchair marathon winner? Because the conditions differ from year to year, and speeds have improved over the years, it seems best to treat these as paired

measurements. Here are summary statistics for the pairwise differences in finishing time (in minutes): (http://www .boston.com/sports/marathon/history/champions/)

Summary of wheelchrF − runM

N = 33

Mean = 3.00

SD = 35.4674

a) Comment on the assumptions and conditions.

b) Assuming that these times are representative of such races and the differences appeared acceptable for inference, construct and interpret a 95% confidence interval for the mean difference in finishing times.

c) Would a hypothesis test at $\alpha = 0.05$ reject the null hypothesis of no difference? What conclusion would you draw?

30. Marathon startup years 2009. When we considered the Boston Marathon in Exercise 29, we were unable to check the Nearly Normal Condition. Here's a histogram of the differences:

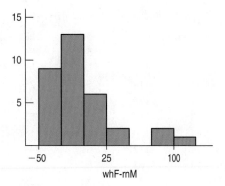

whF-rnM

Those three large differences are the first three years of wheelchair competition: 1977, 1978, and 1979. Often the start-up years of new events are different; later on, more athletes train and compete. If we omit those three years, the summary statistics change as follows:

Summary of wheelchrF - runM

N = 30

Mean = −12.4767

SD = 18.8731

a) Comment on the assumptions and conditions.

b) Assuming that these times are representative of such races, construct and interpret a 95% confidence interval for the mean difference in finishing time.

c) Would a hypothesis test at $\alpha = 0.05$ reject the null hypothesis of no difference? What conclusion would you draw?

31. BST. Many dairy cows now receive injections of BST, a hormone intended to spur greater milk production. After the first injection, a test herd of 60 Ayrshire cows increased their mean daily production from 47 pounds to 61 pounds of milk. The standard deviation of the increases was 5.2 pounds. We want to estimate the mean increase a farmer could expect in his own cows.

a) Check the assumptions and conditions for inference.

b) Write a 95% confidence interval.

c) Explain what your interval means in this context.

d) Given the cost of BST, a farmer believes he cannot afford to use it unless he is sure of attaining at least a 25% increase in milk production. Based on your confidence interval, what advice would you give him?

32. BST II. In the experiment about hormone injections in cows described in Exercise 31, a group of 52 Jersey cows increased average milk production from 43 pounds to 52 pounds per day, with a standard deviation of 4.8 pounds. Is this evidence that the hormone may be more effective in one breed than the other? Test an appropriate hypothesis and state your conclusion. Be sure to discuss any assumptions you make.

JUST CHECKING

ANSWERS

1. These are independent groups sampled at random, so use a two-sample *t* confidence interval to estimate the size of the difference.

2. There is only one sample. Use a one-sample *t*-interval.

3. A brother and sister from the same family represent a matched pair. The question calls for a paired *t*-test.

4. **a)** A before-and-after study calls for paired *t*-methods. To estimate the loss, find a confidence interval for the before–after differences.

   b) The two treatment groups were assigned randomly, so they are independent. Use a two-sample *t*-test to assess whether the mean weight losses differ.

5. Sometimes it just isn't clear. Most likely, couples would discuss the band or even decide to go to the club because they both like a particular band. If we think that's likely, then these data are paired. But maybe not. If we asked them their opinions of, say, the decor or furnishings at the club, the fact that they were couples might not affect the independence of their answers.

Comparing Counts

Where are we going?

Is your favorite color related to how much education you've had? A survey found a higher percentage of those naming blue, and a lower percentage saying red among adults with only a high school education compared with adults with more education. Could this be just random fluctuation, or is the distribution of color preference different for different education levels? We saw tables of counts and percentages in Chapter 3. In this chapter we'll see how to test the strength of the patterns we saw in those tables.

WHO	Executives of Fortune 400 companies
WHAT	Zodiac birth sign
WHY	Maybe the researcher was a Gemini and naturally curious?

A S   **Activity: Children at Risk.** See how a contingency table helps us understand the different risks to which an incident exposed children.

D oes your zodiac sign predict how successful you will be later in life? *Fortune* magazine collected the zodiac signs of 256 heads of the largest 400 companies. The table shows the number of births for each sign.

We can see some variation in the number of births per sign, and there *are* more Pisces, but is that enough to claim that successful people are more likely to be born under some signs than others?

Births	Sign
23	Aries
20	Taurus
18	Gemini
23	Cancer
20	Leo
19	Virgo
18	Libra
21	Scorpio
19	Sagittarius
22	Capricorn
24	Aquarius
29	Pisces

Birth totals by sign for 256 Fortune 400 executives.

Goodness-of-Fit

If births were distributed uniformly across the year, we would expect about 1/12 of them to occur under each sign of the zodiac. That suggests 256/12, or about 21.3 births per sign. How closely do the observed numbers of births per sign fit this simple "null" model?

A hypothesis test to address this question is called a test of **"goodness-of-fit."** The name suggests a certain badness-of-grammar, but it is quite standard. After all, we are asking whether the model that births are uniformly distributed over

"All creatures have their determined time for giving birth and carrying fetus, only a man is born all year long, not in determined time, one in the seventh month, the other in the eighth, and so on till the beginning of the eleventh month."

–Aristotle

the signs fits the data good, . . . er, well. Goodness-of-fit involves testing a hypothesis. We have specified a model for the distribution and want to know whether it fits. There is no single parameter to estimate, so a confidence interval wouldn't make much sense.

If the question were about only one astrological sign (for example, "Are executives more likely to be Pisces?"[1]), we could use a one-proportion z-test and ask if the true proportion of executives with that sign is equal to 1/12. However, here we have 12 hypothesized proportions, one for each sign. We need a test that considers all of them together and gives an overall idea of whether the observed distribution differs from the hypothesized one.

///// **FOR EXAMPLE** Finding Expected Counts

Birth month may not be related to success as a CEO, but what about on the ball field? It has been proposed by some researchers that children who are the older ones in their class at school naturally perform better in sports and that these children then get more coaching and encouragement. Could that make a difference in who makes it to the professional level in sports?

Baseball is a remarkable sport, in part because so much data are available. We have the birth dates of every one of the 16,804 players who ever played in a major league game. Since the effect we're suspecting may be due to relatively recent policies (and to keep the sample size moderate), we'll consider the birth months of the 1478 major league players born since 1975 and who have played through 2006. We can also look up the national demographic statistics to find what percentage of people were born in each month. Let's test whether the observed distribution of ballplayers' birth months shows just random fluctuations or whether it represents a real deviation from the national pattern.

Month	Ballplayer count	National birth %	Month	Ballplayer count	National birth %
1	137	8%	7	102	9%
2	121	7%	8	165	9%
3	116	8%	9	134	9%
4	121	8%	10	115	9%
5	126	8%	11	105	8%
6	114	8%	12	122	9%
			Total	1478	100%

QUESTION: How can we find the expected counts?

There are 1478 players in this set of data. I'd expect 8% of them to have been born in January, and $1478(0.08) = 118.24$. I won't round off, because expected "counts" needn't be integers. Multiplying 1478 by each of the birth percentages gives the expected counts shown in the table.

Month	Expected	Month	Expected
1	118.24	7	133.02
2	103.46	8	133.02
3	118.24	9	133.02
4	118.24	10	133.02
5	118.24	11	118.24
6	118.24	12	133.02

Assumptions and Conditions

These data are organized in tables as we saw in Chapter 3, and the assumptions and conditions reflect that. Rather than having an observation for each individual, we typically work with summary counts in categories. In our example, we don't see the birth signs of each of the 256 executives, only the totals for each sign.

Counted Data Condition: The data must be *counts* for the categories of a categorical variable. This might seem a simplistic, even silly condition. But many kinds of values can be assigned to categories, and it is unfortunately common to find the methods of this chapter applied incorrectly to proportions, percentages, or measurements just because they happen to be organized in a table. So check to be sure the values in each **cell** really are counts.

---

[1] A question actually asked us by someone who was undoubtedly a Pisces.

Independence Assumption

Independence Assumption: The counts in the cells should be independent of each other. The easiest case is when the individuals who are counted in the cells are sampled independently from some population. That's what we'd like to have if we want to draw conclusions about that population. Randomness can arise in other ways, though. For example, these Fortune 400 executives are not a random sample, but we might still think that their birth dates are randomly distributed throughout the year. If we want to generalize to a large population, we should check the Randomization Condition.

Randomization Condition: The individuals who have been counted should be a random sample from the population of interest.

Sample Size Assumption

We must have enough data for the methods to work. We usually check the following:

Expected Cell Frequency Condition: We should expect to see at least 5 individuals in each cell.

The Expected Cell Frequency Condition sounds like—and is, in fact, quite similar to—the condition that np and nq be at least 10 when we tested proportions. In our astrology example, assuming equal births in each month leads us to expect 21.3 births per month, so the condition is easily met here.

FOR EXAMPLE | **Checking Assumptions and Conditions**

RECAP: Are professional baseball players more likely to be born in some months than in others? We have observed and expected counts for the 1478 players born since 1975.

QUESTION: Are the assumptions and conditions met for performing a goodness-of-fit test?

✔ **Counted Data Condition:** I have month-by-month counts of ballplayer births.

✔ **Independence Assumption:** These births were independent.

✔ **Randomization Condition:** Although they are not a random sample, we can take these players to be representative of players past and future.

✔ **Expected Cell Frequency Condition:** The expected counts range from 103.46 to 133.02, all much greater than 5.

✔ **10% Condition:** These 1478 players are less than 10% of the population of 16,804 players who have ever played (or will play) major league baseball.

It's okay to use these data for a goodness-of-fit test.

Calculations

We compare the counts *observed* in each cell with the counts we *expect* to find. The usual notation uses O's and E's or abbreviations such as those we've used here. The method for finding the expected counts depends on the model.

We have observed a count in each category from the data, and have an expected count for each category from the hypothesized proportions. Are the differences just natural sampling variability, or are they so large that they indicate something important? It's natural to look at the *differences* between these observed and expected counts, denoted $(Obs - Exp)$. We'd like to think about the total of the differences, but just adding them won't work because some differences are positive, others negative. We've been in this predicament before—once when we looked at deviations from the mean and again when we dealt with residuals. In fact, these *are* residuals. They're just the differences between the observed data and the counts given by the (null) model. We handle these residuals in essentially the same way we did in regression: We square them. That gives us positive values and focuses attention on any cells with large differences from what we expected. Because the differences between observed and expected counts generally get larger the more data we have, we also need to get an idea of the *relative* sizes of the differences. To do that, we divide each squared difference by the expected count for that cell.

The test statistic, called the **chi-square** (or chi-squared) **statistic,** is found by adding up the sum of the squares of the deviations between the observed and expected counts divided by the expected counts:

$$\chi^2 = \sum_{all\ cells} \frac{(Obs - Exp)^2}{Exp}.$$

The chi-square statistic is denoted χ^2, where χ is the Greek letter chi (pronounced "ky" as in "sky"). It refers to a family of sampling distribution models we have not seen before called (remarkably enough) the **chi-square models.**

This family of models, like the Student's *t*-models, differ only in the number of degrees of freedom. The number of degrees of freedom for a goodness-of-fit test is $n - 1$. Here, however, n is *not* the sample size, but instead is the number of categories. For the zodiac example, we have 12 signs, so our χ^2 statistic has 11 degrees of freedom.

Chi-Square P-Values

The chi-square statistic is used only for testing hypotheses, not for constructing confidence intervals. If the observed counts don't match the expected, the statistic will be large. It can't be "too small." That would just mean that our model *really* fit the data well. So the chi-square test is always one-sided. If the calculated statistic value is large enough, we'll reject the null hypothesis. What could be simpler?

If you don't have technology handy, it's easy to read the χ^2 table (Table X in Appendix D).

A portion of Table X.

Right-Tail Probability		0.10	0.05	0.025	0.01	0.005
	df					
	1	2.706	3.841	5.024	6.635	7.879
Values of χ^2_α	2	4.605	5.991	7.378	9.210	10.597
	3	6.251	7.815	9.348	11.345	12.838
	4	7.779	9.488	11.143	13.277	14.860
	5	9.236	11.070	12.833	15.086	16.750
	6	10.645	12.592	14.449	16.812	18.548
	7	12.017	14.067	16.013	18.475	20.278
	8	13.362	15.507	17.535	20.090	21.955
	9	14.684	16.919	19.023	21.666	23.589
	10	15.987	18.307	20.483	23.209	25.188
	11	17.275	19.675	21.920	24.725	26.757
	12	18.549	21.026	23.337	26.217	28.300
	13	19.812	23.362	24.736	27.688	29.819
	14	21.064	23.685	26.119	29.141	31.319

The usual selected P-values are at the top of the columns. As with the *t*-tables, we have only selected probabilities, so the best we can do is to trap a P-value between two of the values in the table. Just find the row for the correct number of degrees of freedom and read across to find where your calculated χ^2 value falls. Of course, technology can find an exact P-value, and that's usually what we'll see.

Even though its mechanics work like a one-sided test, the interpretation of a chi-square test is in some sense *many*-sided. With more than two proportions, there are many ways the null hypothesis can be wrong. By squaring the differences, we made all the deviations positive, whether our observed counts were higher or lower than expected. There's no direction to the rejection of the null model. All we know is that it doesn't fit.

FOR EXAMPLE   Doing a Goodness-of-Fit Test

RECAP: We're looking at data on the birth months of major league baseball players. We've checked the assumptions and conditions for performing a χ^2 test.

QUESTIONS: What are the hypotheses, and what does the test show?

H_0: The distribution of birth months for major league ballplayers is the same as that for the general population.

H_A: The distribution of birth months for major league ballplayers differs from that of the rest of the population.

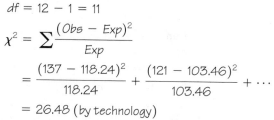

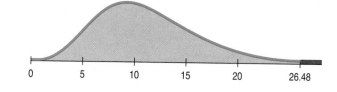

$$df = 12 - 1 = 11$$

$$\chi^2 = \sum \frac{(Obs - Exp)^2}{Exp}$$

$$= \frac{(137 - 118.24)^2}{118.24} + \frac{(121 - 103.46)^2}{103.46} + \cdots$$

$$= 26.48 \text{ (by technology)}$$

$$\text{P-value} = P(\chi^2_{11} \geq 26.48) = 0.0055 \text{ (by technology)}$$

Because of the small P-value, I reject H_0; there's evidence that birth months of major league ballplayers have a different distribution from the rest of us.

STEP-BY-STEP EXAMPLE   A Chi-Square Test for Goodness-of-Fit

We have counts of 256 executives in 12 zodiac sign categories. The natural null hypothesis is that birth dates of executives are divided equally among all the zodiac signs. The test statistic looks at how closely the observed data match this idealized situation.

Question: Are zodiac signs of CEOs distributed uniformly?

THINK **Plan** State what you want to know.	I want to know whether births of successful people are uniformly distributed across the signs of the zodiac. I have counts of 256 Fortune 400 executives, categorized by their birth sign.
Identify the variables and check the W's.	
Hypotheses State the null and alternative hypotheses. For χ^2 tests, it's usually easier to do that in words than in symbols.	H_0: Births are uniformly distributed over zodiac signs.[2]
	H_A: Births are not uniformly distributed over zodiac signs.

---

[2] It may seem that we have broken our rule of thumb that null hypotheses should specify parameter values. If you want to get formal about it, the null hypothesis is that

$$p_{Aries} = p_{Taurus} = \cdots = p_{Pisces}.$$

That is, we hypothesize that the true proportions of births of CEOs under each sign are equal. The role of the null hypothesis is to specify the model so that we can compute the test statistic. That's what this one does.

Model Make a picture. The null hypothesis is that the frequencies are equal, so a bar chart (with a line at the hypothesized "equal" value) is a good display.

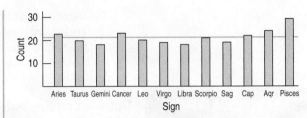

The bar chart shows some variation from sign to sign, and Pisces is the most frequent. But it is hard to tell whether the variation is more than I'd expect from random variation.

Think about the assumptions and check the conditions.

✔ **Counted Data Condition:** I have counts of the number of executives in 12 categories.

✔ **Independence Assumption:** The birth dates of executives should be independent of each other.

✔ **Randomization Condition:** This is a convenience sample of executives, but there's no reason to suspect bias.

✔ **Expected Cell Frequency Condition:** The null hypothesis expects that 1/12 of the 256 births, or 21.333, should occur in each sign. These expected values are all at least 5, so the condition is satisfied.

Specify the sampling distribution model.

Name the test you will use.

The conditions are satisfied, so I'll use a χ^2 model with $12 - 1 = 11$ degrees of freedom and do a **chi-square goodness-of-fit test.**

SHOW

Mechanics Each cell contributes an $\dfrac{(Obs - Exp)^2}{Exp}$ value to the chi-square sum. We add up these components for each zodiac sign. If you do it by hand, it can be helpful to arrange the calculation in a table. We show that after this Step-By-Step.

The expected value for each zodiac sign is 21.333.

$$\chi^2 = \sum \frac{(Obs - Exp)^2}{Exp} = \frac{(23 - 21.333)^2}{21.333}$$
$$+ \frac{(20 - 21.333)^2}{21.333} + \dots$$
$$= 5.094 \text{ for all 12 signs.}$$

The P-value is the area in the upper tail of the χ^2 model above the computed χ^2 value.

The χ^2 models are skewed to the high end, and change shape depending on the degrees of freedom. The P-value considers only the right tail. Large χ^2 statistic values correspond to small P-values, which lead us to reject the null hypothesis.

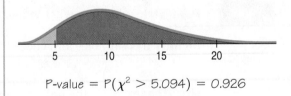

P-value = $P(\chi^2 > 5.094) = 0.926$

Conclusion Link the P-value to your decision. Remember to state your conclusion in terms of what the data mean, rather than just making a statement about the distribution of counts.

> The P-value of 0.926 says that if the zodiac signs of executives were in fact distributed uniformly, an observed chi-square value of 5.09 or higher would occur about 93% of the time. This certainly isn't unusual, so I fail to reject the null hypothesis, and conclude that these data show virtually no evidence of nonuniform distribution of zodiac signs among executives.

The Chi-Square Calculation

A S *Activity:* **Calculating Standardized Residuals.** The incident of the earlier ActivStats activity in which children were placed at risk, also put women at risk. Standardized residuals help us understand the relative risks.

Let's make the chi-square procedure very clear. Here are the steps:

1. **Find the expected values.** These come from the null hypothesis model. Every model gives a hypothesized proportion for each cell. The expected value is the product of the total number of observations times this proportion.

   For our example, the null model hypothesizes *equal* proportions. With 12 signs, 1/12 of the 256 executives should be in each category. The expected number for each sign is 21.333.
2. **Compute the residuals.** Once you have expected values for each cell, find the residuals, *Observed − Expected*.
3. **Square the residuals.**
4. **Compute the components.** Now find the component, $\frac{(Observed - Expected)^2}{Expected}$, for each cell.
5. **Find the sum of the components.** That's the chi-square statistic.
6. **Find the degrees of freedom.** It's equal to the number of cells minus one. For the zodiac signs, that's $12 - 1 = 11$ degrees of freedom.
7. **Test the hypothesis.** Large chi-square values mean lots of deviation from the hypothesized model, so they give small P-values. Look up the critical value from a table of chi-square values, or use technology to find the P-value directly.

A S *Activity:* **The Chi-Square Test.** This animation completes the calculation of the chi-square statistic and the hypothesis test based on it.

The steps of the chi-square calculations are often laid out in tables. Use one row for each category, and columns for observed counts, expected counts, residuals, squared residuals, and the contributions to the chi-square total like this:

Sign	Observed	Expected	Residual = $(Obs - Exp)$	$(Obs - Exp)^2$	Component = $\frac{(Obs - Exp)^2}{Exp}$
Aries	23	21.333	1.667	2.778889	0.130262
Taurus	20	21.333	−1.333	1.776889	0.083293
Gemini	18	21.333	−3.333	11.108889	0.520737
Cancer	23	21.333	1.667	2.778889	0.130262
Leo	20	21.333	−1.333	1.776889	0.083293
Virgo	19	21.333	−2.333	5.442889	0.255139
Libra	18	21.333	−3.333	11.108889	0.520737
Scorpio	21	21.333	−0.333	0.110889	0.005198
Sagittarius	19	21.333	−2.333	5.442889	0.255139
Capricorn	22	21.333	0.667	0.444889	0.020854
Aquarius	24	21.333	2.667	7.112889	0.333422
Pisces	29	21.333	7.667	58.782889	2.755491

$$\sum = 5.094$$

How big is big? When we calculated χ^2 for the zodiac sign example, we got 5.094. That value would have been big for z or t, leading us to reject the null hypothesis. Not here. Were you surprised that $\chi^2 = 5.094$ had a huge P-value of 0.926? What *is* big for a χ^2 statistic, anyway?

Think about how χ^2 is calculated. In every cell, any deviation from the expected count contributes to the sum. Large deviations generally contribute more, but if there are a lot of cells, even small deviations can add up, making the χ^2 value larger. So the more cells there are, the higher the value of χ^2 has to get before it becomes noteworthy. For χ^2, then, the decision about how big is big depends on the number of degrees of freedom.

Unlike the Normal and t families, χ^2 models are skewed. Curves in the χ^2 family change both shape and center as the number of degrees of freedom grows. Here, for example, are the χ^2 curves for 5 and 9 degrees of freedom.

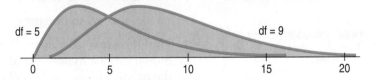

Notice that the value $\chi^2 = 10$ might seem somewhat extreme when there are 5 degrees of freedom, but appears to be rather ordinary for 9 degrees of freedom. Here are two simple facts to help you think about χ^2 models:

▸ The mode is at $\chi^2 = df - 2$. (Look back at the curves; their peaks are at 3 and 7, see?)

▸ The expected value (mean) of a χ^2 model is its number of degrees of freedom. That's a bit to the right of the mode—as we would expect for a skewed distribution.

Our test for zodiac birthdays had 11 df, so the relevant χ^2 curve peaks at 9 and has a mean of 11. Knowing that, we might have easily guessed that the calculated χ^2 value of 5.094 wasn't going to be significant.

But I Believe the Model . . .

Goodness-of-fit tests are likely to be performed by people who have a theory of what the proportions *should* be in each category and who believe their theory to be true. Unfortunately, the only *null* hypothesis available for a goodness-of-fit test is that the theory is true. And as we know, the hypothesis-testing procedure allows us only to *reject* the null or *fail to reject* it. We can never confirm that a theory is in fact true, which is often what people want to do.

Unfortunately, they're stuck. At best, we can point out that the data are consistent with the proposed theory. But this doesn't *prove* the theory. The data *could* be consistent with the model even if the theory were wrong. In that case, we fail to reject the null hypothesis but can't conclude anything for sure about whether the theory is true.

And we can't fix the problem by turning things around. Suppose we try to make our favored hypothesis the alternative. Then it is impossible to pick a single null. For example, suppose, as a doubter of astrology, you want to prove that the distribution of executive births is uniform. If you choose uniform as the null hypothesis, you can only *fail* to reject it. So you'd like uniformity to be your alternative hypothesis. Which particular violation of equally distributed births would you choose as your null? The problem is that the model can be wrong in many, many ways. There's no way to frame a null hypothesis the other way around. There's just no way to prove that a favored model is true.

Why can't we prove the null? A biologist wanted to show that her inheritance theory about fruit flies is valid. It says that 10% of the flies should be type 1, 70% type 2, and 20% type 3. After her students collected data on 100 flies, she did a goodness-of-fit test and found a P-value of 0.07. She started celebrating, since her null hypothesis wasn't rejected–that is, until her students collected data on 100 more flies. With 200 flies, the P-value dropped to 0.02. Although she knew the answer was probably no, she asked the statistician somewhat hopefully if she could just ignore half the data and stick with the original 100. By this reasoning we could always "prove the null" just by not collecting much data. With only a little data, the chances are good that they'll be consistent with almost anything. But they also have little chance of disproving anything either. In this case, the test has no power. Don't be lured into this scientist's reasoning. With data, more is always better. But you can't ever prove that your null hypothesis is true.

Comparing Observed Distributions

Many colleges survey graduating classes to determine the plans of the graduates. We might wonder whether the plans of students are the same at different colleges. Here's a **two-way table** for Class of 2006 graduates from several colleges at one university. Each **cell** of the table shows how many students from a particular college made a certain choice.

WHO	Graduates from 4 colleges at an upstate New York university
WHAT	Post-graduation activities
WHEN	2006
WHY	Survey for general information

	Agriculture	Arts & Sciences	Engineering	Social Science	Total
Employed	379	305	243	125	**1052**
Grad School	186	238	202	96	**722**
Other	104	123	37	58	**322**
Total	**669**	**666**	**482**	**279**	**2096**

TABLE 26.1
Post-graduation activities of the class of 2006 for several colleges of a large university

Because class sizes are so different, we see differences better by examining the proportions for each class rather than the counts:

	Agriculture	Arts & Sciences	Engineering	Social Science	Total
Employed	56.7%	45.8%	50.4%	44.8%	**50.2**
Grad School	27.8	35.7	41.9	34.4	**34.4**
Other	15.5	18.5	7.7	20.8	**15.4**
Total	**100**	**100**	**100**	**100**	**100**

TABLE 26.2
Activities of graduates as a percentage of respondents from each college.

A S *Video:* **The Incident.** You may have guessed which famous incident put women and children at risk. Here you can view the story complete with rare film footage.

We already know how to test whether *two* proportions are the same. For example, we could use a two-proportion z-test to see whether the proportion of students choosing graduate school is the same for Agriculture students as for Engineering students. But now we have more than two groups. We want to test whether the students' choices are the same across all four colleges. The z-test for two proportions generalizes to a **chi-square test of homogeneity.**

Chi-square again? It turns out that the mechanics of this test are *identical* to the chi-square test for goodness-of-fit that we just saw. (How similar can you get?) Why a different name, then? The goodness-of-fit test compared counts with a theoretical model. But here we're asking whether choices are the same among different groups, so we find the expected counts for each category directly from the data. As a result, we count the degrees of freedom slightly differently as well.

The term "homogeneity" means that things are the same. Here, we ask whether the post-graduation choices made by students are the *same* for these four colleges. The homogeneity test comes with a built-in null hypothesis: We hypothesize that the distribution does not change from group to group. The test looks for differences large enough to step beyond what we might expect from random sample-to-sample variation. It can reveal a large deviation in a single category or small, but persistent, differences over all the categories—or anything in between.

Assumptions and Conditions

The assumptions and conditions are the same as for the chi-square test for goodness-of-fit. The **Counted Data Condition** says that these data must be counts. You can't do a test of homogeneity on proportions, so you have to work with the counts of graduates given in the first table. Also, you can't do a chi-square test on measurements. For example, if we had recorded GPAs for these same groups, we wouldn't be able to determine whether the mean GPAs were different using this test.[3]

Often when we test for homogeneity, we aren't interested in some larger population, so we don't really need a random sample. (We would need one if we wanted to draw a more general conclusion—say, about the choices made by all members of the Class of '06.) Don't we need *some* randomness, though? Fortunately, the null hypothesis can be thought of as a model in which the counts in the table are distributed as if each student chose a plan randomly according to the overall proportions of the choices, regardless of the student's class. As long as we don't want to generalize, we don't have to check the **Randomization Condition** or the **10% Condition.**

We still must be sure we have enough data for this method to work. The **Expected Cell Frequency Condition** says that the expected count in each cell must be at least 5. We'll confirm that as we do the calculations.

Calculations

The null hypothesis says that the proportions of graduates choosing each alternative should be the same for all four colleges, so we can estimate those overall proportions by pooling our data from the four colleges together. Within each college, the expected proportion for each choice is just the overall proportion of all students making that choice. The expected counts are those proportions applied to the number of students in each graduating class.

For example, overall, 1052, or about 50.2%, of the 2096 students who responded to the survey were employed. If the distributions are homogeneous (as the null hypothesis asserts), then 50.2% of the 669 Agriculture school graduates (or about 335.8 students) should be employed. Similarly, 50.2% of the 482 Engineering grads (or about 241.96) should be employed.

---

[3] To do that, you'd use a method called Analysis of Variance, as we'll see in Chapters 28 and 29.

Working in this way, we (or, more likely, the computer) can fill in expected values for each cell. Because these are theoretical values, they don't have to be integers. The expected values look like this:

TABLE 26.3
Expected values for the '06 graduates.

	Agriculture	Arts & Sciences	Engineering	Social Science	Total
Employed	335.777	334.271	241.920	140.032	**1052**
Grad School	230.448	229.414	166.032	96.106	**722**
Other	102.776	102.315	74.048	42.862	**322**
Total	**669**	**666**	**482**	**279**	**2096**

Now check the **Expected Cell Frequency Condition.** Indeed, there are at least 5 individuals expected in each cell.

Following the pattern of the goodness-of-fit test, we compute the component for each cell of the table. For the highlighted cell, employed students graduating from the Ag school, that's

$$\frac{(Obs - Exp)^2}{Exp} = \frac{(379 - 335.777)^2}{335.777} = 5.564$$

Summing these components across all cells gives

$$\chi^2 = \sum_{all\ cells} \frac{(Obs - Exp)^2}{Exp} = 54.51$$

How about the degrees of freedom? We don't really need to calculate all the expected values in the table. We know there is a total of 1052 employed students, so once we find the expected values for three of the colleges, we can determine the expected number for the fourth by just subtracting. Similarly, we know how many students graduated from each college, so after filling in two rows, we can find the expected values for the remaining row by subtracting. To fill out the table, we need to know the counts in only $R - 1$ rows and $C - 1$ columns. So the table has $(R - 1)(C - 1)$ degrees of freedom.

In our example, we need to calculate only 2 choices in each column and counts for 3 of the 4 colleges, for a total of $2 \times 3 = 6$ degrees of freedom. We'll need the degrees of freedom to find a P-value for the chi-square statistic.

NOTATION ALERT

For a contingency table, R represents the number of rows and C the number of columns.

STEP-BY-STEP EXAMPLE | **A Chi-Square Test for Homogeneity**

We have reports from four colleges on the post-graduation activities of their 2006 graduating classes.

Question: Are students' choices of post-graduation activities the same across all the colleges?

THINK

Plan State what you want to know.

Identify the variables and check the W's.

I want to know whether post-graduation choices are the same for students from each of four colleges. I have a table of counts classifying each college's Class of 2006 respondents according to their activities.

Hypotheses State the null and alternative hypotheses.

H_0: Students' post-graduation activities are distributed in the same way for all four colleges.
H_A: Students' plans do not have the same distribution.

Model Make a picture: A side-by-side bar chart shows the four distributions of post-graduation activities. Plot column percents to remove the effect of class size differences. A split bar chart would also be an appropriate choice.

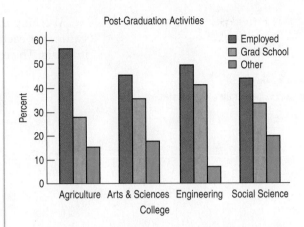

Post-Graduation Activities

A side-by-side bar chart shows how the distributions of choices differ across the four colleges.

Think about the assumptions and check the conditions.

✔ **Counted Data Condition:** I have counts of the number of students in categories.

✔ **Independence Assumption:** Student plans should be largely independent of each other. The occasional friends who decide to join Teach for America together or couples who make grad school decisions together are too rare to affect this analysis.

✔ **Randomization Condition:** I don't want to draw inferences to other colleges or other classes, so there is no need to check for a random sample.

✔ **Expected Cell Frequency Condition:** The expected values (shown below) are all at least 5.

State the sampling distribution model and name the test you will use.

The conditions seem to be met, so I can use a χ^2 model with $(3 - 1) \times (4 - 1) = 6$ degrees of freedom and do a **chi-square test of homogeneity**.

Mechanics Show the expected counts for each cell of the data table. You could make separate tables for the observed and expected counts, or put both counts in each cell as shown here. While observed counts must be whole numbers, expected counts rarely are—don't be tempted to round those off.

	Ag	A&S	Eng	Soc Sci
Empl.	379 / 335.777	305 / 334.271	243 / 241.920	125 / 140.032
Grad sch.	186 / 230.448	238 / 229.414	202 / 166.032	96 / 96.106
Other	104 / 102.776	123 / 102.315	37 / 74.048	58 / 42.862

Calculate χ^2.

$$\chi^2 = \sum_{all\ cells} \frac{(Obs - Exp)^2}{Exp}$$
$$= \frac{(379 - 335.777)^2}{335.777} + \dots$$
$$= 54.52$$

The shape of a χ^2 model depends on the degrees of freedom. A χ^2 model with 6 df is skewed to the high end.

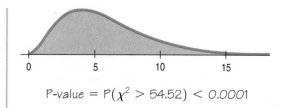

$$\text{P-value} = P(\chi^2 > 54.52) < 0.0001$$

The P-value considers only the right tail. Here, the calculated value of the χ^2 statistic is off the scale, so the P-value is quite small.

Conclusion State your conclusion in the context of the data. You should specifically talk about whether the distributions for the groups appear to be different.

The P-value is very small, so I reject the null hypothesis and conclude that there's evidence that the post-graduation activities of students from these four colleges don't have the same distribution.

If you find that simply rejecting the hypothesis of homogeneity is a bit unsatisfying, you're in good company. Ok, so the post-graduation plans are different. What we'd really like to know is what the differences are, where they're the greatest, and where they're the smallest. The test for homogeneity doesn't answer these interesting questions, but it does provide some evidence that can help us.

Examining the Residuals

Whenever we reject the null hypothesis, it's a good idea to examine residuals. (We don't need to do that when we fail to reject because when the χ^2 value is small, all of its components must have been small.) For chi-square tests, we want to compare residuals for cells that may have very different counts. So we're better off standardizing the residuals. We know the mean residual is zero,[4] but we need to know each residual's standard deviation. When we tested proportions, we saw a link between the expected proportion and its standard deviation. For counts, there's a similar link. To standardize a cell's residual, we just divide by the square root of its expected value:

$$c = \frac{(Obs - Exp)}{\sqrt{Exp}}.$$

Notice that these **standardized residuals** are just the square roots of the **components** we calculated for each cell, and their sign indicates whether we observed more cases than we expected, or fewer.

The standardized residuals give us a chance to think about the underlying patterns and to consider the ways in which the distribution of post-graduation plans may differ from college to college. Now that we've subtracted the mean (zero) and divided by their standard deviations, these are z-scores. If the null hypothesis were true, we could even appeal to the Central Limit Theorem, think of the Normal model, and use the 68–95–99.7 Rule to judge how extraordinary the large ones are.

Here are the standardized residuals for the Class of '06 data:

	Ag	A&S	Eng	Soc Sci
Employed	2.359	−1.601	0.069	−1.270
Grad School	−2.928	0.567	2.791	−0.011
Other	0.121	2.045	−4.305	2.312

TABLE 26.4
Standardized residuals can help show how the table differs from the null hypothesis pattern.

---

[4] Residual = Observed − Expected. Because the total of the expected values is set to be the same as the observed total, the residuals must sum to zero.

The column for Engineering students immediately attracts our attention. It holds both the largest positive and the largest negative standardized residuals. It looks like Engineering college graduates are more likely to go on to graduate work and very unlikely to take time off for "volunteering and travel, among other activities" (as the "Other" category is explained). By contrast, Ag school graduates seem to be readily employed and less likely to pursue graduate work immediately after college.

FOR EXAMPLE | Looking at χ^2 Residuals

RECAP: Some people suggest that schoolchildren who are the older ones in their class naturally perform better in sports and therefore get more coaching and encouragement. To see if there's any evidence for this, we looked at major league baseball players born since 1975. A goodness-of-fit test found their birth months to have a distribution that's significantly different from the rest of us. The table shows the standardized residuals.

Month	Residual	Month	Residual
1	1.73	7	−2.69
2	1.72	8	2.77
3	−0.21	9	0.08
4	0.25	10	−1.56
5	0.71	11	−1.22
6	−0.39	12	−0.96

QUESTION: What's different about the distribution of birth months among major league ballplayers?

It appears that, compared to the general population, fewer ballplayers than expected were born in July and more than expected in August. Either month would make them the younger kids in their grades in school, so these data don't offer support for the conjecture that being older is an advantage in terms of a career as a professional athlete.

Tiny black potato flea beetles can damage potato plants in a vegetable garden. These pests chew holes in the leaves, causing the plants to wither or die. They can be killed with an insecticide, but a canola oil spray has been suggested as a non-chemical "natural" method of controlling the beetles. To conduct an experiment to test the effectiveness of the natural spray, we gather 500 beetles and place them in three Plexiglas® containers. Two hundred beetles go in the first container, where we spray them with the canola oil mixture. Another 200 beetles go in the second container; we spray them with the insecticide. The remaining 100 beetles in the last container serve as a control group; we simply spray them with water. Then we wait 6 hours and count the number of surviving beetles in each container.

1. Why do we need the control group?

2. What would our null hypothesis be?

3. After the experiment is over, we could summarize the results in a table as shown. How many degrees of freedom does our χ^2 test have?

	Natural spray	Insecticide	Water	Total
Survived				
Died				
Total	200	200	100	500

4. Suppose that, all together, 125 beetles survived. (That's the first-row total.) What's the expected count in the first cell—survivors among those sprayed with the natural spray?

5. If it turns out that only 40 of the beetles in the first container survived, what's the calculated component of χ^2 for that cell?

6. If the total calculated value of χ^2 for this table turns out to be around 10, would you expect the P-value of our test to be large or small? Explain.

Independence

A study from the University of Texas Southwestern Medical Center examined whether the risk of hepatitis C was related to whether people had tattoos and to where they got their tattoos. Hepatitis C causes about 10,000 deaths each year in the United States, but often goes undetected for years after infection.

The data from this study can be summarized in a two-way table, as follows:

	Hepatitis C	No Hepatitis C	Total
Tattoo, parlor	17	35	**52**
Tattoo, elsewhere	8	53	**61**
None	22	491	**513**
Total	**47**	**579**	**626**

TABLE 26.5
Counts of patients classified by their hepatitis C test status according to whether they had a tattoo from a tattoo parlor or from another source, or had no tattoo.

WHO	Patients being treated for non–blood-related disorders
WHAT	Tattoo status and hepatitis C status
WHEN	1991, 1992
WHERE	Texas

A S **Activity: Independence and Chi-Square.** This unusual simulation shows how independence arises (and fails) in contingency tables.

These data differ from the kinds of data we've considered before in this chapter because they categorize subjects from a single group on two categorical variables rather than on only one. The categorical variables here are *Hepatitis C Status* ("Hepatitis C" or "No Hepatitis C") and *Tattoo Status* ("Parlor," "Elsewhere," "None"). We've seen counts classified by two categorical variables displayed like this in Chapter 3, so we know such tables are called contingency tables. **Contingency tables** categorize counts on two (or more) variables so that we can see whether the distribution of counts on one variable is contingent on the other.

The natural question to ask of these data is whether the chance of having hepatitis C is *independent* of tattoo status. Recall that for events **A** and **B** to be independent $P(\mathbf{A})$ must equal $P(\mathbf{A} \mid \mathbf{B})$. Here, this means the probability that a randomly selected patient has hepatitis C should not change when we learn the patient's tattoo status. We examined the question of independence in just this way back in Chapter 15, but we lacked a way to test it. The rules for independent events are much too precise and absolute to work well with real data. A **chi-square test for independence** is called for here.

If *Hepatitis Status* is independent of tattoos, we'd expect the proportion of people testing positive for hepatitis to be the same for the three levels of *Tattoo Status*. This sounds a lot like the test of homogeneity. In fact, the mechanics of the calculation are identical.

The only difference between the test for homogeneity and the test for independence is in what you . . .

THINK

The difference is that now we have two categorical variables measured on a single population. For the homogeneity test, we had a single categorical variable measured independently on two or more populations. But now we ask a different question: "Are the variables independent?" rather than "Are the groups homogeneous?" These are subtle differences, but they are important when we state hypotheses and draw conclusions.

FOR EXAMPLE   Which χ^2 test?

Many states and localities now collect data on traffic stops regarding the race of the driver. The initial concern was that Black drivers were being stopped more often (the "crime" ironically called "Driving While Black"). With more data in hand, attention has turned to other issues. For example, data from 2533 traffic stops in Cincinnati[5] report the race of the driver (Black, White, or Other) and whether the traffic stop resulted in a search of the vehicle.

QUESTION: Which test would be appropriate to examine whether race is a factor in vehicle searches? What are the hypotheses?

		Race			
		Black	White	Other	Total
Search	No	787	594	27	1408
	Yes	813	293	19	1125
	Total	1600	887	46	2533

These data represent one group of traffic stops in Cincinnati, categorized on two variables, Race and Search. I'll do a chi-square test of independence.

H_O: Whether or not police search a vehicle is independent of the race of the driver.

H_A: Decisions to search vehicles are not independent of the driver's race.

Assumptions and Conditions

A S

Activity: **Chi-Square Tables.** Work with *ActivStats'* interactive chi-square table to perform a hypothesis test.

Of course, we still need counts and enough data so that the expected values are at least 5 in each cell.

If we're interested in the independence of variables, we usually want to generalize from the data to some population. In that case, we'll need to check that the data are a representative random sample from, and fewer than 10% of, that population.

STEP-BY-STEP EXAMPLE   A Chi-Square Test for Independence

We have counts of 626 individuals categorized according to their "tattoo status" and their "hepatitis status."

Question: Are tattoo status and hepatitis status independent?

THINK

Plan State what you want to know.

Identify the variables and check the W's.

Hypotheses State the null and alternative hypotheses.

We perform a test of independence when we suspect the variables may not be independent. We are on the familiar ground of making a claim (in this case, that knowing *Tattoo Status* will change probabilities for *Hepatitis C Status*) and testing the null hypothesis that it is *not* true.

I want to know whether the categorical variables *Tattoo Status* and *Hepatitis Status* are statistically independent. I have a contingency table of 626 Texas patients with an unrelated disease.

H_O: *Tattoo Status* and *Hepatitis Status* are independent.[6]

H_A: *Tattoo Status* and *Hepatitis Status* are not independent.

---

[5] John E. Eck, Lin Liu, and Lisa Growette Bostaph, Police Vehicle Stops in Cincinnati, Oct. 1, 2003, available at http://www.cincinnati-oh.gov. Data for other localities can be found by searching from http://www.racialprofilinganalysis.neu.edu.

[6] Once again, parameters are hard to express. The hypothesis of independence itself tells us how to find expected values for each cell of the contingency table. That's all we need.

Model Make a picture. Because these are only two categories—Hepatitis C and No Hepatitis C—a simple bar chart of the distribution of tattoo sources for Hep C patients shows all the information.

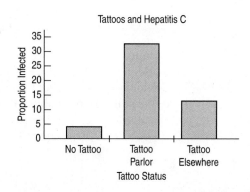

The bar chart suggests strong differences in Hepatitis C risk based on tattoo status.

Think about the assumptions and check the conditions.

✓ **Counted Data Condition:** I have counts of individuals categorized on two variables.

✓ **Independence Assumption:** The people in this study are likely to be independent of each other.

✓ **Randomization Condition:** These data are from a retrospective study of patients being treated for something unrelated to hepatitis. Although they are not an SRS, they were selected to avoid biases.

✓ **10% Condition:** These 626 patients are far fewer than 10% of all those with tattoos or hepatitis.

✗ **Expected Cell Frequency Condition:** The expected values do not meet the condition that all are at least 5.

This table shows both the observed and expected counts for each cell. The expected counts are calculated exactly as they were for a test of homogeneity; in the first cell, for example, we expect $\frac{52}{626}$ (that's 8.3%) of 47.

Warning: Be wary of proceeding when there are small expected counts. If we see expected counts that fall far short of 5, or if many cells violate the condition, we should not use χ^2. (We will soon discuss ways you can fix the problem.) If you do continue, always check the residuals to be sure those cells did not have a major influence on your result.

	Hepatitis C	No Hepatitis C	Total
Tattoo, parlor	17 3.904	35 48.096	52
Tattoo, elsewhere	8 4.580	53 56.420	61
None	22 38.516	491 474.484	513
Total	47	579	626

Although the Expected Cell Frequency Condition is not satisfied, the values are close to 5. I'll go ahead, but I'll check the residuals carefully. I'll use a χ^2 model with $(3 - 1) \times (2 - 1) = 2$ df and do a **chi-square test of independence.**

Specify the model.

Name the test you will use.

Mechanics Calculate χ^2.

The shape of a chi-square model depends on its degrees of freedom. With 2 df, the model looks quite different, as you can see here. We still care only about the right tail.

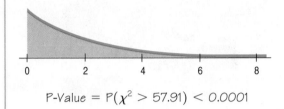

$$\chi^2 = \sum_{\text{all cells}} \frac{(Obs - Exp)^2}{Exp}$$

$$= \frac{(17 - 3.094)^2}{3.094} + \cdots = 57.91$$

P-Value = $P(\chi^2 > 57.91) < 0.0001$

 Conclusion Link the P-value to your decision. State your conclusion about the independence of the two variables.

(We should be wary of this conclusion because of the small expected counts. A complete solution must include the additional analysis, recalculation, and final conclusion discussed in the following section.)

The P-value is very small, so I reject the null hypothesis and conclude that *Hepatitis Status* is not *Independent* of *Tattoo Status*. Because the Expected Cell Frequency Condition was violated, I need to check that the two cells with small expected counts did not influence this result too greatly.

FOR EXAMPLE | Chi-Square Mechanics

RECAP: We have data that allow us to investigate whether police searches of vehicles they stop are independent of the driver's race.

QUESTIONS: What are the degrees of freedom for this test? What is the expected frequency of searches for the Black drivers who were stopped? What's that cell's component in the χ^2 computation? And how is the standardized residual for that cell computed?

		Race		
Search		Black	White	Other	Total
	No	787	594	27	**1408**
	Yes	813	293	19	**1125**
	Total	**1600**	**887**	**46**	**2533**

This is a 2 × 3 contingency table, so $df = (2 - 1)(3 - 1) = 2$.

Overall, 1125 of 2533 vehicles were searched. If searches are conducted independent of race, then I'd expect $\frac{1125}{2533}$ of the 1600 Black drivers to have been searched: $\frac{1125}{2533} \times 1600 \approx 710.62$.

That cell's term in the χ^2 calculation is $\frac{(Obs - Exp)^2}{Exp} = \frac{(813 - 710.62)^2}{710.62} = 14.75$

The standardized residual for that cell is $\frac{Obs - Exp}{\sqrt{Exp}} = \frac{813 - 710.62}{\sqrt{710.62}} = 3.84$

Examine the Residuals

Each cell of the contingency table contributes a term to the chi-square sum. As we did earlier, we should examine the residuals because we have rejected the null hypothesis. In this instance, we have an additional concern that the cells with small expected frequencies not be the ones that make the chi-square statistic large.

Our interest in the data arises from the potential for improving public health. If patients with tattoos are more likely to test positive for hepatitis C, perhaps physicians should be advised to suggest blood tests for such patients.

The standardized residuals look like this:

	Hepatitis C	No Hepatitis C
Tattoo, parlor	6.628	−1.888
Tattoo, elsewhere	1.598	−0.455
None	−2.661	0.758

TABLE 26.6
Standardized residuals for the hepatitis and tattoos data. Are any of them particularly large in magnitude?

The chi-square value of 57.91 is the sum of the squares of these six values. The cell for people with tattoos obtained in a tattoo parlor who have hepatitis C is large and positive, indicating there are more people in that cell than the null hypothesis of independence would predict. Maybe tattoo parlors are a source of infection or maybe those who go to tattoo parlors also engage in risky behavior.

The second-largest component is a negative value for those with no tattoos who test positive for hepatitis C. A negative value says that there are fewer people in this cell than independence would expect. That is, those who have no tattoos are less likely to be infected with hepatitis C than we might expect if the two variables were independent.

What about the cells with small expected counts? The formula for the chi-square standardized residuals divides each residual by the square root of the expected frequency. Too small an expected frequency can arbitrarily inflate the residual and lead to an inflated chi-square statistic. Any expected count close to the arbitrary minimum of 5 calls for checking that cell's standardized residual to be sure it is not particularly large. In this case, the standardized residual for the "Hepatitis C and Tattoo, elsewhere" cell is not particularly large, but the standardized residual for the "Hepatitis C and Tattoo, parlor" cell is large.

We might choose not to report the results because of concern with the small expected frequency. Alternatively, we could include a warning along with our report of the results. Yet another approach is to combine categories to get a larger category total and correspondingly larger expected frequencies, if there are some categories that can be appropriately combined. Here, we might naturally combine the two rows for tattoos, obtaining a 2 × 2 table:

	Hepatitis C	No Hepatitis C	Total
Tattoo	25	88	**113**
None	22	491	**513**
Total	**47**	**579**	**626**

TABLE 26.7
Combining the two tattoo categories gives a table with all expected counts greater than 5.

This table has expected values of at least 5 in every cell, and a chi-square value of 42.42 on 1 degree of freedom. The corresponding P-value is <0.0001.

We conclude that *Tattoo Status* and *Hepatitis C Status* are not independent. The data *suggest* that tattoo parlors may be a particular problem, but we haven't enough data to draw that conclusion.

FOR EXAMPLE | Writing Conclusions for χ^2 Tests

RECAP: We're looking at Cincinnati traffic stop data to see if police decisions about searching cars show evidence of racial bias. With 2 df, technology calculates $\chi^2 = 73.25$, a P-value less than 0.0001, and these standardized residuals:

		Race		
		Black	White	Other
Search	No	−3.43	4.55	0.28
	Yes	3.84	−5.09	−0.31

QUESTION: What's your conclusion?

The very low P-value leads me to reject the null hypothesis. There's strong evidence that police decisions to search cars at traffic stops are associated with the driver's race.

The largest residuals are for White drivers, who are searched less often than independence would predict. It appears that Black drivers' cars are searched more often.

Chi-Square and Causation

Chi-square tests are common. Tests for independence are especially widespread. Unfortunately, many people interpret a small P-value as proof of causation. We know better. Just as correlation between quantitative variables does not demonstrate causation, a failure of independence between two categorical variables does not show a cause-and-effect relationship between them, nor should we say that one variable *depends* on the other.

The chi-square test for independence treats the two variables symmetrically. There is no way to differentiate the direction of any possible causation from one variable to the other. In our example, it is unlikely that having hepatitis causes one to crave a tattoo, but other examples are not so clear.

In this case it's easy to imagine that lurking variables are responsible for the observed lack of independence. Perhaps the lifestyles of some people include both tattoos and behaviors that put them at increased risk of hepatitis C, such as body piercings or even drug use. Even a small subpopulation of people with such a lifestyle among those with tattoos might be enough to create the observed result. After all, we observed only 25 patients with both tattoos and hepatitis.

In some sense, a failure of independence between two categorical variables is less impressive than a strong, consistent, linear association between quantitative variables. Two categorical variables can fail the test of independence in many ways, including ways that show no consistent pattern of failure. Examination of the chi-square standardized residuals can help you think about the underlying patterns.

JUST CHECKING

Which of the three chi-square tests—goodness-of-fit, homogeneity, or independence—would you use in each of the following situations?

7. A restaurant manager wonders whether customers who dine on Friday nights have the same preferences among the four "chef's special" entrées as those who dine on Saturday nights. One weekend he has the wait staff record which entrées were ordered each night. Assuming these customers to be typical of all weekend diners, he'll compare the distributions of meals chosen Friday and Saturday.

8. Company policy calls for parking spaces to be assigned to everyone at random, but you suspect that may not be so. There are three lots of equal size: lot A, next to the building; lot B, a bit farther away; and lot C, on the other side of the highway. You gather data about employees at middle management level and above to see how many were assigned parking in each lot.

9. Is a student's social life affected by where the student lives? A campus survey asked a random sample of students whether they lived in a dormitory, in off-campus housing, or at home, and whether they had been out on a date 0, 1–2, 3–4, or 5 or more times in the past two weeks.

What Can Go Wrong?

- **Don't use chi-square methods unless you have counts.** All three of the chi-square tests apply only to counts. Other kinds of data can be arrayed in two-way tables. Just because numbers are in a two-way table doesn't make them suitable for chi-square analysis. Data reported as proportions or percentages can be suitable for chi-square procedures, *but only after they are converted to counts.* If you try to do the calculations without first finding the counts, your results will be wrong.

A S   *Simulation:* **Sample Size and Chi-Square.** Chi-square statistics have a peculiar problem. They don't respond to increasing the sample size in quite the same way you might expect.

- **Beware large samples.** Beware *large* samples?! That's not the advice you're used to hearing. The chi-square tests, however, are unusual. Be wary of chi-square tests performed on very large samples. No hypothesized distribution fits perfectly, no two groups are exactly homogeneous, and two variables are rarely perfectly independent. The degrees of freedom for chi-square tests don't grow with the sample size. With a sufficiently large sample size, a chi-square test can always reject the null hypothesis. But we have no measure of how far the data are from the null model. There are no confidence intervals to help us judge the effect size.

- **Don't say that one variable "depends" on the other just because they're not independent.** Dependence suggests a pattern and implies causation, but variables can fail to be independent in many different ways. When variables fail the test for independence, you might just say they are "associated."

CONNECTIONS

Chi-square methods relate naturally to inference methods for proportions. We can think of a test of homogeneity as stepping from a comparison of two proportions to a question of whether three or more proportions are equal. The standard deviations of the residuals in each cell are linked to the expected counts much like the standard deviations we found for proportions.

Independence is, of course, a fundamental concept in Statistics. But chi-square tests do not offer a general way to check on independence for all those times when we have had to assume it.

Stacked bar charts or side-by-side pie charts can help us think about patterns in two-way tables. A histogram or boxplot of the standardized residuals can help locate extraordinary values.

WHAT HAVE WE LEARNED?

We've learned how to test hypotheses about categorical variables. We use one of three related methods. All look at counts of data in categories, and all rely on chi-square models, a new family indexed by degrees of freedom.

- ▶ Goodness-of-fit tests compare the observed distribution of a single categorical variable to an expected distribution based on a theory or model.

- ▶ Tests of homogeneity compare the distribution of several groups for the same categorical variable.

- ▶ Tests of independence examine counts from a single group for evidence of an association between two categorical variables.

We've seen that, mechanically, these tests are almost identical. Although the tests appear to be one-sided, we've learned that conceptually they are many-sided, because there are many ways that a table of counts can deviate significantly from what we hypothesized. When that happens and we reject the null hypothesis, we've learned to examine standardized residuals in order to better understand patterns as in the table.

Terms

Cell	A cell is one element of a table corresponding to a specific row and a specific column. Table cells can hold counts, percentages, or measurements on other variables. Or they can hold several values (p. 635).
Chi-square model	Chi-square models are skewed to the right. They are parameterized by their degrees of freedom and become less skewed with increasing degrees of freedom (pp. 636, 640).
Chi-square statistic	The chi-square statistic can be used to test whether the observed counts in a frequency distribution or contingency table match the counts we would expect according to some model. It is calculated as

$$\chi^2 = \sum_{all\ cells} \frac{(Obs - Exp)^2}{Exp}.$$

	Chi-square statistics differ in how expected counts are found, depending on the question asked (p. 636).
Chi-square test of goodness-of-fit	A test of whether the distribution of counts in one categorical variable matches the distribution predicted by a model is called a test of goodness-of-fit. In a chi-square goodness-of-fit test, the expected counts come from the predicting model. The test finds a P-value from a chi-square model with $n - 1$ degrees of freedom, where n is the number of categories in the categorical variable (pp. 633, 637).
Chi-square test of homogeneity	A test comparing the distribution of counts for two or more groups on the same categorical variable. A chi-square test of homogeneity finds expected counts based on the overall frequencies, adjusted for the totals in each group under the (null hypothesis) assumption that the distributions are the same for each group. We find a P-value from a chi-square distribution with $(\#Rows - 1) \times (\#Cols - 1)$ degrees of freedom, where $\#Rows$ gives the number of categories and $\#Cols$ gives the number of independent groups (p. 641).
Chi-square test of independence	A test of whether two categorical variables are independent examines the distribution of counts for one group of individuals classified according to both variables. A chi-square test of *independence* finds expected counts by assuming that knowing the marginal totals tells us the cell frequencies, assuming that there is no association between the variables. This turns out to be the same calculation as a test of homogeneity. We find a P-value from a chi-square distribution with $(\#Rows - 1) \times (\#Cols - 1)$ degrees of freedom, where $\#Rows$ gives the number of categories in one variable and $\#Cols$ gives the number of categories in the other (p. 647).
Chi-square component	The components of a chi-square calculation are

$$\frac{(Observed - Expected)^2}{Expected},$$

	found for each cell of the table (pp. 639, 645).
Standardized residual	In each cell of a two-way table, a standardized residual is the square root of the chi-square component for that cell with the sign of the *Observed − Expected* difference:

$$\frac{(Obs - Exp)}{\sqrt{Exp}}.$$

	When we reject a chi-square test, an examination of the standardized residuals can sometimes reveal more about how the data deviate from the null model (p. 645).
Two-way table	Each *cell* of a two-way table shows counts of individuals. One way classifies a sample according to a categorical variable. The other way can classify different groups of individu-

als according to the same variable or classify the same individuals according to a different categorical variable (pp. 641, 647).

Contingency table   A two-way table that classifies individuals according to two categorical variables (p. 647).

Skills

► Be able to recognize when a test of goodness-of-fit, a test of homogeneity, or a test of independence would be appropriate for a table of counts.

► Understand that the degrees of freedom for a chi-square test depend on the dimensions of the table and not on the sample size. Understand that this means that increasing the sample size increases the ability of chi-square procedures to reject the null hypothesis.

► Be able to display and interpret counts in a two-way table.

► Know how to use the chi-square tables to perform chi-square tests.

► Know how to compute a chi-square test using your statistics software or calculator.

► Be able to examine the standardized residuals to explain the nature of the deviations from the null hypothesis.

► Know how to interpret chi-square as a test of goodness-of-fit in a few sentences.

► Know how to interpret chi-square as a test of homogeneity in a few sentences.

► Know how to interpret chi-square as a test of independence in a few sentences.

CHI-SQUARE ON THE COMPUTER

Most statistics packages associate chi-square tests with contingency tables. Often chi-square is available as an option only when you make a contingency table. This organization can make it hard to locate the chi-square test and may confuse the three different roles that the chi-square test can take. In particular, chi-square tests for goodness-of-fit may be hard to find or missing entirely. Chi-square tests for homogeneity are computationally the same as chi-square tests for independence, so you may have to perform the mechanics as if they were tests of independence and interpret them afterwards as tests of homogeneity.

Most statistics packages work with data on individuals rather than with the summary counts. If the only information you have is the table of counts, you may find it more difficult to get a statistics package to compute chi-square. Some packages offer a way to reconstruct the data from the summary counts so that they can then be passed back through the chi-square calculation, finding the cell counts again. Many packages offer chi-square standardized residuals (although they may be called something else).

DATA DESK

Select variables.
From the **Calc** menu, choose **Contingency Table.** From the table's HyperView menu choose **Table Options.** (Or Choose **Calc** > **Calculation Options** > **Table Options.**) In the dialog, check the boxes for **Chi Square** and for **Standardized Residuals.** Data Desk will display the chi-square and its P-value below the table, and the standardized residuals within the table.

COMMENTS

Data Desk automatically treats variables selected for this command as categorical variables even if their elements are numerals.
The **Compute Counts** command in the table's HyperView menu will make variables that hold the table contents (as selected in the Table Options dialog), including the standardized residuals.

EXCEL

Excel offers the function **CHITEST(actual_range, expected_range),** which computes a chi-square value for homogeneity. Both ranges are of the form UpperLeftcell:LowerRightCell, specifying two rectangular tables that must hold counts (although Excel will not check for integer values). The two tables must be of the same size and shape.

COMMENTS

Excel's documentation claims this is a test for independence and labels the input ranges accordingly, but Excel offers no way to find expected counts, so the function is not particularly useful for testing independence. You can use this function only if you already know both tables of counts or are willing to program additional calculations.

JMP

From the **Analyze** menu, select **Fit Y by X.** Select variables: a Y, Response variable that holds responses for one variable, and an X, Factor variable that holds responses for the other. Both selected variables must be Nominal or Ordinal. JMP will make a plot and a contingency table. Below the contingency table, **JMP** offers a **Tests** panel. In that panel the chi square for independence is called a **Pearson ChiSquare.** The table also offers the P-value.

Click on the Contingency Table title bar to drop down a menu that offers to include a **Deviation** and **Cell Chi square** in each cell of the table.

COMMENTS

JMP will choose a chi-square analysis for a **Fit Y by X** if both variables are nominal or ordinal (marked with an N or O), but not otherwise. Be sure the variables have the right type.

Deviations are the observed—expected differences in counts. Cell chi-squares are the squares of the standardized residuals. Refer to the deviations for the sign of the difference.

Look under **Distributions** in the **Analyze** menu to find a chi-square test for goodness-of-fit.

MINITAB

From the **Stat** menu choose the **Tables** submenu. From that menu, choose **Chi Square Test. . . .** In the dialog, identify the columns that make up the table. Minitab will display the table and print the chi-square value and its P-value.

COMMENTS

Alternatively, select the **Cross Tabulation . . .** command to see more options for the table, including expected counts and standardized residuals.

SPSS

From the **Analyze** menu, choose the **Descriptive Statistics** submenu. From that submenu, choose **Crosstabs. . . .** In the Crosstabs dialog, assign the row and column variables from the variable list. Both variables must be categorical. Click the **Cells** button to specify that standardized residuals should be displayed. Click the **Statistics** button to specify a chi-square test.

COMMENTS

SPSS offers only variables that it knows to be categorical in the variable list for the Crosstabs dialog. If the variables you want are missing, check that they have the right type.

TI-83/84 PLUS

The TI-83 does not have a routine for the chi-square goodness-of-fit test.

To test hyphothesis of homogeneity or independence, enter the data as a matrix. Push the **MATRIX** button, and choose **EDIT** matrix. Specify the dimensions of the table, rows $\times$ columns, then enter the appropriate counts. To do the test, choose **C: χ^2-Test** from the **STAT TESTS** menu. Note that the calculator automatically stores the expected counts in a matrix you specify.

TI-89

To test goodness-of-fit, enter the observed counts in a list and the expected counts in another list. Expected counts can be entered as **n*p**, and the calculator will compute them for you.

From the **STAT Tests** menu, select **7:Chi2 GOF**. Enter the list names using VAR-LINK and the degrees of freedom, $k - 1$, where k is the number of categories. Select whether to simply calculate or display the result with the area corresponding to the P-value highlighted.

To test a hypothesis of homogeneity or independence, enter the data as a matrix. From the home screen, press [APPS] and select **6:Data/Matrix Editor,** then select **3:New.** Specify type as Matrix and name the matrix in the **Variable** box. Specify the number of rows and columns. Type the entries, pressing [ENTER] after each. Press [2nd] [ESC] to leave the editor.

To do the test, choose **8:Chi2 2-way** from the **STAT Tests** menu.

EXERCISES

1. **Which test?** For each of the following situations, state whether you'd use a chi-square goodness-of-fit test, a chi-square test of homogeneity, a chi-square test of independence, or some other statistical test:

   a) A brokerage firm wants to see whether the type of account a customer has (Silver, Gold, or Platinum) affects the type of trades that customer makes (in person, by phone, or on the Internet). It collects a random sample of trades made for its customers over the past year and performs a test.

   b) That brokerage firm also wants to know if the type of account affects the size of the account (in dollars). It performs a test to see if the mean size of the account is the same for the three account types.

   c) The academic research office at a large community college wants to see whether the distribution of courses chosen (Humanities, Social Science, or Science) is different for its residential and nonresidential students. It assembles last semester's data and performs a test.

2. **Which test, again?** For each of the following situations, state whether you'd use a chi-square goodness-of-fit test, a chi-square test of homogeneity, a chi-square test of independence, or some other statistical test:

   a) Is the quality of a car affected by what day it was built? A car manufacturer examines a random sample of the warranty claims filed over the past two years to test whether defects are randomly distributed across days of the workweek.

   b) A medical researcher wants to know if blood cholesterol level is related to heart disease. She examines a database of 10,000 patients, testing whether the cholesterol level (in milligrams) is related to whether or not a person has heart disease.

   c) A student wants to find out whether political leaning (liberal, moderate, or conservative) is related to choice of major. He surveys 500 randomly chosen students and performs a test.

3. **Dice.** After getting trounced by your little brother in a children's game, you suspect the die he gave you to roll may be unfair. To check, you roll it 60 times, recording the number of times each face appears. Do these results cast doubt on the die's fairness?

   a) If the die is fair, how many times would you expect each face to show?

   b) To see if these results are unusual, will you test goodness-of-fit, homogeneity, or independence?

   c) State your hypotheses.

   d) Check the conditions.

   e) How many degrees of freedom are there?

   f) Find χ^2 and the P-value.

   g) State your conclusion.

Face	Count
1	11
2	7
3	9
4	15
5	12
6	6

4. **M&M's.** As noted in an earlier chapter, the Masterfoods Company says that until very recently yellow candies made up 20% of its milk chocolate M&M's, red another 20%, and orange, blue, and green 10% each. The rest are brown. On his way home from work the day he was writing these exercises, one of the authors bought a bag of plain M&M's. He got 29 yellow ones, 23 red, 12 orange, 14 blue, 8 green, and 20 brown. Is this sample consistent with the company's stated proportions? Test an appropriate hypothesis and state your conclusion.

   a) If the M&M's are packaged in the stated proportions, how many of each color should the author have expected to get in his bag?

   b) To see if his bag was unusual, should he test goodness-of-fit, homogeneity, or independence?

   c) State the hypotheses.

   d) Check the conditions.

   e) How many degrees of freedom are there?

   f) Find χ^2 and the P-value.

   g) State a conclusion.

5. **Nuts.** A company says its premium mixture of nuts contains 10% Brazil nuts, 20% cashews, 20% almonds, and 10% hazelnuts, and the rest are peanuts. You buy a large can and separate the various kinds of nuts. Upon weighing them, you find there are 112 grams of Brazil nuts, 183 grams of cashews, 207 grams of almonds,

71 grams of hazelnuts, and 446 grams of peanuts. You wonder whether your mix is significantly different from what the company advertises.

a) Explain why the chi-square goodness-of-fit test is not an appropriate way to find out.

b) What might you do instead of weighing the nuts in order to use a χ^2 test?

6. **Mileage.** A salesman who is on the road visiting clients thinks that, on average, he drives the same distance each day of the week. He keeps track of his mileage for several weeks and discovers that he averages 122 miles on Mondays, 203 miles on Tuesdays, 176 miles on Wednesdays, 181 miles on Thursdays, and 108 miles on Fridays. He wonders if this evidence contradicts his belief in a uniform distribution of miles across the days of the week. Explain why it is not appropriate to test his hypothesis using the chi-square goodness-of-fit test.

7. **NYPD and race.** Census data for New York City indicate that 29.2% of the under-18 population is white, 28.2% black, 31.5% Latino, 9.1% Asian, and 2% other ethnicities. The New York Civil Liberties Union points out that, of 26,181 police officers, 64.8% are white, 14.5% black, 19.1% Hispanic, and 1.4% Asian. Do the police officers reflect the ethnic composition of the city's youth? Test an appropriate hypothesis and state your conclusion.

8. **Violence against women 2006.** In its study *When Men Murder Women: An Analysis of 2006 Homicide Data*, Sept. 2008, the Violence Policy Center (www.vpc.org) reported that 1836 women were murdered by men in 2006. Of these victims, a weapon could be identified for 1670 of them. Of those for whom a weapon could be identified, 909 were killed by guns, 334 by knives or other cutting instruments, 227 by other weapons, and 200 by personal attack (battery, strangulation, etc.). The FBI's Uniform Crime Report says that, among all murders nationwide, the weapon use rates were as follows: guns 63.4%, knives 13.1%, other weapons 16.8%, personal attack 6.7%. Is there evidence that violence against women involves different weapons than other violent attacks in the United States?

9. **Fruit flies.** Offspring of certain fruit flies may have yellow or ebony bodies and normal wings or short wings. Genetic theory predicts that these traits will appear in the ratio 9:3:3:1 (9 yellow, normal: 3 yellow, short: 3 ebony, normal: 1 ebony, short). A researcher checks 100 such flies and finds the distribution of the traits to be 59, 20, 11, and 10, respectively.

a) Are the results this researcher observed consistent with the theoretical distribution predicted by the genetic model?

b) If the researcher had examined 200 flies and counted exactly twice as many in each category—118, 40, 22, 20—what conclusion would he have reached?

c) Why is there a discrepancy between the two conclusions?

T 10. **Pi.** Many people know the mathematical constant π is approximately 3.14. But that's not exact. To be more precise, here are 20 decimal places: 3.14159265358979323846. Still not exact, though. In fact, the actual value is irrational, a decimal that goes on forever without any repeating pattern. But notice that there are no 0's and only one 7 in the 20 decimal places above. Does that pattern persist, or do all the digits show up with equal frequency? The table shows the number of times each digit appears in the first million digits. Test the hypothesis that the digits 0 through 9 are uniformly distributed in the decimal representation of π.

The first million digits of π	
Digit	Count
0	99,959
1	99,758
2	100,026
3	100,229
4	100,230
5	100,359
6	99,548
7	99,800
8	99,985
9	100,106

T 11. **Hurricane frequencies.** The National Hurricane Center provides data that list the numbers of large (category 3, 4, or 5) hurricanes that have struck the United States, by decade since 1851 (http://www.nhc.noaa.gov/Deadliest_Costliest.shtml). The data are given below.

Decade	Count	Decade	Count
1851–1860	6	1931–1940	8
1861–1870	1	1941–1950	10
1871–1880	7	1951–1960	9
1881–1890	5	1961–1970	6
1891–1900	8	1971–1980	4
1901–1910	4	1981–1990	4
1911–1920	7	1991–2000	5
1921–1930	5	2001–2006	7

Recently, there's been some concern that perhaps the number of large hurricanes has been increasing. The natural null hypothesis would be that the frequency of such hurricanes has remained constant.

a) With 96 large hurricanes observed over the 16 periods, what are the expected value(s) for each cell?

b) What kind of chi-square test would be appropriate?

c) State the null and alternative hypotheses.

d) How many degrees of freedom are there?

e) The value of χ^2 is 12.67. What's the P-value?

f) State your conclusion.

g) Look again at the definition of the last "decade." Does that alter your conclusion at all?

T 12. **Lottery numbers.** The fairness of the South African lottery was recently challenged by one of the country's political parties. The lottery publishes historical statistics at its website (http://www.nationallottery.co.za/lotto/statistics.aspx). Here is a table of the number of times

lottery and as the "bonus ball" number as of June 2007:

Number	Count	Bonus	Number	Count	Bonus
1	81	14	26	78	12
2	91	16	27	83	16
3	78	14	28	76	7
4	77	12	29	76	12
5	67	16	30	99	16
6	87	12	31	78	10
7	88	15	32	73	15
8	90	16	33	81	14
9	80	9	34	81	13
10	77	19	35	77	15
11	84	12	36	73	8
12	68	14	37	64	17
13	79	9	38	70	11
14	90	12	39	67	14
15	82	9	40	75	13
16	103	15	41	84	11
17	78	14	42	79	8
18	85	14	43	74	14
19	67	18	44	87	14
20	90	13	45	82	19
21	77	13	46	91	10
22	78	17	47	86	16
23	90	14	48	88	21
24	80	8	49	76	13
25	65	11			

We wonder if all the numbers are equally likely to be the "bonus ball."
a) What kind of test should we perform?
b) There are 655 bonus ball observations. What are the appropriate expected value(s) for the test?
c) State the null and alternative hypotheses.
d) How many degrees of freedom are there?
e) The value of χ^2 is 34.5. What's the P-value?
f) State your conclusion.

13. **Childbirth, part 1.** There is some concern that if a woman has an epidural to reduce pain during childbirth, the drug can get into the baby's bloodstream, making the baby sleepier and less willing to breastfeed. In December 2006, the *International Breastfeeding Journal* published results of a study conducted at Sydney University. Researchers followed up on 1178 births, noting whether the mother had an epidural and whether the baby was still nursing after 6 months. Below are their results.
a) What kind of test would be appropriate?
b) State the null and alternative hypotheses.

		Epidural?		
		Yes	No	Total
Breastfeeding @ 6 months?	Yes	206	498	**704**
	No	190	284	**474**
	Total	**396**	**782**	**1178**

14. **Does your doctor know?** A survey[7] of articles from the *New England Journal of Medicine (NEJM)* classified them according to the principal statistics methods used. The articles recorded were all noneditorial articles appearing during the indicated years. Let's just look at whether these articles used statistics at all.

	Publication Year			
	1978–79	1989	2004–05	Total
No stats	90	14	40	**144**
Stats	242	101	271	**614**
Total	**332**	**115**	**311**	**758**

Has there been a change in the use of Statistics?
a) What kind of test would be appropriate?
b) State the null and alternative hypotheses.

15. **Childbirth, part 2.** In Exercise 13, the table shows results of a study investigating whether aftereffects of epidurals administered during childbirth might interfere with successful breastfeeding. We're planning to do a chi-square test.
a) How many degrees of freedom are there?
b) The smallest expected count will be in the epidural/ no breastfeeding cell. What is it?
c) Check the assumptions and conditions for inference.

16. **Does your doctor know? (part 2).** The table in Exercise 14 shows whether *NEJM* medical articles during various time periods included statistics or not. We're planning to do a chi-square test.
a) How many degrees of freedom are there?
b) The smallest expected count will be in the 1989/No cell. What is it?
c) Check the assumptions and conditions for inference.

17. **Childbirth, part 3.** In Exercises 13 and 15, we've begun to examine the possible impact of epidurals on successful breastfeeding.
a) Calculate the component of chi-square for the epidural/no breastfeeding cell.
b) For this test, $\chi^2 = 14.87$. What's the P-value?
c) State your conclusion.

18. **Does your doctor know? (part 3).** In Exercises 14 and 16, we've begun to examine whether the use of statistics in *NEJM* medical articles has changed over time.
a) Calculate the component of chi-square for the 1989/No cell.
b) For this test, $\chi^2 = 25.28$. What's the P-value?
c) State your conclusion.

19. **Childbirth, part 4.** In Exercises 13, 15, and 17, we've tested a hypothesis about the impact of epidurals on successful breastfeeding. The following table shows the test's residuals.

---

[7] Suzanne S. Switzer and Nicholas J. Horton, "What Your Doctor Should Know about Statistics (but Perhaps Doesn't)," *Chance*, 20:1, 2007.

		Epidural?	
		Yes	**No**
Breastfeeding at 6 months?	**Yes**	−1.99	1.42
	No	2.43	−1.73

a) Show how the residual for the epidural/no breastfeeding cell was calculated.
b) What can you conclude from the standardized residuals?

20. **Does your doctor know? (part 4).** In Exercises 14, 16, and 18, we've tested a hypothesis about whether the use of statistics in *NEJM* medical articles has changed over time. The table shows the test's residuals.

	1978–79	1989	2004–05
No stats	3.39	−1.68	−2.48
Stats	−1.64	0.81	1.20

a) Show how the residual for the 1989/No cell was calculated.
b) What can you conclude from the patterns in the standardized residuals?

21. **Childbirth, part 5.** In Exercises 13, 15, 17, and 19, we've looked at a study examining epidurals as one factor that might inhibit successful breastfeeding of newborn babies. Suppose a broader study included several additional issues, including whether the mother drank alcohol, whether this was a first child, and whether the parents occasionally supplemented breastfeeding with bottled formula. Why would it not be appropriate to use chi-square methods on the 2 × 8 table with yes/no columns for each potential factor?

22. **Does your doctor know? (part 5).** In Exercises 14, 16, 18, and 20, we considered data on articles in the *NEJM*. The original study listed 23 different Statistics methods. (The list read: *t*-tests, contingency tables, linear regression, ...) Why would it not be appropriate to use a chi-square test on the 23 × 3 table with a row for each method?

T 23. *Titanic.* Here is a table we first saw in Chapter 3 showing who survived the sinking of the *Titanic* based on whether they were crew members, or passengers booked in first-, second-, or third-class staterooms:

	Crew	First	Second	Third	Total
Alive	212	202	118	178	**710**
Dead	673	123	167	528	**1491**
Total	**885**	**325**	**285**	**706**	**2201**

a) If we draw an individual at random, what's the probability that we will draw a member of the crew?
b) What's the probability of randomly selecting a third-class passenger who survived?

c) What's the probability of a randomly selected passenger surviving, given that the passenger was a first-class passenger?
d) If someone's chances of surviving were the same regardless of their status on the ship, how many members of the crew would you expect to have lived?
e) State the null and alternative hypotheses.
f) Give the degrees of freedom for the test.
g) The chi-square value for the table is 187.8, and the corresponding P-value is barely greater than 0. State your conclusions about the hypotheses.

T 24. **NYPD and sex discrimination.** The table below shows the rank attained by male and female officers in the New York City Police Department (NYPD). Do these data indicate that men and women are equitably represented at all levels of the department?

		Male	Female
Rank	**Officer**	21,900	4,281
	Detective	4,058	806
	Sergeant	3,898	415
	Lieutenant	1,333	89
	Captain	359	12
	Higher ranks	218	10

a) What's the probability that a person selected at random from the NYPD is a female?
b) What's the probability that a person selected at random from the NYPD is a detective?
c) Assuming no bias in promotions, how many female detectives would you expect the NYPD to have?
d) To see if there is evidence of differences in ranks attained by males and females, will you test goodness-of-fit, homogeneity, or independence?
e) State the hypotheses.
f) Check the conditions.
g) How many degrees of freedom are there?
h) The chi-square value for the table is 290.1 and the P-value is less than 0.0001. State your conclusion about the hypotheses.

25. *Titanic, again.* Examine and comment on this table of the standardized residuals for the chi-square test you looked at in Exercise 23.

	Crew	First	Second	Third
Alive	−4.35	9.49	2.72	−3.30
Dead	3.00	−6.55	−1.88	2.27

26. **NYPD again.** Examine and comment on this table of the standardized residuals for the chi-square test you looked at in Exercise 24.

	Male	Female
Officer	−2.34	5.57
Detective	−1.18	2.80
Sergeant	3.84	−9.14
Lieutenant	3.58	−8.52
Captain	2.46	−5.86
Higher ranks	1.74	−4.14

27. **Cranberry juice.** It's common folk wisdom that drinking cranberry juice can help prevent urinary tract infections in women. In 2001 the *British Medical Journal* reported the results of a Finnish study in which three groups of 50 women were monitored for these infections over 6 months. One group drank cranberry juice daily, another group drank a lactobacillus drink, and the third drank neither of those beverages, serving as a control group. In the control group, 18 women developed at least one infection, compared to 20 of those who consumed the lactobacillus drink and only 8 of those who drank cranberry juice. Does this study provide supporting evidence for the value of cranberry juice in warding off urinary tract infections?

a) Is this a survey, a retrospective study, a prospective study, or an experiment? Explain.
b) Will you test goodness-of-fit, homogeneity, or independence?
c) State the hypotheses.
d) Check the conditions.
e) How many degrees of freedom are there?
f) Find χ^2 and the P-value.
g) State your conclusion.
h) If you concluded that the groups are not the same, analyze the differences using the standardized residuals of your calculations.

28. **Cars.** A random survey of autos parked in the student lot and the staff lot at a large university classified the brands by country of origin, as seen in the table. Are there differences in the national origins of cars driven by students and staff?

		Driver	
		Student	Staff
	American	107	105
Origin	European	33	12
	Asian	55	47

a) Is this a test of independence or homogeneity?
b) Write appropriate hypotheses.
c) Check the necessary assumptions and conditions.
d) Find the P-value of your test.
e) State your conclusion and analysis.

29. **Montana.** A poll conducted by the University of Montana classified respondents by whether they were male or female and political party, as shown in the table. We wonder if there is evidence of an association between being male or female and party affiliation.

	Democrat	Republican	Independent
Male	36	45	24
Female	48	33	16

a) Is this a test of homogeneity or independence?
b) Write an appropriate hypothesis.
c) Are the conditions for inference satisfied?
d) Find the P-value for your test.
e) State a complete conclusion.

30. **Fish diet.** Medical researchers followed 6272 Swedish men for 30 years to see if there was any association between the amount of fish in their diet and prostate cancer. ("Fatty Fish Consumption and Risk of Prostate Cancer," *Lancet*, June 2001)

Fish Consumption	Total Subjects	Prostate Cancers
Never/seldom	124	14
Small part of diet	2621	201
Moderate part	2978	209
Large part	549	42

a) Is this a survey, a retrospective study, a prospective study, or an experiment? Explain.
b) Is this a test of homogeneity or independence?
c) Do you see evidence of an association between the amount of fish in a man's diet and his risk of developing prostate cancer?
d) Does this study prove that eating fish does not prevent prostate cancer? Explain.

31. **Montana, revisited.** The poll described in Exercise 29 also investigated the respondents' party affiliations based on what area of the state they lived in. Test an appropriate hypothesis about this table and state your conclusions.

	Democrat	Republican	Independent
West	39	17	12
Northeast	15	30	12
Southeast	30	31	16

32. **Working parents.** In July 1991 and again in April 2001, the Gallup Poll asked random samples of 1015 adults about their opinions on working parents. The table summarizes responses to the question "Considering the needs of both parents and children, which of the following do you see as the ideal family in today's society?"

	1991	2001
Both work full time	142	131
One works full time, other part time	274	244
One works, other works at home	152	173
One works, other stays home for kids	396	416
No opinion	51	51

a) Is this a survey, a retrospective study, a prospective study, or an experiment?
b) Will you test goodness-of-fit, homogeneity, or independence?
c) Based on these results, do you think there was a change in people's attitudes during the 10 years between these polls?

33. **Grades.** Two different professors teach an introductory Statistics course. The table shows the distribution of final grades they reported. We wonder whether one of these professors is an "easier" grader.

	Prof. Alpha	Prof. Beta
A	3	9
B	11	12
C	14	8
D	9	2
F	3	1

a) Will you test goodness-of-fit, homogeneity, or independence?
b) Write appropriate hypotheses.
c) Find the expected counts for each cell, and explain why the chi-square procedures are not appropriate.

34. **Full moon.** Some people believe that a full moon elicits unusual behavior in people. The table shows the number of arrests made in a small town during weeks of six full moons and six other randomly selected weeks in the same year. We wonder if there is evidence of a difference in the types of illegal activity that take place.

	Full Moon	Not Full
Violent (murder, assault, rape, etc.)	2	3
Property (burglary, vandalism, etc.)	17	21
Drugs/Alcohol	27	19
Domestic abuse	11	14
Other offenses	9	6

a) Will you test goodness-of-fit, homogeneity, or independence?
b) Write appropriate hypotheses.
c) Find the expected counts for each cell, and explain why the chi-square procedures are not appropriate.

35. **Grades, again.** In some situations where the expected cell counts are too small, as in the case of the grades given by Professors Alpha and Beta in Exercise 33, we can complete an analysis anyway. We can often proceed after combining cells in some way that makes sense and also produces a table in which the conditions are satisfied. Here we create a new table displaying the same data, but calling D's and F's "Below C":

	Prof. Alpha	Prof. Beta
A	3	9
B	11	12
C	14	8
Below C	12	3

a) Find the expected counts for each cell in this new table, and explain why a chi-square procedure is now appropriate.
b) With this change in the table, what has happened to the number of degrees of freedom?
c) Test your hypothesis about the two professors, and state an appropriate conclusion.

36. **Full moon, next phase.** In Exercise 34 you found that the expected cell counts failed to satisfy the conditions for inference.

a) Find a sensible way to combine some cells that will make the expected counts acceptable.
b) Test a hypothesis about the full moon and state your conclusion.

37. **Racial steering.** A subtle form of racial discrimination in housing is "racial steering." Racial steering occurs when real estate agents show prospective buyers only homes in neighborhoods already dominated by that family's race. This violates the Fair Housing Act of 1968. According to an article in *Chance* magazine (Vol. 14, no. 2 [2001]), tenants at a large apartment complex recently filed a lawsuit alleging racial steering. The complex is divided into two parts: Section A and Section B. The plaintiffs claimed that white potential renters were steered to Section A, while African-Americans were steered to Section B. The table displays the data that were presented in court to show the locations of recently rented apartments. Do you think there is evidence of racial steering?

	New Renters		
	White	Black	Total
Section A	87	8	95
Section B	83	34	117
Total	170	42	212

38. **Titanic, redux.** Newspaper headlines at the time, and traditional wisdom in the succeeding decades, have held that women and children escaped the *Titanic* in greater proportions than men. Here's a table with the relevant

data. Do you think that survival was independent of whether the person was male or female? Explain.

	Female	Male	Total
Alive	343	367	**710**
Dead	127	1364	**1491**
Total	**470**	**1731**	**2201**

39. **Steering, revisited.** You could have checked the data in Exercise 37 for evidence of racial steering using two-proportion z procedures.
    a) Find the z-value for this approach, and show that when you square your z-value, you get the value of χ^2 you calculated in Exercise 37.
    b) Show that the resulting P-values are the same.

T 40. **Survival on the *Titanic*, one more time.** In Exercise 38 you could have checked for a difference in the chances of survival for men and women using two-proportion z procedures.
    a) Find the z-value for this approach.
    b) Show that the square of your calculated value of z is the value of χ^2 you calculated in Exercise 38.
    c) Show that the resulting P-values are the same.

T 41. **Pregnancies.** Most pregnancies result in live births, but some end in miscarriages or stillbirths. A June 2001

National Vital Statistics Report examined those outcomes in the United States during 1997, broken down by the age of the mother. The table shows counts consistent with that report. Is there evidence that the outcomes are not independent of age group?

		Live Births	Fetal Losses
	Under 20	49	13
Age of Mother	20–29	201	41
	30–34	88	21
	35 or over	49	21

T 42. **Education by age.** Use the survey results in the table to investigate differences in education level attained among different age groups in the United States.

		Age Group				
		25–34	35–44	45–54	55–64	≥65
	Not HS grad	27	50	52	71	101
Education Level	HS	82	19	88	83	59
	1–3 years college	43	56	26	20	20
	≥4 years college	48	75	34	26	20

JUST CHECKING

ANSWERS

1. We need to know how well beetles can survive 6 hours in a Plexiglas® box so that we have a baseline to compare the treatments.

2. There's no difference in survival rate in the three groups.

3. $(2 - 1)(3 - 1) = 2\,\text{df}$

4. 50

5. 2

6. The mean value for a χ^2 with 2 df is 2, so 10 seems pretty large. The P-value is probably small.

7. This is a test of homogeneity. The clue is that the question asks whether the distributions are alike.

8. This is a test of goodness-of-fit. We want to test the model of equal assignment to all lots against what actually happened.

9. This is a test of independence. We have responses on two variables for the same individuals.

Learning About the World

Quick Review

We continue to explore how to answer questions about the statistics we get from samples and experiments. In this part, those questions have been about means—means of one sample, two independent samples, or matched pairs—and about proportions in several categories and in relationships between categorical variables. Here's a brief summary of the key concepts and skills:

▶ A confidence interval uses a sample statistic to estimate a range of possible values for the parameter of a population model.

▶ A hypothesis test proposes a model for the population, then examines the observed statistics to see if the model is plausible.

▶ Statistical inference procedures for proportions are based on the Central Limit Theorem. We can make inferences about a single proportion or the difference of two proportions using Normal models.

▶ Statistical inference procedures for means are also based on the Central Limit Theorem, but we don't usually know the population standard deviation. Student's *t*-models take the additional uncertainty of independently estimating the standard deviation into account.

- We can make inferences about one mean, the difference of two independent means, or the mean of paired differences using *t*-models.
- No inference procedure is valid unless the underlying assumptions are true. Always think about the assumptions and check the conditions before proceeding.
- Because *t*-models assume that samples are drawn from Normal populations, data in the sample should appear to be nearly Normal. Skewness and outliers are particularly problematic.
- When there are two variables, you must think carefully about how the data were collected. You may use two-sample *t* procedures only if the groups are independent.
- Unless there is some obvious reason to suspect that two independent populations have the same standard deviation, you should not pool the variances. It is never wrong to use unpooled *t* procedures.

- If two groups are somehow paired, the data are *not* from independent groups. You must use matched-pairs *t* procedures and test the mean difference rather than the difference in the means.

▶ Not all sampling distributions are unimodal, symmetric, or bell-shaped. Inferences about distributions of counts use chi-square models, which are unimodal but skewed to the high end. Nevertheless, the sampling distribution plays the same role in inference, helping us to translate between probabilities and values based on data.

- To see if an observed distribution is consistent with a proposed model, use a chi-square goodness-of-fit test.
- To see if two or more observed distributions could have arisen from populations with the same model, use a test of homogeneity.
- To see if two categorical variables are independent, perform a chi-square test of independence.

▶ You can now use statistical inference to answer questions about means, proportions, distributions, and associations.

- No inference procedure is valid unless the underlying assumptions are true. Always check the conditions before proceeding.
- You can make inferences about a single proportion or about the difference between two proportions using Normal models.
- You can make inferences about one mean, about the difference between two independent means, or about the mean of paired differences using *t*-models.
- You can make inferences about distributions using chi-square models.
- You can make inferences about association between categorical variables using chi-square models.

Now for some opportunities to review these concepts. Be careful. You have a lot of thinking to do. These review exercises mix questions about proportions and means. You have to determine which of our inference procedures is appropriate in each situation. Then you have to check the proper assumptions and conditions. Keeping track of those can be difficult, so first we summarize the many procedures with their corresponding assumptions and conditions on the next page. Look them over carefully . . . then, on to the Exercises!

Assumptions for Inference	And the Conditions That Support or Override Them
Proportions (z)	
• **One sample**	
1. Individuals are independent.	1. SRS and < 10% of the population.
2. Sample is sufficiently large.	2. Successes and failures ≥ 10.
• **Two sample**	
1. Samples are independent.	1. (Think about how the data were collected.)
2. Data in each sample are independent.	2. Both are SRSs and < 10% of populations OR random allocation.
3. Both samples are sufficiently large.	3. Successes and failures ≥ 10 for both.
Means (t)	
• **One sample** (df = $n - 1$)	
1. Individuals are independent.	1. SRS
2. Population has a Normal model.	2. Histogram is unimodal and symmetric.*
• **Matched pairs** (df = $n - 1$)	
1. Data are paired.	1. (Think about the design.)
2. Individuals are independent.	2. SRS OR random allocation.
3. Population of differences is Normal.	3. Histogram of differences is unimodal and symmetric.
• **Two independent groups** (df from technology)	
1. Groups are independent.	1. (Think about the design.)
2. Data in each group are independent.	2. SRSs OR random allocation.
3. Both populations are Normal.	3. Both histograms are unimodal and symmetric.*
Distributions/Association (χ^2)	
• **Goodness-of-fit** (df = # of cells −1; one variable, one sample compared with population model)	
1. Data are counts.	1. (Are they?)
2. Data in sample are independent.	2. SRS and < 10% of the population.
3. Group is sufficiently large.	3. All expected counts ≥ 5.
• **Homogeneity** [df = $(r - 1)(c - 1)$; samples from many populations compared on one variable]	
1. Data are counts.	1. (Are they?)
2. Data in samples are independent.	2. SRSs and < 10% OR random allocation.
3. Groups are sufficiently large.	3. All expected counts ≥ 5.
• **Independence** [df = $(r - 1)(c - 1)$; sample from one population classified on two variables]	
1. Data are counts.	1. (Are they?)
2. Data are independent.	2. SRSs and < 10% of the population.
3. Group is sufficiently large.	3. All expected counts ≥ 5.

(*less critical as n increases)

REVIEW EXERCISES

1. **Crawling.** A study published in 1993 found that babies born at different times of the year may develop the ability to crawl at different ages! The author of the study suggested that these differences may be related to the temperature at the time the infant is 6 months old. (Benson and Janette, *Infant Behavior and Development*, 1993)

   a) The study found that 32 babies born in January crawled at an average age of 29.84 weeks, with a standard deviation of 7.08 weeks. Among 21 July babies, crawling ages averaged 33.64 weeks with a standard deviation of 6.91 weeks. Is this difference significant?

   b) For 26 babies born in April, the mean and standard deviation were 31.84 and 6.21 weeks, while for 44 October babies the mean and standard deviation of crawling ages were 33.35 and 7.29 weeks. Is this difference significant?

   c) Are these results consistent with the researcher's claim? (We'll examine these data in more detail in a later chapter.)

T 2. **Mazes and smells.** Can pleasant smells improve learning? Researchers timed 21 subjects as they tried to complete paper-and-pencil mazes. Each subject attempted a maze both with and without the presence of a floral aroma. Subjects were randomized with respect to whether they did the scented trial first or second. Some of the data collected are shown in the table. Is there any evidence that the floral scent improved the subjects' ability to complete the mazes? (A. R. Hirsch and L. H. Johnston, "Odors and Learning." Chicago: Smell and Taste Treatment and Research Foundation)

Time to Complete the Maze (sec)	
Unscented	Scented
25.7	30.2
41.9	56.7
51.9	42.4
32.2	34.4
64.7	44.8
31.4	42.9
40.1	42.7
43.2	24.8
33.9	25.1
40.4	59.2
58.0	42.2
61.5	48.4
44.6	32.0
35.3	48.1
37.2	33.7
39.4	42.6
77.4	54.9
52.8	64.5
63.6	43.1
56.6	52.8
58.9	44.3

3. **Women.** The U.S. Census Bureau reports that 26% of all U.S. businesses are owned by women. A Colorado consulting firm surveys a random sample of 410 businesses in the Denver area and finds that 115 of them have women owners. Should the firm conclude that its area is unusual? Test an appropriate hypothesis and state your conclusion.

T 4. **Drugs.** In a full-page ad that ran in many U.S. newspapers in August 2002, a Canadian discount pharmacy listed costs of drugs that could be ordered from a website in Canada. The table compares prices (in US$) for commonly prescribed drugs.

Drug Name	Cost per 100 Pills		
	United States	Canada	Percent savings
Cardizem	131	83	37
Celebrex	136	72	47
Cipro	374	219	41
Pravachol	370	166	55
Premarin	61	17	72
Prevacid	252	214	15
Prozac	263	112	57
Tamoxifen	349	50	86
Vioxx	243	134	45
Zantac	166	42	75
Zocor	365	200	45
Zoloft	216	105	51

a) Give a 95% confidence interval for the average savings in dollars.

b) Give a 95% confidence interval for the average savings in percent.

c) Which analysis do you think is more appropriate? Why?

d) In small print the newspaper ad says, "Complete list of all 1500 drugs available on request." How does this comment affect your conclusions above?

T 5. **Pottery.** Archaeologists can use the chemical composition of clay found in pottery artifacts to determine whether different sites were populated by the same ancient people. They collected five samples of Romano-British pottery from each of two sites in Great Britain—the Ashley Rails site and the New Forest site—and measured the percentage of aluminum oxide in each. Based on these data, do you think the same people used these two kiln sites? Base your conclusion on a 95% confidence interval for the difference in aluminum oxide content of pottery made at the sites. (A. Tubb, A. J. Parker, and G. Nickless, "The Analysis of Romano-British Pottery by Atomic Absorption Spectrophotometry." *Archaeometry*, 22[1980]: 153–171)

Ashley Rails	19.1	14.8	16.7	18.3	17.7
New Forest	20.8	18.0	18.0	15.8	18.3

T 6. **Diet.** Thirteen overweight women volunteered for a study to determine whether eating specially prepared crackers before a meal could help them lose weight. The subjects were randomly assigned to eat crackers with different types of fiber (bran fiber, gum fiber, both, and a control cracker). Unfortunately, some of the women developed uncomfortable bloating and upset stomachs. Researchers suspected that some of the crackers might be at fault. The contingency table of *Cracker* versus *Bloat* shows the relationship between the four different types of crackers and the reported bloating. The study was paid for by the manufacturers of the gum fiber. What would you recommend to them about the prospects for marketing their new diet cracker?

		Bloat	
		Little/None	Moderate/Severe
Cracker	Bran	11	2
	Gum	4	9
	Combo	7	6
	Control	8	4

7. **Gehrig.** Ever since Lou Gehrig developed amyotrophic lateral sclerosis (ALS), this deadly condition has been commonly known as Lou Gehrig's disease. Some believe that ALS is more likely to strike athletes or the very fit. Columbia University neurologist Lewis P. Rowland recorded personal histories of 431 patients he examined between 1992 and 2002. He diagnosed 280 as having ALS; 38% of them had been varsity athletes. The other 151 had other neurological disorders, and only 26% of them had been varsity athletes. (*Science News*, Sept. 28, 2002)

a) Is there evidence that ALS is more common among athletes?

b) What kind of study is this? How does that affect the inference you made in part a?

T 8. **Teen drinking.** A study of the health behavior of school-aged children asked a sample of 15-year-olds in several different countries if they had been drunk at least twice. The results are shown in the following table. Give a 95%

confidence interval for the difference in the rates for males and females. Be sure to check the assumptions that support your chosen procedure, and explain what your interval means. (*Health and Health Behavior Among Young People.* Copenhagen: World Health Organization, 2000)

Country	Percent of 15-Year-Olds Drunk at Least Twice	
	Female	Male
Denmark	63	71
Wales	63	72
Greenland	59	58
England	62	51
Finland	58	52
Scotland	56	53
No. Ireland	44	53
Slovakia	31	49
Austria	36	49
Canada	42	42
Sweden	40	40
Norway	41	37
Ireland	29	42
Germany	31	36
Latvia	23	47
Estonia	23	44
Hungary	22	43
Poland	21	39
USA	29	34
Czech Rep.	22	36
Belgium	22	36
Russia	25	32
Lithuania	20	32
France	20	29
Greece	21	24
Switzerland	16	25
Israel	10	18

9. **Genetics.** Two human traits controlled by a single gene are the ability to roll one's tongue and whether one's ear lobes are free or attached to the neck. Genetic theory says that people will have neither, one, or both of these traits in the ratio 1:3:3:9 (1—attached, noncurling; 3—attached, curling; 3—free, noncurling; 9—free, curling). An Introductory Biology class of 122 students collected the data shown. Are they consistent with the genetic theory? Test an appropriate hypothesis and state your conclusion.

	Trait			
	Attached, noncurling	Attached, curling	Free, noncurling	Free, curling
Count	10	22	31	59

10. **Speeding.** A newspaper report in August 2002 raised the issue of racial bias in the issuance of speeding tickets. The following facts were noted:
    - 16% of drivers registered in New Jersey are black.

- Of the 324 speeding tickets issued in one month on a 65 mph section of the New Jersey Turnpike, 25% went to black drivers.

   a) Is the percentage of speeding tickets issued to blacks unusually high compared with the state's registration information?
   b) Does this prove that racial profiling was used?
   c) What other statistics would you like to know about this situation?

11. **Babies.** The National Perinatal Statistics Unit of the Sydney Children's Hospital reports that the mean birth weight of all babies born in Australia in 1999 was 3360 grams—about 7.41 pounds. A Missouri hospital reports that the average weight of 112 babies born there last year was 7.68 pounds, with a standard deviation of 1.31 pounds. If we believe the Missouri babies fairly represent American newborns, is there any evidence that U.S. babies and Australian babies do not weigh the same amount at birth?

12. **Petitions.** To get a voter initiative on a state ballot, petitions that contain at least 250,000 valid voter signatures must be filed with the Elections Commission. The board then has 60 days to certify the petitions. A group wanting to create a statewide system of universal health insurance has just filed petitions with a total of 304,266 signatures. As a first step in the process, the Board selects an SRS of 2000 signatures and checks them against local voter lists. Only 1772 of them turn out to be valid.

   a) What percent of the sample signatures were valid?
   b) What percent of the petition signatures submitted must be valid in order to have the initiative certified by the Elections Commission?
   c) What will happen if the Elections Commission commits a Type I error?
   d) What will happen if the Elections Commission commits a Type II error?
   e) Does the sample provide evidence in support of certification? Explain.
   f) What could the Elections Commission do to increase the power of the test?

13. **Feeding fish.** In the midwestern United States, a large aquaculture industry raises largemouth bass. Researchers wanted to know whether the fish would grow better if fed a natural diet of fathead minnows or an artificial diet of food pellets. They stocked six ponds with bass fingerlings weighing about 8 grams. For one year, the fish in three of the ponds were fed minnows, and the others were fed the commercially prepared pellets. The fish were then harvested, weighed, and measured. The bass fed a natural food source had a higher average length (19.6 cm) and weight (95.9 g) than those fed the commercial fish food (17.3 cm and 72.0 g, respectively). The researchers reported P-values for both measurements to be less than 0.001.

   a) Explain to someone who has not studied Statistics what the P-values mean here.
   b) What advice should the researchers give the people who raise largemouth bass?
   c) If that advice turns out to be incorrect, what type of error occurred?

14. Risk. A study of auto safety determined the number of driver deaths per million vehicle sales for the model years 1995–1999, classified by type of vehicle. The data below are for 6 midsize models and 6 SUVs. We wonder if there is evidence that drivers of SUVs are less likely to die. We hope to create a 95% confidence interval for the difference in driver death rates for the two types of vehicles. Are these data appropriate for this inference? Explain. (Ross and Wenzel, *An Analysis of Traffic Deaths by Vehicle Type and Model,* March 2002)

Midsize	47	54	64	76	88	97
SUV	55	60	62	76	91	109

15. Twins. In 2000 *The Journal of the American Medical Association* published a study that examined a sample of pregnancies that resulted in the birth of twins. Births were classified as preterm with intervention (induced labor or cesarean), preterm without such procedures, or term or postterm. Researchers also classified the pregnancies by the level of prenatal medical care the mother received (inadequate, adequate, or intensive). The data, from the years 1995–1997, are summarized in the table below. Figures are in thousands of births. (*JAMA,* 284[2002]:335–341)

		Preterm (induced or cesarean)	Preterm (without procedures)	Term or postterm	Total
Level of Prenatal Care	Intensive	18	15	28	61
	Adequate	46	43	65	154
	Inadequate	12	13	38	63
	Total	76	71	131	278

Twin Births 1995–1997 (in thousands)

Is there evidence of an association between the duration of the pregnancy and the level of care received by the mother?

16. Twins, again. After reading of the *JAMA* study in Exercise 15, a large city hospital examined their records of twin births for several years, and found the data summarized in the following table. Is there evidence that the way the hospital deals with pregnancies involving twins may have changed?

		1990	1995	2000
Outcome of Pregnancy	Preterm (induced or cesarean)	11	13	19
	Preterm (without procedures)	13	14	18
	Term or postterm	27	26	32

17. Age. In a study of how depression may impact one's ability to survive a heart attack, the researchers reported the ages of the two groups they examined. The mean age of 2397 patients without cardiac disease was 69.8 years (SD = 8.7 years), while for the 450 patients with cardiac disease the mean and standard deviation of the ages were 74.0 and 7.9, respectively.
a) Create a 95% confidence interval for the difference in mean ages of the two groups.
b) How might an age difference confound these research findings about the relationship between depression and ability to survive a heart attack?

18. Smoking. In the depression and heart attack research described in Exercise 17, 32% of the diseased group were smokers, compared with only 23.7% of those free of heart disease.
a) Create a 95% confidence interval for the difference in the proportions of smokers in the two groups.
b) Is this evidence that the two groups in the study were different? Explain.
c) Could this be a problem in analyzing the results of the study? Explain.

19. Back to Montana. The respondents to the Montana poll described in the exercises in Chapter 26 were also classified by income level: low (under $20,000), middle ($20,000–$35,000), or high (over $35,000). Is there any evidence that party enrollment there is associated with income? Test an appropriate hypothesis about this table, and state your conclusions.

	Democrat	Republican	Independent
Low	30	16	12
Middle	28	24	22
High	26	38	6

20. Hearing. Fitting someone for a hearing aid requires assessing the patient's hearing ability. In one method of assessment, the patient listens to a tape of 50 English words. The tape is played at low volume, and the patient is asked to repeat the words. The patient's hearing ability score is the number of words perceived correctly. Four tapes of equivalent difficulty are available so that each ear can be tested with more than one hearing aid. These lists were created to be equally difficult to perceive in silence, but hearing aids must work in the presence of background noise. Researchers had 24 subjects with normal hearing compare two of the tapes when a background noise was present, with the order of the tapes randomized. Is it reasonable to assume that the two lists are still equivalent for purposes of the hearing test when there is background noise? Base your decision on a confidence interval for the mean difference in the number of words people might misunderstand. (Faith Loven, *A Study of the Interlist Equivalency of the CID W-22 Word List Presented in Quiet and in Noise*. University of Iowa, 1981)

Subject	List A	List B	Subject	List A	List B
1	24	26	13	36	32
2	32	24	14	32	34
3	20	22	15	38	32
4	14	18	16	14	18
5	32	24	17	26	20
6	22	30	18	14	20
7	20	22	19	38	40
8	26	28	20	20	26
9	26	30	21	14	14
10	38	16	22	18	14
11	30	18	23	22	30
12	16	34	24	34	42

21. **Cesareans.** Some people fear that differences in insurance coverage can affect healthcare decisions. A survey of several randomly selected hospitals found that 16.6% of 223 recent births in Vermont involved cesarean deliveries, compared with 18.8% of 186 births in New Hampshire. Is this evidence that the rate of cesarean births in the two states is different?

22. **Newspapers.** Who reads the newspaper more, men or women? Eurostat, an agency of the European Union (EU), conducts surveys on several aspects of daily life in EU countries. Recently, the agency asked samples of 1000 respondents in each of 14 European countries whether they read the newspaper on a daily basis. Below are the data by country and sex.

% Reading a Newspaper Daily		
Country	Men	Women
Belgium	56.3	45.5
Denmark	76.8	70.3
Germany	79.9	76.8
Greece	22.5	17.2
Spain	46.2	24.8
Ireland	58.0	54.0
Italy	50.2	29.8
Luxembourg	71.0	67.0
Netherlands	71.3	63.0
Austria	78.2	74.1
Portugal	58.3	24.1
Finland	93.0	90.0
Sweden	89.0	88.0
UK	32.6	30.4

a) Examine the differences in the percentages for each country. Which of these countries seem to be outliers? What do they have in common?

b) After eliminating the outliers, is there evidence that in Europe men are more likely than women to read the newspaper?

23. **Meals.** A college student is on a "meal program." His budget allows him to spend an average of $10 per day for the semester. He keeps track of his daily food expenses for 2 weeks; the data are given in the table. Is there strong evidence that he will overspend his food allowance? Explain.

Date	Cost ($)
7/29	15.20
7/30	23.20
7/31	3.20
8/1	9.80
8/2	19.53
8/3	6.25
8/4	0
8/5	8.55
8/6	20.05
8/7	14.95
8/8	23.45
8/9	6.75
8/10	0
8/11	9.01

24. **Wall Street.** In February of 2009, the Harris Poll organization asked 1010 randomly sampled American adults whether they agreed or disagreed with the following statement:

> Most people on Wall Street would be willing to break the law if they believed they could make a lot of money and get away with it.

Of those asked, 71% said they agreed with this statement. We know that if we could ask the entire population of American adults, we would not find that exactly 71% think that Wall Street workers would be willing to break the law to make money. Construct a 95% confidence interval for the true percentage of American adults who agree with the statement.

25. **Teach for America.** Several programs attempt to address the shortage of qualified teachers by placing uncertified instructors in schools with acute needs—often in inner cities. A 1999–2000 study compared students taught by certified teachers with others taught by uncertified teachers in the same schools. Reading scores of the students of certified teachers averaged 35.62 points with standard deviation 9.31. The scores of students instructed by uncertified teachers had mean 32.48 points with standard deviation 9.43 points on the same test. There were 44 students in each group. The appropriate t procedure has 86 degrees of freedom. Is there evidence of lower scores with uncertified teachers? Discuss. (*The Effectiveness of "Teach for America" and Other Under-certified Teachers on Student Academic Achievement: A Case of Harmful Public Policy*. Education Policy Analysis Archives, 2002)

26. **Streams.** Researchers in the Adirondack Mountains collect data on a random sample of streams each year. One of the variables recorded is the substrate of the stream—the type of soil and rock over which they flow. The researchers found that 69 of the 172 sampled streams had a substrate of shale. Construct a 95% confidence interval for the proportion of Adirondack streams with a shale substrate. Clearly interpret your interval in this context.

27. **Legionnaires' disease.** In 1974, the Bellevue-Stratford Hotel in Philadelphia was the scene of an outbreak of what later became known as Legionnaires' disease. The cause of the disease was finally discovered to be bacteria that thrived in the air-conditioning units of the hotel. Owners of the Rip Van Winkle Motel, hearing about the Bellevue-Stratford, replace their air-conditioning system. The following data are the bacteria counts, in the air of eight rooms, before and after a new air-conditioning system was installed (measured in colonies per cubic foot of air). The objective is to find out whether the new system has succeeded in lowering the bacterial count. You are the statistician assigned to report to the hotel whether the strategy has worked. Base your analysis on a confidence interval. Be sure to list all your assumptions, methods, and conclusions.

Room Number	Before	After
121	11.8	10.1
163	8.2	7.2
125	7.1	3.8
264	14	12
233	10.8	8.3
218	10.1	10.5
324	14.6	12.1
325	14	13.7

28. **Teach for America, part II.** The study described in Exercise 25 also looked at scores in mathematics and language. Here are software outputs for the appropriate tests. Explain what they show.

Mathematics
T-TEST OF Mu(1) − Mu(2) = 0
Mu(Cert) − Mu(NoCert) = 4.53 t (86) = 2.95 p = 0.002

Language
T-TEST OF Mu(1) − Mu(2) = 0
Mu(Cert) − Mu(NoCert) = 2.13 t (84) = 1.71 p = 0.045

29. **Bipolar kids.** The June 2002 *American Journal of Psychiatry* reported that researchers used medication and psychotherapy to treat children aged 7 to 16 who exhibit bipolar symptoms. After 2 years, symptoms had cleared up in only 26 of the 89 children involved in the study.
   a) Write a 95% confidence interval for the proportion helped by the treatment, and interpret it in this context.
   b) If researchers subsequently hope to produce an estimate (with 95% confidence) of treatment effectiveness for bipolar disorder that has a margin of error of only 6%, how many patients should they study?

30. **Online testing.** The Educational Testing Service is now administering several of its standardized tests online, the CLEP and GMAT exams, for example. Since taking a test on a computer is different from taking a test with pencil and paper, one wonders if the scores will be the same. To investigate this question, researchers created two versions of an SAT-type test and got 20 volunteers to participate in an experiment. Each volunteer took both versions of the test, one with pencil and paper and the other online. Subjects were randomized with respect to the order in which they sat for the tests (online/paper) and which form they took (Test A, Test B) in which environment. The scores (out of a possible 20) are summarized in the table.

Subject	Paper	Online
	Test A	Test B
1	14	13
2	10	13
3	16	8
4	15	14
5	17	16
6	14	11
7	9	12
8	12	12
9	16	16
10	7	14
	Test B	Test A
11	8	13
12	11	13
13	15	17
14	11	13
15	13	14
16	9	9
17	15	9
18	14	15
19	16	12
20	8	10

   a) Were the two forms (A/B) of the test equivalent in terms of difficulty? Test an appropriate hypothesis and state your conclusion.
   b) Is there evidence that testing environment (paper/online) matters? Test an appropriate hypothesis and state your conclusion.

31. **Bread.** Clarksburg Bakery is trying to predict how many loaves to bake. In the last 100 days, the bakery has sold between 95 and 140 loaves per day. Here are a histogram and the summary statistics for the number of loaves sold for the last 100 days.

Summary of sales	
Mean	103
Median	100
StdDev	9.000
Min	95
Max	140
Lower 25th %tile	97
Upper 25th %tile	105.5

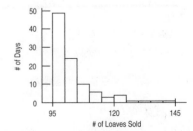

a) Can you use these data to estimate the number of loaves sold on the busiest 10% of all days? Explain.

b) Explain why you can use these data to construct a 95% confidence interval for the mean number of loaves sold per day.

c) Calculate a 95% confidence interval and carefully interpret what that confidence interval means.

d) If the bakery would have been satisfied with a confidence interval whose margin of error was twice as wide, how many days' data could they have used?

e) When the bakery opened, the owners estimated that they would sell an average of 100 loaves per day. Does your confidence interval provide strong evidence that this estimate was incorrect? Explain.

T 32. **Irises.** Can measurements of the petal length of flowers be of value when you need to determine the species of a certain flower? Here are the summary statistics from measurements of the petals of two species of irises. (R. A. Fisher, "The Use of Multiple Measurements in Axonomic Problems." *Annals of Eugenics* 7 [1936]:179–188)

	Species	
	Versicolor	**Virginica**
Count	50	50
Mean	55.52	43.22
Median	55.50	44
StdDev	5.519	5.362
Min	45	30
Max	69	56
Lower quartile	51	40
Upper quartile	59	47

a) Make parallel boxplots of petal lengths for the two species.

b) Describe the differences seen in the boxplots.

c) Write a 95% confidence interval for this difference.

d) Explain what your interval means.

e) Based on your confidence interval, is there evidence of a difference in petal length? Explain.

33. **Insulin and diet.** A study published in the *Journal of the American Medical Association* examined people to see if they showed any signs of IRS (insulin resistance syndrome) involving major risk factors for Type 2 diabetes and heart disease. Among 102 subjects who consumed dairy products more than 35 times per week, 24 were identified with IRS. In comparison, IRS was identified in 85 of 190 individuals with the lowest dairy consumption, fewer than 10 times per week.

a) Is this strong evidence that IRS risk is different in people who frequently consume dairy products than in those who do not?

b) Does this prove that dairy consumption influences the development of IRS? Explain.

34. **World Series.** If the two teams playing in the World Series are evenly matched, the probability that each team wins any game is 0.5. Then the probability that the Series ends with one of the teams sweeping four straight games would be $2(0.5)^4 = 0.125$. Further probability calculations indicate that 25% of all World Series should last five games, 31.25% should last six games, and the other 31.25% should last the full seven games. The table shows the number of games it took to decide all the World Series from 1922 (when the 7-game format was set) through 2003. Do these results indicate that the teams are usually equally matched? Give statistical evidence to support your conclusion.

Length of series	4 games	5 games	6 games	7 games
Number of times	15	15	18	32

T 35. **Rainmakers?** In an experiment to determine whether seeding clouds with silver iodide increases rainfall, 52 clouds were randomly assigned to be seeded or not. The amount of rain they generated was then measured (in acre-feet). Create a 95% confidence interval for the average amount of additional rain created by seeding clouds. Explain what your interval means.

	Unseeded Clouds	**Seeded Clouds**
Count	26	26
Mean	164.588	441.985
Median	44.200	221.600
SD	278.426	650.787
IntQRange	138.600	337.600
25 %ile	24.400	92.400
75 %ile	163	430

36. **Fritos®.** As a project for an Introductory Statistics course, students checked 6 bags of Fritos marked with a net weight of 35.4 grams. They carefully weighed the contents of each bag, recording the following weights (in grams): 35.5, 35.3, 35.1, 36.4, 35.4, 35.5. Is there evidence that the mean weight of bags of Fritos is less than advertised?

a) Write appropriate hypotheses.

b) Do these data satisfy the assumptions for inference? Explain.

c) Test your hypothesis using all 6 weights.

d) Retest your hypothesis with the one unusually high weight removed.

e) What would you conclude about the stated net weight?

T 37. **Color or text?** In an experiment, 32 volunteer subjects are briefly shown seven cards, each displaying the name of a color printed in a different color (example: red, blue, and so on). The subject is asked to perform one of two tasks: memorize the order of the words or memorize the order of the colors. Researchers record the number of cards remembered correctly. Then the cards are shuffled and the subject is asked to perform the other task. The tables display the results for each subject. Is there any evidence that either the color or the written word dominates perception?

a) What role does randomization play in this experiment?
b) State appropriate hypotheses.
c) Are the assumptions necessary for inference reasonable here?
d) Perform the test.
e) State your conclusion.

Subject	Color	Word	Subject	Color	Word
1	4	7	17	4	3
2	1	4	18	7	4
3	5	6	19	4	3
4	1	6	20	0	6
5	6	4	21	3	3
6	4	5	22	3	5
7	7	3	23	7	3
8	2	5	24	3	7
9	7	5	25	5	6
10	4	3	26	3	4
11	2	0	27	3	5
12	5	4	28	1	4
13	6	7	29	2	3
14	3	6	30	5	3
15	4	6	31	3	4
16	4	7	32	6	7

38. And it means? Every statement about a confidence interval contains two parts—the level of confidence and the interval. Suppose that an insurance agent estimating the mean loss claimed by clients after home burglaries created the 95% confidence interval ($1644, $2391).

a) What's the margin of error for this estimate?
b) Carefully explain what the interval means.
c) Carefully explain what the 95% confidence level means.

39. Batteries. We work for the "Watchdog for the Consumer" consumer advocacy group. We've been asked to look at a battery company that claims its batteries last an average of 100 hours under normal use. There have been several complaints that the batteries don't last that long, so we decide to test them. To do this we select 16 batteries and run them until they die. They lasted a mean of 97 hours, with a standard deviation of 12 hours.

a) One of the editors of our newsletter (who does not know statistics) says that 97 hours is a lot less than the advertised 100 hours, so we should reject the company's claim. Explain to him the problem with doing that.
b) What are the null and alternative hypotheses?

c) What assumptions must we make in order to proceed with inference?
d) At a 5% level of significance, what do you conclude?
e) Suppose that, in fact, the average life of the company's batteries is only 98 hours. Has an error been made in part d? If so, what kind?

40. Hamsters. How large are hamster litters? Among 47 golden hamster litters recorded, there were an average of 7.72 baby hamsters, with a standard deviation of 2.5 hamsters per litter.

a) Create and interpret a 90% confidence interval.
b) Would a 98% confidence interval have a larger or smaller margin of error? Explain.
c) How many litters must be used to estimate the average litter size to within 1 baby hamster with 95% confidence?

41. Family planning. Before the introduction of birth control pills, many young women experienced unplanned pregnancies. A 1954 study of 1438 pregnant women examined the association with the women's education levels, producing these data:

	Education Level		
	<3 yr HS	3 + yr HS	Some college
Number of pregnancies	591	608	239
% unplanned	66.2%	55.4%	42.7%

Do these data provide evidence of an association between family planning and education level? (*Fertility Planning and Fertility Rates by Socio-Economic Status*, Social and Psychological Factors Affecting Fertility, 1954)

42. Recruiting. In September 2002, CNN reported on a method of grad student recruiting by the Haas School of Business at U.C.-Berkeley. The school notifies applicants by formal letter that they have been admitted, and also e-mails the accepted students a link to a website that greets them with personalized balloons, cheering, and applause. The director of admissions says this extra effort at recruiting has really worked well. The school accepts 500 applicants each year, and the percentage who actually choose to enroll at Berkeley has increased from 52% the year before the Web greeting to 54% this year.

a) Create a 95% confidence interval for the change in enrollment rates.
b) Based on your confidence interval, are you convinced that this new form of recruiting has been effective? Explain.

Inferences for Regression

Where are we going?

A scatterplot of IQ vs. brain size shows a mildly positive association. Could this just be due to chance? A hypothesis test is clearly what we need. We can estimate the slope, but how reliable is our estimate? In this chapter we'll step from what we know about tests and confidence intervals for means to regression.

WHO	250 male subjects
WHAT	Body fat and waist size
UNITS	% Body fat and inches
WHEN	1990s
WHERE	United States
WHY	Scientific research

Three percent of a man's body is essential fat. (For a woman, the percentage is closer to 12.5%.) As the name implies, essential fat is necessary for a normal, healthy body. Fat is stored in small amounts throughout your body. Too much body fat, however, can be dangerous to your health. For men between 18 and 39 years old, a healthy percent body fat ranges from 8% to 19%. (For women of the same age, it's 21% to 32%.)

Measuring body fat can be tedious and expensive. The "standard reference" measurement is by dual-energy X-ray absorptiometry (DEXA), which involves two low-dose X-ray generators and takes from 10 to 20 minutes.

How close can we get to a useable prediction of body fat from easily measurable variables such as *Height*, *Weight*, or *Waist* size? Here's a scatterplot of *%Body Fat* plotted against *Waist* size for a sample of 250 males of various ages.

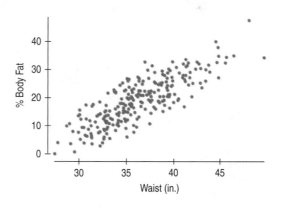

FIGURE 27.1
Percent Body Fat vs. *Waist* size for 250 men of various ages. The scatterplot shows a strong, positive, linear relationship.

Back in Chapter 8 we modeled relationships like this by fitting a least squares line. The plot is clearly straight, so we can find that line. The equation of the least squares line for these data is

$$\widehat{\%Body\ Fat} = -42.7 + 1.7\ Waist.$$

The slope says that, on average, %*Body Fat* is greater by 1.7 percent for each additional inch around the waist.

How useful is this model? When we fit linear models before, we used them to describe the relationship between the variables and we interpreted the slope and intercept as descriptions of the data. Now we'd like to know what the regression model can tell us beyond the 250 men in this study. To do that, we'll want to make confidence intervals and test hypotheses about the slope and intercept of the regression line.

The Population and the Sample

When we found a confidence interval for a mean, we could imagine a single, true underlying value for the mean. When we tested whether two means or two proportions were equal, we imagined a true underlying difference. But what does it mean to do inference for regression? We know better than to think that even if we knew every population value, the data would line up perfectly on a straight line. After all, even in our sample, not all men who have 38-inch waists have the same %*Body Fat*. In fact, there's a whole distribution of %*Body Fat* for these men:

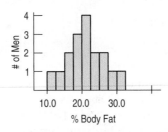

FIGURE 27.2
The distribution of %*Body Fat* for men with a *Waist* size of 38 inches is unimodal and symmetric.

This is true at each *Waist* size. In fact, we could depict the distribution of %*Body Fat* at different *Waist* sizes like this:

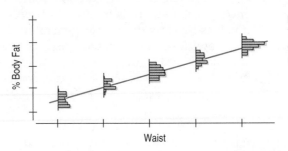

FIGURE 27.3
There's a distribution of %*Body Fat* for each value of *Waist* size. We'd like the means of these distributions to line up.

But we want to *model* the relationship between %*Body Fat* and *Waist* size for all men. To do that, we imagine an idealized regression line. The model assumes that the *means* of the distributions of %*Body Fat* for each *Waist* size fall along the line, even though the individuals are scattered around it. We know that this model is not a perfect description of how the variables are associated, but it may be useful for predicting %*Body Fat* and for understanding how it's related to *Waist* size.

If only we had all the values in the population, we could find the slope and intercept of this *idealized regression line* explicitly by using least squares. Following our usual conventions, we write the idealized line with Greek letters and consider the coefficients (the slope and intercept) to be *parameters:* β_0 is the intercept and β_1 is the slope. Corresponding to our fitted line of $\hat{y} = b_0 + b_1 x$, we write

$$\mu_y = \beta_0 + \beta_1 x.$$

Why μ_y instead of $\hat{y}$? Because this is a model. There is a distribution of *%Body Fat* for each *Waist* size. The model places the *means* of the distributions of *%Body Fat* for each *Waist* size on the same straight line.

Of course, not all the individual y's are at these means. (In fact, the line will miss most—and quite possibly all—of the plotted points.) Some individuals lie above and some below the line, so, like all models, this one makes **errors.** Lots of them. In fact, one at each point. These errors are random and, of course, can be positive or negative. They are model errors, so we use a Greek letter and denote them by ε.

When we put the errors into the equation, we can account for each individual y:

$$y = \beta_0 + \beta_1 x + \varepsilon.$$

This equation is now true for each data point (since there is an ε to soak up the deviation), so the model gives a value of y for any value of x.

For the body fat data, an idealized model such as this provides a summary of the relationship between *%Body Fat* and *Waist* size. Like all models, it simplifies the real situation. We know there is more to predicting body fat than waist size alone. But the advantage of a model is that the simplification might help us to think about the situation and assess how well *%Body Fat* can be predicted from simpler measurements.

We estimate the β's by finding a regression line, $\hat{y} = b_0 + b_1 x$, as we did in Chapter 8. The residuals, $e = y - \hat{y}$, are the sample-based versions of the errors, ε. We'll use them to help us assess the regression model.

We know that least squares regression will give reasonable estimates of the parameters of this model from a random sample of data. Our challenge is to account for our uncertainty in how well they do. For that, we need to make some assumptions about the model and the errors.

Assumptions and Conditions

A S *Activity:* **Conditions for Regression Inference.** View an illustrated discussion of the conditions for regression inference.

Back in Chapter 8 when we fit lines to data, we needed to check only the Straight Enough Condition. Now, when we want to make inferences about the coefficients of the line, we'll have to make more assumptions. Fortunately, we can check conditions to help us judge whether these assumptions are reasonable for our data. And as we've done before, we'll make some checks *after* we find the regression equation.

Also, we need to be careful about the order in which we check conditions. If our initial assumptions are not true, it makes no sense to check the later ones. So now we number the assumptions to keep them in order.

CHECK THE SCATTERPLOT.

The shape must be linear or we can't use linear regression at all.

1. Linearity Assumption

If the true relationship is far from linear and we use a straight line to fit the data, our entire analysis will be useless, so we always check this first.

The **Straight Enough Condition** is satisfied if a scatterplot looks straight. It's generally not a good idea to draw a line through the scatterplot when checking. That can fool your eyes into seeing the plot as more straight. Sometimes it's easier to see violations of the Straight Enough Condition by looking at a scatterplot of the residuals against x or against the predicted values, $\hat{y}$. That plot will have a horizontal direction and should have no pattern if the condition is satisfied.

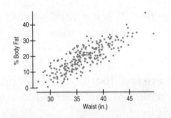

CHECK THE RESIDUALS PLOT (1).

The residuals should appear to be randomly scattered.

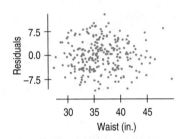

FIGURE 27.4
The residuals show only random scatter when plotted against *Waist* size.

CHECK THE RESIDUALS PLOT (2).

The vertical spread of the residuals should be roughly the same everywhere.

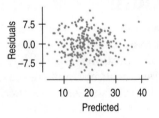

FIGURE 27.5
A scatterplot of residuals against predicted values can help check for plot thickening. Note that this plot looks identical to the plot of residuals against *Waist* size. For a regression of one response variable on one predictor, these plots differ only in the labels on the x-axis.

If the scatterplot is straight enough, we can go on to some assumptions about the errors. If not, stop here, or consider re-expressing the data (see Chapter 10) to make the scatterplot more nearly linear. For the *%Body Fat* data, the scatterplot is beautifully linear. Of course, the data must be quantitative for this to make sense. Check the **Quantitative Data Condition.**

2. Independence Assumption

Independence Assumption: The errors in the true underlying regression model (the ε's) must be mutually independent. As usual, there's no way to be sure that the Independence Assumption is true.

Usually when we care about inference for the regression parameters, it's because we think our regression model might apply to a larger population. In such cases, we can check a **Randomization Condition** that the individuals are a representative sample from that population.

We can also check displays of the regression residuals for evidence of patterns, trends, or clumping, any of which would suggest a failure of independence. In the special case when the x-variable is related to time, a common violation of the Independence Assumption is for the errors to be correlated. (The error our model makes today may be similar to the one it made for yesterday.) This violation can be checked by plotting the residuals against the x-variable and looking for patterns.

The *%Body Fat* data were collected on a sample of men taken to be representative. The subjects were not related in any way, so we can be pretty sure that their measurements are independent. The residuals plot shows no pattern.

3. Equal Variance Assumption

The variability of y should be about the same for all values of x. In Chapter 8 we looked at the standard deviation of the residuals (s_e) to measure the size of the scatter. Now we'll need this standard deviation to build confidence intervals and test hypotheses. The standard deviation of the residuals is the building block for the standard errors of all the regression parameters. But it makes sense only if the scatter of the residuals is the same everywhere. In effect, the standard deviation of the residuals "pools" information across all of the individual distributions at each x-value, and pooled estimates are appropriate only when they combine information for groups with the same variance.

Practically, what we can check is the **Does the Plot Thicken? Condition.** A scatterplot of y against x offers a visual check. Fortunately, we've already made one. Make sure the spread around the line is nearly constant. Be alert for a "fan" shape or other tendency for the variation to grow or shrink in one part of the scatterplot. Often it is better to look at the residuals plotted against the predicted values, $\hat{y}$. With the slope of the line removed, it's easier to see patterns left behind. For the body fat data, the spread of *%Body Fat* around the line is remarkably constant across *Waist* sizes from 30 inches to about 45 inches.

If the plot is straight enough, the data are independent, and the plot doesn't thicken, you can now move on to the final assumption.

4. Normal Population Assumption

We assume the errors around the idealized regression line at each value of x follow a Normal model. We need this assumption so that we can use a Student's t-model for inference.

CHECK A HISTOGRAM OF THE RESIDUALS.

The distribution of the residuals should be unimodal and symmetric.

As we have at other times when we've used Student's t, we'll settle for the residuals satisfying the **Nearly Normal Condition** and the **Outlier Condition.** Look at a histogram or Normal probability plot of the residuals.[1]

The histogram of residuals in the *%Body Fat* regression certainly looks nearly Normal. As we have noted before, the Normality Assumption becomes less important as the sample size grows, because the model is about means and the Central Limit Theorem takes over.

If all four assumptions were true, the idealized regression model would look like this:

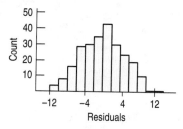

FIGURE 27.6
A histogram of the residuals is one way to check whether they are Nearly Normal. Alternatively, we can look at a Normal probability plot.

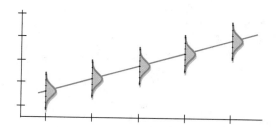

FIGURE 27.7
The regression model has a distribution of y-values for each x-value. These distributions follow a Normal model with means lined up along the line and with the same standard deviations.

At each value of x there is a distribution of y-values that follows a Normal model, and each of these Normal models is centered on the line and has the same standard deviation. Of course, we don't expect the assumptions to be exactly true, and we know that all models are wrong, but the linear model is often close enough to be very useful.

FOR EXAMPLE Checking Assumptions and Conditions

Look at the moon with binoculars or a telescope, and you'll see craters formed by thousands of impacts. The earth, being larger, has been hit even more often. Meteor Crater in Arizona was the first recognized impact crater and was identified as such only in the 1920s. With the help of satellite images, more and more craters have been identified; now more than 180 are known. These, of course, are only a small sample of all the impacts the earth has experienced: Only 29% of earth's surface is land, and many craters have been covered or eroded away. Astronomers have recognized a roughly 35 million-year cycle in the frequency of cratering, although the cause of this cycle is not fully understood. Here's a scatterplot of the known impact craters from the most recent 35 million years.[2] We've taken logs of both age (in millions of years ago) and diameter (km) to make the relationship simpler. (See Chapter 10.)

WHO	39 impact craters
WHAT	Diameter and age
UNITS	km and millions of years ago
WHEN	Past 35 million years
WHERE	Worldwide
WHY	Scientific research

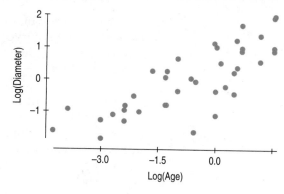

(continued)

---

[1] *This* is why we have to check the conditions in order. We have to check that the residuals are independent and that the variation is the same for all x's so that we can lump all the residuals together for a single check of the Nearly Normal Condition.
[2] Data, pictures, and much more information at the Earth Impact Database found at http://www.unb.ca/passc/ImpactDatabase/

QUESTION: Are the assumptions and conditions satisfied for fitting a linear regression model to these data?

✔ **Linearity Assumption:** The scatterplot satisfies the Straight Enough Condition.

✔ **Independence Assumption:** Sizes of impact craters are likely to be generally independent.

✔ **Randomization Condition:** These are the only known craters, and may differ from others that have disappeared or not yet been found. I'll need to be careful not to generalize my conclusions too broadly.

✔ **Does the Plot Thicken? Condition:** After fitting a linear model, I find the residuals shown.

Two points seem to give the impression that the residuals may be more variable for higher predicted values than for lower ones, but this doesn't seem to be a serious violation of the Equal Variance Assumption.

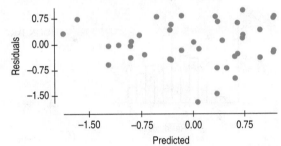

✔ **Nearly Normal Condition:** A Normal probability plot suggests a bit of skewness in the distribution of residuals, and the histogram confirms that.

There are no violations severe enough to stop my regression analysis, but I'll be cautious about my conclusions.

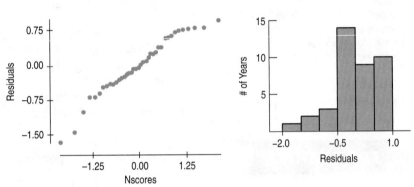

Which Come First: The Conditions or the Residuals?

In regression, there's a little catch. The best way to check many of the conditions is with the residuals, but we get the residuals only *after* we compute the regression. Before we compute the regression, however, we should check at least one of the conditions.

So we work in this order:

1. Make a scatterplot of the data to check the Straight Enough Condition. (If the relationship is curved, try re-expressing the data. Or stop.)
2. If the data are straight enough, fit a regression and find the residuals, *e*, and predicted values, $\hat{y}$.
3. Make a scatterplot of the residuals against *x* or against the predicted values. This plot should have no pattern. Check in particular for any bend (which would suggest that the data weren't all that straight after all), for any thickening (or thinning), and, of course, for any outliers. (If there are outliers, and you can correct them or justify removing them, do so and go back to step 1, or consider performing two regressions—one with and one without the outliers.)
4. If the data are measured over time, plot the residuals against time to check for evidence of patterns that might suggest they are not independent.
5. If the scatterplots look OK, then make a histogram and Normal probability plot of the residuals to check the Nearly Normal Condition.
6. If all the conditions seem to be reasonably satisfied, go ahead with inference.

| STEP-BY-STEP EXAMPLE | Regression Inference |

If our data can jump through all these hoops, we're ready to do regression inference. Let's see how much more we can learn about body fat and waist size from a regression model.

Questions: What is the relationship between *%Body Fat* and *Waist* size in men?
What model best predicts body fat from waist size, and how well does it do the job?

THINK

Plan Specify the question of interest.

Name the variables and report the W's.

Identify the parameters you want to estimate.

Model Think about the assumptions and check the conditions.

Make pictures. For regression inference, you'll need a scatterplot, a residuals plot, and either a histogram or a Normal probability plot of the residuals.

(We've seen plots of the residuals already. See Figures 27.5 and 27.6.)

I have quantitative body measurements on 250 adult males from the BYU Human Performance Research Center. I want to understand the relationship between *%Body Fat* and *Waist* size.

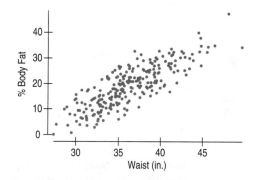

✔ **Straight Enough Condition:** There's no obvious bend in the original scatterplot of the data or in the plot of residuals against predicted values.

✔ **Independence Assumption:** These data are not collected over time, and there's no reason to think that the *%Body Fat* of one man influences the *%Body Fat* of another.

✔ **Does the Plot Thicken? Condition:** Neither the original scatterplot nor the residual scatterplot shows any changes in the spread about the line.

✔ **Nearly Normal Condition, Outlier Condition:** A histogram of the residuals is unimodal and symmetric. The Normal probability plot of the residuals is quite straight, indicating that the Normal model is reasonable for the errors.

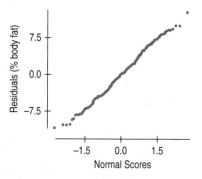

(continued)

Choose your method.

Under these conditions a **regression model** is appropriate.

Mechanics Let's just "push the button" and see what the regression looks like.

The formula for the regression equation can be found in Chapter 8, and the standard error formulas will be shown a bit later, but regressions are almost always computed with a computer program or calculator.

Write the regression equation.

Here's the computer output for this regression:

Dependent variable is: %BF
R-squared = 67.8%
s = 4.713 with 250 − 2 = 248 degrees of freedom

Variable	Coeff	SE(Coeff)	t-ratio	P-value
Intercept	−42.734	2.717	−15.7	<0.0001
Waist	1.70	0.0743	22.9	<0.0001

The estimated regression equation is

$$\widehat{\%Body\ Fat} = -42.73 + 1.70\ Waist.$$

Conclusion Interpret your results in context.

The R^2 for the regression is 67.8%. Waist size seems to account for about 2/3 of the %Body Fat variation in men. The slope of the regression says that %Body Fat increases by about 1.7 percentage points per inch of Waist size, on average.

More Interpretation We haven't worked it out in detail yet, but the output gives us numbers labeled as *t*-statistics and corresponding P-values, and we have a general idea of what those mean.

(Now it's time to learn more about regression inference so we can figure out what the rest of the output means.)

The standard error of 0.07 for the slope is much smaller than the slope itself, so it looks like the estimate is reasonably precise. And there are a couple of t-ratios and P-values given. Because the P-values are small, it appears that some null hypotheses can be rejected.

Intuition About Regression Inference

A S *Simulation: Simulate the Sampling Distribution of a Regression Slope.* Draw samples repeatedly to see for yourself how slope can vary from sample to sample. This simulation experiment lets you build up a histogram to see the sampling distribution.

Wait a minute! We've just pulled a fast one. We've pushed the "regression button" on our computer or calculator but haven't discussed where the standard errors for the slope or intercept come from. We know that if we had collected similar data on a different random sample of men, the slope and intercept would be different. Each sample would have produced its own regression line, with slightly different b_0's and b_1's. This sample-to-sample variation is what generates the sampling distributions for the coefficients.

There's only one regression model; each sample regression is trying to estimate the same parameters, β_0 and β_1. We expect any sample to produce a b_1 whose expected value is the true slope, β_1. What about its standard deviation? What aspects of the data affect how much the slope (and intercept) vary from sample to sample?

- **Spread around the line.** Here are two situations in which we might do regression. Which situation would yield the more consistent slope? That is, if we were to sample over and over from the two underlying populations that these samples come from and compute all the slopes, which group of slopes would vary less?

FIGURE 27.8
Which of these scatterplots shows a situation that would give the more consistent regression slope estimate if we were to sample repeatedly from its underlying population?

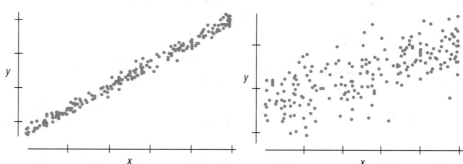

Clearly, data like those in the left plot give more consistent slopes.

Less scatter around the line means the slope will be more consistent from sample to sample. The spread around the line is measured with the **residual standard deviation,** s_e. You can always find s_e in the regression output, often just labeled s. Sometimes it is called the "standard error" although, as we know, that's not quite right—probably a misinterpretation of the s_e notation. You're not likely to calculate the residual standard deviation by hand. When we first saw this formula in Chapter 8, we said that it looks a lot like the standard deviation of y, only subtracting the predicted values rather than the mean and dividing by $n - 2$ instead of $n - 1$:

$$s_e = \sqrt{\frac{\sum (y - \hat{y})^2}{n - 2}}.$$

The less scatter around the line, the smaller the residual standard deviation and the stronger the relationship between x and y.

Some people prefer to assess the strength of a regression by looking at s_e rather than R^2. After all, s_e has the same units as y, and because it's the standard deviation of the errors around the line, it tells you how close the data are to our model. By contrast, R^2 is the proportion of the variation of y accounted for by x. We say, why not look at both?

- **Spread of the x's:** Here are two more situations. Which of these would yield more consistent slopes?

---

n − 2?

For standard deviation (in Chapter 4), we divided by $n - 1$ because we didn't know the true mean and had to estimate it. Now it's later in the course and there's even more we don't know. Here we don't know *two* things: the slope and the intercept. If we knew them both, we'd divide by n and have n degrees of freedom. When we *estimate* both, however, we adjust by subtracting 2, so we divide by $n - 2$ and (as we will see soon) have 2 fewer degrees of freedom.

FIGURE 27.9
Which of these scatterplots shows a situation that would give the more consistent regression slope estimate if we were to sample repeatedly from the underlying population?

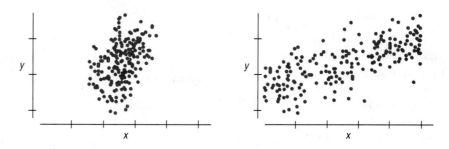

A plot like the one on the right has a broader range of x-values, so it gives a more stable base for the slope. We'd expect the slopes of samples from situations like that to vary less from sample to sample. If s_x, the standard deviation of x is large, it provides a more stable regression.

- **Sample size.** What about these two?

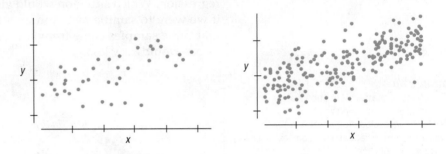

It shouldn't be a surprise that having a larger sample size, n, gives more consistent estimates from sample to sample.

Standard Error for the Slope

Three aspects of the scatterplot, then, affect the standard error of the regression slope:

- Spread around the line: s_e
- Spread of x values: s_x
- Sample size: n

These are in fact the *only* things that affect the standard error of the slope. Although you'll probably never have to calculate it by hand, the formula for the standard error is

$$SE(b_1) = \frac{s_e}{\sqrt{n-1}\, s_x}.$$

A S *Activity: Regression Slope Standard Error.* See how $SE(b_1)$ is constructed and where the values used in the formula are found in the regression output table.

The error standard deviation, s_e, is in the *numerator*, since spread around the line *increases* the slope's standard error. The denominator has both a sample size term, $\sqrt{n-1}$, and s_x, because increasing either of these *decreases* the slope's standard error.

We know the b_1's vary from sample to sample. As you'd expect, their sampling distribution model is centered at β_1, the slope of the idealized regression line. Now we can estimate its standard deviation with $SE(b_1)$. What about its shape? Here the Central Limit Theorem and "Wild Bill" Gosset come to the rescue again. When we standardize the slopes by subtracting the model mean and dividing by their standard error, we get a Student's t-model, this time with $n-2$ degrees of freedom:

A S *Simulation: x-Variance and Slope Variance.* You don't have to just imagine how the variability of the slope depends on the spread of the x's.

$$\frac{b_1 - \beta_1}{SE(b_1)} \sim t_{n-2}.$$

A SAMPLING DISTRIBUTION FOR REGRESSION SLOPES
When the conditions are met, the standardized estimated regression slope,

$$t = \frac{b_1 - \beta_1}{SE(b_1)},$$

follows a Student's t-model with $n-2$ degrees of freedom. We estimate the standard error with

$$SE(b_1) = \frac{s_e}{\sqrt{n-1}\, s_x},$$

$$\text{where } s_e = \sqrt{\frac{\sum(y - \hat{y})^2}{n - 2}},$$

n is the number of data values, and s_x is the ordinary standard deviation of the x-values.

FOR EXAMPLE | **Finding Standard Errors**

RECAP: Recent terrestrial impact craters seem to show a relationship between age and size that is linear when re-expressed using logarithms (see Chapter 10).

Here are summary statistics and regression output:

Variable	Count	Mean	StdDev
LogAge	39	−0.656310	1.57682
LogDiam	39	0.012600	1.04104

Dependent variable is: LogDiam
R-squared = 63.6%
s = 0.6362 with 39 − 2 = 37 degrees of freedom

Variable	Coefficient	Se(coeff)	t-ratio	P-value
Intercept	0.358262	0.1106	3.24	0.0025
LogAge	0.526674	0.0655	8.05	≤0.0001

QUESTIONS: How are the standard error of the slope and the t-ratio for the slope calculated? (And aren't you glad the software does this for you?)

$$SE(b_1) = \frac{s_e}{\sqrt{n-1} \times s_x} = \frac{0.6362}{\sqrt{39 - 1} \times 1.57682} = 0.0655$$

To test the hypothesis of no linear association ($\beta_1 = 0$), $t_{37} = \dfrac{b_1 - \beta_1}{SE(b_1)} = \dfrac{0.526674 - 0}{0.0655} = 8.05$

What About the Intercept?

The same reasoning applies for the intercept. We could write

$$\frac{b_0 - \beta_0}{SE(b_0)} \sim t_{n-2}$$

and use it to construct confidence intervals and test hypotheses, but often the value of the intercept isn't something we care about. The intercept usually isn't interesting. Most hypothesis tests and confidence intervals for regression are about the slope.

Regression Inference

Now that we have the standard error of the slope and its sampling distribution, we can test a hypothesis about it and make confidence intervals. The usual null hypothesis about the slope is that it's equal to 0. Why? Well, a slope of zero would say that y doesn't tend to change linearly when x changes—in other words, that there is no linear association between the two variables. If the slope were zero, there wouldn't be much left of our regression equation.

So a null hypothesis of a zero slope questions the entire claim of a linear relationship between the two variables—and often that's just what we want to know. In fact, every software package or calculator that does regression simply assumes that you want to test the null hypothesis that the slope is really zero.

WHAT IF THE SLOPE WERE 0?

If $b_1 = 0$, our prediction is $\hat{y} = b_0 + 0x$. The equation collapses to just $\hat{y} = b_0$. Now x is nowhere in sight, so y doesn't depend on x at all. And b_0 would turn out to be $\bar{y}$. Why? We know that $b_0 = \bar{y} - b_1\bar{x}$, but when $b_1 = 0$, that becomes simply $b_0 = \bar{y}$. It turns out, then, that when the slope is 0, the equation is just $\hat{y} = \bar{y}$; at every value of x, we always predict the mean value for y.

To test H_0: $\beta_1 = 0$, we find

$$t_{n-2} = \frac{b_1 - 0}{SE(b_1)}.$$

This is just like every t-test we've seen: a difference between the statistic and its hypothesized value, divided by its standard error.

For our body fat data, the computer found the slope (1.7), its standard error (0.0743), and the ratio of the two: $\frac{1.7 - 0}{0.0743} = 22.9$ (see p. 680). Nearly 23 standard errors from the hypothesized value certainly seems big. The P-value (<0.0001) confirms that a t-ratio this large would be very unlikely to occur if the true slope were zero.

Maybe the standard null hypothesis isn't all that interesting here. Did you have any doubts that *%Body Fat* is related to *Waist* size? A more sensible use of these same values might be to make a confidence interval for the slope instead.

We can build a confidence interval in the usual way, as an estimate plus or minus a margin of error. As always, the margin of error is just the product of the standard error and a critical value. Here the critical value comes from the t-distribution with $n - 2$ degrees of freedom, so a 95% **confidence interval for β** is

$$b_1 \pm t^*_{n-2} \times SE(b_1).$$

For the body fat data, $t^*_{248} = 1.970$, so that comes to $1.7 \pm 1.97 \times 0.074$, or an interval from 1.55 to 1.85 *%Body Fat* per inch of *Waist* size.

FOR EXAMPLE | **Interpreting a Regression Model**

RECAP: On a log scale, there seems to be a linear relationship between the diameter and the age of recent terrestrial impact craters. We have regression output from statistics software:

QUESTION: What's the regression model, and what can it tell us?

Dependent variable is: LogDiam
R-squared = 63.6%
s = 0.6362 with 39 − 2 = 37 degrees of freedom

Variable	Coefficient	Se(coeff)	t-ratio	P-value
Intercept	0.358262	0.1106	3.24	0.0025
LogAge	0.526674	0.0655	8.05	≤0.0001

For terrestrial impact craters younger than 35 million years, the logarithm of *Diameter* grows linearly with the logarithm of *Age*: $\overline{\log Diam} = 0.358 + 0.527 \log Age$. The P-value for each coefficient's t-statistic is very small, so I'm quite confident that neither coefficient is zero. Based on my model, I conclude that, on average, the older a crater is, the larger it tends to be. This model accounts for 63.6% of the variation in *logDiam*.

Although it is possible that impacts (and their craters) are getting smaller, it is more likely that I'm seeing the effects of age on craters. Small craters are probably more likely to erode or become buried or otherwise be difficult to find as they age. Larger craters may survive the huge expanses of geologic time more successfully.

JUST CHECKING

Researchers in Food Science studied how big people's mouths tend to be. They measured mouth volume by pouring water into the mouths of subjects who lay on their backs. Unless this is your idea of a good time, it would be helpful to have a model to estimate mouth volume more simply. Fortunately, mouth volume is related to height. (Mouth volume is measured in cubic centimeters and height in meters.)

The data were checked and deemed suitable for regression. Take a look at the computer output.

1. What does the t-ratio of 3.27 for the slope tell about this relationship? How does the P-value help your understanding?

2. Would you say that measuring a person's height could reliably be used as a substitute for the wetter method of determining how big a person's mouth is? What numbers in the output helped you reach that conclusion?

3. What does the value of s_e add to this discussion?

Summary of	Mouth Volume
Mean	60.2704
StdDev	16.8777

Dependent variable is: Mouth Volume
R-squared = 15.3%
s = 15.66 with 61 − 2 = 59 degrees of freedom

Variable	Coefficient	SE(coeff)	t-ratio	P-value
Intercept	−44.7113	32.16	−1.39	0.1697
Height	61.3787	18.77	3.27	0.0018

Another Example

Every spring, Nenana, Alaska, hosts a contest in which participants try to guess the exact minute that a wooden tripod placed on the frozen Tanana River will fall through the breaking ice. The contest started in 1917 as a diversion for railroad engineers, with a jackpot of $800 for the closest guess. It has grown into an event in which hundreds of thousands of entrants enter their guesses on the Internet[3] and vie for as much as $300,000.

Because so much money and interest depends on the time of breakup, it has been recorded to the nearest minute with great accuracy ever since 1917. And because a standard measure of breakup has been used throughout this time, the data are consistent. An article in *Science*[4] used the data to investigate global warming—whether greenhouse gasses and other human actions have been making the planet warmer. Others might just want to make a good prediction of next year's breakup time.

Of course, we can't use regression to tell the *causes* of any change. But we can estimate the *rate* of change (if any) and use it to make better predictions.

Here are some of the data:

A S *Activity:* **A Hypothesis Test for the Regression Slope.** View an animated discussion of testing the standard null hypothesis for slope.

WHO	Years
WHAT	Year, day, and hour of ice breakup
UNITS	x is in years since 1900. y is in days after midnight Dec. 31.
WHEN	1917–present
WHERE	Nenana, Alaska
WHY	Wagering, but proposed to look at global warming

Year (since 1900)	Breakup Date (days after Jan. 1)	Year (since 1900)	Breakup Date (days after Jan. 1)
17	119.4792	30	127.7938
18	130.3979	31	129.3910
19	122.6063	32	121.4271
20	131.4479	33	127.8125
21	130.2792	34	119.5882
22	131.5556	35	134.5639
23	128.0833	36	120.5403
24	131.6319	37	131.8361
25	126.7722	38	125.8431
26	115.6688	39	118.5597
27	131.2375	40	110.6437
28	126.6840	41	122.0764
29	124.6535	⋮	⋮

---

[3] http://www.nenanaakiceclassic.com
[4] "Climate Change in Nontraditional Data Sets." *Science* 294 [26 October 2001]: 811.

STEP-BY-STEP EXAMPLE | A Regression Slope *t*-Test

The slope of the regression gives the change in Nenana ice breakup date per year.

Questions: Is there sufficient evidence to claim that ice breakup times are changing? If so, how rapid is the change?

THINK

Plan State what you want to know.

Identify the *parameter* you wish to estimate. Here our parameter is the slope.

Identify the variables and review the W's.

Hypotheses Write your null and alternative hypotheses.

Model Think about the assumptions and check the conditions.

Make pictures. Because the scatterplot seems straight enough, we can find and plot the residuals.

I wonder whether the date of ice breakup in Nenana has changed over time. The slope of that change might indicate climate change. I have the date of ice breakup annually for 93 years starting in 1917, recorded as the number of days and fractions of a day until the ice breakup.

H_0: There is no change in the date of ice breakup: $\beta_1 = 0$
H_A: Yes, there is: $\beta_1 \neq 0$

✔ **Straight Enough Condition:** I have quantitative data with no obvious bend in the scatterplot.

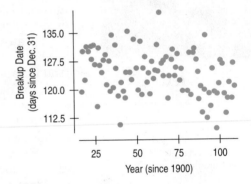

✔ **Independence Assumption:** These data are a time series, which raises my suspicions that they may not be independent. To check, here's a plot of the residuals against time, the x-variable of the regression:

Usually, we check for suggestions that the Independence Assumption fails by plotting the residuals against the predicted values. Patterns and clusters in that plot raise our suspicions. But when the data are measured over time, it is always a good idea to plot residuals against time to look for trends and oscillations.

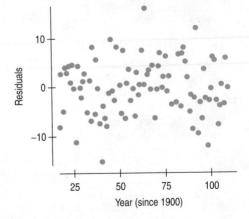

I see a hint that the data oscillate up and down, which suggests some failure of independence, but not so strongly that I can't proceed with the analysis. These data are not a random sample, so I'm reluctant to extend my conclusions beyond this river and these years.

✔ **Does the Plot Thicken? Condition:** The residuals plot shows no obvious trends in the spread.

✔ **Nearly Normal Condition, Outlier Condition:** A histogram of the residuals is unimodal and symmetric.

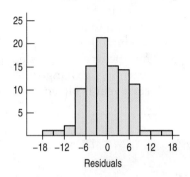

State the sampling distribution model.

Under these conditions, the sampling distribution of the regression slope can be modeled by a Student's t-model with $(n - 2) = 91$ degrees of freedom.

Choose your method.

I'll do a **regression slope t-test.**

 Mechanics The regression equation can be found from the formulas in Chapter 8, but regressions are almost always found from a computer program or calculator.

The P-values given in the regression output table are from the Student's t-distribution on $(n - 2) = 91$ degrees of freedom. They are appropriate for two-sided alternatives.

Here's the computer output for this regression:

Dependent variable is: Breakup Date
R-squared = 10.6%
s = 5.647 with 93 − 2 = 91 degrees of freedom

Variable	Coefficient	SE(Coeff)	t-ratio	P-value
Intercept	128.747	1.494	86.2	≤0.0001
Year Since 1900	−0.071750	0.0218	−3.29	0.0014

The estimated regression equation is

$$\widehat{Date} = 128.747 - 0.072 \ YearSince1900 .$$

 Conclusion Link the P-value to your decision and state your conclusion in the proper context.

The P-value of 0.0014 means that the association we see in the data is unlikely to have occurred by chance. I reject the null hypothesis, and conclude that there is strong evidence that, on average, the ice breakup is occurring earlier each year. But the oscillation pattern in the residuals raises concerns.

SHOW
MORE

Create a confidence interval for the true slope

A 95% confidence interval for β_1 is

$$b_1 \pm t^*_{91} \times SE(b_1)$$
$$-0.072 \pm (1.987)(0.0218)$$
$$\text{or } (-0.11, -0.03) \text{ days per year.}$$

 TELL
MORE

Interpret the interval Simply rejecting the standard null hypothesis doesn't guarantee that the size of the effect is large enough to be important. Whether we want to know the breakup time to the nearest minute or are interested in global warming, a change measured in hours each year is big enough to be interesting.

I am 95% confident that the ice has been breaking up, on average, between 0.03 days (about 40 minutes) and 0.11 days (about 3 hours) earlier each year since 1900.

> **But is it global warming?** So the ice is breaking up earlier. Temperatures are higher. Must be global warming, right?
>
> Maybe.
>
> An article challenging the original analysis of the Nenana data proposed a possible confounding variable. It noted that the city of Fairbanks is upstream from Nenana and suggested that the growth of Fairbanks could have warmed the river. So maybe it's not global warming.
>
> Or maybe global warming is a lurking variable, leading more people to move to a now balmier Fairbanks and also leading to generally earlier ice breakup in Nenana.
>
> Or maybe there's some other variable or combination of variables at work. We can't set up an experiment, so we may never really know.
>
> Only one thing is for sure. When you try to explain an association by claiming cause and effect, you're bound to be on thin ice.[5]

Standard Errors for Predicted Values

Once we have a useful regression, how can we indulge our natural desire to predict, without being irresponsible? We know how to compute predicted values of y for any value of x. We first did that in Chapter 8. This predicted value would be our best estimate, but it's still just an informed guess.

Now, however, we have standard errors. We can use those to construct a confidence interval for the predictions and to report our uncertainty honestly.

From our model of *%Body Fat* and *Waist* size, we might want to use *Waist* size to get a reasonable estimate of *%Body Fat*. A confidence interval can tell us how precise that prediction will be. The precision depends on the question we ask, however, and there are two questions: Do we want to know the mean *%Body Fat* for *all* men with a *Waist* size of, say, 38 inches? Or do we want to estimate the *%Body Fat* for a particular man with a 38-inch *Waist* without making him climb onto the X-ray table?

---

[5] How *do* scientists sort out such messy situations? Even though they can't conduct an experiment, they *can* look for replications elsewhere. A number of studies of ice on other bodies of water have also shown earlier ice breakup times in recent years. That suggests they need an explanation that's more comprehensive than just Fairbanks and Nenana.

What's the difference between the two questions? The predicted *%Body Fat* is the same, but one question leads to an answer much more precise than the other. We can predict the *mean %Body Fat* for *all* men whose *Waist* size is 38 inches with a lot more precision than we can predict the *%Body Fat* of a *particular individual* whose *Waist* size happens to be 38 inches. Both are interesting questions.

We start with the same prediction in both cases. We are predicting the value for a new individual, one that was not part of the original data set. To emphasize this, we'll call his x-value "x sub new" and write it x_ν.[6] Here, x_ν is 38 inches. The regression equation predicts *%Body Fat* as $\hat{y}_\nu = b_0 + b_1 x_\nu$.

Now that we have the predicted value, we construct both intervals around this same number. Both intervals take the form

$$\hat{y}_\nu \pm t^*_{n-2} \times SE.$$

Even the *t** value is the same for both. It's the critical value (from Table T or technology) for $n - 2$ degrees of freedom and the specified confidence level. The intervals differ because they have different standard errors. Our choice of ruler depends on which interval we want.

The standard errors for prediction depend on the same kinds of things as the coefficients' standard errors. If there is more spread around the line, we'll be less certain when we try to predict the response. Of course, if we're less certain of the slope, we'll be less certain of our prediction. If we have more data, our estimate will be more precise. And there's one more piece: If we're farther from the center of our data, our prediction will be less precise. This last factor is new but makes intuitive sense: It's a lot easier to predict a data point near the middle of the data set than far from the center.

Each of these factors contributes uncertainty—that is, variability—to the estimate. Because the factors are independent of each other, we can add their variances to find the total variability. The resulting formula for the standard error of the predicted *mean* value explicitly takes into account each of the factors:

$$SE(\hat{\mu}_\nu) = \sqrt{SE^2(b_1) \cdot (x_\nu - \bar{x})^2 + \frac{s_e^2}{n}}.$$

Individual values vary more than means, so the standard error for a single predicted value has to be larger than the standard error for the mean. In fact, the standard error of a single predicted value has an *extra* source of variability: the variation of individuals around the predicted mean. That appears as the extra variance term, s_e^2, at the end under the square root:

$$SE(\hat{y}_\nu) = \sqrt{SE^2(b_1) \cdot (x_\nu - \bar{x})^2 + \frac{s_e^2}{n} + s_e^2}.$$

Remember to keep this distinction between the two kinds of standard errors when looking at computer output. The smaller one is for the predicted *mean* value, and the larger one is for a predicted *individual* value.[7]

For the Nenana Ice Classic, someone who planned to place a bet would want to predict this year's breakup time. By contrast, scientists studying global warming are likely to be more interested in the mean breakup time. Unfortunately if you want to gamble, the variability is greater for predicting for a single year.

---

[6] Yes, this is a bilingual pun. The Greek letter ν is called "nu." Don't blame me; my co-author suggested this.

[7] You may see the standard error expressions written in other, equivalent ways. The most common alternatives are

$$SE(\hat{\mu}_\nu) = s_e\sqrt{\frac{1}{n} + \frac{(x_\nu - \bar{x})^2}{\sum(x - \bar{x})^2}} \quad \text{and} \quad SE(\hat{y}_\nu) = s_e\sqrt{1 + \frac{1}{n} + \frac{(x_\nu - \bar{x})^2}{\sum(x - \bar{x})^2}}.$$

Confidence Intervals for Predicted Values

Now that we have standard errors, we can ask how well our analysis can predict the mean *%Body Fat* for men with 38-inch *Waists*. The regression output table (still on page 680) provides most of the numbers we need:

$$s_e = 4.713$$
$$n = 250$$
$$SE(b_1) = 0.074, \text{and from the data we need to know that}$$
$$\bar{x} = 36.3$$

The regression model gives a predicted value at $x_\nu = 38$ of

$$\hat{y}_\nu = -42.7 + 1.7(38) = 21.9\%.$$

Let's find the 95% confidence interval for the mean *%Body Fat* for all men with 38-inch *Waists*. We find the standard error from the formula:

$$SE(\hat{\mu}_\nu) = \sqrt{0.074^2(38 - 36.3)^2 + \frac{4.713^2}{250}} = 0.32\%.$$

The t^* value that excludes 2.5% in either tail with $250 - 2 = 248$ df is (according to the tables) 1.97.

Putting it all together, we find the margin of error as

$$ME = 1.97(0.32) = 0.63\%.$$

So, we are 95% confident that the mean *%Body Fat* for men with 38-inch *Waists* is

$$21.9\% \pm 0.63\%.$$

Suppose, instead, we want to predict the *%Body Fat* for an individual man with a 38-inch *Waist*. We need the larger standard error:

$$SE(\hat{y}_\nu) = \sqrt{0.074^2(38 - 36.3)^2 + \frac{4.713^2}{250} + 4.713^2} = 4.72\%.$$

The corresponding margin of error is

$$ME = 1.97(4.72) = 9.30\%,$$

so the **prediction interval** is

$$21.9\% \pm 9.30\%.$$

We can think of this interval as having a 95% chance of capturing the true *%Body Fat* of a randomly selected man whose *Waist* is 38 inches.[8] Notice how much wider this interval is than the previous one. As we've known since Chapter 18, the mean is much less variable than a randomly selected individual value.

Keep in mind this distinction between the two kinds of confidence intervals: The narrower interval is a **confidence interval for the predicted mean value** at x_ν, and the wider interval is a **prediction interval for an individual** with that x-value.

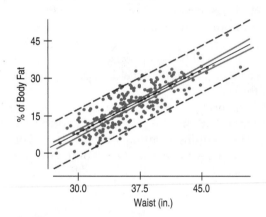

FIGURE 27.11

A scatterplot of %Body Fat vs. Waist size with a least squares regression line. The solid green lines near the regression line show the extent of the 95% confidence intervals for mean %Body Fat at each Waist size. The dashed red lines show the prediction intervals. Most of the points are contained within the prediction intervals, but not within the confidence intervals.

---

[8] Technically, it's a little more complicated, but it's very close to this.

Intervals for Predictions

RECAP: We've found a linear model for the relationship between the *logDiam* and *logAge* for terrestrial impact craters:
$\widehat{logDiam} = 0.358 + 0.527logAge$.

Let's look at some confidence intervals, keeping in mind that the model is in terms of the logs of these variables, so we must work with logs and then transform back to original units when we are done. (Many of the values substituted into the formulas are from the regression tables shown in the For Examples on pages 683 and 684.)

QUESTION #1: If we wish to hunt for new craters that are 5 million years old, what's a 95% confidence interval for the mean size we'd expect them to be?

$$\widehat{logDiam} = 0.358 + 0.527logAge$$
$$= 0.358 + 0.527log(5)$$
$$= 0.726$$

$$SE(\hat{\mu}) = \sqrt{SE^2(b_1)\cdot(x-\bar{x})^2 + \frac{s_e^2}{n}}$$
$$= \sqrt{0.0655^2(log(5)-(-0.65631))^2 + \frac{0.6362^2}{39}}$$
$$= 0.1351$$

From Table T, $t^*_{37} = 2.030$, so a 95% CI is $0.726 \pm 2.030(0.1351)$, or $(0.452, 1.00)$.

Since these represent logs of the diameters, I'll find the original units:

$$10^{0.452} = 2.8 \text{ km and } 10^{1.00} = 10 \text{ km}.$$

I'm 95% confident that terrestrial craters 5 million years old average between 2.8 and 10 kilometers in diameter.

QUESTION #2: There's just been a news report announcing the discovery of a new crater that is 10 million years old. How large do you expect it to be? What's a 95% prediction interval for your estimate?

For a crater 10 million years old, $\widehat{logDiam} = 0.358 + 0.527log(10) = 0.885$ and $10^{0.885} = 7.674$. I expect the new crater to be about 7.7 km in diameter.

$$SE(\hat{y}) = \sqrt{SE^2(b_1)\cdot(x-\bar{x})^2 + \frac{s_e^2}{n} + s_e^2}$$
$$= \sqrt{0.0655^2(log(10)-(-0.65631))^2 + \frac{0.6362^2}{39} + 0.6362^2}$$
$$= 0.653$$

A 95% CI is $0.885 \pm 2.030(0.653) = (-0.441, 2.211)$; in original units, that's $(0.36, 162.55)$.

I'm 95% confident that the newly discovered 10 million-year-old crater is between 0.36 and 162.55 kilometers in diameter. (Without a lot of data and a very strong association, prediction intervals often aren't very precise!)

MATH BOX

So where do those messy formulas for standard errors of predicted values come from? They're based on many of the ideas we've studied so far. Start with regression, add random variables, then throw in the Pythagorean Theorem, the Central Limit Theorem, and a dose of algebra. Mix well. . . .

(continued)

We begin our quest with an equation of the regression line. Usually we write the line in the form $\hat{y} = b_0 + b_1 x$. Mathematicians call that the "slope-intercept" form; in your algebra class you wrote it as $y = mx + b$. In that algebra class you also learned another way to write equations of lines. When you know that a line with slope m passes through the point (x_1, y_1), the "point-slope" form of its equation is $y - y_1 = m(x - x_1)$.

We know the regression line passes through the mean-mean point $(\bar{x}, \bar{y})$ with slope b_1, so we can write its equation in point-slope form as $\hat{y} - \bar{y} = b_1(x - \bar{x})$. Solving for $\hat{y}$ yields $\hat{y} = b_1(x - \bar{x}) + \bar{y}$. This equation predicts the mean y-value for a specific x_ν:

$$\hat{\mu}_y = b_1(x_\nu - \bar{x}) + \bar{y}.$$

To create a confidence interval for the mean value we need to measure the variability in this prediction:

$$Var(\hat{\mu}_y) = Var(b_1(x_\nu - \bar{x}) + \bar{y}).$$

We now call on the Pythagorean Theorem of Statistics once more: the slope, b_1, and mean, $\bar{y}$, should be independent, so their variances add:

$$Var(\hat{\mu}_y) = Var(b_1(x_\nu - \bar{x})) + Var(\bar{y}).$$

The horizontal distance from our specific x-value to the mean, $x_\nu - \bar{x}$, is a constant:

$$Var(\hat{\mu}_y) = (Var(b_1))(x_\nu - \bar{x})^2 + Var(\bar{y}).$$

Let's write that equation in terms of standard deviations:

$$SD(\hat{\mu}_y) = \sqrt{(SD^2(b_1))(x_\nu - \bar{x})^2 + SD^2(\bar{y})}.$$

Because we'll need to estimate these standard deviations using samples statistics, we're really dealing with standard errors:

$$SE(\hat{\mu}_y) = \sqrt{(SE^2(b_1))(x_\nu - \bar{x})^2 + SE^2(\bar{y})}.$$

The Central Limit Theorem tells us that the standard deviation of $\bar{y}$ is $\dfrac{\sigma}{\sqrt{n}}$. Here we'll estimate σ using s_e, which describes the variability in how far the line we drew through our sample mean may lie above or below the true mean:

$$SE(\hat{\mu}_y) = \sqrt{(SE^2(b_1))(x_\nu - \bar{x})^2 + \left(\frac{s_e}{\sqrt{n}}\right)^2}$$

$$= \sqrt{(SE^2(b_1))(x_\nu - \bar{x})^2 + \frac{s_e^2}{n}}.$$

And there it is—the standard error we need to create a confidence interval for a predicted mean value.

When we try to predict an individual value of y, we must also worry about how far the true point may lie above or below the regression line. We represent that uncertainty by adding another term, e, to the original equation:

$$y = b_1(x_\nu - \bar{x}) + \bar{y} + e.$$

To make a long story short (and the equation a wee bit longer), that additional term simply adds one more standard error to the sum of the variances:

$$SE(\hat{y}_\nu) = \sqrt{SE^2(b_1)(x_\nu - \bar{x})^2 + \frac{s_e^2}{n} + s_e^2}.$$

Logistic Regression

The Pima Indians of southern Arizona are a unique community. Their ancestors were among the first people to cross over into the Americas some 30,000 years ago. For at least two millennia, they have lived in the Sonoran Desert near the Gila River. Known throughout history as a generous people, they have given of themselves for the past 30 years helping researchers at the National Institutes of Health study certain diseases like diabetes and obesity. Young Pima Indians often marry other Pimas, making them an ideal group for genetic researchers to study. Pimas also have an extremely high incidence of diabetes.

Researchers investigating factors for increased risk of diabetes examined data on 768 adult women of Pima Indian heritage. One possible predictor is the body mass index, *BMI*, calculated as weight/height^2, where weight is measured in kilograms and height in meters. We are interested in the relationship between *BMI* and the incidence of diabetes. We might start by looking at boxplots of *BMI* for each group:

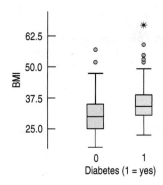

FIGURE 27.12
Side-by-side boxplots for the two *Diabetes* groups (1 = has diabetes; 0 = doesn't have) show significantly elevated body mass index (BMI) for the women who have diabetes.

From the boxplots, we see that the group with diabetes has a higher mean *BMI*. (A *t*-test would show the difference to be more than 9 SEs from 0 with a P-value < 0.0001.) There is clearly a relationship. Here we've displayed *BMI* as the response and *Diabetes* as the predictor. But the researchers are interested in predicting the increased risk of *Diabetes* due to increased *BMI*, not the other way around.

Reversing the roles, we could code having *Diabetes* as "1" and *No Diabetes* as "0" and make a scatterplot:

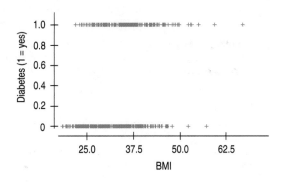

FIGURE 27.13
A scatterplot of *Diabetes* by *BMI* shows a shift in *BMI* for the two groups, but is not easy to interpret because *Diabetes* is a dichotomous (two-valued) variable.

Diabetes is usually treated as a categorical variable, but what if we treat it as a quantitative variable and fit a linear regression to these data? We would get the following regression line:

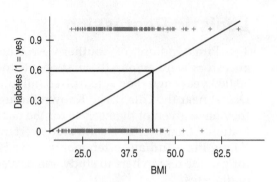

FIGURE 27.14
A regression of *Diabetes* on *BMI* shows an increasing likelihood of having diabetes with increasing *BMI*. Of course, a linear regression is not strictly appropriate unless *Diabetes* is treated as a quantitative variable.

The equation says: Predicted *Diabetes* = −0.351 + 0.022 *BMI*. Does this make *any* sense? Suppose someone had a *BMI* of 44. The equation predicts 0.60. What would you guess is the chance that she will have diabetes? If you said about 60%, then you're using the line to model the *probability* of having *Diabetes*.

There are some obvious problems with this model, though. What's the probability that someone with a *BMI* of 10 has diabetes? It's low, but the equation predicts −0.13, obviously an impossible probability. And if we imagined someone with a *BMI* of 70, we might suspect that the probability of her having diabetes is pretty high, but not the predicted value of 1.16.

We can fix this problem just by setting all negative probabilities to 0 and all probabilities greater than 1 to 1. That would give a model that looked like this:

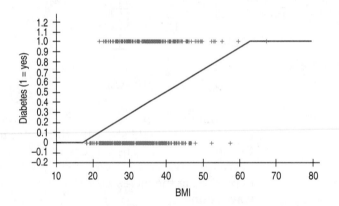

FIGURE 27.15
This model eliminates the problem of probability values that are negative or that exceed 1, but the corners neither are aesthetically pleasing nor do they make scientific sense.

This avoids one problem, but it can't really be correct. The occurrence of *Diabetes* can't be either certain ($p = 1$) or impossible ($p = 0$) based only on *BMI*. And it's not aesthetically pleasing. Real-world changes are likely to be smooth, so we prefer models with smooth transitions rather than corners. That makes good predictive sense, too. There's no reason to expect sharp changes at certain *BMI* values. So instead, we can use a smooth curve to model the probability of having *Diabetes*, like this:

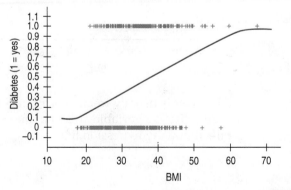

FIGURE 27.16
The smooth curve models the probability of having *Diabetes* as a function of *BMI* in a sensible way. The logistic curve shown here is just one of a number of choices for the form of the curve, but all are fairly similar in shape and in the resulting predicted probabilities.

This smoother version is a sensible way to model the probability of having *Diabetes* as a function of *BMI*. There are many curves in mathematics with shapes like this that we might use for our model. One of the most common is the *logistic curve*. The regression based on this curve is called **logistic regression.**

The equation for logistic regression can be written like an ordinary regression on a transformed y-variable:

$$\ln\left(\frac{\hat{p}}{1 - \hat{p}}\right) = b_0 + b_1 x.$$

The expression on the left-hand side gives the *logistic* curve. The logarithm used is the natural, or base e, log, although that doesn't really matter to the shape of the model.

It turns out that the logistic curve has a particularly useful interpretation. Racetrack enthusiasts know that when p is a probability, $\frac{p}{1 - p}$ is the *odds* in favor of a success. For example, when the probability of success, p, $= 1/3$, we'd get the ratio $\frac{1/3}{2/3} = \frac{1}{2}$. We'd say that the odds in favor of success are 1:2 (or we'd probably say the odds *against* it are 2:1). Logistic regression models the *logarithm* of the odds as a linear function of x. In fact, nobody really thinks in terms of the log of the odds ratio. But it's the combination of that ratio and the logarithm that gets us the nice **S**-curved shape. What is important is that we can work backward from a log odds ratio to get the probability—which is often easier to think about.

Because we're not fitting a straight line to the data directly, we can't use ordinary least squares to estimate the parameters b_0 and b_1.[9] Instead, we use special *nonlinear* methods that require a good deal more computation—but the computers take care of that for us, and can work backward through the fitted equation to give predictions in terms of probabilities.

For the Pima Indians data set, the equation is

$$\ln\left(\frac{\hat{p}}{1 - \hat{p}}\right) = -4.0 + 0.102\ BMI.$$

A computer output will typically provide a table like the following:

Term	Estimate	Std Error	ChiSquare	P-value
Intercept	−3.9967	0.4288	86.8576	<0.00001
BMI	0.1025	0.0126	66.0872	<0.00001

We usually don't interpret the slope itself (unless you happen to think naturally in log odds), but we can perform a test on whether the slope is 0, similar to the t-test we did for linear regression. Unlike linear regression, the ratio of the estimate to its standard error does not have a t distribution, but the square of that ratio has a χ^2 distribution.[10] The P-values for both the slope and the intercept clearly indicate that neither is 0.

Once we have decided that the slope is not 0, we can use the model to predict the probabilities. Solving the equation $\ln\left(\frac{\hat{p}}{1 - \hat{p}}\right) = b_0 + b_1 x$ for $\hat{p}$ gives

$$\hat{p} = \frac{1}{1 + e^{-(b_0 + b_1 x)}}$$

---

[9] This is a tricky point. We've fit regressions with transformed y's before without special methods. If our data consisted of observed *proportions* of people, we could transform the data using $\log(p/(1 - p))$ and use linear regression to fit the equation. But all we have are individual 0's and 1's. To be able to fit this equation with our raw data, we need special nonlinear methods.

[10] Here's that nonlinear fitting popping up again. Don't let it bother you. You've seen χ^2 models before, and all you really need to look at for a test is the P-value.

and the logistic equation guarantees that the estimate of $\hat{p}$ will be between 0 and 1. Fortunately, technology can provide these probability estimates, producing the curve shown previously and giving an estimate for the probability at any *BMI* value.

Response variables that are dichotomous (having only two possible values) like the variable *Diabetes* are common, so there's a widespread need to model data like these. Phone companies want to predict who will switch carriers, credit card companies want to know which transactions are likely to be fraudulent, and loan companies want to know who is most likely to declare bankruptcy. All of these are potential application areas for logistic regression. It's not surprising that logistic regression has become an important modeling tool in the toolbox of analysts in science and industry in the past decade. By understanding the basics of what logistic regression can do, you can expand your ability to apply regression to many other important applications.

What Can Go Wrong?

In this chapter we've added inference to the regression explorations that we did in Chapters 8 and 9. Everything covered in those chapters that could go wrong with regression can still go wrong. It's probably a good time to review Chapter 9. Take your time; we'll wait.

With inference, we've put numbers on our estimates and predictions, but these numbers are only as good as the model. Here are the main things to watch out for:

■ **Don't fit a linear regression to data that aren't straight.** The linearity assumption is the most fundamental assumption. If the relationship between x and y isn't approximately linear, there's no sense in fitting a straight line to it.

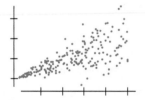

■ **Watch out for the plot thickening.** The common part of confidence and prediction intervals is the estimate of the error standard deviation, the spread around the line. If it changes with x, the estimate won't make sense. Imagine making a prediction interval for these data.

When x is small, we can predict y precisely, but as x gets larger, it's much harder to pin y down. Unfortunately, if the spread changes, the single value of s_e won't pick that up. The prediction interval will use the average spread around the line, with the result that we'll be too pessimistic about our precision for low x-values and too optimistic for high x-values. A re-expression of y is often a good fix for changing spread.

■ **Make sure the errors are Normal.** When we make a prediction interval for an individual, the Central Limit Theorem can't come to our rescue. For us to believe the prediction interval, the errors must be from the Normal model. Check the histogram and Normal probability plot of the residuals to see if this assumption looks reasonable.

■ **Watch out for extrapolation.** It's tempting to think that because we have prediction *intervals*, they'll take care of all our uncertainty so we don't have to worry about extrapolating. Wrong. The interval is only as good as the model. The uncertainty our intervals predict is correct only if our model is true. There's no way to adjust for wrong models. That's why it's always dangerous to predict for x-values that lie far from the center of the data.

■ **Watch out for influential points and outliers.** We always have to be on the lookout for a few points that have undue influence on our estimated model—and regression is certainly no exception.

■ **Watch out for one-tailed tests.** Because tests of hypotheses about regression coefficients are usually two-tailed, software packages report two-tailed P-values. If you are using software to conduct a one-tailed test about slope, you'll need to divide the reported P-value in half.

Regression inference is connected to almost everything we've done so far. Scatterplots are essential for checking linearity and whether the plot thickens. Histograms and normal probability plots come into play to check the Nearly Normal condition. And we're still thinking about the same attributes of the data in these plots as we were back in the first part of the book.

Regression inference is also connected to just about every inference method we have seen for measured data. The assumption that the spread of data about the line is constant is essentially the same as the assumption of equal variances required for the pooled-t methods. Our use of all the residuals together to estimate their standard deviation is a form of pooling.

Inference for regression is closely related to inference for means, so your understanding of means transfers directly to your understanding of regression. Here's a table that displays the similarities:

	Means	Regression Slope
Parameter	μ	β_1
Statistic	$\bar{y}$	b_1
Population spread estimate	$s_y = \sqrt{\dfrac{\Sigma(y - \bar{y})^2}{n - 1}}$	$s_e = \sqrt{\dfrac{\Sigma(y - \hat{y})^2}{n - 2}}$
Standard error of the statistic	$SE(\bar{y}) = \dfrac{s_y}{\sqrt{n}}$	$SE(b_1) = \dfrac{s_e}{s_x\sqrt{n - 1}}$
Test statistic	$\dfrac{\bar{y} - \mu_0}{SE(\bar{y})} \sim t_{n-1}$	$\dfrac{b_1 - \beta_1}{SE(b_1)} \sim t_{n-2}$
Margin of error	$ME = t^*_{n-1} \times SE(\bar{y})$	$ME = t^*_{n-2} \times SE(b_1)$

WHAT HAVE WE LEARNED?

In Chapters 7, 8, and 9, we learned to examine the relationship between two quantitative variables in a scatterplot, to summarize its strength with correlation, and to fit linear relationships by least squares regression. And we saw that these methods are particularly powerful and effective for modeling, predicting, and understanding these relationships.

Now we have completed our study of inference methods by applying them to these regression models. We've found that the same methods we used for means—Student's t-models—work for regression in much the same way as they did for means. And we've seen that although this makes the mechanics familiar, there are new conditions to check and a need for care in describing the hypotheses we test and the confidence intervals we construct.

▶ We've learned that under certain assumptions, the sampling distribution for the slope of a regression line can be modeled by a Student's t-model with $n - 2$ degrees of freedom.

▶ We've learned to check four conditions to verify those assumptions before we proceed with inference. We've learned the importance of checking these conditions in order, and we've seen that most of the checks can be made by graphing the data and the residuals with the methods we learned in Chapters 4, 5, and 8.

▶ We've learned to use the appropriate t-model to test a hypothesis about the slope. If the slope of our regression line is significantly different from zero, we have strong evidence that there is an association between the two variables.

▶ We've also learned to create and interpret a confidence interval for the true slope.

▶ And we've been reminded yet again never to mistake the presence of an association for proof of causation.

Terms

Conditions for inference in regression (and checks for some of them)

▶ **Straight Enough Condition** for linearity. (Check that the scatterplot of y against x has linear form and that the scatterplot of residuals against predicted values has no obvious pattern.) (p. 675)

▶ **Independence Assumption.** (Think about the nature of the data. Check a residuals plot.) (p. 676)

▶ **Does the Plot Thicken? Condition** for constant variance. (Check that the scatterplot shows consistent spread across the range of the x-variable, and that the residuals plot has constant variance, too. A common problem is increasing spread with increasing predicted values—the *plot thickens!*) (p. 676)

▶ **Nearly Normal Condition** for Normality of the residuals. (Check a histogram of the residuals.) (p. 677)

Residual standard deviation

The spread of the data around the regression line is measured with the residual standard deviation, s_e:

$$s_e = \sqrt{\frac{\sum (y - \hat{y})^2}{n - 2}} = \sqrt{\frac{\sum e^2}{n - 2}} \quad \text{(p. 681)}.$$

t-test for the regression slope

When the assumptions are satisfied, we can perform a test for the slope coefficient. We usually test the null hypothesis that the true value of the slope is zero against the alternative that it is not. A zero slope would indicate a complete absence of linear relationship between y and x.

To test $H_0: \beta_1 = 0$, we find

$$t = \frac{b_1 - 0}{SE(b_1)}$$

where

$$SE(b_1) = \frac{s_e}{\sqrt{n - 1}\, s_x}, \quad s_e = \sqrt{\frac{\sum (y - \hat{y})^2}{n - 2}},$$

n is the number of cases, and s_x is the standard deviation of the x-values. We find the P-value from the Student's t-model with $n - 2$ degrees of freedom (p. 682).

Confidence interval for the regression slope

When the assumptions are satisfied, we can find a confidence interval for the slope parameter from $b_1 \pm t^*_{n-2} \times SE(b_1)$. The critical value, t^*_{n-2}, depends on the confidence level specified and on Student's t-model with $n - 2$ degrees of freedom (p. 684).

Confidence interval for a predicted mean value

Different samples will give different estimates of the regression model and, so, different predicted values for the same value of x. We find a confidence interval for the mean of these predicted values at a specified x-value, x_ν, as $\hat{y}_\nu \pm t^*_{n-2} \times SE(\hat{\mu}_\nu)$, where

$$SE(\hat{\mu}_\nu) = \sqrt{SE^2(b_1) \cdot (x_\nu - \bar{x})^2 + \frac{s_e^2}{n}}.$$

The critical value, t^*_{n-2}, depends on the specified confidence level and the Student's t-model with $n - 2$ degrees of freedom (p. 690).

Prediction interval for an individual

Different samples will give different estimates of the regression model and, so, different predicted values for the same value of x. We can make a confidence interval to capture a certain percentage of the entire distribution of predicted values. This makes it much wider than the corresponding confidence interval for the mean. The confidence interval takes the form $\hat{y}_\nu \pm t^*_{n-2} \times SE(\hat{y}_\nu)$, where

$$SE(\hat{y}_\nu) = \sqrt{SE^2(b_1) \cdot (x_\nu - \bar{x})^2 + \frac{s_e^2}{n} + s_e^2}.$$

The critical value, t^*_{n-2}, depends on the specified confidence level and the Student's t-model with $n - 2$ degrees of freedom (p. 690).

Skills

▶ Understand that the "true" regression line does not fit the population data perfectly, but rather is an idealized summary of that data.

▶ Know how to examine your data and a scatterplot of y vs. x for violations of assumptions that would make inference for regression unwise or invalid.

▶ Know how to examine displays of the residuals from a regression to double-check that the conditions required for regression have been met. In particular, know how to judge linearity and constant variance from a scatterplot of residuals against predicted values. Know how to judge Normality from a histogram and Normal probability plot.

▶ Remember to be especially careful to check for failures of the Independence Assumption when working with data recorded over time. To search for patterns, examine scatterplots both of x against time and of the residuals against time.

SHOW

▶ Know how to test the standard hypothesis that the true regression slope is zero. Be able to state the null and alternative hypotheses. Know where to find the relevant numbers in standard computer regression output.

▶ Be able to find a confidence interval for the slope of a regression based on the values reported in a standard regression output table.

TELL

▶ Be able to summarize a regression in words. In particular, be able to state the meaning of the true regression slope, the standard error of the estimated slope, and the standard deviation of the errors.

▶ Be able to interpret the P-value of the t-statistic for the slope to test the standard null hypothesis.

▶ Be able to interpret a confidence interval for the slope of a regression.

REGRESSION ANALYSIS ON THE COMPUTER

All statistics packages make a table of results for a regression. These tables differ slightly from one package to another, but all are essentially the same. We've seen two examples of such tables already.

All packages offer analyses of the residuals. With some, you must request plots of the residuals as you request the regression. Others let you find the regression first and then analyze the residuals afterward. Either way, your analysis is not complete if you don't check the residuals with a histogram or Normal probability plot and a scatterplot of the residuals against x or the predicted values.

You should, of course, always look at the scatterplot of your two variables before computing a regression.

Regressions are almost always found with a computer or calculator. The calculations are too long to do conveniently by hand for data sets of any reasonable size. No matter how the regression is computed, the results are usually presented in a table that has a standard form. Here's a portion of a typical regression results table, along with annotations showing where the numbers come from:

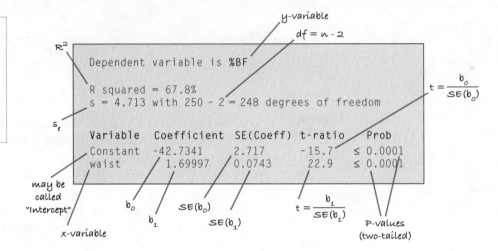

The regression table gives the coefficients (once you find them in the middle of all this other information), so we can see that the regression equation is

$$\widehat{\%BF} = -42.73 + 1.7\,Waist$$

and that the R^2 for the regression is 67.8%. (Is accounting for 68% of the variation in %Body Fat good enough to be useful? Is a prediction ME of more than 9% good enough? Health professionals might not be satisfied.)

The column of t-ratios gives the test statistics for the respective null hypotheses that the true values of the coefficients are zero. The corresponding P-values are also usually reported.

DATA DESK

- Select Y- and X-variables.
- From the **Calc** menu, choose **Regression.**
- Data Desk displays the regression table.
- Select plots of residuals from the Regression table's HyperView menu.

COMMENTS

You can change the regression by dragging the icon of another variable over either the Y- or X-variable name in the table and dropping it there. The regression will recompute automatically.

EXCEL

- In Excel 2003 and earlier, select Data Analysis from the **Tools** menu. In Excel 2007, select Data Analysis from the **Analysis Group** on the Data Tab.
- Select Regression from the **Analysis Tools** list.
- Click the **OK** button.
- Enter the data range holding the Y-variable in the box labeled "Y-range."
- Enter the range of cells holding the X-variable in the box labeled "X-range."
- Select the **New Worksheet Ply** option.
- Select **Residuals** options. Click the **OK** button.

COMMENTS

The Y and X ranges do not need to be in the same rows of the spreadsheet, although they must cover the same number of cells. But it is a good idea to arrange your data in parallel columns as in a data table.

Although the dialog offers a Normal probability plot of the residuals, the data analysis add-in does not make a correct probability plot, so don't use this option.

Excel calls the standard deviation of the residuals the "Standard Error." This is a common error. Don't be confused; it is not $SE(y)$, but rather s_e.

JMP

- From the **Analyze** menu, select **Fit Y by X.**
- Select variables: a Y, Response variable, and an X, Factor variable. Both must be continuous (quantitative).
- JMP makes a scatterplot.
- Click on the red triangle beside the heading labeled **Bivariate Fit . . .** and choose **Fit Line.** JMP draws the least squares regression line on the scatterplot and displays the results of the regression in tables below the plot.
- The portion of the table labeled "Parameter Estimates" gives the coefficients and their standard errors, t-ratios, and P-values.

COMMENTS

JMP chooses a regression analysis when both variables are "Continuous." If you get a different analysis, check the variable types.

The Parameter table does not include the residual standard deviation s_e. You can find that as Root Mean Square Error in the Summary of Fit panel of the output.

MINITAB

- Choose **Regression** from the **Stat** menu.
- Choose **Regression . . .** from the **Regression** submenu.
- In the Regression dialog, assign the Y-variable to the Response box and assign the X-variable to the Predictors box.
- Click the **Graphs** button.
- In the Regression-Graphs dialog, select **Standardized residuals,** and check **Normal plot of residuals** and **Residuals versus fits.**
- Click the **OK** button to return to the Regression dialog.
- Click the **OK** button to compute the regression.

COMMENTS

You can also start by choosing a Fitted Line plot from the **Regression** submenu to see the scatterplot first—usually good practice.

SPSS

- Choose **Regression** from the **Analyze** menu.
- Choose **Linear** from the **Regression** submenu.
- In the Linear Regression dialog that appears, select the Y-variable and move it to the dependent target. Then move the X-variable to the independent target.
- Click the **Plots** button.

- In the Linear Regression Plots dialog, choose to plot the *SRESIDs against the *ZPRED values.
- Click the **Continue** button to return to the Linear Regression dialog.
- Click the **OK** button to compute the regression.

TI-83/84 PLUS

Under **STAT TESTS** choose **LinRegTTest.** Specify the two lists where the data are stored and (usually) choose the two-tail option. In addition to reporting the calculated value of t and the P-value, the calculator will tell you the coefficients of the regression equation (a and b), the values of r^2 and r, and the value of s you need for confidence or prediction intervals.

TI-89

Under **STAT Tests** choose **A:LinRegTTest.** Specify the two lists where the data are stored and (usually) choose the two-tail option. Select an equation name to store the resulting line. In addition to reporting the calculated value of t and the P-value, the calculator will tell you the coefficients of the regression equation (a and b), the values of r^2 and r, the value of s used in prediction and confidence intervals, and the standard error of the slope. For 95% prediction and confidence intervals, choose **7:LinRegTint** from the **STAT Ints** menu. Specify the two lists where the data are stored, and select an equation name to store the resulting line. Select for an interval for the slope or for a response. If for a response, enter the x-value.

EXERCISES

T 1. **Hurricane predictions.** In Chapter 7 we looked at data from the National Oceanic and Atmospheric Administration about their success in predicting hurricane tracks. Here is a scatterplot of the error (in nautical miles) for predicting hurricane locations 72 hours in the future vs. the year in which the prediction (and the hurricane) occurred:

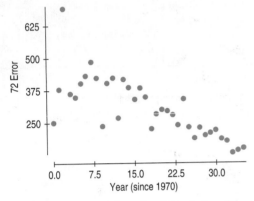

In Chapter 7 we could describe this relationship only in general terms. Now we can learn more. Here is the regression analysis:

Dependent variable is: 72Error
R squared = 58.5%
s = 75.38 with 36 − 2 = 34 degrees of freedom

Variable	Coefficient	SE(Coeff)	t-ratio	P-value
Intercept	453.223	24.61	18.4	≤0.0001
Year since 1970	−8.37084	1.209	−6.92	≤0.0001

a) Explain in words and numbers what the regression says.
b) State the hypothesis about the slope (both numerically and in words) that describes how hurricane prediction quality has changed.
c) Assuming that the assumptions for inference are satisfied, perform the hypothesis test and state your conclusion. Be sure to state it in terms of prediction errors and years.
d) Explain what the R-squared means in terms of this regression.

T 2. **Drug use.** The *European School Study Project on Alcohol and Other Drugs*, published in 1995, investigated the use of marijuana and other drugs. Data from 11 countries are summarized in the following scatterplot and regression analysis. They show the association between the percentage of a country's ninth graders who report having smoked marijuana and who have used other drugs such as LSD, amphetamines, and cocaine.

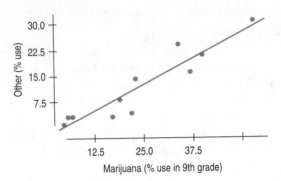

Dependent variable is: Other
R-squared = 87.3%
s = 3.853 with 11 − 2 = 9 degrees of freedom

Variable	Coefficient	SE(Coeff)	t-ratio	P-value
Intercept	−3.06780	2.204	−1.39	0.1974
Marijuana	0.615003	0.0784	7.85	<0.0001

a) Explain in context what the regression says.
b) State the hypothesis about the slope (both numerically and in words) that describes how use of marijuana is associated with other drugs.
c) Assuming that the assumptions for inference are satisfied, perform the hypothesis test and state your conclusion in context.
d) Explain what R-squared means in context.
e) Do these results indicate that marijuana use leads to the use of harder drugs? Explain.

T 3. **Movie budgets.**[11] How does the cost of a movie depend on its length? Data on the cost (millions of dollars) and the running time (minutes) for major release films of 2005 are summarized in these plots and computer output:

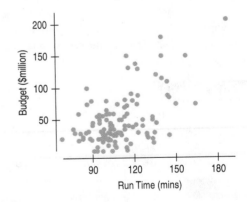

Dependent variable is: Budget($million)
R squared = 27.3%
s = 32.95 with 120 − 2 = 118 degrees of freedom

Variable	Coefficient	SE(Coeff)	t-ratio	P-value
Intercept	−63.9981	17.12	−3.74	0.0003
Run Time	1.02648	0.1540	6.66	≤0.0001

---

[11] Data have been corrected since previous editions.

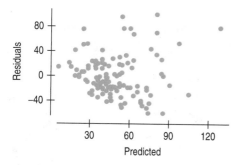

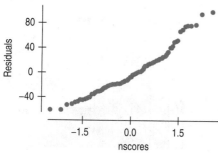

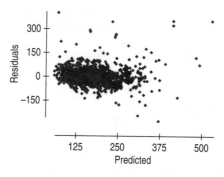

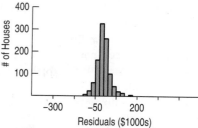

a) Explain in context what the regression says.
b) The intercept is negative. Discuss its value.
c) The output reports s = 32.95. Explain what that means in this context.
d) What's the value of the standard error of the slope of the regression line?
e) Explain what that means in this context.

T 4. **Saratoga house prices.** How does the price of a house depend on its size? Data from Saratoga, New York, on 1064 randomly selected houses that had been sold include data on price ($1000's) and size ($1000's ft^2), producing the following graphs and computer output:

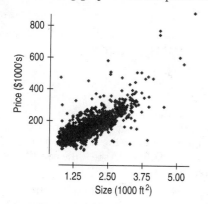

Dependent variable is: Price
R squared = 59.5%
s = 53.79 with 1064 − 2 = 1062 degrees of freedom

Variable	Coefficient	SE(Coeff)	t-ratio	P-value
Intercept	−3.11686	4.688	−0.665	0.5063
Size	94.4539	2.393	39.5	≤0.0001

a) Explain in context what the regression says.
b) The intercept is negative. Discuss its value, taking note of its P-value.
c) The output reports s = 53.79. Explain what that means in this context.
d) What's the value of the standard error of the slope of the regression line?
e) Explain what that means in this context.

T 5. **Movie budgets: the sequel.** Exercise 3 shows computer output examining the association between the length of a movie and its cost.

a) Check the assumptions and conditions for inference.
b) Find a 95% confidence interval for the slope and interpret it in context.

T 6. **Second home.** Exercise 4 shows computer output examining the association between the sizes of houses and their sale prices.

a) Check the assumptions and conditions for inference.
b) Find a 95% confidence interval for the slope and interpret it in context.

T 7. **Hot dogs.** Healthy eating probably doesn't include hot dogs, but if you are going to have one, you'd probably hope it's low in both calories and sodium. In its July 2007 issue, *Consumer Reports* listed the number of calories and sodium content (in milligrams) for 13 brands of all-beef hot dogs it tested. Examine the association, assuming that the data satisfy the conditions for inference.

Dependent variable is: Sodium
R squared = 60.5%
s = 59.66 with 13 − 2 = 11 degrees of freedom

Variable	Coefficient	SE(Coeff)	t-ratio	P-value
Constant	90.9783	77.69	1.17	0.2663
Calories	2.29959	0.5607	4.10	0.0018

a) State the appropriate hypotheses about the slope.
b) Test your hypotheses and state your conclusion in the proper context.

T 8. Cholesterol 2007. Does a person's cholesterol level tend to change with age? Data collected from 1406 adults aged 45 to 62 produced the regression analysis shown. Assuming that the data satisfy the conditions for inference, examine the association between age and cholesterol level.

Dependent variable is: Chol
s = 46.16

Variable	Coefficient	SE(Coeff)	t-ratio	P-value
Intercept	194.232	13.55	14.3	≤0.0001
Age	0.771639	0.2574	3.00	0.0056

a) State the appropriate hypothesis for the slope.
b) Test your hypothesis and state your conclusion in the proper context.

T 9. Second frank. Look again at Exercise 7's regression output for the calorie and sodium content of hot dogs.
a) The output reports s = 59.66. Explain what that means in this context.
b) What's the value of the standard error of the slope of the regression line?
c) Explain what that means in this context.

T 10. More cholesterol. Look again at Exercise 8's regression output for age and cholesterol level.
a) The output reports s = 46.16. Explain what that means in this context.
b) What's the value of the standard error of the slope of the regression line?
c) Explain what that means in this context.

T 11. Last dog. Based on the regression output seen in Exercise 7, create a 95% confidence interval for the slope of the regression line and interpret your interval in context.

T 12. Cholesterol, finis. Based on the regression output seen in Exercise 8, create a 95% confidence interval for the slope of the regression line and interpret it in context.

T 13. Marriage age 2007. The scatterplot suggests a decrease in the difference in ages at first marriage for men and women since 1975. We want to examine the regression to see if this decrease is significant.

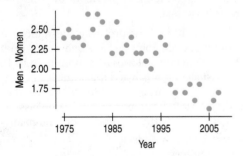

Dependent variable is: Men-Wmn
R squared = 74.1%
s = 0.1824 with 31 − 2 = 29 degrees of freedom

Variable	Coefficient	SE(Coeff)	t-ratio	P-value
Intercept	65.3103	6.931	9.42	≤0.0001
Year	−0.031725	0.0035	−9.11	≤0.0001

a) Write appropriate hypotheses.

b) Here are the residuals plot and a histogram of the residuals. Do you think the conditions for inference are satisfied? Explain.

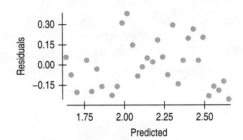

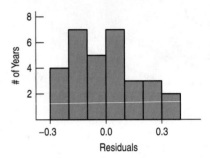

c) Test the hypothesis and state your conclusion about the trend in age at first marriage.

T 14. Used cars 2007. Classified ads in a newspaper offered several used Toyota Corollas for sale. Listed below are the ages of the cars and the advertised prices.

Age (yr)	Advertised Price ($)	Age (yr)	Advertised Price ($)
1	13990	7	6950
1	13495	7	7850
3	12999	8	6999
4	9500	8	5995
4	10495	10	4950
5	8995	10	4495
5	9495	13	2850
6	6999		

a) Make a scatterplot for these data.
b) Do you think a linear model is appropriate? Explain.
c) Find the equation of the regression line.
d) Check the residuals to see if the conditions for inference are met.

T 15. Marriage age 2007, again. Based on the analysis of marriage ages since 1975 given in Exercise 13, give a 95% confidence interval for the rate at which the age gap is closing. Explain what your confidence interval means.

T 16. Used cars 2007, again. Based on the analysis of used car prices you did for Exercise 14, create a 95% confidence interval for the slope of the regression line and explain what your interval means in context.

T 17. Fuel economy. A consumer organization has reported test data for 50 car models. We will examine the association between the weight of the car (in thousands of pounds) and the fuel efficiency (in miles per gallon). Here are the scatterplot, summary statistics, and regression analysis:

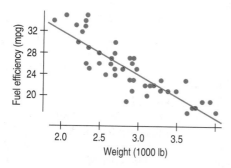

Variable	Count	Mean	StdDev
MPG	50	25.0200	4.83394
wt/1000	50	2.88780	0.511656

Dependent variable is: MPG
R-squared = 75.6%
s = 2.413 with 50 − 2 = 48 df

Variable	Coefficient	SE(Coeff)	t-ratio	P-value
Intercept	48.7393	1.976	24.7	≤0.0001
Weight	−8.21362	0.6738	−12.2	≤0.0001

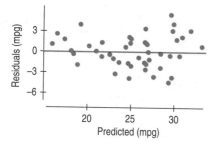

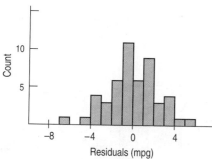

a) Is there strong evidence of an association between the weight of a car and its gas mileage? Write an appropriate hypothesis.
b) Are the assumptions for regression satisfied?
c) Test your hypothesis and state your conclusion.

T 18. SAT scores. How strong was the association between student scores on the Math and Verbal sections of the old SAT? Scores on each ranged from 200 to 800 and were widely used by college admissions offices. Here are

summaries and plots of the scores for a graduating class at Ithaca High School:

Variable	Count	Mean	Median	StdDev	Range	IntQRange
Verbal	162	596.296	610	99.5199	490	140
Math	162	612.099	630	98.1343	440	150

Dependent variable is: Math
R-squared = 46.9%
s = 71.75 with 162 − 2 = 160 df

Variable	Coefficient	SE(Coeff)	t-ratio	P-value
Intercept	209.554	34.35	6.10	≤0.0001
Verbal	0.675075	0.0568	11.9	≤0.0001

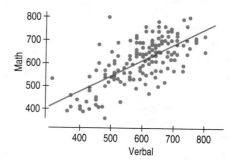

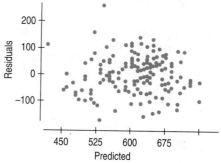

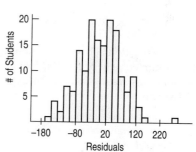

a) Is there evidence of an association between Math and Verbal scores? Write an appropriate hypothesis.
b) Discuss the assumptions for inference.
c) Test your hypothesis and state an appropriate conclusion.

T 19. Fuel economy, part II. Consider again the data in Exercise 17 about the gas mileage and weights of cars.
a) Create a 95% confidence interval for the slope of the regression line.
b) Explain in this context what your confidence interval means.

20. SATs, part II. Consider the high school SAT scores data from Exercise 18.
   a) Find a 90% confidence interval for the slope of the true line describing the association between Math and Verbal scores.
   b) Explain in this context what your confidence interval means.

21. Fuel economy, part III. Consider again the data in Exercise 17 about the gas mileage and weights of cars.
   a) Create a 95% confidence interval for the average fuel efficiency among cars weighing 2500 pounds, and explain what your interval means.
   b) Create a 95% prediction interval for the gas mileage you might get driving your new 3450-pound SUV, and explain what that interval means.

22. SATs, again. Consider the high school SAT scores data from Exercise 18 once more.
   a) Find a 90% confidence interval for the mean SAT-Math score for all students with an SAT-Verbal score of 500.
   b) Find a 90% prediction interval for the Math score of the senior class president if you know she scored 710 on the Verbal section.

23. Cereals. A healthy cereal should be low in both calories and sodium. Data for 77 cereals were examined and judged acceptable for inference. The 77 cereals had between 50 and 160 calories per serving and between 0 and 320 mg of sodium per serving. Here's the regression analysis:

Dependent variable is: Sodium
R-squared = 9.0%
s = 80.49 with 77 − 2 = 75 degrees of freedom

Variable	Coefficient	SE(Coeff)	t-ratio	P-value
Intercept	21.4143	51.47	0.416	0.6786
Calories	1.29357	0.4738	2.73	0.0079

   a) Is there an association between the number of calories and the sodium content of cereals? Explain.
   b) Do you think this association is strong enough to be useful? Explain.

24. Brain size. Does your IQ depend on the size of your brain? A group of female college students took a test that measured their verbal IQs and also underwent an MRI scan to measure the size of their brains (in 1000s of pixels). The scatterplot and regression analysis are shown, and the assumptions for inference were satisfied.

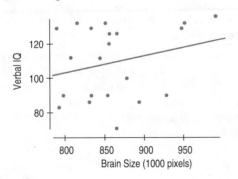

Dependent variable is: IQ_Verbal
R-squared = 6.5%

Variable	Coefficient	SE(Coeff)
Intercept	24.1835	76.38
Size	0.098842	0.0884

   a) Test an appropriate hypothesis about the association between brain size and IQ.
   b) State your conclusion about the strength of this association.

25. Cereals, part 2. Further analysis of the data for the breakfast cereals in Exercise 23 looked for an association between *Fiber* content and *Calories* by attempting to construct a linear model. Here are several graphs. Which of the assumptions for inference are violated? Explain.

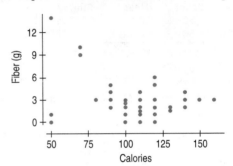

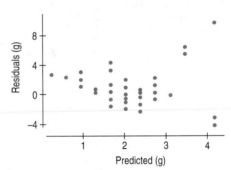

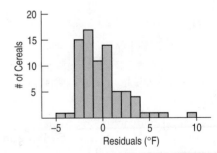

26. Winter. The output shows an attempt to model the association between average *January Temperature* (in degrees Fahrenheit) and *Latitude* (in degrees north of the equator)

for 59 U.S. cities. Which of the assumptions for inference do you think are violated? Explain.

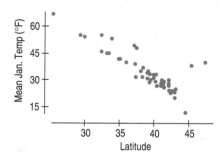

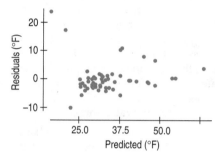

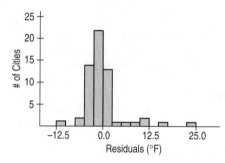

T 27. Streams. Biologists studying the effects of acid rain on wildlife collected data from 163 streams in the Adirondack Mountains. They recorded the *pH* (acidity) of the water and the *BCI*, a measure of biological diversity. Here's a scatterplot of *BCI* against *pH*:

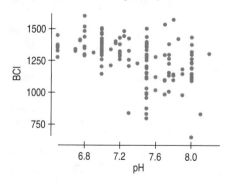

And here is part of the regression analysis:

Dependent variable is: BCI
R-squared = 27.1%
s = 140.4 with 163 − 2 = 161 degrees of freedom

Variable	Coefficient	SE(Coeff)
Intercept	2733.37	187.9
pH	−197.694	25.57

a) State the null and alternative hypotheses under investigation.
b) Assuming that the assumptions for regression inference are reasonable, find the *t*- and P-values.
c) State your conclusion.

T 28. Climate change and CO_2. Concern over the weather associated with El Niño has increased interest in the possibility that the climate on earth is getting warmer. The most common theory relates an increase in atmospheric levels of carbon dioxide (CO_2), a greenhouse gas, to increases in temperature. Here is part of a regression analysis of the mean annual CO_2 concentration in the atmosphere, measured in parts per million (ppm), at the top of Mauna Loa in Hawaii and the mean annual air temperature over both land and sea across the globe, in degrees Celsius. The scatterplots and residuals plots indicated that the data were appropriate for inference.

Dependent variable is: Annual Temp
R-squared = 67.8%
s = 0.0985 with 29 − 2 = 27 degrees of freedom

Variable	Coefficient	SE(Coeff)
Intercept	10.7071	0.4810
CO2	0.010062	0.0013

a) Write the equation of the regression line.
b) Is there evidence of an association between CO_2 level and global temperature?
c) Do you think predictions made by this regression will be very accurate? Explain.

29. Ozone. The Environmental Protection Agency is examining the relationship between the ozone level (in parts per million) and the population (in millions) of U.S. cities. Part of the regression analysis is shown.

Dependent variable is: Ozone
R-squared = 84.4%
s = 5.454 with 16 − 2 = 14 df

Variable	Coefficient	SE(Coeff)
Intercept	18.892	2.395
Pop	6.650	1.910

a) We suspect that the greater the population of a city, the higher its ozone level. Is the relationship significant? Assuming the conditions for inference are satisfied, test an appropriate hypothesis and state your conclusion in context.
b) Do you think that the population of a city is a useful predictor of ozone level? Use the values of both R^2 and *s* in your explanation.

T 30. Sales and profits. A business analyst was interested in the relationship between a company's sales and its profits. She collected data (in millions of dollars) from a random sample of Fortune 500 companies and created the regression analysis and summary statistics shown. The assumptions for regression inference appeared to be satisfied.

	Profits	Sales
Count	79	79
Mean	209.839	4178.29
Variance	635,172	49,163,000
Std Dev	796.977	7011.63

Dependent variable is: Profits
R-squared = 66.2%   s = 466.2

Variable	Coefficient	SE(Coeff)
Intercept	−176.644	61.16
Sales	0.092498	0.0075

a) Is there a significant association between sales and profits? Test an appropriate hypothesis and state your conclusion in context.

b) Do you think that a company's sales serve as a useful predictor of its profits? Use the values of both R^2 and s in your explanation.

31. **Ozone again.** Consider again the relationship between the population and ozone level of U.S. cities that you analyzed in Exercise 29.

a) Give a 90% confidence interval for the approximate increase in ozone level associated with each additional million city inhabitants.

b) For the cities studied, the mean population was 1.7 million people. The population of Boston is approximately 0.6 million people. Predict the mean ozone level for cities of that size with an interval in which you have 90% confidence.

32. **More sales and profits.** Consider again the relationship between the sales and profits of Fortune 500 companies that you analyzed in Exercise 30.

a) Find a 95% confidence interval for the slope of the regression line. Interpret your interval in context.

b) Last year the drug manufacturer Eli Lilly, Inc., reported gross sales of $9 billion (that's $9000 million). Create a 95% prediction interval for the company's profits, and interpret your interval in context.

33. **Start the car!** In October 2002, *Consumer Reports* listed the price (in dollars) and power (in cold cranking amps) of auto batteries. We want to know if more expensive batteries are generally better in terms of starting power. Here are several software displays:

Dependent variable is: Power
R-squared = 25.2%
s = 116.0 with 33 − 2 = 31 degrees of freedom

Variable	Coefficient	SE(Coeff)	t-ratio	P-value
Intercept	384.594	93.55	4.11	0.0003
Cost	4.14649	1.282	3.23	0.0029

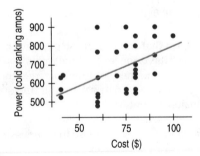

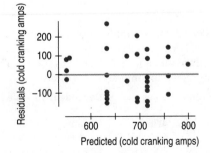

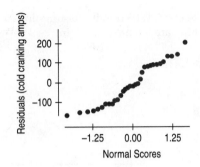

a) How many batteries were tested?

b) Are the conditions for inference satisfied? Explain.

c) Is there evidence of an association between the cost and cranking power of auto batteries? Test an appropriate hypothesis and state your conclusion.

d) Is the association strong? Explain.

e) What is the equation of the regression line?

f) Create a 90% confidence interval for the slope of the true line.

g) Interpret your interval in this context.

34. **Crawling.** Researchers at the University of Denver Infant Study Center wondered whether temperature might influence the age at which babies learn to crawl. Perhaps the extra clothing that babies wear in cold weather would restrict movement and delay the age at which they started crawling. Data were collected on 208 boys and 206 girls. Parents reported the month of the baby's birth and the age (in weeks) at which their child first crawled. The table gives the average *Temperature* (°F) when the babies were 6 months old and average *Crawling Age* (in weeks) for each month of the year. Make the plots and compute the analyses necessary to answer the following questions.

Birth Month	6-Month Temperature	Average Crawling Age
Jan.	66	29.84
Feb.	73	30.52
Mar.	72	29.70
April	63	31.84
May	52	28.58
June	39	31.44
July	33	33.64
Aug.	30	32.82
Sept.	33	33.83
Oct.	37	33.35
Nov.	48	33.38
Dec.	57	32.32

a) Would this association appear to be weaker, stronger, or the same if data had been plotted for individual babies instead of using monthly averages? Explain.

b) Is there evidence of an association between *Temperature* and *Crawling Age*? Test an appropriate hypothesis and state your conclusion. Don't forget to check the assumptions.

c) Create and interpret a 95% confidence interval for the slope of the true relationship.

T 35. Body fat. Do the data shown in the table below indicate an association between *Waist* size and *%Body Fat*?

a) Test an appropriate hypothesis and state your conclusion.

b) Give a 95% confidence interval for the mean *%Body Fat* found in people with 40-inch *Waists*.

Waist (in.)	Weight (lb)	Body Fat (%)	Waist (in.)	Weight (lb)	Body Fat (%)
32	175	6	33	188	10
36	181	21	40	240	20
38	200	15	36	175	22
33	159	6	32	168	9
39	196	22	44	246	38
40	192	31	33	160	10
41	205	32	41	215	27
35	173	21	34	159	12
38	187	25	34	146	10
38	188	30	44	219	28

T 36. Body fat again. Use the data from Exercise 35 to examine the association between *Weight* and *%Body Fat*.

a) Find a 90% confidence interval for the slope of the regression line of *%Body Fat* on *Weight*.

b) Interpret your interval in context.

c) Give a 95% prediction interval for the *%Body Fat* of an individual who weighs 165 pounds.

T 37. Grades. The data set below shows midterm scores from an Introductory Statistics course.

First Name	Midterm 1	Midterm 2	Homework
Timothy	82	30	61
Karen	96	68	72
Verena	57	82	69
Jonathan	89	92	84
Elizabeth	88	86	84
Patrick	93	81	71
Julia	90	83	79
Thomas	83	21	51
Marshall	59	62	58
Justin	89	57	79
Alexandra	83	86	78
Christopher	95	75	77
Justin	81	66	66
Miguel	86	63	74
Brian	81	86	76
Gregory	81	87	75
Kristina	98	96	84
Timothy	50	27	20
Jason	91	83	71
Whitney	87	89	85
Alexis	90	91	68
Nicholas	95	82	68
Amandeep	91	37	54
Irena	93	81	82

First Name	Midterm 1	Midterm 2	Homework
Yvon	88	66	82
Sara	99	90	77
Annie	89	92	68
Benjamin	87	62	72
David	92	66	78
Josef	62	43	56
Rebecca	93	87	80
Joshua	95	93	87
Ian	93	65	66
Katharine	92	98	77
Emily	91	95	83
Brian	92	80	82
Shad	61	58	65
Michael	55	65	51
Israel	76	88	67
Iris	63	62	67
Mark	89	66	72
Peter	91	42	66
Catherine	90	85	78
Christina	75	62	72
Enrique	75	46	72
Sarah	91	65	77
Thomas	84	70	70
Sonya	94	92	81
Michael	93	78	72
Wesley	91	58	66
Mark	91	61	79
Adam	89	86	62
Jared	98	92	83
Michael	96	51	83
Kathryn	95	95	87
Nicole	98	89	77
Wayne	89	79	44
Elizabeth	93	89	73
John	74	64	72
Valentin	97	96	80
David	94	90	88
Marc	81	89	62
Samuel	94	85	76
Brooke	92	90	86

a) Fit a model predicting the second midterm score from the first.

b) Comment on the model you found, including a discussion of the assumptions and conditions for regression. Is the coefficient for the slope statistically significant?

c) A student comments that because the P-value for the slope is very small, Midterm 2 is very well predicted from Midterm 1. So, he reasons, next term the professor can give just one midterm. What do you think?

38. Grades? The professor teaching the Introductory Statistics class discussed in Exercise 37 wonders whether performance on homework can accurately predict midterm scores.

a) To investigate it, she fits a regression of the sum of the two midterms scores on homework scores. Fit the regression model.

b) Comment on the model including a discussion of the assumptions and conditions for regression. Is the coefficient for the slope "statistically significant"?

c) Do you think she can accurately judge a student's performance without giving the midterms? Explain.

39. Strike two. Remember the Little League instructional video discussed in Chapter 25? Ads claimed it would improve the performances of Little League pitchers. To test this claim, 20 Little Leaguers threw 50 pitches each, and we recorded the number of strikes. After the players participated in the training program, we repeated the test. The table shows the number of strikes each player threw before and after the training. A test of paired differences failed to show that this training improves ability to throw strikes. Is there any evidence that the effectiveness of the video (*After – Before*) depends on the player's initial ability to throw strikes (*Before*)? Test an appropriate hypothesis and state your conclusion. Propose an explanation for what you find.

Number of Strikes (out of 50)			
Before	After	Before	After
28	35	33	33
29	36	33	35
30	32	34	32
32	28	34	30
32	30	34	33
32	31	35	34
32	32	36	37
32	34	36	33
32	35	37	35
33	36	37	32

40. All the efficiency money can buy. A sample of 84 model-2004 cars from an online information service was examined to see how fuel efficiency (as highway mpg) relates to the cost (Manufacturer's Suggested Retail Price in dollars) of cars. Here are displays and computer output:

Dependent variable is: Highway MPG
R squared = 30.1%
s = 5.298 with 84 − 2 = 82 degrees of freedom

Variable	Coefficient	SE(Coeff)	t-ratio	P-value
Constant	33.0581	1.299	25.5	≤0.0001
MSRP	−2.16543e−4	0.0000	−5.95	≤0.0001

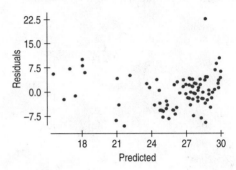

a) State what you want to know, identify the variables, and give the appropriate hypotheses.

b) Check the assumptions and conditions.

c) If the conditions are met, complete the analysis.

41. Education and mortality. The following software output is based on the mortality rate (deaths per 100,000 people) and the education level (average number of years in school) for 58 U.S. cities.

Variable	Count	Mean	StdDev
Mortality	58	942.501	61.8490
Education	58	11.0328	0.793480

Dependent variable is: Mortality
R-squared = 41.0%
s = 47.92 with 58 − 2 = 56 degrees of freedom

Variable	Coefficient	SE(Coeff)
Intercept	1493.26	88.48
Education	−49.9202	8.000

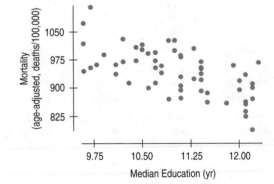

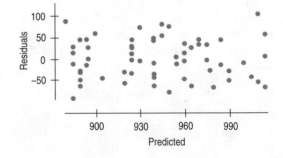

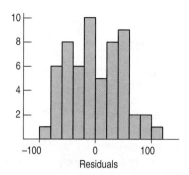

a) Comment on the assumptions for inference.
b) Is there evidence of a strong association between the level of *Education* in a city and the *Mortality* rate? Test an appropriate hypothesis and state your conclusion.
c) Can we conclude that getting more education is likely (on average) to prolong your life? Why or why not?
d) Find a 95% confidence interval for the slope of the true relationship.
e) Explain what your interval means.
f) Find a 95% confidence interval for the average *Mortality* rate in cities where the adult population completed an average of 12 years of school.

T 42. **Property assessments.** The following software outputs provide information about the *Size* (in square feet) of 18 homes in Ithaca, New York, and the city's assessed *Value* of those homes.

Variable	Count	Mean	StdDev	Range
Size	18	2003.39	264.727	890
Value	18	60946.7	5527.62	19710

Dependent variable is: Value
R-squared = 32.5%
s = 4682 with 18 − 2 = 16 degrees of freedom

Variable	Coefficient	SE(Coeff)
Intercept	37108.8	8664
Size	11.8987	4.290

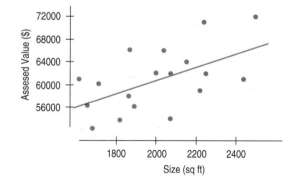

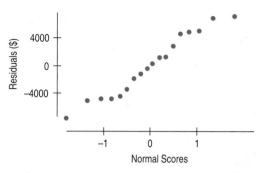

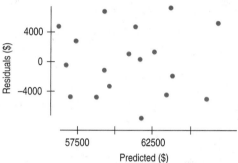

a) Explain why inference for linear regression is appropriate with these data.
b) Is there a significant association between the *Size* of a home and its assessed *Value*? Test an appropriate hypothesis and state your conclusion.
c) What percentage of the variability in assessed *Value* is explained by this regression?
d) Give a 90% confidence interval for the slope of the true regression line, and explain its meaning in the proper context.
e) From this analysis, can we conclude that adding a room to your house will increase its assessed *Value*? Why or why not?
f) The owner of a home measuring 2100 square feet files an appeal, claiming that the $70,200 assessed *Value* is too high. Do you agree? Explain your reasoning.

43. **Right-to-work laws.** Are state right-to-work laws related to the percent of public sector employees in unions and the percent of private sector employees in unions? This data set looks at these percentages for the states in the United States in 1982. The dependent variable is whether the state had a right-to-work law or not. The computer output for the logistic regression is given here. (Source: N. M. Meltz, "Interstate and Interprovincial Differences in Union Density," *Industrial Relations*, 28:2 [Spring 1989], 142–158 by way of DASL.)

Logistic Regression Table

Predictor	Coeff	SE(Coeff)	z	P
Intercept	6.19951	1.78724	3.47	0.001
publ	−0.106155	0.0474897	−2.24	0.025
pvt	−0.222957	0.0811253	−2.75	0.006

a) Write out the estimated regression equation.

b) The following are scatterplots of the response variable against each of the explanatory variables. Examine them for the conditions required by logistic regression. Does logistic regression seem appropriate here? Explain.

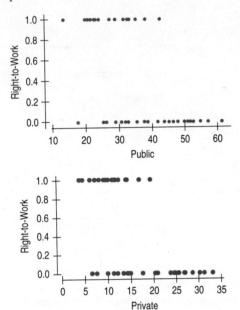

ANSWERS

1. A high t-ratio of 3.27 indicates that the slope is different from zero—that is, that there is a linear relationship between height and mouth size. The small P-value says that a slope this large would be very unlikely to occur by chance if, in fact, there was no linear relationship between the variables.

2. Not really. The R^2 for this regression is only 15.3%, so height doesn't account for very much of the variability in mouth size.

3. The value of s tells the standard deviation of the residuals. Mouth sizes have a mean of 60.3 cubic centimeters. A standard deviation of 15.7 in the residuals indicates that the errors made by this regression model can be quite large relative to what we are estimating. Errors of 15 to 30 cubic centimeters would be common.

44. **Cost of higher education.** Are there fundamental differences between liberal arts colleges and universities? In this case, we have information on the top 25 liberal arts colleges and the top 25 universities in the Unites States. We will consider the type of school as our response variable and will use the percent of students who were in the top 10% of their high school class and the amount of money spent per student by the college or university as our explanatory variables. The output from this logistic regression is given here.

Logistic Regression Table

Predictor	Coeff	SE(Coeff)	z	P
Intercept	−13.1461	3.98629	−3.30	0.001
Top 10%	0.0845469	0.0396345	2.13	0.033
$/Student	0.0002594	0.0000860	3.02	0.003

a) Write out the estimated regression equation.
b) Is percent of students in the top 10% of their high school class statistically significant in predicting whether or not the school is a university? Explain.
c) Is the amount of money spent per student statistically significant in predicting whether or not the school is a university? Explain.

Analysis of Variance

Where are we going?

In Chapter 24 we compared the mean lifetimes of generic and brand-name batteries. But our supermarket carries four different "name" brands of batteries and two cheaper generic brands. Are all these brands equally good? How can we compare them all? We could run a *t*-test for each of the 15 head-to-head comparisons, but we'll learn a better way to compare more than two groups in this chapter.

WHO	Hand washings by four different methods, assigned randomly and replicated 8 times each
WHAT	Number of bacteria colonies
HOW	Sterile media plates incubated at 36°C for 2 days

D id you wash your hands with soap before eating? You've undoubtedly been asked that question a few times in your life. Mom knows that washing with soap eliminates most of the germs you've managed to collect on your hands. Or does it? A student decided to investigate just how effective washing with soap is in eliminating bacteria. To do this she tested four different methods—washing with water only, washing with regular soap, washing with antibacterial soap (ABS), and spraying hands with antibacterial spray (AS) (containing 65% ethanol as an active ingredient). Her experiment consisted of one experimental factor, the washing *Method*, at four levels.

She suspected that the number of bacteria on her hands before washing might vary considerably from day to day. To help even out the effects of those changes, she generated random numbers to determine the order of the four treatments. Each morning she washed her hands according to the treatment randomly chosen. Then she placed her right hand on a sterile media plate designed to encourage bacteria growth. She incubated each plate for 2 days at 36°C, after which she counted the bacteria colonies. She replicated this procedure 8 times for each of the four treatments.

A side-by-side boxplot of the numbers of colonies seems to show some differences among the treatments:

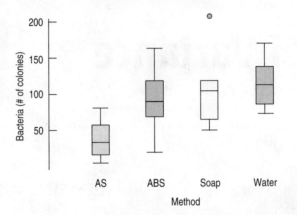

FIGURE 28.1
Boxplots of the bacteria colony counts for the four different washing methods suggest some differences between treatments.

When we first looked at a quantitative variable measured for each of several groups in Chapter 5, we displayed the data this way with side-by-side boxplots. And when we compared the boxes, we asked whether the centers seemed to differ, using the spreads of the boxes to judge the size of the differences. Now we want to make this more formal by testing a hypothesis. We'll make the same kind of comparison, comparing the variability among the means with the spreads of the boxes. It looks like the alcohol spray has lower bacteria counts, but as always, we're skeptical. Could it be that the four methods really have the same mean counts and we just *happened* to get a difference like this because of natural sampling variability?

What is the null hypothesis here? It seems natural to start with the hypothesis that *all the group means are equal.* That would say it doesn't matter what method you use to wash your hands because the mean bacteria count will be the same. We know that even if there were no differences at all in the *means* (for example, if someone replaced all the solutions with water) there would still be sample-to-sample differences. We want to see, statistically, whether differences as large as those observed in the experiment could naturally occur by chance in groups that have equal means. If we find that the differences in washing *Methods* are so large that they would occur only very infrequently in groups that actually have the same mean, then, as we've done with other hypothesis tests, we'll reject the null hypothesis and conclude that the washing *Methods* really have different means.[1]

Contrast baths are a treatment commonly used in hand clinics to reduce swelling and stiffness after surgery. Patients' hands are immersed alternately in warm and cool water. (That's the *contrast* in the name.) Sometimes, the treatment is combined with mild exercise. Although the treatment is widely used, it had never been verified that it would accomplish the stated outcome goal of reducing swelling.

---

[1] The alternative hypothesis is that "the means are not *all* equal." Be careful not to confuse that with "all the means are different." With 11 groups we could have 10 means equal to each other and 1 different. The null hypothesis would still be false.

Researchers[2] randomly assigned 59 patients who had received carpal tunnel release surgery to one of three treatments: contrast bath, contrast bath with exercise, and (as a control) exercise alone. Hand therapists who did not know how the subjects had been treated measured hand volumes before and after treatments in milliliters by measuring how much water the hand displaced when submerged. The change in hand volume (after treatment minus before) was reported as the outcome.

QUESTION: Specify the details of the experiment's design. Identify the subjects, the sample size, the experiment factor, the treatment levels, and the response. What is the null hypothesis? Was randomization employed? Was the experiment blinded? Was it double-blinded?

Subjects were patients who received carpal tunnel release surgery. Sample size is 59 patients. The factor was contrast bath treatment with three levels: contrast baths alone, contrast baths with exercise, and exercise alone. The response variable is the change in hand volume. The null hypothesis is that the mean changes in hand volume will be the same for the three treatment levels. Patients were randomly assigned to treatments. The study was single-blind because the evaluators were blind to the treatments. It was not (and could not be) double-blind because the patients had to be aware of their treatments.

Are the Means of Several Groups Equal?

We saw in Chapter 24 how to use a *t*-test to see whether two groups have equal means. We compared the difference in the means to a standard error estimated from all the data. And when we were willing to assume that the underlying group variances were equal, we pooled the data from the two groups to find the standard error.

Now we have more groups, so we can't just look at differences in the means.[3] But all is not lost. Even if the null hypothesis were true, and the means of the populations underlying the groups were equal, we'd still expect the sample means to vary a bit. We could measure that variation by finding the variance of the means. How much should they vary? Well, if we look at how much the data themselves vary, we can get a good idea of how much the means should vary. And if the underlying means are actually different, we'd expect that variation to be larger.

It turns out that we can build a hypothesis test to check whether the variation in the means is bigger than we'd expect it to be just from random fluctuations. We'll need a new sampling distribution model, called the *F*-model, but that's just a different table to look at (Table F, remarkably enough, found in Appendix D).

---

[2] Janssen, Robert G., Schwartz, Deborah A., and Velleman, Paul F., "A Randomized Controlled Study of Contrast Baths on Patients with Carpal Tunnel Syndrome," *Journal of Hand Therapy,* **22**:3, pp. 200–207. The data reported here differ slightly from those in the original paper because they include some additional subjects and exclude some outliers.

[3] You might think of testing all pairs, but that method generates too many Type I errors. We'll see more about this later in the chapter.

To get an idea of how it works, let's start by looking at the following two sets of boxplots:

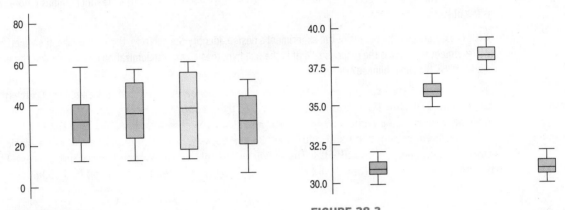

FIGURE 28.2
It's hard to see the difference in the means in these boxplots because the spreads are large relative to the differences in the means.

FIGURE 28.3
In contrast with Figure 28.2, the smaller variation makes it much easier to see the differences among the group means.

We're trying to decide if the means are different enough for us to reject the null hypothesis. If they're close, we'll attribute the differences to natural sampling variability. What do you think? It's easy to see that the means in the second set differ. It's hard to imagine that the means could be that far apart just from natural sampling variability alone. How about the first set? It looks like these observations *could* have occurred from treatments with the same means.[4] This much variation among groups does seem consistent with equal group means.

Believe it or not, the two sets of treatment means in both figures are the same. (They are 31, 36, 38, and 31, respectively.) Then why do the figures look so different? In the second figure, the variation *within* each group is so small that the differences *between* the means stand out. This is what we looked for when we compared boxplots by eye back in Chapter 5. And it's the central idea of the *F*-test. We compare the differences *between* the means of the groups with the variation *within* the groups. When the differences between means are large compared with the variation within the groups, we reject the null hypothesis and conclude that the means are not equal. In the first figure, the differences among the means look as though they could have arisen just from natural sampling variability from groups with equal means, so there's not enough evidence to reject H_0.

How can we make this comparison more precise statistically? All the tests we've seen have compared differences of some kind with a ruler based on an estimate of variation. And we've always done that by looking at the ratio of the statistic to that variation estimate. Here, the differences among the means will show up in the numerator, and the ruler we compare them with will be based on the underlying standard deviation—that is, on the variability *within* the treatment groups.

---

[4]Of course, with a large enough sample, we can detect any differences that we like. For experiments with the same sample size, it's easier to detect the differences when the variation *within* each box is smaller.

// FOR EXAMPLE

RECAP: Fifty-nine postsurgery patients were randomly assigned to one of three treatment levels. Changes in hand volume were measured. Here are the boxplots. The recorded values are volume after treatment–volume before treatment, so positive values indicate swelling. Some swelling is to be expected.

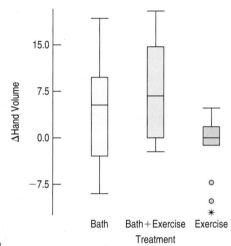

QUESTION: What do the boxplots say about the results?

There doesn't seem to be much difference between the two contrast bath treatments. The exercise only treatment may result in less swelling.

How Different Are They?

> Why variances? We've usually measured variability with standard deviations. Standard deviations have the advantage that they're in the same units as the data. Variances have the advantage that for independent variables, the variances add. Because we're talking about sums of variables, we'll stay with variances before we get back to standard deviations.

The challenge here is that we can't take a simple difference as we did when comparing two groups. In the hand-washing experiment, we have differences in mean bacteria counts across *four* treatments. How should we measure how different the four group means are? With only two groups, we naturally took the difference between their means as the numerator for the *t*-test. It's hard to imagine what else we could have done. How can we generalize that to more than two groups? When we've wanted to know how different many observations were, we measured how much they vary, and that's what we do here.

How much natural variation should we expect among the means if the null hypothesis were true? If the null hypothesis were *true*, then each of the treatment means would estimate the *same* underlying mean. If the washing methods are all the same, it's as if we're just estimating the mean bacteria count on hands that have been washed with plain water. And we have several (in our experiment, four) different, independent estimates of this mean. Here comes the clever part. We can treat these estimated means as if they were observations and simply calculate their (sample) variance. This variance is the measure we'll use to assess how different the group means are from each other. It's the generalization of the difference between means for only two groups.

The more the group means resemble each other, the smaller this variance will be. The more they differ (perhaps because the treatments actually have an effect), the larger this variance will be.

For the bacteria counts, the four means are listed in the table to the left. If you took those four values, treated them as observations, and found their sample variance, you'd get 1245.08. That's fine, but how can we tell whether it is a big value? Now we need a model, and the model is based on our null hypothesis that all the group means are equal. Here, the null hypothesis is that it doesn't matter what washing method you use; the mean bacteria count will be about the same:

$$H_0: \mu_1 = \mu_2 = \mu_3 = \mu_4 = \mu.$$

As always when testing a null hypothesis, we'll start by assuming that it is true. And if the group means are equal, then there's an overall mean, μ—the

Level	n	Mean
Alcohol spray	8	37.5
Antibacterial soap	8	92.5
Soap	8	106.0
Water	8	117.0

bacteria count you'd expect all the time after washing your hands in the morning. And each of the observed group means is just a sample-based estimate of that underlying mean.

We know how sample means vary. The variance of a sample mean is σ^2/n. With eight observations in a group, that would be $\sigma^2/8$. The estimate that we've just calculated, 1245.08, should estimate this quantity. If we want to get back to the variance of the *observations*, σ^2, we need to multiply it by 8. So $8 \times 1245.08 = 9960.64$ should estimate σ^2.

Is 9960.64 large for this variance? How can we tell? We'll need a hypothesis test. You won't be surprised to learn that there is just such a test. The details of the test, due to Sir Ronald Fisher in the early 20th century, are truly ingenious, and may be the most amazing statistical result of that century.

The Ruler Within

We need a suitable ruler for comparison—one based on the underlying variability in our measurements. That variability is due to the day-to-day differences in the bacteria count even when the same soap is used. Why would those counts be different? Maybe the experimenter's hands were not equally dirty, or she washed less well some days, or the plate incubation conditions varied. We randomized just so we could see past such things.

We need an independent estimate of σ^2, one that doesn't depend on the null hypothesis being true, one that won't change if the groups have different means. As in many quests, the secret is to look "within." We could look in *any* of the treatment groups and find its variance. But which one should we use? The answer is, *all* of them!

At the start of the experiment (when we randomly assigned experimental units to treatment groups), the units were drawn randomly from the same pool, so each treatment group had a sample variance that estimated the same σ^2. If the null hypothesis is true, then not much has happened to the experimental units—or at least, their means have not moved apart. It's not much of a stretch to believe that their variances haven't moved apart much either. (If the washing methods are equivalent, then the choice of method would not affect the mean *or* the variability.) So each group variance still estimates a common σ^2.

We're assuming that the null hypothesis is true. If the group variances are equal, then the common variance they all estimate is just what we've been looking for. Since all the group variances estimate the same σ^2, we can pool them to get an overall estimate of σ^2. Recall that we pooled to estimate variances when we tested the null hypothesis that two proportions were equal—and for the same reason. It's also exactly what we did in a pooled t-test. The variance estimate we get by pooling we'll denote, as before, by s_p^2.

Level	n	Mean	Std Dev	Variance
Alcohol spray	8	37.5	26.56	705.43
Antibacterial soap	8	92.5	41.96	1760.64
Soap	8	106.0	46.96	2205.24
Water	8	117.0	31.13	969.08

For the bacteria counts, the standard deviations and variances are listed to the left. If we pool the four variances (here we can just average them because all the sample sizes are equal), we'd get $s_p^2 = 1410.10$. In the pooled variance, each variance is taken around its *own* treatment mean, so the pooled estimate doesn't depend on the treatment means being equal. But the estimate in which we took the four means as observations and took their variance does. That estimate gave 9960.64. That seems a lot bigger than 1410.10. Might this be evidence that the four means are not equal?

Let's see what we have. We have an estimate of σ^2 from the variation *within* groups of 1410.10. That's just the variance of the *residuals* pooled across all groups. Because it's a pooled variance, we could write it as s_p^2. Traditionally this quantity is also called the **Error Mean Square,** or sometimes the *Within Mean Square* and denoted by MS_E. These names date back to the early 20th century when the methods were developed. If you think about it, the names do make sense—variances are means of squared differences.[5]

But we also have a *separate* estimate of σ^2 from the variation *between* the groups because we know how much means ought to vary. For the hand-washing data, when we took the variance of the four means and multiplied it by *n* we got 9960.64. We expect this to estimate σ^2 too, *as long as we assume the null hypothesis is true.* We call this quantity the **Treatment Mean Square** (or sometimes the *Between Mean Square*[6]) and denote by MS_T.

The *F*-statistic

Now we have two different estimates of the underlying variance. The first one, the MS_T, is based on the differences *between* the group means. If the group means are equal, as the null hypothesis asserts, it will estimate σ^2. But, if they are not, it will give some bigger value. The other estimate, the MS_E, is based only on the variation *within* the groups around each of their own means, and doesn't depend at all on the null hypothesis being true.

So, how do we test the null hypothesis? When the null hypothesis is true, the treatment means are equal, and both MS_E and MS_T estimate σ^2. Their ratio, then, should be close to 1.0. But, when the null hypothesis is false, the MS_T will be *larger* because the treatment means are not equal. The MS_E is a pooled estimate in which the variation within each group is found around its own group mean, so differing means won't inflate it. That makes the ratio MS_T/MS_E perfect for testing the null hypothesis. When the null hypothesis is true, the ratio should be near 1. If the treatment means really are different, the numerator will tend to be larger than the denominator, and the ratio will tend to be bigger than 1.

Of course, even when the null hypothesis *is* true, the ratio will vary around 1 just due to natural sampling variability. How can we tell when it's big enough to reject the null hypothesis? To be able to tell, we need a sampling distribution model for the ratio. Sir Ronald Fisher found the sampling distribution model of the ratio in the early 20th century. In his honor we call the distribution of MS_T/MS_E the **F-distribution.** And we call the ratio MS_T/MS_E the **F-statistic.** By comparing this statistic with the appropriate *F*-distribution we (or the computer) can get a P-value.

The **F-test** is simple. It is one-tailed because any differences in the means make the *F*-statistic larger. Larger differences in the treatments' effects lead to the means being more variable, making the MS_T bigger. That makes the *F*-ratio grow. So the test is significant if the *F*-ratio is big enough. In practice, we find a P-value, and big *F*-statistic values go with small P-values.

The entire analysis is called the Analysis of Variance, commonly abbreviated **ANOVA** (and pronounced uh-NŌ-va). You might think that it should be called the analysis of means, since it's the equality of the means we're testing. But we use the *variances* within and between the groups for the test.

Like Student's *t*-models, the *F*-models are a family. *F*-models depend on not one, but two, degrees of freedom parameters. The degrees of freedom come from the two variance estimates and are sometimes called the *numerator df* and

[5] Well, actually, they're sums of squared differences divided by their degrees of freedom—$n-1$ for the first variance we saw back in Chapter 4, and other degrees of freedom for each of the others we've seen. But even back in Chapter 4 we said this was a "kind of" mean, and indeed, it still is.
[6] Grammarians would probably insist on calling it the *Among Mean Square*, since the variation is among *all* the group means. Traditionally, though, it's called the *Between Mean Square* and we have to talk about the variation between all the groups (as bad as that sounds).

the *denominator df.* The *Treatment Mean Square,* MS_T, is the sample variance of the observed treatment means. If we think of them as observations, then since there are k groups, this variance has $k - 1$ degrees of freedom. The *Error Mean Square,* MS_E, is the pooled estimate of the variance within the groups. If there are n observations in each group, then we get $n - 1$ degrees of freedom from each, for a total of $k(n - 1)$ degrees of freedom.

A simpler way of tracking the degrees of freedom is to start with all the cases. We'll call that N. Each group has its own mean, costing us a degree of freedom—k in all. So we have $N - k$ degrees of freedom for the error. When the groups all have equal sample size, that's the same as $k(n - 1)$, but this way works even if the group sizes differ.

We say that the F-statistic, MS_T/MS_E, has $k - 1$ and $N - k$ degrees of freedom.

Back to Bacteria

For the hand-washing experiment, the $MS_T = 9960.64$. The $MS_E = 1410.14$. If the treatment means were equal, the *Treatment Mean Square* should be about the same size as the *Error Mean Square,* about 1410. But it's 9960.64, which is 7.06 times bigger. In other words, $F = 7.06$. This F-statistic has $4 - 1 = 3$ and $32 - 4 = 28$ degrees of freedom.

An F-value of 7.06 is bigger than 1, but we can't tell for sure whether it's big enough to reject the null hypothesis until we check the $F_{3,28}$ model to find its P-value. (Usually, that's most easily done with technology, but we can use printed tables.) It turns out the P-value is 0.0011. In other words, if the treatment means were actually equal, we would expect the ratio MS_T/MS_E to be 7.06 or larger about 11 times out of 10,000, just from natural sampling variability. That's not very likely, so we reject the null hypothesis and conclude that the means are different. We have strong evidence that the four different methods of hand washing are not equally effective at eliminating germs.

The ANOVA Table

You'll often see the mean squares and other information put into a table called the **ANOVA table**. Here's the table for the washing experiment:

Analysis of Variance Table

Source	Sum of Squares	DF	Mean Square	F-ratio	P-value
Method	29882	3	9960.64	7.0636	0.0011
Error	39484	28	1410.14		
Total	69366	31			

The ANOVA table was originally designed to organize the calculations. With technology, we have much less use for that. We'll show how to calculate the sums of squares later in the chapter, but the most important quantities in the table are the F-statistic and its associated P-value. When the F-statistic is large, the Treatment (here *Method*) Mean Square is large compared to the Error Mean Square (MS_E), and provides evidence that in fact the means of the groups are not all equal.

You'll almost always see ANOVA results presented in a table like this. After nearly a century of writing the table this way, statisticians (and their technology) aren't going to change. Even though the table was designed to facilitate hand calculation, computer programs that compute ANOVAs still present the results in this form. Usually the P-value is found next to the F-ratio. The P-value column may be labeled with a title such as "Prob > F," "sig," or "Prob." Don't let that confuse you; it's just the P-value.

You'll sometimes see the two mean squares referred to as the *Mean Square Between* and the *Mean Square Within*—especially when we test data from observational studies rather than experiments. ANOVA is often used for such observational data, and as long as certain conditions are satisfied, there's no problem with using it in that context.

FOR EXAMPLE

RECAP: An experiment to determine the effect of contrast bath treatments on swelling in postsurgical patients recorded hand volume changes for patients who had been randomly assigned to one of three treatments.

Here is the Analysis of Variance for these data:

Analysis of Variance for Hand Volume Change

Source	df	Sum of Squares	Mean Square	F-ratio	P-value
Treatment	2	716.159	358.080	7.4148	0.0014
Error	56	2704.38	48.2926		
Total	58	3420.54			

QUESTION: What does the ANOVA say about the results of the experiment? Specifically, what does it say about the null hypothesis?

The F-ratio of 7.4148 has a P-value that is quite small. We can reject the null hypothesis that the mean change in hand volume is the same for all three treatments.

The *F*-table

Usually, you'll get the P-value for the *F*-statistic from technology. Any software program performing an ANOVA will automatically "look up" the appropriate one-sided P-value for the *F*-statistic. If you want to do it yourself, you'll need an *F*-table. *F*-tables are usually printed only for a few values of α, often 0.05, 0.01, and 0.001. They give the critical value of the *F*-statistic with the appropriate number of degrees of freedom determined by your data, for the α level that you select. If your *F*-statistic is greater than that value, you know that its P-value is less than that α level. So, you'll be able to tell whether the P-value is greater or less than 0.05, 0.01, or 0.001, but to be more precise, you'll need technology (or an interactive table like the one in the *ActivStats* program on the DVD).

Here's an excerpt from an *F*-table for $\alpha = 0.05$:

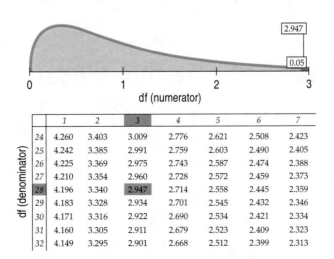

	1	2	3	4	5	6	7
24	4.260	3.403	3.009	2.776	2.621	2.508	2.423
25	4.242	3.385	2.991	2.759	2.603	2.490	2.405
26	4.225	3.369	2.975	2.743	2.587	2.474	2.388
27	4.210	3.354	2.960	2.728	2.572	2.459	2.373
28	4.196	3.340	2.947	2.714	2.558	2.445	2.359
29	4.183	3.328	2.934	2.701	2.545	2.432	2.346
30	4.171	3.316	2.922	2.690	2.534	2.421	2.334
31	4.160	3.305	2.911	2.679	2.523	2.409	2.323
32	4.149	3.295	2.901	2.668	2.512	2.399	2.313

FIGURE 28.4
Part of an *F*-table showing critical values for $\alpha = 0.05$ and highlighting the critical value, 2.947, for 3 and 28 degrees of freedom. We can see that only 5% of the values will be greater than 2.947 with this combination of degrees of freedom.

Notice that the critical value for 3 and 28 degrees of freedom at $\alpha = 0.05$ is 2.947. Since our F-statistic of 7.06 is larger than this critical value, we know that the P-value is less than 0.05. We could also look up the critical value for $\alpha = 0.01$ and find that it's 4.568 and the critical value for $\alpha = 0.001$ is 7.193. So our F-statistic sits between the two critical values 0.01 and 0.001, and our P-value is slightly *greater* than 0.0001. Technology can find the value precisely. It turns out to be 0.011.

JUST CHECKING

A student conducted an experiment to see which, if any, of four different paper airplane designs results in the longest flights (measured in inches). The boxplots look like this:

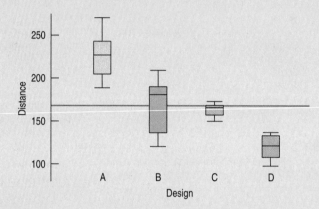

The ANOVA table shows:

Analysis of Variance

Source	DF	Sum of Squares	Mean Square	F Ratio	Prob > F
Design	3	51991.778	17330.6	37.4255	<.0001
Error	32	14818.222	463.1		
C. Total	35	66810.000			

1. What is the null hypothesis?

2. From the boxplots, do you think that there is evidence that the mean flight distances of the four designs differ?

3. Does the F-test in the ANOVA table support your preliminary conclusion in (2)?

4. The researcher concluded that "there is substantial evidence that all four of the designs result in different mean flight distances." Do you agree?

The ANOVA Model

To understand the ANOVA table, let's start by writing a model for what we observe. We start with the simplest interesting model: one that says that the only differences of interest among the groups are the differences in their means. So we'll characterize each group in terms of its mean and assume that any variation around that mean is just random error:

$$y_{ij} = \mu_j + \varepsilon_{ij}.$$

That is, each observation is the sum of the mean for the treatment it received plus a random error. Our null hypothesis is that the treatments made no difference—that is, that the means are all equal:

$$H_0: \mu_1 = \mu_2 = \cdots = \mu_k.$$

It will help our discussion if we think of the overall mean of the experiment and consider the treatments as adding or subtracting an effect to this overall mean. Thinking in this way, we could write μ for the overall mean and τ_j for the deviation from this mean to get to the jth treatment mean—the *effect* of the treatment (if any) in moving that group away from the overall mean:

$$y_{ij} = \mu + \tau_j + \varepsilon_{ij}.$$

Thinking in terms of the effects, we could also write the null hypothesis in terms of these treatment *effects* instead of the means:

$$H_0: \tau_1 = \tau_2 = \cdots = \tau_k = 0.$$

We now have three different kinds of parameters: the overall mean, the treatment effects, and the errors. We'll want to estimate them from the data. Fortunately, we can do that in a straightforward way.

To estimate the overall mean, μ, we use the mean of all the observations: $\bar{\bar{y}}$, (called the "grand mean."[7]) To estimate each treatment effect, we find the difference between the mean of that particular treatment and the grand mean:

$$\hat{\tau}_j = \bar{y}_j - \bar{\bar{y}}.$$

There's an error, ε_{ij}, for each observation. We estimate those with the residuals from the treatment means: $e_{ij} = y_{ij} - \bar{y}_j$.

Now we can write each observation as the sum of three quantities that correspond to our model:

$$y_{ij} = \bar{\bar{y}} + (\bar{y}_j - \bar{\bar{y}}) + (y_{ij} - \bar{y}_j).$$

What this says is simply that we can write each observation as the sum of

- the grand mean,
- the effect of the treatment it received, and
- the residual

Or:

Observations = Grand mean + Treatment effect + Residual.

If we look at the equivalent equation

$$y_{ij} = \bar{\bar{y}} + (\bar{y}_j - \bar{\bar{y}}) + (y_{ij} - \bar{y}_j)$$

closely, it doesn't really seem like we've done anything. In fact, collecting terms on the right-hand side will give back just the observation, y_{ij} again. But this decomposition is actually the secret of the Analysis of Variance. We've split each observation into "sources"—the grand mean, the treatment effect, and error.

Where does the residual term come from? Think of the annual report from any Fortune 500 company. The company spends billions of dollars each year and at the end of the year, the accountants show where each penny goes. How do they do it? After accounting for salaries, bonuses, supplies, taxes, etc., etc., etc., what's the last line? It's always labeled "other" or miscellaneous. Using "other" as the difference between all the sources they know and the total they start with, they can always make it add up perfectly. The residual is just the statisticians' "other." It takes care of all the other sources we didn't think of or don't want to consider, and makes the decomposition work by adding (or subtracting) back in just what we need.

---

[7] The father of your father is your grandfather. The mean of the group means should probably be the grandmean, but we usually spell it as two words.

Let's see what this looks like for our hand-washing data. Here are the data again, displayed a little differently:

	Alcohol	AB Soap	Soap	Water
	51	70	84	74
	5	164	51	135
	19	88	110	102
	18	111	67	124
	58	73	119	105
	50	119	108	139
	82	20	207	170
	17	95	102	87
Treatment Means	37.5	92.5	106	117

The grand mean of all observations is 88.25. Let's put that into a similar table:

Alcohol	AB Soap	Soap	Water
88.25	88.25	88.25	88.25
88.25	88.25	88.25	88.25
88.25	88.25	88.25	88.25
88.25	88.25	88.25	88.25
88.25	88.25	88.25	88.25
88.25	88.25	88.25	88.25
88.25	88.25	88.25	88.25
88.25	88.25	88.25	88.25

The treatment *means* are 37.5, 92.5, 106, and 117, respectively, so the treatment *effects* are those minus the grand mean (88.25). Let's put the treatment effects into their table:

Alcohol	AB Soap	Soap	Water
−50.75	4.25	17.75	28.75
−50.75	4.25	17.75	28.75
−50.75	4.25	17.75	28.75
−50.75	4.25	17.75	28.75
−50.75	4.25	17.75	28.75
−50.75	4.25	17.75	28.75
−50.75	4.25	17.75	28.75
−50.75	4.25	17.75	28.75

Finally, we compute the residuals as the differences between each observation and its treatment mean:

Alcohol	AB Soap	Soap	Water
13.5	−22.5	−22	−43
−32.5	71.5	−55	18
−18.5	−4.5	4	−15
−19.5	18.5	−39	7
20.5	−19.5	13	−12
12.5	26.5	2	22
44.5	−72.5	101	53
−20.5	2.5	−4	−30

Now we have four tables for which

$$Observations = Grand\ Mean\ +\ Treatment\ Effect\ +\ Residual.$$

(You can verify, for example, that the first observation, $51 = 88.25 + (-50.75) + 13.5$).

Why do we want to think in this way? Think back to the boxplots in Figures 28.2 and 28.3. To *test* the hypothesis that the treatment effects are zero, we want to see whether the treatment effects are large *compared* to the errors. Our eye looks at the variation *between* the treatment means and compares it to the variation *within* each group.

The ANOVA separates those two quantities into the Treatment Effects and the Residuals. Sir Ronald Fisher's insight was how to turn those quantities into a statistical test. We want to see if the Treatment Effects are large compared with the Residuals. To do that, we first compute the Sums of Squares of each table. Fisher's insight was that dividing these sums of squares by their respective degrees of freedom lets us test their ratio by a distribution that he found (which was later named the F in his honor). When we divide a sum of squares by its degrees of freedom we get the associated *mean square*.

When the Treatment Mean Square is *large* compared to the Error Mean Square, this provides evidence that the treatment means are different. And we can use the F-distribution to see how large "large" is.

The sums of squares for each table are easy to calculate. Just take every value in the table, square it, and add them all up. For the *Methods*, the Treatment Sum of Squares, $SS_T = (-50.75)^2 + (-50.75)^2 + \cdots + (28.75)^2 = 29882$. There are four treatments, and so there are 3 degrees of freedom. So,

$$MS_T = SS_T/3 = 29882/3 = 9960.64$$

In general, we could write the Treatment Sum of Squares as

$$SS_T = \sum \sum (\bar{y}_j - \bar{\bar{y}})^2.$$

Be careful to note that the summation is over the *whole* table, rows and columns. That's why there are two summation signs.

And,

$$MS_T = SS_T/(k - 1).$$

The table of residuals shows the variation that remains after we remove the overall mean and the treatment effects. These are what's left over after we account for what we're interested in—in this case the treatments. Their variance is the variance *within* each group that we see in the boxplots of the four groups. To find its value, we first compute the Error Sum of Squares, SS_E, by summing up the squares of every element in the residuals table. To get the Mean Square (the variance) we have to divide it by $N - k$ rather than by $N - 1$ because we found them by subtracting each of the k treatment means. So,

$$SS_E = (13.5)^2 + (-32.5)^2 + \cdots + (-30)^2 = 39484$$

and

$$MS_E = SS_E/(32 - 4) = 1410.14$$

As equations:

$$SS_E = \sum \sum (y_{ij} - \bar{y}_j)^2,$$

and

$$MS_E = SS_E/(N - k).$$

Now where are we? To test the null hypothesis that the treatment means are all equal we find the *F*-statistic:

$$F_{k-1, N-k} = MS_T/MS_E$$

and compare that value to the *F*-distribution with $k - 1$ and $N - k$ degrees of freedom. When the *F*-statistic is large enough (and its associated P-value small) we reject the null hypothesis and conclude that at least one mean is different.

There's another amazing result hiding in these tables. If we take each of these tables, square every observation, and add them up, the sums add as well!

$$SS_{Observations} = SS_{Grand\,Mean} + SS_T + SS_E$$

The $SS_{Observations}$ is usually very large compared to SS_T and SS_E, so when ANOVA was originally done by hand, or even by calculator, it was hard to check the calculations using this fact. The first sum of squares was just too big. So, usually the ANOVA table uses the "Corrected Total" sum of squares. If we write

Observations = Grand Mean + Treatment Effect + Residual,

we can naturally write

Observations − Grand Mean = Treatment Effect + Residual.

Mathematically, this is the same statement, but numerically this is more stable. What's amazing is that the sums of the squares still add up. That is, if you make the first table of observations with the grand mean subtracted from each, square those, and add them up, you'll have the SS_{Total} and

$$SS_{Total} = SS_T + SS_E.$$

That's what the ANOVA table shows. If you find this surprising, you must be following along. The tables add up, so sums of their elements must add up. But it is not at all obvious that the sums of the *squares* of their elements should add up, and this is another great insight of the Analysis of Variance.

Back to Standard Deviations

We've been using the variances because they're easier to work with. But when it's time to think about the data, we'd really rather have a standard deviation because it's in the units of the response variable. The natural standard deviation to think about is the standard deviation of the residuals.

The variance of the residuals is staring us in the face. It's the MS_E. All we have to do to get the **residual standard deviation** is take the square root of MS_E:

$$s_p = \sqrt{MS_E} = \sqrt{\frac{\Sigma e^2}{(N-k)}}.$$

The *p* subscript is to remind us that this is a *pooled* standard deviation, combining residuals across all *k* groups. The denominator in the fraction shows that finding a mean for each of the *k* groups cost us one degree of freedom for each.

This standard deviation should "feel" right. That is, it should reflect the kind of variation you expect to find in any of the experimental groups. For the hand-washing data, $s_p = \sqrt{1410.14} = 37.6$ bacteria colonies. Looking back at the boxplots of the groups, we see that 37.6 seems to be a reasonable compromise standard deviation for all four groups.

Plot the Data . . .

Just as you would never find a linear regression without looking at the scatterplot of y vs. x, you should never embark on an ANOVA without first examining side-by-side boxplots of the data comparing the responses for all of the groups. You already know what to look for—we talked about that back in Chapter 5. Check for outliers within any of the groups and correct them if there are errors in the data. Get an idea of whether the groups have similar spreads (as we'll need) and whether the centers seem to be alike (as the null hypothesis claims) or different. If the spreads of the groups are very different—and especially if they seem to grow consistently as the means grow—consider re-expressing the response variable to make the spreads more nearly equal. Doing so is likely to make the analysis more powerful and more correct. Likewise, if the boxplots are skewed in the same direction, you may be able to make the distributions more symmetric with a re-expression.

Don't ever carry out an Analysis of Variance without looking at the side-by-side boxplots first. The chance of missing an important pattern or violation is just too great.

Assumptions and Conditions

When we checked assumptions and conditions for regression we had to take care to perform our checks in order. Here we have a similar concern. For regression we found that displays of the residuals were often a good way to check the corresponding conditions. That's true for ANOVA as well.

Independence Assumptions

The groups must be independent of each other. No test can verify this assumption. You have to think about how the data were collected. The assumption would be violated, for example, if we measured subjects' performance before some treatment, again in the middle of the treatment period, and then again at the end.[8]

The data *within* each treatment group must be independent as well. The data must be drawn independently and at random from a homogeneous population, or generated by a randomized comparative experiment.

We check the **Randomization Condition:** Were the data collected with suitable randomization? For surveys, are the data drawn from each group a representative random sample of that group? For experiments, were the treatments assigned to the experimental units at random?

We were told that the hand-washing experiment was randomized.

Equal Variance Assumption

The ANOVA requires that the variances of the treatment groups be equal. After all, we need to find a pooled variance for the MS_E. To check this assumption, we can check that the groups have similar variances:

Similar Spread Condition: There are some ways to see whether the variation in the treatment groups seems roughly equal:

- Look at side-by-side boxplots of the groups to see whether they have roughly the same spread. It can be easier to compare spreads across groups when they have the same center, so consider making side-by-side boxplots

---

[8] There is a modification of ANOVA, called *repeated measures* ANOVA, that deals with such data. (If the design reminds you of a paired-t situation, you're on the right track, and the lack of independence is the same kind of issue we discussed in Chapter 25.)

of the residuals. If the groups have differing spreads, it can make the pooled variance—the MS_E—larger, reducing the F-statistic value and making it less likely that we can reject the null hypothesis. So the ANOVA will usually fail on the "safe side," rejecting H_0 less often than it should. Because of this, we usually require the spreads to be quite different from each other before we become concerned about the condition failing. If you've rejected the null hypothesis, this is especially true.

- Look at the original boxplots of the response values again. In general, do the spreads seem to change *systematically* with the centers? One common pattern is for the boxes with bigger centers to have bigger spreads. This kind of systematic trend in the variances is more of a problem than random differences in spread among the groups and should not be ignored. Fortunately, such systematic violations are often helped by re-expressing the data. (If, in addition to spreads that grow with the centers, the boxplots are skewed with the longer tail stretching off to the high end, then the data are pleading for a re-expression. Try taking logs of the dependent variable for a start. You'll likely end up with a much cleaner analysis.)

- Look at the residuals plotted against the predicted values. Often, larger predicted values lead to larger magnitude residuals. This is another sign that the condition is violated. (This may remind you of the **Does the Plot Thicken? Condition** of regression. And it should.) When the plot thickens (to one side or the other), it's usually a good idea to consider re-expressing the response variable. Such a systematic change in the spread is a more serious violation of the equal variance assumption than slight variations of the spreads across groups.

Let's check the conditions for the hand-washing data. Here's a boxplot of residuals by group and residuals by predicted value:

FIGURE 28.5
Boxplots of residuals for the four washing methods and a plot of residuals vs. predicted values. There's no evidence of a systematic change in variance from one group to the other or by predicted value.

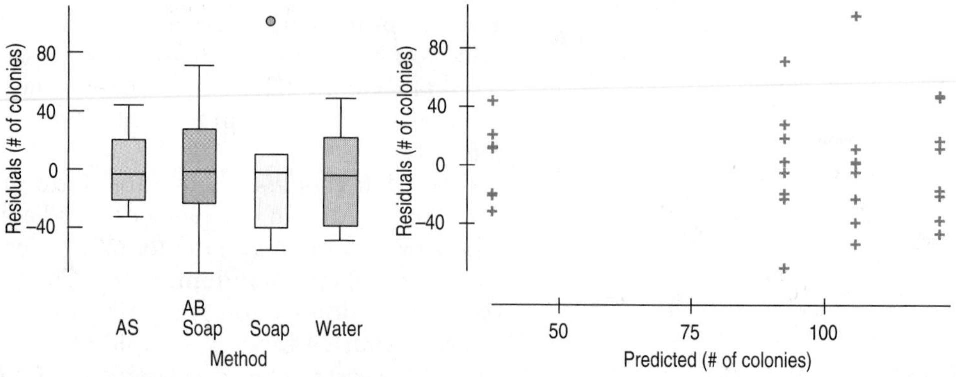

Neither plot shows a violation of the condition. The IQRs (the box heights) are quite similar and the plot of residuals vs. predicted values does not show a pronounced widening to one end. The pooled estimate of 37.6 colonies for the error standard deviation seems reasonable for all four groups.

Normal Population Assumption

Like Student's t-tests, the F-test requires the underlying errors to follow a Normal model. As before when we've faced this assumption, we'll check a corresponding **Nearly Normal Condition.**

Technically, we need to assume that the Normal model is reasonable for the populations underlying *each* treatment group. We can (and should) look at the side-by-side boxplots for indications of skewness. Certainly, if they are all (or mostly) skewed in the same direction, the Nearly Normal Condition fails (and re-expression is likely to help).

In experiments, we often work with fairly small groups for each treatment, and it's nearly impossible to assess whether the distribution of only six or eight numbers is Normal (though sometimes it's so skewed or has such an extreme outlier that we can see that it's not). Here we are saved by the Equal Variance Assumption (which we've already checked). The residuals have their group means subtracted, so the mean residual for each group is 0. If their variances are equal, we can group all the residuals together for the purpose of checking the Nearly Normal Condition.

Check Normality with a histogram or a Normal probability plot of all the residuals together. The hand-washing residuals look nearly Normal in the Normal probability plot, although, as the boxplots showed, there's a possible outlier in the Soap group.

Because we really care about the Normal model within each group, the Normal Population Assumption is violated if there are outliers in any of the groups. Check for outliers in the boxplots of the values for each treatment group. The Soap group of the hand-washing data shows an outlier, so we might want to compute the analysis again without that observation. (For these data, it turns out to make little difference.)

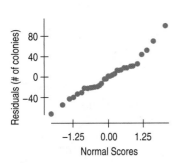

FIGURE 28.6
The hand-washing residuals look nearly Normal in this Normal probability plot.

ONE-WAY ANOVA *F*-TEST

We test the null hypothesis $H_0: \mu_1 = \mu_2 = \cdots = \mu_k$ against the alternative that the group means are not all equal. We test the hypothesis with the *F*-statistic, $F = \dfrac{MS_T}{MS_E}$, where MS_T is the Treatment Mean Square, found from the variance of the means of the treatment groups, and MS_E is the Error Mean Square, found by pooling the variances within each of the treatment groups. If the *F*-statistic is large enough, we reject the null hypothesis.

STEP-BY-STEP EXAMPLE   Analysis of Variance

In Chapter 5 we looked at side-by-side boxplots of four different containers for holding hot beverages. The experimenter wanted to know which type of container would keep his hot beverages hot longest. To test it, he heated water to a temperature of 180°F, placed it in the container, and then measured the temperature of the water again 30 minutes later. He randomized the order of the trials and tested each container 8 times. His response variable was the difference in temperature (in °F) between the initial water temperature and the temperature after 30 minutes.

Question: Do the four containers maintain temperature equally well?

THINK   **Plot** Plot the side-by-side boxplots of the data.

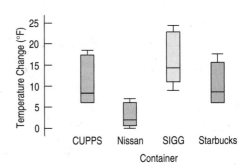

(continued)

Plan State what you want to know and the null hypothesis you wish to test. For ANOVA, the null hypothesis is that all the treatment groups have the same mean. The alternative is that at least one mean is different.

I want to know whether there is any difference among the four containers in their ability to maintain the temperature of a hot liquid for 30 minutes. I'll write μ_k for the mean temperature difference for container k, so the null hypothesis is that these means are all the same:

$$H_0: \mu_1 = \mu_2 = \mu_3 = \mu_4.$$

The alternative is that the group means are not all equal.

Think about the assumptions and check the conditions.

✔ **Independence Assumption:** The "experimental units" in this experiment are cups of heated water. It's easy to believe that one cup of water is independent of another. It also seems reasonable that the performance of one tested cup should be independent of other cups.

✔ **Randomization Condition:** The experimenter performed the trials in random order.

✔ **Similar Spread Condition:** The Nissan mug variation seems to be a bit smaller than the others. I'll look later at the plot of residuals vs. predicted values to see if the plot thickens.

 Mechanics Fit the ANOVA model.

Analysis of Variance

Source	DF	Sum of Squares	Mean Square	F-ratio	P-value
Container	3	714.1875	238.063	10.713	<0.0001
Error	28	622.1875	22.221		
Total	31	1336.3750			

✔ **Nearly Normal Condition, Outlier Condition:** The Normal probability plot is not very straight, but there are no outliers.

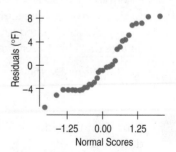

The histogram shows that the distribution of the residuals is skewed to the right:

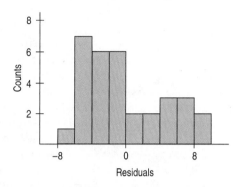

The table of means and SDs (below) shows that the standard deviations grow along with the means. Possibly a re-expression of the data would improve matters.

Under these circumstances, I cautiously find the P-value for the F-statistic from the F-model with 3 and 28 degrees of freedom.

The ratio of the mean squares gives an F-ratio of 10.7134 with a P-value of <0.0001.

 SHOW Show the table of means.

From the ANOVA table, the Error Mean Square, MS_E, is 22.22, which means that the standard deviation of all the errors is estimated to be $\sqrt{22.22} = 4.71$ degrees F.

This seems like a reasonable value for the error standard deviation in the four treatments (with the possible exception of the Nissan mug).

Level	n	Mean	Std Dev
CUPPS	8	10.1875	5.20259
Nissan	8	2.7500	2.50713
SIGG	8	16.0625	5.90059
Starbucks	8	10.2500	4.55129

 TELL **Interpretation** Tell what the F-test means.

An F-ratio this large would be very unlikely if the containers all had the same mean temperature difference.

 THINK State your conclusions.

(You should be more worried about the changing variance if you fail to reject the null hypothesis.) More specific conclusions might require a re-expression of the data.

Conclusions: Even though some of the conditions are mildly violated, I still conclude that the means are not all equal and that the four cups do not maintain temperature equally well.

The Balancing Act

The two examples we've looked at so far share a special feature. Each treatment group has the same number of experimental units. For the hand-washing experiment, each washing method was tested 8 times. For the cups, there were also 8 trials for each cup. This feature (the equal numbers of cases in each group, not the number 8) is called **balance,** and experiments that have equal numbers of experimental units in each treatment are said to be balanced or to have balanced designs.

Balanced designs are a bit easier to analyze because the calculations are simpler, so we usually try for balance. But in the real world we often encounter unbalanced data. Participants drop out or become unsuitable, plants die, or maybe we just can't find enough experimental units to fit a particular criterion.

Everything we've done so far works just fine for unbalanced designs except that the calculations get a bit more complicated. Where once we could write n for the number of experimental units in a treatment, now we have to write n_k and sum more carefully. Where once we could pool variances with a simple average, now we have to adjust for the different n's. Technology clears these hurdles easily, so you're safe thinking about the analysis in terms of the simpler balanced formulas and trusting that the technology will make the necessary adjustments.

FOR EXAMPLE

RECAP: An ANOVA for the contrast baths experiment had a statistically significant F-value.

Here are summary statistics for the three treatment groups:

Group	Count	Mean	StdDev
Bath	22	4.54545	7.76271
Bath+Exercise	23	8	7.03885
Exercise	14	−1.07143	5.18080

QUESTION: What can you conclude about these results?

We can be confident that there is a difference. However, it is the exercise treatment that appears to reduce swelling and not the contrast bath treatments. We might conclude (as the researchers did) that contrast bath treatments are of limited value.

Comparing Means

When we reject H_0, it's natural to ask which means are different. No one would be happy with an experiment to test 10 cancer treatments that concluded only with "We can reject H_0—the treatments are different!" We'd like to know more, but the F-statistic doesn't offer that information.

What can we do? If we can't reject the null, we've got to stop. There's no point in further testing. If we've rejected the simple null hypothesis, however, we *can* do more. In particular, we can test whether any pairs or combinations of group means differ. For example, we might want to compare treatments against a control or a placebo, or against the current standard treatment.

In the hand-washing experiment, we could consider plain water to be a control. Nobody would be impressed with (or want to pay for) a soap that did no better than water alone. A test of whether the antibacterial soap (for example) was different from plain water would be a simple test of the difference between two group means. To be able to perform an ANOVA, we first check the Similar Variance Condition. If things look OK we assume that the variances are equal. If the variances *are* equal then a pooled *t*-test is appropriate. Even better (this is the special part), we already have a pooled estimate of the standard deviation based on *all* of the tested washing methods. That's s_p, which, for the hand-washing experiment, was equal to 37.55 bacteria colonies.

The null hypothesis is that there is no difference between water and the antibacterial soap. As we did in Chapter 24, we'll write that as a hypothesis about the difference in the means:

$$H_0: \mu_W - \mu_{ABS} = 0. \text{ The alternative is}$$
$$H_0: \mu_W - \mu_{ABS} \neq 0.$$

The natural test statistic is $\bar{y}_W - \bar{y}_{ABS}$, and the (pooled) standard error is

$$SE(\mu_W - \mu_{ABS}) = s_p\sqrt{\frac{1}{n_W} + \frac{1}{n_{ABS}}}.$$

Level	n	Mean	Std Dev
Alcohol spray	8	37.5	26.56
Antibacterial soap	8	92.5	41.96
Soap	8	106.0	46.96
Water	8	117.0	31.13

The difference in the observed means is $117.0 - 92.5 = 24.5$ colonies. The standard error comes out to 18.775. The *t*-statistic, then, is $t = \dfrac{24.5}{18.775} = 1.31$. To find the P-value we consult the Student's *t*-distribution on $N - k = 32 - 4 = 28$ degrees of freedom. The P-value is about 0.1—not small enough to impress us. So we can't discern a significant difference between washing with the antibacterial soap and just using water.

Our *t*-test asks about a simple difference. We could also ask a more complicated question about groups of differences. Does the average of the two soaps differ from the average of three sprays, for example? Complex combinations like these are called *contrasts*. Finding the standard errors for contrasts is straightforward but beyond the scope of this book. We'll restrict our attention to the common question of comparing pairs of treatments after H_0 has been rejected.

*Bonferroni Multiple Comparisons

Our hand-washing experimenter *was* pretty sure that alcohol would kill the germs even before she started the experiment. But alcohol dries the skin and leaves an unpleasant smell. She was hoping that one of the antibacterial soaps would work as well as alcohol so she could use that instead. That means she really wanted to compare *each* of the other treatments against the alcohol spray. We know how to compare two of the means with a *t*-test. But now we want to do several tests, and each test poses the risk of a Type I error. As we do more and more tests, the risk that we might make a Type I error grows bigger than the α level of each individual test. With each additional test, the risk of making an error grows. If we do enough tests, we're almost sure to reject one of the null hypotheses by mistake—and we'll never know which one.

There is a defense against this problem. In fact, there are several defenses. As a class, they are called **methods for multiple comparisons.** All multiple comparisons methods require that we first be able to reject the overall null hypothesis with the ANOVA's *F*-test. Once we've rejected the overall null, then we can think about comparing several—or even all—pairs of group means.

Let's look again at our test of the water treatment against the antibacterial soap treatment. This time we'll look at a confidence interval instead of the pooled *t*-test. We did a test at significance level $\alpha = 0.05$. The corresponding confidence level is $1 - \alpha = 95\%$. For *any* pair of means, a confidence interval for their difference is $(\bar{y}_1 - \bar{y}_2) \pm ME$, where the margin of error is

$$ME = t^* \times s_p\sqrt{\frac{1}{n_1} + \frac{1}{n_2}}.$$

As we did in the previous section, we get s_p as the pooled standard deviation found from *all* the groups in our analysis. Because s_p uses the information

Carlo Bonferroni (1892–1960) was a mathematician who taught in Florence. He wrote two papers in 1935 and 1936 setting forth the mathematics behind the method that bears his name.

about the standard deviation from *all* the groups it's a better estimate than we would get by combining the standard deviation of just two of the groups. This uses the **Equal Variance Assumption** and "borrows strength" in estimating the common standard deviation of all the groups. We find the critical value t^* from the Student's t-model corresponding to the specified confidence level found with $N - k$ degrees of freedom, and the n_k's are the number of experimental units in each of the treatments.

To reject the null hypothesis that the two group means are equal, the difference between them must be larger than the ME. That way 0 won't be in the confidence interval for the difference. When we use it in this way, we call the margin of error the **least significant difference** (**LSD** for short). If two group means differ by more than this amount, then they are significantly different at level α *for each individual test.*

For our hand-washing experiment, each group has $n = 8$, $s_p = 37.55$, and df $= 32 - 4 = 28$. From technology or Table T, we can find that t^* with 28 df (for a 95% confidence interval) is 2.048. So

$$\text{LSD} = 2.048 \times 37.55 \times \sqrt{\frac{1}{8} + \frac{1}{8}} = 38.45 \text{ colonies,}$$

and we could use this margin of error to make a 95% confidence interval for any difference between group means. Any two washing methods whose means differ by more than 38.45 colonies could be said to differ at $\alpha = 0.05$ by this method.

Of course, we're still just examining individual pairs. If we want to examine *many* pairs simultaneously, there are several methods that adjust the critical t^*-value so that the resulting confidence intervals provide appropriate tests for all the pairs. And, in spite of making *many* such intervals, the overall Type I error rate stays at (or below) α.

One such method is called the **Bonferroni method.** This method adjusts the LSD to allow for making many comparisons. The result is a wider margin of error called the **minimum significant difference,** or **MSD.** The MSD is found by replacing t^* with a slightly larger number. That makes the confidence intervals wider for each contrast and the corresponding Type I error rates lower for *each* test. And it keeps the *overall* Type I error rate at or below α.

The Bonferroni method distributes the error rate equally among the confidence intervals. It divides the error rate among J confidence intervals, finding each interval at confidence level $1 - \dfrac{\alpha}{J}$ instead of the original $1 - \alpha$. To signal this adjustment, we label the critical value t^{**} rather than t^*. For example, to make the three confidence intervals comparing the alcohol spray with the other three washing methods, and preserve our overall α risk at 5%, we'd construct each with a confidence level of

$$1 - \frac{0.05}{3} = 1 - 0.01667 = 0.98333.$$

The only problem with this is that t-tables don't have a column for 98.33% confidence (or, correspondingly, for $\alpha = 0.01667$). Fortunately, technology has no such constraints.[9] For the hand-washing data, if we want to examine the three confidence intervals comparing each of the other methods with the alcohol spray, the t^{**}-value (on 28 degrees of freedom) turns out to be 2.238. That's somewhat larger than the individual t^*-value of 2.048 that we would have used for a single confidence interval. And the corresponding ME is 42.02 colonies (rather than 38.45 for a single comparison). The larger critical value along with correspondingly wider intervals is the price we pay for making multiple comparisons.

---

[9] The electronic t-tables provided on the DVD in *ActivStats* let you add new columns to the t-table at any alpha level, so you can do the Bonferroni calculation easily.

Many statistics packages assume that you'd like to compare all pairs of means. Some will display the result of these comparisons in a table like this:

Level	n	Mean	Groups
Alcohol spray	8	37.5	A
Antibacterial soap	8	92.5	B
Soap	8	106.0	B
Water	8	117.0	B

This table shows that the alcohol spray is in a class by itself and that the other three hand-washing methods are indistinguishable from one another.

ANOVA on Observational Data

So far we've applied ANOVA only to data from designed experiments. That's natural for several reasons. The primary one is that, as we saw in Chapter 13, randomized comparative experiments are specifically designed to compare the results for different treatments. The overall null hypothesis, and the subsequent tests on pairs of treatments in ANOVA, address such comparisons directly. In addition, as we discussed earlier, the **Equal Variance Assumption** (which we need for all of the ANOVA analyses) is often plausible in a randomized experiment because the treatment groups start out with sample variances that all estimate the same underlying variance of the collection of experimental units.

Sometimes, though, we just can't perform an experiment. When ANOVA is used to test equality of group means from observational data, there's no *a priori* reason to think the group variances might be equal at all. Even if the null hypothesis of equal means were true, the groups might easily have different variances. But if the side-by-side boxplots of responses for each group show roughly equal spreads and symmetric, outlier-free distributions, you can use ANOVA on observational data.

Observational data tend to be messier than experimental data. They are much more likely to be unbalanced. If you aren't assigning subjects to treatment groups, it's harder to guarantee the same number of subjects in each group. And because you are not controlling conditions as you would in an experiment, things tend to be, well, less controlled. The only way we know to avoid the effects of possible lurking variables is with control and randomized assignment to treatment groups, and for observational data, we have neither.

ANOVA is often applied to observational data when an experiment would be impossible or unethical. (We can't randomly break some subjects' legs, but we *can* compare pain perception among those with broken legs, those with sprained ankles, and those with stubbed toes by collecting data on subjects who have already suffered those injuries.) In such data, subjects are already in groups, but not by random assignment.

Be careful; if you have not assigned subjects to treatments randomly, you can't draw *causal* conclusions even when the F-test is significant. You have no way to control for lurking variables or confounding, so you can't be sure whether any differences you see among groups are due to the grouping variable or to some other unobserved variable that may be related to the grouping variable.

Because observational studies often are intended to estimate parameters, there is a temptation to use pooled confidence intervals for the group means for this purpose. Although these confidence intervals are statistically correct, be sure to think carefully about the population that the inference is about. The relatively few subjects that happen to be in a group may not be a simple random sample of any interesting population, so their "true" mean may have only limited meaning.

STEP-BY-STEP EXAMPLE | One More Example

Here's an example that exhibits many of the features we've been discussing. It gives a fair idea of the kinds of challenges often raised by real data.

A study at a liberal arts college attempted to find out who watches more TV at college: men or women? Varsity athletes or non-athletes? Student researchers asked 200 randomly selected students questions about their backgrounds and about their television-viewing habits. The researchers found that men watch, on average, about 2.5 hours per week more TV than women, and that varsity athletes watch about 3.5 hours per week more than those who are not varsity athletes. But is this the whole story? To investigate further, they divided the students into four groups: male athletes (MA), male non-athletes (MNA), female athletes (FA), and female non-athletes (FNA).

Question: Do these four groups of students spend about the same amount of time watching TV?

THINK

Variables Name the variables, report the W's, and specify the questions of interest.

I have the number of hours spent watching TV in a week for 197 randomly selected students. We know their sex and whether they are varsity athletes or not. I wonder whether TV watching differs according to sex and athletic status.

Plot Always start an ANOVA with side-by-side boxplots of the responses in each of the groups. Always.

These data offer a good example why.

Here are the side-by-side boxplots of the data:

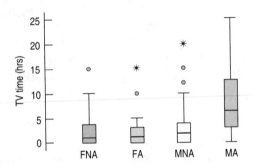

This plot suggests problems with the data. Each box shows a distribution skewed to the high end, and outliers pepper the display, including some extreme outliers. The box with the highest center (MA) also has the largest spread. These data just don't pass our first screening for suitability. This sort of pattern calls for a re-expression.

The responses are counts—numbers of TV hours. You may recall from Chapter 10 that a good re-expression to try first for counts is the square root.

Here are the boxplots for the square root of TV hours.

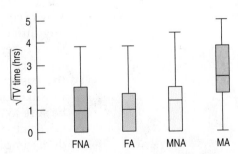

The spreads in the four groups are now more similar and the individual distributions more symmetric. And now there are no outliers.

Think about the assumptions and check the conditions.

✔ **Independence Assumption:** Because this is a random sample, the assumption of independence is reasonable. However, I'll want to check that the sample does not contain, for example, too many athletes on the same sports team.

✔ **Randomization Condition:** The data come from a random sample of students.

✔ **Similar Spread Condition:** The boxplots show similar spreads. I may want to check the residuals later.

Fit the ANOVA model.

The ANOVA table looks like this:

Source	DF	Sum of Squares	Mean Square	F-ratio	P-value
Group	3	47.24733	15.7491	12.8111	<0.0001
Error	193	237.26114	1.2293		
Total	196	284.50847			

Nearly Normal Condition, Outlier Condition:
A histogram of the residuals looks reasonably Normal:

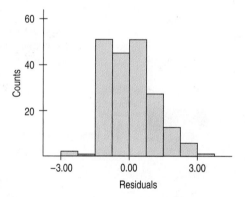

Interestingly, the few cases that seem to stick out on the low end are male athletes who watched no TV, making them different from all the other male athletes.

Under these conditions, it's appropriate to use Analysis of Variance.

 Interpretation

The F-statistic is large and the corresponding P-value small. I conclude that the TV-watching behavior is not the same among these groups.

*So Do Male Athletes Watch More TV?

Here's a Bonferroni comparison of all pairs of groups:

In case you were wondering . . . The standard errors are different because this isn't a balanced design. Differing numbers of experimental units in the groups generate differing standard errors.

	Difference	Std. Err.	P-Value
FA–FNA	0.049	0.270	0.9999
MNA–FNA	0.205	0.182	0.8383
MNA–FA	0.156	0.268	0.9929
MA–FNA	1.497	0.250	<0.0001
MA–FA	1.449	0.318	<0.0001
MA–MNA	1.292	0.248	<0.0001

Three of the differences are very significant. It seems that among women there's little difference in TV watching between varsity athletes and others. Among men, though, the corresponding difference is large. And among varsity athletes, men watch significantly more TV than women.

But wait. How far can we extend the inference that male athletes watch more TV than other groups? The data came from a random sample of students made during the week of March 21. If the students carried out the survey correctly using a simple random sample, we should be able to make the inference that the generalization is true for the entire student body during that week.

Is it true for other colleges? Is it true throughout the year? The students conducting the survey followed up the survey by collecting anecdotal information about TV watching of male athletes. It turned out that during the week of the survey, the NCAA men's basketball tournament was televised. This could explain the increase in TV watching for the male athletes. It could be that the increase extends to other students at other times, but we don't know that. Always be cautious in drawing conclusions too broadly. Don't generalize from one population to another.

What Can Go Wrong?

- **Watch out for outliers.** One outlier in a group can change both the mean and the spread of that group. It will also inflate the Error Mean Square, which can influence the F-test. The good news is that ANOVA fails on the safe side by losing power when there are outliers. That is, you are less likely to reject the overall null hypothesis if you have (and leave) outliers in your data. But they are not likely to cause you to make a Type I error.

- **Watch out for changing variances.** The conclusions of the ANOVA depend crucially on the assumptions of independence and constant variance, and (somewhat less seriously as n increases) on Normality. If the conditions on the residuals are violated, it may be necessary to re-express the response variable to approximate these conditions more closely. ANOVA benefits so greatly from a judiciously chosen re-expression that the choice of a re-expression might be considered a standard part of the analysis.

- **Be wary of drawing conclusions about causality from observational studies.** ANOVA is often applied to data from randomized experiments for which causal conclusions are appropriate. If the data are not from a designed experiment, however, the Analysis of Variance provides no more evidence for causality than any other method we have studied. Don't get into the habit of assuming that ANOVA results have causal interpretations.

- **Be wary of generalizing** to situations other than the one at hand. Think hard about how the data were generated to understand the breadth of conclusions you are entitled to draw.
- **Watch for multiple comparisons.** When rejecting the null hypothesis, you can conclude that the means are not *all* equal. But you can't start comparing every pair of treatments in your study with a *t*-test. You'll run the risk of inflating your Type I error rate. Use a multiple comparisons method when you want to test many pairs.

CONNECTIONS

We first learned about side-by-side boxplots in Chapter 5. There we made general statements about the shape, center, and spread of each group. When we compared groups, we asked whether their centers looked different compared with how spread out the distributions were. Now we've made that kind of thinking precise. We've added confidence intervals for the difference and tests of whether the means are the same.

We pooled data to find a standard deviation when we tested the hypothesis of equal proportions. For that test, the assumption of equal variances was a consequence of the null hypothesis that the proportions were equal, so it didn't require an extra assumption. Means don't have a linkage with their corresponding variances, so to use pooled methods we must make the additional assumption of equal variances. In a randomized experiment, that's a plausible assumption.

Chapter 13 offered a variety of designs for randomized comparative experiments. Each of those designs can be analyzed with a variant or extension of the ANOVA methods discussed in this chapter. Entire books and courses deal with these extensions, but all follow the same fundamental ideas presented here.

ANOVA is closely related to the regression analyses we saw in Chapter 27. (In fact, most statistics packages offer an ANOVA table as part of their regression output.) The assumptions are similar—and for good reason. The analyses are, in fact, related at a deep conceptual (and computational) level, but those details are beyond the scope of this book.

The pooled two-sample *t*-test for means is a special case of the ANOVA *F*-test. If you perform an ANOVA comparing only two groups, you'll find that the P-value of the *F*-statistic is exactly the same as the P-value of the corresponding pooled *t*-statistic. That's because in this special case the *F*-statistic is just the square of the *t*-statistic. The *F*-test is more general. It can test the hypothesis that several group means are equal.

WHAT HAVE WE LEARNED?

We learned, in Chapter 24, how to test whether the means of two groups are equal. Now in this chapter, we've extended that to testing whether the means of *several* groups are equal. We first learned in Chapter 5 that a good first step in looking at the relationship between a quantitative response and a categorical grouping variable is to look at side-by-side boxplots. We've seen that it's still a good first step before formally testing the null hypothesis.

We've learned that the *F*-test is a generalization of the *t*-test that we used for testing two groups. And we've seen that although this makes the mechanics familiar, there are new conditions to check. We've also learned that when the null hypothesis is rejected and we conclude that there are differences, we need to adjust the confidence intervals for the pair-wise differences between means. We also need to adjust the alpha levels of tests we perform once we've rejected the null hypothesis.

▶ We've learned that under certain assumptions, the statistic used to test whether the means of k groups are equal is distributed as an F-statistic with $k - 1$ and $N - k$ degrees of freedom.

▶ We've learned to check four conditions to verify the assumptions before we proceed with inference and we've seen that most of the checks can be made by graphing the data and the residuals with the methods we learned in Chapters 4, 5, and 8.

▶ We've learned that if the F-statistic is large enough we reject the null hypothesis that all the means are equal.

▶ We've also learned to create and interpret confidence intervals for the differences between each pair of group means, recognizing that we need to adjust the confidence interval for the number of comparisons we make.

Terms

Error (or Within) Mean Square (MS_E)	The Error Mean Square (MS_E) is the estimate of the error variance obtained by *pooling* the variances of each treatment group. The square root of the (MS_E) is the estimate of the error standard deviation, s_p (p. 719).
Treatment (or Between) Mean Square (MS_T)	The Treatment Mean Square (MS_T) is the estimate of the error variance under the assumption that the treatment means are all equal. If the (null) assumption is not true, the MS_T will be larger than the error variance (p. 719).
F-distribution	The F-distribution is the sampling distribution of the F-statistic when the null hypothesis that the treatment means are equal is true. It has two degrees of freedom parameters, one for the numerator, $(k - 1)$, and one for the denominator, $(N - k)$, where N is the total number of observations and k is the number of groups (p. 719).
F-statistic	The F-statistic is the ratio MS_T/MS_E. When the F-statistic is sufficiently large, we reject the null hypothesis that the group means are equal (p. 719).
F-test	The F-test tests the null hypothesis that all the group means are equal against the one-sided alternative that they are not all equal. We reject the hypothesis of equal means if the F-statistic exceeds the critical value from the F-distribution corresponding to the specified significance level and degrees of freedom (p. 720).
ANOVA	An analysis method for testing equality of means across treatment groups (p. 719).
ANOVA table	The ANOVA table is convenient for showing the degrees of freedom, the Treatment Mean Square, the Error Mean Square, their ratio, the F-statistic, and its P-value. There are usually other quantities of lesser interest included as well (p. 720).
ANOVA model	The model for a one-way (one response, one factor) ANOVA is $$y_{ij} = \mu_j + \varepsilon_{ij}.$$ Estimating with $y_{ij} = \bar{y}_j + e_{ij}$ gives predicted values $\hat{y}_{ij} = \bar{y}_j$ and residuals $e_{ij} = y_{ij} - \bar{y}_j$ (p. 722).
Residual standard deviation	The residual standard deviation, $$s_p = \sqrt{MS_E} = \sqrt{\frac{\sum e^2}{N - k}}$$ gives an idea of the underlying variability of the response values (p. 726).
Assumptions for ANOVA (and conditions to check)	▶ **Independence Assumption.** Think about the design of the experiment or, if an observational study, how the data were collected (p. 727). ▶ **Equal Variance Assumption. (Similar Spread Condition.)** Look at side-by-side boxplots to check for similar spreads, or look at residuals vs. predicted to see if the plot thickens (p. 727). ▶ **Normal Population Assumption. (Nearly Normal Condition.)** Check a histogram or Normal probability plot of the residuals (p. 728).
Balance	An experiment's design is balanced if each treatment level has the same number of experimental units. Balanced designs make calculations simpler and are generally more powerful (p. 732).

Multiple comparisons	If we reject the null hypothesis of equal means, we often then want to investigate further and compare pairs of treatment group means to see if they differ. If we want to test several such pairs, we must adjust for performing several tests to keep the overall risk of a Type I error from growing too large. Such adjustments are called methods for multiple comparisons (p. 733).
Least significant difference (LSD)	The standard margin of error in the confidence interval for the difference of two means. It has the correct Type I error rate for a single test, but not when performing more than one comparison (p. 734).
*Bonferroni method	One of many methods for adjusting the length of the margin of error when testing the differences between several group means (p. 734).
Minimum significant difference (MSD)	The Bonferroni method's margin of error for the confidence interval for the difference of two group means. This can be used to test differences of several pairs of group means. If their difference exceeds the MSD, they are different at the overall rate (p. 734).

Skills

▶ Recognize situations for which ANOVA is the appropriate analysis.

▶ Know how to examine your data for violations of conditions that would make ANOVA unwise or invalid.

▶ Recognize when a further analysis of differences between group means would be appropriate.

▶ Be able to perform an ANOVA using a statistics package or calculator for one response variable and one factor with any number of levels.

▶ Be able to perform several subsequent tests using a multiple comparisons procedure.

▶ Be able to explain the contents of an ANOVA table, in particular the roles of the MS_T, the MS_E, and the pooled standard deviation, s_p.

▶ Be able to interpret a test of the null hypothesis that the true means of several independent groups are equal. (Your interpretation should include a defense of your assumption of equal variances.)

▶ *Be able to interpret the results of tests that use multiple comparisons methods.

ANOVA ON THE COMPUTER

Most analyses of variance are found with computers. And all statistics packages present the results in an ANOVA table much like the one we discussed. Technology also makes it easy to examine the side-by-side boxplots and check the residuals for violations of the assumptions and conditions.

Statistics packages offer different choices among possible multiple comparisons methods (although Bonferroni is quite common). This is a specialized area. Get advice or read further if you need to choose a multiple comparisons method.

As we saw in Chapter 5, there are two ways to organize data recorded for several groups. We can put all the response values in a single variable and use a second, "factor," variable to hold the group identities. This is sometimes called stacked format. The alternative is to place the data for each group in its own column or variable. Then the variable identities become the group identifiers.

Most statistics packages expect the data to be in stacked format because this form also works for more complicated experimental designs. Some packages can work with either form, and some use one form for some things and the other for others. (Be careful, for example, when you make side-by-side boxplots; be sure to give the appropriate version of the command to correspond to the structure of your data.)

Most packages offer to save residuals and predicted values and make them available for further tests of conditions. In some packages you may have to request them specifically.

DATA DESK

- Select the response variable as Y and the factor variable as X.
- From the **Calc** menu, choose **ANOVA**.
- Data Desk displays the ANOVA table.
- Select plots of residuals from the ANOVA table's HyperView menu.

COMMENTS

Data Desk expects data in "stacked" format. You can change the ANOVA by dragging the icon of another variable over either the Y or X variable name in the table and dropping it there. The analysis will recompute automatically.

EXCEL

- In Excel 2003 and earlier, select **Data Analysis** from the Tools menu.
- In Excel 2007, select **Data Analysis** from the Analysis Group on the Data Tab.
- Select **Anova Single Factor** from the list of analysis tools.
- Click the **OK** button.
- Enter the data range in the box provided.
- Check the **Labels in First Row** box, if applicable.
- Enter an alpha level for the F-test in the box provided.
- Click the **OK** button.

COMMENTS

The data range should include two or more columns of data to compare. Unlike all other statistics packages, Excel expects each column of the data to represent a different level of the factor. However, it offers no way to label these levels. The columns need not have the same number of data values, but the selected cells must make up a rectangle large enough to hold the column with the most data values.

JMP

- From the **Analyze** menu select **Fit Y by X.**
- Select variables: a quantitative Y, Response variable, and a categorical X, Factor variable.
- JMP opens the **Oneway** window.
- Click on the red triangle beside the heading, select **Display Options,** and choose **Boxplots.**
- From the same menu choose the **Means/ANOVA. t-test** command.
- JMP opens the oneway ANOVA output.

COMMENTS

JMP expects data in "stacked" format with one response and one factor variable.

MINITAB

- Choose **ANOVA** from the Stat menu.
- Choose **One-way . . .** from the **ANOVA** submenu.
- In the One-way Anova dialog, assign a quantitative Y variable to the Response box and assign a categorical X variable to the Factor box.
- Check the **Store Residuals** check box.
- Click the **Graphs** button.
- In the ANOVA-Graphs dialog, select **Standardized residuals,** and check **Normal plot of residuals** and **Residuals versus fits.**
- Click the **OK** button to return to the ANOVA dialog.
- Click the **OK** button to compute the ANOVA.

COMMENTS

If your data are in unstacked format, with separate columns for each treatment level, choose **One-way (unstacked)** from the **ANOVA** submenu.

SPSS

- Choose **Compare Means** from the **Analyze** menu.
- Choose **One-way ANOVA** from the **Compare Means** submenu.
- In the One-Way ANOVA dialog, select the Y-variable and move it to the dependent target. Then move the X-variable to the independent target.
- Click the **OK** button.

COMMENTS

SPSS expects data in stacked format. The **Contrasts** and **Post Hoc** buttons offer ways to test contrasts and perform multiple comparisons. See your SPSS manual for details.

TI-89

Under **STAT Tests**, choose **C:ANOVA**

- Specify the input method (Data or Stats) according to whether you have data entered as one list for each group or summary statistics for each group, and specify the number of groups. Press ÷.
- If Data, you will then be asked to supply the name of each list.
- If Stats, you will be asked for the stats for each group. Enter n, x̄, and s for each group separated by commas and within curly braces ({and}).
- Press ÷ to perform the calculations.

COMMENTS

In addition to the ANOVA table output, the calculator creates three new lists—the means for each group (in the order specified) and *individual* 95% confidence interval upper and lower bounds.

EXERCISES

1. **Popcorn.** A student runs an experiment to test four different popcorn brands, recording the number of kernels left unpopped. She pops measured batches of each brand 4 times, using the same popcorn popper and randomizing the order of the brands. After collecting her data and analyzing the results, she reports that the F-ratio is 13.56.
   a) What are the null and alternative hypotheses?
   b) How many degrees of freedom does the treatment sum of squares have? How about the error sum of squares?
   c) Assuming that the conditions required for ANOVA are satisfied, what is the P-value? What would you conclude?
   d) What else about the data would you like to see in order to check the assumptions and conditions?

2. **Skating.** A figure skater tried various approaches to her Salchow jump in a designed experiment using 5 different places for her focus (arms, free leg, midsection, takeoff leg, and free). She tried each jump 6 times in random order, using two of her skating partners to judge the jumps on a scale from 0 to 6. After collecting the data and analyzing the results, she reports that the F-ratio is 7.43.
   a) What are the null and alternative hypotheses?
   b) How many degrees of freedom does the treatment sum of squares have? How about the error sum of squares?

   c) Assuming that the conditions are satisfied, what is the P-value? What would you conclude?
   d) What else about the data would you like to see in order to check the assumptions and conditions?

3. **Gas mileage.** A student runs an experiment to study the effect of three different mufflers on gas mileage. He devises a system so that his Jeep Wagoneer uses gasoline from a one-liter container. He tests each muffler 8 times, carefully recording the number of miles he can go in his Jeep Wagoneer on one liter of gas. After analyzing his data, he reports that the F-ratio is 2.35 with a P-value of 0.1199.
   a) What are the null and alternative hypotheses?
   b) How many degrees of freedom does the treatment sum of squares have? How about the error sum of squares?
   c) What would you conclude?
   d) What else about the data would you like to see in order to check the assumptions and conditions?
   e) If your conclusion in part c is wrong, what type of error have you made?

4. **Darts.** A student interested in improving her dart-throwing technique designs an experiment to test 4 different stances to see whether they affect her accuracy. After warming up for several minutes, she randomizes

the order of the 4 stances, throws a dart at a target using each stance, and measures the distance of the dart in centimeters from the center of the bull's-eye. She replicates this procedure 10 times. After analyzing the data she reports that the *F*-ratio is 1.41.

a) What are the null and alternative hypotheses?

b) How many degrees of freedom does the treatment sum of squares have? How about the error sum of squares?

c) What would you conclude?

d) What else about the data would you like to see in order to check the assumptions and conditions?

e) If your conclusion in part c is wrong, what type of error have you made?

5. Activating baking yeast. To shorten the time it takes him to make his favorite pizza, a student designed an experiment to test the effect of sugar and milk on the activation times for baking yeast. Specifically, he tested four different recipes and measured how many seconds it took for the same amount of dough to rise to the top of a bowl. He randomized the order of the recipes and replicated each treatment 4 times.

Here are the boxplots of activation times from the four recipes:

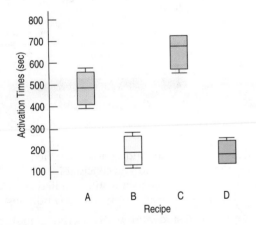

The ANOVA table follows:

Analysis of Variance

Source	DF	Sum of Squares	Mean Square	F-ratio	P-value
Recipe	3	638967.69	212989	44.7392	<0.0001
Error	12	57128.25	4761		
Total	15	696095.94			

a) State the hypotheses about the recipes (both numerically and in words).

b) Assuming that the assumptions for inference are satisfied, perform the hypothesis test and state your conclusion. Be sure to state it in terms of activation times and recipes.

c) Would it be appropriate to follow up this study with multiple comparisons to see which recipes differ in their mean activation times? Explain.

6. Frisbee throws. A student performed an experiment with three different grips to see what effect it might have on the distance of a backhanded Frisbee throw. She tried

it with her normal grip, with one finger out, and with the Frisbee inverted. She measured in paces how far her throw went. The boxplots and the ANOVA table for the three grips are shown below:

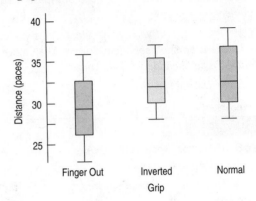

Analysis of Variance

Source	DF	Sum of Squares	Mean Square	F-ratio	P-value
Grip	2	58.58333	29.2917	2.0453	0.1543
Error	21	300.75000	14.3214		
Total	23	359.33333			

a) State the hypotheses about the grips.

b) Assuming that the assumptions for inference are satisfied, perform the hypothesis test and state your conclusion. Be sure to state it in terms of Frisbee grips and distance thrown.

c) Would it be appropriate to follow up this study with multiple comparisons to see which grips differ in their mean distance thrown? Explain.

7. Eye and hair color. In Chapter 5, Exercise 22, we saw a survey of 1021 school-age children conducted by randomly selecting children from several large urban elementary schools. Two of the questions concerned eye and hair color. In the survey, the following codes were used:

Hair color:	Eye color:
1 = Blond	1 = Blue
2 = Brown	2 = Green
3 = Black	3 = Brown
4 = Red	4 = Grey
5 = Other	5 = Other

The students analyzing the data were asked to study the relationship between eye and hair color. They produced this plot:

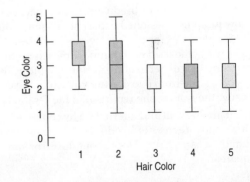

They then ran an Analysis of Variance with *Eye color* as the response and *Hair color* as the factor. The ANOVA table they produced follows:

Analysis of Variance

Source	DF	Sum of Squares	Mean Square	F-ratio	P-value
Hair color	4	1.46946	0.367365	0.4024	0.8070
Error	1016	927.45317	0.912848		
Total	1020	928.92263			

What suggestions do you have for the Statistics students? What alternative analysis might you suggest?

8. **Zip codes, revisited.** The intern from the marketing department at the Holes R Us online piercing salon (Chapter 4, Exercise 49) has recently finished a study of the company's 500 customers. He wanted to know whether people's zip codes vary by the last product they bought. They have 16 different products, and the ANOVA table of zip code by product showed the following:

ANOVA table

Source	DF	Sum of Squares	Mean Square	F-ratio	P-value
Product	15	3.836e10	2.55734e9	4.9422	<0.0001
Error	475	2.45787e11	517445573		
Total	490	2.84147e11			

(Nine customers were not included because of missing zip code or product information.)

What criticisms of the analysis might you make? What alternative analysis might you suggest?

9. **Fuel economy, revisited.** In Chapter 5, Exercise 18, we looked at what these boxplots told us about the relationship between the number of cylinders a car's engine has and the car's fuel economy.

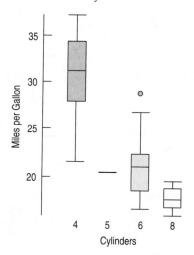

a) State the null and alternative hypotheses.
b) Do the conditions for an ANOVA seem to be met here? Why or why not?

10. **Wines, revisited.** The boxplots we saw in Chapter 5, Exercise 24, display case prices (in dollars) of wines produced by wineries along three of the Finger Lakes.

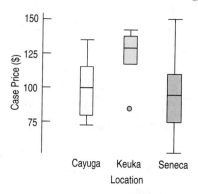

a) What are the null and alternative hypotheses? Talk about prices and location, not symbols.
b) Do the conditions for an ANOVA seem to be met here? Why or why not?

11. **Tellers.** A bank is studying the time that it takes 6 of its tellers to serve an average customer. Customers line up in the queue and then go to the next available teller. Here is a boxplot of the last 140 customers and the times it took each teller:

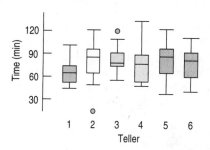

Analysis of Variance

Source	DF	Sum of Squares	Mean Square	F-ratio	P-value
Teller	5	3315.32	663.064	1.508	0.1914
Error	134	58919.1	439.695		
Total	139	62234.4			

a) What are the null and alternative hypotheses?
b) What do you conclude?
c) Would it be appropriate to run a multiple comparisons test (for example, a Bonferroni test) to see which tellers differ from each other? Explain.

12. **Hearing.** A researcher investigated four different word lists for use in hearing assessment. She wanted to know whether the lists were equally difficult to understand in the presence of a noisy background. To find out, she tested 96 subjects with normal hearing randomly assigning 24 to each of the four word lists and measured the number of words perceived correctly in the

presence of background noise. Here are the boxplots of the four lists:

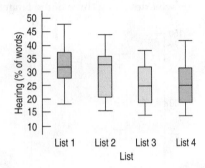

Analysis of Variance

Source	DF	Sum of Squares	Mean Square	F-ratio	P-value
List	3	920.4583	306.819	4.9192	0.0033
Error	92	5738.1667	62.371		
Total	95	6658.6250			

a) What are the null and alternative hypotheses?
b) What do you conclude?
c) Would it be appropriate to run a multiple comparisons test (for example, a Bonferroni test) to see which lists differ from each other in terms of mean percent correct? Explain.

13. **Yogurt.** An experiment to determine the effect of several methods of preparing cultures for use in commercial yogurt was conducted by a food science research group. Three batches of yogurt were prepared using each of three methods: traditional, ultrafiltration, and reverse osmosis. A trained expert then tasted each of the 9 samples, presented in random order, and judged them on a scale from 1 to 10. A partially complete Analysis of Variance table of the data follows.

An incomplete ANOVA Table for the Yogurt Data

Source	Sum of Squares	Degrees of Freedom	Mean Square	F-ratio
Treatment	17.300			
Residual	0.460			
Total	17.769			

a) Calculate the mean square of the treatments and the mean square of the error.
b) Form the F-statistic by dividing the two mean squares.
c) The P-value of this F-statistic turns out to be 0.000017. What does this say about the null hypothesis of equal means?
d) What assumptions have you made in order to answer part c?
e) What would you like to see in order to justify the conclusions of the F-test?
f) What is the average size of the error standard deviation in the judge's assessment?

14. **Smokestack scrubbers.** Particulate matter is a serious form of air pollution often arising from industrial production. One way to reduce the pollution is to put a filter, or scrubber, at the end of the smokestack to trap

the particulates. An experiment to determine which smokestack scrubber design is best was run by placing four scrubbers of different designs on an industrial stack in random order. Each scrubber was tested 5 times. For each run, the same material was produced, and the particulate emissions coming out of the scrubber were measured (in parts per billion). A partially complete Analysis of Variance table of the data is shown below.

An incomplete ANOVA Table for the Smokestack Data

Source	Sum of Squares	Degrees of Freedom	Mean Square	F-ratio
Treatment	81.2			
Residual	30.8			
Total	112.0			

a) Calculate the mean square of the treatments and the mean square of the error.
b) Form the F-statistic by dividing the two mean squares.
c) The P-value of this F-statistic turns out to be 0.0000949. What does this say about the null hypothesis of equal means?
d) What assumptions have you made in order to answer part c?
e) What would you like to see in order to justify the conclusions of the F-test?
f) What is the average size of the error standard deviation in particulate emissions?

15. **Eggs.** A student wants to investigate the effects of real vs. substitute eggs on his favorite brownie recipe. He enlists the help of 10 friends and asks them to rank each of 8 batches on a scale from 1 to 10. Four of the batches were made with real eggs, four with substitute eggs. The judges tasted the brownies in random order. Here is a boxplot of the data:

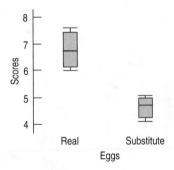

Analysis of Variance

Source	DF	Sum of Squares	Mean Square	F-ratio	P-value
Eggs	1	9.010013	9.01001	31.0712	0.0014
Error	6	1.739875	0.28998		
Total	7	10.749883			

The mean score for the real eggs was 6.78 with a standard deviation of 0.651. The mean score for the substitute eggs was 4.66 with a standard deviation of 0.395.

a) What are the null and alternative hypotheses?
b) What do you conclude from the ANOVA table?
c) Do the assumptions for the test seem to be reasonable?

d) Perform a two-sample pooled *t*-test of the difference. What P-value do you get? Show that the square of the *t*-statistic is the same (to rounding error) as the *F*-ratio.

T 16. Auto noise filters. In a statement to a Senate Public Works Committee, a senior executive of Texaco, Inc., cited a study on the effectiveness of auto filters on reducing noise. Because of concerns about performance, two types of filters were studied, a standard silencer and a new device developed by the Associated Octel Company. Here are the boxplots from the data on noise reduction (in decibels) of the two filters. Type 1 = standard; Type 2 = Octel.

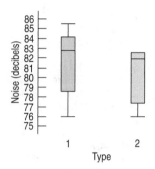

Analysis of Variance

Source	DF	Sum of Squares	Mean Square	F-ratio	P-value
Type	1	6.31	6.31	0.7673	0.3874
Error	33	271.47	8.22		
Total	34	2.77			

Level	n	Mean	StdDev
Standard	18	81.5556	3.2166
Octel	17	80.7059	2.43708

a) What are the null and alternative hypotheses?
b) What do you conclude from the ANOVA table?
c) Do the assumptions for the test seem to be reasonable?
d) Perform a two-sample pooled *t*-test of the difference. What P-value do you get? Show that the square of the *t*-statistic is the same (to rounding error) as the *F*-ratio.

T 17. School system. A school district superintendent wants to test a new method of teaching arithmetic in the fourth grade at his 15 schools. He plans to select 8 students from each school to take part in the experiment, but to make sure they are roughly of the same ability, he first gives a test to all 120 students. Here are the scores of the test by school:

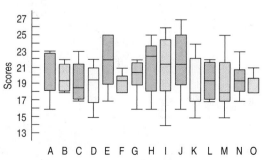

The ANOVA table shows:

Analysis of Variance

Source	DF	Sum of Squares	Mean Square	F-ratio	P-value
School	14	108.800	7.7714	1.0735	0.3899
Error	105	760.125	7.2392		
Total	119	868.925			

a) What are the null and alternative hypotheses?
b) What does the ANOVA table say about the null hypothesis? (Be sure to report this in terms of scores and schools.)
c) An intern reports that he has done *t*-tests of every school against every other school and finds that several of the schools seem to differ in mean score. Does this match your finding in part b? Give an explanation for the difference, if any, of the two results.

T 18. Fertilizers. A biology student is studying the effect of 10 different fertilizers on the growth of mung bean sprouts. She sprouts 12 beans in each of 10 different petri dishes, and adds the same amount of fertilizer to each dish. After one week she measures the heights of the 120 sprouts in millimeters. Here are boxplots and an ANOVA table of the data:

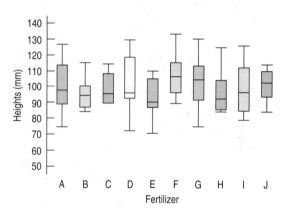

Analysis of Variance

Source	DF	Sum of Squares	Mean Square	F-ratio	P-value
Fertilizer	9	2073.708	230.412	1.1882	0.3097
Error	110	21331.083	193.919		
Total	119	23404.791			

a) What are the null and alternative hypotheses?
b) What does the ANOVA table say about the null hypothesis? (Be sure to report this in terms of heights and fertilizers).
c) Her lab partner looks at the same data and says that he did *t*-tests of every fertilizer against every other fertilizer and finds that several of the fertilizers seem to have significantly higher mean heights. Does this match your finding in part b? Give an explanation for the difference, if any, between the two results.

T 19. Cereals. Supermarkets often place similar types of cereal on the same supermarket shelf. The same data set we met in the Step-By-Step of Chapter 8 keeps track of the shelf as well as the sugar, sodium, and calorie content of

77 cereals. Does sugar content vary by shelf? Here is a boxplot and an ANOVA table for the 77 cereals:

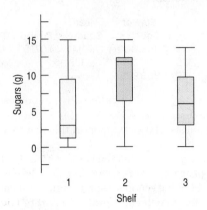

Analysis of Variance

Source	DF	Sum of Squares	Mean Square	F-ratio	P-value
Shelf	2	248.4079	124.204	7.3345	0.0012
Error	74	1253.1246	16.934		
Total	76	1501.5325			

Level	n	Mean	StdDev
1	20	4.80000	4.57223
2	21	9.61905	4.12888
3	36	6.52778	3.83582

a) What are the null and alternative hypotheses?
b) What does the ANOVA table say about the null hypothesis? (Be sure to report this in terms of *Sugars* and *Shelves*.)
c) Can we conclude that cereals on shelf 2 have a higher mean sugar content than cereals on shelf 3? Can we conclude that cereals on shelf 2 have a higher mean sugar content than cereals on shelf 1? What *can* we conclude?
d) To check for significant differences between the shelf means, we can use a Bonferroni test, whose results are shown below. For each pair of shelves, the difference is shown along with its standard error and significance level. What does it say about the questions in part c?

Dependent Variable: SUGARS

	(I) SHELF	(J) SHELF	Mean Difference (I-J)	Std. Error	P-value	95% Confidence Interval Lower Bound	Upper Bound
Bonferroni							
	1	2	−4.819(*)	1.2857	0.001	−7.969	−1.670
		3	−1.728	1.1476	0.409	−4.539	1.084
	2	1	4.819(*)	1.2857	0.001	1.670	7.969
		3	3.091(*)	1.1299	0.023	0.323	5.859
	3	1	1.728	1.1476	0.409	−1.084	4.539
		2	−3.091(*)	1.1299	0.023	−5.859	−0.323

*The mean difference is significant at the 0.05 level.

T 20. Cereals, redux. We also have data on the protein content of cereals by their shelf number. Here are the boxplot and ANOVA table:

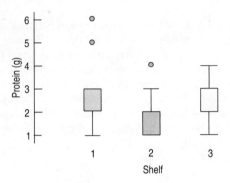

Analysis of Variance

Source	DF	Sum of Squares	Mean Square	F-ratio	P-value
Shelf	2	12.4258	6.2129	5.8445	0.0044
Error	74	78.6650	1.0630		
Total	76	91.0909			

Means and Std Deviations

Level	n	Mean	StdDev
1	20	2.65000	1.46089
2	21	1.90476	0.99523
3	36	2.86111	0.72320

a) What are the null and alternative hypotheses?
b) What does the ANOVA table say about the null hypothesis? (Be sure to report this in terms of protein content and shelves.)
c) Can we conclude that cereals on shelf 2 have a lower mean protein content than cereals on shelf 3? Can we conclude that cereals on shelf 2 have a lower mean protein content than cereals on shelf 1? What *can* we conclude?
d) To check for significant differences between the shelf means we can use a Bonferroni test, whose results are shown below. For each pair of shelves, the difference is shown along with its standard error and significance level. What does it say about the questions in part c?

Dependent Variable: PROTEIN Bonferroni

(I) SHELF	(J) SHELF	Mean Difference (I-J)	Std. Error	P-value	95% Confidence Interval Lower Bound	Upper Bound
1	2	0.75	0.322	0.070	−0.04	1.53
	3	−0.21	0.288	1.000	−0.92	0.49
2	1	−0.75	0.322	0.070	−1.53	0.04
	3	−0.96(*)	0.283	0.004	−1.65	−0.26
3	1	0.21	0.288	1.000	−0.49	0.92
	2	0.96(*)	0.283	0.004	0.26	1.65

*The mean difference is significant at the 0.05 level.

21. Downloading. To see how much of a difference time of day made on the speed at which he could download files, a college sophomore performed an experiment. He placed a file on a remote server and then proceeded to download it at three different time periods of the day. He downloaded the file 48 times in all, 16 times at each *Time of Day*, and recorded the *Time* in seconds that the download took.

Early (7 a.m.) Time (sec)	Evening (5 p.m.) Time (sec)	Late night (12 a.m.) Time (sec)
68	299	216
138	367	175
75	331	274
186	257	171
68	260	187
217	269	213
93	252	221
90	200	139
71	296	226
154	204	128
166	190	236
130	240	128
72	350	217
81	256	196
76	282	201
129	320	161

a) State the null and alternative hypotheses, being careful to talk about download *Time* and *Time of Day* as well as parameters.
b) Perform an ANOVA on these data. What can you conclude?
c) Check the assumptions and conditions for an ANOVA. Do you have any concerns about the experimental design or the analysis?
*d) Perform a multiple comparisons test to determine which times of day differ in terms of mean download time.

22. Analgesics. A pharmaceutical company tested three formulations of a pain relief medicine for migraine headache sufferers. For the experiment, 27 volunteers were selected and 9 were randomly assigned to one of three drug formulations. The subjects were instructed to take the drug during their next migraine headache episode and to report their pain on a scale of 1 = no pain to 10 = extreme pain 30 minutes after taking the drug.

Drug	Pain	Drug	Pain	Drug	Pain
A	4	B	6	C	6
A	5	B	8	C	7
A	4	B	4	C	6
A	3	B	5	C	6
A	2	B	4	C	7
A	4	B	6	C	5
A	3	B	5	C	6
A	4	B	8	C	5
A	4	B	6	C	5

a) State the null and alternative hypotheses, being careful to talk about *Drug* and *Pain* levels as well as parameters.
b) Perform an ANOVA on these data. What can you conclude?
c) Check the assumptions and conditions for an ANOVA. Do you have any concerns about the experimental design or the analysis?
*d) Perform a multiple comparisons test to determine which drugs differ in terms of mean pain level reported.

JUST CHECKING

ANSWERS

1. The null hypothesis is that the mean flight distance for all four designs is the same.

2. Yes, it looks as if the variation between the means is greater than the variation within each boxplot.

3. Yes, the *F*-test rejects the null hypothesis with a P-value < 0.0001.

4. No. The alternative hypothesis is that *at least* one mean is different from the other three. Rejecting the null hypothesis does not imply that all four means are different.

Multifactor Analysis of Variance

Where are we going?

When production problems forced a computer manufacturer to halt operations, teams of engineers, scientists, production workers, and managers were formed to solve the problem. They found themselves overwhelmed with many potential causes. But experiments that test one factor at a time would have been far too inefficient. And they might have missed important combinations of factors. In this chapter, we'll see how to look at two factors at a time, and we'll discover why these two-factor designs are so much more interesting than one-factor experiments.

WHO	Attempts at hitting a bull's-eye in a dartboard
WHAT	Distance from the center of the target
UNITS	Inches
HOW	Two-factor experiment, randomized order of runs

How accurately can you throw a dart? It probably depends on which hand you use and on how far from the target you're standing. A student designed an experiment to test the effects of both factors. He used three levels of the factor *Distance: Near, Middle,* and *Far*—and the two natural levels *Left* and *Right* for *Hand*. (He was right-handed.) In random order, he threw six darts at a dartboard under each of the six treatment conditions, and measured the distance each dart landed from the bull's-eye in inches.[1]

It makes sense that dart throws would be more accurate from nearer distances, especially when using his regular throwing hand. Side-by-side boxplots of the distance from the bull's-eye by each factor seem to support our intuition:

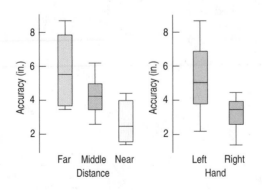

FIGURE 29.1
Boxplots of *Accuracy* (distance in inches from the bull's-eye) plotted for each experimental factor, *Distance* and *Hand*.

---

[1] He called the response variable *Accuracy,* although it measures inches *away* from the target. That means that bigger values correspond to *less* accuracy, not more.

It appears that both the distance and the hand he used affected the accuracy of his throw, and by about the same amount. How can we measure the effect of each variable? And how can we test whether the observed differences are larger than we'd expect just from sampling variation?

Two Factors at Once?!

Unlike the experiments we've seen so far, this experiment has not one but two factors. You might expect this to confuse things, but the surprising fact is that it actually improves the experiment and its analysis. It also lets us see something we'd never be able to see if we ran two different one-factor experiments.

True, with two factors, we have two hypothesis tests. Each of those hypotheses asks whether the mean of the response variable is the same for each of the treatment levels. In comparing means, we know that any excess variation just gets in the way. Back in Chapter 13, we considered ways to remove or avoid extra variation in designing experiments.

A two-factor experiment does just that. If both factors do in fact have an effect on the response (as we'd expect hand choice and distance to do for dart accuracy), then removing the effects of one factor should make it easier to see and assess the effects of the other. We'll have removed variability due to an identifiable cause.

FOR EXAMPLE

Students set up an experiment to investigate factors that determine how far a paper airplane will fly. Among the factors are whether the plane is constructed of copier paper or heavier construction paper, whether it is a "dart" or a glider, and whether the ends of the wings are folded up into "flaps." We'll combine the paper and shape factor into a single *Design* factor with four levels; *Dart/Copier, Dart/Construction, Glider/Copier,* and *Glider/Construction*. The second factor, *Flaps*, has two levels: *Flaps* and *None*.

Four planes were constructed of each of the four designs. For each design, two planes were given flaps and two were not. Each plane was flown 4 times and the distance of the flight carefully measured.

QUESTION: What are the details of the experiment design?

We have two factors: Design and Flaps. Design has 4 levels, Dart/Copier, Dart/Construction, Glider/Copier and Glider/Construction. Flaps has 2: Flaps and None. There are thus 4 × 2 = 8 treatment groups. Each treatment group has 2 planes and each plane was flown 4 times for a total of 8 observations. The response variable is the distance flown.

An ANOVA Model

The model for one-way ANOVA represented each observation as the sum of simple components. It broke each observation into the sum of three effects: the grand mean, the treatment effect, and an error:

$$y_{ij} = \mu + \tau_j + \varepsilon_{ij}.$$

Now we have two factors, *Hand* and *Distance*. So each treatment is a combination of a *Hand* assignment and a *Distance*. For example, throwing with the left hand from a near distance would be one treatment. Our model should reflect the effects of both factors. We've already used τ for the treatment effects for the first factor, so for the second factor, we'll denote the effects by γ. Now, for each observation, we write:

$$y_{ijk} = \mu + \tau_j + \gamma_k + \varepsilon_{ijk}.$$

Here we see the i-th observation at level j of factor A and level k of factor B. Why the i subscript? We're told that the student recorded six trials at each treatment, so the subscript i denotes the different observations at each combination of the two factors. The error has three subscripts because we associate a possibly different error with each observation. (If there were only one observation at each treatment, we wouldn't need the extra subscript.)

When you write out a model, you'll usually find it clearer to name the factors rather than using Greek letters. So the model will look like this:

$$y_{ijk} = \mu + Hand\ effect_j + Distance\ effect_k + Error_{ijk}.$$

To estimate the effect of each treatment level for each factor, we take the *difference* between the mean response of that treatment and the grand mean, $\bar{\bar{y}}$. So, to find the effect of throwing with the left hand, we'd find the mean accuracy of *all* runs using the left hand and subtract the grand average from that.

In general, we'd like to know whether the mean dart accuracy changes with changes in the levels of either factor. The model helps us to think about this question. Our null hypothesis on each factor is that the effects of that treatment are all zero. So we can write two null hypotheses we'd like to test as

$$H_0: \tau_1 = \tau_2 \quad \text{and} \quad H_0: \gamma_1 = \gamma_2 = \gamma_3,$$

where the τ's (taus) denote the effects due to the two levels of *Hand* and the γ's (gammas) denote the effects due to the three levels of *Distance*.

The alternative hypotheses are that the treatment effects are not all equal. The details of the analysis are pretty much the same as for a one-factor ANOVA. We want to compare the differences across the treatment effects with the underlying variability within the treatment. But now with two factors, that underlying variability has the effect of both factors removed from it. So the error term holds the variability that's left after removing the effects of both factors. Here's the ANOVA table for the dart experiment:

Analysis of Variance for Accuracy

Source	df	Sum of Squares	Mean Square	F-Ratio	Prob
Distance	2	51.0439	25.5219	28.561	≤0.0001
Hand	1	39.6900	39.6900	44.416	≤0.0001
Error	32	28.5950	0.893594		
Total	35	119.329			

This analysis splits up the total variation into three sources: the variation due to changes in *Distance*, the variation due to changes in *Hand*, and the error variation. The square root of the Error Mean Square, $\sqrt{0.894} = 0.946$ inches, estimates the standard deviation we'd expect if we made repeated observations at any treatment condition. Both F-ratios are large. Looking back at the boxplots, we shouldn't be surprised. It was clear that there were differences in accuracy among *near*, *middle*, and *far* even without accounting for the fact that each level contained both hands. The F-ratio for *Distance* is 28.56. The probability of an F-ratio that big occurring by chance is very small (<0.0001). So we can reject the hypothesis that the mean effects of *near*, *middle*, and *far* distances are equal.

The means of *left* and *right* don't look the same either. The F-ratio of 44.42 with a P-value of <0.0001 leads us to reject the null hypothesis about *Hand* effects as well.

Of course, there are assumptions and conditions to check. The good news is that most are the same as for one-way ANOVA.

Assumptions and Conditions

Even before checking any assumptions, you should have already plotted the data. Boxplots give a first look that draws our attention to the aspects tested by the ANOVA.

Plot the Data …

You could start by examining side-by-side boxplots of the data across levels of *each factor* as we did earlier, looking for outliers within any of the groups and correcting them if they are errors, or consider omitting them if we can't correct them.[2] (There don't seem to be any.) We'd like to get an idea of whether the groups have similar spreads (as we'll want) and whether the centers seem to be alike (as the null hypotheses claim) or different.

The problem of just looking at the boxplots is that the responses at each level of one factor contain *all levels of the other factor*. For example, the responses for the *left* hand were measured at all three distances. As a result, the boxplots show more variability than they need to. Our model will deal with both factors, so it will take care of this extra variability but the simple side-by-side boxplots can't do that.

On the other hand, if you do see a difference here, that's a difference *in spite of this extra variation*, so it's very likely that the *F*-test will reject the null hypothesis for that factor. Just be careful not to rule out the possibility of differences that aren't apparent.

A better alternative might be to make boxplots for each factor level *after removing the effects of the other factor*. How would you do that? You could compute a one-way ANOVA on one factor and find the residuals. Then make boxplots of those *residuals* for each level of the other factor. We might call such a display a **partial boxplot**.[3] By removing the variability due to the other factor, you should see differences in the responses among the levels of the factor you are graphing more readily. For example, here are the boxplots of dart toss *Accuracy* plotted for the two levels of *Hand* both in the original data and after removing the effects of *Distance*:

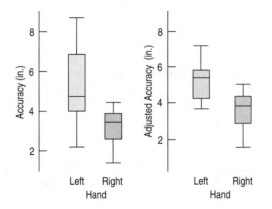

FIGURE 29.2

Boxplots of *Accuracy* by *Hand* (as in Figure 29.1) show the effect of changing *Hand* less clearly than the corresponding partial boxplots on the right, which show the effects of *Hand* after the effect of *Distance* has been removed. The effect of changing hands is much easier to see without the unwanted variation caused by changing the other factor, *Distance*.

Removing the variation due to *Distance* makes the differences between the two *Hands* in the partial boxplots much clearer. We'll want to do this for each factor, removing the effects of the other factor by computing the one-way ANOVA and finding the residuals. On the following page are the

---

[2] Unlike regression coefficients, which can be changed drastically by outliers, ANOVA tests tend to fail on the "safe side." Outliers generally inflate the Error Mean Square, which reduces the *F*-statistics and increases the P-values. That makes it harder to reject the null hypotheses, so we are more likely to make Type II errors. That's generally thought to be safer than making Type I errors. For example, we are not likely to be mistakenly led to approve a drug that isn't really effective. But it also means that we might fail to approve a drug that really is effective. Generally, it is best to remove outliers and report them separately.

[3] This is not a standard term. You are not likely to find it in a statistics package. But it is a natural term because we've removed *part* of the variability.

side-by-side and the partial boxplots for *Distance* removing the variability due to *Hand:*

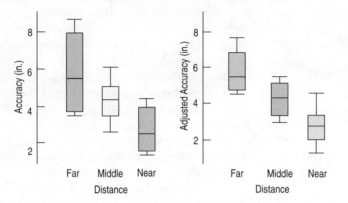

FIGURE 29.3
Partial boxplots after removing the effect of *Hand* (on the right) show the differences in *Accuracy* among the three distances more clearly than the original boxplots of *Accuracy* by *Distance* (on the left).

We'll use the partial boxplots for diagnostic purposes as we used the side-by-side boxplots in a one-way analysis. If the partial boxplots were all skewed in the same direction, or if their spreads changed systematically from level to level across a factor, we would have considered re-expressing the response variable to make them more symmetric. Re-expression would be likely to make the analysis more powerful and more correct.

Additivity

Each observation is assigned a treatment that combines two factor levels, one from each factor. Our model assumes that we can just *add* the effects of these two factor levels together:

$$y_{ijk} = \mu + Hand\ effect_j + Distance\ effect_k + Error_{ijk}.$$

That's a nice simple model, but it may not be a good one for our data. Just like linearity, **additivity** is an assumption. Like all assumptions, it helps us by making the model simpler, but it may be wrong for our data. Of course, we can't know for sure, but we can check the **Additive Enough Condition.** You won't be surprised that we check it with displays.

For the effects of *Hand* and *Distance* to be additive, changing hands must make the *same* difference in accuracy no matter what distance you throw from (or *vice versa*). Is this reasonable for this experiment? Alternatively, we could ask whether moving away from the target makes a greater difference to accuracy when throwing with the weaker hand. That's a *conditional* question, so we look at how *Accuracy* changes by *Distance* for each *Hand* separately.

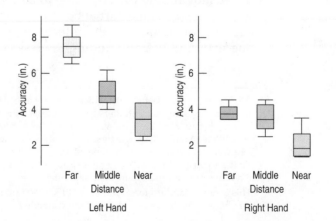

FIGURE 29.4
Boxplots of *Accuracy* by *Distance* conditional on each *Hand*. Changing the *Distance* seems to have a greater effect on left *Hand* accuracies.

The changes in accuracy due to *Distance* do look larger for the *left* hand, especially at the *far* distance.

When the effects of one factor *change* for different levels of another factor, we say that there's an **interaction**. To show the interaction, we often plot the treatment means in a single display, called an **interaction plot**. This plot shows the average of the observations at *each* level of one factor broken up by the levels of the other factor.

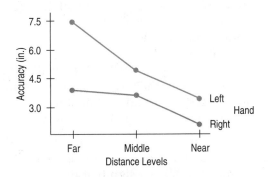

FIGURE 29.5
An interaction plot showing the mean accuracy at the six combinations of *Distance* and *Hand*. Connecting lines for right and left hands shows the greater effect of the far *Distance* on left-hand *Accuracy* than on right-hand *Accuracy*.

Here it's easier to see that throwing from the Far distance has more of an effect for the Left hand. If the effect of *Distance* were constant, the lines in this plot would be parallel. If the lines aren't parallel, that's evidence that the effects are not additive. But how parallel is enough? Are these lines sufficiently far from parallel for us to doubt whether our data satisfy the **Additive Enough Condition?** A graphical check like this is good enough to proceed. Later in the chapter, we'll see a way to include a term for non-additivity in our model and test for it.

Independence Assumptions

"Without randomization, your inferences haven't got a chance."

—George Cobb,
famous statistician

The **Independence Assumption** for the two-factor model is the same as in one-way ANOVA. That is, the observations *within each treatment group* must be independent of each other. As usual, no test can verify this assumption. Ideally, the data should be generated by a randomized comparative experiment or at least be drawn independently and at random from a homogeneous population. Without randomization of some sort, our P-values are meaningless. We have no sampling distribution models and no valid inference.

Check the **Randomization Condition.** Were the data collected with suitable randomization? For surveys, are the data for each group a representative random sample of that group? For experiments, was the experiment randomized?

We were told that the order of treatment conditions in the dart experiment was randomized. This is especially crucial for an experiment like this because of the possibility of either learning or tiring. Your accuracy might increase after a few throws as you warm up, but after an hour or two, your concentration—and your aim—might start to wander.

If we are satisfied with the Additivity and Independence Assumptions, then we can look at the ANOVA table. For the dart data, we might have some doubts about the additivity, but we'll go ahead for now.

		Analysis of Variance for Accuracy			
Source	df	Sum of Squares	Mean Square	F-Ratio	Prob
Distance	2	51.0439	25.5219	28.561	≤0.0001
Hand	1	39.600	39.6900	44.416	≤0.0001
Error	32	28.5950	0.893594		
Total	35	119.329			

These may look like small P-values, but inference requires additional assumptions. Before we believe the *F*-tests and use the P-values to make decisions, we have other assumptions to think about.

Equal Variance Assumption

Like the one-way ANOVA, the two-factor ANOVA requires that the variances of all treatment groups be equal. It's the residuals after fitting *both* effects that we'll pool for the Error Mean Square, so it's their variance whose stability we care about most. We check this assumption by checking the **Similar Variance Condition.**

In one-way ANOVA, we looked at the side-by-side boxplots of the groups to see whether they had roughly the same spread. But here, we need to check for equal spread across all treatment groups. The easiest way to check this is to plot the residuals from the two-way ANOVA model.

- Look at the residuals plotted against the predicted values. If the variance is constant, the plot should be patternless. If the plot thickens (to one side or the other), it's a sign that the variance is changing *systematically*. Then it's usually a good idea to consider re-expressing the response variable.

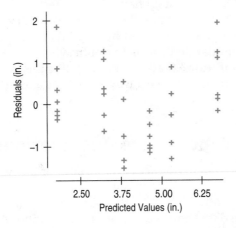

FIGURE 29.6
A scatterplot of residuals vs. predicted values from the two-way ANOVA model for *Accuracy* shows a U-shaped pattern. This suggests that a condition is violated.

This plot doesn't seem to show changing variance, but it's certainly not patternless. Residuals at both ends of the predicted values are higher than those in the middle. This could mean that something is missing from our model. Could that nagging suspicion we had about the additivity be resurfacing? Stay tuned.

- We can also plot the residuals grouped by each factor. Here's a boxplot of residuals by treatment level for each factor in the dart experiment:

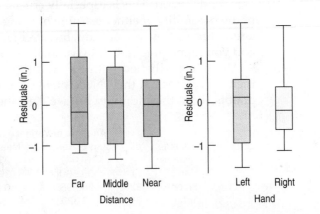

FIGURE 29.7
Boxplots of the residuals from the two-way ANOVA of the dart accuracy experiment show roughly equal variability when plotted for each factor.

Remember, we can't use the original boxplots to check changing spread. From those it seemed that the spread of the *near* level was smaller than that of the *far* level. But much of that difference was due to changes in the way the factor *Hand* behaves at different distances rather than a change in the underlying variance. By contrast, these boxplots of residuals look fine.

Normal Error Assumption

As with one-way ANOVA, the *F*-tests require that the underlying errors follow a Normal model. We'll check a corresponding **Nearly Normal Condition** with a Normal probability plot or histogram of the residuals.

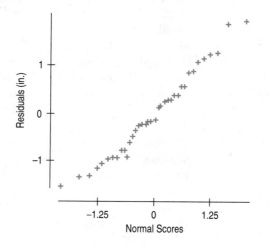

FIGURE 29.8
A Normal probability plot of the residuals from the two-way ANOVA model for dart accuracy seems reasonably straight.

This condition appears to be satisfied.

What Have We Learned so Far?

The ANOVA table confirmed what we saw in the boxplots. Both *Distance* and *Hand* affect accuracy of throwing a dart, at least at the levels chosen in this experiment. The small P-values give us confidence that the differences we see are due to the factors and not to chance. We did worry that the Additive Enough Condition might not be satisfied, and there seems to be something amiss in the plot of residuals versus predicted values. Before we examine that more, let's look at a different example with two factors, Step-By-Step.

STEP-BY-STEP EXAMPLE   Two-Factor Analysis of Variance

Another student, who prefers the great outdoors to damp pub basements, wonders whether leaving her tennis balls in the trunk of her car for several days after the can is opened affects their performance, especially in the winter when it can get quite cold. She also wonders if the more expensive brand might retain its bounce better. To investigate, she performed a two-factor experiment on *Brand* and *Temperature*, using two *Brands* and three levels of *Temperature*. She bounced three balls under each of the six treatment conditions by first randomly selecting a *Brand* and for that ball, randomly selecting whether to leave it at room temperature or to put it in the refrigerator or the freezer. After 8 hours she dropped the balls from a height of 1 meter to a concrete floor, recording the *Height* of the bounce in centimeters.

(continued)

This is a completely randomized replicated design in two factors. The factor *Brand* has two levels: *Premium* and *Standard*. The factor *Temperature* has three levels: *Room, Fridge,* and *Freezer*. The null hypotheses are that *Brand* has no effect and that *Temperature* has no effect.

WHO	Tennis balls of two *Brands* allocated to three *Temperatures*.	
WHAT	*Height* of the ball after it bounces from an initial height of 100 centimeters	
UNITS	Centimeters	
HOW	Two-factor experiment, randomized order of runs	

Brand	Temperature	Bounce Height
Standard	Freezer	37
Standard	Fridge	59
Standard	Room	59
Premium	Freezer	45
Premium	Fridge	60
Premium	Room	63
Standard	Freezer	37
Standard	Fridge	58
Standard	Room	60
Premium	Freezer	39
Premium	Fridge	64
Premium	Room	62
Standard	Freezer	41
Standard	Fridge	60
Standard	Room	61
Premium	Freezer	37
Premium	Fridge	63
Premium	Room	61

THINK

State what we want to know and the null hypotheses we wish to test. For two-factor ANOVA, the null hypotheses are that all the treatment groups have the same mean for each factor. The alternatives are that the means are not all equal.

I want to know whether the storage *Temperature* affects the mean *Bounce Height* of tennis balls. I also want to know whether the two *Brands* have different mean *Bounce Heights*. Writing τ_j for the effect of *Brand j*, the null hypothesis is that the two brand effects are the same:

$$H_0: \tau_P = \tau_S.$$

Writing γ_k for the effect of *Temperature* level k, then the other null hypothesis is

$$H_0: \gamma_{room} = \gamma_{fridge} = \gamma_{freezer}.$$

The alternative for the first hypothesis is that the effects of the two *Brands* on the *Bounce Height* are different. The alternative for the second is that the effects of the *Temperature* levels are not all equal.

Plot Examine the side-by-side partial boxplots of the data.

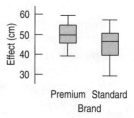

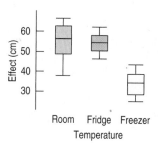

It's pretty clear that *Temperature* has an effect, but we can't be sure about the effect of *Brand* just from the plots.

✔ **Independence Assumption:** Because the order was randomized, I can assume the observations are independent. However, I might want to check to see if the balls of each brand came from the same box. It might be better to have selected balls of the same brand from different boxes.

✔ **Randomization Condition:** The experimenter performed the trials in random order.

✔ **Additive Enough Condition:** Here's the plot of the mean *Bounce Heights* at each treatment condition:

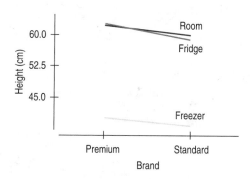

The lines look reasonably parallel. Now I can fit the additive model in two factors:

$$y_{ijk} = \mu + Brand\ effect_j + Temp\ effect_k + Error_{ijk}.$$

Under these conditions it's appropriate to fit the two-way Analysis of Variance model.

Plan Think about the assumptions and check the appropriate conditions.

Show the ANOVA table.

Analysis of Variance For Height

Source	DF	Sum of Squares	Mean Square	F-ratio	P-value
Temp	2	1849.33	924.665	209.55	<0.0001
Brand	1	26.8889	26.8889	6.0935	0.0271
Error	14	61.7778	4.41270		
Total	17	1938.00			

(continued)

The Error Mean Square is 4.41, which means that the common error standard deviation is estimated to be $\sqrt{4.41}$, a bit more than 2 centimeters. This seems like a reasonable accuracy for measuring *Bounce Heights* of tennis balls dropped from a height of 1 meter.

Before testing the hypotheses, I need to check some more conditions:

✔ **Similar Variance Condition:** A plot of residuals vs. predicted values shows some increased spread on the left, but with only 5 points it's hard to say.

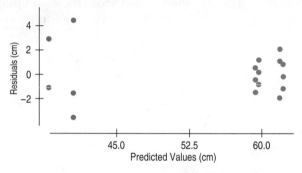

✔ **Nearly Normal Condition, Outlier Condition:** A Normal probability plot of the residuals looks OK too:

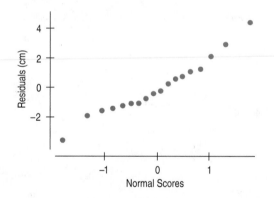

Under these conditions, it is appropriate to perform and interpret the F-test.

 Mechanics Discuss the tests.

Display the effects for each level of the significant factors. In the next section, we'll show how they're calculated. Remember, significance does not guarantee that the differences are meaningful.

The F-ratio for *Brand* is 6.09, with a P-value of 0.0271. The F-ratio for *Temperature* is 209.55, with a P-value less than 0.0001.

The overall mean *Bounce Height* is 53.67 cm.

The *Brand* effects are

Premium	1.22
Standard	−1.22

The *Temperature* effects are

Room	7.33
Fridge	7.00
Freezer	−14.33

Interpretation Tell what the *F*-tests mean.

F-ratios this large would be very unlikely if the factors had no effect. I conclude that the means of the two *Brand* levels are not equal. I also conclude that the means of the three *Temperature* levels are not all equal.

Overall, the premium brand bounces about 2.44 cm higher than the standard brand, averaging across the temperature ranges. Generally, there seems to be little difference between the average *Bounce Height* for room and fridge temperatures, but tennis balls from the freezer bounce almost 22 cm less on average.

Conclusions: It appears that the premium brand outperforms the standard brand overall. It looks from the plots like our student needn't be too worried about playing tennis on a cold day, but if the temperature drops below freezing, she should try to warm the tennis balls first.[4]

FOR EXAMPLE

RECAP: In an experiment to investigate the effects of paper airplane design on flight distance two factors were tested. Here are boxplots of distance by each of the factors:

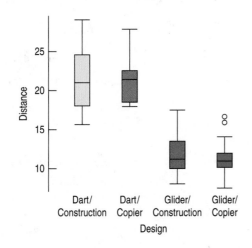
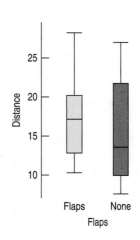

QUESTION: What can we learn from the boxplots?

It appears that the Darts travel farther than the Gliders. There does not appear to be an effect for flaps. The spreads of all the groups are comparable. There are two observations plotted as outliers for the Glider/Copier design, but they don't seem to be very extreme and are probably not a problem.

---

[4] It's quite obvious from the plots and the tables of effects that there's a difference between *freezer* and the other temperatures, but little difference between *room* and *fridge*. A Bonferroni test would confirm this.

*How ANOVA Works—the Gory Details

You'll probably never carry out a two-way ANOVA by hand. But to understand the model it can help to see exactly how it decomposes each observation into components. Our tennis ball model says

$$y_{ijk} = \mu + Brand\ effect_j + Temp\ effect_k + Error_{ijk}.$$

To apply this model to data, we need to *estimate* each of these quantities:

$$y_{ijk} = \overline{\overline{y}} + \widehat{Brand\ effect_j} + \widehat{Temp\ effect_k} + Residual_{ijk}.$$

Here are the data again, displayed a little differently:

		Temperature		
		Room	Fridge	Freezer
Brand	Premium	63	60	45
		62	64	39
		61	63	37
	Standard	59	59	37
		60	58	37
		61	60	41

The first term in the model is μ, the overall mean. It's reasonable to estimate that with the grand mean, $\overline{\overline{y}}$, which is just the mean of all the observations. For the tennis balls, the average bounce was 53.67 cm. The grand mean has the same value for every observation, which we can lay out as a table like this:

		Temperature		
		Room	Fridge	Freezer
Brand	Premium	53.67	53.67	53.67
		53.67	53.67	53.67
		53.67	53.67	53.67
	Standard	53.67	53.67	53.67
		53.67	53.67	53.67
		53.67	53.67	53.67

Now we turn to the *Brand* effects. To estimate the effect of, say, the *premium Brand*, we find the mean of all *premium* balls and subtract the grand mean. That's 55.89 cm − 53.67 cm = 1.22 cm. In other words, the *premium Brand* balls bounced an average of 1.22 cm higher than the overall average. The *standard Brand* balls bounced an average of 51.45 cm, so their effect is 51.45 cm − 53.67 cm = −1.22 cm. Of course, with only two levels, the fact that these two effects are equal and opposite is no coincidence. Effects *always sum to* 0.

We can put the *Brand* effects into a table like this:

		Temperature		
		Room	Fridge	Freezer
Brand	Premium	1.22	1.22	1.22
		1.22	1.22	1.22
		1.22	1.22	1.22
	Standard	−1.22	−1.22	−1.22
		−1.22	−1.22	−1.22
		−1.22	−1.22	−1.22

We can find a *Temperature* effect table in a similar way. We take the treatment average at each level of *Temperature* and subtract the grand mean. Here's the table:

		Temperature		
		Room	Fridge	Freezer
Brand	Premium	7.33	7.00	−14.33
		7.33	7.00	−14.33
		7.33	7.00	−14.33
	Standard	7.33	7.00	−14.33
		7.33	7.00	−14.33
		7.33	7.00	−14.33

We're missing only one table, the *Residuals.* Finding the residuals is easy, too. The additive model represents each observation as the sum of these three components. As always, the residuals are the differences between the observed values and those given by the model. Look at the first ball in the table. It was a *premium Brand* ball at *room Temperature.* From the effects, we'd predict it to bounce 53.67 (grand mean) + 1.22 (from being premium) + 7.33 (from being at room temperature) = 62.22 cm. Because it actually bounced 63 cm, its residual is 63 − 62.22 = 0.78 cm. If we compute all the residuals this way (or have the computer do it), we'd find the residual table to be

		Temperature		
		Room	Fridge	Freezer
Brand	Premium	0.78	−1.89	4.44
		−0.22	2.11	−1.56
		−1.22	1.11	−3.56
	Standard	−0.78	−0.44	−1.11
		0.22	−1.44	−1.11
		1.22	0.56	2.89

What we have done with all these tables is to decompose each original bounce height into the three estimated components plus a residual:

Bounce Height = mean Bounce Height + Brand effect + Temperature effect + Residual.

These tables actually display the workings of the ANOVA. Each term in the ANOVA model corresponds to one of these tables. Because the overall mean doesn't vary, we usually leave it out of the ANOVA table. Each Sum of Squares term in the ANOVA is simply the sum of the squares of all the values in its table.

The degrees of freedom are easy to find too. Look at the grand mean table. Because it has the same value for all observations, it has 1 degree of freedom. The *Brand* effect table seems to have 2 degrees of freedom, but we know that the two effects always sum to 0. So it really has only 1. In general, the degrees of freedom for a factor is one less than its number of levels. So *Temperature* has 2 degrees of freedom.

The original table has 18 observations and so 18 degrees of freedom. But what about the residuals? Remarkably enough, the degrees of freedom from the tables add. That means that we can find the degrees of freedom for the residuals by subtraction. They must have $18 - (1 + 1 + 2) = 14$ degrees of freedom.

The mean squares are just the sum of squares for each effect divided by the degrees of freedom. The *F*-ratios are the ratios of each Treatment Mean Square to the Error Mean Square. You can check the degrees of freedom and the mean squares in the step-by-step ANOVA table. The original table has 18 observations and so 18 degrees of freedom, but most programs choose not to display the overall mean and subtract out the degree of freedom associated with it to give what's sometimes called a "corrected" total or just "total." That's why our ANOVA table has only 17 degrees of freedom in "total."

Blocks

Back in Chapter 13 we thought about a *randomized block design*.[5] We imagined obtaining tomato plants from two stores and randomly assigning plants from each store to one of three fertilizer treatments. We can view the *Store* as a factor in this design. We now have a two-factor model, with *Store* and *Fertilizer* as factors. The ANOVA looks just the same, but now we're not likely to be interested in testing the *Store* effect. We *assumed* that the plants might be different from the different stores—that's why we used a blocked design.

The two-way ANOVA takes advantage of the blocking to remove the unwanted variation due to *Store*. As in a completely randomized two-factor design, removing the unwanted variation makes the effects of the other factor easier to see.

We wouldn't call a blocked design *completely randomized*, because we can't assign the plants at random to which store they come from. We must randomize *within* the blocks when we assign plants to levels of the *Fertilizer* factor.

Interactions

Up to now, we have assumed that whatever effects the two factors have on the response can be modeled by simply adding the separate effects together. What if that's not good enough? In the dart-throwing experiment, we worried that the experimenter's accuracy when throwing from farther away may deteriorate *even more* when he uses his nondominant hand. Looking at the interaction plot added evidence to that suspicion.

---

[5]You may have noticed in the tennis ball example that balls couldn't be randomly assigned to a *Brand*. *Brand* is actually a blocking factor, and the experiment is a randomized block design. But it doesn't change the analysis at all.

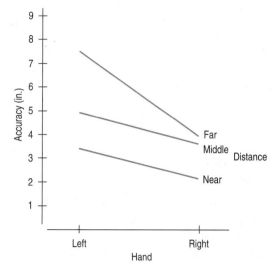

FIGURE 29.9
Interaction plot of *Accuracy* by *Hand* and *Distance*. The effect of *Distance* appears to be greater for the left hand.

It doesn't look like the effect of *Distance* is constant for the two hands. In particular, moving from *middle* to *far* has a much larger effect when using the *left Hand*. It's not unusual for the effect of a factor's level to depend on the level of the other factor. Unfortunately, our model doesn't take this into account. So, we need a new model.

When the effect of one factor *changes* depending on the levels of the other factor, the factors are said to *interact*. We can model this interaction by adding a term to our model and testing whether the adjustment is worthwhile with an *F*-test.

We include a new term to our model, adding an interaction effect at every *combination* of the levels of the other two factors. The model can now be written:

$$y_{ijk} = \mu + \tau_j + \gamma_k + \omega_{jk} + \varepsilon_{ijk}.$$

The new term, ω_{jk}, represents the *interaction* effect at level *j* of factor 1 and level *k* of factor 2.

We add a corresponding line to the ANOVA table. Of course, we'll have to spend some degrees of freedom on this new term. We'll have to take them from the error term where we parked all the leftover degrees of freedom after fitting the two factors. We'll need the *product* of each term's degrees of freedom. For the darts, this is only 2 degrees of freedom (1 from *Hand* and 2 from *Distance*), but for factors with more levels, this could get expensive.

Here's the ANOVA table for the dart experiment:

ANOVA Table for Accuracy

Source	DF	Sum of Squares	Mean Square	F-ratio	P-value
Distance	2	51.044	25.522	41.977	<0.0001
Hand	1	39.690	39.690	65.28	<0.0001
Distance × Hand	2	10.355	5.178	8.516	0.0012
Error	30	18.240	0.608		
Total	35	119.329			

What has changed? Nearly everything. (Of course, the Total Sum of Squares remains the same because it was calculated from the observations minus the overall mean, and the overall mean is still the same.) The interaction term is significant. Because we've now removed a significant source of variability, our Error Mean Square is smaller in spite of the loss of the 2 degrees of freedom devoted to the interaction term.

FOR EXAMPLE

RECAP: An experiment on paper airplanes considered two factors; *Design* and *Flaps*. Here is the two-factor ANOVA:

Source	df	Sum of Squares	Mean Square	F-ratio	P-value
Design	3	1549.50	516.499	56.610	≤0.0001
Flap	1	27.0140	27.0140	2.9608	0.0905
Error	59	538.303	9.12378		
Total	63	2114.81			

State and test the null hypotheses. What can you conclude?

The first null hypothesis is that the mean distance flown will be the same for each design, after allowing for any effects of flaps. That hypothesis is rejected. As we saw in the boxplots, the darts fly farther.

The second null hypothesis is that the mean distance will be the same whether or not flaps are present, after allowing for the differences due to design. The P-value for that factor is too large to allow us to reject the null hypothesis.

Inference When Variables Are Related

The *F*-test for the interaction leads us to reject the null hypothesis of no interaction effect and indicates that the effect of *Distance* really does change depending on which *Hand* is used. (Or, equivalently the effect of *Hand* changes depending on the *Distance*.)

What about the main effects? When a significant interaction term is present, it's difficult to *Tell* about the main effect of each factor. How much does switching from *right* hand to *left* hand matter? It depends on the *Distance*. It no longer makes sense to talk about the main effects of each factor separately. The main effects are just averages of the effects over all the levels of the other factor. As we saw with Simpson's paradox (back in Chapter 3), it can be misleading to average over very different situations.

When a significant interaction is present, the best way to display the results is with an interaction plot.

Now that we have the interaction term in our model, we recalculate the residuals. They're the values that are left over after accounting for the overall mean, the effect of each factor, and the interaction effect. Here's a plot of residuals versus predicted values for the two-factor model with interaction:

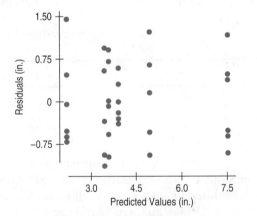

FIGURE 29.10
Residuals for the two-way ANOVA of dart accuracy with an interaction term included. Now, unlike Figure 29.6, there is no U-shaped pattern.

How has the interaction term helped? These residuals have less structure. Fitting the interaction term succeeded in removing structure from the error. This new model seems to satisfy the assumptions more successfully, and so our inferences are likely to be closer to the truth.

Why Not Always Start by Including an Interaction Term?

With two factors in the model, there's always the possibility that they interact. We checked the Additive Enough Condition with an interaction plot. Now we know we can include the interaction term and use the F-test for added information. If the interaction plot shows parallel lines and the F-test for the interaction term is not significant, you can comfortably proceed with a two-way ANOVA model without interaction. This will put back into the error term the degrees of freedom you used for the interaction.

On the other hand, if the interaction is significant, the model *must* contain more than just the main effects. In this case, the interpretation of the analysis depends crucially on the interaction term, and the interaction plot shows most of what to *Tell* about the data. When there's a significant interaction term, the effect of each factor depends on the level of the other factor, so it may not make sense to talk about the effects of a factor by itself. However, we may still be able to talk about the main effects. For example, we can see that the *right* hand is better *on average* than the *left* at any distance, and the closer you are to the target, the better the accuracy, no matter which hand you use. So both main effects are significant, and we can discuss them. A significant interaction effect tells us more. It says that *how much* better the performance is at near distances *depends* on which hand you use.

But if the lines in the interaction plot cross, you need to be careful. Here's an interaction plot from an experiment on fuels and alternative engine designs:

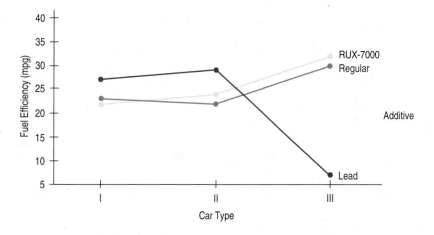

FIGURE 29.11
An interaction plot of gas mileage by *Car Type* and *Gas Additive*. If the lines cross, as they do here, be careful of interpreting the main effects.

It looks like the fuel with lead is great for two of the engine designs, but a disaster for the third. Suppose someone asked you to summarize the effect of adding lead. Nothing simpler than saying it *depends* on the engine type makes sense. *On average*, it lowers gas mileage, but is that relevant? The real message is that sometimes it improves mileage and sometimes it hurts. You won't want to *Tell anything* about the main effect. Its F-test is irrelevant.

Whether the lines cross or not, including an interaction term in our model is always a good way to start. But sometimes we just can't. Suppose we had run the dart experiment with only one dart at each treatment. A design like that is called an **unreplicated two-factor design.** Imagine that the throw at the combination of left hand and far distance missed the bull's-eye by a mile. Would this be evidence of an interaction between *Distance* and *Hand*, or was it just a bad throw? How could we tell? The obvious thing to do would be to *repeat* the treatment. If accuracy was consistently poor at this combination of factors, we'd want to model the interaction. If the other throws were no worse than we'd expect from the two main effects, then we might set the outlier aside. We can't distinguish an interaction effect from error unless we replicate the experiment. In fact, the design tells us that. Without replication, if we try to fit an interaction term, there are exactly 0 degrees of freedom left for the error. That makes any inference or testing impossible. If we are willing to *assume* that the additive model is adequate, then we can reduce the number of runs required and use an unreplicated design. In that case, although the residuals are indistinguishable from the interaction effects, the residual plot may help reveal whether the assumption made sense.

When you do replicate, it's best to replicate *all* treatment conditions equally. (Unbalanced and other more complicated designs can be analyzed, but they are beyond the scope of this book.)

The student conducting the paper airplane experiment decides to do a follow-up experiment to investigate two other factors in more detail. Using his favorite design, he uses one of three different size weights and places the weight either on the nose or on the tail of the plane. At each combination, he performs four test flights (randomizing the order of the flights). An interaction plot looks like this:

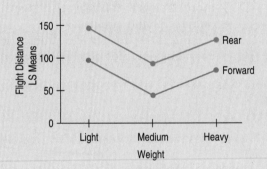

The ANOVA table shows:

			Analysis of Variance		
Source	DF	Sum of Squares	Mean Square	F-Ratio	P-value
Position	1	13371.9314	13371.9314	53.3868719	<0.0001
Weight	2	12625.5486	6312.7743	25.2034851	<0.0001
Position × Weight	2	10.6208	5.3104	0.02120155	0.9790
Error	18	4508.501	250.47		
Total	23	30516.602			

1. What are the null hypotheses?

2. From the interaction plot, do the effects appear to be additive?

3. Does the *F*-test in the ANOVA table support your conclusion in (2)?

4. Does the position of the weight seem to matter?

5. Does the amount of weight appear to affect flight distance?

6. What would you recommend to the student to increase flight distance?

STEP-BY-STEP EXAMPLE   Two-Factor ANOVA with Interaction

In Chapter 28 we looked at how much TV four groups of students watched on average. Let's look at their grade point averages (GPAs). Back in that chapter, we treated the four groups (male athletes, female athletes, male non-athletes, and female non-athletes) as four levels of the factor *Group*. Now we recognize that there are really two factors: the factor *Sex* with levels *male* and *female* and the factor *Varsity* with levels *yes* and *no*. Let's analyze the GPA data with a two-factor ANOVA.

As with many Social Science studies, this isn't an experiment. (Some fans might object if we assigned students at random to varsity teams.[6]) Instead, it is an application of ANOVA to an observational study.

THINK

State what we want to know and the null hypotheses we wish to test. For two-factor ANOVA with interaction, the null hypotheses are that all the treatment groups have the same mean for each factor and that the interaction effect is 0. The alternatives are that at least one effect is not 0.

I want to know whether the mean GPA is the same for men and women, whether the mean GPA is the same for athletes and non-athletes, and whether the factors interact. Writing τ_j for the effect of Sex level j, the null hypothesis is that the effects are the same:

$$H_0: \tau_M = \tau_F$$

Writing γ_k for the effect of Varsity level k, the other null hypothesis is

$$H_0: \gamma_Y = \gamma_N$$

Finally, the null hypothesis for interaction is that the interaction terms are all 0.

The alternative for the first hypothesis is that the mean GPA is different for men and women. The alternative for the second is that the mean GPA is different for Varsity athletes than for other students. The alternative for the last is that there is an interaction effect.

Plot Examine the side-by-side partial boxplots of the data.

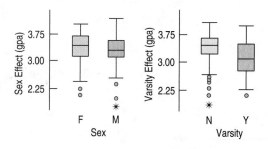

Plan Think about the assumptions and check the conditions.

It appears that there is no difference in mean GPA between men and women (after accounting for varsity status), but there may be a difference between varsity athletes and non-athletes (after accounting for sex).

(continued)

---

[6]And randomly assigning the *other* factor would be problematic.

The study is a random sample, so students' GPAs should be independent of each other.

✔ **Independence Assumption, Randomization Condition:** The study was based on a random sample of students.

✔ **Additive Enough Condition:** Here's the interaction plot of the means at each treatment condition:

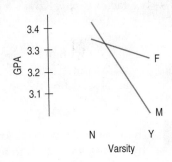

The lines cross, so I'll add an interaction term to the model and fit the two-way ANOVA with interaction:

$$y_{ijk} = \mu + Sex\ effect_j + Varsity\ effect_k + Interaction_{jk} + Error_{ijk}.$$

Show the ANOVA table.

Analysis of Variance for gpa

Source	DF	Sum of Squares	Mean Square	F-ratio	P-value
Sex	1	0.3040	0.3040	1.7681	0.1852
Varsity	1	2.4345	2.4345	14.1871	0.0002
Sex × Varsity	1	1.0678	1.0678	6.2226	0.0134
Error	196	33.6397	0.1716		
Total	199	37.4898			

✔ **Similar Variance Condition:** A plot of residuals vs. predicted shows no thickening or other patterns:

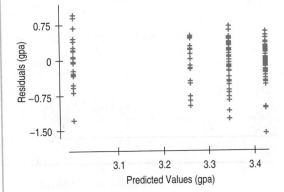

✔ **Nearly Normal Condition, Outlier Condition:** The Normal probability plot of the residuals is reasonable as well.

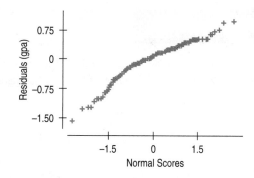

Under these conditions it is appropriate to interpret the F-tests and their P-values.

Mechanics	The error mean square is 0.172 points, which means that the common error standard deviation is estimated to be $\sqrt{0.172} = 0.415$ points (nearly half a letter grade).
	The F-ratio for the interaction term is 6.223, with a P-value of 0.0134. Based on this, I reject the hypothesis of no interaction. Because there is interaction, I am cautious in testing the main effects, but will examine the means further to understand the factor effects.
	Overall mean GPA = 3.32.

Level of Varsity	Effect
N	0.1281
Y	−0.1281

The main effect of Varsity is significant. Varsity athletes have lower GPA by about a quarter of a grade point averaged over all, but this effect is much greater for men than for women. Here are the four group mean GPAs:

Show the table of means.

Group	Mean GPA
Female, non-athlete	3.35
Female, varsity	3.26
Male, non-athlete	3.43
Male, varsity	3.00

TELL ▷ **Interpretation** Tell what the *F*-test means.

The interaction term is significant, and it appears that most of the effect we see is due to a difference of 0.43 in mean GPAs between male athletes and non-athletes. It doesn't look like any of the other differences is large enough to be important. Probably none of them is statistically significant either.

FOR EXAMPLE

RECAP: The ANOVA for paper airplane distance on two factors, *Design* and *Flaps*, showed a significant effect for *Design*, but not for *Flaps*. But maybe there is an interaction. Here is the ANOVA with the interaction term and an interaction plot.

Source	df	Sum of Squares	Mean Square	F-ratio	P-value
Design	3	1549.50	516.499	81.065	≤ 0.0001
Flaps	1	27.0140	27.0140	4.2399	0.0441
Design × Flaps	3	181.503	60.5009	9.4956	≤ 0.0001
Error	56	356.800	6.37144		
Total	63	2114.81			

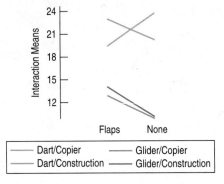

| Dart/Copier | Glider/Copier |
| Dart/Construction | Glider/Construction |

QUESTION: Interpret the interactions. What do they say about paper airplane design?

Including the interaction term reveals that the *Flaps* factor is significant. But it matters in different ways for different designs. Flaps help improve the distance of both kinds of gliders. But flaps hinder the performance of darts made of lighter copier paper, although they help darts made of construction paper.

What Next?

This chapter introduces some of the issues that arise in advanced experimental design and analysis. We have scratched the surface of this large and important area by showing the analysis of a balanced two-way design with interaction.

Of course, there's no reason to stop with two factors. And if we have more than two, there are many ways in which they may interact. As you can imagine, real-life experiments and observational studies may pose many interesting challenges. Anyone who is serious about analyzing or designing multifactor studies should consult a professional statistician. You now know the vocabulary and principles, so you are well equipped to do that.

What Can Go Wrong?

- **Beware of unreplicated designs unless you are sure there is no interaction.** Without replicating the experiment for each treatment combination, there is no way to distinguish the interaction terms from the residuals. If you are designing a two-factor experiment, you must be willing to assume that there is no interaction if you choose not to replicate. In such a case, you can fit an additive model only in the two factors. If there is an interaction, it will show up in the error term. You should examine a residual plot to help reveal a possible interaction effect. You must be prepared to defend the assumption of no interaction and the decision not to replicate.

- **Don't attempt to fit an interaction term to an unreplicated two-factor design.** If you have an unreplicated two-factor experiment or observational study, you'll find that if you try to fit an interaction term, you'll get a strange ANOVA table. The design exhausts the degrees of freedom for error, so fitting the interactions leaves no degrees of freedom for residuals. That wipes out the mean square errors, F-ratios, and P-values, which may appear in the computer output as dots, dashes, or some other indication of things gone wrong. Remove the interaction term from the model and try again.

- **Be sure to fit an interaction term when it exists.** When the design is replicated, it is always a good idea to fit an interaction term. If it turns out not to be statistically significant, you can then fit a simpler two-factor main effects model instead.

- **When the interaction effect is significant, don't interpret the main-effects.** Main effects can be very misleading in the presence of interaction terms. Look at this interaction plot.

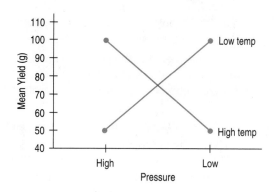

FIGURE 29.12

An interaction plot of *Yield* by *Temperature* and *Pressure*. The main effects are misleading. There is no (main) effect of *Pressure* because the average *Yield* at the two pressures is the same. That doesn't mean that *Pressure* has no effect on the *Yield*. In the presence of an interaction effect, be careful when interpreting the main effects.

The experiment was run at two temperatures and two pressure levels. High amounts of material were produced at high pressure with high temperature and at low pressure with low temperature. What's the effect of *Temperature*? Of *Pressure*? Both main effects are 0, but it would be silly (and wrong) to say that neither *Temperature* nor *Pressure* was important. The real story is in the interaction.

- **Always check for outliers.** As in any analysis, outliers can distort your conclusions. An outlier can inflate the Error Mean Square so much that it may be hard to discern any effect, whether it exists or not. Use the partial boxplots to search for outliers. Consider setting outliers aside and re-analyzing the results. An outlier can make an interaction term appear significant. For example, a single male varsity athlete with a very low GPA could account for the results we saw.

- **Check for skewness.** If the underlying data distributions are skewed, you should consider a transformation to make them more symmetric.

- **Beware of unbalanced designs and designs with empty cells.** We've been assuming that the data are *balanced* over the design—that we have equal numbers of observations in each cell. There are methods that will easily compensate for small amounts of imbalance, but empty cells and other more serious violations of balance require different methods and additional assumptions.

CONNECTIONS

We first discussed designing experiments in Chapter 13. Now we know how to analyze a two-factor design. In Chapter 28, we saw that the one-way ANOVA generalized the pooled *t*-test to more than two levels of one factor. Now we've added a second factor.

Of course, we're still relying on boxplots, Normal probability plots, and scatterplots of residuals to help us check conditions and understand relationships. What we first said we'd look for in these displays back in Chapters 4, 5, 6, and 7 is still what we're concerned with here.

WHAT HAVE WE LEARNED?

In Chapter 28, we learned that the Analysis of Variance is a natural way to extend the *t*-test for testing the means of two groups to compare several groups. Often, those groups are different levels of a single factor. Now we've learned that we can extend the Analysis of Variance to designs with more than one factor. We've learned that partial boxplots are a good way to examine the effect of each factor on the response. We've seen that separate tests of hypotheses can be carried out for each of the factors, testing whether the factor has any discernible effect on the response across any of the levels tested.

We've also learned that sometimes factors can interact with each other. When we have at least two observations at each combination of factor levels, we can add an interaction term to our model to account for the possible interaction.

Of course, we still need to check the appropriate assumptions and conditions as we did for simple ANOVA. All the methods we learned for displaying data in the early chapters are still important. And we've begun to see the deep connection between the design concepts we learned in Chapter 13, the exploratory methods we learned in the early chapters, and the inference methods of the latter part of the book.

Terms

Two-way ANOVA model (main effects model)	The model for a two-way (one response, two factors) ANOVA is $$y_{ijk} = \mu + \tau_j + \gamma_k + \varepsilon_{ijk},$$ where τ_j represents the effect of (level *j*) of factor one and γ_k represents the effect of (level *k*) of factor two. The subscript *i* designates the *i*th replication of the treatment combination (p. 752).
Partial boxplot	Partial boxplots are found by plotting the response at each of the levels of one factor with the effects of the second factor removed. A simple way to generate partial boxplots of factor A is to first save the residuals from a one-way ANOVA of the response on factor B. Add the overall mean to the residuals and make boxplots of the resulting data against factor A (p. 753).
Additivity, additive model	A model in two factors is additive if it simply adds the effects of the two factors and doesn't include an interaction term (p. 754).
Interaction	When the effects of the levels of one factor change depending on the level of the other factor, the two factors are said to interact. When interaction terms are present, it is misleading to talk about the main effect of one factor because how large it is *depends* on the level of the other factor (p. 755).
Interaction plot	A plot that shows the means at each treatment combination, highlighting the factor effects and their behavior at all the combinations (p. 755).

Assumptions for the two-way ANOVA model (and conditions to check)	▶ **Additivity Assumption:** We assume that we can *add* the effects of the two factors together. (Check the **Additive Enough Condition** that the effects of one factor are about the same across all levels of the other factor. An interaction plot should show roughly parallel lines.) Of course, this assumption (and associated condition) is not required when an interaction term is included in the model (p. 754).

▶ **Independence Assumptions:** The observations within each treatment group must be independent of each other. (Think about the nature of the data and the **Randomization Condition**. Check a residuals plot.) (p. 755)

▶ **Equal Variance Assumption:** (Check the **Similar Variance Condition** by looking at side-by-side partial boxplots of the groups and a plot of residuals against predicted values. A common problem is *increasing* spread with increasing predicted values—the *plot thickens!*) (p. 756)

▶ **Normal Error Assumption:** (Check the **Nearly Normal Condition** by looking at a Normal probability plot of the residuals.) (p. 757)

Two-way ANOVA model with interaction (main effects model)

For a replicated two-way design, an interaction term can also be fit. The resulting model is

$$y_{ijk} = \mu + \tau_j + \gamma_k + \omega_{jk} + \varepsilon_{ijk},$$

where τ_j represents the effect of (level j) of factor 1 and γ_k represents the effect of (level k) of factor 2. Now ω_{jk} represents the effect of the interaction at levels j of factor one and k of factor two. The subscript i designates the ith replication of the treatment combination (p. 765).

Skills

▶ Understand the advantages of an experiment in two factors.

▶ Know how to set up an additive model in two factors.

▶ Know how to examine the Additivity Assumption and when to consider an interaction term.

▶ Know how to make partial boxplots.

▶ Be able to use a statistics package to compute a two-way ANOVA.

▶ Know how to make an interaction plot for replicated data.

▶ Be able to interpret main effects in a two-way ANOVA.

▶ Be able to use an interaction plot to explain an interaction effect.

▶ Be able to distinguish when a discussion of main effects is appropriate in the presence of a significant interaction.

TWO-WAY ANOVA ON THE COMPUTER

Some statistics packages distinguish between models with one factor and those with two or more factors. You must be alert to these differences when analyzing a two-factor ANOVA. It's not unusual to find ANOVA models in several different places in the same package.

Usually, you must specify the interaction term yourself. That's because these features of statistics packages typically are designed for models with three or more factors where the number of interactions can explode unless the data analyst carefully selects the interactions to be included in the model.

(continued)

DATA DESK

- Select the response variable as Y and the factor variable as X.
- From the **Calc** menu, choose **ANOVA > ANOVA** or **ANOVA > ANOVA with Interactions.**
- Data Desk displays the ANOVA table.
- Select plots of residuals from the ANOVA table's HyperView menu.

COMMENTS

Data Desk expects data in "stacked" format. You can change the ANOVA by dragging the icon of another variable over either the Y or the X variable name in the table and dropping it there. The analysis will recompute automatically.

EXCEL

- The Excel Data Analysis Add-in offers a two-way ANOVA "with and without replication."

COMMENTS

Excel requires that the data be in a special format and cannot deal with unbalanced data, estimate interactions, or make boxplots.

JMP

- From the **Analyze** menu select **Fit Model.**
- Select variables: and **Add** them to the **Construct Model Effects** box.
- To specify an interaction, select both factors and press the **Cross** button.
- Click **Run Model.**
- JMP opens a **Fit Least Squares** window.
- Click on the red triangle beside each effect to see the means plots for that factor. For the interaction term, this is the interaction plot.
- Consult the JMP documentation for information about other features.

COMMENTS

JMP expects data in "stacked" format with one continuous response and two nominal factor variables.

MINITAB

- Choose **ANOVA** from the **Stat** menu.
- Choose **Two-way...** from the **ANOVA** submenu.
- In the Two-way ANOVA dialog, assign a quantitative Y variable to the Response box and assign the categorical X factors to the Factor box.
- Specify interactions.
- Check the **Store Residuals** check box.
- Click the **Graphs** button.
- In the ANOVA-Graphs dialog, select **Standardized residuals,** and check **Normal plot of residuals** and **Residuals versus fits.**

COMMENTS

Minitab expects "stacked" format data for two-way designs.

SPSS

- Choose **Analyze > General Linear Model > Univariate.**
- Assign the response variable to the **Dependent Variable** box.
- Assign the two factors to the **Fixed Factor(s)** box. This will fit the model with interactions by default.
- To omit interactions, click on **Model.** Select **Custom.** Highlight the factors. Select **Main Effects** under the **Build Terms** arrow and click the arrow.
- Click **Continue** and **OK** to compute the model.

TI-83/84 PLUS

COMMENTS

You need a special program to compute a two-way ANOVA on the TI-83.

TI-89

Under **STAT Tests** *choose* **D:ANOVA2-Way.**

To compute the analysis for a complete 2-way design (with interaction):

- *Enter the data into lists (one for each level of factor A the column factor) with data for each level of factor B (row factor) grouped together.*

- *Select* **2 Factor, Eq Reps** *for the Design, and enter the number of column and row factors.*

- *Specify the lists containing the column factors, and the number of row factor levels. Press* $\div$ *to perform the calculations.*

COMMENTS

- The calculator cannot fit a model without interactions unless the second factor is considered as blocking, in which each row of the input columns is considered as a block.

- The calculator gives only the components of the ANOVA table as output. It does not give residuals or factor level means.

EXERCISES

1. **Popcorn revisited.** A student runs a two-factor experiment to test how microwave power and temperature affect popping. She chooses 3 levels of *Power* (*low, medium,* and *high*) and 3 *Times* (*3 minutes, 4 minutes,* and *5 minutes*), running one bag at each condition. She counts the number of uncooked kernels as the response variable.
   a) What are the null and alternative hypotheses for the main effects?
   b) How many degrees of freedom does each factor sum of squares have? How about the error sum of squares?
   c) Should she consider fitting an interaction term to the model? Why or why not?

2. **Gas mileage revisited.** A student runs an experiment to study the effect of *Tire Pressure* and *Acceleration* on gas mileage. He devises a system so that his Jeep Wagoneer uses gasoline from a one-liter container. He uses 3 levels of *Tire Pressure* (*low, medium,* and *full*) and 2 levels of *Acceleration,* either holding the pedal *steady* or *pumping* it every few seconds. He randomizes the trials, performing 4 runs under each treatment condition, carefully recording the number of miles he can go in his Jeep Wagoneer on one liter of gas.
   a) What are the null and alternative hypotheses for the main effects?
   b) How many degrees of freedom does each treatment sum of squares have? How about the error sum of squares?
   c) Should he consider fitting an interaction term to the model? Why might it be a good idea?
   d) If he fits an interaction term, how many degrees of freedom would it have?

3. **Popcorn again.** Refer to the experiment in Exercise 1. After collecting her data and analyzing the results, the student reports that the *F*-ratio for *Power* is 13.56 and the *F*-ratio for *Time* is 9.36.
   a) What are the P-values?
   b) What would you conclude?
   c) What else about the data would you like to see in order to check the assumptions and conditions?

4. **Gas mileage again.** Refer to the experiment in Exercise 2. After analyzing his data the student reports that the *F*-ratio for *Tire Pressure* is 4.29 with a P-value of 0.030, the *F*-ratio for *Acceleration* is 2.35 with a P-value of 0.143, and the *F*-ratio for the *Interaction effect* is 1.54 with a P-value of 0.241.
   a) What would you conclude?
   b) What else about the data would you like to see in order to check the assumptions and conditions?
   c) If your conclusion about the *Acceleration* factor in part a is wrong, what type of error have you made?

5. **Crash analysis.** The National Highway Transportation Safety Administration runs crash tests in which stock automobiles are crashed into a wall at 35 mph with dummies in both the passenger and the driver's seats. The THOR Alpha crash dummy is capable of recording 134 channels of data on the impact of the crash at various sites on the dummy. In this test 335 cars are crashed. The response variable is a measure of head injury. Researchers want to know if which seat the dummy is in affects head injury severity, as well as whether the type of car affects severity.

Here are partial boxplots for the 2 different *Seat* (*driver, passenger*) and the 6 different *Size* classifications (*compact, light, medium, mini, pickup, van*):

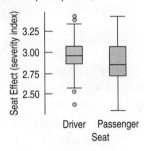

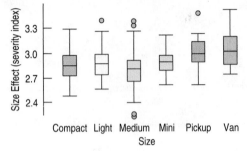

An interaction plot shows:

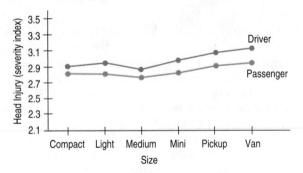

A scatterplot of residuals vs. predicted values shows:

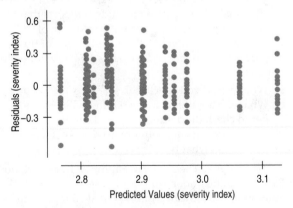

The ANOVA table follows:

Analysis of Variance for Head Severity: 293 cases

Source	DF	Sum of Squares	Mean Square	F-ratio	P-value
Seat	1	0.88713	0.88713	25.501	<0.0001
Size	5	1.49253	0.29851	8.581	<0.0001
Seat × Size	5	0.07224	0.01445	0.415	0.838
Error	282	9.8101	0.03479		
Total	293	12.3853			

a) State the hypotheses about the main effects (both numerically and in words).
b) Are the conditions for two-way ANOVA met?
c) If so, perform the hypothesis tests and state your conclusion. Be sure to state it in terms of head injury severity, seats, and vehicle types.

6. **Sprouts.** An experiment on mung beans was performed to investigate the environmental effects of salinity and water temperature on sprouting. Forty beans were randomly allocated to each of 36 petri dishes that were subject to one of four levels of *Salinity* (0, 4, 8 and 12 *ppm*) and one of three *Temperatures* (32°, 34°, or 36° C). After 48 hours, the biomass of the sprouts was measured.

Here are partial boxplots of *Biomass* on *Salinity* and *Temperature*:

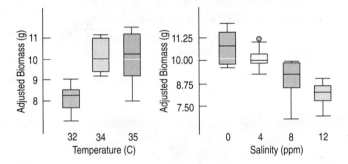

The interaction plot shows:

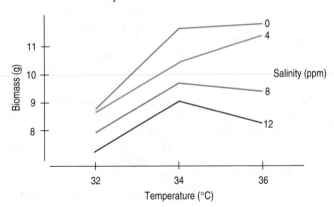

A two-way ANOVA model is fit, and the following ANOVA table results:

Analysis of Variance for Biomass (g)

Source	DF	Sum of Squares	Mean Square	F-ratio	P-value
Salinity	3	36.4701	12.1567	16.981	<0.0001
Temp	2	34.7168	17.3584	24.247	<0.0001
Salinity × Temp	6	5.2972	0.8829	1.233	0.3244
Error	24	17.1816	0.7159		
Total	35	93.6656			

The plot of residuals vs. predicted values shows:

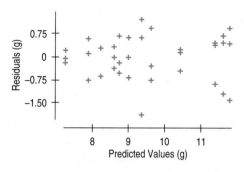

a) State the hypotheses about the factors (both numerically and in words).
b) Are the conditions for two-way ANOVA met?
c) Perform the hypothesis tests and state your conclusions. Be sure to state your conclusions in terms of biomass, salinity, and water temperature.

7. **Baldness and heart disease.** A retrospective study examined the link between baldness and the incidence of heart disease. In the study, 1435 middle-aged men were selected at random and examined to see whether they showed signs of *Heart Disease* (or not) and what amount of *Baldness* they exhibited (*none, little, some,* or *much*). A student runs a two-factor ANOVA on these data and finds the following ANOVA table:

Source	DF	Sum of Squares	Mean Square	F-Ratio	P-value
Baldness	3	62441.375	20813.792	17.956	0.020
Heart disease	1	1485.125	1485.125	1.281	0.340
Error	3	3477.375	1159.125		

a) Comment on her analysis. What problems, if any, do you find with the analysis?
b) What sort of analysis might you do instead?

Baldness	Heart Disease	Number of Men
None	Yes	251
None	No	331
Little	Yes	165
Little	No	221
Some	Yes	195
Some	No	185
Much	Yes	52
Much	No	35

8. **Fish and prostate.** In the Chapter 3 Step-By-Step, we looked at a Swedish study that asked 6272 men how much fish they ate and whether or not they had prostate cancer. Here are the data:

		Prostate Cancer?	
		No	Yes
	Never/seldom	110	14
Eat Fish?	Small part of diet	2420	201
	Moderate part	2769	209
	Large part	507	42

Armed with the methods of this chapter, a student performs a two-way ANOVA on the data. Here is her ANOVA table:

Source	DF	Sum of Squares	Mean Square	F-ratio	P-value
Fish	3	3110203.0	1036734.3	1.3599	0.4033
Prostate cancer	1	3564450.0	3564450.0	4.6756	0.1193
Error	3	2287051.0	762350.0		

a) Comment on her analysis. What problems, if any, do you find with the analysis?
b) What sort of analysis might you do instead?

9. **Baldness and heart disease again.** Refer back to Exercise 7. Perform your own anlysis of the data to see if baldness and heart disease are related. Do your conclusions support the claim that baldness is a cause of heart disease? Explain.

10. **Fish and prostate.** Refer back to Exercise 8. Perform your own analysis of the data to see if eating fish and contracting prostate cancer are related.

T 11. **Basketball shots.** A student performed an experiment to see if her favorite sneakers and the time of day might affect her free throw percentage. She tried shooting with and without her favorite sneakers and in the early morning and at night. For each treatment combination, she shot 50 baskets on 4 different occasions, recording the number of shots made each time. She randomized the treatment conditions by drawing numbers out of a hat. Here are her data:

Time of Day	Shoes	Shots Made
Morning	Others	25
Morning	Others	26
Night	Others	27
Night	Others	27
Morning	Favorite	32
Morning	Favorite	22
Night	Favorite	30
Night	Favorite	34
Morning	Others	35
Morning	Others	34
Night	Others	33
Night	Others	30
Morning	Favorite	33
Morning	Favorite	37
Night	Favorite	36
Night	Favorite	38

a) What are the null and alternative hypotheses?
b) Write a short report on your findings, being sure to include diagnostic analysis as well as practical conclusions.

T 12. **Washing.** For his final project, Jonathan examined the effects of two factors on how well stains are removed when washing clothes. On each of 16 new white handkerchiefs, he spread a teaspoon of dirty motor oil (obtained from a local garage). He chose 4 *Temperature* settings (each of which is a combination of wash and rinse: *cold-cold, cold-warm, warm-hot,* and *hot-hot*) and 4 *Cycle* lengths (*short, med short, med long,* and *long*). After its washing, each handkerchief was dried in a dryer for 20 minutes and hung up. He rounded up 10 family members to judge the handkerchiefs for cleanliness on a scale of 1 to 10 and used the average score as his response. Here are the data:

Temp	Cycle	Score
Cold-cold	Med long	3.7
Warm-hot	Med long	6.5
Cold-warm	Med long	4.9
Hot-hot	Med long	6.5
Cold-cold	Long	4.6
Warm-hot	Long	8.3
Cold-warm	Long	4.7
Hot-hot	Long	9.1
Cold-cold	Short	3.4
Warm-hot	Short	5.6
Cold-warm	Short	3.8
Hot-hot	Short	7.1
Cold-cold	Med short	3.1
Warm-hot	Med short	6.3
Cold-warm	Med short	5
Hot-hot	Med short	6.1

You may assume, as Jonathan did, that interactions between *Temperature* and *Cycle* are negligible. Write a report showing what you found about washing factors and stain removal.

13. **Sprouts again.** The students running the sprouts experiment (Exercise 6) also kept track of the number of beans sprouted (out of 40) for each of the 36 dishes. Here are the partial boxplots of *Sprouts* plotted against *Salinity* and *Temperature:*

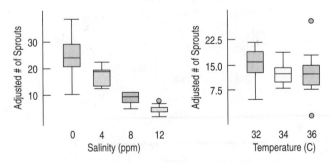

An ANOVA table shows:

Source	DF	Sum of Squares	Mean Square	F-ratio	P-value
Salinity	3	2014.750	671.583	23.657	<0.0001
Temp	2	57.556	28.778	1.014	0.3779
Salinity × Temp	6	96.000	16.000	0.564	0.7549
Error	24	681.333	28.389		
Total	35	2849.639			

An interaction plot shows:

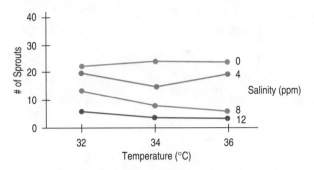

A plot of residuals vs. predicted values shows:

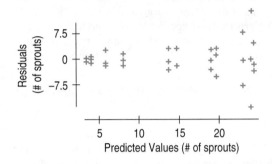

a) State the hypotheses about the factors (both numerically and in words).
b) Perform the hypothesis tests and state your conclusions. Be sure to check conditions.

T 14. **Containers revisited.** Building on the cup experiment of the Chapter 4 Step-By-Step, a student selects one type of container and designs an experiment to see whether the type of *Liquid* stored and the outside *Environment* affect the ability of a cup to maintain temperature. He randomly chooses an experimental condition and runs each twice:

Liquid	Environment	Change in Temperature
Water	Room	13
Water	Room	14
Water	Outside	31
Water	Outside	31
Coffee	Room	11
Coffee	Room	11
Coffee	Outside	27
Coffee	Outside	29

After fitting a two-way ANOVA model, he obtains the following interaction plot, ANOVA table, and effects table:

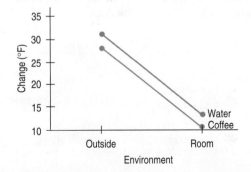

Source	DF	Sum of Squares	Mean Square	F-ratio	P-value
Liquid	1	15.125	15.125	24.2	0.0079
Envirn	1	595.125	595.125	952.2	<0.0001
Interaction	1	0.125	0.125	0.2	0.6779
Error	4	2.500	0.625		
Total	7	612.875			

Term	Estimate
Overall mean	20.875
Liquid II [Coffee]	−1.375
Liquid II [Water]	1.375
Environment II [Outside]	8.625
Environment II [Room]	−8.625

a) State the null and alternative hypotheses.
b) Test the hypotheses at $\alpha = 0.05$.
c) Perform a residual analysis.
d) Summarize your findings.

15. **Gas additives.** An experiment to test a new gasoline additive, *Gasplus*, was performed on three different cars: a sports car, a minivan, and a hybrid. Each car was tested with both *Gasplus* and regular gas on 10 different occasions and their gas mileage was recorded. Here are the partial boxplots:

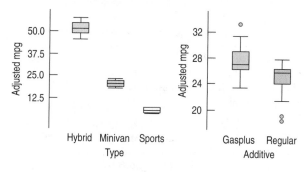

A two-way ANOVA with interaction model was run, and the following ANOVA table resulted:

Source	DF	Sum of Squares	Mean Square	F-ratio	P-value
Type	2	23175.4	11587.7	2712.2	<0.0001
Additive	1	92.1568	92.1568	21.57	<0.0001
Type × Additive	2	51.8976	25.9488	6.0736	0.0042
Error	54	230.711	4.27242		
Total	59	23550.2			

A plot of the residuals vs. predicted values showed:

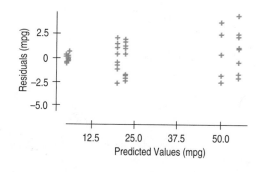

What conclusions about the additive and car types do you draw? Do you see any potential problems with the analysis?

T 16. **Chromatography.** A gas chromatograph is an instrument that measures the amounts of various compounds contained in a sample by separating the various constituents. Because different components are flushed through the system at different rates, chromatographers are able to both measure and distinguish the various constituents of the sample. A counter is placed somewhere along the instrument that records how much material is passing at various times. By looking at the counts at various times, the chemist is able to reconstruct the amounts of various compounds present. The total number of counts is proportional to the amount of the compound present.

An experiment was performed to see whether slowing down the flow rate would increase total counts. A mixture was produced with three different *Concentration* levels: *low, medium,* and *high*. The two *Flow Rates* used were *slow* and *fast*. Each mixture was run 5 times and the total counts recorded each time. Partial boxplots for *Concentration* and *Flow Rate* show:

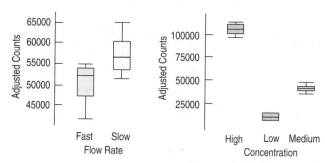

A two-way ANOVA with interaction model was run, and the following ANOVA table resulted:

Source	DF	Sum of Squares	Mean Square	F-ratio	P-value
Conc	2	483655E5	241828E5	1969.44	<0.0001
Flow rate	1	364008E3	364008E3	29.65	<0.0001
Interaction	2	203032E3	101516E3	8.27	0.0019
Error	24	294698E3	122791E3		
Total	29	492272E5			

A plot of residuals vs. predicted values showed:

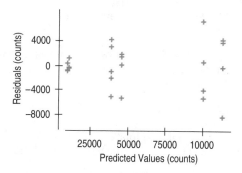

What conclusions about the effect of flow rate do you draw? Do you see any potential problems with the analysis?

17. **Gas additives again.** Refer back to the experiment in Exercise 15. Instead of *mpg* redo the analysis using log (*mpg*) as the response. Do your conclusions change? How? Are the assumptions of the model better satisfied?

T 18. **Chromatography again.** Refer back to the experiment in Exercise 16. Instead of *Total counts*, redo the analysis using log(*Total counts*) as the response. Do your conclusions change? How? Are the assumptions of the model better satisfied?

19. **Batteries again.** A student experiment was run to test the performance of 4 brands of batteries under 2 different *Environments* (*room temperature* and *cold*). For each of the 8 treatments, 2 batteries of a particular brand were put into a flashlight. The flashlight was then turned on and allowed to run until the light went out. The number of minutes the flashlight stayed on was recorded. Each treatment condition was run twice.
   Partial boxplots showed:

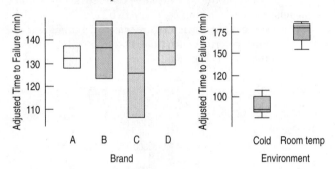

An interaction plot showed:

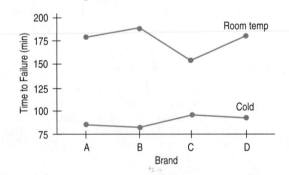

An ANOVA table showed:

Source	DF	Sum of Squares	Mean Square	F-ratio	P-value
Envir	1	30363.1	30363.1	789.93	<0.0001
Brand	3	338.187	112.729	2.9328	0.0994
Interaction	3	1278.19	426.063	11.085	0.0032
Error	8	307.5	38.4375		
Total	15	32286.9			

a) What are the main effect null and alternative hypotheses?
b) From the partial boxplots, do you think that the *Brand* has an effect on the time the batteries last? How about the condition?
c) Do the conclusions of the ANOVA table match your intuition?

d) What does the interaction plot say about the performance of the brands?
e) Why might you be uncomfortable with a recommendation to go with the cheapest battery (brand C)?

20. **Peas.** In an experiment on growing sweet peas, a team of students selected 2 factors at 4 levels each and recorded *Weight*, *Stem Length*, and *Root Length* after $6\frac{1}{2}$ days of growth. They grew plants using various amounts of *Water* and *Quickgrow* solution, a fertilizer designed to help plants grow faster. Each factor was run at 4 levels: little, some, moderate, and full. They grew 2 plants under each of the 16 conditions.
   An interaction plot of *Weight* in mg (*x*-axis is *Quickgrow*—levels are water) shows:

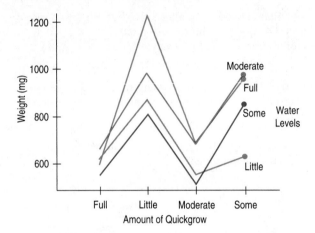

Because of this, a two-way ANOVA with interaction model was fit to the data, resulting in the following ANOVA table:

Analysis of Variance for Weight

Source	DF	Sum of Squares	Mean Square	F-ratio	P-value
Water	3	246255	82084.8	2.5195	0.0948
QS	3	827552	275851	8.4669	0.0013
Water × QS	8	176441	22055.1	0.6770	0.7130
Error	16	521275	32579.7		
Total	31	1771523			

Residuals plots showed no violations of the variance or Normality conditions. A table of effects for *Quickgrow* shows:

Level of Quickgrow	Effect
Little	213.8
Some	97.1
Moderate	−155.5
Full	−155.5

A table of effects for *Water* shows:

Level of Water	Effect
Little	−91.1
Some	−81.9
Moderate	66.4
Full	106.0

Write a report summarizing what the students have learned about the effects of *Water* and *Quickgrow* solution on the early stages of sweet pea growth as measured by *Weight*.

21. **Batteries once more.** Another student analyzed the battery data from Exercise 19, using a one-way ANOVA. He considered the experimental factor to be an 8-level factor consisting of the 8 possible combinations of *Brand* and *Environment*. Here are the boxplots for the 8 treatments and a one-way ANOVA:

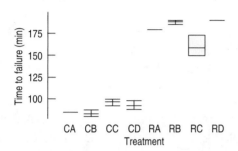

Source	DF	Sum of Squares	Mean Square	F-ratio	P-value
Treatment	7	31895.8	4556.54	93.189	<0.0001
Error	8	391.167	48.8958		
Total	15	32286.9			

Compare this analysis with the one performed in Exercise 19. Which one provides a better understanding of the data? Explain.

22. **Containers one more time.** Another student performs a one-way ANOVA on the container data of Exercise 14, using the 4 treatments *water room, water outside, coffee room,* and *coffee outside.* Perform this analysis and comment on the differences between this analysis and the one in Exercise 14.

JUST CHECKING

ANSWERS

1. The null hypotheses are that the mean flight distance for all three *Weights* is the same, that the mean flight distance for both *Positions* is the same, and that the interaction effects are all 0.

2. Yes, the effects appear to be additive. The lines are nearly parallel.

3. Yes, the *F*-test provides no evidence to reject the null hypothesis with a P-value of >0.9.

4. Yes, we reject the null hypothesis with a P-value <0.0001.

5. Yes, we reject the null hypothesis with a P-value <0.0001.

6. Because both factors are significant, it appears that using the light *Weight* in the rear *Position* may result in the longest mean flight distance.

CHAPTER

30

Multiple Regression

Where are we going?

We've seen that the top wind speed in a hurricane depends on the central barometric pressure. But what about the sea surface temperature? Can we include other variables in our model? Linear models are often useful, but the world is usually not so simple that a two-variable model does the trick. For a more realistic understanding, we need models with several variables.

WHO	250 Male subjects
WHAT	Body fat and waist size
UNITS	%Body fat and inches
WHEN	1990s
WHERE	United States
WHY	Scientific research

In Chapter 27 we tried to predict the percent body fat of male subjects from their waist size, and we did pretty well. The R^2 of 67.8% says that we accounted for almost 68% of the variability in *%Body Fat* by knowing only the *Waist* size. We completed the analysis by performing hypothesis tests on the coefficients and looking at the residuals.

But that remaining 32% of the variance has been bugging us. Couldn't we do a better job of accounting for *%Body Fat* if we weren't limited to a single predictor? In the full data set there were 15 other measurements on the 250 men. We might be able to use other predictor variables to help us account for the left-over variation that wasn't accounted for by waist size.

What about *Height*? Does *Height* help to predict *%Body Fat*? Men with the same *Waist* size can vary from short and corpulent to tall and emaciated. Knowing a man has a 50-inch waist suggests that he's likely to carry a lot of body fat. If we found out that he was 7 feet tall, that might change our impression of his body type. Knowing his *Height* as well as his *Waist* size might help us to make a more accurate prediction.

Just Do It

Does a regression with *two* predictors even make sense? It does—and that's fortunate because the world is too complex a place for simple linear regression alone to model it. A regression with two or more predictor variables is called a **multiple regression.** (When we need to note the difference, a regression on a single predictor is called a *simple* regression.) We'd never try to find a regression by hand, and even calculators aren't really up to the task. This is a job for a statistics program on a computer. If you know how to find the regression of

%Body Fat on *Waist* size with a statistics package, you can usually just add *Height* to the list of predictors without having to think hard about how to do it.

For simple regression, we found the **Least Squares** solution, the one whose coefficients made the sum of the squared residuals as small as possible. For multiple regression, we'll do the same thing but this time with more coefficients. Remarkably enough, we can still solve this problem. Even better, a statistics package can find the coefficients of the least squares model easily.

Here's a typical example of a multiple regression table:

Dependent variable is: %Body Fat
R-squared = 71.3%   R-squared (adjusted) = 71.1%
s = 4.460 with 250 − 3 = 247 degrees of freedom

Variable	Coefficient	SE(Coeff)	t-ratio	P-value
Intercept	−3.10088	7.686	−0.403	0.6870
Waist	1.77309	0.0716	24.8	≤0.0001
Height	−0.60154	0.1099	−5.47	≤0.0001

You should recognize most of the numbers in this table. Most of them mean what you expect them to.

R^2 gives the fraction of the variability of *%Body Fat* accounted for by the *multiple* regression model. (With *Waist* alone predicting *%Body Fat*, the R^2 was 67.8%.) The multiple regression model accounts for 71.3% of the variability in *%Body Fat*. We shouldn't be surprised that R^2 has gone up. It was the hope of accounting for some of that leftover variability that led us to try a second predictor.

The standard deviation of the residuals is still denoted s (or sometimes s_e to distinguish it from the standard deviation of y).

The degrees of freedom calculation follows our rule of thumb: The degrees of freedom is the number of observations (250) minus 1 for each coefficient estimated—for this model, 3.

For each predictor we have a coefficient, its standard error, a *t*-ratio, and the corresponding P-value. As with simple regression, the *t*-ratio measures how many standard errors the coefficient is away from 0. So, we can find a P-value from a Student's *t*-model to test the null hypothesis that the true value of the coefficient is 0.

Using the coefficients from this table, we can write the regression model:

$$\widehat{\%Body\ Fat} = -3.10 + 1.77\ Waist - 0.60\ Height.$$

As before, we define the residuals as

$$Residuals = \%Body\ Fat - \widehat{\%\ Body\ Fat}.$$

We've fit this model with the same least squares principle: The sum of the squared residuals is as small as possible for any choice of coefficients.

So, What's New?

So what's different? With so much of the multiple regression looking just like simple regression, why devote an entire chapter (or two) to the subject?

There are several answers to this question. First—and most important—the *meaning* of the coefficients in the regression model has changed in a subtle but important way. Because that change is not obvious, multiple regression coefficients are often misinterpreted. This chapter will show some examples to help make the meaning clear.

Second, multiple regression is an extraordinarily versatile calculation, underlying many widely used Statistics methods. A sound understanding of the multiple regression model will help you to understand these other applications.

Third, multiple regression offers our first glimpse into statistical models that use more than two quantitative variables. The real world is complex. Simple models of the kind we've seen so far are a great start, but often they're just not detailed enough to be useful for understanding, predicting, and decision making. Models that use several variables can be a big step toward realistic and useful modeling of complex phenomena and relationships.

Real Estate

As a class project, students in a large Statistics class collected publicly available information on recent home sales in their hometowns. There are 894 properties. These are not a random sample, but they may be representative of home sales during a short period of time, nationwide.

Variables available include the price paid, the size of the living area (sq ft), the number of bedrooms, the number of bathrooms, the year of construction, the lot size (acres), and a coding of the location as urban, suburban, or rural made by the student who collected the data.

Here's a regression to model the sale price from the living area (sq ft) and the number of bedrooms.

Dependent variable is: Price
R squared = 14.6%   R squared (adjusted) = 14.4%
s = 266899 with 894 − 3 = 891 degrees of freedom

Variable	Coefficient	SE(Coeff)	t-ratio	P-value
Intercept	308100	41148	7.49	≤0.0001
Living area	135.089	11.48	11.8	≤0.0001
bedrooms	−43346.8	12844	−3.37	0.0008

QUESTION: How should we interpret the regression output?

The model is

$$\widehat{Price} = 308{,}100 + 135\ Living\ Area - 43{,}346\ Bedrooms$$

The R-squared says that this model accounts for 14.6% of the variation in *Price*. But the value of *s* leads us to doubt that this model would provide very good predictions because the standard deviation of the residuals is more than $266,000. Nevertheless, we may be able to learn about home prices because the P-values of the coefficients are all very small, so we can be quite confident that none of them is really zero.

What Multiple Regression Coefficients Mean

We said that height might be important in predicting body fat in men. What's the relationship between *%Body Fat* and *Height* in men? We know how to approach this question; we follow the three rules. Here's the scatterplot:

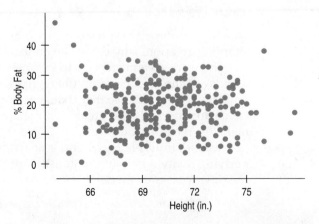

FIGURE 30.1
The scatterplot of *%Body Fat* against *Height* seems to say that there is little relationship between these variables.

It doesn't look like *Height* tells us much about *%Body Fat*. You just can't tell much about a man's *%Body Fat* from his *Height*. Or can you? Remember, in the multiple regression model, the coefficient of *Height* was −0.60, had a *t*-ratio of −5.47, and had a very small P-value. So it *did* contribute to the *multiple* regression model. How could that be?

The answer is that the multiple regression coefficient of *Height* takes account of the other predictor, *Waist* size, in the regression model.

To understand the difference, let's think about all men whose waist size is about 37 inches—right in the middle of our sample. If we think only about *these* men, what do we expect the relationship between *Height* and *%Body Fat* to be? Now a negative association makes sense because taller men probably have less body fat than shorter men *who have the same waist size*. Let's look at the plot:

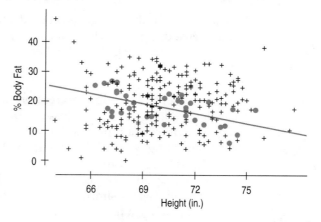

FIGURE 30.2
When we restrict our attention to men with waist sizes between 36 and 38 inches (points in blue), we can see a relationship between *%Body Fat* and *Height*.

Here we've highlighted the men with waist sizes between 36 and 38 inches. Overall, there's little relationship between *%Body Fat* and *Height*, as we can see from the full set of points. But when we focus on *particular* waist sizes, there *is* a relationship between body fat and height. This relationship is *conditional* because we've restricted our set to only those men within a certain range of waist size. For men with that waist size, an extra inch of height is associated with a decrease of about 0.60% in body fat. If that relationship is consistent for each *Waist* size, then the multiple regression coefficient will estimate it. The simple regression coefficient simply couldn't see it.

We've picked one particular *Waist* size to highlight. How could we look at the relationship between *%Body Fat* and height conditioned on *all waist* sizes *at the same time*? Once again, residuals come to the rescue.

We plot the residuals of *%Body Fat* after a regression on *Waist* size against the residuals of *Height* after regressing *it* on *Waist* size. This display is called a *partial regression plot*. It shows us just what we asked for: the relationship of *%Body Fat* to *Height* after removing the linear effects of *Waist* size.

> As their name reminds us, residuals are what's left over after we fit a model. That lets us remove the effects of some variables. The residuals are what's left.

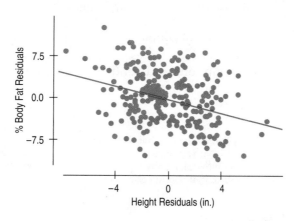

FIGURE 30.3
A partial regression plot for the coefficient of *Height* in the regression model has a slope equal to the coefficient value in the multiple regression model.

A **partial regression plot** for a particular predictor has a slope that is the same as the *multiple* regression coefficient for that predictor. Here, it's −0.60. It also has the same residuals as the full multiple regression, so you can spot any outliers or influential points and tell whether they've affected the estimation of this particular coefficient.

Many modern statistics packages offer partial regression plots as an option for any coefficient of a multiple regression. For the same reasons that we always look at a scatterplot before interpreting a simple regression coefficient, it's a good idea to make a partial regression plot for any multiple regression coefficient that you hope to understand or interpret.

The Multiple Regression Model

We can write a multiple regression model like this, numbering the predictors arbitrarily (we don't care which one is x_1), writing β's for the model coefficients (which we will estimate from the data), and including the errors in the model:

$$y = \beta_0 + \beta_1 x_1 + \beta_2 x_2 + \varepsilon.$$

Of course, the multiple regression model is not limited to two predictor variables, and regression model equations are often written to indicate summing any number (a typical letter to use is k) of predictors. That doesn't really change anything, so we'll often stick with the two-predictor version just for simplicity. But don't forget that we can have many predictors.

The assumptions and conditions for the multiple regression model sound nearly the same as for simple regression, but with more variables in the model, we'll have to make a few changes.

Assumptions and Conditions
Linearity Assumption

We are fitting a linear model.[1] For that to be the right kind of model, we need an underlying linear relationship. But now we're thinking about several predictors. To see whether the assumption is reasonable, we'll check the Straight Enough Condition for *each* of the predictors.

Straight Enough Condition: Scatterplots of y against each of the predictors are reasonably straight. As we have seen with *Height* in the body fat example, the scatterplots need not show a strong (or any!) slope; we just check that there isn't a bend or other nonlinearity. For the body fat data, the scatterplot is beautifully linear in *Waist* as we saw in Chapter 27. For *Height*, we saw no relationship at all, but at least there was no bend.

As we did in simple regression, it's a good idea to check the residuals for linearity after we fit the model. It's good practice to plot the residuals against the predicted values and check for patterns, especially bends or other nonlinearities. (We'll watch for other things in this plot as well.)

If we're willing to assume that the multiple regression model is reasonable, we can fit the regression model by least squares. But we must check the other assumptions and conditions before we can interpret the model or test any hypotheses.

CHECK THE RESIDUAL PLOT (PART 1)

The residuals should appear to have no pattern with respect to the predicted values.

---

[1] By *linear* we mean that each x appears simply multiplied by its coefficient and added to the model. No x appears in an exponent or some other more complicated function. That means that as we move along any x-variable, our prediction for y will change at a constant rate (given by the coefficient) if nothing else changes.

Independence Assumption

As with simple regression, the errors in the true underlying regression model must be independent of each other. As usual, there's no way to be sure that the Independence Assumption is true. Fortunately, even though there can be many predictor variables, there is only one response variable and only one set of errors. The Independence Assumption concerns the errors, so you should check the corresponding conditions on the residuals.

Randomization Condition: The data should arise from a random sample or randomized experiment. Randomization assures us that the data are representative of some identifiable population. If you can't identify the population, you can interpret the regression model only as a description of the data you have, and you can't interpret the hypothesis tests at all because they are about a regression model for that population. Regression methods are often applied to data that were not collected with randomization. Regression models fit to such data may still do a good job of modeling the data at hand, but without some reason to believe that the data are representative of a particular population, you should be reluctant to believe that the model generalizes to other situations.

You should also check displays of the regression residuals for evidence of patterns, trends, or clumping, any of which would suggest a failure of independence. In the special case when one of the *x*-variables is related to time, be sure that the residuals do not have a pattern when plotted against that variable or against *Time*.

The body fat data were collected on a sample of men. The men were not related in any way, so we can be pretty sure that their measurements are independent.

Equal Variance Assumption

The variability of the errors should be about the same for all values of *each* predictor. To see if this is reasonable, we look at scatterplots.

Does the Plot Thicken? Condition: Scatterplots of the regression residuals against each *x* or against the predicted values, $\hat{y}$, offer a visual check. The spread around the line should be nearly constant. Be alert for a "fan" shape or other tendency for the variability to grow or shrink in one part of the scatterplot.

Here are the residuals plotted against *Waist* and *Height*. Neither plot shows patterns that might indicate a problem.

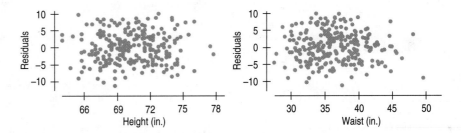

If residual plots show no pattern, if the data are plausibly independent, and if the plots don't thicken, we can feel good about interpreting the regression model. Before we test hypotheses, however, we must check one final assumption.

Normality Assumption

We assume that the errors around the idealized regression model at any specified values of the *x*-variables follow a Normal model. We need this assumption so that we can use a Student's *t*-model for inference. As with other times when we've used Student's *t*, we'll settle for the residuals satisfying the Nearly Normal Condition.

CHECK THE RESIDUAL PLOT (PART 2)

The residuals should appear to be randomly scattered and show no patterns or clumps when plotted against the predicted values.

CHECK THE RESIDUAL PLOT (PART 3)

The spread of the residuals should be uniform when plotted against any of the *x*'s or against the predicted values.

FIGURE 30.4
Residuals plotted against each predictor show no pattern. That's a good indication that the Straight Enough Condition and the Does the Plot Thicken? Condition are satisfied.

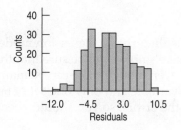

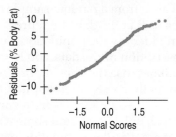

FIGURE 30.5
Check a histogram of the residuals. The distribution of the residuals should be unimodal and symmetric. Or check a Normal probability plot to see whether it is straight.

Nearly Normal Condition: Because we have only one set of residuals, this is the same set of conditions we had for simple regression. Look at a histogram or Normal probability plot of the residuals. The histogram of residuals in the body fat regression certainly looks Nearly Normal, and the Normal probability plot is fairly straight. And, as we have said before, the Normality Assumption becomes less important as the sample size grows.

Let's summarize all the checks of conditions that we've made and the order that we've made them:

1. Check the Straight Enough Condition with scatterplots of the y-variable against each x-variable.
2. If the scatterplots are straight enough (that is, if it looks like the regression model is plausible), fit a multiple regression model to the data. (Otherwise, either stop or consider re-expressing an x- or the y-variable.)
3. Find the residuals and predicted values.
4. Make a scatterplot of the residuals against the predicted values.[2] This plot should look patternless. Check in particular for any bend (which would suggest that the data weren't all that straight after all) and for any thickening. If there's a bend and especially if the plot thickens, consider re-expressing the y-variable and starting over.
5. Think about how the data were collected. Was suitable randomization used? Are the data representative of some identifiable population? If the data are measured over time, check for evidence of patterns that might suggest they're not independent by plotting the residuals against time to look for patterns.
6. If the conditions check out this far, feel free to interpret the regression model and use it for prediction. If you want to investigate a particular coefficient, make a partial regression plot for that coefficient.
7. If you wish to test hypotheses about the coefficients or about the overall regression, then make a histogram and Normal probability plot of the residuals to check the Nearly Normal Condition.

STEP-BY-STEP EXAMPLE   **Multiple Regression**

Question: How should we model *%Body Fat* in terms of *Height* and *Waist* size?

 THINK

Variables Name the variables, report the W's, and specify the questions of interest.

I have quantitative body measurements on 250 adult males from the BYU Human Performance Research Center. I want to understand the relationship between *%Body Fat*, *Height*, and *Waist* size.

Plan Think about the assumptions and check the conditions.

✔ **Straight Enough Condition:** There is no obvious bend in the scatterplots of *%Body Fat* against either x-variable. The scatterplot of residuals against predicted values below shows no patterns that would suggest nonlinearity.

---

[2] In Chapter 27 we noted that a scatterplot of residuals against the predicted values looked just like the plot of residuals against x. But for a multiple regression, there are several x's. Now the predicted values, $\hat{y}$, are a combination of the x's—in fact, they're the combination given by the regression equation we have computed. So they combine the effects of all the x's in a way that makes sense for our particular regression model. That makes them a good choice to plot against.

✔ **Independence Assumption:** These data are not collected over time, and there's no reason to think that the *%Body Fat* of one man influences that of another. I don't know whether the men measured were sampled randomly, but the data are presented as being representative of the male population of the United States.

✔ **Does the Plot Thicken? Condition:** The scatterplot of residuals against predicted values shows no obvious changes in the spread about the line.

Now we can find the regression and examine the residuals.

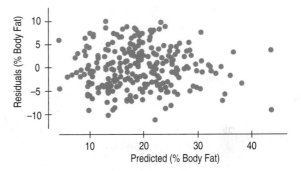

Actually, we need the Nearly Normal Condition only if we want to do inference.

✔ **Nearly Normal Condition, Outlier Condition:** A histogram of the residuals is unimodal and symmetric.

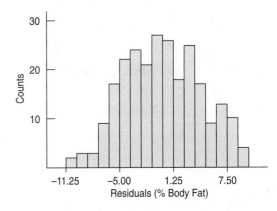

The Normal probability plot of the residuals is reasonably straight:

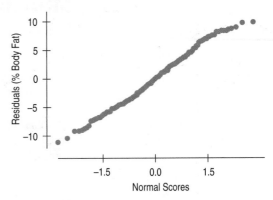

(continued)

Choose your method.	Under these conditions a full multiple regression analysis is appropriate.

SHOW **Mechanics**	Here is the computer output for the regression:

Dependent variable is: %Body Fat

R-squared = 71.3%   R-squared (adjusted) = 71.1%

s = 4.460 with 250 − 3 = 247 degrees of freedom

Source	Sum of Squares	DF	Mean Square	F-ratio	P-value
Regression	12216.6	2	6108.28	307	<0.0001
Residual	4912.26	247	19.8877		

Variable	Coefficient	SE(Coeff)	t-ratio	P-value
Intercept	−3.10088	7.686	−0.403	0.6870
Waist	1.77309	0.0716	24.8	<0.0001
Height	−0.60154	0.1099	−5.47	<0.0001

The estimated regression equation is

$$\widehat{\%Body\ Fat} = -3.10 + 1.77\ Waist - 0.60\ Height.$$

TELL **Interpretation**	The R^2 for the regression is 71.3%. Waist size and Height together account for about 71% of the variation in %Body Fat among men. The regression equation indicates that each inch in Waist size is associated with about a 1.77 increase in %Body Fat among men who are of a particular Height. Each inch of Height is associated with a decrease in %Body Fat of about 0.60 among men with a particular Waist size.
More Interpretation	The standard errors for the slopes of 0.07 (Waist) and 0.11 (Height) are both small compared with the slopes themselves, so it looks like the coefficient estimates are fairly precise. The residuals have a standard deviation of 4.46%, which gives an indication of how precisely we can predict %Body Fat with this model.

Multiple Regression Inference I: I Thought I Saw an ANOVA Table . . .

There are several hypothesis tests in the multiple regression output, but all of them talk about the same thing. Each is concerned with whether the underlying model parameters are actually zero.

The first of these hypotheses is one we skipped over for simple regression (for reasons that will be clear in a minute). Now that we've looked at ANOVA (in Chapter 28),[3] we can recognize the ANOVA table sitting in the middle of the regression output. Where'd that come from?

---

[3] If you skipped over Chapter 28, you can just take our word for this and read on.

The answer is that now that we have more than one predictor, there's an overall test we should consider before we do more inference on the coefficients. We ask the global question "Is this multiple regression model any good at all?" That is, would we do as well using just $\bar{y}$ to model y? What would that mean in terms of the regression? Well, if all the coefficients (except the intercept) were zero, we'd have

$$\hat{y} = b_0 + 0x_1 + \cdots + 0x_k$$

and we'd just set $b_0 = \bar{y}$.

To address the overall question, we'll test

$$H_0: \beta_1 = \beta_2 = \cdots = \beta_k = 0.$$

(That null hypothesis looks very much like the null hypothesis we tested with an F-test in the Analysis of Variance in Chapter 28.)

We can test this hypothesis with a statistic that is labeled with the letter F (in honor of Sir Ronald Fisher, the developer of Analysis of Variance). In our example, the F-value is 307 on 2 and 247 degrees of freedom. The alternative hypothesis is just that the slope coefficients aren't all equal to zero, and the test is one-sided—bigger F-values mean smaller P-values. If the null hypothesis were true, the F-statistic would be near 1. The F-statistic here is quite large, so we can easily reject the null hypothesis and conclude that the multiple regression model is better than just using the mean.[4]

Why didn't we do this for simple regression? Because the null hypothesis would have just been that the lone model slope coefficient was zero, and we were already testing that with the t-statistic for the slope. In fact, the *square* of that t-statistic is equal to the F-statistic for the simple regression, so it really was the identical test.

Multiple Regression Inference II: Testing the Coefficients

Once we check the F-test and reject the null hypothesis—and, if we are being careful, *only* if we reject that hypothesis—we can move on to checking the test statistics for the individual coefficients. Those tests look like what we did for the slope of a simple regression in Chapter 27. For each coefficient, we test

$$H_0: \beta_j = 0$$

against the (two-sided) alternative that it isn't zero. The regression table gives a standard error for each coefficient and the ratio of the estimated coefficient to its standard error. If the assumptions and conditions are met (and now we need the Nearly Normal Condition), these ratios follow a Student's t-distribution.

$$t_{n-k-1} = \frac{b_j - 0}{SE(b_j)}$$

How many degrees of freedom? We have a rule of thumb and it works here. The degrees of freedom is the number of data values minus the number of predictors (counting the intercept term). For our regression on two predictors, that's $n - 3$. You shouldn't have to look up the t-values. Almost every regression report includes the corresponding P-values.

---

[4] There are F tables in Table F at the end of the book, and they work pretty much as you'd expect. Most regression tables include a P-value for the F-statistic, but there's almost never a need to perform this particular test in a multiple regression. Usually we just glance at the F-statistic to see that it's reasonably far from 1.0, the value it would have if the true coefficients were really all zero.

We can build a confidence interval in the usual way, as an estimate ± a margin of error. As always, the margin of error is just the product of the standard error and a critical value. Here the critical value comes from the t-distribution on $n - k - 1$ degrees of freedom. So a confidence interval for β_j is

$$b_j \pm t^*_{n-k-1} \, SE(b_j).$$

The tricky parts of these tests are that the standard errors of the coefficients now require harder calculations (so we leave it to the technology) and the meaning of a coefficient, as we have seen, depends on all the *other* predictors in the multiple regression model.

That last bit is important. If we fail to reject the null hypothesis for a multiple regression coefficient, it does **not** mean that the corresponding predictor variable has no linear relationship to y. It means that the corresponding predictor contributes nothing to modeling y *after allowing for all the other predictors.*

How's That, Again?

This last point bears repeating. The multiple regression model looks so simple and straightforward:

$$y = \beta_0 + \beta_1 x_1 + \cdots + \beta_k x_k + \varepsilon.$$

It *looks* like each β_j tells us the effect of its associated predictor, x_j, on the response variable, y. But that is not so. This is, without a doubt, the most common error that people make with multiple regression:

- It is possible for there to be no simple relationship between y and x_j, and yet β_j in a *multiple* regression can be significantly different from 0. We saw this happen for the coefficient of *Height* in our example.
- It is also possible for there to be a strong two-variable relationship between y and x_j, and yet β_j in a multiple regression can be almost 0 with a large P-value so that we cannot reject the null hypothesis that the true coefficient is zero. If we're trying to model the horsepower of a car, using both its weight and its engine size, it may turn out that the coefficient for *Engine Size* is nearly 0. That *doesn't* mean that engine size isn't important for understanding horsepower. It simply means that after allowing for the weight of the car, the engine size doesn't give much *additional* information.
- It is even possible for there to be a significant linear relationship between y and x_j in one direction, and yet β_j can be of the *opposite* sign and strongly significant in a multiple regression. More expensive cars tend to be bigger, and since bigger cars have worse fuel efficiency, the price of a car has a slightly negative association with fuel efficiency. But in a multiple regression of *Fuel Efficiency* on *Weight* and *Price*, the coefficient of *Price* may be positive. If so, it means that *among cars of the same weight*, more expensive cars have better fuel efficiency. The simple regression on *Price*, though, has the opposite direction because, *overall*, more expensive cars are bigger. This switch in sign may seem a little strange at first, but it's not really a contradiction at all. It's due to the change in the *meaning* of the coefficient of *Price* when it is in a multiple regression rather than a simple regression.

So we'll say it once more: The coefficient of x_j in a multiple regression depends as much on the *other* predictors as it does on x_j. Remember that when you interpret a multiple regression model.

FOR EXAMPLE   **Interpreting Coefficients**

We looked at a multiple regression to predict the price of a house from its living area and the number of bedrooms. We found the model

$$\widehat{Price} = 308,100 + 135\,Living\,Area - 43,346\,Bedrooms.$$

However, common sense says that houses with more bedrooms are usually worth more. And, in fact, the simple regression of *Price* on *Bedrooms* finds the model

$$\widehat{Price} = 33,897 + 40,234\,Bedrooms$$

and the P-value for the slope coefficient is 0.0005.

QUESTION: How should we understand the coefficient of *Bedrooms* in the multiple regression?

The coefficient of Bedrooms in the multiple regression does not mean that houses with more bedrooms are generally worth less. It must be interpreted taking account of the other predictor (Living area) in the regression. If we consider houses with a given amount of living area, those that devote more of that area to bedrooms either must have smaller bedrooms or less living area for other parts of the house. Those differences could result in reducing the home's value.

JUST CHECKING

Recall the regression example in Chapter 8 to predict hurricane maximum wind speed from central barometric pressure. Another researcher, interested in the possibility that global warming was causing hurricanes to become stronger, added the variable Year as a predictor and obtained the following regression:

Dependent variable is: Max. Winds (kn)
275 total cases of which 113 are missing

R squared = 77.9%   R squared (adjusted) = 77.6%
s = 7.727 with 162 − 3 = 159 degrees of freedom

Source	Sum of Squares	df	Mean Square	F-ratio
Regression	33446.2	2	16723.1	280
Residual	9493.45	159	59.7072	

Variable	Coefficient	SE(Coeff)	t-ratio	P-value
Intercept	1009.99	46.53	21.7	≤0.0001
Central Pressure	−0.933491	0.0395	−23.6	≤0.0001
Year	−0.010084	0.0123	−0.821	0.4128

1. Interpret the R^2 of this regression.

2. Interpret the coefficient of *Central Pressure*.

3. The researcher concluded that "There has been no change over time in the strength of Atlantic hurricanes." Is this conclusion a sound interpretation of the regression model?

Another Example: Modeling Infant Mortality

Infant Mortality is often used as a general measure of the quality of health care for children and mothers. It is reported as the rate of deaths of newborns per 1000 live births. Data recorded for each of the 50 states of the United States may allow us to build regression models to help understand or predict infant mortality. The variables available for our model are *Child Deaths* (deaths per 100,000 children aged 1–14), percent of teens (ages 16–19) who drop out of high school (*HS Drop%*), percent of low-birth-weight babies (*Low BW%*), *Teen Births* (births per 100,000 females aged 15–17), and *Teen Deaths* by accident, homicide, and suicide (deaths per 100,000 teens ages 15–19).[5]

WHO	U.S. states
WHAT	Various measures relating to children and teens
WHEN	1999
WHY	Research and policy

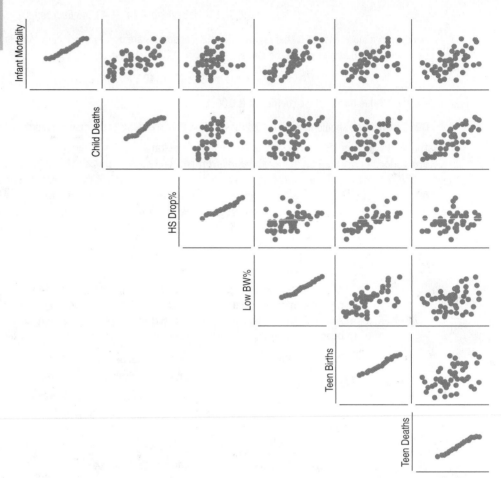

FIGURE 30.6
A scatterplot matrix shows a scatterplot of each pair of variables arrayed so that the vertical and horizontal axes are consistent across rows and down columns. You can tell which variable is plotted on the *x*-axis of any plot by reading down to the diagonal and looking to the left. The diagonal cells may hold Normal probability plots (as they do here), histograms, or just the names of the variables. These are a great way to check the Straight Enough Condition and to check for simple outliers.

All of these variables were displayed and found to have no outliers and Nearly Normal distributions.[6] One useful way to check many of our conditions is with a **scatterplot matrix.** Figure 30.6 shows an array of scatterplots set up so that the plots in each row have the same variable on their *y*-axis and those in each column have the same variable on their *x*-axis. This way every pair of variables is graphed. On the diagonal, rather than plotting a variable against itself, you'll usually find either a Normal probability plot or a histogram of the variable to help us assess the **Nearly Normal Condition.**

The individual scatterplots show at a glance that each of the relationships is straight enough for regression. There are no obvious bends, clumping, or outliers. And the plots don't thicken. So it looks like we can examine some multiple regression models with inference.

---

[5]The data are available from the Kids Count section of the Annie E. Casey Foundation (http://datacenter.kidscount.org/), and are all for 1999.

[6]In the interest of complete honesty, we should point out that the original data include the District of Columbia, but it proved to be an outlier on several of the variables, so we've restricted attention to the 50 states here.

STEP-BY-STEP EXAMPLE | Inference for Multiple Regression

Question: How should we model *Infant Mortality* using the available predictors?

THINK

Hypotheses State what we want to know.

I wonder whether all or some of these predictors contribute to a useful model for *Infant Mortality*.

First, I'll check the overall null hypothesis that asks whether the entire model is better than just modeling y with its mean:

(Hypotheses on the intercept are not particularly interesting for these data.)

H_O: The model itself contributes nothing useful, and all the slope coefficients are zero:

$$\beta_1 = \beta_2 = \cdots = \beta_k = O.$$

H_A: At least one of the β_j is not O.

If I reject this hypothesis, then I'll test a null hypothesis for each of the coefficients of the form:

Plan State the null model.

H_O: The j-th variable contributes nothing useful, after allowing for the other predictors in the model: $\beta_j = O$.

H_A: The j-th variable makes a useful contribution to the model: $\beta_j \neq O$.

Think about the assumptions and check the conditions.

✓ **Straight Enough Condition, Outlier Condition:** The scatterplot matrix shows no bends, clumping, or outliers.

✓ **Independence Assumption:** These data are based on random samples and can be considered independent.

These conditions are enough to compute the regression model and find residuals.

✓ **Does the Plot Thicken? Condition:** The residual plot shows no obvious trends in the spread:

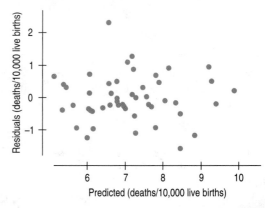

(continued)

✔ **Nearly Normal Condition:** A histogram of the residuals is unimodal and symmetric.

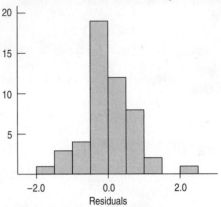

The one possible outlier is South Dakota. I may want to repeat the analysis after removing South Dakota to see whether it changes substantially.

Choose your method.

Under these conditions I can continue with the multiple regression analysis.

 Mechanics Multiple regressions are always found from a computer program.

Computer output for this regression looks like this:

Dependent variable is: Infant mort
R-squared = 71.3%   R-squared (adjusted) 68.0 %
s = 0.7520 with 50 − 6 = 44 degrees of freedom

Source	Sum of Squares	DF	Mean Square	F-ratio
Regression	61.7319	5	12.3464	21.8
Residual	24.8843	44	0.565553	

The P-values given in the regression output table are from the Student's t-distribution on $(n − 6) = 44$ degrees of freedom. They are appropriate for two-sided alternatives.

Variable	Coefficient	SE(Coeff)	t-ratio	P-value
Intercept	1.63168	0.9124	1.79	0.0806
Child Deaths	0.03123	0.0139	2.25	0.0292
HS Drop%	−0.09971	0.0610	−1.63	0.1096
Low BW%	0.66103	0.1189	5.56	<0.0001
Teen Births	0.01357	0.0238	0.57	0.5713
Teen Deaths	0.00556	0.0113	0.49	0.6245

Consider the hypothesis tests.

The F-ratio of 21.8 on 5 and 44 degrees of freedom is certainly large enough to reject the default null hypothesis that the regression model is no better than using the mean infant mortality rate. So I will examine the individual coefficients.

Under the assumptions we're willing to accept, and considering the conditions we've checked, the individual coefficients follow Student's t-distributions on 44 degrees of freedom.

Most of these coefficients have relatively small t-ratios, so I can't be sure that their underlying values are not zero. Two of the coefficients, *Child Deaths* and *Low BW%*, have P-values less than 5%. So I can be confident that in this model both of these variables are unlikely to really have zero coefficients.

 Interpretation

Overall the R^2 indicates that more than 71% of the variability in *Infant Mortality* can be accounted for with this regression model.

After allowing for the linear effects of the other variables in the model, an increase in *Child Deaths* of 1 death per 100,000 is associated with an increase of 0.03 deaths per 1000 live births in the *Infant Mortality* rate. And an increase of 1% in the percentage of live births that are low birth weight is associated with an increase of 0.66 deaths per 1000 live births.

Comparing Multiple Regression Models

There may be even more variables available to model *Infant Mortality*. Moreover, several of those we tried don't seem to contribute to the model. How do we know that some other choice of predictors might not provide a better model? What exactly *would* make an alternative model better?

These are not easy questions. There is no simple measure of the success of a multiple regression model. Many people look at the R^2 value, and certainly we are not likely to be happy with a model that accounts for only a small fraction of the variability of y. But that's not enough. You can always drive the R^2 up by piling on more and more predictors, but models with many predictors are hard to understand. Keep in mind that the meaning of a regression coefficient depends on all the *other* predictors in the model, so it is best to keep the number of predictors as small as possible.

Regression models should make sense. Predictors that are easy to understand are usually better choices than obscure variables. Similarly, if there is a known mechanism by which a predictor has an effect on the response variable, that predictor is usually a good choice for the regression model.

How can we know whether we have the best possible model? The simple answer is that we can't. There's always the chance that some other predictors might bring an improvement (in higher R^2 or fewer predictors or simpler interpretation).

Adjusted R^2

You may have noticed that the full regression tables shown in this chapter include another statistic we haven't discussed. It is called adjusted R^2 and sometimes appears in computer output as R^2 (adjusted). The **adjusted R^2** statistic is a rough attempt to adjust for the simple fact that when we add another predictor to a multiple regression, the R^2 can't go down and will most likely go up. Only if we were to add a predictor whose coefficient turned out to be exactly zero would the R^2 remain the same. This fact complicates the comparison of alternative regression models that have different numbers of predictors.

We can write a formula for R^2 using the sums of squares in the ANOVA table portion of the regression output table:

$$R^2 = \frac{SS_{Regression}}{SS_{Regression} + SS_{Residual}} = 1 - \frac{SS_{Residual}}{SS_{Total}}.$$

Adjusted R^2 simply substitutes the corresponding *Mean Squares* for the SS's:[7]

$$R^2_{adj} = 1 - \frac{MS_{Residual}}{MS_{Total}}.$$

---

[7] We learned about Mean Squares in Chapter 28. A Mean Square is just a Sum of Squares divided by its appropriate degrees of freedom. Mean Squares are variances.

Because the Mean Squares are Sums of Squares divided by degrees of freedom, they are adjusted for the number of predictors in the model. As a result, the adjusted R^2 value won't necessarily increase when a new predictor is added to the multiple regression model. That's fine. But adjusted R^2 no longer tells the fraction of variability accounted for by the model, and it isn't even bounded by 0 and 100%, so it can be awkward to interpret.

Comparing alternative regression models is a challenge, especially when they have different numbers of predictors. The search for a summary statistic to help us choose among models is the subject of much contemporary research in Statistics. Adjusted R^2 is one common—but not necessarily the best—choice often found in computer regression output tables. Don't use it as the sole decision criterion when you compare different regression models.

What Can Go Wrong?

Interpreting Coefficients

■ **Don't claim to "hold everything else constant" for a single individual.** It's often meaningless to say that a regression coefficient says what we expect to happen if all variables but one were held constant for an individual and the predictor in question changed. While it's mathematically correct, it often just doesn't make any sense. We can't gain a year of experience or have another child without getting a year older. Instead, we *can* think about all those who fit given criteria on some predictors and ask about the conditional relationship between y and one x for those individuals. The coefficient -0.60 of *Height* for predicting *%Body Fat* says that among men of the same *Waist* size, those who are one inch taller in *Height* tend to be, on average, 0.60% lower in *%Body Fat*. The multiple regression coefficient measures that average conditional relationship.

■ **Don't interpret regression causally.** Regressions are usually applied to observational data. Without deliberately assigned treatments, randomization, and control, we can't draw conclusions about causes and effects. We can never be certain that there are no variables lurking in the background, causing everything we've seen. Don't interpret b_1, the coefficient of x_1 in the multiple regression, by saying, "If we were to change an individual's x_1 by 1 unit (holding the other x's constant) it would change his y by b_1 units." We have no way of knowing what applying a change to an individual would do.

■ **Be cautious about interpreting a regression model as predictive.** Yes, we do call the x's predictors, and you can certainly plug in values for each of the x's and find a corresponding *predicted value, $\hat{y}$*. But the term "prediction" suggests extrapolation into the future or beyond the data, and we know that we can get into trouble when we use models to estimate $\hat{y}$ values for x's not in the range of the data. Be careful not to extrapolate very far from the span of your data. In simple regression it was easy to tell when you extrapolated. With many predictor variables, it's often harder to know when you are outside the bounds of your original data.[8] We usually think of fitting models to the data more as modeling than as prediction, so that's often a more appropriate term.

---

[8] With several predictors it is easy to wander beyond the data because of the *combination* of values even when individual values are not extraordinary. For example, both 28-inch waists and 76-inch heights can be found in men in the body fat study, but a single individual with both these measurements would not be at all typical. The model we fit is probably not appropriate for predicting the *%Body Fat* for such a tall and skinny individual.

- **Don't think that the sign of a coefficient is special.** Sometimes our primary interest in a predictor is whether it has a positive or negative association with y. As we have seen, though, the sign of the coefficient also depends on the other predictors in the model. Don't look at the sign in isolation and conclude that "the direction of the relationship is positive (or negative)." Just like the value of the coefficient, the sign is about the relationship after allowing for the linear effects of the other predictors. The sign of a variable can change depending on which other predictors are in or out of the model. For example, in the regression model for infant mortality, the coefficient of *HS Drop%* was negative and its P-value was fairly small, but the simple association between *Dropout Rate* and *Infant Mortality* is positive. (Check the plot matrix.)

- **If a coefficient's *t*-statistic is not significant, don't interpret it at all.** You can't be sure that the value of the corresponding parameter in the underlying regression model isn't really zero.

What *Else* Can Go Wrong?

- **Don't fit a linear regression to data that aren't straight.** This is the most fundamental regression assumption. If the relationship between y and the x's isn't approximately linear, there's no sense in fitting a linear model to it. What we mean by "linear" is a model of the form we have been writing for the regression. When we have two predictors, this is the equation of a plane, which is linear in the sense of being flat in all directions. With more predictors, the geometry is harder to visualize, but the simple structure of the model is consistent; the predicted values change consistently with equal size changes in any predictor.

  Usually we're satisfied when plots of y against each of the x's are straight enough. We'll also check a scatterplot of the residuals against the predicted values for signs of nonlinearity.

- **Watch out for the plot thickening.** The estimate of the error standard deviation shows up in all the inference formulas. But that estimate assumes that the error standard deviation is the same throughout the range of the x's so that we can combine (pool, actually) all the residuals when we estimate it. If s_e changes with any x, these estimates won't make sense. The most common check is a plot of the residuals against the predicted values. If plots of residuals against several of the predictors all show a thickening, and especially if they also show a bend, then consider re-expressing y. If the scatterplot against only one predictor shows thickening, consider re-expressing that predictor.

- **Make sure the errors are nearly Normal.** All of our inferences require that the true errors be modeled well by a Normal model. Check the histogram and Normal probability plot of the residuals to see whether this assumption looks reasonable.

- **Watch out for high-influence points and outliers.** We always have to be on the lookout for a few points that have undue influence on our model, and regression is certainly no exception. Partial regression plots are a good place to look for influential points and to understand how they affect each of the coefficients.

We would never consider a regression analysis without first making scatterplots. The aspects of scatterplots that we always look for—their direction, form, and strength—relate directly to regression, and we assess the nearly normal condition by examining the shape of a residual histogram or with a normal probability plot.

Regression inference is connected to just about every inference method we have seen for measured data. The assumption that the spread of data about the line is constant is essentially the same as the assumption of equal variances required for the pooled-*t* methods. Our use of all the residuals together to estimate their standard deviation is a form of pooling.

Of course, the ANOVA table in the regression output connects to our consideration of ANOVA in Chapter 28. This, too, is not coincidental. Multiple Regression, ANOVA, pooled *t*-tests, and inference for means are all part of a more general statistical model known as the General Linear Model (often just called the GLM).

WHAT HAVE WE LEARNED?

In Chapter 27 we learned to apply our inference methods to linear regression models. Now we've seen that much of what we know about those models is also true for multiple regression:

▶ The assumptions and conditions are the same: linearity (checked now with scatterplots of *y* against each *x*), independence (think about it), constant variance (checked with the scatterplot of residuals against predicted values), and Nearly Normal residuals (checked with a histogram or probability plot).

▶ R^2 is still the fraction of the variation in *y* accounted for by the regression model.

▶ s_e is still the standard deviation of the residuals—a good indication of the precision of the model.

▶ The degrees of freedom (in the denominator of s_e and for each of the *t*-tests) follows the same rule: *n* minus the *number of parameters estimated*.

▶ The regression table produced by any statistics package shows a row for each coefficient, giving its estimate, a standard error, a *t*-statistic, and a P-value.

▶ If all the conditions are met, we can test each coefficient against the null hypothesis that its parameter value is zero with a Student's *t*-test.

And we've learned some new things that are useful now that we have multiple predictors:

▶ We can perform an overall test of whether the multiple regression model provides a better summary for *y* than its mean by using the *F*-distribution we saw in the previous chapters.

▶ We learned that R^2 may not be appropriate for comparing multiple regression models with different numbers of predictors. Adjusted R^2 is one approach to this problem.

Finally, we've learned that multiple regression models extend our ability to model the world to many more situations, but that we must take great care when we interpret its coefficients. To interpret a coefficient of a multiple regression model, remember that it estimates the linear relationship between *y* and that predictor *after allowing for the linear effects of all the other predictors on both y and that x.*

Terms

Multiple regression
A linear regression with two or more predictors whose coefficients are found to minimize the sum of the squared residuals is a least squares linear multiple regression. But it is usually just called a multiple regression. When the distinction is needed, a least squares linear regression with a single predictor is called a simple regression. The multiple regression model is (p. 784)

$$y = \beta_0 + \beta_1 x_1 + \cdots + \beta_k x_k + \varepsilon.$$

Least Squares
We still fit multiple regression models by choosing the coefficients that make the sum of the squared residuals as small as possible, the method of least squares (p. 785).

Partial regression plot
The partial regression plot for a specified coefficient is a display that helps in understanding the meaning of that coefficient in a multiple regression. It has a slope equal to the coefficient value and shows the influences of each case on that value. Partial regression plots display the residuals when y is regressed on the *other* predictors against the residuals when the specified x is regressed on the other predictors (p. 788).

Assumptions for inference in regression (and conditions to check for some of them)
- **Linearity Assumption.** Check that the scatterplots of y against each x are straight enough and that the scatterplot of residuals against predicted values has no obvious pattern. (If we find the relationships straight enough, we may fit the regression model to find residuals for further checking.) (p. 788)

- **Independence Assumption.** Think about the nature of the data. Check a residual plot. Any evident pattern in the residuals can call the assumption of independence into question (p. 789).

- **Equal Variance Assumption.** Check that the scatterplots show consistent spread across the ranges of the x-variables and that the residual plot has constant variance, too. A common problem is increasing spread of the residuals with increasing predicted values—*the plot thickens!* (p. 789)

- **Normality Assumption.** Check a histogram or a Normal probability plot of the residuals (pp. 789-790).

ANOVA
The Analysis of Variance table that is ordinarily part of the multiple regression results offers an F-test to test the null hypothesis that the overall regression is no improvement over just modeling y with its mean:

$$H_0: \beta_1 = \beta_2 = \cdots = \beta_k = 0.$$

If this null hypothesis is not rejected, then do not proceed to test the individual coefficients (p. 792).

t-ratios for the coefficients
The t-ratios for the coefficients can be used to test the null hypotheses that the true value of each coefficient is zero against the alternative that it is not (p. 793).

Scatterplot matrix
A scatterplot matrix displays scatterplots for all pairs of a collection of variables, arranged so that all the plots in a row have the same variable displayed on their y-axis and all plots in a column have the same variable on their x-axis. Usually, the diagonal holds a display of a single variable such as a histogram or Normal probability plot, and identifies the variable in its row and column (p. 796).

Adjusted R^2
An adjustment to the R^2 statistic that attempts to allow for the number of predictors in the model. It is sometimes used when comparing regression models with different numbers of predictors (p. 799).

$$R_{adj}^2 = 1 - \frac{MS_{Residual}}{MS_{Total}}$$

Skills

THINK
- Understand that the "true" regression model is an idealized summary of the data.

- Know how to examine scatterplots of y vs. each x for violations of assumptions that would make inference for regression unwise or invalid.

> ▶ Know how to examine displays of the residuals from a multiple regression to check that the conditions have been satisfied. In particular, know how to judge linearity and constant variance from a scatterplot of residuals against predicted values. Know how to judge Normality from a histogram and Normal probability plot.

> ▶ Remember to be especially careful to check for failures of the independence assumption when working with data recorded over time. Examine scatterplots of the residuals against time and look for patterns.

> ▶ Be able to use a statistics package to perform the calculations and make the displays for multiple regression, including a scatterplot matrix of the variables, a scatterplot of residuals vs. predicted values, and partial regression plots for each coefficient.

> ▶ Know how to use the ANOVA F-test to check that the overall regression model is better than just using the mean of y.

> ▶ Know how to test the standard hypotheses that each regression coefficient is really zero. Be able to state the null and alternative hypotheses. Know where to find the relevant numbers in standard computer regression output.

> ▶ Be able to summarize a regression in words. In particular, be able to state the meaning of the regression coefficients, taking full account of the effects of the other predictors in the model.

> ▶ Be able to interpret the F-statistic for the overall regression.

> ▶ Be able to interpret the P-value of the t-statistics for the coefficients to test the standard null hypotheses.

REGRESSION ANALYSIS ON THE COMPUTER

All statistics packages make a table of results for a regression. If you can read a package's regression output table for simple regression, then you can read its table for a multiple regression. You'll want to look at the ANOVA table, and you'll see information for each of the coefficients, not just for a single slope.

Most packages offer to plot residuals against predicted values. Some will also plot residuals against the x's. With some packages you must request plots of the residuals when you request the regression. Others let you find the regression first and then analyze the residuals afterward. Either way, your analysis is not complete if you don't check the residuals with a histogram or Normal probability plot and a scatterplot of the residuals against the x's or the predicted values.

One good way to check assumptions before embarking on a multiple regression analysis is with a scatterplot matrix. This is sometimes abbreviated SPLOM in commands.

Multiple regressions are always found with a computer or programmable calculator. Before computers were available, a full multiple regression analysis could take months or even years of work.

DATA DESK

- Select Y- and X-variable icons.
- From the **Calc** menu, choose **Regression.**
- Data Desk displays the regression table.
- Select plots of residuals from the Regression table's HyperView menu.

COMMENTS

You can change the regression by dragging the icon of another variable over either the Y- or an X-variable name in the table and dropping it there. You can add a predictor by dragging its icon into that part of the table. The regression, predicted values, residuals, and diagnostic statistics will recompute automatically.

EXCEL

- In Excel 2003 and earlier, select **Data Analysis** from the **Tools** menu.
- In Excel 2007, select **Data Analysis** from the **Analysis Group** on the Data Tab.
- Select **Regression** from the **Analysis Tools** list.
- Click the **OK** button.
- Enter the data range holding the Y-variable in the box labeled "Y-range."
- Enter the range of cells holding the X-variables in the box labeled "X-range."
- Select the **New Worksheet Ply** option.
- Select **Residuals** options. Click the **OK** button.

COMMENTS

The Y and X ranges do not need to be in the same rows of the spreadsheet, although they must cover the same number of cells. But it is a good idea to arrange your data in parallel columns as in a data table. The X-variables must be in adjacent columns. No cells in the data range may hold non-numeric values.

Although the dialog offers a Normal probability plot of the residuals, the data analysis add-in does not make a correct probability plot, so don't use this option.

JMP

- From the **Analyze** menu select **Fit Model.**
- Specify the response, Y. Assign the predictors, X, in the **Construct Model Effects** dialog box.
- Click on **Run Model.**

COMMENTS

JMP chooses a regression analysis when the response variable is "Continuous." The predictors can be any combination of quantitative or categorical. If you get a different analysis, check the variable types.

MINITAB

- Choose **Regression** from the **Stat** menu.
- Choose **Regression...** from the **Regression** submenu.
- In the Regression dialog, assign the Y-variable to the Response box and assign the X-variables to the Predictors box.
- Click the **Graphs** button.
- In the Regression-Graphs dialog, select **Standardized residuals,** and check **Normal plot of residuals** and **Residuals versus fits.**
- Click the **OK** button to return to the Regression dialog.
- Click the **OK** button to compute the regression.

SPSS

- Choose **Regression** from the **Analyze** menu.
- Choose **Linear** from the **Regression** submenu.
- When the Linear Regression dialog appears, select the Y-variable and move it to the dependent target. Then move the X-variables to the independent target.
- Click the **Plots** button.
- In the Linear Regression Plots dialog, choose to plot the *SRESIDs against the *ZPRED values.
- Click the **Continue** button to return to the Linear Regression dialog.
- Click the **OK** button to compute the regression.

(continued)

TI-83/84 PLUS	COMMENTS
	You need a special program to compute a multiple regression on the TI-83.

TI-89	COMMENTS

Under **STAT Tests** choose **B:MultREg Tests**

- Specify the number of predictor variables, and which lists contain the response variable and predictor variables.
- Press ÷ to perform the calculations.

- The first portion of the output gives the *F*-statistic and its P-value as well as the values of R^2, $Adj'R^2$, the standard deviation of the residuals (s), and the Durbin-Watson statistic, which measures correlation among the residuals.
- The rest of the main output gives the components of the *F*-test, as well as values of the coefficients, their standard errors, and associated *t*-statistics along with P-values. You can use the right arrow to scroll through these lists (if desired).
- The calculator creates several new lists that can be used for assessing the model and its conditions: Yhatlist, resid, sresid (standardized residuals), leverage, and cookd, as well as lists of the coefficients, standard errors, *t*'s, and P-values.

EXERCISES

1. **Interpretations.** A regression performed to predict the selling price of houses found the equation

   $$\widehat{Price} = 169,328 + 35.3\,Area + 0.718\,Lotsize - 6543\,Age$$

   where *Price* is in dollars, *Area* is in square feet, *Lotsize* is in square feet, and *Age* is in years. The R^2 is 92%. One of the interpretations below is correct. Which is it? Explain what's wrong with the others.

   a) Each year a house *Age*s it is worth $6543 less.
   b) Every extra square foot of *Area* is associated with an additional $35.30 in average price, for houses with a given *Lotsize* and *Age*.
   c) Every dollar in price means *Lotsize* increases 0.718 square feet.
   d) This model fits 92% of the data points exactly.

2. **More interpretations.** A household appliance manufacturer wants to analyze the relationship between total sales and the company's three primary means of advertising (television, magazines, and radio). All values were in millions of dollars. They found the regression equation

   $$\widehat{Sales} = 250 + 6.75\,TV + 3.5\,Radio + 2.3\,Magazines.$$

   One of the interpretations below is correct. Which is it? Explain what's wrong with the others.

   a) If they did no advertising, their income would be $250 million.
   b) Every million dollars spent on radio makes sales increase $3.5 million, all other things being equal.
   c) Every million dollars spent on magazines increases TV spending $2.3 million.
   d) Sales increase on average about $6.75 million for each million spent on TV, after allowing for the effects of the other kinds of advertising.

3. **Predicting final exams.** How well do exams given during the semester predict performance on the final? One class had three tests during the semester. Computer output of the regression gives

   Dependent variable is Final
   s = 13.46   R-Sq = 77.7%   R-Sq(adj) = 74.1%

Predictor	Coeff	SE(Coeff)	t	P-value
Intercept	−6.72	14.00	−0.48	0.636
Test1	0.2560	0.2274	1.13	0.274
Test2	0.3912	0.2198	1.78	0.091
Test3	0.9015	0.2086	4.32	<0.0001

   Analysis of Variance

Source	DF	SS	MS	F	P-value
Regression	3	11961.8	3987.3	22.02	<0.0001
Error	19	3440.8	181.1		
Total	22	15402.6			

   a) Write the equation of the regression model.
   b) How much of the variation in final exam scores is accounted for by the regression model?
   c) Explain in context what the coefficient of *Test3* scores means.
   d) A student argues that clearly the first exam doesn't help to predict final performance. She suggests that this exam not be given at all. Does Test1 have no effect on the final exam score? Can you tell from this model? (*Hint:* Do you think test scores are related to each other?)

4. **Scottish hill races 2008.** Hill running—races up and down hills—has a written history in Scotland dating back to the year 1040. Races are held throughout the year

at different locations around Scotland. A recent compilation of information for 91 races (for which full information was available and omitting two unusual races) includes the *Distance* (km), the *Climb* (m), and the *Record Time* (minutes). A regression to predict the men's records as of 2008 looks like this:

Dependent variable is: Time (mins)
R-squared = 98.0%   R-squared (adjusted) = 98.0%
s = 6.623 with 90 − 3 = 87 degrees of freedom

Source	Sum of Squares	df	Mean Square	F-ratio
Regression	189204	2	94602.1	2157
Residual	3815.92	87	43.8612	

Variable	Coefficient	SE(Coeff)	t-ratio	P-value
Intercept	−10.3723	1.245	−8.33	≤0.0001
Climb (m)	0.034227	0.0022	15.7	≤0.0001
Distance (km)	4.04204	0.1448	27.9	≤0.0001

a) Write the regression equation. Give a brief report on what it says about men's record times in hill races.
b) Interpret the value of R^2 in this regression.
c) What does the coefficient of *Climb* mean in this regression?

5. **Home prices.** Many variables have an impact on determining the price of a house. A few of these are *Size* of the house (square feet), *Lotsize,* and number of *Bathrooms.* Information for a random sample of homes for sale in the Statesboro, GA, area was obtained from the Internet. Regression output modeling the *Asking Price* with *Square Footage* and number of *Bathrooms* gave the following result:

Dependent Variable is: Asking Price
s = 67013   R-Sq = 71.1%   R-Sq (adj) = 64.6%

Predictor	Coeff	SE(Coeff)	t-ratio	P-value
Intercept	−152037	85619	−1.78	0.110
Baths	9530	40826	0.23	0.821
Sq ft	139.87	46.67	3.00	0.015

Analysis of Variance

Source	DF	SS	MS	F	P-value
Regression	2	99303550067	49651775033	11.06	0.004
Residual	9	40416679100	4490742122		
Total	11	1.39720E+11			

a) Write the regression equation.
b) How much of the variation in home asking prices is accounted for by the model?
c) Explain in context what the coefficient of *Square Footage* means.
d) The owner of a construction firm, upon seeing this model, objects because the model says that the number of bathrooms has no effect on the price of the home. He says that when *he* adds another bathroom, it increases the value. Is it true that the number of bathrooms is unrelated to house price? (*Hint:* Do you think bigger houses have more bathrooms?)

T 6. **More hill races 2008.** Here is the regression for the women's records for the same Scottish hill races we considered in Exercise 4:

Dependent variable is: Women's Time (mins)
R-squared = 96.7%   R-squared (adjusted) = 96.7%
s = 10.06 with 90 − 3 = 87 degrees of freedom

Source	Sum of Squares	df	Mean Square	F-ratio
Regression	261029	2	130515	1288
Residual	8813.02	87	101.299	

Variable	Coefficient	SE(Coeff)	t-ratio	P-value
Intercept	−11.6545	1.891	−6.16	<0.0001
Climb (m)	0.045195	0.0033	13.7	<0.0001
Distance	4.43427	0.2200	20.2	<0.0001

a) Compare the regression model for the women's records with that found for the men's records in Exercise 4.

Here's a scatterplot of the residuals for this regression:

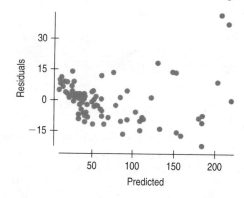

b) Discuss the residuals and what they say about the assumptions and conditions for this regression.

7. **Predicting finals II.** Here are some diagnostic plots for the final exam data from Exercise 3. These were generated by a computer package and may look different from the plots generated by the packages you use. (In particular, note that the axes of the Normal probability plot are swapped relative to the plots we've made in the text. We only care about the pattern of this plot, so it shouldn't affect your interpretation.) Examine these plots and discuss whether the assumptions and conditions for the multiple regression seem reasonable.

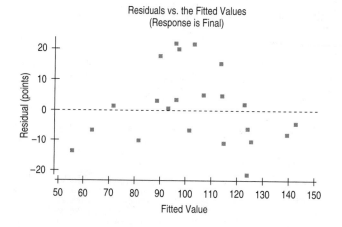

Residuals vs. the Fitted Values
(Response is Final)

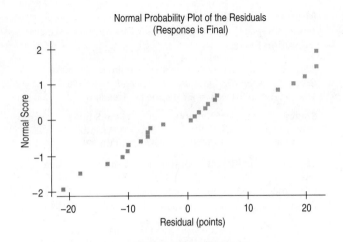

Normal Probability Plot of the Residuals
(Response is Final)

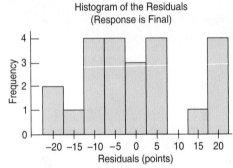

Histogram of the Residuals
(Response is Final)

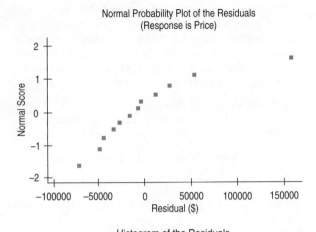

Normal Probability Plot of the Residuals
(Response is Price)

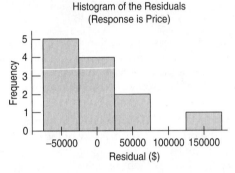

Histogram of the Residuals
(Response is Price)

8. **Home prices II.** Here are some diagnostic plots for the home prices data from Exercise 5. These were generated by a computer package and may look different from the plots generated by the packages you use. (In particular, note that the axes of the Normal probability plot are swapped relative to the plots we've made in the text. We only care about the pattern of this plot, so it shouldn't affect your interpretation.) Examine these plots and discuss whether the assumptions and conditions for the multiple regression seem reasonable.

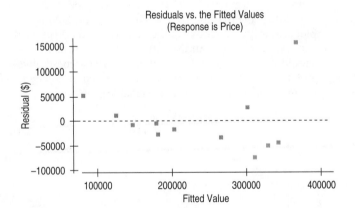

Residuals vs. the Fitted Values
(Response is Price)

9. **Secretary performance.** The AFL-CIO has undertaken a study of 30 secretaries' yearly salaries (in thousands of dollars). The organization wants to predict salaries from several other variables.

The variables considered to be potential predictors of salary are:

X1 = months of service
X2 = years of education
X3 = score on standardized test
X4 = words per minute (wpm) typing speed
X5 = ability to take dictation in words per minute

A multiple regression model with all five variables was run on a computer package, resulting in the following output:

Variable	Coefficient	Std. Error	t-value
Intercept	9.788	0.377	25.960
X1	0.110	0.019	5.178
X2	0.053	0.038	1.369
X3	0.071	0.064	1.119
X4	0.004	0.307	0.013
X5	0.065	0.038	1.734

$s = 0.430$ $R^2 = 0.863$

Assume that the residual plots show no violations of the conditions for using a linear regression model.

a) What is the regression equation?
b) From this model, what is the predicted *Salary* (in thousands of dollars) of a secretary with 10 years (120 months) of experience, 9th grade education (9 years of education), a 50 on the standardized test, 60 wpm typing speed, and the ability to take 30 wpm dictation?

c) Test whether the coefficient for words per minute of typing speed (*X4*) is significantly different from zero at $\alpha = 0.05$.

d) How might this model be improved?

e) A correlation of *Age* with *Salary* finds $r = 0.682$, and the scatterplot shows a moderately strong positive linear association. However, if X6 = *Age* is added to the multiple regression, the estimated coefficient of *Age* turns out to be $b_6 = -0.154$. Explain some possible causes for this apparent change of direction in the relationship between age and salary.

10. **GPA and SATs.** A large section of Stat 101 was asked to fill out a survey on grade point average and SAT scores. A regression was run to find out how well Math and Verbal SAT scores could predict academic performance as measured by GPA. The regression was run on a computer package with the following output:

Response: GPA

Variable	Coefficient	SE(Coeff)	t-ratio	P-value
Intercept	0.574968	0.253874	2.26	0.0249
SAT Verbal	0.001394	0.000519	2.69	0.0080
SAT Math	0.001978	0.000526	3.76	0.0002

a) What is the regression equation?

b) From this model, what is the predicted GPA of a student with an SAT Verbal score of 500 and an SAT Math score of 550?

c) What else would you want to know about this regression before writing a report about the relationship between SAT scores and grade point averages? Why would these be important to know?

T 11. **Body fat revisited.** The data set on body fat contains 15 body measurements on 250 men from 22 to 81 years old. Is average *%Body Fat* related to *Weight*? Here's a scatterplot:

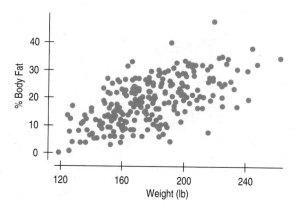

And here's the simple regression:

Dependent variable is: Pct BF
R-squared = 38.1%   R-squared (adjusted) = 37.9%
s = 6.538 with 250 − 2 = 248 degrees of freedom

Variable	Coefficient	SE(Coeff)	t-ratio	P-value
Intercept	−14.6931	2.760	−5.32	<0.0001
Weight	0.18937	0.0153	12.4	<0.0001

a) Is the coefficient of *%Body Fat* on *Weight* statistically distinguishable from 0? (Perform a hypothesis test.)

b) What does the slope coefficient mean in this regression?

We saw before that the slopes of both *Waist* size and *Height* are statistically significant when entered into a multiple regression equation. What happens if we add *Weight* to that regression? Recall that we've already checked the assumptions and conditions for regression on *Waist* size and *Height* in the chapter. Here is the output from a regression on all three variables:

Dependent variable is: Pct BF
R-squared = 72.5%   R-squared (adjusted) = 72.2%
s = 4.376 with 250 − 4 = 246 degrees of freedom

Source	Sum of Squares	df	Mean Square	F-ratio
Regression	12418.7	3	4139.57	216
Residual	4710.11	246	19.1468	

Variable	Coefficient	SE(Coeff)	t-ratio	P-value
Intercept	−31.4830	11.54	−2.73	0.0068
Waist	2.31848	0.1820	12.7	<0.0001
Height	−0.224932	0.1583	−1.42	0.1567
Weight	−0.100572	0.0310	−3.25	0.0013

c) Interpret the slope for *Weight*. How can the coefficient for *Weight* in this model be negative when its coefficient was positive in the simple regression model?

d) What does the P-value for *Height* mean in this regression? (Perform the hypothesis test.)

T 12. **Breakfast cereals.** We saw in Chapter 8 that the calorie content of a breakfast cereal is linearly associated with its sugar content. Is that the whole story? Here's the output of a regression model that regresses *Calories* for each serving on its *Protein(g)*, *Fat(g)*, *Fiber(g)*, *Carbohydrate(g)*, and *Sugars(g)* content.

Dependent variable is: Calories
R-squared = 93.6%   R-squared (adjusted) = 93.1%
s = 5.113 with 77 − 6 = 71 degrees of freedom

Source	Sum of Squares	df	Mean Square	F-ratio
Regression	26995.9	5	5399.18	207
Residual	1856.03	71	26.1412	

Variable	Coefficient	SE(Coeff)	t-ratio	P-value
Intercept	−0.879994	4.383	−0.201	0.8414
Protein	3.60495	0.6977	5.17	≤0.0001
Fat	8.56877	0.6625	12.9	≤0.0001
Fiber	0.309180	0.3337	0.927	3.572
Carbo	4.13996	0.2049	20.2	≤0.0001
Sugars	4.00677	0.1719	23.3	≤0.0001

Assuming that the conditions for multiple regression are met,

a) What is the regression equation?

b) Do you think this model would do a reasonably good job at predicting calories? Explain.

c) To check the conditions, what plots of the data might you want to examine?

d) What does the coefficient of *Fat* mean in this model?

13. Body fat again. Chest size might be a good predictor of body fat. Here's a scatterplot of *%Body Fat* vs. *Chest Size.*

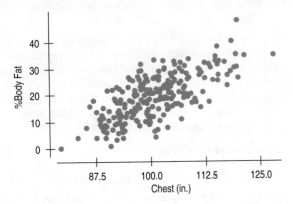

A regression of *%Body Fat* on *Chest Size* gives the following equation:

Dependent variable is: Pct BF
R-squared = 49.1%   R-squared (adjusted) = 48.9%
S = 5.930 with 250 − 2 = 248 degrees of freedom

Variable	Coefficient	SE(Coeff)	t-ratio	P-value
Intercept	−52.7122	4.654	−11.3	<0.0001
Chest Size	0.712720	0.0461	15.5	<0.0001

a) Is the slope of *%Body Fat* on *Chest Size* statistically distinguishable from 0? (Perform a hypothesis test.)
b) What does the answer in part a mean about the relationship between *%Body Fat* and *Chest Size*?

We saw before that the slopes of both *Waist* size and *Height* are statistically significant when entered into a multiple regression equation. What happens if we add *Chest Size* to that regression? Here is the output from a regression on all three variables:

Dependent variable is: Pct BF
R-squared = 72.2%   R-squared (adjusted) = 71.9%
s = 4.399 with 250 − 4 = 246 degrees of freedom

Source	Sum of Squares	df	Mean Square	F-ratio	P
Regression	12368.9	3	4122.98	213	<0.0001
Residual	4759.87	246	19.3491		

Variable	Coefficient	SE(Coeff)	t-ratio	P-value
Intercept	2.07220	7.802	0.266	0.7908
Waist	2.19939	0.1675	13.1	<0.0001
Height	−0.561058	0.1094	−5.13	<0.0001
Chest Size	−0.233531	0.0832	−2.81	0.0054

c) Interpret the coefficient for *Chest Size.*
d) Would you consider removing any of the variables from this regression model? Why or why not?

14. Grades. The following table shows the five scores from an Introductory Statistics course. Find a model for final exam score by trying all possible models with two predictor variables. Which model would you choose? Be sure to check the conditions for multiple regression.

Name	Final	Midterm 1	Midterm 2	Project	Home-work
Timothy F.	117	82	30	10.5	61
Karen E.	183	96	68	11.3	72
Verena Z.	124	57	82	11.3	69
Jonathan A.	177	89	92	10.5	84
Elizabeth L.	169	88	86	10.6	84
Patrick M.	164	93	81	10	71
Julia E.	134	90	83	11.3	79
Thomas A.	98	83	21	11.2	51
Marshall K.	136	59	62	9.1	58
Justin E.	183	89	57	10.7	79
Alexandra E.	171	83	86	11.5	78
Christopher B.	173	95	75	8	77
Justin C.	164	81	66	10.7	66
Miguel A.	150	86	63	8	74
Brian J.	153	81	86	9.2	76
Gregory J.	149	81	87	9.2	75
Kristina G.	178	98	96	9.3	84
Timothy B.	75	50	27	10	20
Jason C.	159	91	83	10.6	71
Whitney E.	157	87	89	10.5	85
Alexis P.	158	90	91	11.3	68
Nicholas T.	171	95	82	10.5	68
Amandeep S.	173	91	37	10.6	54
Irena R.	165	93	81	9.3	82
Yvon T.	168	88	66	10.5	82
Sara M.	186	99	90	7.5	77
Annie P.	157	89	92	10.3	68
Benjamin S.	177	87	62	10	72
David W.	170	92	66	11.5	78
Josef H.	78	62	43	9.1	56
Rebecca S.	191	93	87	11.2	80
Joshua D.	169	95	93	9.1	87
Ian M.	170	93	65	9.5	66
Katharine A.	172	92	98	10	77
Emily R.	168	91	95	10.7	83
Brian M.	179	92	80	11.5	82
Shad M.	148	61	58	10.5	65
Michael R.	103	55	65	10.3	51
Israel M.	144	76	88	9.2	67
Iris J.	155	63	62	7.5	67
Mark G.	141	89	66	8	72
Peter H.	138	91	42	11.5	66
Catherine R.M.	180	90	85	11.2	78
Christina M.	120	75	62	9.1	72
Enrique J.	86	75	46	10.3	72
Sarah K.	151	91	65	9.3	77
Thomas J.	149	84	70	8	70
Sonya P.	163	94	92	10.5	81
Michael B.	153	93	78	10.3	72
Wesley M.	172	91	58	10.5	66
Mark R.	165	91	61	10.5	79
Adam J.	155	89	86	9.1	62

(continued)

Name	Final	Midterm 1	Midterm 2	Project	Home-work
Jared A.	181	98	92	11.2	83
Michael T.	172	96	51	9.1	83
Kathryn D.	177	95	95	10	87
Nicole M.	189	98	89	7.5	77
Wayne E.	161	89	79	9.5	44
Elizabeth S.	146	93	89	10.7	73
John R.	147	74	64	9.1	72
Valentin A.	160	97	96	9.1	80
David T.O.	159	94	90	10.6	88
Marc I.	101	81	89	9.5	62
Samuel E.	154	94	85	10.5	76
Brooke S.	183	92	90	9.5	86

T 15. Fifty states 2009. Here is a data set on various measures of the 50 United States. The *Murder* rate is per 100,000, *HS Graduation* rate is in %, *Income* is per capita income in dollars, *Illiteracy* rate is per 1000, and *Life Expectancy* is in years. Find a regression model for *Life Expectancy* with three predictor variables by trying all four of the possible models.

a) Which model appears to do the best?
b) Would you leave all three predictors in this model?
c) Does this model mean that by changing the levels of the predictors in this equation, we could affect life expectancy in that state? Explain.
d) Be sure to check the conditions for multiple regression. What do you conclude?

State	Life expectancy	Murder rate08	Graduation Rate	Income/cap07	Illiteracy
Alabama	74.4	7.6	62%	27557	0.15
Alaska	77.1	4.1	67	34316	0.09
Arizona	77.5	6.3	59	28088	0.13
Arkansas	75.2	5.7	72	25563	0.14
California	78.2	5.8	68	35352	0.23
Colorado	78.2	3.2	68	34902	0.10
Connecticut	78.7	3.5	75	46021	0.09
Delaware	76.8	6.5	73	34533	0.11
Florida	77.5	6.4	59	32693	0.20
Georgia	75.3	6.6	54	28452	0.17
Hawaii	80.0	1.9	69	33369	0.16
Idaho	77.9	1.5	78	26530	0.11
Illinois	76.4	6.1	78	34290	0.13
Indiana	76.1	5.1	74	28587	0.08
Iowa	78.3	2.5	93	29784	0.07
Kansas	77.3	4	76	31268	0.08
Kentucky	75.2	4.6	71	26457	0.12
Louisiana	74.2	11.9	69	29557	0.16
Maine	77.6	2.4	78	28677	0.07
Maryland	76.3	8.8	75	39136	0.11
Massachusetts	78.4	2.6	75	41740	0.10
Michigan	76.3	5.4	75	29837	0.08
Minnesota	78.8	2.1	82	34896	0.06
Mississippi	73.6	8.1	62	24530	0.16

State	Life expectancy	Murder rate08	Graduation Rate	Income/cap07	Illiteracy
Missouri	75.9	7.7	75	29245	0.07
Montana	77.2	2.4	83	27602	0.09
Nebraska	77.8	3.8	85	31015	0.07
Nevada	75.8	6.3	58	34424	0.16
New Hampshire	78.3	1	71	35302	0.06
New Jersey	77.5	4.3	75	41835	0.17
New Mexico	77.0	7.2	65	26766	0.16
New York	77.7	4.3	70	40296	0.22
North Carolina	75.8	6.5	63	28604	0.14
North Dakota	78.3	0.5	88	29633	0.06
Ohio	76.2	4.7	77	29657	0.09
Oklahoma	75.2	5.8	74	29044	0.12
Oregon	77.8	2.2	67	29580	0.10
Pennsylvania	76.7	5.6	82	32986	0.13
Rhode Island	78.3	2.8	72	33560	0.08
South Carolina	74.8	6.8	62	26374	0.15
South Dakota	77.7	3.2	80	28833	0.07
Tennessee	75.1	6.6	60	28301	0.13
Texas	76.7	5.6	67	31624	0.19
Utah	78.7	1.4	81	26523	0.09
Vermont	78.2	2.7	84	31184	0.07
Virginia	76.8	4.7	74	35162	0.12
Washington	78.2	2.9	70	34368	0.10
West Virginia	75.1	3.3	82	25118	0.13
Wisconsin	77.9	2.6	85	30655	0.07
Wyoming	76.7	1.9	81	36760	0.09

T 16. Breakfast cereals again. We saw in Chapter 8 that the calorie count of a breakfast cereal is linearly associated with its sugar content. Can we predict the calories of a serving from its vitamin and mineral content? Here's a multiple regression model of *Calories* per serving on its *Sodium (mg)*, *Potassium (mg)*, and *Sugars (g)*:

Dependent variable is: Calories
R-squared = 38.4%   R-squared (adjusted) = 35.9%
s = 15.60 with 77 − 4 = 73 degrees of freedom

Source	Sum of Squares	dF	Mean Square	F-ratio	P-value
Regression	11091.8	3	3697.28	15.2	<0.0001
Residual	17760.1	73	243.289		

Variable	Coefficient	SE(Coeff)	t-ratio	P-value
Intercept	83.0469	5.198	16.0	<0.0001
Sodium	0.05721	0.0215	2.67	0.0094
Potass	−0.01933	0.0251	−0.769	0.4441
Sugars	2.38757	0.4066	5.87	<0.0001

Assuming that the conditions for multiple regression are met,

a) What is the regression equation?
b) Do you think this model would do a reasonably good job at predicting calories? Explain.
c) Would you consider removing any of these predictor variables from the model? Why or why not?
d) To check the conditions, what plots of the data might you want to examine?

17. Burger King. Recall the Burger King menu data from Chapter 8. BK's nutrition sheet lists many variables. Here's a multiple regression to predict calories for Burger King foods from *Protein* content *(g)*, *Total Fat (g)*, *Carbohydrate (g)*, and *Sodium (mg)* per serving:

Dependent variable is: Calories
R-squared = 100.0%   R-squared (adjusted) = 100.0%
s = 3.140 with 31 − 5 = 26 degrees of freedom

Source	Sum of Squares	df	Mean Square	F-ratio
Regression	1419311	4	354828	35994
Residual	256.307	26	9.85796	

Variable	Coefficient	SE(Coeff)	t-ratio	P-value
Intercept	6.53412	2.425	2.69	0.0122
Protein	3.83855	0.0859	44.7	<0.0001
Total fat	9.14121	0.0779	117	<0.0001
Carbs	3.94033	0.0336	117	<0.0001
Na/Serv.	−0.69155	0.2970	−2.33	0.0279

a) Do you think this model would do a good job of predicting calories for a new BK menu item? Why or why not?

b) The mean of *Calories* is 455.5 with a standard deviation of 217.5. Discuss what the value of *s* in the regression means about how well the model fits the data.

c) Does the R^2 value of 100.0% mean that the residuals are all actually equal to zero?

JUST CHECKING

ANSWERS

1. 77.9% of the variation in *Maximum Wind Speed* can be accounted for by multiple regression on *Central Pressure* and *Year*.

2. In any given year, hurricanes with a *Central Pressure* that is 1 mb lower can be expected to have, on average, winds that are 0.933 kn faster.

3. First, the researcher is trying to prove his null hypothesis for this coefficient and, as we know, statistical inference won't permit that. Beyond that problem, we can't even be sure we understand the relationship of *Wind Speed* to *Year* from this analysis. For example, both *Central Pressure* and *Wind Speed* might be changing over time, but their relationship might well stay the same during any given year.

Multiple Regression Wisdom

Where have we been?

We've looked ahead in each of the preceding chapters, but this is a good time to take stock. Wisdom in building and interpreting multiple regressions uses all that we've discussed throughout this book–even histograms and scatterplots. But most important is to keep in mind that we use models to help us understand the world with data. This chapter is about building that understanding even when we use powerful, complex methods. And that's been our purpose all along.

R oller coasters are an old thrill that continues to grow in popularity. Engineers and designers compete to make them bigger and faster. For a two-minute ride on the best roller coasters, fans will wait hours. Can we learn what makes a roller coaster fast? Or how long the ride will last? Here are data on some of the fastest roller coasters in the world:

Name	Park	Country	Type	Duration (sec)	Speed (mph)	Height (ft)	Drop (ft)	Length (ft)	Inversion?
New Mexico Rattler	Cliff's Amusement Park	USA	Wooden	75	47	80	75	2750	No
Fujiyama	Fuji-Q Highlands	Japan	Steel	216	80.8	259.2	229.7	6708.67	No
Goliath	Six Flags Magic Mountain	USA	Steel	180	85	235	255	4500	No
Great American Scream Machine	Six Flags Great Adventure	USA	Steel	140	68	173	155	3800	Yes
Hangman	Wild Adventures	USA	Steel	125	55	115	95	2170	Yes
Hayabusa	Tokyo SummerLand	Japan	Steel	108	60.3	137.8	124.67	2559.1	No
Hercules	Dorney Park	USA	Wooden	135	65	95	151	4000	No
Hurricane	Myrtle Beach Pavilion	USA	Wooden	120	55	101.5	100	3800	No

TABLE 31.1
A small selection of coasters from the larger data set available on the DVD.

WHO	Roller coasters
WHAT	See Table 31.1. (For multiple regression we have to know "What" and the units for each variable.)
WHERE	Worldwide
WHEN	All were in operation in 2003.
SOURCE	The Roller Coaster DataBase, www.rcdb.com

Here are the variables and their units:

- *Type* indicates what kind of track the roller coaster has. The possible values are "wooden" and "steel." (The frame usually is of the same construction as the track, but doesn't have to be.)
- *Duration* is the duration of the ride in seconds.
- *Speed* is top speed in miles per hour.
- *Height* is maximum height above ground level in feet.
- *Drop* is greatest drop in feet.
- *Length* is total length of the track in feet.
- *Inversions* reports whether riders are turned upside down during the ride. It has the values "yes" or "no."

It's always a good idea to explore the data before starting to build a model. Let's first consider the ride's *Duration*. We have that information for only 63 of the 80 coasters in our data set, but there's no reason to believe that the data are missing in any patterned way so we'll look at those 63 coasters. The average *Duration* for these coasters is 142 seconds, but one ride is as short as 28 seconds and another as long as 240 seconds. It makes sense that the duration of the ride should depend on the length of the track. Here's the scatterplot of *Duration* against *Length* and the regression:

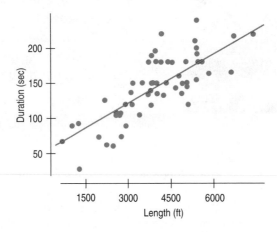

FIGURE 31.1
Duration *of the ride appears to be linearly related to the* Length *of the track.*

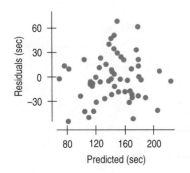

Dependent variable is: Duration
R-squared = 62.0%   R-squared (adjusted) = 61.4%
s = 27.23 with 63 − 2 = 61 degrees of freedom

Source	Sum of Squares	DF	Mean Square	F-ratio
Regression	73901.7	1	73901.7	99.6
Residual	45243.7	61	741.700	

Variable	Coefficient	SE(Coeff)	t-ratio	P-value
Intercept	53.9348	9.488	5.68	≤0.0001
Length	0.0231	0.0023	9.98	≤0.0001

The regression conditions seem to be met, and the regression makes sense. We'd expect longer tracks to give longer rides. From a base of 53.9 seconds, the duration of the ride increases by about 0.0231 seconds per foot of track—about 23 seconds more for each 1000 additional feet of track.

Indicators

Of course, there's more to these data. One interesting variable might not be one you'd naturally think of. Many modern coasters have "inversions." That's a nice way of saying that they turn riders upside down, with loops, corkscrews, or other devices. These inversions add excitement, but they must be carefully engineered, and that enforces some speed limits on that portion of the ride.

We'd like to add the information of whether the roller coaster has an inversion to our model. Until now, all our predictor variables have been quantitative. Whether or not a roller coaster has any inversions is a categorical variable ("yes" or "no"). Can we introduce the categorical variable *Inversions* as a predictor in our regression model? What would it mean if we did?

Let's start with a plot. Figure 31.2 shows the same scatterplot of duration against length, but now with the roller coasters that have inversions shown as red x's and a separate regression line drawn for each type of roller coaster.

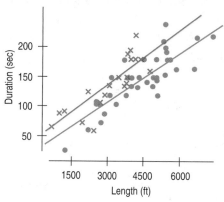

FIGURE 31.2
The two lines fit to coasters with inversions and without are roughly parallel.

It's easy to see that, for a given length, the roller coasters with inversions take a bit longer, and that for each type of roller coaster, the slopes of the relationship between duration and length are not quite equal but are similar.

We could split the data into two groups—coasters without inversions and those with inversions—and compute the regression for each group. That would look like this:

Dependent variable is: Duration

Cases selected according to: No inversions

R-squared = 69.4%   R-squared (adjusted) = 68.5%

s = 25.12 with 38 − 2 = 36 degrees of freedom

Variable	Coefficient	SE(Coeff)	t-ratio	P-value
Intercept	25.9961	14.10	1.84	0.0734
Length	0.0274	0.003	9.03	≤0.0001

Dependent variable is: Duration

Cases selected according to: Inversions

R-squared = 70.5%   R-squared (adjusted) = 69.2%

s = 23.20 with 25 − 2 = 23 degrees of freedom

Variable	Coefficient	SE(Coeff)	t-ratio	P-value
Intercept	47.6454	12.50	3.81	0.0009
Length	0.0299	0.004	7.41	≤0.0001

As the scatterplot showed, the slopes are very similar, but the intercepts are different.

When we have a situation like this with roughly parallel regressions for each group,[1] there's an easy way to add the group information to a single regression model. We make up a special variable that *indicates* what type of roller coaster we have, giving it the value 1 for roller coasters that have inversions and the value 0 for those that don't. (We could have reversed the coding; it's an

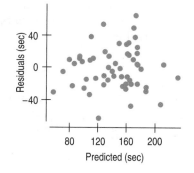

---

[1] The fact that the individual regression lines are nearly parallel is really a part of the Straight Enough Condition. You should check that the lines are nearly parallel before using this method. Or read on to see what to do if they are not parallel enough.

arbitrary choice.[2]) Such variables are called **indicator variables** or *indicators* because they indicate which of two categories each case is in.[3]

When we add our new indicator, *Inversions*, to the regression model, the model looks like this:

Dependent variable is: Duration
R-squared = 70.4%   R-squared (adjusted) = 69.4%
s = 24.24 with 63 − 3 = 60 degrees of freedom

Variable	Coefficient	SE(Coeff)	t-ratio	P-value
Intercept	22.3909	11.39	1.97	0.0539
Length	0.028239	0.0024	11.7	<0.0001
Inversions	30.0824	7.290	4.13	<0.0001

This looks like a better model than the simple regression for all the data. The R^2 is larger, the *t*-ratios of both coefficients are large, and the residuals look reasonable. But what does the coefficient for *Inversions* mean?

Let's see how an indicator variable works when we calculate predicted values for two of the roller coasters given at the start of the chapter:

Name	Park	Country	Type	Duration	Speed	Height	Drop	Length	Inversion?
Hangman	Wild Adventures	USA	Steel	125	55	115	95	2170	Yes
Hayabusa	Tokyo SummerLand	Japan	Steel	108	60.3	137.8	124.67	2559.1	No

The model says that for all coasters, the predicted *Duration* is

$$22.39 + 0.028 \times Length + 30.08 \times Inversions.$$

For *Hayabusa*, the length is 2559.1 feet and the value of *Inversions* is 0, so the model predicts a duration of[4]

$$22.3909 + 0.028239 \times 2559.1 + 30.0824 \times 0 = 94.66 \text{ seconds.}$$

That's not far from the actual duration of 108 seconds.

For the *Hangman*, the length is 2170 feet. It has an inversion, so the value of *Inversions* is 1, and the model predicts a duration of

$$22.3909 + 0.028239 \times 2170 + 30.0824 \times 1 = 113.75 \text{ seconds.}$$

That compares well with the actual duration of 125 seconds.

Notice how the indicator works in the model. When there is an inversion (as in *Hangman*), the value 1 for the indicator causes the amount of the indicator's coefficient, 30.0824, to be *added* to the prediction. When there is no inversion (as in *Hayabusa*), the indicator is zero, so nothing is added. Looking back at the scatterplot, we can see that this is exactly what we need. The difference between the two lines is a vertical shift of about 30 seconds.

This may seem a bit confusing at first. We usually think of the coefficients in a multiple regression as slopes. For indicator variables, however, they act differently. They're vertical shifts that keep the slopes for the other variables apart.

---

[2]Some implementations of indicator variables use −1 and 1 for the levels of the categories.
[3]They are also commonly called *dummies* or *dummy variables*. But this sounds like an insult, so the more politically correct term is *indicator variable.*
[4]We round coefficient values when we write the model but calculate with the full precision, rounding at the end of the calculation.

//// **FOR EXAMPLE** | **Using Indicator Variables**

As a class project, students in a large Statistics class collected publicly available information on recent home sales in their home-towns. There are 894 properties. These are not a random sample, but they may be representative of home sales during a short period of time, nationwide. In Chapter 30 we looked at these data and constructed a multiple regression model. Let's look further. Among the variables available is an indication of whether the home was in an urban, suburban, or rural setting.

QUESTION: How can we incorporate information such as this in a multiple regression model?

We might suspect that homes in rural communities might differ in price from similar homes in urban or suburban settings. We can define an indicator (dummy) variable to be 1 for homes in rural communities and 0 otherwise. A scatterplot shows that rural homes have, on average, lower prices for a given living area:

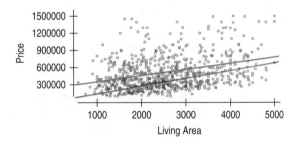

The multiple regression model is

Dependent variable is: Price
R-squared = 18.4%   R-squared (adjusted) = 18.2%
s = 260996 with 894 − 3 = 891 degrees of freedom

Variable	Coefficient	SE(Coeff)	t-ratio	P-value
Intercept	230945	25706	8.98	≤0.0001
Living area	112.534	9.353	12.0	≤0.0001
Rural	−172359	23749	−7.26	≤0.0001

The coefficient of Rural indicates that, for a given living area, rural homes sell for on average about $172,000 less than comparable homes in urban or suburban settings.

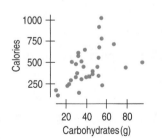

FIGURE 31.3
Calories of Burger King foods plotted against Carbohydrates *seems to fan out.*

Adjusting for Different Slopes

What if the lines aren't parallel? An indicator variable that is 0 or 1 can only shift the line up and down. It can't change the slope, so it works only when we have lines with the same slope and different intercepts.

Let's return to the Burger King data we looked at in Chapter 8 and look at how *Calories* are related to *Carbohydrates* (*Carbs* for short). Figure 31.3 shows the scatterplot.

It's not surprising to see that more *Carbs* goes with more *Calories*, but the plot seems to thicken as we move from left to right. Could there be something else going on?[5]

Burger King foods can be divided into two groups: those with meat (including chicken and fish) and those without. When we color the plot (red for

---

[5]Would we even ask if there weren't?

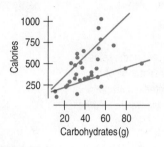

FIGURE 31.4
Plotting the meat-based and non-meat items separately, we see two distinct linear patterns.

meat, blue for non-meat) and look at the regressions for each group, we see a different picture.

Clearly, meat-based dishes have more calories for each gram of carbohydrate than do other Burger King foods. But the regression model can't account for the kind of difference we see here by just including an indicator variable. It isn't just the height of the lines that is different; they have entirely different slopes. How can we deal with that in our regression model?

The trick is to adjust the slopes with another constructed variable. This one is the *product* of an indicator for one group and the predictor variable. The co-efficient of this constructed **interaction term** in a multiple regression gives an adjustment to the slope, b_1, to be made for the individuals in the indicated group.[6] Here we have the indicator variable *Meat*, which is 1 for meat-containing foods and 0 for the others. We then construct an interaction variable, *Carbs*Meat*, which is just the product of those two variables. That's right; just multiply them. The resulting variable has the value of *Carbs* for foods containing meat (those coded 1 in the *Meat* indicator) and the value 0 for the others. By including the interaction variable in the model, we can adjust the slope of the line fit to the meat-containing foods. Here's the resulting analysis:

Dependent variable is: Calories
R-squared = 78.1%   R-squared (adjusted) = 75.7%
s = 106.0 with 32 − 4 = 28 degrees of freedom

Source	Sum of Squares	DF	Mean Square	F-ratio
Regression	1119979	3	373326	33.2
Residual	314843	28	11244.4	

Variable	Coefficient	SE(Coeff)	t-ratio	P-value
Intercept	137.395	58.72	2.34	0.0267
Carbs(g)	3.93317	1.113	3.53	0.0014
Meat	−26.1567	98.48	−0.266	0.7925
Carbs*Meat	7.87530	2.179	3.61	0.0012

What does the coefficient for the indicator *Meat* mean? It provides a different intercept to separate the meat and non-meat items at the origin (where *Carbs* = 0). For these data, there is a different slope, but the two lines nearly meet at the origin, so there seems to be no need for an additional adjustment. The estimated difference of 26.16 calories is small. That's why the coefficient for *Meat* has a small *t*-statistic.

By contrast, the coefficient of the interaction term, *Carbs*Meat*, says that the slope relating calories to carbohydrates is steeper by 7.875 calories per carbohydrate gram for meat-containing foods than for meat-free foods. Its small P-value suggests that this difference is real.

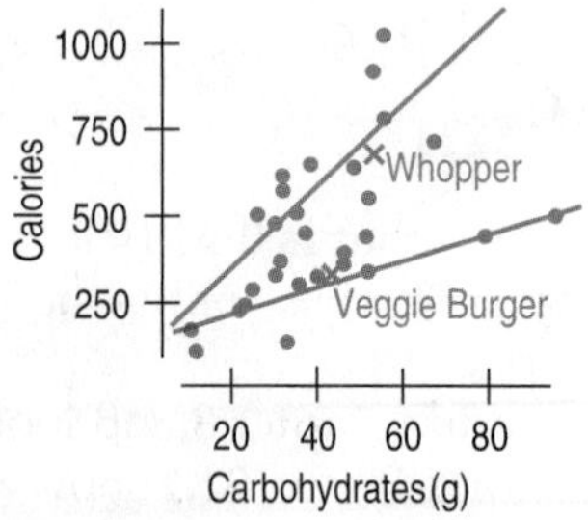

$$137.40 + 3.93\,Carbs - 26.16\,Meat + 7.88\,Carbs*Meat$$

Let's see how these adjustments work. A BK Whopper has 53g of *Carbohydrates* and is a meat dish. The model predicts its *Calories* as

$$137.395 + 3.93317 \times 53 - 26.1567 \times 1 + 7.8753 \times 53 \times 1 = 737.1,$$

not far from the measured calorie count of 680. By contrast, the Veggie Burger, with 43g of *Carbohydrates*, is predicted to have

$$137.395 + 3.93317 \times 43 - 26.1567 \times 0 + 7.87530 \times 0 \times 43 = 306.5\,calories,$$

not far from the 330 measured officially. The last two terms in the equation for the Veggie Burger are just zero because the indicator for *Meat* is 0 for the Veggie Burger.

FIGURE 31.5
The Whopper and Veggie Burger belong to different groups.

---

[6]Chapter 29 discussed interaction effects in two-way ANOVA. Interaction terms such as these are exactly the same idea.

Diagnosing Regression Models: Looking at the Cases

We often use regression analyses to try to understand the world. By working with the data and creating models, we can learn a great deal about the relationships among variables. As we saw with simple regression, sometimes we can learn as much from the cases that *don't* fit the model as from the bulk of cases that do. Extraordinary cases often tell us more about the world simply by the ways in which they fail to conform and the reasons we can discover for those deviations.

If a case doesn't conform to the others, we should identify it and, if possible, understand why it is different. As in simple regression, a case can be extraordinary by standing away from the model in the *y* direction or by having unusual values in an *x*-variable. In multiple regression it can also be extraordinary by having an unusual *combination* of values in the *x*-variables. Deviations in the *y* direction show up in the residuals. Deviations in the *x*'s show up as *leverage*.

Leverage

Recent events have focused attention on airport screening of passengers. But screening has a longer history. The *Sourcebook of Criminal Justice Statistics Online* lists the numbers of various violations found by airport screeners for each of several types of violations in each year from 1977 to 1999. Here's a regression of the number of long guns (rifles and the like) found vs. the number of times false information was discovered.

Dependent variable is: Long guns
R-squared = 7.8%   R-squared (adjusted) = 3.4%
s = 38.67 with 23 − 2 = 21 degrees of freedom

Variable	Coefficient	SE(Coeff)	t-ratio	P-value
Intercept	78.9069	13.60	5.80	≤0.0001
False info	0.242679	0.1823	1.33	0.1975

That summary doesn't look like it's a particularly successful regression. The R^2 is only 7.8%, and the P-value for *False Info* is large. But a look at the scatterplot tells us more.

The unusual case is from 1988 when, for some reason, the number of false information reports jumped over 200. The resulting case has high leverage because it is so far from the *x*-values of the other points. It's easy to see the influence of that one high-leverage case if we look at the regression lines with and without that case (Figure 31.7).

The **leverage** of a case measures its ability to move the regression model all by itself by just moving in the *y* direction. In Chapter 9, when we had only one predictor variable, we could *see* high leverage points in a scatterplot because they stood far from the mean of *x*. But now, with several predictors, we can't count on seeing them in our plots.

Fortunately, we can put a number on the leverage. If we keep everything else the same, change the *y*-value of a case by 1.0, and find a new regression, the leverage of that case is the amount by which the *predicted* value at that case would change. Leverage can never be greater than 1.0—we wouldn't expect the line to move *farther* than we move the case, only to try to keep up. Nor can it be less than 0.0—we'd hardly expect the line to move in the *opposite* direction. A point with zero leverage has no effect at all on the regression model, although it does participate in the calculations of R^2, s, and the *F*- and *t*-statistics.

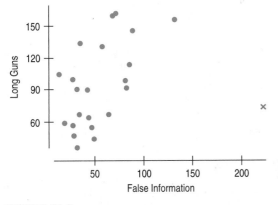

FIGURE 31.6

A high-leverage point can hide a strong relationship, so that you can't see it in the regression. Make a plot.

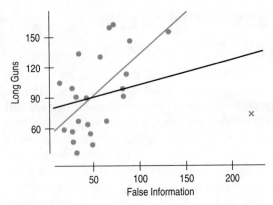

FIGURE 31.7
A single high-leverage point can change the regression slope quite a bit. The line omitting the point for 1988 is quite different from the line that includes the outlier.

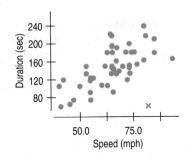

For the airport inspections, the leverage of 1988 is 0.63. That's quite high. If there had been even one fewer long gun discovered that year (decreasing the *observed* y-value by 1), the *predicted* y-value for 1988 would have decreased by 0.63, dragging the regression line down still farther. For comparison, the next highest leverage of any other case is only 0.155.

The leverage of a case is a measure of how far that case is from the center of the x's. As always in Statistics, we expect to measure that distance with a ruler based on a standard deviation—here, the standard deviation of the x's. And that's really all the leverage is: an indication of how far each point is away from the center of all the x-values, measured in standard deviations. Fortunately, there's a less tedious way to calculate leverage than moving each case in turn, but it's beyond the scope of this book and you'd never want to do it by hand anyway. So just let the computer do the computing and think about what the result *means*. Most statistics programs calculate leverage values, and you should examine them.

A case can have large leverage in two different ways:

- It might be extraordinary in one or more individual variables. For example, the fastest or slowest roller coaster may stand out.
- It may be extraordinary in a *combination* of variables. For example, one roller coaster stands out in the scatterplot of *Duration* against *Speed*. It isn't extraordinarily fast and others have shorter duration, but the combination of high speed and short duration is unusual. Looking at leverage values can be a very effective way to discover cases that are extraordinary on a combination of x-variables.

There are no tests for whether the leverage of a case is too large. The average leverage value among all cases in a regression is $1/n$, but that doesn't give us much of a guide. One common approach is to just make a histogram of the leverages. Any case whose leverage stands out in a histogram of leverages probably deserves special attention. You may decide to leave the case in the regression or to see how the regression model changes when you delete the case, but you should be aware of its potential to influence the regression.

FOR EXAMPLE | Diagnosing a Regression

Here's another regression model for the real estate data we looked at in the previous For Example.

Dependent variable is: Price
R squared = 23.1%   R squared (adjusted) = 22.8%
s = 253709 with 893 − 5 = 888 degrees of freedom

Variable	Coefficient	SE(Coeff)	t-ratio	P-value
Intercept	322470	40192	8.02	≤0.0001
Living area	92.6272	13.09	7.08	≤0.0001
Bedrooms	−69720.6	12764	−5.46	≤0.0001
Bathrooms	82577.6	13410	6.16	≤0.0001
Rural	−161575	23313	−6.93	≤0.0001

QUESTION: What do diagnostic statistics tell us about these data and this model?

A boxplot of the leverage values shows one extraordinarily large leverage:

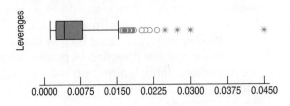

Investigation of that case reveals it to be a home that sold for $489,900 that has 8 bedrooms and only 2.5 baths. This is a particularly unusual combination, especially for a home with that value. If we were pursuing this analysis further, we'd want to check the records for this house to be sure that the number of bedrooms and bathrooms were recorded accurately.

Residuals and Standardized Residuals

Residuals are not all alike. Consider a point with leverage 1.0. That's the highest a leverage can be, and it means that the line follows the point perfectly. So, a point like that must have a zero residual. And since we know the residual exactly, that residual has zero standard deviation. This tendency is true in general: The larger the leverage, the *smaller* the standard deviation of its residual.[7]

When we want to compare values that have differing standard deviations, it's a good idea to standardize them.[8] We can do that with the regression residuals, dividing each one by an estimate of its own standard deviation. When we do that, the resulting values follow a Student's *t*-distribution. In fact, these standardized residuals are called **Studentized residuals.** It's a good idea to examine the Studentized residuals (rather than the simple residuals) to assess the Nearly Normal Condition and the **Does the Plot Thicken? Condition.** Any Studentized residual that stands out from the others deserves our attention.[9]

It may occur to you that we've always plotted the *unstandardized* residuals when we made regression models. And we've treated them as if they all had the same standard deviation when we checked the **Nearly Normal Condition.** It turns out that this was a simplification. It didn't matter much for simple regression, but for multiple regression models, it's a better idea to use the Studentized residuals when checking the Nearly Normal Condition. (Of course, Student's *t* isn't exactly Normal either—that's why we say "nearly" Normal.)

IT ALL FITS TOGETHER DEPARTMENT

Make an indicator variable for a single case—that is, construct a variable that is 0 everywhere except that it is 1 just for the case in question. When you include that indicator in the regression model, its *t*-ratio will be what that case's externally Studentized residual was in the original model without the indicator. That tells us that an externally Studentized residual can be used to perform a *t*-test of the null hypothesis that a case is *not* an outlier. If we reject that null hypothesis, we can call the point an outlier.[10]

---

[7]Technically, $SD(e_i) = \sigma\sqrt{1 - h_i}$ where h_i is the leverage of the *i*-th case, e_i is its residual, and σ is the standard deviation of the regression model errors.

[8] Be cautious when you encounter the term "standardized residual." It is used in different books and by different statistics packages to mean quite different things. Be sure to check the meaning.

[9] There's more than one way to Studentize residuals according to how you estimate σ. You may find statistics packages referring to *externally Studentized residuals* and *internally Studentized residuals*. It is the *externally* Studentized version that follows a *t*-distribution, so those are the ones we recommend.

[10]Finally we have a test to decide whether a case is an outlier. Up until now, all we've had was our judgment based on how the plots looked. But you must still use your common sense and understanding of the data to decide *why* the case is extraordinary and whether it should be corrected or removed from the analysis. That important decision is *still* a judgment call.

Influential Cases

A case that has *both* high leverage and large Studentized residuals is likely to have changed the regression model substantially all by itself. Such a case is said to be **influential.** An influential case cries out for special attention because removing it is likely to give a very different regression model.

The surest way to tell whether a case is influential is to omit it[11] and see how much the regression model changes. You should call a case "influential" if omitting it changes the regression model by enough to matter for *your* purposes. To identify possibly influential cases, check the leverage and Studentized residuals. Two statistics that combine leverage and Studentized residuals into a single measure of influence, Cook's distance and DFFITs, are offered by many statistics programs. If either measure is unusually large for a case, that case should be checked as a possibly influential point.

When a regression analysis has cases that have both high leverage and large Studentized residuals, it would be irresponsible to report only the regression on all the data. You should also compute and discuss the regression found with such cases removed, and discuss the extraordinary cases individually if they offer additional insight. If your interest is to understand the world, the extraordinary cases may well tell you more than the rest of the model. If your only interest is in the model (for example, because you hope to use it for prediction), then you'll want to be certain that the model wasn't determined by only a few influential cases, but instead was built on the broader base of the bulk of your data.

STEP-BY-STEP EXAMPLE   **Diagnosing a Multiple Regression**

Let's consider what makes a roller coaster fast and then diagnose the model to understand more. Roller coasters get their *Speed* from gravity (the "coaster" part), so we'd naturally look to such variables as the *Height* and largest *Drop* as predictors. Let's make and diagnose that multiple regression.

THINK	**Variables** Name the variables, report the W's, and specify the questions of interest. **Plot**	I have data on 75 roller coasters that give their top *Speed* (mph), maximum *Height*, and largest *Drop* (both in feet).

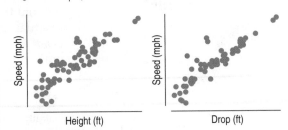

	Plan Think about the assumptions and check the conditions.	✔ **Straight Enough Condition:** The plots look reasonably straight. ✔ **Independence Assumption:** There are only a few manufacturers of roller coasters worldwide, and coasters made by the same

---

[11]Or, equivalently, include an indicator variable that selects only for that case.

company may be similar in some respects, but each roller coaster in our data is individualized for its site, so the coasters are likely to be independent.

Because these conditions are met I computed the regression model and found the Studentized residuals.

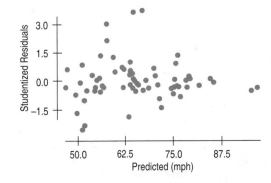

✔ **Straight Enough Condition (2):** The values for one roller coaster don't seem to affect the values for the others in any systematic fashion. This makes the independence assumption more plausible.

✔ **Does the Plot Thicken? Condition:** The scatterplot of Studentized residuals against predicted values shows no obvious changes in the spread about the line. There do seem to be some large residuals that might be outliers.

✔ **Nearly Normal Condition:** A histogram of the Studentized residuals is unimodal and symmetric.

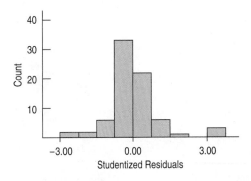

✖ **Outlier Condition:** The histogram does show three residuals separated from the others at the high end. I'll want to look at those.

Under these conditions, the multiple regression model is appropriate.

Actually, we need the Nearly Normal Condition only if we want to do inference, but it's hard not to look at the P-values, so we usually check it out. In a multiple regression, it's best to check the Studentized residuals, although the difference is rarely large enough to change our assessment of the normality.

Choose your method.

(continued)

Mechanics

Here is the computer output for the regression:

Dependent variable is: Speed
R-squared = 85.2%   R-squared (adjusted) = 84.8%
s = 4.633 with 75 − 3 = 72 degrees of freedom

Source	Sum of Squares	DF	Mean Square	F-ratio
Regression	8902.58	2	4451.29	207
Residual	1545.50	72	21.4652	

Variable	Coefficient	SE(Coeff)	t-ratio	P-value
Intercept	36.9098	1.502	24.6	≤0.0001
Height	0.067218	0.0195	3.46	0.0009
Drop	0.124870	0.0190	6.57	≤0.0001

The estimated regression equation is

$$\widehat{Speed} = 36.91 + 0.067\ Height + 0.125\ Drop.$$

Interpretation

The R^2 for the regression is 85.2%. *Height* and *Drop* account for 85% of the variation in *Speed* in roller coasters like these. Both *Height* and *Drop* contribute significantly to the *Speed* of a roller coaster.

Diagnosis

A histogram of the leverages shows one roller coaster with a rather high leverage of more than 0.24.

Leverage Most computer regression programs will calculate leverages. There is a leverage value for each case.

It may not be necessary to remove high leverage points from the model, but it's certainly wise to know where they are and, if possible, why they are unusual.

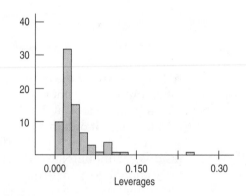

This high-leverage point is the *Oblivion* coaster in Alton, England. Neither the *Height* nor the *Drop* is extraordinary. To see what's going on, I made a scatterplot of *Drop* against *Height* with *Oblivion* shown as a red x.

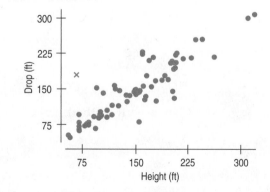

Although *Oblivion's* maximum height is a modest 65 feet, it has a surprisingly long drop of 180 feet. Looking it up, I discovered that the unusual feature of the *Oblivion* coaster is that it plunges riders down a deep hole below the ground.

Residuals At this point, we might consider recomputing the regression model after removing these three coasters. That's what we do in the next section.

The histogram of the Studentized residuals (above) also nominates some cases for special attention. That bar on the right of the histogram holds three roller coasters with large positive residuals: the *Xcelerator, Hypersonic XCL,* and *Volcano, the Blast Coaster.* New technologies such as hydraulics or compressed air are used to launch all three roller coasters. These three coasters are different in that their speed doesn't depend only on gravity.

Diagnosis Wrap-Up

What have we learned from diagnosing the regression? We've discovered four roller coasters that may be influencing the model. And for each of them, we've been able to understand why and how they differed from the others. The oddness of *Oblivion* in plunging into a hole in the ground may cause us to value *Drop* as a predictor of speed more than *Height.*

The three influential cases turned out to be different from the other roller coasters because they are "blast coasters" that don't rely only on gravity for their acceleration. Although we can't count on always discovering why influential cases are special, diagnosing influential cases raises the question of what about them might be different. Understanding influential cases can help us understand our data better.

When there are influential cases, we always want to consider the regression model without them:

The *Oblivion* roller coaster plunges into a hole in the ground.

Dependent variable is: Speed
R-squared = 92.7%   R-squared (adjusted) = 92.1%
s = 3.331 with 72 − 3 = 69 degrees of freedom

Variable	Coefficient	SE(Coeff)	t-ratio	P-value
Intercept	36.4715	1.084	33.7	≤0.0001
Drop	0.175161	0.0151	11.6	≤0.0001
Height	0.016047	0.0154	1.04	0.3013

Without the three blast coasters, *Height* no longer appears to be important in the model, so we might try omitting it:

Dependent variable is: Speed
R-squared = 92.1%   R-squared (adjusted) = 92.0%
s = 3.333 with 72 − 2 = 70 degrees of freedom

Variable	Coefficient	SE(Coeff)	t-ratio	P-value
Intercept	36.7620	1.048	35.1	≤0.0001
Drop	0.189301	0.0066	28.7	≤0.0001

That looks like a good model. It seems that our diagnosis has led us back to a simple regression.

INDICATORS FOR INFLUENCE

One good way to examine the effect of an extraordinary case on a regression is to construct a special indicator variable that is zero for all cases *except* the one we want to isolate. Including such an indicator in the regression model has the same effect as removing the case from the data, but it has two special advantages. First, it makes it clear to anyone looking at the regression model that we have treated that case specially. Second, the *t*-statistic for the indicator variable's coefficient can be used as a test of whether the case is influential. If the P-value is small, then that case really didn't fit well with the rest of the data. Typically, we name such an indicator with the identifier of the case we want to remove. Here's the last roller coaster model in which we have removed the influence of the three blast coasters by constructing indicators for them instead of by removing them from the data. Notice that the coefficients for the other predictors are just the same as the ones we found by omitting the cases.

Dependent variable is: Speed
R-squared = 92.1%   R-squared (adjusted) = 92.0%
s = 3.333 with 74 − 5 = 69 degrees of freedom

Variable	Coefficient	SE(Coeff)	t-ratio	P-value
Intercept	36.7620	1.048	35.1	≤0.0001
Drop	0.189301	0.0066	28.7	≤0.0001
Xcelerator	19.4678	3.536	5.51	≤0.0001
HyperSonic	17.5842	3.386	5.19	≤0.0001
Volcano, . . .	17.0282	3.534	4.82	≤0.0001

The P-values for the three indicator variables confirm that each of these roller coasters doesn't fit with the others.

The Best Multiple Regression Model

When many possible predictors are available, we will naturally want to select only a few of them for a regression model. But which ones? The first and most important thing to realize is that often there is no such thing as the "best" regression model. (After all, all models are wrong.) Several alternative models may be useful or insightful. The "best" for one purpose may not be best for another. And the one with the highest R^2 may well not be best for many purposes. There is nothing wrong with continuing to work with several models without choosing among them.

Multiple regressions are subtle. The coefficients often don't mean what they at first appear to mean. The choice of which predictors to use determines almost everything about the regression.

Predictors interact with each other, which complicates interpretation and understanding. So it is usually best to build a *parsimonious* model, using as few predictors as you can. On the other hand, we don't want to leave out predictors that are theoretically or practically important. Making this trade-off is

"It is the mark of an educated mind to be able to entertain a thought without accepting it."

—Aristotle

the heart of the challenge of selecting a good model. The best regression models, in addition to satisfying the assumptions and conditions of multiple regression, have:

- Relatively few predictors to keep the model simple.
- A relatively high R^2 indicating that much of the variability in y is accounted for by the regression model.
- A relatively small value of s, the standard deviation of the residuals, indicating that the magnitude of the errors is small.
- Relatively small P-values for their F- and t-statistics, showing that the overall model is better than a simple summary with the mean, and that the individual coefficients are reliably different from zero.
- No cases with extraordinarily high leverage that might dominate and alter the model.
- No cases with extraordinarily large residuals, and Studentized residuals that appear to be Nearly Normal. Outliers can alter the model and certainly weaken the power of any test statistics. And the Nearly Normal Condition is required for inference.
- Predictors that are reliably measured and relatively unrelated to each other.

The term "relatively" in this list is meant to suggest that you should favor models with these attributes over others that satisfy them less, but, of course, there are many trade-offs and no absolute rules.

Cases with high leverage or large residuals can be dealt with by introducing indicator variables.

In addition to favoring predictors that can be measured reliably, you may want to favor those that are less expensive to measure, especially if your model is intended for prediction with values not yet measured.

Seeking Multiple Regression Models Automatically

How can we find the best multiple regression model? The list of desirable features we just looked at should make it clear that there is no simple definition of the "best" model. A computer can try all combinations of the predictors to find the regression model with the highest R^2, or optimize some other criterion,[12] but models found that way are not best for all purposes, and may not even be particularly good for many purposes.

Another alternative is to have the computer build a regression "stepwise." In a **stepwise regression,** at each step, a predictor is either added to or removed from the model. The predictor chosen to add is the one whose addition increases the R^2 the most (or similarly improves some other measure). The predictor chosen to remove is the one whose removal reduces the R^2 least (or similarly loses the least on some other measure). The hope is that by following this path, the computer can settle on a good model. The model will gain or lose a predictor only if that change in the model makes a big enough improvement in the performance measure. The changes stop when no more changes pass the criterion.

---

[12]This is literally true. Even for many variables and a moderately large number of cases, it is computationally possible to find the "best subset" of predictors that maximizes R^2. Many statistics programs offer this capability.

STEPPING IN THE WRONG DIRECTION

Here's an example of how stepwise regression can go astray. We might want to find a regression to model *Horsepower* in a sample of cars from the car's engine size (*Displacement*) and its *Weight*. The simple correlations are as follows:

	HP	Disp	Wt
Horsepower	1.000		
Displacement	0.872	1.000	
Weight	0.917	0.951	1.000

Because *Weight* has a slightly higher correlation with *Horsepower*, stepwise regression will choose it first. Then, because *Weight* and engine size (*Displacement*) are so highly correlated, once *Weight* is in the model, *Displacement* won't be added to the model. But *Weight* is, at best, a lurking variable leading to both the need for more horsepower and a larger engine. Don't try to tell an engineer that the best way to increase horsepower is to add weight to the car and that the engine size isn't important! From an engineering standpoint, *Displacement* is a far more appropriate predictor of *Horsepower*, but stepwise regression for these data doesn't find that model.

Stepwise methods can be valuable when there are hundreds or thousands of potential predictors, as can happen in data mining applications. They can build models that are useful for prediction or as starting points in the search for better models. Because they do each step *automatically*, however, stepwise methods are inevitably affected by influential points and nonlinear relationships. A better strategy might be to mimic the stepwise procedure yourself, but more carefully. You could consider adding or removing a variable yourself with a careful look at the assumptions and conditions each time a variable is considered. That kind of guided stepwise method is still not guaranteed to find a good model, but it may be a sensible way to search among the potential candidates.

Building Multiple Regression Models

You can build a regression model by adding variables to a growing regression. Each time you add a predictor, you hope to account for a little more of the variation in the response. What's left over is the residuals. At each step, consider the predictors still available to you. Those that are most highly correlated with the current residuals are the ones that are most likely to improve the model.

If you see a variable with a high correlation at this stage and it is *not* among those that you thought were important, stop and think about it. Is it correlated with another predictor or with several other predictors? Don't let a variable that doesn't make sense enter the model just because it has a high correlation, but at the same time, don't exclude a predictor just because you didn't initially think it was important. (That would be a good way to make sure that you never learn anything new.) Finding the balance between these two choices underlies the art of successful model building.

Alternatively, you can start with all available predictors in the model and remove those with small *t*-ratios. At each step make a plot of the residuals to check for outliers, and check the leverages (say, with a histogram of the leverage values) to be sure there are no high-leverage points. Influential cases can strongly affect which variables appear to be good or poor predictors in the model. It's also a good idea to check that a predictor doesn't appear to be unimportant in the model only because it's correlated with other predictors in the model. It may (as is true of *Displacement* in the example of predicting *Horsepower*) actually be a more useful or meaningful predictor than some of those in the model.

In either method, adding or removing a predictor will usually change *all* of the coefficients, sometimes by quite a bit.

STEP-BY-STEP EXAMPLE   Building Multiple Regression Models

Let's return to the Kids Count infant mortality data. In Chapter 30, we fit a large multiple regression model in which several of the *t*-ratios for coefficients were too small to be discernibly different from zero. Maybe we can build a more parsimonious model. Which model should we build?

The most important thing to do is to think about the data. Regression models can and should make sense. Many factors can influence your choice of a model, including the cost of measuring particular predictors, the reliability or possible biases in some predictors, and even the political costs or advantages to selecting predictors.

THINK

Variables Name the available variables, report the W's, and specify the question of interest or the purpose of finding the regression model.

I have data on the 50 states. The available variables are (all for 1999):

Infant Mortality (deaths per 1000 live births)

Low Birth Weight (Low BW%—%babies with low birth weight)

Child Deaths (deaths per 100,000 children ages 1–14)

%Poverty (percent of children in *poverty* in the previous year)

HS Drop% (percent of teens who are high school *dropouts*, ages 16–19)

Teen Births (births per 100,000 females ages 15–17)

We've examined a scatterplot matrix and the regression with all potential predictors in Chapter 30.

Teen Deaths (by accident, homicide, and suicide; deaths per 100,000 teens ages 15–19)

I hope to gain a better understanding of factors that affect infant mortality.

Plan Think about the assumptions and check the conditions.

✔ **Straight Enough Condition:** The scatterplot matrix shows no bends, clumping, or outliers in any of the scatterplots.

✔ **Independence Assumption:** These data are based on random samples.

With this assumption and condition satisfied, I can compute the regression model and find residuals.

(continued)

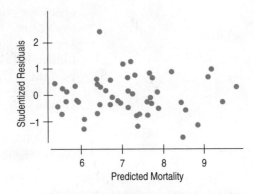

Remember that in a multiple regression, rather than plotting residuals against each of the predictors, we usually plot Studentized residuals against the predicted values.

✔ **Does the Plot Thicken? Condition:** This scatterplot of Studentized residuals vs. predicted values for the full model (all predictors) shows no obvious trends in the spread.

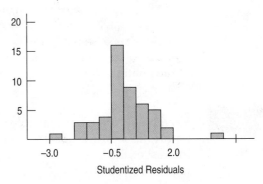

✘ **Nearly Normal Condition, Outlier Condition:** A histogram of the Studentized residuals from the full model is unimodal and symmetric, but it seems to have an outlier. The unusual state is South Dakota. I'll test whether it really is an outlier by making an indicator variable for South Dakota and including it in the predictors.

 Mechanics Multiple regressions are always found from a computer program.

I'll start with the full regression and work backward:

Dependent variable is: Infant mort
R-squared = 78.7%   R-squared (adjusted) = 75.2%
s = 0.6627 with 50 − 8 = 42 degrees of freedom

Variable	Coefficient	SE(Coeff)	t-ratio	P-value
Intercept	1.31183	0.8639	1.52	0.1364
Low BW%	0.73272	0.1067	6.87	≤0.0001
Child Deaths	0.02857	0.0123	2.31	0.0256
%Poverty	−5.3026e-3	0.0332	−0.160	0.8737
HS Drop%	−0.10754	0.0540	−1.99	0.0531
Teen Births	0.02402	0.0234	1.03	0.3111
Teen Deaths	−1.5516e-4	0.0101	−0.015	0.9878
S. Dakota	2.74813	0.7175	3.83	0.0004

For model building, look at the P-values only as general indicators of how much a predictor contributes to the model.

You shouldn't remove more than one predictor at a time from the model because each predictor can influence how the others contribute to the model. If removing a predictor from the model doesn't change the remaining coefficients very much (or reduce the R^2 by very much), that predictor wasn't contributing very much to the model.

Adjusted R^2 can increase when you remove a predictor if that predictor wasn't contributing very much to the regression model.

The coefficient for the S. Dakota indicator variable has a very small P-value, so that case is an outlier in this regression model. Teen Births, Teen Deaths, and %Poverty have large P-values and look like they are less successful predictors in this model.

I'll remove Teen Deaths first:

Dependent variable is: Infant mort
R-squared = 78.7%   R-squared (adjusted) = 75.7%
s = 0.6549 with 50 − 7 = 43 degrees of freedom

Variable	Coefficient	SE(Coeff)	t-ratio	P-value
Intercept	1.30595	0.7652	1.71	0.0951
Low BW%	0.73283	0.1052	6.97	≤0.0001
Child Deaths	0.02844	0.0085	3.34	0.0018
%Poverty	−5.3548e-3	0.0326	−0.164	0.8703
HS Drop%	−0.10749	0.0533	−2.02	0.0501
Teen Births	0.02402	0.0231	1.04	0.3053
S. Dakota	2.74651	0.7014	3.92	0.0003

Removing Teen Births and %Poverty, in turn, gives this model:

Dependent variable is: Infant mort
R-squared = 78.1%   R-squared (adjusted) = 76.2%
s = 0.6489 with 50 − 5 = 45 degrees of freedom

Variable	Coefficient	SE(Coeff)	t-ratio	P-value
Intercept	1.03782	0.6512	1.59	0.1180
Low BW%	0.78334	0.0934	8.38	≤0.0001
Child Deaths	0.03104	0.0075	4.12	0.0002
HS Drop%	−0.06732	0.0381	−1.77	0.0837
S. Dakota	2.66150	0.6899	3.86	0.0004

Compared with the full model, the R^2 has come down only very slightly, and the adjusted R^2 has actually increased. The P-value for HS Drop% is bigger than the standard .05 level, but more to the point, Child Deaths and Low Birth Weight are both variables that look at birth and early childhood. HS Drop% seems not to belong with them. When I take that variable out, the model looks like this:

Dependent variable is: Infant mort
R-squared = 76.6%   R-squared (adjusted) = 75.1%
s = 0.6638 with 50 − 4 = 46 degrees of freedom

Variable	Coefficient	SE(Coeff)	t-ratio	P-value
Intercept	0.760145	0.6465	1.18	0.2457
Child Deaths	0.026988	0.0073	3.67	0.0006
Low BW%	0.750461	0.0937	8.01	≤0.0001
S.Dakota	2.74057	0.7042	3.89	0.0003

This looks like a good model. It has a reasonably high R^2 and small P-values for each of the coefficients.

(continued)

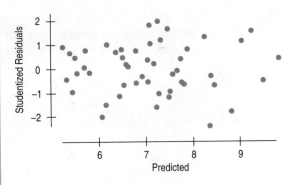

Before deciding that any regression model is a "keeper," remember to check the residuals.

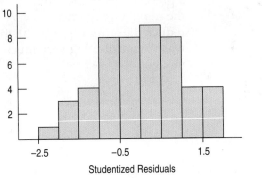

Summarize the features of this model.

Here's an example of an outlier that might help us learn something about the data or the world. Whatever makes South Dakota's infant mortality rate so much higher than the model predicts, it might be something we could address with new policies or interventions.

The scatterplot of Studentized residuals against predicted values shows no structure, and the histogram of Studentized residuals is Nearly Normal. So this looks like a good model for *Infant Mortality*. The coefficient for *S. Dakota* is still very significant, so I'd prefer to keep South Dakota separate and look into why its *Infant Mortality* rate is so much higher (2.74 deaths per 1000 live births) than we would otherwise expect from its *Child Death Rate* and *Low Birth Weight* percent.

Let's try the other way and build a regression model "forward" by selecting variables to add to the model.

The data include variables that concern young adults: *Teen Births*, *Teen Deaths*, and the *HS Drop%*.

Both *Teen Births* and *Teen Deaths* are promising predictors, but births to teens seem more directly relevant. Here's the regression model:

Dependent variable is: Infant mort
R-squared = 29.3%   R-squared (adjusted) = 27.9%
s = 1.129 with 50 − 2 = 48 degrees of freedom

Variable	Coefficient	SE(Coeff)	t-ratio	P-value
Intercept	4.96399	0.5098	9.74	≤0.0001
Teen Births	0.081217	0.0182	4.47	≤0.0001

One way to select variables to add to a growing regression model is to find the correlation of the residuals of the current state of the model with the potential new predictors. Predictors with higher correlations can be expected to account for more of the remaining residual variation if we include them in the regression model.

Notice that adding a predictor that does not contribute to the model can reduce the adjusted R^2.

The regression that models *Infant Mortality* on *Teen Births, Teen Deaths,* and *HS Drop%* may be worth keeping as well. But, of course, we're not finished until we check the residuals:

The correlations of the residuals with other predictors look like this:

	Resids
HS Drop%	−0.188
Teen Deaths	0.333
%Poverty	0.105

Teen Deaths looks like a good choice to add to the model:

Dependent variable is: Infant mort
R-squared = 39.1%   R-squared (adjusted) = 36.5%
s = 1.059 with 50 − 3 = 47 degrees of freedom

Variable	Coefficient	SE(Coeff)	t-ratio	P-value
Intercept	3.98643	0.5960	6.69	≤0.0001
Teen Births	0.057880	0.0191	3.04	0.0039
Teen Deaths	0.028228	0.0103	2.75	0.0085

Finally, I'll try adding HS Drop% to the model:

Dependent variable is: Infant mort
R-squared = 44.0%   R-squared (adjusted) = 40.4%
s = 1.027 with 50 − 4 = 46 degrees of freedom

Variable	Coefficient	SE(Coeff)	t-ratio	P-value
Intercept	4.51922	0.6358	7.11	≤0.0001
Teen Births	0.097855	0.0272	3.60	0.0008
Teen Deaths	0.026844	0.0100	2.69	0.0099
HS Drop%	−0.164347	0.0819	−2.01	0.0506

Here is one more step, adding %Poverty to the model:

Dependent variable is: Infant mort
R-squared = 44.0%   R-squared (adjusted) = 39.1%
s = 1.038 with 50 − 5 = 45 degrees of freedom

Variable	Coefficient	SE(Coeff)	t-ratio	P-value
Intercept	4.49810	0.7314	6.15	≤0.0001
Teen Births	0.09690	0.0317	3.06	0.0038
Teen Deaths	0.02664	0.0106	2.50	0.0160
HS Drop%	−0.16397	0.0830	−1.98	0.0544
%Poverty	3.1053e-3	0.0513	0.061	0.9520

The P-value for %Poverty is quite high, so I prefer the previous model.

Here are the residuals:

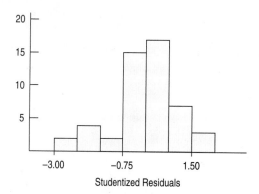

Studentized Residuals

(continued)

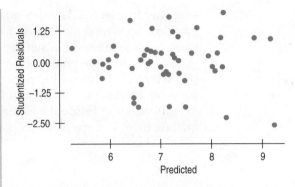

This histogram hints of a low mode holding some large negative residuals, and the scatterplot shows two in particular that trail off at the bottom right corner of the plot. They are Texas and New Mexico. These states are neighbors and may share some regional attributes. To be careful, I'll try removing them from the model. I'll construct two indicator variables that are 1 for the named state and 0 for all others:

Dependent variable is: Infant mort
R-squared = 58.9%   R-squared (adjusted) = 54.2%
s = 0.8997 with 50 − 6 = 44 degrees of freedom

Variable	Coefficient	SE(Coeff)	t-ratio	P-value
Intercept	4.15748	0.5673	7.33	≤0.0001
Teen Births	0.13823	0.0259	5.33	≤0.0001
Teen Deaths	0.02669	0.0090	2.97	0.0048
HS Drop%	−0.22808	0.0735	−3.10	0.0033
New Mexico	−3.01412	0.9755	−3.09	0.0035
Texas	−2.74363	0.9748	−2.81	0.0073

Removing the two outlying states has improved the model noticeably. The indicators for both states have small P-values, so I conclude that they were in fact outliers for this model. The R^2 has improved to 58.9%, and the P-values of all the other coefficients have been reduced.

A final check on the residuals from this model shows that they satisfy the regression conditions:

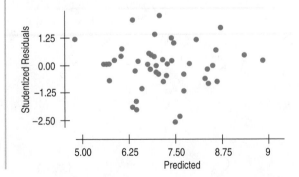

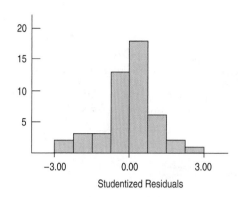

This model is an alternative to the first one I found. It has a smaller R^2 (58.9%) and larger s value, but it might be useful for understanding the relationships between these variables and infant mortality.

| Compare and contrast the models. | I have found two reasonable regression models for infant mortality. The first finds that Infant Mortality can be modeled by Child Deaths and %Low Birth Weight, removing the influence of South Dakota: |

$$\widehat{Infant\ Mortally} = 0.76 + 0.027\ Child\ Deaths + 0.75\ LowBW\%.$$

It may be worthwhile to look into why South Dakota is so different from the other states. The other model focused on teen behavior, modeling Infant Mortality by Teen Births, Teen Deaths, and HS Drop%, removing the influence of Texas and New Mexico:

$$\widehat{Infant\ Mortally} = 4.16 + 0.138\ Teen\ Births + 0.027\ Teen\ Deaths - 0.228\ HS\ Drop\%$$

For a more complete understanding of infant mortality, we should look into South Dakota's early childhood variables and the teen-related variables in New Mexico and Texas. We might well learn as much about infant mortality by understanding why these states stand out—and how they differ from each other—as we would from the regression models themselves.

The coefficient of HS Drop% is the opposite sign of the simple relationship between Infant Deaths and HS Drop%.

Each model has nominated different states as outliers. For a more complete understanding of infant mortality, it might be worthwhile to look into why these states are outliers in these models.

Which model is better? That depends on what you want to know. Remember—all models are wrong. But both may offer useful information and insights about infant mortality and its relationship with other variables and about the states that stood out and why they differ from the others.

1. Give two ways that we use histograms to support the construction, inference, and understanding of multiple regression models.

2. Give two ways that we use scatterplots to support the construction, inference, and understanding of multiple regression models.

3. What role does the Normal model play in the construction, inference, and understanding of multiple regression models?

Regression Roles

We build regression models for a number of reasons. One reason is to model how variables are related to each other in the hope of understanding the relationships. Another is to build a model that might be used to predict values for a response variable when given values for the predictor variables.

When we hope to understand, we are often particularly interested in simple, straightforward models in which predictors are as unrelated to each other as possible. We are especially happy when the *t*-statistics are large, indicating that the predictors each contribute to the model. We are likely to want to look at partial regression plots to understand the coefficients and to check that no outliers or influential points are affecting them.

When prediction is our goal, we are more likely to care about the overall R^2. Good prediction occurs when much of the variability in y is accounted for by the model. We might be willing to keep variables in our model that have relatively small *t*-statistics simply for the stability that having several predictors can provide. We care less whether the predictors are related to each other because we don't intend to interpret the coefficients anyway.

In both roles, we may include some predictors to "get them out of the way." Regression offers a way to approximately control for factors when we have observational data because each coefficient measures effects *after removing the effects* of the other predictors. Of course, it would be better to control for factors in a randomized experiment, but often that's just not possible.

*Indicators for Three or More Levels

It's easy to construct indicators for a variable with two levels; we just assign 0 to one level and 1 to the other. But variables like *Month* or *Class* often have several levels. You can construct indicators for a categorical variable with several levels by constructing a separate indicator for each of these levels. There's just one trick: You have to choose one of the categories as a "baseline" and *leave out* its indicator. Then the coefficients of the other indicators can be interpreted as the amount by which their categories differ from the baseline, after allowing for the linear effects of the other variables in the model.[13]

Make sure your collection of indicators doesn't exhaust all the categories. One category must be left out to serve as a baseline or the regression model can't be found. For the two-category variable *Inversions*, we used "no inversion" as the baseline and coasters with an inversion got a 1. We needed only one variable for two levels. If we wished to represent *Month* with indicators, we would need 11 of them. We might, for example, define *January* as the baseline, and make indicators for *February, March, . . . , November*, and *December*. Each of these

---

[13]There are alternative coding schemes that compare all the levels with the mean. Make sure you know how the indicators are coded.

indicators would be 0 for all cases except for the ones that had that value for the variable *Month.* Why not just a single variable with "1" for *January,* "2" for *February,* and so on? That might work. But it would impose the pretty strict assumption that the responses to the months are ordered and equally spaced—that is, that the change in our response variable from January to February is the same in both direction and amount as the change from July to August. That's a pretty severe restriction and may not be true for many kinds of data. Using 11 indicators releases the model from that restriction, but, of course, at the expense of having 10 fewer degrees of freedom for all of our *t*-tests.

Collinearity

Let's look at the infant mortality data one more time. One good predictor of *Infant Mortality* is *Teen Deaths.*

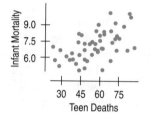

Dependent variable is: Infant mort
R-squared = 27.2%   R-squared (adjusted) = 25.7%
s = 1.146 with 50 − 2 = 48 degrees of freedom

Variable	Coefficient	SE(Coeff)	t-ratio	P-value
Intercept	4.73979	0.5866	8.08	≤0.0001
Teen Deaths	0.042129	0.0100	4.23	0.0001

Teen Deaths has a positive coefficient (as we might expect) and a very small P-value. Suppose we now add *Child Deaths Rate (CDR)* to the regression model:

Dependent variable is: Infant mort
R-squared = 42.6%   R-squared (adjusted) = 40.1%
s = 1.029 with 50 − 3 = 47 degrees of freedom

Variable	Coefficient	SE(Coeff)	t-ratio	P-value
Intercept	5.79561	0.6049	9.58	≤0.0001
Teen Deaths	−1.86877e-3	0.0153	−0.122	0.9032
Child Deaths	0.059398	0.0168	3.55	0.0009

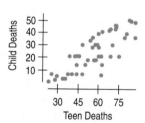

FIGURE 31.8
Child Deaths *and* Teen Deaths *are linearly related.*

Suddenly *Teen Deaths* has a small negative coefficient and a very large P-value. What happened? The problem is that *Teen Deaths* and *Child Deaths* are closely associated. The coefficient of *Teen Deaths* now reports how *Infant Mortality* is related to *Teen Deaths after allowing for the linear effects of Child Deaths* on both variables.

When we have several predictors, we must think about how the predictors are related to each other. When predictors are unrelated to each other, each provides new information to help account for more of the variation in *y.* Just as we need a predictor to have a large enough variability to provide a stable base for simple regression, when we have several predictors, we need for them to vary in different directions for the multiple regression to have a stable base. If you wanted to build a deck on the back of your house, you wouldn't build it with supports placed just along one diagonal. Instead, you'd want the supports spread out in different directions as much as possible to make the deck stable. We're in a similar situation with multiple regression. When predictors are highly correlated, they line up together, which makes the regression they support balance precariously. Even small variations can rock it one way or the other. A more stable model can be built when predictors have low correlation and the points are spread out.

When two or more predictors are linearly related, they are said to be **collinear.** The general problem of predictors with close (but perhaps not perfect) linear relationships is called the problem of **collinearity.**

Fortunately, there's an easy way to assess collinearity. To measure how much one predictor is linearly related to the others, just find the regression of

MULTICOLLINEARITY?

You may find this problem referred to as "multicollinearity." But there is no such thing as "unicollinearity"—we need at least two predictors for there to be a linear association between them—so there is no need for the extra two syllables.

WHY NOT JUST LOOK AT THE CORRELATIONS?

It's sometimes suggested that we examine the table of correlations of all the predictors to search for collinearity. But this will find only associations among *pairs* of predictors. Collinearity can—and does—occur among *several* predictors working together. You won't find that more subtle collinearity with a correlation table.

that predictor on the others[14] and look at the R^2. That R^2 gives the fraction of the variability of the predictor in question that is accounted for by the other predictors. So $1 - R^2$ is the amount of the predictor's variance that is left after we allow for the effects of the other predictors. That's what the predictor has left to bring to the regression model. And we know that a predictor with little variance can't do a good job of predicting.[15]

Collinearity can hurt our analysis in yet another way. We've seen that the variance of a predictor plays a role in the standard error of its associated coefficient. Small variance leads to a larger SE. In fact, it's exactly this leftover variance that shows up in the formula for the SE of the coefficient. That's what happened in the infant mortality example.

As a final blow, when a predictor is collinear with the other predictors, it's often difficult to figure out what its coefficient means in the multiple regression. We've blithely talked about "removing the effects of the other predictors," but now when we do that, there may not be much left. What is left is not likely to be about the original predictor, but more about the fractional part of that predictor not associated with the others. In a regression of horsepower on weight and engine size, once we've removed the effect of weight on horsepower, engine size doesn't tell us anything *more* about horsepower. That's certainly not the same as saying that engine size doesn't tell us anything about horsepower. It's just that most cars with big engines also weigh a lot.

When a predictor is collinear with the other predictors in the model, two things can happen:

1. Its coefficient can be surprising, taking on an unanticipated sign or being unexpectedly large or small.
2. The standard error of its coefficient can be large, leading to a smaller *t*-statistic and correspondingly large P-value.

One telltale sign of collinearity is the paradoxical situation in which the overall *F*-test for the multiple regression model is significant, showing that at least one of the coefficients is discernably different from zero, and yet most or all of the individual coefficients have small *t*-values, each in effect, denying that *it* is the significant one.

What should you do about a collinear regression model? The simplest cure is to remove some of the predictors. That both simplifies the model and generally improves the *t*-statistics. And, if several predictors give pretty much the same information, removing some of them won't hurt the model. Which should you remove? Keep the predictors that are most reliably measured, least expensive to find, or even those that are politically important.

CHOOSING A SENSIBLE MODEL

The mathematics department at a large university built a regression model to help them predict success in graduate study. They were shocked when the coefficient for Mathematics GRE score was not significant. But the Math GRE was collinear with some of the other predictors, such as math course GPA and Verbal GRE, which made its slope not significant. They decided to omit some of the other predictors and retain Math GRE as a predictor because that model seemed more appropriate—even though it predicted no better (and no worse) than others without

---

[14] The residuals from this regression are plotted as the *x*-axis of the partial regression plot for this variable. So if they have a very small variance, you can see it by looking at the *x*-axis labels of the partial regression plot, and get a sense of how precarious a line fit to the partial regression plot—and its corresponding multiple regression coefficient—may be.

[15] The statistic $1 - R^2$ for the R^2 found from the regression of one predictor on the other predictors in the model is also called the *Variance Inflation Factor*, or *VIF*, in some computer programs and books.

What Can Go Wrong?

In the Oscar-winning movie *The Bridge on the River Kwai* and in the book on which it is based,[16] the character Colonel Green famously says, "As I've told you before, in a job like yours, even when it's finished, there's always one more thing to do." It is wise to keep Colonel Green's advice in mind when building, analyzing, and understanding multiple regression models.

■ *Beware of collinearities.* When the predictors are linearly related to each other, they add little to the regression model after allowing for the contributions of the other predictors. Check the R^2s when each predictor is regressed on the others. If these are high, consider omitting some of the predictors.

■ *Don't check for collinearity only by looking at pairwise correlations.* Collinearity is a relationship among any number of the predictors. Pairwise correlations can't always show that. (Of course, a high pairwise correlation between two predictors does indicate collinearity of a special kind.)

■ *Don't be fooled when high-influence points and collinearity show up together.* A single high-influence point can be the difference between your predictors being collinear and seeming not to be collinear. (Picture that deck supported only along its diagonal and with a single additional post in another corner. Supported in this way, the deck is stable, but the height of that single post completely determines the tilt of the deck, so it's very influential.) Removing a high-influence point may surprise you with unexpected collinearity. Alternatively, a single value that is extreme on several predictors can make them appear to be collinear when in fact they would not be if you removed that point. Removing that point may make apparent collinearities disappear (and would probably result in a more useful regression model).

■ *Beware missing data.* Values may be missing or unavailable for any case in any variable. In simple regression, when the cases are missing for reasons that are unrelated to the variable we're trying to predict, that's not a problem. We just analyze the cases for which we have data. But when several variables participate in a multiple regression, any case with data missing on any of the variables will be omitted from the analysis. You can unexpectedly find yourself with a much smaller set of data than you started with. Be especially careful, when comparing regression models with different predictors, that the cases participating in the models are the same.

■ *Remember linearity.* The **Linearity Assumption** (and the **Straight Enough Condition**) require linear relationships among the variables in a regression model. As you build and compare regression models, be sure to plot the data to check that it is straight. Violations of this assumption make everything else about a regression model invalid.

■ *Check for parallel regression lines.* When you introduce an indicator variable for a category, check the underlying assumption that the other coefficients in the model are essentially the same for both groups. If not, consider adding an interaction term.

---

[16]The author of the book, Pierre Boulle, also wrote the book and script for *Planet of the Apes.* The director, David Lean, also directed *Lawrence of Arabia.*

Now that we understand indicator variables, we can see that multiple regression and ANOVA are really the same analysis. If the only predictor in a regression is an indicator variable that is 1 for one group and 0 for the other, the *t*-test for its coefficient is just the pooled *t*-test for the difference in the means of those groups. In fact, most of the Student's *t*–based methods in this book can be seen as part of a more general statistical model known as the General Linear Model (GLM). That accounts for why they seem to be so connected, using the same general ideas and approaches.[17] We've generalized the concept of leverage that we first saw in Chapter 9. Everything we said about how to think about these ideas back in Chapters 9 and 27 still applies to the multiple regression model.

Don't forget that the Straight Enough Condition is essential to all of regression. At any stage in developing a model, if the scatterplot that you check is not straight, consider re-expressing the variables to make the relationship straighter. The topics of Chapter 10 will help you with that.

WHAT HAVE WE LEARNED?

In Chapter 30, we learned that multiple regression is a natural way to extend what we knew about linear regression models to include several predictors. Now we've learned that multiple regression is both more powerful and more complex than it may appear at first. As with other chapters in this book whose titles spoke of greater "wisdom," this chapter has drawn us deeper into the uses and cautions of multiple regression.

We've glimpsed the power of the multiple regression model. We can incorporate categorical data by using indicator variables, modeling relationships that have parallel slopes but at different levels for different groups. With interaction terms, we can allow for different slopes as well. We can create identifier variables that isolate individual cases to remove their influence from the model while exhibiting how they differ from the other points and testing whether that difference is statistically significant.

We've learned to beware unusual cases. A single case can have high leverage, allowing it to influence the entire regression. Such cases should be treated specially, possibly by fitting the model both with and without them or by including indicator variables to isolate their influence.

We've learned that in complex models one has to be careful in interpreting the coefficients. Associations among the predictors can change the coefficients to values that can be quite different from the coefficient in the simple regression of a predictor and the response, even changing the sign.

And we've learned that building multiple regression models is an art that speaks to the central goal of statistical analysis: understanding the world with data. We've learned that there is no right model. We've seen that the same response variable can be modeled with several alternative models, each showing us different aspects of the data and of the relationships among the variables and nominating different cases as special and deserving of our attention.

We've seen that everything we've discussed throughout this book fits together to help us understand the world. The graphical methods are the same ones we learned in the early chapters, and the inference methods are those we originally developed for means. In short, there's been a consistent tale of how we understand data to which we've added more and more detail and richness, but which has been consistent throughout.

---

[17]It has been wistfully observed that if only we could start the course by teaching multiple regression, everything else would just be simplifications of the general method. Now that you're here, you might try reading the book backward, contradicting the White King's advice to Alice, which we quoted in Chapter 1.

WHAT ELSE HAVE WE LEARNED?

We, the authors, hope that you've also learned to see the world differently, to understand what has been measured and about whom, to be skeptical of untested claims and curious about patterns and relationships. We hope that you find the world a more interesting, more nuanced place that can be understood and appreciated with the tools of Statistics and Science.

Finally, we hope you'll consider further study in Statistics. Whatever your field, whatever your job, whatever your interests, you can use Statistics to understand the world better.

Terms

Indicator variable
A variable constructed to indicate for each case whether it is in a designated group or not. A common way to assign values to indicator variables is to let them take on the values 0 and 1, where 1 indicates group membership (p. 816).

Interaction term
A constructed variable found as the product of a predictor and an indicator variable. An interaction term adjusts the *slope* of the cases identified by the indicator against the predictor (p. 818).

Leverage
The leverage of a case measures how far its *x*-values are from the center of the *x*'s and, consequently, how much influence it can exert on the regression model. Points with high leverage can determine a regression model and should, therefore, be examined carefully (p. 819).

Studentized residual
When a residual is divided by an independent estimate of its standard deviation, the result is a Studentized residual. The type of Studentized residual that has a *t*-distribution is an *externally* Studentized residual (p. 821).

Influential case
A case is *influential* on a multiple regression model if, when it is omitted, the model changes by enough to matter for your purposes. (There is no specific amount of change defined to declare a case influential.) Cases with high leverage and large Studentized residual are likely to be influential (p. 822).

Stepwise regression
An automated method of building regression models in which predictors are added to or removed from the model one at a time in an attempt to optimize a measure of the success of the regression. Stepwise methods rarely find the best model and are easily influenced by influential cases, but they can be valuable in winnowing down a large collection of candidate predictors (p. 827).

Collinearity
When one (or more) of the predictors can be fit closely by a multiple regression on the other predictors, we have collinearity. When collinear predictors are in a regression model, they may have unexpected coefficients and often have inflated standard errors (and correspondingly small *t*-statistics) (p. 837).

Skills

When you complete this lesson you should:

▶ Understand how individual cases can influence a regression model.

▶ Know how to define and use indicator variables to introduce categorical variables as predictors in a multiple regression model.

▶ Know how to examine histograms of leverages and of Studentized residuals to identify extraordinary cases that deserve special attention.

▶ Know how to recognize when a regression model may suffer from collinearity.

▶ Know how to check for high-leverage cases by identifying cases whose leverage stands apart from the others.

▶ Know how to check for cases with large Studentized residuals.

▶ Be able to use a statistics package to diagnose a multiple regression model.

▶ Know how to build a multiple regression model, selecting predictors from a larger collection of potential predictors.

▶ Be able to interpret the coefficients found for indicator variables in a multiple regression.

▶ Be able to discuss the influence that a case with high leverage or a large Studentized residual may have in a regression.

▶ Be able to recognize when collinearity among the predictors may be present. Be able to check for it and discuss its consequences.

▶ Be careful in interpreting regression coefficients when the predictors are collinear. Avoid the pitfalls of interpreting the sign of the coefficient as if it were different from the coefficient itself.[18]

REGRESSION ANALYSIS ON THE COMPUTER

Statistics packages differ in how much information they provide to diagnose a multiple regression. Most packages provide leverage values. Many provide far more, including statistics that we have not discussed. But for all, the principle is the same. We hope to discover any cases that don't behave like the others in the context of the regression model and then to understand why they are special.

Many of the ideas in this chapter rely on the concept of examining a regression model and then finding a new one based on your growing understanding of the model and the data. Regression diagnosis is meant to provide steps along that road. A thorough regression analysis may involve finding and diagnosing several models.

DATA DESK

Request diagnostic statistics and graphs from the HyperView menus in the regression output table. Most will update and can be set to update automatically when the model or data change.

COMMENTS

You can add a predictor to the regression by dragging its icon into the table, or replace variables by dragging the icon over their name in the table. Click on a predictor's name to drop down a menu that lets you remove it from the model.

EXCEL

Excel does not offer diagnostic statistics with its regression function.

COMMENTS

Although the dialog offers a Normal probability plot of the residuals, the data analysis add-in does not make a correct probability plot, so don't use this option. The "standardized residuals" are just the residuals divided by their standard deviation (with the wrong df), so they too should be ignored.

JMP

- From the **Analyze** menu select **Fit Model.**
- Specify the response, Y. Assign the predictors, X, in the **Construct Model Effects** dialog box.
- Click on **Run Model.**
- Click on the red triangle in the title of the Model output to find a variety of plots and diagnostics available.

COMMENTS

JMP chooses a regression analysis when the response variable is "Continuous."

---

[18]Now stop. (See page 2 margin.)

MINITAB

- Choose **Regression** from the **Stat** menu.
- Choose **Regression...** from the **Regression** submenu.
- In the Regression dialog, assign the Y variable to the Response box and assign the X-variables to the Predictors box.
- Click the **Storage** button.
- In the Regression Storage dialog, you can select a variety of diagnostic statistics. They will be stored in columns of your worksheet.
- Click the **OK** button to return to the Regression dialog.
- To specify displays, click **Graphs,** and check the displays you want.
- Click the **OK** button to return to the Regression dialog.
- Click the **OK** button to compute the regression.

COMMENTS

You will probably want to make displays of the stored diagnostic statistics. Use the usual Minitab methods for creating displays.

SPSS

- Choose **Regression** from the **Analyze** menu.
- Choose **Linear** from the **Regression** submenu.
- When the Linear Regression dialog appears, select the Y-variable and move it to the dependent target. Then move the X-variables to the independent target.
- Click the **Save** button.
- In the Linear Regression Save dialog, choose diagnostic statistics. These will be saved in your worksheet along with your data.
- Click the **Continue** button to return to the Linear Regression dialog.
- Click the **OK** button to compute the regression.

TI-83/84 PLUS

COMMENTS

You need a special program to compute a multiple regression on the TI-83.

TI-89

Under **STAT** Tests choose **B:MultREg** Tests

- Specify the number of predictor variables, and which lists contain the response variable and predictor variables.
- Press ÷ to perform the calculations.
- When finished, you will want to plot residuals (either raw or standardized) against predicted values.
- Examine the list of leverages for any influential points.

COMMENTS

- The first portion of the output gives the F-statistic and its P-value as well as the values of R^2, $AdjR^2$, the standard deviation of the residuals (s), and the Durbin-Watson statistic, which measures correlation among the residuals.
- The rest of the main output gives the components of the F-test, as well as values of the coefficients, their standard errors, and associated t-statistics along with P-values. You can use the right arrow to scroll through these lists (if desired).
- The calculator creates several new lists that can be used for assessing the model and its conditions. Yhatlist, resid, sresid (standardized residuals), leverage, and cookd, as well as lists of the coefficients, standard errors, t's, and P-values.

EXERCISES

1. Climate change 2009. Recent concern with the rise in global temperatures has focused attention on the level of carbon dioxide (CO_2) in the atmosphere. The National Oceanic and Atmospheric Administration (NOAA) records the CO_2 levels in the atmosphere atop the Mauna Loa volcano in Hawaii, far from any industrial contamination, and calculates the annual overall temperature of the atmosphere and the oceans using an established method. (See *http://www.esrl.noaa.gov/gmd/ccgg/trends/* for the CO_2 levels and *http://www.ncdc.noaa.gov/oa/climate/research/anomalies/index.php#means* for the temperatures.) Here is a regression predicting Mean Annual Temperature Anomaly (°C away from the 20th-century mean) from annual CO_2 levels (parts per million). We'll examine the data from 1959 to 2009.

Dependent variable is: Global Temperature Anomaly
R-squared = 85.3%   R-squared (adjusted) = 84.7%
s = 0.0850 with 51 − 3 = 48 degrees of freedom

Source	Sum of Squares	df	Mean Square	F-ratio
Regression	2.00798	2	1.00399	139
Residual	0.346811	48	0.007225	

Variable	Coefficient	SE(Coeff)	t-ratio	P-value
Intercept	30.9595	12.52	2.47	0.0170
Year	−0.019404	0.0072	−2.71	0.0094
CO_2	0.022413	0.0049	4.55	≤0.0001

A histogram of the externally Studentized residuals looks like this:

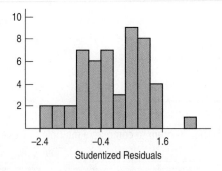

Studentized Residuals

a) Comment on the distribution of the Studentized residuals.
b) It is widely understood that global temperatures have been rising consistently during this period. But the coefficient of *Year* is negative and its P-value is small. Does this contradict the common wisdom?

2. Pizza. Consumers' Union rated frozen pizzas. Their report includes the number of *Calories, Fat* content, and *Type* (cheese or pepperoni, represented here as an indicator variable that is 1 for cheese and 0 for pepperoni).

Here's a regression model to predict the *"Score"* awarded each pizza from these variables:

Dependent variable is: Score
R-squared = 28.7%
R-squared (adjusted) = 20.2%
s = 19.79 with 29 − 4 = 25 degrees of freedom

Source	Sum of Squares	DF	Mean Square	F-ratio
Regression	3947.34	3	1315.78	3.36
Residual	9791.35	25	391.654	

Variable	Coefficient	SE(Coeff)	t-ratio	P-value
Intercept	−148.817	77.99	−1.91	0.0679
Calories	0.743023	0.3066	2.42	0.0229
Fat	−3.89135	2.138	−1.82	0.0807
Type	15.6344	8.103	1.93	0.0651

a) What is the interpretation of the coefficient of cheese in this regression?
b) What displays would you like to see to check assumptions and conditions for this model?

3. Healthy breakfast, sick data. A regression model for data on breakfast cereals originally looked like this:

Dependent variable is: Calories
R-squared = 84.5%
R-squared (adjusted) = 83.4%
s = 7.947 with 77 − 6 = 71 degrees of freedom

Source	Sum of Squares	DF	Mean Square	F-ratio
Regression	24367.5	5	4873.50	77.2
Residual	4484.45	71	63.1613	

Variable	Coefficient	SE(Coeff)	t-ratio	P-value
Intercept	20.2454	5.984	3.38	0.0012
Protein	5.69540	1.072	5.32	≤0.0001
Fat	8.35958	1.033	8.09	≤0.0001
Fiber	−1.02018	0.4835	−2.11	0.0384
Carbo	2.93570	0.2601	11.3	≤0.0001
Sugars	3.31849	0.2501	13.3	≤0.0001

Let's take a closer look at the coefficient for *Fiber*. Here's the partial regression plot for *Fiber* in that regression model:

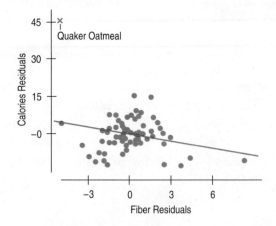

a) The line on the plot is the least squares line fit to this plot. What is its slope? (You may need to look back at the facts about partial regression plots in Chapter 30.)

b) One point is labeled as corresponding to Quaker Oatmeal. What effect does this point have on the slope of the line? (Does it make it larger, smaller, or have no effect at all?)

Here is the same regression with Quaker Oatmeal removed from the data:

Dependent variable is: Calories
77 total cases of which 1 is missing
R-squared = 93.9%   R-squared (adjusted) = 93.5%
s = 5.002 with 76 − 6 = 70 degrees of freedom

Source	Sum of Squares	DF	Mean Square	F-ratio	P-value
Regression	27052.4	5	5410.49	216	
Residual	1751.51	70	25.0216		

Variable	Coefficient	SE(Coeff)	t-ratio	P-value
Intercept	−1.25891	4.292	−0.293	0.7701
Protein	3.88601	0.6963	5.58	≤0.0001
Fat	8.69834	0.6512	13.4	≤0.0001
Fiber	0.250140	0.3277	0.763	0.4478
Carbo	4.14458	0.2005	20.7	≤0.0001
Sugars	3.96806	0.1692	23.4	≤0.0001

c) Compare this regression with the previous one. In particular, which model is likely to make the best predictions of calories? Which seems to fit the data better?

d) How would you interpret the coefficient of *Fiber* in this model? Does *Fiber* contribute significantly to modeling calories?

(In fact, the data for Quaker Oatmeal was determined to be in error and was corrected for the subsequent analyses seen elsewhere in this book.)

4. **Fifty states.** In Exercise 15 of Chapter 30 we looked at data from the 50 states. Here's an analysis of the same data from a few years earlier. The *Murder* rate is per 100,000, *HS Graduation* rate is in %, *Income* is per capita income in dollars, *Illiteracy* rate is per 1000, and *Life Expectancy* is in years. We are trying to find a regression model for *Life Expectancy*.

Here's the result of a regression on all the available predictors:

Dependent variable is: Lifeexp
R-squared = 67.0%   R-squared (adjusted) = 64.0%
s = 0.8049 with 50 − 5 = 45 degrees of freedom

Source	Sum of Squares	DF	Mean Square	F-ratio
Regression	59.1430	4	14.7858	22.8
Residual	29.1560	45	0.6479	

Variable	Coefficient	SE(Coeff)	t-ratio	P-value
Intercept	69.4833	1.325	52.4	≤0.0001
Murder	−0.261940	0.0445	−5.89	≤0.0001
HS grad	0.046144	0.0218	2.11	0.0403
Income	1.24948e-4	0.0002	0.516	0.6084
Illiteracy	0.276077	0.3105	0.889	0.3787

Here's a histogram of the leverages and a scatterplot of the externally Studentized residuals against the leverages:

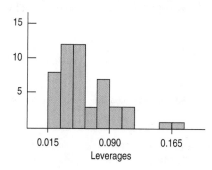

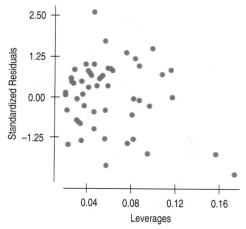

a) The two states with high leverages and large (negative) Studentized residuals are Nevada and Alaska. Do you think they are likely to be influential in the regression? From just the information you have here, why or why not?

Here's the regression with indicator variables for Alaska and Nevada added to the model to remove those states from affecting the model:

Dependent variable is: Lifeexp
R-squared = 74.1%   R-squared (adjusted) = 70.4%
s = 0.7299 with 50 − 7 = 43 degrees of freedom

Variable	Coefficient	SE(Coeff)	t-ratio	P-value
Intercept	66.9280	1.442	46.4	≤0.0001
Murder	−0.207019	0.0446	−4.64	≤0.0001
HS grad	0.065474	0.0206	3.18	0.0027
Income	3.91600e-4	0.0002	1.63	0.1105
Illiteracy	0.302803	0.2984	1.01	0.3159
Alaska	−2.57295	0.9039	−2.85	0.0067
Nevada	−1.95392	0.8355	−2.34	0.0241

b) What evidence do you have that Nevada and Alaska are outliers with respect to this model? Do you think they should continue to be treated specially? Why?

c) Would you consider removing any of the predictors from this model? Why or why not?

5. **Cereals, part 2.** In Exercise 16 of Chapter 30, we considered a multiple regression model for predicting calories in breakfast cereals. The regression looked like this:

Dependent variable is: Calories
R-squared = 38.4%   R-squared (adjusted) = 35.9%
s = 15.60 with 77 − 4 = 73 degrees of freedom

Source	Sum of Squares	DF	Mean Square	F-ratio	P-value
Regression	11091.8	3	3697.28	15.2	≤0.0001
Residual	17760.1	73	243.289		

Variable	Coefficient	SE(Coeff)	t-ratio	P-value
Intercept	83.0469	5.198	16.0	≤0.0001
Sodium	0.057211	0.0215	2.67	0.0094
Potassium	−0.019328	0.0251	−0.769	0.4441
Sugars	2.38757	0.4066	5.87	≤0.0001

Here's a histogram of the leverages and a partial regression plot for *Potassium* in which the three high-leverage points are plotted with red x's. (They are *All-Bran, 100% Bran,* and *All-Bran with Extra Fiber.*)

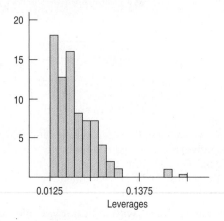

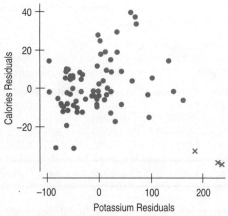

With this additional information, answer the following:

a) How would you interpret the coefficient of *Potassium* in the multiple regression?

b) *Without doing any calculating,* how would you expect the coefficient and *t*-statistic for *Potassium* to change if we were to omit the three high-leverage points?

Here's a histogram of the externally Studentized residuals. The selected bar, holding the two most negative residuals, holds the two bran cereals that had the largest leverages.

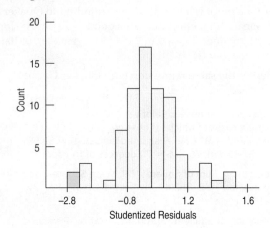

With this additional information, answer the following:

c) What term would you apply to these two cases? Why?

d) Do you think they should be omitted from this analysis? Why or why not? (*Note:* There is no correct choice. What matters is your reasons.)

T 6. **Scottish hill races 2008.** In Chapter 30, Exercises 4 and 6, we considered data on hill races in Scotland. These are overland races that climb and descend hills—sometimes several hills in the course of one race. Here is a regression analysis to predict the *Women's Record* times from the *Distance* and total vertical *Climb* of the races:

Dependent variable is: Women's record
R-squared = 96.7%   R-squared (adjusted) = 96.7%
s = 10.06 with 90 − 3 = 87 degrees of freedom

Source	Sum of Squares	dF	Mean Square	F-ratio
Regression	261029	2	130515	1288
Residual	8813.02	87	101.299	

Variable	Coefficient	SE(Coeff)	t-ratio	P-value
Intercept	−11.6545	1.891	−6.16	≤0.0001
Distance	4.43427	0.2200	20.2	≤0.0001
Climb	0.045195	0.0033	13.7	≤0.0001

Here is the scatterplot of externally Studentized residuals against predicted values, as well as a histogram of leverages for this regression:

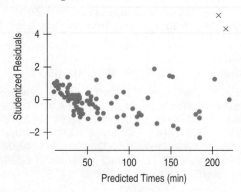

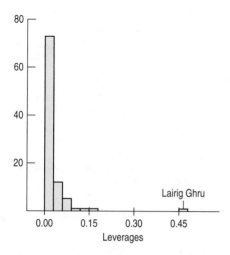

Leverages

a) Comment on what these diagnostic displays indicate.
b) The two races with the largest Studentized residuals are the Arochar Alps race and the Glenshee 9. Both are relatively new races, having been run only one or two times with relatively few participants. What effects can you be reasonably sure they have had on the regression? What displays would you want to see to investigate other effects? Explain.
c) If you have access to a suitable statistics package, make the diagnostic plots you would like to see and discuss what you find.

T 7. Traffic delays. The Texas Transportation Institute studies traffic delays. Data the institute published for the year 2001 include information on the *Total Delay per Person* (hours per year spent delayed by traffic), the *Average Arterial Road Speed* (mph), the *Average Highway Road Speed* (mph), and the *Size* of the city (small, medium, large, very large). The regression model based on these variables looks like this. The variables *Small, Large*, and *Very Large* are indicators constructed to be 1 for cities of the named size and 0 otherwise.

Dependent variable is: Delay/person
R-squared = 79.1%   R-squared (adjusted) = 77.4%
s = 6.474 with 68 − 6 = 62 degrees of freedom

Source	Sum of Squares	DF	Mean Square	F-ratio
Regression	9808.23	5	1961.65	46.8
Residual	2598.64	62	41.9135	

Variable	Coefficient	SE(Coeff)	t-ratio	P-value
Intercept	139.104	16.69	8.33	≤0.0001
Arterial mph	−2.04836	0.6672	−3.07	0.0032
HiWay mph	−1.07347	0.2474	−4.34	≤0.0001
Small	−3.58970	2.953	−1.22	0.2287
Large	5.00967	2.104	2.38	0.0203
Very large	3.41058	3.230	1.06	0.2951

a) Explain how the coefficients of *Small, Large,* and *Very Large* account for the size of the city in the model. Why is there no coefficient for *Medium*?
b) What is the interpretation of the coefficient of *Large* in this regression model?

T 8. Gourmet pizza. Here's a plot of the Studentized residuals against the predicted values for the regression model found in Exercise 2:

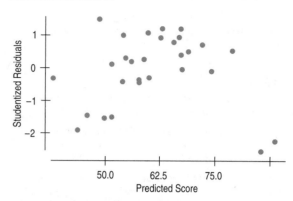

The two extraordinary cases in the plot of residuals are Reggio's and Michelina's, two gourmet pizzas.

a) Interpret these residuals. What do they say about these two brands of frozen pizza? Be specific—that is, talk about the *Scores* they received and might have been expected to receive.

We can create indicator variables to isolate these cases. Adding them to the model results in the following model:

Dependent variable is: Score
R-squared = 65.2%   R-squared (adjusted) = 57.7%
s = 14.41 with 29 − 6 = 23 degrees of freedom

Source	Sum of Squares	DF	Mean Square	F-ratio
Regression	8964.13	5	1792.83	8.64
Residual	4774.56	23	207.590	

Variable	Coefficient	SE(Coeff)	t-ratio	P-value
Intercept	−363.109	72.15	−5.03	≤0.0001
Calories	1.56772	0.2824	5.55	≤0.0001
Fat	−8.82748	1.887	−4.68	0.0001
Cheese	25.1540	6.214	4.05	0.0005
Reggio's	−67.6401	17.86	−3.79	0.0010
Michelina's	−67.0036	16.62	−4.03	0.0005

b) What does the coefficient of *Michelina's* mean in this regression model? Do you think that Michelina's pizza is an outlier for this model for these data? Explain.

T 9. More traffic. Here's a plot of Studentized residuals against *Arterial mph* for the model of Exercise 7. The plot is colored according to *City Size*, and regression lines are fit for each size.

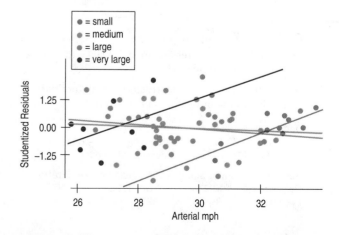

a) The model of Exercise 7 includes indicators for *City Size*. Considering this display, have these indicator variables accomplished what is needed for the regression model? Explain.

We constructed additional indicators as the product of *Small* with *Arterial mph* and the product of *Very Large* with *Arterial mph*. Here's the resulting model:

Dependent variable is: Delay/person
R-squared = 80.7%   R-squared (adjusted) = 78.5%
s = 6.316 with 68 − 8 = 60 degrees of freedom

Source	Sum of Squares	DF	Mean Square	F-ratio
Regression	10013.0	7	1430.44	35.9
Residual	2393.82	60	39.8970	

Variable	Coefficient	SE(Coeff)	t-ratio	P-value
Intercept	153.110	17.42	8.79	≤0.0001
Arterial mph	−2.60848	0.6967	−3.74	0.0004
HiWay mph	−1.02104	0.2426	−4.21	≤0.0001
Small	−125.979	66.92	−1.88	0.0646
Large	4.89837	2.053	2.39	0.0202
Very large	−89.4993	63.25	−1.41	0.1623
AM*sml	3.81461	2.077	1.84	0.0712
AM*VLg	3.38139	2.314	1.46	0.1491

b) What does the predictor *AM*sml* (*Arterial mph* by *Small*) do in this model? Interpret the coefficient.

c) Does this appear to be a good regression model? Would you consider removing any predictors? Why or why not?

10. **Another slice of pizza.** A plot of Studentized residuals against predicted values for the regression model found in Exercise 8 now looks like this. It has been colored according to *Type* of pizza and separate regression lines fitted for each type:

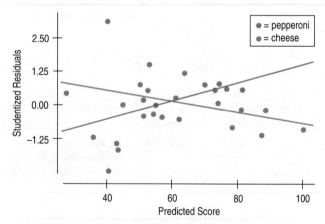

a) Comment on this diagnostic plot. What does it say about how the regression model deals with cheese and pepperoni pizzas?

Based on this plot, we constructed yet another variable consisting of the indicator *cheese* multiplied by *Calories*:

Dependent variable is: Score
R-squared = 73.7%   R-squared (adjusted) = 66.5%
s = 12.82 with 29 − 7 = 22 degrees of freedom

Source	Sum of Squares	DF	Mean Square	F-ratio
Regression	10121.4	6	1686.90	10.3
Residual	3617.32	22	164.424	

Variable	Coefficient	SE(Coeff)	t-ratio	P-value
Intercept	−464.498	74.73	−6.22	≤0.0001
Calories	1.92005	0.2842	6.76	≤0.0001
Fat	−10.3847	1.779	−5.84	≤0.0001
Cheese	183.634	59.99	3.06	0.0057
Cheese*cals	−0.461496	0.1740	−2.65	0.0145
Reggio's	−64.4237	15.94	−4.04	0.0005
Michelina's	−51.4966	15.90	−3.24	0.0038

b) Interpret the coefficient of *Cheese*cals* in this regression model.

c) Would you prefer this regression model to the model of Exercise 8? Explain.

11. **Influential traffic?** Here are histograms of the leverage and Studentized residuals for the regression model of Exercise 9.

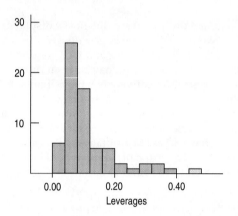

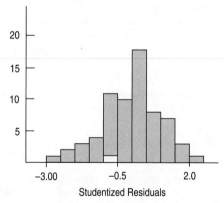

The city with the highest leverage is Colorado Springs, CO. It's highlighted in both displays.

Do you think Colorado Springs is an influential case? Explain your reasoning.

12. **The final slice.** Here's the residual plot corresponding to the regression model of Exercise 10:

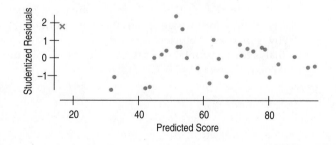

The extreme case this time is *Weight Watchers Pepperoni* (makes sense, doesn't it?). We can make one more indicator for *Weight Watchers*. Here's the model:

Dependent variable is: Score
R-squared = 77.1%   R-squared (adjusted) = 69.4%
s = 12.25 with 29 − 8 = 21 degrees of freedom

Source	Sum of Squares	DF	Mean Square	F-ratio
Regression	10586.8	7	1512.41	10.1
Residual	3151.85	21	150.088	

Variable	Coefficient	SE(Coeff)	t-ratio	P-value
Intercept	−525.063	79.25	−6.63	≤0.0001
Calories	2.10223	0.2906	7.23	≤0.0001
Fat	−10.8658	1.721	−6.31	≤0.0001
Cheese	231.335	63.40	3.65	0.0015
Cheese*cals	−0.586007	0.1806	−3.24	0.0039
Reggio's	−66.4706	15.27	−4.35	0.0003
Michelina's	−52.2137	15.20	−3.44	0.0025
Weight W...	28.3265	16.09	1.76	0.0928

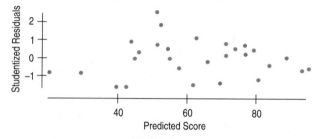

a) Compare this model with the others we've seen for these data. In what ways does this model seem better or worse than the others?

b) Do you think the indicator for *Weight Watchers* should be in the model? (Consider the effect that including it has had on the other coefficients also.)
c) What do the Consumers' Union tasters seem to think makes for a really good pizza?

JUST CHECKING

ANSWERS

1. Histograms are used to examine the shapes of distributions of individual variables. We check especially for multiple modes, outliers, and skewness. They are also used to check the shape of the distribution of the residuals for the Nearly Normal Condition.

2. Scatterplots are used to check the Straight Enough Condition in plots of *y vs.* any of the *x*'s. They are used to check plots of the residuals or Studentized residuals against the predicted values, against any of the predictors, or against *Time* to check for patterns. Scatterplots are also the display used in partial regression plots, where we check for influential points and unexpected subgroups.

3. The Normal model is needed only when we use inference; it isn't needed for computing a regression model. We check the Nearly Normal Condition on the residuals.

REVIEW OF PART

PART VII

Inference When Variables Are Related

Quick Review

This part introduces inference for the most widely used class of statistical models: linear models. You may have read one, two, or all of the chapters in this part. They have a consistent central theme, and it is one you'll likely see if you study more Statistics.

▶ Linear models predict a single quantitative variable from one or more other variables. The predictor variables can be quantitative or categorical. Linear models require only that the effects of the predictor variables be simply added together (rather than, for example, being multiplied or exponential).

▶ When the predictors are categorical, we are finding an Analysis of Variance (ANOVA).

▶ When the predictors are quantitative, we are finding a regression model. With two or more predictors, we have a *multiple* regression model.

▶ The overall fit of a linear model is tested with an *F*-test. The null hypothesis for this test is that we'd do better just modeling *y* with its mean.

▶ If the overall *F*-test is significant, we can move on to consider the individual coefficients. For regression, this means considering the *t*-statistics. For ANOVA, it means first considering the *F*-statistics for

Quick Guide to Inference

	Think						Tell?
Inference about?	**One sample or two?**	**Procedure**	**Model**	**Parameter**	**Estimate**	**SE**	**Chapter**
Proportions	One sample	1-Proportion z-Interval	z	p	$\hat{p}$	$\sqrt{\dfrac{\hat{p}\hat{q}}{n}}$	19
		1-Proportion z-Test				$\sqrt{\dfrac{p_0 q_0}{n}}$	20, 21
	Two independent groups	2-Proportion z-Interval	z	$p_1 - p_2$	$\hat{p}_1 - \hat{p}_2$	$\sqrt{\dfrac{\hat{p}_1\hat{q}_1}{n_1} + \dfrac{\hat{p}_2\hat{q}_2}{n_2}}$	22
		2-Proportion z-Test				$\sqrt{\dfrac{\hat{p}\hat{q}}{n_1} + \dfrac{\hat{p}\hat{q}}{n_2}}, \hat{p} = \dfrac{y_1 + y_2}{n_1 + n_2}$	22
Means	One sample	t-Interval t-Test	t df $= n - 1$	μ	$\bar{y}$	$\dfrac{s}{\sqrt{n}}$	23
	Two independent groups	2-Sample t-Test 2-Sample t-Interval	t df from technology	$\mu_1 - \mu_2$	$\bar{y}_1 - \bar{y}_2$	$\sqrt{\dfrac{s_1^2}{n_1} + \dfrac{s_2^2}{n_2}}$	24
	n Matched pairs	Paired t-Test Paired t-Interval	t df $= n - 1$	μ_d	$\bar{d}$	$\dfrac{s_d}{\sqrt{n}}$	25
Distributions (one categorical variable)	One sample	Goodness of fit	χ^2 df $= cells - 1$			$\displaystyle\sum \dfrac{(\text{obs} - \text{exp})^2}{\text{exp}}$	26
	Many independent samples	Homogeneity χ^2 Test					
Independence (two categorical variables)	One sample	Independence χ^2 Test	χ^2 df $= (r - 1)(c - 1)$				
Association (two quantitative variables)	One sample	Linear Regression t-Test or Confidence Interval for β	t df $= n - 2$	β_1	b_1	$\dfrac{s_e}{s_x \sqrt{n - 1}}$ (compute with technology)	27
		Confidence Interval for μ_v		μ_v	$\hat{y}_v$	$\sqrt{SE^2(b_1) \cdot (x_v - \bar{x})^2 + \dfrac{s_e^2}{n}}$	
		Prediction Interval for y_v		y_v	$\hat{y}_v$	$\sqrt{SE^2(b_1) \cdot (x_n - \bar{x})^2 + \dfrac{s_e^2}{n} + s_e^2}$	
Inference about?	**One sample or two?**	**Procedure**	**Model**	**Parameter**	**Estimate**	**SE**	**Chapter**

individual factors. We may then move on to consider *contrasts* among different levels of significant factors.

▶ Models with two or more predictors can fail conditions in even more ways than those with only one, and some of these can be quite subtle. These models require care in checking assumptions and conditions, and especially in checking for outliers and high-leverage points.

▶ If the response variable is categorical but dichotomous (has only two values) and the single predictor is quantitative, then you can use a logistic regression.

▶ If the response variable is categorical and the predictor is categorical as well, then use the chi-squared methods from Chapter 26.

Remember (have we said this often enough yet?): Never use any inference procedure without first checking the assumptions and conditions. We summarized those in the last unit; have another look.

Here come more opportunities to practice using these concepts and skills, but we've also sprinkled in some exercises that call for methods learned in earlier chapters, so keep alert.

REVIEW EXERCISES

T 1. **Tableware.** Nambe Mills manufactures plates, bowls, and other tableware made from an alloy of several metals. Each item must go through several steps, including polishing. To better understand the production process and its impact on pricing, the company checked the polishing *Time* (in minutes) and the retail *Price* (in US$) of these items. The regression analysis is shown below. The scatterplot showed a linear pattern, and residuals were deemed suitable for inference.

Dependent variable is: Price
R-squared = 84.5%
s = 20.50 with 59 − 2 = 57 degrees of freedom

Variable	Coefficient	SE(Coeff)
Intercept	−2.89054	5.730
Time	2.49244	0.1416

a) How many different products were included in this analysis?
b) What percentage of the variation in retail price is explained by the polishing time?
c) Create a 95% confidence interval for the slope of this relationship.
d) Interpret your interval in this context.

T 2. **Hard water.** In an investigation of environmental causes of disease, data were collected on the annual *Mortality* rate (deaths per 100,000) for males in 61 large towns in England and Wales. In addition, the water hardness was recorded as the *Calcium* concentration (parts per million, or ppm) in the drinking water. Here are the scatterplot and regression analysis of the relationship between mortality and calcium concentration.

Dependent variable is: Mortality
R-squared = 43%
s = 143.0 with 61 − 2 = 59 degrees of freedom

Variable	Coefficient	SE(Coeff)
Intercept	1676	29.30
Calcium	−3.23	0.48

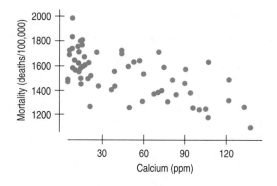

a) Is there an association between the hardness of the water and the mortality rate? Write the appropriate hypothesis.
b) Assuming the assumptions for regression inference are met, what do you conclude?
c) Create a 95% confidence interval for the slope of the true line relating *Calcium* concentration and *Mortality*.
d) Interpret your interval in context.

T 3. **Mutual funds.** In March 2002, *Consumer Reports* listed the rate of return for several large cap mutual funds over the previous 3-year and 5-year periods. ("Large cap" refers to companies worth over $10 billion.)

a) Create a 95% confidence interval for the difference in rate of return for the 3- and 5-year periods covered by these data. Clearly explain what your interval means.
b) It's common for advertisements to carry the disclaimer that "past returns may not be indicative of future performance," but do these data indicate that there was an association between 3-year and 5-year rates of return?

Annualized Returns (%)		
Fund name	3-year	5-year
Ameristock	7.9	17.1
Clipper	14.1	18.2
Credit Suisse Strategic Value	5.5	11.5
Dodge & Cox Stock	15.2	15.7
Excelsior Value	13.1	16.4
Harbor Large Cap Value	6.3	11.5
ICAP Discretionary Equity	6.6	11.4
ICAP Equity	7.6	12.4
Neuberger Berman Focus	9.8	13.2
PBHG Large Cap Value	10.7	18.1
Pelican	7.7	12.1
Price Equity Income	6.1	10.9
USAA Cornerstone Strategy	2.5	4.9
Vanguard Equity Income	3.5	11.3
Vanguard Windsor	11.0	11.0

T 4. **Polling.** How accurate are pollsters in predicting the outcomes of Congressional elections? The table shows the actual number of Democrat seats in the House of Representatives and the number predicted by the Gallup organization for nonpresidential election years since World War II.

a) Is there a significant difference between the number of seats predicted for the Democrats and the number they actually held? Test an appropriate hypothesis and state your conclusions.

b) Is there a strong association between the pollsters' predictions and the outcomes of the elections? Test an appropriate hypothesis and state your conclusions.

Democrat Congressmen		
Year	Predicted	Actual
1946	190	188
1950	235	234
1954	232	232
1958	272	283
1962	259	258
1966	247	248
1970	260	255
1974	292	291
1978	277	277
1982	275	269
1986	264	258
1990	260	267
1994	201	204
1998	211	211

5. **Football.** A student runs an experiment to test four different grips on his football throwing distance, recording the distance in yards that he can throw the football using each grip. He randomizes the grip used each time by drawing numbers out of a hat until each grip has been used 5 times. After collecting his data and analyzing the results, he reports that the P-value of his test is 0.0032.

a) What kind of test should he have performed?
b) What are the null and alternative hypotheses?
c) Assuming that the conditions required for the test are satisfied, what would you conclude?
d) What else about the data would you like to see in order to check the assumptions and conditions?
e) What might he want to test next?

6. **Golf.** A student runs an experiment to test four different clubs on her putting accuracy, recording the distance in centimeters from a small target that she places on the green. She randomizes the club used each time by drawing numbers out of a hat until each club has been used 6 times. After collecting her data and analyzing the results, she reports that the P-value of her test is 0.0245.

a) What kind of test should she have performed?
b) What are the null and alternative hypotheses?
c) Assuming that the conditions required for the test are satisfied, what would you conclude?
d) What else about the data would you like to see in order to check the assumptions and conditions?
e) What might she want to test next?

T 7. **Wild horses.** Large herds of wild horses can become a problem on some federal lands in the West. Researchers hoping to improve the management of these herds collected data to see if they could predict the number of foals that would be born based on the size of the current herd. Their attempt to model this herd growth is summarized in the output shown.

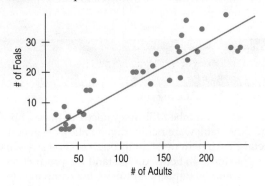

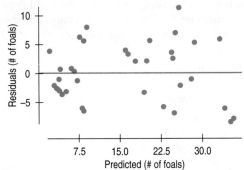

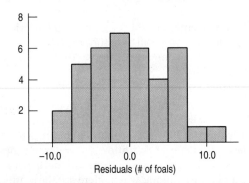

Variable	Count	Mean	StdDev
Adults	38	110.237	71.1809
Foals	38	15.3947	11.9945

Dependent variable is: Foals
R-squared = 83.5%
s = 4.941 with 38 − 2 = 36 degrees of freedom

Variable	Coefficient	SE(Coeff)	t-ratio	Prob
Intercept	−1.57835	1.492	−1.06	0.2970
Adults	0.153969	0.0114	13.5	≤0.0001

a) How many herds of wild horses were studied?
b) Are the conditions necessary for inference satisfied? Explain.
c) Create a 95% confidence interval for the slope of this relationship.
d) Explain in this context what that slope means.
e) Suppose that a new herd with 80 adult horses is located. Estimate, with a 90% prediction interval, the number of foals that may be born.

8. North and hard. The data on water hardness and mortality considered in Exercise 2 also included information on whether the town was north or south of Derby. When that information is coded in a variable as 1 for "north of Derby" and 0 for "south of Derby," and that variable is included in the regression, the resulting table looks like this:

Dependent variable is: Mortality
R-squared = 56.2%   R-squared (adjusted) = 54.7%
s = 126.4 with 61 − 3 = 58 degrees of freedom

Variable	Coefficient	SE(Coeff)	t-ratio	P-value
Intercept	1537.50	42.02	36.6	≤0.0001
Calcium	−2.16011	0.4979	−4.34	≤0.0001
North of Derby	158.892	37.87	4.20	≤0.0001

a) What is the name for the kind of variable that the 0/1 variable is for being north of Derby?
b) What does the coefficient of that variable mean in this regression?
c) Would you prefer this regression to the one of Exercise 2? Explain why or why not.

9. Horses again, a bit less wild. In an attempt to control the growth of the herds of wild horses discussed in Exercise 7, managers sterilized some of the stallions. The variable *Sterilized* is coded 0 for herds and years in which no stallions were sterilized and 1 for herds and years in which some stallions were sterilized. The resulting regression looks like this:

Dependent variable is: Foals
R-squared = 84.8%   R-squared (adjusted) = 83.9%
s = 4.814 with 38 − 3 = 35 degrees of freedom

Source	Sum of squares	DF	Mean square	F-ratio
Regression	4511.85	2	2255.92	97.3
Residual	811.229	35	23.1780	

Variable	Coefficient	SE(Coeff)	t-ratio	P-value
Intercept	6.23667	4.801	1.30	0.2024
Adults	0.112271	0.0268	4.18	0.0002
Sterilized	−6.43657	3.769	−1.71	0.0965

a) What is the name for the kind of variable used here to represent sterilizing?
b) What is the interpretation of the coefficient of *Sterilized?*
c) Does sterilizing appear to work? That is, do these data show a statistically significant effect of sterilizing stallions on the number of foals born?

10. Bowling. A student runs an experiment to test how different factors affect her bowling performance. She uses 3 levels for the weight of the ball (low, medium, and high) and 2 approaches (standing and walking), throwing 4 balls at each condition and choosing the conditions in random order. She counts the number of pins knocked down as the response variable.

a) What are the null and alternative hypotheses for the main effects?
b) How many degrees of freedom does each factor sum of squares have? How about the error sum of squares?
c) Should she consider fitting an interaction term to the model? Why or why not? How many degrees of freedom would it take?

11. Video racing. A student runs an experiment to test how different factors affect his reaction time while playing video games. He uses a specified race course with random hazards and times how long it takes him (in seconds) to finish 4 laps under a variety of experimental conditions. As factors, he uses three different types of *Mouse* (cordless ergonomic, cordless regular, or corded) and keeps the *Lights* on or off. He measures his time once at each condition. To avoid a learning effect, he runs the race 5 times before selecting the conditions in random order.

a) What are the null and alternative hypotheses for the main effects?
b) How many degrees of freedom does each factor sum of squares have? How about the error sum of squares?
c) Should he consider fitting an interaction term to the model? Why or why not? How many degrees of freedom would it take?

12. Resume fraud. In 2002 the Veritas Software company found out that its chief financial officer did not actually have the MBA he had listed on his resume. They fired him, and the value of the company's stock dropped 19%. Kroll, Inc., a firm that specializes in investigating such matters, said that they believe as many as 25 percent of background checks might reveal false information. How many such random checks would they have to do to estimate the true percentage of people who misrepresent their backgrounds to within ±5% with 98% confidence?

13. Paper planes. In preparation for a regional paper airplane competition, a student tried out her latest design. The distances her plane traveled (in feet) in 11 trial flights are given here. (The world record is an astounding 193.01 feet!) The data were 62, 52, 68, 23, 34, 45, 27, 42, 83, 56, and 40 feet.

Here are some summaries:

Count	11
Mean	48.3636
Median	45
StdDev	18.0846
StdErr	5.45273
IntQRange	25
25th %tile	35.5000
75th %tile	60.5000

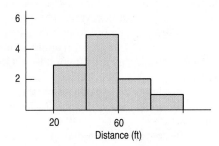

a) Construct a 95% confidence interval for the mean distance.
b) Based on your confidence interval, is it plausible that the mean distance is 40 ft? Explain.

c) How would a 99% confidence interval for the mean distance differ from your answer in part a? Explain briefly, without actually calculating a new interval.

d) How large a sample size would the student need to get a confidence interval half as wide as the one you got in part a, at the same confidence level?

T 14. Nuclear plants. Here are data on 32 light water nuclear power plants. The variables are:

Cost: In $100,000, adjusted to 1976 base.

Date: Date that construction permit was issued in years after 1900. Thus, 68.58 is roughly halfway through 1968.

Mwatts: Power plant net capacity in megawatts.

We are interested in the *Cost* of the plants as a function of *Date* and *Mwatts.*

Cost	Mwatts	Date	Cost	Mwatts	Date
345.39	514	67.92	712.27	845	69.50
460.05	687	68.58	289.66	530	68.42
452.99	1065	67.33	881.24	1090	69.17
443.22	1065	67.33	490.88	1050	68.92
652.32	1065	68.00	567.79	913	68.75
642.23	1065	68.00	665.99	828	70.92
272.37	822	68.17	621.45	786	69.67
317.21	457	68.42	608.80	821	70.08
457.12	822	68.42	473.64	538	70.42
690.19	792	68.33	697.14	1130	71.08
350.63	560	68.58	207.51	745	67.25
402.59	790	68.75	288.48	821	67.17
412.18	530	68.42	284.88	886	67.83
495.58	1050	68.92	280.36	886	67.83
394.36	850	68.92	217.38	745	67.25
423.32	778	68.42	270.71	886	67.83

a) Examine the relationships between *Cost* and *Mwatts* and between *Cost* and *Date*. Make appropriate displays and interpret them with a sentence or two.

b) Find the regression of *Cost* on *Mwatts*. Write a sentence that explains the relationship as described by the regression.

c) Make a scatterplot of residuals vs. predicted values and discuss what it shows. Make a Normal probability plot or histogram of the residuals. Discuss the four assumptions needed for regression analysis and indicate whether you think they are satisfied here. Give your reasons.

d) State the standard null hypothesis for the slope coefficient and complete the *t*-test at the 5% level. State your conclusion.

e) Estimate the cost of a 1000-mwatt plant. Show your work.

f) Compute the residuals for this regression. Discuss the meaning of the *R*-squared in this regression. Plot the residuals against *Date*. Does it appear that *Date* can account for some of the remaining variability?

g) Compute the multiple regression of *Cost* on both *Mwatts* and *Date*. Compare the coefficient in this regression with those you have found for each of these predictors.

h) Would you expect *Mwatts* and *Date* to be correlated? Why or why not? Examine the relationship between *Mwatts* and *Date*. Make a scatterplot and find the correlation coefficient, for example. It's only because of the extraordinary nature of this relationship that the relationships you saw at earlier steps were this simple.

T 15. Barbershop music. At a barbershop music singing competition, choruses are judged on three scales: *Music* (quality of the arrangement, etc.), *Performance*, and *Singing*. The scales are supposed to be independent of each other, and each is scored by a different judge, but a friend claims that he can predict a chorus's singing score from the other two scores. He offers the following regression based on the scores of all 34 choruses in a recent competition:

Dependent variable is: Singing
R-squared = 90.9%   R-squared (adjusted) = 90.3%
s = 6.483 with 34 − 3 = 31 degrees of freedom

Variable	Coefficient	SE(Coeff)	t-ratio	P-value
Intercept	2.08926	7.973	0.262	0.7950
Performance	0.793407	0.0976	8.13	≤0.0001
Music	0.219100	0.1196	1.83	0.0766

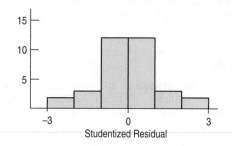

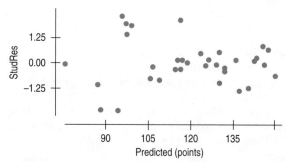

a) What do you think of your friend's claim? Can he predict singing scores? Explain.

b) State the standard null hypothesis for the coefficient of *Performance* and complete the *t*-test at the 5% level. State your conclusion.

c) Complete the analysis. Check assumptions and conditions to the extent you can with the information provided.

16. **Sleep.** Using a simple random sample, a student group asked 450 students about their sleep and study habits and received about 200 responses. The group wanted to know

if the average amount of sleep varied by sex (F or M) or by year (Freshman, Sophomore, Junior, or Senior) of the respondent. Partial boxplots of the amount of sleep last night by the two factors are shown below with the interaction plot:

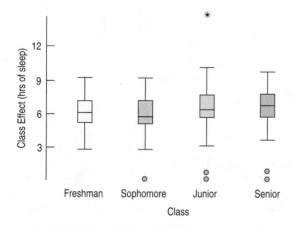

The ANOVA table shows:

Source	DF	Sum of squares	F-ratio	P-value
Sex	1	8.0658	2.2821	0.1325
Year	3	8.8990	0.8393	0.4739
Sex × Year	3	19.4075	1.8303	0.1431
Total	196	707.0180		

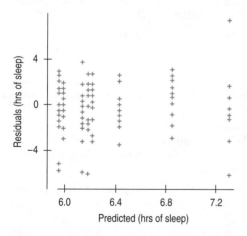

a) What are the null and alternative hypotheses for the main effects?
b) Should the group consider fitting an interaction term to the model? Why or why not?
c) What effects appear to be significant?
d) What, if any, reservations do you have about the conclusions?

17. **Study habits.** The survey in Exercise 16 also asked students about the number of hours they studied. Those doing the survey wanted to know if the average amount of studying varied by sex or by class of the respondent. Partial boxplots of the *Hours Studied* last night by the two factors are shown below:

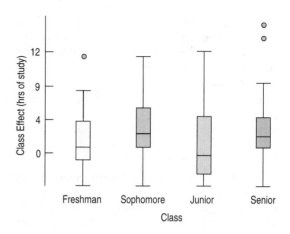

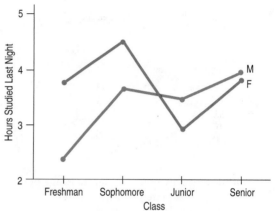

Source	DF	Sum of squares	F-ratio	P-value
Sex	1	7.1861	1.0744	0.3013
Year	3	34.9922	1.7440	0.1595
Sex × Year	3	27.8188	1.3865	0.2483
Total	196	1329.8020		

A plot of residuals vs. predicted shows:

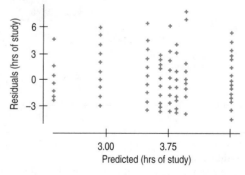

a) What are the null and alternative hypotheses for the main effects?
b) Should those doing the survey consider fitting an interaction term to the model? Why or why not?
c) What effects appear to be significant?
d) What, if any, reservations do you have about the conclusions?

18. **Pregnancy.** In 1998 a San Diego reproductive clinic reported 42 live births to 157 women under the age of 38, but only 7 successes for 89 clients aged 38 and older. Is this evidence of a difference in the effectiveness of the clinic's methods for older women?

   a) Test the appropriate hypotheses using the 2-proportion z-procedure.
   b) Repeat the analysis using an appropriate chi-square procedure.
   c) Explain how the two results are equivalent.

19. **Lost baggage.** The Bureau of Transportation Statistics of the U.S. Department of Transportation reports statistics about airline performance. For 2005 they report the following number of bags lost per 1000 passengers.

Airline	Lost Bags
Continental	4.12
Southwest	4.25
United	4.26
America West	4.33
Northwest	4.84
American	5.92
Delta	7.09
USAir	9.62

   Are the airlines roughly equal in their baggage performance? Perform a chi-square goodness-of-fit test, or explain why that would not be appropriate.

20. **Old Faithful.** As you saw in an earlier chapter, Old Faithful isn't all that faithful. Eruptions do not occur at uniform intervals, and may vary greatly. Can we improve our chances of predicting the time *Interval* until the next eruption if we know the *Duration* of the previous eruption?

   a) Describe what you see in this scatterplot.

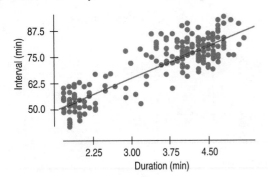

   b) Write an appropriate hypothesis.
   c) Here are a histogram of the residuals and the residuals plot. Do you think the assumptions for inference are met? Explain.

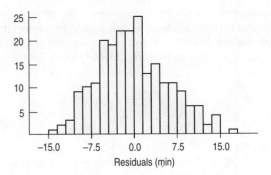

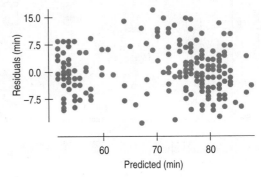

   d) State a conclusion based on this regression analysis:

Dependent variable is: Interval
R-squared = 77.0%
s = 6.159 with 222 − 2 = 220 degrees of freedom

Variable	Coefficient	SE(Coeff)	t-ratio	P-value
Intercept	33.9668	1.428	23.8	≤0.0001
Duration	10.3582	0.3822	27.1	≤0.0001

Variable	Mean	StdDev
Duration	3.57613	1.08395
Interval	71.0090	12.7992

   e) The second table shows the summary statistics for the two variables. Create a 95% confidence interval for the mean *Interval* following a 2-minute eruption.
   f) You arrive at Old Faithful just as an eruption ends. Witnesses say it lasted 4 minutes. Create a 95% prediction interval for the length of time you will wait to see the next eruption.

21. **Togetherness.** Are good grades in high school associated with family togetherness? A simple random sample of 142 high school students was asked how many meals per week their families ate together. Their responses produced a mean of 3.78 meals per week, with a standard deviation of 2.2. Researchers then matched these responses against the students' grade point averages. The scatterplot appeared to be reasonably linear, so they went ahead with the regression analysis, seen below. No apparent pattern emerged in the residuals plot.

Dependent variable: GPA
R-squared = 11.0%
s = 0.6682 with 142 − 2 = 140 df

Variable	Coefficient	SE(Coeff)
Intercept	2.7288	0.1148
Meals/wk	0.1093	0.0263

a) Is there evidence of an association? Test an appropriate hypothesis and state your conclusion.
b) Do you think this association would be useful in predicting a student's grade point average? Explain.
c) Are your answers to parts a and b contradictory? Explain.

T 22. Is Old Faithful getting older? The data on Old Faithful eruptions that we saw in Exercise 20 includes another variable recording the *Day* on which the eruption occurred (where 1 is the first day and each successive day just counts one more). The correlation of *Interval* (minutes until the next eruption) with *Day* is −0.004. But when we include *Day* in a multiple regression along with the *Duration* (in minutes) of the previous eruption, we get the following model:

Dependent variable is: Interval
R-squared = 77.6%   R-squared (adjusted) = 77.3%
s = 6.092 with 222 − 3 = 219 degrees of freedom

Variable	Coefficient	SE(Coeff)	t-ratio	P-value
Intercept	35.2463	1.509	23.4	≤0.0001
Duration	10.4348	0.3794	27.5	≤0.0001
Day	−0.126316	0.0523	−2.42	0.0166

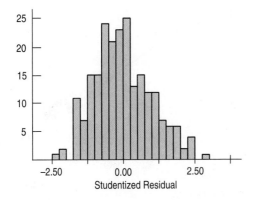

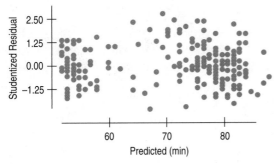

a) What is the model fit by this regression?
b) Is the *Interval* changing over time? Perform a formal test of the relevant hypothesis.

c) Doesn't the small P-value for *Day* contradict the correlation between *Interval* and *Day* being virtually zero? Explain.
d) Is the amount of change in *Interval* due to *Day* meaningful?

23. **Lefties and music.** In an experiment to see if left- and right-handed people have different abilities in music, subjects heard a tone and were then asked to identify which of several other tones matched the first. Of 76 right-handed subjects, 38 were successful in completing this test, compared with 33 of 53 lefties. Is this strong evidence of a difference in musical abilities based on handedness?

24. **Preemies.** Do the effects of being born prematurely linger into adulthood? Researchers examined 242 Cleveland area children born prematurely between 1977 and 1979, and compared them with 233 children of normal birth weight; 24 of the "preemies" and 12 of the other children were described as being of "subnormal height" as adults. Is this evidence that babies born with a very low birth weight are more likely to be smaller than normal adults? ("Outcomes in Young Adulthood for Very-Low-Birth-Weight Infants," *New England Journal of Medicine*, 346, no. 3 [January 2002])

T 25. Teen traffic deaths 2007. The Insurance Institute for Highway Safety publishes data on a variety of traffic-related risks. One report gives the numbers of male and female teenagers killed in highway accidents during each year from 1975 to 2002. Here is a regression predicting *Female Deaths* by *Year*:

Dependent variable is: Female deaths
R-squared = 57.4%   R-squared (adjusted) = 56.1%
s = 182.6 with 33 − 2 = 31 degrees of freedom

Source	Sum of squares	df	Mean square	F-ratio
Regression	1396025	1	1396025	41.9
Residual	1033964	31	33353.7	

Variable	Coefficient	SE(Coeff)	t-ratio	P-value
Intercept	45074.0	6648	6.78	≤0.0001
Year	−21.6006	3.339	−6.47	≤0.0001

a) Here's a scatterplot of residuals vs. predicted values for this regression. Discuss the assumptions and conditions of the regression.

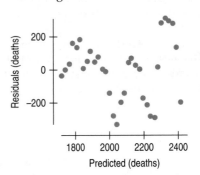

b) What is the meaning of the R^2 value of 52.3% for this regression?

c) What is the regression equation?

d) Give an interpretation of the coefficient of *Year* in this regression.

T 26. More teen traffic 2007. The data discussed in Exercise 25 included the numbers of male teen traffic deaths as well. We can add that as a predictor to obtain the following model:

Dependent variable is: Female deaths
R-squared = 89.5%   R-squared (adjusted) = 88.8%
s = 92.05 with 33 − 3 = 30 degrees of freedom

Source	Sum of squares	df	Mean square	F-ratio
Regression	2175799	2	1087899	128
Residual	254191	30	8473.02	

Variable	Coefficient	SE(Coeff)	t-ratio	P-value
Intercept	−20186.2	7583	−2.66	0.0124
Year	10.5383	3.749	2.81	0.0086
male	0.271321	0.0283	9.59	≤0.0001
Year	20.3628	4.326	4.71	≤0.0001

a) How does this regression compare with the regression of Exercise 25? Which would you prefer to use? Why?

b) What does the coefficient of *Year* mean in this regression?

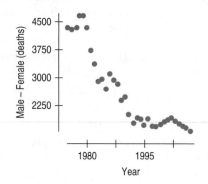

c) Considering both the regression model of Exercise 25 and this one, would you say that female teen traffic fatalities have been increasing or decreasing over time? How does the scatterplot above help to explain what is happening?

27. **Typing.** For a class project, Nick M. designed and carried out an experiment to see if the room *Temperature* and the wearing of *Gloves* affected his typing speed. He ran each combination of hot and cold temperature and gloves on and off 8 times, recording the net number of words typed (words typed minus mistakes). Partial boxplots and interaction plots show:

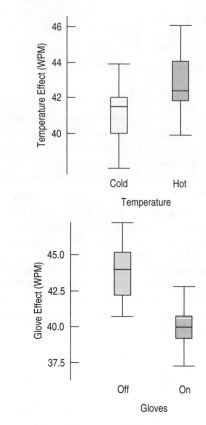

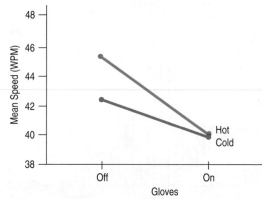

The ANOVA table shows:

Source	DF	Sum of squares	F-ratio	P-value
Gloves	1	120.1250	57.2511	<0.0001
Temperature	1	18.0000	8.5787	0.0067
Gloves*Temp	1	15.1250	7.2085	0.0121
Total	28	58.75		

A scatterplot of residuals vs. predicted values shows:

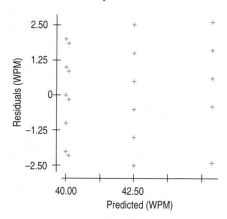

a) What are the null and alternative hypotheses for the main effects?
b) Given the partial boxplots, what do you suspect the *F*-test will say about the two main effects?
c) Should he consider fitting an interaction term to the model? Why or why not?
d) If he does fit an interaction term, do you suspect it will be significant on the basis of the interaction plot? Explain.
e) Which effects appear to be significant?
f) Describe the effects of the factors.
g) What is the size of the estimated standard deviation of the errors? Does this seem reasonable given the partial boxplots? Explain.
h) If Nick wants to increase his typing speed, what recommendations would you give him?
i) What reservations, if any, do you have about the conclusions?

28. **Typing again.** Nick (see Exercise 27) designed a follow-up experiment to see if having *Music* or *Television* on would affect his typing speed. In particular, he'd like to know if he can type just as effectively with the music and/or the TV on while he types. He ran each combination of *Music* and *Television* on or off 8 times, recording the net number of words typed (words typed minus mistakes). Partial boxplots and interaction plots show:

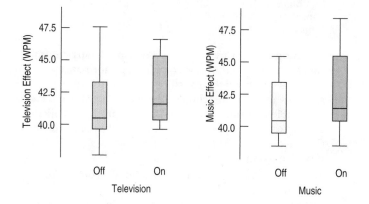

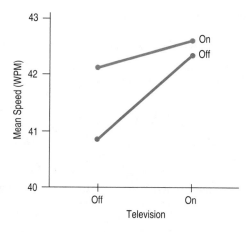

The ANOVA table shows:

Source	DF	Sum of squares	F-ratio	P-value
Music	1	4.5000	0.6380	0.4312
Television	1	8.0000	1.1342	0.2960
Interaction	1	2.0000	0.2835	0.5986
Total	28	197.5000		

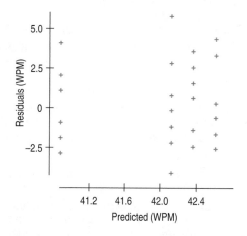

a) What are the null and alternative hypotheses for the main effects?
b) Given the partial boxplots, what do you suspect the *F*-test will say about the two main effects?
c) Should he consider fitting an interaction term to the model? Why or why not?
d) If he does fit an interaction term, do you suspect it will be significant on the basis of the interaction plot? Explain.
e) What effects appear to be significant?
f) Describe the effects of the factors.
g) What is the size of the estimated standard deviation of the errors? Does this seem reasonable given the partial boxplots? Explain.
h) If Nick wants to increase his typing speed, what recommendations would you give him?
i) What reservations, if any, do you have about the conclusions?

T 29. **NY Marathon.** The *New York Times* reported the results of the 2003 NY Marathon by listing time brackets and the number of racers who finished within that bracket. Because the brackets are of different sizes, we look at the number of racers finishing *per minute* against the time at the middle of the time bracket. The resulting regression looks like this:

Dependent variable is: #/minute
R-squared = 10.9%   R-squared (adjusted) = 9.5%
s = 78.45 with 63 − 2 = 61 degrees of freedom

Variable	Coefficient	SE(Coeff)	t-ratio	P-value
Intercept	45.1407	51.74	0.873	0.3863
Mid time ...	0.519037	0.1899	2.73	0.0082

a) How would you interpret the coefficient of *Mid Time*?

Here's a scatterplot of the Studentized residuals against the predicted values for this regression:

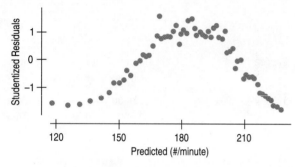

b) Comment on the regression in the light of this plot.

30. Births. The National Vital Statistics Report provides information on live *Births* (per 1000 women), according to the age of the woman (in 5-year brackets—*Age* used here is the midpoint of the bracket) and the *Year* from 1990 to 1999. The report isolates births to women younger than 20 as a separate category. Looking only at women 20 years old and older, we find the following regression:

Dependent variable is: Births
R-squared = 98.1%   R-squared (adjusted) = 98.0%
s = 15.55 with 50 − 3 = 47 degrees of freedom

Variable	Coefficient	SE(Coeff)	t-ratio	P-value
Intercept	4422.98	1527	2.90	0.0057
Age	−15.1898	0.3110	−48.8	≤0.0001
Year	−1.89830	0.7656	−2.48	0.0168

a) Write out the regression model.
b) How would you interpret the coefficient of *Year* in this regression? What happened to pregnancy rates during the decade of the 1990s?

Here's a scatterplot of the Studentized residuals against the predicted values.

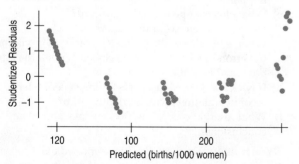

c) What might you do to improve the regression model?

31. Depression and the Internet. The September 1998 issue of the *American Psychologist* published an article reporting on an experiment examining "the social and psychological impact of the Internet on 169 people in 73 households during their first 1 to 2 years on-line." In the experiment, a sample of households was offered free Internet access for one or two years in return for allowing their time and activity online to be tracked. The members of the households who participated in the study were also given a battery of tests at the beginning and again at the end of the study. One of the tests measured the subjects' levels of depression on a 4-point scale, with higher numbers meaning the person was more depressed. Internet usage was measured in average number of hours per week. The regression analysis examines the association between the subjects' depression levels and the amounts of Internet use. The conditions for inference were satisfied.

Dependent variable is: Depression after
R-squared = 4.6%
s = 0.4563 with 162 − 2 = 160 degrees of freedom

Variable	Coefficient	SE(coeff)	t-ratio	Probe
Constant	0.565485	0.0399	14.2	≤0.0001
Intr_use	0.019948	0.0072	2.76	0.0064

a) Do these data indicate that there is an association between Internet use and depression? Test an appropriate hypothesis and state your conclusion clearly.
b) One conclusion of the study was that those who spent more time online tended to be more depressed at the end of the experiment. News headlines said that too much time on the Internet can lead to depression. Does the study support this conclusion? Explain.
c) As noted, the subjects' depression levels were tested at both the beginning and the end of this study; higher scores indicated the person was more depressed. Results are summarized in the table. Is there evidence that the depression level of the subjects changed during this study?

Depression Level
162 subjects

Variable	Mean	StdDev
DeprBfore	0.730370	0.487817
DeprAfter	0.611914	0.461932
Difference	−0.118457	0.552417

32. Learning math. Developers of a new math curriculum called "Accelerated Math" compared performances of students taught by their system with control groups of students in the same schools who were taught using traditional instructional methods and materials. Statistics about pretest and posttest scores are shown in the table. (J. Ysseldyke and S. Tardrew, *Differentiating Math Instruction*, Renaissance Learning, 2002)

a) Did the groups differ in average math score at the start of this study?
b) Did the group taught using the Accelerated Math program show a significant improvement in test scores?
c) Did the control group show significant improvement in test scores?

d) Were gains significantly higher for the Accelerated Math group than for the control group?

		Instructional Method	
		Acc. math	Control
Number of students		231	245
Pretest	Mean	560.01	549.65
	St. Dev	84.29	74.68
Posttest	Mean	637.55	588.76
	St. Dev	82.9	83.24
Individual gain	Mean	77.53	39.11
	St. Dev.	78.01	66.25

33. **Pesticides.** A study published in 2002 in the journal *Environmental Health Perspectives* examined the sex ratios of children born to workers exposed to dioxin in Russian pesticide factories. The data covered the years 1961 to 1988 in the city of Ufa, Bashkortostan, Russia. Of 227 children born to workers exposed to dioxin, only 40% were male. Overall in the city of Ufa the proportion of males was 51.2%. Is this evidence that human exposure to dioxin may results in the birth of more girls? (An interesting note: It appeared that paternal exposure was most critical; 51% of babies born to mothers exposed to the chemical were boys.)

T 34. **Dairy sales.** Peninsula Creameries sells both cottage cheese and ice cream. The CEO recently noticed that in months when the company sells more cottage cheese, it seems to sell more ice cream as well. Two of his aides were assigned to test whether this is true or not. The first aide's plot and analysis of sales data for the past 12 months (in millions of pounds for cottage cheese and for ice cream) appears below.

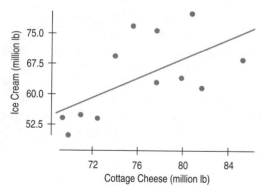

Dependent variable is: Ice cream
R-squared = 36.9%
s = 8.320 with 12 − 2 = 10 degrees of freedom

Variable	Coefficient	SE(Coeff)	t-ratio	Probe
Constant	−26.5306	37.68	−0.704	0.4975
Cottage C ...	1.19334	0.4936	2.42	0.0362

The other aide looked at the differences in sales of ice cream and cottage cheese for each month, and created the following output:

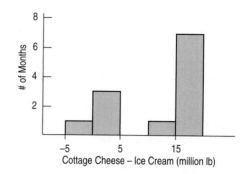

Cottage Cheese–Ice Cream
Count	12
Mean	11.8000
Median	15.3500
StdDev	7.99386
IntQRange	14.3000
25th %tile	3.20000
75th %tile	17.5000

Test HO: $\mu(CC - IC) = 0$ vs Ha: $\mu(CC - IC) \neq 0$
Sample Mean = 11.800000 t-Statistic = 5.113 w/11 df
Prob = 0.0003
Lower 95% bound = 6.7209429
Upper 95% bound = 16.879057

a) Which analysis would you use to answer the CEO's question? Why?
b) What would you tell the CEO?
c) Which analysis would you use to test whether the company sells more cottage cheese or ice cream in a typical year? Why?
d) What would you tell the CEO about this other result?
e) What assumptions are you making in the analysis you chose in part a? What assumptions are you making in the analysis in part c?
f) Next month's cottage cheese sales are 82 million pounds. Ice cream sales are not yet available. How much ice cream do you predict Peninsula Creameries will sell?
g) Give a 95% confidence interval for the true slope of the regression equation of ice cream sales by cottage cheese sales.
h) Explain what your interval means.

35. **Video pinball.** A student runs an experiment to test how different factors affect his score while playing video pinball. Here are the results of 16 runs of an experiment performed in random order. Factor *Eyes* has two levels: *both* open and *right eye* closed. Factor *Tilt* has two levels: tilt *on* and tilt *off*. The response is the score of one ball at a combination of factors. Each combination was repeated 4 times in random order.

a) What are the null and alternative hypotheses for the main effects?
b) Analyze the data and write a short report on your findings. Include appropriate graphics and diagnostic plots.

Eyes	Tilt	Score
Both	On	67059
Right eye	Off	21036
Both	On	59520
Right eye	Off	3100
Both	Off	61272
Right eye	On	55957
Both	Off	72472
Right eye	On	18460
Right eye	On	16556
Both	Off	89553
Right eye	On	37950
Both	Off	74336
Right eye	Off	700
Both	On	79037
Right eye	Off	36591
Both	On	74610

36. Javelin. Brianna, a member of the track team, runs an experiment to test how different factors affect her javelin throw. She wants to know if the more expensive (*premium*) javelin is worth the extra price and is curious to know how much warming up helps her distance. She tries all four combinations of the two *Javelins* (*standard* and *premium*) and *Preparation* (no warm-up—*cold*—and *warm-up*), repeating each combination twice in random order. She measures the distance of her throw to the nearest meter. Here are the results of the 8 runs.

Javelin	Preparation	Distance (meters)
Premium	Warm-up	46
Standard	Warm-up	39
Premium	Cold	37
Standard	Cold	30
Premium	Warm-up	45
Standard	Warm-up	40
Premium	Cold	35
Standard	Cold	32

a) What are the null and alternative hypotheses for the main effects?

b) Analyze the data and write a short report on your findings. Include appropriate graphics and diagnostic plots and a recommendation for Brianna to optimize her javelin distance.

T 37. Eye and hair color. A survey of 1021 school-age children was conducted by randomly selecting children from several large urban elementary schools. Two of the questions concerned eye and hair color. In the survey, the following codes were used:

Hair Color	Eye Color
1 = Blond	1 = Blue
2 = Brown	2 = Green
3 = Black	3 = Brown
4 = Red	4 = Grey
5 = Other	5 = Other

The Statistics students analyzing the data were asked to study the relationship between eye and hair color.

a) One group of students produced the output shown below. What kind of analysis is this? What are the null and alternative hypotheses? Is the analysis appropriate? If so, summarize the findings, being sure to include any assumptions you've made and/or limitations to the analysis. If it's not an appropriate analysis, explicitly state why not.

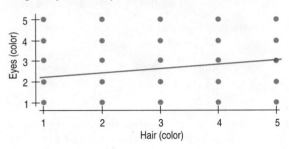

Dependent variable is: Eye color
R-squared = 3.7%
s = 1.112 with 1021 − 2 = 1019 degrees of freedom

Variable	Coefficient	SE(Coeff)	t-ratio	Probe
Constant	1.99541	0.08346	23.9	≤0.0001
Hair Color	0.211809	0.03372	0.28	≤0.0001

b) A second group of students used the same data to produce the output shown below. What kind of analysis is this? What are the null and alternative hypotheses? Is the analysis appropriate? If so, summarize the findings, being sure to include any assumptions you've made and/or limitations to the analysis. If it's not an appropriate analysis, explicitly state why not.

Table contents: Counts
Standardized Residuals

		Eye Color				
		1	2	3	4	5
Hair Color	1	143	30	58	15	12
		7.6754	0.417988		−0.63925	−0.314506
	2	90	45	215	30	20
		−2.57141	0.290189	1.72235	0.491885	−0.0824592
	3	28	15	190	10	10
		−5.39425	−2.3478	6.28154	−1.76376	−0.803818
	4	30	15	10	10	5
		2.06116	2.71589	−4.0554	2.37402	0.759931
	5	10	5	15	5	5
		−0.521945	0.332621	−0.941918	1.36326	2.07578

$$\sum \frac{(Observed - Expected)^2}{Expected} = 223.6 \text{ P-value} < 0.00001$$

38. Infliximab. In an article appearing in the journal *Lancet* in 2002, medical researchers reported on the experimental use of the arthritis drug infliximab in treating Crohn's disease. In a trial, 573 patients were given initial 5-mg injections of the drug. Two weeks later, 335 had responded positively. These patients were then randomly assigned to three groups. Group I received continued injections of a placebo, Group II continued with 5 mg of infliximab, and Group III received 10 mg of the drug. After 30 weeks, 23 of 110 Group I patients were in remission, compared with 44 of 113 Group II and 50 of 112 Group III patients. Do these data indicate that continued treatment with infliximab is of value for Crohn's disease patients who exhibit a positive initial response to the drug?

T 39. LA rainfall. The Los Angeles Almanac website reports recent annual rainfall (in inches), as shown in the table.

a) Create a 90% confidence interval for the mean annual rainfall in LA.

b) If you wanted to estimate the mean annual rainfall with a margin of error of only 2 inches, how many years' data would you need?

c) Do these data suggest any change in annual rainfall as time passes? Check for an association between rainfall and year.

Year	Rain (in.)	Year	Rain (in.)
1980	8.96	1991	21.00
1981	10.71	1992	27.36
1982	31.28	1993	8.14
1983	10.43	1994	24.35
1984	12.82	1995	12.46
1985	17.86	1996	12.40
1986	7.66	1997	31.01
1987	12.48	1998	9.09
1988	8.08	1999	11.57
1989	7.35	2000	17.94
1990	11.99	2001	4.42

40. TV and athletics. Using a simple random sample, a student asked 200 students questions about their study and workout habits, and received 124 responses. One of the questions asked, "Do you participate in intramural (IM) athletics, varsity athletics, or no athletics?" while the second asked, "How many hours of television did you watch last week?" The student wants to see if participation in athletics is associated with amount of TV watched. Here are the boxplots of *TV Watching* by *Athletic Participation*:

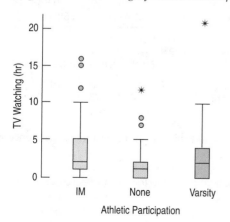

Analysis of Variance

Source	DF	Sum of squares	Mean square	F-ratio	P-value
Athl. Part.	2	128.271	64.1353	4.2343	0.0167
Error	121	1832.72	15.1465		
Total	123	1960.99			

a) State the hypothesis about the students (both numerically and in words).

b) Do the conditions for ANOVA appear to be satisfied? What concerns, if any, do you have?

c) Assuming that the assumptions for inference are satisfied, perform the hypothesis test and state your conclusion. Be sure to state it in terms of television watching and athletic participation.

d) An analysis of the data with all the points highlighted as outliers removed was performed. The F-test showed a P-value of 0.0049. How does this affect your answer to part b?

41. Weight and athletics. Using the same survey as in Exercise 40, the student examined the relationship between *Athletic Participation* and *Weight*. Here are the boxplots of *Weight* by *Athletic Participation*:

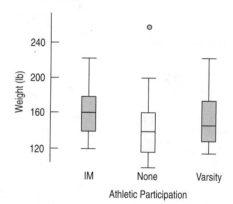

Analysis of Variance

Source	DF	Sum of squares	Mean square	F-ratio	P-value
Athl. Part.	2	9025.55	4512.78	5.7368	0.0042
Error	121	95183.2	786.638		
Total	123	104209			

a) State the null hypothesis about the students (both numerically and in words).

b) Do the conditions for ANOVA appear to be satisfied? What concerns, if any, do you have?

c) Assuming that the assumptions for inference are satisfied, perform the hypothesis test and state your conclusion. Be sure to state it in terms of *Weight* and *Athletic Participation*. What might explain the apparent relationship?

d) An analysis of the data with the points highlighted as an outlier removed was performed. The F-test showed a P-value of 0.0030. How does this affect your answer to part b?

42. Weight loss. A weight loss clinic advertises that its program of diet and exercise will allow clients to lose 10 pounds in one month. A local reporter investigating weight reduction gets permission to interview a randomly selected sample of clients who report the given weight losses during their first month in this program. Create a confidence interval to test the clinic's claim that typical weight loss is 10 pounds.

Pounds Lost	
9.5	9.5
13	9
9	8
10	7.5
11	10
9	7
5	8
9	10.5
12.5	10.5
6	9

43. Cramming. Students in two basic Spanish classes were required to learn 50 new vocabulary words. One group of 45 students received the list on Monday and studied the words all week. Statistics summarizing this group's scores on Friday's quiz are given. The other group of 25 students did not get the vocabulary list until Thursday. They also took the quiz on Friday after "cramming" Thursday night. Then, when they returned to class the following Monday they were retested—without advance warning. Both sets of test scores for these students are shown.

Group 1

Fri.

Number of students = 45

Mean = 43.2 (of 50)

StDev = 3.4

Students passing (score ≥ 40) = 33%

a) Did the week-long study group have a mean score significantly higher than that of the overnight crammers?
b) Was there a significant difference in the percentages of students who passed the quiz on Friday?
c) Is there any evidence that when students cram for a test their "learning" does not last for 3 days?
d) Use a 95% confidence interval to estimate the mean number of words that might be forgotten by crammers.
e) Is there any evidence that how much students forget depends on how much they "learned" to begin with?

Group 2			
Fri.	Mon.	Fri.	Mon.
42	36	50	47
44	44	34	34
45	46	38	31
48	38	43	40
44	40	39	41
43	38	46	32
41	37	37	36
35	31	40	31
43	32	41	32
48	37	48	39
43	41	37	31
45	32	36	41
47	44		

44. Education vs. income. The following information examines the *Median Income* and *Median Education* level (years in school) for several U.S. cities.

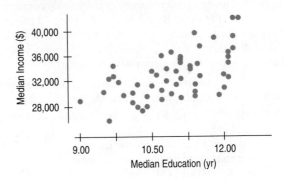

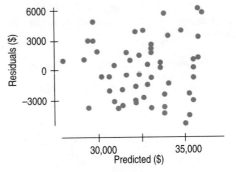

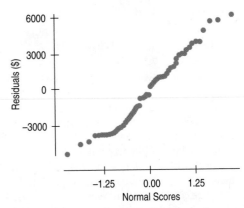

Variable	Count	Mean	StdDev
Education	57	10.9509	0.848344
Income	57	32742.6	3618.01

Dependent variable is: Income

R-squared = 32.9%

s = 2991 with 57 − 2 = 55 degrees of freedom

Variable	Coefficient	SE(Coeff)	t-ratio	Probe
Constant	5970.05	5175	1.15	0.2537
Education	2444.79	471.2	5.19	≤0.0001

a) Do you think the assumptions for inference are met? Explain.
b) Does there appear to be an association between education and income levels in these cities?
c) Would this association appear to be weaker, stronger, or the same if data were plotted for individual people rather than for cities in aggregate? Explain.
d) Create and interpret a 95% confidence interval for the slope of the true line that describes the association between income and education.

e) Predict the *Median Income* for cities where residents spent an average of 11 years in school. Describe your estimate with a 90% confidence interval, and interpret that result.

45. Airport screening. Concern with terrorism leads us to look at the records of airport screening for the years between 1977 and 2000 as provided by the *Sourcebook of Criminal Justice Statistics Online*. We find the following regression predicting the incidence of *False Information* from other problems discovered while screening passengers:

Dependent variable is: False info
R-squared = 8.7%   R-squared (adjusted) = −9.5%
s = 48.41 with 19 − 4 = 15 degrees of freedom

Source	Sum of squares	DF	Mean square	F-ratio
Regression	3364.40	3	1121.47	0.479
Residual	35150.2	15	2343.35	

Variable	Coefficient	SE(Coeff)	t-ratio	P-value
Intercept	27.8181	65.24	0.426	0.6759
Long guns	0.545680	0.6010	0.908	0.3783
Handguns	−4.25650e-3	0.0357	−0.119	0.9066
Explosives	−0.093442	0.0952	−0.982	0.3417

a) Does this appear to be a successful model for incidence of false information? Why or why not?

Here's a scatterplot of the Studentized residuals vs. the predicted values:

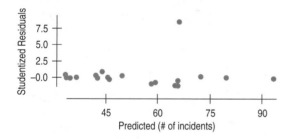

The outlying value is the year 1988. We created an indicator variable for 1988 and included it in the model with the following result:

Dependent variable is: False info
R-squared = 86.8%   R-squared (adjusted) = 83.0%
s = 19.05 with 19 − 5 = 14 degrees of freedom

Source	Sum of squares	DF	Mean square	F-ratio
Regression	33436.2	4	8359.05	23.0
Residual	5078.44	14	362.746	

Variable	Coefficient	SE(Coeff)	t-ratio	P-value
Intercept	47.2152	25.76	1.83	0.0882
Long guns	0.750819	0.2375	3.16	0.0069
Handguns	−0.023583	0.0142	−1.66	0.1190
Explosives	−0.089368	0.0374	−2.39	0.0316
1988	181.310	19.91	9.10	≤0.0001

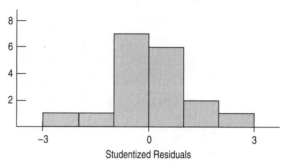

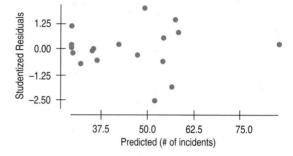

b) Complete the analysis. Check the assumptions and conditions, so far as you can with the information provided. Summarize the model. Discuss any concerns you may still have about the data or the model.
c) Would you remove any of the predictors from the model? If so, which one and why?

Answers

Here are the "answers" to the exercises for the chapters and the unit reviews. As we said in Chapter 1, the answers are outlines of the complete solution. Your solution should follow the model of the Step-By-Step examples, where appropriate. You should explain the context, show your reasoning and calculations, and draw conclusions. For some problems, what you decide to include in an argument may differ somewhat from the answers here. But, of course, the numerical part of your answer should match the numbers in the answers shown.

Chapter 2

1. Categorical

3. Quantitative

5. Answers will vary.

7. *Who*—2500 cars; *What*—Distance from car to bicycle; *Population*—All cars passing bicyclists

9. *Who*—Coffee drinkers at a Newcastle University coffee station; *What*—Amount of money contributed; *Population*—All people in honor system payment situations

11. *Who*—25,892 men aged 30 to 87; *What*—Fitness level and cause of death; *Population*—All men

13. *Who*—54 bears; *Cases*—Each bear is a case; *What*—Weight, neck size, length, and sex; *When*—Not specified; *Where*—Not specified; *Why*—To estimate weight from easier-to-measure variables; *How*—Researchers collected data on 54 bears they were able to catch. *Variable*—Weight; *Type*—Quantitative; *Units*—Not specified; *Variable*—Neck size; *Type*—Quantitative; *Units*—Not specified; *Variable*—Length; *Type*—Quantitative; *Units*—Not specified; *Variable*—Sex; *Type*—Categorical

15. *Who*—Arby's sandwiches; *Cases*—Each sandwich is a case; *What*—Type of meat, number of calories, and serving size; *When*—Not specified; *Where*—Arby's restaurants; *Why*—To assess nutritional value of sandwiches; *How*—Report by Arby's restaurants; *Variable*—Type of meat; *Type*—Categorical; *Variable*—Number of calories; *Type*—Quantitative; *Units*—Calories; *Variable*—Serving size; *Type*—Quantitative; *Units*—Ounces

17. *Who*—882 births; *Cases*—Each of the 882 births is a case; *What*—Mother's age, length of pregnancy, type of birth, level of prenatal care, birth weight of baby, sex of baby, and baby's health problems; *When*—1998–2000; *Where*—Large city hospital; *Why*—Researchers were investigating the impact of prenatal care on newborn health; *How*—Not specified exactly, but probably from hospital records; *Variable*—Mother's age; *Type*—Quantitative; *Units*—Not specified; probably years; *Variable*—Length of pregnancy; *Type*—Quantitative; *Units*—Weeks; *Variable*—Birth weight of baby; *Type*—Quantitative; *Units*—Not specified, probably pounds and ounces; *Variable*—Type of birth;

Type—Categorical; *Variable*—Level of prenatal care; *Type*—Categorical; *Variable*—Sex; *Type*—Categorical; *Variable*—Baby's health problems; *Type*—Categorical

19. *Who*—Experiment subjects; *Cases*—Each subject is a case; *What*—Treatment (herbal cold remedy or sugar solution) and cold severity; *When*—Not specified; *Where*—Not specified; *Why*—To test efficacy of herbal remedy on common cold; *How*—The scientists set up an experiment; *Variable*—Treatment; *Type*—Categorical; *Variable*—Cold severity rating; *Type*—Quantitative (perhaps ordinal categorical); *Units*—Scale from 0 to 5; *Concerns*—The severity of a cold seems subjective and difficult to quantify. Scientists may feel pressure to report negative findings of herbal product.

21. *Who*—Streams; *Cases*—Each stream is a case; *What*—Name of stream, substrate of the stream, acidity of the water, temperature, BCI; *When*—Not specified; *Where*—Upstate New York; *Why*—To study ecology of streams; *How*—Not specified; *Variable*—Stream name; *Type*—Identifier; *Variable*—Substrate; *Type*—Categorical; *Variable*—Acidity of water; *Type*—Quantitative; *Units*—pH; *Variable*—Temperature; *Type*—Quantitative; *Units*—Degrees Celsius; *Variable*—BCI; *Type*—Quantitative; *Units*—Not specified

23. *Who*—41 refrigerator models; *Cases*—Each of the 41 refrigerator models is a case; *What*—Brand, cost, size, type, estimated annual energy cost, overall rating, and repair history; *When*—2006; *Where*—United States; *Why*—To provide information to the readers of *Consumer Reports*; *How*—Not specified; *Variable*—Brand; *Type*—Categorical; *Variable*—Cost; *Type*—Quantitative; *Units*—Not specified (dollars); *Variable*—Size; *Type*—Quantitative; *Units*—Cubic feet; *Variable*—Type; *Type*—Categorical; *Variable*—Estimated annual energy cost; *Type*—Quantitative; *Units*—Not specified (dollars); *Variable*—Overall rating; *Type*—Categorical (ordinal); *Variable*—Percent requiring repair in last 5 years; *Type*—Quantitative; *Units*—Percent

25. *Who*—Kentucky Derby races; *What*—Date, winner, margin, jockey, net proceed to winner, duration, track condition; *When*—1875 to 2008; *Where*—Churchill Downs, Louisville, Kentucky; *Why*—Not specified (To see trends in horse racing?); *How*—Official statistics collected at race

Variable—Year; *Type*—Quantitative; *Units*—Day and year; *Variable*—Winner; *Type*—Identifier; *Variable*—Margin; *Type*—Quantitative; *Units*—Horse lengths; *Variable*—Jockey; *Type*—Categorical; *Variable*—Net proceeds to winner; *Type*—Quantitative; *Units*—Dollars; *Variable*—Duration; *Type*—Quantitative; *Units*—Minutes and seconds; *Variable*—Track condition; *Type*—Categorical

Chapter 3

1. Answers will vary.

3. Answers will vary.

5. a) Yes; each is categorized in a single genre.
   b) Thriller/Horror

7. a) Comedy
   b) It is easier to tell from the bar chart; slices of the pie chart are too close in size.

9. 1755 students applied for admission to the magnet schools program. 53% were accepted, 17% were waitlisted, and the other 30% were turned away.

11. a) Yes. We can add because these categories do not overlap. (Each person is assigned only one cause of death.)
    b) $100 - (26.6 + 22.8 + 5.9 + 5.3 + 4.8) = 34.6\%$
    c) Either a bar chart or pie chart with "other" added would be appropriate. A bar chart is shown.

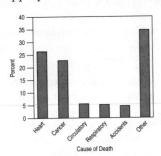

13. a) The bar chart shows that grounding and collision are the most frequent causes of oil spills. Very few have unknown causes.
    b) A pie chart seems appropriate as well.

15. There's no title, the percentages total only 92%, and the three-dimensional display distorts the sizes of the regions.

17. In both the South and West, about 58% of the eighth-grade smokers preferred Marlboro. Newport was the next most popular brand, but was far more popular in the South than in the West, where Camel was cited nearly 3 times as often as in the South. Nearly twice as many smokers in the West as in the South indicated that they had no usual brand (12.9% to 6.7%).

19. a) The column totals are 100%.
    b) 31.7%
    c) 60%
    d) i. 35.7%; ii. can't tell; iii. 0%; iv. can't tell

21. a) 82.5%   b) 12.9%   c) 11.1%
    d) 13.4%   e) 85.7%

23. a) 73.9% 4-yr college, 13.4% 2-year college, 1.5% military, 5.2% employment, 6.0% other
    b) 77.2% 4-yr college, 10.5% 2-year college, 1.8% military, 5.3% employment, 5.3% other
    c) Many charts are possible. Here is a side-by-side bar chart.

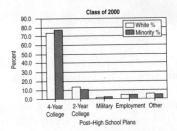

d) The white and minority students' plans are very similar. The small differences should be interpreted with caution because the total number of minority students is small. There is little evidence of an association between race and plans.

25. a) 16.6%   b) 11.8%   c) 37.7%   d) 53.0%

27. 1755 students applied for admission to the magnet schools program: 53% were accepted, 17% were waitlisted, and the other 30% were turned away. While the overall acceptance rate was 53%, 93.8% of blacks and Hispanics were accepted, compared to only 37.7% of Asians and 35.5% of whites. Overall, 29.5% of applicants were black or Hispanic, but only 6% of those turned away were. Asians accounted for 16.6% of all applicants, but 25.4% of those were turned away. Whites were 54% of the applicants and 68.5% of those who were turned away. It appears that the admissions decisions were not independent of the applicant's ethnicity.

29. a) 9.3%   b) 24.7%   c) 80.8%
    d) No, there appears to be no association between weather and ability to forecast weather. On days it rained, his forecast was correct 79.4% of the time. When there was no rain, his forecast was correct 81.0% of the time.

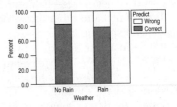

31. a) Low 20.0%, Normal 48.9%, High 31.0%
    b)

		Under 30	30–49	Over 50
Blood Pressure	Low	27.6%	20.7%	15.7%
	Normal	49.0%	50.8%	47.2%
	High	23.5%	28.5%	37.1%

c)

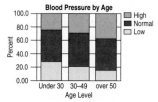

d) As age increases, the percent of adults with high blood pressure increases. By contrast, the percent of adults with low blood pressure decreases.

e) No, but it gives an indication that it might. There might be additional reasons that explain the differences in blood pressures.

33. No, there's no evidence that Prozac is effective. The relapse rates were nearly identical: 28.6% among the people treated with Prozac, compared to 27.3% among those who took the placebo.

35. a) 4.8% b) 50.0%

c) There are about 50% of each sex in each age group. Below age 35, there are more male drivers in each age group. Above age 35, there are more female drivers in each group.

d) There is a slight association. As the age increases, there is a small increase in the percentage of female drivers.

37. a) 160 of 1300, or 12.3%

b) Yes. Major surgery: 15.3% vs. minor surgery: 6.7%

c) Large hospital: 13%; small hospital: 10%

d) Large hospital: Major 15% vs. minor 5%
Small hospital: Major 20% vs. minor 8%.

e) No. Smaller hospitals have a higher rate for both kinds of surgery, even though it's lower "overall."

f) The small hospital has a larger percentage of minor surgeries (83.3%) than the large hospital (20%). Minor surgeries have a lower delay rate, so the small hospital looks better "overall."

39. a) 42.6%

b) A higher percentage of males than females were admitted: Males: 47.2% to females: 30.9%

c) Program 1: Males 61.9%, females 82.4%
Program 2: Males 62.9%, females 68.0%
Program 3: Males 33.7%, females 35.2%
Program 4: Males 5.9%, females 7.0%

d) The comparisons in part c show that males have a lower admittance rate in every program, even though the overall rate shows males with a higher rate of admittance. This is an example of Simpson's paradox.

Chapter 4

1. Answers will vary.

3. Answers will vary.

5. a) Unimodal (near 0) and skewed to the right. Many seniors will have 0 or 1 speeding tickets. Some may have several, and a few may have more than that.

b) Probably unimodal and slightly skewed to the right. It is easier to score 15 strokes over the mean than 15 strokes under the mean.

c) Probably unimodal and symmetric. Weights may be equally likely to be over or under the average.

d) Probably bimodal. Men's and women's distributions may have different modes. It may also be skewed to the right, since it is possible to have very long hair, but hair length can't be negative.

7. a) Bimodal. Looks like two groups. Modes are near 6% and 46%. No real outliers.

b) Looks like two groups of cereals, a low-sugar and a high-sugar group.

9. a) 78%

b) Skewed to the right with at least one high outlier. Most of the vineyards are less than 90 acres with a few high ones. The mode is between 0 and 30 acres.

11. a) Because the distribution is skewed to the right, we expect the mean to be larger.

b) Bimodal and skewed to the right. Center mode near 8 days. Another mode at 1 day (may represent patients who didn't survive). Most of the patients stay between 1 and 15 days. There are some extremely high values above 25 days.

c) The median and IQR, because the distribution is strongly skewed.

13. a) 45 points b) 37 points and 54.5 (or 55) points

c) In the Super Bowl teams typically score a total of about 45 points, with half the games totaling between 37 and 55 points. In only one fourth of the games have the teams scored fewer than 37 points, and they once totaled 75.

15. a) $1001.50 b) 1025, 850, 1200 c) 835, 350

17. a) Median will probably be unaffected. The mean will be larger.

b) The range and standard deviation will increase; IQR will be unaffected.

19. a) The standard deviation will be larger for set 2, since the values are more spread out. SD(set 1) = 2.2, SD(set 2) = 3.2.

b) The standard deviation will be larger for set 2, since 11 and 19 are farther from 15 than are 14 and 16. Other numbers are the same. SD(set 1) = 3.6, SD(set 2) = 4.5.

c) The standard deviation will be the same for both sets, since the values in the second data set are just the values in the first data set +80. The spread has not changed. SD(set 1) = 4.2, SD(set 2) = 4.2.

21. The mean and standard deviation because the distribution is unimodal and symmetric.

23. a) The mean is closest to $2.60 because that's the balancing point of the histogram.

b) The standard deviation is closest to $0.15 since that's a typical distance from the mean. There are no prices as far as $0.50 or $1.00 from the mean.

25. a) About 100 minutes

b) Yes, only 4 of these movies run that long.

c) The mean would be higher. The distribution is skewed high.

27. a) i. The middle 50% of movies ran between 97 and 119 minutes.

ii. On average, movie lengths varied from the mean run time by 19.6 minutes.

b) We should be cautious in using the standard deviation because the distribution of run times is skewed to the right.

29. The publication is using the median; the watchdog group is using the mean, pulled higher by the several very expensive movies in the long right tail.

31. a) Mean $525, median $450

b) 2 employees earn more than the mean.

c) The median because of the outlier.

d) The IQR will be least sensitive to the outlier of $1200, so it would be the best to report.

33. a)

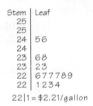

```
Stem | Leaf
  25 |
  25 |
  24 | 5 6
  24 |
  23 | 6 8
  23 | 2 3
  22 | 6 7 7 7 8 9
  22 | 1 2 3 4
```

22|1 = $2.21/gallon

b) The distribution of gas prices is unimodal and skewed to the right (upward), centered around $2.27, with most stations charging between $2.26 and $2.33 per gallon. The lowest and highest prices were $2.21 and $2.46.

c) There are two high prices separated from the other gas stations by a gap.

35. a) Since these data are strongly skewed, the median and IQR are the best statistics to report.

b) The mean will be larger than the median because the data are skewed to the high end.

c) The median is 4 million. The IQR is 4.5 million (Q3 = 6 million, Q1 = 1.5 million).

d) The distribution of populations of the states and Washington, DC, is unimodal and skewed to the right. The median population is 4 million. One state is an outlier, with a population of 34 million.

37. Reasonably symmetric, except for 2 low outliers, median at 41.

39. a)

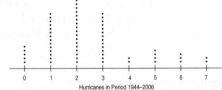

Hurricanes in Period 1944–2006

b) Slightly skewed to the right. Unimodal, mode near 2. Possibly a second mode near 5. No outliers.

41. a) This is not a histogram. The horizontal axis should split the number of home runs hit in each year into bins. The vertical axis should show the number of years in each bin.

b)

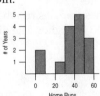

Home Runs

43. Skewed to the right, possibly bimodal with one fairly symmetric group near 4.4, another at 5.6. Two outliers in middle seem not to belong to either group.

```
Stem | Leaf
  57 | 8
  56 | 2 7
  55 | 1
  54 |
  53 |
  52 | 9
  51 |
  50 | 8
  49 |
  48 | 2
  47 | 3
  46 | 0 3 4
  45 | 2 6 7
  44 | 0 1 5
  43 | 0 1 9 9
  42 | 6 6 9
  41 | 2 2
```

41|2 = 4.12 pH

45. Histogram bins are too wide to be useful.

47. Neither appropriate nor useful. Zip codes are categorical data, not quantitative. But they do contain *some* information. The leading digit gives a rough East-to-West placement in the United States. So, we see that they have almost no customers in the Northeast, but a bar chart by leading digit would be more appropriate.

49. a) Median 239, IQR 9, Mean 237.6, SD 5.7

b) Because it's skewed to the left, probably better to report Median and IQR.

c) Skewed to the left; may be bimodal. The center is around 239. The middle 50% of states scored between 233 and 242. Alabama, Mississippi, and New Mexico scores were much lower than other states' scores.

51. In the year 2004, per capita gasoline use by state in the United States averaged around 500 gallons per person (mean 488.8, median 500.5). States varied in per capita consumption, with a standard deviation of 68.7 gallons. The only outlier is New York. The IQR of 96.9 gallons shows that 50% of the states had per capita consumption of between 447.5 and 544.4 gallons. The data appear to be bimodal, so the median and IQR are better choices of summary statistics.

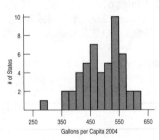

Gallons per Capita 2004

Chapter 5

1. Answers will vary.   **3.** Answers will vary.

5. a) Prices appear to be both higher on average and more variable in Baltimore than in the other three cities. Prices in Chicago may be slightly higher than in Dallas and Denver, but the difference is very small.

b) There are outliers on the low end in Baltimore and Chicago and one high outlier in Dallas, but these do not affect the overall conclusions reached in part a.

7. a) Essentially symmetric, very slightly skewed to the right with two high outliers at 36 and 48. Most victims are between the ages of 16 and 24.

b) The slight increase between ages 22 and 24 is apparent in the histogram but not in the boxplot. It may be a second mode.

c) The median would be the most appropriate measure of center because of the slight skew and the extreme outliers.

d) The IQR would be the most appropriate measure of spread because of the slight skew and the extreme outliers.

9. a) About 59%   b) Bimodal

c) Some cereals are very sugary; others are healthier low-sugar brands.

d) Yes

e) Although the ranges appear to be comparable for both groups (about 28%), the IQR is larger for the adult cereals, indicating that there's more variability in the sugar content of the middle 50% of adult cereals.

11. a)

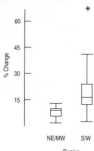

b) Growth rates in NE/MW states are tightly clustered near 5%. S/W states are more variable, and bimodal with modes near 14 and 22. The S/W states have an outlier as well. Around all the modes, the distributions are fairly symmetric.

13. a) They should be put on the same scale, from 0 to 20 days.

b) Lengths of men's stays appear to vary more than for women. Men have a mode at 1 day and then taper off from there. Women have a mode near 5 days, with a sharp drop afterward.

c) A possible reason is childbirth.

15. a) Both girls have a median score of about 17 points per game, but Scyrine is much more consistent. Her IQR is about 2 points, while Alexandra's is over 10.

b) If the coach wants a consistent performer, she should take Scyrine. She'll almost certainly deliver somewhere between 15 and 20 points. But if she wants to take a chance and needs a "big game," she should take Alexandra. Alex scores over 24 points about a quarter of the time. (On the other hand, she scores under 11 points as often.)

17. Women appear to marry about 3 years younger than men, but the two distributions are very similar in shape and spread.

19. (*Note:* Numerical details may vary.) In general, fuel economy is higher in cars than in either SUVs or vans. There are numerous outliers on both ends for cars and a few high outliers for SUVs. The top 50% of cars gets higher fuel economy than 75% of SUVs and nearly all vans. On average, SUVs and vans get about the same fuel economy, although the distribution for vans shows less spread. The range for vans is about 40 mpg, while for SUVs it is nearly 30 mpg.

21. The class A is 1, class B is 2, and class C is 3.

23. a) Probably slightly left skewed. The mean is slightly below the median, and the 25th percentile is farther from the median than the 75th percentile.

b) No, all data are within the fences.

c)

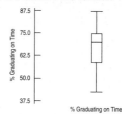

d) The 48 universities graduate, on average, about 68% of freshmen "on time," with percents ranging from 43% to 87%. The middle 50% of these universities graduate between 59% and 75% of their freshmen in 4 years.

25. a) *Who:* Student volunteers
What: Memory test
Where, when: Not specified
How: Students took memory test 2 hours after drinking caffeine-free, half-dose caffeine, or high-caffeine soda.
Why: To see if caffeine makes you more alert and aids memory retention.

b) *Drink:* categorical; Test score: quantitative.

c)

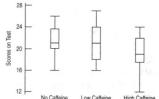

d) The participants scored about the same with no caffeine and low caffeine. The medians for both were 21 points, with slightly more variation for the low-caffeine group. The high-caffeine group generally scored lower than the other two groups on all measures of the 5-number summary: min, lower quartile, median, upper quartile, and max.

27. a) About 36 mph

b) Q_1 about 35 mph and Q_3 about 37 mph

c) The range appears to be about 7 mph, from about 31 to 38 mph. The IQR is about 2 mph.

d) We can't know exactly, but the boxplot may look something like this:

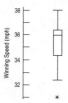

e) The median winning speed has been about 36 mph, with a max of about 38 and a min of about 31 mph. Half have run between about 35 and 37 mph, for an IQR of 2 mph.

29. a) Boys　　　b) Boys　　　c) Girls
　　d) The boys appeared to have more skew, as their scores were less symmetric between quartiles. The girls' quartiles are the same distance from the median, although the left tail stretches a bit farther to the left.
　　e) Girls. Their median and upper quartiles are larger. The lower quartile is slightly lower, but close.
　　f) $[14(4.2) + 11(4.6)]/25 = 4.38$

31.

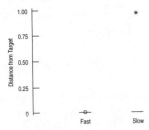

There appears to be an outlier! This point should be investigated. We'll proceed by redoing the plots with the outlier omitted:

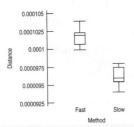

It appears that slow speed provides much greater accuracy. But the outlier should be investigated. It is possible that slow speed can induce an infrequent very large distance.

33. a)

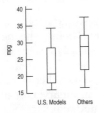

　　b) Mileage for U.S. models is typically lower, although the variability is about the same as for cars made elsewhere. The median for U.S. models is around 21 mpg, compared to 28 for the others. Half of U.S. models fall below the first quartile of others. (Other answers possible.)

35. a) Day 16 (but any estimate near 20 is okay).
　　b) Day 65 (but anything around 60 is okay).
　　c) Around day 50

37. a) Most of the data are found in the far left of this histogram. The distribution is very skewed to the right.
　　b) Re-expressing the data by, for example, logs or square roots might help make the distribution more nearly symmetric.

39. a) The logarithm makes the histogram more symmetric. It is easy to see that the center is around 3.5 in log assets.
　　b) That has a value of around 2500 million dollars.
　　c) That has a value of around 1000 million dollars.

41. a) Fusion time and group.
　　b) Fusion time is quantitative (units = seconds). Group is categorical.
　　c) Both distributions are skewed to the right with high outliers. The boxplot indicates that visual information may reduce fusion time. The median for the Verbal/Visual group seems to be about the same as the lower quartile of the No/Verbal group.

Chapter 6

1. a) 72 oz., 40 oz.　　　b) 4.5 lb, 2.5 lb

3. a) Skewed to the right; mean is higher than median.
　　b) $350 and $950.
　　c) Minimum $350. Mean $750. Median $550. Range $1200. IQR $600. Q1 $400. SD $400.
　　d) Minimum $330. Mean $770. Median $550. Range $1320. IQR $660. Q1 $385. SD $440.

5. Lowest score = 910. Mean = 1230. SD = 120. Q3 = 1350. Median = 1270. IQR = 240.

7. Your score was 2.2 standard deviations higher than the mean score in the class.

9. 65

11. In January, a high of 55 is not quite 2 standard deviations above the mean, whereas in July a high of 55 is more than 2 standard deviations lower than the mean. So it's less likely to happen in July.

13. The z-scores, which account for the difference in the distributions of the two tests, are 1.5 and 0 for Derrick and 0.5 and 2 for Julie. Derrick's total is 1.5, which is less than Julie's 2.5.

15. a) Megan　　　　b) Anna

17. a) About 1.81 standard deviations below the mean.
　　b) 1000 ($z = 1.81$) is more unusual than 1250 ($z = 1.17$).

19. a) Mean = $1152 - 1000 = 152$ pounds; SD is unchanged at 84 pounds.
　　b) Mean = $0.40(1152) = 460.80; SD = $0.40(84) = $33.60.

21. Min = $0.40(980) - 20 = 372;
　　median = $0.40(1140) - 20 = 436;
　　SD = $0.40(84) = 33.60; IQR = $0.40(102) = 40.80.

23. College professors can have between 0 and maybe 40 (or possibly 50) years' experience. A standard deviation of 1/2 year is impossible, because many professors would be 10 or 20 SDs away from the mean, whatever it is. An SD of 16 years would mean that 2 SDs on either side of the mean is plus or minus 32, for a range of 64 years. That's too high. So, the SD must be 6 years.

25. a)

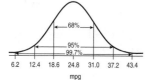

b) 18.6 to 31.0 mpg   **c)** 16%
d) 13.5%   **e)** less than 12.4 mpg

27. Any weight more than 2 standard deviations below the mean, or less than $1152 - 2(84) = 984$ pounds, is unusually low. We expect to see a steer below $1152 - 3(84) = 900$ pounds only rarely.

29. a)

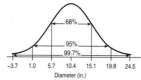

b) Between 1.0 and 19.8 inches   **c)** 2.5%
d) 34%   **e)** 16%

31. Since the histogram is not unimodal and symmetric, it is not wise to have faith in numbers from the Normal model.

33. a) 16%   **b)** 3.8%
c) Because the Normal model doesn't fit well.
d) Distribution is skewed to the right.

35. a) 2.5%
b) 2.5% of the receivers should gain less than −333 yards, but that's impossible, so the model doesn't fit well.
c) Data are strongly skewed to the right, not symmetric.

37. a) 12.2%   **b)** 71.6%   **c)** 23.3%

39. a) 1259.7 lb   **b)** 1081.3 lb   **c)** 1108 lb to 1196 lb

41. a) 1130.7 lb   **b)** 1347.4 lb   **c)** 113.3 lb

43. a)

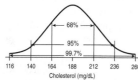

b) 30.85%   **c)** 17.00%   **d)** 32 points   **e)** 212.9 points

45. a) 11.1%   **b)** (35.9, 40.5) inches   **c)** 40.5 inches

47. a) 5.3 grams   **b)** 6.4 grams
c) Younger because SD is smaller.

Part I Review

1. a)

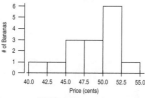

b) Median 49 cents, IQR 6 cents.
c) The distribution is unimodal and left skewed. The center is near 50 cents; values range from 42 cents to 53 cents.

3. a) If enough sopranos have a height of 65 inches, this can happen.
b) The distribution of heights for each voice part is roughly symmetric. The basses are slightly taller than the tenors. The sopranos and altos have about the same median height. Heights of basses and sopranos are more consistent than those of altos and tenors.

5. a) It means their heights are also more variable.
b) The z-score for women to qualify is 2.40, compared with 1.75 for men, so it is harder for women to qualify.

7. a) *Who*—People who live near State University
What—Age, attended college? Favorable opinion of State?
When—Not stated
Where—Region around State U.
Why—To report to the university's directors
How—Sampled and phoned 850 local residents
b) Age—Quantitative (years); attended college?—categorical; favorable opinion?—categorical.
c) The fact that the respondents know they are being interviewed by the university's staff may influence answers.

9. a) These are categorical data, so mean and standard deviation are meaningless.
b) Not appropriate. Even if it fits well, the Normal model is meaningless for categorical data.

11. a)

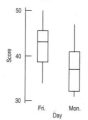

b) The scores on Friday were higher by about 5 points on average. This is a drop of more than 10% off the average score and shows that students fared worse on Monday after preparing for the test on Friday. The spreads are about the same, but the scores on Monday are a bit skewed to the right.
c)

d) The changes (Friday–Monday) are unimodal and centered near 4 points, with a spread of about 5 (SD). They are fairly symmetric, but slightly skewed to the right. Only 3 students did better on Monday (had a negative difference).

13. a) Categorical
b) Go fish. All you need to do is match the denomination. The denominations are not ordered. (Answers will vary.)
c) Gin rummy. All cards are worth their value in points (face cards are 10 points). (Answers will vary.)

15. a) Annual mortality rate for males (quantitative) in deaths per 100,000 and water hardness (quantitative) in parts per million.
b) Calcium is skewed right, possibly bimodal. There looks to be a mode down near 12 ppm that is the center of a fairly tight symmetric distribution and another mode near 62.5 ppm that is the center of a much more spread out, symmetric (almost uniform) distribution. Mortality, however, appears unimodal and symmetric with the mode near 1500 deaths per 100,000.

17. a) They are on different scales.
b) January's values are lower and more spread out.
c) Roughly symmetric but slightly skewed to the left. There are more low outliers than high ones. Center is around 40 degrees with an IQR of around 7.5 degrees.

19. a) Bimodal with modes near 2 and 4.5 minutes. Fairly symmetric around each mode.
b) Because there are two modes, which probably correspond to two different groups of eruptions, an average might not make sense.
c) The intervals between eruptions are longer for long eruptions. There is very little overlap. More than 75% of the short eruptions had intervals less than about an hour (62.5 minutes), while more than 75% of the long eruptions had intervals longer than about 75 minutes. Perhaps the interval could even be used to predict whether the next eruption will be long or short.

21. a)

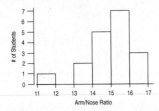

The distribution is left skewed with a center of about 15. It has an outlier between 11 and 12.
b) Even though the distribution is somewhat skewed, the mean and median are close. The mean is 15.0 and the SD is 1.25.
c) Yes. 11.8 is already an outlier. 9.3 is more than 4.5 SDs below the mean. It is a very low outlier.

23. If we look only at the overall statistics, it appears that the follow-up group is insured at a much lower rate than those not traced (11.1% of the time compared with 16.6%). But most of the follow-up group were black, who have a lower rate of being insured. When broken down by race, the follow-up group actually has a higher rate of being insured for both blacks and whites. So the overall statistic is misleading and is attributable to the difference in race makeup of the two groups.

25. a)

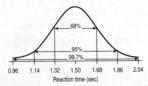

b) According to the model, reaction times are symmetric with center at 1.5 seconds. About 95% of all reaction times are between 1.14 and 1.86 seconds.
c) 8.2%   **d)** 24.1%

e) Quartiles are 1.38 and 1.62 seconds, so the IQR is 0.24 seconds.
f) The slowest 1/3 of all drivers have reaction times of 1.58 seconds or more.

27. a)

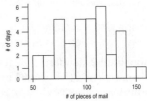

b) Mean 100.25, SD 25.54 pieces of mail.
c) The distribution is somewhat symmetric and unimodal, but the center is rather flat, almost uniform.
d) 64%. The Normal model seems to work reasonably well, since it predicts 68%.

29. a) *Who*—100 health food store customers
What—Have you taken a cold remedy?, and Effectiveness (scale 1 to 10)
When—Not stated
Where—Not stated
Why—Promotion of herbal medicine
How—In-person interviews
b) Have you taken a cold remedy?—categorical. Effectiveness—categorical or ordinal.
c) No. Customers are not necessarily representative, and the Council had an interest in promoting the herbal remedy.

31. a) 38 cars
b) Possibly because the distribution is skewed to the right.
c) Center—median is 148.5 cubic inches. Spread—IQR is 126 cubic inches.
d) No. It's bigger than average, but smaller than more than 25% of cars. The upper quartile is at 231 inches.
e) No. 1.5 IQR is 189, and 105 − 189 is negative, so there can't be any low outliers. 231 + 189 = 420. There aren't any cars with engines bigger than this, since the maximum has to be at most 105 (the lower quartile) + 275 (the range) = 380.
f) Because the distribution is skewed to the right, this is probably not a good approximation.
g) Mean, median, range, quartiles, IQR, and SD all get multiplied by 16.4.

33. a) 30.4%
b) If this were a random sample of all voters, yes.
c) 36.6%   **d)** 8.8%
e) 23.1%   **f)** 47.0%

35. a) Republican—16,535, Democrat—17,183, Other— 20,666; or Republican—30.4%, Democrat—31.6%, Other—38.0%.
b)

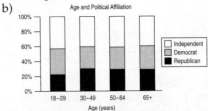

c) Among voters over 30, political affiliation appears to be largely unrelated to age. However, there is some evidence that younger voters are less likely to be Republican.
d) Voters who identified themselves as "Other" seem to be generally younger than Democrats or Republicans.

37. a) 0.43 hour.
  b) 1.4 hours.
  c) 0.89 hour (or 53.4 minutes).
  d) Survey results vary, and the mean and the SD may have changed.

Chapter 7

1. a) Weight in ounces: explanatory; Weight in grams: response. (Could be other way around.) To predict the weight in grams based on ounces. Scatterplot: positive, straight, strong (perfectly linear relationship).
  b) Circumference: explanatory. Weight: response. To predict the weight based on the circumference. Scatterplot: positive, linear, moderately strong.
  c) Shoe size: explanatory; GPA: response. To try to predict GPA from shoe size. Scatterplot: no direction, no form, very weak.
  d) Miles driven: explanatory; Gallons remaining: response. To predict the gallons remaining in the tank based on the miles driven since filling up. Scatterplot: negative, straight, moderate.

3. a) Altitude: explanatory; Temperature: response. (Other way around possible as well.) To predict the temperature based on the altitude. Scatterplot: negative, possibly straight, weak to moderate.
  b) Ice cream cone sales: explanatory. Air conditioner sales: response—although the other direction would work as well. To predict one from the other. Scatterplot: positive, straight, moderate.
  c) Age: explanatory; Grip strength: response. To predict the grip strength based on age. Scatterplot: curved down, moderate. Very young and elderly would have grip strength less than that of adults.
  d) Reaction time: explanatory; Blood alcohol level: response. To predict blood alcohol level from reaction time test. (Other way around is possible.) Scatterplot: positive, nonlinear, moderately strong.

5. a) None   b) 3 and 4   c) 2, 3, and 4
  d) 1 and 2   e) 3 and possibly 1

7. There seems to be a very weak—or possibly no—relation between brain size and performance IQ.

9. a)
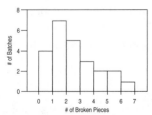
  b) Unimodal, skewed to the right. The skew.
  c) The positive, somewhat linear relation between batch number and broken pieces.

11. a) 0.006   b) 0.777   c) −0.923   d) −0.487

13. There may be an association, but not a correlation unless the variables are quantitative. There could be a correlation between average number of hours of TV watched per week per person and number of crimes committed per year. Even if there is a relationship, it doesn't mean one causes the other.

15. a) Yes. It shows a linear form and no outliers.
  b) There is a strong, positive, linear association between drop and speed; the greater the coaster's initial drop, the higher the top speed.

17. a) The scatterplot is not linear; correlation is not appropriate.
  b) The scatterplot does show a steady increase. Kendall's tau would be an appropriate measure of this.

19. The correlation may be near 0. We expect nighttime temperatures to be low in January, increase through spring and into the summer months, then decrease again in the fall and winter. The relationship is not linear.

21. The correlation coefficient won't change, because it's based on z-scores. The z-scores of the prediction errors are the same whether they are expressed in nautical miles or miles.

23. a) Assuming the relation is linear, a correlation of −0.772 shows a strong relation in a negative direction.
  b) Continent is a categorical variable. Correlation does not apply.

25. a) Actually, yes, taller children will tend to have higher reading scores, but this doesn't imply causation.
  b) Older children are generally both taller and are better readers. Age is the lurking variable.

27. a) No. We don't know this from the correlation alone. There may be a nonlinear relationship or outliers.
  b) No. We can't tell from the correlation what the form of the relationship is.
  c) No. We don't know from the correlation coefficient.
  d) Yes, the correlation doesn't depend on the units used to measure the variables.

29. This is categorical data even though it is represented by numbers. The correlation is meaningless.

31. a) The association is positive, moderately strong, and roughly straight, with several states whose HCI seems high for their median income and one state whose HCI appears low given its median income.
  b) The correlation would still be 0.65.
  c) The correlation wouldn't change.
  d) DC would be a moderate outlier whose HCI is high for its median income. It would lower the correlation slightly.
  e) No. We can only say that higher median incomes are associated with higher housing costs, but we don't know why. There may be other economic variables at work.
  f) No. We can say that there is a consistent monotone pattern, but correlation—even nonparametric correlation—does not demonstrate causation.

33. a)
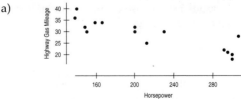
  b) Negative, linear, strong.
  c) −0.869
  d) There is a strong linear relation in a negative direction between horsepower and highway gas mileage. Lower fuel efficiency is associated with higher horsepower.

35.

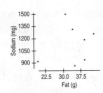

(Plot could have explanatory and predictor variables swapped.)
a) Correlation is 0.199. There does not appear to be a relation between sodium and fat content in burgers, especially without the low-fat, low-sodium item. The correlation of 0.199 shows a weak relationship, even with the outlier included.
b) Spearman's rho is slightly negative. Using ranks doesn't allow the outlier to have as strong an influence and the remaining points have little or no association.

37. a) Yes, the scatterplot appears to be somewhat linear.
b) As the number of runs increases, the attendance also increases.
c) There is a positive association, but it does not *prove* that more fans will come if the number of runs increases. Association does not indicate causality.

39. A scatterplot shows a generally straight scattered pattern with no outliers. The correlation between *Drop* and *Duration* is 0.35, indicating that rides on coasters with greater initial drops generally last somewhat longer, but the association is weak.

41. a)

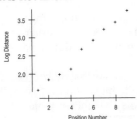

The relation between position and distance is nonlinear, with a positive direction. There is very little scatter from the trend.
b) The relation is not linear.
c)

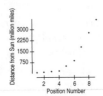

The relation between position number and log of distance appears to be roughly linear.
*d) Kendall's tau is 1.0 because the points are perfectly monotonically increasing.

Chapter 8

Note: Answers assume that full precision has been maintained throughout the calculation. If you round, for example, slope and intercept values to those given in one part of the answer rather than keeping full precision in your calculator or program, your answers may vary somewhat from those given here.

1. 281 milligrams

3. The potassium content is actually lower than the model predicts for a cereal with that much fiber.

5. The model predicts that cereals will have approximately 27 more milligrams of potassium for every additional gram of fiber.

7. 81.5%

9. The true potassium contents of cereals vary from the predicted amounts with a standard deviation of 30.77 milligrams.

11.

	$\bar{x}$	s_x	$\bar{y}$	s_y	r	$\hat{y} = b_0 + b_1x$
a)	10	2	20	3	0.5	$\hat{y} = 12.5 + 0.75x$
b)	2	0.06	7.2	1.2	−0.4	$\hat{y} = 23.2 − 8x$
c)	12	6	152	30	−0.8	$\hat{y} = 200 − 4x$
d)	2.5	1.2	25	100	0.6	$\hat{y} = −100 + 50x$

13. a) Model is appropriate.
b) Model is not appropriate. Relationship is nonlinear.
c) Model may not be appropriate. Spread is changing.

15. a) *Price* (in thousands of dollars) is y and *Size* (in square feet) is x.
b) Slope is thousands of $ per square foot.
c) Positive. Larger homes should cost more.

17. 300 pounds/foot. If a "typical" car is 15 feet long, all of 3, 30, and 3000 would give ridiculous weights.

19. A linear model on *Size* accounts for 71.4% of the variation in home *Price*.

21. a) R^2 does not tell whether the model is appropriate, but measures the strength of the linear relationship. High R^2 could also be due to an outlier.
b) Predictions based on a regression line are for average values of y for a given x. The actual wingspan will vary around the prediction.

23. a) 0.845
b) Price should be 0.845 SDs above the mean in price.
c) Price should be 1.690 SDs below the mean in price.

25. a) Probably not. Your score is better than about 97.5% of people, assuming scores follow the Normal model. Your next score is likely to be closer to the mean.
b) The friend probably should retake the test. His score is better than only about 16% of people. His score is likely to be closer to the mean.

27. a) *Price* increases by about $0.061/1000 ft², or $61.00, per square foot.
b) 230.82 thousand, or $230,820.
c) $115,020; $6000 is the residual.

29. a) Probably. The residuals show some initially low points, but there is no clear curvature.
b) The linear model on *Tar* content accounts for 92.4% of the variability in *Nicotine*.

31. a) $r = 0.961$
b) Nicotine should be 1.922 SDs below average.
c) Tar should be 0.961 SDs above average.

33. a) $\widehat{Nicotine} = 0.15403 + 0.065052\ Tar$.
   b) 0.414 mg
   c) Nicotine content increases by 0.065 mg of nicotine per milligram of tar.
   d) We'd expect a cigarette with no tar to have 0.154 mg of nicotine.
   e) 0.1094 mg

35. a) Yes. The relationship is straight enough, with a few outliers. The spread increases a bit for states with large median incomes, but we can still fit a regression line.
   b) From summary statistics: $\widehat{HCI} = -156.50 + 0.0107\ MFI$; from original data: $\widehat{HCI} = -157.64 + 0.0107\ MFI$
   c) From summary statistics: predicted HCI = 324.93; from original data: 323.79.
   d) 223.09
   e) $\widehat{z_{HCI}} = 0.65 z_{MFI}$
   f) $\widehat{z_{MFI}} = 0.65 z_{HCI}$

37. a) $\widehat{Total} = 539.803 + 1.103\ Age$.
   b) Yes. Both variables are quantitative; the plot is straight (although flat); there are no apparent outliers; the plot does not appear to change spread throughout the range of Age.
   c) $559.65; $594.94
   d) 0.14%
   e) No. The plot is nearly flat. The model explains almost none of the variation in Total Yearly Purchases.

39. a) Moderately strong, fairly straight, and positive. Possibly some outliers (higher-than-expected math scores).
   b) The student with 500 verbal and 800 math.
   c) Positive, fairly strong linear relationship. 46.9% of variation in math scores is explained by verbal scores.
   d) $\widehat{Math} = 217.7 + 0.662 \times Verbal$.
   e) Every point of verbal score adds 0.662 points to the predicted average math score.
   f) 548.5 points.
   g) 53.0 points.

41. a) 0.685
   b) $\widehat{Verbal} = 162.1 + 0.71 \times Math$.
   c) The observed verbal score is higher than predicted from the math score.
   d) 516.7 points.
   e) 559.6 points.
   f) Regression to the mean. Someone whose math score is below average is predicted to have a verbal score below average, but not as far (in SDs). So if we use *that* verbal score to predict math, they will be even closer to the mean in predicted math score than their observed math score. If we kept cycling back and forth, eventually we would predict the mean of each and stay there.

43. a) The relationship is straight enough, but very weak. In fact, there may be no relationship at all between these variables.
   b) The number of wildfires has been increasing by about 297 per year.
   c) Yes, the intercept estimates the number of wildfires in 1985 as about 73,790.

   d) The residuals are distributed around zero with a standard deviation of 12,323 fires. Compared to the observed values, most of which are between 50,000 and 80,000 fires, this amount of errors in our model's predictions is quite large.
   e) Only 2.9% of the variation in the number wildfires can be accounted for by the linear model on Year. This confirms the impression from the scatterplot that there is very little association between these variables—that is, that there has been little change in the number of wildfires during this period.

45. a)

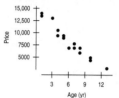

   b) Negative, linear, strong.
   c) Yes.
   d) −0.972
   e) Age accounts for 94.4% of the variation in Advertised Price.
   f) Other factors contribute—options, condition, mileage, etc.

47. a) $\widehat{Price} = 14{,}286 - 959 \times Years$.
   b) Every extra year of age decreases average value by $959.
   c) The average new Corolla costs $14,286.
   d) $7573
   e) Negative residual. Its price is below the predicted value for its age.
   f) −$1195
   g) No. After age 14, the model predicts negative prices. The relationship is no longer linear.

49. a)

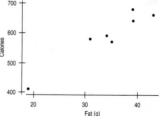

   b) 92.3% of the variation in calories can be accounted for by the fat content.
   c) $\widehat{Calories} = 211.0 + 11.06 \times Fat$.
   d)

Residuals vs. the Fitted Values
(response is Calories)

Residuals show no clear pattern, so the model seems appropriate.

e) Could say a fat-free burger still has 211.0 calories, but this is extrapolation (no data close to 0).

f) Every gram of fat adds 11.06 calories, on average.

g) 553.5 calories.

51. a) The regression was for predicting calories from fat, not the other way around.

b) $\widehat{Fat} = -15.0 + 0.083 \times Calories$.
Predict 34.8 grams of fat.

53. a) This model predicts that every bridge will be deficient. The intercept is less than 5 and the slope is negative, so all predicted conditions are less than 5.

b) This model says bridges in New York City are decreasing in condition at only 0.004 per year—less rapidly than bridges in Tompkins County.

c) The R^2 for this model is only 2.6%. I don't think it has much predictive value.

55. a) 0.823

b) CO_2 levels account for 67.8% of the variation in mean temperature.

c) $\overline{Mean\ Temperature} = 10.71 + 0.010 \times CO_2$.

d) The predicted mean temperature has been increasing at an average rate of 0.01 degrees (C)/ppm of CO_2.

e) One *could* say that with no CO_2 in the atmosphere, there would be a temperature of 10.71 degrees Celsius, but this is extrapolation to a nonsensical point.

f) No.

g) Predicted 14.65 degrees C.

57. a) $\overline{\%\ Body\ Fat} = -27.4 + 0.25 \times Weight$.

b) Residuals look randomly scattered around 0, so conditions are satisfied.

c) *% Body Fat* increases, on average, by 0.25 percent per pound of *Weight*.

d) Reliable is relative. R^2 is 48.5%, but residuals have a standard deviation of 7%, so variation around the line is large.

e) 0.9 percent.

59. a) $\overline{Highjump} = 2.681 - 0.00671 \times 800m\ Time$. High-jump height is lower, on average, by 0.00671 meters per second of 800-m race time.

b) 16.4%

c) Yes, the slope is negative. Faster runners tend to jump higher as well.

d) There is a slight tendency for less variation in high-jump height among the slower runners than among the faster runners.

e) Not especially. The residual standard deviation is 0.060 meters, which is not much smaller than the SD of all high jumps (0.066 meters). The model doesn't appear to do a very good job of predicting.

61. a) As calcium levels increase, mortality rate decreases. Relationship is fairly strong, negative, and linear.

b) $\overline{Mortality} = 1676.0 - 3.23 \times Calcium$.

c) Mortality decreases 3.23 deaths per 100,000, on average, for each part per million of calcium. The intercept indicates a baseline mortality of 1676 deaths per 100,000 with no calcium, but this is extrapolation.

d) Exeter has 348.6 fewer deaths per 100,000 than the model predicts.

e) 1353 deaths per 100,000.

f) Calcium concentration accounts for 43.0% of the variation in death rate per 100,000 people.

Chapter 9

1. a) The trend appears to be somewhat linear up to about 1940, but from 1940 to about 1970 the trend appears to be nonlinear. From 1975 or so to the present, the trend appears to be linear.

b) Relatively strong for certain periods.

c) No, as a whole the graph is clearly nonlinear. Within certain periods (ex: 1975 to the present) the correlation is high.

d) Overall, no. You could fit a linear model to the period from 1975 to 2003, but why? You don't need to interpolate, since every year is reported, and extrapolation seems dangerous.

3. a) The relationship is not straight.

b) It will be curved downward.

c) No. The relationship will still be curved.

5. a) No. We need to see the scatterplot first to see if the conditions are satisfied, and models are always wrong.

b) No, the linear model might not fit the data everywhere.

7. a) Millions of dollars per minute of run time.

b) Costs for movies increase at the same rate per minute.

c) On average dramas cost about $20 million less for the same runtime.

9. a) The use of the Oakland airport has been growing at about 59,700 passengers/year, starting from about 282,000 in 1990.

b) 71% of the variation in passengers is accounted for by this model.

c) Errors in predictions based on this model have a standard deviation of 104,330 passengers.

d) No, that would extrapolate too far from the years we've observed.

e) The negative residual is September 2001. Air traffic was artificially low following the attacks on 9/11.

11. a) 1) High leverage, small residual.

2) No, not influential for the slope.

3) Correlation would decrease because outlier has large z_x and z_y, increasing correlation.

4) Slope wouldn't change much because the outlier is in line with other points.

b) 1) High leverage, probably small residual.

2) Yes, influential.

3) Correlation would weaken, increasing toward zero.

4) Slope would increase toward 0, since outlier makes it negative.

c) 1) Some leverage, large residual.
   2) Yes, somewhat influential.
   3) Correlation would increase, since scatter would decrease.
   4) Slope would increase slightly.
d) 1) Little leverage, large residual.
   2) No, not influential.
   3) Correlation would become stronger and become more negative because scatter would decrease.
   4) Slope would change very little.

13. 1) e   2) d   3) c   4) b   5) a

15. Perhaps high blood pressure causes high body fat, high body fat causes high blood pressure, or both could be caused by a lurking variable such as a genetic or lifestyle issue.

17. a) The graph shows that, on average, students progress at about one reading level per year. This graph shows averages for each grade. The linear trend has been enhanced by using averages.
b) Very close to 1.
c) The individual data points would show much more scatter, and the correlation would be lower.
d) A slope of 1 would indicate that for each 1-year grade level increase, the average reading level is increasing by 1 year.

19. a) *Cost* decreases by $2.13 per degree of average daily *Temp*. So warmer temperatures indicate lower costs.
b) For an avg. monthly temperature of 0°F, the cost is predicted to be $133.
c) Too high; the residuals (observed − predicted) around 32°F are negative, showing that the model overestimates the costs.
d) $111.70
e) About $105.70
f) No, the residuals show a definite curved pattern. The data are probably not linear.
g) No, there would be no difference. The relationship does not depend on the units.

21. a) 0.88
b) Interest rates during this period grew at about 0.25% per year, starting from an interest rate of about 0.64%.
c) Substituting 50 in the model yields a predicted of about 13%.
d) Not really. Extrapolating 20 years beyond the end of these data would be dangerous and unlikely to be accurate.

23. a) The two models fit comparably well, but they have very different slopes.
b) This model predicts the interest rate in 2000 to be 3.24%, much lower than the other model predicts.
c) We can trust the new predicted value because it is in the middle of the data used for the regression.
d) The best answer is "I can't predict that."

25. a) Stronger. Both slope and correlation would increase.
b) Restricting the study to nonhuman animals would justify it.
c) Moderately strong.
d) For every year increase in life expectancy, the gestation period increases by about 15.5 days, on average.
e) About 270.5 days.

27. a) Removing hippos would make the association stronger, since hippos are more of a departure from the pattern.
b) Increase.
c) No, there must be a good reason for removing data points.
d) Yes, removing it lowered the slope from 15.5 to 11.6 days per year.

29. a) Answers may vary. Using the data for 1955–2005 results in a scatterplot that is relatively linear with some curvature. The residuals plot shows a definite trend, indicating that the data are not linear. You might use the data after 1955 only to predict 2010, but that would still call for extrapolation and would not be safe.
b) Not much, since the data are not truly linear and 2015 is 10 years from the last data point (extrapolating is risky).
c) No, that extrapolation of more than 50 years would be absurd. There's no reason to believe the trend from 1955 to 2005 will continue.

31.

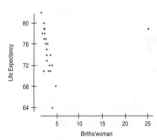

a) Except for the outlier, Costa Rica, the data appear to have a linear form in a negative direction.
b) The outlier is Costa Rica, whose data appear to be wrong, with 25 births per woman. That's impossible.
c) With Costa Rica, $r = 0.168$ and R-squared $= 2.8\%$, indicating that 2.8% of the variation in *Life Expectancy* is explained by the variation in *Births per Woman*. Without Costa Rica, $r = -0.796$ and R-squared $= 63.3\%$, indicating that 63.3% of the variation in *Life Expectancy* is explained by the variation in *Births/Woman*.
d) With Costa Rica, $\overline{Life\ Expectancy} = 72.6 + 0.15\ Births$; without Costa Rica, $\overline{Life\ Expectancy} = 84.5 - 4.44\ Births$.
e) The model with Costa Rica is not appropriate. The residuals plot shows a distinct outlier, which is Costa Rica. Removing Costa Rica gives a better residuals plot, suggesting that the linear equation is more appropriate.
f) With Costa Rica, the slope is near 0, suggesting that the linear model is not very useful. The y-intercept suggests that with no births, the life expectancy is about 72.6 years. Without Costa Rica, the slope is −4.44, indicating that an average increase of one child per woman predicts a lower life expectancy of 4.44 years, on average. The y-intercept indicates that a country with a birth rate of zero would have a life expectancy of 84.5 years. This is extrapolation.
g) While there is an association, there is no reason to expect causality. Lurking variables may be involved.

33. a) The scatterplot is clearly nonlinear; however, the last few years—say, from 1970 on—do appear to be linear.
 b) Using the data from 1970 to 2006 gives $r = 0.997$ and $\widehat{CPI} = -9052.42 + 4.61\ Year$. Predicted CPI in 2016 = 241.34 (an extrapolation of doubtful accuracy).

Chapter 10

1. a) No re-expression needed.
 b) Re-express to straighten the relationship.
 c) Re-express to equalize spread.

3. a) There's an annual pattern in when people fly, so the residuals cycle up and down.
 b) No, this kind of pattern can't be helped by re-expression.

5. a) 16.44 **b)** 7.84 **c)** 0.36 **d)** 1.75 **e)** 27.59

7. a) Fairly linear, negative, strong.
 b) Gas mileage decreases an average 7.652 mpg for each thousand pounds of weight.
 c) No. Residuals show a curved pattern.

9. a) Residuals are more randomly spread around 0, with some low outliers.
 b) $\overline{Fuel\ Consumption} = 0.625 + 1.178 \times Weight$.
 c) For each additional 1000 pounds of *Weight*, an additional 1.178 gallons will be needed to drive 100 miles.
 d) 21.06 miles per gallon.

11. a) Although more than 97% of the variation in GDP can be accounted for by this model, we should examine a scatterplot of the residuals to see if it's appropriate.
 b) No. The residuals show clear curvature.

13. Yes, the pattern in the residuals is somewhat weaker.

15. a)

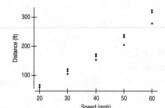

$\widehat{Distance} = -65.9 + 5.98\ Speed$.
But residuals have a curved shape, so linear model is not appropriate.

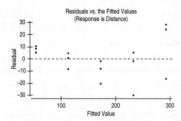

b)

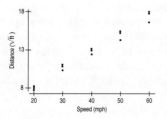

$\sqrt{Distance}$ linearizes the plot.

c) Predicted $\sqrt{Distance} = 3.30 + 0.235 \times Speed$.
 d) 263.25 feet.
 e) 390.1 feet (an extrapolation)
 f) Fairly confident, since $R^2 = 98.4\%$, and s is small.

17. a) The plot looks fairly straight. (It is okay to see a bend in the plot; there's one there.)

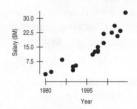

b) $\widehat{Salary} = -1952.77 + 0.985\ Year$

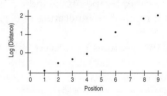

The residuals plot shows a strong bend.
 c) $\log(Salary)$ works well.
 d) $\log(\widehat{Salary}) = -109.133 + 0.05516\ Year$

19. a)

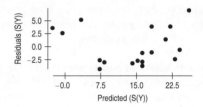

$\log(Distance)$ against position works pretty well.
$\overline{\log(Distance)} = 1.245 + 0.271 \times Position\ number$.
 b) Pluto's residual is not especially large in the log scale. However, a model without Pluto predicts the 9th planet should be 5741 million miles. Pluto, at "only" 3707 million miles, doesn't fit very well, giving support to the argument that Pluto doesn't behave like a planet.

21. The predicted $\log(Distance)$ of Eris is 3.956, corresponding to a distance of 9046 million miles. That's more than the actual average distance of 6300 million miles.

22. Both models have high R^2, but the Exercise 18 model has $R^2 = 1$. That indicates that perhaps we have found a physical law. If we find another system with the same pattern, it would add evidence for Titius-Bode "law"; otherwise, we would discount the law.

23. a)

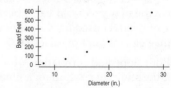

$\widehat{\sqrt{Bdft}} = -4 + diam$
The model is exact.
 b) 36 board feet.
 c) 1024 board feet.

25.

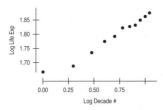

$$\widehat{\log Life} = 1.647 + 0.2125 \log Decade.$$

27. The relationship cannot be made straight by the methods of this chapter.

29. a) $\widehat{\sqrt{Left}} = 8.465 - 0.06926(Age)$.
   b) 52.10 years
   c) No; the residuals plot still shows a pattern.

Part II Review

1. % over 50, 0.69.
   % under 20, −0.71.
   % Graduating on time, −0.51.
   % Full-time Faculty, 0.09

3. a) There does not appear to be a linear relationship.
   b) Nothing, there is no reason to believe that the results for the Finger Lakes region are representative of the vineyards of the world.
   c) $\widehat{CasePrice} = 92.77 + 0.567 \times Years$.
   d) Only 2.7% of the variation in case price is accounted for by the ages of vineyards. Most of that is due to two outliers. We are better off using the mean price rather than this model.

5. a) $\widehat{TwinBirths} = -5235191 + 2676.4 \times Year$.
   b) Each year, the number of twins born in a year increases, on average, by approximately 2676.
   c) 144,371.57 births. The scatterplot appears to be somewhat linear, but there is some curvature in the pattern. There is no reason to believe that the increase will continue to be linear 5 years beyond the data.
   d) The residuals plot shows a definite curved pattern, so the relation is not linear.

7. a) −0.520
   b) Negative, not strong, somewhat linear, but with more variation as pH increases.
   c) The BCI would also be average.
   d) The predicted BCI will be 1.56 SDs of BCI below the mean BCI.

9. a) $\widehat{Manatee\ Deaths} = -45.67 \times 0.1315\ Powerboat\ Registrations$ (in 1000s).
   b) According to the model, for each increase of 10,000 motorboat registrations, the number of manatees killed increases by approximately 1.315 on average.
   c) If there were 0 motorboat registrations, the number of manatee deaths would be −45.67. This is obviously a silly extrapolation.
   d) The predicted number is 82.41 deaths. The actual number of deaths was 79. The residual is 79 − 82.41 = −3.41. The model overestimated the number of deaths by 3.41.
   e) Negative residuals would suggest that the actual number of deaths was lower than the predicted number.

f) Over time, the number of motorboat registrations has increased and the number of manatee kills has increased. The trend may continue. Extrapolation is risky, however, because the government may enact legislation to protect the manatee.

11. a) −0.984   b) 96.9%   c) 32.95 mph   d) 1.66 mph
   e) Slope will increase.
   f) Correlation will weaken (become less negative).
   g) Correlation is the same, regardless of units.

13. a) Weight (but unable to verify linearity).
   b) As weight increases, mileage decreases.
   c) *Weight* accounts for 81.5% of the variation in *Fuel Efficiency*.

15. a) $\widehat{Horsepower} = 3.50 + 34.314 \times Weight$.
   b) Thousands. For the equation to have predicted values between 60 and 160, the X values would have to be in thousands of pounds.
   c) Yes. The residual plot does not show any pattern.
   d) 115.0 horsepower.

17. a) The scatterplot shows a fairly strong linear relation in a positive direction. There seem to be two distinct clusters of data.
   b) $\widehat{Interval} = 33.967 + 10.358 \times Duration$.
   c) The time between eruptions increases by about 10.4 minutes per minute of *Duration* on average.
   d) Since 77% of the variation in *Interval* is accounted for by *Duration* and the error standard deviation is 6.16 minutes, the prediction will be relatively accurate.
   e) 75.4 minutes.
   f) A residual is the observed value minus the predicted value. So the residual = 79 − 75.4 = 3.6 minutes, indicating that the model underestimated the interval in this case.

19. a) $r = 0.888$. Although r is high, you must look at the scatterplot and verify that the relation is linear in form.
   b)

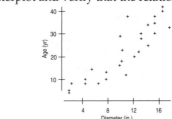

The association between diameter and age appears to be strong, somewhat linear, and positive.
   c) $\widehat{Age} = -0.97 + 2.21 \times Diameter$.
   d)

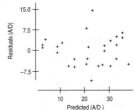

The residuals show a curved pattern (and two outliers).
   e) The residuals for five of the seven largest trees (15 in. or larger) are positive, indicating that the predicted values underestimate the age.

21. Most houses have areas between 1000 and 5000 square feet. Increasing 1000 square feet would result in either $1000(.008) = 8$ thousand dollars, $1000(.08) = 80$ thousand dollars, $1000(.8) = 800$ thousand dollars, or $1000(8) = 8000$ thousand dollars. Only $80,000 is reasonable, so the slope must be 0.08.

23. a) The model predicts % smoking from year, not the other way around.
    b) $\widehat{Year} = 2027.91 - 202.74 \times \% Smoking$.
    c) The smallest % smoking given is 12.7, and an extrapolation to $x = 0$ is probably too far from the given data. The prediction is not very reliable in spite of the strong correlation.

25. The relation shows a negative direction, with a somewhat linear form, but perhaps with some slight curvature. There are several model outliers.

27. a) 71.9%
    b) As latitude increases, the average January temperature decreases.
    c) $\widehat{January\ Temperature} = 108.80 - 2.111 \times Latitude$.
    d) As the latitude increases by 1 degree, the average January temperature drops by about 2.11 degrees, on average.
    e) The y-intercept would indicate that the average January temperature is 108.8 when the latitude is 0. However, this is extrapolation and may not be meaningful.
    f) 24.4 degrees.
    g) The equation underestimates the average January temperature.

29. a) The scatterplot shows a strong, linear, positive association.
    b) There is an association, but it is likely that training and technique have increased over time and affected both jump performances.
    c) Neither; the change in units does not affect the correlation.
    d) I would predict the winning long jump to be 0.920 SDs above the mean long jump.

31. a) No relation; the correlation would probably be close to 0.
    b) The relation would have a positive direction and the correlation would be strong, assuming that students were studying French in each grade level. Otherwise, no correlation.
    c) No relation; correlation close to 0.
    d) The relation would have a positive direction and the correlation would be strong, since vocabulary would increase with each grade level.

33. $\widehat{Calories} = 560.7 - 3.08 \times Time$.
    Each minute extra at the table results in 3.08 fewer calories being consumed, on average. Perhaps the hungry children eat fast and eat more.

35. There seems to be a strong, positive, linear relationship with one high-leverage point (Northern Ireland) that makes the overall R^2 quite low. Without that point, the R^2 increases to 61.5%. Of course, these data are averaged across thousands of households, so the correlation appears to be higher than it would be for individuals. Any conclusions about individuals would be suspect.

37. a) 3.842   b) 501.187   c) 4.0

39. a) 30,818 pounds.
    b) 1302 pounds.
    c) 31,187.6 pounds.
    d) I would be concerned about using this relation if we needed accuracy closer than 1000 pounds or so, as the residuals are more than ± 1000 pounds.
    e) Negative residuals will be more of a problem, as the predicted weight would overestimate the weight of the truck; trucking companies might be inclined to take the ticket to court.

41. The original data are nonlinear, with a significant curvature. Using reciprocal square root of diameter gave a scatterplot that is nearly linear:
    $\widehat{1/\sqrt{Drain\ Time}} = 0.0024 + 0.219\ Diameter$.

Chapter 11

1. Yes. You cannot predict the outcome beforehand.

3. A machine pops up numbered balls. If it were truly random, the outcome could not be predicted and the outcomes would be equally likely. It is random only if the balls generate numbers in equal frequencies.

5. Use two-digit numbers 00–99; let 00–02 = defect, 03–99 = no defect

7. a) 45, 10   b) 17, 22

9. If the lottery is random, it doesn't matter which number you play; all are equally likely to win.

11. a) The outcomes are not equally likely; for example, tossing 5 heads does not have the same probability as tossing 0 or 9 heads, but the simulation assumes they are equally likely.
    b) The even-odd assignment assumes that the player is equally likely to score or miss the shot. In reality, the likelihood of making the shot depends on the player's skill.
    c) The likelihood for the first ace in the hand is not the same as for the second or third or fourth. But with this simulation, the likelihood is the same for each. (And it allows you to get 5 aces, which could get you in trouble in a real poker game!)

13. The conclusion should indicate that the simulation *suggests* that the average length of the line would be 3.2 people. Future results might not match the simulated results exactly.

15. a) The component is one voter voting. An outcome is a vote for our candidate or not. Use two random digits, giving 00–54 a vote for your candidate and 55–99 for the underdog.
    b) A trial is 100 votes. Examine 100 two-digit random numbers, and count how many people voted for each candidate. Whoever gets the majority of votes wins that trial.
    c) The response variable is whether the underdog wins or not.

17. Answers will vary, but average answer will be about 51%.

19. Answers will vary, but average answer will be about 26%.

21. a) Answers will vary, but you should win about 10% of the time.
b) You should win at the same rate with any number.

23. Answers will vary, but you should win about 10% of the time.

25. Answers will vary, but average answer will be about 1.9 tests.

27. Answers will vary, but average answer will be about 1.24 points.

29. Do the simulation in two steps. First simulate the payoffs. Then count until $500 is reached. Answers will vary, but average should be near 10.2 customers.

31. Answers will vary, but average answer will be about 3 children.

33. Answers will vary, but average answer will be about 7.5 rolls.

35. No, it will happen about 40% of the time.

37. Answers will vary, but average answer will be about 37.5%.

39. Three women will be selected about 7.8% of the time.

Chapter 12

1. a) No. It would be nearly impossible to get exactly 500 males and 500 females from every country by random chance.
b) A stratified sample, stratified by whether the respondent is male or female.

3. a) Voluntary response.
b) We have no confidence at all in estimates from such studies.

5. a) The population of interest is all adults in the United States aged 18 and older.
b) The sampling frame is U.S. adults with telephones.
c) Some members of the population (e.g., many college students) don't have landline phones, which could create a bias.

7. a) Population—All U.S. adults.
b) Parameter—Proportion who have used and benefited from alternative medicine.
c) Sampling Frame—All Consumers Union subscribers.
d) Sample—Those who responded.
e) Method—Questionnaire to all (nonrandom).
f) Bias—Nonresponse. Those who respond may have strong feelings one way or another.

9. a) Population—Adults.
b) Parameter—Proportion who think drinking and driving is a serious problem.
c) Sampling Frame—Bar patrons.
d) Sample—Every 10th person leaving the bar.
e) Method—Systematic sampling (may be random).
f) Bias—Those interviewed had just left a bar. They may think drinking and driving is less of a problem than do other adults.

11. a) Population—Soil around a former waste dump.
b) Parameter—Concentrations of toxic chemicals.
c) Sampling Frame—Accessible soil around the dump.
d) Sample—16 soil samples.
e) Method—Not clear.
f) Bias—Don't know if soil samples were randomly chosen. If not, may be biased toward more or less polluted soil.

13. a) Population—Snack food bags.
b) Parameter—Weight of bags, proportion passing inspection.
c) Sampling Frame—All bags produced each day.
d) Sample—Bags in 10 randomly selected cases, 1 bag from each case for inspection.
e) Method—Multistage random sampling.
f) Bias—Should be unbiased.

15. Bias. Only people watching the news will respond, and their preference may differ from that of other voters. The sampling method may systematically produce samples that don't represent the population of interest.

17. a) Voluntary response. Only those who see the ad, have Internet access, *and* feel strongly enough will respond.
b) Cluster sampling. One school may not be typical of all.
c) Attempted census. Will have nonresponse bias.
d) Stratified sampling with follow-up. Should be unbiased.

19. a) This is a multistage design, with a cluster sample at the first stage and a simple random sample for each cluster.
b) If any of the three churches you pick at random is not representative of all churches, then you'll introduce sampling error by the choice of that church.

21. a) This is a systematic sample.
b) The sampling frame is patrons willing to wait for the roller coaster on that day at that time. It should be representative of the people in line, but not of all people at the amusement park.
c) It is likely to be representative of those waiting for the roller coaster. Indeed, it may do quite well if those at the front of the line respond differently (after their long wait) than those at the back of the line.

23. a) Answers will definitely differ. Question 1 will probably get many "No" answers, while Question 2 will get many "Yes" answers. This is response bias.
b) "Do you think standardized tests are appropriate for deciding whether a student should be promoted to the next grade?" (Other answers will vary.)

25. a) Biased toward yes because of "pollute." "Should companies be responsible for any costs of environmental cleanup?"
b) Biased toward no because of "old enough to serve in the military." "Do you think the drinking age should be lowered from 21?"

27. a) Not everyone has an equal chance. Misses people with unlisted numbers, or without landline phones, or at work.
b) Generate random numbers and call at random times.

c) Under the original plan, those families in which one person stays home are more likely to be included. Under the second plan, many more are included. People without landline phones are still excluded.

d) It improves the chance of selected households being included.

e) This takes care of phone numbers. Time of day may be an issue. People without landline phones are still excluded.

29. a) Answers will vary.

b) Your own arm length. Parameter is your own arm length; population is all possible measurements of it.

c) Population is now the arm lengths of you and your friends. The average estimates the mean of these lengths.

d) Probably not. Friends are likely to be of the same age and not very diverse or representative of the larger population.

31. a) Assign numbers 001 to 120 to each order. Use random numbers to select 10 transactions to examine.

b) Sample proportionately within each type. (Do a stratified random sample.)

33. a) Select three cases at random; then select one jar randomly from each case.

b) Use random numbers to choose 3 cases from numbers 61 through 80; then use random numbers between 1 and 12 to select the jar from each case.

c) No. Multistage sampling.

35. a) Depends on the Yellow Pages listings used. If from regular (line) listings, this is fair if all doctors are listed. If from ads, probably not, as those doctors may not be typical.

b) Not appropriate. This cluster sample will probably contain listings for only one or two business types.

Chapter 13

1. a) No. There are no manipulated factors. Observational study.

b) There may be lurking variables that are associated with both parental income and performance on the SAT.

3. a) This is a retrospective observational study.

b) That's appropriate because MS is a relatively rare disease.

c) The subjects were U.S. military personnel, some of whom had developed MS.

d) The variables were the vitamin D blood levels and whether or not the subject developed MS.

5. a) This was a randomized, placebo-controlled experiment.

b) Yes, such an experiment is the right way to determine whether black cohosh has an effect.

c) 351 women aged 45 to 55 who reported at least two hot flashes a day.

d) The treatments were black cohosh, a multiherb supplement, a multiherb supplement plus advice, estrogen, and a placebo. The response was the women's symptoms (presumably frequency of hot flashes).

7. a) Experiment.

b) Bipolar disorder patients.

c) Omega-3 fats from fish oil, two levels.

d) 2 treatments.

e) Improvement (fewer symptoms?).

f) Design not specified.

g) Blind (due to placebo), unknown if double-blind.

h) Individuals with bipolar disease improve with high-dose omega-3 fats from fish oil.

9. a) Observational study.

b) Prospective.

c) Men and women with moderately high blood pressure and normal blood pressure, unknown selection process.

d) Memory and reaction time.

e) As there is no random assignment, there is no way to know that high blood pressure *caused* subjects to do worse on memory and reaction-time tests. A lurking variable may also be the cause.

11. a) Experiment.

b) Postmenopausal women.

c) Alcohol—2 levels; blocking variable—estrogen supplements (2 levels).

d) 1 factor (alcohol) at 2 levels = 2 treatments.

e) Increase in estrogen levels.

f) Blocked.

g) Not blind.

h) Indicates that alcohol consumption *for those taking estrogen supplements* may increase estrogen levels.

13. a) Observational study.

b) Retrospective.

c) Women in Finland, unknown selection process with data from church records.

d) Women's lifespans.

e) As there is no random assignment, there is no way to know that having sons or daughters shortens or lengthens the life span of mothers.

15. a) Observational study.

b) Prospective.

c) People with or without depression, unknown selection process.

d) Frequency of crying in response to sad situations.

e) There is no apparent difference in crying response (to sad movies) for depressed and nondepressed groups.

17. a) Experiment.

b) People experiencing migraines.

c) 2 factors (pain reliever and water temperature), 2 levels each.

d) 4 treatments.

e) Level of pain relief.

f) Completely randomized over 2 factors.

g) Blind, as subjects did not know if they received the pain medication or the placebo, but not blind, as the subjects will know if they are drinking regular or ice water.

h) It may indicate whether pain reliever alone or in combination with ice water gives pain relief, but patients are not blinded to ice water, so placebo effect may also be the cause of any relief seen caused by ice water.

19. a) Experiment.

b) Athletes with hamstring injuries.

c) 1 factor: type of exercise program (2 levels).
d) 2 treatments.
e) Time to return to sports.
f) Completely randomized.
g) No blinding—subjects must know what kind of exercise they do.
h) Can determine which of the two exercise programs is more effective.

21. They need to compare omega-3 results to something. Perhaps bipolarity is seasonal and would have improved during the experiment anyway.

23. a) Subjects' responses might be related to many other factors (diet, exercise, genetics, etc). Randomization should equalize the two groups with respect to unknown factors.
  b) More subjects would minimize the impact of individual variability in the responses, but the experiment would become more costly and time consuming.

25. People who engage in regular exercise might differ from others with respect to bipolar disorder, and that additional variability could obscure the effectiveness of this treatment.

27. Answers may vary. Use a random-number generator to randomly select 24 numbers from 01 to 24 without replication. Assign the first 8 numbers to the first group, the second 8 numbers to the second group, and the third 8 numbers to the third group.

29. a) First, they are using athletes who have a vested interest in the success of the shoe by virtue of their sponsorship. They should choose other athletes. Second, they should randomize the order of the runs, not run all the races with their shoes second. They should blind the athletes by disguising the shoes if possible, so they don't know which is which. The timers shouldn't know which athletes are running with which shoes, either. Finally, they should replicate several times, since times will vary under both shoe conditions.
  b) Because of the problems in part a, the results they obtain may favor their shoes. In addition, the results obtained for Olympic athletes may not be the same as for the general runner.

31. a) Allowing athletes to self-select treatments could confound the results. Other issues such as severity of injury, diet, age, etc., could also affect time to heal; randomization should equalize the treatment groups with respect to any such variables.
  b) A control group could have revealed whether either exercise program was better (or worse) than just letting the injury heal.
  c) Doctors who evaluated the athletes to approve their return to sports should not know which treatment the subject had.
  d) It's hard to tell. The difference of 15 days seems large, but the standard deviations indicate that there was a great deal of variability in the times.

33. a) The differences among the Mozart and quiet groups were more than would have been expected from sampling variation.

b)

c) The Mozart group seems to have the smallest median difference and thus the *least* improvement, but there does not appear to be a significant difference.
d) No, if anything, there is less improvement, but the difference does not seem significant compared with the usual variation.

35. a) Observational, prospective study.
  b) The supposed relation between health and wine consumption might be explained by the confounding variables of income and education.
  c) None of these. While the variables have a relation, there is no causality indicated for the relation.

37. a) Arrange the 20 containers in 20 separate locations. Use a random-number generator to identify the 10 containers that should be filled with water.
  b) Guessing, the dowser should be correct about 50% of the time. A record of 60% (12 out of 20) does not appear to be significantly different.
  c) Answers may vary. You would need to see a high level of success—say, 90% to 100%, that is, 18 to 20 correct.

39. Randomly assign half the reading teachers in the district to use each method. Students should be randomly assigned to teachers as well. Make sure to block both by school and grade (or control grade by using only one grade). Construct an appropriate reading test to be used at the end of the year, and compare scores.

41. a) They mean that the difference is higher than they would expect from normal sampling variability.
  b) An observational study.
  c) No. Perhaps the differences are attributable to some confounding variable (e.g., people are more likely to engage in riskier behaviors on the weekend) rather than the day of admission.
  d) Perhaps people have more serious accidents and traumas on weekends and are thus more likely to die as a result.

43. Answers may vary. This experiment has 1 factor (pesticide), at 3 levels (pesticide A, pesticide B, no pesticide), resulting in 3 treatments. The response variable is the number of beetle larvae found on each plant. Randomly select a third of the plots to be sprayed with pesticide A, a third with pesticide B, and a third with no pesticide (since the researcher also wants to know whether the pesticides even work at all). To control the experiment, the plots of land should be as similar as possible with regard to amount of sunlight, water, proximity to other plants, etc. If not, plots with similar characteristics should be blocked together. If possible, use some inert substance as a placebo pesticide on the control group, and do not tell the counters of the beetle larvae which plants have been treated with pesticides. After a given period of time, count the number of beetle larvae on each plant and compare the results.

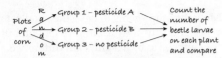

45. Answers may vary. Find a group of volunteers. Each volunteer will be required to shut off the machine with his or her left hand and right hand. Randomly assign the left or right hand to be used first. Complete the first attempt for the whole group. Now repeat the experiment with the alternate hand. Check the differences in time for the left and right hands.

47. a) Jumping with or without a parachute.
  b) Volunteer skydivers (the dimwitted ones).
  c) A parachute that looks real but doesn't work.
  d) A good parachute and a placebo parachute.
  e) Whether parachutist survives the jump (or extent of injuries).
  f) All should jump from the same altitude in similar weather conditions and land on similar surfaces.
  g) Randomly assign people the parachutes.
  h) The skydivers (and the people involved in distributing the parachute packs) shouldn't know who got a working chute. And the people evaluating the subjects after the jumps should not be told who had a real parachute either!

Part III Review

1. Observational prospective study. Indications of behavior differences can be seen in the two groups. May show a link between premature birth and behavior, but there may be lurking variables involved.

3. Experiment, matched by gender and weight, randomization within blocks of two pups of same gender and weight. Factor: type of diet. Treatments: low-calorie diet and allowing the dog to eat all it wants. Response variable: length of life. Can conclude that, on average, dogs with a lower-calorie diet live longer.

5. Observational prospective study. Indicates folate *may* help in reducing colon cancer for those with family histories of the disease.

7. Sampling. Probably a simple random sample, although may be stratified by type of firework. Population is all fireworks produced each day. Parameter is proportion of duds. Can determine if the day's production is ready for sale.

9. Observational retrospective study. Living near strong electromagnetic fields may be associated with more leukemia than normal. May be lurking variables, such as socioeconomic level.

11. Experiment. Blocked by sex of rat. Randomization is not specified. Factor is type of hormone given. Treatments are leptin and insulin. Response variable is lost weight. Can conclude that hormones can help suppress appetites in rats, and the type of hormone varies by gender.

13. Experiment. Factor is gene therapy. Hamsters were randomized to treatments. Treatments were gene therapy or not.

Response variable is heart muscle condition. Can conclude that gene therapy is beneficial (at least in hamsters).

15. Sampling. Population is all oranges on the truck. Parameter is proportion of unsuitable oranges. Procedure is probably simple random sampling. Can conclude whether or not to accept the truckload.

17. Observational prospective study. Physically fit men may have a lower risk of death from cancer.

19. Answers will vary. This is a simulation problem. Using a random digits table or software, call 0–4 a loss and 5–9 a win for the gambler on a game. Use blocks of 5 digits to simulate a week's pick.

21. Answers will vary.

23. a) Experiment. Actively manipulated candy giving, diners were randomly assigned treatments, control group was those with no candy, lots of dining parties.
  b) It depends on when the decision was made. If early in the meal, the server may give better treatment to those who will receive candy—biasing the results.
  c) A difference in response so large it cannot be attributed to natural sampling variability.

25. a) Voluntary response. Only those who feel strongly will pay for the 900 phone call.
  b) "If it would help future generations live a longer, healthier life, would you be in favor of human cloning?"

27. a) Simulation results will vary. Average will be around 5.8 points.
  b) Simulation results will vary. Average will also be around 5.8 points.
  c) Answers will vary.

29. a) Yes.
  b) No. Residences without phones are excluded. Residences with more than one phone number had a higher chance.
  c) No. People who respond to the survey may be of age but not registered voters.
  d) No. Households who answered the phone may be more likely to have someone at home when the phone call was generated. These may not be representative of all households.

31. a) Does not prove it. There may be other confounding variables. Only way to prove this would be to do a controlled experiment.
  b) Alzheimer's usually shows up late in life. Perhaps smokers have died of other causes before Alzheimer's is evident.
  c) An experiment would be unethical. One could design a prospective study in which groups of smokers and nonsmokers are followed for many years and the incidence of Alzheimer's is tracked.

33.

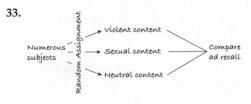

Numerous subjects will be randomly assigned to see shows with violent, sexual, or neutral content. They will see the same commercials. After the show, they will be interviewed for their recall of brand names in the commercials.

35. a) May have been a simple random sample, but given the relative equality in age groups, may have been stratified.
    b) 35.1%.
    c) We don't know. Perhaps cell phones or unlisted numbers were excluded, and Democrats have more (or fewer) of those. Probably OK, though.
    d) Do party affiliations differ for different age groups?

37. The factor in the experiment will be type of bird control. I will have three treatments: scarecrow, netting, and no control. I will randomly assign several different areas in the vineyard to one of the treatments, taking care that there is sufficient separation that the possible effect of the scarecrow will not be confounded. At the end of the season, the response variable will be the proportion of bird-damaged grapes.

39. a) We want all subjects treated as alike as possible. If there were no "placebo surgery," subjects would know this and perhaps behave differently.
    b) The experiment looked for a difference in the effectiveness of the two treatments. (If we wanted to generalize, we would need to assume that the results for these volunteers are the same as on all patients who might need this operation.)
    c) "Not statistically significant" means the difference in results were small enough that it could be explained by natural sampling variability.

41. a) Use stratified sampling to select 2 first-class passengers and 12 from coach.
    b) Number passengers alphabetically, 01 = Bergman to 20 = Testut. Read in blocks of two, ignoring any numbers more than 20. This gives 65, 43, 67, 11 (selects Fontana), 27, 04 (selects Castillo).
    c) Number passengers alphabetically from 001 to 120. Use the random-number table to find three-digit numbers in this range until 12 different values have been selected.

43. Simulation results will vary. (Use integers 00 to 99 as a basis. Use integers 00 to 69 to represent a tee shot on the fairway. If on the fairway, use digits 00 to 79 to represent on the green. If off the fairway, use 00 to 39 to represent getting on the green. If not on the green, use digits 00 to 89 to represent landing on the green. For the first putt, use digits 00 to 9 to represent making the shot. For subsequent putts, use digits 00 to 89 to represent making the shot.)

Chapter 14

1. S = {HH, HT, TH, TT}, equally likely.
   S = {0, 1, 2, 3}, not equally likely.
   S = {H, TH, TTH, TTT}, not equally likely.
   S = {1, 2, 3, 4, 5, 6}, not equally likely.

3. This context "truly random" should mean that every number is equally likely to occur.

5. There is no "Law of Averages." She would be wrong to think they are "due" for a harsh winter.

7. There is no "Law of Averages." If at bats are independent, his chance for a hit does not change based on recent successes or failures.

9. a) There is some chance you would have to pay out much more than the $300.
   b) Many customers pay for insurance. The small risk for any one customer is spread among all.

11. a) Legitimate.
    b) Legitimate.
    c) Not legitimate (sum more than 1).
    d) Legitimate.
    e) Not legitimate (can't have negatives or values more than 1).

13. A family may own both a car and an SUV. The events are not disjoint, so the Addition Rule does not apply.

15. When cars are traveling close together, their speeds are not independent, so the Multiplication Rule does not apply.

17. a) He has multiplied the two probabilities.
    b) He assumes that being accepted at the colleges are independent events.
    c) No. Colleges use similar criteria for acceptance, so the decisions are not independent.

19. a) 0.72   b) 0.89   c) 0.28

21. a) 0.5184   b) 0.0784   c) 0.4816

23. a) Repair needs for the two cars must be independent.
    b) Maybe not. An owner may treat the two cars similarly, taking good (or poor) care of both. This may decrease (or increase) the likelihood that each needs to be repaired.

25. a) 342/1005 = 0.340.
    b) 30/1005 + 50/1005 = 80/1005 = 0.080.

27. a) 0.195   b) 0.913
    c) Responses are independent.
    d) People were polled at random.

29. a) 0.2888   b) 0.7112
    c) (1 − 0.76) + 0.76(1 − 0.38) or 1 − (0.76)(0.38)

31. a) 1) 0.30   2) 0.30   3) 0.90   4) 0.0
    b) 1) 0.027   2) 0.128   3) 0.512   4) 0.271

33. a) Disjoint (can't be both red and orange).
    b) Independent (unless you're drawing from a small bag).
    c) No. Once you know that one of a pair of disjoint events has occurred, the other is impossible.

35. a) 0.0046   b) 0.125   c) 0.296   d) 0.421   e) 0.995

37. a) 0.027   b) 0.063   c) 0.973   d) 0.014

39. a) 0.024   b) 0.250   c) 0.543

41. 0.078

43. a) For any day with a valid three-digit date, the chance is 0.001, or 1 in 1000. For many dates in October through December, the probability is 0. (No three digits will make 10/15, for example.)
    b) There are 65 days when the chance to match is 0. (Oct. 10–31, Nov. 10–30, and Dec. 10–31.) The chance for no matches on the remaining 300 days is 0.741.
    c) 0.259   d) 0.049

Chapter 15

1. a) 0.68   b) 0.32   c) 0.04

3. a) 0.31   b) 0.48   c) 0.31

5. a) 0.2025   b) 0.6965   c) 0.2404   d) 0.0402

7. a) 0.50   b) 1.00   c) 0.077   d) 0.333

9. a) 0.11   b) 0.27   c) 0.407   d) 0.344

11. a) 0.011   b) 0.222   c) 0.054   d) 0.337   e) 0.436

13. 0.21

15. a) 0.145   b) 0.118   c) 0.414   d) 0.217

17. a) 0.318   b) 0.955   c) 0.071   d) 0.009

19. a) 32%   b) 0.135
c) No, 7% of juniors have taken both.
d) No, the probability that a junior has taken a computer course is 0.23. The probability that a junior has taken a computer course *given* he or she has taken a Statistics course is 0.135.

21. a) 0.266
b) No, 26.6% of homes with garages have pools; 21% of homes overall have pools.
c) No, 17% of homes have both.

23. Yes, $P(\text{Ace}) = 4/52$. $P(\text{Ace}|\text{any suit}) = 1/13$.

25. a) 0.17
b) No; 13% of the chickens had both contaminants.
c) No; $P(C \mid S) = 0.87 \neq P(C)$. If a chicken is contaminated with salmonella, it's more likely also to have campylobacter.

27. No, only 32% of all men have high cholesterol, but 40.7% of those with high blood pressure do.

29. a) 95.6%
b) Probably. 95.4% of people with cell phones had landlines, and 95.6% of all people did.

31. No. Only 34% of men were Democrats, but over 41% of all voters were.

33. a) No, the probability that the luggage arrives on time depends on whether the flight is on time. The probability is 95% if the flight is on time and only 65% if not.
b) 0.695

35. 0.975

37. a) No, the probability of missing work for day-shift employees is 0.01. It is 0.02 for night-shift employees. The probability depends on whether they work the day or night shift.
b) 1.4%

39. 57.1%

41. a) 0.20   b) 0.272   c) 0.353   d) 0.033

43. 0.563

45. Over 0.999

Chapter 16

1. a) 19   b) 4.2

3. a)

Amount won	$0	$5	$10	$30
P(Amount won)	$\frac{26}{52}$	$\frac{13}{52}$	$\frac{12}{52}$	$\frac{1}{52}$

b) $4.13   c) $4 or less (answers may vary)

5. a)

Children	1	2	3
P(Children)	0.5	0.25	0.25

b) 1.75 children   c) 0.875 boys

Boys	0	1	2	3
P(Boys)	0.5	0.25	0.125	0.125

7. $27,000

9. a) 7   b) 1.89

11. $5.44

13. 0.83

15. a) 1.7   b) 0.9

17. $\mu = 0.64, \sigma = 0.93$

19. a) $50   b) $100

21. a) No. The probability of winning the second depends on the outcome of the first.
b) 0.42   c) 0.08
d)

Games won	0	1	2
P(Games won)	0.42	0.50	0.08

e) $\mu = 0.66, \sigma = 0.62$

23. a)

Number good	0	1	2
P(Number good)	0.067	0.467	0.467

b) 1.40   c) 0.61

25. a) $\mu = 30, \sigma = 6$ b) $\mu = 26, \sigma = 5$ c) $\mu = 30, \sigma = 9$
d) $\mu = -10, \sigma = 5.39$ e) $\mu = 20, \sigma = 2.83$

27. a) $\mu = 240, \sigma = 12.80$ b) $\mu = 140, \sigma = 24$
c) $\mu = 720, \sigma = 34.18$ d) $\mu = 60, \sigma = 39.40$
e) $\mu = 600, \sigma = 22.63$

29. a) 1.8   b) 0.87
c) Cartons are independent of each other.

31. $\mu = 13.6, \sigma = 2.55$ (assuming the hours are independent of each other).

33. a) $\mu = 23.4, \sigma = 2.97$
b) We assume each truck gets tickets independently.

35. a) There will be many gains of $150 with a few large loss.
b) $\mu = \$300, \sigma = \$8485.28$
c) $\mu = \$1,500,000, \sigma = \$600,000$
d) Yes. $0 is 2.5 SDs below the mean for 10,000 policie
e) Losses are independent of each other. A major catastrophe with many policies in an area would violate the assumption.

37. a) 1 oz   b) 0.5 oz   c) 0.023
d) $\mu = 4$ oz, $\sigma = 0.5$ oz
e) 0.159
f) $\mu = 12.3$ oz, $\sigma = 0.54$ oz

39. a) 12.2 oz   b) 0.51 oz   c) 0.058

41. a) $\mu = 200.57$ sec, $\sigma = 0.46$ sec

b) No, $z = \dfrac{199.48 - 200.57}{0.461} = -2.36$. There is only 0.009 probability of swimming that fast or faster.

43. a) A = price of a pound of apples; P = price of a pound of potatoes; Profit $= 100A + 50P - 2$

b) $63.00   c) $20.62

d) Mean—no; SD—yes (independent sales prices).

45. a) $\mu = 1920, \sigma = 48.99$; $P(T > 2000) = 0.051$

b) $\mu = 220, \sigma = 11.09$; No—$300 is more than 7 SDs above the mean.

c) $P(D - \frac{1}{2}C > 0) \approx 0.26$

Chapter 17

1. a) No. More than two outcomes are possible.

b) Yes, assuming the people are unrelated to each other.

c) No. The chance of a heart changes as cards are dealt.

d) No, 500 is more than 10% of 3000.

e) If packages in a case are independent of each other, yes; otherwise, no.

3. a) Use single random digits. Let 0, 1 = LeBron. Count the number of random numbers until a 0 or 1 occurs.

c) Results will vary.

d)

x	1	2	3	4	5	6	7	8	≥9
$P(x)$	0.2	0.16	0.128	0.102	0.082	0.066	0.052	0.042	0.168

5. a) Use single random digits. Let 0, 1 = LeBron. Examine random digits in groups of five, counting the number of 0's and 1's.

c) Results will vary.

d)

x	0	1	2	3	4	5
$P(x)$	0.33	0.41	0.20	0.05	0.01	0.0

7. Departures from the same airport during a 2-hour interval may not be independent. All could be delayed by weather, for example.

9. a) 0.0819   b) 0.0064   c) 0.992

11. 5

13. 20 calls

15. a) 25   b) 0.185   c) 0.217   d) 0.693

17. a) 50   b) $E(X) = np = 100(0.5) = 50$.

19. a) 0.0745   b) 0.502   c) 0.211

d) 0.0166   e) 0.0179   f) 0.9987

21. a) 0.65   b) 0.75   c) 7.69 picks

23. a) $\mu = 10.44, \sigma = 1.16$

b) 0.812   c) 0.475   d) 0.00193   e) 0.998

25. $\mu = 20.28, \sigma = 4.22$

27. a) 0.118   b) 0.324   c) 0.744   d) 0.580

29. $\mu = 56, \sigma = 4.10$

b) Yes, $np = 56 \geq 10, nq = 24 \geq 10$, serves are independent.

c) In a match with 80 serves, approximately 68% of the time she will have between 51.9 and 60.1 good serves, approximately 95% of the time she will have between 47.8 and 64.2 good serves, and approximately 99.7% of the time she will have between 43.7 and 68.3 good serves.

d) Normal, approx.: 0.014; Binomial, exact: 0.016

31. a) Assuming apples fall and become blemished independently of each other, Binom(300, 0.06) is appropriate. Since $np \geq 10$ and $nq \geq 10$, $N(18, 4.11)$ is also appropriate.

b) Normal, approx.: 0.072; Binomial, exact: 0.085

c) No, 50 is 7.8 SDs above the mean.

33. Normal, approx.: 0.053; Binomial, exact: 0.061

35. The mean number of sales should be 24 with SD 4.60. Ten sales is more than 3.0 SDs below the mean. He was probably misled.

37. a) 0.0869   b) 0.0364

39. a) 4 cases   b) 0.9817

41. a) 5   b) 0.066   c) 0.107   d) $\mu = 24, \sigma = 2.19$

e) Normal, approx.: 0.819; Binomial, exact: 0.848

43. $\mu = 20, \sigma = 4$. I'd want *at least* 32 (3 SDs above the mean). (Answers will vary.)

45. Probably not. There's a more than 9% chance that he could hit 4 shots in a row, so he can expect this to happen nearly once in every 10 sets of 4 shots he takes. That does not seem unusual.

47. Yes. We'd expect him to make 22 shots, with a standard deviation of 3.15 shots. 32 shots is more than 3 standard deviations above the expected value, an unusually high rate of success.

49. a) The Poission model

b) 0.9502   c) 0.0025

51. a) The exponential model

b) 1/3 minutes   c) 0.0473

Part IV Review

1. a) 0.34   b) 0.27   c) 0.069

d) No, 2% of cars have both types of defects.

e) Of all cars with cosmetic defects, 6.9% have functional defects. Overall, 7.0% of cars have functional defects. The probabilities here are estimates, so these are probably close enough to say the defects are independent.

3. a) C = Price to China; F = Price to France; Total $= 3C + 5F$

b) $\mu = 5500, \sigma = 672.68$

c) $\mu = 500, \sigma = 180.28$

d) Means—no. Standard deviations—yes; ticket prices must be independent of each other for different countries, but all tickets to the same country are at the same price.

5. a) $\mu = -\$0.20, \sigma = \$1.89$
   b) $\mu = -\$0.40, \sigma = \$2.67$

7. a) 0.106   b) 0.651   c) 0.442

9. a) 0.590   b) 0.328   c) 0.00856

11. a) $\mu = 15.2, \sigma = 3.70$
    b) Yes, $np \geq 10$ and $nq \geq 10$
    c) Normal, approx.: 0.080; Binomial, exact: 0.097

13. a) 0.0173   b) 0.591
    c) Left: 960; right: 120; both: 120
    d) $\mu = 120, \sigma = 10.39$
    e) About 68% chance of between 110 and 130; about 95% between 99 and 141; about 99.7% between 89 and 151.

15. a) Men's heights are more variable than women's.
    b) Men
    c) M = Man's height; W = Woman's height; $M - W$ is how much taller the man is.
    d) 5.1"   e) 3.75"   f) 0.913
    g) If independent, it should be about 91.3%. We are told 92%. This difference seems small and may be due to natural sampling variability.

17. a) The chance is 1.6×10^{-7}.
    b) 0.952   c) 0.063

19. −$2080.00

21. a) 0.717   b) 0.588

23. a) $\mu = 100, \sigma = 8$
    b) $\mu = 1000, \sigma = 60$
    c) $\mu = 100, \sigma = 8.54$
    d) $\mu = -50, \sigma = 10$
    e) $\mu = 100, \sigma = 11.31$

25. a) Many do both, so the two categories can total more than 100%.
    b) No. They can't be disjoint. If they were, the total would be 100% or less.
    c) No. Probabilities are different for boys and girls.
    d) 0.0524

27. a) 21 days
    b) 1649.73 som
    c) 3300 som extra. About 157-som "cushion" each day.

29. No, you'd expect 541.2 homeowners, with an SD of 13.56. 523 is 1.34 SDs below the mean; not unusual.

31. a) 0.018   b) 0.300   c) 0.26

33. a) 6   b) 15   c) 0.402

35. a) 34%   b) 35%   c) 31.4%
    d) 31.4% of classes that used calculators used computer assignments, while in classes that didn't use calculators, 30.6% used computer assignments. These are close enough to think the choice is probably independent.

37. a) A Poisson model   b) 0.1 failures
    c) 0.090   d) 0.095

39. a) 1/11   b) 7/22   c) 5/11   d) 0   e) 19/66

41. a) Expected number of stars with planets.
    b) Expected number of planets with intelligent life.
    c) Probability of a planet with a suitable environment having intelligent life.

d) f_l: If a planet has a suitable environment, the probability that life develops.
   f_i: If a planet develops life, the probability that the life evolves intelligence.
   f_c: If a planet has intelligent life, the probability that it develops radio communication.

43. 0.991

Chapter 18

1. All the histograms are centered near 0.05. As n gets larger, the histograms approach the Normal shape, and the variability in the sample proportions decreases.

3. a)

n	Observed mean	Theoretical mean	Observed st. dev.	Theoretical st. dev.
20	0.0497	0.05	0.0479	0.0487
50	0.0516	0.05	0.0309	0.0308
100	0.0497	0.05	0.0215	0.0218
200	0.0501	0.05	0.0152	0.0154

   b) They are all quite close to what we expect from the theory.
   c) The histogram is unimodal and symmetric for $n = 200$.
   d) The Success/Failure Condition says that np and nq should both be at least 10, which is not satisfied until $n = 200$ for $p = 0.05$. The theory predicted my choice.

5. a) Symmetric, because probability of heads and tails is equal.
   b) 0.5   c) 0.125   d) $np = 8 < 10$

7. a) About 68% should have proportions between 0.4 and 0.6, about 95% between 0.3 and 0.7, and about 99.7% between 0.2 and 0.8.
   b) $np = 12.5, nq = 12.5$; both are ≥ 10.
   c)

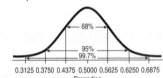

0.3125 0.3750 0.4375 0.5000 0.5625 0.6250 0.6875
Proportion

   $np = nq = 32$; both are ≥ 10.
   d) Becomes narrower (less spread around 0.5).

9. This is a fairly unusual result: about 2.26 SDs below the mean. The probability of that is about 0.012. So, in a class of 100 this is certainly a reasonable possibility.

11. a)

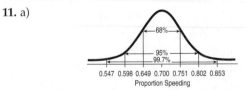

0.547 0.598 0.649 0.700 0.751 0.802 0.853
Proportion Speeding

   b) Both $np = 56$ and $nq = 24 \geq 10$. Drivers *may* be independent of each other, but if flow of traffic is very fast, they may not be. Or weather conditions may affect all drivers. In these cases they may get more or fewer speeders than they expect.

13. a) Assume that these children are typical of the population. They represent fewer than 10% of all children. We expect 20.4 nearsighted and 149.6 not; both are at least 10.

b)

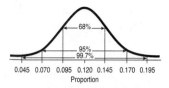

c) Probably between 12 and 29.

15. a) $\mu = 7\%, \sigma = 1.8\%$

b) Assume that clients pay independently of each other, that we have a random sample of all possible clients, and that these represent less than 10% of all possible clients. $np = 14$ and $nq = 186$ are both at least 10.

c) 0.048

17.

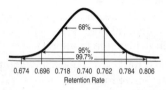

These are not random samples, and not all colleges may be typical (representative). $np = 296, nq = 104$ are both at least 10.

19. Yes; if their students were typical, a retention rate of 522/603 = 86.6% would be over 7 standard deviations above the expected rate of 74%.

21. 0.212. Reasonable that those polled are independent of each other and represent less than 10% of all potential voters. We assume the sample was selected at random. Success/Failure Condition met: $np = 208, nq = 192$. Both ≥ 10.

23 0.088 using $N(0.08, 0.022)$ model.

25. Answers will vary. Using $\mu + 3\sigma$ for "very sure," the restaurant should have 89 nonsmoking seats. Assumes customers at any time are independent of each other, a random sample, and represent less than 10% of all potential customers. $np = 72, nq = 48$, so Normal model is reasonable ($\mu = 0.60, \sigma = 0.045$).

27. a) Normal, center at μ, standard deviation $\sigma/\sqrt{n}$.

b) Standard deviation will be smaller. Center will remain the same.

29. a) The histogram is unimodal and slightly skewed to the right, centered at 36 inches with a standard deviation near 4 inches.

b) All the histograms are centered near 36 inches. As n gets larger, the histograms approach the Normal shape and the variability in the sample means decreases. The histograms are fairly normal by the time the sample reaches size 5.

31. a)

n	Observed mean	Theoretical mean	Observed st. dev.	Theoretical st. dev.
2	36.314	36.33	2.855	2.842
5	36.314	36.33	1.805	1.797
10	36.341	36.33	1.276	1.271
20	36.339	36.33	0.895	0.899

b) They are all very close to what we would expect.

c) For samples as small as 5, the sampling distribution of sample means is unimodal and very symmetric.

d) The distribution of the original data is nearly unimodal and symmetric, so it doesn't take a very large sample size for the distribution of sample means to be approximately Normal.

33.

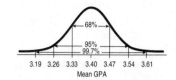

Normal, $\mu = 3.4, \sigma = 0.07$. We assume that the students are randomly assigned to the seminars and represent less than 10% of all possible students, and that individuals' GPAs are independent of one another.

35. a) As the CLT predicts, there is more variability in the smaller outlets.

b) If the lottery is random, all outlets are equally likely to sell winning tickets.

37. a) 21.1%   b) 276.8 days or more

c) $N(266, 2.07)$ d) 0.002

39. a) There are more premature births than very long pregnancies. Modern practice of medicine stops pregnancies at about 2 weeks past normal due date.

b) Parts (a) and (b)—yes—we can't use Normal model if it's very skewed. Part (c)—no—CLT guarantees a Normal model for this large sample size.

41. a) $\mu = \$2.00, \sigma = \$3.61$

b) $\mu = \$4.00, \sigma = \$5.10$

c) 0.191. Model is $N(80, 22.83)$.

43. a) $\mu = 2.859, \sigma = 1.324$

b) No. The score distribution in the sample should resemble that in the population, somewhat uniform for scores 1–4 and about half as many 5's.

c) Approximately $N\left(2.859, \dfrac{1.324}{\sqrt{40}}\right)$.

45. About 20%, based on $N(2.859, 0.167)$.

47. a) $N(2.9, 0.045)$ b) 0.0131   c) 2.97 gm/mi

49. a) Can't use a Normal model to estimate probabilities. The distribution is skewed right—not Normal.

b) 4 is probably not a large enough sample to say the average follows the Normal model.

c) No. This is 3.16 SDs above the mean.

51. a) 0.0003. Model is $N(384, 34.15)$.

b) $427.77 or more.

53. a) 0.734

b) 0.652. Model is $N(10, 12.81)$.

c) 0.193. Model is $N(120, 5.774)$.

d) 0.751. Model is $N(10, 7.394)$.

Chapter 19

1. She believes the true proportion is within 4% of her estimate, with some (probably 95%) degree of confidence.

3. a) Population—all cars; sample—those actually stopped at the checkpoint; p—proportion of all cars with safety problems; $\hat{p}$—proportion actually seen with safety problems (10.4%); if sample (a cluster sample) is representative, then the methods of this chapter will apply.

b) Population—general public; sample—those who logged onto the website; p—population proportion of those who favor prayer in school; $\hat{p}$—proportion of those who voted in the poll who favored prayer in school (81.1%); can't use methods of this chapter—sample is biased and nonrandom.

c) Population—parents at the school; sample—those who returned the questionnaire; p—proportion of all parents who favor uniforms; $\hat{p}$—proportion of respondents who favor uniforms (60%); should not use methods of this chapter, since not SRS. (Possible non-response bias.)

d) Population—students at the college; sample—the 1632 students who entered that year; p—proportion of all students who will graduate on time; $\hat{p}$—proportion of that year's students who graduate on time (85.0%); can use methods of this chapter if that year's students (a cluster sample) are viewed as a representative sample of all possible students at the school.

5. a) Not correct. This implies certainty.

b) Not correct. Different samples will give different results. Many fewer than 95% will have 88% on-time orders.

c) Not correct. The interval is about the population proportion, not the sample proportion in different samples.

d) Not correct. In this sample, we *know* 88% arrived on time.

e) Not correct. The interval is about the parameter, not the days.

7. a) False   b) True   c) True   d) False

9. On the basis of this sample, we are 90% confident that the proportion of Japanese cars is between 29.9% and 47.0%.

11. a) (0.798, 0.863)

b) We're 95% confident that between 80% and 86% of all broiler chicken sold in U.S. food stores is infected with *Campylobacter*.

c) The size of the population is irrelevant. If *Consumer Reports* had a random sample, 95% of intervals generated by studies like this will capture the true contamination level.

13. a) 0.025

b) We're 90% confident that this poll's estimate is within $\pm 2.5\%$ of the true proportion of people who are baseball fans.

c) Larger. To be more certain, we must be less precise.

d) 0.039

e) less confidence

f) No evidence of change; given the margin of error, 0.37 is a plausible value for 2007 as well.

15. a) (0.0465, 0.0491). The assumptions and conditions for constructing a confidence interval are satisfied.

b) The confidence interval gives the set of plausible values (with 95% confidence). Since 0.05 is outside the interval, that seems to be a bit too optimistic.

17. a) (12.7%, 18.6%)

b) We are 95% confident, based on this sample, that the proportion of all auto accidents that involve teenage drivers is between 12.7% and 18.6%.

c) About 95% of all random samples will produce confidence intervals that contain the true population proportion.

d) Contradicts. The interval is completely below 20%.

19. Probably nothing. Those who bothered to fill out the survey may be a biased sample.

21. a) Response bias (wording)

b) (54%, 60%)

c) Smaller—the sample size was larger.

23. a) (18.2%, 21.8%)

b) We are 98% confident, based on the sample, that between 18.2% and 21.8% of English children are deficient in vitamin D.

c) About 98% of all random samples will produce a confidence interval that contains the true proportion of children deficient in vitamin D.

25. a) Wider. The sample size is probably about one fourth of the sample size for all adults, so we'd expect the confidence interval to be about twice as wide.

b) Smaller. The second poll used a slightly larger sample size.

27. a) (15.5%, 26.3%)   b) 612

c) Sample may not be random or representative. Deer that are legally hunted may not represent all sexes and ages.

29. a) 141   b) 318   c) 564

31. 1801

33. 384 total, using $p = 0.15$

35. 90%

Chapter 20

1. a) $H_0: p = 0.30; H_A: p < 0.30$

b) $H_0: p = 0.50; H_A: p \neq 0.50$

c) $H_0: p = 0.20; H_A: p > 0.20$

3. Statement d is correct.

5. No, we can say only that there is a 27% chance of seeing the observed effectiveness just from natural sampling variation. There is no *evidence* that the new formula is more effective, but we can't conclude that they are equally effective.

7. a) No. There's a 25% chance of losing twice in a row. That's not unusual.

b) 0.125

c) No, we expect that to happen 1 time in 8.

d) Maybe 5? The chance of 5 losses in a row is only 1 in 32 which seems unusual.

9. 1) Use p, not $\hat{p}$, in hypotheses.

2) The question was about failing to meet the goal, so H_A should be $p < 0.96$.

3) Did not check $0.04(200) = 8$. Since $nq < 10$, the Success/Failure Condition is violated. Didn't check 10% Condition.

4) $188/200 = 0.94; SD(\hat{p}) = \sqrt{\dfrac{(0.96)(0.04)}{200}} = 0.014$

5) z is incorrect; should be $z = \dfrac{0.94 - 0.96}{0.014} = -1.43$

6) $P = P(z < -1.43) = 0.076$

7) There is only weak evidence that the new instructions do not work.

11. a) $H_0: p = 0.30; H_A: p > 0.30$

b) Possibly an SRS; we don't know if the sample is less than 10% of his customers, but it could be viewed as less than 10% of all possible customers; $(0.3)(80) \geq 10$ and $(0.7)(80) \geq 10$. Wells are independent only if customers don't have farms on the same underground springs.

c) $z = 0.73$; P-value $= 0.232$

d) If his dowsing is no different from standard methods, there is more than a 23% chance of seeing results as good as those of the dowser's, or better, by natural sampling variation.

e) These data provide no evidence that the dowser's chance of finding water is any better than normal drilling.

13. a) $H_0: p_{2000} = 0.34; H_A: p_{2000} \neq 0.34$

b) Students were randomly sampled and should be independent. 34% and 66% of 8302 are greater than 10. 8302 students is less than 10% of the entire student population of the United States.

c) $P = 0.058$

d) The P-value provides weak evidence against the null hypothesis.

e) No. A difference this small, although statistically significant, is not meaningful. We might look at new data in a few years.

15. a) $H_0: p = 0.05$ vs. $H_A: p < 0.05$

b) We assume the whole mailing list has over 1,000,000 names. This is a random sample, and we expect 5000 successes and 95,000 failures.

c) $z = -3.178$; P-value $= 0.00074$, so we reject H_0; there is strong evidence that the donation rate would be below 5%.

17. a) $H_0: p = 0.63, H_A: p > 0.63$

b) The sample is representative. $240 < 10\%$ of all law school applicants. We expect $240(0.63) = 151.2$ to be admitted and $240(0.37) = 88.8$ not to be, both at least 10. $z = 1.58$; P-value $= 0.057$

c) Although the evidence is weak, there is some indication that the program may be successful. Candidates should decide whether they can afford the time and expense.

19. $H_0: p = 0.20; H_A: p > 0.20$. SRS (not clear from information provided); 22 is more than 10% of the population of 150; $(0.20)(22) < 10$. Do not proceed with a test.

21. $H_0: p = 0.03; p \neq 0.03. \hat{p} = 0.015$. One mother having twins will not affect another, so observations are independent; not an SRS; sample is less than 10% of all births. However, the mothers at this hospital may not be representative of all teenagers; $(0.03)(469) = 14.07 \geq 10$; $(0.97)(469) \geq 10. z = -1.91$; P-value $= 0.0556$. With a P-value this low, reject H_0. These data show some evidence that the rate of twins born to teenage girls at this hospital is less than the national rate of 3%. It is not clear whether this can be generalized to all teenagers.

23. $H_0: p = 0.25; H_A: p > 0.25$. SRS; sample is less than 10% of all potential subscribers; $(0.25)(500) \geq 10; (0.75)(500) \geq 10. z = 1.24$; P-value $= 0.1076$. The P-value is high, so do not reject H_0. These data do not show that more than 25% of current readers would subscribe; the company should not go ahead with the WebZine on the basis of these data.

25. $H_0: p = 0.40; H_A: p < 0.40$. Data are for all executives in this company and may not be able to be generalized to all companies; $(0.40)(43) \geq 10; (0.60)(43) \geq 10. z = -1.31$; P-value $= 0.0955$. Because the P-value is high, we fail to reject H_0. These data do not show that the proportion of women executives is less than the 40% of women in the company in general.

27. $H_0: p = 0.103; H_A: p > 0.103. \hat{p} = 0.118; z = 2.06$; P-value $= 0.02$. Because the P-value is low, we reject H_0. These data provide evidence that the dropout rate has increased.

29. $H_0: p = 0.90; H_A: p < 0.90. \hat{p} = 0.844; z = -2.05$; P-value $= 0.0201$. Because the P-value is so low, we reject H_0. There is strong evidence that the actual rate at which passengers with lost luggage are reunited with it within 24 hours is less than the 90% claimed by the airline.

31. a) Yes; assuming this sample to be a typical group of people, $P = 0.0008$. This cancer rate is very unusual.

b) No, this group of people may be atypical for reasons that have nothing to do with the radiation.

Chapter 21

1. a) Two sided. Let p be the percentage of students who prefer Diet Pepsi. $H_0: p = 0.5$ vs. $H_A: p \neq 0.5$

b) One sided. Let p be the percentage of teenagers who prefer the new formulation. $H_0: p = 0.5$ vs. $H_A: p > 0.5$

c) One sided. Let p be the percentage of people who intend to vote for the override. $H_0: p = 2/3$ vs. $H_A: p > 2/3$

d) Two sided. Let p be the percentage of days that the market goes up. $H_0: p = 0.5$ vs. $H_A: p \neq 0.5$

3. If there is no difference in effectiveness, the chance of seeing an observed difference this large or larger is 4.7% by natural sampling variation.

5. $\alpha = 0.10$: Yes. The P-value is less than 0.05, so it's less than 0.10. But to reject H_0 at $\alpha = 0.01$, the P-value must be below 0.01, which isn't necessarily the case.

7. a) There is only a 1.1% chance of seeing a sample proportion as low as 89.4% vaccinated by natural sampling variation if 90% have really been vaccinated.

b) We conclude that p is below 0.9, but a 95% confidence interval would suggest that the true proportion is between (0.889, 0.899). Most likely, a decrease from 90% to 89.9% would not be considered important. On the other hand, with 1,000,000 children a year vaccinated, even 0.1% represents about 1000 kids—so this may very well be important.

9. a) (1.9%, 4.1%)

b) Because 5% is not in the interval, there is strong evidence that fewer than 5% of all men use work as their primary measure of success.

c) $\alpha = 0.01$; it's a lower-tail test based on a 98% confidence interval.

11. a) (0.273, 0.327)
   b) Since 0.27 is not in the confidence interval, we reject the hypothesis that $p = 0.27$

13. a) The Success/Failure Condition is violated: only 5 pups had dysplasia.
   b) We are 95% confident that between 5% and 26% of puppies will show signs of hip dysplasia at the age of 6 months.

15. a) Type II error
   b) Type I error
   c) By making it easier to get the loan, the bank has reduced the alpha level.
   d) The risk of a Type I error is decreased and the risk of a Type II error is increased.

17. a) Power is the probability that the bank denies a loan that would not have been repaid.
   b) Raise the cutoff score.
   c) A larger number of trustworthy people would be denied credit, and the bank would miss the opportunity to collect interest on those loans.

19. a) The null is that the level of home ownership remains the same. The alternative is that it rises.
   b) The city concludes that home ownership is on the rise, but in fact the tax breaks don't help.
   c) The city abandons the tax breaks, but they were helping.
   d) A Type I error causes the city to forego tax revenue, while a Type II error withdraws help from those who might have otherwise been able to buy a home.
   e) The power of the test is the city's ability to detect an actual increase in home ownership.

21. a) It is decided that the shop is not meeting standards when it is.
   b) The shop is certified as meeting standards when it is not.
   c) Type I
   d) Type II

23. a) The probability of detecting a shop that is not meeting standards.
   b) 40 cars. Larger n.
   c) 10%. More chance to reject H_0.
   d) A lot. Larger differences are easier to detect.

25. a) One-tailed. The company wouldn't be sued if "too many" minorities were hired.
   b) Deciding the company is discriminating when it is not.
   c) Deciding the company is not discriminating when it is.
   d) The probability of correctly detecting actual discrimination.
   e) Increases power.
   f) Lower, since n is smaller.

27. a) One-tailed. Software is supposed to decrease the dropout rate.
   b) $H_0: p = 0.13$; $H_A: p < 0.13$
   c) He buys the software when it doesn't help students.
   d) He doesn't buy the software when it does help students.
   e) The probability of correctly deciding the software is helpful.

29. a) $z = -3.21, p = 0.0007$. The change is statistically significant. A 95% confidence interval is (2.3%, 8.5%). This is clearly lower than 13%. If the cost of the software justifies it, the professor should consider buying the software.
   b) The chance of observing 11 or fewer dropouts in a class of 203 is only 0.07% if the dropout rate is really 13%.

31. a) $H_A: p = 0.30$, where p is the probability of heads.
   b) Reject the null hypothesis if the coin comes up tails—otherwise fail to reject.
   c) P(tails given the null hypothesis) $= 0.1 = \alpha$.
   d) P(tails given the alternative hypothesis) $=$ power $= 0.70$
   e) Spin the coin more than once and base the decision on the sample proportion of heads.

33. a) 0.0464   b) Type I   c) 37.6%
   d) Increase the number of shots. Or keep the number of shots at 10, but increase alpha by declaring that 8, 9, or 10 will be deemed as having improved.

Chapter 22

1. It's very unlikely that samples would show an observed difference this large if in fact there is no real difference in the proportions of boys and girls who have used online social networks.

3. The ads may be working. If there had been no real change in name recognition, there'd be only about a 3% chance the percentage of voters who heard of this candidate would be at least this much higher in a different sample.

5. The responses are not from two independent groups, but are from the same individuals.

7. a) Stratified   b) 6% higher among males   c) 4%
   d)

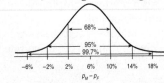

   e) Yes; a poll result showing little difference is only 1–2 standard deviations below the expected outcome.

9. a) Yes. Random sample; less than 10% of the population; samples are independent; more than 10 successes and failures in each sample.
   b) (0.055, 0.140)
   c) We are 95% confident, based on these samples, that the proportion of American women age 65 and older who suffer from arthritis is between 5.5% and 14.0% more than the proportion of American men of the same age who suffer from arthritis.
   d) Yes; the entire interval lies above 0.

11. a) 0.035   b) (0.356, 0.495)
   c) We are 95% confident, based on these data, that the proportion of pets with a malignant lymphoma in homes where herbicides are used is between 35.6% and 49.5% higher than the proportion of pets with lymphoma in homes where no pesticides are used.

13. a) Yes, subjects were randomly divided into independent groups, and more than 10 successes and failures were observed in each group.

  b) (4.7%, 8.9%)

  c) Yes, we're 95% confident that the rate of infection is 5–9 percentage points lower. That's a meaningful reduction, considering the 20% infection rate among the unvaccinated kids.

15. a) $H_0: p_V - p_{NV} = 0, H_A: p_V - p_{NV} < 0$.

  b) Because 0 is not in the confidence interval, reject the null. There's evidence that the vaccine reduces the rate of ear infections.

  c) 2.5%   d) Type I

  e) Babies would be given ineffective vaccinations.

17. a) Prospective study

  b) $H_0: p_1 - p_2 = 0; H_A: p_1 - p_2 \neq 0$ where p_1 is the proportion of students whose parents disapproved of smoking who became smokers and p_2 is the proportion of students whose parents are lenient about smoking who became smokers.

  c) Yes. We assume the students were randomly selected; they are less than 10% of the population; samples are independent; at least 10 successes and failures in each sample.

  d) $z = -1.17$, P-value $= 0.2422$. These samples do not show evidence that parental attitudes influence teens' decisions to smoke.

  e) If there is no difference in the proportions, there is about a 24% chance of seeing the observed difference or larger by natural sampling variation.

  f) Type II

19. a) $(-0.065, 0.221)$

  b) We are 95% confident that the proportion of teens whose parents disapprove of smoking who will eventually smoke is between 22.1% less and 6.5% more than for teens with parents who are lenient about smoking.

  c) 95% of all random samples will produce intervals that contain the true difference.

21. a) No; subjects weren't assigned to treatment groups. It's an observational study.

  b) $H_0: p_1 - p_2 = 0; H_A: p_1 - p_2 \neq 0$. $z = 3.56$, P-value $= 0.0004$. With a P-value this low, we reject H_0. There is a significant difference in the clinic's effectiveness. Younger mothers have a higher birth rate than older mothers. Note that the Success/Failure Condition is met based on the pooled estimate of p.

  c) We are 95% confident, based on these data, that the proportion of successful live births at the clinic is between 10.0% and 27.8% higher for mothers under 38 than in those 38 and older. However, the Success/Failure Condition is not met for the older women, since # Successes < 10. We should be cautious in trusting this confidence interval.

23. a) $H_0: p_1 - p_2 = 0; H_A: p_1 - p_2 > 0$. $z = 1.18$, P-value $= 0.118$. With a P-value this high, we fail to reject H_0. These data do not show evidence of a decrease in the voter support for the candidate.

  b) Type II

  c) No need to do anything.

25. a) $H_0: p_1 - p_2 = 0; H_A: p_1 - p_2 \neq 0$. $z = -0.39$, P-value $= 0.6951$. With a P-value this high, we fail to reject H_0. There is no evidence of racial differences in the likelihood of multiple births, based on these data.

  b) Type II

27. a) We are 95% confident, that between 67.0% and 83.0% of patients with joint pain will find medication A effective.

  b) We are 95% confident, that between 51.9% and 70.3% of patients with joint pain will find medication B effective.

  c) Yes, they overlap. This might indicate no difference in the effectiveness of the medications. (Not a proper test.)

  d) We are 95% confident that the proportion of patients with joint pain who will find medication A effective is between 1.7% and 26.1% higher than the proportion who will find medication B effective.

  e) No. There appears to be a difference in the effectiveness of the medications.

  f) To estimate the variability in the difference of proportions, we must add variances. The two one-sample intervals do not. The two-sample method is the correct approach.

29. The conditions are satisfied to test $H_0: p_{young} = p_{old}$ against $H_A: p_{young} > p_{old}$. The one-sided P-value is 0.0619, so we may reject the null hypothesis. Although the evidence is not strong, *Time* may be justified in saying that younger men are more comfortable discussing personal problems.

31. Yes. With a low P-value of 0.003, reject the null hypothesis of no difference. There's evidence of an increase in the proportion of parents checking the websites visited by their teens.

Part V Review

1. H_0: There is no difference in cancer rates, $p_1 - p_2 = 0$. H_A: The cancer rate in those who use the herb is higher, $p_1 - p_2 > 0$.

3. a) 10.29

  b) Not really. The z-score is -1.11. Not any evidence to suggest that the proportion for Monday is low.

  c) Yes. The z-score is 2.26 with a P-value of 0.024 (two-sided).

  d) Some births are scheduled for the convenience of the doctor and/or the mother.

5. a) $H_0: p_1 = 0.40; H_A: p_1 < 0.40$

  b) Random sample; less than 10% of all California gas stations, $0.4(27) = 10.8, 0.6(27) = 16.2$. Assumptions and conditions are met.

  c) $z = -1.49$, P-value $= 0.0677$

  d) With a P-value this high, we fail to reject H_0. These data do not provide sufficient evidence that the proportion of leaking gas tanks is less than 40% (or that the new program is effective in decreasing the proportion).

  e) Yes, Type II.

  f) Increase α, increase the sample size.

  g) Increasing α—increases power, lowers chance of Type II error, but increases chance of Type I error. Increasing sample size—increases power, costs more time and money.

7. a) The researcher believes that the true proportion of "A's" is within 10% of the estimated 54%, namely, between 44% and 64%.
   b) Small sample
   c) No, 63% is contained in the interval.

9. a) Pew uses a 95% confidence level. We can be 95% confident that the true proportion is within 2% of 13%—that is, that it is between 11% and 15%.
   b) The cell phone group would have the larger ME because its sample size is smaller.
   c) CI = (78.5%, 85.5%)
   d) The ME is 0.035, which is larger than the 2% ME in part a, largely because of the smaller sample size. It is larger than the ME in part b, mostly because 0.82 is smaller than 0.87 and proportions closer to 0.50 have larger MEs.

11. a) Bimodal!
    b) μ, the population mean. Sample size does not matter.
    c) $\sigma/\sqrt{n}$; sample size does matter.
    d) It becomes closer to a Normal model and narrower as the sample size increases.

13. a) $\mu = 0.80, \sigma = 0.028$
    b) Yes. $0.8(200) = 160, 0.2(200) = 40$. Both ≥ 10.
    c)

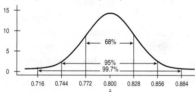

    d) 0.039

15. H_0: There is no difference, $p_1 - p_2 = 0$. H_A: Early births have increased, $p_1 - p_2 < 0$. $z = -0.729$, P-value = 0.2329. Because the P-value is so high, we do not reject H_0. These data do not show an increase in the incidence of early birth of twins.

17. a) H_0: There is no difference, $p_1 - p_2 = 0$. H_A: Treatment prevents deaths from eclampsia, $p_1 - p_2 < 0$.
    b) Samples are random and independent; less than 10% of all pregnancies (or eclampsia cases); more than 10 successes and failures in each group.
    c) 0.8008
    d) There is insufficient evidence to conclude that magnesium sulfate is effective in preventing eclampsia deaths.
    e) Type II   f) Increase the sample size, increase α.
    g) Increasing sample size: decreases variation in the sampling distribution, is costly. Increasing α: Increases likelihood of rejecting H_0, increases chance of Type I error.

19. a) It is not clear what the pollster asked. Otherwise they did fine.
    b) Stratified sampling.   c) 4%
    d) 95%   e) Smaller sample size.
    f) Wording and order of questions (response bias).

21. a) H_0: There is no difference, $p = 0.143$. H_A: The fatal accident rate is lower in girls, $p < 0.143$. $z = -1.67$, P-value = 0.0479. Because the P-value is low, we reject H_0. These data give some evidence that the fatal accident rate is lower for girls than for teens in general.
    b) If the proportion is really 14.3%, we will see the observed proportion (11.3%) or lower 4.8% of the time by sampling variation.

23. a) One would expect many small fish, with a few large ones.
    b) We don't know the exact distribution, but we know it's not Normal.
    c) Probably not. With a skewed distribution, a sample size of five is not a large enough sample to say the sampling model for the mean is approximately Normal.
    d) 0.961

25. a) Yes. $0.8(60) = 48, 0.2(60) = 12$. Both are ≥ 10.
    b) 0.834
    c) Higher. Bigger sample means smaller standard deviation for $\hat{p}$.
    d) Answers will vary. For $n = 500$, the probability is 0.997.

27. a) 54.4% to 62.5%
    b) Based on this study, with 95% confidence the proportion of Crohn's disease patients who will respond favorably to infliximab is between 54.4% and 62.5%.
    c) 95% of all such random samples will produce confidence intervals that contain the true proportion of patients who respond favorably.

29. At least 423, assuming that p is near 50%.

31. a) Random sample (?); certainly less than 10% of all preemies and normal babies; more than 10 failures and successes in each group. 1.7% to 16.3% greater for normal-birth-weight children.
    b) Since 0 is not in the interval, there is evidence that preemies have a lower high school graduation rate than children of normal birth weight.
    c) Type I, since we rejected the null hypothesis.

33. a) H_0: The computer is undamaged. H_A: The computer is damaged.
    b) 20% of good PCs will be classified as damaged (bad), while all damaged PCs will be detected (good).
    c) 3 or more.   d) 20%
    e) By switching to two or more as the rejection criterion, 7% of the good PCs will be misclassified, but only 10% of the bad ones will, increasing the power from 20% to 90%.

35. The null hypothesis is that Bush's disapproval proportion is 66%—the Nixon benchmark. The one-tailed test has a z-value of −2.00, so the P-value is 0.0228. It looks like Bush's May 2007 ratings were better than the Nixon benchmark low.

37. a) The company is interested only in confirming that the athlete is well known.
    b) Type I: the company concludes that the athlete is well known, but that's not true. It offers an endorsement contract to someone who lacks name recognition. Type II: the company overlooks a well-known athlete, missing the opportunity to sign a potentially effective spokesperson.
    c) Type I would be more likely, Type II less likely.

39. I am 95% confident that the proportion of U.S. adults who favor nuclear energy is between 7 and 19 percentage points higher than the proportion who would accept a nuclear plant near their area.

Chapter 23

1. a) 1.74   b) 2.37   c) 0.0524   d) 0.0889

3. Shape becomes closer to Normal; center does not change; spread becomes narrower.

5. a) The confidence interval is for the population mean, not the individual cows in the study.
   b) The confidence interval is not for individual cows.
   c) We *know* the average gain in this study was 56 pounds!
   d) The average weight gain of all cows does not vary. It's what we're trying to estimate.
   e) No. There is not a 95% chance for another sample to have an average weight gain between 45 and 67 pounds. There is a 95% chance that another sample will have its average weight gain within two standard errors of the true mean.

7. a) No. A confidence interval is not about individuals in the population.
   b) No. It's not about individuals in the sample, either.
   c) No. We know the mean cost for students in the sample was $1196.
   d) No. A confidence interval is not about other sample means.
   e) Yes. A confidence interval estimates a population parameter.

9. a) Based on this sample, we can say, with 95% confidence, that the mean pulse rate of adults is between 70.9 and 74.5 beats per minute.
   b) 1.8 beats per minute
   c) Larger

11. The assumptions and conditions for a *t*-interval are not met. The distribution is highly skewed to the right and there is a large outlier.

13. a) Yes. Randomly selected group; less than 10% of the population; the histogram is not unimodal and symmetric, but it is not highly skewed and there are no outliers, so with a sample size of 52, the CLT says $\bar{y}$ is approximately Normal.
   b) (98.06, 98.51) degrees F
   c) We are 98% confident, based on the data, that the average body temperature for an adult is between 98.06°F and 98.51°F.
   d) 98% of all such random samples will produce intervals containing the true mean temperature.
   e) These data suggest that the true normal temperature is somewhat less than 98.6°F.

15. a) Narrower. A smaller margin of error, so less confident.
   b) Advantage: more chance of including the true value. Disadvantage: wider interval.
   c) Narrower; due to the larger sample, the SE will be smaller.
   d) About 252 (256 if you rounded up at each step).

17. a) (709.90, 802.54)
   b) With 95% confidence, based on these data, the speed of light is between 299,709.9 and 299,802.5 km/sec.
   c) Normal model for the distribution, independent measurements. These seem reasonable here, but it would be nice to see if the Nearly Normal Condition held for the data.

19. a) Given no time trend, the monthly on-time departure rates should be independent. Though not a random sample, these months should be representative, and they're fewer than 10% of all months. The histogram looks unimodal, but slightly left-skewed; not a concern with this large sample.
   b) $80.17 < \mu(\text{OT Departure\%}) < 81.35$
   c) We can be 90% confident that the interval from 80.17% to 81.35% holds the true mean monthly percentage of on-time flight departures.

21. The 95% confidence interval lies entirely above the 0.08 ppm limit, evidence that mirex contamination is too high and consistent with rejecting the null. We used an upper-tail test, so the P-value should therefore be smaller than $\frac{1}{2}(1 - 0.95) = 0.025$, and it was.

23. If in fact the mean cholesterol of pizza eaters does not indicate a health risk, then only 7 of every 100 samples would have mean cholesterol levels as high (or higher) as observed in this sample.

25. a) Upper-tail. We want to show it will hold 500 pounds (or more) easily.
   b) They will decide the stands are safe when they're not.
   c) They will decide the stands are unsafe when they are in fact safe.

27. a) Decrease α. This means a smaller chance of declaring the stands safe if they are not.
   b) The probability of correctly detecting that the stands are capable of holding more than 500 pounds.
   c) Decrease the standard deviation—probably costly. Increase the sample size—takes more time for testing and is costly. Increase α—more Type I errors. Increase the "design load" to be well above 500 pounds—again, costly.

29. a) $H_0: \mu = 23.3$; $H_A: \mu > 23.3$
   b) We have a random sample of the population. Population may not be normally distributed, as it would be easier to have a few much older men at their first marriage than some very young men. However, with a sample size of 40, $\bar{y}$ should be approximately Normal. We should check the histogram for severity of skewness and possible outliers.
   c) $(\bar{y} - 23.3)/(s/\sqrt{40}) \sim t_{39}$
   d) 0.1447
   e) If the average age at first marriage is still 23.3 years, there is a 14.5% chance of getting a sample mean of 24.2 years or older simply from natural sampling variation.
   f) We lack evidence that the average age at first marriage has increased from the mean of 23.3 years.

31. a) Probably a representative sample; the Nearly Normal Condition seems reasonable. (Show a Normal probability plot or histogram.) The histogram is nearly uniform, with no outliers or skewness.
   b) $\bar{y} = 28.78, s = 0.40$
   c) (28.36, 29.21) grams
   d) Based on this sample, we are 95% confident the average weight of the content of Ruffles bags is between 28.36 and 29.21 grams.

e) The company is erring on the safe side, as it appears that, on average, it is putting in slightly more chips than stated.

33. a) Type I; he mistakenly rejected the null hypothesis that $p = 0.10$ (or worse).
b) Yes. These are a random sample of bags and the Nearly Normal Condition is met (Show a Normal probability plot or histogram.); $t = -2.51$ with 7 df for a one-sided P-value of 0.0203.

35. a) Random sample; the Nearly Normal Condition seems reasonable from a Normal probability plot. The histogram is roughly unimodal and symmetric with no outliers. (Show plot.)
b) (1187.9, 1288.4) chips
c) Based on this sample, the mean number of chips in an 18-ounce bag is between 1187.9 and 1288.4, with 95% confidence. The *mean* number of chips is clearly greater than 1000. However, if the claim is about individual bags, then it's not necessarily true. If the mean is 1188 and the SD deviation is near 94, then 2.5% of the bags will have fewer than 1000 chips, using the Normal model. If in fact the mean is 1288, the proportion below 1000 will be about 0.1%, but the claim is still false.

37. a) The Normal probability plot is relatively straight, with one outlier at 93.8 sec. Without the outlier, the conditions seem to be met. The histogram is roughly unimodal and symmetric with no other outliers. (Show your plot.)
b) $t = -2.63$, P-value $= 0.0160$. With the outlier included, we might conclude that the mean completion time for the maze is not 60 seconds; in fact, it is less.
c) $t = -4.46$, P-value $= 0.0003$. Because the P-value is so small, we reject H_0. Without the outlier, we see strong evidence that the average completion time for the maze is less than 60 seconds. The outlier here did not change the conclusion.
d) The maze does not meet the "one-minute average" requirement. Both tests rejected a null hypothesis of a mean of 60 seconds.

39. a) $285.1 < \mu(\text{Drive Distance}) < 287.6$
b) These data are not a random sample of golfers. The top professionals are (unfortunately) not representative and were not selected at random. We might consider the 2009 data to represent the population of all professional golfers, past, present, and future.
c) The data are means for each golfer, so they are less variable than if we looked at all the separate drives.

Chapter 24

1. Yes. The high P-value means that we lack evidence of a difference, so 0 is a possible value for $\mu_{Meat} - \mu_{Beef}$.

3. a) Plausible values of $\mu_{Meat} - \mu_{Beef}$ are all negative, so the mean fat content is probably higher for beef hot dogs.
b) The difference is significant.
c) 10%

5. a) False. The confidence interval is about means, not about individual hot dogs.

b) False. The confidence interval is about means, not about individual hot dogs.
c) True.
d) False. CI's based on other samples will also try to estimate the true difference in population means; there's no reason to expect other samples to conform to this result.
e) True.

7. a) 2.927   b) Larger
c) Based on this sample, we are 95% confident that students who learn Math using the CPMP method will score, on average, between 5.57 and 11.43 points better on a test solving applied Algebra problems with a calculator than students who learn by traditional methods.
d) Yes; 0 is not in the interval.

9. a) H_0: $\mu_C - \mu_T = 0$ vs. H_A: $\mu_C - \mu_T \neq 0$
b) Yes. Groups are independent, though we don't know if students were randomly assigned to the programs. Sample sizes are large, so CLT applies.
c) If the means for the two programs are really equal, there is less than a 1 in 10,000 chance of seeing a difference as large as or larger than the observed difference just from natural sampling variation.
d) On average, students who learn with the CPMP method do significantly worse on Algebra tests that do not allow them to use calculators than students who learn by traditional methods.

11. a) (1.36, 4.64)
b) No; 5 minutes is beyond the high end of the interval.

13.

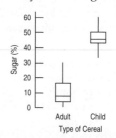

Random sample—questionable, but probably representative, independent samples, less than 10% of all cereals; boxplot shows no outliers—not exactly symmetric, but these are reasonable sample sizes. Based on these samples, with 95% confidence, children's cereals average between 32.15% and 40.82% more sugar content than adult's cereals.

15. H_0: $\mu_N - \mu_C = 0$ vs. H_A: $\mu_N - \mu_C > 0$; $t = 2.207$; P-value $= 0.0168$; df $= 33.4$. Because of the small P-value, we reject H_0. These data do suggest that new activities are better. The mean reading comprehension score for the group with new activities is significantly (at $\alpha = 0.05$) higher than the mean score for the control group. (*Tukey's test isn't appropriate because one group holds both the largest and smallest elements. The rank sum test has a $z = 3.7333$, $P = 0.0002$ and rejects the null hypothesis of equal medians.)

17. a)

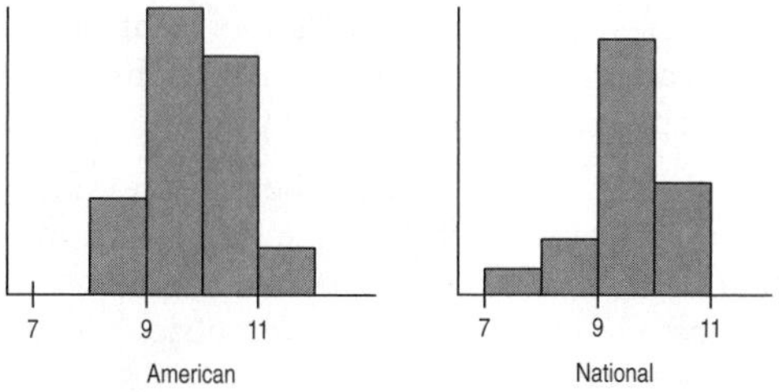

The histograms are labeled "American" (x-axis: 7, 9, 11) and "National" (x-axis: 7, 9, 11).

Both are unimodal and reasonably symmetric.
b) Based on these data, the average number of runs in an American League stadium is between 9.36 and 10.23, with 95% confidence.
c) No. The boxplot indicates it isn't an outlier.
d) We want to work directly with the average difference. The two separate confidence intervals do not answer questions about the difference. The difference has a different standard deviation, found by adding variances.

19. a) $(-0.18, 0.89)$
b) Based on these data, with 95% confidence, American League stadiums average between 0.18 fewer runs and 0.89 more runs per game than National League stadiums.
c) No; 0 is in the interval.

21. These are not two independent samples. These are before and after scores for the same individuals.

23. a) These data do not provide evidence of a difference in ad recall between shows with sexual content and violent content.
b) $H_0: \mu_S - \mu_N = 0$ vs. $H_A: \mu_S - \mu_N \neq 0$. $t = -6.08$, df $= 213.99$, P-value $= 5.5 \times 10^{-9}$. Because the P-value is low, we reject H_0. These data suggest that ad recall between shows with sexual and neutral content is different; those who saw shows with neutral content had higher average recall.

25. a) $H_0: \mu_V - \mu_N = 0$ vs. $H_A: \mu_V - \mu_N \neq 0$. $t = -7.21$, df $= 201.96$, P-value $= 1.1 \times 10^{-11}$. Because of the very small P-value, we reject H_0. There is a significant difference in mean ad recall between shows with violent content and neutral content; viewers of shows with neutral content remember more brand names, on average.
b) With 95% confidence, the average number of brand names remembered 24 hours later is between 1.45 and 2.41 higher for viewers of neutral content shows than for viewers of sexual content shows, based on these data.

27. $H_0: \mu_{big} - \mu_{small} = 0$ vs. $H_A: \mu_{big} - \mu_{small} \neq 0$; bowl size was assigned randomly; amount scooped by individuals and by the two groups should be independent. With 34.3 df, $t = 2.104$ and P-value $= 0.0428$. The low P-value leads us to reject the null hypothesis. There is evidence of a difference in the average amount of ice cream that people scoop when given a bigger bowl.

29. a) The 95% confidence interval for the difference is $(0.61, 5.39)$. 0 is not in the interval, so scores in 1996 were significantly higher. (Or the t, with more than 7500 df, is 2.459 for a P-value of 0.0070.)
b) Since both samples were very large, there shouldn't be a difference in how certain you are, assuming conditions are met.

31. Independent Groups Assumption: The runners are different women, so the groups are independent. The Randomization Condition is satisfied since the runners are selected at random for these heats.

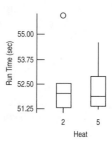

Nearly Normal Condition: The boxplots show an outlier, but we will proceed and then redo the analysis with the outlier deleted. When we include the outlier, $t = 0.035$ with a two-sided P-value of 0.97. With the outlier deleted, $t = -1.14$, with P $= 0.2837$. Either P-value is so large that we fail to reject the null hypothesis of equal means and conclude that there is no evidence of a difference in the mean times for runners in unseeded heats.

33. With $t = -4.57$ and a very low P-value of 0.0013, we reject the null hypothesis of equal mean velocities. There is strong evidence that golf balls hit off Stinger tees will have a higher mean initial velocity.

35. a) We can be 95% confident that the interval 74.8 ± 178.05 minutes includes the true difference in mean crossing times between men and women. Because the interval includes zero, we cannot be confident that there is any difference at all.
b) Independence Assumption: There is no reason to believe that the swims are not independent or that the two groups are not independent of each other.

Randomization Condition: The swimmers are not a random sample from any identifiable population, but they may be representative of swimmers who tackle challenges such as this.

Nearly Normal Condition: The boxplots show no outliers. The histograms are unimodal; the histogram for men is somewhat skewed to the right. (Show your graphs.)

37. a) $H_0: \mu_R - \mu_N = 0$ vs. $H_A: \mu_R - \mu_N < 0$. $t = -1.36$, df $= 20.00$, P-value $= 0.0945$. Because the P-value is large, we fail to reject H_0. These data show no evidence of a difference in mean number of objects recalled between listening to rap or no music at all.
b) Didn't conclude any difference.

Chapter 25

1. a) Randomly assign 50 hens to each of the two kinds of feed. Compare production at the end of the month.
b) Give all 100 hens the new feed for 2 weeks and the old feed for 2 weeks, randomly selecting which feed the hens get first. Analyze the differences in production for all 100 hens.
c) Matched pairs. Because hens vary in egg production, the matched-pairs design will control for that.

3. a) Show the same people ads with and without sexual images, and record how many products they remember in each group. Randomly decide which ads a person sees first. Examine the differences for each person.

b) Randomly divide volunteers into two groups. Show one group ads with sexual images and the other group ads without. Compare how many products each group remembers.

5. a) Matched pairs—same cities in different periods.

b) There is a significant difference (P-value = 0.0244) in the labor force participation rate for women in these cities; women's participation seems to have increased between 1968 and 1972.

7. a) Use the paired t-test because we have pairs of Fridays in 5 different months. Data from adjacent Fridays within a month may be more similar than data from randomly chosen Fridays.

b) We conclude that there is evidence (P-value 0.0212) that the mean number of cars found on the M25 motorway on Friday the 13th is less than on the previous Friday.

c) We don't know if these Friday pairs were selected at random. If these are the Fridays with the largest differences, this will affect our conclusion. The Nearly Normal Condition appears to be met by the differences, but the sample size is small.

9. Adding variances requires that the variables be independent. These price quotes are for the same cars, so they are paired. Drivers quoted high insurance premiums by the local company will be likely to get a high rate from the online company, too.

11. a) The histogram—we care about differences in price.

b) Insurance cost is based on risk, so drivers are likely to see similar quotes from each company, making the differences relatively smaller.

c) The price quotes are paired; they were for a random sample of fewer than 10% of the agent's customers; the histogram of differences looks approximately Normal.

13. $H_0: \mu(Local - Online) = 0$ vs. $H_A: \mu(Local - Online) > 0$; with 9 df, $t = 0.83$. With a high P-value of 0.215, we don't reject the null hypothesis. These data don't provide evidence that online premiums are lower, on average.

15.

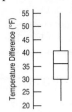

Data are paired for each city; cities are independent of each other; boxplot shows the temperature differences are reasonably symmetric, with no outliers. This is probably not a random sample, so we might be wary of inferring that this difference applies to all European cities. Based on these data, we are 90% confident that the average temperature in European cities in July is between 32.3°F and 41.3°F higher than in January.

17. Based on these data, we are 90% confident that boys, on average, can do between 1.6 and 13.0 more push-ups than girls (independent samples—not paired).

19. a) Paired sample test. Data are before/after for the same workers; workers randomly selected; assume fewer than 10% of all this company's workers; boxplot of differences shows them to be symmetric, with no outliers.

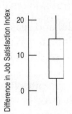

b) $H_0: \mu_D = 0$ vs. $H_A: \mu_D > 0$. $t = 3.60$, P-value = 0.0029. Because P < 0.01, reject H_0. These data show evidence that average job satisfaction has increased after implementation of the program.

c) Type I

*d) H_0: Median difference = 0, H_A Median difference > 0. P = 0.0547. This is sufficiently different from part b that we should examine the data for anomalies.

21. $H_0: \mu_D = 0$ vs. $H_A: \mu_D \neq 0$. Data are paired by brand; brands are independent of each other; fewer than 10% of all yogurts (questionable); boxplot of differences shows an outlier (100) for Great Value:

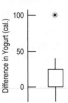

With the outlier included, the mean difference (Strawberry − Vanilla) is 12.5 calories with a t-stat of 1.332, with 11 df, for a P-value of 0.2098. Deleting the outlier, the difference is even smaller, 4.55 calories with a t-stat of only 0.833 and a P-value of 0.4241. With P-values so large, we do not reject H_0. We conclude that the data do not provide evidence of a difference in mean calories.

23. a) Cars were probably not a simple random sample, but may be representative in terms of stopping distance; boxplot does not show outliers, but does indicate right skewness. A 95% confidence interval for the mean stopping distance on dry pavement is (131.8, 145.6) feet.

b) Data are paired by car; cars were probably not randomly chosen, but representative; boxplot shows an outlier (car 4) with a difference of 12. With deletion of that car, a Normal probability plot of the differences is relatively straight.

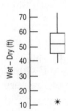

Retaining the outlier, we estimate with 95% confidence that the average braking distance is between 38.8 and 62.6 feet more on wet pavement than on dry, based on this sample. (Without the outlier, the confidence interval is 47.2 to 62.8 feet.)

25. a) Paired Data Assumption: Data are paired by college. Randomization Condition: This was a random sample of public colleges and universities. 10% Condition: these are fewer than 10% of all public colleges and universities.

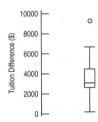

Normal Population Assumption: U.C. Irvine seems to be an outlier; we might consider removing it.

b) Having deleted the observation for U.C.-Irvine, whose difference of $9300 was an outlier, we are 90% confident, based on the remaining data, that nonresidents pay, on average, between $2615.31 and $3918.02 more than residents. If we retain the outlier, the interval is ($2759, $4409).

c) Assertion is reasonable; with or without the outlier, $3500 is in the confidence interval.

27. a) 60% is 30 strikes; $H_0: \mu = 30$ vs. $H_A: \mu > 30$. $t = 6.07$, P-value $= 3.92 \times 10^{-6}$. With a very small P-value, we reject H_0. There is very strong evidence that players can throw more than 60% strikes after training, based on this sample.

b) $H_0: \mu_D = 0$ vs. $H_A: \mu_D > 0$. $t = 0.135$, P-value $= 0.4472$. With such a high P-value, we do not reject H_0. These data provide no evidence that the program has improved pitching in these Little League players.

c) H_0: Median difference $= 0$, H_A: Median difference > 0. $P = 0.4073$.

29. a) The data are clearly paired. Both races may have improved over time, but the pairwise differences are likely to be independent. We can only check the Nearly Normal Condition by using the computer files.

b) With 95% confidence we can say the mean time difference is between -15.58 minutes (men are faster) and $+9.57$ minutes (women are faster).

c) The interval contains 0, so we would not reject the hypothesis of no mean difference at $\alpha = 0.05$. We can't discern a difference between the female wheelchair times and the male running times.

31. a) Same cows before and after injection; the cows should be representative of others of their breed; cows are independent of each other; less than 10% of all cows; don't know about Nearly Normal differences.

b) (12.66, 15.34)

c) Based on this sample, with 95% confidence, the average increase in milk production for Ayrshire cows given BST is between 12.66 and 15.34 pounds per day.

d) $0.25(47) = 11.75$. The average increase is much more than this, so we would recommend he incur the extra expense.

Chapter 26

1. a) Chi-square test of independence. We have one sample and two variables. We want to see if the variable *Account Type* is independent of the variable *Trade Type*.

b) Other test. *Account Size* is quantitative, not counts.

c) Chi-square test of homogeneity. We want to see if the distribution of one variable, *Courses*, is the same for two groups (resident and nonresident students).

3. a) 10   b) Goodness-of-fit

c) H_0: The die is fair (all faces have $p = 1/6$). H_A: The die is not fair.

d) Count data; rolls are random and independent; expected frequencies are all bigger than 5.

e) 5

f) $\chi^2 = 5.600$, P-value $= 0.3471$

g) Because the P-value is high, do not reject H_0. The data show no evidence that the die is unfair.

5. a) Weights are quantitative, not counts.

b) Count the number of each kind of nut, assuming the company's percentages are based on counts rather than weights.

7. H_0: The police force represents the population (29.2% white, 28.2% black, etc.). H_A: The police force is not representative of the population. $\chi^2 = 16516.88$, df $= 4$, P-value $= 0.0000$. Because the P-value is so low, we reject H_0. These data show that the police force is not representative of the population. In particular, there are too many white officers in relationship to their membership in the community.

9. a) $\chi^2 = 5.671$, df $= 3$, P-value $= 0.1288$. With a P-value this high, we fail to reject H_0. Yes, these data are consistent with those predicted by genetic theory.

b) $\chi^2 = 11.342$, df $= 3$, P-value $= 0.0100$. Because of the low P-value, we reject H_0. These data provide evidence that the distribution is not as specified by genetic theory.

c) With small samples, many more data sets will be consistent with the null hypothesis. With larger samples, small discrepancies will show evidence against the null hypothesis.

11. a) $96/16 = 6$

b) Goodness-of-fit

c) H_0: The number of large hurricanes remains constant over decades. H_A: The number of large hurricanes has changed.

d) 15

e) P-value $= 0.63$

f) The very high P-value means these data offer no evidence that the numbers of large hurricanes has changed.

g) The final period is only 6 years rather than 10 and already 7 large hurricanes have been observed. Perhaps this decade will have an unusually large number of such hurricanes.

13. a) Independence

b) H_0: Breastfeeding success is independent of having an epidural. H_A: There's an association between breastfeeding success and having an epidural.

15. a) 1   b) 159.34
   c) Breastfeeding behavior should be independent for these babies. They are fewer than 10% of all babies; we assume they are representative. We have counts, and all the expected counts are at least 5.

17. a) 5.90
   b) P-value < 0.005
   c) The P-value is very low, so reject the null. There's evidence of an association between having an epidural and subsequent success in breastfeeding.

19. a) $\dfrac{(190 - 159.34)}{\sqrt{159.34}} = 2.43$
   b) It appears that babies whose mothers had epidurals during childbirth are much less likely to be breastfeeding 6 months later.

21. These factors would not be mutually exclusive. There would be yes or no responses for every baby for each.

23. a) 40.2%   b) 8.1%   c) 62.2%   d) 285.48
   e) H_0: Survival was independent of status on the ship. H_A: Survival was not independent of the status.
   f) 3
   g) We reject the null hypothesis. Survival depended on status.
      We can see that first-class passengers were more likely to survive than passengers of any other class.

25. First class passengers were most likely to survive, while third-class passengers and crew were underrepresented among the survivors.

27. a) Experiment—actively imposed treatments (different drinks)
   b) Homogeneity
   c) H_0: The rate of urinary tract infection is the same for all three groups. H_A: The rate of urinary tract infection is different among the groups.
   d) Count data; random assignment to treatments; all expected frequencies larger than 5.
   e) 2
   f) $\chi^2 = 7.776$, P-value $= 0.020$.
   g) With a P-value this low, we reject H_0. These data provide reasonably strong evidence that there is a difference in urinary tract infection rates between cranberry juice drinkers, lactobacillus drinkers, and the control group.
   h) The standardized residuals are

	Cranberry	Lactobacillus	Control
Infection	−1.87276	1.19176	0.68100
No Infection	1.24550	−0.79259	−0.45291

   From the standardized residuals (and the sign of the residuals), it appears those who drank cranberry juice were less likely to develop urinary tract infections; those who drank lactobacillus were more likely to have infections.

29. a) Independence
   b) H_0: *Political Affiliation* is independent of *Sex*. H_A: There is a relationship between *Political Affiliation* and *Sex*.

c) Counted data; probably a random sample, but can't extend results to other states; all expected frequencies greater than 5.
   d) $\chi^2 = 4.851$, df $= 2$, P-value $= 0.0884$.
   e) Because of the high P-value, we do not reject H_0. These data do not provide evidence of a relationship between *Political Affiliation* and *Sex*.

31. H_0: *Political Affiliation* is independent of *Region*. H_A: There is a relationship between *Political Affiliation* and *Region*. $\chi^2 = 13.849$, df $= 4$, P-value $= 0.0078$. With a P-value this low, we reject H_0. *Political Affiliation* and *Region* are related. Examination of the residuals shows that those in the West are more likely to be Democrat than Republican; those in the Northeast are more likely to be Republican than Democrat.

33. a) Homogeneity
   b) H_0: The grade distribution is the same for both professors.
      H_A: The grade distributions are different.
   c)

	Prof. Alpha	Prof. Beta
A	6.667	5.333
B	12.778	10.222
C	12.222	9.778
D	6.111	4.889
F	2.222	1.778

   Three cells have expected frequencies less than 5.

35. a)

	Prof. Alpha	Prof. Beta
A	6.667	5.333
B	12.778	10.222
C	12.222	9.778
Below C	8.333	6.667

   All expected frequencies are now larger than 5.
   b) Decreased from 4 to 3.
   c) $\chi^2 = 9.306$, P-value $= 0.0255$. Because the P-value is so low, we reject H_0. The grade distributions for the two professors are different. Dr. Alpha gives fewer A's and more grades below C than Dr. Beta.

37. $\chi^2 = 14.058$, df $= 1$, P-value $= 0.0002$. With a P-value this low, we reject H_0. There is evidence of racial steering. Blacks are much less likely to rent in Section A than Section B.

39. a) $z = 3.74936$, $z^2 = 14.058$.
   b) P-value $(z) = 0.0002$ (same as in Exercise 25).

41. $\chi^2 = 5.89$, df $= 3$, P $= 0.117$. Because the P-value is >0.05, these data show no evidence of an association between the mother's age group and the outcome of the pregnancy.

Part VI Review

1. a) H_0: $\mu_{Jan} - \mu_{Jul} = 0$; H_A: $\mu_{Jan} - \mu_{Jul} \neq 0$. $t = -1.94$, df $= 43.68$, P-value $= 0.0590$. Since P-value is fairly low, reject the null. These data show some evidence of a difference in mean *Age* to crawl between January and July babies.
   b) H_0: $\mu_{Apr} - \mu_{Oct} = 0$; H_A: $\mu_{Apr} - \mu_{Oct} \neq 0$. $t = -0.92$; df $= 59.40$; P-value $= 0.3610$. Since P-value is high, do

not reject the null; these data do not provide evidence of a significant difference between April and October with regard to the mean age at which crawling begins.

c) These results are not consistent with the claim. (But note that temperatures are generally different between January and July, but similar in April and October.)

3. $H_0: p = 0.26$; $H_A: p \neq 0.26$. $z = 0.946$; P-value $= 0.3443$. Because the P-value is high, we do not reject H_0. These data do not provide evidence that the Denver area rate is different from the national rate in the proportion of businesses with women owners.

5. Based on these data we are 95% confident that the mean difference in aluminum oxide content is between -3.37 and 1.65. The means in aluminum oxide content of the pottery made at the two sites could reasonably be the same.

7. a) $H_0: P_{\text{Varsity} | \text{ALS}} = P_{\text{nonVarsity} | \text{ALS}}$; $H_A: P_{\text{Varsity} | \text{ALS}} > P_{\text{nonVarsity} | \text{ALS}}$. $z = 2.52$; P-value $= 0.0058$. With such a low P-value, we reject H_0. This is strong evidence that there is a higher proportion of varsity athletes among ALS patients than among non-ALS patients.

b) Observational retrospective study. To make the inference one must assume the patients studied are representative.

9. H_0: The proportions are as specified by the ratio 1:3:3:9; H_A: The proportions are not as stated. $\chi^2 = 5.01$; df $= 3$; P-value $= 0.1711$. Since P > 0.05, we fail to reject H_0. These data do not provide evidence to indicate that the proportions are other than 1:3:3:9.

11. $H_0: \mu = 7.41$; $H_A: \mu \neq 7.41$. $t = 2.18$; df $= 111$; P-value $= 0.0313$. With such a low P-value, we reject H_0. Assuming that Missouri babies fairly represent the United States, these data show that American babies are different from Australian babies in birth weight; American babies are heavier, on average.

13. a) If there is no difference in the average fish sizes, the chance of seeing an observed difference this large just by natural sampling variation is 0.1%.

b) If cost is justified, feed them a natural diet.

c) Type I

15. $\chi^2 = 6.14$; P-value $= 0.1887$. Since P > 0.05, we do not reject H_0. These data do not provide evidence of an association between *Duration* of pregnancy and *Level* of care.

17. a) Assuming the conditions are met, from these data we are 95% confident that patients with cardiac disease average between 3.39 and 5.01 years older than those without cardiac disease.

b) Older patients are at greater risk from a variety of other health issues, and perhaps more depressed.

19. H_0: *Income* and *Party* are independent. H_A: *Income* and *Party* are not independent. $\chi^2 = 17.19$ with 4 df; P-value $= 0.0018$. With such a small P-value, we reject H_0. These data provide evidence that income level and party are not independent. Examination of components shows Democrats are most likely to have low incomes, Independents are most likely to have middle incomes, and Republicans most likely to have high incomes.

21. $H_0: p_{\text{VT}} - p_{\text{NH}} = 0$; $H_A: p_{\text{VT}} - p_{\text{NH}} \neq 0$. $z = -0.59$; P-value $= 0.5563$. With such a high P-value, we do not reject H_0. These data show no evidence of a difference in the rates of cesarean deliveries between Vermont and New Hampshire.

23. $H_0: \mu = \$10$; $H_A: \mu > \$10$. $t = 0.66$; df $= 13$; P-value $= 0.26$. With such a high P-value, we do not reject H_0. These data do not provide evidence that he is likely to overspend his budget of $10 per day, provided that these days are representative.

25. $H_0: \mu_{\text{Cert}} - \mu_{\text{UC}} = 0$; $H_A: \mu_{\text{Cert}} - \mu_{\text{UC}} > 0$. $t = 1.57$; df $= 86$; P-value $= 0.0598$. The P-value is just greater than 0.05. Although there may be some indication that students of certified teachers achieve higher mean reading scores than students of uncertified teachers, we cannot reject the null hypothesis at the 5% level.

27. Data are matched pairs (before and after for the same rooms); less than 10% of all rooms in a large hotel; uncertain how these rooms were selected (are they representative?). Histogram shows that differences are roughly unimodal and symmetric with no outliers. A 95% confidence interval for the difference, before $-$ after is (0.58, 2.65) counts. Since the entire interval is above 0, these data show that the new air-conditioning system was effective in reducing average bacteria counts.

29. a) We are 95% confident that between 19.77% and 38.66% of children with bipolar symptoms will be helped with medication and psychotherapy, based on this study.

b) 221 children

31. a) From this histogram, about 115 loaves or more. (Not Normal.)

b) Large sample size; CLT says $\bar{y}$ will be approximately Normal.

c) From the data, we are 95% confident that the bakery will sell between 101.2 and 104.8 loaves of bread on an average day.

d) 25

e) Yes, 100 loaves per day is too low—the entire confidence interval is above that.

33. a) $H_0: p_{\text{High}} - p_{\text{Low}} = 0$; $H_A: p_{\text{High}} - p_{\text{Low}} \neq 0$. $z = -3.57$; P-value $= 0.0004$. Because the P-value is so low, we reject H_0. These data show the IRS risk is different in the two groups; people who consume dairy products often have a lower risk on average.

b) Doesn't prove it. Association does not demonstrate causation.

35. Based on these data, we are 95% confident that seeded clouds will produce an average of between -4.76 and 559.56 more acre-feet of rain than unseeded clouds. Since the interval contains negative values, it may be that seeding is unproductive.

37. a) Randomizing order of the tasks helps avoid bias and memory effects. Randomizing the cards helps avoid bias as well.

b) $H_0: \mu_D = 0$; $H_A: \mu_D \neq 0$

c)

Boxplot of the differences looks symmetric with no outliers.

d) $t = -1.70$ on 31 df; P-value $= 0.0999$; do not reject H_0 because the P-value is high.

e) The data do not provide evidence that the color or written word dominates.

39. a) Different samples give different means; this is a fairly small sample. The difference may be due to natural sampling variation.

b) $H_0: \mu = 100$; $H_A: \mu < 100$

c) Batteries selected are a SRS (representative); less than 10% of the company's batteries; lifetimes are approximately Normal.

d) $t = -1.0$ on 15 df; P-value $= 0.1666$; do not reject H_0. This sample does not show that the average life of the batteries is significantly less than 100 hours.

e) Type II.

41. $\chi^2 = 40.715$ on 2df; P-value < 0.0001. These data do indicate an association between *Education* and *Family Planning*. More educated women have fewer unplanned pregnancies.

Chapter 27

1. a) $\widehat{Error} = 453.22 - 8.37\ YearSince1970$; according to the model, the error made in predicting a hurricane's path was about 453 nautical miles, on average, in 1970. It has been declining at a rate of about 8.37 nautical miles per year.

b) $H_0: \beta_1 = 0$; there has been no change in prediction accuracy. $H_A: \beta_1 \neq 0$; there has been a change in prediction accuracy.

c) With a P-value < 0.001, I reject the null hypothesis and conclude that prediction accuracies have in fact been changing during this period.

d) 58.5% of the variation in hurricane prediction accuracy is accounted for by this linear model on time.

3. a) $\widehat{Budget} = -63.998 + 1.026\ RunTime$. The model suggests that movies cost about $1 million per minute to make.

b) A negative starting value makes no sense, but it can be a starting value for the model.

c) Amounts by which movie costs differ from predictions made by this model vary, with a standard deviation of about $33 million.

d) 0.154 $m/min

e) If we constructed other models based on different samples of movies, we'd expect the slopes of the regression lines to vary, with a standard deviation of about $154,000 per minute.

5. a) The scatterplot looks straight enough, the residuals look random and nearly normal, and the residuals don't display any clear change in variability.

b) I'm 95% confident that the cost of making longer movies increases at a rate of between 0.72 and 1.325 million dollars per additional minute.

7. a) $H_0: \beta_1 = 0$; there's no association between calories and sodium content in all-beef hot dogs. $H_A: \beta_1 \neq 0$: there is an association.

b) Based on the low P-value (0.0018), I reject the null. There is evidence of an association between the number of calories in all-beef hot dogs and their sodium contents.

9. a) Among all-beef hot dogs with the same number of calories, the sodium content varies, with a standard deviation of about 60 mg.

b) 0.561 mg/cal

c) If we tested many other samples of all-beef hot dogs, the slopes of the resulting regression lines would be expected to vary, with a standard deviation of about 0.56 mg of sodium per calorie.

11. I'm 95% confident that for every additional calorie, all-beef hot dogs have, on average, between 1.07 and 3.53 mg more sodium.

13. a) H_0: Difference in age at first marriage has not been changing, $\beta_1 = 0$. H_A: Difference in age at first marriage has been changing, $\beta_1 \neq 0$.

b) Residual plot shows possible slight bend; histogram is unimodal and a bit skewed, unimodal and a bit skewed, but shows no obvious skewness or outliers.

c) $t = -9.11$, P-value < 0.0001. With such a low P-value, we reject H_0. These data show evidence that difference in age at first marriage is decreasing.

15. Based on these data, we are 95% confident that the average difference in age at first marriage is decreasing at a rate between 0.024 and 0.039 years per year.

17. a) H_0: *Fuel Economy* and *Weight* are not (linearly) related, $\beta_1 = 0$. H_A: *Fuel Economy* changes with *Weight*, $\beta_1 \neq 0$. P-value < 0.0001, indicating strong evidence of an association.

b) Yes, the conditions seem satisfied. Histogram of residuals is unimodal and symmetric; residual plot looks OK, but some "thickening" of the plot with increasing values.

c) $t = -12.2$, P-value < 0.0001. These data show evidence that *Fuel Economy* decreases with the *Weight* of the car.

19. a) $(-9.57, -6.86)$ mpg per 1000 pounds.

b) Based on these data, we are 95% confident that *Fuel Efficiency* decreases between 6.86 and 9.57 miles per gallon, on average, for each additional 1000 pounds of *Weight*.

21. a) We are 95% confident that 2500-pound cars will average between 27.34 and 29.07 miles per gallon.

b) Based on the regression, a 3450-pound car will get between 15.44 and 25.36 miles per gallon, with 95% confidence.

23. a) Yes. $t = 2.73$, P-value $= 0.0079$. With a P-value so low, we reject H_0. There is a positive relationship between *Calories* and *Sodium* content.

b) No. $R^2 = 9\%$ and s appears to be large, although without seeing the data, it is a bit hard to tell.

25. Plot of *Calories* against *Fiber* does not look linear; the residuals plot also shows increasing variance as predicted values get large. The histogram of residuals is right skewed.

27. a) H_0: No (linear) relationship between *BCI* and *pH*, $\beta_1 = 0$. H_A: There seems to be a relationship, $\beta_1 \neq 0$.
  b) $t = -7.73$ with 161 df; P-value < 0.0001
  c) There seems to be a negative relationship; *BCI* decreases as *pH* increases at an average of 197.7 *BCI* units per increase of 1 *pH*.

29. a) H_0: No linear relationship between *Population* and *Ozone*, $\beta_1 = 0$. H_A: *Ozone* increases with *Population*, $\beta_1 > 0$. $t = 3.48$, P-value $= 0.0018$. With a P-value so low, we reject H_0. These data show evidence that *Ozone* increases with *Population*.
  b) Yes, *Population* accounts for 84% of the variability in *Ozone* level, and *s* is just over 5 parts per million.

31. a) Based on this regression, each additional million residents corresponds to an increase in average ozone level of between 3.29 and 10.01 ppm, with 90% confidence.
  b) The mean *Ozone* level for cities with 600,000 people is between 18.47 and 27.29 ppm, with 90% confidence.

33. a) 33 batteries.
  b) Yes. The scatterplot is roughly linear with lots of scatter; plot of residuals vs. predicted values shows no overt patterns; Normal probability plot of residuals is reasonably straight.
  c) H_0: No linear relationship between *Cost* and *Cranking Amps*, $\beta_1 = 0$. H_A: *Cranking Amps* increase with cost, $\beta_1 > 0$. $t = 3.23$; P-value $= \frac{1}{2}(0.0029) = 0.00145$. With a P-value so low, we reject H_0. These data provide evidence that more expensive batteries do have more cranking amps.
  d) No. $R^2 = 25.2\%$ and $s = 116$ amps. Since the range of amperage is only about 400 amps, an *s* of 116 is not very useful.
  e) *Cranking amps* $= 384.59 + 4.15 \times$ *Cost*.
  f) $(1.97, 6.32)$ cold cranking amps per dollar.
  g) *Cranking amps* increase, on average, between 1.97 and 6.32 per dollar of battery *Cost* increase, with 90% confidence.

35. a) H_0: No linear relationship between *Waist* size and *%Body Fat*, $\beta_1 = 0$. H_A: *%Body Fat* changes with *Waist* size, $\beta_1 \neq 0$. $t = 8.14$; P-value < 0.0001. There's evidence that *%Body Fat* seems to increase with *Waist* size.
  b) With 95% confidence, mean *%Body Fat* for people with 40-inch waists is between 23.58 and 29.02, based on this regression.

37. a) The regression model is $\widehat{Midterm2} = 12.005 + 0.721$ *Midterm1*

	Estimate	Std Error	t-ratio	P-value
Intercept	12.00543	15.9553	0.752442	0.454633
Slope	0.72099	0.183716	3.924477	0.000221

RSquare	0.198982
s	16.78107
n	64

b) The scatterplot shows a weak, somewhat linear, positive relationship. There are several outlying points, but removing them only makes the relationship slightly stronger. There is no obvious pattern in the residual plot. The regression model appears appropriate. The small P-value for the slope shows that the slope is statistically distinguishable from 0 even though the R^2 value of 0.199 suggests that the overall relationship is weak.
  c) No. The R^2 value is only 0.199 and the value of *s* of 16.8 points indicates that he would not be able to predict performance on *Midterm2* very accurately.

39. H_0: Slope of *Effectiveness* vs. *Initial Ability* $= 0$; H_A: Slope $\neq 0$

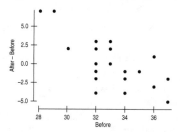

Scatterplot is straight enough. Regression conditions appear to be met. $t = -4.34$, df $= 19$, *P-value* $= 0.0004$. With a P-value this small, we reject the null hypothesis. There is strong evidence that the effectiveness of the video depends on the player's initial ability. The negative slope observed that the method is more effective for those whose initial performance was poorest and less so for those whose initial performance was better. This looks like a case of regression to the mean. Those who were above average initially tended to be worse after training. Those who were below average initially tended to improve.

41. a) Data plot looks linear; no overt pattern in residuals; histogram of residuals roughly symmetric and unimodal.
  b) H_0: No linear relationship between *Education* and *Mortality*, $\beta_1 = 0$. H_A: $\beta_1 \neq 0$. $t = -6.24$; P-value < 0.001. There is evidence that cities in which the mean education level is higher also tend to have a lower mortality rate.
  c) No. Data are on cities, not individuals. Also, these are observational data. We cannot predict causal consequences from them.
  d) $(-65.95, -33.89)$ deaths per 100,000 people.
  e) *Mortality* decreases, on average, between 33.89 and 65.95 deaths per 100,000 for each extra year of average *Education*.
  f) Based on the regression, the average *Mortality* for cities with an average of 12 years of *Education* will be between 874.239 and 914.196 deaths per 100,000 people.

43. a) $\widehat{Logit(Right\text{-}to\text{-}work)} = 6.19951 - 0.106155$ *publ* $- 0.222957$ *pvt*.
  b) 1.847
  c) 0.864
  d) -2.506
  e) 0.075

Chapter 28

1. a) $H_0: \mu_1 = \mu_2 = \mu_3 = \mu_4$ vs. the alternative that not all means are equal. Here μ_k refers to the mean number of popcorn kernels left unpopped for brand k.

b) MS_T has 3 df; MS_E has 12.

c) An $F_{3,12}$ of 13.56 has a P-value of 0.00037. The data provide strong evidence that the means are not all equal. The brands do not produce the same mean number of unpopped kernels.

d) I would like to see side-by-side boxplots of the treatment groups, a Normal probability plot of the residuals, a histogram of the residuals, and a plot of the residuals vs. the predicted values.

3. a) The null hypothesis is that the mean mileage driven on one liter of gas using each muffler is the same. The alternative is that not all means are equal.

b) MS_T has 2 df; MS_E has 21 df.

c) We would not reject the null hypothesis of equal means with a P-value of 0.1199. The data from this experiment provide no evidence that the muffler type affects gas mileage.

d) I would like to see side-by-side boxplots of the treatment groups, a normal probability plot of the residuals, a histogram of the residuals, and a plot of the residuals vs. the predicted values.

e) A Type II error.

5. a) $H_0: \mu_1 = \mu_2 = \mu_3 = \mu_4$ vs. the alternative that not all means are equal. Here μ_k refers to the mean activation time using recipe k.

b) The data provide strong evidence with a P-value < 0.0001 to reject the null hypothesis and conclude that the means are not all equal. This experiment provides strong evidence that the mean activation times differ among the recipes.

c) Yes, because we have rejected the null hypothesis, we can proceed with a multiple comparisons method to compare all the groups.

7. An ANOVA is not appropriate because *Eye Color* is a categorical variable. The students should consider analyzing the data with a χ^2 test of independence between eye and hair color.

9. a) $H_0: \mu_1 = \mu_2 = \mu_3 = \mu_4$ vs. the alternative that not all means are equal. Here μ_k refers to the mean mileage using cylinder level k.

b) The spreads of the four groups look very different. The similar variance condition is not met. The response variable might need to be re-expressed before proceeding with an Analysis of Variance.

11. a) $H_0: \mu_1 = \mu_2 = \mu_3 = \mu_4 = \mu_5 = \mu_6$ vs. the alternative that not all means are equal. Here μ_k refers to the mean *Time* it takes to serve a customer by *Teller k*.

b) The data do not provide evidence that the mean *Time* it takes to serve a customer differs by *Teller*.

c) No, we do not reject the null hypothesis of equal means, so we cannot perform multiple comparisons.

13. a) $MS_T = 8.65$; $MS_E = 0.0767$.

b) $F = 112.78$.

c) The data provide very strong evidence that the means are not equal.

d) We have assumed that the experimental runs were performed in random order, that the variances of the treatment groups are similar, and that the residuals are nearly Normal.

e) A boxplot of the *Scores* by *Method*, a plot of residuals vs. predicted values, a Normal probability plot, and a histogram of the residuals.

f) $s_p = \sqrt{0.0767} = 0.277$ points.

15. a) The null hypothesis is that the mean *Scores* are the same for both *Types* of eggs. The alternative is that they are different.

b) The data provide strong evidence that the means are different. The real eggs have a higher mean score.

c) Yes, the conditions look reasonable. It would be good to examine the residuals as well.

d) The pooled estimate of the variance is equal to MS_E. The t-ratio is 5.5727 and $5.5727^2 = 31.05499$, which is approximately the F-ratio. The two-sided P-value for the t with 6 df is 0.00141, which agrees with the P-value for the F-statistic.

17. a) The null hypothesis is that the mean *Test Scores* from all the schools are equal. The alternative is that not all the means are equal.

b) The data provide no evidence that the mean *Test Scores* differ.

c) The intern's Type I error rate is higher because he is performing multiple t-tests rather than an Analysis of Variance or a multiple comparisons method.

19. a) The null hypothesis is that the mean *Sugar Content* is the same for the cereals on each *Shelf*. The alternative is that not all the means are equal.

b) The P-value of 0.0012 provides strong evidence that the mean *Sugar Content* is not the same for each *Shelf*.

c) We cannot conclude that cereals on *Shelf* 2 have a higher mean *Sugar Content* than cereals on *Shelf* 3 or that cereals on *Shelf* 2 have a higher mean *Sugar Content* than cereals on *Shelf* 1. We can conclude only that the means are not all equal.

d) The Bonferroni test shows that at $\alpha = .05$. Now we can conclude that the mean *Sugar Content* of cereals on *Shelf* 2 is *not* equal to the mean *Sugar Content* on *Shelf* 1 and is *not* equal to the mean *Sugar Content* on *Shelf* 3. In other words, we can conclude from this test what we wanted to conclude in part c.

21. a) The null hypothesis states that the mean *Download Time* (DT) is the same at each *Time of Day*. The alternative is that not all the means are equal.

b)

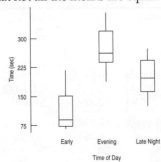

Analysis of Variance for Time (in sec)

Source	DF	Sum of Squares	Mean Square	F-ratio	P-value
DT	2	204641	102320	46.035	<0.0001
Error	45	100020	2222.67		
Total	47	304661			

The data provide very strong evidence that mean *Download Time* is not the same at each *Time of Day*.

c) The similar variance condition seems to be met. The residual plots below show no evidence of violating the nearly Normal or similar variance condition. The runs were not randomized because they were in specific time periods. It would be good to check a timeplot of residuals by run order to see if there is dependence from one run to the next.

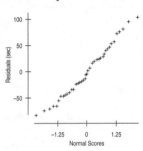

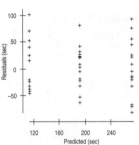

d) A Bonferroni test shows that all three pairs can be distinguished from each other at $\alpha = .05$.

Chapter 29

1. a) $H_0: \gamma_1 = \gamma_2 = \gamma_3$ where γ represents the effect of the *Power* level vs. H_A: Not all of the *Power* levels have the same effect on the response.
$H_0: \tau_1 = \tau_2 = \tau_3$ where τ represents the effect of *Time* level vs. H_A: Not all of the *Time* levels have the same effect on the response.

b) The *Power* sum of squares has 2 df; the *Time* sum of squares has 2 df; the error sum of squares has 4 df.

c) There are no degrees of freedom left for the interaction term. She must assume that the interaction effects are negligible.

3. a) The P-value for *Power* is 0.0165. The P-value for *Time* is 0.0310.

b) We reject the null hypothesis that *Power* has no effect and conclude that the mean number of uncooked kernels is not equal across all 3 *Power* levels. We also reject the null hypothesis that *Time* has no effect and conclude that the mean number of uncooked kernels is not the same across all 3 *Times*.

c) Partial boxplots, scatterplots of residuals vs. predicted values, and a Normal probability plot of the residuals to check the assumptions.

5. a) $H_0: \gamma_1 = \gamma_2$ where γ represents the effect of the *Seat* level vs. H_A: One of the *Seat* levels has a different effect. The null hypothesis states that *seat* choice (driver vs. passenger) has no effect on mean head injury sustained. The alternative states that it does.
$H_0: \tau_1 = \tau_2 = \tau_3 = \tau_4 = \tau_5 = \tau_6$ where τ represents the effect of the *Size* level vs. H_A: The *Size* level effects are not all equal. The null hypothesis states that *Size* of vehicle (the 6 represented) has no effect on mean *Head Injury* sustained. The alternative states that it does.

b) The conditions appear to be met. The effects are additive enough, the data we assume were collected independently, the boxplots show that the variance is roughly constant, and there are no patterns in the scatterplot of residuals vs. predicted values.

c) There is no significant interaction. The P-values for both *Seat* and *Size* are <0.0001. Thus, we reject the null hypotheses and conclude that both the *Seat* and the *Size* of car affect the severity of head injury. From the partial boxplots we see that the mean *Head Injury* level is higher for the driver's side. The effect of driver's seat seems to be roughly the same for all 6 car sizes.

7. a) A two-factor ANOVA must have a quantitative response variable. Here the response is whether they exhibited baldness or not, which is a categorical variable. A two-factor ANOVA is not appropriate.

b) We could use a chi-square analysis to test whether *Baldness* and *Heart Disease* are independent.

9. a) A chi-square test of independence gives a chi-square statistic of 14.510 with a P-value of 0.0023. We reject the hypothesis that *Baldness* and *Heart Disease* are independent.

b) No, the fact that these are not independent does NOT mean that one causes the other. There could be a lurking variable (such as age) that influences both.

11. a) $H_0: \gamma_1 = \gamma_2$ where γ represents the effect of the *Time of Day* on *Shots Made*. H_A: The means of *Shots Made* differ across the two *Time of Day* levels. The null hypothesis states that *Time of Day* has no effect on *Shots Made*. The alternative states that it does.
$H_0: \tau_1 = \tau_2$ where τ represents the kind of *Shoes* worn vs. H_A: The means of *Shots Made* differ across the two types of *Shoes*. The null hypothesis states that type of *Shoes* worn has no effect on *Shots Made*. The alternative states that it does.

b) Partial boxplots show little effect of either *Time of Day* or *Shoes* on *Shots Made*.

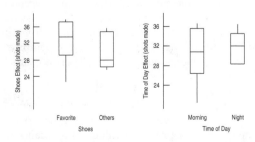

Neither the scatterplot of residuals vs. predicted values nor the Normal probability plots of residuals show any conditions that aren't met. We assume the number of shots made were independent from one treatment condition to the next.

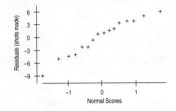

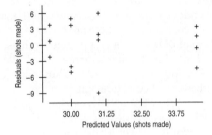

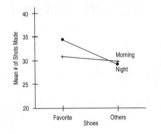

An interaction plot shows a possible interaction effect. It looks as though the favorite shoes may make more of a difference at night. However, the *F*-test shows that the null hypothesis of no interaction effect is not rejected. None of the effects appears to be significant. It looks as though she cannot conclude that either *Shoes* or *Time of Day* affect her mean *Shots Made*.

13. a) $H_0: \gamma_1 = \gamma_2 = \gamma_3$ where γ represents the effect of the *Temperature* level vs. H_A: Not all the γ are equal. The null hypothesis states that *Temperature* level (32°, 34°, and 36° C) has no effect on the *Number of Sprouts*. The alternative states that it does.
$H_0: \tau_1 = \tau_2 = \tau_3 = \tau_4$ where τ represents the effect of the *Salinity* level vs. H_A: Not all the τ are equal. The null hypothesis states that *Salinity* level (0, 4, 8, and 12 ppm) has no effect on the *Number of Sprouts*. The alternative states that it does.

b) The partial boxplots show that *Salinity* appears to have an effect on the *Number of Sprouts*, but *Temperature* does not. The ANOVA supports this. The P-value for *Salinity* is <.0001, but the P-value for *Temperature* is 0.3779, providing no evidence of a temperature effect over this range of temperatures. There appears to be no interaction as well.
There appears to be more spread in *Number of Sprouts* for the lower *Salinity* levels (where the response is higher). This is evident in both the partial boxplot and the residual vs. predicted values. This is cause for some concern, but most likely does not affect the conclusions.

15. The ANOVA table shows that both *Car Type* and *Additive* affect *Gas Mileage*, with P-values <0.0001. There is a significant interaction effect as well that makes interpretation of the main effects problematic. However, the residual plot shows a strong increase in variance, which makes the whole analysis suspect.

17. The ANOVA table now shows only main effects to be significant:

Source	DF	Sum of Squares	Mean Square	F-ratio	P-value
Type	2	10.1254	5.06268	5923.1	<0.0001
Additive	1	0.026092	0.0260915	30.526	<0.0001
Interaction	2	7.57E-05	3.78E-05	0.044265	0.9567
Error	54	0.046156	8.55E-04		
Total	59	10.1977			

The residuals now show *no* violation of the conditions:

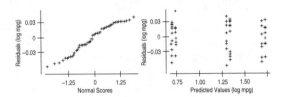

We can conclude that both the *Car Type* and *Additive* have an effect on mileage and that the effects are constant (in log10 *Mpg*) over the values of the various levels of the other factor.

19. a) $H_0: \gamma_1 = \gamma_2$ where γ represents the effect of the *Environment* level vs. H_A: The mean *Time* the battery lasted is not the same for the two conditions.
$H_0: \tau_1 = \tau_2 = \tau_3 = \tau_4$ where τ represents the effect of the *Brand* vs. H_A: Not all the τ are equal.

b) From the partial boxplot, it is not clear that *Brand* has an effect. The condition clearly has an effect.

c) Yes, the *Brand* effect has a P-value of 0.099, while the *Environment* effect is clearly significant with a P-value <0.0001.

d) There is also an interaction, however, which makes the statement about *Brands* problematic. Not all brands are affected by the environment in the same way. Brand C, which works best in the cold, performs worst at room temperature.

e) Because it performs the worst of the four at room temperature.

21. In this one-way ANOVA, we can see that the means vary across treatments. (However, boxplots with only 2 observations are not appropriate.) By looking closely, it seems obvious that the four flashlights at room temperature lasted much longer than the ones in the cold. But it is much harder to see whether the means of the four brands are different, or whether they differ by the same amounts across both environmental conditions. The two-way ANOVA with interaction makes these distinctions clear.

Chapter 30

1. a) Doesn't mention the other predictors.
   b) This is correct.
   c) Can't predict from y to x.
   d) R^2 is about the fraction of vkariability accounted for by the regression model, not the fraction of data values.

3. a) $\widehat{Final} = -6.7210 + 0.2560\ Test1 + 0.3912\ Test2 + 0.9015\ Test3$
   b) R^2 is 77.7%, so 77.7% of the variation in final grade is accounted for by the regression model.
   c) After allowing for the linear effects of the other predictors, each point on *Test3* is associated with an average increase of 0.9015 points on the final.
   d) Test scores are probably collinear. If all we are concerned about is predicting final exam score, *Test1* may not add much to the regression. However, we'd expect it to be associated with final exam score.

5. a) $\widehat{Price} = -152,037 + 9530\ Baths + 139.87\ Sqft$
   b) 71.1% of the variation in asking price is accounted for by this regression model.
   c) For homes with the same number of bathrooms, asking price increases, on average, by about $139.87 per square foot.
   d) The number of bathrooms is probably associated with the size of the house (even after considering the square footage of the bathroom itself). That association may account for the coefficient of *Baths* not being discernibly different from zero. Moreover, the regression model does not predict what will happen when a house is modified (for example, by converting existing space into a bathroom).

7. The plot of residuals vs. predicted values looks bent, rising in the middle and falling at both ends. This violates the Straight Enough Condition. The Normal probability plot and the histogram of the residuals suggest that the highest five residuals (which we know are in the middle of the predicted value range) are extraordinarily high. These data may benefit from a re-expression.

9. a) $\widehat{Salary} = 9.788 + 0.11\ Service + 0.053\ Educ + 0.071\ Score + 0.004\ Speed + 0.065\ Dictation$
   b) 29.2 thousand dollars
   c) b_4 has a t-statistic of 0.013. That is not significant at any reasonable alpha level.
   d) Omitting X4 (typing speed) would simplify the model and might result in one that was almost as good. Other predictors might also be omitted, but we can't tell that from what we know here.
   e) *Age* may be collinear with other predictors in the model. In particular, it is likely to be highly associated with X1, months of service.

11. a) $H_0: \beta = 0$; $H_A: \beta \neq 0$; $t = 12.4$; $P \leq 0.0001$, so we reject the null hypothesis. The coefficient of *Weight* is statistically discernible from zero.
   b) Each pound of weight is associated, on average, with an increase of 0.189 in *%Body Fat*.
   c) After removing the linear effects of *Waist* and *Height*, each pound of *Weight* is associated on average with a decrease of 0.10 in *%Body Fat*. Alternatively, for men with the same *Waist* and *Height*, each pound of *Weight* is associated, on average, with a decrease of 0.10 in *%Body Fat*. The change in coefficient and sign is a result of including the other predictors. We expect *Weight* to be correlated with both *Waist* and *Height*. It may be collinear with them.
   d) The P-value of 0.1567 says that if the coefficient of *Height* in this model is truly zero, we could expect to observe a sample regression coefficient as far from zero as the one we have here about 15.6% of the time.

13. a) P-value is ≤ 0.0001, so we can reject the null hypothesis that the true slope is zero.
   b) It says that there is a clear relationship between *%Body Fat* and *Chest Size*. In fact, *%Body Fat* grows, on average, by 0.71272% per inch of *Chest Size*.
   c) After allowing for the linear effects of *Waist* and *Height*, each inch of *Chest Size* is associated, on average, with a *%Body Fat* that is on average 0.233531 lower. This coefficient is statistically discernible from zero (P = 0.0054).
   d) Each of the variables appears to contribute to the model. There appears to be no advantage in removing one.

15. a) The only model that seems to do poorly is the one that omits *Murder*. The other three are hard to choose among.
   b) Each of the models has at least one coefficient with a large P-value. This one could be omitted to simplify the model without degrading it much.
   c) No. A regression model can't be used to predict what effects changes in the predictors might cause.
   d) Plots of the residuals highlight some states that may be outliers. You may want to consider setting them aside to see if the model changes.

17. a) With an R^2 of 100%, it looks like the model would do a good job of prediction.
   b) The value of s, 3.140 calories, is very small compared with the initial standard deviation of *Calories*. This indicates that the model fits the data quite well, leaving very little variation unaccounted for.
   c) No, the residuals are not all zero. Indeed, we know that their standard deviation, s, is 3.140. But they are very small compared with the original values. The true value of R^2 was rounded up to 100%.

Chapter 31

1. a) There may be an outlier. We should examine the data to check for that.
   b) That would not be a correct interpretation. If both temperature and CO_2 were changing together, then they may be collinear.

3. a) The slope of a partial regression plot is the coefficient of the corresponding predictor—in this case, -1.020.
   b) Quaker oatmeal makes the slope more strongly negative. It appears to have substantial influence on this slope.

c) Not surprisingly, omitting Quaker oatmeal changes the coefficient of *Fiber*. It is now positive (although not significantly different from 0). This second regression model has a higher R^2, suggesting that it fits the data better. Without the influential point, the second regression is probably the better model.

d) The coefficient of *Fiber* is not discernibly different from zero. We have no evidence that it contributes significantly to *Calories* after allowing for the other predictors in the model.

5. a) After allowing for the effects of *Sodium* and *Sugars*, each gram of *Potassium* is associated with a 0.019-calorie decrease in *Calories*.

b) Those points pull the slope of the relationship down. Omitting them should increase the value of the coefficient of *Potassium*. It would likely become positive, since the remaining points show a positive slope in the partial regression plot.

c) These appear to be influential points. They have both high leverage and large residual, and the partial regression plot shows their influence.

d) If our goal is to understand the relationships among these variables, then it might be best to omit these cereals because they seem to behave as if they are from a separate subgroup.

7. a) This set of indicators uses *medium* size as its base. The coefficients of *Small*, *Large*, and *Very Large* estimate the average change in amount of *Delay/Person* relative to the amount of *Delay/Person* for *medium* size cities found for each of the other three sizes. If there were an indicator variable for *medium* as well, the four indicators would be collinear, so the coefficients could not be estimated.

b) The delays per person in *Large* cities are, on average, about 5 hours per year longer than in *medium* size cities, after allowing for the effects of highway mph and arterial mph.

9. a) An assumption required for indicator variables to be useful is that the regression model's fit for the different groups identified by the indicators be parallel. These lines are not parallel.

b) The coefficient of *Am*Sml* adjusts the slope of the regression model fit for the small cities. We would say that the slope of *Delay/Person* on *Arterial Mph* for *small* cities (after allowing for the linear effects of the other variables in the model) is $-2.60848 + 3.81461 = 1.20613$.

c) The regression model seems to do a good job. The R^2 shows that 80.7% of the variability in *Delay/Person* is accounted for by the model. Most of the P-values for the coefficients are small. The coefficients concerning the very large cities have larger P-values, but that may be due to having a relatively small number of such cities. It may still be wise to keep those predictors in the model.

11. Colorado Springs has high leverage, but it does not have a particularly high Studentized residual. It appears that the point has influence but doesn't exert it. Removing this case from the regression may not result in a large change in the model, so the case is probably not influential.

Part VII Review

1. a) 59 products
   b) 84.5%
   c) (2.21, 2.78) dollars per minute
   d) Based on this regression, average *Price* increases between $2.21 and $2.78 for each minute of polishing *Time*, with 95% confidence.

3. a) Based on these data, with 95% confidence 5-year yields are between 3.15% and 5.93% higher than 3-year yields on average (paired data).
   b) Yes (at least for this data set). The regression line is 5-year $= 6.93 + 0.719 \times$ 3-year. H_0: $\beta_1 = 0$ against H_A: $\beta_1 \neq 0$ has $t = 4.27$; P-value $= 0.0009$. Since P is so small, we reject H_0. There is evidence of an association. (But we don't know that this was an SRS or even a representative sample of large cap funds.)

5. a) An ANOVA and an *F*-test.
   b) The null hypothesis is that the mean *Distance* thrown is the same for all 4 grips. The alternative is that it is not.
   c) Conclude that the mean *Distance* thrown is not the same for all the grips.
   d) Boxplots of *Distance* by *Grip*, residual vs. predicted value scatterplot, and a Normal probability plot of the residuals.
   e) A multiple comparison test to see which grip is best.

7. a) 38
   b) Data plot looks linear; residuals plot shows random scatter; histogram of residuals approximately Normal.
   c) (0.131, 0.177) foals per adult.
   d) Based on this regression, we have 95% confidence that for every *Adult* horse an average of between 0.131 and 0.177 *Foals* will be born.
   e) A herd with 80 *Adult* horses will have between 2.27 and 19.21 *Foals*, with 90% confidence.

9. a) Indicator variable.
   b) There were, on average, 6.4 fewer foals in herds in which some stallions were sterilized, after allowing for the number of adults in the herd.
   c) The P-value for *Sterilized* is 0.096. That's not significant at the 0.05 level.

11. a) The null hypotheses are that the mean *Time* is the same for all three *Mice* and for *Lights* on and off. The alternatives are that the means are not all the same for the three *Mice* or for the two lighting conditions.
    b) *Mouse* has 2 df; *Light* has 1. The error has 1 df.
    c) No, the interaction would use 2 df. There are none left, so the interaction can't be estimated.

13. a) (36.21, 60.51) feet.
    b) Yes, 40 is in the interval.
    c) Wider; we'd need to have a wider interval to be more confident.
    d) Roughly 4 times as big—44 flights.

15. a) With an R^2 of 90.9%, your friend is right about being able to predict.
    b) H_0: $\beta_{PRS} = 0$, $t = 8.13$ on 31 df. P is <0.0001, so we reject the null hypothesis.
    c) The plots show evidence that all 4 conditions are met.

17. a) Null hypotheses are that mean hours studied are the same for both sexes and that the mean hours studied are the same for all four classes. Alternatives are that the mean hours studied are not the same for both sexes and that not all four classes have the same mean hours studied.

b) Yes, an interaction term should be fit because the Additive Enough Condition does not seem to be met.

c) None appear to be significant.

d) There are a few outliers. The Constant Variance Condition appears to be met, but we do not have Normal probability plots of residuals. The main concern is that we may not have enough power to detect differences between groups. We would need more data to increase the power.

19. These are not counts, so they fail the Counted Data Condition. We cannot use chi-square methods.

21. a) $H_0: \beta_1 = 0; H_A: \beta_1 \neq 0. t = 4.16$, P-value < 0.0001. These data show evidence of a positive relationship between number of meals eaten together and grades.

b) No. R^2 is small and $s = 0.66$ points. So we could predict only to within 1.32 grade points at best.

c) No. The slope is clearly not 0, but that doesn't mean the relationship is strong or the predictions are useful.

23. $H_0: p_L - p_R = 0; H_A: p_L - p_R \neq 0. z = 1.38$; P-value $= 0.1683$. Since $P > 0.05$, we do not reject H_0. These data do not provide evidence of a difference in musical abilities between right- and left-handed people.

25. a) Linearity: There's no evidence of nonlinearity in the residual scatterplot.
Independence: These data are measured over time. We should plot the data and the residuals against time to look for failures of independence.
Constant variance: The scatterplot of the residuals shows no evidence of thickening.
Normality: Without a histogram or Normal probability plot of the Studentized residuals, we can't confirm whether the residuals are Nearly Normal.

b) 57.4% of the variation in female teen traffic deaths is accounted for by the linear trend over time.

c) Predicted *Deaths* $= 45074 - 21.6\ Year$.

d) *Deaths* have been declining at the rate of 21.6 per *Year*. (With a standard deviation of the errors estimated to be 182.6, this doesn't seem like a very meaningful decrease.)

27. a) Null hypotheses are that the mean number of *Words Typed* is the same for both *Temperatures* and for both *Gloves* on and off. The alternatives are that the mean number of *Words Typed* is different for the two *Temperatures* and for *Gloves* on and off.

b) Based on the partial boxplots, both effects appear to be significant.

c) Yes, an interaction term should be fit because the Additive Enough Condition does not seem to be met.

d) The interaction term appears to be real. The temperature effect is only active with gloves off.

e) All three effects are significant.

f) *Gloves* decrease *Words Typed* at both *Temperatures*. Hot temperature increases words typed with gloves off, but has little effect with gloves on.

g) $s = \sqrt{\dfrac{58.75}{28}} = 1.45$ words per minute. Yes, this seems consistent with the size of the variation shown in the partial boxplots.

h) Tell him to type in a warm room with his gloves off.

i) Even though we haven't seen a histogram or Normal probability plot of residuals, the other conditions seem to be satisfied. For the levels of the factors that he used, it seems clear that a warm room and not wearing gloves are beneficial for his typing.

29. a) The number of finishers per minute appears to grow at about 0.523 finisher per minute for each additional minute of race time.

b) Oops! Clearly, the Linearity Assumption is violated. We can't interpret this regression model at all, although there seems to be a strong—but not linear—association.

31. a) H_0: There is no (linear) relationship between *Depression* and *Internet Use*. $\beta_1 = 0$. H_A: There is a linear relationship. $\beta_1 \neq 0$. $t = 2.76$; P-value $= 0.0064$. With such a small P-value, we reject H_0. These data provide strong evidence of a relationship between *Internet Use* and *Depression*.

b) The study says nothing about causality. Many other factors may be involved.

c) $H_0: \mu_D = 0; H_A: \mu_D > 0. t = -2.73$; P-value $= 0.0071$. With such a small P-value, we reject H_0. These data indicate that mean *Depression* level actually got better (decreased) during the study.

33. $H_0: p = 0.512; H_A: p < 0.512. z = -3.35$; P-value $= 0.0004$. With such a small P-value, we reject H_0. These data provide evidence that exposure to dioxin is related to a reduced rate of male births.

35. a) H_0: There is no effect of *Tilt* on average score.
H_A: There is an effect.
H_0: There is no effect of using one or two *Eyes* on average score.
H_A: There is an effect.

b) Partial boxplots show that *Eye* effect is strong, *Tilt* effect is much weaker:

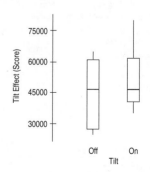

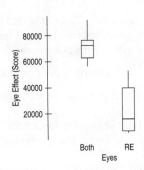

There is a possible interaction effect:

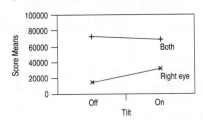

Conditions appear to be met.
The ANOVA table shows that only the *Eye* effect is significant:

Source	DF	Sum of squares	F-ratio	P-value
Eye	1	9385201568	44.9369	<0.0001
Tilt	1	156806745	0.7508	0.4032
Interaction	1	450532463	2.1572	0.1676
Error	12	2506236979		

Residual plots show no evidence of non-Normality or changing variance:

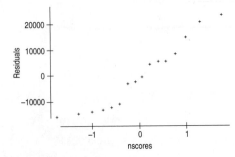

Conclude that keeping both eyes open improves score but that using the tilt does not.

37. a) This is a linear regression that is meaningless—the data are categorical.
   b) This is a two-way table, that is appropriate. H_0: *Eye Color* and *Hair Color* are independent; H_A: *Eye Color* and *Hair Color* are not independent. However, four cells have expected counts less than 5, so the χ^2 analysis is not valid unless cells are merged. However with a χ^2 value of 223.6 with 16 df and a P-value of <0.0001, the results are not likely to change if we merge appropriate eye colors.

39. a) Based on these data, the average annual rainfall in LA is between 11.65 and 17.39 inches, with 90% confidence.
   b) About 46 years
   c) No. The regression equation is *Rain* = −51.684 + 0.033 × *Year*. R^2 = 0.1%. The *t*-stat for the slope is 0.12 with a P-value of 0.9029.

41. a) The null hypothesis is that mean *Weight* is the same for all three groups. The alternative is that it is not.
   b) The variance for the three groups appears to be about the same. There is one outlier. We don't have a Normal probability plot of the residuals, but we suspect that the data may be Normal enough.
   c) The *F*-test indicates that the mean *Weight* is not the same for the three groups. It may be that more men are involved with athletics, which might explain the weight differences. On the other hand, it may simply be that those who weigh more are more likely to be involved with sports.
   d) The differences are evident even when the outlier is removed. It seems that the conclusion was valid.

43. a) $H_0: \mu_1 - \mu_2 = 0; H_A: \mu_1 - \mu_2 > 0$. $t = 0.90$; P-value = 0.1864. With a t that small we do not reject H_0. Week-long study scores were not significantly higher.
   b) $H_0: p_1 - p_2 = 0; H_A: p_1 - p_2 \neq 0$. $z = -3.10$, P-value = 0.0019. With such a small P-value, we reject H_0. There is evidence of a difference in proportion passing on Friday; it appears that cramming may be more effective.
   c) $H_0: \mu_D = 0; H_A: \mu_D > 0$. $t = 5.17$, P-value < 0.0001. These data show evidence that learning does not last for 3 days because mean score declined.
   d) Based on these data, the average number of words forgotten by crammers is between 3.03 and 7.05 with 95% confidence.
   e) Yes. Regression equation is *Monday* = 14.6 + 0.536 × *Friday*. $t = 2.57$; P-value = 0.0170.

45. a) No, the R^2 is only 8.7%.
   b) Isolating the outlier with an indicator variable improves the model dramatically. We can predict incidents of false information well (R^2 = 86.8%) as 47.21 + 0.75 *long gun discoveries* − 0.02358 *handgun discoveries* − 0.0894 *explosive discoveries* with 1988 removed from the data.
   c) The coefficient for *Handgun Discoveries* has a large P-value and might be removed from the model to simplify it.

APPENDIX B

Photo Acknowledgments

Page 1 Leland Bobbe/Digital Image/Getty Images   **Page 2** *Frazz* copyright © Jef Mallett/Distributed by United Feature Syndicate, Inc.   **Page 6** Andrew McKinney/Dorling Kindersley Media Library   **Page 9** Johner Images Royalty Free/Getty Images   **Page 10** (C) Fancy/Alamy **Page 12** Eric Gaillard/Corbis   **Page 13** Andrew McKinney/Dorling Kindersley   **Page 18** ©20th Century Fox Film Corp./Courtesy Everett Collection   **Page 19 (Florence Nightingale)** Schlesinger Library, Radcliffe Institute, Harvard University/The Bridgeman Art Library   **Page 19 (Diagram of the Causes of Mortality)** Florence Nightingale Museum, London/The Bridgeman Art Library   **Page 22** Stephen J. Carrera/AP Images   **Page 29** Pete Atkinson/Stone Allstock/Getty Images   **Page 34** ©20th Century Fox Film Corp./Courtesy Everett Collection   **Page 44** Photographer's Mate 3rd Class Jennifer Rivera/US Navy   **Page 50** Sygma/Corbis   **Page 51** MGM/Photofest **Page 55** Digital Vision/Getty Images   **Page 67** Photographer's Mate 3rd Class Jennifer Rivera/US Navy **Page 80** PhotoDisc/Getty Images   **Page 84** NRA/Shutterstock   **Page 87** Beth Anderson/Pearson **Page 94** Courtesy of author   **Page 95** PhotoDisc/Getty Images   **Page 109** Xinhua/Newscom **Page 111** Herbert Kratky/Shutterstock   **Page 115** Sportsphotographer.eu/Shutterstock **Page 126** Olga Nayashkova/Shutterstock   **Page 132** Xinhua/Newscom   **Page 150** National Oceanic and Atmospheric Administration   (NOAA)   **Page 162** International Association for Statistical Computing   **Page 164** Department of Psychology, University College London
Page 166 AP Images   **Page 167** National Oceanic and Atmospheric Administration (NOAA) **Page 178** Nayashkova Olga/Shutterstock   **Page 181** National Oceanic and Atmospheric Administration (NOAA)   **Page 186** Mary Evans Picture Library/The Image Works
Page 193 Mikeledray/Shutterstock   **Page 197** Nayashkova Olga/Shutterstock   **Page 213** Courtesy of the author   **Page 216** *Foxtrot* copyright © 2010 Bill Amend/Distributed by Universal Uclick **Page 219 (Ralph Nader)** Shawn Thew/AFP/Getty Images   **Page 219 (Pat Buchanan)** Jim Cole/AP Images   **Page 220** Ann Ronan Picture Library/Art Resource   **Page 224** PhotoDisc/Getty Images   **Page 226** Courtesy of the author   **Page 237** Joel Saget/AFP/Getty Images
Page 250 Joel Saget/AFP/Getty Images   **Page 267** Oleg Svyatoslavsky/PhotoDisc/Getty Images **Page 268** *Dilbert* copyright © 2010 by Scott Adams/Distributed by United Feature Syndicate, Inc. **Page 269 (LeBron James)** Ed Suba, Jr./Newscom   **Page 269 (David Beckham)** Sandra Behne/Bongarts/Allsport/Getty Images   **Page 269 (Serena Williams)** Tobias Schwarz/Corbis **Page 275** Oleg Svyatoslavsky/PhotoDisc/Getty Images   **Page 281** Stockbyte/Getty Images **Page 282 (Literary Digest)** Courtesy of author   **Page 282 (George Gallup)** AP Images
Page 289 Jamie Duplass/Shutterstock   **Page 292** *Calvin and Hobbes* copyright © 2010 Bill Watterson/Distributed by Universal Uclick   **Page 294** *Wizard of Id* copyright © 2010 by Parker and Hart/Distributed by Creators Syndicate, Inc.   **Page 298** Stockbyte/Getty Images
Page 305 Beth Anderson/Pearson   **Page 306** Grant Shimmin/iStockphoto   **Page 307** Photo Researchers/Alamy   **Page 309 (C. S. Peirce)** National Oceanic and Atmospheric Administration (NOAA)   **Page 309 (Dilbert)** *Dilbert* copyright © 2010 by Scott Adams/Distributed by United Feature Syndicate, Inc.   **Page 310 (puppies)** PhotoDisc/Getty Images   **Page 310 (tomatoes)** F. Schussler/PhotoDisc/Getty Images   **Page 316** *Wizard of Id* copyright © 2010 by Parker and Hart/Distributed by Creators Syndicate, Inc.   **Page 318** Vuk Nenezic/Shutterstock
Page 320 iStockphoto   **Page 322** Beth Anderson/Pearson   **Page 336** Jessica Bethke/Shutterstock **Page 338** Interfoto/Alamy   **Page 339** SpxChrome/iStockphoto   **Page 340** PhotoDisc/Getty Images **Page 341** Royal Society   **Page 349** Jessica Bethke/Shutterstock   **Page 354** CNN/Getty Images **Page 355 (stones)** Corbis   **Page 355 (money)** US Department of Treasury   **Page 371** Courtesy of St Andrew's University   **Page 373** Photos.com/Thinkstock   **Page 374** Corbis   **Page 381** Digital Vision/Getty Images   **Page 391** US Central Intelligence Agency (CIA), map #802421
Page 393 PhotoDisc/Getty Images   **Page 397** Digital Vision/Getty Images   **Page 404 (inspector)** Juice Images/Alamy   **Page 404 (LeBron James)** Chris Keane/Newscom   **Page 405 (Daniel Bernoulli)** Pearson   **Page 405 (Calvin and Hobbes)** *Calvin and Hobbes* copyright © 2010 Bill Watterson/Distributed by Universal Uclick   **Page 412** Keith Brofsky/PhotoDisc/Getty Images   **Page 414** David James/Touchstone/Kobal Collection/The Picture Desk

APPENDIX C

Index

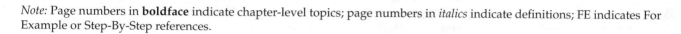

Note: Page numbers in **boldface** indicate chapter-level topics; page numbers in *italics* indicate definitions; FE indicates For Example or Step-By-Step references.

APPENDIX D

Tables and Selected Formulas

Row					TABLE OF RANDOM DIGITS					
1	96299	07196	98642	20639	23185	56282	69929	14125	38872	94168
2	71622	35940	81807	59225	18192	08710	80777	84395	69563	86280
3	03272	41230	81739	74797	70406	18564	69273	72532	78340	36699
4	46376	58596	14365	63685	56555	42974	72944	96463	63533	24152
5	47352	42853	42903	97504	56655	70355	88606	61406	38757	70657
6	20064	04266	74017	79319	70170	96572	08523	56025	89077	57678
7	73184	95907	05179	51002	83374	52297	07769	99792	78365	93487
8	72753	36216	07230	35793	71907	65571	66784	25548	91861	15725
9	03939	30763	06138	80062	02537	23561	93136	61260	77935	93159
10	75998	37203	07959	38264	78120	77525	86481	54986	33042	70648
11	94435	97441	90998	25104	49761	14967	70724	67030	53887	81293
12	04362	40989	69167	38894	00172	02999	97377	33305	60782	29810
13	89059	43528	10547	40115	82234	86902	04121	83889	76208	31076
14	87736	04666	75145	49175	76754	07884	92564	80793	22573	67902
15	76488	88899	15860	07370	13431	84041	69202	18912	83173	11983
16	36460	53772	66634	25045	79007	78518	73580	14191	50353	32064
17	13205	69237	21820	20952	16635	58867	97650	82983	64865	93298
18	51242	12215	90739	36812	00436	31609	80333	96606	30430	31803
19	67819	00354	91439	91073	49258	15992	41277	75111	67496	68430
20	09875	08990	27656	15871	23637	00952	97818	64234	50199	05715
21	18192	95308	72975	01191	29958	09275	89141	19558	50524	32041
22	02763	33701	66188	50226	35813	72951	11638	01876	93664	37001
23	13349	46328	01856	29935	80563	03742	49470	67749	08578	21956
24	69238	92878	80067	80807	45096	22936	64325	19265	37755	69794
25	92207	63527	59398	29818	24789	94309	88380	57000	50171	17891
26	66679	99100	37072	30593	29665	84286	44458	60180	81451	58273
27	31087	42430	60322	34765	15757	53300	97392	98035	05228	68970
28	84432	04916	52949	78533	31666	62350	20584	56367	19701	60584
29	72042	12287	21081	48426	44321	58765	41760	43304	13399	02043
30	94534	73559	82135	70260	87936	85162	11937	18263	54138	69564
31	63971	97198	40974	45301	60177	35604	21580	68107	25184	42810
32	11227	58474	17272	37619	69517	62964	67962	34510	12607	52255
33	28541	02029	08068	96656	17795	21484	57722	76511	27849	61738
34	11282	43632	49531	78981	81980	08530	08629	32279	29478	50228
35	42907	15137	21918	13248	39129	49559	94540	24070	88151	36782
36	47119	76651	21732	32364	58545	50277	57558	30390	18771	72703
37	11232	99884	05087	76839	65142	19994	91397	29350	83852	04905
38	64725	06719	86262	53356	57999	50193	79936	97230	52073	94467
39	77007	26962	55466	12521	48125	12280	54985	26239	76044	54398
40	18375	19310	59796	89832	59417	18553	17238	05474	33259	50595

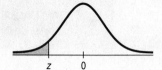

Table Z
Areas under the
standard Normal curve

0.09	0.08	0.07	0.06	0.05	0.04	0.03	0.02	0.01	0.00	z
									0.0000†	−3.9
0.0001	0.0001	0.0001	0.0001	0.0001	0.0001	0.0001	0.0001	0.0001	0.0001	−3.8
0.0001	0.0001	0.0001	0.0001	0.0001	0.0001	0.0001	0.0001	0.0001	0.0001	−3.7
0.0001	0.0001	0.0001	0.0001	0.0001	0.0001	0.0001	0.0001	0.0002	0.0002	−3.6
0.0002	0.0002	0.0002	0.0002	0.0002	0.0002	0.0002	0.0002	0.0002	0.0002	−3.5
0.0002	0.0003	0.0003	0.0003	0.0003	0.0003	0.0003	0.0003	0.0003	0.0003	−3.4
0.0003	0.0004	0.0004	0.0004	0.0004	0.0004	0.0004	0.0005	0.0005	0.0005	−3.3
0.0005	0.0005	0.0005	0.0006	0.0006	0.0006	0.0006	0.0006	0.0007	0.0007	−3.2
0.0007	0.0007	0.0008	0.0008	0.0008	0.0008	0.0009	0.0009	0.0009	0.0010	−3.1
0.0010	0.0010	0.0011	0.0011	0.0011	0.0012	0.0012	0.0013	0.0013	0.0013	−3.0
0.0014	0.0014	0.0015	0.0015	0.0016	0.0016	0.0017	0.0018	0.0018	0.0019	−2.9
0.0019	0.0020	0.0021	0.0021	0.0022	0.0023	0.0023	0.0024	0.0025	0.0026	−2.8
0.0026	0.0027	0.0028	0.0029	0.0030	0.0031	0.0032	0.0033	0.0034	0.0035	−2.7
0.0036	0.0037	0.0038	0.0039	0.0040	0.0041	0.0043	0.0044	0.0045	0.0047	−2.6
0.0048	0.0049	0.0051	0.0052	0.0054	0.0055	0.0057	0.0059	0.0060	0.0062	−2.5
0.0064	0.0066	0.0068	0.0069	0.0071	0.0073	0.0075	0.0078	0.0080	0.0082	−2.4
0.0084	0.0087	0.0089	0.0091	0.0094	0.0096	0.0099	0.0102	0.0104	0.0107	−2.3
0.0110	0.0113	0.0116	0.0119	0.0122	0.0125	0.0129	0.0132	0.0136	0.0139	−2.2
0.0143	0.0146	0.0150	0.0154	0.0158	0.0162	0.0166	0.0170	0.0174	0.0179	−2.1
0.0183	0.0188	0.0192	0.0197	0.0202	0.0207	0.0212	0.0217	0.0222	0.0228	−2.0
0.0233	0.0239	0.0244	0.0250	0.0256	0.0262	0.0268	0.0274	0.0281	0.0287	−1.9
0.0294	0.0301	0.0307	0.0314	0.0322	0.0329	0.0336	0.0344	0.0351	0.0359	−1.8
0.0367	0.0375	0.0384	0.0392	0.0401	0.0409	0.0418	0.0427	0.0436	0.0446	−1.7
0.0455	0.0465	0.0475	0.0485	0.0495	0.0505	0.0516	0.0526	0.0537	0.0548	−1.6
0.0559	0.0571	0.0582	0.0594	0.0606	0.0618	0.0630	0.0643	0.0655	0.0668	−1.5
0.0681	0.0694	0.0708	0.0721	0.0735	0.0749	0.0764	0.0778	0.0793	0.0808	−1.4
0.0823	0.0838	0.0853	0.0869	0.0885	0.0901	0.0918	0.0934	0.0951	0.0968	−1.3
0.0985	0.1003	0.1020	0.1038	0.1056	0.1075	0.1093	0.1112	0.1131	0.1151	−1.2
0.1170	0.1190	0.1210	0.1230	0.1251	0.1271	0.1292	0.1314	0.1335	0.1357	−1.1
0.1379	0.1401	0.1423	0.1446	0.1469	0.1492	0.1515	0.1539	0.1562	0.1587	−1.0
0.1611	0.1635	0.1660	0.1685	0.1711	0.1736	0.1762	0.1788	0.1814	0.1841	−0.9
0.1867	0.1894	0.1922	0.1949	0.1977	0.2005	0.2033	0.2061	0.2090	0.2119	−0.8
0.2148	0.2177	0.2206	0.2236	0.2266	0.2296	0.2327	0.2358	0.2389	0.2420	−0.7
0.2451	0.2483	0.2514	0.2546	0.2578	0.2611	0.2643	0.2676	0.2709	0.2743	−0.6
0.2776	0.2810	0.2843	0.2877	0.2912	0.2946	0.2981	0.3015	0.3050	0.3085	−0.5
0.3121	0.3156	0.3192	0.3228	0.3264	0.3300	0.3336	0.3372	0.3409	0.3446	−0.4
0.3483	0.3520	0.3557	0.3594	0.3632	0.3669	0.3707	0.3745	0.3783	0.3821	−0.3
0.3859	0.3897	0.3936	0.3974	0.4013	0.4052	0.4090	0.4129	0.4168	0.4207	−0.2
0.4247	0.4286	0.4325	0.4364	0.4404	0.4443	0.4483	0.4522	0.4562	0.4602	−0.1
0.4641	0.4681	0.4721	0.4761	0.4801	0.4840	0.4880	0.4920	0.4960	0.5000	−0.0

Second decimal place in z

†For $z \leq -3.90$, the areas are 0.0000 to four decimal places.

Table Z (cont.)

Areas under the standard Normal curve

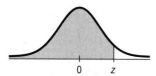

0   z

z	0.00	0.01	0.02	0.03	0.04	0.05	0.06	0.07	0.08	0.09
					Second decimal place in z					
0.0	0.5000	0.5040	0.5080	0.5120	0.5160	0.5199	0.5239	0.5279	0.5319	0.5359
0.1	0.5398	0.5438	0.5478	0.5517	0.5557	0.5596	0.5636	0.5675	0.5714	0.5753
0.2	0.5793	0.5832	0.5871	0.5910	0.5948	0.5987	0.6026	0.6064	0.6103	0.6141
0.3	0.6179	0.6217	0.6255	0.6293	0.6331	0.6368	0.6406	0.6443	0.6480	0.6517
0.4	0.6554	0.6591	0.6628	0.6664	0.6700	0.6736	0.6772	0.6808	0.6844	0.6879
0.5	0.6915	0.6950	0.6985	0.7019	0.7054	0.7088	0.7123	0.7157	0.7190	0.7224
0.6	0.7257	0.7291	0.7324	0.7357	0.7389	0.7422	0.7454	0.7486	0.7517	0.7549
0.7	0.7580	0.7611	0.7642	0.7673	0.7704	0.7734	0.7764	0.7794	0.7823	0.7852
0.8	0.7881	0.7910	0.7939	0.7967	0.7995	0.8023	0.8051	0.8078	0.8106	0.8133
0.9	0.8159	0.8186	0.8212	0.8238	0.8264	0.8289	0.8315	0.8340	0.8365	0.8389
1.0	0.8413	0.8438	0.8461	0.8485	0.8508	0.8531	0.8554	0.8577	0.8599	0.8621
1.1	0.8643	0.8665	0.8686	0.8708	0.8729	0.8749	0.8770	0.8790	0.8810	0.8830
1.2	0.8849	0.8869	0.8888	0.8907	0.8925	0.8944	0.8962	0.8980	0.8997	0.9015
1.3	0.9032	0.9049	0.9066	0.9082	0.9099	0.9115	0.9131	0.9147	0.9162	0.9177
1.4	0.9192	0.9207	0.9222	0.9236	0.9251	0.9265	0.9279	0.9292	0.9306	0.9319
1.5	0.9332	0.9345	0.9357	0.9370	0.9382	0.9394	0.9406	0.9418	0.9429	0.9441
1.6	0.9452	0.9463	0.9474	0.9484	0.9495	0.9505	0.9515	0.9525	0.9535	0.9545
1.7	0.9554	0.9564	0.9573	0.9582	0.9591	0.9599	0.9608	0.9616	0.9625	0.9633
1.8	0.9641	0.9649	0.9656	0.9664	0.9671	0.9678	0.9686	0.9693	0.9699	0.9706
1.9	0.9713	0.9719	0.9726	0.9732	0.9738	0.9744	0.9750	0.9756	0.9761	0.9767
2.0	0.9772	0.9778	0.9783	0.9788	0.9793	0.9798	0.9803	0.9808	0.9812	0.9817
2.1	0.9821	0.9826	0.9830	0.9834	0.9838	0.9842	0.9846	0.9850	0.9854	0.9857
2.2	0.9861	0.9864	0.9868	0.9871	0.9875	0.9878	0.9881	0.9884	0.9887	0.9890
2.3	0.9893	0.9896	0.9898	0.9901	0.9904	0.9906	0.9909	0.9911	0.9913	0.9916
2.4	0.9918	0.9920	0.9922	0.9925	0.9927	0.9929	0.9931	0.9932	0.9934	0.9936
2.5	0.9938	0.9940	0.9941	0.9943	0.9945	0.9946	0.9948	0.9949	0.9951	0.9952
2.6	0.9953	0.9955	0.9956	0.9957	0.9959	0.9960	0.9961	0.9962	0.9963	0.9964
2.7	0.9965	0.9966	0.9967	0.9968	0.9969	0.9970	0.9971	0.9972	0.9973	0.9974
2.8	0.9974	0.9975	0.9976	0.9977	0.9977	0.9978	0.9979	0.9979	0.9980	0.9981
2.9	0.9981	0.9982	0.9982	0.9983	0.9984	0.9984	0.9985	0.9985	0.9986	0.9986
3.0	0.9987	0.9987	0.9987	0.9988	0.9988	0.9989	0.9989	0.9989	0.9990	0.9990
3.1	0.9990	0.9991	0.9991	0.9991	0.9992	0.9992	0.9992	0.9992	0.9993	0.9993
3.2	0.9993	0.9993	0.9994	0.9994	0.9994	0.9994	0.9994	0.9995	0.9995	0.9995
3.3	0.9995	0.9995	0.9995	0.9996	0.9996	0.9996	0.9996	0.9996	0.9996	0.9997
3.4	0.9997	0.9997	0.9997	0.9997	0.9997	0.9997	0.9997	0.9997	0.9997	0.9998
3.5	0.9998	0.9998	0.9998	0.9998	0.9998	0.9998	0.9998	0.9998	0.9998	0.9998
3.6	0.9998	0.9998	0.9999	0.9999	0.9999	0.9999	0.9999	0.9999	0.9999	0.9999
3.7	0.9999	0.9999	0.9999	0.9999	0.9999	0.9999	0.9999	0.9999	0.9999	0.9999
3.8	0.9999	0.9999	0.9999	0.9999	0.9999	0.9999	0.9999	0.9999	0.9999	0.9999
3.9	1.0000[†]									

[†]For $z \geq 3.90$, the areas are 1.0000 to four decimal places.

Two-tail probability	0.20	0.10	0.05	0.02	0.01	
One-tail probability	0.10	0.05	0.025	0.01	0.005	
Table T df						df
Values of t_α 1	3.078	6.314	12.706	31.821	63.657	1
2	1.886	2.920	4.303	6.965	9.925	2
3	1.638	2.353	3.182	4.541	5.841	3
4	1.533	2.132	2.776	3.747	4.604	4
5	1.476	2.015	2.571	3.365	4.032	5
6	1.440	1.943	2.447	3.143	3.707	6
7	1.415	1.895	2.365	2.998	3.499	7
8	1.397	1.860	2.306	2.896	3.355	8
9	1.383	1.833	2.262	2.821	3.250	9
10	1.372	1.812	2.228	2.764	3.169	10
11	1.363	1.796	2.201	2.718	3.106	11
12	1.356	1.782	2.179	2.681	3.055	12
13	1.350	1.771	2.160	2.650	3.012	13
14	1.345	1.761	2.145	2.624	2.977	14
15	1.341	1.753	2.131	2.602	2.947	15
16	1.337	1.746	2.120	2.583	2.921	16
17	1.333	1.740	2.110	2.567	2.898	17
18	1.330	1.734	2.101	2.552	2.878	18
19	1.328	1.729	2.093	2.539	2.861	19
20	1.325	1.725	2.086	2.528	2.845	20
21	1.323	1.721	2.080	2.518	2.831	21
22	1.321	1.717	2.074	2.508	2.819	22
23	1.319	1.714	2.069	2.500	2.807	23
24	1.318	1.711	2.064	2.492	2.797	24
25	1.316	1.708	2.060	2.485	2.787	25
26	1.315	1.706	2.056	2.479	2.779	26
27	1.314	1.703	2.052	2.473	2.771	27
28	1.313	1.701	2.048	2.467	2.763	28
29	1.311	1.699	2.045	2.462	2.756	29
30	1.310	1.697	2.042	2.457	2.750	30
32	1.309	1.694	2.037	2.449	2.738	32
35	1.306	1.690	2.030	2.438	2.725	35
40	1.303	1.684	2.021	2.423	2.704	40
45	1.301	1.679	2.014	2.412	2.690	45
50	1.299	1.676	2.009	2.403	2.678	50
60	1.296	1.671	2.000	2.390	2.660	60
75	1.293	1.665	1.992	2.377	2.643	75
100	1.290	1.660	1.984	2.364	2.626	100
120	1.289	1.658	1.980	2.358	2.617	120
140	1.288	1.656	1.977	2.353	2.611	140
180	1.286	1.653	1.973	2.347	2.603	180
250	1.285	1.651	1.969	2.341	2.596	250
400	1.284	1.649	1.966	2.336	2.588	400
1000	1.282	1.646	1.962	2.330	2.581	1000
∞	1.282	1.645	1.960	2.326	2.576	∞
Confidence levels	80%	90%	95%	98%	99%	

Two tails

One tail

Right-tail probability	0.10	0.05	0.025	0.01	0.005
df					

Table X
Values of χ_α^2

df	0.10	0.05	0.025	0.01	0.005
1	2.706	3.841	5.024	6.635	7.879
2	4.605	5.991	7.378	9.210	10.597
3	6.251	7.815	9.348	11.345	12.838
4	7.779	9.488	11.143	13.277	14.860
5	9.236	11.070	12.833	15.086	16.750
6	10.645	12.592	14.449	16.812	18.548
7	12.017	14.067	16.013	18.475	20.278
8	13.362	15.507	17.535	20.090	21.955
9	14.684	16.919	19.023	21.666	23.589
10	15.987	18.307	20.483	23.209	25.188
11	17.275	19.675	21.920	24.725	26.757
12	18.549	21.026	23.337	26.217	28.300
13	19.812	22.362	24.736	27.688	29.819
14	21.064	23.685	26.119	29.141	31.319
15	22.307	24.996	27.488	30.578	32.801
16	23.542	26.296	28.845	32.000	34.267
17	24.769	27.587	30.191	33.409	35.718
18	25.989	28.869	31.526	34.805	37.156
19	27.204	30.143	32.852	36.191	38.582
20	28.412	31.410	34.170	37.566	39.997
21	29.615	32.671	35.479	38.932	41.401
22	30.813	33.924	36.781	40.290	42.796
23	32.007	35.172	38.076	41.638	44.181
24	33.196	36.415	39.364	42.980	45.559
25	34.382	37.653	40.647	44.314	46.928
26	35.563	38.885	41.923	45.642	48.290
27	36.741	40.113	43.195	46.963	49.645
28	37.916	41.337	44.461	48.278	50.994
29	39.087	42.557	45.722	59.588	52.336
30	40.256	43.773	46.979	50.892	53.672
40	51.805	55.759	59.342	63.691	66.767
50	63.167	67.505	71.420	76.154	79.490
60	74.397	79.082	83.298	88.381	91.955
70	85.527	90.531	95.023	100.424	104.213
80	96.578	101.879	106.628	112.328	116.320
90	107.565	113.145	118.135	124.115	128.296
100	118.499	124.343	129.563	135.811	140.177

Table F

Denominator df \ Numerator df	1	2	3	4	5	6	7	8	9	10	11	12	13	14	15	16	17	18	19	20	21	22
1	4052.2	4999.3	5403.5	5624.3	5764.0	5859.0	5928.3	5981.0	6022.4	6055.9	6083.4	6106.7	6125.8	6143.0	6157.0	6170.0	6181.2	6191.4	6200.7	6208.7	6216.1	6223.1
2	98.50	99.00	99.16	99.25	99.30	99.33	99.36	99.38	99.39	99.40	99.41	99.42	99.42	99.43	99.43	99.44	99.44	99.44	99.45	99.45	99.45	99.46
3	34.12	30.82	29.46	28.71	28.24	27.91	27.67	27.49	27.34	27.23	27.13	27.05	26.98	26.92	26.87	26.83	26.79	26.75	26.72	26.69	26.66	26.64
4	21.20	18.00	16.69	15.98	15.52	15.21	14.98	14.80	14.66	14.55	14.45	14.37	14.31	14.25	14.20	14.15	14.11	14.08	14.05	14.02	13.99	13.97
5	16.26	13.27	12.06	11.39	10.97	10.67	10.46	10.29	10.16	10.05	9.96	9.89	9.82	9.77	9.72	9.68	9.64	9.61	9.58	9.55	9.53	9.51
6	13.75	10.92	9.78	9.15	8.75	8.47	8.26	8.10	7.98	7.87	7.79	7.72	7.66	7.60	7.56	7.52	7.48	7.45	7.42	7.40	7.37	7.35
7	12.25	9.55	8.45	7.85	7.46	7.19	6.99	6.84	6.72	6.62	6.54	6.47	6.41	6.36	6.31	6.28	6.24	6.21	6.18	6.16	6.13	6.11
8	11.26	8.65	7.59	7.01	6.63	6.37	6.18	6.03	5.91	5.81	5.73	5.67	5.61	5.56	5.52	5.48	5.44	5.41	5.38	5.36	5.34	5.32
9	10.56	8.02	6.99	6.42	6.06	5.80	5.61	5.47	5.35	5.26	5.18	5.11	5.05	5.01	4.96	4.92	4.89	4.86	4.83	4.81	4.79	4.77
10	10.04	7.56	6.55	5.99	5.64	5.39	5.20	5.06	4.94	4.85	4.77	4.71	4.65	4.60	4.56	4.52	4.49	4.46	4.43	4.41	4.38	4.36
11	9.65	7.21	6.22	5.67	5.32	5.07	4.89	4.74	4.63	4.54	4.46	4.40	4.34	4.29	4.25	4.21	4.18	4.15	4.12	4.10	4.08	4.06
12	9.33	6.93	5.95	5.41	5.06	4.82	4.64	4.50	4.39	4.30	4.22	4.16	4.10	4.05	4.01	3.97	3.94	3.91	3.88	3.86	3.84	3.82
13	9.07	6.70	5.74	5.21	4.86	4.62	4.44	4.30	4.19	4.10	4.02	3.96	3.91	3.86	3.82	3.78	3.75	3.72	3.69	3.66	3.64	3.62
14	8.86	6.51	5.56	5.04	4.69	4.46	4.28	4.14	4.03	3.94	3.86	3.80	3.75	3.70	3.66	3.62	3.59	3.56	3.53	3.51	3.48	3.46
15	8.68	6.36	5.42	4.89	4.56	4.32	4.14	4.00	3.89	3.80	3.73	3.67	3.61	3.56	3.52	3.49	3.45	3.42	3.40	3.37	3.35	3.33
16	8.53	6.23	5.29	4.77	4.44	4.20	4.03	3.89	3.78	3.69	3.62	3.55	3.50	3.45	3.41	3.37	3.34	3.31	3.28	3.26	3.24	3.22
17	8.40	6.11	5.19	4.67	4.34	4.10	3.93	3.79	3.68	3.59	3.52	3.46	3.40	3.35	3.31	3.27	3.24	3.21	3.19	3.16	3.14	3.12
18	8.29	6.01	5.09	4.58	4.25	4.01	3.84	3.71	3.60	3.51	3.43	3.37	3.32	3.27	3.23	3.19	3.16	3.13	3.10	3.08	3.05	3.03
19	8.18	5.93	5.01	4.50	4.17	3.94	3.77	3.63	3.52	3.43	3.36	3.30	3.24	3.19	3.15	3.12	3.08	3.05	3.03	3.00	2.98	2.96
20	8.10	5.85	4.94	4.43	4.10	3.87	3.70	3.56	3.46	3.37	3.29	3.23	3.18	3.13	3.09	3.05	3.02	2.99	2.96	2.94	2.92	2.90
21	8.02	5.78	4.87	4.37	4.04	3.81	3.64	3.51	3.40	3.31	3.24	3.17	3.12	3.07	3.03	2.99	2.96	2.93	2.90	2.88	2.86	2.84
22	7.95	5.72	4.82	4.31	3.99	3.76	3.59	3.45	3.35	3.26	3.18	3.12	3.07	3.02	2.98	2.94	2.91	2.88	2.85	2.83	2.81	2.78
23	7.88	5.66	4.76	4.26	3.94	3.71	3.54	3.41	3.30	3.21	3.14	3.07	3.02	2.97	2.93	2.89	2.86	2.83	2.80	2.78	2.76	2.74
24	7.82	5.61	4.72	4.22	3.90	3.67	3.50	3.36	3.26	3.17	3.09	3.03	2.98	2.93	2.89	2.85	2.82	2.79	2.76	2.74	2.72	2.70
25	7.77	5.57	4.68	4.18	3.85	3.63	3.46	3.32	3.22	3.13	3.06	2.99	2.94	2.89	2.85	2.81	2.78	2.75	2.72	2.70	2.68	2.66
26	7.72	5.53	4.64	4.14	3.82	3.59	3.42	3.29	3.18	3.09	3.02	2.96	2.90	2.86	2.81	2.78	2.75	2.72	2.69	2.66	2.64	2.62
27	7.68	5.49	4.60	4.11	3.78	3.56	3.39	3.26	3.15	3.06	2.99	2.93	2.87	2.82	2.78	2.75	2.71	2.68	2.66	2.63	2.61	2.59
28	7.64	5.45	4.57	4.07	3.75	3.53	3.36	3.23	3.12	3.03	2.96	2.90	2.84	2.79	2.75	2.72	2.68	2.65	2.63	2.60	2.58	2.56
29	7.60	5.42	4.54	4.04	3.73	3.50	3.33	3.20	3.09	3.00	2.93	2.87	2.81	2.77	2.73	2.69	2.66	2.63	2.60	2.57	2.55	2.53
30	7.56	5.39	4.51	4.02	3.70	3.47	3.30	3.17	3.07	2.98	2.91	2.84	2.79	2.74	2.70	2.66	2.63	2.60	2.57	2.55	2.53	2.51
32	7.50	5.34	4.46	3.97	3.65	3.43	3.26	3.13	3.02	2.93	2.86	2.80	2.74	2.70	2.65	2.62	2.58	2.55	2.53	2.50	2.48	2.46
35	7.42	5.27	4.40	3.91	3.59	3.37	3.20	3.07	2.96	2.88	2.80	2.74	2.69	2.64	2.60	2.56	2.53	2.50	2.47	2.44	2.42	2.40
40	7.31	5.18	4.31	3.83	3.51	3.29	3.12	2.99	2.89	2.80	2.73	2.66	2.61	2.56	2.52	2.48	2.45	2.42	2.39	2.37	2.35	2.33
45	7.23	5.11	4.25	3.77	3.45	3.23	3.07	2.94	2.83	2.74	2.67	2.61	2.55	2.51	2.46	2.43	2.39	2.36	2.34	2.31	2.29	2.27
50	7.17	5.06	4.20	3.72	3.41	3.19	3.02	2.89	2.78	2.70	2.63	2.56	2.51	2.46	2.42	2.38	2.35	2.32	2.29	2.27	2.24	2.22
60	7.08	4.98	4.13	3.65	3.34	3.12	2.95	2.82	2.72	2.63	2.56	2.50	2.44	2.39	2.35	2.31	2.28	2.25	2.22	2.20	2.17	2.15
75	6.99	4.90	4.05	3.58	3.27	3.05	2.89	2.76	2.65	2.57	2.49	2.43	2.38	2.33	2.29	2.25	2.22	2.18	2.16	2.13	2.11	2.09
100	6.90	4.82	3.98	3.51	3.21	2.99	2.82	2.69	2.59	2.50	2.43	2.37	2.31	2.27	2.22	2.19	2.15	2.12	2.09	2.07	2.04	2.02
120	6.85	4.79	3.95	3.48	3.17	2.96	2.79	2.66	2.56	2.47	2.40	2.34	2.28	2.23	2.19	2.15	2.12	2.09	2.06	2.03	2.01	1.99
140	6.82	4.76	3.92	3.46	3.15	2.93	2.77	2.64	2.54	2.45	2.38	2.31	2.26	2.21	2.17	2.13	2.10	2.07	2.04	2.01	1.99	1.97
180	6.78	4.73	3.89	3.43	3.12	2.90	2.74	2.61	2.51	2.42	2.35	2.28	2.23	2.18	2.14	2.10	2.07	2.04	2.01	1.98	1.96	1.94
250	6.74	4.69	3.86	3.40	3.09	2.87	2.71	2.58	2.48	2.39	2.32	2.26	2.20	2.15	2.11	2.07	2.04	2.01	1.98	1.95	1.93	1.91
400	6.70	4.66	3.83	3.37	3.06	2.85	2.68	2.56	2.45	2.37	2.29	2.23	2.17	2.13	2.08	2.05	2.01	1.98	1.95	1.92	1.90	1.88
1000	6.66	4.63	3.80	3.34	3.04	2.82	2.66	2.53	2.43	2.34	2.27	2.20	2.15	2.10	2.06	2.02	1.98	1.95	1.92	1.90	1.87	1.85

Table F (cont.)

Numerator df

α = .01	23	24	25	26	27	28	29	30	32	35	40	45	50	60	75	100	120	140	180	250	400	1000
1	6228.7	6234.3	6239.9	6244.5	6249.2	6252.9	6257.1	6260.4	6266.9	6275.3	6286.4	6295.7	6302.3	6313.0	6323.7	6333.9	6339.5	6343.2	6347.9	6353.5	6358.1	6362.8
2	99.46	99.46	99.46	99.46	99.46	99.46	99.46	99.47	99.47	99.47	99.48	99.48	99.48	99.48	99.48	99.49	99.49	99.49	99.49	99.50	99.50	99.50
3	26.62	26.60	26.58	26.56	26.55	26.53	26.52	26.50	26.48	26.45	26.41	26.38	26.35	26.32	26.28	26.24	26.22	26.21	26.19	26.17	26.15	26.14
4	13.95	13.93	13.91	13.89	13.88	13.86	13.85	13.84	13.81	13.79	13.75	13.71	13.69	13.65	13.61	13.58	13.56	13.54	13.53	13.51	13.49	13.47
5	9.49	9.47	9.45	9.43	9.42	9.40	9.39	9.38	9.36	9.33	9.29	9.26	9.24	9.20	9.17	9.13	9.11	9.10	9.08	9.06	9.05	9.03
6	7.33	7.31	7.30	7.28	7.27	7.25	7.24	7.23	7.21	7.18	7.14	7.11	7.09	7.06	7.02	6.99	6.97	6.96	6.94	6.92	6.91	6.89
7	6.09	6.07	6.06	6.04	6.03	6.02	6.00	5.99	5.97	5.94	5.91	5.88	5.86	5.82	5.79	5.75	5.74	5.72	5.71	5.69	5.68	5.66
8	5.30	5.28	5.26	5.25	5.23	5.22	5.21	5.20	5.18	5.15	5.12	5.09	5.07	5.03	5.00	4.96	4.95	4.93	4.92	4.90	4.89	4.87
9	4.75	4.73	4.71	4.70	4.68	4.67	4.66	4.65	4.63	4.60	4.57	4.54	4.52	4.48	4.45	4.41	4.40	4.39	4.37	4.35	4.34	4.32
10	4.34	4.33	4.31	4.30	4.28	4.27	4.26	4.25	4.23	4.20	4.17	4.14	4.12	4.08	4.05	4.01	4.00	3.98	3.97	3.95	3.94	3.92
11	4.04	4.02	4.01	3.99	3.98	3.96	3.95	3.94	3.92	3.89	3.86	3.83	3.81	3.78	3.74	3.71	3.69	3.68	3.66	3.64	3.63	3.61
12	3.80	3.78	3.76	3.75	3.74	3.72	3.71	3.70	3.68	3.65	3.62	3.59	3.57	3.54	3.50	3.47	3.45	3.44	3.42	3.40	3.39	3.37
13	3.60	3.59	3.57	3.56	3.54	3.53	3.52	3.51	3.49	3.46	3.43	3.40	3.38	3.34	3.31	3.27	3.25	3.24	3.23	3.21	3.19	3.18
14	3.44	3.43	3.41	3.40	3.38	3.37	3.36	3.35	3.33	3.30	3.27	3.24	3.22	3.18	3.15	3.11	3.09	3.08	3.06	3.05	3.03	3.02
15	3.31	3.29	3.28	3.26	3.25	3.24	3.23	3.21	3.19	3.17	3.13	3.10	3.08	3.05	3.01	2.98	2.96	2.95	2.93	2.91	2.90	2.88
16	3.20	3.18	3.16	3.15	3.14	3.12	3.11	3.10	3.08	3.05	3.02	2.99	2.97	2.93	2.90	2.86	2.84	2.83	2.81	2.80	2.78	2.76
17	3.10	3.08	3.07	3.05	3.04	3.03	3.01	3.00	2.98	2.96	2.92	2.89	2.87	2.83	2.80	2.76	2.75	2.73	2.72	2.70	2.68	2.66
18	3.02	3.00	2.98	2.97	2.95	2.94	2.93	2.92	2.90	2.87	2.84	2.81	2.78	2.75	2.71	2.68	2.66	2.65	2.63	2.61	2.59	2.58
19	2.94	2.92	2.91	2.89	2.88	2.87	2.86	2.84	2.82	2.80	2.76	2.73	2.71	2.67	2.64	2.60	2.58	2.57	2.55	2.54	2.52	2.50
20	2.88	2.86	2.84	2.83	2.81	2.80	2.79	2.78	2.76	2.73	2.69	2.67	2.64	2.61	2.57	2.54	2.52	2.50	2.49	2.47	2.45	2.43
21	2.82	2.80	2.79	2.77	2.76	2.74	2.73	2.72	2.70	2.67	2.64	2.61	2.58	2.55	2.51	2.48	2.46	2.44	2.43	2.41	2.39	2.37
22	2.77	2.75	2.73	2.72	2.70	2.69	2.68	2.67	2.65	2.62	2.58	2.55	2.53	2.50	2.46	2.42	2.40	2.39	2.37	2.35	2.34	2.32
23	2.72	2.70	2.69	2.67	2.66	2.64	2.63	2.62	2.60	2.57	2.54	2.51	2.48	2.45	2.41	2.37	2.35	2.34	2.32	2.30	2.29	2.27
24	2.68	2.66	2.64	2.63	2.61	2.60	2.59	2.58	2.56	2.53	2.49	2.46	2.44	2.40	2.37	2.33	2.31	2.30	2.28	2.26	2.24	2.22
25	2.64	2.62	2.60	2.59	2.58	2.56	2.55	2.54	2.52	2.49	2.45	2.42	2.40	2.36	2.33	2.29	2.27	2.26	2.24	2.22	2.20	2.18
26	2.60	2.58	2.57	2.55	2.54	2.53	2.51	2.50	2.48	2.45	2.42	2.39	2.36	2.33	2.29	2.25	2.23	2.22	2.20	2.18	2.16	2.14
27	2.57	2.55	2.54	2.52	2.51	2.49	2.48	2.47	2.45	2.42	2.38	2.35	2.33	2.29	2.26	2.22	2.20	2.18	2.17	2.15	2.13	2.11
28	2.54	2.52	2.51	2.49	2.48	2.46	2.45	2.44	2.42	2.39	2.35	2.32	2.30	2.26	2.23	2.19	2.17	2.15	2.13	2.11	2.10	2.08
29	2.51	2.49	2.48	2.46	2.45	2.44	2.42	2.41	2.39	2.36	2.33	2.30	2.27	2.23	2.20	2.16	2.14	2.12	2.10	2.08	2.07	2.05
30	2.49	2.47	2.45	2.44	2.42	2.41	2.40	2.39	2.36	2.34	2.30	2.27	2.25	2.21	2.17	2.13	2.11	2.10	2.08	2.06	2.04	2.02
32	2.44	2.42	2.41	2.39	2.38	2.36	2.35	2.34	2.32	2.29	2.25	2.22	2.20	2.16	2.12	2.08	2.06	2.05	2.03	2.01	1.99	1.97
35	2.38	2.36	2.35	2.33	2.32	2.30	2.29	2.28	2.26	2.23	2.19	2.16	2.14	2.10	2.06	2.02	2.00	1.98	1.96	1.94	1.92	1.90
40	2.31	2.29	2.27	2.26	2.24	2.23	2.22	2.20	2.18	2.15	2.11	2.08	2.06	2.02	1.98	1.94	1.92	1.90	1.88	1.86	1.84	1.82
45	2.25	2.23	2.21	2.20	2.18	2.17	2.16	2.14	2.12	2.09	2.05	2.02	2.00	1.96	1.92	1.88	1.85	1.84	1.82	1.79	1.77	1.75
50	2.20	2.18	2.17	2.15	2.14	2.12	2.11	2.10	2.08	2.05	2.01	1.97	1.95	1.91	1.87	1.82	1.80	1.79	1.76	1.74	1.72	1.70
60	2.13	2.12	2.10	2.08	2.07	2.05	2.04	2.03	2.01	1.98	1.94	1.90	1.88	1.84	1.79	1.75	1.73	1.71	1.69	1.66	1.64	1.62
75	2.07	2.05	2.03	2.02	2.00	1.99	1.97	1.96	1.94	1.91	1.87	1.83	1.81	1.76	1.72	1.67	1.65	1.63	1.61	1.58	1.56	1.53
100	2.00	1.98	1.97	1.95	1.93	1.92	1.91	1.89	1.87	1.84	1.80	1.76	1.74	1.69	1.65	1.60	1.57	1.55	1.53	1.50	1.47	1.45
120	1.97	1.95	1.93	1.92	1.90	1.89	1.87	1.86	1.84	1.81	1.76	1.73	1.70	1.66	1.61	1.56	1.53	1.51	1.49	1.46	1.43	1.40
140	1.95	1.93	1.91	1.89	1.88	1.86	1.85	1.84	1.81	1.78	1.74	1.70	1.67	1.63	1.58	1.53	1.50	1.48	1.46	1.43	1.40	1.37
180	1.92	1.90	1.88	1.86	1.85	1.83	1.82	1.81	1.78	1.75	1.71	1.67	1.64	1.60	1.55	1.49	1.47	1.45	1.42	1.39	1.35	1.32
250	1.89	1.87	1.85	1.83	1.82	1.80	1.79	1.77	1.75	1.72	1.67	1.64	1.61	1.56	1.51	1.46	1.43	1.41	1.38	1.34	1.31	1.27
400	1.86	1.84	1.82	1.80	1.79	1.77	1.76	1.75	1.72	1.69	1.64	1.61	1.58	1.53	1.48	1.42	1.39	1.37	1.33	1.30	1.26	1.22
1000	1.83	1.81	1.79	1.77	1.76	1.74	1.73	1.72	1.69	1.66	1.61	1.58	1.54	1.50	1.44	1.38	1.35	1.33	1.29	1.25	1.21	1.16

Denominator df

Table F (cont.)

Numerator df

α = .05 Denominator df	1	2	3	4	5	6	7	8	9	10	11	12	13	14	15	16	17	18	19	20	21	22
1	161.4	199.5	215.7	224.6	230.2	234.0	236.8	238.9	240.5	241.9	243.0	243.9	244.7	245.4	245.9	246.5	246.9	247.3	247.7	248.0	248.3	248.6
2	18.51	19.00	19.16	19.25	19.30	19.33	19.35	19.37	19.38	19.40	19.40	19.41	19.42	19.42	19.43	19.43	19.44	19.44	19.44	19.45	19.45	19.45
3	10.13	9.55	9.28	9.12	9.01	8.94	8.89	8.85	8.81	8.79	8.76	8.74	8.73	8.71	8.70	8.69	8.68	8.67	8.67	8.66	8.65	8.65
4	7.71	6.94	6.59	6.39	6.26	6.16	6.09	6.04	6.00	5.96	5.94	5.91	5.89	5.87	5.86	5.84	5.83	5.82	5.81	5.80	5.79	5.79
5	6.61	5.79	5.41	5.19	5.05	4.95	4.88	4.82	4.77	4.74	4.70	4.68	4.66	4.64	4.62	4.60	4.59	4.58	4.57	4.56	4.55	4.54
6	5.99	5.14	4.76	4.53	4.39	4.28	4.21	4.15	4.10	4.06	4.03	4.00	3.98	3.96	3.94	3.92	3.91	3.90	3.88	3.87	3.86	3.86
7	5.59	4.74	4.35	4.12	3.97	3.87	3.79	3.73	3.68	3.64	3.60	3.57	3.55	3.53	3.51	3.49	3.48	3.47	3.46	3.44	3.43	3.43
8	5.32	4.46	4.07	3.84	3.69	3.58	3.50	3.44	3.39	3.35	3.31	3.28	3.26	3.24	3.22	3.20	3.19	3.17	3.16	3.15	3.14	3.13
9	5.12	4.26	3.86	3.63	3.48	3.37	3.29	3.23	3.18	3.14	3.10	3.07	3.05	3.03	3.01	2.99	2.97	2.96	2.95	2.94	2.93	2.92
10	4.96	4.10	3.71	3.48	3.33	3.22	3.14	3.07	3.02	2.98	2.94	2.91	2.89	2.86	2.85	2.83	2.81	2.80	2.79	2.77	2.76	2.75
11	4.84	3.98	3.59	3.36	3.20	3.09	3.01	2.95	2.90	2.85	2.82	2.79	2.76	2.74	2.72	2.70	2.69	2.67	2.66	2.65	2.64	2.63
12	4.75	3.89	3.49	3.26	3.11	3.00	2.91	2.85	2.80	2.75	2.72	2.69	2.66	2.64	2.62	2.60	2.58	2.57	2.56	2.54	2.53	2.52
13	4.67	3.81	3.41	3.18	3.03	2.92	2.83	2.77	2.71	2.67	2.63	2.60	2.58	2.55	2.53	2.51	2.50	2.48	2.47	2.46	2.45	2.44
14	4.60	3.74	3.34	3.11	2.96	2.85	2.76	2.70	2.65	2.60	2.57	2.53	2.51	2.48	2.46	2.44	2.43	2.41	2.40	2.39	2.38	2.37
15	4.54	3.68	3.29	3.06	2.90	2.79	2.71	2.64	2.59	2.54	2.51	2.48	2.45	2.42	2.40	2.38	2.37	2.35	2.34	2.33	2.32	2.31
16	4.49	3.63	3.24	3.01	2.85	2.74	2.66	2.59	2.54	2.49	2.46	2.42	2.40	2.37	2.35	2.33	2.32	2.30	2.29	2.28	2.26	2.25
17	4.45	3.59	3.20	2.96	2.81	2.70	2.61	2.55	2.49	2.45	2.41	2.38	2.35	2.33	2.31	2.29	2.27	2.26	2.24	2.23	2.22	2.21
18	4.41	3.55	3.16	2.93	2.77	2.66	2.58	2.51	2.46	2.41	2.37	2.34	2.31	2.29	2.27	2.25	2.23	2.22	2.20	2.19	2.18	2.17
19	4.38	3.52	3.13	2.90	2.74	2.63	2.54	2.48	2.42	2.38	2.34	2.31	2.28	2.26	2.23	2.21	2.20	2.18	2.17	2.16	2.14	2.13
20	4.35	3.49	3.10	2.87	2.71	2.60	2.51	2.45	2.39	2.35	2.31	2.28	2.25	2.22	2.20	2.18	2.17	2.15	2.14	2.12	2.11	2.10
21	4.32	3.47	3.07	2.84	2.68	2.57	2.49	2.42	2.37	2.32	2.28	2.25	2.22	2.20	2.18	2.16	2.14	2.12	2.11	2.10	2.08	2.07
22	4.30	3.44	3.05	2.82	2.66	2.55	2.46	2.40	2.34	2.30	2.26	2.23	2.20	2.17	2.15	2.13	2.11	2.10	2.08	2.07	2.06	2.05
23	4.28	3.42	3.03	2.80	2.64	2.53	2.44	2.37	2.32	2.27	2.24	2.20	2.18	2.15	2.13	2.11	2.09	2.08	2.06	2.05	2.04	2.02
24	4.26	3.40	3.01	2.78	2.62	2.51	2.42	2.36	2.30	2.25	2.22	2.18	2.15	2.13	2.11	2.09	2.07	2.05	2.04	2.03	2.01	2.00
25	4.24	3.39	2.99	2.76	2.60	2.49	2.40	2.34	2.28	2.24	2.20	2.16	2.14	2.11	2.09	2.07	2.05	2.04	2.02	2.01	2.00	1.98
26	4.23	3.37	2.98	2.74	2.59	2.47	2.39	2.32	2.27	2.22	2.18	2.15	2.12	2.09	2.07	2.05	2.03	2.02	2.00	1.99	1.98	1.97
27	4.21	3.35	2.96	2.73	2.57	2.46	2.37	2.31	2.25	2.20	2.17	2.13	2.10	2.08	2.06	2.04	2.02	2.00	1.99	1.97	1.96	1.95
28	4.20	3.34	2.95	2.71	2.56	2.45	2.36	2.29	2.24	2.19	2.15	2.12	2.09	2.06	2.04	2.02	2.00	1.99	1.97	1.96	1.95	1.93
29	4.18	3.33	2.93	2.70	2.55	2.43	2.35	2.28	2.22	2.18	2.14	2.10	2.08	2.05	2.03	2.01	1.99	1.97	1.96	1.94	1.93	1.92
30	4.17	3.32	2.92	2.69	2.53	2.42	2.33	2.27	2.21	2.16	2.13	2.09	2.06	2.04	2.01	1.99	1.98	1.96	1.95	1.93	1.92	1.91
32	4.15	3.29	2.90	2.67	2.51	2.40	2.31	2.24	2.19	2.14	2.10	2.07	2.04	2.01	1.99	1.97	1.95	1.94	1.92	1.91	1.90	1.88
35	4.12	3.27	2.87	2.64	2.49	2.37	2.29	2.22	2.16	2.11	2.07	2.04	2.01	1.99	1.96	1.94	1.92	1.91	1.89	1.88	1.87	1.85
40	4.08	3.23	2.84	2.61	2.45	2.34	2.25	2.18	2.12	2.08	2.04	2.00	1.97	1.95	1.92	1.90	1.89	1.87	1.85	1.84	1.83	1.81
45	4.06	3.20	2.81	2.58	2.42	2.31	2.22	2.15	2.10	2.05	2.01	1.97	1.94	1.92	1.89	1.87	1.86	1.84	1.82	1.81	1.80	1.78
50	4.03	3.18	2.79	2.56	2.40	2.29	2.20	2.13	2.07	2.03	1.99	1.95	1.92	1.89	1.87	1.85	1.83	1.81	1.80	1.78	1.77	1.76
60	4.00	3.15	2.76	2.53	2.37	2.25	2.17	2.10	2.04	1.99	1.95	1.92	1.89	1.86	1.84	1.82	1.80	1.78	1.76	1.75	1.73	1.72
75	3.97	3.12	2.73	2.49	2.34	2.22	2.13	2.06	2.01	1.96	1.92	1.88	1.85	1.83	1.80	1.78	1.76	1.74	1.73	1.71	1.70	1.69
100	3.94	3.09	2.70	2.46	2.31	2.19	2.10	2.03	1.97	1.93	1.89	1.85	1.82	1.79	1.77	1.75	1.73	1.71	1.69	1.68	1.66	1.65
120	3.92	3.07	2.68	2.45	2.29	2.18	2.09	2.02	1.96	1.91	1.87	1.83	1.80	1.78	1.75	1.73	1.71	1.69	1.67	1.66	1.64	1.63
140	3.91	3.06	2.67	2.44	2.28	2.16	2.08	2.01	1.95	1.90	1.86	1.82	1.79	1.76	1.74	1.72	1.70	1.68	1.66	1.65	1.63	1.62
180	3.89	3.05	2.65	2.42	2.26	2.15	2.06	1.99	1.93	1.88	1.84	1.81	1.77	1.75	1.72	1.70	1.68	1.66	1.64	1.63	1.61	1.60
250	3.88	3.03	2.64	2.41	2.25	2.13	2.05	1.98	1.92	1.87	1.83	1.79	1.76	1.73	1.71	1.68	1.66	1.65	1.63	1.61	1.60	1.58
400	3.86	3.02	2.63	2.39	2.24	2.12	2.03	1.96	1.90	1.85	1.81	1.78	1.74	1.72	1.69	1.67	1.65	1.63	1.61	1.60	1.58	1.57
1000	3.85	3.00	2.61	2.38	2.22	2.11	2.02	1.95	1.89	1.84	1.80	1.76	1.73	1.70	1.68	1.65	1.63	1.61	1.60	1.58	1.57	1.55

Table F (cont.)

Numerator df

Denominator df	23	24	25	26	27	28	29	30	32	35	40	45	50	60	75	100	120	140	180	250	400	1000
α = .05																						
1	248.8	249.1	249.3	249.5	249.6	249.8	250.0	250.1	250.4	250.7	251.1	251.5	251.8	252.2	252.6	253.0	253.3	253.4	253.6	253.8	254.0	254.2
2	19.45	19.45	19.46	19.46	19.46	19.46	19.46	19.46	19.46	19.47	19.47	19.47	19.48	19.48	19.48	19.49	19.49	19.49	19.49	19.49	19.49	19.49
3	8.64	8.64	8.63	8.63	8.63	8.62	8.62	8.62	8.61	8.60	8.59	8.59	8.58	8.57	8.56	8.55	8.55	8.55	8.54	8.54	8.53	8.53
4	5.78	5.77	5.77	5.76	5.76	5.75	5.75	5.75	5.74	5.73	5.72	5.71	5.70	5.69	5.68	5.66	5.66	5.65	5.65	5.64	5.64	5.63
5	4.53	4.53	4.52	4.52	4.51	4.50	4.50	4.50	4.49	4.48	4.46	4.45	4.44	4.43	4.42	4.41	4.40	4.39	4.39	4.38	4.38	4.37
6	3.85	3.84	3.83	3.83	3.82	3.82	3.81	3.81	3.80	3.79	3.77	3.76	3.75	3.74	3.73	3.71	3.70	3.70	3.69	3.69	3.68	3.67
7	3.42	3.41	3.40	3.40	3.39	3.39	3.38	3.38	3.37	3.36	3.34	3.33	3.32	3.30	3.29	3.27	3.27	3.26	3.25	3.25	3.24	3.23
8	3.12	3.12	3.11	3.10	3.10	3.09	3.08	3.08	3.07	3.06	3.04	3.03	3.02	3.01	2.99	2.97	2.97	2.96	2.95	2.95	2.94	2.93
9	2.91	2.90	2.89	2.89	2.88	2.87	2.87	2.86	2.85	2.84	2.83	2.81	2.80	2.79	2.77	2.76	2.75	2.74	2.73	2.73	2.72	2.71
10	2.75	2.74	2.73	2.72	2.72	2.71	2.70	2.70	2.69	2.68	2.66	2.65	2.64	2.62	2.60	2.59	2.58	2.57	2.57	2.56	2.55	2.54
11	2.62	2.61	2.60	2.59	2.59	2.58	2.58	2.57	2.56	2.55	2.53	2.52	2.51	2.49	2.47	2.46	2.45	2.44	2.43	2.43	2.42	2.41
12	2.51	2.51	2.50	2.49	2.48	2.48	2.47	2.47	2.46	2.44	2.43	2.41	2.40	2.38	2.37	2.35	2.34	2.33	2.33	2.32	2.31	2.30
13	2.43	2.42	2.41	2.41	2.40	2.39	2.39	2.38	2.37	2.36	2.34	2.33	2.31	2.30	2.28	2.26	2.25	2.25	2.24	2.23	2.22	2.21
14	2.36	2.35	2.34	2.33	2.33	2.32	2.31	2.31	2.30	2.28	2.27	2.25	2.24	2.22	2.21	2.19	2.18	2.17	2.16	2.15	2.15	2.14
15	2.30	2.29	2.28	2.27	2.27	2.26	2.25	2.25	2.24	2.22	2.20	2.19	2.18	2.16	2.14	2.12	2.11	2.11	2.10	2.09	2.08	2.07
16	2.24	2.24	2.23	2.22	2.21	2.21	2.20	2.19	2.18	2.17	2.15	2.14	2.12	2.11	2.09	2.07	2.06	2.05	2.04	2.03	2.02	2.02
17	2.20	2.19	2.18	2.17	2.17	2.16	2.15	2.15	2.14	2.12	2.10	2.09	2.08	2.06	2.04	2.02	2.01	2.00	1.99	1.98	1.98	1.97
18	2.16	2.15	2.14	2.13	2.13	2.12	2.11	2.11	2.10	2.08	2.06	2.05	2.04	2.02	2.00	1.98	1.97	1.96	1.95	1.94	1.93	1.92
19	2.12	2.11	2.11	2.10	2.09	2.08	2.08	2.07	2.06	2.05	2.03	2.01	2.00	1.98	1.96	1.94	1.93	1.92	1.91	1.90	1.89	1.88
20	2.09	2.08	2.07	2.07	2.06	2.05	2.05	2.04	2.03	2.01	1.99	1.98	1.97	1.95	1.93	1.91	1.90	1.89	1.88	1.87	1.86	1.85
21	2.06	2.05	2.05	2.04	2.03	2.02	2.02	2.01	2.00	1.98	1.96	1.95	1.94	1.92	1.90	1.88	1.87	1.86	1.85	1.84	1.83	1.82
22	2.04	2.03	2.02	2.01	2.00	2.00	1.99	1.98	1.97	1.96	1.94	1.92	1.91	1.89	1.87	1.85	1.84	1.83	1.82	1.81	1.80	1.79
23	2.01	2.01	2.00	1.99	1.98	1.97	1.97	1.96	1.95	1.93	1.91	1.90	1.88	1.86	1.84	1.82	1.81	1.80	1.79	1.78	1.77	1.76
24	1.99	1.98	1.97	1.97	1.96	1.95	1.95	1.94	1.93	1.91	1.89	1.88	1.86	1.84	1.82	1.80	1.79	1.78	1.77	1.76	1.75	1.74
25	1.97	1.96	1.96	1.95	1.94	1.93	1.93	1.92	1.91	1.89	1.87	1.86	1.84	1.82	1.80	1.78	1.77	1.76	1.75	1.74	1.73	1.72
26	1.96	1.95	1.94	1.93	1.92	1.91	1.91	1.90	1.89	1.87	1.85	1.84	1.82	1.80	1.78	1.76	1.75	1.74	1.73	1.72	1.71	1.70
27	1.94	1.93	1.92	1.91	1.90	1.90	1.89	1.88	1.87	1.86	1.84	1.82	1.81	1.79	1.76	1.74	1.73	1.72	1.71	1.70	1.69	1.68
28	1.92	1.91	1.91	1.90	1.89	1.88	1.88	1.87	1.86	1.84	1.82	1.80	1.79	1.77	1.75	1.73	1.71	1.70	1.69	1.68	1.67	1.66
29	1.91	1.90	1.89	1.88	1.88	1.87	1.86	1.85	1.84	1.83	1.81	1.79	1.77	1.75	1.73	1.71	1.70	1.69	1.68	1.67	1.66	1.65
30	1.90	1.89	1.88	1.87	1.86	1.85	1.85	1.84	1.83	1.81	1.79	1.77	1.76	1.74	1.72	1.70	1.68	1.68	1.66	1.65	1.64	1.63
32	1.87	1.86	1.85	1.85	1.84	1.83	1.82	1.82	1.80	1.79	1.77	1.75	1.74	1.71	1.69	1.67	1.66	1.65	1.64	1.63	1.61	1.60
35	1.84	1.83	1.82	1.82	1.81	1.80	1.79	1.79	1.77	1.76	1.74	1.72	1.70	1.68	1.66	1.63	1.62	1.61	1.60	1.59	1.58	1.57
40	1.80	1.79	1.78	1.77	1.77	1.76	1.75	1.74	1.73	1.72	1.69	1.67	1.66	1.64	1.61	1.59	1.58	1.57	1.55	1.54	1.53	1.52
45	1.77	1.76	1.75	1.74	1.73	1.73	1.72	1.71	1.70	1.68	1.66	1.64	1.63	1.60	1.58	1.55	1.54	1.53	1.52	1.51	1.49	1.48
50	1.75	1.74	1.73	1.72	1.71	1.70	1.69	1.69	1.67	1.66	1.63	1.61	1.60	1.58	1.55	1.52	1.51	1.50	1.49	1.47	1.46	1.45
60	1.71	1.70	1.69	1.68	1.67	1.66	1.66	1.65	1.64	1.62	1.59	1.57	1.56	1.53	1.51	1.48	1.47	1.46	1.44	1.43	1.41	1.40
75	1.67	1.66	1.65	1.64	1.63	1.63	1.62	1.61	1.60	1.58	1.55	1.53	1.52	1.49	1.47	1.44	1.42	1.41	1.40	1.38	1.37	1.35
100	1.64	1.63	1.62	1.61	1.60	1.59	1.58	1.57	1.56	1.54	1.52	1.49	1.48	1.45	1.42	1.39	1.38	1.36	1.35	1.33	1.31	1.30
120	1.62	1.61	1.60	1.59	1.58	1.57	1.56	1.55	1.54	1.52	1.50	1.47	1.46	1.43	1.40	1.37	1.35	1.34	1.32	1.30	1.29	1.27
140	1.61	1.60	1.59	1.57	1.56	1.55	1.55	1.54	1.53	1.51	1.49	1.46	1.44	1.41	1.38	1.35	1.33	1.32	1.30	1.29	1.27	1.25
180	1.59	1.58	1.57	1.56	1.55	1.54	1.53	1.52	1.51	1.49	1.46	1.44	1.42	1.39	1.36	1.33	1.31	1.30	1.28	1.26	1.24	1.22
250	1.57	1.56	1.55	1.54	1.53	1.52	1.51	1.50	1.49	1.47	1.44	1.42	1.40	1.37	1.34	1.31	1.29	1.27	1.25	1.23	1.21	1.18
400	1.56	1.54	1.53	1.52	1.51	1.50	1.50	1.49	1.47	1.45	1.42	1.40	1.38	1.35	1.32	1.28	1.26	1.25	1.23	1.20	1.18	1.15
1000	1.54	1.53	1.52	1.51	1.50	1.49	1.48	1.47	1.46	1.43	1.41	1.38	1.36	1.33	1.30	1.26	1.24	1.22	1.20	1.17	1.14	1.11

Denominator df

Table F (cont.)

Numerator df

α = .1	1	2	3	4	5	6	7	8	9	10	11	12	13	14	15	16	17	18	19	20	21	22
1	39.9	49.5	53.6	55.8	57.2	58.2	58.9	59.4	59.9	60.2	60.5	60.7	60.9	61.1	61.2	61.3	61.5	61.6	61.7	61.7	61.8	61.9
2	8.53	9.00	9.16	9.24	9.29	9.33	9.35	9.37	9.38	9.39	9.40	9.41	9.41	9.42	9.42	9.43	9.43	9.44	9.44	9.44	9.44	9.45
3	5.54	5.46	5.39	5.34	5.31	5.28	5.27	5.25	5.24	5.23	5.22	5.22	5.21	5.20	5.20	5.20	5.19	5.19	5.19	5.18	5.18	5.18
4	4.54	4.32	4.19	4.11	4.05	4.01	3.98	3.95	3.94	3.92	3.91	3.90	3.89	3.88	3.87	3.86	3.86	3.85	3.85	3.84	3.84	3.84
5	4.06	3.78	3.62	3.52	3.45	3.40	3.37	3.34	3.32	3.30	3.28	3.27	3.26	3.25	3.24	3.23	3.22	3.22	3.21	3.21	3.20	3.20
6	3.78	3.46	3.29	3.18	3.11	3.05	3.01	2.98	2.96	2.94	2.92	2.90	2.89	2.88	2.87	2.86	2.85	2.85	2.84	2.84	2.83	2.83
7	3.59	3.26	3.07	2.96	2.88	2.83	2.78	2.75	2.72	2.70	2.68	2.67	2.65	2.64	2.63	2.62	2.61	2.61	2.60	2.59	2.59	2.58
8	3.46	3.11	2.92	2.81	2.73	2.67	2.62	2.59	2.56	2.54	2.52	2.50	2.49	2.48	2.46	2.45	2.45	2.44	2.43	2.42	2.42	2.41
9	3.36	3.01	2.81	2.69	2.61	2.55	2.51	2.47	2.44	2.42	2.40	2.38	2.36	2.35	2.34	2.33	2.32	2.31	2.30	2.30	2.29	2.29
10	3.29	2.92	2.73	2.61	2.52	2.46	2.41	2.38	2.35	2.32	2.30	2.28	2.27	2.26	2.24	2.23	2.22	2.22	2.21	2.20	2.19	2.19
11	3.23	2.86	2.66	2.54	2.45	2.39	2.34	2.30	2.27	2.25	2.23	2.21	2.19	2.18	2.17	2.16	2.15	2.14	2.13	2.12	2.12	2.11
12	3.18	2.81	2.61	2.48	2.39	2.33	2.28	2.24	2.21	2.19	2.17	2.15	2.13	2.12	2.10	2.09	2.08	2.08	2.07	2.06	2.05	2.05
13	3.14	2.76	2.56	2.43	2.35	2.28	2.23	2.20	2.16	2.14	2.12	2.10	2.08	2.07	2.05	2.04	2.03	2.02	2.01	2.01	2.00	1.99
14	3.10	2.73	2.52	2.39	2.31	2.24	2.19	2.15	2.12	2.10	2.07	2.05	2.04	2.02	2.01	2.00	1.99	1.98	1.97	1.96	1.96	1.95
15	3.07	2.70	2.49	2.36	2.27	2.21	2.16	2.12	2.09	2.06	2.04	2.02	2.00	1.99	1.97	1.96	1.95	1.94	1.93	1.92	1.92	1.91
16	3.05	2.67	2.46	2.33	2.24	2.18	2.13	2.09	2.06	2.03	2.01	1.99	1.97	1.95	1.94	1.93	1.92	1.91	1.90	1.89	1.88	1.88
17	3.03	2.64	2.44	2.31	2.22	2.15	2.10	2.06	2.03	2.00	1.98	1.96	1.94	1.93	1.91	1.90	1.89	1.88	1.87	1.86	1.86	1.85
18	3.01	2.62	2.42	2.29	2.20	2.13	2.08	2.04	2.00	1.98	1.95	1.93	1.92	1.90	1.89	1.87	1.86	1.85	1.84	1.84	1.83	1.82
19	2.99	2.61	2.40	2.27	2.18	2.11	2.06	2.02	1.98	1.96	1.93	1.91	1.89	1.88	1.86	1.85	1.84	1.83	1.82	1.81	1.81	1.80
20	2.97	2.59	2.38	2.25	2.16	2.09	2.04	2.00	1.96	1.94	1.91	1.89	1.87	1.86	1.84	1.83	1.82	1.81	1.80	1.79	1.79	1.78
21	2.96	2.57	2.36	2.23	2.14	2.08	2.02	1.98	1.95	1.92	1.90	1.87	1.86	1.84	1.83	1.81	1.80	1.79	1.78	1.78	1.77	1.76
22	2.95	2.56	2.35	2.22	2.13	2.06	2.01	1.97	1.93	1.90	1.88	1.86	1.84	1.83	1.81	1.80	1.79	1.78	1.77	1.76	1.75	1.74
23	2.94	2.55	2.34	2.21	2.11	2.05	1.99	1.95	1.92	1.89	1.87	1.84	1.83	1.81	1.80	1.78	1.77	1.76	1.75	1.74	1.74	1.73
24	2.93	2.54	2.33	2.19	2.10	2.04	1.98	1.94	1.91	1.88	1.85	1.83	1.81	1.80	1.78	1.77	1.76	1.75	1.74	1.73	1.72	1.71
25	2.92	2.53	2.32	2.18	2.09	2.02	1.97	1.93	1.89	1.87	1.84	1.82	1.80	1.79	1.77	1.76	1.75	1.74	1.73	1.72	1.71	1.70
26	2.91	2.52	2.31	2.17	2.08	2.01	1.96	1.92	1.88	1.86	1.83	1.81	1.79	1.77	1.76	1.75	1.73	1.72	1.71	1.71	1.70	1.69
27	2.90	2.51	2.30	2.17	2.07	2.00	1.95	1.91	1.87	1.85	1.82	1.80	1.78	1.76	1.75	1.74	1.72	1.71	1.70	1.70	1.69	1.68
28	2.89	2.50	2.29	2.16	2.06	2.00	1.94	1.90	1.87	1.84	1.81	1.79	1.77	1.75	1.74	1.73	1.71	1.70	1.69	1.69	1.68	1.67
29	2.89	2.50	2.28	2.15	2.06	1.99	1.93	1.89	1.86	1.83	1.80	1.78	1.76	1.75	1.73	1.72	1.71	1.69	1.68	1.68	1.67	1.66
30	2.88	2.49	2.28	2.14	2.05	1.98	1.93	1.88	1.85	1.82	1.79	1.77	1.75	1.74	1.72	1.71	1.70	1.69	1.68	1.67	1.66	1.65
32	2.87	2.48	2.26	2.13	2.04	1.97	1.91	1.87	1.83	1.81	1.78	1.76	1.74	1.72	1.71	1.69	1.68	1.67	1.66	1.65	1.64	1.64
35	2.85	2.46	2.25	2.11	2.02	1.95	1.90	1.85	1.82	1.79	1.76	1.74	1.72	1.70	1.69	1.67	1.66	1.65	1.64	1.63	1.62	1.62
40	2.84	2.44	2.23	2.09	2.00	1.93	1.87	1.83	1.79	1.76	1.74	1.71	1.70	1.68	1.66	1.65	1.64	1.62	1.61	1.61	1.60	1.59
45	2.82	2.42	2.21	2.07	1.98	1.91	1.85	1.81	1.77	1.74	1.72	1.70	1.68	1.66	1.64	1.63	1.62	1.60	1.59	1.58	1.58	1.57
50	2.81	2.41	2.20	2.06	1.97	1.90	1.84	1.80	1.76	1.73	1.70	1.68	1.66	1.64	1.63	1.61	1.60	1.59	1.58	1.57	1.56	1.55
60	2.79	2.39	2.18	2.04	1.95	1.87	1.82	1.77	1.74	1.71	1.68	1.66	1.64	1.62	1.60	1.59	1.58	1.56	1.55	1.54	1.53	1.53
75	2.77	2.37	2.16	2.02	1.93	1.85	1.80	1.75	1.72	1.69	1.66	1.63	1.61	1.60	1.58	1.57	1.55	1.54	1.53	1.52	1.51	1.50
100	2.76	2.36	2.14	2.00	1.91	1.83	1.78	1.73	1.69	1.66	1.64	1.61	1.59	1.57	1.56	1.54	1.53	1.52	1.50	1.49	1.48	1.48
120	2.75	2.35	2.13	1.99	1.90	1.82	1.77	1.72	1.68	1.65	1.63	1.60	1.58	1.56	1.55	1.53	1.52	1.50	1.49	1.48	1.47	1.46
140	2.74	2.34	2.12	1.99	1.89	1.82	1.76	1.71	1.68	1.64	1.62	1.59	1.57	1.55	1.54	1.52	1.51	1.50	1.48	1.47	1.46	1.45
180	2.73	2.33	2.11	1.98	1.88	1.81	1.75	1.70	1.67	1.63	1.61	1.58	1.56	1.54	1.53	1.51	1.50	1.48	1.47	1.46	1.45	1.44
250	2.73	2.32	2.11	1.97	1.87	1.80	1.74	1.69	1.66	1.62	1.60	1.57	1.55	1.53	1.52	1.50	1.49	1.47	1.46	1.45	1.44	1.43
400	2.72	2.32	2.10	1.96	1.86	1.79	1.73	1.69	1.65	1.61	1.59	1.56	1.54	1.52	1.50	1.49	1.47	1.46	1.45	1.44	1.43	1.42
1000	2.71	2.31	2.09	1.95	1.85	1.78	1.72	1.68	1.64	1.61	1.58	1.55	1.53	1.51	1.49	1.48	1.46	1.45	1.44	1.43	1.42	1.41

Denominator df

Table F (cont.)

Numerator df

α = .1	23	24	25	26	27	28	29	30	32	35	40	45	50	60	75	100	120	140	180	250	400	1000
1	61.9	62.0	62.1	62.1	62.1	62.2	62.2	62.3	62.3	62.4	62.5	62.6	62.7	62.8	62.9	63.0	63.1	63.1	63.1	63.2	63.2	63.3
2	9.45	9.45	9.45	9.45	9.45	9.46	9.46	9.46	9.46	9.46	9.47	9.47	9.47	9.47	9.48	9.48	9.48	9.48	9.49	9.49	9.49	9.49
3	5.18	5.18	5.17	5.17	5.17	5.17	5.17	5.17	5.17	5.16	5.16	5.16	5.15	5.15	5.15	5.14	5.14	5.14	5.14	5.14	5.14	5.1
4	3.83	3.83	3.83	3.83	3.82	3.82	3.82	3.82	3.81	3.81	3.80	3.80	3.80	3.79	3.78	3.78	3.78	3.77	3.77	3.77	3.77	3.76
5	3.19	3.19	3.19	3.18	3.18	3.18	3.18	3.17	3.17	3.16	3.16	3.15	3.15	3.14	3.13	3.13	3.12	3.12	3.12	3.11	3.11	3.11
6	2.82	2.82	2.81	2.81	2.81	2.81	2.80	2.80	2.80	2.79	2.78	2.77	2.77	2.76	2.75	2.75	2.74	2.74	2.74	2.73	2.73	2.72
7	2.58	2.58	2.57	2.57	2.56	2.56	2.56	2.56	2.55	2.54	2.54	2.53	2.52	2.51	2.51	2.50	2.49	2.49	2.49	2.48	2.48	2.47
8	2.41	2.40	2.40	2.40	2.39	2.39	2.39	2.38	2.38	2.37	2.36	2.35	2.35	2.34	2.33	2.32	2.32	2.31	2.31	2.30	2.30	2.30
9	2.28	2.28	2.27	2.27	2.26	2.26	2.26	2.25	2.25	2.24	2.23	2.22	2.22	2.21	2.20	2.19	2.18	2.18	2.18	2.17	2.17	2.16
10	2.18	2.18	2.17	2.17	2.17	2.16	2.16	2.16	2.15	2.14	2.13	2.12	2.12	2.11	2.10	2.09	2.08	2.08	2.07	2.07	2.06	2.06
11	2.11	2.10	2.10	2.09	2.09	2.08	2.08	2.08	2.07	2.06	2.05	2.04	2.04	2.03	2.02	2.01	2.00	2.00	1.99	1.99	1.98	1.98
12	2.04	2.04	2.03	2.03	2.02	2.02	2.01	2.01	2.01	2.00	1.99	1.98	1.97	1.96	1.95	1.94	1.93	1.93	1.92	1.92	1.91	1.91
13	1.99	1.98	1.98	1.97	1.97	1.96	1.96	1.96	1.95	1.94	1.93	1.92	1.92	1.90	1.89	1.88	1.88	1.87	1.87	1.86	1.86	1.85
14	1.94	1.94	1.93	1.93	1.92	1.92	1.92	1.91	1.91	1.90	1.89	1.88	1.87	1.86	1.85	1.83	1.83	1.82	1.82	1.81	1.81	1.80
15	1.90	1.90	1.89	1.89	1.88	1.88	1.88	1.87	1.87	1.86	1.85	1.84	1.83	1.82	1.80	1.79	1.79	1.78	1.78	1.77	1.76	1.76
16	1.87	1.87	1.86	1.86	1.85	1.85	1.84	1.84	1.83	1.82	1.81	1.80	1.79	1.78	1.77	1.76	1.75	1.75	1.74	1.73	1.73	1.72
17	1.84	1.84	1.83	1.83	1.82	1.82	1.81	1.81	1.80	1.79	1.78	1.77	1.76	1.75	1.74	1.73	1.72	1.71	1.71	1.70	1.70	1.69
18	1.82	1.81	1.80	1.80	1.80	1.79	1.79	1.78	1.78	1.77	1.75	1.74	1.74	1.72	1.71	1.70	1.69	1.69	1.68	1.67	1.67	1.66
19	1.79	1.79	1.78	1.78	1.77	1.77	1.76	1.76	1.75	1.74	1.73	1.72	1.71	1.70	1.69	1.67	1.67	1.66	1.65	1.65	1.64	1.64
20	1.77	1.77	1.76	1.76	1.75	1.75	1.74	1.74	1.73	1.72	1.71	1.70	1.69	1.68	1.66	1.65	1.64	1.64	1.63	1.62	1.62	1.61
21	1.75	1.75	1.74	1.74	1.73	1.73	1.72	1.72	1.71	1.70	1.69	1.68	1.67	1.66	1.64	1.63	1.62	1.62	1.61	1.60	1.60	1.59
22	1.74	1.73	1.73	1.72	1.72	1.71	1.71	1.70	1.69	1.68	1.67	1.66	1.65	1.64	1.63	1.61	1.60	1.60	1.59	1.59	1.58	1.57
23	1.72	1.72	1.71	1.70	1.70	1.69	1.69	1.69	1.68	1.67	1.66	1.64	1.64	1.62	1.61	1.59	1.59	1.58	1.57	1.57	1.56	1.55
24	1.71	1.70	1.70	1.69	1.69	1.68	1.68	1.67	1.66	1.65	1.64	1.63	1.62	1.61	1.59	1.58	1.57	1.57	1.56	1.55	1.54	1.54
25	1.70	1.69	1.68	1.68	1.67	1.67	1.66	1.66	1.65	1.64	1.63	1.62	1.61	1.59	1.58	1.56	1.56	1.55	1.54	1.54	1.53	1.52
26	1.68	1.68	1.67	1.67	1.66	1.66	1.65	1.65	1.64	1.63	1.61	1.60	1.59	1.58	1.57	1.55	1.54	1.54	1.53	1.52	1.52	1.51
27	1.67	1.67	1.66	1.65	1.65	1.64	1.64	1.64	1.63	1.62	1.60	1.59	1.58	1.57	1.55	1.54	1.53	1.53	1.52	1.51	1.50	1.50
28	1.66	1.66	1.65	1.64	1.64	1.63	1.63	1.63	1.62	1.61	1.59	1.58	1.57	1.56	1.54	1.53	1.52	1.51	1.51	1.50	1.49	1.48
29	1.65	1.65	1.64	1.63	1.63	1.62	1.62	1.62	1.61	1.60	1.58	1.57	1.56	1.55	1.53	1.52	1.51	1.50	1.50	1.49	1.48	1.47
30	1.64	1.64	1.63	1.63	1.62	1.62	1.61	1.61	1.60	1.59	1.57	1.56	1.55	1.54	1.52	1.51	1.50	1.49	1.49	1.48	1.47	1.46
32	1.63	1.62	1.62	1.61	1.60	1.60	1.59	1.59	1.58	1.57	1.56	1.54	1.53	1.52	1.50	1.49	1.48	1.47	1.47	1.46	1.45	1.44
35	1.61	1.60	1.60	1.59	1.58	1.58	1.57	1.57	1.56	1.55	1.53	1.52	1.51	1.50	1.48	1.47	1.46	1.45	1.44	1.43	1.43	1.42
40	1.58	1.57	1.57	1.56	1.56	1.55	1.55	1.54	1.53	1.52	1.51	1.49	1.48	1.47	1.45	1.43	1.42	1.42	1.41	1.40	1.39	1.38
45	1.56	1.55	1.55	1.54	1.53	1.53	1.52	1.52	1.51	1.50	1.48	1.47	1.46	1.44	1.43	1.41	1.40	1.39	1.38	1.37	1.37	1.36
50	1.54	1.54	1.53	1.52	1.52	1.51	1.51	1.50	1.49	1.48	1.46	1.45	1.44	1.42	1.41	1.39	1.38	1.37	1.36	1.35	1.34	1.33
60	1.52	1.51	1.50	1.50	1.49	1.49	1.48	1.48	1.47	1.45	1.44	1.42	1.41	1.40	1.38	1.36	1.35	1.34	1.33	1.32	1.31	1.30
75	1.49	1.49	1.48	1.47	1.47	1.46	1.45	1.45	1.44	1.43	1.41	1.40	1.38	1.37	1.35	1.33	1.32	1.31	1.30	1.29	1.27	1.26
100	1.47	1.46	1.45	1.45	1.44	1.43	1.43	1.42	1.41	1.40	1.38	1.37	1.35	1.34	1.32	1.29	1.28	1.27	1.26	1.25	1.24	1.22
120	1.46	1.45	1.44	1.43	1.43	1.42	1.41	1.41	1.40	1.39	1.37	1.35	1.34	1.32	1.30	1.28	1.26	1.26	1.24	1.23	1.22	1.20
140	1.45	1.44	1.43	1.42	1.42	1.41	1.41	1.40	1.39	1.38	1.36	1.34	1.33	1.31	1.29	1.26	1.25	1.24	1.23	1.22	1.20	1.19
180	1.43	1.43	1.42	1.41	1.40	1.40	1.39	1.39	1.38	1.36	1.34	1.33	1.32	1.29	1.27	1.25	1.23	1.22	1.21	1.20	1.18	1.16
250	1.42	1.41	1.41	1.40	1.39	1.39	1.38	1.37	1.36	1.35	1.33	1.31	1.30	1.28	1.26	1.23	1.22	1.21	1.19	1.18	1.16	1.14
400	1.41	1.40	1.39	1.39	1.38	1.37	1.37	1.36	1.35	1.34	1.32	1.30	1.29	1.26	1.24	1.21	1.20	1.19	1.17	1.16	1.14	1.12
1000	1.40	1.39	1.38	1.38	1.37	1.36	1.36	1.35	1.34	1.32	1.30	1.29	1.27	1.25	1.23	1.20	1.18	1.17	1.15	1.13	1.11	1.08

Denominator df

Selected Formulas

$Range = Max - Min$

$IQR = Q3 - Q1$

Outlier Rule-of-Thumb: $y < Q1 - 1.5 \times IQR$ or $y > Q3 + 1.5 \times IQR$

$$\bar{y} = \frac{\Sigma y}{n}$$

$$s = \sqrt{\frac{\Sigma (y - \bar{y})^2}{n - 1}}$$

$$z = \frac{y - \mu}{\sigma} \text{ (model based)}$$

$$z = \frac{y - \bar{y}}{s} \text{ (data based)}$$

$$r = \frac{\Sigma z_x z_y}{n - 1}$$

$$\hat{y} = b_0 + b_1 x \qquad \text{where } b_1 = r\frac{s_y}{s_x} \text{ and } b_0 = \bar{y} - b_1\bar{x}$$

$$P(\mathbf{A}) = 1 - P(\mathbf{A}^C)$$

$$P(\mathbf{A} \text{ or } \mathbf{B}) = P(\mathbf{A}) + P(\mathbf{B}) - P(\mathbf{A} \text{ and } \mathbf{B})$$

$$P(\mathbf{A} \text{ and } \mathbf{B}) = P(\mathbf{A}) \times P(\mathbf{B} \mid \mathbf{A})$$

$$P(\mathbf{B} \mid \mathbf{A}) = \frac{P(\mathbf{A} \text{ and } \mathbf{B})}{P(\mathbf{A})}$$

If $\mathbf{A}$ and $\mathbf{B}$ are independent, $P(\mathbf{B} \mid \mathbf{A}) = P(\mathbf{B})$

$E(X) = \mu = \Sigma x \cdot P(x)$ $Var(X) = \sigma^2 = \Sigma (x - \mu)^2 P(x)$

$E(X \pm c) = E(X) \pm c$ $Var(X \pm c) = Var(X)$

$E(aX) = aE(X)$ $Var(aX) = a^2 Var(X)$

$E(X \pm Y) = E(X) \pm E(Y)$ $Var(X \pm Y) = Var(X) + Var(Y)$

                                         if X and Y are independent

Geometric: $P(x) = q^{x-1}p$ $\mu = \dfrac{1}{p}$ $\sigma = \sqrt{\dfrac{q}{p^2}}$

Binomial: $P(x) = {}_nC_x p^x q^{n-x}$ $\mu = np$ $\sigma = \sqrt{npq}$

$$\hat{p} = \frac{x}{n} \qquad \mu(\hat{p}) = p \qquad SD(\hat{p}) = \sqrt{\frac{pq}{n}}$$

Sampling distribution of $\bar{y}$:

(CLT) As n grows, the sampling distribution approaches the Normal model with

$$\mu(\bar{y}) = \mu_y \qquad SD(\bar{y}) = \frac{\sigma}{\sqrt{n}}$$

Inference:

Confidence interval for parameter = *statistic ± critical value × SE(statistic)*

$$\text{Test statistic} = \frac{statistic - parameter}{SD(statistic)}$$

Parameter	Statistic	SD(statistic)	SE(statistic)
p	$\hat{p}$	$\sqrt{\dfrac{pq}{n}}$	$\sqrt{\dfrac{\hat{p}\hat{q}}{n}}$
$p_1 - p_2$	$\hat{p}_1 - \hat{p}_2$	$\sqrt{\dfrac{p_1 q_1}{n_1} + \dfrac{p_2 q_2}{n_2}}$	$\sqrt{\dfrac{\hat{p}_1 \hat{q}_1}{n_1} + \dfrac{\hat{p}_2 \hat{q}_2}{n_2}}$
μ	$\bar{y}$	$\dfrac{\sigma}{\sqrt{n}}$	$\dfrac{s}{\sqrt{n}}$
$\mu_1 - \mu_2$	$\bar{y}_1 - \bar{y}_2$	$\sqrt{\dfrac{\sigma_1^2}{n_1} + \dfrac{\sigma_2^2}{n_2}}$	$\sqrt{\dfrac{s_1^2}{n_1} + \dfrac{s_2^2}{n_2}}$
μ_d	$\bar{d}$	$\dfrac{\sigma_d}{\sqrt{n}}$	$\dfrac{s_d}{\sqrt{n}}$
σ_ε	$s_e = \sqrt{\dfrac{\Sigma(y - \hat{y})^2}{n - 2}}$	(dividing $n - k - 1$ in multiple regression)	
β_1	b_1	(in simple regression)	$\dfrac{s_e}{s_x \sqrt{n - 1}}$
μ_ν	$\hat{y}_\nu$	(in simple regression)	$\sqrt{SE^2(b_1) \cdot (x_\nu - \bar{x})^2 + \dfrac{s_e^2}{n}}$
y_ν	$\hat{y}_\nu$	(in simple regression)	$\sqrt{SE^2(b_1) \cdot (x_\nu - \bar{x})^2 + \dfrac{s_e^2}{n} + s_e^2}$

Pooling: For testing difference between proportions: $\hat{p}_{pooled} = \dfrac{y_1 + y_2}{n_1 + n_2}$

For testing difference between means: $s_p = \sqrt{\dfrac{(n_1 - 1)s_1^2 + (n_2 - 1)s_2^2}{n_1 + n_2 - 2}}$

Substitute these pooled estimates in the respective SE formulas for both groups when assumptions and conditions are met.

Chi-square: $\chi^2 = \Sigma \dfrac{(Obs - Exp)^2}{Exp}$

ANOVA: $SS_T = \Sigma \Sigma (\bar{y}_j - \bar{\bar{y}})^2$; $MS_T = SS_T/(k - 1)$

$SS_E = \Sigma \Sigma (y_{ij} - \bar{y}_j)^2$; $MS_E = SS_E/(N - k)$

$F_{k-1, N-k} = MS_T/MS_E$